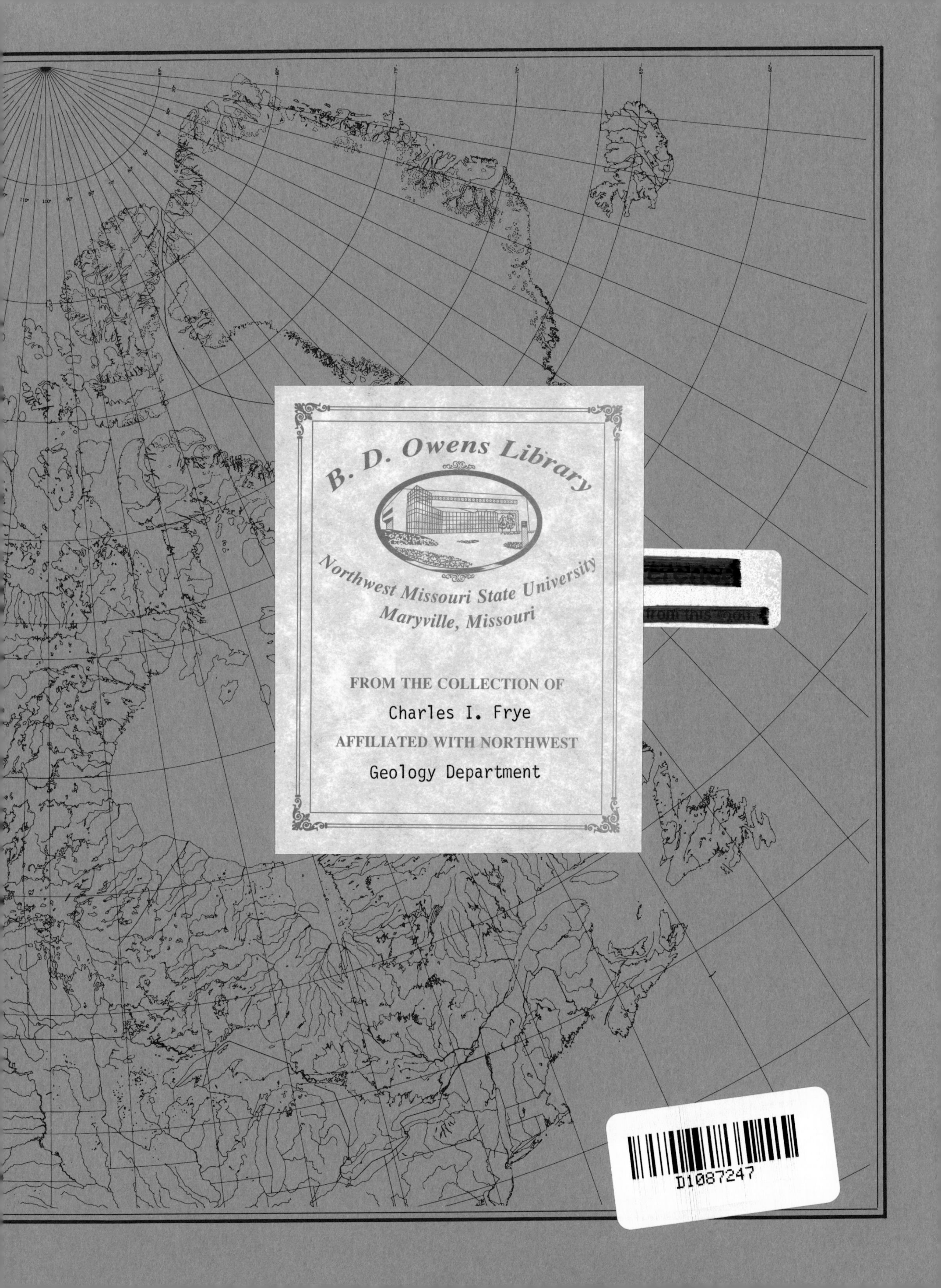

Northeastern Section
of the
Geological Society of America

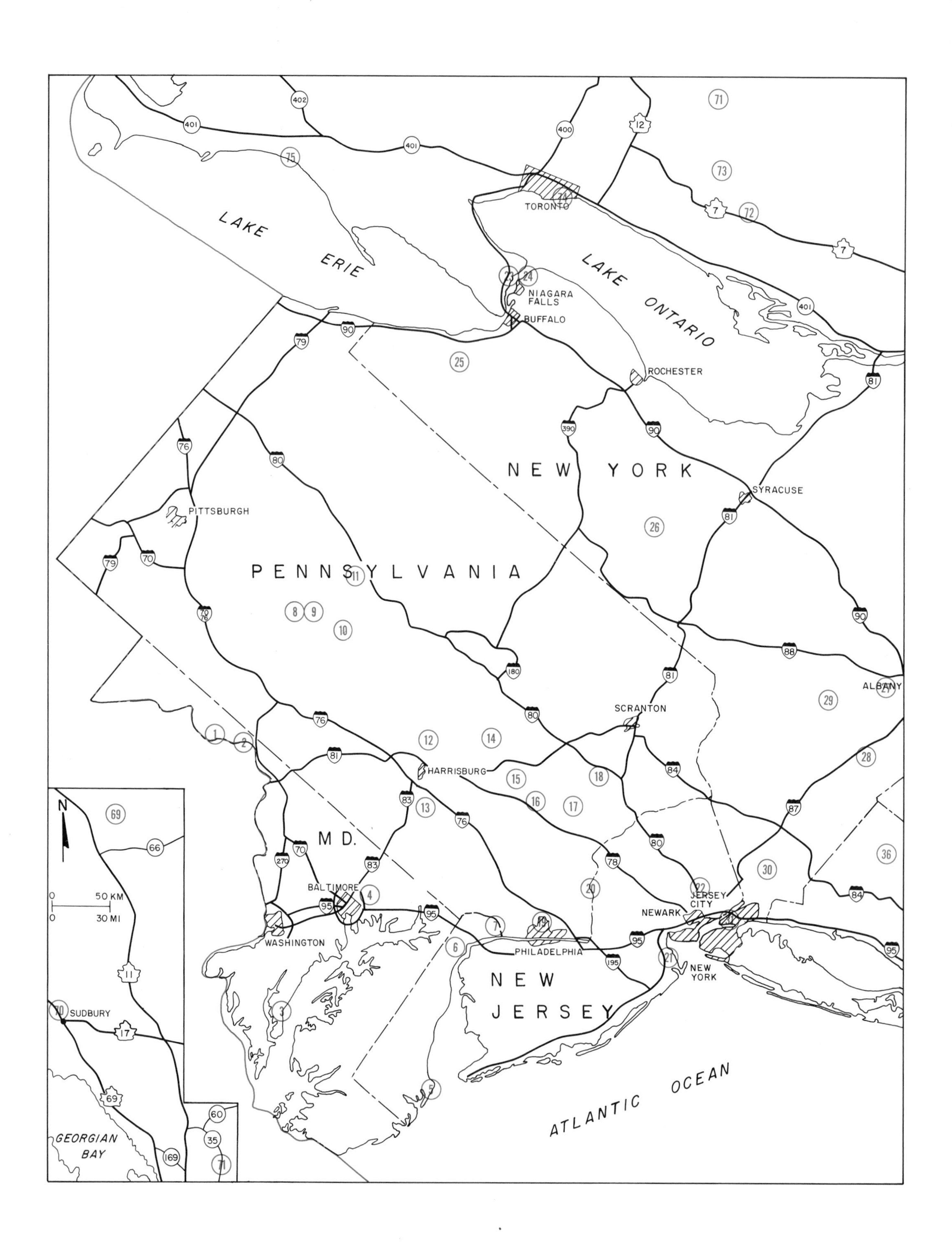

LAKE ERIE
LAKE ONTARIO
TORONTO
NIAGARA FALLS
BUFFALO
ROCHESTER
SYRACUSE
NEW YORK
PENNSYLVANIA
PITTSBURGH
ALBANY
SCRANTON
HARRISBURG
MD.
BALTIMORE
WASHINGTON
PHILADELPHIA
NEWARK
JERSEY CITY
NEW YORK
NEW JERSEY
ATLANTIC OCEAN
SUDBURY
GEORGIAN BAY
50 KM
30 MI
N

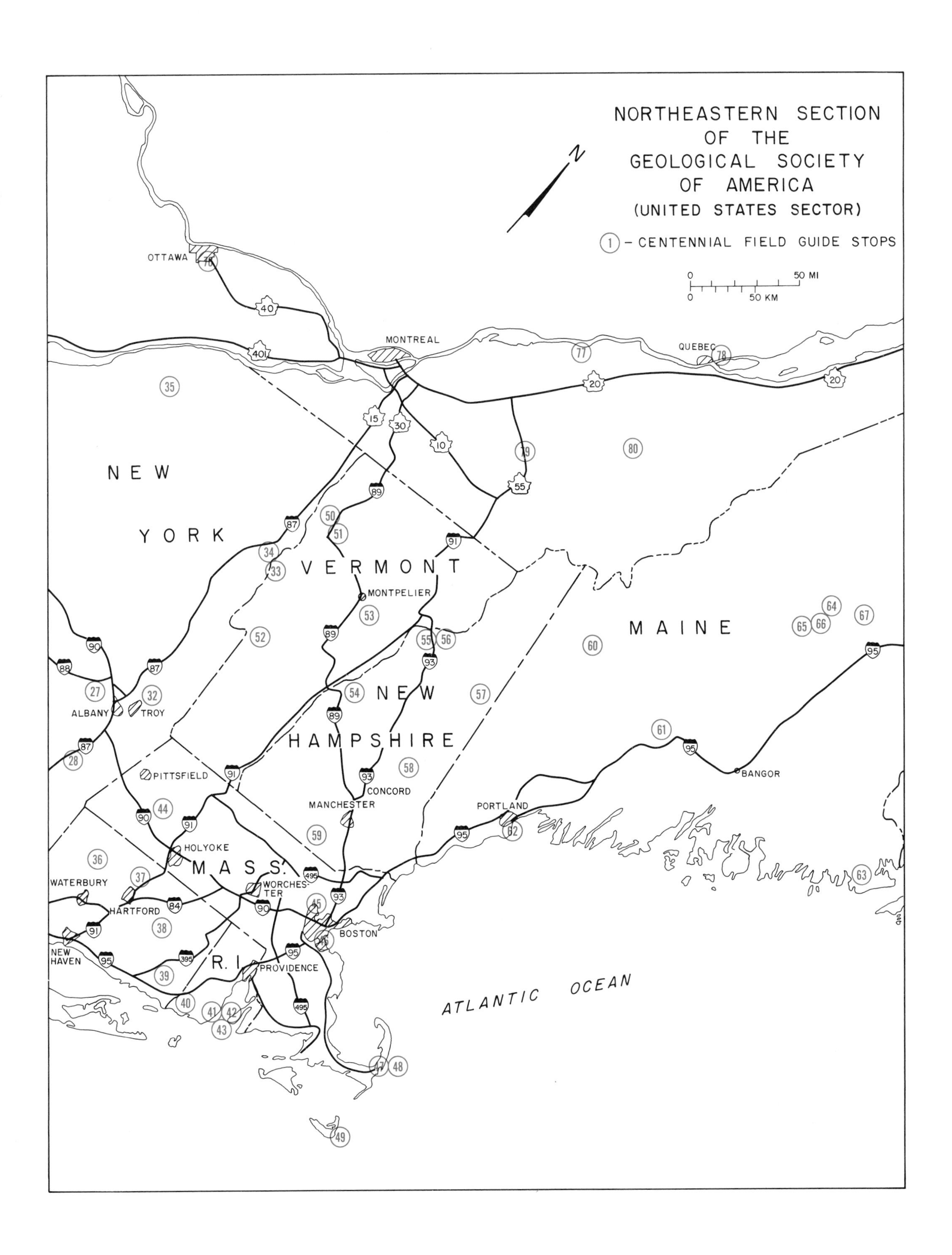
NORTHEASTERN SECTION
OF THE
GEOLOGICAL SOCIETY
OF AMERICA
(UNITED STATES SECTOR)
1 - CENTENNIAL FIELD GUIDE STOPS
0 50 MI
0 50 KM
N
OTTAWA
MONTREAL
QUEBEC
NEW YORK
VERMONT
MONTPELIER
MAINE
NEW HAMPSHIRE
ALBANY
TROY
PITTSFIELD
CONCORD
MANCHESTER
PORTLAND
BANGOR
HOLYOKE
MASS.
WATERBURY
HARTFORD
WORCHESTER
BOSTON
NEW HAVEN
R.I.
PROVIDENCE
ATLANTIC OCEAN

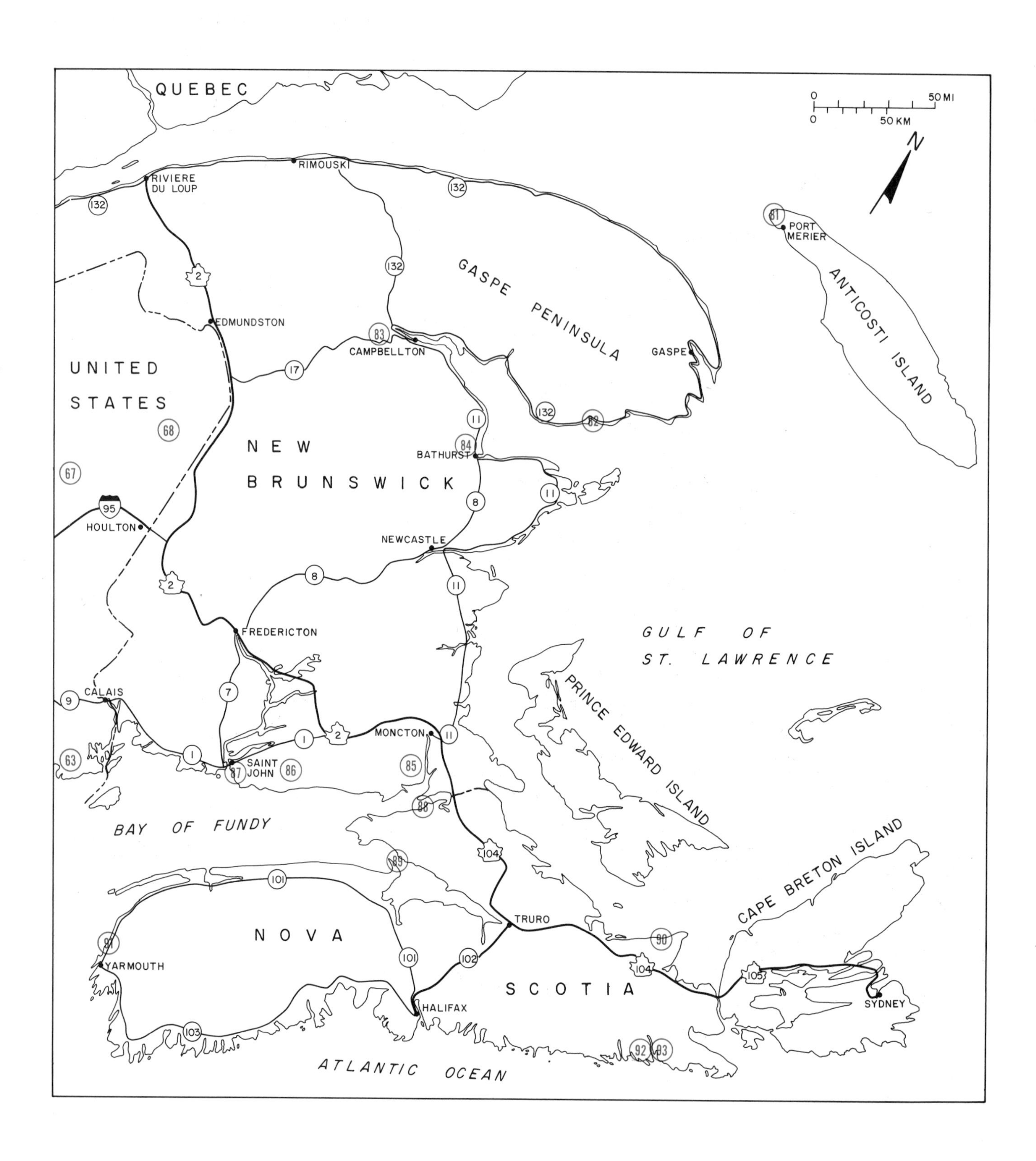
QUEBEC
0
50 MI
0
50 KM
N
RIMOUSKI
RIVIERE DU LOUP
132
132
132
GASPE PENINSULA
81
PORT MERIER
ANTICOSTI ISLAND
2
EDMUNDSTON
83
CAMPBELLTON
GASPE
UNITED STATES
17
132
82
11
68
NEW BRUNSWICK
84
BATHURST
67
8
11
95
HOULTON
NEWCASTLE
2
8
11
FREDERICTON
GULF OF ST. LAWRENCE
PRINCE EDWARD ISLAND
7
CALAIS
9
1
2
MONCTON
11
63
1
SAINT JOHN
87
86
85
88
BAY OF FUNDY
104
CAPE BRETON ISLAND
89
101
TRURO
NOVA SCOTIA
81
90
YARMOUTH
101
102
104
105
HALIFAX
SYDNEY
103
92
93
ATLANTIC OCEAN

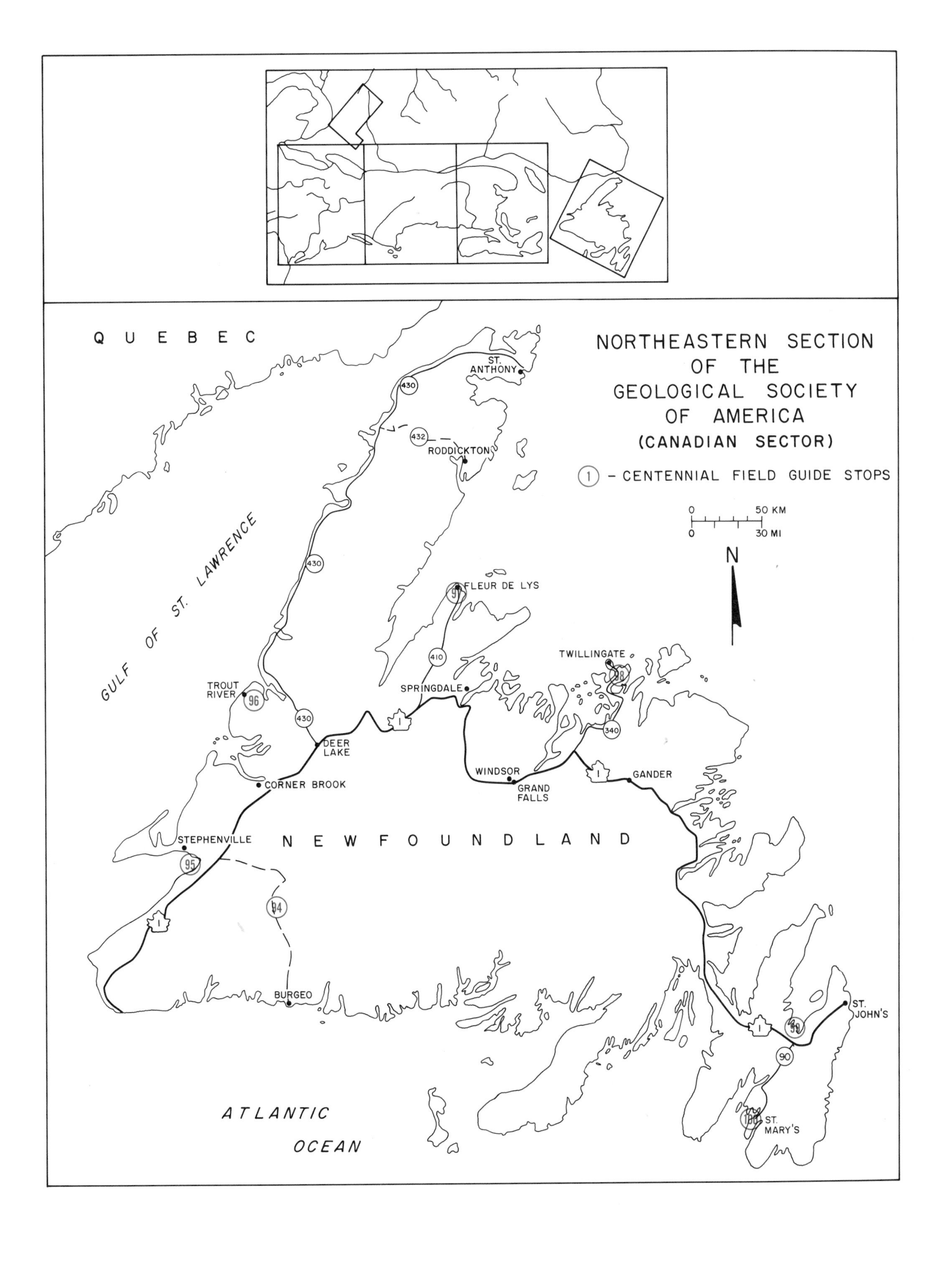
NORTHEASTERN SECTION
OF THE
GEOLOGICAL SOCIETY
OF AMERICA
(CANADIAN SECTOR)
1 - CENTENNIAL FIELD GUIDE STOPS
0 50 KM
0 30 MI
N
QUEBEC
GULF OF ST. LAWRENCE
NEWFOUNDLAND
ATLANTIC
OCEAN
ST. ANTHONY
430
432
RODDICKTON
FLEUR DE LYS
97
410
TWILLINGATE
98
SPRINGDALE
TROUT RIVER
96
430
1
340
DEER LAKE
WINDSOR
GRAND FALLS
GANDER
CORNER BROOK
STEPHENVILLE
95
94
1
BURGEO
ST. JOHN'S
1
99
90
100
ST. MARY'S

Centennial Field Guide Volume 5

Northeastern Section of the Geological Society of America

Edited by

David C. Roy
Department of Geology and Geophysics
Boston College
Chestnut Hill, Massachusetts 02167

1987

Acknowledgment

Publication of this volume, one of the Centennial Field Guide Volumes of *The Decade of North American Geology Project* series, has been made possible by members and friends of the Geological Society of America, corporations, and government agencies through contributions to the Decade of North American Geology fund of the Geological Society of America Foundation.

Following is a list of individuals, corporations, and government agencies giving and/or pledging more than $50,000 in support of the DNAG Project:

ARCO Exploration Company
Chevron Corporation
Cities Service Company
Conoco, Inc.
Diamond Shamrock Exploration Corporation
Exxon Production Research Company
Getty Oil Company
Gulf Oil Exploration and Production Company
Paul V. Hoovler
Kennecott Minerals Company
Kerr McGee Corporation
Marathon Oil Company
McMoRan Oil and Gas Company
Mobil Oil Corporation
Pennzoil Exploration and Production Company
Phillips Petroleum Company
Shell Oil Company
Caswell Silver
Sohio Petroleum Corporation
Standard Oil Company of Indiana
Sun Exploration and Production Company
Superior Oil Company
Tenneco Oil Company
Texaco, Inc.
Union Oil Company of California
Union Pacific Corporation and its operating companies:
  Champlin Petroleum Company
  Missouri Pacific Railroad Companies
  Rocky Mountain Energy Company
  Union Pacific Railroad Companies
  Upland Industries Corporation
U.S. Department of Energy

Published by the Geological Society of America, Inc.
3300 Penrose Place, P.O. Box 9140, Boulder, Colorado 80301

Printed in U.S.A.

Library of Congress Cataloging-in-Publication Data
(Revised for vol. 5)

Centennial field guide.

"Prepared under the auspices of the regional Sections of the Geological Society of America as a part of the Decade of North American Geology (DNAG) Project"—Vol. 6, pref.
Vol. : maps on lining papers.
Includes bibliographies and indexes.
Contents: —v. 5. Northeastern Section of the Geological Society of America / edited by David C. Roy. —v. 6. Southeastern Section of the Geological Society of America / edited by Thornton L. Neathery.
1. Geology—United States—Guide-books. 2. Geology—Canada—Guide-books. 3. United States—Description and travel—1981- —Guide-books. 4. Canada—Description and travel—1981- —Guide-books. I. Title: Geological Society of America.
QE77.C46 86-11986
ISBN 0-8137-5405-4 (v. 5)

Front Cover: Fall color highlights the banks of South Branch Ponds Brook where it flows over exposures of the Black Cat Member of the Traveler Rhyolite in Baxter State Park, Maine (Site 64). Here gently dipping surfaces on the right bank of the stream are roughly parallel to a compaction foliation (eutaxitic texture) in the ash-flow tuff. Crude columnar joints are perpendicular to the foliation. (Photo by D. C. Roy)

Contents

TOPICAL AND GEOGRAPHIC CROSS-REFERENCES

		Topic	EASTERN STATES					NEW ENGLAND					
			Md.	Del	Pa.	N.J.	N.Y.	Conn.	R.I.	Mass.	Vt.	N.H.	Me.
General		Coastal Geology		5						48			
General		Economic Geology			11,14								
General		Geomorphology	1		12		23					57	66
General		Glacial Geology			18		23,25	39		47,49		57	66
General		Igneous Petrology		7		22	30,34	37	40		53	54,57, 58	63-65
General		Metamorphic Petrology	4	7	19		30,31, 34,35	36,38	41,42	44,45		57,59	61
General		Stratigraphy and Sedimentation	2,3	6	8-11, 13-17, 19	20-22	24,26-29, 33	37	42,43	46	50-52	54-57	60-63, 67,68
General		Structural Geology	2		10,13, 14,16		26,28, 31,32, 34	36,38	41-43	44,45	50	55,56	61,62
Conti-nental		Precambrian of Canada											
Conti-nental		Adirondack Precambrian					34,35						
Conti-nental		Craton Cover Stratigraphy											
Appalachian Orogen	Regional Domains	Valley and Ridge/Plateau	2		8-12, 14,15, 17		26						
Appalachian Orogen	Regional Domains	Hudson Valley/ Logan's Line					28,32				50		
Appalachian Orogen	Regional Domains	Miogeocline					33				51		
Appalachian Orogen	Regional Domains	Taconic Klippen			16		32				52		
Appalachian Orogen	Regional Domains	Piedmont	4		13,19								
Appalachian Orogen	Regional Domains	Green Mt.-Sutton Mt. Anticlinorium											
Appalachian Orogen	Regional Domains	Cameron/Baie Verte-Brompton lines					31	36		44			
Appalachian Orogen	Regional Domains	Bronson Hill-Boundary Mt. Anticlinorium										54,55	
Appalachian Orogen	Regional Domains	Kearsarge-Central Maine Synclinorium										56	60,61
Appalachian Orogen	Regional Domains	Miramichi Anticlinorium											
Appalachian Orogen	Regional Domains	Maine Coastal Belt											62,63
Appalachian Orogen	Regional Domains	Avalon/Meguma Terranes							43	46			
Appalachian Orogen	General Topics	Fault Zones					32,34	36,38		45	50		
Appalachian Orogen	General Topics	Interior Precambrian/ Cambrian	4				31					59	62,67
Appalachian Orogen	General Topics	Ophiolite Sequences and Mafic/Ultramafic Complexes		7			30			44			
Appalachian Orogen	General Topics	Post Taconian-Pre Acadian Stratigraphy					27					56	60,61
Appalachian Orogen	General Topics	Acadian Plutonic Rocks									53		65
Appalachian Orogen	General Topics	Post Acadian Molasse					29						68
Appalachian Orogen	General Topics	Carboniferous Stratigraphy			15,17				42				
Appalachian Orogen	General Topics	Alleghanian Deformation and Plutonism			14		26		40-42			59	
Mesozoic and Cenozoic		Juro-Triassic Basins				20,22		37					
Mesozoic and Cenozoic		White Mt. Magma Series										57,58	
Mesozoic and Cenozoic		Coastal Plain	3	6		21							

TOPICAL AND GEOGRAPHIC CROSS-REFERENCES (CONTINUED)

		Topic	EASTERN CANADA				
			Ont.	Que.	N.B.	N.S.	Nfld.
General		Coastal Geology				91	95
General		Economic Geology	70,72	80	85	93	
General		Geomorphology	23				
General		Glacial Geology	74,75	77		91	95
General		Igneous Petrology	69-72	80	84	89	94,96
General		Metamorphic Petrology	71,72	79	87	90	97
General		Stratigraphy and Sedimentation	24,72 73,76	78,79, 81,82	83-87	88-90 92	94,98-100
General		Structural Geology	71,76	78-80	87	92	97,98
Continental Interior		Precambrian of Canada	69-73				
Continental Interior		Adirondack Precambrian					
Continental Interior		Craton Cover Stratigraphy	73,76	78			
Appalachian Orogen	Regional Domains	Valley and Ridge/Plateau					
Appalachian Orogen	Regional Domains	Hudson Valley/Logan's Line					
Appalachian Orogen	Regional Domains	Miogeocline					
Appalachian Orogen	Regional Domains	Taconic Klippen					96
Appalachian Orogen	Regional Domains	Piedmont					
Appalachian Orogen	Regional Domains	Green Mt.-Sutton Mt. Anticlinorium		79			
Appalachian Orogen	Regional Domains	Cameron/Baie Verte-Brompton lines		80			94,97
Appalachian Orogen	Regional Domains	Bronson Hill-Boundary Mt. Anticlinorium					
Appalachian Orogen	Regional Domains	Kearsarge-Central Maine Synclinorium					
Appalachian Orogen	Regional Domains	Miramichi Anticlinorium			84		
Appalachian Orogen	Regional Domains	Maine Coastal Belt					
Appalachian Orogen	Regional Domains	Avalon/Meguma Terranes			86,87	90, 92,93	99,100
Appalachian Orogen	General Topics	Fault Zones					
Appalachian Orogen	General Topics	Interior Precambrian/Cambrian		79			
Appalachian Orogen	General Topics	Ophiolite Sequences and Mafic/Ultramafic Complexes	69,70	80	84		94,96, 97
Appalachian Orogen	General Topics	Post Taconian-Pre Acadian Stratigraphy		81,82	83		98
Appalachian Orogen	General Topics	Acadian Plutonic Rocks					
Appalachian Orogen	General Topics	Post Acadian Molasse					
Appalachian Orogen	General Topics	Carboniferous Stratigraphy			85,87	88	
Appalachian Orogen	General Topics	Alleghanian Deformation and Plutonism					
Mesozoic and Cenozoic		Juro-Triassic Basins				89	
Mesozoic and Cenozoic		White Mt. Magma Series					
Mesozoic and Cenozoic		Coastal Plain					

Preface

This volume is one of a six–volume set of Centennial Field Guides prepared under the auspices of the regional Sections of the Geological Society of America as a part of the Decade of North American Geology (DNAG) Project. The intent of this volume is to highlight the best and most accessible geologic localities of the northeastern United States and eastern Canada for the geologic traveler and for students and professional geologists interested in major geologic features of regional significance. The leadership provided by the editor, David C. Roy, and the support provided to him by the Northeastern Section of the Geological Society of America and by the Department of Geology and Geophysics of Boston College are greatly appreciated.

Drafting services were offered by the DNAG Project to those authors of field guide texts who did not have access to drafting facilities. Particular thanks are given here to Ms. Karen Canfield of Louisville, Colorado, who prepared final drafted copy of many figures from copy provided by the authors.

In addition to Centennial Field Guides, the DNAG Project includes a 29–volume set of syntheses that constitute *The Geology of North America,* and 8 wall maps at a scale of 1:5,000,000 that summarize the geology, tectonics, magnetic and gravity anomaly patterns, regional stress fields, thermal aspects, seismicity, and neotectonics of North America and its surroundings. Together, the synthesis volumes and maps are the first coordinated effort to integrate all available knowledge about the geology and geophysics of a crustal plate on a regional scale. They are supplemented, as a part of the DNAG project, by 23 Continent–Ocean Transects providing strip maps and both geologic and tectonic cross–sections strategically sited around the margins of the continent, and by several related topical volumes.

The products of the DNAG Project have been prepared as a part of the celebration of the Centennial of the Geological Society of America. They present the state of knowledge of the geology and geophysics of North America in the 1980s, and they point the way toward work to be done in the decades ahead.

Allison R. Palmer
Centennial Science Program Coordinator

Foreword

The region of the Northeastern Section of the Geological Society of America includes the northeastern states of the United States and the eastern provinces of Canada, including most of Ontario. In many ways the science of geology in North America was born and nurtured in this region with the pioneer work of such people as James Hall, in the United States, and Sir William Logan, in Canada, just a decade or so after publication of Charles Lyell's *Principles of Geology.* Lyell himself visited the region in the early 1840s and visited some of the sites described in this volume.

With approximately 200 colleges and universities offering programs in the geological sciences and the very active work of the United States Geological Survey, the Geological Survey of Canada, and the several state and provincial surveys, it is understandable why the region of the Northeastern Section is one of the most studied and restudied terranes on the continent. Despite this long history of intense study the region is still the source of new information, new ideas, and new problems.

This field guide project began with selection of the editor by the Management Board of the Section in the spring of 1982. The commission given the editor was to produce a field guide to the Northeastern Section, containing a hundred "classic" field sites, to be a part of a Decade of North American Geology (DNAG) field guide series. This sort of field guide had never been attempted by the Society before but it seemed like a good idea since no other DNAG volumes were planned that would document the field evidence on which the regional syntheses are necessarily based. Although there is a rich library of guidebooks by local and regional societies, associations, and conferences, most of those guidebooks are not generally available. The Management Board suggested that everyone who has, or is, working in the region be given an opportunity to nominate sites and that selection be made by a panel of geologists familiar with the region. This was accomplished by mailing nomination materials to all members of the Section, as well as by promotions at Section meetings and notices by the Society to all members. Individuals familiar with the geology of each state/province reviewed the nominations and made recommendations both as to which nominated sites should be included and which other sites should be sought. The Management Board also considered it important that the guide be regionally and topically "balanced" insofar as possible with the limited number of sites.

Although the "sites" were originally thought of as single outcrops, it early became clear that individual exposures rarely provide sufficient evidence to establish major conclusions about the geologic history of a region. It is the integration of observations from several, sometimes many, outcrops that usually convince us of the extent and importance of geologic events. In much of this wooded region, one cannot usually describe a stratigraphic section, delineate a major structure, or explore a moraine without using information from numerous exposures. Therefore most of the sites described in this volume are small areas containing diverse geologic information of regional importance. Individual exposures are designated as numbered "stops" (sometimes with "stations" within them), but the descriptions are written so that users can examine the stops in any order. Although collecting of fossils, minerals, and rocks is possible and encouraged at many of the sites described here, sites that are important primarily as collecting localities were not included because of the limited audience to which they would appeal and also to prevent excess "people pressure" on them. Localities featuring archetypical examples of geologic features without also providing information of widespread significance to our understanding of the geologic history of the region were also not given priority. In some cases, authors have requested that visitors to the site, or to some stops, not collect or mar the outcrop with hammers or other tools. Please respect these cautions so that future visitors can see all of the things that make the site important.

The limitation on the number of sites makes it, of course, impossible to include all important sites in such a large region. This field guide, and the others in the series, can however provide a foundation upon which to build field trips or study particular aspects of the geology for geologists unfamiliar with the geology of the region. Many of the sites have been described, at least in part, in guidebooks produced by regional organizations. These previous guidebook descriptions have been cited in the site articles to aid further investigation. The value of this field guide lies, to no small extent, in its availablity to a wide audience and its presence in libraries across the continent. A special attempt was made to ensure that most major urban areas in the region have sites in or near them so that visitors with short stays can explore at least some aspect of the local geology.

Topically, 15 of the sites deal with Quaternary Geology (glacial geology, geomorphology, and coastal geology) with the remainder treating bedrock. Most of the bedrock sites illustrate some aspect of the Precambrian-through-Mesozoic history of the region; three bedrock sites are significant in large part for their economic geology. As shown in the index maps at the front of this volume, the areal distribution of sites is good. They are arranged by state and province, from south to north, with ordering within each state or province from west to east. This arrangement should assist users coming from the rest of the continent to locate sites as they travel by automobile into the Northeastern Section. One effect of this arrangement is the separation of the United States portion of the guide from that for eastern Canada, which should not be interpreted as suggesting that the geology of the two countries is somehow divided. The New England states are also together in the middle of the book. Those interested in tracing some geologic feature (such as the ophiolitic suture within the Appalachian Mountains) or examining sites dealing with a particular general topic (e.g. igneous rocks, glacial geology, etc.) will find the Topical Cross-references for Field Guide Sites in the Contents useful. Unfortunately, some bedrock domains, such as the allochthonous Precambrian massifs along the western margin of the Appalachian orogen in the Hudson Highlands, the Berkshire and Green Mountains in New England, and the Long Range of Newfoundland, are not represented or are under-represented in the guide. In most cases the absence of sites within certain terranes is usually the result of a lack of volunteers despite considerable effort; in a few cases, it results from the failure of an author to produce an article on a site that was accepted for the book.

In addition to obvious contributions by the 120 authors and coauthors, many people have helped to make this field guide come together before the end of the Decade of North American Geology. The following were particularly helpful in selection of the sites from the almost 200 nominations: Gail M. Ashley, Michael Bell, Harold W. Borns, Leslie R. Fyffe, Richard Goldsmith, Douglas R. Grant, David Harper, O. Donald Hermes, Donald M. Hoskins, Yngvar Isachsen, Robert R. Jordan, Paul F. Karrow, J. Duncan Keppie, John B. Lyons, Allan Ludman, Rolfe S. Stanley, Walter E. Trzcienski, Kenneth N. Weaver, and Harold Williams. The work of Robert D. Collins was essential in the processing of manuscripts that soon began overwhelming the editor. The Department of Geology and Geophysics of Boston College provided important support without which the work of the editor would have been impossible. The Management Board of the Section provided funds to support the Field Guide Project.

A special acknowledgment is here made to the many individuals who nominated sites that were not selected. The support of the Section's DNAG effort indicated by the efforts put into these nominations is greatly appreciated.

A final word of caution. Most of the sites contain stops that are on private property. Publication of a description of the stop in this field guide in no way constitutes approval by owners for access by visitors. Users must obtain permission on their own for each visit. In addition, the stops of each site present certain levels of risk that are borne by the visitor and not by land owners, authors, or the Society.

Visit the sites with care and enjoy!

David C. Roy
Chestnut Hill, Massachusetts
December 29, 1986

Geological Society of America Centennial Field Guide—Northeastern Section, 1987

Active and abandoned incised meanders of the Potomac River, south of Little Orleans, Maryland

Michael F. Fitzpatrick, Bureau of Land Management, P.O. Box 631, Milwaukee, Wisconsin 53201

LOCATION

The incised meanders of the Potomac River are located between Paw Paw, West Virginia, and Little Orleans, Maryland, approximately 17 mi (27 km) southeast of Cumberland, Maryland, and 20 mi (32 km) southwest of Hancock, Maryland (Fig. 1). These meanders are shown on the Paw Paw, West Virginia, Topographic Quadrangle. The northernmost meander is accessible in dry weather by automobile from U.S. 40, about 6 mi (10 km) south on Orleans Road to Little Orleans, then 2.7 mi (4.4 km) west on Oldtown Road to Orleans overlook (Stop 1). An abandoned channel of Fifteen Mile Creek is present immediately south of Little Orleans. The Reckley Flat abandoned meander now occupied by Purslane Run is accessible via Maryland 51 or West Virginia 9. The Reckley Flat meander is best seen from Maryland 51 (Stop 2). The high terrace cut by the Little Cacapon can be seen from Moser Avenue 0.7 mi (1 km) southwest of its junction with West Virginia 9 in Paw Paw (Stop 3). Caution is required, however, since this road is narrow and has few places to stop. Orleans overlook and the Reckley Flat meander are also accessible to each other in dry weather via Oldtown Road and Thomas Road. A panoramic view of the Potomac valley meanders is visible from Bannecs overlook (Stop 4) atop Town Hill halfway between Little Orleans and Maryland 51.

The valley between Town Hill and Sideling Hill is a broad anticlinorium with minor anticlines and synclines, all striking northeast. The 3,172-ft-long (967 m) Paw Paw tunnel on the C & O Canal, built between 1848 and 1850, cuts diagonally through the axis of one of these smaller anticlines. The anticlinal structure is best seen at the north portal of the tunnel and is accessible by walking north from the parking area off Maryland 51, along the canal, and through the tunnel (Stop 5).

SIGNIFICANCE

Near Paw Paw, West Virginia, the Potomac River flows parallel to the strike of the regional geologic structure of the Appalachian Mountains and passes through a series of exceptionally large, well-formed incised meanders with sub-parallel limbs. These meanders are a good example of the elongate rectilinear meanders that are typical of many Appalachian streams, and are structurally controlled at least in part by weaknesses in the bedrock. The meander farthest upstream in this series has been cut off and abandoned, not by lateral corrasion of the meander neck but by meander elongation breaking into the neighboring Little Cacapon River valley to provide a more direct route.

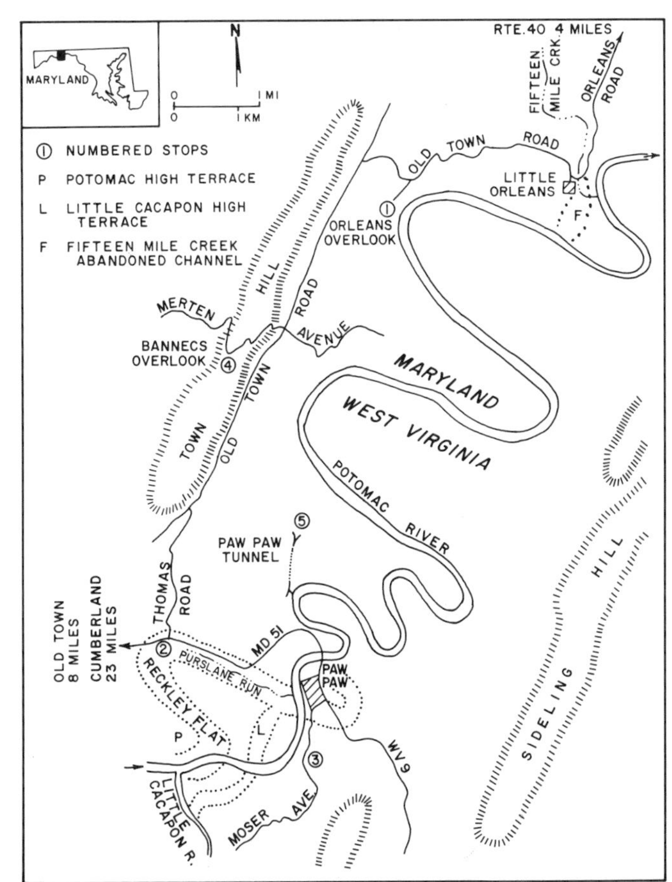

Figure 1. Sketch map showing the meanders of the Potomac River between Paw Paw and Little Orleans. Note also the NE–SW trend of the major ridges.

SITE INFORMATION

Incised meanders are common throughout the Appalachian Valley and Ridge Province. Typically these valley meanders are elongate and deeply incised in the bedrock, have parallel to sub-parallel limbs, nearly regular wave lengths, a rock bed, and relatively cohesive alluvial banks with a narrow flood plain (Hack and Young, 1959; Braun, 1976). A good example of these meanders is located near Little Orleans, Maryland (Fig. 1), where the Potomac passes through a series of well-formed rectilinear meanders that are incised in bedrock composed primarily of shales and thin platy sandstones of the Hampshire, Chemung, Parkhead, and Woodmont formations of upper Devonian age.

The origin and significance of river meanders incised in bedrock has been debated by various authors for nearly a century. It had been believed by some that the incised meanders were

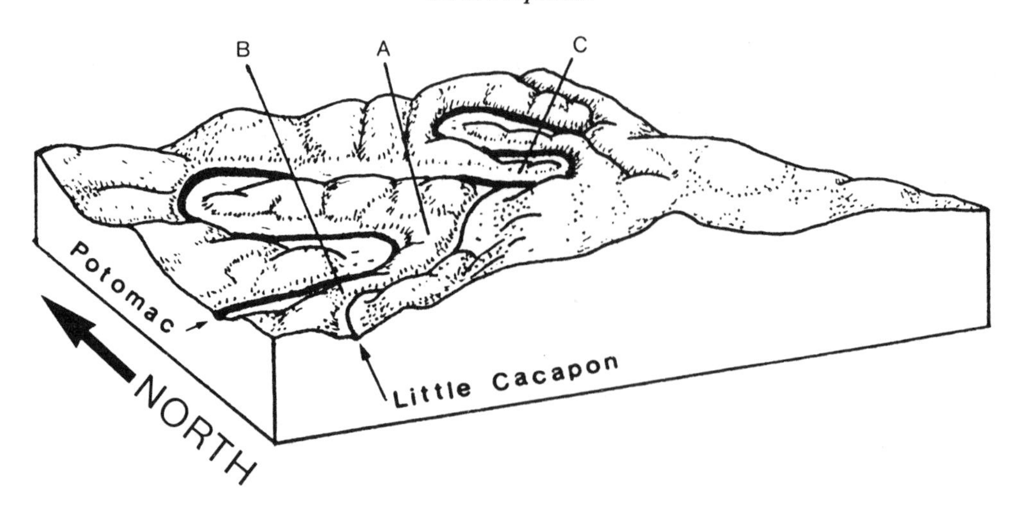

Figure 2. Block diagram showing postulated positions of the Potomac and Little Cacapon Rivers prior to the Reckley Flat meander abandonment. Points A, B, and C are locations where the rivers have broken through their divides. At A the Potomac breaks into the valley of the Little Cacapon and abandons Reckley Flat; B, the Little Cacapon cuts into the Potomac; and C, the combined Potomac–Little Cacapon cuts off the Paw Paw meander (from Fitzpatrick, 1980).

inherited from alluvial floodplain meanders that became entrenched by renewed downcutting of the stream (Davis, 1906; Butts, 1940). Whether this was accomplished by superposition or antecedence, the result was the same (Leopold and others, 1964). An alternative view was that the meanders developed as the stream cut its valley. In this case the river is thought to cut laterally as well as vertically so that a meandering pattern evolves as the valley is cut. According to this second hypothesis, asymmetric valley walls, steeply sloping on the outside of meander bends and gently sloping inside, are thought to indicate developing first-cycle meanders, whereas symmetrical and usually steep valley walls on both sides of meander bends are thought to indicate second-cycle, inherited meanders (Thornbury, 1969; Morisawa, 1968).

Hack and Young (1959) believed that incised rectilinear meanders along the Shenandoah River in Virginia are controlled by structural weakness in the shale bedrock and are not relict forms inherited from a previous erosion cycle. They also believed that the regular wavelength was a function of the discharge or width of the stream. Carlston (1964) expanded upon this idea, stating that "original discharges of entrenched meandering streams can be reconstructed by comparing their wave lengths with the discharges associated with equivalent wave lengths of modern free meanders." His calculations, based on streams throughout the central and northern Appalachians, indicated a decrease of 80% in runoff and discharge since the Pleistocene.

Dury (1964, 1965) also considered that the large scale incised meanders were cut by larger discharges earlier in the Pleistocene. He calculated that incised meanders on the nearby North Fork Shenandoah River are five times larger than alluvial meanders and require a discharge 25 times that of present to cut them.

Tinkler (1972) on the other hand believed that the lithological environment of a channel exerts a controlling influence on channel meander wave lengths and concludes that "it is increasingly evident that meander parameters are not unambiguous indicators of Pleistocene climates." Braun (1982) supports this finding and states that "this lithologically controlled variation in incised meander dimensions indicates that such dimensions cannot be used to uniquely determine paleohydrologic or paleoclimatologic conditions and that bedrock meanders and alluvial meanders need not be geometric or hydraulic homologies of each other." Prior to this, Braun (1976) found an excellent correlation between rock structure and meander planform. Although this correlation holds for a variety of rock types, it is most perfectly developed on rock types with a thin-platy structure. He also noted a "progression in the perfection of the rectilinear form from alluvial meanders to the incised meanders of the Valley and Ridge," and concludes, "The effect of rock structure in creating the elongate rectilinear shape is secondary rather than primary, even though an excellent correlation of structural and meander geometries exists."

Throughout the Valley and Ridge, incised meandering reaches of streams are common, especially where streams flow parallel to the regional structure and have cut into thin-platy rock. Abandoned or cut off meanders are also abundant, although not as common as such features in alluvial rivers. Incised meander loops are often more than twice the size of those along alluvial meanders on streams of equivalent discharges. Nevertheless, they appear to cut off in a similar way. This is accomplished by the migration and intersection of the upstream and downstream limbs of a meander. However, the cutoff of a meander loop by lateral corrasion of the meander neck is understandably slower for a meander incised into bedrock than for an alluvial one.

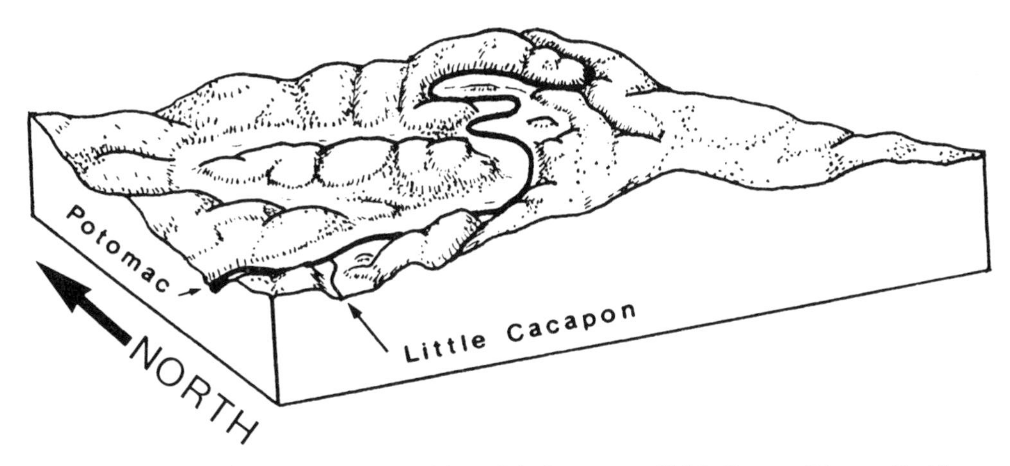

Figure 3. Block diagram of the present positions of the Potomac and Little Cacapon Rivers at Reckley Flat (from Fitzpatrick, 1980).

Indeed, the very process of incision may retard cutoffs and allow more time for meanders to elongate. Alluvial rivers often cut off when overbank flood water crosses the neck of the meander. Incised meanders, because of the height of the meander necks, cannot be cut off by this method (Braun, 1976).

The Reckley Flat abandoned meander (Fig. 1) is geometrically anomalous in that the positions of the limbs of this meander suggest that lateral corrasion of the meander neck is not the method by which the Potomac River changed its course here. The strath or bedrock terrace at the 700-ft (213-m) elevation across the former meander neck shows that a valley had been cut across the neck before the river abandoned the lower elevation through Reckley Flat (Fitzpatrick, 1980). Shaw (1911) studied the terraces and abandoned valley segments along the Allegheny River in western Pennsylvania. The terraces lie approximately 300 ft (91 m) above the present river throughout the length of the Allegheny and are composed of as much as 125 ft (38 m) of alluvium over bedrock. Not all of the abandoned valley segments are meanders with constricted necks; in places there are multiple abandoned channels, all at an elevation approximately 300 ft (91 m) above the present Allegheny. Shaw concluded that due to continental glaciation, the Allegheny River aggraded its channel. As deposition took place, the level of the channel rose, and the river overflowed low places in former divides and split into multiple channels. When the glacier retreated and the river renewed downcutting, in most cases it became established along the shortest route whether or not this route was its original course. In this manner it succeeded in superimposing itself across old divides and abandoning former courses. Referring to a specific abandoned meander Shaw states, "The length, depth, and narrowness of the rock channel through which the river now flows across the neck of the oxbow suggests that the oxbow was not cut off in the way that streams ordinarily cut off their meander, but points rather to superimposition."

Stose and Swartz (1912) noticed the anomalous geometry of the Reckley Flat meander and assumed that it formed in the same manner that Shaw suggested. This supposition did not account for the fact that the Potomac was not affected by debris supplied by the Pleistocene glaciers, and the terraces along the Potomac are strath or bedrock terraces, not constructional alluvial terraces.

While studying this meander, Braun (1971) and Fitzpatrick (1980) concluded that the Potomac was actually pirated by a tributary stream, notably the Little Cacapon River. This smaller stream had previously established its valley across the meander neck (Fig. 2). It was the Little Cacapon that cut the 700 ft (213 m) terrace across the meander neck while the Potomac was still occupying Reckley Flat (Fig. 1). Gravel on this terrace contained only lithologies carried by the Little Cacapon and not the wider variety of lithologies carried by the Potomac. Through lateral corrasion of its valley wall, the Potomac first broke through into the Little Cacapon valley at point A of Figure 2. The Little Cacapon and Potomac intersected later at point B. Once the Reckley Flat meander was abandoned, the full force of the Potomac began to undercut the Paw Paw meander neck at point C, leading to accelerated neck narrowing and abandonment of this smaller meander (Fig. 3).

Thus the Reckley Flat meander was abandoned by valley intersection, the Paw Paw meander by the pinching off of the meander neck. The abandonment of the Fifteen Mile Creek channel south of Little Orleans also occurred from the lateral intersection of the incised meanders of the Potomac River and the creek (Fig. 1). Though the Reckley Flat and Little Orleans channels are similar in form to several abandoned valleys in western Pennsylvania, their origin is quite different.

REFERENCES CITED

Braun, D. D., 1971, Elongate incised meanders in the Appalachian Ridge and Valley Province [M.S. thesis]: Baltimore, Maryland, The Johns Hopkins University, 50 p.

——, 1976, The relation of rock resistance to incised meander form in the Appalachian Valley and Ridge Province [Ph.D. thesis]: Baltimore, Maryland, The Johns Hopkins University, 203 p.

——, 1982, Incised meanders in the Appalachian Valley and Ridge Province; Underfit, misfit, or misunderstood?: Geological Society of America Abstracts with Programs, v. 14, nos. 1–2, p. 7.

Butts, C., 1940, Geology of the Appalachian Valley in Virginia: Virginia Geological Survey Bulletin 52, 568 p.

Carlston, C. W., 1964, Free and incised meanders in the United States and their geomorphic and paleoclimatic implications [abs.]: Geological Society of America Special Paper 76, p. 28–29.

Davis, W. M., 1906, Incised meandering valleys: Geological Society of Philadelphia Bulletin, v. 4, p. 182–192.

Dury, G. H., 1964, Principles of underfit streams: General theory of meandering valleys: U.S. Geological Survey Professional Paper 452-A, 76 p.

——, 1965, Theoretical implications of underfit streams; General theory of meandering valleys: U.S. Geological Survey Professional Paper 452-C, 43 p.

Fitzpatrick, M. F., 1980, Origin of the present course of the Potomac River near Paw Paw, West Virginia [M.S. thesis]: Terre Haute, Indiana State University, 73 p.

Hack, J. T., and Young, R. S., 1959, Intrenched meanders of the North Fork of the Shenandoah River, Virginia: U.S. Geological Survey Professional Paper 354-A, 10 p.

Leopold, L. B., Wolman, M. G., and Miller, J. P., 1964, Fluvial Processes in Geomorphology: San Francisco, W. H. Freeman and Company, 522 p.

Morisawa, M., 1968, Streams; Their dynamics and morphology: New York, McGraw-Hill Book Company, 175 p.

Shaw, E. W., 1911, High terraces and abandoned valleys in western Pennsylvania: Journal of Geology, v. 19, p. 140–156.

Stose, G. W., and Swartz, C. K., 1912, Paw Paw–Hancock Folio, Maryland, West Virginia, Pennsylvania: U.S. Geological Survey Geologic Atlas of the United States, Folio 179, 24 p.

Thornbury, W. D., 1969, Principles of geomorphology (second edition): New York, John Wiley and Sons, Incorporated, 594 p.

Tinkler, K. J., 1972, The superimposition hypotheses for incised meanders; A general rejection and specific test: Area, v. 4, no. 2, p. 86–91.

The Silurian section at Roundtop Hill near Hancock, Maryland

John D. Glaser, Maryland Geological Survey, 2300 St. Paul Street, Baltimore, Maryland 21218

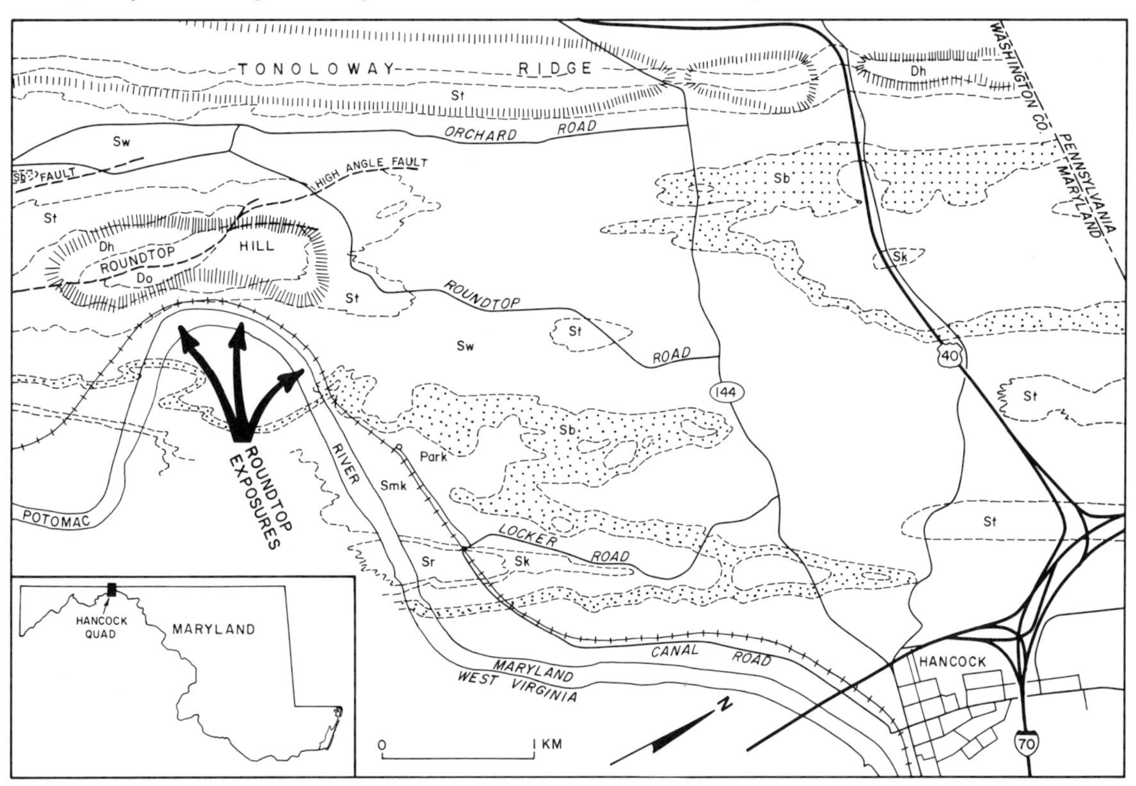

Figure 1. Index map showing location of Roundtop exposures. Sr = Rochester Shale; Sk = Keefer Sandstone; Smk = McKenzie Formation; Sb = Bloomsburg Formation; Sw = Wills Creek Formation; St = Tonoloway Limestone; Dh = Helderberg Limestone; Do = Oriskany Formation.

LOCATION

The Western Maryland Railway cut at Roundtop Hill lies in the Valley and Ridge Province of Maryland, midway between Cumberland and Hagerstown, in Washington County (Fig. 1). The cut can be conveniently reached from I-70 if approaching from the east or north, or from U.S. 40 if approaching from the west. In either case, one must take the Hancock exit and drive south into the town of Hancock, a distance of less than 1 mi (1.6 km). Once in town, two choices of a route to the outcrop are presented. One can elect Canal Road, reached from the center of town by turning south at the traffic light and crossing the railroad tracks (Fig. 1), or one can choose Maryland 144 west to Locker Road and turn south. Either route brings one to a short, unimproved lane that runs west along the railroad tracks for about 0.4 mi (0.7 km) to a small parking area maintained for hunters by the Maryland Department of Natural Resources. The outcrop begins about 0.2 mi (0.4 km) farther west along the tracks.

At the present time, the tracks of the Western Maryland Railway in the vicinity of Hancock are not in use, and there are plans for their eventual removal. Therefore, no permission is required. However, in the unlikely event that these plans are changed and the tracks become active, permission to view the cuts should be sought from the Chessie System offices in Baltimore.

SIGNIFICANCE

The Silurian outcrops at Roundtop, comprising the Bloomsburg and Wills Creek formations, provide the best section of folded rocks in Maryland and are certainly among the finest in the entire Appalachian region. Cloos (1951, 1958) spoke in glowing terms of the Roundtop exposures as a veritable laboratory of the mechanics of folding, formation of cleavage, thrusting, and the deformation of bedded rocks. Specifically, Geiser (1974) pointed to Roundtop as an excellent illustration of the role of pressure solution in the generation of cleavage. Much of the value of the section lies in the fact that deformation at this locality has progressed only far enough to show early structures, but not to the point of obliterating sedimentary features. The importance, then, of the Roundtop section is its prominence as an example of typical Valley and Ridge structural style, and the general excellence of the minor structures displayed.

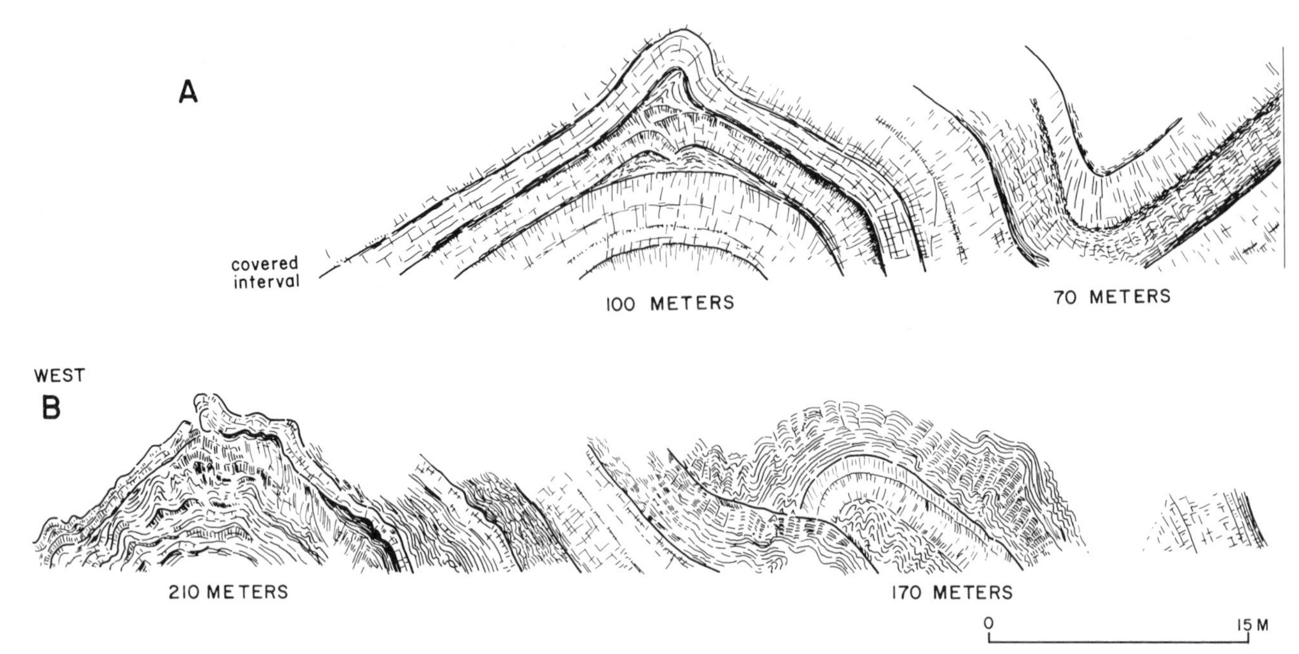

Figure 2. Sketches showing details of exposures in cuts along the Western Maryland Railway at Roundtop Hill. A, easternmost cut, this and facing page; B, western cut.

SITE INFORMATION

The Valley and Ridge Province of the central Appalachians displays a characteristic structural signature, marked by closely spaced parallel folds and faults with remarkable uniformity along strike. The Cacapon Mountain anticlinorium is one such fold, and it is within the eastern limb of this fold that the Roundtop section is located. Deformation here has been guided by local conditions such as the relative competence of beds, the frequency of failure through low-angle thrusting, and lithologic control of cleavage. This deformation plan is in sharp contrast both with that of the South Mountain anticlinorium, 37 mi (60 km) eastward where a regional penetrative deformation has been imposed over a broad area, and with that of the gently deformed to nearly flat-lying beds typical of the Appalachian Plateau to the west.

The Silurian rocks in the Cacapon Mountain anticline are exposed in a belt about 2.5 mi (4 km) wide between Cove and Tonoloway Ridges, and comprise nearly 1,900 ft (600 m) of strata of which only about 500 ft (160 m) can be seen in the Roundtop cuts. The Wills Creek Formation is exposed in its entirety and dominates the section, but the underlying Bloomsburg, although much thinner, is structurally significant and displays many tectonic features not seen in the other rocks.

The Cacapon Mountain structure, a surface reflection of an underlying ramp anticline involving the Cambro-Ordovician carbonates, is an asymmetric box fold with a gentle northward plunge. Bedding in the west limb rotates from nearly horizontal to steeply dipping over a relatively short distance. The crest of the anticline has collapsed into a large faulted syncline at Roundtop Hill, and the outcrops here display some of the most complex structure in the entire Cacapon Mountain fold. There are at least ten well-exposed third-order folds visible in outcrop along the hill. For the most part, the folds are open, and competent beds typically have flexure folds showing constant bed thickness and bedding-plane slippage. This type of folding is characteristic of sandstones and coarse siltstones in the Bloomsburg redbeds. In contrast, the weaker shales and thin-bedded limestones of the Wills Creek Formation are abundantly sheared and thickened into the fold crests and display intense subparallel flow cleavage.

Cleavage, as a persistent plane of parting, is present in nearly all of the rocks of the Roundtop section, but it is most conspicuous in the shales and siltstones of the Bloomsburg and lower Wills Creek formations. In these rocks, one can see an array of prominent subparallel irregular surfaces with a spacing of 1 to 2 cm, generally normal to bedding, and fanning through a large angle around the axes of folds. Geiser (1970, 1974) views this structure, historically identified as "fracture cleavage," as the product of pressure solution and not the result of brittle failure. He examined minor structures in considerable detail in both the Bloomsburg and Wills Creek rocks and found two generations of cleavage. The first (S_a) is uncommon, has a constant orientation with the plane of bedding (always an acute angle), and is selectively developed in shales bracketed by more competent beds in both the Bloomsburg and Wills Creek formations. S_a is a regional structure imposed prior to flexural slip folding. The second (S_p), formed by pressure solution subsequent to lithification as a product of bedding-parallel compression, is more widespread. S_p predates finite amplitude folding and is locally penetrative.

The Roundtop area is crossed by several large-scale north-south oriented high-angle faults (Fig. 1), but none of these are

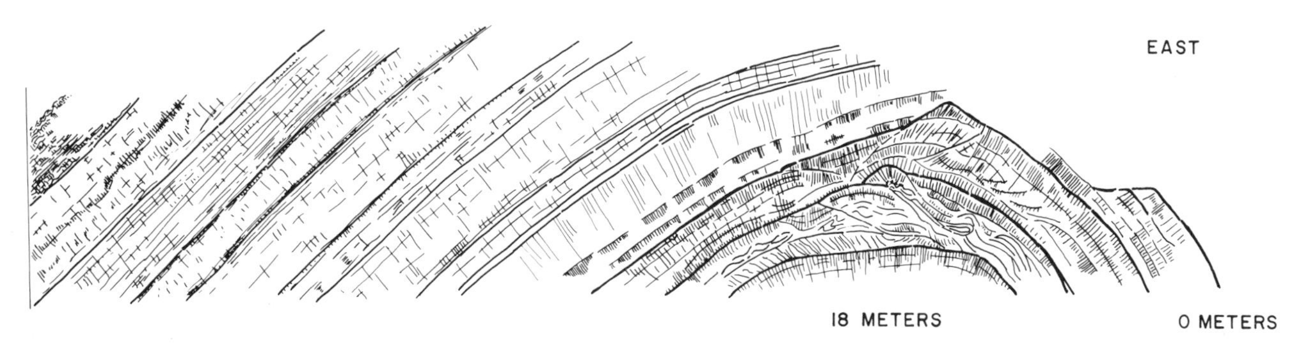

exposed in the railway cuts. Small-scale curviplanar thrust faults are quite common and conspicuous in outcrop. Bedding planes were usually the surfaces along which movement occurred, but in many instances, the thrust surfaces cut through beds at very low angles. Cloos (1964) coined the term "wedges" to describe the wedge-shaped blocks bounded by bedding planes and the low-angle curviplanar thrust surfaces so common in the Roundtop section. The wedges occur at all scales, in beds 6 to 10 ft (2 to 3 m) thick as well as those only a few cm thick. The host rocks are nearly always massive siltstones or coarse shales in the Bloomsburg Formation, but some wedges occur in the Tonoloway limestones. Careful measurements and stereo net plots by Geiser (1974) showed that the wedges in most cases have a westward sense of transport and occur on both the limbs and crests of folds. This evidence tends to support the earlier conclusion of Cloos that wedging took place in response to regional bedding-parallel compression prior to folding. Geiser (1974) also hypothesized that the wedging postdated the formation of S_p cleavage since cleavage is rotated by motion on the wedges. The majority of "wedge slips" examined resulted in thickening of the faulted beds, but exceptionally, thinning can be observed. Thinned beds were termed "lags" by Geiser and were ascribed to local movements concurrent with folding. Cloos visualized the overall process as a crowding of material outward from the limbs of the folds with a multitude of small thrusts serving as movement planes.

Approaching the exposure from the east, the first cut reveals a complexly faulted anticline (Fig. 2A, at 18 m) in Bloomsburg red shale and sandstone. Of major interest here is the behavior of the thin sandstone bed in the lower portion of the section that has been telescoped and repeated by thrusting as much as six times. The associated shales show excellent fanning cleavage. Some of these shale beds, particularly the laminated and crenulated greenish bed above the sandstone, have clearly served as planes of slippage. It is also worth noting that the degree of deformation in this outcrop is too severe to be related only to the anticline in this cut, and moreover, the telescoping of beds is asymmetric with respect to the axial plane of the fold. Rather, such movements may be accommodations within the much larger Cacapon Mountain fold, or even be due to bedding plane slippage caused by gravity sliding or slumping prior to lithification and folding.

Proceeding westward, one passes steadily upward through evenly dipping beds of soft, poorly-consolidated mudstone and shale of the Wills Creek Formation, and about 164 ft (50 m) beyond the first anticline encounters a tight syncline with conspicuous axial-plane cleavage, some rotated blocks on antithetic faults, and low-angle thrust faults in the west limb (Fig. 2A, at 70 m). About 100 ft (30 m) farther west and downsection is a second anticline (300 ft; 100 m) which shows a textbook example of the contrast in response between competent and incompetent beds during tight folding. The less competent shales have flowed into the fold crest, whereas the more competent red siltstone has thickened into the crest by repeated low-angle thrust faults. The sliding of competent beds past one another along bedding planes to accommodate the growing fold produced well-defined bedding plane striations. Beyond the second anticline, and covered at track level, are about 160 ft (50 m) of west-dipping calcareous siltstone, mudstone, and limestone of the Wills Creek Formation. These rocks show cleavage normal to bedding, and, in part, at a low angle to bedding as a result of rotation due to bedding plane slippage. Visible at about this point, several meters above track level, is a partially collapsed tunnel excavated in thin-bedded argillaceous limestone that has the approximate composition of cement. Other tunnels were diven here at canal level, and the site is marked by a ruined cement kiln in operation at the turn of the century. Just west of the tunnel is a tight syncline followed by a short covered interval.

About 50 ft (15 m) farther along, the section resumes with an asymmetric, faulted anticline that has been thrust westward at a low angle over the broken west limb of the fold (Fig. 2B at 170 m). The thrust plane is clearly shown in the outcrop, as is small-scale crumpling of the thin-bedded limestones involved in the anticline. As noted before, the siltstones in the section are the least deformed lithology but show the strongest cleavage. Walking west along the tracks for another 130 ft (40 m) and downsection through east-dipping Wills Creek strata, we encounter another anticline (Fig. 2B, at 210 m) involving thin-bedded limestones and calcareous siltstones. This fold is essentially symmetrical but the limbs are crowded with minor folds, developed partly in response to flexural slip. Some earlier workers termed these folds "drag folds," but Geiser (1974) believes that some of them are really a type of conjugate kink fold formed during the final stages of deformation. These kink folds are most apparent in limbs that show reversal of movement sense in consecutive beds.

A second tunnel can be seen high on the slope from a point 100 ft (30 m) farther west along the tracks, beyond the exposures shown in Figure 2B. The rocks here have a steep westward dip. Over the succeeding 200 ft (60 m), the dip remains westward but becomes less steep and passes into a very tight syncline in which beds were conspicuously crowded toward the fold axis with attendant crumpling and small-scale thrusting.

Another 60 ft (20 m) brings one to a paired anticline and syncline, both with steep limbs marked by thin crumpled limestone beds and thrust displacement toward the axes of the folds. The thick calcareous shale at track level displays excellent fan cleavage. A few paces farther west, about midway in the railway cut, is an excellent example of desiccation cracks in a light greenish gray shaly limestone. Just beyond the desiccation cracks is the entrance to the third and largest cement mine, in this case excavated into the core of a large tight anticline. The tunnel is steeply inclined and opens into a large chamber.

The next 300 ft (90 m) of the exposure consist of uniformly dipping mudstone, shale, and thin-bedded limestone and sandstone. In places, the limestone beds are selectively folded. At the 300-ft (90-m) point is a small anticline marked by small-scale folding and telescoping in the thin rusty sandstone beds that make up the core of the fold. About 25 ft (8 m) beyond the anticline is a thick bed of green silty shale with conspicuous cleavage at an acute angle to bedding. This is a good example of Geiser's S_a cleavage. Thirty ft (10 m) farther along, S_a cleavage is very well developed in shaly beds involved in a fairly open syncline.

The dip reversal on the west limb of the syncline repeats the foregoing section for the next 700 ft (60 m) of traverse. The beds dip eastward, steeply at first, then more gently as the next anticline is approached. Between these two folds, further examples of S_a cleavage at an angle to bedding can be seen in a thick steeply-dipping green shale bed a few feet beyond the syncline. At this site, the S_a cleavage has been folded. A few paces farther along, a sequence of thin fine-grained sandstone beds shows considerable minor folding with many beds broken and thickened by wedges. Similar structures are present in places over the next 130 ft (40 m) of otherwise homoclinal section until a large open anticline, with prominent fan cleavage in the thick shaly beds, is encountered.

The section can be followed for about another 1,400 ft (440 m) before the exposure ends, but the beds dip uniformly west, and no further structures of interest are to be seen. All of the rocks in this homoclinal sequence are within the Wills Creek Formation. Limestone visibly increases in abundance as the contact with the overlying Tonoloway Formation is approached.

REFERENCES CITED

Cloos, E., 1951, Stratigraphy and structural geology of Washington County, Maryland, *in* Physical features of Washington County: Maryland Department of Geology, Mines, and Water Resources, p. 124–161.

—— , 1958, Structural geology of South Mountain and Appalachians in Maryland (Guidebook 4 and 5, Johns Hopkins University Studies in Geology no. 17): Baltimore, Maryland, Johns Hopkins Press, p. 1–76.

—— , 1964, Wedging, bedding plane slips, and gravity tectonics in the Appalachians: Virginia Polytechnic Institute Department of Geological Sciences, Memoir 1, p. 63–70.

Geiser, P. A., 1970, Deformation of the Bloomsburg Formation in the Cacapon Mountain Anticline, Hancock, Maryland [Ph.D. thesis]: Baltimore, Maryland, Johns Hopkins University, 189 p.

—— , 1974, Cleavage in some sedimentary rocks of the central Valley and Ridge Province, Maryland: Geological Society of America, Bulletin, v. 85, p. 1399–1412.

Maryland's Cliffs of Calvert: A fossiliferous record of mid-Miocene inner shelf and coastal environments

Peter R. Vogt and Ralph Eshelman, *Calvert Marine Museum, P.O. Box 97, Solomons, Maryland 20688*

LOCATION

The Cliffs of Calvert (Calvert Cliffs) form a series of wave-cut bluffs up to 100 to 130 ft (30 to 40 m) high, which extend about 25 mi (40 km) along the western shore of the Chesapeake Bay (Fig. 1). For brevity, we refer to sites by distance in miles (km) either + (north) or – (south) of Governors Run, and keyed to Figure 2.

Maryland 2/4 is the main route of access and runs parallel to the cliffs. Most of the Calvert Cliffs have been residentially developed so public access from the land is limited to commercial marinas (CM), public parking (PP), and simple road access (RA; Fig. 1). There are two parks along the cliffs: the Flag Ponds Natural Area and the Calvert Cliffs State Park. The cliffs behind Flag Ponds lack good exposures: those at the Calvert Cliffs State Park can be reached only by a 0.25 mi (0.4 km) footpath. Permission must be obtained for access through private property. Digging into or climbing on the cliffs is dangerous and prohibited. Significant fossil finds such as vertebrate remains should be reported to the Calvert Marine Museum. The best geologic overview of the cliffs is by boat, with landings at selected sites (public beach access in Maryland extends up to Mean High Water).

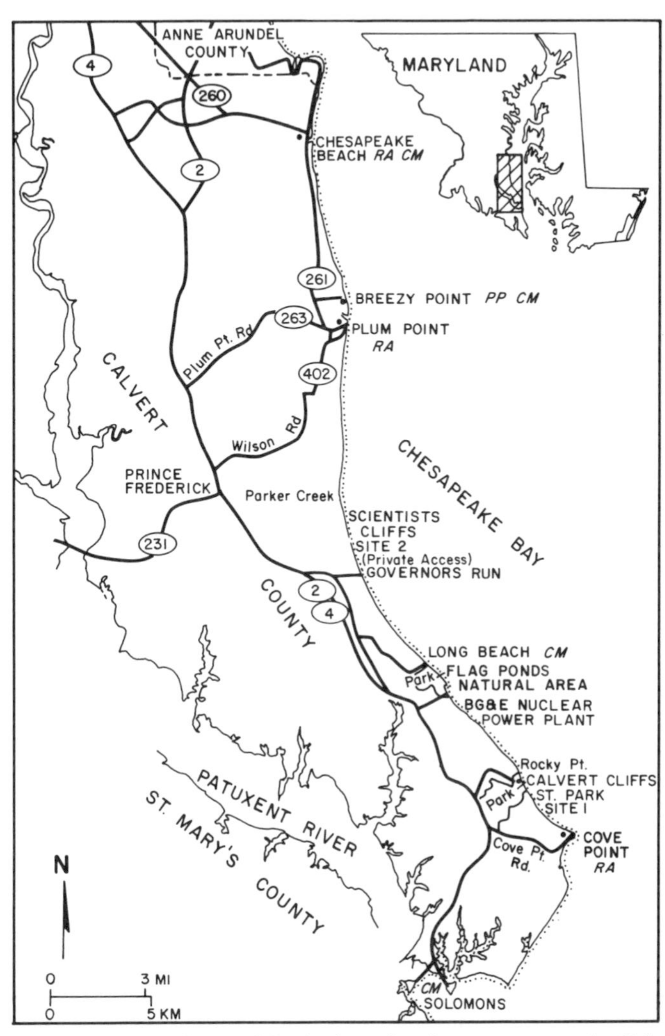

Figure 1. Calvert County, Maryland, road and water access to the Calvert Cliffs: RA, road access: CM, commercial marina: PP, public parking; and PARK, public park. Main road access points are marked by dots. The entire line of cliffs is most easily examined by boat. Upper right: location of the Calvert Cliffs area in southern Maryland.

SIGNIFICANCE

The Calvert Cliffs expose inner shelf to marginal-marine and littoral deposits of the Chesapeake Group (late early to early late Miocene) that are marginal to a major mid-Atlantic depocenter, the Salisbury Embayment. Three formations of the Chesapeake Group are exposed in the cliffs: from oldest to youngest, they are the Calvert, Choptank, and St. Mary's formations, together traditionally subdivided into 24 lithostratigraphic "zones" of which Zones 3 through 23, and the time-equivalent of Zone 24 are exposed in the cliffs. Except for diatomite in the lower Calvert Formation and littoral deposits in the upper St. Mary's Formation, the sediments are variably fossiliferous fine sands, silts, and clays. A slight (ca. 1 or 2:1000) southward dip exposes successively younger strata down dip, offering beach-level examination of all "zones" present. The Miocene marine sediments are overlain by late Miocene or Pliocene fluvial or littoral "Upland Deposits" (UD in Fig. 2) and Pleistocene "Lowland Deposits" (LD) largely of estuarine origin. In southern Calvert County the "Upland Deposits" are littoral facies near the top of the Chesapeake Group. Unconformities within the Chesapeake Group are subtle in outcrop expression in the cliffs.

The Calvert Cliffs have yielded a rich fossil record of local marine and terrestrial organisms, comprising the best available record of mid-Miocene life in the northeastern United States. Biota represented include diatoms, terrestrial plants, and invertebrates, particularly molluscs; and vertebrates including sharks, rays, bony fish, reptiles, birds, cetaceans, seals, and sea cows. Abundant remains of juvenile cetaceans suggest the area was a calving ground. Occasional bones and teeth of land mammals, as well as carbonized wood probably were floated as carcasses or tree trunks into the sea by rivers.

Several 6- to 12-ft-thick (2 to 4 m) mollusc beds outcrop

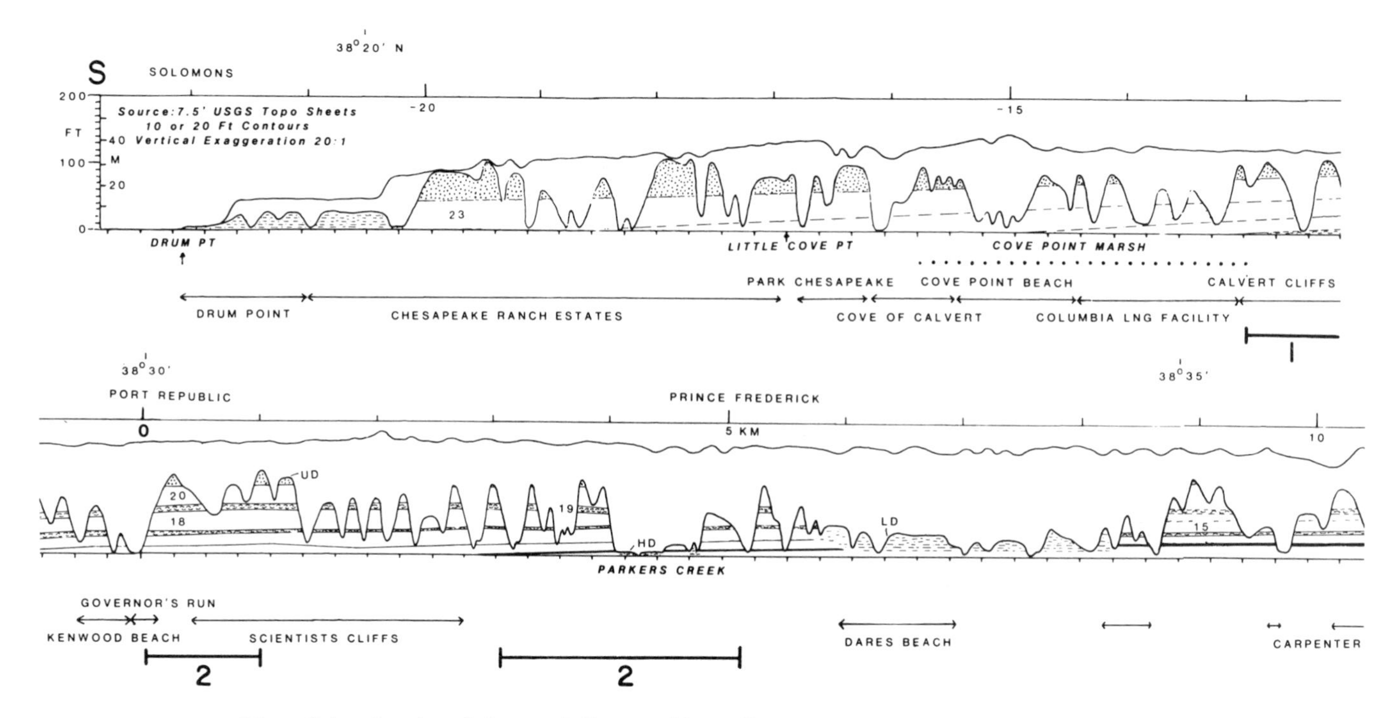

Figure 2 (continued on facing page). Topographic profile constructed along cliff crest from 1:24,000 quadrangle sheets, with generalized geology: Shattuck's "zones" and "post-Miocene" deposits added (UD, upland deposits (stippled); LD, lowland deposits; HD, Holocene deposits). Note that the "UD" deposits at least in southern Calvert County are now considered (McCartan and others, 1985) of Zone 24 age (early late Miocene). The lateral and bottom contacts of the Lowland Deposits have not been mapped in detail. No fault contacts have been identified along Calvert Cliffs. Geology generalized from Shattuck (1904), Dryden (unpublished sections, dated 1965), Gernant (1970, unpublished manuscript), Gernant and others (1971), Glaser (1971), Blackwelder and Ward (1976), and Abbott (1978, 1982). Horizontal scale (top) is in kilometers from Governors Run (0), and close-spaced marks along bottom are 1,000 ft apart. The curve at top shows the highest elevations of Calvert County due west of the cliffs, as derived from quadrangle sheets.

continuously over long distances along the cliffs. These shell concentrations have been variously attributed to storm-generated currents, hydraulic effects of passing swells, abnormally high productivity, or slower sedimentation. The Calvert Cliffs exposures illustrate the interrelationships among sediment substrate (fine sand versus mud), infauna and epifauna, and allochthonous fossil debris. However, the processes ultimately responsible for the quasi-periodic vertical variations in lithology and fossil content remain unknown. The gradual upward shoaling and climatic cooling recorded by the Chesapeake Group sediments and biota may reflect mid-Miocene cooling and high-latitude glaciation. Concurrent uplift of the Appalachian Mountains, uneven subsidence of the Coastal Plain, and eustatic sea level changes may all be recorded in the Chesapeake Group. The Calvert Cliffs offer a single slice through a complex sediment package representing a first-order transgressive/regressive cycle modulated by numerous lesser sea-level perturbations. A full understanding of this package will depend on drill cores and outcrops elsewhere in the region.

INTRODUCTION

The Calvert Cliffs, and a shorter stretch of cliffs (Nomini Cliffs) along the Potomac River to the west are a topographic rarity along the mostly flat and low-lying tidewater shoreline of the U.S. East Coast south of New England. Systematic geologic studies of the Calvert Cliffs that began early in the nineteenth century culminated with Shattuck (1904), who described the Calvert, Choptank, and St. Mary's formations, and divided them into the 24 lithostratigraphic "zones" informally used here. The Baltimore Gas and Electric Company (BG&E) site (km: –8) was examined by the Maryland Academy of Sciences and Smithsonian Institution in great detail in 1968 to 1969 but many results remain unpublished. Gibson (1983), Newell and Rader (1982), and Ward and Strickland (1985) present regional syntheses and references to earlier studies. A new geological synthesis, more complete than is possible in this chapter, is in preparation (Vogt and Eshelman, 1987).

In the formational contact areas, the exact assignment of Shattuck's zones to particular formations has been the subject of much discussion. The assignments in Figure 3 reflect the most recent opinions (McCartan and others, 1985).

Unfortunately only a small fraction of the Calvert Cliffs remain in a natural state, mainly in the State Park, Flag Ponds,

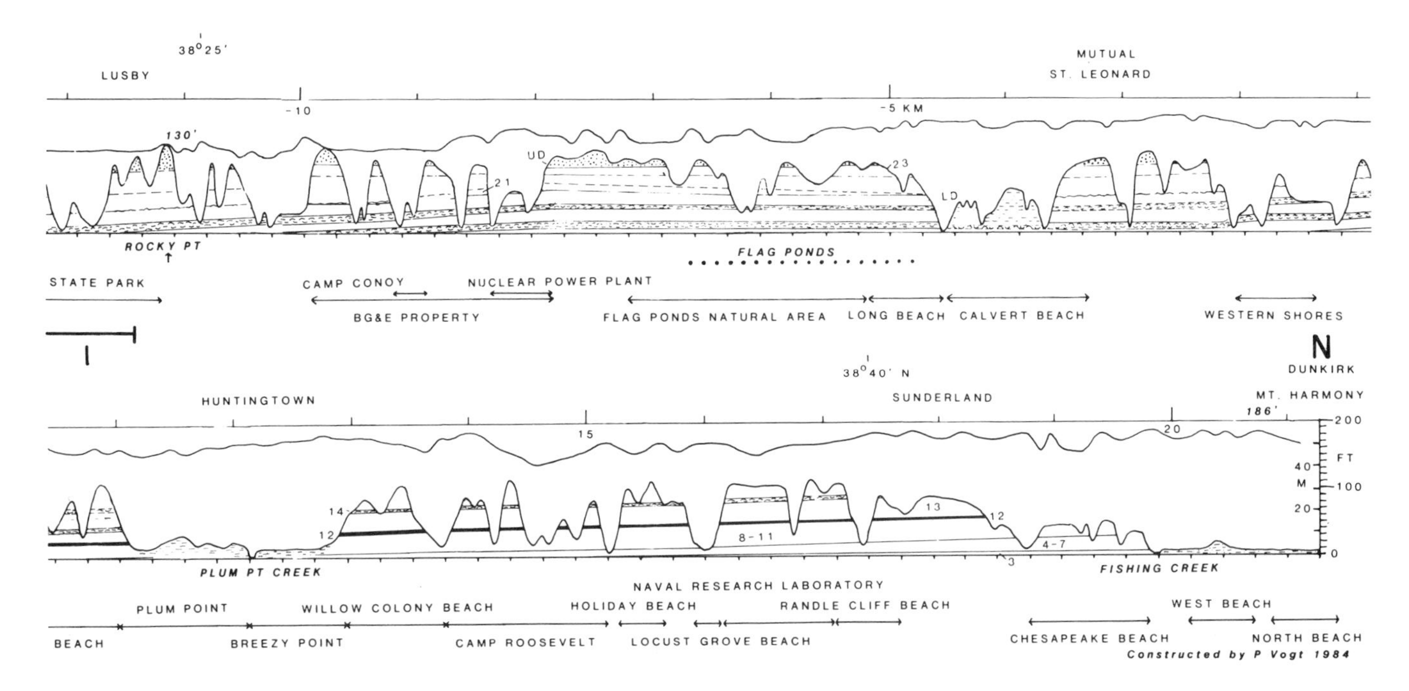

and Parker Green areas, the last of which is privately owned. The Calvert Cliffs are dynamic outcrops in which exposures change from one year to the next: it therefore seems pointless to discuss specific outcrops in great detail.

AGE OF DEPOSITS

The formations and members of the Chesapeake Group are lithostratigraphic units which, especially in slowly transgressing/regressing situations, may change significantly in average age or age range, particularly in landward/seaward directions (e.g., Fig. 2 of Newell and Rader, 1982). The ages given here refer to the Calvert County area, which is small enough for rough litho/chronostratigraphic equivalence. Generally the formations and intra-group contacts young landward, toward the Fall Line.

The age of the Chesapeake Group has long been known as Miocene. The strata were originally subdivided into formations on the basis of macrofossils—specifically, mollusc assemblages. Although global microfossil zonations for marine sediments have improved since the mid-1960s, the Chesapeake Group comprises only shallow water sediments; planktonic foraminifera are sparse in the Calvert and Choptank Formations, and virtually absent in the St. Mary's Formation (Gibson, 1983). However, planktonic foraminifera in Zone 10 allowed Gibson (in Gernant and others, 1971) to assign it to foraminiferal zone N8 or N9, i.e., latest Early Miocene to earliest Middle Miocene. The only other precise age date on the Chesapeake Group is 12.0 ± 0.5 Ma, determined from glauconite recovered at about the Zone 22–23 level (Blackwelder and Ward, 1976). If this date is correct, these zones were deposited in the middle to later parts of the Middle Miocene. Blackwelder and Ward (1976) also observed that "mollusks in the lower part of the Calvert Formation are closely related to those in 'zone' 10," and concluded that the "lower part of the Calvert Formation is not much older than 'zone' 10, despite unconformities. . .". Until rather recently (e.g., Gibson, 1983), the Calvert Formation was considered equivalent to the Langhian Stage (foraminiferal zones N8 and N9). However, subsequent studies of diatoms, found throughout the Calvert and Choptank Formations, indicate that Zone 3 extends from about mid-Langhian (i.e., about the Early/Middle Miocene boundary) down into the Burdigalian Stage (Abbott, 1978, 1982; Andrews, 1978). The lower parts of Zone 3, as well as Zones 1 and 2, are distinctly older, suggesting an unconformity somewhere within Zone 3 (Gibson, 1982; Abbott, 1982). A recent correlation chart for the mid-Atlantic Neogene thus shows the Calvert Formation extending down into the middle of the Early Miocene, or about 20 Ma (Gibson, 1982). However, only the very top of Zone 3 is exposed in the Calvert Cliffs, and this is unlikely to be older than early Langhian. The Langhian/Burdigalian boundary (16.6 Ma) is one of the "anchor points" for the DNAG time scale (Kent and Gradstein, 1986). Thus the absolute age of the oldest sediments exposed in the Calvert Cliffs near Chesapeake Beach is probably about 16–17 Ma. The Calvert/Choptank Formation boundary is about mid-Serravalian (12.5–13 Ma), or perhaps early Serravalian (Gibson, 1982). The Choptank formation was thus deposited during the mid-to-late Serravalian, approximately equivalent in age to foraminiferal zone N12 (based on diatoms; Abbott, 1982). The St. Mary's Formation has not yielded any "datable" microfossils. However, if the one radiometric date (Blackwelder and Ward, 1976) is accepted as correct, the St. Mary's formation (Zones 20–24) was deposited during the late Serravalian to early Tortonian. The absolute age range of the Chesapeake Gorup is then 10–20 Ma, of which the period 10–16 Ma is represented in the Calvert Cliffs. The lengths of possible

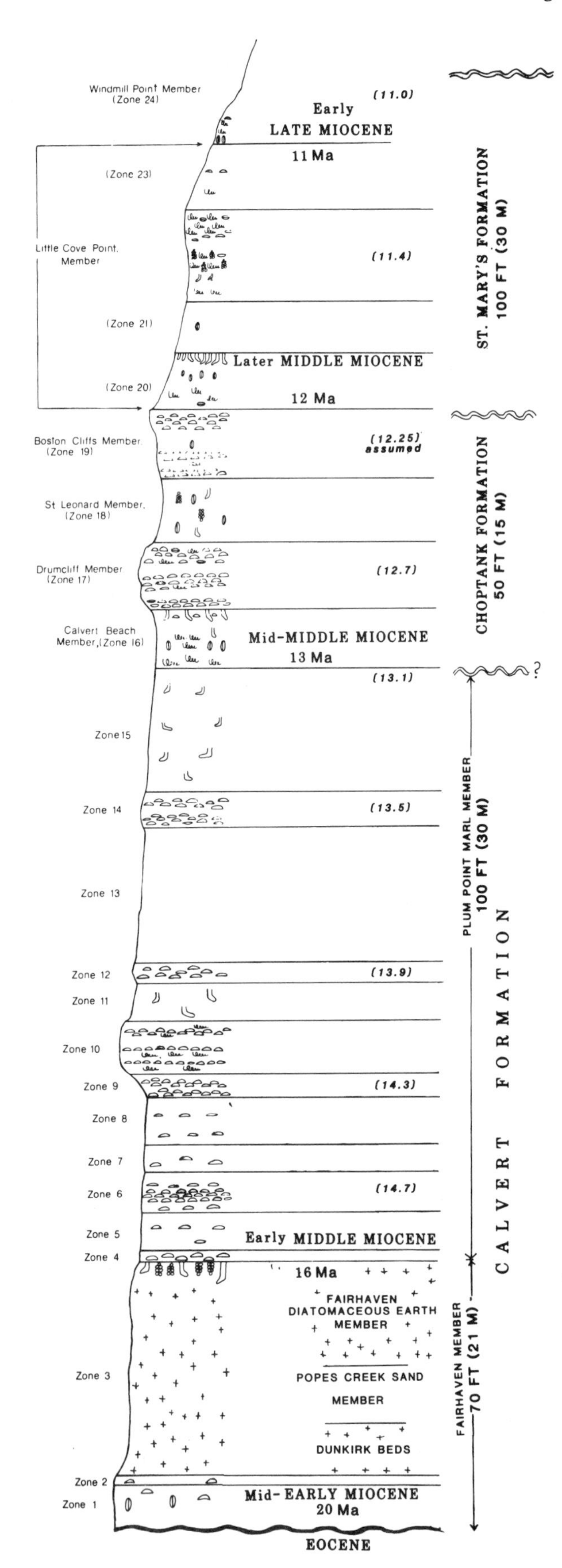

Figure 3. Composite section of beds exposed in Calvert Cliffs (plus Zone 24, not exposed) adapted from Gernant and others (1971). "Zone" designation is from Shattuck (1904). Member designations of "Zones" 16 to 19 are from Gernant and others (1971); and designations of "Windmill Point Member" (Zone 24), "Little Cove Point Member," and "St. Mary's Formation" are from Blackwelder and Ward (1976). Traditionally, "Zones" 21 to 24 were assigned to the St. Mary's Formation (Shattuck, 1904). Recent work (McCartan and others, 1985) eliminates the Little Cove Point Member by extending the St. Mary's Formation down to the Zone 19/20 contact. Stratigraphic and absolute ages are present authors' estimate based on DNAG time scale (Kent and Gradstein, 1986) and available literature. Ages in parentheses are based on hypothesis that conditions favorable for mollusc growth recurred quasi-periodically at about the 413,000-year-long eccentricity cycle and that one shell bed equivalent is missing at each of the two unconformities.

missing intervals due to erosion or nondeposition cannot be reliably estimated. Correlation charts (e.g., Gibson, 1982; Blackwelder and Ward, 1976) show gaps in the range of 0.1 to 1 m.y. at the unconformities. In the cliffs, gaps in some cases could be much smaller but are unlikely to exceed 1 m.y. The long-term average accumulation rate for the Chesapeake Group in this area was about 15 m/m.y.

UNCONFORMITIES

The Chesapeake Group is bounded above and below, and less obviously within, by unconformities. In general there are more unconformities and/or more time "missing" in the up-dip direction (west or northwest, toward the Fall Line). The major unconformity between the Eocene Nanjemoy Formation and the lower Miocene base of the Calvert Formation is not exposed along the Calvert Cliffs, but lies about –6 m (below sea level) at km: 19, and –65 m at km: –8. The diatom biostratigraphy suggests an unconformity within Zone 3 (Abbott, 1982). Several unconformities (essentially disconformities) have been postulated within the Calvert Cliffs exposures, for example, at the boundaries between Zones 3/4, 15/16, 19/20, and 20/21. Identifying and tracing them within the cliffs and elsewhere is a challenge for field investigation. Some may not exist at all, and others may be local phenomena. It is also not always clear whether the unconformities are associated with complete marine regressions followed by transgressions or whether they were formed by submarine erosion.

The Zone 3/4 contact is a slight angular unconformity, the diatomite of Zone 3 having been tilted slightly prior to deposition of Zone 4. Reworked "deep" water oysters (*Pyncnodonte percrassa*), once considered part of Zone 3 (Fairhaven member), form the lowest bed of the Plum Point Marl (Gernant, unpublished ms.). Shattuck (1904) postulated a Calvert/Choptank (Zone 15/16) unconformity largely on the basis of exposures in the type area just south of Parker Creek. Dryden (unpublished measured sections, 1965) traced this unconformity southward almost to Governors Run. His interpretation suggests that Zone 14 thins, and Zone 15 pinches out updip against the unconformity. Gernant (1970) looked for the Zone 15/16 unconformity and remained unconvinced, as have some other investigators. He suggested that Shattuck may have been misled by a leaching-induced color change prominent near the Zone 15/16 boundary. Whereas Shattuck considered the Choptank/St. Mary's contact (Zone 20/21) conformable, Gernant (1970) identified an unconformity there, best revealed near Camp Conoy. Based on sedimentological studies at the BG&E site (km: –8), Shideler and Swift (unpublished ms.) first suggested that the Choptank/St. Mary's boundary should be placed at the Zone 19/20 boundary since the Zone 20/21 boundary is gradational. However, Gernant (unpublished ms.) considered the BG&E site anomalous and therefore placed the Choptank/St. Mary's boundary at the Zone 20/21 boundary. The St. Mary's formation contains extra beds (Zone 21) not present to the north and south (Fig. 2). The youngest intra-Miocene unconformity separates Zone 23, the top of Blackwelder and Ward's "Little Cove Point Member" (Zones 20–23), from the overlying "Windmill Point Member" (Zone 24); since Zone 24 is not exposed in the Calvert Cliffs, neither is the supposed unconformity. However, McCartan and others (1985) considered Zone 24 to be merely the downdip equivalent of the thick unfossiliferous sand—now recognized as Miocene—above Zone 23, near Little Cove Point (km: –16 to –18). This study has reestablished the St. Mary's Formation extending down to a Zone 19/20 unconformity and eliminated the unconformity long thought to separate Zone 23 from the overlying sands.

SHELL BEDS

Some of the more conspicuous features of the Calvert Cliffs are the zones in which shells of mostly disarticulated, flat-lying bivalves are concentrated (Zones 6, 9, 10, 12, 14, 16, 17, 19, and 22; Fig. 3). The finer-grained parts of the Calvert Formation were perhaps inhospitable substrates for invertebrates, which are practically non-existent in these lithologies (Blackwelder and Ward, 1976). Zone 16 includes some macrofauna dominated by articulated burrowers in life position and other accumulations apparently concentrated by currents into tidal channels (Gernant, unpublished manuscript). The conical form of *Turritella* allowed it to roll easily on the sea floor and concentrate in small depressions. The prominent shell concentrations in Zones 10, 14, 17, and 19 have been attributed to high mollusc productivity (Blackwelder and Ward, 1976). However, the disarticulated character of most shells led Gernant (1970, and unpublished manuscript) to suggest they were repositioned and concentrated in the substrate by the hydraulic action of passing swells. Shideler and Swift (unpublished manuscript) attribute the shell concentrations to transient currents set up by tropical storms. However, a block removed from Zone 10 had only two-thirds of the shells in the convex-up position expected from strong currents (Gernant, unpublished manuscript).

Kidwell (1982a, 1982b) recognized shell enrichments on ecologic (1 to 100 years), subevolutionary (10^3 to 10^5 years), and evolutionary (1 m.y. and more) scales. She considered some shell concentrations to result from low dilution by sediments. Temporal changes in morphology of eight Chesapeake Group molluscs are dominated by non-directional fluctuations rather than long-term trends (Kelley, 1983).

The major shell beds are spaced at intervals of the order of 0.5 m.y. (Fig. 3), suggesting that conditions favorable to mollusc proliferation may have recurred quasi-periodically at about the 413,000-year (Berger, 1977) eccentricity cycle. It is possible to construct a plausible time scale on this premise (Fig. 3). Perhaps the cycles involve ice volume fluctuations in East Antarctica, and therefore eustatic sea level fluctuations.

SITE DESCRIPTIONS

Two areas are described here in more detail: the Calvert Cliffs State Park (Site 1, km: –11.5 to –13; public access) and from Governors Run to Parker Creek (Site 2, km: 0 to +5). Other sites are described in Gernant and others (1971), Blackwelder and Ward (1976), and Newell and Rader (1982).

Site 1 extends south about 0.6 mi (1 km) from Rocky Point (+131 ft; +40 m), the highest point along the Cliffs of Calvert. Exposed here are the upper Choptank Formation and all but the topmost St. Mary's Formation. At Rocky Point, 3 ft (1 m) of exposed Zone 18 is overlain by Zone 19, 11.5 ft (3.5 m) thick, consisting of nine distinct subunits of variably muddy fine sand and variable shell concentration, species diversity, and leaching. The lowest subunit includes worn and fragmented shells, suggesting a high-energy environment. The Zone 19/20 contact—the Choptank/St. Mary's unconformity (e.g., McCartan and others, 1985)—is distinct and undulating. Zone 20, 16 ft (5 m) thick, is a gray-green, very muddy, very fine sand with occasional sand and pebbles in the 1.5-ft-thick (0.5 m) basal unit. The Zone 20/21 contact is a distinct, burrowed surface (the Choptank/St. Mary's boundary according to Shattuck, 1904). Higher in the cliffs is seen a regressive sequence topped by Miocene tidal flat and littoral facies of Zone 24 equivalent.

Site 2 is the section of the cliffs just north of Governors Run (km 0 to +1) and on either side of Parker Creek valley (km: +3 to +5, of which +4.1 to +4.6 is the valley floor, lacking Miocene outcrops). The Site 2 cliffs are privately owned. The presence of good exposures indicates rapid shoreline erosion and a slide hazard. The exposed section extends from Zone 11 up to Zone 20. Where not hidden by slide masses, zone 14 can be inspected on the beach from km 0 to +2; zone 13 from km +2 to +4, zone 12 from km +3 to +4, and zone 11 from km +3.5 to +5. The two most identical 6- to 10-ft-thick (2 to 3 m) sandy shell beds, Zones 17 and 19, are conspicuous at heights of +33 and 66 ft (+10 and +20 m) in the high cliffs about km +0.5. Loose slide material and blocks up to 3 ft (1 m) size, as well as abundant reworked mollusc remains along the beach allow examination of the composition of Zones 17 and 19 without climbing, which is forbidden. Zone 14 is a complex of shelly horizons set in a greenish bluish gray fine muddy sand.

Occasional reptilian and bird bones (more commonly marine mammals) have been recovered from Zone 14. The much-debated Calvert/Choptank unconformity should exist between Zone 15 and Zone 16 at ca. 20 ft (6 m) height. Farther north (km +2 to 3.5) the unconformity(?) in the area of its type locality (Shattuck, 1904; Gernant, 1970) appears to truncate Zone 15 and upper Zone 14 (L. Dryden, unpublished sections dated 1965). Below it, the 20-ft-thick (6 m) massive, blue-gray fine muddy sand Zone 13 is generally unfossiliferous, although complete marine mammal skeletons occasionally come to light. The underlying "Parker Creek Bone Bed" (Zone 12) is a distinctive brown, biogenically mottled, 1.7-ft-thick (0.5 m) bed containing occasional shark teeth, vertebrae, and mollusc fragments. Although the exact Calvert/Choptank boundary is debated, the two formations are easily distinguished in the area of Site 1: The Calvert Formation tends to be bluish gray, with more clay; the overlying brownish Choptank is sandier and more permeable to groundwater. The systematic subvertical jointing in the Calvert Formation may attest to regional stresses and basement structural control (e.g., Newell and Rader, 1982).

REFERENCES CITED

Abbott, W. H., 1978, Correlation and zonation of Miocene strata along the Atlantic margin of North America using diatoms and silicoflagellates: Marine Micropaleontology, v. 3, p. 15–34.

—— , 1982, Diatom biostratigraphy of the Chesapeake Group, Virginia and Maryland, *in* Scott, T. H., and Upchurch, S. B., eds., Miocene of the southeastern United States: Florida, Bureau of Geology, Special Publication 25, p. 23–34.

Andrews, G. W., 1978, Marine diatom sequence in Miocene strata of the Chesapeake Bay region, Maryland: Micropaleontology, v. 24, p. 371–406.

Berger, A. L., 1977, Support for the astronomical theory of climatic change: Nature, v. 269, p. 44–45.

Blackwelder, B. W., and Ward, L. W., 1976, Stratigraphy of the Chesapeake Group of Maryland and Virginia: Guidebook 7b (Northeast-Southeast Sections Joint Meeting 1976) Geological Society of America, Arlington, Virginia, 52 p.

Gernant, R. E., 1970, Paleoecology of the Choptank Formation (Miocene) of Maryland and Virginia: Maryland Geological Survey Report of Investigation no. 12, 90 p.

Gernant, R. E., Gibson, T. G., and Whitmore, F. C., Jr., 1971, Environmental history of Maryland Miocene: Guidebook no. 3, Maryland Geological Survey, 58 p.

Gibson, T. G., 1983, Stratigraphy of Miocene through lower Pleistocene strata of the United States central Atlantic Coastal Plain, *in* Ray, C. E., ed., Geology and paleontology of the Lee Creek Mine, North Carolina, I: Smithsonian Contributions to Paleobiology, no. 53, p. 35–80.

Glaser, J. D., 1971, Geology and mineral resources of southern Maryland: Maryland Geological Survey, Report of Investigation no. 15, 84 p.

Kelley, P. H., 1983, Evolutionary patterns of eight Chesapeake Group molluscs; Evidence for the model of punctuated equilibria: Journal of Paleontology, v. 57, p. 581–598.

Kent, D. V., and Gradstein, F. M., 1986, A Jurassic to Recent chronology, *in* Vogt, P. R., and Tucholke, B. E., eds., The western North Atlantic region: Geological Society of America, Geology of North America, v. M, p. 45–50.

Kidwell, S. M., 1982a, Stratigraphy, invertebrate taphonomy, and depositional history of the Calvert and Choptank formations (Miocene), Atlantic Coastal Plain [Ph.D. thesis]: Yale University, 531 p.

—— , 1982b, Time scales of fossil accumulation; Patterns from Miocene benthic assemblages: Proceedings of the Third North American Paleontological Convention, v. I, p. 295–300.

McCartan, L., Blackwelder, B. W., and Lemon, E. M., Jr., 1985, Stratigraphic section through the St. Mary's Formation, Miocene, at Little Cove Point, Maryland: Southeastern Geology, v. 25, p. 123–139.

Newell, W. L., and Rader, E. K., 1982, Tectonic control of cyclic sedimentation in the Chesapeake Group of Virginia and Maryland, *in* Lyttle, P. E., ed., central Appalachian geology: Northeast-Southeast Geological Society of America Field Trip Guidebook, p. 1–27.

Shattuck, G. B., 1904, Geologic and paleontological relations, with a review of earlier investigations, *in* Clark, W. B., Shattuck, G. B., and Dall, W. H., eds., The Miocene deposits of Maryland: Maryland Geological Survey, Miocene Volume, p. 33–94.

Vogt, P. R., and Eshelman, R., 1987, The Cliffs of Calvert. A geological synthesis, Calvert Marine Museum Special Publication, in preparation.

Ward, L. W., and Strickland, G. L., 1985, Outline of Tertiary stratigraphy and depositional history of the U.S. Atlantic Coastal Plain, *in* Poag, C. W., ed., Geologic evolution of the United States Atlantic Margin: New York, Van Nostrand Reinhold Company, p. 87–123.

Baltimore gneiss and the Glenarm Supergroup, northeast of Baltimore, Maryland

Jonathan Edwards, Jr., Maryland Geological Survey, 2300 St. Paul Street, Baltimore, Maryland 21218

LOCATION

The section described herein extends across the northwestern flank of the Towson anticline in the eastern Piedmont region of Maryland, about 3.5 mi (6 km) northeast of Baltimore City (Fig. 1). Although better individual localities exist for separately examining the Baltimore Gneiss and units of the Glenarm Supergroup, the most accessible area to see a more-or-less continuous sequence lies along Big Gunpowder Falls in east-central Baltimore County, Maryland (Fig. 2). To reach this site from I-695 (Baltimore Beltway), take Exit 31 north onto Maryland 147 (Harford Road) and go 1.8 mi (2.9 km), turn left onto Cub Hill Road and proceed 1.3 mi (2.1 km) to the bottom of the hill. Where the road makes a sharp turn to the left, park in the dead-end, abandoned entrance road of the Charles H. Hickey School (formerly the Maryland Training School for Boys). This is Stop 1.

At this locality the Baltimore Gneiss is exposed on both sides of Cub Hill Road. On the south side, the land is private property and posted NO TRESPASSING. It may be possible to obtain permission at the large stone house in the elbow of the creek in order to examine the fresh outcrops at close hand, but it is important to remember that NO HAMMERS are to be used and that the rocks are not to be defaced in any manner. Exposures of the gneiss on the north side of the road and on the hillside above, as well as outcrops of the Setters Formation downstream from the small bridge, are on property of the State of Maryland and samples may be collected.

To reach Stop 2, continue west on Cub Hill Road for 0.1 mi (0.2 km) and turn left onto Cromwell Bridge Road. Proceed southwest for 0.5 mi (0.8 km), turn right onto Loch Raven Drive, go about 0.3 mi (0.5 km), and park beside the large, abandoned quarry just below the lower dam. This is Stop 2, where two units of the Cockeysville Formation may be seen. For large groups, it may be advisable to check in at Reservoir Maintenance Headquarters, 9800 Loch Raven Drive, Towson, Maryland 21204, about 0.2 mi (0.4 km) farther along the road.

Stop 3 is about 0.6 mi (1.0 km) farther on Loch Raven Drive, just below the large parking lot at the upper dam. Exposures of the Rush Brook Member of the Loch Raven Schist are in the roadbank along Loch Raven Drive down the hill from the parking area.

From the parking lot, continue west and north along Loch Raven Drive to Providence Road. A few poor exposures of the Loch Raven Schist may be seen at several places along the way. Parking along this winding road is not permitted and large groups should obtain permission to park from the Reservoir Maintenance Headquarters. At the intersection with Providence Road,

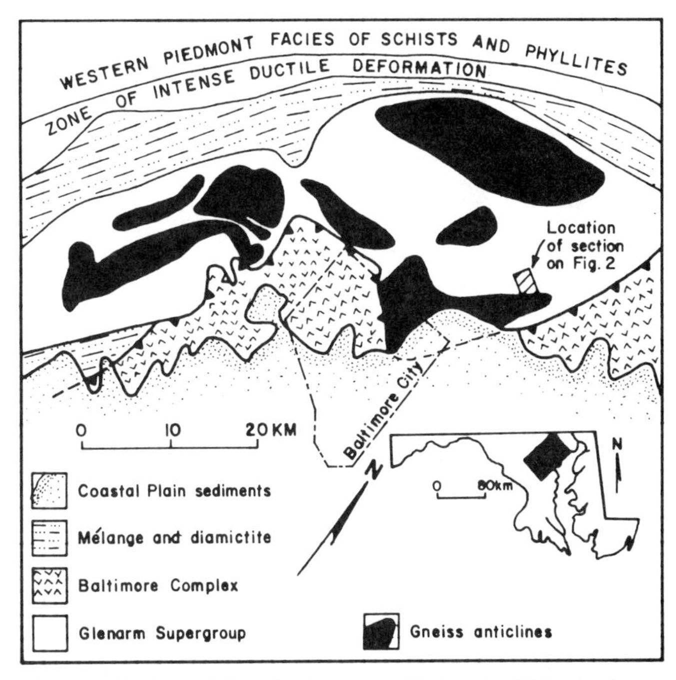

Figure 1. Gneiss anticlines in the eastern Piedmont of Maryland near Baltimore, with location of section described. (Modified from Cleaves and others, 1968; and Muller and Chapin, 1984).

turn left and travel for 2.3 mi (3.7 km) to return to I-695 at Exit 28.

Other localities where these lithologies may be seen are described in Crowley and others (1970, 1971) and Crowley (1976).

SIGNIFICANCE

In the vicinity of Baltimore, Maryland, gneissic rocks of the Precambrian basement complex and metasedimentary rocks of the overlying Glenarm Supergroup are exposed in seven anticlinal or antiformal structures (Fig. 1). Initially studied by G. H. Williams of the Johns Hopkins University in the final decades of the nineteenth century, these rocks have provided a classic stratigraphic framework upon which all geologic interpretations of the Piedmont in the mid-Atlantic region have been made. For many years these folds were considered to be mantled gneiss domes or diapirs (Hopson, 1964); current models interpret these structures as allochthonous gneiss-cored thrust nappes that have been subsequently refolded into complex patterns (Crowley, 1976; Fisher and others, 1979; Muller and Chapin, 1984).

SITE INFORMATION

Stop 1. At Stop 1, both the Baltimore Gneiss and the over-

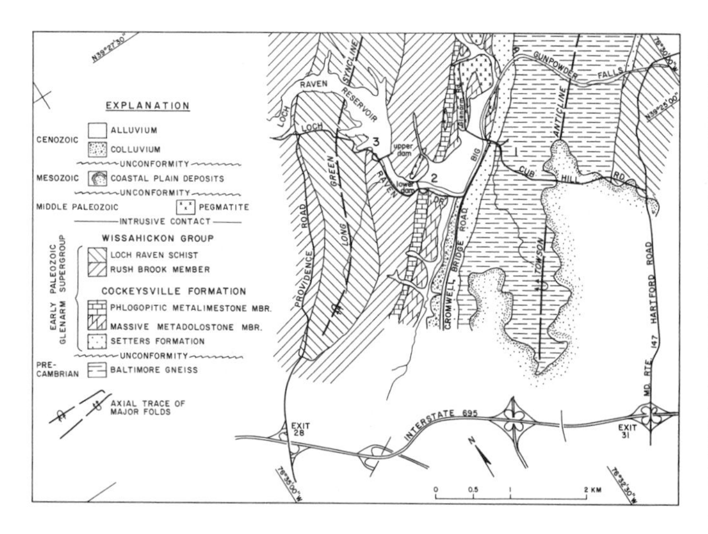

Figure 2. Geologic map of Baltimore Gneiss and units of the Glenarm Supergroup near Baltimore, Maryland, showing route of trip and stops. (Modified from Crowley and Cleaves, 1974.)

Figure 3. Augen gneiss in outcrop at Stop 1: A, quartz-microcline augen; f, foliation planes; length of pen 5½ in (14 cm), width ⅜ in (1.0 cm).

lying Setters Formation can be seen (Fig. 2). Throughout its area of occurrence (Fig. 1), the Baltimore Gneiss is primarily banded or layered quartzofeldspathic gneiss with lesser amounts of biotite-hornblende gneiss, amphibolite, augen gneiss, and locally, granitic leucogneiss or granofels (Hopson, 1964; Crowley, 1976; Muller and Chapin, 1984). However, the most conspicuous rock type present in the Baltimore Gneiss at Stop 1 is augen gneiss, a fine- to medium-grained biotite-microcline-plagioclase-quartz gneiss in which the augen occur as aggregates of medium-grained quartz and microcline (Fig. 3). Foliation planes are about ⅛ to ¼ in (2 to 5 mm) apart and are emphasized by concentrations of biotite. Layers in the gneiss range between 1 in and 1 ft (2.5 to 30 cm) in thickness. The augen have been modified by cataclasis into ovoid or lozenge-shaped masses ¼ by ⅜ in (5 by 7 mm) in size, and in places have been sheared out into flattened lenses or disks that "tail out" into the foliation.

The gneiss is streaked and banded with lighter-colored, more felsic layers that show little or no obvious foliation. Sporadic layers of darker gray biotite-plagioclase-quartz gneiss, up to 1 ft (30 cm) thick, also occur. Fine-grained alaskite dikes, 2 to 4 in (5 to 10 cm) thick, cut across the gneissic banding at low to moderate angles, and pegmatite dikes up to 1 ft (30 cm) thick and composed of large crystals of quartz and pink to pale orange-red microcline are also present. The gneiss becomes more felsic toward the top of the section (west), and just below the contact with the overlying Setters Formation a large mass of alaskite or granofels is exposed in the creek bed.

Radiometric age determinations on zircons from the Baltimore Gneiss (Tilton and others, 1970; Grauert, 1973a) indicate that the rock has been subjected to at least two distinct thermal events. The earliest, ranging from 1,000 to 1,200 Ma, represents the Grenville period of deformation when the gneiss was formed from the original sedimentary and volcanic rocks. The later event, between 420 and 450 Ma, corresponds to the period of uplift and intrusion in Late Ordovician time (Taconic orogeny) when the overlying sedimentary rocks of the Glenarm Group were deformed, metamorphosed, and intruded by remobilized subjacent continental crust (Hopson, 1964). According to Hopson (1964), the augen developed as porphyroblasts in the Baltimore Gneiss during this later metamorphic event, and are probably associated with the emplacement of the anatectic Gunpowder Granite into the east flank of the Towson Anticline just east of the area shown on Figure 2. However, zircons from this granite have yielded a radiometric age of 330 Ma (Grauert, 1973b), indicating a Late Devonian age.

The Setters Formation in the vicinity of Stop 1 (Fig. 2) consists of three informal members (Crowley and Cleaves, 1974). The lower member is medium-grained biotite-microcline-quartz-muscovite schist, locally coarse-grained and feldspathic, which ranges in thickness from zero to 300 ft (90 m) but is not everywhere present at the base of the formation. The middle member is fine- to medium-grained tourmaline-bearing quartzite and microcline-bearing schistose muscovitic quartzite in layers 2 to 8 in (5 to 20 cm) thick. The thickness of this member ranges from zero to 600 ft (180 m). The upper member is thinly layered microcline-quartz gneiss banded with biotite and muscovite and containing conspicuous black needles of tourmaline. The thickness of this member ranges up to 1,150 ft (350 m).

The basal schist member of the Setters Formation is exposed in the cut on the south side of the road at the bridge and in the cliff just upstream. This rock is somewhat feldspathized, which may be due either to an original arkosic composition or to granitization associated with the intrusion of the Gunpowder Granite. The middle quartzite member crops out in ledges on the hillside north of Cub Hill Run, downstream from the small bridge. The upper gneiss member is not well exposed at Stop 1, but forms the western dip slope of the prominent ridge that extends both northeast and southwest.

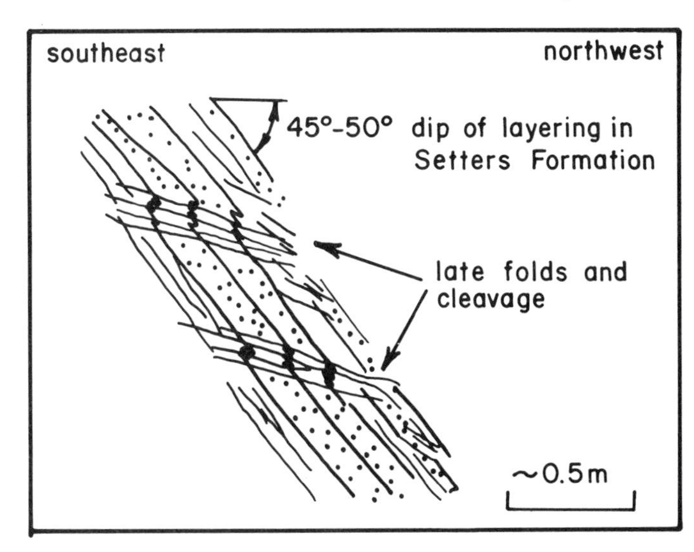

Figure 4. Sketch of minor folds in the Setters Formation at Stop 1, looking southwest.

Figure 5. Unconformity at top of Baltimore Gneiss at Stop 1, looking northwest.

In the roadcut at Stop 1, small folds in the lower schist member have a rotational sense opposite to that expected for drag folds on an anticline (Fig. 4). The axial planes of these small folds are subhorizontal to gently northwest-dipping and result from late vertical uplift of the Towson nappe-anticline (Muller and Chapin, 1984). Broad, low-amplitude folds with steep, northwest-plunging axes can also be seen in the layers of the Setters Formation exposed in the cliff face south of Cub Hill Road.

The units in the Setters Formation are considered to represent a transgressive clastic sequence deposited upon a subsiding shelf or platform of Baltimore Gneiss (Crowley, 1976; Fisher and others, 1979). Although described in the geological literature as an unconformity or nonconformity (Hopson, 1964), the actual contact between the Baltimore Gneiss and the overlying Setters Formation exposed at Stop 1 is obscure because of the feldspathization of the Setters. The position of this unconformity lies about 1.5 ft (0.5 m) below the thin, 4 to 8 in (10 to 20 cm), pegmatite dike which can be seen near the base of the cliff beside the bridge (Fig. 5).

The terrain between the section just described at Stop 1 and the exposures at Stops 2 and 3 in the Loch Raven Reservoir area illustrates the control of topography by bedrock lithology. Cromwell Bridge Road traverses a valley carved by Big Gunpowder Falls into the basal part of the Cockeysville Formation, a carbonate unit very susceptible to chemical weathering and erosion in Maryland's humid climate. The steep-sided ridge to the southeast is underlain by the middle quartzite member of the Setters Formation and the hills to the west are composed of Loch Raven Schist.

Stop 2. In the vicinity of the lower dam, in the Loch Raven Reservoir area (Fig. 2), the Cockeysville Formation is made up of three members, each composed of carbonate rocks (Crowley and Cleaves, 1974). The lowest member is layered metadolostone, but in this area is entirely covered by alluvium in the valley of Big Gunpowder Falls (Crowley and Cleaves, 1974) and no exposures can be observed. The overlying massive metadolostone member may be seen in the small quarry on Loch Raven Drive about 300 ft (100 m) southeast of the parking area at the large quarry. The rock is white, massive, medium- to coarse-grained, and composed predominantly of dolomite. Upon weathering, it breaks down into coarse dolomitic sand such as can be seen on the quarry floor. Compositional layering, emphasized by concentrations of minor amounts of diopside, tremolite, phlogopite, and quartz, dips about 40° to the northwest. Overlying this unit is the phlogopitic metalimestone member that can be seen in the large quarry at the lower dam. Good exposures of this member are accessible only in the upper quarry walls as the lower faces have been covered with rock debris, soil, and vegetation. The rock is composed of fine- to medium-grained white to bluish-gray calcitic marble interlayered on a scale of ¼ to 2 in (0.5 to 5 cm) with fine- to medium-grained gray to purplish-gray phlogopitic marble or calcareous phlogopitic schist. The overall dip of the layering is about 35° to the northwest.

The carbonate rocks of the Cockeysville Formation are believed to have originated as limy sediments deposited on a stable, shallow-water bank or platform (Crowley, 1976; Fisher and others, 1979).

At the northern end of the quarry and extending for about 300 ft (100 m) up the hillside away from the lower dam is a large body of white, coarse-grained quartz-albite pegmatite. Minor amounts of muscovite and sporadic small garnets are also present.

Stop 3. The Loch Raven Schist (Crowley and Cleaves, 1974; Crowley, 1976) is the lowest of the series of schistose clastic rocks that make up the Wissahickon Group (Crowley, 1976). The Rush Brook Member, the basal unit of the Loch Raven, is exposed in the roadbank along Loch Raven Drive between the parking lot at the upper dam and the bottom of the hill, about 0.2 mi (0.3 km) to the south (Fig. 2). At the base of the hill is an exposure of biotite-quartz-hornblende schist or amphibolite containing small pods or layers of epidote. Farther up the hill are muscovite-biotite-feldspar-quartz schist and fine-grained biotite-feldspar-quartz gneiss interlayered with fine-grained, flaggy, micaceous quartzites. Veins and pods of massive, gray to light-brown quartz occur sporadically within these lithologies.

Some authors (Rodgers, 1970; Rankin, 1975) interpret the heterogeneous lithology of the Rush Brook Member as a tectonic unit or sheared mélange at the base of an allochthonous sheet of the Wissahickon Group. However, Crowley (1976) and Fisher and others (1979) consider the Rush Brook to be in stratigraphic conformity with the Cockeysville and to represent a transition from a shallow-water to a deep marine environment. Muller and Chapin (1984) describe an exposure where this contact appears to be gradational and conformable.

The Loch Raven Schist (Fig. 2), formerly mapped as Wissahickon Schist, is the highest formation of the Glenarm Supergroup along the route of this trip. According to Crowley (1976), the lithology of the Loch Raven is remarkably uniform throughout its area of occurrence in the Maryland Piedmont, although it may be divided into several facies based on the presence or absence of various mineral porphyproblasts. The unit represents deep-water deposition of pelitic materials and suggests foundering of the former shelf into a marine basin (Crowley, 1976; Fisher and others, 1979).

The formation is poorly exposed along the route described and no formal stop was prepared. However, several small outcrops are present in the roadbank along Loch Raven Drive about 0.6 mi (1.0 km) northwest of the parking lot at the upper dam (Stop 3). These exposures are of medium-grained biotite-plagioclase-muscovite-quartz schist in which the muscovite flakes may be several times the size of other mineral grains. Scattered small crystals of garnet are also present. Veins and pods of quartz, about 2 to 2.5 in (5 to 6 cm) thick, occur in places parallel to the foliation, which dips about 45° to the northwest. In contrast to the schists of the Rush Brook Member exposed at Stop 3, the Loch Raven Schist displays a pervasive, tight crinkling of the foliation along subhorizontal to gently northwest-dipping cleavage planes. This cleavage may have the same origin as the gently northwest-dipping axial-plane cleavage seen in the Setters Formation at Stop 1.

The Glenarm Supergroup is considered by many workers to be correlative with rocks of known early Paleozoic age in the Piedmont of Pennsylvania (Fisher and others, 1979). Conclusive proof of this, however, is lacking. Recent investigations (Drake and Lyttle, 1981; Edwards, 1984; Muller and Edwards, 1985) show that the Piedmont is made up of multiple thrust sheets and such a correlation may not be justified.

REFERENCES CITED

Cleaves, E. T., Edwards, J., Jr., Glaser, J. D., compilers, 1968, Geological map of Maryland: Maryland Geological Survey, scale 1:250,000.

Crowley, W. P., 1976, The geology of the crystalline rocks near Baltimore and its bearing on the evolution of the eastern Maryland Piedmont: Maryland Geological Survey Report of Investigations 27, 40 p.

Crowley, W. P., and Cleaves, E. T., 1974, Geologic map of the Towson Quadrangle, Maryland: Maryland Geological Survey Quadrangle Atlas 2, map 1, scale 1:24,000.

Crowley, W. P., Higgins, M. W., Bastian, T., and Olsen, S., 1970, New interpretations of the Piedmont geology of Maryland: Field Conference of Pennsylvania Geologists Guidebook, 35th annual field conference, 60 p.

——, 1971, New interpretations of the Piedmont geology of Maryland: Maryland Geological Survey Guidebook 2, 43 p.

Drake, A. A., and Lyttle, P. T., 1981, The Accotink Schist, Lake Barcroft Metasandstone, and Popes Head Formation; Keys to an understanding of the tectonic evolution of the northern Virginia Piedmont: U.S. Geological Survey Professional Paper 1205, 16 p.

Edwards, J., Jr., 1984, The Linganore Fault; Key to age of Sams Creek Formation in the Piedmont of Maryland: Geological Society of America Abstracts with Programs, v. 16, no. 3, p. 135.

Fisher, G. W., Higgins, M. W., and Zietz, I., 1979, Geological interpretation of aeromagnetic maps of the crystalline rocks in the Appalachians, northern Virginia to New Jersey: Maryland Geological Survey Report of Investigations 32, 43 p.

Grauert, B. W., 1973a, U-Pb isotopic studies of zircons from the Baltimore Gneiss of the Towson Dome, Maryland: Carnegie Institute of Washington Yearbook 72, p. 285–288.

——, 1973b, U-Pb isotopic studies of zircons from the Gunpowder Granite, Baltimore County, Maryland: Carnegie Institute of Washington Yearbook 72, p. 289–290.

Hopson, C. A., 1964, The crystalline rocks of Howard and Montgomery counties, *in* The geology of Howard and Montgomery counties: Maryland Geological Survey County Report, p. 27–215.

Muller, P. D., and Chapin, D. A., 1984, Tectonic evolution of the Baltimore Gneiss anticlines, Maryland, *in* Bartholomew, M. J., ed., The Grenville event in the Appalachians and related topics: Geological Society of America Special paper 194, p. 127–148.

Muller, P. D., and Edwards, J., Jr., 1985, Tectonostratigraphic relationships in the central Maryland Piedmont: Geological Society of America Abstracts with Programs, v. 17, no. 1, p. 55.

Rankin, D. W., 1975, The continental margin of eastern North America in the southern Appalachians; The opening and closing of the proto-Atlantic Ocean: American Journal of Science, v. 275-A (Rodgers Volume), p. 298–336.

Rodgers, J., 1970, The tectonics of the Appalachians: New York, Wiley-Interscience, 271 p.

Tilton, G. R., Doe, B. R., and Hopson, C. A., 1970, Zircon age measurements in the Maryland Piedmont, with special reference to Baltimore Gneiss problems, *in* Fisher, G. W., and others, eds., Studies of Appalachian geology; Central and southern: New York, Wiley-Interscience, p. 429–434.

Cape Henlopen spit: A late Holocene sand accretion complex at the mouth of Delaware Bay, Delaware

John C. Kraft and Anne V. Hiller, Department of Geology, University of Delaware, Newark, Delaware 19716

LOCATION

The Cape Henlopen spit and associated dunes, beaches, marshes, and other coastal landforms are located at the confluence of Delaware Bay and the Atlantic Ocean near Lewes, Delaware. Much of this area is included in the State of Delaware's Cape Henlopen State Park. The state park is open all year; however, entry fees are charged from Memorial Day to Labor Day as the park is a popular recreation area. In general, the beaches, berm, tidal flats, and large portions of the dunes and barriers are open to foot traffic. There are restrictions in bird-nesting areas and in several small U.S. military zones. Trails inland are good, but we suggest care regarding excess heat in the summer and local patches of poison ivy. Should you wish to dig beach trenches, we suggest that they be shallow and broad to avoid collapse and that all trenches be refilled at the end of the study. Much of the park is included and protected in the Cape Henlopen Archaeological District. The staff at the park headquarters or at the nature center can provide further information on available facilities.

Once in southern Delaware, take U.S. 9 east towards Lewes, following signs for the Cape May–Lewes Ferry and the Cape Henlopen State Park (Fig. 1). The park entrance is approximately 1 mi (1.6 km) east from the ferry terminal. During the summer months this is a busy resort area, and reservations are suggested for local motels. Cape Henlopen State Park has an excellent campground, but again reservations may be needed.

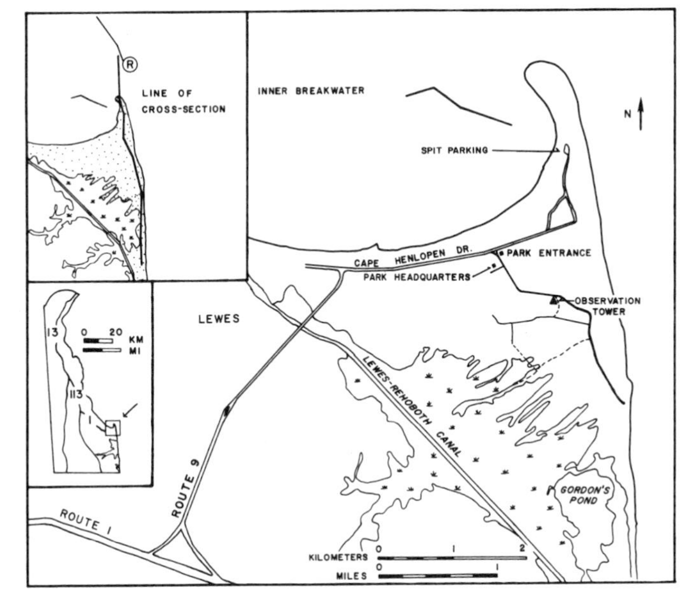

Figure 1. Location and access to the Cape Henlopen complex, Cape Henlopen Quadrangle, Sussex County, Delaware. Line of section for Figure 3 is shown.

SIGNIFICANCE

Cape Henlopen spit complex is an area of many separate and distinct coastal sedimentary depositional and erosional environments constantly changing in response to wind and wave processes of the Atlantic Ocean and Delaware Bay as well as the many intrusions into the natural environments by people. Air photos and topographic maps show excellent evidence of coastal changes in the late Holocene Epoch as the Atlantic coastal transgression continues landward at rates of approximately 3+ ft/yr (1+ m/yr) during the past 2,000 years. Surface lineaments clearly show the remnants of recurved spit tips, cuspate forelands, beach accretion plains, and the present-day coastal dune systems that fringe the beach-berm and tidal flat areas. The sands of the Cape Henlopen spit system have been mainly derived from erosion of the shoreface and headlands to the south, transported north to the Cape via littoral transport and deposited in the various spit systems over the past 2,000+ years. A large portion of sand and pebbles has also been deposited in the nearshore areas of the lower Delaware Bay and by ebbtide flow on Hen and Chickens Shoal. While these sandy coastal environments were eroded on the Atlantic shore and deposited into the Delaware estuary, a large embayed area to the west was surrounded against a previous (3,500 years B.P.) shoreline of Delaware Bay. With completion of the formation of a cuspate foreland, this embayed area filled in with silt and is presently Lewes Creek Marsh. At the tip of Cape Henlopen, the simple spit is presently arcing northwesterly between the two breakwaters. The Harbor of Refuge is rapidly shoaling with sand. Breakwater Harbor has shoaled to less than 10 ft (3 m; except in dredge channels) in an environment of black muds. The various coastal sedimentary environmental lithosomes show a complex lateral and vertical relationship of facies change as the separate elements of the depositional systems continue to accrete, aggrade, or erode back into the sediment transport system along the Atlantic coastline. Rapid deposition rates are common. The tip of the spit will likely join the outer side of the inner breakwater in several decades.

SITE INFORMATION

The triangular-shaped Cape Henlopen spit complex, bounded by Roosevelt Inlet to the west, Whiskey Beach to the south, and the present Cape Henlopen, includes a large number of coastal sedimentary environments with a very complex lateral and vertical facies relationship. The majority of the sands and gravels in the complex were derived from coastal erosion of

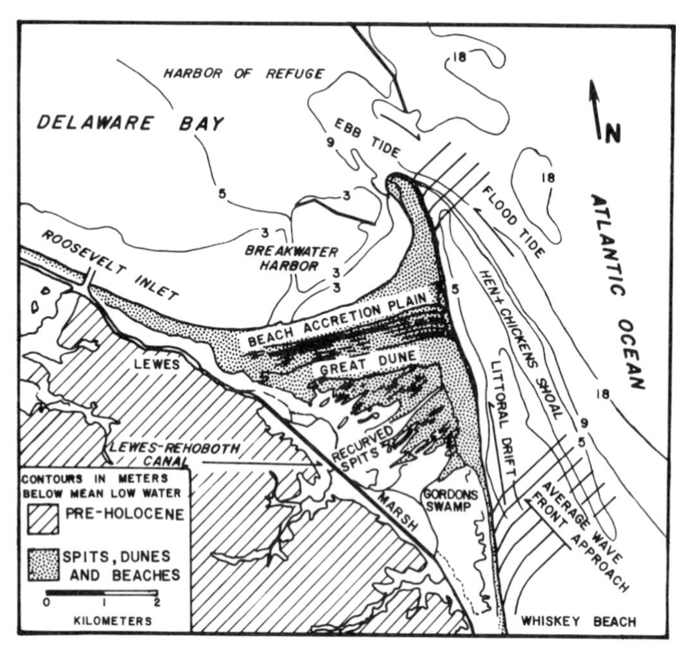

Figure 2. Geomorphic map of the Cape Henlopen area and adjacent Delaware Bay and Atlantic shelf (modified from Kraft and Caulk, 1972).

Pleistocene sediments in headlands to the south, transported northward as a barrier-spit, much as originally proposed by G. K. Gilbert (Kraft, 1980). Lesser amounts of sediment were transported southeasterly along the shoreline of Delaware Bay and in turbid suspension in the coastal waters of Delaware Bay. The recycling of silts and clays, like coastal erosion in the shoreface to the south, from Holocene and Pleistocene coastal lagoon and estuarine sediments, is also important as a source of the sediments in the spit-marsh complex (Belknap and Kraft, 1985).

As along many of the world's coastlines, the Cape Henlopen area and Atlantic coast to the south are undergoing a marine transgression. The rate of sea level rise along the Delaware coast averaged approximately 7 in/100 yr (18 cm/100 yr) over the past 2,000 years (Belknap and Kraft, 1977). As a consequence of this erosion and drowning of the coast, the environments present in the Cape Henlopen region are continually evolving and moving northwestward and upward through time.

Since the area of Lewes was first settled by the Dutch in 1631, the historical record of this region is long and well detailed and may be easily equated with changes in coastal landforms (Kraft, 1977). The Cape Henlopen Lighthouse, built on the cuspate foreland-type cape in 1765 approximately 0.25 mi (0.4 km) from the Atlantic Ocean, eroded into the sea in 1926 (Beach, 1970). Thus, the lighthouse serves as an excellent marker for monitoring historic coastal change in the area (Kraft, 1971).

As shown in Figure 2, the major geomorphic elements of the spit complex are: the simple spit, Cape Henlopen; the shallow estuarine shoaling areas of Breakwater Harbor and the Harbor of Refuge; the Atlantic beach and berm; the beach accretion plain of a former cuspate foreland; high and low coast-parallel dunes of the Atlantic coast; the Great Dune—a dune 2–2.5 mi (3–4 km) long perpendicular to the Atlantic coast; more ancient recurved spit tips in the present Lewes Creek Marsh; the offshore submerged Hen and Chickens Shoal; and the Whiskey Beach washover barrier.

Cape Henlopen spit is an ideal area to observe coastal processes. Following the road through the park entrance and taking the second left turn will bring you to the parking lot on the spit at an old gun emplacement (Fig. 1). Access around the edges of the spit to the beach, berm, and tidal flats is excellent. However, the interior of the spit is a bird sanctuary and entrance is restricted.

The present Cape Henlopen is a simple spit, rapidly accreting to the northwest between the inner and outer breakwaters at a rate of approximately 100 ft/yr (30 m/yr; Maurmeyer, 1974). A high beach and berm forms the Atlantic Ocean edge of the spit in continuity with the Atlantic beaches to the south. When the beach and berm are full (summer beach), the berm elevation is approximately 3.3 to 5 ft (1 to 1.5 m) above the lower tidal flats and swash bars on the western side of the spit. A series of ridges and runnels arc around the tip of the spit as it continues to build in a northwesterly direction. The interior of the spit is covered by an extensive, partially vegetated dune field that is progressively younger toward the tip of the spit. The dune sands, with a slight parallel skew and long, sweeping foresets, may be well exposed near the parking lot where one can cross the spit to the Atlantic Ocean. The spit is composed of Holocene sands and gravels from the berm and tidal flat surface downward to the base of the depositional spit, approximately 60 ft (20 m) below sea level. Beneath the spit sediments are estuarine and marine sediments.

Hen and Chickens Shoal is a submerged ebbtidal shoal extending southeast from Cape Henlopen approximately 6.2 mi (10 km). Sand diverted by ebbtidal current movement is transported along the shoal. It is believed that the shoal has always been associated with Cape Henlopen (Kraft and others, 1976). From the spit parking lot, the position of the shoal can be seen by several bell buoys trending to the southeast and by breaking waves at low tide.

Coast-parallel dunes of the Cape Henlopen spit complex extend from the tip of Cape Henlopen spit southward to Whiskey Beach. Those from south of the spit parking lot to the Army Recreation Area are the most accessible. Access onto the dunes farther south is restricted by either the military or the park service. The dunes are of varying heights to a maximum of approximately 45 ft (13.7 m). These vegetated dunes serve as a buffer, protecting the landward environments from wind and wave activity along the Atlantic shore. A large amount of sand-size sediment is eroded from these exposed dunes and blown landward, continuing the evolution of the coast-parallel dunes. Sand is also removed during coastal storm events and transported in a net northerly direction to the tip of the accreting Cape Henlopen. The coast in this area is presently eroding approximately 10 ft/yr (3 m/yr) but has eroded at rates up to 32 ft/yr (10 m/yr), as shown by shoreline changes on maps over the longer term. As the transgression continues, these coast-parallel dunes are moving landward, burying other coastal geomorphic elements in the spit complex, such as the recurved spit tips, Lewes Creek Marsh, and beach accretion ridges.

Some areas of beach ridge lineaments are visible in aerial photographs of the beach accretion plain. However, much of the beach accretion plain has been altered by the development of Fort Miles. Remnants of the ridges can be seen by walking the nature trail that begins south of the nature center. The parallel beach ridges formed when Cape Henlopen was a cuspate foreland, from approximately 500 years B.P. to the early nineteenth century, when the inner breakwater was constructed.

An anomalous feature in the Cape Henlopen spit complex is the Great Dune, sometimes called the "galloping dune of Lewes." This sand ridge is 2 to 2.5 mi (3 to 4 km) long and 90 ft (27 m) high and is oriented perpendicular to the Atlantic coast. Presently the dune is divided into three sections: a stabilized and built-upon eastern section, a relatively unvegetated and still migrating middle section, and a quarried and reworked western section. Kraft and Caulk (1972) believe that the Great Dune formed from the deforestation of the Cape Henlopen area during the construction of the inner breakwater, which allowed northerly winds to pick up sediment from the beach and create the ridge. The dune was first seen fully developed on an 1831 map but was not present on any previous map. In the time since its formation, the Great Dune has migrated at an average rate of 5.5 ft/yr (1.7 m/yr; Kraft and others, 1978) and is presently burying the forest growing on ancient recurved spit tips to the south. This dune and most of the rest of the spit complex can be studied from an excellent viewpoint on the top of a gun-sighting tower maintained by the park service approximately 0.8 mi (1.2 km) south from the entrance along the park headquarters road. One can also examine the slip face of the dune by walking westward from the parking lot at the campground along the crest of the dune.

To the west and south in the Cape Henlopen spit complex lie the Lewes Creek Marsh and the more ancient recurved spit tips of Cape Henlopen that extend into and under the marsh. Based on archaeological evidence and radiocarbon dates, these spit tips range from 2,500 years B.P. for the southernmost spit to 500 years B.P. for the northernmost one. These recurved spit tips formed when Cape Henlopen was located far to the southeast of its present position. The recurving of the spits created a shallow estuarine lagoonal area separated from the open estuarine Delaware Bay of several thousand years ago. With the continuing transgression of the sea, the recurved spits moved upward and landward through time. As the area behind the spits became more and more cut off from the Delaware estuary, the area silted in and eventually was covered by the present tidal Lewes Creek Marsh. As Lewes Creek Marsh evolved and the arm of the estuary filled in, the sands of the Cape Henlopen spit of approximately A.D. 1600 to 1700 moved many miles (kilometers) to the west of Lewes (as documented by Maurmeyer, 1974). Access to trails maintained by the park service along the crests of old recurved spit tips is available from the park road going southeast from the park headquarters toward the Atlantic coast at appropriate parking ares (Fig. 1).

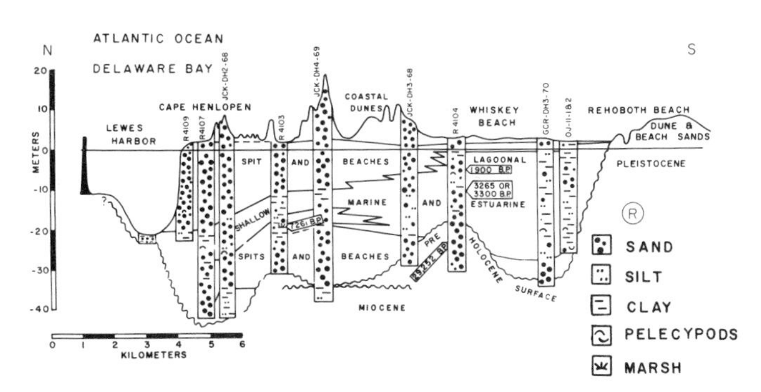

Figure 3. Coast-parallel cross section from Lewes Harbor south to Rehoboth Beach (Kraft and others, 1978). Radiocarbon dates corrected to 5,730 years half-life.

The southernmost section of the spit complex is the Whiskey Beach washover barrier east of Gordons Pond. This area can be reached by going south on the park headquarters road to the parking lot at the end and then walking along the beach approximately 1 mi (1.5 km) south. Whiskey Beach is a washover beach with low dunes, washover fans, and it is rapidly eroding in the beachface-shoreface. This low-lying area is subject to frequent overwash during storms. Gordons Pond is a brackish swamp, pond, and the site of a former colonial salt works.

Figure 3 is a coast-parallel cross section from the tip of Cape Henlopen in the north extending southward to the Pleistocene highlands at Rehoboth Beach. The cross section shows the interrelationships of the sedimentary geomorphologic features discussed above. The line of cross section is shown in Figure 1.

Much of the Holocene Epoch stratigraphic section in the southern part of the spit complex is comprised of estuarine and lagoonal sands and silts of the past three millennia and greater, deposited when Cape Henlopen (assuming it existed) lay to the east and south of the present coastal zone. As coastal erosion and landward transgression of the coastal environments continued over the last several thousand years, increasing amounts of sand and gravel appear in the Holocene section to the north as shown in Figure 3. Thus, the coastal shallow marine sands and gravels are progressively younger in age towards the north as progradation or accretion of this spit continued to present time. As this is a coast-parallel section, a surficial cover of coast-parallel dune sands covers the section.

Figure 4 is a paleogeographic map showing the positions of the Cape Henlopen shoreline from 2,500 years B.P. (the oldest spit tip in the coastal zone) through 1,600 years B.P., 1,000 years B.P., A.D. 1631, A.D. 1765, A.D. 1831, and present. The coastal zone in the area of study has evolved from a series of recurved spit tips to a cuspate foreland and thence to the present configuration of a simple spit. These projections can be correlated with the subsurface continuation of the sediments of the coastal environments as shown in Figure 3. In this case, the dating of the shoreline positions can be made with precision. The A.D. 1631 shoreline position is based on coastal morphologic information, stratigraphic information, and an early sketch map produced by some of the first settlers of the Lewes area. The 1765 shoreline is based on surveys made of the area when the Cape Henlopen Lighthouse was built. With the construction of the inner breakwater in A.D. 1831 and additions in the mid-nineteenth century, tidal currents in the area were altered. Information available shows that the inner harbor is rapidly infilling with a dark gray-black silt in the deeper areas and a silty sand in the tidal flat areas toward the shoreline. The Harbor of Refuge between the two breakwaters is infilling with sand shoals at a fairly rapid rate. It now seems evident that the tip of Cape Henlopen will join with a sand shoal on the northern side of the inner breakwater at some time in the near future, closing the eastern end of Breakwater Harbor. The harbor will then rapidly silt in with an expansion of the tidal flat areas. Possibly the beginnings of a coastal marsh will form as a new recurve of Cape Henlopen appears along the inner breakwater (Demarest and Kraft, 1979). Long-term projections into the future are fraught with many problems. However, a simple linear projection of coastal erosion with relative sea level rise—as observed over the past several millennia—suggests that an Atlantic coastal position several miles (kilometers) landward might be anticipated at a peak interglacial position 2,000 to 6,000 years into the future (Kraft and Caulk, 1972; Belknap and Kraft, 1985).

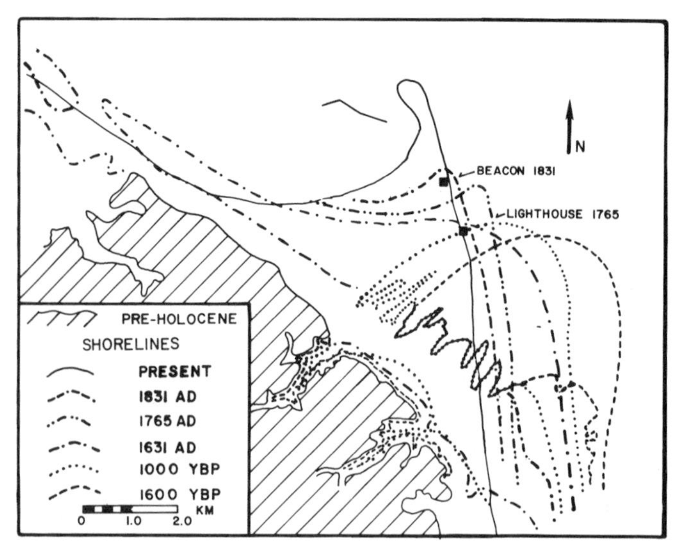

Figure 4. Paleogeographic reconstructions of Cape Henlopen showing the evolution from a recurved spit to a cuspate foreland and thence to a simple spit. Based in part on Kraft and Caulk, 1972; Allen, 1974; and Maurmeyer, 1974.

REFERENCES CITED

Allen, E. A., 1974, Identification of Gramineae fragments in salt marsh peats and their use in late Holocene environmental reconstructions [M.S. thesis]: Newark, University of Delaware, 141 p.

Beach, J. W., 1970, The Cape Henlopen Lighthouse: Dover, Delaware, Henlopen Publishing Company, 92 p.

Belknap, D. F., and Kraft, J. C., 1977, Holocene relative sea level changes and coastal stratigraphic units on the northwest flank of the Baltimore Canyon Trough geosyncline: Journal of Sedimentary Petrology, v. 47, no. 2, p. 610–629.

—— , 1985, Influence of antecedent geology on stratigraphic preservation potential and evolution of Delaware's barrier systems: Marine Geology, v. 62, p. 235–262.

Demarest, J. M., II, and Kraft, J. C., 1979, Projection of sedimentation patterns in Breakwater Harbor, Delaware: Shore and Beach, v. 47, p. 17–24.

Kraft, J. C., 1971, A guide to the geology of Delaware's coastal environments: Newark, Delaware, College of Marine Studies publication no. 2GL-039, 220 p.

—— , 1977, Late Quaternary paleogeographic changes in the coastal environments of Delaware, Mid-Atlantic Bight, related to archaeological settings, *in* Newman, W. S., and Salwen, B., eds., Amerinds and their paleoenvironments in northeastern North America: Annals of the New York Academy of Science, v. 288, p. 35–69.

—— , 1980, Grove Karl Gilbert and the origin of barrier shorelines, *in* Yochelson, E. L., ed., The scientific ideas of G. K. Gilbert: Geological Society of America Special Paper 183, p. 105–113.

Kraft, J. C., and Caulk, R. L., 1972, The evolution of Lewes Harbor: Transactions of the Delaware Academy of Science, v. 2, p. 79–125.

Kraft, J. C., Allen, E. A., Belknap, D. F., John, C. J., and Maurmeyer, E. M., 1976, Geologic processes and the geology of Delaware's coastal zone: Dover, State of Delaware, Executive Department, Planning Office, published as Delaware's Coastal Management Program, 319 p.

Kraft, J. C., Allen, E. A., and Maurmeyer, E. M., 1978, The geological and paleogeomorphological evolution of a spit system and its associated coastal environments; Cape Henlopen spit, Delaware: Journal of Sedimentary Petrology, v. 48, no. 1, p. 211–226.

Maurmeyer, E. M., 1974, Analysis and short- and long-term elements of coastal change in a simple spit system; Cape Henlopen, Delaware [M.S. thesis]: Newark, University of Delaware, 150 p.

Upper Cretaceous and Quaternary stratigraphy of the Chesapeake and Delaware Canal

Thomas E. Pickett, Delaware Geological Survey, University of Delaware, Newark, Delaware 19716

LOCATION

The Chesapeake and Delaware Canal (C & D Canal) connects the Delaware River and upper Chesapeake Bay (Elk River) (Fig. 1). The best geologic exposures are located east of Delaware 896 "Deep Cut," the bluff on the north side of the canal just east of the bridge crossing at Summit, Delaware. The bluff is reached on dirt roads that are located parallel to the canal banks as indicated in Figure 1. Another good exposure, seen only at low tide, is about 1 mi (1.6 km) east of U.S. 13 near St. Georges, along dirt road on the south bank of the canal (Fig. 1). This is the Biggs Farm locality where more than 200 species of fossils have been collected from the Mount Laurel Formation. The Biggs Farm locality is marked by a square bulkheaded area at the edge of the water.

The U.S. Army Corps of Engineers has jurisdiction over the C & D Canal Reservation. Part of the reservation is managed by the Delaware Department of Natural Resources and Environmental Control. No permits are needed for access. It is against federal law to collect fossils here for later sale; however, small-scale personal collecting is permitted. Deep Cut is an officially designated "Delaware Natural Area."

Avoid wading in the water along the canal because there is a 35-ft (10-m)-deep ship channel just offshore. Because the rocks in the outcrop are unconsolidated, they can slump without warning. Visitors should stay away from overhanging cliffs. If fossil collecting only is desired, the Corps prefers that this be done in dredge spoils areas. The best ones can be reached near the town of North St. Georges (Fig. 1) by driving 0.5 mi (0.8 km) west from North St. Georges along the dirt road adjacent to the canal, and walking to the top of the spoil area. The area 0.25 mi (0.4 km) east of town may be reached along a paved road (Fig. 1).

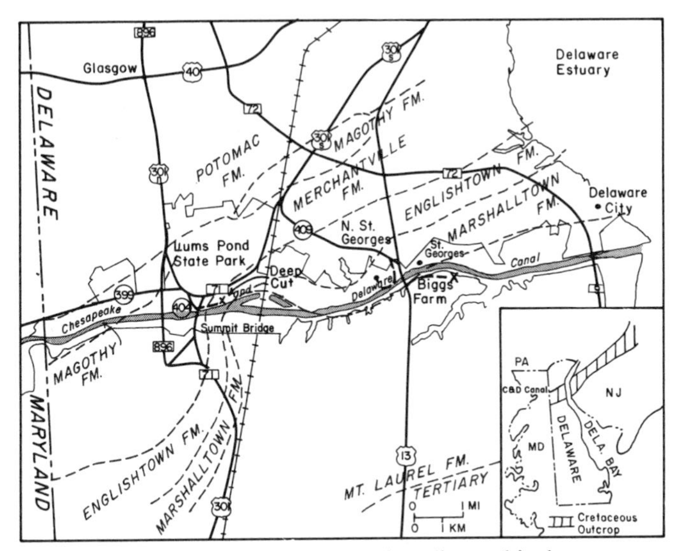

Figure 1. Index map showing locations discussed in the text.

SIGNIFICANCE

The C & D Canal outcrops are some of the best exposures of the Cretaceous formations in the Atlantic Coastal Plain between Cape Fear, North Carolina, and northern New Jersey.

Until canal-widening operations in the last 20 years, the exposures were much more continuous (Groot and others, 1954). The bluff exposures near Summit Bridge are the best remaining and rival all others in the northern Atlantic Coastal Plain in height (60 ft; 18 m), breadth (0.75 m; 1.2 km), formations exposed (5), and accessibility. They are keys to understanding Cretaceous stratigraphy in the mid-Atlantic states and are of great regional importance for correlation and historic interest.

SITE INFORMATION

Figure 2 shows the upper Cretaceous and Quaternary stratigraphic succession at the C & D Canal. It is generalized for the entire canal area.

The oldest coastal plain unit in Delaware, the Potomac Formation, was deposited on ancient crystalline rocks of the basement complex from the latter part of Early Cretaceous time into Late Cretaceous time. Streams transported clays and sands from the Appalachians in the northwest and the sediments were deposited here, probably in a deltaic environment.

The overlying white sands and lignitic black silts of the Magothy Formation are separated from the Potomac Formation by an unconformity. The Magothy indicates the transition from older nonmarine sediments to the later marine deposits. Magothy sediments were deposited in a shoreline environment containing elements of strand line, barrier island, and lagoonal conditions. Neither the Potomac nor the Magothy formations are exposed at sites shown in Figure 1.

A sequence of varied marine sedimentary rocks was deposited essentially continuously from Late Cretaceous to at least Middle Eocene time. The oldest Cretaceous sediments above the Magothy form the Matawan Group, which is exposed at Deep Cut, consisting of the Merchantville, Englishtown, and Marshalltown formations. None of these units persist very far into the subsurface, so the Matawan is assigned formational status in the subsurface a few miles south of the C & D Canal. The Mount Laurel Formation is found above the Matawan Group. The Merchantville, Marshalltown, and Mount Laurel sedimentary rocks were probably deposited in fairly shallow, open marine, perhaps embayed areas as evidenced by the presence of the mineral glauconite and marine fossils. However, lithology and fossil burrows indicate that the Englishtown represents a shoreline environment in which sea level was dropping. Paleocene and Eocene sediments are found several miles south of the C & D Canal. There is no sedimentary record of the geologic history in the canal area between Eocene and Pleistocene time.

Much later, during Pleistocene time, the advance and retreat of the continental glaciers brought about changes in sea level and in the streams draining into Delaware. The Pleistocene Columbia Formation, making up the uppermost part of the sequence exposed at the site, consisting mostly of coarse sand and gravel, was deposited on the stream-channeled surface of the truncated Cretaceous and Tertiary beds.

At Deep Cut the Merchantville Formation forms the beach to about 10 ft (3 m) up the bluff. Ammonites may be found. The alternating gray and oxidized sands of the overlying Englishtown Formation comprise the midway area up the bluff. *Ophiomorpha* and other trace fossils are found in abundance in the upper Englishtown Formation and are accessible with care in slumped areas. The more massive gray Marshalltown Formation near the top of the bluff is also burrowed and is fossiliferous. The top of the bluff is capped by a few feet of oxidized Mount Laurel Formation and fluvial Columbia Formation.

The Mount Laurel Formation exposed at low tide at the Biggs Farm locality is very fossiliferous. One should examine the formation exposed on the beach around the steel bulkheaded area. The bulkheads were erected in the 1970s in a futile attempt to preserve the low bluff here.

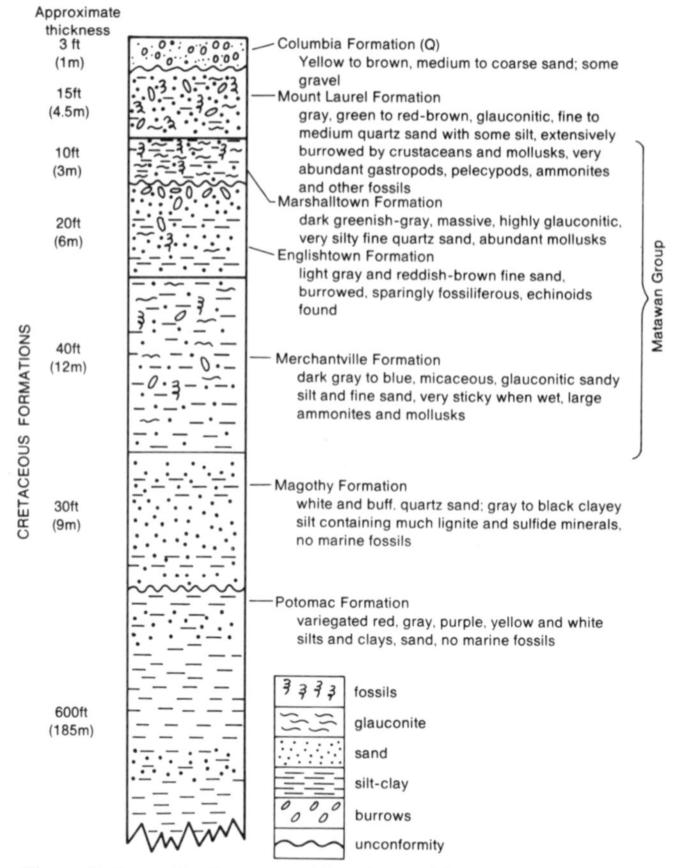

Figure 2. Generalized stratigraphic column, Chesapeake and Delaware Canal.

REFERENCES CITED

Groot, J. J. and Penny, J. S., 1960, Plant microfossils and age of nonmarine Cretaceous sediments of Maryland and Delaware: Micropaleontology, v. 6, no. 2, p. 225–236.

Groot, J. J., Organist, D. M., and Richards, H. G., 1954, Marine Upper Cretaceous formations of the Chesapeake and Delaware Canal: Delaware Geological Survey Bulletin 3, 64 p.

Jordan, R. R., 1962, Stratigraphy of the sedimentary rocks of Delaware: Delaware Geological Survey Bulletin 9, 51 p.

Lauginiger, E. M., and Hartstein, J.E.F., 1983, A guide to fossil sharks, skates, and rays from the Chesapeake and Delaware Canal Area, Delaware: Delaware Geological Survey Open-File Report 21, 64 p.

Morton, S. C., 1829, Description of the fossil shells which characterize the Atlantic Secondary Formation of New Jersey and Delaware: Academy of Natural Science of Philadelphia Journal, v. 6, p. 72–100.

Mumby, J. L., 1961, Upper Cretaceous Foraminifera from the marine formations along the Chesapeake and Delaware Canal [Ph.D. thesis]: Bryn Mawr, Bryn Mawr College, 174 p.

Owens, J. P., Minard, J. P., Sohl, N. F., and others, 1970, Stratigraphy of the outcropping post-Magothy Upper Cretaceous formations in southern New Jersey and northern Delmarva Peninsula, Delaware and Maryland: U.S. Geological Survey Professional Paper 674, 60 p.

Pickett, T. E., 1970, Geology of the Chesapeake and Delaware Canal area: Delaware Geological Survey Geologic Map Series 1.

——, 1972, Guide to common Cretaceous fossils of Delaware: Delaware Geological Survey Report of Investigations 21, 28 p.

Pickett, T. E., Kraft, J. C., and Smith, K., 1971, Cretaceous burrows, Chesapeake and Delaware Canal, Delaware: Journal of Paleontology, v. 45, no. 2, p. 209–211.

Richards, H. G., Howell, B. F., Wells, J. W., and Cooke, C. W., 1958, The Cretaceous fossils of New Jersey: New Jersey Geological Survey Bulletin 61, Part 1, 266 p.

Richards, H. G., and Shapiro, E., 1963. An invertebrate macrofauna from the Upper Cretaceous of Delaware: Delaware Geological Survey Report of Investigations 7, 37 p.

Spoljaric, N., and Jordan, R. R., 1966, Generalized geologic map of Delaware: Delaware Geological Survey, revised 1976.

Banded gneiss and Bringhurst gabbro of the Wilmington Complex in Bringhurst Woods Park, northern Delaware

Mary Emma Wagner, Lee Ann Srogi, and Elayne Brick**, *Department of Geology, University of Pennsylvania, Philadelphia, Pennsylvania 19104*

LOCATION

The Wilmington Complex underlies the eastern part of northern Delaware and extends into adjacent Pennsylvania. The site described here is in the Wilmington North 7½-minute Quadrangle, Delaware-Pennsylvania (Fig. 1), in New Castle County, Delaware, just northeast of the city of Wilmington. The outcrops are in Bringhurst Woods Park, which is easily accessible by car. Because the road into the park and the parking areas are not paved and can be muddy, caution is urged in wet weather. The area is not accessible for buses because the turning area is very limited, and bus parking on adjacent Carr Road is presently not feasible. Buses might park on the main thoroughfare (Washington Street Extension) about 0.3 mi (0.5 km) southwest of the park. In the spring of 1986, the woods across Carr Road from the park were cleared for the construction of office buildings, so additional parking may become available there.

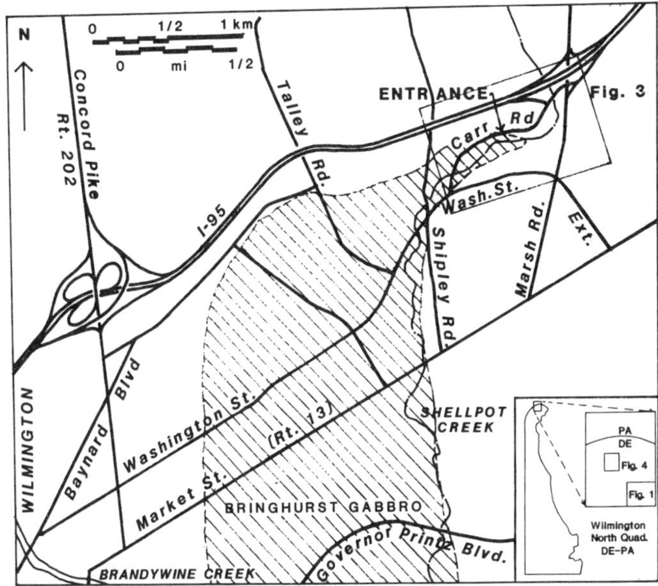

Figure 1. Map of northeastern Wilmington and outlying areas. Bringhurst gabbro shown by diagonal lines; rest of area is underlain by granulite facies gneisses.

SIGNIFICANCE

The Wilmington Complex consists of granulite facies gneisses intruded by plutonic igneous rocks. The largest of the plutons (Fig. 2) are the Arden (ranging from charnockite through norite, anorthositic norite, and gabbro) and the Bringhurst (gabbro, troctolite, norite, and anorthositic norite). Numerous other, smaller gabbroid bodies also intrude the gneisses. The age of the granulite facies metamorphism is early Paleozoic, making this one of very few Paleozoic granulite facies terranes in the Appalachians. The presence of abundant plutonic igneous rocks suggests that the granulite facies metamorphism may have been caused by a regional geothermal gradient that was higher than normal (Bohlen, 1983). This area may represent the deep infrastructure of an early Paleozoic volcanic arc that was subsequently thrust onto the North American continent (Crawford and Mark, 1982; Wagner, 1982; Wagner and Srogi, unpublished manuscript).

SITE INFORMATION

Geologic Setting. Bringhurst Woods Park lies at the northeastern tip of the Bringhurst gabbro (Fig. 1). Very coarse-grained gabbro with subophitic texture (including a pegmatitic phase) and both mafic and felsic granulite-facies gneisses can be seen here in the valley of Shellpot Creek. The contact between the gabbro and the gneisses is exposed in at least one place.

The rocks of the Wilmington Complex were first described by Bascom (Bascom and Miller, 1920; Bascom and Stose, 1932). In these two folios the entire Wilmington Complex is shown as gabbro. Ward (1959) recognized that granulite-facies gneisses underlie much of the area. He gave the name Wilmington Complex to the granulites and their associated plutonic igneous rocks and named the largest pluton the Arden "granite." He also described a small gabbro pluton in Bringhurst Woods Park. Woodruff and Thompson (1975) recognized that the Bringhurst pluton is much larger than was indicated by Ward. Its full extent is not known as it disappears below Cretaceous coastal plain sediments to the south. A low aeromagnetic anomaly directly overlies the gabbro and does not extend far into the coastal plain, suggesting that most of the gabbro is exposed. The aeromagnetic low is caused by the scarcity of magnetite in the igneous rocks, whereas magnetite is abundant in the surrounding gneisses.

The Wilmington Complex is surrounded on all sides (except the south and its northern tip) by gneisses of the Wissahickon Group, part of the Glenarm Supergroup of latest Precambrian or earliest Paleozoic age (Higgins, 1972). Along the northwestern edge of the Wilmington Complex, in the area of Brandywine Creek State Park, mapping by Srogi (1982) suggests that the Wilmington Complex overlies Wissahickon gneisses along a fault contact that strikes ~N45°E and dips ~5°SE. Hager and Thomp-

*Present addresses: Srogi, Department of Geology, Smith College, Northampton, Massachusetts 01060; Brick, 412 Linden St., Jenkintown, Pennsylvania 19046.

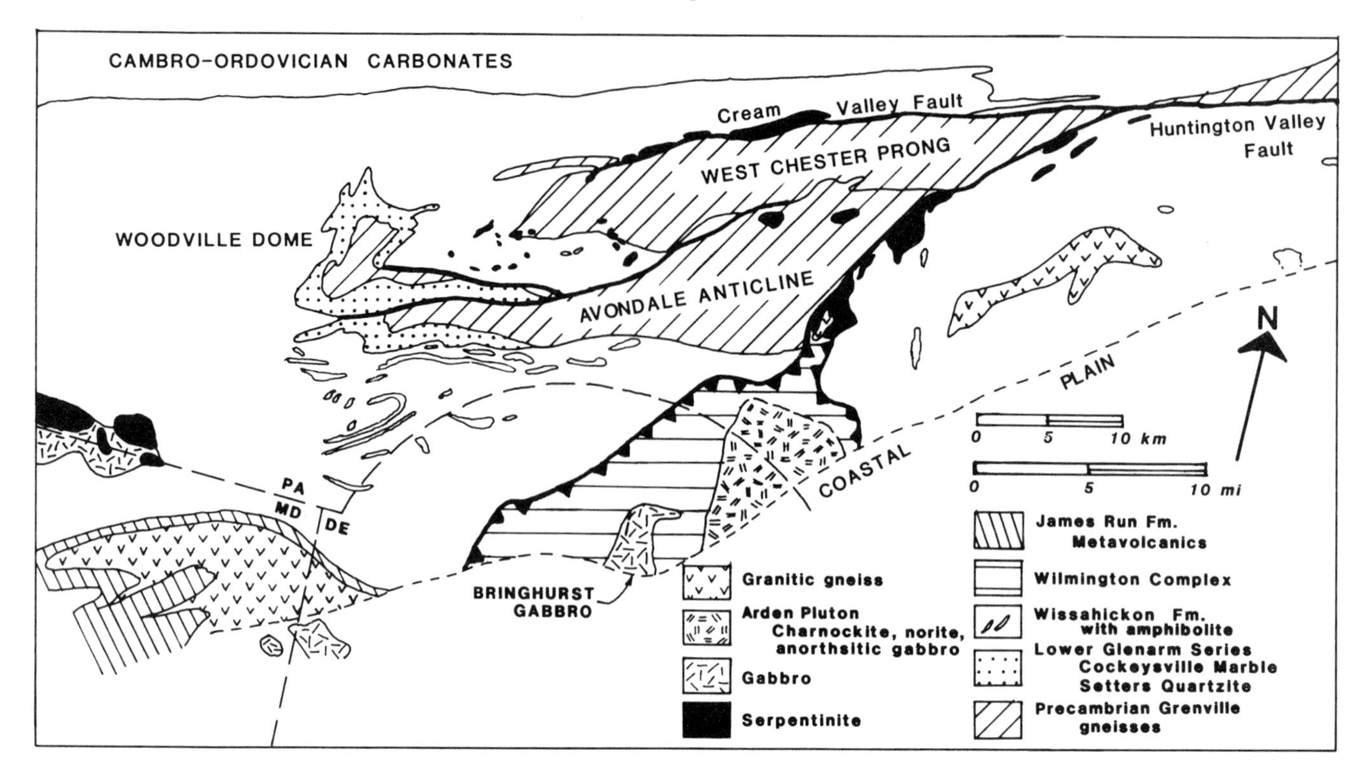

Figure 2. Geologic map of northern Delaware, southeastern Pennsylvania and part of eastern Maryland.

son (1975) also found evidence for a fault contact between the Wilmington Complex and the Wissahickon Group. The Wissahickon is typically above the second sillimanite isograd adjacent to the Wilmington Complex and falls off in grade a few kilometers away both to the east and west, suggesting that the metamorphism of the Wissahickon was caused, at least in part, by the high temperature of the Wilmington Complex at the time of its emplacement, and that the isograds in the Wissahickon are possibly inverted under the Wilmington Complex. The rocks of the Wilmington Complex are unconformably overlain by Cretaceous coastal plain sediments to the south. The northern tip of the Wilmington Complex is in a geologically complex area where poor exposure has impeded deciphering the relationships among granite, serpentinite, Wilmington Complex, Grenville gneisses, Wissahickon Group, and amphibolite.

Age Relationships. U/Pb isotopic analyses of zircons from one sample of felsic gneiss of the Wilmington Complex have yielded a nearly concordant age of 441 Ma. (Grauert and Wagner, 1975), which has been interpreted as the age of the granulite facies metamorphism. Foland and Muessig (1978) determined a whole rock Rb/Sr isochron of 502 ± 20 Ma for the Arden pluton, interpreted by those authors as the age of intrusion. Minerals separated from one sample gave a possible age consistent with resetting during the 441-Ma metamorphic event, although a precisely defined isochron did not emerge. There are some problems with the interpretation of the concordant 441-Ma zircon age as the age of metamorphism, primarily that the granulite-facies foliation of the felsic gneisses is deformed adjacent to the plutons in places. We plan further work on the U/Pb, Rb/Sr, and Sm/Nd systematics of the gneisses and plutons of the Wilmington Complex. Although one of us (LAS) has made Sm/Nd analyses of the Bringhurst gabbro, these analyses have not yielded a good isochron and the age remains undetermined.

Tectonic Setting. Crawford and Mark (1982) have proposed a model involving imbricate thrusting of the Wissahickon Group and Wilmington Complex, with a low-pressure metamorphism followed by a higher pressure metamorphism due to crustal thickening (both early Paleozoic in age), in order to explain overprinting relationships in the Wissahickon Group to the east of the Wilmington Complex. Wagner (1982) and Wagner and Srogi (unpublished manuscript) have suggested a model involving thrusting of a thick plate with hot Wilmington Complex at its base over cold North American continental rocks, producing moderate-pressure Paleozoic metamorphism of the Wilmington Complex and a high-pressure Paleozoic overprint on Precambrian granulite-facies gneisses and associated diabase dikes of the West Chester prong in Pennsylvania (Fig. 2).

The rocks that can be seen in Bringhurst Woods Park therefore represent deep-seated gneisses intruded by plutonic igneous rocks. The gneisses were subjected to granulite facies metamorphism during the early Paleozoic, probably because of a high geothermal gradient associated with the pervasive intrusion of mantle-derived magmas. The rocks may have formed in the deep infrastructure of a volcanic arc, overlying a subduction zone, and were thrust over the North American continent when the continental plate entered the east-dipping subduction zone.

Features of the Rocks within the Park. Most of the Bringhurst gabbro, including that found in Bringhurst Woods Park, is very coarse-grained with subophitic texture, but variations in grain size exist over a scale of only a few centimeters. Typical grains measure about 1 cm in diameter. Cumulate textures are commonly seen in thin section. The primary minerals of the gabbro are plagioclase, olivine, clinopyroxene, and orthopyroxene. Olivine is less common than the other minerals and when present is always surrounded by double coronas, with an inner corona of orthopyroxene, often in symplectites with spinel, and an outer corona of pargasitic hornblende, also in symplectites with spinel. Clinopyroxene is usually exsolved and also replaced in patches by hornblende as well as locally being surrounded by coronas of hornblende. The primary orthopyroxene is usually bronzite or hypersthene. The plagioclase, which in most of the gabbro ranges in composition from An_{55} to An_{80}, often contains small inclusions of orthopyroxene and hornblende. The plagioclase displays reverse zoning around orthopyroxene inclusions, e.g., in one place from An_{69} to An_{79}.

A pegmatitic phase of the gabbro, in which plagioclase laths measure up to 1 ft (30 cm) in length, is also found in the park. In this pegmatitic phase, orthopyroxene grains, weathered to a rusty color, are surrounded by coronas of black hornblende.

Typical Wilmington Complex gneiss was described by Ward (1959) as banded gneiss with alternating mafic and felsic bands. However, the banding is not always present and large areas consist of only felsic gneiss or mafic gneiss. The felsic gneiss is typically coarser grained than the mafic gneiss, and both the mafic and felsic gneisses are dark colored on fresh surfaces. A foliation, visible only on weathered surfaces, helps to distinguish the gneisses from the igneous rocks. The mineralogy of the felsic gneiss is orthopyroxene + clinopyroxene + plagioclase + quartz + magnetite ± biotite, whereas the assemblage in the mafic gneiss is orthopyroxene + clinopyroxene + plagioclase + magnetite ± hornblende. The quartz of the felsic gneisses is highly variable in abundance, often occurring in clots that stand out in relief on weathered surfaces. There is evidence of only one metamorphism and no evidence of disequilibrium in any of the pyroxene gneisses. At a few localities in the Wilmington Complex, of which Bringhurst Woods Park is one, garnet and/or cordierite have been found in aluminous gneisses close to contacts with gabbro. We believe these unusual lithologies, which have complex disequilibrium textures, are restites left from partial melting of narrow layers of pelitic composition adjacent to the hot intrusions (Srogi and others, 1985).

The equilibrium textures in the pyroxene gneisses are in contrast to those in the gabbros, in which igneous minerals have been partially replaced by metamorphic minerals and disequilibrium textures abound, due to subsolidus recrystallization during cooling under granulite facies conditions. They are also in sharp contrast to the Grenville gneisses of the West Chester prong in Pennsylvania a few kilometers away (Fig. 2), where Grenville granulite assemblages have been partially overprinted by high-pressure coronas (~650 to 700°C, 9 to 10 kb) during the Taconic orogeny (Wagner and Crawford, 1975; Wagner, 1982; Wagner and Srogi, unpublished manuscript). The granulite facies assemblages in the Wilmington Complex indicate P-T conditions of ~800 to 850°C and 6 to 6.5 kb during the early Paleozoic metamorphism (Srogi and others, 1983; Wagner and Srogi, unpublished manuscript).

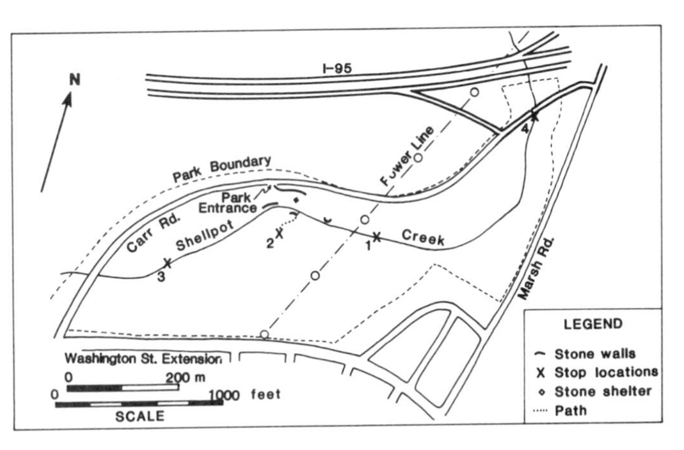

Figure 3. Map of Bringhurst Woods Park, showing location of features described in the text.

There are moderately abundant xenoliths of mafic granulite facies gneiss in the gabbro. Although felsic gneiss is much more abundant than mafic gneiss in the Wilmington Complex, no xenoliths of felsic gneiss have been found. We believe this is because any inclusions of felsic gneiss melted completely in the hot gabbroic magma.

Stop 1. About 150 ft (45 m) up Shellpot Creek from directly under the power line on the near side of the stream, an outcrop or large boulder in the bank of the stream shows the contact between the gabbro and mafic granulite facies gneiss (Fig. 3). (Please refrain from hammering in the vicinity of the contact.) The gneissic banding in some felsic gneisses near the contact is contorted and some rocks appear migmatitic. Detailed mapping of the gabbro (Brick, 1980) has shown that this is often the case around the margins of the pluton. Clots of quartz that have weathered in relief are conspicuous features of the felsic gneiss upstream from the contact. A few small offshoots of the gabbro can be seen upstream.

Stop 2. Boulders of a pegmatitic phase of the gabbro can be seen on the south side of Shellpot Creek (Fig. 3), about 165 ft (50 m) south-southwest of a very large boulder just east of a stone wall on the south bank. Since the boulders of the pegmatitic phase are rapidly disappearing due to indiscriminate hammering, please do not use your hammers on these rocks!

Stop 3. Additional boulders of the pegmatitic phase can be seen in the stream bed about 800 to 900 ft (250 to 275 m) southwest of the stone shelter near the parking area, where the creek's course changes from ~S60°W to ~S35°W. Near these pegmatitic boulders, a finer grained, olivine-bearing lithology shows flow structure.

Stop 4. One of the rare garnet-bearing lithologies can be seen in an outcrop in the middle of the stream at the easternmost extremity of the park, just south of the bridge that carries Carr Road over Shellpot Creek, near the I-95 offramp. This is one of the *very few* examples of this lithology; please leave it for others to see. At this same stop there are abundant foliated felsic gneisses and a dike of the gabbro that crosses the stream, particularly noticeable on the west bank because of its different weathering properties. Note that the foliation in the gneiss is contorted close to the contact with the dike and that flow structure within the dike is discordant to the foliation of the gneiss.

OTHER SITES TO VIEW THE WILMINGTON COMPLEX

Another area where rocks are well-exposed is north of Wilmington in Brandywine Creek State Park (Fig. 4). The contact between Wilmington Complex and Wissahickon Group runs through the park, striking ~N45°E, nearly parallel to the trend of Brandywine Creek to the southwest and its tributary, Rocky Run, to the northeast. Along the east side of the Brandywine from Thompson Bridge on Delaware 92 to Rocky Run, high-grade pelitic gneisses (above second sillimanite isograd) of the Wissahickon Group are exposed along the side of the valley. Granulite facies gneisses of the Wilmington Complex appear about 600 ft (200 m) beyond Rocky Run. Since the contact between Wilmington Complex and Wissahickon Group dips very shallowly (~5°SE), its exposure is strongly affected by the topography (Srogi, 1982). On the west side of the Brandywine, southeast of the park headquarters in the side of the valley below Marsh Overlook (Fig. 4), migmatitic-appearing Wissahickon outcrops are found, whereas hilltops in the park northwest of the valley are underlain by mafic Wilmington Complex (Srogi, 1982).

Boulders of the Bringhurst gabbro, including olivine-bearing lithologies, occur in a small park south of Market Street, in Wilmington, between 28th and 30th Streets. Most of the gabbro here is considerably more mafic than that in Bringhurst Woods Park.

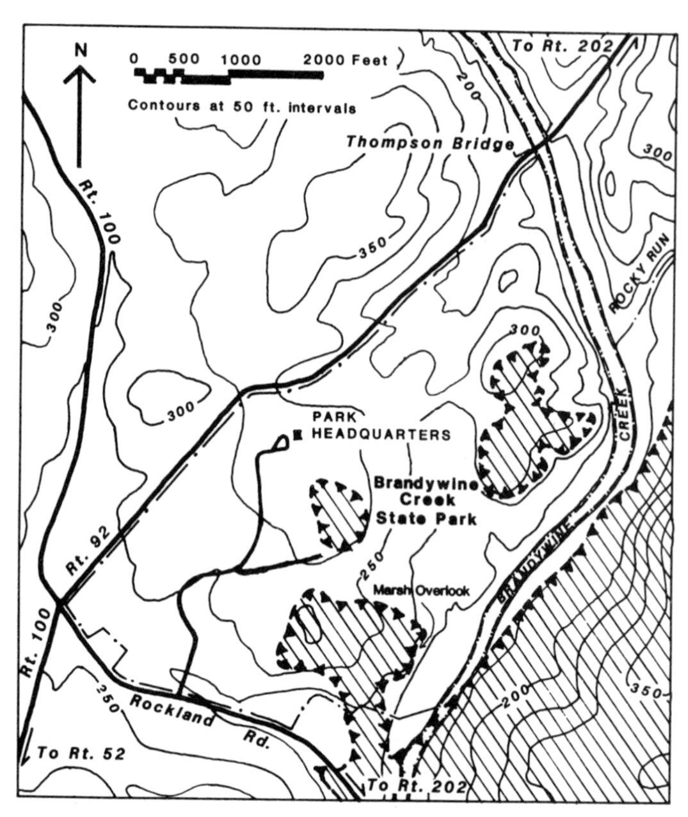

Figure 4. Map of Brandywine Creek State Park and surrounding areas. Contact shown is intersection with topography of hypothetical thrust striking N45°E and dipping 5°SE, which places Wilmington Complex (diagonal ruling) over Wissahickon Group metasediments (no pattern). Exposures in park correspond closely to suggested contact.

REFERENCES CITED

Bascom, F., and Miller, B. L., 1920, Description of the Elkton and Wilmington Quadrangles: U.S. Geological Survey Folio No. 211.

Bascom, F., and Stose, G. W., 1932, Description of the Coatesville and West Chester Quadrangles: U.S. Geological Survey Folio No. 223.

Bohlen, S. R., 1983, Retrograde P-T paths for granulites: EOS American Geophysical Union Transactions, v. 64, p. 878.

Brick, E., 1980, Field and petrographic study of the Bringhurst gabbro, northern Delaware [Senior thesis]: Philadelphia University of Pennsylvania.

Crawford, M. L., and Mark, L., 1982, Evidence from metamorphic rocks for overthrusting, Pennsylvania Piedmont: Canadian Mineralogist, v. 20, p. 333–347.

Foland, K. A., and Muessig, K. W., 1978, A Paleozoic age for some charnockitic-anorthositic rocks: Geology, v. 6, p. 143–146.

Grauert, B., and Wagner, M. E., 1975, Age of the granulite facies metamorphism of the Wilmington Complex, Delaware-Pennsylvania piedmont: American Journal of Science, v. 275, p. 683–691.

Hager, G. M., and Thompson, A. M., 1975, Fault origin of major ltihologic boundary in Delaware Piedmont: Geological Society of America Abstracts with Programs, v. 7, p. 68–69.

Higgins, M. W., 1972, Age, origin, regional relations, and nomenclature of the Glenarm Series, Central Appalachian Piedmont; A reinterpretation: Geological Society of America Bulletin, v. 83, p. 989–1026.

Srogi, L. A., 1982, A new interpretation of contact relationships and early Paleozoic history of the Delaware Piedmont: Geological Society of America Abstracts with Programs, v. 14, p. 85.

Srogi, E. A., Wagner, M. E., Lutz, T. L., and Hamre, J., 1983, Metamorphic and tectonic history of a Paleozoic granulite facies terrane, Delaware-Pennsylvania piedmont: Geological Society of America Abstracts with Programs, v. 15, p. 694.

Srogi, E. A., Wagner, M. E., and Lutz, T. L., 1985, Dehydration partial melting in the granulite facies; An example from the Wilmington complex, Delaware-Pennsylvania Piedmont; EOS American Geophysical Union Transactions, v. 66, p. 398.

Wagner, M. E., 1982, Tectonic metamorphism at two crustal levels and a tectonic model for the Pennsylvania-Delaware Piedmont: Geological Society of America Abstracts with Programs, v. 14, p. 640.

Wagner, M. E., and Crawford, M. L., 1975, Polymetamorphism of the Precambrian Baltimore gneiss in southeastern Pennsylvania: American Journal of Science, v. 275, p. 653–682.

Ward, R. F., 1959, Petrology and metamorphism of the Wilmington Complex, Delaware, Pennsylvania, and Maryland: Geological Society of America Bulletin, v. 70, p. 1425–1458.

Woodruff, K. D., and Thompson, A. M., 1975, Geology of the Wilmington area, Delaware: Delaware Geological Survey, Geologic Map Series No. 4.

Upper Paleozoic stratigraphy along the Allegheny topographic front at the Horseshoe Curve, west-central Pennsylvania

***Jon D. Inners**, Pennsylvania Geological Survey, P.O. Box 2357, Harrisburg, Pennsylvania 17120*

LOCATION

The Horseshoe Curve, a world-famous engineering landmark on the main line of the old Pennsylvania Railroad, is located 87 mi (140 km) west-northwest of Harrisburg and 3.7 mi (6 km) west of Altoona in Blair County, Pennsylvania (Fig. 1). Extensive exposures of Upper Devonian to Upper Pennsylvanian rocks occur along approximately 8.7 mi (14 km) of the railroad right-of-way (now Conrail) from a point just west of Altoona through the Horseshoe Curve to the Gallitzin Tunnels. Highway cuts on nearby U.S. 22 supplement the Lower Mississippian portion of the railroad section. All exposures are within the Hollidaysburg and Cresson 7½-minute quadrangles.

The long railroad traverse can be conveniently divided into two legs (Fig. 2). The lower part of the section (Leg 1: Upper Devonian and Lower Mississippian) crops out on the north side of Burgoon Run for a distance of 2.9 mi (4.6 km) from near the hamlet of Coburn to the head of Horseshoe Curve below Kittanning Point. This part of the cut can be reached from several places along Legislative Route 07023 (40th Street in Altoona). The upper part of the section (Leg 2: Upper Devonian to lowermost Upper Pennsylvanian) is exposed on the north side of the steep-sided valley of Sugar Run for 3 mi (5 km) from a sharp bend of the railroad 1.6 mi (2.6 km) south-southeast of the head of Horseshoe Curve to the Gallitzin Tunnels. Best access to both the east and west ends of this exposure is from Legislative Route 07108 (Sugar Run Road). Although the part of the traverse between the Burgoon Run and Sugar Run legs can be omitted (exposures are poor and repeat section better seen elsewhere), a fine view of the curve itself can be obtained on the south side of Burgoon Run valley at the southeast end of the great "U" (Fig. 3). Please note that the railroad exposures are on private property: permission to enter the right-of-way should be obtained from the Superintendent of the Altoona Division of Conrail at 1331 12th Street, Altoona, Pennsylvania 16601 (phone 814-949-1003).

The highway-cut exposures on U.S. 22 lie on the south side of Sugar Run valley 4.3 mi (7 km) northwest of the Duncansville interchange and 2.5 mi (4 km) east of the Gallitzin interchange. Parking on the shoulder of the road is permitted, but be wary of high-speed traffic on this 4-lane, divided highway.

SIGNIFICANCE

The Horseshoe Curve section provides the finest display of Upper Devonian to Pennsylvanian rocks in west-central Pennsylvania. Its continuity of exposure and ease of access have long made it a reference section for bedrock geologic mapping in this part of the state (Butts, 1905, 1945; Swartz, 1965). Of particular stratigraphic interest are the following: type section of the Burgoon Sandstone (Lower Mississippian), excellent exposures of Upper Devonian–Lower Mississippian transition strata, exposures that provide a transect across a distal portion of the Catskill Delta (Upper Devonian), and rocks representative of a wide range of Mississippian and Pennsylvanian deltaic and marginal marine environments, including several Allegheny coals (Middle Pennsylvanian). Good opportunities for fossil collecting (plants, invertebrates, and traces) are also available.

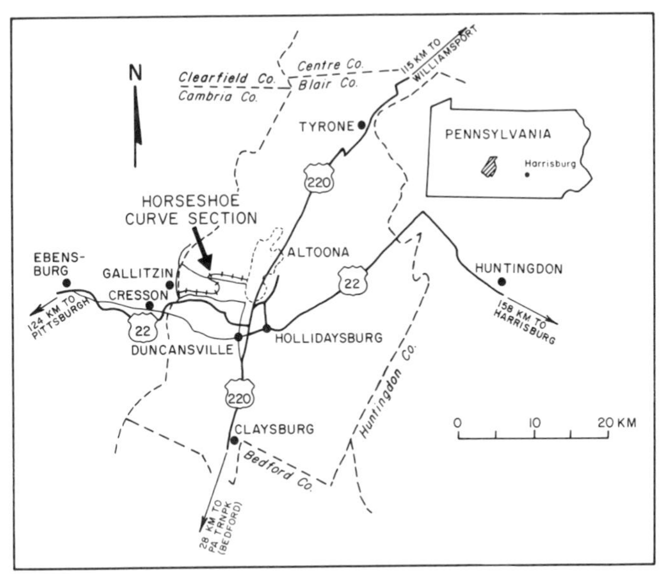

Figure 1. Location map.

SITE INFORMATION

Figure 4 shows the litho- and chronostratigraphic framework of the bedrock formations in the Horseshoe Curve area.

Upper Devonian. Except for the lower half of the Rockwell Formation, all Upper Devonian rocks in the Horseshoe Curve section are part of the Catskill Delta complex. This great eastward- and southeastward-thickening wedge of marine and continental sediments was shed from a tectonic highland that was pushed up along the eastern edge of the Laurentian continent during the Late Devonian Acadian orogeny, approximately 373 Ma (Kent, 1985; Faill, 1985). In west-central Pennsylvania, the submarine portion of the Catskill sediment wedge is represented by the Lock Haven Formation (as well as the underlying Brallier Formation), the shoreface portion by the Irish Valley Member of the Catskill Formation, and the subaerial part by the Sherman Creek and Duncannon members of the Catskill Formation. These

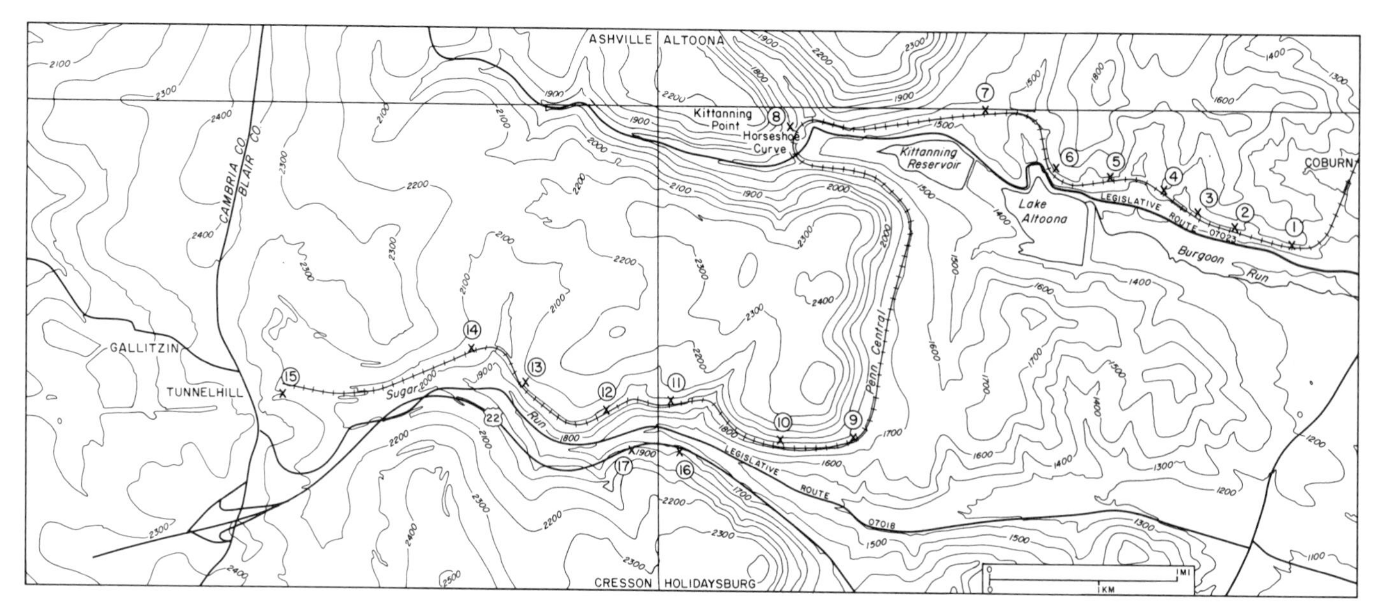

Figure 2. Index map for Horseshoe Curve section, showing location of stops. (Site 8: Railroad Section Leg 1, Stops 1 to 8; Site 9: Railroad Section Leg 2, Stops 9 to 15; U.S. 22 Section, Stops 16 and 17).

sedimentary units formed in a variety of environmental settings generated by a northwestward-prograding coastline (Rahmanian, 1979; Williams, 1985). The predominantly subaerial part of the delta (Catskill Formation *sensu stricto*) increases in thickness from about 1,970 ft (600 m) in the Horseshoe Curve area to about 7,550 ft (2,300 m) in the central Susquehanna Valley, concomitant with a decrease in the thickness of the submarine part from 4,590 to 1,970 ft (1,400 to 600 m; Swartz, 1965; Dyson, 1963).

Sedimentation in the Catskill Formation is typically cyclic, characterized by marine-nonmarine motifs in the Irish Valley Member (Walker and Harms, 1971; Rahmanian, 1979) and fining-upward alluvial cycles in the Sherman Creek and Duncannon members (Allen and Friend, 1968; Sevon, 1985). The cycles are usually 6 to 60 ft (2 to 20 m) thick in the Irish Valley, 5 to 30 ft (1.5 to 9 m) thick in the Sherman Creek, and 10 to 60 ft (3 to 20 m) thick in the Duncannon (Rahmanian, 1979; see Fig. 6). Differences between the fining-upward cycles of the Sherman Creek and Duncannon members are largely the result of stream regimen and paleogeographic setting: the thinner, finer-grained Sherman Creek cycles reflect sedimentation by shallow low-gradient, high-sinuosity meandering streams on broad, relatively inactive flood basins, whereas the thicker, coarser-grained Duncannon cycles were deposited by deeper, moderate- to high-sinuosity, meandering streams on an active alluvial plain (Rahmanian, 1979; Sevon, 1985).

Rahmanian (1979) has documented the existence of a major sediment-input center along the Catskill shoreline in the vicinity of Port Matilda, 33 mi (55 km) northeast of Horseshoe Curve (Fig. 5). In contrast to the active delta progradation in the Port Matilda area, the shoreline at Horseshoe Curve was built-up mainly as a tidal-flat/chenier-plain complex supplied with sediment from this delta lobe (and possibly an unidentified one to the south) by longshore currents.

The Upper Devonian portion of the Rockwell Formation was presumably deposited after northwestward progradation on the Catskill Delta had ceased (Sevon, 1985). These rocks (predominantly gray, fine- to medium-grained sandstones) are near-shore and coastal plain equivalents of the transgressive, latest Devonian Oswayo Formation of northwestern Pennsylvania (Berg and Edmunds, 1979; Berg and others, 1983). Like the younger Burgoon (Lower Mississippian) and Pottsville (Lower Pennsylvanian) sandstones, the Rockwell sediments were probably derived from stream cannibalization of the proximal portions of the former Catskill alluvial plain rather than from erosion of the highlands (Sevon, 1985).

Mississippian. The Mississippian succession at Horseshoe Curve consists predominantly of quartz-rich sandstone in the lower three-fourths (upper Rockwell Formation, Burgoon Sandstone, and Loyalhanna Formation) and variegated coarse and fine clastic rock in the upper one-fourth (Mauch Chunk Formation). While the conformable Devonian-Mississippian systemic boundary exists within a well-exposed transitional sequence in the Rockwell Formation, the Mississippian-Pennsylvanian boundary falls within a concealed interval (see Swartz, 1965). North of Horseshoe Curve, an erosional hiatus exists between preserved Mississippian and Pennsylvanian rocks (Edmunds, 1968; Glass, 1972). Evidence bearing on the possible southward extension of this disconformity is lacking in the Horseshoe Curve area.

The upper part of the Rockwell Formation presumably formed in shallow marine and coastal alluvial-plain environ-

Figure 3. View of Horseshoe Curve from southeast side of Burgoon Run valley. Kittanning Point and Stop 8 are at head of the curve. (Photo courtesy of Altoona Public Library and Railroaders' Memorial Museum.)

ments. Thick, gray sandstones are predominantly fluvial (meandering streams), and the purplish red and green mudstones are tidal-flat and/or alluvial overbank deposits (see Berg and Glover, 1976).

Overlying the Rockwell above a sharp erosional contact is the Burgoon Sandstone, a sequence of thick-bedded, coarse-grained to conglomeratic sandstones that accumulated on a great braided alluvial plain.

The Loyalhanna Formation—which abruptly succeeds the Burgoon—is a thin tongue of the mid-Mississippian Greenbrier Group limestone that extends across southwestern and central Pennsylvania (Edmunds and others, 1979). Wells (1974) has traced this unit along the Allegheny Front as far northeast as Sullivan County, 37 mi (60 km) east of Williamsport. Although the spectacular trough cross-bedding of the Loyalhanna is highly suggestive of eolian deposition, the most detailed recent study (Adams, 1970) indicates that these rocks were formed by migrating shallow-marine tidal sand waves with transport direction to the northeast and east-northeast.

Nonmarine conditions returned with the deposition of the overlying Mauch Chunk Formation. While this stratigraphic unit is only 165 ft (50 m) thick in the Horseshoe Curve area, it thickens to more than 4,000 ft (1,200 m) in the anthracite region of eastern Pennsylvania (Wood and others, 1969). The Mauch Chunk represents alluvial deposition on a delta that prograded in a northwesterly direction across the area, progressively pushing the marine Mississippian ("Greenbrier") to the west (Edmunds and others, 1979).

Pennsylvanian. Pennsylvanian rocks in the Horseshoe Curve area consist of a highly variable, interbedded sequence of sandstones, siltstones, claystones, shales, fresh-water limestones,

"AGE"				UNIT		THICKNESS M	LITHOLOGY
PENNSYLVANIAN	Upper		MISSOURIAN	CONEMAUGH GROUP GLENSHAW FM.		12 +	Greenish-gray silty clay shale. Poorly exposed.
PENNSYLVANIAN	Middle	WESTPHALIAN	DESMOINESIAN	ALLEGHENY GROUP		105	Irregularly cyclic sequences of clastic rocks and coal beds (to 1.8 meters thick). Some fresh-water limestones. Mostly gray and olive colors. STOPS 14 & 15.
PENNSYLVANIAN	Middle / Lower	WESTPHALIAN / NAMURIAN	ATOKAN / MORROWAN	POTTSVILLE GROUP		32	Light gray quartzose sandstone, clay shale, coal (not exposed), and underclay. Somewhat cyclic. STOPS 13 & 14.
MISSISSIPPIAN	Upper	NAMURIAN / VISEAN	CHESTERIAN / MERE-MECIAN	MAUCH CHUNK FM.		50	Gray quartzose sandstone and conglomerate; red and green silty claystones at top and bottom. STOP 12.
MISSISSIPPIAN	Upper	VISEAN	MERE-MECIAN	LOYALHANNA FM.		14	Gray, strongly cross-bedded calcareous sandstone. STOP 11.
MISSISSIPPIAN	Lower	TOURNAISIAN	OSAGIAN	BURGOON SS.		91	Gray, cross-bedded quarzitic sandstone, pitted at top. Siltstone, shale, and coal in middle. STOPS 11, 16 & 17.
MISSISSIPPIAN	Lower	TOURNAISIAN	KINDER-HOOKIAN	ROCKWELL FM.		180	Predominantly gray sandstone, but also containing considerable siltstone and silty clay shale. Fine clastics locally purplish red; Patton "red beds" at top. STOPS 8, 9, 10, & 16.
DEVONIAN	Upper	FAMENNIAN	CONEWANGOAN	CATSKILL FM.	DUNCANNON MBR	30	Fluvial fining-upward cycle(s), consisting of basal gray sandstone grading up into red silty claystone. STOP 9.
DEVONIAN	Upper	FAMENNIAN	CONEWANGOAN / CASADAGAN	CATSKILL FM.	SHERMAN CREEK MBR.	480	Interbedded grayish-red silty claystone, siltstone and sandstone; some greenish-gray beds of these same lithologies. Fining upward cycles common. STOPS 6 & 7.
DEVONIAN	Upper	FAMENNIAN	CASADAGAN	CATSKILL FM.	IRISH VALLEY MBR.	83	Interbedded gray and red sandstone, siltstone, and silty claystone, mostly arranged in marine-nonmarine motifs. STOP 5.
DEVONIAN	Upper	FRASNIAN	CHEMUNGIAN	LOCK HAVEN FM.		820	Interbedded gray silty clay shale, siltstone and sandstone, minor quartz pebble conglomerate. Lower part thinner bedded and finer grained than upper. STOPS 1, 2, 3, & 4.

Figure 4. Upper Paleozoic bedrock units exposed at Horseshoe Curve (keyed to stop descriptions).

and coals (Butts, 1905; Swartz, 1965). Individual lithologic units generally have limited areal extent, commonly occurring as lenses and pods with thicknesses ranging from a few centimeters to 30 ft (10 m) or more. A crude cyclicity characterizes the Pennsylvanian rock succession. Cycles typical of the Allegheny Formation have the following basic pattern (after Edmunds, 1968): (5) fluvial-deltaic sediments (typically sandstones); (4) localized swamp deposits (rider coals); (3) open-water sediments (marine to fresh-water shales and siltstone); (2) swamp deposits (major coals); (1) backswamp sediments (rooted claystones and fresh-water limestones). Such cyclic sedimentation is believed to result from a combination of isostatic or tectonic fluctuations (Edmunds, 1968) and delta-distributary switching and channel avulsion (Skema and others, 1982).

Eight coal seams occur within the Pottsville and Allegheny groups (Fig. 6). The main targets of current mining are the Mercer, Brookville (A), Lower Kittanning (B), and Upper Free-

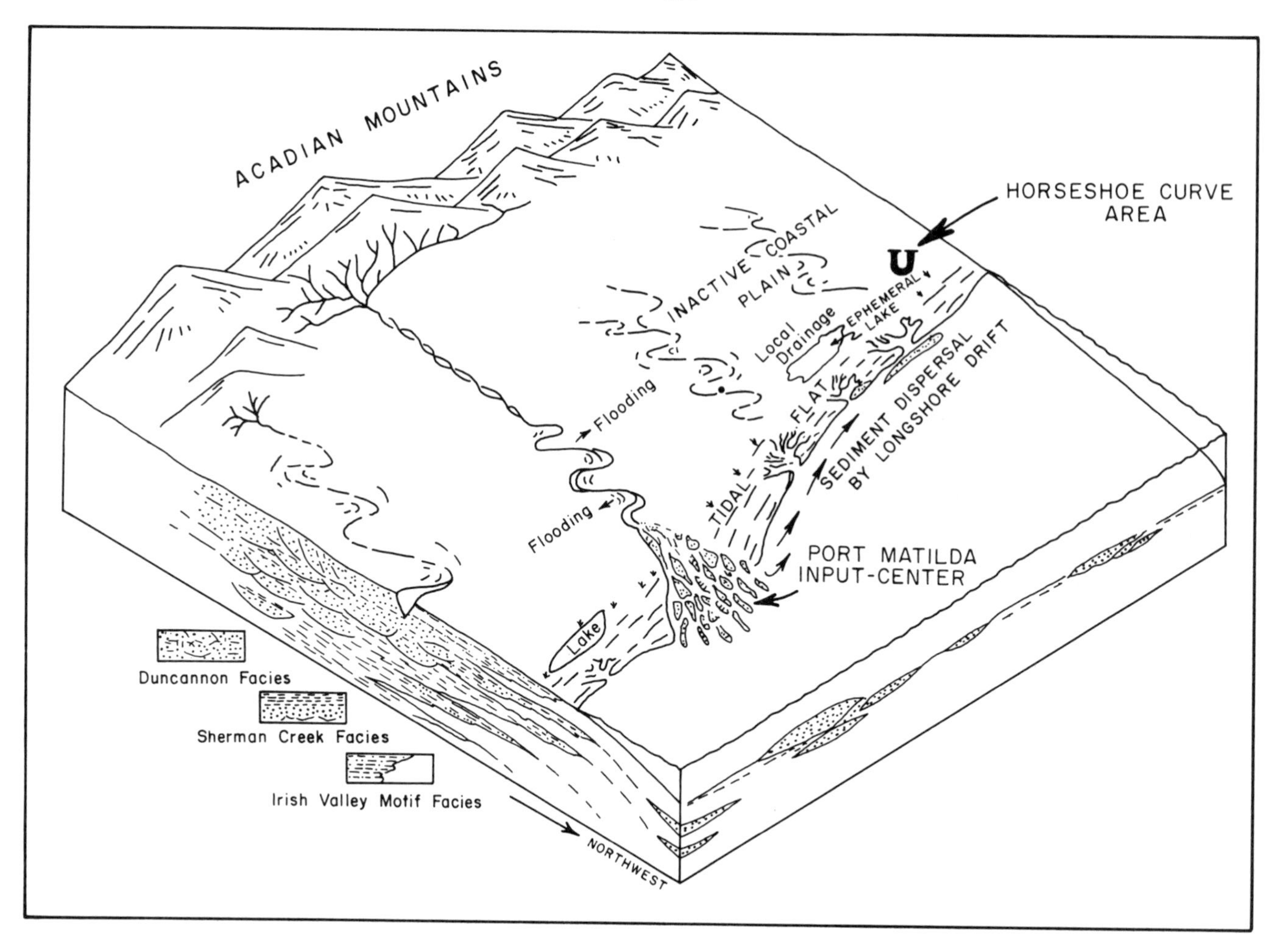

Figure 5. Sedimentation model of Upper Devonian in central Pennsylvania, showing Port Matilda input center. Slightly modified from Rahmanian (1979).

port (E). Although all mining is presently by surface methods, the B and E coals were at one time extensively deep mined, especially in the Sugar Run valley (Platt and Platt, 1877; Platt, 1881; Butts, 1905). Only the A, A′, D, and E coals are exposed along the railroad right-of-way.

SITE DESCRIPTIONS

Stop descriptions for Sites 8 and 9 are given below. Figures 2, 4, and 6 provide the geographic locations, lithologic summaries, and stratigraphic information, respectively, for each stop. The traverses along each leg of the railroad section proceed upsection from the east to the west.

SITE 8

Railroad Section, Leg 1: Burgoon Run Valley (Fig. 6A)

Stop 1. Lock Haven Formation (at Milepost 239). The strata exposed here occur in the basal part of the "Chemung Formation" of Butts (1945) and Swartz (1965). Sedimentary features include thin (0.4 to 6 in; 1 to 15 cm) siltstone-sandstone beds with planar bases and rippled tops; horizontal, fucoidlike burrows on the bases of some beds; and locally intense bioturbation. Common fossils include the brachhiopods *Productella, Ambocoelia, Cariniferella,* and *Atrypa* (Swartz, 1965). These rocks formed from sediments that accumulated on an offshore, shallow-marine shelf subject to occasional storm deposition (Rahmanian, 1979).

Stop 2. Lock Haven Formation (just west of power-line crossing). Many of the fine sandstone beds exposed here exhibit small scale cross-laminations and ball-and-pillow structures. Swartz (1965) noted the occurrence of brachiopods (*Douvillina, Productella, Atrypa, Tylothyris,* and *Ambocoelia*) and bryozoans at several horizons. These sediments represent deposition on a more proximal portion of the marine shelf than those at Stop 1 (see Rahmanian, 1979).

At the east end of this cut is a steeply west-dipping thrust fault containing mesoscopic kink folds (see Swartz, 1965).

Stop 3. Lock Haven Formation (at easternmost rail-

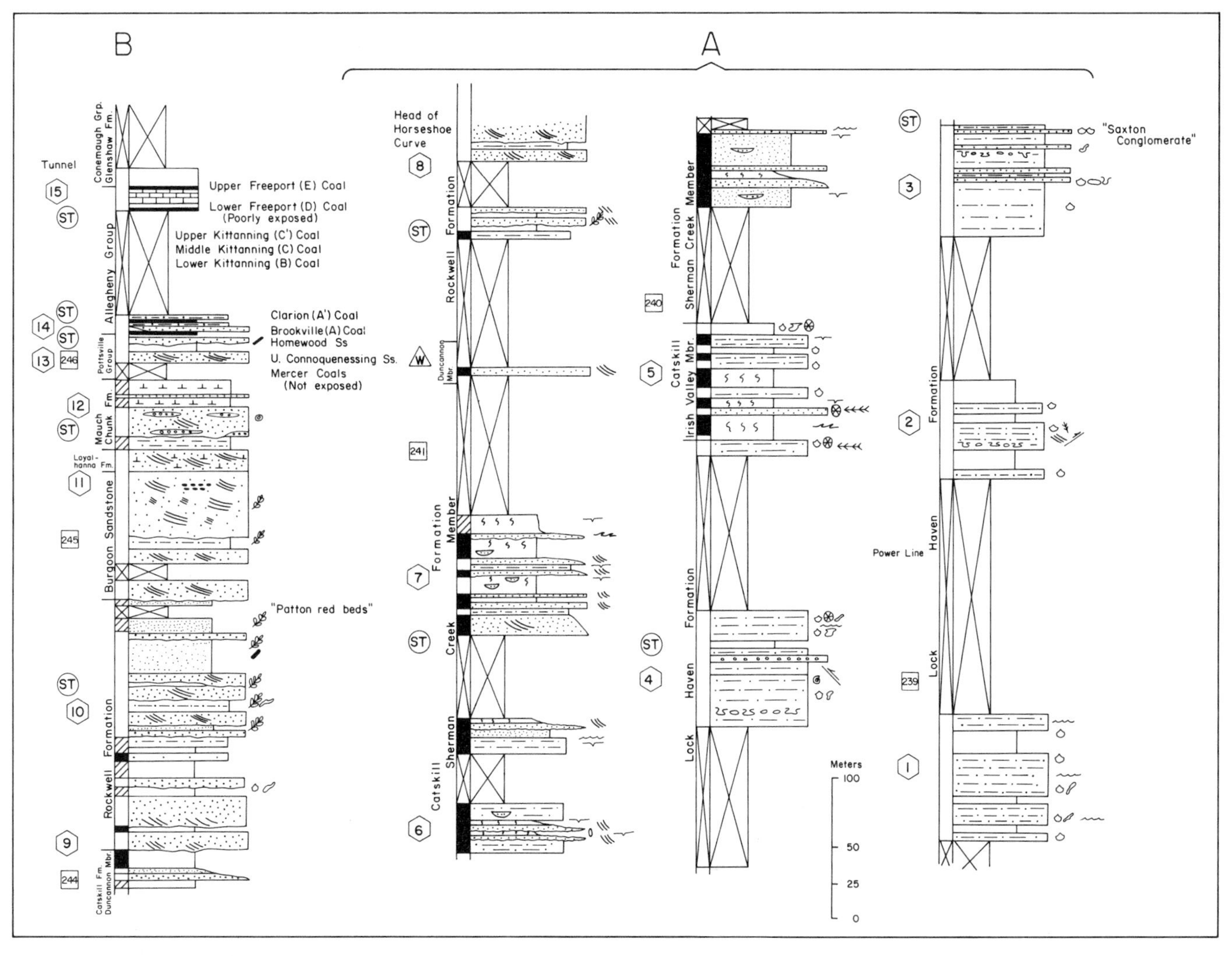

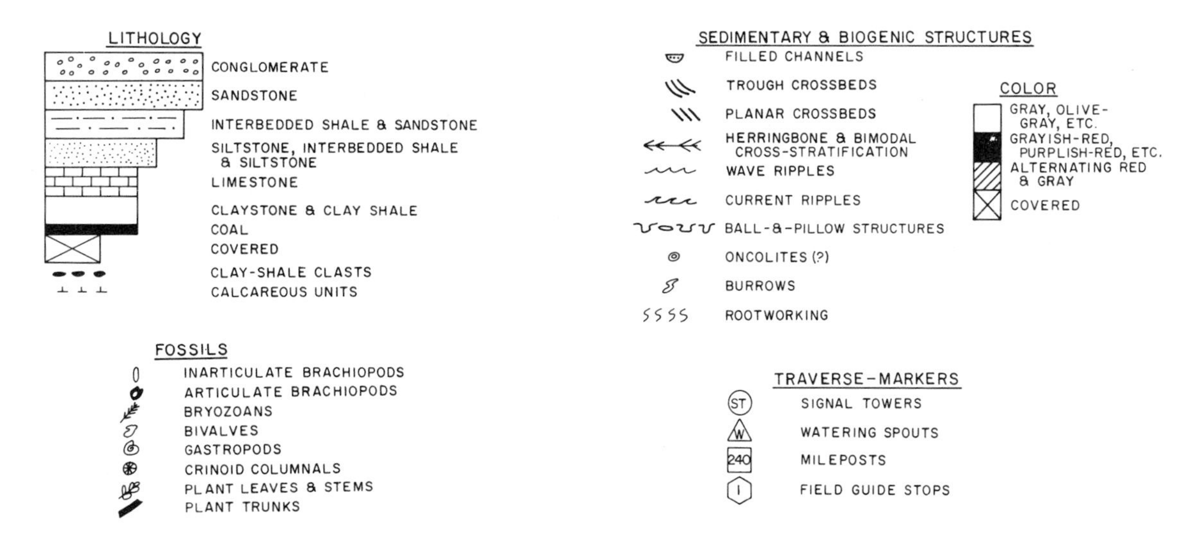

Figure 6. Stratigraphic sections at Horseshoe Curve, showing position of stops (in hexagons). A, north side of Burgoon Run valley (Railroad Section, Leg 1); B, north side of Sugar Run valley (Railroad Section, Leg 2). Modified from Swartz (1965) and Rahmanian (1979).

road signal tower in valley). Five sandstone packets form distinct ledges separated by recessed intervals of silty clay shale. Ball-and-pillow structures, load casts, bioturbation (large fucoid-like burrows), climbing ripples, and symmetrical ripples characterize the sandstone beds. The brachiopods *Orthospirifer, Ambocoelia,* and *Camarotoechia* occur throughout. At the top of the exposure is a coquinite composed of the disarticulated shells of *Orthospirifer.* The sandstones probably formed from migrating tidal sand waves and ridges (Rahmanian, 1979).

Butts (1945) correlated the sandy, brachiopod-coquinite at the west end of the exposure with the "Saxton conglomerate," one of several, coarse offshore-bar or tidal-channel deposits better developed in the Lock Haven to the east of the Horseshoe Curve area.

Stop 4. Lock Haven Formation (at second signal tower). Two distinctive lithologies at this stop are purplish gray to brownish red silty clay shale at the west end (upper part) and a flat-pebble, quartz conglomerate bed about 1 ft (0.3 m) thick at a point 75 ft (23 m) east of the signal tower. Climbing ripples, symmetrical ripples, bioturbation, flaser bedding, herringbone cross-stratification, shallow scour-and-fill structures, and isolated ball-and-pillow structures, occur at various horizons (Rahmanian, 1979). Brachiopods (*Cyrtospirifer, Atrypa,* etc.), bivalves, and crinoid ossicles are common here. The depositional environment is similar to that at Stop 3, but probably somewhat more proximal.

About 50 ft (15 m) east of the signal tower is a thrust fault that dips 50° to the east or southeast and has about 6 ft (2 m) of displacement.

Stop 5. Irish Valley Member of Catskill Formation (east of Milepost 240). The greenish gray and grayish red silty clay shales, siltstones, and sandstones here are organized into fining-upward "motifs" (Walker and Harms, 1971). Four such motifs are exposed (Rahmanian, 1979). Sedimentary structures include symmetrical and asymmetrical ripples, herringbone cross-stratification (in a 16-ft-thick [5 m] sandstone bed 50 ft [15 m] above the base), flaser bedding, mudcracks, bioturbation, and rootworking (Rahmanian, 1979). Brachiopods, bivalves, and crinoid ossicles are diagnostic of the marine portions of the motifs. According to Rahmanian (1979), the sedimentary environment is shallow marine shelf-shoreface (locally containing tidal sand waves or tidal deltas) grading upward into tidal and supratidal mudflats.

Stop 6. Sherman Creek Member of Catskill Formation (north of Lake Altoona). Trough cross-bedding, parting-step lineation, and symmetrical ripples occur in the red sandstones. The red claystones are intensely rootworked and bioturbated. Carbonaceous plant remains are common in the basal parts of a few pale red sandstones. In the red claystone beneath the most prominent sandstone, the inarticulate brachiopod *Lingula* is common. The fluvial fining-upward cycles at this stop are characteristic of meandering-stream deposits on the lower delta plain. The occurrence of *Lingula* suggests a distal setting, subject to brief brackish water incursions.

Stop 7. Sherman Creek Member of Catskill Formation (east of Milepost 241). The fining-upward cycles at this stop differ from those at the previous one in containing a few thin olive-gray sandstone and clay shale beds. Trough cross-bedding is well developed in the basal sandstones of the cycles. The sandstones also exhibit asymmetrical ripples and basal erosional scour. Mudcracks, bioturbation, and rootworking characterizes the red claystones. Carbonaceous plant remains are abundant in the sandstones.

Stop 8. Rockwell Formation (at head of Horseshoe Curve; see Fig. 3). Trough and planar cross-bedding and carbonaceous plant remains are common in the thick-bedded, greenish gray sandstones exposed below Kittanning Point. Presumably these rocks represent a series of stacked, meandering-stream channel deposits on a coastal alluvial plain.

SITE 9

Railroad Section, Leg 2. Sugar Run Valley (Fig. 6B)

Stop 9. Duncannon Member of the Catskill Formation and contact with the overlying Rockwell Formation (at Milepost 244). In the Horseshoe Curve area, the Duncannon Member is less than 98 ft (30 m) thick and comprises only one or two fining-upward cycles. The Rockwell/Duncannon contact occurs about 49 ft (15 m) west of Milepost 244 (see Swartz, 1965). Trough cross-bedding and carbonaceous plant remains occur in the sandstones of both formations. The basal sandstones of the Rockwell contain interference ripples at several horizons. Rootworking is profuse in the red claystones of the Duncannon. At this stop the meandering stream deposits of the Duncannon upper delta plain grade upward into basal meandering stream deposits of the Rockwell coastal alluvial plain.

Stop 10. Rockwell Formation (in vicinity of first, or easternmost, signal tower in valley). The sandstones of the Rockwell here exhibit both trough cross-bedding and planar bedding. Invertebrate fossils (bivalves and rhynchonellid brachiopods) occur in some of the greenish gray silty clay shales interbedded with the dominant sandstones at this stop. Some thin argillaceous sandstone beds contain basal fucoidlike horizontal burrows. Identifiable plant remains reported from these beds include *Triphyllopteris, Rhodea, Adiantites, Lepidodendropsis,* and *Archaeopteris.* Periodic marine transgressions within the alluvial-plain complex formed the fine-grained fossiliferous, estuarine (?) deposits.

From a study of spores collected here, Streel and Traverse (1978) concluded that the Mississippian-Devonian boundary lies within the stratigraphic interval from 246 to 459 ft (75 to 140 m) below the top of the Rockwell Formation. Thick sandstones about 820 ft (250 m) east of the signal tower apparently correlate with the Mississippian Murrysville Sand, an important gas-producing unit in western Pennsylvania (Bayles, 1949). The Patton "red beds" at the top of the Rockwell (poorly exposed here) are a key horizon marker for surface and subsurface mapping along the Allegheny Front and in the eastern portion of the Allegheny Plateau.

Stop 11. Burgoon Sandstone and Loyalhanna Formation (at Milepost 245). The spectacularly trough cross-bedded, calcareous sandstone of the Loyalhanna Formation here abruptly overlies the thick-bedded, coarse-grained, trough and planar cross-bedded sandstones of the Burgoon Sandstone. Planar cross-beds in the Burgoon indicate a west-northwest transport direction. According to Adams (1970), the transport direction in the Loyalhanna was to the northeast and east-northeast. A few plant fossils (*Triphyllopteris, Rhodea*) can be found in the Burgoon (Swartz, 1965); no plant or invertebrate fossils are reported from the Loyalhanna. The Burgoon sandstones are braided river deposits on a great coastal sandplain, whereas the Loyalhanna probably represents shallow-marine reworking of these sediments into tidal sand waves during a transgressive episode (Adams, 1970).

Stop 12. Mauch Chunk Formation (near second signal tower). The thick-bedded, greenish gray, calcareous, locally conglomeratic sandstones at this stop display good trough cross-bedding and locally contain limestone nodules 0.4 to 1.2 in (1 to 3 cm) in diameter that may be of algal origin (Swartz, 1965). These rocks probably formed from deposits of meandering rivers on a coastal alluvial plain.

Stop 13. Upper Connoquenessing Sandstone of Pottsville Group (at Milepost 246). The excellent trough cross-bedding and paucity of finer interbeds indicates that these thick-bedded, medium- to coarse-grained, quartzose sandstones are braided stream deposits, probably formed in an upper delta plain environment.

Stop 14. Upper Pottsville Group (including Homewood Sandstone) and lower Allegheny Group (including Brookville and Clarion coals; in Bennington Cut at third and fourth signal towers). The thin to thick-bedded, channel-fill sandstones display trough cross-bedding and basal cutouts, and the olive-gray underclays are extensively rootworked. *Lepidodendron* casts are common in the Homewood (the lower of the two sandstones exposed in the cut). These rocks represent deposition on a broad delta plain (fluvial and peat-swamp deposits).

The channel sandstone resting directly on top of the Brookville (A) coal displays a common type of superposition in coal-bearing rock sequences. Since compacted peat is difficult to erode, downcutting streams and migrating delta distributaries that are able to remove overlying sediments cannot incise the matted peat.

For various sedimentologic and petrographic reasons, Williams (1960) considered the 5 ft (1.5 m) coal, here called Brookville, to be a Mercer coal (older) and the lower sandstone, here called Homewood, to be the Upper Connoquenessing. However, the stratigraphic interval between this coal and the Lower Kittanning (reported to outcrop in the old Bennington Cemetery by F. Platt and W. Platt, 1877) is only about 52 ft (16 m), far too thin for the lower coal to be a Mercer (see also Butts, 1905). Recent drilling in this area indicates that locally mineable coals occurring approximately 60 ft (20 m) lower in the section actually represent the Mercer seams.

Stop 15. Uppermost Allegheny Formation (including Upper Freeport, or E, coal; at eastbound Gallitzin Tunnel). Well exposed here is a portion of a typical Allegheny sedimentary cycle composed of medium-dark-gray, finely crystalline freshwater limestone (base), greenish gray silty clay shale and fine-grained sandstone, rootworked underclay, and a 5 ft (1.5 m) coal (E; top). These rocks formed from peat swamp and lake deposits on the lower delta plain.

U.S. 22 Section: Sugar Run Valley

Stop 16. Rockwell Formation and Burgoon Sandstone (in eastern cut). The Rockwell Formation is exposed east of the high, subvertical cut and consists of cyclic units of greenish gray sandstone and siltstone and greenish gray to pale red or purplish gray, silty clay shales and claystone; at the top is variegated shale and claystone of the Patton Member. The presplit, subvertical cut exposes the massive sandstones of the Burgoon. The sandstones of both formations exhibit planar bedding and both planar and trough cross-bedding. Pyrite nodules and lenses to 5 ft long and 1 ft thick (1.5 and 0.3 m) occur in the lower part of the Burgoon. The base of the Burgoon is marked by a spectacular erosional cutout, with relief of up to 16.4 ft (5 m). Sandstone casts of plant trunks and stems are common throughout the cut.

Stop 17. Burgoon Sandstone (in western cut). In addition to the thick-bedding and trough and planar cross-bedding typical of braided river deposits, the Burgoon here displays pockets of gray-shale-chip conglomerate containing rounded clasts to 6 in (15 cm) in diameter. A thin coaly layer crops out in the lower middle part of the cut. Fossils include abundant comminuted plant debris in the coal-bearing shaly zone and casts of trunks and stems in the sandstones.

The Burgoon Sandstone at Stops 16 and 17 exhibits two distinct pulses of braided stream alluviation (each represented by sand bodies about 148 ft [45 m] thick) separated by an interval of reduced sedimentation (shaly zone with coal). The coaly layer is 3.6 to 4.6 ft (1.1 to 1.4 m) thick and consists of two high-ash "coal" beds (the upper of which is actually coaly shale) separated by 1 ft (0.3 m) of rootworked claystone (A. D. Glover, Pennsylvania Geological Survey, personal communication, 1985).

REFERENCES CITED

Adams, R. W., 1970, Loyalhanna Limestone; Cross-bedding and provenance, *in* Fisher, G. W., Pettijohn, F. J., Reed, J. C., Jr., eds., Studies of Appalachian geology; Central and southern: New York, Interscience, p. 83–100.

Allen, J.R.L., and Friend, P. F., 1968, Deposition of the Catskill facies, Appalachian region, with notes on some other Old Red Sandstone basins, *in* Klein, G. deV., ed., Late Paleozoic and Mesozoic continental sedimentation, northeastern North America: Geological Society of America Special Paper 160, p. 21–74.

Bayles, R. C., 1949, Subsurface Upper Devonian sections in southwestern Pennsylvania: American Association of Petroleum Geologists Bulletin, v. 33, p. 1682–1703.

Berg, T. M., and Edmunds, W. E., 1979, The Huntley Mountain Formation; Catskill-to-Burgoon transition in north-central Pennsylvania: Pennsylvania Geological Survey, 4th ser., Information Circular 83, 80 p.

Berg, T. M., and Glover, A. D., 1976, Geology and mineral resources of the Sabula and Penfield quadrangles, Clearfield, Elk, and Jefferson counties, Pennsylvania: Pennsylvania Geological Survey, 4th ser., Atlas 74ab, 98 p.

Berg, T. M., McInerney, M. K., Way, J. H., and MacLachlan, D. B., 1983, Stratigraphic correlation chart of Pennsylvania: Pennsylvania Geological Survey, 4th ser., General Geology Report 75.

Butts, C., 1905, Description of the Ebensburg Quadrangle: U.S. Geological Survey Atlas, Folio 8, 9 p.

—— , 1945, Description of the Hollidaysburg and Huntingdon quadrangles: U.S. Geological Survey Atlas, Folio 227, 20 p.

Dyson, J. L., 1963, Geology and mineral resources of the northern half of the New Bloomfield Quadrangle: Pennsylvania Geological Survey, 4th ser., Atlas 137ab, 63 p.

Edmunds, W. E., 1968, Geology and mineral resources of the northern half of the Houtzdale 15-minute Quadrangle, Pennsylvania: Pennsylvania Geological Survey, 4th ser., Atlas 85ab, 150 p.

Edmunds, W. E., Berg, T. M., Sevon, W. D., Piotrowski, R. C., Heyman, L., and Rickard, L. V., 1979, The Mississippian and Pennsylvanian (Carboniferous) systems in the United States, Pennsylvania and New York: U.S. Geological Survey Professional Paper 1110-B, 33 p.

Faill, R. T., 1985, The Acadian orogeny and the Catskill Delta, *in* Woodrow, D. L., and Sevon, W. D., eds., The Catskill Delta: Geological Society of America Special Paper 201, p. 15–37.

Glass, G. B., 1972, Geology and mineral resources of the Phillipsburg 7½-minute Quadrangle, Centre and Clearfield counties, Pennsylvania: Pennsylvania Geological Survey, 4th ser., Atlas 95a, 241 p.

Kent, D. V., 1985, Paleocontinental setting for the Catskill Delta, *in* Woodrow, D. L., and Sevon, W. D., eds., The Catskill Delta: Geological Society of America Special Paper 201, p. 9–13.

Platt, F., 1881, The geology of Blair County: Pennsylvania Geological Survey, 2nd ser., Report T, 311 p.

Platt, F., and Platt, W., 1877, Report of progress in the Cambria and Somerset district of the bituminous coal fields of western Pennsylvania; Part I, Cambria: Pennsylvania Geological Survey, 2nd ser., Report HH, 194 p.

Rahmanian, V. D., 1979, Stratigraphy and sedimentology of the Upper Devonian Catskill and uppermost Trimmers Rock Formation in central Pennsylvania [Ph.D. thesis]: University Park, The Pennsylvania State University, 340 p.

Sevon, W. D., 1985, Nonmarine facies of the Middle and Late Devonian Catskill coastal alluvial plain, *in* Woodrow, D. L., and Sevon, W. D., eds., The Catskill Delta: Geological Society of America Special Paper 201, p. 79–90.

Skema, V. W., Sholes, M. A., and Edmunds, W. E., 1982, The economic geology of the upper Freeport coal in northeastern Greene County, Pennsylvania: Pennsylvania Geological Survey, 4th ser., Mineral Resource Report 76, 51 p.

Streel, M., and Traverse, A., 1978, Spores from the Devonian/Mississippian transition near the Horseshoe Curve section, Altoona, Pennsylvania, U.S.A.: Review of Paleobotany and Palynology, v. 26, p. 21–39.

Swartz, F. M., 1965, Guide to the Horse Shoe Curve section between Altoona and Gallitzin, central Pennsylvania: Pennsylvania Geological Survey, 4th ser., General Geology Report 50, 56 p.

Walker, R. G., and Harms, J. C., 1971, The "Catskill Delta"; A prograding muddy shoreline in central Pennsylvania: Journal of Geology, v. 79, p. 281–299.

Wells, R. B., 1974, Loyalhanna sandstone extended into north-central Pennsylvania: Geological Society of America Abstracts with Programs, v. 5, p. 84–85.

Williams, E. G., 1960, Relationship between the stratigraphy and petrography of Pottsville sandstones and the occurrence of high-alumina Mercer clay: Economic Geology, v. 55, p. 1291–1302.

—— , 1985, Catskill sedimentation in central Pennsylvania, *in* Gold, D. P., and others, leaders, Central Pennsylvania revisited: Guidebook, 50th Annual Field Conference of Pennsylvania Geologists, Pennsylvania Geological Survey, p. 20–32.

Wood, G. H., Jr., Trexler, J. P., and Kehn, T. M., 1969, Geology of the west-central part of the southern Anthracite field and adjoining areas, Pennsylvania: U.S. Geological Survey Professional Paper 602, 150 p.

The Birmingham window; Alleghanian décollement tectonics in the Cambrian-Ordovician succession of the Appalachian Valley and Ridge Province, Birmingham, Pennsylvania

***Rodger T. Faill**, Pennsylvania Geological Survey, P.O. Box 2357, Harrisburg, Pennsylvania 17120*

LOCATION

The Birmingham site is located in central Pennsylvania, 75 mi (120 km) west-northwest of Harrisburg, 20 mi (30 km) northeast of Altoona (Fig. 1). The site comprises 11 outcrops in a traverse along the common boundary of Blair and Huntingdon counties, within the Tyrone and Spruce Creek 7½-minute quadrangles. Ten of the 11 outcrops are along Pennsylvania 453 and the railroad that follows the Little Juniata River from Tyrone southeastward for 5 mi (8 km) toward Water Street. The other outcrop (Stop 3), near Nealmont, is an active quarry off Pennsylvania 453. Descriptions of the specific structures present at each outcrop follow the general discussion; a number of the outcrops have been recently described in some detail (Gold, 1984, 1985).

No permissions are necessary for the outcrops along the highway, but the shoulders are very narrow and alertness and keen listening are strongly advised to warn oneself of the fast, occassionally heavy traffic on this curving road. The three outcrops on the railroad are on private property and permission to visit these exposures, especially for large groups, can be sought from the Superintendent of the Altoona Division of Conrail (currently W. E. Flight, 1331 12th Street, Altoona, Pennsylvania 16601, [814] 949-1003). Visits to the active Narehood quarry of the New Enterprise Stone & Lime Company require permission, which can be obtained from the quarry foreman (currently Jim Moist).

SIGNIFICANCE

The late Paleozoic Alleghanian orogeny produced the foreland folds and faults of the Appalachian Valley and Ridge and Plateau provinces. In Pennsylvania the structures are primarily folds, with the décollements buried at depths of 3 to 6 mi (5 to 10 km) or more. One exception is at Birmingham, where a secondary décollement, the Sinking Valley fault duplex, has thrust the Cambrian and Ordovician sequence over younger Ordovician and Silurian rocks. The windows and associated faults are exposed in the hinge of the Nittany anticline, the northwesternmost, largest, and structurally highest major fold in the Valley and Ridge of Pennsylvania.

SITE INFORMATION

Stratigraphy. The sedimentary rocks exposed along this site range from the Upper Cambrian Gatesburg to the Upper Ordovician Juniata Formation (Table 1). Most of these formations crop out in both limbs of the Nittany anticline, with a percentage of exposure unequaled elsewhere in the central Appalachians.

The Lower Paleozoic stratigraphic units in this part of the Appalachian basin consist overwhelmingly of carbonates (Collie, 1903; Ulrich, 1911; Butts, 1918; Field, 1919; Butts and others, 1939; Kay, 1944a, 1944b; Wilson, 1952; and Wagner, 1966)—only the uppermost part of the section is noncarbonate, represented by the flysch and molasse deposits of the Taconian clastic wedge (Thompson, 1970; Thompson and Sevon, 1982). The carbonates are largely fine-grained, rather featureless dolomites (see also Spelman, 1966). Limestones are confined to specific intervals: a nearly continuous limestone sequence occurs below the Taconian deposits (e.g., Thompson, 1963; Rones, 1969; Chafetz, 1969; and Faill and others, 1987); an interbedded limestone/dolomite (Stonehenge) appears at the base of the Ordovician, another (Pleasant Hill) at the top of the Middle Cambrian, and one locally (Axemann) near the top of the Lower Ordovician (Lees, 1967). The Upper Cambrian Gatesburg Formation contains two quartz sandstone members.

The Birmingham area is important from a mineralogical perspective as well. A number of significant zinc-lead occurrences, particularly the Keystone mine, are in or close by the exposures comprising the site (Zeller, 1949; Smith, 1977, p. 89–102).

Structural Geology. The Nittany anticline affords a rare surface exposure of a complicated structure that is best explained by décollement tectonics (Moebs and Hoy, 1959, Gwinn, 1970). The anticline, 9 mi wide, 112 mi long with a maximum structural relief of 4.0 mi (15 by 180 by 6.5 km), sits above a basal décollement that is located in the shaly Lower Cambrian Elbrook/Waynesboro interval. Throughout the Valley and Ridge, splays from this décollement converted the northwestward movement of the overiding Paleozoic rocks into anticlines. The Nittany anticline probably consists of two splay slices that have been stacked, one above the other. The upper slice, to which most of the present surface exposures belong, rode on the Sinking Valley fault, up and over the back of the lower slice (Fig. 2). The Sinking Valley fault is a large, anastomotic, duplex fault (Fig. 3) containing numerous horses of Silurian–Ordovician clastic rocks and Ordovician limestones that have been scooped mostly from the footwall, although one or more horses of Warrior Formation from the hanging wall may have been carried along as well. The Birmingham fault is a single fault within the duplex, and locally is the roof thrust. North of the Nittany hinge, the Birmingham fault

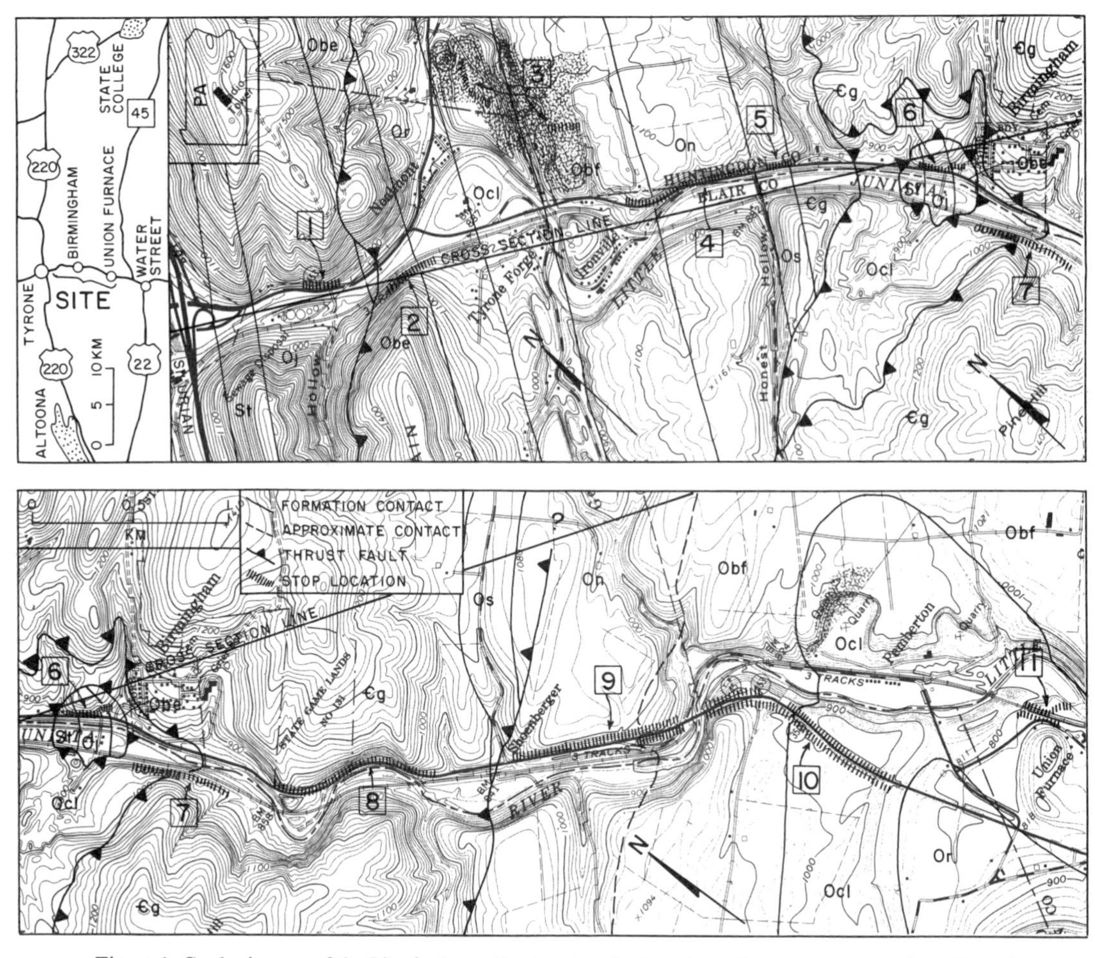

Figure 1. Geologic map of the Birmingham site, showing the locations of the various stops from Stop 1 in the Tyrone Gap to Stop 11 at the Union Furnace railroad cut. Base from the Tyrone and Spruce Creek 7½-minute quadrangle maps. See Figure 3 for key for geologic abbreviations.

splays upward into the overturned rocks of the north limb, cutting off the Honest Hollow fault, itself a minor upward splay from the Sinking Valley duplex.

A number of mesoscopic structures developed prior to the folding, and others were created during the growth of the Nittany anticline. These reveal a progressive change of deformation style as the enclosing bedding dip increased. The structures include (from oldest to youngest) cleavage, disharmonic folds, wedges, ramps and duplexes, kink bands, hybrid fault-kink structures, cross-folds and wrench faults, and conjugate (normal and thrust) faults. Whether or not a particular structure is present at a specific outcrop is dependent not only on the lithology, but also on position within the Nittany anticline, and on the inclination of the enclosing bedding to the deforming stress. Most of these structures are present in the vertical to overturned beds of the northwest limb; in the moderately dipping beds of the southeast limb, wrench faults and thrusts are most common, with occasional kink structures and disharmonic folds, and (rarely) cleavage.

The first deformational event was layer parallel shortening, which took several forms: cleavage, ramps, wedges, disharmonic folds, and kink bands. Cleavage, the earliest structure, tends to be perpendicular to bedding, except where rotated by subsequent fold movements. The presence of wedges, ramps, and kink bands indicate that the mechanical anistotropy of bedding was an important factor while bedding was still largely horizontal and parallel to the presumed horizontal deforming stress. Wedges developed in the more arenaceous beds and seem to be equally distributed between northwest and southeast vergences. Kink bands, steeply dipping relative to the enclosing bedding, are a third form of early layer parallel shortening. The originally southeast dipping (but now dipping to the northwest) kink bands are more common than their southeast-verging conjugates, which suggests that kink banding continued after some small inclination of the enclosing bedding. The hybrid structures, consisting of the northwest-verging kink bands for one member and southeast-verging faults for the other, were formed after the enclosing bedding had rotated further, completely suppressing the southeast-verging kink bands. These hybrid structures gave way to

TABLE 1. STRATIGRAPHIC COLUMN FOR THE BIRMINGHAM VICINITY

System and Series	Stage	Formation/Thickness	Lithic Description	Stop No.
Silurian	Niagaran	Tuscarora/170±15m	White quartzitic sandstone, with shale interbeds	6
Upper Ordovician	Richmondian	Juniata/450±50m	Red quartzitic sandstone, siltstone, and mudstone	1, 6
		Bald Eagle/275±25m	Gray quartzitic sandstone, siltstone, and shale	2
	Edenian and Maysvillian	Reedsville/250±30m	Olive gray shale, with siltstones in the upper part	2
	Trentonian	Antes/75±10m	Black shale, with a few interbeds of limestones	6
		Coburn/90±5m	Fossiliferous and nonfossiliferous limestones and interbeds of shale	10
		Salona/55±2m	Interbedded limestone and black shale with several ash beds	3, 10
		Nealmont/17±2m	Fossiliferous nodular and sparry limestone	3, 10, 11
		Linden Hall/24±3m	Fossiliferous, fine grained limestone	3, 10, 11
		Snyder/55±2m	Conglomeratic, oolitic limestone	3, 10, 11
		Hatter/30±2m	Fossiliferous, magnesian limestone	3, 10, 11
Middle Ordovician	Chazyan and Whiterockian	Loysburg/28±2m	Limestone, dolomite, sublithographic limestone	10, 11
		Bellefonte 300±50m	Fine grained dolomite, minor chert	3, 4, 9, 10
Lower Ordovician	Canadian	Axemann/1-15 m	Fossiliferous, conglomeratic limestone and dolomite	none
		Nittany/380±25m	Very fine grained dolomite, with chert beds and nodules	4, 9
		Stonehenge/150±20m	Limestone, oolitic, with some dolomite interbeds	5
Upper Cambrian	Trempealeauan and Franconian	Gatesburg/600±50m	Coarse dolomite, with two intervals of interbedded quartzitic sandstone	6, 7, 8
	Franconian and Dresbachian	Warrior/480±50m	Argillaceous, oolitic dolomite with limestone interbeds	7(?)

conjugate fault systems of several orientations after the enclosing bedding became moderately steep and the kink-banding mechanism was mechanically untenable.

The wrench faults are transverse structures, 30° to 60° apart in strike, and with moderately (approximately 40°) southeast-plunging intersections, a geometry suggesting conjugacy. Movements began as early as when the bedding was still horizontal and continued through much of the folding, because slickenlines on these faults range in orientation from parallel to as much as 60° to bedding. That some of these faults have slickenlines of several orientations indicates that movements were intermittent and were spread over a significant portion of the folding (but not in the terminal stages). Multiple slickenline orientations are also present on some bedding and wedge-fault surfaces, indicating that the movements were complex during the growth of the Nittany anticline. The relative ages of the cross-fold and the filled tension gash arrays are uncertain. The cross-fold may have been coeval with the wrench faults, absorbing some of the extension parallel to the Nittany fold axis that was produced by the wrench faults.

The Birmingham Locality sits astride the Tyrone–Mount Union lineament (Kowalik and Gold, 1974), which stands out primarily because of the fairly linear course of the Juniata River valley. A number of structures are cited as additional evidence, including the slickensided transverse wrench faults, increased fracture density (Canich, 1976) and termination of second- and third-order folds, among others. This lineament may be linked to a Precambrian basement feature, it probably affected the Paleozoic rocks during the Alleghanian orogeny, and it may still be active (Canich and Gold, 1985; Rodgers and Anderson, 1984).

Stops on the Northwest Limb of the Nittany Anticline
(Bedding mostly vertical to overturned)

Stop 1. Tyrone Gap, Pennsylvania 453 (Swartz, 1955). The Juniata is the only formation exposed here. Structures include northwest-dipping kink bands, hybrid kink-fault pairs, northwest-dipping normal and reverse faults in conjugate pairs, and cleavage. Wedge faults are also present, but apparently wrench faults are not.

Stop 2. Tyrone Gap, Conrail railroad (Butts and others, 1939; Swartz, 1955; Gold and Parizek, 1976; Gold, 1984, 1985). The upper Reedsville and the sandstones of the lower Bald Eagle formations are present. Two large, northwest-dipping kink bands are present, with unusually steep (approximately 20°) southwest-plunging kink axes. The other significant structures are faults, comprising wedges, a steeply southeast-dipping normal fault, duplex faults, and wrench faults. Multiple directions of slickenlines on faults and bedding surfaces are common. The early age of kink

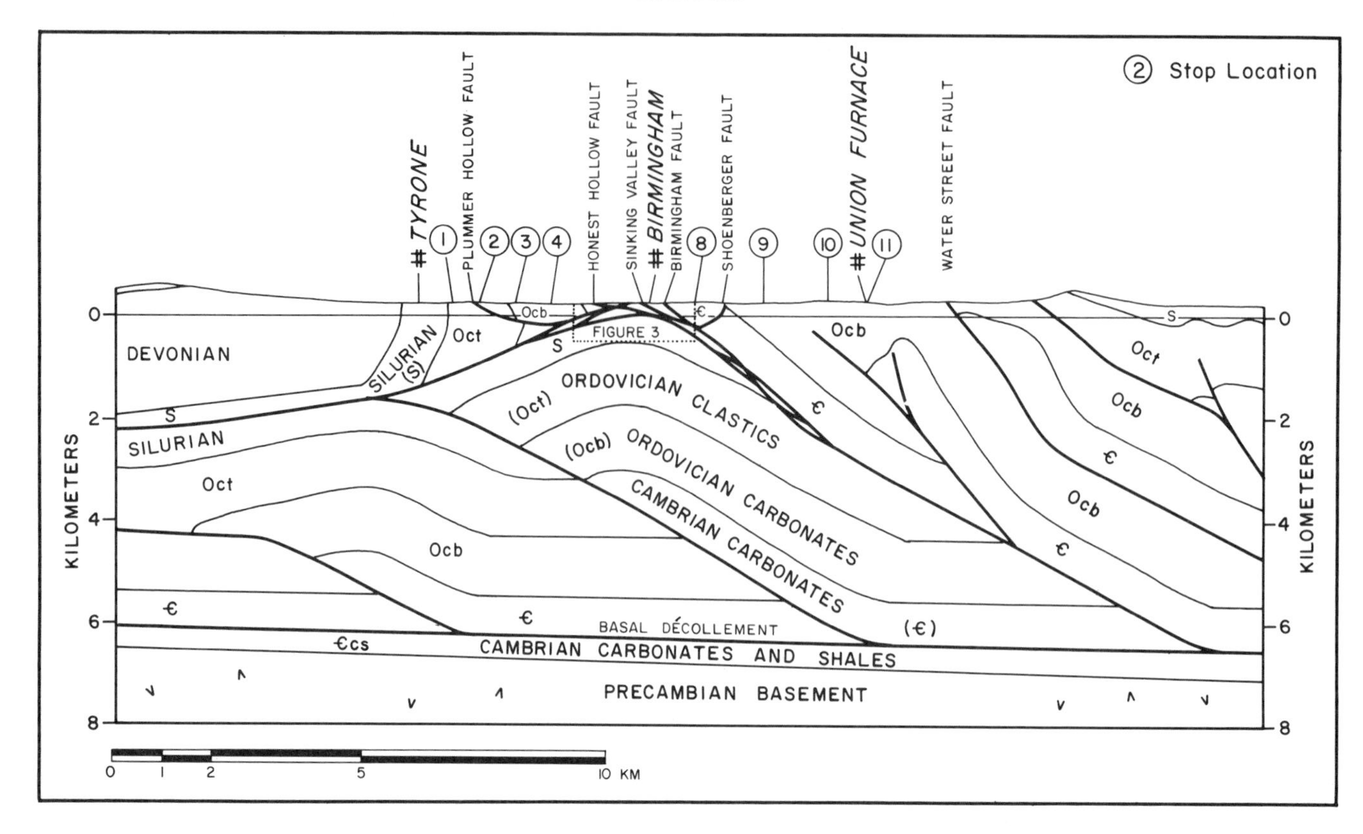

Figure 2. Cross section of the Nittany anticline. A basal décollement within the Lower(?) Cambrian carbonates and shales underlies the entire structure. The lower part of the anticline is formed by a ramp from the décollement to the Silurian strata. The Birmingham duplex is the base of a higher block that ramped over the lower one. The present exposures of the Nittany anticline are entirely from this upper block, and from pieces of the lower block that were incorporated into the duplex.

banding is indicated by the slickenlines on a wrench fault that crosses a kink band with no change in orientation.

Stop 3. Narehood (New Enterprise Co.) Quarry (no published descriptions). Enter from Pennsylvania 453. The Middle Ordovician Bellefonte dolomites are overturned to as much as 45° to the southeast, but despite this severe overturning, these rocks exhibit virtually no mesoscopic structures. The overturned Middle and Upper Ordovician limestones are northwest of the creek. The most prominent structure there is a complicated cross-fold with breccia zones and arrays of en echelon tension gashes. Slickenlines in numerous directions are present on almost every surface. Other structures include several wrench faults, a few symmetric low-amplitude disharmonic folds, a small duplex fault, and a number of wedges.

Stop 4. Ironville, Pennsylvania 453 (no published descriptions). This long outcrop exposes the Bellefonte and the Nittany dolomites; their contact is undefined because the sometimes intervening Axemann limestone is wanting here. Within the vertical to overturned beds, mesoscopic structures include faults (wedges and a hanging-wall ramp halfway through the cut), a kink band, and a northwest-dipping spaced cleavage. The structures are colinear, with their intersections and axes all plunging approximately 031-04.

Stop 5. Honest Hollow, Pennsylvania 453 (Butts and others, 1939; Gold, 1984). The Stonehenge Formation occupies most of this outcrop; part of the Mines Member of the Gatesburg Formation is at the south end. The largest mesoscopic structure is a 82-ft-wide (25 m), southeast-dipping kink band, within which the vertical enclosing bedding has been rotated back to subhorizontal and right-side-up. Some small faults parallel to, and antithetic to, the kink plane are in or near these kink-band boundaries. Also present are conjugate kink bands, one of which contains a faulted fold.

Stops in the Hinge of the Nittany Anticline

Stop 6. Birmingham window, Pennsylvania 453 (Fox, 1950; Swartz, 1955; Gold, 1984, 1985). This outcrop comprises the Birmingham window (Fig. 1a). In the north part of the outcrop are the vertical and overturned beds of the Tuscarora, Juniata, and Bald Eagle formations (on the road up to Birmingham). The Sinking Valley fault is not exposed but occurs just

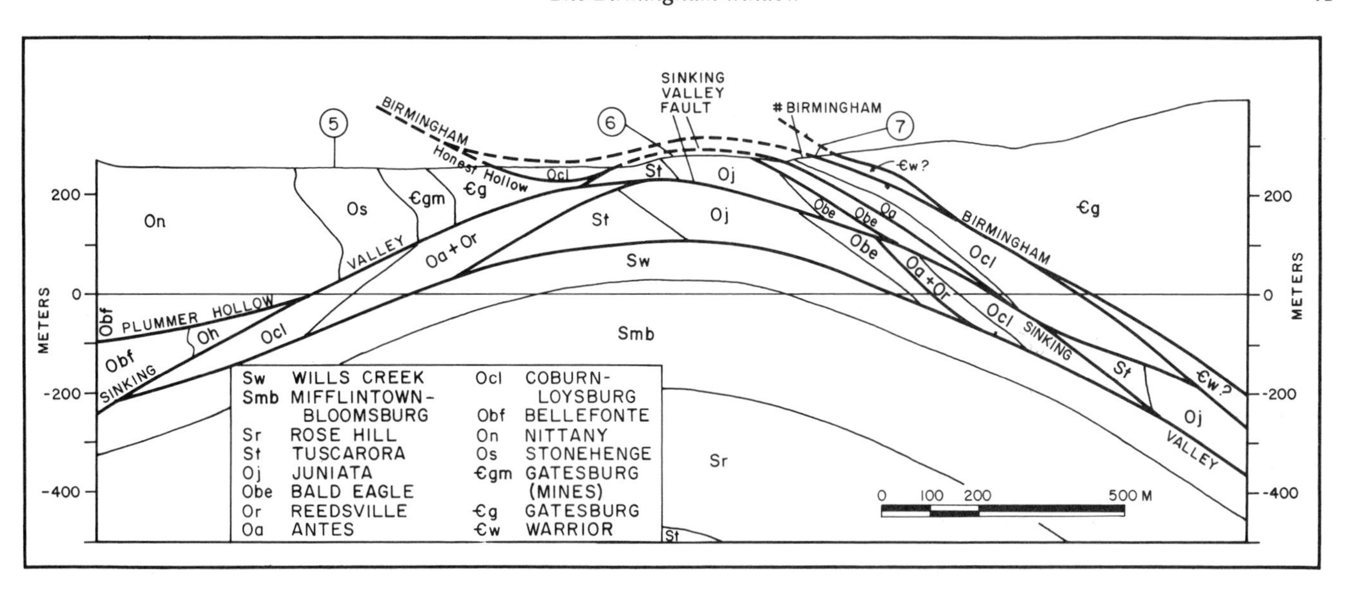

Figure 3. Detailed cross section of the hinge of the Nittany anticline and the Birmingham duplex. The Birmingham fault is the roof of the duplex; the Sinking Valley fault(s) is an anastomic fault within the duplex, and locally constitutes the floor thrust. Lithic units within the duplex are slightly to extremely overturned.

south of the Birmingham Road. Farther south, a cleaved and folded black shale (Antes Formation) extends from the large overgrown borrow pit southward to the Grier School entrance (Weitz, 1979), lying underneath and in fault contact with Ordovician(?) limestones. The Upper Cambrian Gatesburg Formation is on the hill above, but the intervening Birmingham fault is not exposed.

Stop 7. Birmingham, Conrail railroad (Butts, 1918; Butts and others, 1939; Fox, 1950; Nickelsen, 1963; Gold and Parizek, 1976; Gold, 1984, 1985). This exposure of the Birmingham fault is south of and structurally higher than Stop 6. South-dipping Gatesburg(?) dolomite overlies folded and faulted Ordovician(?) limestone. Mesoscopic structures farther south along this outcrop include thrust faults, wedges, small folds, kink structures, and disharmonic folds.

Stops on the Southeast Limb of the Nittany Anticline
(Bedding moderately southeast dipping)

Stop 8. Birmingham South, Pennsylvania 453 (Butts and others, 1939; Wilson, 1952; Nickelsen, 1963). The upper and lower sandy members and the intervening Ore Hill member of the Gatesburg Formation are exposed. The principal structures in this outcrop are rather closely spaced fractures. At highway paddle 0-95 is a 6-ft-thick (2 m) thrust zone (duplex?) dipping moderately to the south. A number of wedges are also present (e.g., at paddle 0-75); all of them seem to be south dipping.

Stop 9. Shoenberger, Pennsylvania 453 (Spelman, 1966). Much of the Nittany and the basal portion of the Bellefonte Formation are well exposed. Structures include wrench faults with slickenlines subparallel to bedding. Also present are an occasional wedge fault, and at least one mullion structure.

Stop 10. Union Furnace, Pennsylvania 453 (Nickelsen, 1963; Berkheiser, 1986; Faill, 1986). Most complete and best exposed section of the Middle and Upper Ordovician carbonates in central Pennsylvania. The most prominent structures comprise the transverse master fractures and the large wrench faults that parallel them. On most of the wrench faults the slickenlines are parallel to bedding, indicating an early, prefolding origin. Other faults include wedges, duplexes, and conjugate extension faults. Mullion structures are present, mostly in the northern end of the roadcut, on the west side. Stylolites are mostly parallel to bedding, although a few, perpendicular to bedding, appear to be of tectonic origin. A poorly developed cleavage is present in some of the beds. Two or three folds overlie apparent detachment faults in a structurally distinct block separated from the rest of the outcrop by a large wrench fault some 500 ft (150 m) from the south end (west side) of the roadcut.

Stop 11. Union Furnace, Conrail railroad (Rosenkranz, 1934; Butts and others, 1939; Kay, 1944a, 1944b; Rones, 1969). Excellent exposure, important reference section, and type locality for several Upper Ordovician limestones, including the Linden Hall, Snyder, Hatter, and Loysburg formations (Kay, 1944a, 1944b). Structures include wrench and oblique faults, and multiple slickenline directions on some bedding surfaces.

REFERENCES CITED

Berkheiser, S. W., Jr., 1986, Middle and Ordovician carbonates at Union Furnace, *in* Sevon, W. D., ed., Selected localities in Bedford and Huntingdon counties, Pennsylvania: Guidebook, 51st Annual Field Conference of Pennsylvania Geologists, p. 7–12, 111–119.

Butts, C., 1918, Geologic section of Blair and Huntingdon counties, central Pennsylvania: American Journal of Science, 4th ser., v. 46, p. 523–537.

Butts, C., Swartz, F. M., and Willard, B., 1939, Geology and mineral resources of the Tyrone Quadrangle, Pennsylvania: Pennsylvania Geological Survey, 4th ser., Geologic Atlas 96, 118 p.

Canich, M. R., 1976, A study of fractures along the western segment of the Tyrone–Mount Union lineament [M.S. thesis]: University Park, Pennsylvania State University, 59 p.

Canich, M. R., and Gold, D. P., 1985, Structural features in the Tyrone–Mount Union lineament, across the Nittany anticlinorium in central Pennsylvania, *in* Gold, D. P., ed., Central Pennsylvania geology revisited: Guidebook, 50th Annual Conference of Pennsylvania Geologists, Pennsylvania Geological Survey, p. 120–137.

Chafetz, H. S., 1969, Carbonates of the Lower and Middle Ordovician in central Pennsylvania: Pennsylvania Geological Survey, 4th ser., General Geology Report 58, 39 p.

Collie, G. L., 1903, Ordovician section near Bellefonte, Pennsylvania: Geological Society of America Bulletin, v. 14, p. 407–420.

Faill, R. T., 1986, Structural Geology, *in* Sevon, W. D., ed., Selected localities in Bedford and Huntingdon counties, Pennsylvania: Guidebook, 51st Annual Field Conference of Pennsylvania Geologists, p. 119–126.

Faill, R. T., Glover, A. D., and Way, J. H., Jr., 1987, Geology and mineral resources of the Blandburg, Tipton, Altoona and Bellwood Quadrangles, Blair, Cambria, Clearfield, and Centre counties, Pennsylvania: Pennsylvania Geological Survey, 4th ser., Geological Atlas 86, 253 p.

Field, R. M., 1919, The Middle Ordovician of central and southcentral Pennsylvania: American Journal of Science, 4th ser., v. 48, p. 403–428.

Fox, H. D., 1950, Structure and origin of two windows exposed on the Nittany arch at Birmingham, Pennsylvania: American Journal of Science, v. 248, p. 153–170.

Gold, D. P., ed., 1984, Strike parallel and cross-strike structures in Tyrone–Mt. Union lineament in central Pennsylvania: Guidebook, Pittsburgh, American Association of Petroleum Geologists, Eastern Section, 71 p.

—— , 1985, Central Pennsylvania geology revisited: Guidebook, 50th Annual Field Conference of Pennsylvania Geologists, Pennsylvania Geological Survey, 290 p.

Gold, D. P., and Parizek, R. R., 1976, Field guide to lineaments and fractures in central Pennsylvania: Guidebook, 2nd International Conference on the New Basement Tectonics, Newark, Delaware, 75 p.

Gwinn, V. E., 1970, Kinematic patterns and estimates of lateral shortening, Valley and Ridge and Great Valley provinces, central Appalachians, south-central Pennsylvania, *in* Fisher, G. W., Pettijohn, F. J., Reed, J. C., Jr., and Weaver, K. N., eds., Studies of Appalachian geology; Central and southern: New York, Interscience–Wiley, p. 127–146.

Kay, G. M., 1944a, Middle Ordovician of central Pennsylvania; Part 1, Chazyan and earlier Mohawkian (Black River) formations: Journal of Geology, v. 52, p. 1–23.

—— , 1944b, Middle Ordovician of central Pennsylvania; Part 2, Later Mohawkian (Trenton) formations: Journal of Geology, v. 52, p. 97–116.

Kowalik, W. S., and Gold, D. P., 1974, The use of Landsat-l imagery in mapping lineaments in Pennsylvania: Proceedings of the First International Conference on the New Basement Tectonics, Salt Lake City, Utah, p. 236–249.

Lees, J. A., 1967, Stratigraphy of the Lower Ordovician Axemann limestone in central Pennsylvania: Pennsylvania Geological Survey, 4th ser., General Geology Report 52, 79 p.

Moebs, N. N., and Hoy, R. B., 1959, Thrust faults in Sinking Valley, Blair and Huntingdon counties, Pennsylvania: Geological Society of America Bulletin, v. 70, p. 1079–1088.

Nickelsen, R. P., 1963, Fold patterns and continuous deformation mechanisms of the central Pennsylvania folded Appalachians, *and* Stops III, IV, and V, *in* Tectonics and Cambrian–Ordovician stratigraphy, central Appalachians of Pennsylvania: Pittsburgh, Pennsylvania, Pittsburgh Geological Society, Guidebook, p. 13–29 and 53–59.

Rodgers, M. R., and Anderson, T. H., 1984, Tyrone–Mt. Union cross-strike lineament of Pennsylvania; A major Paleozoic basement fracture and uplift boundary: American Association of Petroleum Geologists Bulletin, v. 68, p. 92–105.

Rones, M., 1969, A lithostratigraphic, petrographic, and chemical investigation of the lower Middle Ordovician carbonate rocks in Pennsylvania: Pennsylvania Geological Survey, 4th ser., General Geology Report 53, 224 p.

Rosenkrans, R. R., 1934, Correlation studies of the central and south central Pennsylvania bentonite occurrences: American Journal of Science, 5th ser., v. 27, p. 113–134.

Smith, R. C., II, 1977, Zinc and lead occurrences in Pennsylvania: Pennsylvania Geological Survey, 4th ser., Mineral Resource Report 72, 318 p.

Spelman, A. R., 1966, Stratigraphy of Lower Ordovician Nittany dolomite in central Pennsylvania: Pennsylvania Geologcal Survey, 4th ser., General Geology Report 47, 187 p.

Swartz, F. M., 1955, Stratigraphy and structure of Ridge and Valley area from University Park to Tyrone, Mt. Union, and Lewistown: Guidebook, 21st Annual Field Conference of Pennsylvania Geologists, Harrisburg, p. S11–S138.

Thompson, A. M., 1970, Lithofacies and formation nomenclature in Upper Ordovician stratigraphy, central Appalachians: Geological Society of America Bulletin, v. 81, p. 1255–1260.

Thompson, A. M., and Sevon, W. D., 1982, Excursion 19B; Comparative sedimentology of Paleozoic clastic wedges in the central Appalachians, U.S.A.: Field excursion guidebook, 11th International Congress on Sedimentology, Hamilton, Ontario, Canada, International Association of Sedimentologists, 136 p.

Thompson, R. F., 1963, Lithostratigraphy of the Middle Ordovician Salona and Coburn formations in central Pennsylvania: Pennsylvania Geological Survey, 4th ser., General Geology Report 88, 154 p.

Ulrich, E. O., 1911, Revision of the Paleozoic systems: Geological Society of America Bulletin, v. 22, p. 281–680.

Wagner, W. R., 1966, Stratigraphy of the Cambrian to Middle Ordovician rocks of central and western Pennsylvania: Pennsylvania Geological Survey, 4th ser., General Geology Report 49, 156 p.

Weitz, J. H., Jr., 1979, Structural geology of selected areas in the vicinity of Birmingham, Pennsylvania [B.S. thesis]: Lewisburg, Bucknell University, 55 p.

Wilson, J. L., 1952, Upper Cambrian stratigraphy in the central Appalachians: Geological Society of America Bulletin, v. 63, p. 275–322.

Zeller, R. A., Jr., 1949, The structural geology and mineralization of Sinking Valley, Pennsylvania [M.S. thesis]: University Park, Pennsylvania State University, 64 p.

Mississippian-Pennsylvanian boundary and variability of coal-bearing facies at Curwensville Reservoir, Clearfield County, Pennsylvania

***Thomas M. Berg,** Pennsylvania Geological Survey, Harrisburg, Pennsylvania 17120*

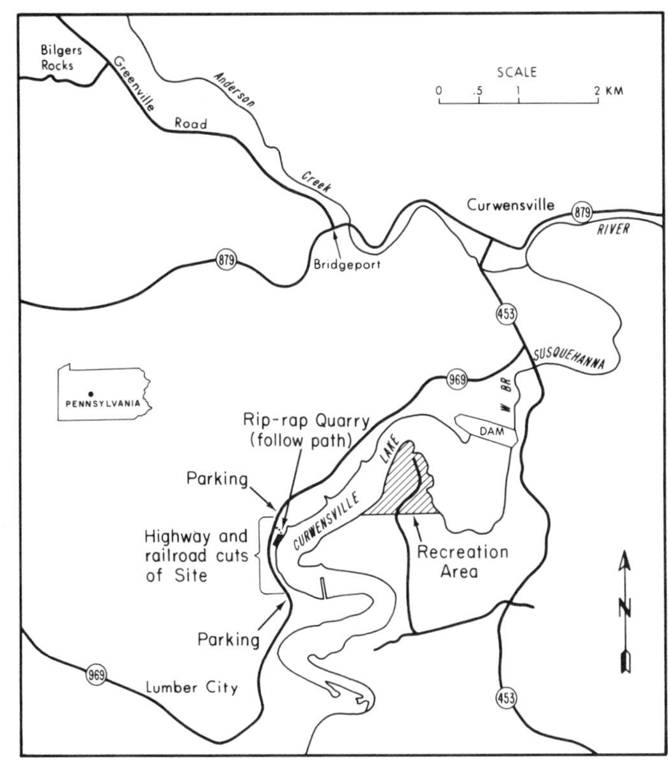

Figure 1. Map of the Curwensville area of Pennsylvania showing the locations of the exposures at this site.

LOCATION

The upper part of the Mississippian Burgoon Sandstone and the Pennsylvanian Pottsville Group and lower Allegheny Group are well-exposed in a quarry and series of exposures along Pennsylvania 969, on the west side of Curwensville Reservoir on the West Branch of the Susquehanna River, approximately 2.3 mi (3.7 km) southwest of the Borough of Curwensville (Fig. 1). Part of the section is also exposed in the railroad cuts parallel to Pennsylvania 969. Parking is limited along Pennsylvania 969; locations having space for one or two automobiles are indicated in Figure 1. Parking for more or larger vehicles will have to be located farther away. Pennsylvania 969 is a well-traveled road; exercise caution and common sense! The railroad tracks are in use, so caution should be exercised there, too. Stay alert! The easiest way to enter the rip-rap quarry is to follow the short path through the woods from the highway, as shown in Figure 1.

Curwensville Dam was built by the U.S. Army Corps of Engineers for flood control purposes. The lake is maintained by the Corps of recreation: Day use facilities and activities include boating, fishing, swimming, picnicking, and hiking.

SIGNIFICANCE

This site is located at the northeastern end of the Pittsburgh-Huntingdon synclinorium, which was a basin during Pennsylvanian time, with rates of subsidence increasing to the southwest. The subtle disconformity (or paraconformity) marking the Mississippian-Pennsylvanian systemic boundary, and the high-alumina hard clay, which formed in response to regional paleotopographic conditions at the disconformity, are well-exposed in the rip-rap quarry of this site. The hard clay exposed here and elsewhere in the Anderson Creek region, together with other areas in Clearfield and Clinton Counties, constitute one of the largest occurrences of high-alumina hard clay in the United States.

The highway and railroad cuts above the quarry (Fig. 2) provide an excellent opportunity to observe some of the lateral thickness and lithologic variations that characterize much of the Pennsylvanian coal-bearing succession of the Allegheny Plateau. Recognition of the exact configuration of the pre-Pennsylvanian erosion surface has greatly facilitated geologic mapping in north-central Pennsylvania. Awareness of relatively rapid lateral variations within coal-bearing sequence has changed approaches to mapping of coal and concomitant resource evaluations.

SITE INFORMATION

In the rip-rap quarry (Fig. 1), the Burgoon Sandstone (the oldest unit exposed) is light gray to very light gray, and weathers to light brown or buff. It is fine- to medium-grained and is medium- to thin-bedded, with the thinner beds near the top. Broad, trough-style crossbedding is the dominant sedimentary structure. Detailed mapping in quadrangles to the east and north has shown that pre-Pottsville erosion removed the Mississippian succession, including the Loyalhanna and Mauch Chunk Formations, from south-central Pennsylvania northward to this area, where the Pottsville rocks rest directly on the Burgoon Sandstone (Edmunds and Berg, 1971, p. 13). Farther north, the Burgoon itself is missing and the Pottsville rests upon the subjacent Huntley Mountain Formation (Berg and Edmunds, 1979, p. 47). Recognition of the great extent of this unconformity has allowed mappers to correct the erroneous widespread mapping of Mauch Chunk redbeds in north-central Pennsylvania (Berg and others, 1980), which had been mistaken for redbeds that are a minor component of the Huntley Mountain Formation. At this site, the unconformity is more properly termed a paraconformity.

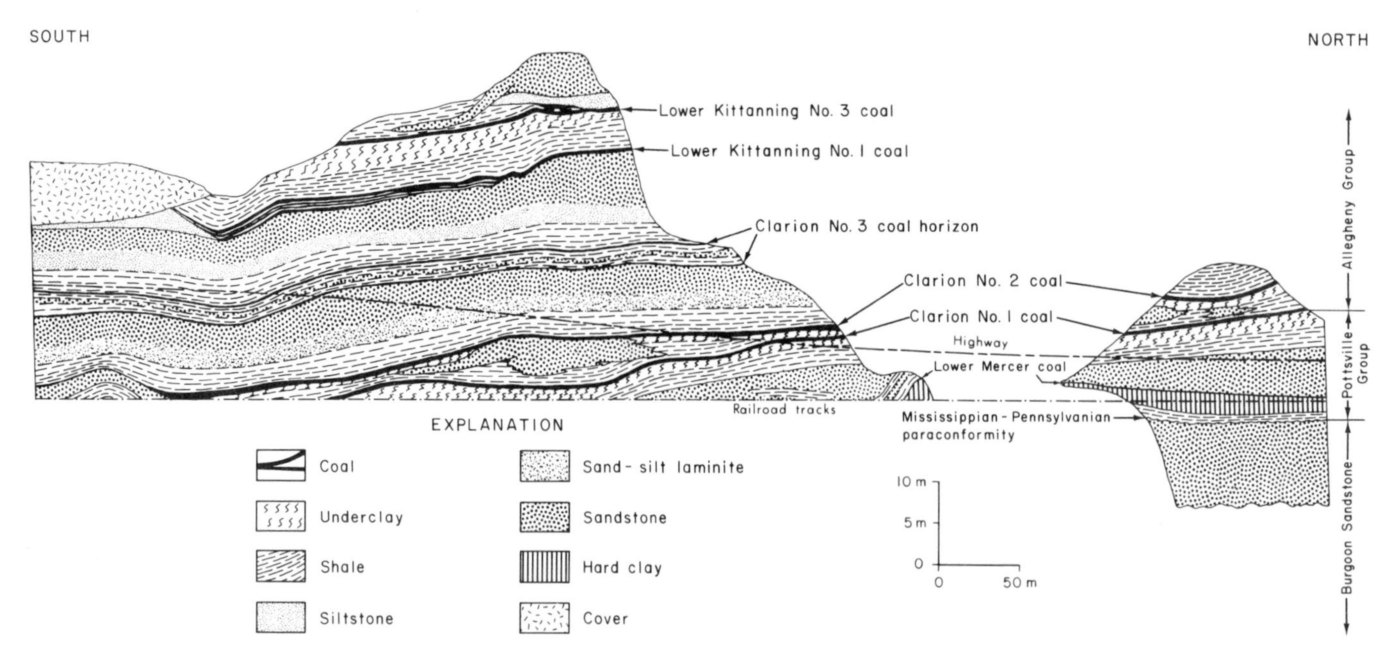

Figure 2. Diagram of the Carboniferous section exposed at this site (Redrawn from Edmunds and Berg, 1971, plate 22).

The overlying Pottsville Group comprises a relatively thin sequence of siltstones, silt shales, fine-grained sandstones, claystones (both hard clay and soft clay), and coal. The Pottsville varies from 35 to 60 ft (11 to 18 m) thick here in marked contrast with the 130 to 180 ft (40 to 55 m) thickness of the formation to the north and east. In the quarry, 3 to 5 ft (1.0 to 1.5 m) of silt shale, siltstone, very fine-grained sandstone, and claystone separate the Burgoon Sandstone from the overlying high-alumina Mercer hard clay. This thin interval is probably the lateral equivalent of the Lower Connoquenessing Sandstone of the Pottsville. The hard clay is more than 8 ft (2.5 m) thick, is medium gray to medium dark gray, and appears to be mostly a high-alumina diaspore block clay (a variety of hard clay that breaks with a blocky fracture). It occurs in individual beds up to 2 ft (0.5 m) thick. Above the hard clay the lower Mercer coal varies from 2 to 5 in (5 to 13 cm) thick.

The hard clay is thought to have developed on paleotopographic highs resulting from differential erosion of the regionally uplifted Mississippian Burgoon Sandstone (Edmunds and Berg, 1971, p. 60). In an alternative interpretation, Williams (1960) concluded that the hard clay developed on paleotopographic highs on the Lower Connoquenessing Sandstone, and that the hard clay was coeval with the Upper Connoquenessing Sandstone. Because of lesser resistance to erosion, the younger Mauch Chunk Formation and older Huntley Mountain Formation are thought to have formed paleotopographically lower areas of the pre-Pottsville erosion surface. Some sandstone units of the Mauch Chunk Formation may have formed similar paleotopographic highs to the south. The hard clay occurs only where the Connoquenessing sandstone units of the Pottsville are absent or very thin (as at this site) and directly or almost directly overlie the Burgoon Sandstone. Where the lower Mercer clay horizon overlies well-developed Connoquenessing sandstones, it is almost always a soft clay. This suggests that the hard clay developed as a Pennsylvanian weathering product on top of the Burgoon. The hard clay apparently developed in areas where drainage conditions prevented surface water from removing weathering products until they were reduced to colloidal size or removed in solution—usually on "positive" topographic features such as the Burgoon "ridge." This would have been an environment where loss of water was greater than supply, and drainage of groundwater produced an environment of leaching. Williams and Bragonier (1974, p. 143) suggested that growing folds may have been a third-order factor controlling erosion patterns, allowing leaching of silica. Bragonier (1970, p. 191–192) earlier indicated that the Mercer high-alumina clay formed in response to a sequence of conditions controlled by abundant vegetation, poor drainage, water table fluctuations, relative paleotectonic stability, and prolonged leaching. Weitz (1954, p. 92–93) cited similar factors controlling formation of the hard clays, including low topographic relief, sluggish streams, poor drainage, dense vegetation, deposition of very fine materials including colloidal gels, long periods of marked stability, little change in base level, and lithification by crystallization and dehydration of gels. Further elaboration of high-alumina clay genesis is given by Williams and Bragonier (1985, p. 204ff). Smith (1978, p. 62) listed kaolinite as the most abundant mineral in the hard clay; diaspore and boehmite are the second- and third-most abundant minerals. The hard clay also contains traces of "limonite," siderite, pyrite, illite, and sparse detrital grains of zircon and tourmaline. The Mercer high-alumina hard clay is used in the production of a variety of refractory products.

In the railroad cuts and exposures along Pennsylvania 969 (Fig. 2), upper portions of the Pottsville Group are exposed along with the Clarion coals and the lower Kittanning coals. Strata occurring between the coals show varying degrees of lateral persistence. For example, in the railroad cut the sandstone between the Clarion no. 1 coal and the Clarion no. 2 coal grades laterally to medium gray and grayish black silt shale, which in turn grades upward to a dark gray underclay. The sandstone is root-worked where directly overlain by the Clarion no. 2 coal. Some internal slumps occur in this sandstone, probably reflecting mass movement of sand toward lower areas of peat accumulation. The interval between the Clarion no. 1 and Clarion no. 2 varies from less than 1 m to as much as 6 m. The Clarion no. 2 coal varies from 0.5 to 2.5 ft (0.15 to 0.76 m) thick. In contrast, the succession of silt shale, laminite, and sandstone above the Clarion no. 2 coal is quite persistent, and the stratigraphic thickness from Clarion no. 2 to the Clarion no. 3 coal horizon varies only from 25 to 30 ft (7.9 m to 8.8 m).

A further example of lateral discontinuity is the lower Kittanning no. 1 coal (Fig. 2), which has up to four distinct benches, the uppermost being most persistent. At its maximum development, where it is 5 ft (1.5 m) thick, the four benches are separated by one sandstone and two shale partings. The entire seam appears to disappear to the south, where it rises over a poorly-exposed succession of siltstone and silt shale. Unusual variations in thickness and lithology of units above the lower Kittanning no. 3 coal may be due to penecontemporaneous subaqueous slumping. Williams and Bragonier (1974, p. 149–150) explain the lateral variations and thickness changes as due to small-scale "fourth order" controls of the paleotopography by differential sedimentation, compaction, and erosion. For example, the thickest coal development probably represents a paleotopographic low; these lows are seen to be transmitted vertically in the section. The greatest amount of compaction occurs where the original peat was thickest. Fossil plants are present at many horizons at this site. Fossil invertebrates are also present, but are less common (E. G. Williams, written communication, 1986). *Anthraconauta* (a freshwater bivalve) occurs above the Clarion no. 2 coal, *Lingula* and *Dunbarella* (brackish water fossils) are present above the Clarion no. 2 coal, a few *Chonetes* (a marine brachiopod) occur above the Clarion no. 3 coal, and *Lingula* occur above the lower Kittanning no. 3 coal. Lateral variation and lateral discontinuity are to be expected in Appalachian Pennsylvanian stratigraphy. An appreciation of these lateral and vertical variations is most important in mapping bituminous coal-bearing sequences and consequent calculation of coal resources.

Those who wish to see the well-developed Connoquenessing sandstones of the Pottsville Group resting directly upon the Burgoon Sandstone with no intervening hard clay may examine the road cut exposures along I-80 between 0.3 mi (0.5 km) and 4.2 mi (6.7 km) east of the Elliott Park (no. 18) interchange (Edmunds and Berg, 1971, Plate 4). While in the Curwensville area, visitors will also come to appreciate important lateral variations by visiting Bilgers Rocks west of Curwensville, an excellent "rock city" developed in the massive Homewood Sandstone (a major channel system in the uppermost Pottsville Group). If the massive Homewood was present at this site, it would occur just below the Clarion no. 1 coal. To get to Bilgers Rocks, follow the Greenville Road northwest from Pennsylvania 879 at Bridgeport, along the west side of Anderson Creek, for 2.3 mi (3.7 km); turn left and follow the dirt road for 0.6 mi (0.9 km) to Bilgers Rocks. The exposures at this present site, as well as other Allegheny Group exposures, are also included in the guidebook descriptions for Field Trip number 5 (Williams and others, 1985) of the 50th Field Conference of Pennsylvania Geologists. Of particular interest is Stop 3, which covers middle Allegheny rocks at the Curwensville Dam spillway.

REFERENCES CITED

Berg, T. M., and Edmunds, W. E., 1979, The Huntley Mountain Formation—Catskill-to-Burgoon transition in north-central Pennsylvania: Pennsylvania Geological Survey, 4th series, Information Circular 83, 80 p.

Berg, T. M., Edmunds, W. E., Geyer, A. R., and others, 1980, Geologic map of Pennsylvania: Pennsylvania Geological Survey, 4th series, Map 1, scale 1:250,000.

Bragonier, W. A., 1970, Genesis and geologic relations of the high-alumina Mercer fireclay, western Pennsylvania [M.S. thesis]: University Park, Pennsylvania State University, 212 p.

Edmunds, W. E., and Berg, T. M., 1971, Geology and mineral resources of the southern half of the Penfield 15-minute Quadrangle, Pennsylvania: Pennsylvania Geological Survey, 4th series, Atlas 74cd, 184 p.

Smith, R. C., II, 1978, The mineralogy of Pennsylvania 1966–1975: Friends of Mineralogy, Pennsylvania Chapter, Inc., 304 p.

Weitz, J. H., 1954, The Mercer fire-clay in Clinton and Centre Counties, Pennsylvania [Ph.D. thesis]: University Park, Pennsylvania State College, 107 p.

Williams, E. G., 1960, Relationship between the stratigraphy and petrography of Pottsville sandstones and the occurrence of high-alumina Mercer clay: Economic Geology, v. 55, p. 1291–1302.

Williams, E. G., and Bragonier, W. A., 1974, Controls of Early Pennsylvanian sedimentation in western Pennsylvania, *in* Briggs, G., Carboniferous of the southeastern United States: Geological Society of America Special Paper 148, p. 135–152.

—— 1985, Origin of the Mercer high-alumina clay, *in* Gold, D. P., Canich, M. R., Cuffey, R. J., and others, Central Pennsylvania geology revisited: Guidebook, 50th Annual Field Conference of Pennsylvania Geologists, State College, 1985, Pennsylvania Geological Survey, p. 204–211.

Williams, E. G., Gardner, T., Davis, A., and others, 1985, Economic and mining aspects of the coal-bearing rocks in western Pennsylvania, *in* Gold, D. P., Canich, M. R., Cuffey, R. J., and others, Central Pennsylvania geology revisited: Guidebook, 50th Annual Field Conference of Pennsylvania Geologists, State College, 1985, Pennsylvania Geological Survey, p. 275–286.

The Susquehanna River Water Gaps near Harrisburg, Pennsylvania

Donald M. Hoskins, Pennsylvania Geological Survey, Box 2357, Harrisburg, Pennsylvania 17120

LOCATION

Watergaps occur north of Harrisburg, Pennsylvania, and appear in stark contrast to the elongate ridges that demarcate the boundary between the Great Valley and Appalachian Mountain sections of the Valley and Ridge Physiographic Province. In order from south to north (Fig. 1), five gaps of 0.5 to 0.95 mi width (0.9 to 1.5 km) occur in Blue Mountain (with its associated Little Mountain) 2.1 mi (3.4 km) north of Harrisburg, and then in Second, Peters, Berry, and Mahantango mountains, over a distance of 17 mi (27 km).

Three of the five gaps are selected for stop descriptions as examples of local geologic structures that have controlled the location of the Susquehanna River gaps. Each of the gaps may be reached by major federal or state highways that parallel both the east and west banks of the Susquehanna River.

Rocks at Stop 1 at the Blue-Little Mountain Gap are best exposed on the east side of the river. Stop 1 is reached by driving north from Harrisburg on U.S. 22-322. Park on the east berm at the south end of Blue Mountain. West bank exposures of Stop 1 may be seen by driving north on U.S. 11-15 from Marysville and parking in available berm areas near the west end of the Rockville Railroad Bridge.

Stop 2 exposures, at Peters Mountain, are best on the east side of the gap and are reached by driving north on U.S. 22-322 to near the south part of the mountain. Visitors to this stop should turn right onto a local road 0.2 mi (0.3 km) north of the intersection of U.S. 22-322 with Pennsylvania 325. This is immediately south of the highway bridge over the Conrail railroad. Safe parking and access to Stop 2 exposures are at the north end of this local road, formerly the main highway.

Stop 3 is reached by continuing north, approximately 13 mi (20 km), from Stop 2 on Pennsylvania 147. Parking is available between the river side of the highway and the railroad. Location maps (Figs. 2 to 4) show convenient parking places for easy access to each stop's outcrops.

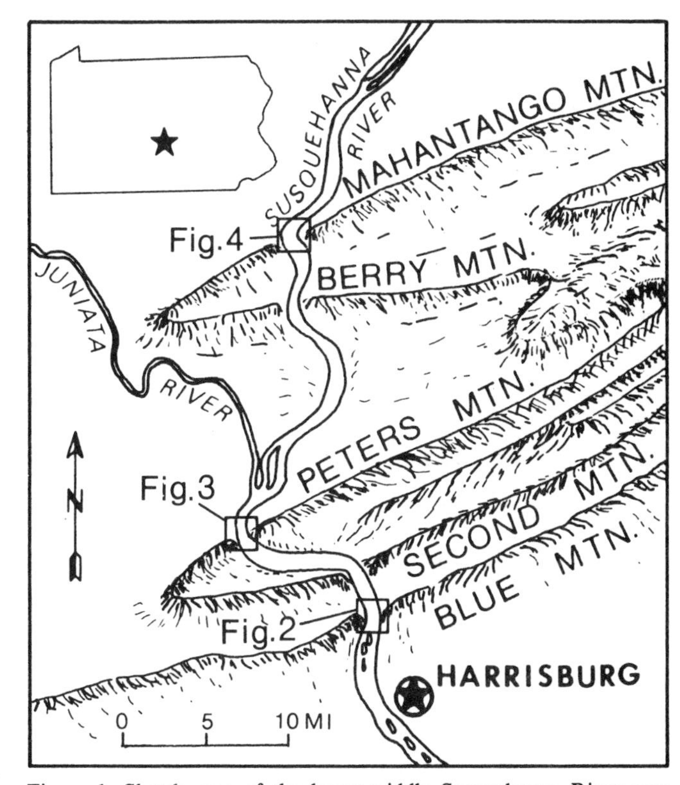

Figure 1. Sketch map of the lower-middle Susquehanna River area showing location of Susquehanna River Gaps and stop location maps.

SIGNIFICANCE

The Susquehanna River Water Gaps, now designated a National Landmark, have been used in geologic education as one of the classic examples that demonstrate the theory of regional superposition of drainage. The five water gaps have been eroded through linear ridges, locally called mountains, supported by Paleozoic sandstones and conglomerates, and seemingly without regard to the resistant lithologies. The ridge surfaces have been interpreted as remnants of a peneplain upon which the drainage began its erosion. Interpretation of the gaps' origin has ranged from erosion along transverse faults, headward erosion, superposition of drainage, and erosion along local fractures.

In addition to the interest generated by the multiplicity of theories of origin, these gaps expose significant sections of Silurian, Devonian, and Mississippian Period formations and are the site of type sections of some members of these formations. Further, two gaps include complete exposures of Triassic diabase dikes penetrating Paleozoic clastic rocks.

SITE INFORMATION

From the first description of the Susquehanna Water Gaps by Rogers (1858, p. 896) and continuing through such authors as Davis (1889), Johnson (1931), Ashley (1935), Strahler (1945), Meyerhoff (1972), Oberlander (1984), and Sevon (1985), the Susquehanna River Gaps and other Appalachian Mountain gaps have been the subject of considerable discussion and speculation about how existing drainage eroded through major resistant ridges. The Susquehanna River, with its five gaps, was a focus of much of this discussion. Early postulations were (1) that the gaps

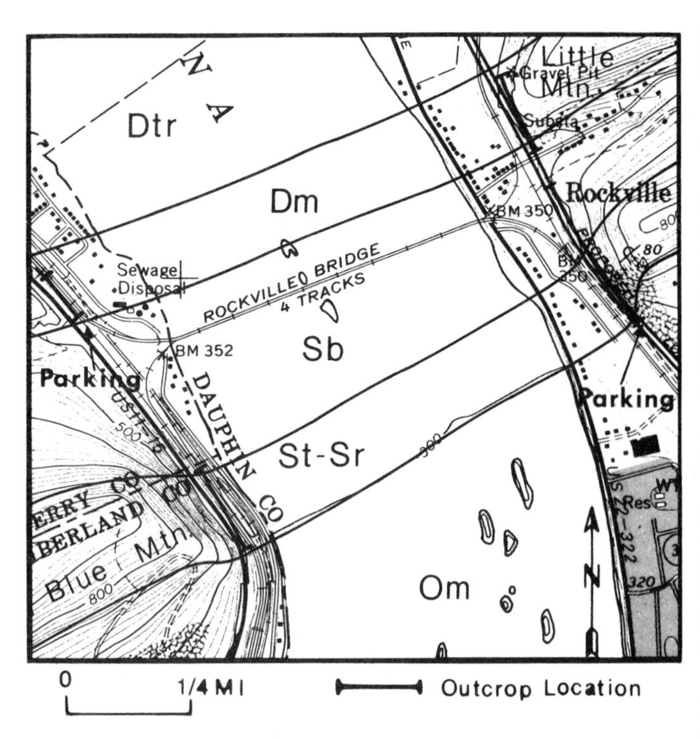

Figure 2. Blue–Little Mountains Gap at Rockville showing Stop 1 outcrops. Om, Martinsburg Formation; St-Sr, Tuscarora-Rose Hill formations; Sb, Bloomsburg Formation; Dm, Montebello [Mahantango] Formation; Dtr, Trimmers Rock Formation.

were locations of transverse faults and the gaps were thus related to these local structures, or (2) that superposition of the rivers from a peneplain or an ancient coastal plain determined the course of the rivers and thus the gaps.

In 1863, Davis (1931) observed and described the three southerly gaps cut through what were seen as three "even crested" ridges. This early visit and description supported his later interpretation (Davis, 1889) of the Susquehanna River as transverse drainage that flowed across a Cretaceous peneplain, and was subsequently incised to its present course. Johnson (1931, p. 65–68) argued that the present river position resulted from direct superposition from an ancient coastal plain.

Thompson (1949, p. 59–61) was the first to describe details of the local Susquehanna River structures and stratigraphy and to suggest that small-scale local faulting and crushing of the rocks controlled the gap locations. He did not find transverse faults.

More recently, Epstein (1966) convincingly showed that in each of the six gaps in the Stroudsburg area of eastern Pennsylvania, including the Delaware River Water Gap, structural features occur that include abrupt changes in strike, local intense folding, and other structures. He concluded that the rocks in this gap were weakened by one or more of these local structural features, leading to easier erodibility of the rocks. Similar structural weaknesses have been suggested as the explanation of transverse drainage across resistant ridges by many authors (for a detailed list see Sevon, 1985, p. 11). This explanation has been challenged by others, as also noted by Sevon (1985).

Recent detailed geologic mapping and detailed descriptions of the structural geology (Dyson, 1963; Hoskins, 1976b; Theisen, 1983) in most of the Susquehanna River Gaps show that complex fracture systems, strike-parallel faults, kinked folds, or overturned bedding occur in each gap. While these gaps have not been examined in the elegant detail provided by Epstein's 1966 report, sufficient structural data have been reported or are readily apparent to the visitor's view, such that each of the five Susquehanna River gap locations can be related to one or more significant local fracture systems, or other structural features interpreted to have locally caused the rocks to be weakened and thus more easily eroded at each gap.

Questions remain as to whether similar structures and thereby weakened areas exist along other parts of the linear ridges. Detailed examination of the ridge crests shows them to be undulatory and not always "even crested," and from this evidence one can postulate the presence of local structures similar to those exposed in the Susquehanna River Gaps. A fundamental question also remains as to whether a major regional feature determined the present course of the Susquehanna River and the specific local weak areas, now gaps, used by the drainage rather than other locally weak areas of the ridges.

Oberlander (1984) suggests that the regional position of the Susquehanna River has had a "close and consistent relation to geologic structure" and further suggests that the current position is controlled by structural slope and erodibility of formations. He used the term "obstinate" stream at the Fifteenth Binghamton Symposium to describe a stream that "seems to follow the most difficult route possible" (Geotimes, 1985). I suggest that a more appropriate term for the Susquehanna River is "opportunistic" because the river has used every opportunity to erode its gaps at structurally weak areas.

Sevon (1986) argues that the initial post-Alleghanian position of the Susquehanna was determined by the western and northern margin of a thrust sheet, the Anthracite Thrust Sheet, hypothesized to have covered the Anthracite Region (MacLachlan, 1985; Levine, 1985). Once the course of the river was established at the margin of the sheet, it became locked in place by the structural weakness of the rocks.

Stop 1. Blue and Little Mountain Gaps (Fig. 2). Excellent stratigraphic sections occur on the east side, and good exposures are present on the west side of the gap along highway cuts. The east side exposures are safer for visitors than are the west side exposures and are reached by walking north from the recommended parking site. Here are exposed many compressional (wedge faults, kink bands, strike-parallel thrust faults) and extensional (conjugate faults and boudinage) features. Details of the east section, from the Martinsburg Formation to the Bloomsburg Formation, are described by Theisen (1983), who includes a detailed structural sketch and measured stratigraphic sections of both east and west side outcrops. The Montebello Formation, also best exposed on the east side of the gap in Little Mountain

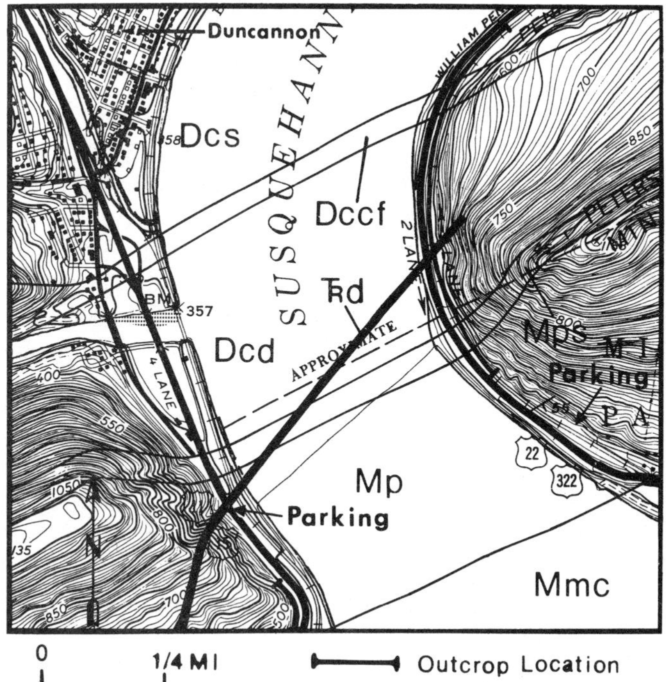

Figure 3. Peters Mountain Gap at Duncannon showing Stop 2 outcrops. Dcd, Duncannon Member; Dccf, Clarks Ferry Member; Dcs, Sherman Creek Member; all of Catskill Formation. Mp, Pocono Formation; Mps, Spechty Kopf Formation; Mmc, Mauch Chunk Formation; Trd, Triassic-Jurassic diabase.

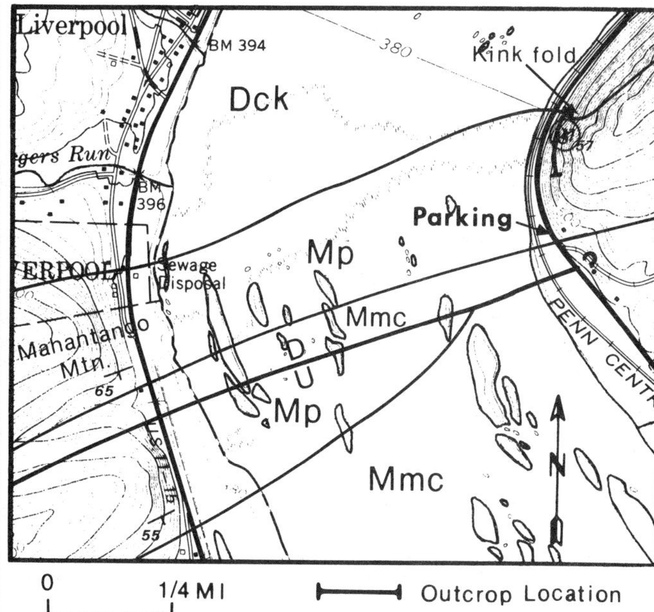

Figure 4. Mahantango Mountain Gap at Liverpool showing Stop 3 outcrops. Dck, Catskill Formation; Mp, Pocono Formation; Mmc, Mauch Chunk Formation.

(see Ellison, 1965, for a detailed stratigraphic section), is reached by continuing north across a narrow valley. A description of extensional faulting in the Montebello rocks on the west side is reported by Cloos and Broedel (1943). Excellent brachiopod collecting in the Montebello on the east side is reported by Inners (1984). The exposures at Stop 1 are the best example of the Susquehanna River Gaps for multiplicity of both compressional and extensional structural features that have weakened the local rocks and made them more susceptible to erosion.

The Second Mountain Gap area, the next gap north of Stop 1, has not yet been mapped in detail. Exposures of 60°S overturned rocks in this gap are present only on the east side but are better viewed from the west side north of the village of Marysville approximately 2 mi (3 km) north of the Stop 1 west side recommended parking area. Aerial photographs show that the strike of the sandstone and conglomerate layers of the Pocono Formation in Second Mountain changes significantly at the west side of the gap. In addition, the steeply overturned Pocono returns to vertical or north dipping within 0.6 mi (1 km) west of the gap and near 3 mi (5 km) east of the gap. The sharp bending of the Pocono Formation, shown by the location of the strike and dip changes, is interpreted to have induced local fractures, thereby producing easier erodibility of the resistant, ridge-forming strata at this location along Second Mountain.

Stop 2. Peters Mountain Gap (Fig. 3). Dyson (1963) mapped this area in detail and demonstrated that a major fracture, now filled with Triassic diabase, traverses the gap. The diabase is much more easily eroded than the quartzites and other sandstone rocks that it abuts. As viewed from the east side of the gap, the diabase occupies a pronounced notch on the south slope of Peters Mountain. Of note is the parallelism of the river to local strike in the immediate upstream reach of the river, just prior to entering the gap.

The diabase filling the fracture is only exposed on the east side of the river. The local rocks adjacent to the diabase have been baked to a hornfels. Dyson (1963, p. 55–63) described in detail the exposed stratigraphic section and later (1967) named the members of the Catskill Formation present. The section includes 125 ft (38 m) of the lower part of the Pocono Formation, and 207 ft (64 m) of the Spechty Kopf Formation. About 1,820 ft (560 m) of the Catskill Formation is also exposed, including all of the type section of the Duncannon Member (1,322 ft; 400 m), the type section of the Clarks Ferry Member (199 ft; 61 m), and the upper part of the Sherman Creek Member (299 ft; 92 m). The Peters Mountain exposure of these members shows the repetitive fining-upward cycles characteristic of the Catskill Formation. The cycles usually begin with a non-red, occasionally conglomeratic sandstone, occupying an erosional channel, and grade upward to a grayish-red claystone. An excellent description of the lithologies and interpretation of the sedimentary history of the Catskill Formation cycles and members in the immediate area of Stop 2 is given by Faill and Wells (1974).

Berry Mountain Gap is similar to Peters Mountain Gap in that the local position of the gap is also controlled by a fracture filled with Triassic diabase. At Berry Mountain the diabase is exposed on the west side of the gap (Hoskins, 1976a, 1976b) along U.S. 11-15. The diabase is hidden at road level by a retaining wall built to control the frequent collapses of loose rock onto the roadway, but the diabase and surrounding rocks can be examined on a retaining level above the highway.

Stop 3. Mahantango Mountain Gap (Fig. 4). Detailed mapping by Hoskins (1976b) revealed two significant structures present in or immediately adjacent to the gap. The more obvious structure is a strike-parallel thrust fault that repeats part of the Pocono and Mauch Chunk formations. Exposures are very limited in the gap; thus it is impossible to observe whether there are other fractures in the gap that relate to the thrust fault and its movement. Less obvious, but perhaps more important to the origin of this gap, are overturned and faulted layers of Pocono sandstone and conglomerate on the east side of the gap. An overall perspective requires viewing from the west bank at leaf-off seasons. Hoskins (1967) includes a photograph of the structure from this vantage and season. R. Faill (personal communication, 1969) suggested that this structure represents a kink-band fold. The local fracturing from the kinking, as in the Second Mountain Gap, may be the primary structural feature controlling the location of the Mahantango Mountain Gap.

REFERENCES CITED

Ashley, G. H., 1935, Studies in Appalachian Mountain sculpture: Geological Society of America Bulletin, v. 46, p. 1398–1436.

Cloos, E., and Broedel, C. H., 1943, Reverse faulting north of Harrisburg, Pennsylvania: Geological Society of America Bulletin, v. 54, p. 1375–1397.

Davis, W. M., 1889, The rivers and valleys of Pennsylvania: The National Geographic Magazine, v. 1, no. 3, p. 183–253.

—— , 1931, Foreword *in* Johnson, D., Stream sculpture on the Atlantic slope: New York, Columbia University Press, 142 p.

Dyson, J. L., 1963, Geology and mineral resources of the northern half of the New Bloomfield Quadrangle: Pennsylvania Geological Survey, 4th series, Atlas 137ab, 63 p.

—— , 1967, Geology and mineral resources of the southern half of the New Bloomfield Quadrangle, Pennsylvania: Pennsylvania Geological Survey, 4th series, Atlas 137cd, 86 p.

Ellison, R. L., 1965, Stratigraphy and paleontology of the Mahantango Formation in south-central Pennsylvania: Pennsylvania Geological Survey, 4th series, General Geology Report 48, 298 p.

Epstein, J. B., 1966, Structural control of wind gaps and water gaps and of stream capture in the Stroudsburg area, Pennsylvania and New Jersey: U.S. Geological Survey Professional Paper 550-B, p. 1380–1386.

Faill, R. T., and Wells, R. B., 1974, Geology and mineral resources of the Millerstown Quadrangle, Perry, Juniata, and Snyder counties, Pennsylvania: Pennsylvania Geological Survey, 4th series, Atlas 136, 276 p.

Geotimes editor, 1985, Explanation of cover illustration: Geotimes, v. 30, no. 3, p. 3.

Hoskins, D. M., 1967, Origin of locally overturned strata in northern Dauphin County, Pennsylvania: Pennsylvania Academy of Science Proceedings, v. 40, no. 2, p. 99–103.

—— , 1976a, Triassic-Jurassic diabase extended: Pennsylvania Geology, v. 7, no. 3, p. 6–8.

—— , 1976b, Geology and mineral resources of the Millersburg 15-minute Quadrangle, Dauphin, Juniata, Northumberland, Perry, and Snyder counties, Pennsylvania: Pennsylvania Geological Survey, 4th series, Atlas 146, 38 p.

Inners, J. D., 1984, Spiral feeding-organ supports preserved in the brachiopod *Athyris spiriferoides* (Eaton), Rockville, Dauphin County, Pennsylvania: Pennsylvania Geology, v. 15, no. 2, p. 9–12.

Johnson, D. W., 1931, Stream sculpture on the Atlantic slope: New York, Columbia University Press, 142 p.

Levine, J. R., 1985, Interpretation of Alleghanian tectonic events in the Anthracite Region, Pennsylvania, using aspects of coal metamorphism: Geological Society of America Abstracts with Programs, v. 17, p. 31.

MacLachlan, D. B., 1985, Pennsylvania anthracite as foreland effect of Alleghanian thrusting: Geological Society of America Abstracts with Programs, v. 17, p. 53.

Meyerhoff, H. A., 1972, Postorogenic development of the Appalachians: Geological Society of America Bulletin, v. 83, p. 1708–1728.

Oberlander, T. M., 1984, Origin of drainage transverse to structures in orogens, *in* Morisawa, M., and Hack, J. T., eds., Tectonic geomorphology, Proceedings, 15th Annual Binghamton Geomorphology Symposium, September, 1984: Boston, Allen and Unwin, p. 155–182.

Rogers, H. D., 1858, The geology of Pennsylvania, a government survey: Philadelphia, J. B. Lippincott & Co., 1045 p.

Sevon, W. D., 1985, Pennsylvania's polygenetic landscape: Fourth Annual Field Trip of the Harrisburg Area Geological Society, Guidebook, 55 p.

—— , 1986, Susquehanna River water gaps; Many years of speculation: Pennsylvania Geology, v. 17, no. 3, p 4–7.

Strahler, A. N., 1945, Hypotheses of stream development in the folded Appalachians of Pennsylvania: Geological Society of America Bulletin, v. 56, p. 45–88.

Theisen, J. P., 1983, Is there a fault in our gap?: Pennsylvania Geology, v. 14, p. 5–11.

Thompson, H. D., 1949, Drainage evolution in the Appalachians of Pennsylvania: Annals of the New York Academy of Science, v. 52, no. 3, p. 31–62.

Rheems quarry; The underside of a Taconian nappe in Lancaster County, Pennsylvania

Rodger T. Faill, Pennsylvania Geological Survey, P.O. Box 2357, Harrisburg, Pennsylvania 17120
Alan R. Geyer, 5334 Huntingwood Court, Sarasota, Florida 33580

LOCATION

The village of Rheems lies in the carbonate valley of northern Lancaster County, in southeastern Pennsylvania, at the edge of the Appalachian Piedmont Province (Fig. 1). Rheems is 2 mi (3.2 km) east of Elizabethtown, in the 7½-minute quadrangle bearing that town's name. The quarry is a short distance west of the village, adjacent to East Harrisburg Avenue in West Donegal Township. If you are arriving from some distance, the quarry can be most easily approached from the Rheems/Elizabethtown exit of Pennsylvania 283 (Fig. 2).

Rheems quarry is currently active; it is accessible only on weekends and after work ceases at 3:30 P.M. on weekdays. Permission to enter may be obtained by letter, phone, or in person from Union Quarries, Incorporated, Old Harrisburg Pike, Rheems, Pennsylvania 17570 (717-367-1080). Permission may also be obtained from the present superintendent, Denny Dupler, whose phone number is 367-4515. The quarry may be entered by foot or by car. Because the quarry is active, hard hats, safety glasses, and hard-toed shoes are necessary.

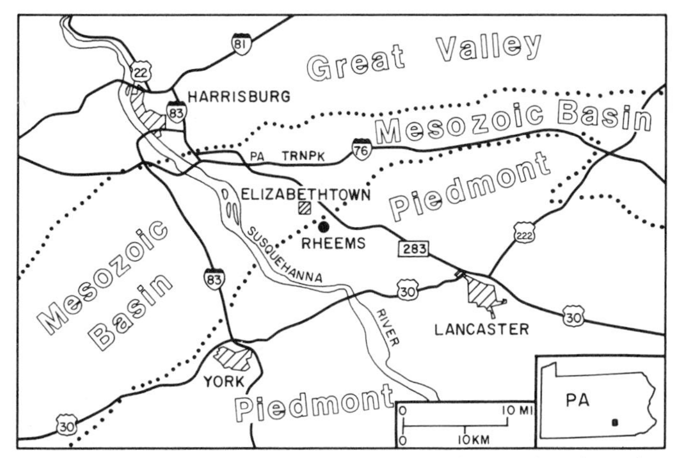

Figure 1. Roadmap of southcentral Pennsylvania, showing the major highways in the Rheems vicinity.

SIGNIFICANCE

The Taconian orogeny produced the alpine-style Lebanon Valley nappe in south-central Pennsylvania. Weathering and erosion have reduced this enormous recumbent anticlinorium to a humble topography, but a few glimpses here and there provide clues to its complexities and grandeur. The quarry at Rheems exhibits recumbent folds, rock flowage, boudinage, faults and fractures, cleavage, and transport indicators in the overturned carbonate beds of the lower limb of the nappe.

SITE INFORMATION

The rocks. Rheems Quarry was opened in the 1920s to supply stone for building purposes, the use that continues to this day. The quarry exposes only the Lower Ordovician Epler Formation, an approximately 2,500 ft (750 m) thick carbonate unit within the Beekmantown Group (Meisler and Becher, 1971) present throughout the entire Lebanon Valley nappe from Reading to Harrisburg.

Perhaps the most distinctive aspect of the Epler is the interbedding of limestone and dolomite, which on weathered surfaces provides striking displays of mesoscale structures. The limestones are mostly medium to medium-light gray (locally light pinkish gray) and weather to a light olive-gray or light gray. The finely crystalline beds contain very fine dark gray laminations.

The medium crystalline dolomite beds are medium gray in color, and weather to a yellowish gray. The dolomites are also laminated, and both lithologies tend to be medium to thick-bedded. Chert, occurring as dark gray to black nodules, lenses, and stringers, is scattered throughout the Epler Formation.

Regional setting. The sedimentary rocks at Rheems quarry were deposited on the carbonate shelf on the edge of the Laurentian continent, the Precambrian core of what is now North America. During the Middle and Late Ordovician, the continental convergence and closing of the proto-Atlantic Ocean created the Taconian orogeny. During this diastrophism, enormous blocks of the sedimentary shelf and large fragments of the underlying Precambrian rocks were forced up and into a younger shale basin to the (then?) northwest, which contained sediments of the present Martinsburg Formation. In the course of this (presently) northward movement, some of the blocks overrode their leading edge, forming recumbent anticlinoria of considerable complexity. The Lebanon Valley nappe is one of these thrust blocks.

The Lebanon Valley nappe contains rocks of Precambrian, Cambrian, and Ordovician age. The nappe is more than 60 mi (100 km) in width, and extends across the regional trend for at least 30 mi (50 km), from the Great Valley north of Lebanon well into the Piedmont terrane south of Lancaster. Although the nappe was modified, mostly by faulting, during the late Paleozoic Alleghanian deformation, the structures in the quarry are almost entirely of Taconian age. The southeastern edge of the Mesozoic basin lies only 1000 ft (300 m) to the north of the quarry, but the extent to which the quarry was affected by the Jurassic(?) deformation, if at all, is unknown.

The quarry exposures. The quarry highwalls provide ex-

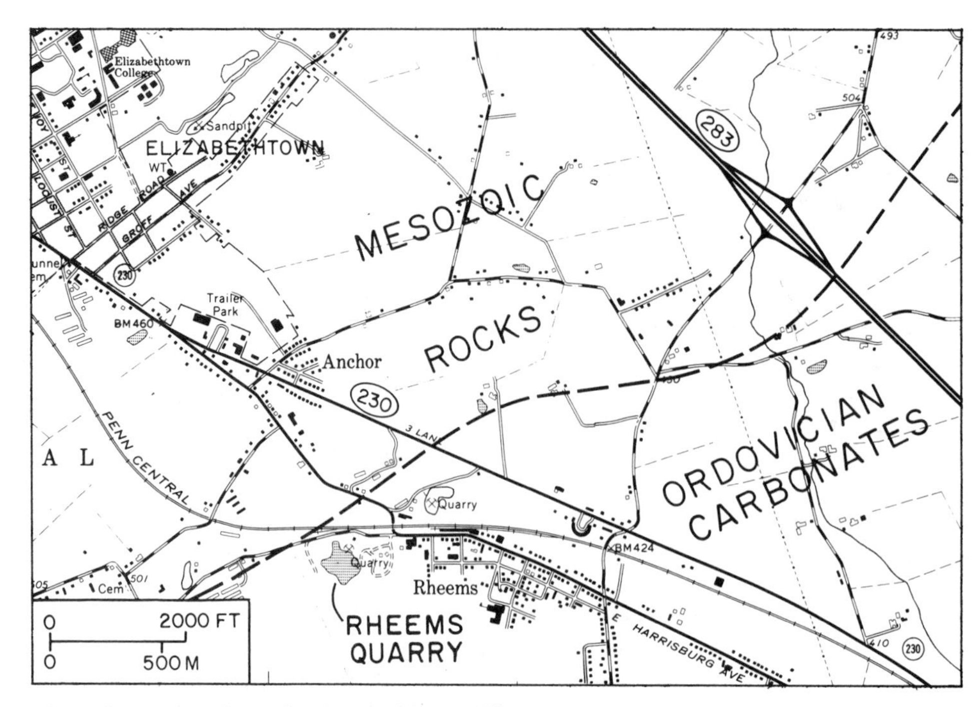

Figure 2. Local roadmap showing the Rheems-Elizabethtown exit from Pennsylvania 283, and the quarry just west of the village of Rheems.

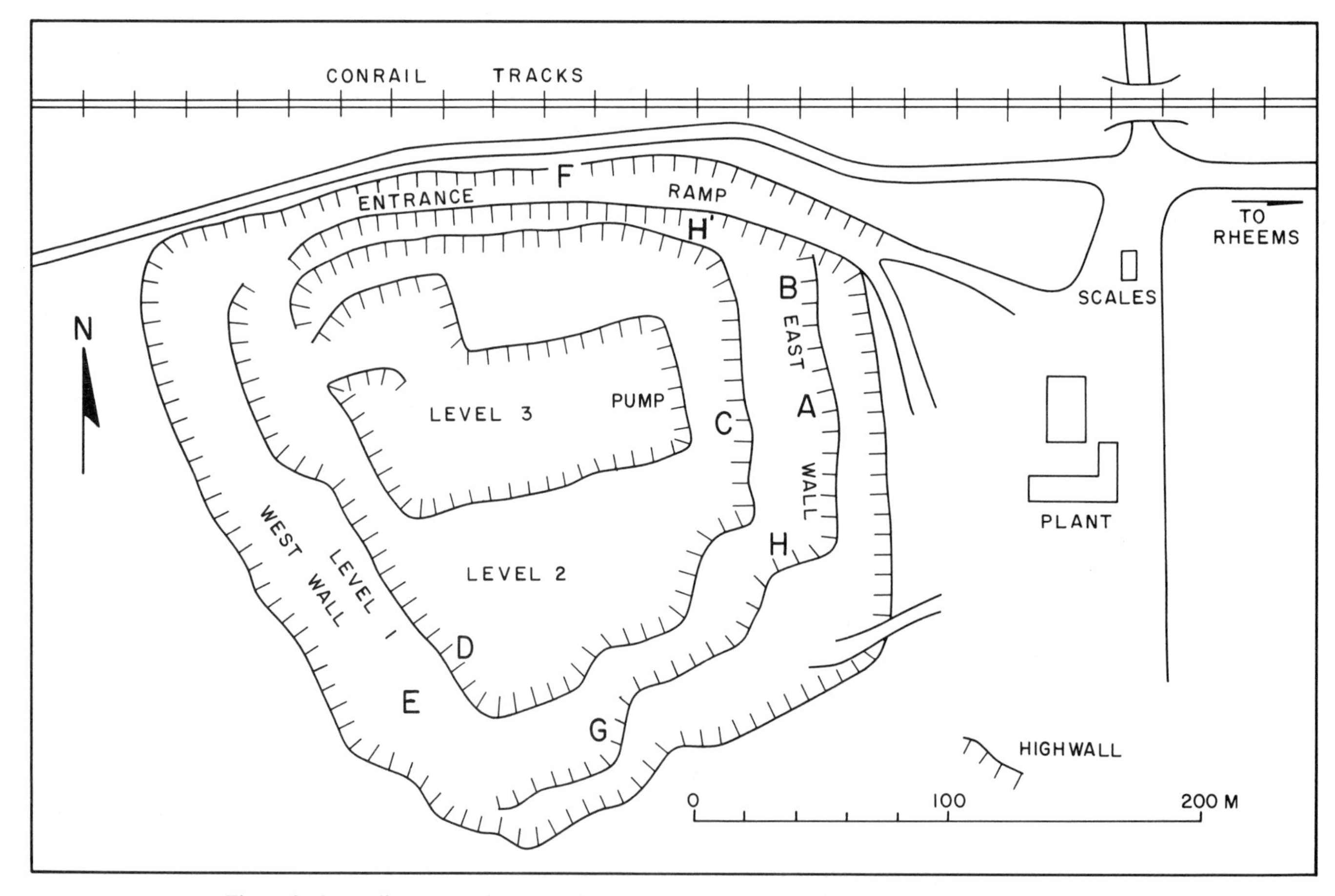

Figure 3. Generalized map of Rheems Quarry, which currently has three levels. Letters indicate the Stations where features described in the text may be seen.

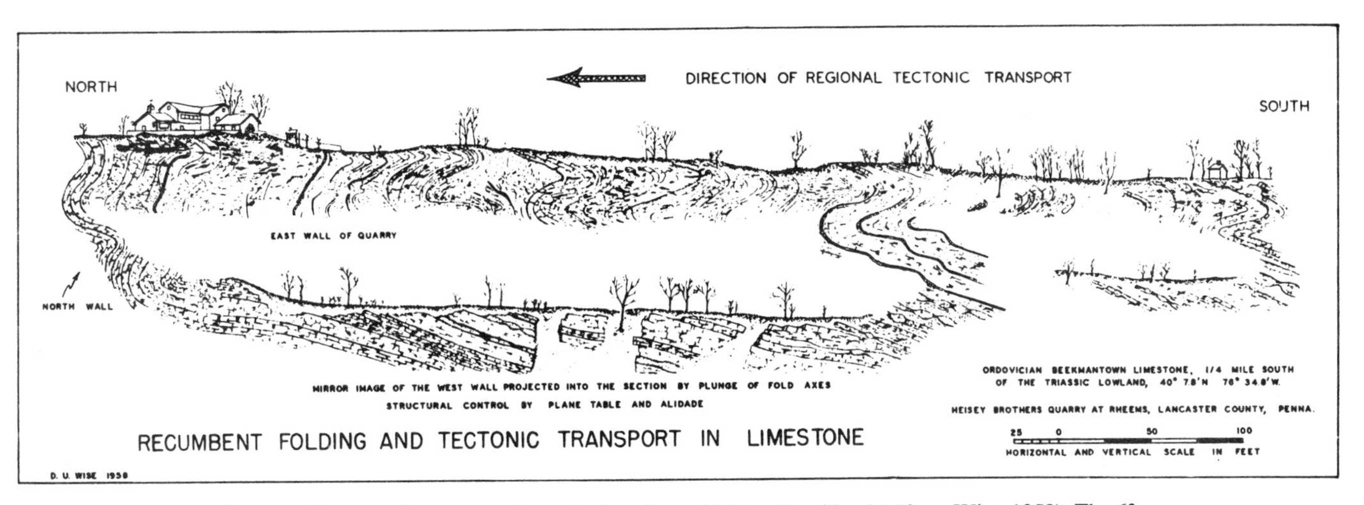

Figure 4. Diagram of the east and west (mirror image) highwalls of level 1 (from Wise, 1958). The 6° northeastward plunge of the structures in the quarry indicates that the west wall is structurally below the folds in the east wall. See text for discussion.

cellent exposures of the mesoscale structures that comprise the local tectonic grain. The north and south highwalls approximately parallel the 075 azimuth trend of the folds and display the changes along the grain; the east and west walls exhibit good cross sections of the structures (Fig. 3).

The dominant tectonic factor was the northward transport of the nappe; all the other Taconian structures in this quarry are derivative from this fundamental movement (Wise, 1958, 1960). A consequence of this movement is that most of the rocks in this underlying limb have undergone, to varying degrees, an extension in the north-south direction. Most of the mesostructures in the quarry amply demonstrate this strain.

The two most dramatic structures are the two recumbent folds in the east wall at Level 1 (Fig. 4). No evidence of depositional tops of beds has been found in this quarry, so which fold is the anticline and which is the syncline cannot be determined from sedimentary criteria. However, because the quarry is in the lower limb of the nappe, one can presume that the stratigraphic section is probably inverted. If so, then the upper fold, the one that is concave toward the north, has younger beds in the core, and thus it is the syncline. Similarly, the lower fold, the one that is convex toward the north, should have older beds in the core, and so it is the anticline. The axial surfaces of both the anticline and the higher syncline persist across the east wall with a 15 to 25 ft (5 to 8 m) separation, extending horizontally through at least 650 ft (200 m) of rock. This exposure also enables one to see that the folds possess a similar geometry.

The highwall exposure in the west wall at Level 1 is in distinct contrast to the east wall; no recumbent folds disturb the gentle south dip of the (presumably) overturned beds (Fig. 4). The relation between the different structures displayed in the two walls can be determined from the folds in the north highwall, above the entrance ramp (Station F in Fig. 3). The 5° to 15° plunge of the folds at Station F to the east-northeast demonstrate that the western highwall exposure is structurally below the folds in the eastern wall (Fig. 4). This can be verified by examining the east highwalls at Levels 2 and 3. No folds are present at these levels; only the fairly constant gentle south dip is present, similar to that in all three levels in the western part of the quarry.

The smaller mesostructures. Calcite crystals possess several slip systems that are not present in dolomite crystals. As a consequence, calcite deforms ductilely at low stress levels, whereas dolomite deforms in a brittle fashion at much higher stresses. This difference in deformational behavior shows up in several ways in Rheems quarry.

The dolomite beds tend to preserve a constant bed-normal thickness around fold hinges because of their greater strength and brittle behavior. The more ductile limestones flowed into the hinges increasing the bed normal thickness there, in some instances by large amounts, and greatly changing the fold profile (e.g., at Station A and southward, Fig. 3). Other types of profile distortion are also present. At Station B (Fig. 3), just north of the beds expressing a concentric fold geometry, the beds in the middle limb (between the syncline and anticline hinges) appear to be anomalously thickened, either by ductile flow or faulting. Nearby, several small recumbent folds in dolomite beds are detached from one another and surrounded by limestone. Although cleavage is not a strongly developed structure in this quarry, a weak axial-plane cleavage is present in many of the limestones in the fold hinges, representing a certain amount of flattening of the folds.

Boudinage, the "necking" or separation of beds, is common in many of the "overturned" layers (particularly at Stations C, D, and E, Fig. 3). The boudins trend 075 azimuth, parallel to the fold axes, suggesting that the boudins and folds are coeval. Again, the contrast in behavior of the rocks led to the boudinage. The stronger, less ductile dolomite beds developed "necks" of enhanced extension during the northward transport. The more duc-

Figure 5. Boudins in a dolomite embedded between limestone beds (at Station D in Fig. 3). The northward tectonic transport pinched and fractured the more brittle dolomite beds; the more ductile limestones extended by flow, and expanded into the boudin necks as well. The fractures in the dolomite were filled with white vein calcite. The white notebook on the left side is 8.3 in (21 cm) long.

tile limestones adjacent to the dolomites "flowed" into the necks between the dolomite boudins, producing the "hour glass" structure typical of boudins (Fig. 5). Interestingly, this boudinage does not occur in the "right-side-up" beds above the anticlinal hinge, because these beds were carried along as a unit by the northward transport of the nappe and did not experience extension.

Fractures filled with white vein calcite (Fig. 5) are another aspect of the brittle extension produced in the dolomite beds. These fractures parallel the east-trending fold axes, tend to be perpendicular to bedding, and are far more common in the overturned beds than in the upright ones. This, in addition to their association with boudin necks (as at Stations D and E, Fig. 3), and their absence in the more ductile limestones indicates that they formed, opened, and were filled during the later part of the northward transport and extension.

Ductile flow was an important, if not dominant, mode of deformation in these rocks, but it was not the only mode. Abundant slickensides on bedding surfaces attest to flexural slip folding being a significant process that contributed to the total deformation. Slickensides are particularly well exposed in the folds above the entrance ramp along the north wall (at Station F in Fig. 3).

The important faults present in this quarry occur in two orientations—parallel and transverse to the fold trends (Faill, 1983). Both types of faults, a steeply north-dipping parallel fault and a steeply west-dipping transverse fault, are present at Station G (Fig. 3). They are both postfolding because they offset parts of a fold. Another steeply west-dipping transverse fault (Station H in Fig. 3) has an apparent offset of 6 ft (2 m) down on the west, but its gently south-plunging slickenlines indicate strike-slip movement. This fault extends northward across the quarry (to Station H′ in Fig. 3), but at Station H′ the slickenlines plunge steeply (obliquely) to the northwest. These complex fault movements may represent a late stage of the Taconian deformation, effects of the late Paleozoic Alleghanian orogeny, or they may even be of Jurassic age and be a consequence of the Mesozoic rifting and opening of the Atlantic Ocean.

It is only in the past few decades that the regional nappe structure of this part of the Appalachians has been unraveled by Gray (1954), Wise (1958, 1960), and others, using the few excellent exposures provided by quarries such as the one at Rheems. However, the complexity of this terrane was recognized as early as the late nineteenth century by geologists, such as J. Peter Lesley, of the Second Pennsylvania Geological Survey. Lesley (1883, p. 54) wrote "Level as the general surface may be, it is the planed-off section of as gnarled and twisted a piece of the earth's crust as can be found in any country. Although these plications are comparatively small, they are of the same nature as the gigantic overthrown anticlinals of the Alps and Apennines."

REFERENCES CITED

Faill, R. T., 1983, Rheems quarry, *in* Mowery, J. R., ed., Geology along the Susquehanna River, southcentral Pennsylvania: Guidebook, 2nd Annual Field Trip of the Harrisburg Area Geological Society, p. 47–52.

Gray, C., 1954, Recumbent folding in the Great Valley: Pennsylvania Academy of Science Proceedings, v. 28, p. 96–101.

Lesley, J. P., 1883, Introduction; *in* Prime, F. P., ed., The Geology of Lehigh and Northampton counties: Pennsylvania Geological Survey, 2nd ser., Report of Progress DDD, vol. 1, p. 1–82.

Wise, D. U., 1958, An example of recumbent folding south of the Great Valley of Pennsylvania: Pennsylvania Academy of Science Proceedings, v. 32, p. 172–176.

——, 1960, Rheems quarry, *in* Wise, D. U., and Kauffman, M. E., eds., Some tectonic and structural problems of the Appalachian Piedmont along the Susquehanna River: Guidebook, 25th Annual Field Conference of Pennsylvania Geologists, p. 76–83.

Sequence of structural stages of the Alleghany orogeny at the Bear Valley Strip Mine, Shamokin, Pennsylvania

***Richard P. Nickelsen,** Department of Geology, Bucknell University, Lewisburg, Pennsylvania 17837*

LOCATION

The abandoned Bear Valley Strip Mine is in the Western Middle Field of the Pennsylvania Anthracite Region in the Valley and Ridge Province (Fig. 1). The intersection of lat. 40°45′50″N and long. 76°35′45″W marks the locality on the Shamokin, Pennsylvania 7½-minute Quadrangle map (1969 edition, photo revised 1975). The mine can be reached from the junction of Pennsylvania 61 and 125 in Shamokin as follows. Proceed south on Pennsylvania 125 for 1.4 mi (2.2 km) to the curve where Pennsylvania 125 turns sharply left; at this curve, continue straight in a southwest direction for 1.45 mi (2.3 km) to the end of the paved road; proceed uphill on a rough dirt road for 0.3 mi (0.5 km) to where the road becomes level and wide; park at a trail/road to the left and walk downhill 0.15 mi (0.24 km) southwest and then south to the "Entrance Road" shown on Figure 2. The site is on property owned by the Reading Anthracite Company; therefore, you are trespassing on private property. However, professional and educational groups have been granted free access for the past 25 years, but no liability is assumed by the company. Strip mine highwalls with loose rock pose a hazard, and one should exercise caution to avoid injury by falling or being fallen upon. Collecting specimens of in-place bedrock is prohibited.

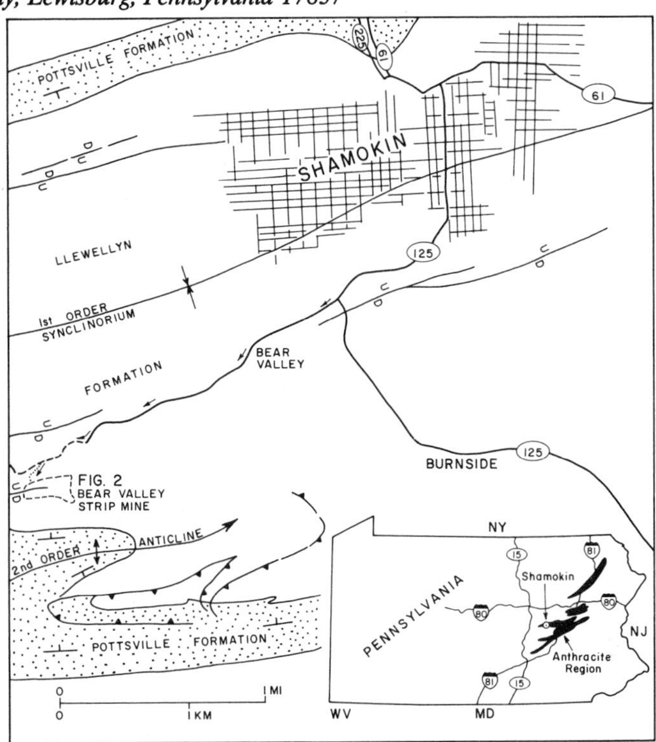

Figure 1. Index map and location map of Bear Valley Strip Mine. Pennsylvania 225, 125, and 61; U.S. 15, and I-80, I-81 are labeled.

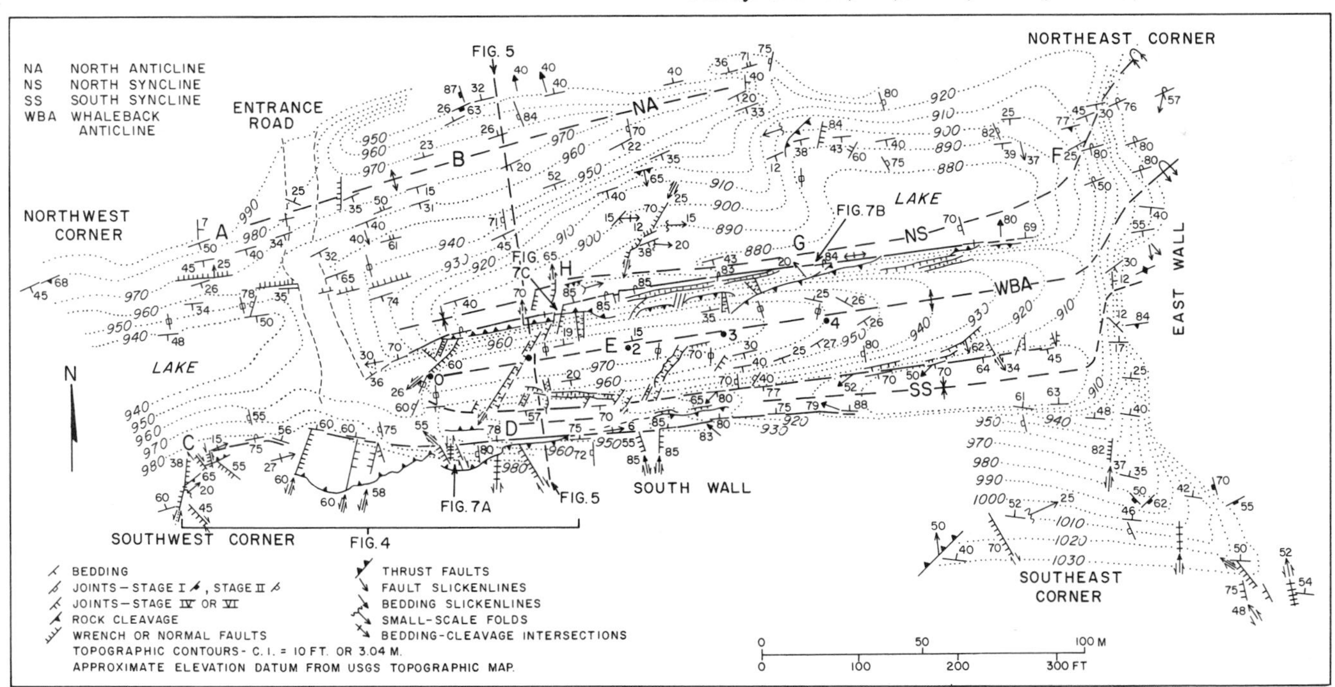

Figure 2. Map showing topography, geology, geologic localities A to H, and figure localities 4, 5, 7, and 8, of the Bear Valley Strip Mine. Location numbers 0, 1, 2, 3, 4 are at 100 ft (30 m) intervals along crest of Whaleback anticline.

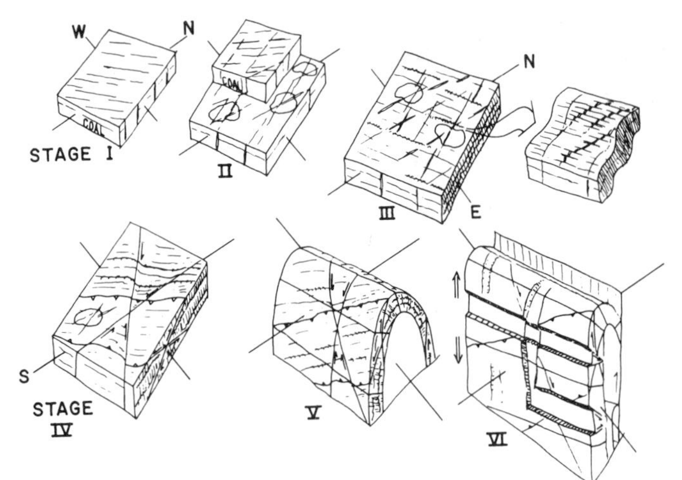

Figure 3. Cartoon depicting the sequence of development of structural stages of the Alleghany orogeny: Stage I. Orthogonal joint sets form in coal; Stage II. Several sets of hydraulic extensional joints form in sandstones and shales; Stage III. Pressure solution and primary crenulation cleavage and small-scale folds form; pressure solution of Stage II joint fillings occurs; Stage IV. Conjugate wrench and wedge faults deform Stage III cleavage; Stage V. Large-scale folding of all previous structures occurs; Stage VI. Extensional joints and faults produce flattening perpendicular to bedding and layer-parallel extension, both parallel and perpendicular to fold hinges.

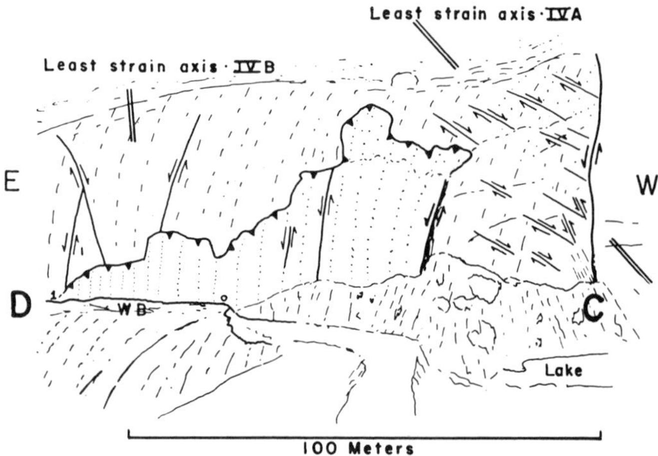

Figure 4. Conjugate wrench fault systems and a thrust fault between Stations C and D on the south wall, viewed from A. Least strain axis, IVA, is overprinted by IVB.

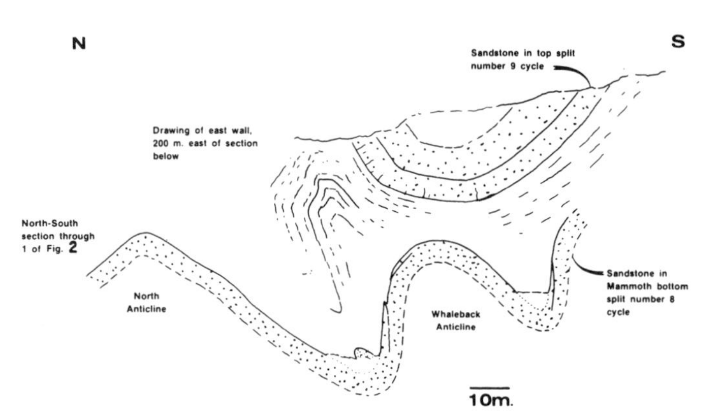

Figure 5. North-south section through location 1 of Figure 2.

SIGNIFICANCE

The Bear Valley Strip Mine offers superb three-dimensional exposures of the structural elements of the northern Valley and Ridge Province, and clear views of the geometric relationships, and the sequential overprinting of structures that elucidate the stages and processes of deformation within one orogeny. All of this is visible within a 320,000 ft^2 (30,000 m^2) strip mine area that contains exposures of two disharmonically-folded, Pennsylvanian-age cycles of sedimentation in cross section along highwalls and on unique bedding plane surfaces. Deformation mechanisms and the structural sequence of layer-parallel shortening (Stages II, III, IV) overprinted first by large-scale folding (Stage V) and later by layer-parallel extension (Stage VI) can be fully demonstrated at this locality (Fig. 3).

The variety of structural elements and deformation mechanisms displayed includes: Stage I joints in coal that are eastern extensions of the pre-Alleghanian Set I joints in coal observed on the Appalachian Plateau by Nickelsen and Hough (1967, Plate 3); hydraulic extensional joints with quartz-fiber fillings in ironstone and sandstone (Stage II); spaced cleavage in shales and silty shales formed by pressure solution, or grain rotation and sliding, or primary crenulation, accompanied by incipient recrystallization (stage III); wrench or wedge (thrust) faults with very obvious slickensides and slickenlines (Stage IV); third (size) order flexural slip folds showing disharmony between adjacent structural lithic units (Stage V); and extensional fault "grabens" or extensional joints formed by buckling or release fracturing (Stage VI). Finally, it is apparent that ductility contrast ranging from stiff ironstone to sandstone to more ductile coal and shale controls the relative structural behavior of rock types.

SITE INFORMATION

The Bear Valley Strip Mine is situated along several faulted second order folds on the south limb of the first order Shamokin Synclinorium (Fig. 1). The local geologic setting is provided by the map of Arndt and others (1973) while the larger anthracite region is described by Wood and others (1969), and Wood and Bergin (1970). The mine area has been previously described by Nickelsen (1979, 1983).

Figure 2 shows the detailed structural geology and topography of the site as well as the location of eight stations (A to H) of particular interest.

Station A. Turn right (west) off the entrance road and climb to the crest of the North anticline at A, a flat-topped box fold. The rocks beneath your feet represent the floor of the coal swamp and have impressions of large lycopsid trees and stigmaria (roots). From A, the entire mine is visible, including third order folds, fold disharmonies, many faults, and ironstone concretions.

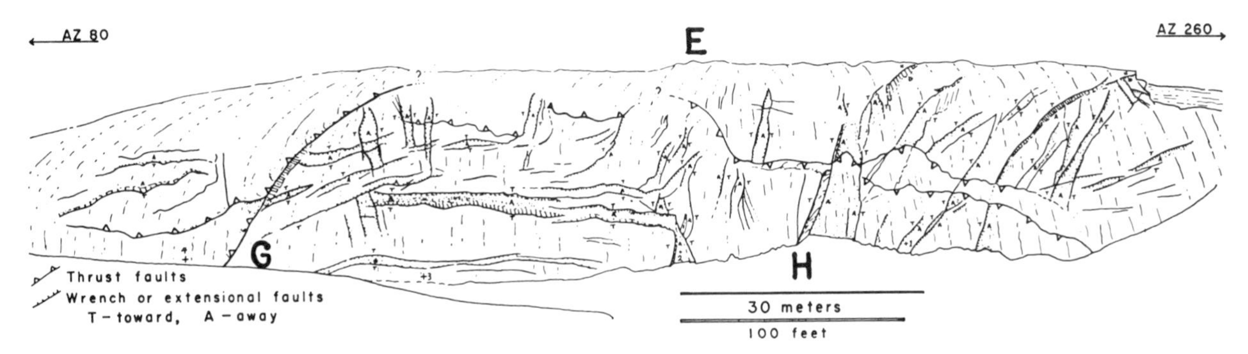

Figure 6. North limb of Whaleback anticline viewed from Station B. Extensional faults (Stage VI) define "grabens" that are both parallel and transverse to the fold hinge. The "grabens" overprint Stage IV wrench and thrust faults. Stations E, G, and H are labeled.

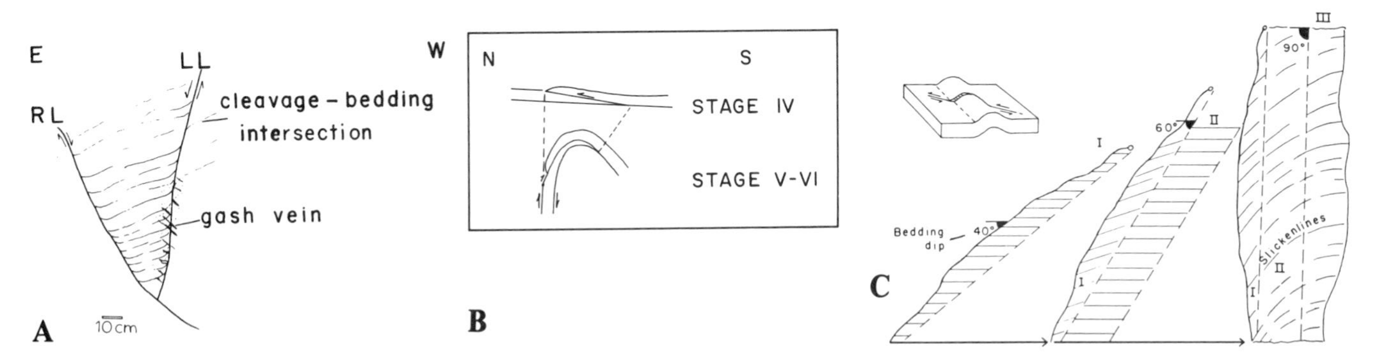

Figure 7(A). Conjugate wrench faults (Stage IV) viewed on the south wall between Stations C and D cause drag of cleavage-bedding intersections. Gash veins along the left lateral wrench fault cut preexisting Stage III cleavage. (B) Drawing to show how thrust faulting followed by flexural-slip folding has caused pull back of hanging wall to expose a fault surface at Station G. (C) Interpretation of curved slickenlines on a traverse fault at Station H. Folding to 40° north dip preceded wrench faulting. The Whaleback anticline and the fault then evolved together during Stage V folding and Stage VI layer-parallel extension.

The view of the southwest corner (Fig. 4) shows overprinted Stage IV conjugate wrench fault systems (dihedral angle 35°) and a Stage IV thrust. To the east the disharmonic third order folding shown in Figure 5 is visible.

Station B. Here, 330 ft (100 m) east of A, is an excellent view to the north limb of the Whaleback anticline showing Stage IV thrusts and conjugate wrench faults, the Stage V anticline plunging east, and Stage VI extensional joints and faults (Fig. 6). The North anticline on which you are standing has a chevron profile, different from that at Station A, because the kink junction axis is inclined to bedding (Faill, 1973, Fig. 20).

Station C. The Stage IV wrench and wedge (thrust) faults of Figure 4 can be studied here. Note that slickenlines on wrench faults are parallel to the fault-bedding intersection, indicating that they formed prior to folding and were later folded to their present attitude during Stage V. Slickenlines on wedge faults bisect the dihedral angle between wrench faults and were formed at the same time. Spaced cleavage (primary crenulation and pressure solution mechanisms) occurs in shales upslope to the southwest of C. Ironstone concretions are the stiffest member of the sedimentary succession, showing only Stage II joints and surface slickenlines that demonstrate how other strata have shortened and flowed around them. The contact between concretions and their enclosing rock are shear planes or boundary zones between different structural lithic units. The sequence of structural stages is established by offset of Stage II joints by Stage III pressure solution cleavage and by drag of Stage III cleavage-bedding intersections against Stage IV wrench faults (Fig. 7A). All of these structures formed by layer-parallel shortening in horizontal beds prior to folding.

While walking eastward between Stations C and D, you see Stage II joints in ironstone concretions, ductility contrast between ironstone and sandstone or shale, and a major wrench fault that is the west boundary of the thrust slice of Figure 4.

Station D. The east end of the thrust slice illustrated in Figure 4 can be observed. On the Whaleback anticline to the north, are wrench faults with intricate slickenline patterns that indicate overlap between Stage IV faulting and Stage V folding. Stage VI strike and transverse extensional faults and "grabens" are well exposed.

Station E. The crest of the Whaleback anticline provides the best view of the fold disharmony to the east (Fig. 5), Stage II

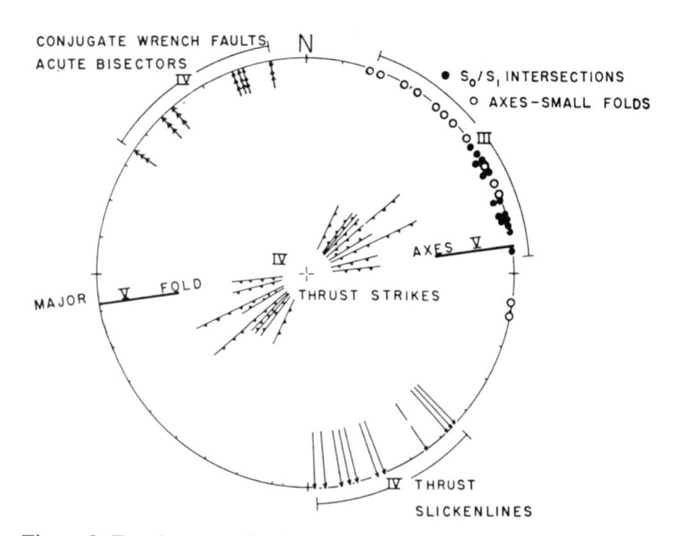

Figure 8. Equal-area projection showing angular relations of structures after they have been rotated with bedding to prefolding attitude. Stage III and IV structures trend counterclockwise of the Stage V fold hinge.

hydraulic joints on the whaleback (at 0 and 1), and the Stage IV wrench and wedge faults and Stage VI strike joints (of release or buckling origin) on the south wall. Note that the Stage VI strike joints are not symmetrical with the acute bisectors and slickenlines of stage IV wrench and wedge faults, indicating that the structural array formed in response to differently oriented strains—Stage IV layer parallel shortening, versus Stage VI extension due to fold buckling.

Station F. The tight hinge of the North syncline is seen here. Upslope in this multilayered sequence of shale and sandstone are excellent bedding-plane slickenlines as well as spaced cleavages in the shales and shaly sandstones.

Station G. Vertical bedding on the north limb of the Whaleback anticline is cut by Stage IV thrust faults and Stage VI extensional faults and "grabens" trending both parallel to and perpendicular to the fold hinge (Fig. 6, above G). Evidence for the relative age of thrusting and folding is illustrated in Figures 7B and 8. Thrust faults preceded folding because flexural-slip associated with folding has reversed their slip sense, causing a pull back of the tip line of the thrust and exposing the thrust surface as illustrated in Figure 7B. Also, thrust faults and their slickenlines as well as the acute bisectors of conjugate wrench faults consistently trend counterclockwise of the third order folds that overprint them (Fig. 8).

About 60 ft (20 m) west of Station G toward Station H is a triangular fault block (photograph in Nickelsen, 1979, Plate 10A) initially defined by Stage IV conjugate wrench faults but later reacting to Stage VI extensional strains to form a "graben." Overprinting of slickenlines on the faults bounding the block establishes the sequence. Stage IV slickenlines parallel the bedding-fault intersection, but Stage VI slickenlines are perpendicular to that intersection.

Station H. Two structural features are of importance here: a fourth order anticline in the trough of the North syncline that displays the evidence for the sequence of structural Stages IV and V, and a wrench fault on the north limb of the Whaleback anticline that remained active from Stage IV through Stage VI. The fourth order anticline illustrated by Nickelsen (1979, Pls. 8A and B) has slickenlines on a left lateral wrench fault (Stage IV) that nearly parallel the bedding-fault intersection and have been folded through 50° during Stage V.

The history of the wrench fault on the north limb is interpreted in Figure 7C. This fault was initiated under compression as a wrench fault, participated in folding (Stage V), and ended movement as an extensional fault, contributing to stretching of the hinge of the Whaleback anticline (Stage VI).

SUMMARY

At the Bear Valley Strip Mine, a sequence of deformational stages of the Alleghany orogeny can be interpreted from overprinted structures in rocks of Pennsylvanian age. All of the stages displayed must be Pennsylvanian or younger but no Mesozoic deformation is thought to have occurred here. Similar sequences of deformation have been recognized elsewhere in the Canadian Rockies (Price, 1967) and the Valley and Ridge Province (Perry, 1975). Rocks of the Bear Valley Strip Mine were deformed at temperatures of 185° to 205°C under an overburden of 3.1 to 5.0 mi (5 to 8 km) (Nickelsen, 1983, p. 137).

REFERENCES

Arndt, H. H., Wood, G. H., Jr., and Schryver, R. F., 1973, Geologic map of the south half of the Shamokin Quadrangle, Northumberland and Columbia counties, Pennsylvania: U.S. Geological Survey Miscellaneous Investigations Series Map 1-734, scale, 1:62,500.

Faill, R. T., 1973, Kink band folding, Valley and Ridge Province, Pennsylvania: Geological Society of America Bulletin, v. 84, p. 1289–1314.

Nickelsen, R. P., 1979, Sequence of structural stages of the Alleghany orogeny, at the Bear Valley Strip Mine, Shamokin, Pennsylvania: American Journal of Science, v. 279, p. 225–271.

—— , 1983, Stop III, Bear Valley, *in* Nickelsen, R. P., and Cotter, E., leaders, Silurian depositional history and Alleghanian deformation in the Pennsylvanian Valley and Ridge: 48th Annual Field Conference of Pennsylvania Geologists, Guidebook, p. 128–140.

Nickelsen, R. P., and Hough, V.N.D., 1967, Jointing in the Appalachian Plateau of Pennsylvania: Geological Society of America Bulletin, v. 78, p. 609–630.

Perry, W. J., 1975, Tectonics of the western Valley and Ridge fold belt, Pendleton County, West Virginia; A summary report: U.S. Geological Survey Journal of Research, v. 3, no. 5, p. 583–588.

Price, R. A., 1967, The tectonic significance of mesoscopic subfabrics in the Southern Rocky Mountains of Alberta and British Columbia: Canadian Journal of Earth Sciences, v. 4, p. 39–70.

Wood, G. H., Jr., and Bergin, M. J., 1970, Structural controls of the anthracite region, Pennsylvania, *in* Fisher, G. W., Pettijohn, F. J., Reed, J. C., Jr., and Weaver, K. N., eds., Studies of Appalachian geology central and southern: New York, Interscience Publishers, p. 147–160.

Wood, G. H., Jr., Trexler, J. P., and Kehn, T. M., 1969, Geology of the west-central part of the Southern Anthracite Field and adjoining areas, Pennsylvania: U.S. Geological Survey Professional Paper 602, 150 p.

Upper Mississippian to Middle Pennsylvanian stratigraphic section Pottsville, Pennsylvania

*Jeffrey Ross Levine**, *Department of Geological Sciences, McGill University, Montréal, Québec HA3 2A7, Canada*
Rudy Slingerland, *Department of Geosciences, The Pennsylvania State University, University Park, Pennsylvania 16802*

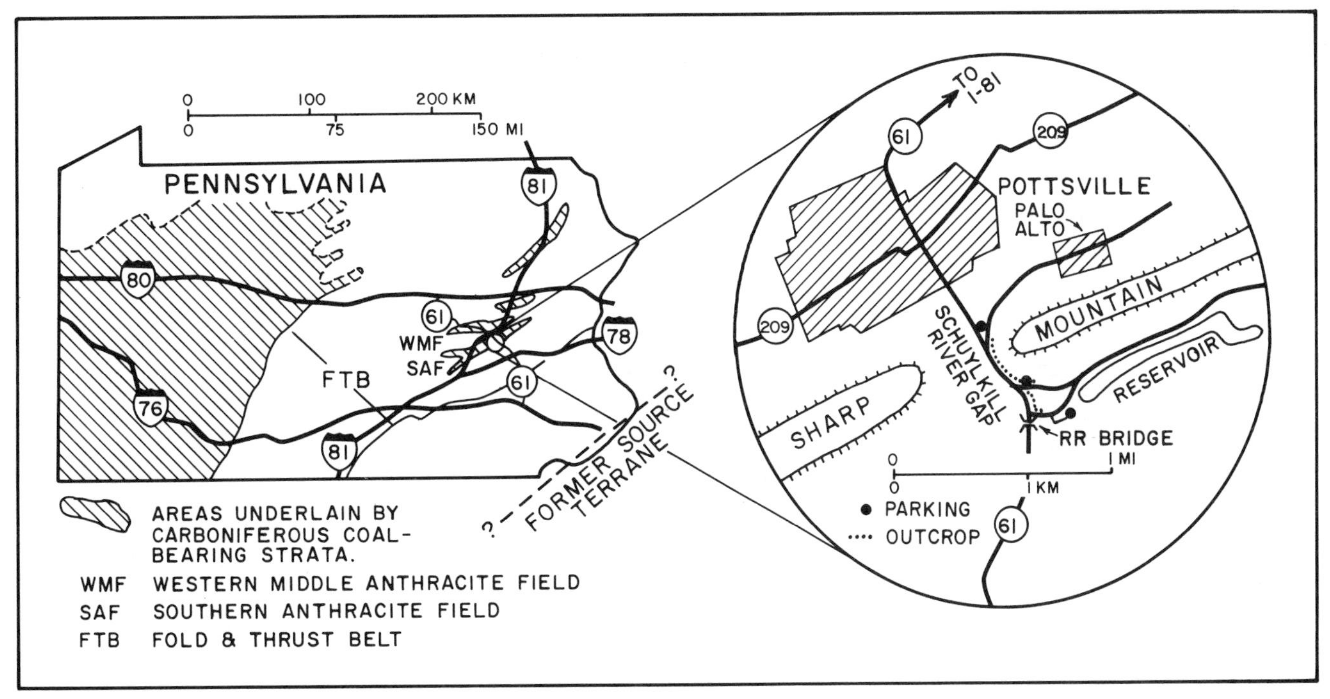

Figure 1. Key map of field locality.

LOCATION AND ACCESSIBILITY

The rocks at this site are exposed along a road cut on the eastern side of Pennsylvania 61, 0.3 to 0.5 mi (0.4 to 0.8 km) south of Pottsville, Pennsylvania (Fig. 1), on the southern margin of the Southern Anthracite field where the Schuylkill River has cut a deep gap in Sharp Mountain. Parking is available at several places, but it is advantageous to begin at the southern end of the outcrop and walk up section.

SIGNIFICANCE OF SITE

The outcrop exposes a 2,000-ft (600+-m)-thick section of upper Carboniferous molasse, representing the northwestward influx of clastic detritus into the Appalachian foreland basin from an orogenic source terrane formerly situated along the present Atlantic Coastal Plain. The alternation of facies (Fig. 2) reflects the gradual but progressive evolution of depositional environments from a semi-arid alluvial plain (Mauch Chunk Formation), to a semi-humid alluvial plain (Pottsville Formation), to a humid alluvial plain dominated by peat swamps (Llewellyn Formation). This transition, documented by dramatic changes in sedimentary facies, facies sequences, and maximum clast sizes, clearly reflects regional (perhaps even world-wide) climatic changes occurring near the end of the Mississippian; however, incipient Alleghanian tectonism and the evolution of many new plant groups occurring at this time may have played an influential role as well.

Subsequent to their deposition, the Carboniferous sediments were deeply buried, metamorphosed, tectonically deformed in the Alleghanian orogeny, uplifted, and largely eroded. The Southern Anthracite field now preserves the thickest, coarsest-grained, most proximal to the source, and most stratigraphically continuous occurrence of upper Carboniferous molasse in the central Appalachians.

STRATIGRAPHIC AND GEOMORPHIC OVERVIEW

Molasse sediments of the Anthracite region are stratigraphically subdivided on the basis of grain size and predominant coloration (Wood and others, 1969). The fine-grained, red Mauch Chunk Formation (Middle to Upper Mississippian) intertongues with and is replaced by the coarse-grained, gray Pottsville Formation (Lower to Middle Pennsylvanian), which in turn gives way

*Present address: School of Mines and Energy Development, The University of Alabama, Drawer AY, Tuscaloosa, Alabama 35487-1489.

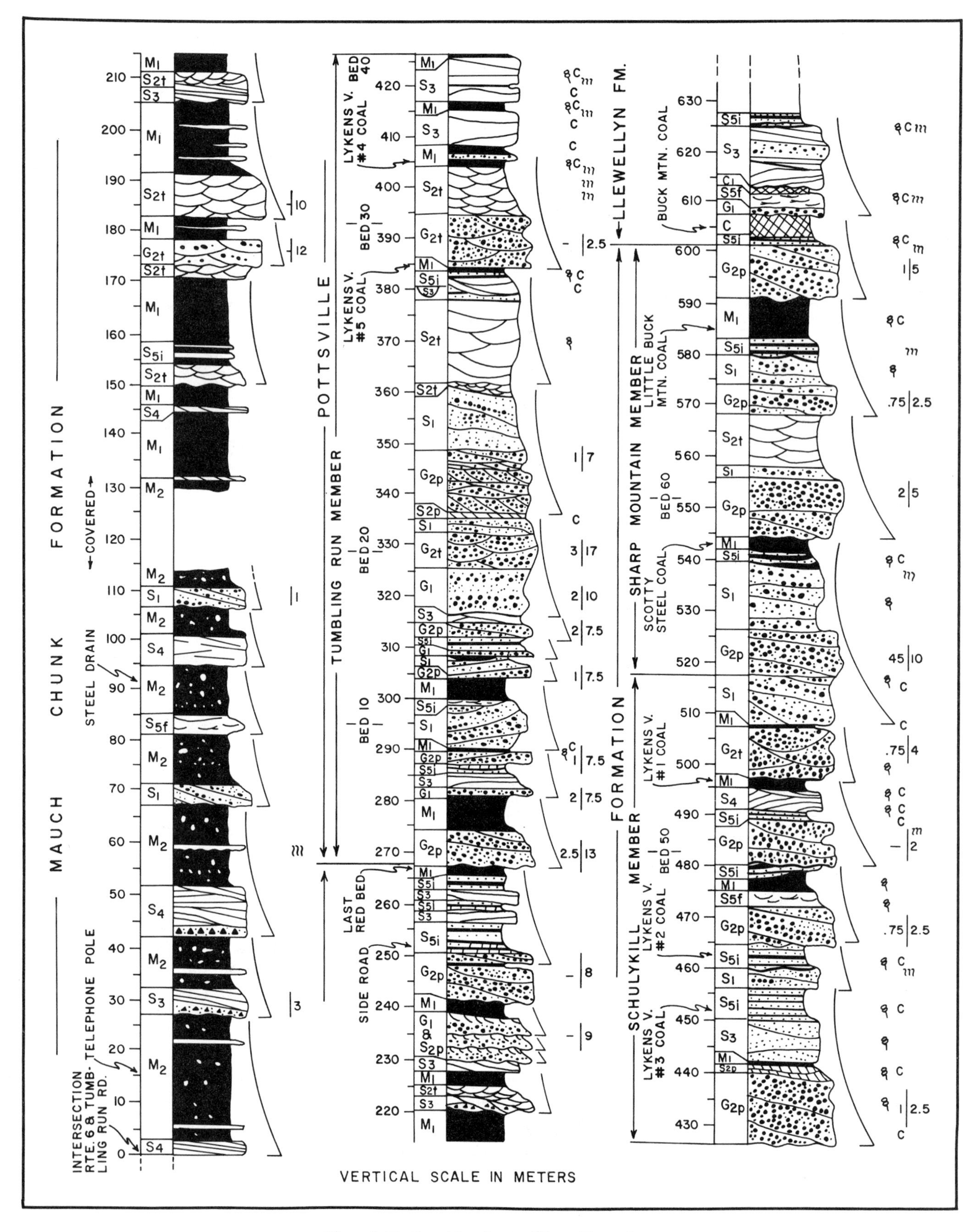

Figure 2. Stratigraphic column of Pottsville section.

TABLE 1 — FACIES STATES, SEQUENCES, COMPOSITION, AND FEATURES OF POTTSVILLE SECTION.

CODE:	G_1	$G_{2T\&P}$	S_1	$S_{2T\&P}$	S_3	S_4	$S_{5F\&I}$	M_1	M_2	C
SYMBOL:		t, p		t, p			f, i			
NAME:	Crudely bedded sandy conglom.	Cross bedded sandy conglom.	Plane bedded pebbly sandst.	Cross bedded pebbly sandst.	Coarse, plane bedded sandst.	Fine, plane bedded sandst.	Flaser or interbedded sandst. & mudst.	Noncalcareous mudst.	Calcareous mudst.	Coal
COLOR:	Pale gray ss. with variegated conglom.	Dusky yellow conglom. with gray sandst.	Light olive gray	Grayish orange to pink	Pale red (Mauch Ch.) Pale olive (Pottsv.)	Grayish to pink	Gray red (Mauch Ch.) or dark gray (Pv.)	Ruddy (MC) to light brown to black (Pv.)	Ruddy to brown	Black as coal
GRAIN SIZE:	Coarse sand to pebble conglom.	Pebble conglom. to very coarse sand	Coarse to granule with pebbly stringers	Coarse to granule	Coarse to very coarse	Very fine to fine sand	Fine sand with interbed. mud	Fine clay to silt	Fine clay & silt with carbonate concretions	Finely macerated plant fragments
INTERNAL BED FORMS:	Subhoriz. interstat. coarse sand & pebble conglom. Lenticular medium - thick beds	Medium to very thick lensatic beds. Matrix supported conglom (Note 1.)	Low angle laterally contin. wedge sets Medium to thick, concave upward	Cross bedding Lg. scale trough (S_{2t}) or Sm. scale planar (S_{2p})	Decimeter - thick tabular or wedge sets. Lateral cont. to 10s of m. Mass. or par. laminated	Tab. or wedge - shape beds. Laterally extensive thin to thick parallel laminae	Small scale cross - strata with mud flasers (S_5f) or laminated w/ mud layers (S_5i)(Note 2.)	Finely laminated or rooted see Note 3.	Stratified or paleosol features (see Note 3.) M. C. only	Finely stratified but sheared during tectonic deform.
COMPOSITION MAUCH CHUNK TYPE:	(No specific data)		Maturity: mature	Monoxl quartz: 50%	Feldspar: <1%	Rock fragments plus mica. Avg. 15% Generally sedimentary or low - grade metamorphic origin; primarily zircon & tourmaline.		90% of M. C. Shales are red beds (org. - void) Relatively Fe - rich. Clays: 80% illite; 20% chlorite; $CaCO_3$ & SiO_2 concretions in M_1.		None
PV.—LLEWELL.—TYPE:	Vein qtz > sandstone, chert, conglom., silt, and shale > low to med grade metamorph. incl: phyllite, slate and schist.		more mature	60%	<1%	(3 - 38%) Med. to high - grade metamorphic minerals first appear in upper Mauch Chunk indicating unroofing of these rocks in source terrane.		All are grey to black org. - bearing Fe - poor, Al - rich. Primarily ill. & chlor. with → 40% kaol. & pyro phyllite. No caliche.		> 92% fixed carbon, dry mineral matter - free.
TYPICAL BASE:	Undulatory, sharp	Undulatory, sharp, erosive	Undulatory, sharp, erosive	Gradational from S_1 or erosive from M_1	Gradational or sharp	Sharp, planar dips to 15°	Gradational from S_1 or S_3	Gradational from S_2 or S_5	Gradational	Gradational from M_1
TOP:	Gradational	Grades to S_{2t}	Sharp, undulatory, fine - gr.	Grad. to S_5i or M_1	Grad. to S_{2t} or S_{5i}	Grad. to M_2	Grad. to S_1 or M_2	Sharp or erosive	Sharp or erosive	Eros. to G_2 Grad. to M_1

NOTES:

1. G_2 bed continuity ranges from 3m to across outcrop. Lenses subparallel to 2nd - order truncation surfaces; either laterally extensive tangential low angle (<15°) planar cross - strata (G_{2p}) or large - scale (2 - 4 m) cut - and - fill troughs (G_{2t})
2. S_5 unit contains abundant dark brown calcareous root traces, desiccation cracks, and raindrop impressions in Mauch Chunk or plant fragments in Pottsville.
3. Paleosol features of M_1 & M_2 : In Mauch Chunk, paleosols are "vertisols" with wetting/drying features, including wedge - shaped peds, blocky joints, mud cracks, raindrop impressions, pretectonic slickensides, & caliche.
   Paleosols in Pottsville Fm. are "underclays" with abundant root impressions, leached clay minerals, & no bedding.

OTHER SYMBOLS:

▲▲▲ Intraformational conglomerate; Calcareous concretions; Root traces; Plant fragments; C Carbonaceous; Erosional contact; 3|10 average (3) and maximum (10) clast size in centimeters; Fining and thinning upward cycle; Disconformity

TRANSITION MATRIX SUMMARY SEQUENCES:

Mauch Chunk No. 1 $S_4 \rightarrow M_2$

Mauch Chunk No. 2 S_3, S_{2t}, S_{5i}, M_1

Pottsville $G_{2p} \rightarrow S_1 \rightarrow S_{2t} \rightarrow M_1$; $S_3 \rightarrow S_{5i}$; S_{5f}

(Compositional information from Meckel, 1967; Wood et al., 1969; Holbrook, 1970; Hosterman et al., 1970)

to the finer-grained, gray to black, coal-rich Llewellyn Formation (Middle Pennsylvanian), representing the youngest extant molasse in the region. The former presence of many miles (kilometers) of overlying rocks is implied by the high coal rank and compaction of the Llewellyn sediments (Paxton, 1983; Levine, 1986).

The Mauch Chunk Formation is informally subdivided into three members (Wood and others, 1969). The middle member represents the 'type' Mauch Chunk red bed lithofacies. The lower and upper members represent respectively the zones of intertonguing with the underlying Pocono Formation and the overlying Pottsville Formation. The upper contact of the Mauch Chunk is defined as the top of the uppermost Mauch Chunk–type red bed (Fig. 2).

The Pottsville Formation is formally subdivided into three members (Wood and others, 1956), each representing a crudely fining-upward megacycle. Of the three, the Tumbling Run and the Sharp Mountain members are the coarser-grained, while the intervening Schuylkill Member is finer-grained and contains a greater proportion of coal. The lower contacts of the Schuylkill and Sharp Mountain members are defined at the base of major conglomeratic units. The base of the Schuylkill Member is by no means obvious at the outcrop, but the "Great White Egg" quartz pebble conglomerate at the base of the Sharp Mountain Member is very distinctive. The contact between the Pottsville and Llewellyn Formations is placed at the base of the lowermost thick, stratigraphically persistent coal horizon, the Buck Mountain (#5), which has been correlated over large areas of the Anthracite fields (Wood and others, 1963).

Chronostratigraphic age designations in the Anthracite region, based upon the 13 upper Paleozoic floral zones defined by Read and Mamay (1964; also see Edmunds and others, 1979, Fig. 11), indicate the Pottsville section is conformable, extending from Zone 3 in the upper Mauch Chunk Formation (Chesterian Series) to Zone 10 in the lower Llewellyn Formation (Des Moinesian/Missourian Series); however, Zones 7 and 8 have not been explicitly recognized at this site. The Mauch Chunk/Pottsville contact, occurring between Zones 3 and 4, corresponds roughly to the Mississippian/Pennsylvanian systemic boundary. In areas of the central Appalachians other than the Southern and

Middle Anthracite fields, Zones 4, 5, and 6 are absent, suggesting the presence of a significant disconformity between the youngest Mississippian and oldest Pennsylvanian strata (see discussion in Edmunds and others, 1979).

The strata exposed at the site are slightly overturned and comprise part of the southern limb of the Minersville Synclinorium, forming the southern margin of the Southern Anthracite field. They attained their present attitude during the late Paleozoic Alleghanian orogeny when northwest-directed tectonic forces produced a progression of deformational phases that migrated northwestward across the foreland basin. At the Pottsville site all structural phases are superposed (Wood and Bergin, 1970; Nickelsen, 1979).

The structure and stratigraphy of the upper Paleozoic molasse sequence are revealed geomorphically by the relative resistance to erosion of the near-vertical component units. The Pocono sandstone, subjacent to the Mauch Chunk Formation, upholds Second Mountain, the major ridge visible to the south of the Pottsville section. The Mauch Chunk Formation underlies the valley between Second and Sharp mountains. The distinctive double ridge of Sharp Mountain is formed by the Tumbling Run and Sharp Mountain members of the Pottsville Formation. The Schuylkill River, which excavated the gap in Sharp Mountain, flows southeasterly across the Valley and Ridge Province on its course to the Chesapeake Bay, opposite to the streams that originally deposited the Pottsville sediments.

SEDIMENTOLOGY OF THE POTTSVILLE SECTION —FACIES STATES AND COMPOSITION

Sedimentary bed forms, sediment composition, facies sequences, and paleobotany reveal a significant alteration in paleoclimatic conditions across the Pottsville section, ranging from generally semi-arid, poorly vegetated conditions at the base to perennially humid, lush conditions at the top. Ten general facies have been defined at this site and are described in Table 1. Transition matrix analysis reveals two repeating motifs, one characteristic of the Mauch Chunk and one of the Pottsville. When compared to facies sequences from modern environments of deposition, the Mauch Chunk sequence is similar to that of Bijou Creek, Colorado, a sandy, braided, ephemeral stream subject to catastrophic floods (Miall, 1977). Facies S_3 and S_4 probably comprised sand flats or shallow channel deposits; S_{5i} and S_{2t} comprised waning flow deposits or overbank deposits more removed from the active channel. M_1 represents intra-channel, slack water deposits and M_2 represents overbank soils.

The Pottsville sequence is similar to that produced by the Donjek River, Yukon Territory, a gravel-sand mixed bedload, perennial braided stream (Miall, 1977). Facies G_2, S_1, and S_3 formed in the lower parts of the active channels by longitudinal braid bar migration. Facies S_{2t} and S_{5t} formed in the upper parts of active channels or minor channels and on the tops of braid bars. Facies S_{5i} and M_1 formed on bar tops, abandoned channels, and overbank areas, and facies C was deposited in inter-channel swamps. The channels forming the Pottsville Formation were deeper with greater cross-sectional areas, and lower width/depth ratios than those forming the Mauch Chunk Formation. In consequence, maximum clast size is greater as is the thickness of cross-bed sets.

Sandstone petrology, organic matter content, clay mineralogy, and features of the paleosols (Table 1) all show a progressive trend to more highly leaching, less oxidizing (i.e., more humid) conditions higher in the section. Sandstones are compositionally mature throughout the section but become even more mature up section. The Tumbling Run Member of the Pottsville Formation contains the highest variety and proportion of non-quartzose fragments while the Sharp Mountain Member contains the highest proportion of vein quartz (Meckel, 1967). Preservation of organic matter in the upper part of the section implies conditions of low Eh, maintained by continuous saturation by stagnant or slowly moving water. Clay minerals are enriched in alumina and depleted in iron higher in the section indicating a greater degree of chemical and biological leaching.

Paleosols occurring throughout the section are particularly useful in revealing paleo-environmental conditions. Most paleosols of the Pottsville and Llewellyn Formations formed as underclays beneath peat swamps and, therefore, must have been water-saturated during most of their development. In contrast, paleosols of the Mauch Chunk Formation, classified as vertisols by Holbrook (1970), exhibit a variety of features indicating episodic wetting/drying cycles (Table 1).

Caliche, occurring as thin, bed-parallel laminae or in nodular layers less than 3 ft (1 m) in thickness is common in the middle member of the Mauch Chunk (Fig. 2) and occurs occasionally in the upper member. Caliche forms in seasonally arid conditions when surface evaporation produces supersaturation of dissolved salts, especially calcium carbonate and silica. The laminar caliche is interpreted to have formed at the sediment surface in shallow ponds during evaporative cycles (Holbrook, 1970). A surface or near-surface origin is indicated for the nodular caliche as well (Holbrook, 1970) based on: (1) sedimentary laminations that pass from the surrounding sediment into the concretions, (2) nodules occurring as intraformational clasts in conglomerates, (3) the presence of carbonate as nodules in the shales but not as cement in the adjacent sandstones, and (4) ball and pillow structures occurring between the nodules and the underlying (but not the overlying) sediments.

The composition of the organic matter and clay minerals has been strongly influenced by diagenetic conditions during burial. The coal has been elevated to anthracite rank. Expandable layer clays are not present and illite is of the highly ordered 2-M form, representing "anchizone" alteration. Pyrophyllite is an anchizone alteration product of kaolinite that forms only in Fe-depleted rocks (cf., Hosterman and others, 1970, Table 1). Ammonium illite is thought to form at high coal rank in organic matter–rich sediments by nitrogen released during late stages of coalification (Paxton, 1983). These transformations imply temperatures of ca. 225–275°C and 4 to 6 mi (6 to 9 km) of burial.

TECTONIC SIGNIFICANCE OF THE POTTSVILLE SECTION

During deposition of the Pottsville section the depositional margin of the basin lay in the vicinity of Philadelphia as indicated by paleocurrent directions and regional trends in maximum grain size (Pelletier, 1958; Meckel, 1967; Wood and others, 1969). Northeast-flowing streams carried sediments toward the basin axis, which trended northeast-southwest across western Pennsylvania. Time equivalent upper Carboniferous rocks are alluvial in eastern Pennsylvania and deltaic and shallow marine to the west (Edmunds and others, 1979). The Mauch Chunk Formation documents a relatively quiescent interval represented variously by fine-grained sedimentation and soil development in the east, an erosional disconformity toward the west, and shallow marine carbonate sedimentation along the basin axis. The influx of coarse clastics in the Pottsville interval has traditionally been ascribed to tectonic uplift in the source (e.g., Meckel, 1967), but while this might be partly true, it is neither a necessary nor sufficient explanation. The simplest explanation is that the change to more humid climatic conditions in the Pennsylvanian produced larger sediment yields and stream discharges. The continued influx of clastic sediments would represent isostatic unloading of the Acadian source terrane.

An additional factor influencing the stratigraphic succession may have been the diversification and proliferation of terrestrial plants during the middle Carboniferous. Plant evolution could have helped to stabilize stream banks, allowed peat accumulation rates to equal or exceed basin subsidence, and influenced climatic patterns.

The intertonguing of Mauch Chunk and Pottsville facies in the upper member of the Mauch Chunk clearly indicates an alternation of depositional environments, but it is problematical whether this represents the lateral migration of two co-existing subenvironments in the sense of Walther's Law, or the sedimentological adjustment of an entire depositional system to cyclic climatic changes. In the former case, the Pottsville Formation would represent a higher elevation, proximal, more humid facies and the Mauch Chunk a more distal, flood basin facies, subject to wetting more by flooding than by rainfall.

The interpreted tectonic and paleoenvironmental setting during Mauch Chunk deposition would have resembled in many respects the current alluvial plain extending from the Zagros Mountains to the Persian Gulf where arid conditions produce little clastic influx from the tectonically active mountain belt. The adjacent foreland basin axis—lying parallel to the mountain belt—receives primarily carbonate sedimentation. Were a future global climatic change to transform the Middle East into a humid region, the margins of the Persian Gulf could perhaps evolve into a broad peat-forming environment such as existed in the Appalachian basin during Pottsville and Llewellyn times.

REFERENCES CITED

Edmunds, W. E., Berg, T. M., Sevon, W. D., Piotrowski, R. C., Heyman, L., and Rickard, L. V., 1979, The Mississippian and Pennsylvanian (Carboniferous) systems in the United States; Pennsylvania and New York: U.S. Geological Survey Professional Paper 1110-B, 33 p.

Holbrook, P. W., 1970, The sedimentology and pedology of the Mauch Chunk Formation at Pottsville, Pennsylvania, and their climatic implications [M.S. thesis]: Lancaster, Pennsylvania, Franklin and Marshall College, 111 p.

Hosterman, J. W., Wood, G. H., Jr., and Bergin, M. J., 1970, Mineralogy of underclays in the Pennsylvania Anthracite region: U.S. Geological Survey Professional Paper 700-C, p. C89–C97.

Levine, J. R., 1986, Deep burial of coal-bearing strata, Anthracite region, Pennsylvania—Sedimentation or tectonics?: Geology, v. 14, p. 577–580.

Meckel, L. D., 1967, Origin of Pottsville conglomerates (Pennsylvanian) in the central Appalachians: Geological Society of America Bulletin, v. 78, p. 223–258.

Miall, A. D., 1977, A review of the braided river depositional environment: Earth Science Review, v. 13, p. 1–62.

Nickelsen, R. P., 1979, Sequence of structural stages of the Alleghany orogeny at the Bear Valley strip mine, Shamokin, Pennsylvania: American Journal of Science, v. 279, p. 225–271.

Paxton, S. T., 1983, Relationships between Pennsylvanian-age lithic sandstone and mudrock diagenesis and coal rank in the central Appalachians [Ph.D. thesis]: University Park, Pennsylvania State University, 503 p.

Pelletier, B. R., 1958, Pocono paleocurrents in Pennsylvania and Maryland: Geological Society of America Bulletin, v. 69, p. 1033–1064.

Read, C. B., and Mamay, S. H., 1964, Upper Paleozoic floral zones and floral provinces of the United States: U.S. Geological Survey Professional Paper 454-K, 35 p.

Wood, G. H., Jr., and Bergin, M. J., 1970, Structural controls of the Anthracite region, Pennsylvania, *in* Fisher, G. W., Pettijohn, F. J., Reed, J. C., Jr., and Weaver, K. N., eds., Studies of Appalachian tectonics; Central and southern: New York, John Wiley Interscience, p. 161–173.

Wood, G. H., Jr., Arndt, H. H., Yelonsky, A., and Soren, J., 1956, Subdivision of the Pottsville Formation in Southern Anthracite field, Pennsylvania: American Association of Petroleum Geologists Bulletin, v. 40, p. 2669–2688.

Wood, G. H., Jr., Arndt, H. H., and Hoskins, D. M., 1963, Geology of the southern part of the Pennsylvania anthracite region: Annual Meeting of the Geological Society of America, Field Trip Guide Book 4, U.S. and Pennsylvania Geologic Surveys, 84 p.

Wood, G. H., Jr., Trexler, J. P., and Kehn, T. M., 1969, Geology of the west-central part of the Southern Anthracite field and adjoining areas, Pennsylvania: U.S. Geological Survey Professional Paper 602, 150 p.

Structural and sedimentologic characteristics of the Greenwich slice of the Hamburg klippe: An ancient subduction complex in eastern Pennsylvania

Gary G. Lash, Department of Geology, State University College, Fredonia, New York 14063

LOCATION

This site within the Greenwich slice of the Hamburg klippe is located in the northwest corner of the Kutztown 7½-minute Quadrangle (Lash, 1985a) in Berks County, Pennsylvania, and includes two large fault bounded exposures approximately 1000 ft (300 m) apart on Pennsylvania 143 (Fig. 1). The exposures can be reached by proceeding about 4 mi (6 km) north on Pennsylvania 143 from the Lenhartsville exit of U.S. 78. Parking of buses and automobiles is possible at the south end of the northern exposure (Fig. 1) and both exposures are easily accessible by foot from the parking area. Visitors are warned that shoulders on Pennsylvania 143 are narrow and the road is heavily traveled.

SIGNIFICANCE

Exposures at this site illustrate the major sedimentologic and structural characteristics of the Greenwich slice, the most extensive tectonic slice of the Hamburg klippe, and part of a proposed subduction complex. Visitors to this site will have the opportunity to study numerous structural and stratigraphic features that are analogous to those documented from modern convergent margins. These rocks are not only extremely important to the geologic synthesis of central Appalachian tectonic history, but they offer a great deal of information about sedimentation patterns and sediment behavior at convergent margins.

The southern exposure illustrates a coarsening-upward sequence composed of red and light-green pelagic mudstone and white dolomitic chert of Early to Middle Ordovician age overlain by a transitional zone of mixed pelagic and hemipelagic mudstone and minor turbidite sandstone. These rocks are in turn overlain by a transitional zone of mixed pelagic and hemipelagic mudstone and minor turbidite sandstone. These rocks are in turn overlain by a thick (possibly greater than 0.6 mi or 1 km) sequence of interbedded sandstone and olive-green hemipelagic mudstone of Middle Ordovician age. The sequence exposed here accumulated over a period of about 25 m.y. and is inferred to record advance of a site on oceanic lithosphere from a region of pelagic sedimentation in Early Ordovician time through an area of mixed pelagic-hemipelagic sedimentation and finally into the trench axis in Middle Ordovician time.

The northern exposure, which is tectonically overlain by the southern exposure, best illustrates the style of deformation of the Greenwich slice. Rocks here include interbedded turbidite sandstone and hemipelagic mudstone. Individual sandstone beds have been disrupted to various degrees and are surrounded by scaly cleaved mudstone. Pinch-and-swell and complete boudinage of sandstone beds are common. Stratal disruption of this sort is restricted to specific zones or horizons that are found adjacent to mesoscopically undeformed zones of interbedded sandstone and mudstone. This alternation of coherent and disrupted zones is quite similar to deformational styles in accreted oceanic sediment in many younger orogens around the world.

SITE INFORMATION

Rogers (1858) was the first to recognize the peculiar nature of the rocks that constitute the Hamburg klippe. Subsequent studies by Kay (1941) and Stose (1946, 1950a, 1950b) pointed out that these rocks, anomalous in age and lithology with respect to surrounding rocks of the Pennsylvania Great Valley, are similar to rocks of the Taconic sequence of New York and New England. Stose (1946) referred to the gray, black, red, and green shale and graywacke, limestone, chert, and volcanic rocks that extend between the Lehigh and Susquehanna rivers as the Hamburg klippe (Fig. 1). He maintained that these rocks were thrust westward from an eastern ocean or oceans onto the Martinsburg Formation and underlying older carbonate rocks.

Despite the arguments of Stose and Kay, Gray and Willard (1955) asserted that these rocks were facies equivalents of the Martinsburg Formation and not allochthonous. McBride (1962) in his classic study of the Martinsburg throughout the central Appalachians did not separate the klippe rocks from the Martinsburg.

In the late 1960s and early 1970s, a number of geologists reviewed the problem of the Hamburg klippe. In particular, studies by Platt and others (1972) and Alterman (1972) showed that there are, indeed, allochthonous elements within the Pennsylvania Great Valley. Moreover, conodont-based biostratigraphic studies of carbonate rocks within the klippe (Bergström and others, 1972) illustrated that not only are these rocks older than the surrounding Martinsburg, but that they also contain a deep-water North Atlantic conodont fauna rather than the shallow-water North American Midcontinent fauna typical of the partly coeval and parautochthonous Beekmantown Group. In addition, graptolite-based biostratigraphic studies by Wright and Stephens (1978) and Wright and others (1979) illustrated that the sandstone and shale within the klippe is significantly older than lithologically similar rocks of the Martinsburg.

Despite the fact that most geologists now agree on the al-

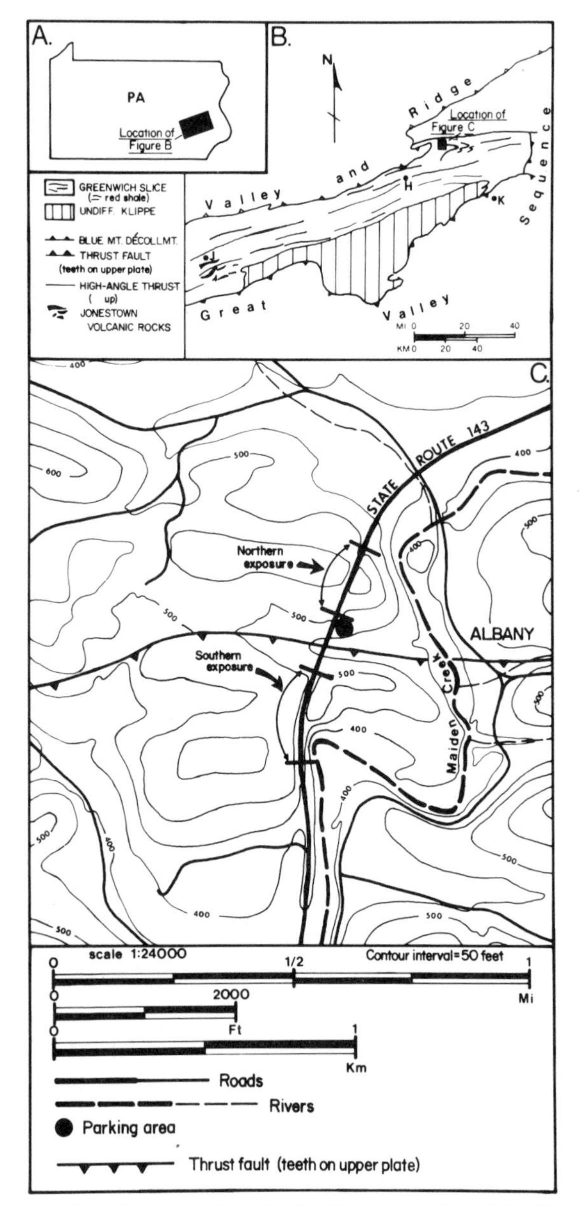

Figure 1. Location map of site in the Greenwich slice of the Hamburg klippe (J, Jonestown; K, Kutztown; H, Hamburg).

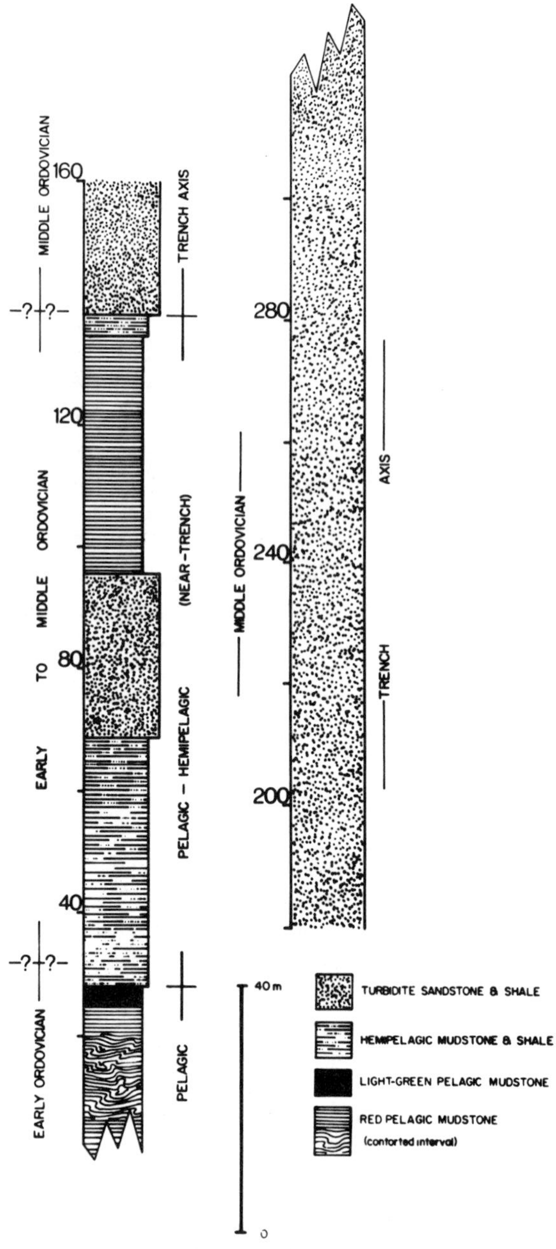

Figure 2. Measured section of a coarsening-upward sequence present in the southern exposure illustrating inferred tectono-depositional environments.

lochthonous nature of these rocks, the internal stratigraphy and structure of the klippe is not fully understood. Indeed, numerous field studies (for example, Wood and MacLachlan, 1978; Alterman, 1972; Lash, 1980) have emphasized the problems of tracing individual lithologic units along strike. Platt and others (1972) suggested that instead of the klippe being a single plate it may be a jumble of smaller parts that were emplaced into the Martinsburg basin by gravity sliding. Recent investigations (for example, Lash, 1980, 1985a; Lash and Drake, 1984; Stephens and others, 1982; MacLachlan, 1979), however, have illustrated that certain lithologic associations can be defined within the klippe suggesting that an internal stratigraphy does exist. An internal, mappable stratigraphy is more consistent with an allochthon composed of stacked thrust slices (for example, Alterman, 1972; Lash and Drake, 1984; Root and MacLachlan, 1978) rather than of a chaotic melange.

Mapping at the eastern end of the allochthon by Lash (1980; Lash and Drake, 1984) has led to the recognition of two lithotectonic units or sequences of rocks bounded by thrust faults. The tectonically lowest and most extensive slice at the eastern end of the klippe, the Greenwich slice, is a tectonically thickened (20,000 ft; 6,200 m) unit that includes sandstone and siltstone, olive-green mudstone and shale, red, purple, and light-green pelagic mudstone and shale, deep-water limestone, chert, and minor

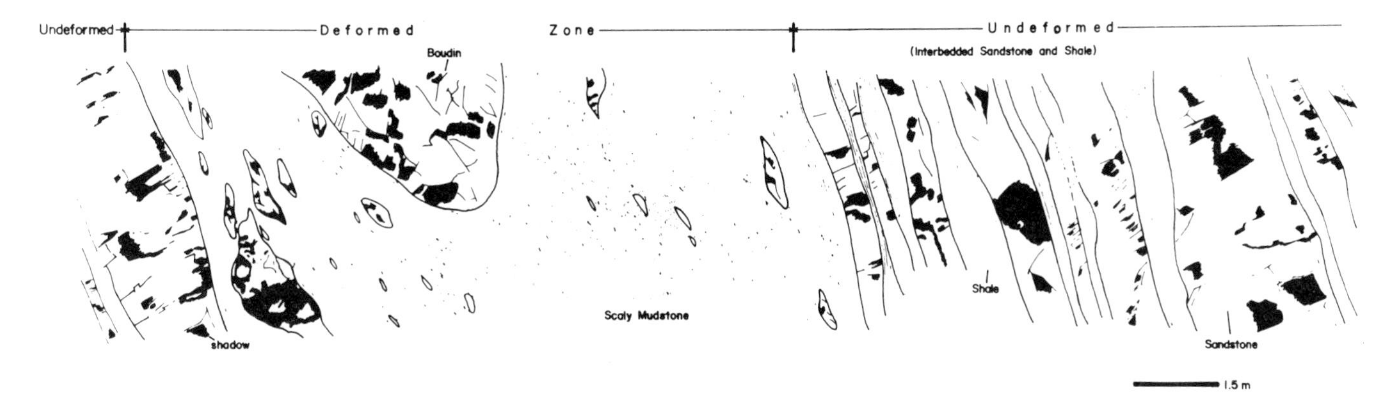

Figure 3. Line drawing of a photograph illustrating abrupt variations in mesoscopic style of deformation between deformed and undeformed horizons exposed in the northern exposure (shaded areas are shadows; view to east).

proportions of boulder conglomerate and mafic intrusive and extrusive volcanic rocks (Lash, 1980; Lash and Drake, 1984; Lash and others, 1984). The dominant lithologies of the Greenwich slice are arranged in coarsening-upward sequences similar to that exposed in the southern exposure of this site (Figs. 1 and 2; Lash and others, 1984).

Southern Exposure. Starting at the north end on the east side of the exposure and proceeding south or up-section (see Fig. 2), bedding is right-side-up and moderately inclined (65°) to the southeast. The lower 82 ft (25 m) of the section consists of pelagic strata that include red mudstone and white dolomitic chert. Some of these beds are characterized by a high degree of folding and boudinage that is probably the result of soft-sediment deformation (Lash and others, 1984). Early Ordovician conodonts have been collected from red mudstone at this level of the section. The red shale passes into about 10 ft (3 m) of laminated light-green mudstone and shale. These deposits are compositionally similar to the red shale except for lower Fe^{3+}/Fe^{+2} ratios and lower manganese contents suggestive of sedimentation under reducing conditions and at greater sedimentation rates than underlying red mudstone (Lash, 1986a). The light-green shale grades into about 130 ft (40 m) of olive-green mudstone, silty mudstone, and thin-bedded siltstone. The hemipelagic deposits grade into approximately 80 ft (25 m) of mudstone and sandstone, some of which is highly disrupted. These deposits are overlain by approximately 130 ft (40 m) of red and light-green, locally folded, pelagic mudstone. These rocks are finally overlain by interbedded sandstone and hemipelagic mudstone. Mapping of this area (Lash, 1985a, 1986b) suggests that the sandstone-hemipelagic mudstone sequence may be greater than 0.6 mi (1 km) thick. The presence of graded bedding and a vertical arrangement of primary sedimentary structures similar to the Bouma sequence (Bouma, 1962) argues for deposition of the sandstone from turbidity currents. Moreover, turbidite facies analysis of this part of the section argues for sedimentation in submarine channels and interchannel areas (Lash and others, 1984; Lash, 1986c). Although graptolites were not collected from the interbedded sandstone and mudstone, they have been collected from lithologically similar rocks in other parts of the Greenwich slice (Wright and Stephens, 1978; Stephens and others, 1982; Lash, 1980; Lash and Drake, 1984) and suggest that these rocks, the most prevalent lithology of the Greenwich slice (Lash and others, 1984), accumulated during Middle Ordovician time (Fig. 2). The stratigraphic relations illustrated at this exposure are similar to sequences recovered from active convergent margins around the world (Schweller and Kulm, 1978). Additionally, Moore and Karig (1976) maintain that recognition of coarsening-upward sequences like that exposed at the southern exposure within orogenic belts provides the best single indicator of accreted pelagic and trench axis sediment. The stratigraphic sequence exposed at the southern exposure, then, is interpreted to record the 25 m.y. advance of oceanic lithosphere from an area of pelagic sedimentation in Early Ordovician time through a near-trench area characterized by pelagic-hemipelagic sedimentation followed by movement into the trench in Middle Ordovician time after which it was scraped off the descending plate and accreted to the overriding plate.

Northern Exposure. The northern exposure is comprised solely of interbedded turbidite sandstone and hemipelagic mudstone of Middle Ordovician age and tectonically underlies the Early Ordovician pelagic sedimentary rocks of the southern exposure (Fig. 1). Visitors should start at the south end of the exposure and move north or down-section. Beds dip about 60° southeast and are right-side-up. Immediately one is impressed with the intensity of deformation illustrated at this part of the exposure. Sandstone beds of any thickness have been disrupted and are surrounded by scaly cleaved mudstone characterized by pervasive anastomosing, commonly polished and striated surfaces 2 to 6 cm in diameter. Pinch-and-swell to complete boudinage of sandstone beds is common. Boudins range from a few cm to at least 16 ft (5 m) long and are oriented parallel to a plane defined by scaly cleavage. Zones of deformed sandstone beds and scaly cleaved mudstone are adjacent to coherent zones of mesoscopically undeformed sandstone and mudstone characterized by relatively high sandstone:mudstone ratios (Fig. 3). Locally, mudstone

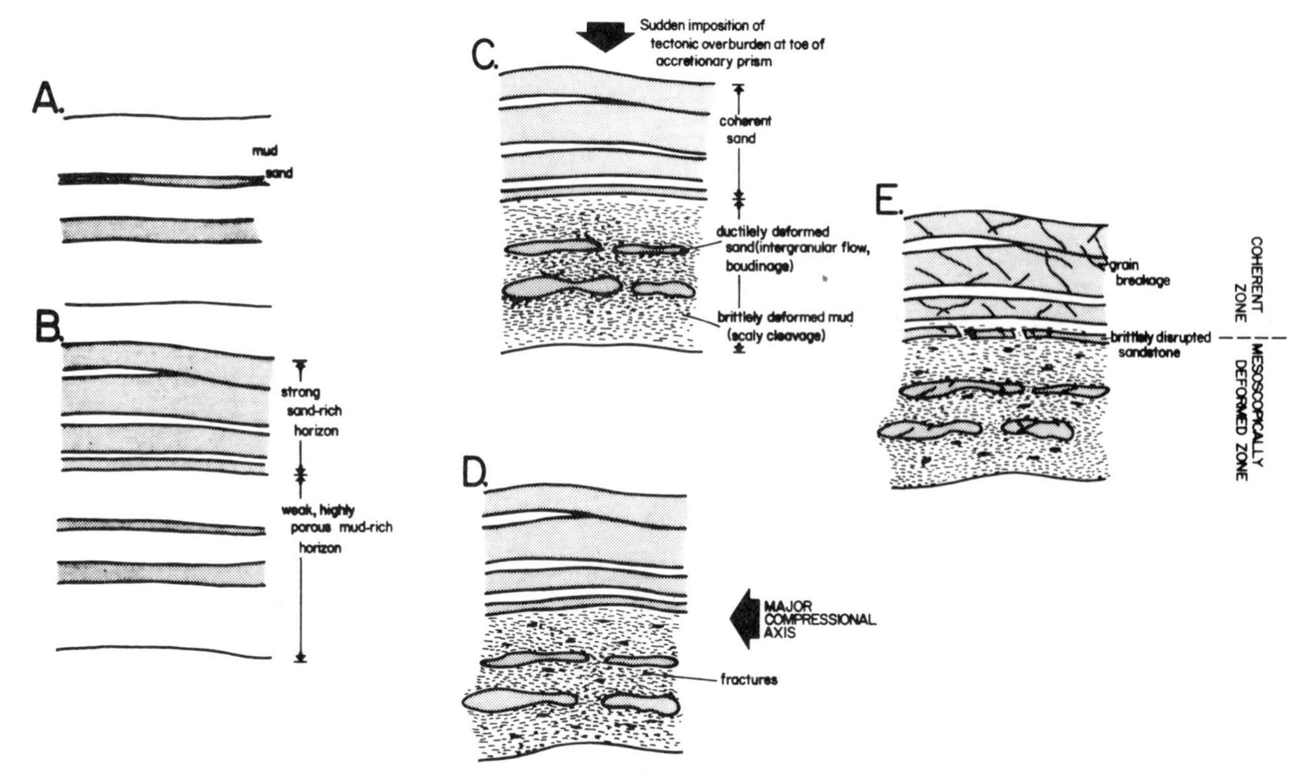

Figure 4. Proposed structural evolution of the Greenwich slice trench-fill deposits. This model stresses the importance of vertical variations in pore fluid content on the manner in which sediment deforms in the early stages of accretion-related deformation. Refer to text for discussion.

within coherent zones displays a poorly-developed slaty cleavage that dips to the southeast at a lesser angle than associated bedding and is axial planar to folds that have folded some of the mesoscopically deformed (scaly cleaved mudstone, boudined sandstone) zones. These relations indicate that scaly cleavage formed well before folding and slaty cleavage formation (Lash, 1985a, 1985b; Lash and others, 1984).

Mesoscopically deformed and undeformed sandstones show various degrees of cataclastic movement or grain breakage such as microgouge zones (0.5 to 1.0 mm thick) composed of dark opaque material and granulated quartz and feldspar grains (Lash, 1985b, Fig. 14). The presence of corroded and sutured grain contacts suggests that pressure solution acted in conjunction with mechanical grain breakage in these zones. The grain-granulation zones in stratally disrupted sandstones are oblique or perpendicular to boudin edges, suggesting that they were not responsible for disruption. Based on this, it is likely that stratal disruption was accommodated by high fluid pressures that fostered intergranular grain movement rather than shearing. The fact that almost all boudins lack internal primary sedimentary structures and some have irregular bottom and top surfaces similar to load casts suggests that the sand was deformed while water-rich (for example, Cowan, 1982).

Minor folds are rare but can be found within some deformed horizons. Petrographic examination indicates that the scaly cleavage is axial planar to the folds. Although the vast majority of boudins at this exposure, as well as others of the Greenwich slice, are tapered and internally structureless, there are several that can be found toward the northern end of the exposure that display excellent primary sedimentary structures such as graded bedding and laminations. The boudins are characterized by square, blunt ends. Such features suggest that the sediment composing these boudins were strong enough to retain its internal integrity while deforming. Disruption of these beds probably occurred during the late stages of deformation as part of a continuum initiated in unconsolidated sediment and eventually ended with brittle failure (for example, Nelson, 1982). Indeed, Carson (1977) pointed out that most deformation of accreted sediment occurs at the time of offscraping and emplacement of a tectonic overburden when the sediment is water-rich and decreases with time.

The meso- and microfabric of the Greenwich slice suggests that its style of deformation, in particular, the vertical juxtaposition of coherent and deformed zones, reflects local variations in lithology and porosity or state of compaction of the accreting sediment (Lash, 1985b). Initial accretion-related deformation and dewatering were localized along weak mud-rich horizons, whereas stronger sand-rich zones remained coherent. Increased pore fluid pressures of sand beds within mud-rich zones resulted in intergranular flow and stratal disruption of the sand. The zones of intense stratal deformation, then, may be taken as indirect evidence that sedimentation rates of inferred trench sediments of the Greenwich slice were, at times, relatively high (Lash, 1985b). Rapid sedimentation of sand over a water-rich hemipelagic mud horizon would tend to trap pore fluid in the mud whereas slower accumulation of sediment over the water-rich mud would allow for natural dewatering and strengthening of the mud (Faas, 1982). The mudstone-rich deformation zones of the Greenwich slice are interpreted as horizons in which water was trapped by rapid accumulation of overlying sand-rich horizons (Fig. 4). The water-rich horizons were, upon gravitational- and accretion-related compaction, areas in which interbedded sand was fluidized and liquified resulting in a reduction of the internal strength of the sediment. Deformation of these beds was accommodated by increased pore fluid pressure. A sequence of events that may have produced the structural variations seen at this exposure and others like it in the Greenwich slice is as follows:

1. Accumulation of hemipelagic mud and lesser sand beds in a near-trench or trench axis location (Fig. 4A);

2. Rapid sedimentation of a thick sand sequence over a water-rich mud horizon resulted in entrapment of pore fluid and increased pre-tectonic (i.e., pre-accretion) pore fluid pressure in the mud-rich horizon (Fig. 4B);

3. Upon accretion and the instantaneous imposition of a tectonic load, increased pore fluid pressures in sand beds within the mud-rich horizons reduced their coefficients of internal and sliding friction, resulting in ductile flow of sand and stratal disruption. Formation of scaly cleavage probably accelerated dewatering and strengthening of mud (Fig. 4C).

4. As deformation continued pore-fluid pressures within deforming mud-rich horizons became sufficient enough to generate a micro-fracture porosity (Lash, 1985b; Fig. 4D);

5. Eventually strength differences of the mud-rich and sand-rich horizons were reduced and both deformed and undeformed sand failed brittlely (Fig. 4E).

The subduction complex origin of the Greenwich slice suggested by the sedimentologic characteristics best illustrated in the southern exposure is substantiated by the style of deformation seen in the northern exposure. This style of deformation is similar to other ancient subduction complexes (for example, Coastal Belt of the Franciscan Complex; Kodiak Islands of coastal southern Alaska) and illustrates the importance of pre-tectonic (i.e., pre-accretion) variations in pore fluid content and lithology of incoming sediment in determining the style of deformation of offscraped and accreted sediment.

REFERENCES CITED

Alterman, I. B., 1972, Structure and history of the Taconic allochthon and surrounding autochthon, east-central Pennsylvania [Ph.D. thesis]: New York, Columbia University, 287 p.

Bergström, S. M., Epstein, A. G., and Epstein, J. B., 1972, Early Ordovician North Atlantic Province conodonts in eastern Pennsylvania: U.S. Geological Survey Professional Paper 800-D, p. D37–D44.

Bouma, A. H., 1962, Sedimentology of some flysch deposits, a graphic approach to facies interpretations: Amsterdam, Elsevier Publishing Company, 168 p.

Carson, B., 1977, Tectonically induced deformation of deep-sea sediments off Washington and northern Oregon; Mechanical consolidation: Marine Geology, v. 24, p. 289–307.

Cowan, D. S., 1982, Deformation of partly dewatered and consolidated Franciscan sediments near Piedras Point, California, *in* Leggett, J. K., ed., Trench–forearc geology: Geological Society of London Special Publication 10, p. 439–457.

Faas, R. W., 1982, Gravitational compaction patterns determined from sediment cores recovered during the Deep Sea Drilling Project Leg 67, Guatemala transect; Continental slope, Middle America Trench, and Cocos Plate, *in* von Huene, R., Aubouin, J., eds., Initial reports of the Deep Sea Drilling Project: Washington, D.C., U.S. Government Printing Office, v. 67, p. 617–638.

Gray, C., and Willard, B., 1955, Stratigraphy and structure of lower Paleozoic rocks in eastern Pennsylvania, *in* Field guidebook of Appalachian geology, Pittsburgh to New York: Pittsburgh Geological Society, p. 87–92.

Kay, G. M., 1941, Taconic allochthon and the Martic thrust: Science, v. 44, p. 73.

Lash, G. G., 1980, Structural geology and stratigraphy of the allochthonous and autochthonous rocks of the Kutztown and Hamburg 7½-minute Quadrangles, eastern Pennsylvania [Ph.D. thesis]: Bethlehem, Pennsylvania, Lehigh University, 240 p.

——, 1985a, Geologic map and sections of the Kutztown 7½-minute Quadrangle: U.S. Geological Survey Quadrangle Map GQ-1577, scale 1:24,000.

——, 1985b, Accretion-related deformation of an ancient (early Paleozoic) trench-fill deposit, central Appalachian orogen: Geological Society of America Bulletin, v. 96, p. 1167–1178.

——, 1986a, Possible significance of early Paleozoic fluctuations in bottom current intensity, northwest Iapetus Ocean: Paleoceanography, v. 1, p. 119–135.

——, 1986b, Geologic map and sections of the Hamburg 7½-minute Quadrangle, Pennsylvania: U.S. Geological Survey Quadrangle Map GQ-1637, scale 1:24,000 (in press).

——, 1986c, Sedimentology of channelized turbidite deposits in an ancient (early Paleozoic) subduction complex, central Appalachians: Geological Society of America Bulletin, v. 97, p. 703–710.

Lash, G. G., and Drake, A. A., Jr., 1984, The Richmond and Greenwich slices of the Hamburg klippe in eastern Pennsylvania; Stratigraphy, structure, and plate tectonic implications: U.S. Geological Survey Professional Paper 1312, 40 p.

Lash, G. G., Lyttle, P. T., and Epstein, J. B., 1984, Geology of an accreted terrane; The Hamburg klippe and surrounding rocks, eastern Pennsylvania: Annual Field Conference of Pennsylvania Geologists, 49th, Guidebook, 152 p.

MacLachlan, D. B., 1979, Geology and mineral resources of the Temple and Fleetwood quadrangles, Berks County, Pennsylvania: Pennsylvania Geological Survey, 4th series, Atlas 187ab, 71 p.

McBride, E. F., 1962, Flysch and associated beds of the Martinsburg Formation (Ordovician), central Appalachians: Journal of Sedimentology Petrology, v. 32, p. 39–91.

Moore, J. C., and Karig, D. E., 1976, Sedimentology, structural geology, and tectonics of the Shikoku subduction zone, southwestern Japan: Geological Society of America Bulletin, v. 87, p. 1259–1268.

Nelson, K. D., 1982, A suggestion for the origin of microscopic fabric in accretionary melange, based on features observed in the Chrystalls Beach Complex, South Island, New Zealand: Geological Society of America Bulletin, v. 93, p. 625–634.

Platt, L. B., Loring, R. B., Papaspyros, A., and Stephens, G. C., 1972, The Hamburg klippe reconsidered: American Journal of Science, v. 272, p. 305–318.

Rogers, H. D., 1858, The geology of Pennsylvania, Volume 1: Philadelphia, J. B. Lippincott and Company, 586 p.

Root, S. I., and MacLachlan, D. B., 1978, Western limit of Taconic allochthons in Pennsylvania: Geological Society of America Bulletin, v. 89, p. 1515–1528.

Schweller, W. J., and Kulm, L. D., 1978, Depositional patterns and channelized sedimentation in active eastern Pacific trenches, *in* Stanley, D. J., and Kulm, L. D., eds., Sedimentation in submarine canyons, fans, and trenches:

Stroudsburg, Pennsylvania, Dowden, Hutchinson, and Ross, p. 311–324.
Stephens, G. C., Wright, T. O., and Platt, L. B., 1982, Geology of Middle Ordovician Martinsburg Formation and related rocks in Pennsylvania: Annual Field Conference of Pennsylvania Geologists, 47th, Guidebook, 87 p.
Stose, G. W., 1946, The Taconic sequence in Pennsylvania: American Journal of Science, v. 244, p. 665–696.
—— , 1950a, Evidence of the Taconic sequence in the vicinity of Lehigh River, Pennsylvania: American Journal of Science, v. 248, p. 815–819.
—— , 1950b, Comments on the Taconic sequence in Pennsylvania: Geological Society of America Bulletin, v. 61, p. 133–135.
Wood, C. R., and MacLachlan, D. B., 1978, Ground-water resources of northern Berks County, Pennsylvania: Pennsylvania Geological Survey, 4th series, Water Resources Map 44, 91 p.
Wright, T. O., and Stephens, G. C., 1978, Regional implications of the stratigraphy and structure of Shochary Ridge, Berks and Lehigh Counties, Pennsylvania: American Journal of Science, v. 278, p. 1000–1017.
Wright, T. O., Stephens, G. C., and Wright, E. K., 1979, A revised stratigraphy of the Martinsburg Formation of eastern Pennsylvania and paleogeographic consequences: American Journal of Science, v. 279, p. 1176–1186.

Ordovician through Mississippian rocks, Lehigh River, Carbon County, Pennsylvania

W. D. Sevon, Pennsylvania Geological Survey, Box 2357, Harrisburg, Pennsylvania 17120

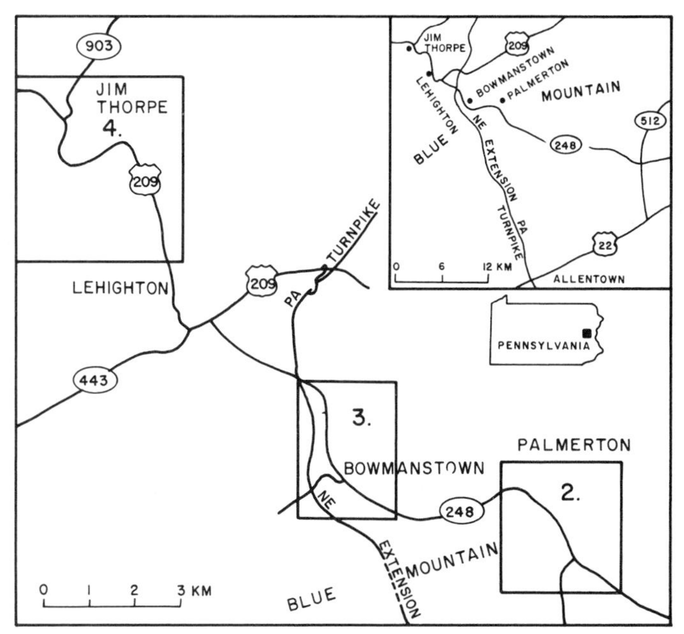

Figure 1. General location map for the Lehigh River, Ordovician through Mississippian exposures. Numbered boxes are locations of subsequent figures.

Figure 2. Location of Ordovician and Silurian exposures in the Lehigh Gap area. See Figure 1 for location of area.

LOCATION

Rocks of Ordovician through Mississippian age are exposed along roads and railroads paralleling the Lehigh River between Palmerton and Jim Thorpe in Carbon County, Pennsylvania (Fig. 1). Rocks of the Ordovician Martinsburg Formation and the Silurian Shawangunk Formation are well exposed on the east side of Lehigh Gap along an abandoned railroad bed (Fig. 2). An unmaintained access road along the railroad bed is reached from Pennsylvania 248 just east of the former railroad overpass or by climbing the slope near the east end of the bridge across the Lehigh River (Fig. 2). Part of the Silurian Bloomsburg Formation is exposed along a small side road paralleling Pennsylvania 248 (Fig. 2). This road is accessed from Delaware Avenue in Palmerton or Pennsylvania 248 just south of Aquashicola Creek.

Several Lower, Middle, and Upper Devonian rock units are beautifully exposed at Bowmanstown (Fig. 2). These exposures are reached by exiting from Pennsylvania 248 to Bowmanstown. The Devonian Ridgeley (Oriskany), Schoharie-Esopus, Palmerton, Buttermilk Falls, and Marcellus Formations (Fig. 3) are exposed in a deep roadcut along the Northeast Extension Pennsylvania Turnpike at West Bowmans. Proceed west on Pennsylvania 895 from Pennsylvania 248 and park in a pull-off area on the north side of the road at the east end of the bridge over the turnpike. It is possible to go over the southeast side of the bridge onto a terrace above road level and walk south through the Marcellus Formation to the lower units. Warning! This is a narrow roadcut with high-speed traffic. Examining the rocks at road level is dangerous and requires special permission from the Turnpike Commission. Stay on the terrace level and do not take large groups to the outcrop.

The Centerfield fossil zone in the Mahantango Formation is reached by parking to the southwest of the railroad overpass over Pennsylvania 895 and walking along the railroad tracks to the fossil zone.

To reach the outcrops north of Bowmanstown, turn north on the main street in Bowmanstown and follow the road north to Pennsylvania 248. Just before this road joins Pennsylvania 248 N (divided highway) there is a large pull-off area on the east side of the road (Fig. 3). The Mahantango Formation is exposed for many meters south of this pull-off and the Tully fossil zone is the first good roadside exposure (poison ivy–covered) south of the pull-off. North of the pull-off is continuous exposure through the Trimmers Rock Formation, the Towamensing Member of the Catskill Formation, and part of the Walcksville Member of the Catskill Formation. This roadcut has a narrow berm and the road is heavily traveled. The first terrace level above the road has somewhat poorer exposure, but is traffic free and less noisy.

The remainder of the rocks are exposed north of Lehighton, mainly along U.S. 209 (Fig. 4). The Beaverdam Run Member of the Catskill Formation is exposed on the west side of U.S. 209 north of Jamestown. Park in the small area on the east side of the road at a curve and the start of a small downgrade. This roadcut, the type section for this unit, is very close to the road and dangerous. The Long Run, Packerton, Sawmill Run, and Berry Run Members of the Catskill Formation are exposed along the former New Jersey Central Railroad bed north of the mouth of Beaverdam Run, and described sections of these rocks were done from these outcrops. However, since cessation of the railroad line, these outcrops have deteriorated and are not as good now as some along U.S. 209. The railroad outcrops are accessed by a road which intersects U.S. 209 from the east a few tens of meters north of the road coming down Beaverdam Run Valley. Parking is available, and the former railroad bed is easily located and walked.

The Long Run Member is well exposed along U.S. 209 just north of Beaverdam Run (Fig. 4), but the berm is very narrow and dangerous. Parking is available north of Packerton at a drive-in (eatery) where the Packerton, Sawmill Run, and Berry Run Members can be viewed along the west side of U.S. 209.

The Clarks Ferry and Duncannon Members are spectacularly exposed on the southwest side of U.S. 209 south of Jim Thorpe. For these exposures either park at the drive-in and walk down the section or park in a small lot on the west side of the road at the top of the section (near a beverage distributor) and walk up and then back (the whole section is on a downgrade). The berm here is narrow and caution is required.

Exposures of the Spechty Kopf and Pocono Formations are reached as follows: north of the Jim Thorpe railroad terminal (used as a tourist center) turn onto Pennsylvania 903 N from U.S. 209 and cross the bridge over the Lehigh River (Fig. 4). At the east end of the bridge, turn right on a road leading down to a supermarket (Pennsylvania 903 turns left). Either park in the parking lot and walk south along a gravel road paralleling the Lehigh Valley railroad tracks or drive the road, which leads to the Jim Thorpe sewage treatment plant, and park on the left after crossing the railroad at the treatment plant. Warning! Be alert for trains.

The Mauch Chunk Formation type locality is exposed in part along the west side of U.S. 209 north of the Jim Thorpe railroad terminal and in part along former railroad tracks below road level (Fig. 4). Park in Jim Thorpe and walk north to the exposures. Warning! The road has a very narrow berm and is on curve.

The base of the transition zone between the Pennsylvanian Pottsville and the Mississippian Mauch Chunk Formations is exposed along the south side of U.S. 209 about 0.6 mi (1 km) northwest of the junction of U.S. 209 and Pennsylvania 903. Parking is available on the north side of the road some distance west of the exposures.

On a clear day an excellent view of the countryside, including the Pocono Plateau, Lehigh River Gorge, Jim Thorpe (the Switzerland of America), and the Lehigh Gap is available from the lookout at Flagstaff Mountain Park on top of Mauch Chunk Ridge south of Jim Thorpe (Fig. 4).

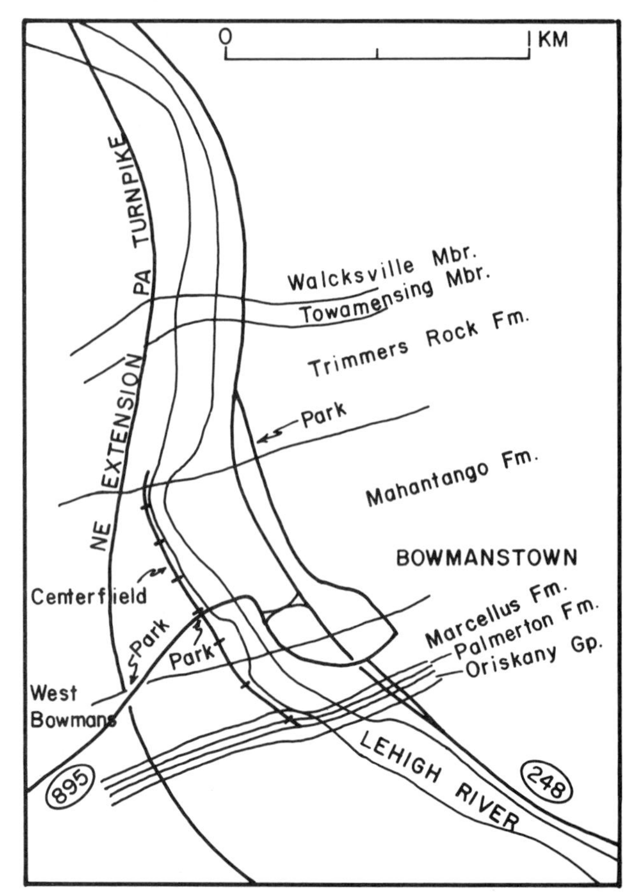

Figure 3. Location of Lower, Middle, and Upper Devonian exposures in the Bowmanstown area. See Figure 1 for location of area.

SIGNIFICANCE

Excellent rock exposures along the Lehigh River between Lehigh Gap and Jim Thorpe provide the most complete and readily accessible stratigraphic sequence in northeastern Pennsylvania for rocks of Ordovician through Mississippian age. Rocks in most exposures are tilted to a near-vertical attitude so that a maximum stratigraphic interval is covered in a minimum distance. These exposures, which include several type localities and a great variety of marine and nonmarine facies, are the foundation of stratigraphic and sedimentologic understanding of northeastern Pennsylvania.

SITE INFORMATION

Study of the various members of the Catskill Formation along the Lehigh River combined with mapping and study in adjacent areas has led to the idea that the Lehigh River area was

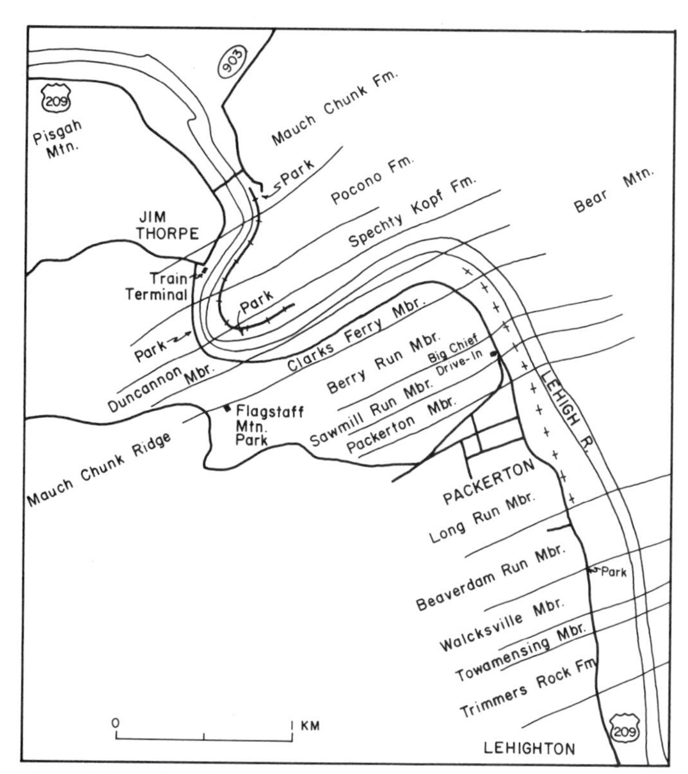

Figure 4. Location of Upper Devonian and Mississippian exposures in the Jim Thorpe area. See Figure 1 for location of area.

the center of a major sediment-dispersal system during the Upper Devonian (Sevon, 1979a; Thompson and Sevon, 1982; Sevon, 1985). Evidence supporting this concept includes: 1) the local concentration of conglomeratic facies (Duncannon Member), 2) the local abundance of braided stream facies (Packerton, Berry Run, Clarks Ferry Members), 3) lateral facies changes in the previously mentioned members, and 4) localized occurrence of the basal diamictite facies of the Spechty Kopf Formation. The work of Epstein and Epstein (1972) on the distribution of conglomeratic facies in the Shawangunk Formation suggests that a similar sediment-dispersal system may have existed during part of the Silurian.

At the Shawangunk and Martinsburg locality (Fig. 2), the contrasting rock types are well shown. The contact is well exposed and lends itself well to discussion regarding the presence or absence of a fault (Epstein and Epstein, 1972; Thompson and Sevon, 1982). The exposures of Shawangunk include the type localities of the Weiders (lower) and Lizard Creek (upper) Members (Epstein and Epstein, 1972) as well as the intervening Minsi Member. The rocks are thought to represent a variety of subaerial (braided stream) to subtidal depositional environments (Epstein and others, 1984). The excellent Bloomsburg exposure is one of the few of a unit normally poorly exposed. It is interesting for discussions of meandering stream environments and the origin of red beds. The Lehigh Gap is one of several structurally controlled wind and water gaps discussed by Epstein (1966).

The exposures along the Northeast Extension Pennsylvania Turnpike (Ridgeley through Marcellus; Fig. 3) are important because some of these units are rarely exposed and the units are part of a complex sequence of marine deposits (Epstein, 1984). The ridge-forming Palmerton Sandstone is a valuable economic resource, a key to deciphering some of the complex folding in lithotectonic unit 3 (Epstein and others, 1974, Fig. 157, p. 280), and of problematic marginal-marine origin. Several layers of Tioga bentonite occur within the Buttermilk Falls. The Marcellus is complexly faulted (Epstein and others, 1974, Fig. 161, p. 288) and probably represents the deepest water facies in the sequence.

The exposures north of Bowmanstown (Mahantango through Walcksville; Fig. 3) offer an opportunity to examine a complete transition from offshore marine environments (Mahantango) through proximal delta front (Trimmers Rock) to nonmarine (Towamensing) and lower delta plain (Walcksville). Specific environments of deposition are not clear in some of the sequence, and the work of Walker (1971, 1972) is a good starting point for discussion. The Mahantango Formation is a superb example of a unit so dominated by cleavage that bedding is obscured. The whole sequence is exposed on the south limb of the Weir Mountain syncline. Some wedge faulting and bedding-plane parallel faulting occurs near the syncline axis as does some folded cleavage (Epstein and others, 1974, Fig. 160, p. 286 and Fig. 164, p. 293).

The Beaverdam Run Member (Fig. 4) represents a return to marine environments. The unit apparently merges with the Trimmers Rock Formation to the southwest and pinches out totally more than 30 mi (50 km) to the northeast. The upper part of the Catskill Formation starts with a thick, lower-delta plain sequence (Long Run Member), goes upward into a very thick, non-red, dominantly braided stream sequence (Packerton, Sawmill Run, Berry Run, and Clarks Ferry Members), and culminates with a sequence of 12 fluvial fining-upward cycles (Duncannon Member). The exposures of the Clarks Ferry Member are spectacular and show the epitome of braided stream development found in the Catskill Formation in Pennsylvania.

The base of the Spechty Kopf Formation has a sequence of polymictic diamictite, laminite, and planar-bedded sandstone which marks an interval that is locally restricted but regionally widespread (Sevon, 1979b; 1985). The sequence probably reflects a significant tectonic event (end of the Acadian or start of the Alleghanian). The remainder of the Spechty Kopf and all of the Pocono Formation reflect a return to braided stream deposition and an enrichment in quartz that dominates the remainder of the Paleozoic sequence in Pennsylvania. The red Mauch Chunk Formation is very calcareous, in contrast to almost all of the underlying Mississippian and Devonian rocks. It presumably is all fluvial in origin, but specific environments of deposition are not clear. The transition zone into the Pennsylvanian Pottsville Formation heralds the change to coarse, quartz-rich, clastic sedimentation which dominated northeastern Pennsylvania throughout the Pennsylvanian Period.

REFERENCES CITED

Epstein, J. B., 1966, Structural control of wind gaps and water gaps and of stream capture in the Stroudsburg area, Pennsylvania and New Jersey: U.S. Geological Survey Professional Paper 550-B, p. B80–B86.

—— , 1984, Onesquethawan stratigraphy (Lower and Middle Devonian) of northeastern Pennsylvania: U.S. Geological Survey Professional Paper 1337, 35 p.

Epstein, J. B., and Epstein, A. G., 1972, The Shawangunk Formation (Upper Ordovician[?] to Middle Silurian) in eastern Pennsylvania: U.S. Geological Survey Professional Paper 744, 49 p.

Epstein, J. B., Sevon, W. D., and Glaeser, J. D., 1974, Geology and mineral resources of the Lehighton and Palmerton Quadrangles, Carbon and Northampton Counties, Pennsylvania: Pennsylvania Geological Survey, 4th ser., Atlas 195cd, 460 p.

Sevon, W. D., 1979a, Devonian sediment-dispersal systems in Pennsylvania: Geological Society of America Abstracts with Programs, v. 11, p. 53.

—— , 1979b, Polymictic diamictites in the Spechty Kopf and Rockwell Formations, *in* Dennison, J. M., and others, eds., Devonian shales in south central Pennsylvania and Maryland: Pennsylvania Geological Survey, 44th Annual Field Conference of Pennsylvania Geologists, Guidebook, p. 61–66; 107–110.

—— , 1985, Nonmarine facies of the Middle and Late Devonian Catskill coastal alluvial plain, *in* Woodrow, D. L., and Sevon, W. D., eds., The Catskill Delta: Geological Society of America Special Paper 201, p. 79–90.

Thompson, A. M., and Sevon, W. D., 1982, Excursion 19B; Comparative sedimentology of Paleozoic clastic wedges in the central Appalachians, U.S.A.: International Association of Sedimentologists, 11th International Congress on Sedimentology, Guidebook, 136 p.

Walker, R. G., 1971, Nondeltaic depositional environments in the Catskill clastic wedge (Upper Devonian) of central Pennsylvania: Geological Society of America Bulletin, v. 82, p. 1309–1326.

—— , 1972, Upper Devonian marine-nonmarine transition, southern Pennsylvania: Pennsylvania Geological Survey, 4th ser., General Geology Report 62, 25 p.

The Hickory Run boulder field, a periglacial relict, Carbon County, Pennsylvania

W. D. Sevon, Pennsylvania Geological Survey, Box 2357, Harrisburg, Pennsylvania 17120

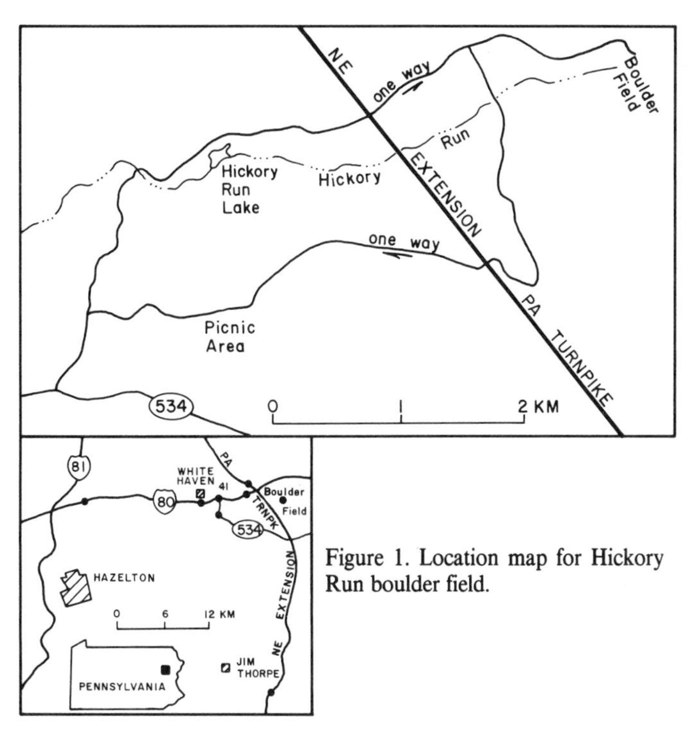

Figure 1. Location map for Hickory Run boulder field.

Figure 2. Hickory Run boulder field. View is up-field from near the parking lot.

LOCATION

The Hickory Run boulder field is located within Hickory Run State Park, Carbon County, Pennsylvania. The park is reached by Pennsylvania 534, which intersects I-80 at Interchange 41, a few miles northwest of the park. The boulder field is reached via a park road from a well-marked intersection with Pennsylvania 534 (Fig. 1). The road is both paved and gravel. It can accommodate all vehicles, including buses, and ends at a large parking lot adjacent to the boulder field. Although the road is not plowed during snow months, it is accessible by snowmobiles. Hickory Run State Park, which has camping facilities, is open year-round from 8:00 a.m. to sundown; no special permission is necessary to visit the boulder field.

SIGNIFICANCE

The Hickory Run boulder field is probably the best example of felsenmeer or block field in the eastern United States. The boulder field is exceptional because of its gradient, size of exposed area, diversity of features, demonstrable evidence of movement, and accessibility. It is a spectacular example of a deposit formed by processes associated with Pleistocene periglacial climates. In 1967 the field was designated as a Registered National Landmark.

SITE INFORMATION

Hickory Run boulder field (Fig. 2) is a deposit of boulders derived from the Duncannon Member of the Catskill Formation (Sevon, 1975). The open expanse of the field consists of an east-west–trending area about 1,600 ft (500 m) long and 400 ft (125 m) wide (area A, Fig. 3) and a smaller, open expanse (area B, Fig. 3) separated from the larger area by a narrow corridor of trees and low shrubs. The larger expanse has a gradient of 1° to the west. The boulder deposit is continuous into the surrounding vegetated area for many tens of meters. The maximum thickness of the deposit is unknown, but it is presumably much greater than the 6 ft (2 m) known from small excavations.

The boulders are dominantly reddish gray and vary from angular to well rounded. Boulder size ranges from rounded boulders a few centimeters in diameter to angular blocks more than 30 ft (10 m) long (area B, Fig. 3), with larger sizes concentrated on the surface. There is a general downfield (east-to-west) decrease in maximum boulder size, accompanied by an increase in boulder roundness. Undisturbed boulders on the field's surface are usually etched and bleached by weathering on their upper exposed side, but are smooth-surfaced on their lower protected side. Rounded boulders below the surface layer are usually smooth surfaced, sometimes highly polished, and usually coated with a reddish iron stain. The boulders rest in point contact with each other and are sometimes unstable at the surface. Interstitial material is absent to a depth of up to 6 ft (2 m), beyond which the boulders are surrounded by sand.

The boulder field has a surface morphology (Fig. 4) consist-

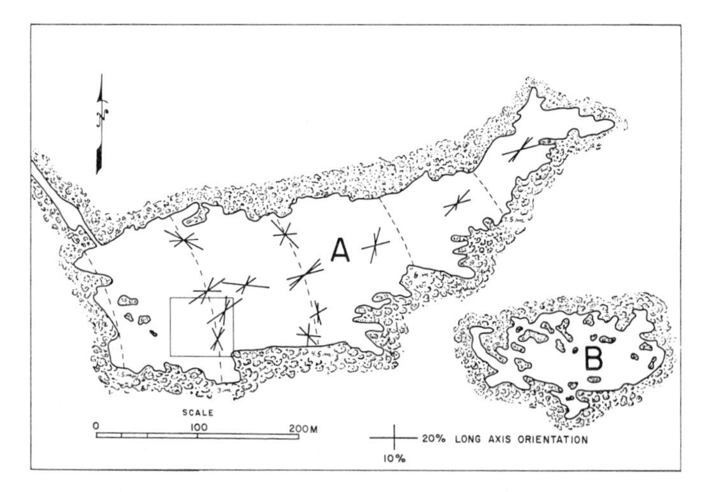

Figure 3. Map of Hickory Run boulder field showing open areas, gradient, and fabric orientations. Fabric data collected from boulders with 2:1 axis ratio by Lehigh University students in 1974. Base and gradient contours from Smith (1953, Fig. 1, Pl. 1, 1939 air photo). Box indicates location of Figure 4.

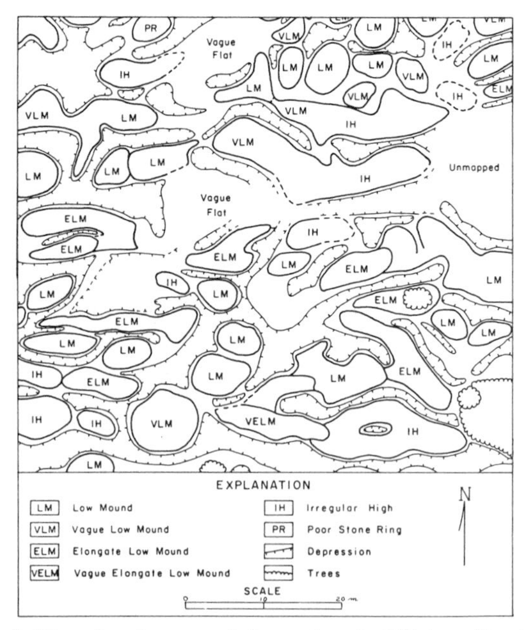

Figure 4. Map of surface morphology of part of Hickory Run boulder field; see Figure 3 for location of area. Mapping from unpublished work by Alan Adler.

ing of a variety of mounds and depressions, but overall the field appears flat. Some of the low mounds are true stone polygons with size-differentiated boulders. Limited study indicates a definite fabric (Fig. 3), and detailed lithologic studies show distinctive zones (Fig. 5) interpreted to be discrete flows. Some boulders have "fitted surfaces" (mutual form-fitting surfaces on adjacent boulders) thought to have resulted from grinding during or after transport to their present position.

The Hickory Run boulder field, as well as numerous other fields—less accessible but similar—in northeastern Pennsylvania, is interpreted to be the result of transportation and deposition of rock split by frost action from outcrops marginal to the field. The exact mechanism of movement is not known, but a mixture of sand, ice, and boulders was probably involved. The boulder field is assumed to be post-Illinoian in age (Sevon and others, 1975) and presumably formed during the climatic extremes accompanying the two Wisconsinan ice advances that stopped within 0.6 mi (1 km) of the boulder field (Crowl and Sevon, 1980).

Smith (1953) suggested that the open expanse of the boulder field was originally filled with interstitial sand, which has subsequently been removed by marginal sapping, thus enlarging the field. I have suggested (Sevon, 1969; Sevon and others, 1975) that the interstitial sand never existed (except at depth). I believe that marginal deposition of sediment, combined with accumulations of organic material, is encroaching onto the field and is thus making it smaller.

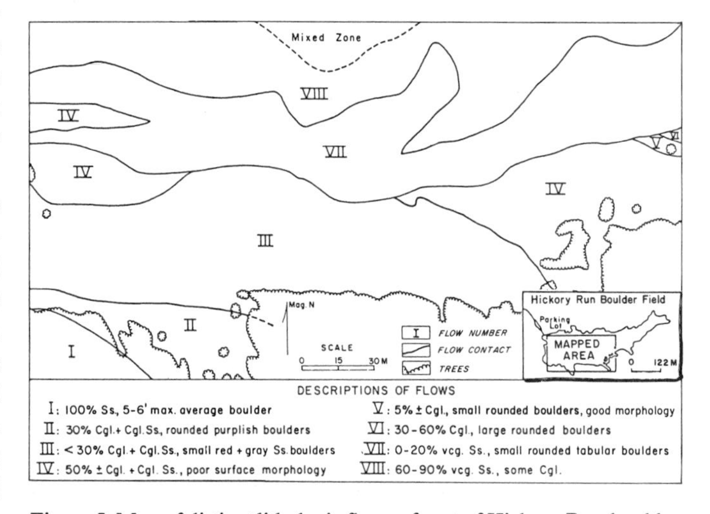

Figure 5. Map of distinct lithologic flows of part of Hickory Run boulder field. A tentative sequence of flow is: V (oldest), VI, IV, II, III, VII, and I or VIII (youngest). Mapping from unpublished work by Alan Adler.

REFERENCES CITED

Crowl, G. H., and Sevon, W. D., 1980, Glacial border deposits of Late Wisconsinan age in northeastern Pennsylvania: Pennsylvania Geological Survey, 4th ser., General Geology Report 71, 68 p.

Sevon, W. D., 1969, Sedimentology of some Mississippian and Pleistocene deposits of northeastern Pennsylvania, *in* Subitsky, S., ed., Geology of selected areas in New Jersey and eastern Pennsylvania and guidebook of excursions: New Brunswick, New Jersey, Rutgers University Press, p. 214–234.

——, 1975, Geology and Mineral Resources of the Hickory Run and Blakeslee Quadrangles, Carbon and Monroe Counties, Pennsylvania: Pennsylvania Geological Survey, 4th ser., Atlas 194 cd.

Sevon, W. D., Crowl, G. H., and Berg, T. M., 1975, The Late Wisconsinan drift border in northeastern Pennsylvania: Guidebook, 40th Annual Field Conference of Pennsylvania Geologists, Pennsylvania Geological Survey, 108 p.

Smith, H.T.U., 1953, The Hickory Run boulder field, Carbon County, Pennsylvania: American Journal of Science, v. 251, p. 625–642.

The Wissahickon Schist type section, Wissahickon Creek, Philadelphia, Pennsylvania

Maria Luisa Crawford, Department of Geology, Bryn Mawr College, Bryn Mawr, Pennsylvania 19010

LOCATION

Wissahickon Creek is located in the western half of the Germantown 7½-minute Quadrangle. The creek flows south from Ambler, enters the Wissahickon Schist just south of the Philadelphia County line, and joins the Schuylkill River at Manayunk. Access to Wissahickon Creek and the Wissahickon Creek section of Fairmount Park, is available from Bells Mill Road at the northern end of the section, at Valley Green accessible from Wise's Mill Road in the center, and from Ridge Avenue and Lincoln Drive at the southern end (Fig. 1). Hiking trails are developed along both sides of the creek, providing easy access along its entire length. Parking lots are provided both at Bells Mill Road and Valley Green, where there is also a small snack bar and restrooms.

SIGNIFICANCE

The exposure along Wissahickon Creek is an almost continuous section that shows much of the lithologic, structural, and metamorphic evidence for the origin and emplacement of the Wissahickon Formation. The Wissahickon is one of the major units of the northern part of the southern Appalachian Piedmont, probably correlative with the Manhattan Schist of New York and the Chilhowee Group in Virginia. This greenschist and amphibolite facies quartzose and micaceous schist, is the thickest unit in the Glenarm Series that underlies much of southeastern Pennsylvania and eastern Maryland. The Glenarm Series unconformably overlies Grenville age gneisses and records the latest Precambrian and lowermost Paleozoic sedimentary and tectonic history in this portion of the Appalachians. The metamorphic and structural features in the Wissahickon Formation and other units of the Glenarm Series are interpreted to have formed during the Taconic collisional tectonic event (Crawford and Crawford, 1980; Crawford and Mark, 1982; Wagner, 1982).

SITE INFORMATION

The age of the Wissahickon Formation is not firmly established. The entire Glenarm Series has been correlated with the Cambrian and Ordovician sequence in the Mine Ridge anticlinorium immediately to the northwest (e.g., Fisher and others, 1979). This suggests that the Wissahickon Formation is lower Paleozoic, possibly as young as Early Ordovician. The metamorphism, probably Taconian in age (Lapham and Root, 1971; Sutter and others, 1980) is of the intermediate pressure, or Barrovian, facies series; metamorphic grade increases from northwest to southeast (Weiss, 1949; Wyckoff, 1952). Here at its type locality, where it

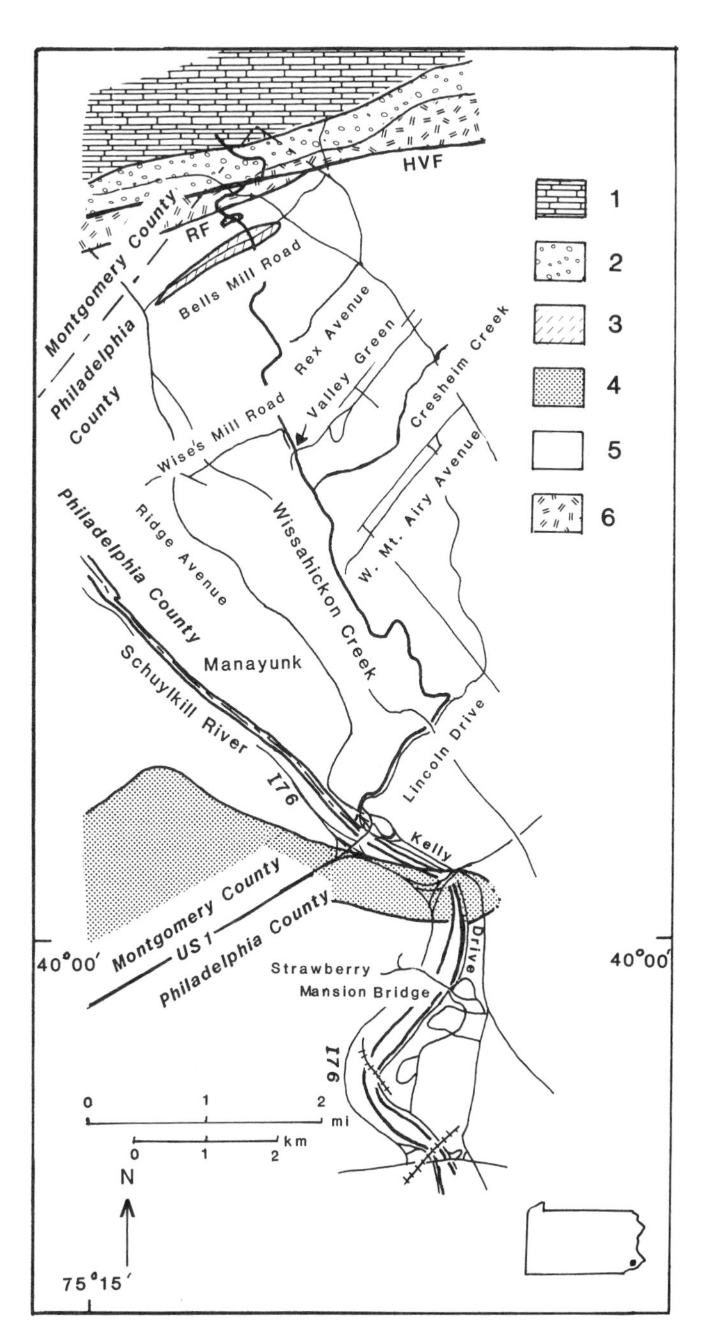

Figure 1. Area surrounding Wissahickon Creek, southeastern Germantown 7½-minute Quadrangle and northeastern Philadelphia 7½-minute Quadrangle. 1, Cambro-Ordovician limestone; 2, Cambrian quartzite; 3, Ultramafic rocks; 4, Granodiorite; 5, Wissahickon schist; 6, Grenville gneiss; RF, Rosemont Fault; HVF, Huntingdon Valley Fault.

was first described by Bascom (1905), the Wissahickon schist consists of alternating psammitic and pelitic layers. At the northern end of the Wissahickon Creek section the Huntingdon Valley and Rosemont faults juxtapose the Wissahickon schist against Grenville age Baltimore gneiss. The faults are not exposed. South of the faults the schists are in the lowermost amphibolite facies (staurolite grade). Metamorphic grade increases downstream to upper amphibolite facies (sillimanite-muscovite grade).

Locally along Wissahickon Creek, and elsewhere in the Wissahickon Formation, lenticular bodies of ultramafic rock lie surrounded by the schist (Figs. 1, 2). A band of these ultramafic lenses can be traced to the southwest along and south of the Hungtingdon Valley and Rosemont faults that separate the schist from the Grenville gneiss. The mineral assemblages and structures in the ultramafic bodies suggest they were metamorphosed with the enclosing schist (Roberts, 1969).

The dominant structures consist of 3-ft-scale (meter-scale) folds overturned to the southeast and a northeast-trending nearly horizontal lineation. Careful examination suggests that the mesoscopic folds correspond to the second of three episodes of deformation (Weiss, 1949; Amenta, 1974; Tearpock and Bischke, 1980). The earliest foliation coincides with lithologic layering except in the hinges of first-generation folds. Metamorphism coincides with the second deformation episode while the faults that juxtapose the schist against Grenville age gneiss to the north postdate the metamorphic culmination. Additional information and detailed descriptions of specific localities along this traverse can be found in Crawford and Crawford (1974) and Goodwin (1964).

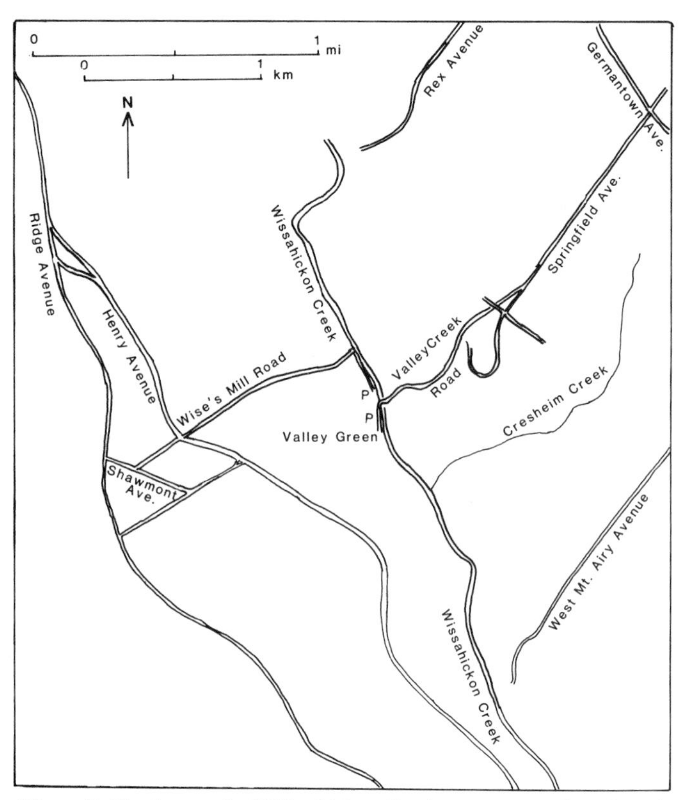

Figure 2. Northern end of Wissahickon Creek traverse, north and south of Bells Mill Road. P, location of parking lots.

Description of rocks along Wissahickon Creek. (A). Northern limit of outcrop to Bells Mill Road (0.3 mi; 0.5 km; Fig. 2). The northernmost outcrops along this section in the bed of the creek and along the bank, accessible if the water is not too high, consist of banded gray Baltimore gneiss. South of the gneiss, outcrops of the Wissahickon schist consist of dominantly psammitic quartz-muscovite-biotite schist locally containing small garnets and showing a marked vertical foliation. The steep foliation in the schist and the associated upright tight isoclinal folds occur only in this northern section of the traverse and are assumed to be associated with the faults that juxtapose the Grenville age gneiss against the schist. A set of 3-ft-scale (meter-scale) open folds and, in micaceous units, of centimeter-sized crenulations, both with horizontal axial surfaces and fold axes, deform the foliation. A small quarry along the west side of the Creek 0.2 mi (0.3 km) north of Bells Mill Road exposes a slightly foliated, strongly lineated dike of granodiorite approximately 50 ft (15 m) wide. The deformation has converted this dike to augen gneiss. A lens of ultramafic rock, best seen on the eastern side of the creek, outcrops immediately north of Bells Mill Road. Here the ultramafic schist consists of three units: (1) a talc-serpentine rock in which centimeter-sized patches of serpentine, apparently replacing olivine, occur in a talc-magnesite matrix; (2) a tremolite-anthophyllite schist composed of radial clusters of amphibole; and (3) a foliated chlorite-magnetite schist. Just south of Bells Mill Road, on the east side of the creek, the first large outcrop of aluminous pelitic schist is encountered. Throughout the section, pelitic units similar to this contain the metamorphic index minerals used to document the southward increase in metamorphic grade. At this outcrop the schist is highly garnetiferous; rare small staurolite crystals also occur. A few thin layers of calcareous quartzite outline upright isoclinal folds. These calcareous quartzite layers are commonly associated with the pelitic units.

(B). Bells Mill Road to Valley Green (Figs. 2, 3). The section described lies along the east bank of the creek, which is less congested by joggers, cyclists, and equestrians. The main features displayed along this section include: a change from vertical to more gently-dipping orientation of the lithologic layers, foliation, and fold axial surfaces; more open folds; an increase in the number and size of staurolite porphyroblasts; and the appearance of kyanite in the aluminous layers. Locally staurolite and kyanite are oriented randomly relative to the foliation; however, observations in thin section suggest the prominent subhorizontal crenulation lineation postdates porphyroblast growth.

Approximately 0.6 mi (0.9 km) south of Bells Mill Road, the first of several coarse-grained granite pegmatites intrude the schist. These pegmatite bodies are approximately concordant to the foliation but are clearly intrusive; some contain schist xenoliths. At 0.9 mi (1.5 km), just north of the extension of Rex

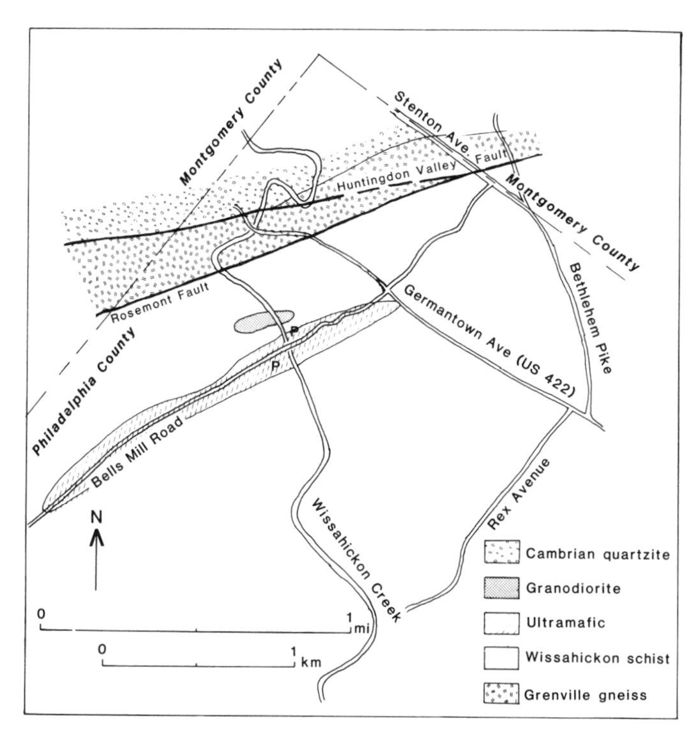

Figure 3. Middle portion of Wissahickon Creek traverse, north and south of Valley Green. The entire area of this map is underlain by Wissahickon schist. P, location of parking lots.

Avenue, several large outcrops above the trail to the east contain the first kyanite porphyroblasts. The location can be recognized by a large cement statue of an Indian atop the highest ledge. Some outcrops show traces of early isoclinal folds refolded by the dominant 3-ft-scale (meter-scale) open-to-close folds that characterize this section. The rare, early folds are tight with long limbs and a narrow hinge zone. Kyanite and staurolite porphyroblasts also can be found cutting across the fabric of the garnet schist at the small outcrop below the stone shelter just above and northeast of the bridge that crosses the creek at Valley Green.

(C). Valley Green to the Schuylkill River (Fig. 3). One of the notable features along this section occurs on the east bank, 0.3 mi (0.5 km) south of the bridge at Valley Green. At this locality, a few hundred feet (meters) north of Cresheim Creek, there is a 20 ft (6 m) concordant layer of ultramafic anthophyllite/tremolite schist with the same rosette texture of amphibole clusters seen in the ultramafic rocks farther north. The next relatively large, low outcrop lying between the trail and the creek shows a complexly contorted fabric locally defined by thin (~0.5 in, 1 cm) pink layers that consist of quartz, minute garnet crystals, and scattered biotite flakes. These bands may represent metamorphosed thin cherty layers. The rest of the outcrop is semi-pelitic and exhibits quartzo-feldspathic segregations apparently derived from the enclosing schists. These segregations contain the same minerals as the host schist. They are distinguished from the pegmatite dikes described earlier by the absence of potassium feldspar, which is abundant in the pegmatite. Another easily accessible spot to observe this type of schist is the last outcrop on the west side of the creek, just before it crosses under Ridge Avenue and flows into the Schuylkill River.

Continuing downstream from Cresheim Creek, the relative amount of staurolite in pelitic units decreases and the amount of kyanite increases. A staurolite-out isograd is difficult to locate exactly; occasional staurolite grains have been observed at least as far as 1.25 mi (2 km) downstream from Valley Green. The quartzo-feldspathic segregations become increasingly abundant and more well developed, eventually forming in more micaceous layers as well. At 1.1 mi (1.75 km) downstream from the Valley Green bridge, on the east side of the creek at the end of West Mt. Airy Avenue, the first outcrop of amphibolite is encountered. The outcrop lies at the end of a short path slightly above the creek. Downstream from this locality, fine-grained dark amphibolite with well-lineated hornblende needles occurs occasionally, in layers up to tens of feet (meters) thick. The appearance of amphibolite in the section coincides approximately with the disappearance of the most aluminous layers, making it difficult to locate diagnostic assemblages to infer metamorphic grade.

Also at West Mt. Airy Avenue, fibrolite appears, initially associated with quartz segregations and subsequently as mats and bundles of fine fibers in the schist matrix. This fibrolite is extremely difficult to detect in hand specimen. In this lower section of the creek, kyanite and fibrolite can occasionally be found in the same specimen, but they normally occur separately and it is not possible to tell whether one formed before the other.

(D). Along the east bank of the Schuylkill River. A good exposure of fibrolite-bearing schist is found 1.8 mi (2.9 km) south of the mouth of Wissahickon Creek just south of Strawberry Mansion Bridge (Fig. 1). This outcrop is just south of the small creek, across Kelly Drive from a large parking lot. The schist contains well-developed quartz-fibrolite nodules that weather out on the dip surface of the foliation. The nodules are flattened in the plane of the schistosity and elongated in a north-northeasterly direction approximately parallel to axes of recumbent isoclinal folds that can be seen on the southwestern side of the outcrop. Just to the north, across the creek and along the road coming down from the east, there is a thick amphibolite unit with hornblende lineation parallel to that of the quartz-fibrolite nodules.

In summary, this section presents a fairly continuous traverse through the amphibolite facies of the Wissahickon Formation. The rocks along the traverse are representative of the Wissahickon Formation that underlies the area between the Grenville gneiss to the north and the Coastal Plain sediments to the south, from Trenton to the western border of Philadelphia County. It demonstrates a continuous increase in metamorphic grade of the Barrovian facies series, and the accompanying southeasterly verging structures that formed approximately contemporaneously with the metamorphism. Lack of good exposure and structural complexities associated with the emplacement of the Wilmington Complex thrust sheet (Delaware and adjacent southeasternmost Pennsylvania) and the State Line Mafic-ultramafic Complex (Maryland and Pennsylvania) preclude correlation of the schists in the Wissahickon Creek section with the subdivisions of the Wissahickon established in Maryland (Higgins, 1972). All evidence thus far suggests the original stratigraphic sequence throughout the northern Piedmont has been thoroughly disrupted by tectonic telescoping during the Taconic orogeny (e.g., Wise, 1970; Fisher and others, 1979).

REFERENCES CITED

Amenta, R. V., 1974, Multiple deformation and metamorphism from structural analysis in the eastern Pennsylvania Piedmont: Geological Society of America Bulletin, v. 85, p. 1647–1660.

Bascom, F., 1905, The Piedmont district of Pennsylvania: Geological Society of America Bulletin, v. 16, p. 289–328.

Crawford, M. L., and Crawford, W. A., 1980, Metamorphic and tectonic history of the Pennsylvania Piedmont: Journal of the Geological Society of London, v. 137, p. 311–332.

Crawford, M. L., and Mark, L. E., 1982, Evidence from metamorphic rocks for overthrusting, Pennsylvania Piedmont, U.S.A.: Canadian Mineralogist, v. 20, p. 333–347.

Crawford, W. A., and Crawford, M. L., 1974, Geology of the Piedmont of southeastern Pennsylvania: Guidebook of 39th Field Conference of Pennsylvania Geologists, 104 p.

Fisher, G. W., Higgins, M. W., and Zietz, I., 1979, Geological interpretations of aeromagnetic maps of the crystalline rocks in the Appalachians, northern Virginia to New Jersey: Maryland Geological Survey Report of Investigations 32, 43 p.

Goodwin, B. K., 1964, Guidebook to the geology of the Philadelphia area: Pennsylvania Geological Survey, General Geology Report G41, 189 p.

Higgins, M. W., 1972, Age, origin, regional relations, and nomenclature of the Glenarm Series, central Appalachian Piedmont; A reinterpretation: Geological Society of America Bulletin, v. 83, p. 989–1026.

Johnson, R. C., and Meyer, G. H., 1975, Sillimanite nodules in the Wissahickon schist, Philadelphia: Pennsylvania Academy of Science Proceedings, v. 49, p. 34–36.

Lapham, D. L., and Root, S., 1971, Summary of isotopic age determinations in Pennsylvania: Pennsylvania Geological Survey Information Circular 7, 29 p.

Roberts, F. H., 1969, Ultramafic rocks along the Precambrian axis of southeastern Pennsylvania [Ph.D. thesis]: Bryn Mawr, Pennsylvania, Bryn Mawr College, 39 p.

Sutter, J. F., Crawford, M. L., and Crawford, W. A., 1980, $^{40}Ar/^{39}Ar$ age spectra of coexisting hornblende and biotite from the Piedmont of SE Pennsylvania; Their bearing on the metamorphic and tectonic history: Geological Society of America Abstracts with Programs, v. 12, p. 85.

Tearpock, D. J., and Bischke, R., 1980, The structural analysis of the Wissahickon schist near Philadelphia, Pa.: Geological Society of America Bulletin, Part II, v. 91, p. 2432–2456.

Wagner, M. E., 1982, Taconic metamorphism at two crustal levels and a tectonic model for the Pennsylvania-Delaware Piedmont: Geological Society of America Abstracts with Programs, v. 14, p. 640.

Weiss, J. V., 1949, Wissahickon schist at Philadelphia, Pennsylvania: Geological Society of America Bulletin, v. 60, p. 1689–1726.

Wise, D. U., 1970, Multiple deformation, geosynclinal transitions, and the Martic problem in Pennsylvania, *in* Fisher, G. W., Pettijohn, F. J., Reed, J. C., Jr., and Weaver, K. N., eds., Studies of Appalachian geology; Central and southern: New York, Wiley Interscience, p. 317–334.

Wyckoff, D., 1952, Metamorphic facies in the Wissahickon schist near Philadelphia: Geological Society of America Bulletin, v. 63, p. 25–57.

Late Triassic cyclic sedimentation: Upper Lockatong and lower Passaic Formations (Newark Supergroup), Delaware Valley, west-central New Jersey

Franklyn B. Van Houten, *Department of Geologic and Geophysical Sciences, Princeton University, Princeton, New Jersey 08544*

LOCATION

This section of Triassic rocks along New Jersey 29 extends from Byram, 8 mi (13 km) north of Lambertville, 2.8 mi (4.5 km) northward to the Devils Tea Table, 5 mi (8 km) south of Frenchtown. The site, in the Lumberville (Pa.-N.J.) Quadrangle, lat. 40° 25′20″–27′45″N, long. 75°3′30″W, is reached by car from the junction of U.S. 202/New Jersey 29 to the south or of New Jersey 12/29 to the north (Fig. 1). The roadcuts on the east side of New Jersey 29 are easily accessible. Stops 1 to 4 (Figs. 1 to 5) display the significant features, but there are numerous additional outcrops along the continuous exposure between Stops 1 and 3 as well as north and south of Stop 4.

SIGNIFICANCE

Triassic lacustrine (Lockatong) and mudflat (Passaic) deposits exposed here record alternations of climate operating at Milankovitch intervals (Figs. 2, 3). Small-scale detrital and chemical cycles (6 to 20 ft or 2 to 6 m thick) mark the repeated expansion and waning of the lake in 21,000- to 23,000-year periods. Clusters of several of these record 100,000-year spans, and large-scale repetitions of dark gray and reddish-brown deposits (300 to 400 ft or 100 to 125 m thick) reflect 400,000- to 500,000-year fluctuations.

A diabase (dolerite) sill at the base of the traverse (Figs. 1, 4) produced a distinctive suite of soda-rich hornfels minerals in metamorphosed chemical cycles.

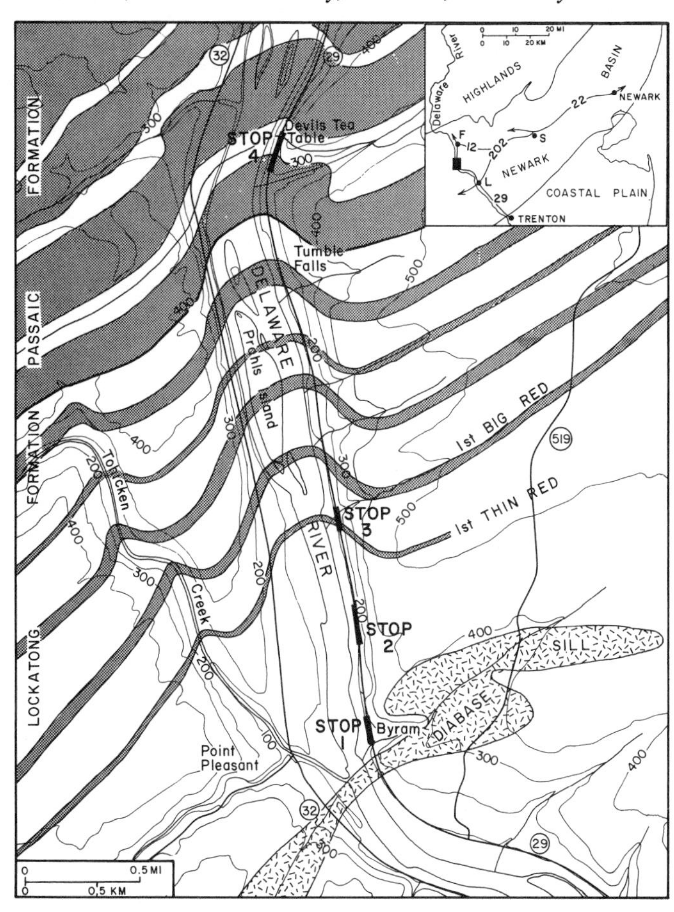

Figure 1. Geologic sketch map of upper Lockatong and lower Passaic formations, Delaware Valley section, west-central New Jersey, with location of Stops 1 to 4 along New Jersey 29. Map shows Byram diabase sill, and subordinate grayish-red to reddish-brown units (stippled) in dark gray Lockatong and dark gray units in reddish-brown (stippled) Passaic deposits (see Fig. 2). Base is USGS 7½-minute Lumberville (Pa.-N.J.) Quadrangle. Inset indicates access to site; F, Frenchtown, L, Lambertville, S, Somerville.

SITE INFORMATION

The Late Triassic part of the Newark Supergroup comprises as much as 3 mi (5 km) of nonmarine deposits that accumulated in a rift valley during an early phase of the opening of the Atlantic Basin. Along its northwestern margin the valley is bounded by Precambrian and Paleozoic rocks of the Highlands (Fig. 1). Within it, Newark strata lie on Paleozoic and Precambrian rocks, and along the southeastern margin Triassic rocks are overlapped by the Cretaceous Coastal Plain mantle.

In central New Jersey the Newark Supergroup (Fig. 2A) includes a lower, locally conglomeratic Stockton Arkose (5,050 ft or 1,540 m thick), a middle dark gray to reddish-brown Lockatong Formation (as much as 3,870 ft or 1,180 m thick), and an upper reddish-brown Passaic ("Brunswick") Mudstone (more than 3,300 ft or 1,000 m thick) that interfingers with the Hammer Creek Conglomerate along the northwestern border of the basin; for details see Olsen, 1980a; Van Houten, 1969, 1980. The site comprises the upper 2,800 ft (850 m) of the Lockatong Formation and the lower 300 ft (100 m) of the Passaic Formation (Figs. 1, 2).

The Lockatong Formtion is a huge lacustrine lens that thins northeastward and southwestward along the axis of the basin, and it grades upward into the Passaic Formation through a succession of alternating large-scale dark gray and reddish-brown units (Fig. 2). Lockatong deposits are arranged in small-scale detrital and chemical cycles averaging several yd (m) thick (Fig.

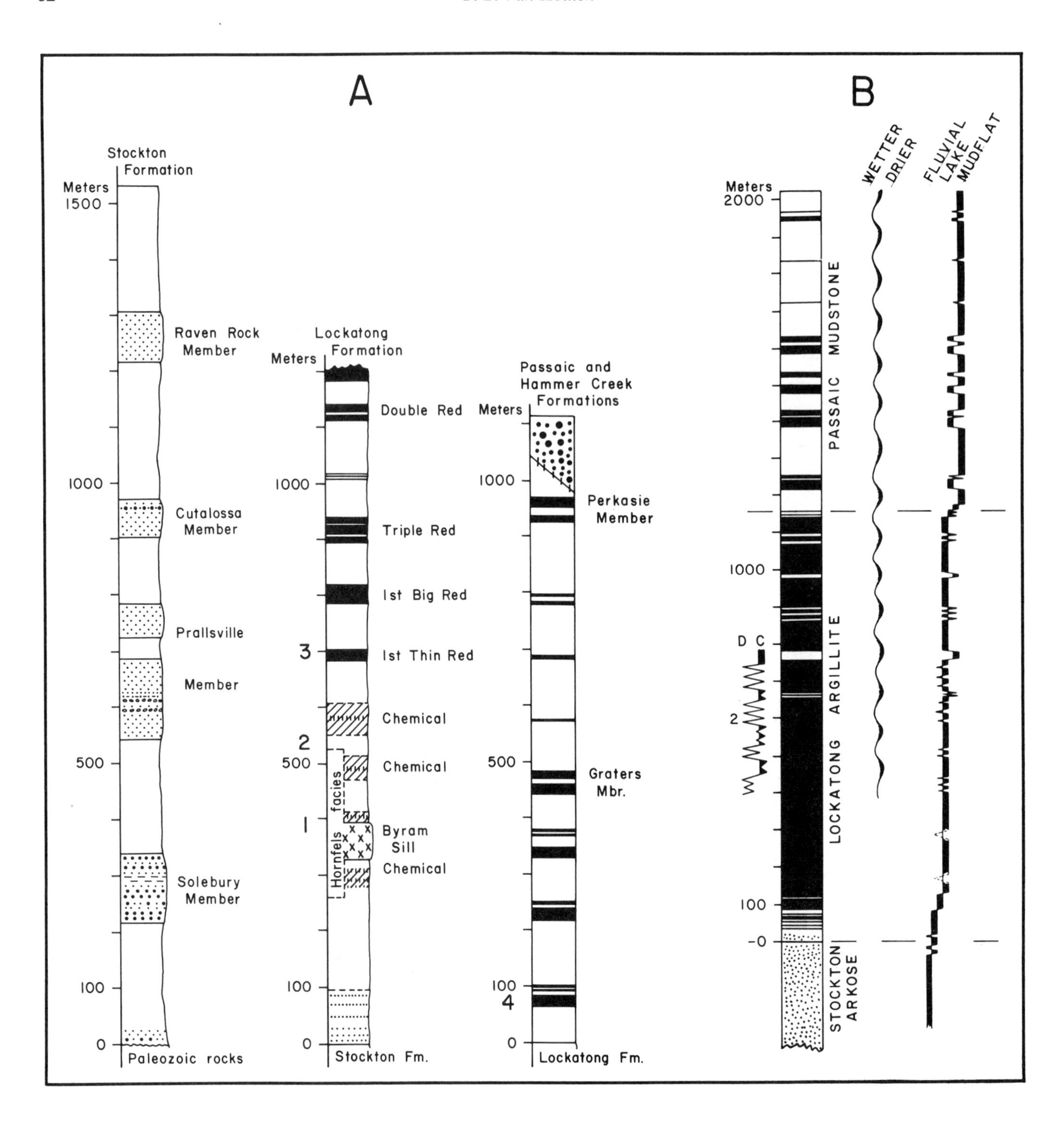

Figure 2. Stratigraphic section of Newark Supergroup along Delaware Valley. A: Subordinate lithic units in the formations. Black units in Lockatong Formation are grayish-red to reddish-brown analcime-rich chemical cycles (Fig. 3); those in Passaic Formation are dark gray detrital deposits. Intervals of gray chemical cycles and hornfels facies in Lockatong Formation and location of Stops 1 to 4 are indicated. B: Patterns of cycles in Lockatong and Passaic formations. Black, dark gray deposits; white, grayish-red to reddish-brown deposits. Left column indicates distribution of short chemical (C) detrital (D) 21,000- to 23,000-year cycles in patterns of intermediate (100,000-year) and long (400,000- to 500,000-year) cycles. Sequence of changing environments sketched in columns to right of stratigraphic section.

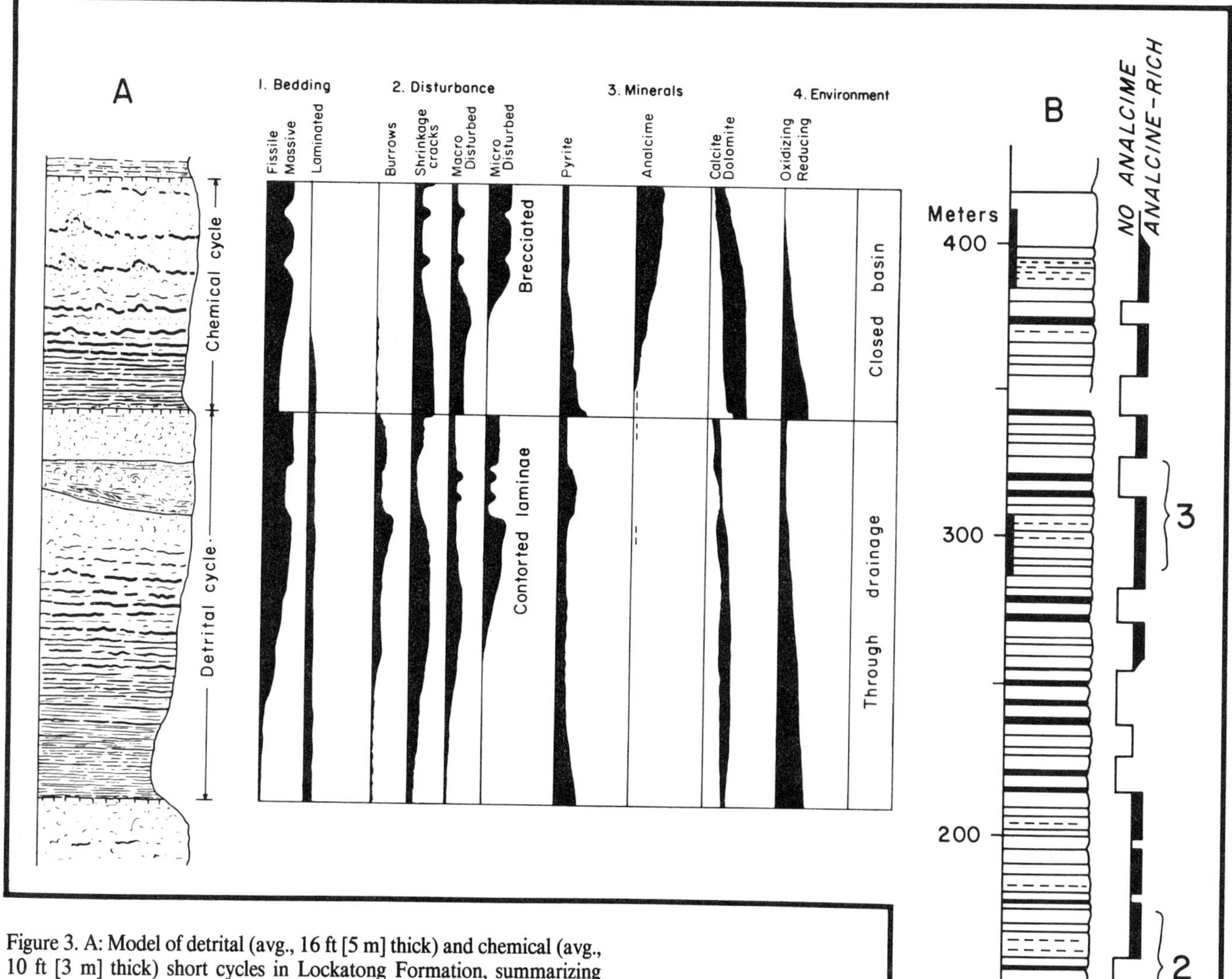

Figure 3. A: Model of detrital (avg., 16 ft [5 m] thick) and chemical (avg., 10 ft [3 m] thick) short cycles in Lockatong Formation, summarizing distribution of sedimentary features. B: Measured section of short cycles above Byram Sill along east side of New Jersey 29, with location of Stops 1 to 3. Bundles of detrital and chemical cycles are indicated at right of column. Thick lines along left margin span First Thin Red and First Big Red units in pattern of long cycles (Fig. 2). 0 mark is base of measured section marked in 5-ft (1.5-m) intervals.

3; Stops 1-3). Detrital cycles are more common in long gray sequences; chemical cycles are best developed in the grayish-red to reddish-brown sequences that recur at about 300- to 400-ft (100- to 125-m) intervals, as seen at Stop 2 and 3.

Short detrital cycles (Fig. 3A), about 13 to 20 ft (4 to 6 m) thick, consist of a lower black pyritic shale succeeded by platy dark gray carbonate-rich mudstone in the lower part, and tough, massive dark gray calcareous mudstone (argillite) in the upper. The argillite has a very small-scale contorted and disrupted fabric produced largely by crumpled shrinkage crack and burrow casts (Van Houten, 1969). Some thicker detrital cycles contain a 1.6- to 5-ft(0.5- to 1.5-m)-thick lens of thin- and ripple-bedded siltstone and fine-grained feldspathic sandstone with small-scale

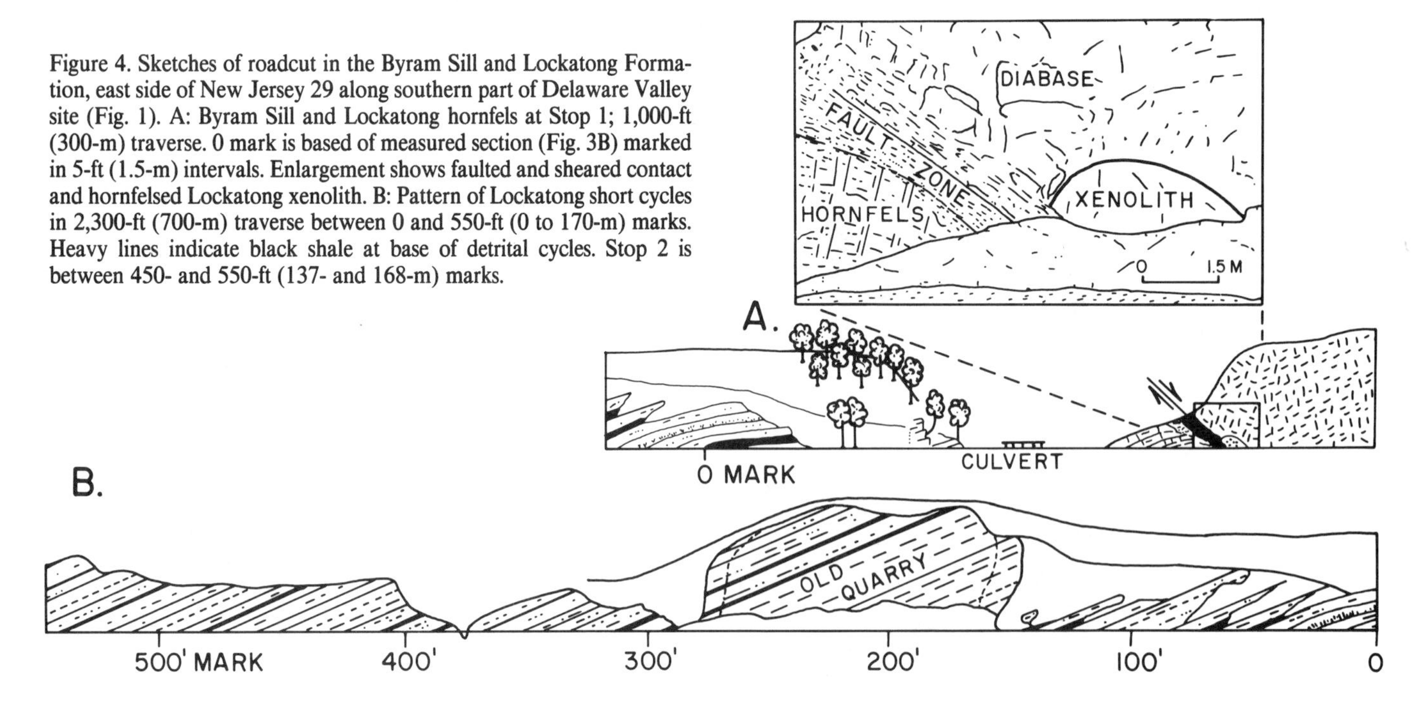

Figure 4. Sketches of roadcut in the Byram Sill and Lockatong Formation, east side of New Jersey 29 along southern part of Delaware Valley site (Fig. 1). A: Byram Sill and Lockatong hornfels at Stop 1; 1,000-ft (300-m) traverse. 0 mark is based of measured section (Fig. 3B) marked in 5-ft (1.5-m) intervals. Enlargement shows faulted and sheared contact and hornfelsed Lockatong xenolith. B: Pattern of Lockatong short cycles in 2,300-ft (700-m) traverse between 0 and 550-ft (0 to 170-m) marks. Heavy lines indicate black shale at base of detrital cycles. Stop 2 is between 450- and 550-ft (137- and 168-m) marks.

disturbed bedding. On average these deposits contain abundant Na-feldspar, illite and muscovite, some K-feldspar, chlorite and calcite, and a little quartz.

Short chemical cycles (Fig. 3A), about 6 to 12 ft (2 to 4 m) thick, are most common in the upper part of the formation, and are limited to the central 60 mi (100 km) along the axis of the basin. Lower beds 1 to 8 cm thick are alternating dark gray to black platy dolomitic mudstone and tan-weathering marlstone disrupted by shrinkage cracks. Locally, basal beds contain thin lenses of dolomite and pyrite. The middle, more massive argillite encloses layers of tan-weathering dolomitic marlstone extensively disrupted by shrinkage cracking. The middle part of many chemical cycles, as well as of reddish-brown ones in the lowest part of the Passaic Formation, also exhibits a pattern of upward-concave surfaces and thin zones of shearing in tent-like structures 6 to 12 in (15 to 30 cm) high recurring laterally in wave lengths of 1.5 to 3 ft (0.5 to 1 m). The beds involved contain numerous small, crumpled shrinkage cracks and are fractured and brecciated (Van Houten, 1969). The upper part of a chemical cycle is tough gray analcime- and dolomite-rich argillite brecciated on a microscopic scale. Tiny, slender, crumpled shrinkage cracks filled with dolomite and analcime produce a "birdseye" fabric.

Some thinner chemical cycles are grayish-red to reddish-brown. In these, lower platy dark red layers are disrupted by shrinkage cracks, with patches of analcime and dolomite in the cracks. Many of the upper, thicker, massive beds are speckled with tiny lozenge-shaped pseudomorphs of dolomite and analcime after gypsum or glauberite that are locally arranged in rosettes of radiating skeletal crystals. The argillite is also marked by long, intricately crumpled crack filling. The upper part of chemical cycles contains as much as 7 percent Na_2O and as little as 47 percent SiO_2; it is composed of a maximum of 35 to 40 percent analcime, together with albite, dolomite and calcite, and illite and minor chlorite.

The Lockatong Formation was converted to aphanitic to very fine grained varieties of calcitic biotite and Na-feldspar hornfels marked by an absence of quartz (Van Houten, 1971). Metamorphosed carbonate-rich deposits in the middle of detrital cycles and in the lower part of chemical cycles are banded calc-silicate hornfels and thin layers of marble. The upper massive argillite of detrital cycles within 100 ft (30 m) of the Byram Sill (Fig. 4) contains scapolite, aegirine, diopside, clinozoisite, K-feldspar, Na-feldspar, calcite, chlorite, and biotite. Upper argillite of chemical cycles, contains nepheline, sodalite, cancrinite, thompsonite, pyroxene and amphibole, calcite, biotite, and albite within 100 ft (30 m) of the sill; 100 to 425 ft (30 m to 130 m) from it cancrinite, thompsonite, and increasing unaltered analcime predominate.

Throughout the central part of the basin, the Passaic Formation consists of a rather uniform succession of reddish-brown mudstone and siltstone with subordinate claystone and fine-grained feldspathic sandstone. These strata commonly preserve large and small tracks and trails. Molds of glauberite now filled with calcite or barite are present locally in 2- to 8-in(5- to 20-cm)-thick lenses. Widespread units of dark gray pyritic mudstone and marlstone (Fig. 5) recur in the formation at 300- to 400-ft (100- to 125-m) intervals, matching those in the upper part of the Lockatong Formation (Fig. 2). The common mudstone and siltstone in the Passaic Formation contain abundant illite and subordinate chlorite, abundant quartz (50 to 75 percent in siltstone, 10 to 30 percent in mudstone), less than 15 percent feldspar (Na-feldspar normally predominates), and relatively rare lithic fragments except near the northwest border. Hematite is the pigment mineral coating grains and staining the clay fraction; it is also the common opaque mineral grain.

CONDITIONS OF DEPOSITION

Accumulation of Stockton sandy deposits ended with waning of detrital influx and widespread ponding along the axial drainageway of the rift valley. This produced a huge Lockatong lake (Olsen, 1980b) with narrow marginal Passaic mudflats and small deltas supplied largely from the distant southeastern upland. Along its northern margin the lake was fringed by fan deltas. The ponding may have been caused partly by the especially active building of alluvial fans at both ends of the basin (Van Houten, 1969) combined with continued slow subsidence. Once established, conditions in the long lake varied only within narrow limits for several million years, and cyclic variation in climate (Fig. 2B) exerted a major control on the lacustrine sedimentation. This produced short detrital cycles during times of throughflowing drainage, and chemical cycles when the lake was closed. In its late stage the lake became a carbonate-clay playa with salts crystallizing in the mud and repeated wetting and drying disrupting the beds. Based on counts of assumed varves, Lockatong short cycles apparently resulted from expansion and waning of the lake during 21,000- to 23,000-year precession cycles. Clusters of several cycles record 100,000-year periods, and groups of 20 to 25 cycles (Figs. 2, 3), either predominantly detrital or chemical, suggest a large-scale climate control duration of 400,000 to 500,000 years (Olsen, 1980b).

The Lockatong lacustrine facies gradually gave way to oxygenated Passaic mudflats and ponds where large and small reptiles and small crustaceans left their tracks and trails. In the eastern source area, weathering was intense enough to convert crystalline basement in the upland into a continuing supply of ferric oxide–rich feldspathic mud. Along the northwestern border of the basin, Passic mud interfingered directly with the fringing alluvial fans. Much of the basin regime may have been that of ephemeral clay flat playas where sedimentation was largely due to suspension. Here and there glauberite formed as scattered crystals in dried-out mud extensively broken by shrinkage cracks. Long (400,000 to 500,000-year) climatic cycles (Figs. 2, 4), in phase with those recorded in upper Lockatong deposits, produced prolonged periods of dry, oxidizing environment and thick sequences of ferric oxide–rich mud alternating with briefer, moist intervals and accumulation of thin units of pyritic dark gray mud and minor carbonate.

STOP 1. At southern end of site (Fig. 1). Park along New Jersey 29 at the north end of Byram Sill. Observe the following significant features in a northward traverse (Figs. 2 to 4).

a. Fractured coarse-grained diabase, with common "joint minerals" including amphibole, calcite, epidote, prehnite, and tourmaline.

b. Faulted and sheared Lockatong-diabase contact, and essentially undisturbed contact with Lockatong xenolith (Fig. 4A). Chemical cycles are metamorphosed soda-rich hornfels.

Walk northward beyond culvert to beginning of continuous outcrop, with a vestige of 5-ft (1.5-m) marks (Fig. 4A).

c. Indistinct pattern of hornfelsed Lockatong short cycles. At the 21-ft (6.4-m) mark note the white sprays of cancrinite, albite, and calcite in hornfels developed from analcime-dolomite–rich chemical cycle like those in the 1st Thin Red unit at Stop 3 (Figs. 2, 3).

Drive northward. A large abandoned quarry is in a thick

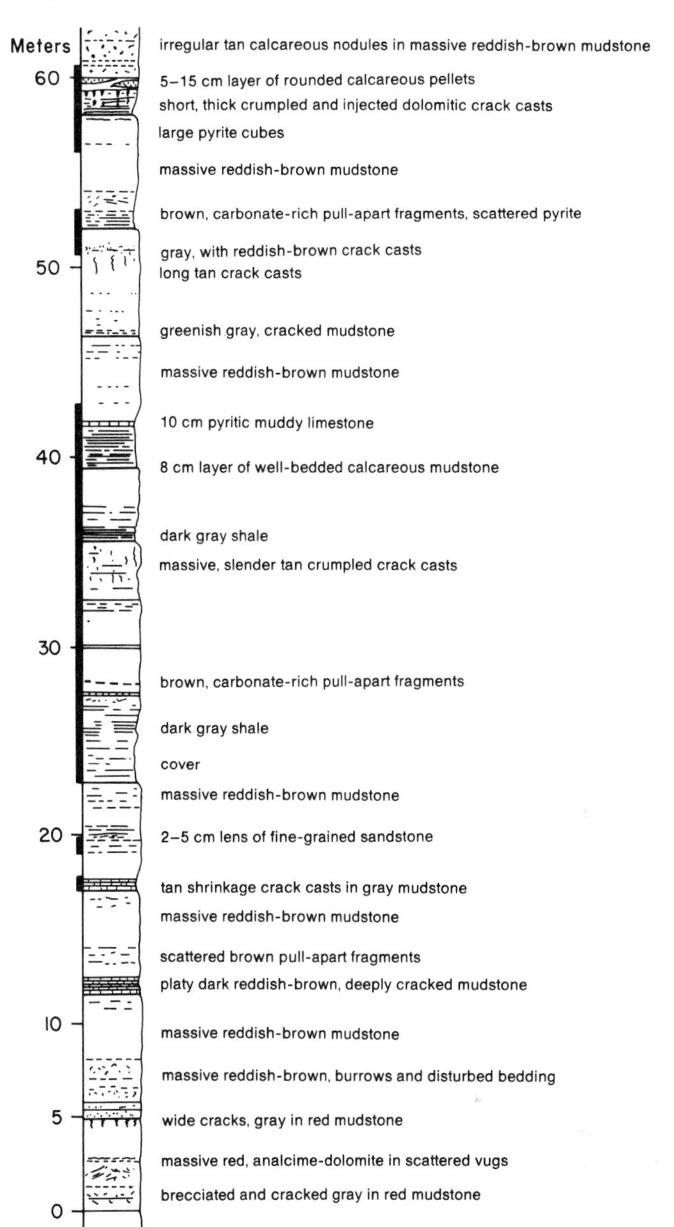

Figure 5. Measured section of lower part of Passaic Formation, Stop 4 at north end of the Delaware Valley site (Figs. 1, 2). Diagram shows cyclic pattern in upper 75 ft (23 m) of the lowest reddish-brown mudstone unit and in overlying unit of dark gray mudstone and shale (heavy lines along left margin).

succession of hornfelsed chemical cycles. Two detrital cycles cross the top of the quarry wall (Fig. 4B).

STOP 2. Between the 450- and 550-ft marks. Well-displayed short chemical cycles exhibit the significant features shown in Figure 3A.

STOP 3. Park along New Jersey 29 near vertical roadcut about the 850-ft (259-m) mark.

a. 1st Thin Red unit of chemical cycles between the 850- and 900-ft (259- and 274-m) marks. Note shrinkage cracks and "birdseye" fabric with patches of analcime and dolomite.

b. At the 953- and 970-ft (290- and 296-m) marks: short detrital cycle with lens of feldspathic sandstone in the upper part. Identify features shown in Figure 3A.

c. 1st Big Red unit (Figs. 1, 2A, 3B) of chemical cycles is about 250 ft (80 m) stratigraphically above the top of 1st Thin Red unit.

STOP 4. At the northern end of the site. Park near the Tumble Falls (Fig. 1) where stopping is permitted. Walk northward to vertical roadcut below the Devils Tea Table.

a. Four cycles of the upper part of the lowest reddish-brown mudstone and siltstone unit of the Passaic Formation (Figs. 1, 2) exhibit small-scale disruption features (Fig. 5).

b. The upper 10 ft (3 m) of the lowest, 125-ft (38-m)-thick, cyclic dark gray unit (Figs. 1, 2, 5) includes (1) black pyritic mudstone with crumpled, composite shrinkage crack casts composed of dolomitic and calcitic mudstone; and (2) lenses of pyritic peloidal dolomite 4 to 18 in (10 to 20 cm) thick, with silty burrow casts and fragments of thin shells. Overlying massive reddish-brown mudstone contains irregular calcareous nodules 0.5 to 2 in (1 to 5 cm) long that may be algal structures, as well as scattered flecks of pyrite.

REFERENCES CITED

Olsen, P. E., 1980a, Triassic and Jurassic Formation of the Newark Basin, *in* Manspeizer, W., ed., Field Studies of New Jersey Geology and Guide to Field Trips, 52nd Annual Meeting of the New York State Geological Association: Newark, New Jersey, Rutgers University, p. 2–39.

—— , 1980b, Fossil great lakes of the Newark Supergroup in New Jersey, *in* Manspeizer, W., ed., Field Studies of New Jersey Geology and Guide to Field Trips, 52nd Annual Meeting of the New York State Geological Association: Newark, New Jersey, Rutgers University, p. 352–398.

Van Houten, F. B., 1969, Late Triassic Newark Group, north central New Jersey and adjacent Pennsylvania and New York, *in* Subitsky, S. S., ed., Geology of Selected Areas in New Jersey and Pennsylvania and Guidebook: New Brunswick, New Jersey, Rutgers University Press, p. 314–347.

—— , 1971, Contact metamorphic mineral assemblages, Late Triassic Newark Group, New Jersey: Contributions to Mineralogy and Petrology, v. 30, p. 1–14.

—— , 1980, Late Triassic part of Newark Supergroup, Delaware River section, west-central New Jersey, *in* Manspeizer, W., ed., Field Studies of New Jersey Geology and Guide to Field Trips, 52nd Annual Meeting of the New York State Geological Association: Newark, New Jersey, Rutgers University, p. 246–276.

Cretaceous stratigraphy of the Atlantic Coastal Plain, Atlantic Highlands of New Jersey

Richard K. Olsson, *Department of Geological Sciences, Rutgers University, New Brunswick, New Jersey 08903*

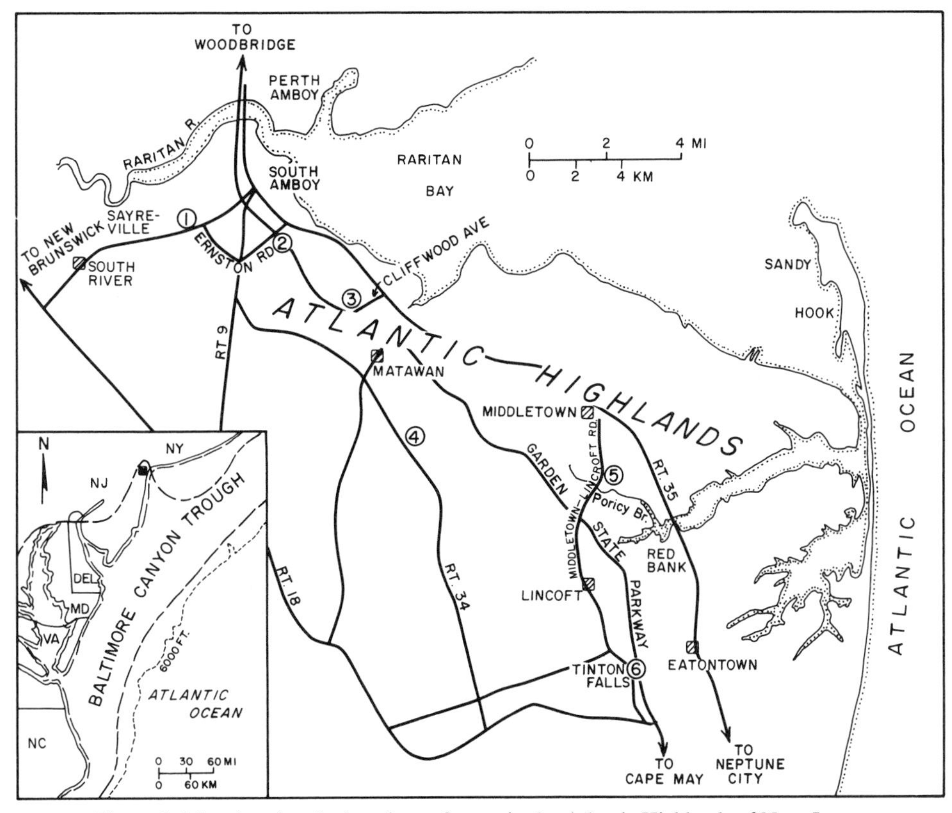

Figure 1. Map showing the locations of stops in the Atlantic Highlands of New Jersey.

LOCATION

The Upper Cretaceous section of the Atlantic Highlands is located in the coastal plain of northern Monmouth County, New Jersey (Fig. 1). The stops described here can be reached by passenger car from major streets and roadways. Accesses to the stops are not restricted, although one should check in with appropriate people where a site is located on commercial property. Stop 1 is exposured on slopes below Washington Road, Sayreville, in John F. Kennedy Memorial Park, entrance just opposite intersection with Ernston Road. Stop 2 is approximately 0.25 mi (0.42 km) southeast of Dwight D. Eisenhower School on Ernston Road, Sayreville, in slopes facing southeast above Cheesequake Creek and parallel to the Garden State Parkway. For Stop 3, proceed north on County Road off of Cliffwood Avenue (access from New Jersey 35 at Cliffwood) just north of overpass of Garden State Parkway (GSP); then go north on Biondi Street 150 ft (46 m) to slopes of old pit facing southwest towards GSP. Stop 4 is a series of excavations on the north side of New Jersey 34 behind Bunkhouse Furniture Store, Gilberts' Gymnastic School, Ern Construction Company, and beyond toward the southeast. Stop 5 is in the banks of Poricy Brook in the Poricy Park Fossil Beds, Middletown Township, on Middletown-Lincroft Road. Stop 6 is at Tinton Falls behind the old mill house (Milldam Restaurant) in the center of town.

Useful references on the geology of the Atlantic Highland sector of the coastal plain include Minard (1969), Owens and Sohl (1969), Owens and others (1970, 1971), Olsson (1980), and Olsson and others, 1986.

SIGNIFICANCE

The Upper Cretaceous section of the Atlantic Highlands is located in the northern part of the coastal plain of New Jersey. It lies along the western edge of the Baltimore Canyon Trough, a large sedimentary basin that extends along the United States middle Atlantic states (Fig. 1). The sediments in the Baltimore Canyon Trough were initially deposited when North America and Africa separated during Jurassic time. Deposition in the coastal plain began in the Early Cretaceous as fluvial sedimentation in an upper delta plain environment. Marine incursions did not fully extend over the coastal plain until Late Cretaceous time. Marine deposition in the coastal plain is related to the well-known Late Cretaceous rise in sea level, which inundated and flooded widely onto the continents of the world. The Upper Cretaceous stratigraphy of the Atlantic Highlands reflects the overall rise in sea level during the Late Cretaceous, but it also records shorter term changes in sea level.

SITE INFORMATION

Sea-level change has been long recognized in the New Jersey Coastal Plain. Weller (1907) distinguished two general molluscan assemblages, a *Cucullea* and a *Lucina* assemblage, which he associated, respectively, with formations deposited during transgression and regression. The concept of "depositional sequence" has, subsequently, been applied to the coastal plain following the work of Vail and others (1977) and Mitchum and others (1977) on interpreting seismic reflection profiles on continental margins. A depositional sequence is a chronostratigraphic unit that is unconformity-bounded. Owens and Gohn (1985) first described depositional sequences in the Cretaceous of the Atlantic Coastal Plain. Each depositional sequence is physically and biostratigraphically constrained and can be traced throughout the coastal plain. Owens and Gohn suggest that large-scale processes such as eustatic sea-level change or continental-scale tectonic events controlled the initial distribution of each sequence. Olsson (1986) modified the depositional sequence stratigraphy of Owens and Gohn for the New Jersey Coastal Plain (Fig. 2). He regards, in agreement with Vail and others (1977), eustatic sea-level change as the primary process in the development of depositional sequences in the New Jersey Coastal Plain.

Each depositional sequence in the New Jersey Coastal Plain corresponds to a short-term cycle of sea-level change (Olsson, 1986). Each sequence begins with a transgressive unit and ends with a regressive unit. A sequence records a cycle of sea-level rise and fall (Fig. 2).

The transgressive part of each depositional sequence is characterized by ubiquitous glauconite in clayey, silty units. Regressive units are characterized by sands and silts. The environments of deposition range from middle to outer shelf in transgressive units to inner shelf, shoreface, and delta facies in the regressive units. In the Atlantic Highlands of the New Jersey Coastal Plain, parts of five sequences can be observed.

Stop 1. Raritan Formation overlain by the Pensauken Formation (Pleistocene?). The sands and clays of the Raritan represent regressive continental sediments of depositional sequence KCE1 (Fig. 2). The transgressive marine facies of this sequence, which is called the Bass River Formation, is encountered in coastal plain wells.

The Pensauken Formation consists of 5 to 10 ft (1.5 to 3 m) of dark yellowish brown sand and gravel. Large-scale steeply dipping cross-beds represent slip faces of fluvial bars. Approximately 150 ft (46 m) to the west, a point-bar fining-upward sequence is present. A gravel layer marks the contact with the Raritan Formation.

Raritan Formation (Woodbridge Member) is here 65 ft (20 m) thick. The upper 30 to 40 ft (9 to 12 m) consists of interbedded sand and clay with numerous siderite and iron-oxide–cemented sand layers. Easterly and westerly directions of planar cross-bedding can be observed. The cross-bedding probably represents sand waves of a meandering stream or of intertidal currents. Marine fossils are present in some of the sand layers. Abundant carbonized wood can be found in a 5-ft (1.5 m) interval below the interbedded section. Below this, in sharp contact, is a 20-ft (6 m) section of dark gray, finely laminated, clay and silt overbank deposits.

Plant fossils and palynological remains in the Raritan indicate a Cenomanian to Turonian (Washita to Eagle Ford) age. Marine fossils (planktonic foraminifera, coccoliths) of a similar age occur in the Bass River marine facies in New Jersey Coastal Plain Wells.

Stop 2. Magothy Formation overlain by Pleistocene? sediments. The Magothy represents estuarine sediments of depositional sequence KS1. This exposure represents an intertidal sequence of tidal delta sands and lagoonal clays.

The Pleistocene sediments consist of 1.5 ft (0.5 m) of sand and gravel. The Magothy Formation consists of an upper 29 ft (9 m) of tidal delta deposits, which contain dark gray sands and clays with carbonaceous-rich layers, uniformly cross-bedded sands, intermixed flaser bedding, and layers of rip-up clasts. Below this, 16 ft (5 m) of lagoonal deposits occur. They consist of dark gray laminated clay, dark gray alternating sands and clays, and carbonaceous-rich layers.

Palynological studies indicate that the Magothy is Santonian in age. This suggests that the Coniacian is absent due to a disconformity. This is confirmed in the subsurface by the absence of marine assemblages of Coniacian age.

Stop 3. Magothy, Merchantville, and Woodbury formations of depositional sequences KS1 and KC1, respectively. The Merchantville glauconitic sediments were deposited during a major rise in sea level, a rise that established marine environments over the entire coastal plain for the first time. The Woodbury is a regressive facies of sequence KC1.

The Woodbury Formation, which is 5 to 10 ft (1.5 to 3 m) thick, is a dark gray micaceous finely bedded clayey silt. Molluscs and foraminifera indicate that the Woodbury was deposited in an inner to middle shelf environment.

The Merchantville Formation is a dark gray to greenish black, uniformly bedded sandy silt and clay to sandy clay. Glauconite, mica, and siderite are abundant throughout the 12 ft (4 m) of outcrop. Many molds of molluscs are also present in the outcrop. The Merchantville was deposited in a middle to outer shelf environment.

The Magothy Formation consists of 8 ft (2.4 m) of the Cliffwood Member. It is a dark gray alternating fine sand, silt, and silty clay. Rare *Ophiomorpha* burrows indicate a probable shoreface environment of deposition.

Ammonite species (*Scaphites hippocrepis* and others) found in the Merchantville in outcrop date this formation as early Campanian (Taylor) in age. In the subsurface marine assemblages (planktonic foraminifera) of Santonian (Austin) to early Campanian age are present in the Merchantville.

Stop 4. Depositional sequences KM1 and KC2. The base of section I is exposed in the lower excavation (Stop 4A) and the

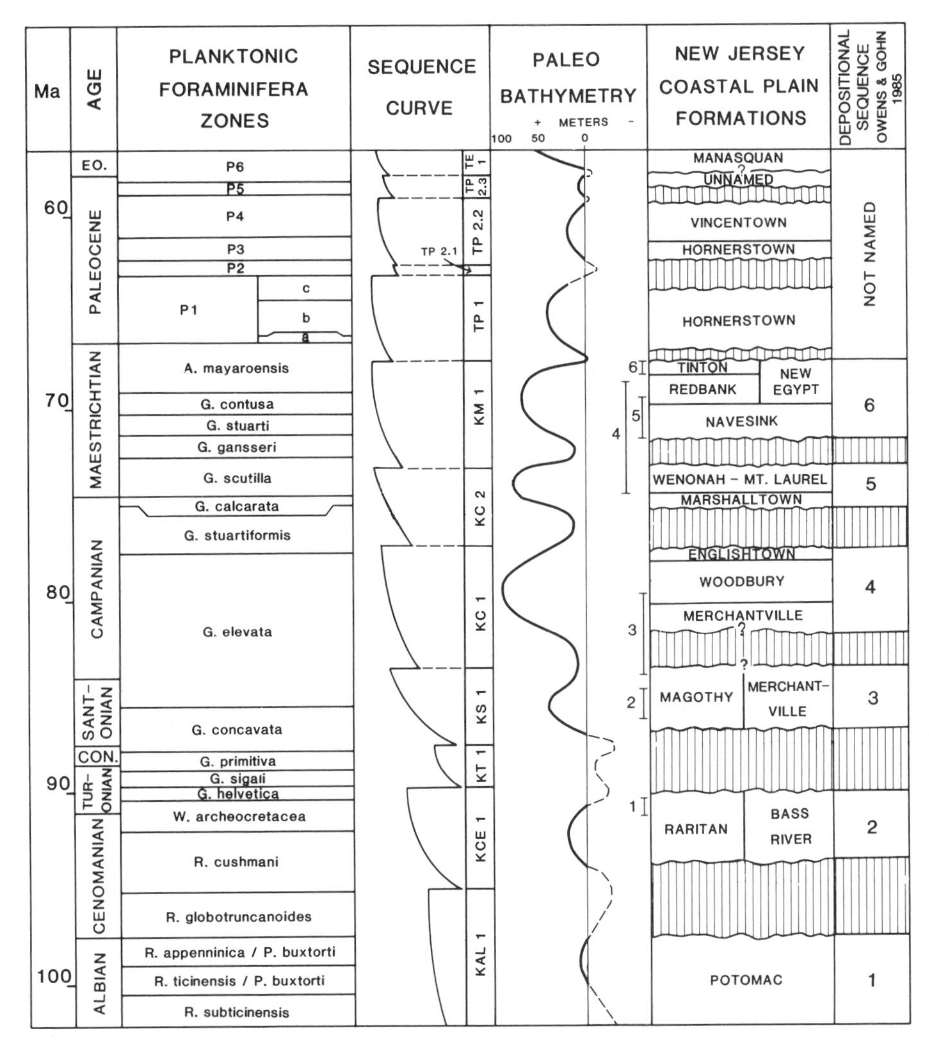

Figure 2. Upper Cretaceous and Paleocene depositional sequences in New Jersey Coastal Plain (after Olsson, 1986). The sequence designations KAL1, KCE1, . . . refer to Cretaceous Albian sequence 1, Cretaceous Cenomanian sequence 1, etc. Vertical numbered bars refer to the stops.

upper part is exposed in the higher excavation behind the gymnasium (Stop 4B). The sedimentary units at this section and the next, exhibit the typical glauconitic (transgressive) and sandy (regressive) facies of depositional sequences in the coastal plain.

The lower section at Stop 4A displays depositional sequences KM1 and KC2. The section begins with the Wenonah Formation (8 ft; 2.4 m) at the base succeeded upward by the Mount Laurel (21 to 27 ft; 6.4 to 8.2 m), Navesink (26 ft; 7.9 m), and Redbank (15 ft; 4.6 m) formations. The Wenonah Formation represents an inner shelf regressive facies of depositional sequence KC2. The Wenonah consists of a gray clayey, slightly glauconitic, micaceous, fine quartz sand, which is burrow-mottled with indistinguishable bedding. Occasional large clay-filled subvertical burrows (*Asterosoma*) and *Zoophycus* are present.

The Mount Laurel Sand was deposited during a sea-level fall that ended the KC2 cycle of deposition. The lower 10 ft (3.0 m) of the Mount Laurel was deposited in a transitional zone from offshore (innermost shelf) to shoreface (just below the surf zone). The interval from 10 to 20 ft (3.0 to 6.1 m) above the base represents lower shoreface deposition. The Mount Laurel is a thin- to medium-bedded light gray to white, fine to medium sand with thin chocolate brown silt and clay layers. The sand is well sorted, angular, and slightly glauconitic with dark micaceous laminae. Tabular planar cross-bedding is common in the sand beds. The beds range in thickness from 2 to 3 in (5 to 8 cm) at the base of the section to 4 to 6 in (10 to 15 cm) at the top. Maximum thickness is 15 in (38 cm). Common broad shallow trough cross-bedding suggests shoreward migration of lunate megaripples. The paleocurrent direction is bimodal, being northwest and southeast. The predominate direction is to the northwest (onshore). Vertical cylindrical burrows of the trace fossil *Ophiomorpha* and cylindrical to rod-shaped burrows of the trace fossil *Asterosoma* are present in the lower 20 ft (6 m) of the section. From about 15 to 20 ft (4.6 to 6.1 m) above the base, large robust vertical cylindrical and ellipsoidal horizontal burrows of *Ophiomorpha* predominate. *Thalassinoides* also occurs within this interval.

The Navesink Formation was deposited during sea level rise in the KM1 cycle. It shows the typical transgressive glauconitic shelf facies of the Cretaceous of New Jersey. Paleobathymetric indicators (various fossil groups and sediment characteristics) suggest deposition under middle-shelf environments. The Navesink is a dark gray clayey, silty, glauconite sand that is burrow-mottled. Contact with the overlying Sandy Hook is gradational. The Navesink contains abundant molds of megafossils and rounded pebbles. Megafossils include large bivalves, gastropods, and phragmacones of belemites. The basal 6 ft (1.8 m) is interpreted as a lag deposit formed during the Navesink transgression. It consists of poorly sorted bioturbated clayey sand containing rounded pebbles, abraded fossil molds, and glauconite. Glauconite infills burrows in the upper 3 ft (1 m).

The Sandy Hook Member of the Redbank Formation, which was deposited in an inner-shelf environment, is the initial regressive facies of the KM1 depositional sequence. The Sandy Hook is a dark gray, very micaceous, argillaceous, feldspathic, fine quartz sand. It is thick-bedded with a mottled texture, but it also contains disrupted sandy laminae and occasional small-scale cross-laminae. Sand-filled burrow tubes and the trace fossil *Zoophycus* are scattered throughout the exposure.

The upper section at Stop 4B displays the regressive Redbank Formation in depositional sequence KM1. The basal part of the section is exposed behind the Ern Construction Company, and the upper part is exposed above and just beyond in a long excavation. The section includes the lower Sandy Hook Member (15 ft; 4.6 m) and the upper Shrewsbury Member (25+ ft; 7.6+ m). The Sandy Hook Member is a dark gray, very micaceous, argillaceous, feldspathic, fine quartz sand. It has a mottled texture with sand-filled burrow tubes and the trace fossil *Zoophycus.* The large light-dark mottled shapes are caused by weathering-related permeability. The trace fossil assemblage includes vertical cylindrical and horizontal ellipsoidal *Ophiomorpha,* rod-shaped *Asterosoma,* and *Chondrites.* The Shrewsbury Member consists of light gray to white, micaceous, feldspathic, mostly well-sorted fine to medium quartz sand. The lower 10 ft (3 m) is slightly silty. Occasional thin clay horizons define inclined bedding, which represents foreset beds of a prograding sand body. Planar cross-stratification occurs within the foreset beds. Both dip to the southwest.

Stop 5. Poricy Park Fossil Beds. Navesink Formation (15 ft; 4.5 m) overlain by 25 ft (7.6 m) of the Redbank Formation (Sandy Hook Member). In contrast to the Navesink at Stop 4, the formation at this locality is richly fossiliferous with skeletal material. It is a classic collecting locality. The Sandy Hook is a dark gray, micaceous, feldspathic, fine to medium quartz sand. It is glauconitic in the basal part, which contains microfossils and small megafossils. The Navesink Formation consists of greenish black, clayey glauconite sand. Several shell layers are present. The oysters *Exogyra, Pycnodonte,* and *Ostrea*; the brachiopod *Choristothyris*; and the belemnite *Belemnitella* are well preserved and common. Molds of various molluscs are common and an extensive well-preserved epibiont bryozoan fauna is present. Microfossils include foraminifera, ostracodes, coccoliths, and dinoflagellates.

Stop 6. At Tinton Falls. This is the type locality of the Tinton Formation, the only indurated unit within the Upper Cretaceous section and it represents the uppermost part of depositional sequence KM1. The formation (22 ft; 6.7 m) is very limited in its geographic extent, disappearing within a short distance along strike to the southwest and in a downdip direction. The Tinton represents an inner shelf facies related to the Redbank regressive facies. It is separated from the overlying Hornerstown Formation (depositional sequence TP1) by an unconformity. The Tinton consists of brownish green, argillaceous, quartz, and glauconite sandstone interbedded with layers and lenses of gray claystone. Molds of megafossils of gastropods and pelecypods are common. Rare specimens of the ammonite *Sphenodiscus* occur. A well-developed dinoflagellate flora is also present.

REFERENCES CITED

Minard, J. P., 1969, Geology of the Sandy Hook Quadrangle in Monmouth County, New Jersey: U.S. Geological Survey Bulletin 1276, 43 p.

Mitchum, R. M., Jr., Vail, P. R., and Thompson, S., III, 1977, The depositional sequence as a basic unit for stratigraphic analysis, *in* Payton, C. E., ed., Seismic stratigraphy; Applications to hydrocarbon exploration: American Association Petroleum Geologists Memoir 26, p. 53–62.

Owens, J. P., and Gohn, G. S., 1985, Depositional history of the Cretaceous Series in the U.S. Atlantic Coastal Plain; Stratigraphy, paleoenvironments, and tectonic controls of sedimentation, *in* Poag, C. W., ed., Geological evolution of the United States Atlantic margin, New York, Van Nostrand Reinhold, p. 25–86.

Owens, J. P., and Sohl, N. F., 1969, Shelf and deltaic paleoenvironments in the Cretaceous-Tertiary formations of the New Jersey Coastal Plain, *in* Subitsky, S., ed., Geology of selected areas in New Jersey and eastern Pennsylvania and guidebook of excursions: New Brunswick, New Jersey, Rutgers University Press, p. 235–278.

Owens, J. P., Minard, J. P., Sohl, N. F., and Mello, J., 1970, Stratigraphy of the outcropping of post-Magothy Upper Cretaceous formations in New Jersey and northern Delmarva Peninsula, Delaware and Maryland: U.S. Geological Survey Professional Paper 674, 60 p.

Owens, J. P., Sohl, N. F., and Minard, J. P., 1977, A field guide to Cretaceous and lower Tertiary beds of the Raritan and Salisbury embayments, New Jersey, Delaware and Maryland: Washington, D.C., American Association Petroleum Geologists—Society of Economic Paleontologists and Mineralogists, 113 p.

Olsson, R. K., 1980, The New Jersey Coastal Plain and its relationship with the Baltimore Canyon Trough: New York Geological Association 52nd Annual Meeting, Guidebook, p. 116–129.

—— , 1986, Foraminiferal modeling of sea-level change in the Late Cretaceous of New Jersey, *in* Sea level changes; An integrated approach: Society of Economic Paleontologists and Mineralogists Special publication (in press).

Olsson, R. K., Gibson, T. G., Hansen, H. J., and Owens, J. P., 1986, Geology of the northern Atlantic Coastal Plain; Long Island to Virginia, *in* Grow, J. A., and Sheridan, R. E., eds., The Atlantic continental margin: Boulder, Colorado, Geological Society of America, Geology of North America, v. I-2 (in press).

Vail, P. R., Mitchum, R. M., Jr., and Thompson, S., III, 1977, Seismic stratigraphy and global changes of sea level; Part 3, Relative changes of sea level from coastal onlap, *in* Payton, C. E., ed., Seismic stratigraphy; Applications to hydrocarbon exploration: American Association Petroleum Geologists Memoir 26, p. 63–81.

Weller, S., 1907, A report on the Cretaceous paleontology of New Jersey: New Jersey Geological Survey, Paleontology series, v. 4, 871 p.

The Palisades Sill and Watchung basalt flows, northern New Jersey

John H. Puffer, Geology Department, Rutgers University, Newark, New Jersey 07102

LOCATION

The early Jurassic Palisades Sill and the three Watchung Basalts (the Orange Mountain, Preakness, and Hook Mountain Basalts) of northeastern New Jersey are best examined at the nine stops listed below and shown in Figure 1.

Stop 1. Palisades Sill at Ross Dock, Palisades State Park. Road cut through park, minor traffic, large parking lot, picnic tables. From I-95 eastbound, take Exit 67—Fort Lee (last exit in New Jersey), and continue east toward the Hudson River on Bridge Plaza. Turn right onto River Road, proceed 0.4 mi (0.6 km) then turn left into entrance of Palisades Park and follow the park road 0.9 mi (1.4 km) east and north under the George Washington Bridge. At the first traffic circle, turn to the right and down the hill, to the parking lot at Ross Dock.

Stop 2. Palisades Sill at Nyack Beach State Park. Foot path along Hudson River escarpment, large parking lot, picnic tables. From downtown Nyack, New York, proceed 2.0 mi (3.2 km) north on Broadway along the Hudson River. Turn right into the parking lot of Nyack Beach State Park.

Stop 3. Ladentown Basalt, Ladentown, New York. Exposures along major highway, heavy traffic. From New Jersey 306 northbound, turn left onto New Jersey 202 and proceed 1.0 mi (1.6 km) southwest. The Ladentown Basalt is exposed along New Jersey 202 on the left, directly across from a fire hydrant.

Stop 4. Orange Mountain Basalt and Preakness Basalt (Type Sections). Road cuts along major interstate, very heavy traffic, dangerous and noisy, no parking. From I-280 westbound, at the Garden State Parkway interchange, proceed 3.2 mi (5.1 km) west. Extensive 1-mi-long (1.6 km) exposures of Orange Mountain Basalt (Stop 4a) are located on both sides of I-280. A 0.3 mi (0.5 km) long exposure of Preakness Basalt (Stop 4b) is located 0.8 mi (1.3 km) west of the Orange Mountain exposures.

Stop 5. Hook Mountain Basalt at Walter Kidde Dinosaur Park. From I-280 westbound take Exit 4A onto Eisenhower Parkway, proceed 1.1 mi (1.9 km) south, then turn left onto Beaufort Avenue; proceed another 0.4 mi (0.6 km) and turn left again into entrance road for Riker Hill Park. Follow the park road up hill and follow signs to Geology Museum.

Stop 6. Orange Mountain Basalt at Great Falls, Paterson, New Jersey. Falls escarpment at park in downtown Paterson, New Jersey. From I-80 westbound, take the downtown Paterson Exit (Exit 57); turn left onto Grand Street. Proceed 0.3 mi (0.5 km) on Grand Street, then turn right onto Spruce Street at the light. Proceed 0.5 mi (0.8 km) on Spruce Street, then turn right onto Market Street, and immediately turn right into the Paterson Museum parking lot.

Stop 7. Preakness Basalt at High Mountain, North Haledon, New Jersey. Alpine meadow at end of gravel road (four-wheel drive). From New Jersey 208 southbound through Wyckoff, New Jersey, turn right onto Russell Avenue. Proceed 0.6 mi (1 km) on Russel Avenue, then turn left onto Sicomac Road and proceed four blocks. Turn right onto Mountain Avenue, proceed 1.4 mi (2.2 km), then turn left (south) onto High Mountain Road. Bear right at the fork on High Mountain Road, then turn right onto Glen Plaza, proceed 0.2 mi (0.3 km), and park at the base of the jeep trail. Hike or take four-wheel drive vehicle up the 0.6 mi (1 km) gravel road to the summit of High Mountain in the High Mountain Wilderness Park.

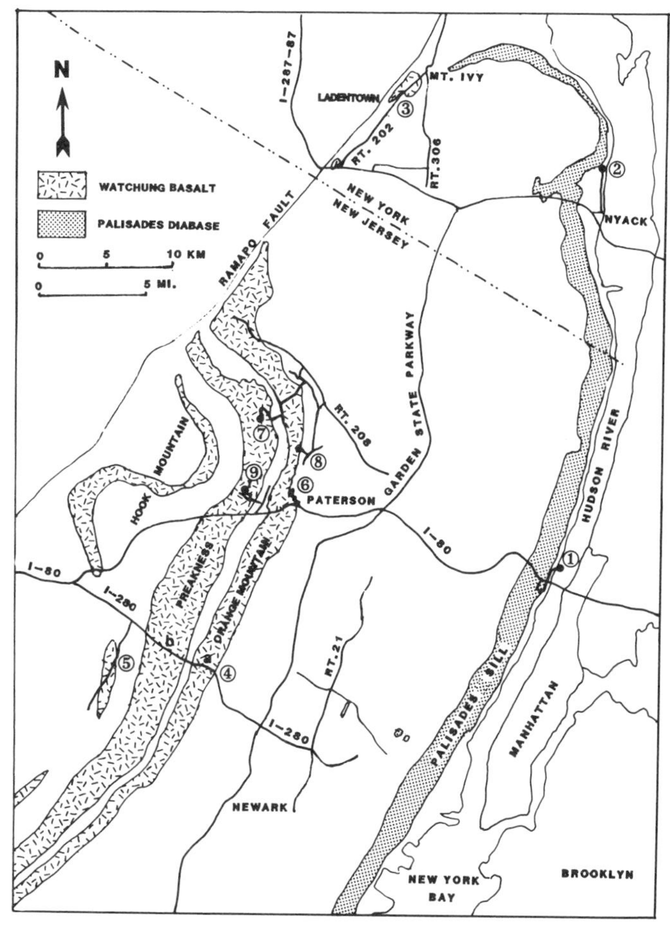

Figure 1. Geologic map of Watchung basalt and Palisades diabase within the northeastern New Jersey and southeastern New York area showing the distribution of localities 1-10.

Stop 8. Orange Mountain pillow basalt at Prospect Park trap-rock quarry, Prospect Park. Active quarry, access by permission only. From New Jersey 208 southbound through Hawthorne, New Jersey, turn right onto Goffle Road. Proceed 2.1 mi (3.4 km) on Goffle Road, then turn right at fork in road onto North 8th Street. Proceed 0.5 mi (0.8 km) on North 8th Street, then turn right onto Planten Road and drive up hill to the entrance of the Prospect Park Quarry.

Stop 9. Preakness pillow basalt at Totowa, New Jersey. Road along suburban street, minor traffic. From I-80 westbound, take Exit 55 onto Union Avenue. Proceed 0.6 mi (1 km) north on Union Avenue, then turn left onto Crews Street at Washington Park School and merge with Totowa Avenue. Turn right onto Greene Avenue, then turn left onto Claremont Avenue. Proceed 0.2 mi (0.3 km) north on Claremont, then park along the exposure of Preakness Basalt on the west side of Claremont Avenue.

SIGNIFICANCE

The Palisades diabase sill and the three Watchung basalt flows record an early rifting, pre-spreading tectonic stage. The Palisades-Watchung igneous activity was preceeded and followed by continental red-bed sedimentation. The igneous activity of the Newark basin, New Jersey is part of a widespread network of early Jurassic eastern North American basaltic rock exposures extending from Nova Scotia to Georgia. The Palisades-Watchung rock is light REE enriched (Kay and Hubbard, 1978) and is compositionally very similar to the Karroo basalts of Basutoland (Cox and Hornung, 1966) and some other "continental" quartz tholeiites. Most of the Palisades-Watchung rocks are good examples of the high-Ti type of eastern North American quartz tholeiite (Weigand and Ragland, 1970) and the high-Fe type that fractionated out of high-Ti magma (Ragland and Whittington, 1983). The second of the three Watchung basalts, however, may have fractionated out of a low-Ti magma type (Puffer and Philpotts, 1986). Recent research activity has been focusing on determining magma sources and fractionation patterns of early Jurassic eastern North American basaltic magmas (Gottfried and others, 1983; Ragland and Whittington, 1983; Philpotts and Martello, 1986; Puffer and Philpotts, 1986), and deciphering the tectonic setting and structural controls on magma emplacement (deBoer and others, 1987; Manspeizer and others, 1987; McHone and Butler, 1984; and Whittington and Cummins, 1986).

The Palisades Diabase

The Palisades diabase intruded 186 to 192 m.y. ago, based on the Ar^{40}-Ar^{39} dates of Dallmeyer (1975). The Palisades magma intruded as a sill throughout most of its exposed length but is somewhat discordant north of Nyack, New York. Contact metamorphic effects on the Lockatong Argillite and Passaic Formation host-rocks, including pervasive hornfels development have been described by Van Houten (1969). Some local melting of Lockatong hornfels and the generation of some pyroxene trondhjemite has been described by Benimoff and Sclar (1984).

The Palisades intrusion has been interpreted by Walker (1969) as having involved two pulses of magma but is interpreted by Shirley and Warren (1985) as having involved three separate magma pulses, each characterized by chemical distinctions.

The Watchung Basalts

Paleontological, stratigraphic, and geochemical evidence, (Olsen, 1984; Puffer and others, 1981) indicate that the three Watchung basalts extruded in rapid succession during the Hettangian age of the early Jurassic period. Olsen (1984) suggests that the time interval separating the stratigraphically equivalent flows of the Hartford basin, Connecticut was only 475,000 years. Subsequent burial has apparently induced zeolite facies metamorphic effects (Puffer and Laskovich, 1984) that may have included mobilization of copper mineralizing solutions (Puffer, 1984). Several contrasting fractionation patterns relating the three Watchung basalts have been proposed, but there is a general consensus that the uppermost flow (the Hook Mountain basalt) is not a fractionation product of the central unit (the Preakness basalt). The chemical composition of the first flow unit (the Orange Mountain) is equivalent to the chill zone of the Palisades sill, and the composition of the uppermost flow unit closely resembles some of the fractionated interior portion of the Palisades. The second flow unit, however, either fractionated out of Orange Mountain magma as suggested by Puffer and others (1981) and Philpotts and Reichenbach (1985), or it fractionated out of a low-Ti type magma as suggested by Puffer and Philpotts (1986).

SITE INFORMATION

Palisades Sill and Ladentown basalt (Stops 1, 2, and 3)

The Palisades diabase was intruded largely as a composite sill throughout most of its length, but it is discordant near its northern end, particularly northwest of Nyack, New York (Fig. 1). The Palisades magma probably broke through to the surface near Mount Ivy, New York and flowed south during early Jurassic time to form the Ladentown basalt (Stop 3).

Stop 1. Lower contact of Palisades sill with Lockatong hornfels at Ross Dock. Walk south from the parking lot along the road to the boat launch, then up the stone steps, through the tunnel under the road, and out onto road level. Walk north along the contact of the lower chill zone with the Lockatong hornfels.

The fine-grained diabase chill zone consists of about 1% olivine (Fo_{80}), 49% plagioclase (An_{64}), and 50% pyroxene (augite and hypersthene). Tholeiitic-trend fractionation of the diabase magma culminated in the development of a granophyre layer about 100 ft (33 m) thick near the upper contact of the 1,100 ft (330 m)-thick sill. Some thin white veins of late quartzofeldspathic granophyre cut the diabase along the road.

A 10 to 13 ft (3 to 4 m) thick olivine enriched zone is exposed 52 ft (16 m) above the road level and is exposed at road level 0.1 mi (0.2 km) east of the entrance to Palisades Park. The olivine zone was an important part of Palisades fractionation series and has been attributed by some to be the result of gravitational crystal settling. Walker (1969), however, has shown that most of the olivine in the olivine zone is too iron-rich (Fo_{65}) to have formed as an early cumulate phase and too fine-grained to have settled very far through the viscous magma. Instead, most of the olivine probably formed by fractional crystallization at the base of a second pulse of magma that intruded into the composite sill.

While walking north along the road from the stone path, observe metamorphosed buff arkose exposed 0 to 250 ft (0 to 75 m) and 600 to 800 ft (180 to 250 m) north of the chill-zone contact, and see the metamorphosed platey and laminated siltstone 250 to 600 ft (75 to 180 m) north of the contact. These meta-sediments have been described by Olsen (1980) and correlate with other Lockatong exposures to the south. The pelitic layers have been particularly affected by contact metamorphism and have been converted into black hornfels consisting of biotite and albite with minor analcime, diopside, and calcite, or to green hornfels consisting of diopside, grossularite, chlorite, and calcite with minor biotite, feldspar, amphibole, and prehnite (Van Houten, 1969).

Porphyroblasts of pinite after cordierite commonly occur as small green spots in the black biotite-albite hornfels (Miller and Puffer, 1972). Large tourmaline porphyroblasts and even larger green spherical structures up to 1.5 in (4 cm) across composed largely of clinozoisite, are less common in the hornfels but have been found in several contact-metamorphosed rocks located throughout the Newark and Culpeper basins.

Stop 2. Palisades Sill at Nyack Beach State Park. From the parking lot next to the snack bar, walk north on the trail along the Hudson River. The Palisades diabase intruded into the Passaic Formation, which overlies the Lockatong Formation. The contact is approximately parallel to the Hudson River escarpment and is exposed about 30 ft (10 m) above the level of the trail. Bright reddish-brown fluvial deposits of micaceous mudstone interbedded with arkose displaying cross-bedding, channels, and burrows are well exposed along the trail at the base of the spectacular escarpment formed by the Palisades sill. The Palisades diabase along the southern part of the trail can be examined by climbing the lower slope of the escarpment; then 1.1 mi (1.85 km) north of the parking lot the lower contact is exposed at path level. Still farther north, the contact rises again above the underlying Passaic Formation. The sedimentary rocks of the Passaic Formation have not been metamorphosed beyond a few meters from the sill. The arkosic sandstones are particularly resistant to metamorphic effects unlike the shales of the Lockatong Formation (Stop 1), which have been pervasively metamorphosed.

The olivine-rich zone is not exposed anywhere along the Nyack Beach escarpment, which led Sanders (1974) to suggest that the olivine zone is displaced below the surface by a nearly vertical normal fault along the base of the sill. Close examination of the diabase near the contact, however, indicates a distinctly fine grain size that suggests it is a chill zone; there is also an absence of any recognizable evidence for a fault. The absence of the olivine zone is therefore probably a result of the discordant emplacement of the northern Palisades sill, an absence of a gravitational field directed perpendicular to a horizontal intrusive floor, and the relatively shallow and therefore rapid crystallization of a solid front at a faster rate than diffusion and convection rates.

Stop 3. Ladentown Basalt at Ladentown, New York. The closely-spaced curved cooling columns displayed by the Ladentown basalt along New Jersey 202 contrast with the thick massive columns typical of the Palisades sill that underlies the basalt and probably was its source of magma. The basalt exposed along New Jersey 202 is fine grained, contains large plagioclase phenocrysts, and is slightly vesicular. Chemical analyses of the Ladentown Basalt (Puffer and others, 1982) compare closely (particularly the Cu and Sr content) with the fractionated interior (second magma pulse) of the Palisades sill. The Ladentown Basalt, therefore, appears to have extruded onto sediments of the Passaic Formation as the second magma pulse of the Palisades sill was emplaced, before the Feltville Formation was deposited, and before the Preakness Basalt was extruded. A physical connection between the Ladentown flow and the western end of the Palisades sill at Mount Ivy is suggested by magnetic (Kodama and others, 1981) and gravity data (Koutsomitis, 1980) but is not visible at the surface.

Good exposures of coarse boulder conglomerates are exposed another 0.4 mi (0.6 km) south along New Jersey 202. The coarse clast size of the sediment coincides with close proximity to the western border fault in the valley parallel to New Jersey 202. Note the distinct change in the topography on the opposite side of the border fault.

Watchung Basalt Flows (Stops 4–9)

The basalt flows of the Newark Basin are exposed as three N-E trending ridges. The three basalt units dip to the west at about 15° and were known as the First, Second, and Third Watchungs but they were recently renamed by Olsen (1980) as the Orange Mountain, Preakness, and Hook Mountain basalts respectively. One deeply altered exposure of basalt near the Ramapo Fault may represent a flow younger than the Hook Mountain Basalt but preliminary chemical analyses by Puffer suggests that it is probably a displaced fault slice of Hook Mountain or possibly Preakness Basalt.

Stop 4. Type section of Orange Mountain and Preakness Basalt Flows, West Orange, New Jersey. The type section of the Orange Mountain Basalt (Stop 4a) displays the westward dipping lower flow unit and a structural sequence within the flow comparable to that of the Giant's Causeway of Northern Ireland, as described by Tomkeieff (1940). Tomkeieff's use of the classical architectural terms "colonnade" (a series of columns) and "entablature" (a series of arches) to describe the joint patterns also apply to the Watchung basalts. The lower colonnade is exposed as a fine-grained columnar jointed, but otherwise massive, layer about 20 ft (6 m) thick. The overlying entablature is a medium- to fine-grained layer, about 115 ft (35 m) thick, characterized by radiating and curved joints. The joints of the entablature are characteristically more closely spaced than those of the colonnades. Distinct vertical left-lateral faults and low-angle faults with little apparent effect commonly cut through the entablature.

The upper colonnade is the uppermost layer of the lower flow unit exposed along I-280. The top of the upper colonnade contains a frothy amygdular crust that was weathered prior to the deposition of less than 3 ft (1 m) of red volcanoclastic sediment and before the upper flow unit was extruded.

Only about half of the total thickness of the Orange Mountain Basalt is exposed along I-280. The upper flow unit is best exposed at Site 8 (Fig. 1). The combined thickness of the two flow units is about 410 ft (125 m) along I-280 but varies from about 300 to 600 ft (100 to 200 m) throughout the Watchung Syncline (Olsen, 1980). Chemical analyses of samples from along the I-280 roadcut indicate no discernible differences in the composition of the three Tomkeieff structural portions of the lower flow unit (Puffer and Lechler, 1980). The petrology of the Orange Mountain Basalt is indistinguishable from that of the Talcott Basalt of Connecticut as described by Philpotts and Reichenbach (1985).

The type section of the Preakness Basalt along New Jersey 280 (Stop 4b) exposes almost the entire, unusually thick, lower flow unit, and a nearly complete Tomkeieff structural sequence. The lower colonnade is exposed as a medium-to-fine–grained and massive layer that is about 15 ft (5 m) thick. The entablature is an unusually coarse-grained central layer, about 260 ft (80 m) thick, characterized by closely-spaced columnar joints. The joints of the Preakness Basalt are more closely spaced and are less commonly curved or radiating than those of the Orange Mountain Basalt. The splintery nature of the basalt is a characteristic of the Preakness. The overlying two flows of the Preakness Basalt are separated by thin red layers of siltstone but are not exposed along I-280. The combined thickness of the Preakness flows is about 700 ft (215 m).

The petrology of the Preakness Basalt is indistinguishable from that of the Holyoke Basalt of Connecticut. Philpotts (1979) found good evidence of silicate liquid immiscibility in the Rattlesnake Basalt (an outlier of Holyoke Basalt) that takes the form of iron-rich glass. The immiscible globules are absent from the lowest 18 ft (5.5 m) of the flow but are abundant in the basalt above 30 ft (9 m) from the base. Similar globules are found in the entablature of the Preakness Basalt but are absent from the lower colonnade. The development of an immiscible liquid phase is a characteristic of the entablature that apparently crystallized under lower oxygen pressures than the lower colonnade. Philpotts and Doyle (1983) have shown that the lower colonnade crystallized at oxygen pressures close to that of the NNO buffer, unlike the upper portion that crystallized near the QFM buffer.

Site 5. Hook Mountain Basalt at Walter Kidde Dinosaur Park, Roseland, New Jersey. Take the path from the Geology Museum parking lot over the crest of Riker Hill, past the abandoned NIKE missile silo, down through the woods into Dinosaur Park. Jurassic red-beds are exposed along the lower contact of the Hook Mountain Basalt. The red-beds consist of fining upwards cycles of Towaco Formation. Olsen (1980) has described reptile footprints, coprolites, fish scales, calcareous concretions, chert nodules, salt casts, rill marks, rain-drop impressions, and a black coaly siltstone layer in the Towaco sediments of Dinosaur Park. A badly weathered volcanoclastic layer is exposed near the base of the Hook Mountain basalt, and some bleaching and evidence of low-grade thermal metamorphism is seen at the lower contact.

In contrast to the Preakness and Orange Mountain basalts, the Hook Mountain Basalt is relatively enriched in amygdules and vesicles. The high stratigraphic position of the Hook Mountain Basalt compared with the underlying basalts has resulted in less pervasive zeolite facies mineralization, but prehnite is abundant at this site and is easily collected, particularly near the north end of the exposure.

Pipe vesicles are common near the base of the flow and have been interpreted by Manspeizer (1980) as indicating southerly flow in contrast to the northeasterly flow of the underlying basalt units. The Hook Mountain exposure displays columnar joints, but a well defined Tomkeieff structured sequence is not apparent. The basalt averages 300 ft (91 m) in thickness (Faust, 1975) and consists of at least two flows.

Stop 6. Orange Mountain Basalt at Great Falls, Paterson, New Jersey. At the exposure along the north edge of the parking lot by the statue of Alexander Hamilton, observe the lower contact of the Orange Mountain Basalt with the upper Passaic Formation. Southwest plunging pipe-amygdules and vesicles are exposed in the basalt near the contact, but consistently occur within the basalt above the basal contact.

From the statue of Alexander Hamilton, view: (1) vertical strike-slip fault plains through the lower flow unit of the Orange mountain basalt; and (2) the contact between the lower colonnade and the overlying entablature. Glacial activity has removed the overlying friable pillowed flow unit. The "S.U.M." over the door of the historic building near the river at the base of the view area stands for "Society of Useful Manufacturers" an organization founded in 1789 to promote local trade. The hydroelectric plant operated from 1914 through 1968 and was closed in 1969 because of flood damage.

From the statue of Alexander Hamilton, walk north up Spruce Street to the path leading to the foot bridge over the falls. Cross the foot bridge to the park and observe the columnar jointing in the basalt and some large convex upward amygdules. Climb into the narrow notches in the basalt eroded along strike-slip faults for close inspection at Orange Mountain Basalt.

From the falls, walk south along Spruce Street, cross McBride Avenue, and continue one-half block to the Paterson Museum. Some of the best examples of the secondary minerals found in the Paterson area trap-rock quarries are on display at the museum. Prehnite (the State Mineral of New Jersey), heulandite, datolite, chabazite, stilbite, and amethest crystal clusters are particularly outstanding.

Stop 7. Preakness Mountain Basalt at High Mountain, North Haledon, New Jersey. High Mountain is the highest point (970 ft; 296 m) in the Preakness Mountain chain. The upper portion of the hiking or jeep trail to the top of High Mountain cuts through closely-spaced columnar-jointed basalt typical of the entablature of the Preakness Basalt. At the top of the mountain is a beautiful alpine meadow that on a clear day affords a spectacular view of most of northern New Jersey. The basalt exposed in the meadow is very coarse-grained; it is typical of Preakness Basalt and resembles diabase.

The view to the southeast includes the New York City skyline in the background behind the ridge formed by the Palisades Sill. (A copy of the 1981 Paterson 7½-minute Quadrangle is recommended.) The city of Paterson is in the foreground where the Passaic River cuts through the First Watchung Mountain ridge formed by the westward dipping Orange Mountain Basalt. The New Street trap-rock quarries and Garrett Mountain are clearly visible just south of downtown Paterson. The Prospect Park trap-rock quarry is clearly visible just north of Paterson. The view directly to the south includes the city of Newark in the distant background on the other side of the First Watchung Mountain, Montclair State College southeast of Paterson, and Paterson State College in the foreground. The view to the southwest includes the Precambrian New Jersey Highlands province west of the Ramapo Fault scarp in the background and the curved "inverted S" shaped Hook Mountain ridge formed by the Hook Mountain Basalt in the foreground.

Stop 8. Prospect Park Trap-rock Quarry (also known as Vandermade and Sowerbutt Quarry). Several large trap-rock quarries occur throughout the Paterson area, some of which are currently in operation. Although permission to enter the quarries is difficult to obtain, the quarries constitute some of the best exposures of Watchung Basalt. The Prospect Park Quarry has been in operation since 1901 (Peters, 1984). Recent mining has removed much of the upper flow unit of the Orange Mountain Basalt, but portions of the upper flow unit still remain. The upper flow unit is made up of (1) columnar basalt, (2) subaqueous flow lobes, and (3) bedded pillow lava (Manspeizer, 1980).

Some of the upper flow unit formed as massive and columnar basalt, particularly as thick lenses, close to the source of volcanic activity and as the central portion of flow channels or as the proximal end of flow lobes. These lenses display each of the three Tomkeieff structures more typical of the lower flow unit. The subaqueous flow lobes of the Orange Mountain Basalt overlie the vesicular upper surface of the lower flow unit and are typically composed of three zones (Manspeizer, 1980): (1) a proximal zone of massive and columnar basalt, (2) a transitional zone of columnar basalt with isolated pillows, and (3) a distal steeply dipping zone containing ellipsoidal flow lobe pillows and pahoehoe toes. Pahoehoe toes are characterized by an ellipsoidal cross section with an aspect ratio of 3 or 4, concentric structures, and common lava tubes resulting in central cavities that in the Prospect Park Quarry are typically mineralized with quartz and zeolites. A layer consisting of pillow lava overlying subaqueous flow lobes is exposed at the Prospect Park Quarry. Secondary minerals are typically abundant in the interstices between pillows but are not confined to the pillow lava.

A sequence of eight events is probably responsible for most of the mineralogical development of the basalts. (1) Primary igneous crystallization—plagioclase, pyroxene, olivine, and others. (2) Deuteric alteration—particularly boron later recrystallized as datolite. (3) Spilitization—secondary albite, chlorite, calcite, epidote, and chalcedony. (4) Diagenesis—widespread analcime, heulandite mineralization. (5) Zeolite facies Metamorphism. Zeolite facies mineral assemblages developed only where appropriate structural and chemical conditions existed. Coombs and others (1959) have observed that zeolite facies assemblages are not found in recent volcanic rocks and suggest that wherever such minerals are found in volcanic rocks they are probably the result of metamorphism. Armstrong and Besancon (1970) suggest that zeolite facies metamorphism is widespread throughout the Newark Supergroup from Nova Scotia through New Jersey and has affected K/Ar dating of the basalts. Hay (1966) suggests a laumintite-albite assemblage as representative of zeolite facies metamorphism. Laumontite and albite is common throughout the Prospect Park Quarry and probably equilibrated under zeolite facies conditions. (6) Prehnite-pumpellyite facies metamorphism. The assemblage prehnite-pumpellyite-albite is common throughout the Prospect Park Quarry and probably equilibrated under prehnite-pumpellynite facies conditions. Lausonite, however, has not been reported within the Paterson area suggesting that confining pressures did not exceed about 3 kb (Coombs and others, 1959). (7) Hydrothermal alteration. Some hydrothermal water driven out of underlying sediments and released by subsequent magmas probably complicated concurrent metamorphic activity. (8) Weathering—such as, clay after feldspar, fibrous rutile after ilmenite.

Stop 9. Preakness Pillow Basalt, Claremont Avenue, Totowa. The lower portion of a Preakness Mountain flow unit, the second of three Preakness Mountain flows, is exposed along the west side of Claremont Avenue. The base of the flow is a subaqueous flow lobe containing ellipsoidal flow lobe pillows and pahoehoe toes and is mineralized with calcite, quartz, and minor heulandite in small stretched amygdules. The pillowed base of the flow rests on a previously undescribed thin layer of red siltstone exposed at road level. The flow grades upward into massive columnar basalt. Further north on Claremont Avenue, the upper part of the flow is exposed. The upper part of the flow, in contrast to the bottom, is not pillowed, and contains large spherical amygdules mineralized with prehnite and pectolite.

REFERENCES CITED

Armstrong, R. L., and Besancon, J., 1970, A Triassic time scale dilemma; K-Ar dating of Upper Triassic mafic igneous rocks, eastern U.S.A. and Canada, and post-Upper Triassic plutons, western Idaho, U.S.A.: Eclogae Geologicas Helvetiae, v. 63, p. 15–28.

Benimoff, A. I., and Sclar, C. B., 1984, Coexisting silicic and mafic melts resulting from marginal fusion of a xenolith of Lockatong argillite in the Palisades Sill, Graniteville, Staten Island, New York: American Mineralogist, v. 69, p. 1005–1014.

Coombs, D. S., Ellis, A. J., Fyfe, W. S., and Taylor, A. M., 1959, The zeolite

facies, with comments on the interpretation of hydrothermal syntheses: Geochimica et Cosmochimica Acta, v. 17, p. 53–107.
Cox, K. G., and Hornung, G., 1966, The petrology of the Karroo basalts of Basutoland: American Mineralogist, v. 51, p. 1414–1432.
Dallmeyer, R. D., 1975, The Palisades Sill; A Jurassic intrusion? Evidence from $^{40}Ar/^{39}Ar$ incremental release ages: Geology, v. 3, p. 242–245.
deBoer, J. Z., McHone, J. G., Puffer, J. H., Ragland, P. C., and Whittington, D., 1987, Mesozoic and Cenozoic magmatism, *in* Hatcher, R. D., Jr., Thomas, W. A., and Viele, G. W., eds., The Appalachian/Ouachita regions; U.S.: Boulder, Colorado, Geological Society of America, The Geology of North America, v. F-2 (in press).
Faust, G. T., 1975, A review and interpretation of the geologic setting of the Watchung Basalt flows, New Jersey: U.S. Geological Survey Professional Paper 864-A, 40 p.
Gottfried, D., Annell, C. S., and Byerly, G. R., 1983, Geochemistry and tectonic significance of subsurface basalts from Charleston, South Carolina—Clubhouse Crossroads test holes #2 and #3, *in* Gohn, G. S., ed., Studies related to the Charleston, South Carolina earthquake of 1886: U.S. Geological Survey Professional Paper 1313, p. A1–A19.
Hay, R. L., 1966, Zeolite and zeolitic reactions in sedimentary rocks: Geological Society of America Special Paper 85, 130 p.
Kay, R. W., and Hubbard, N. J., 1978, Trace elements in ocean ridge basalts: Earth and Planetary Science Letters, v. 38, p. 95–116.
Kodama, K. P., Evans, R., and Davis, J. P., 1981, Magnetics and gravity evidence for a subsurface connection between the Palisades Sill and Ladentown basalts, Mount Ivy, New York: Geological Society of America Abstracts with Programs, v. 13, p. 141.
Koutsomitis, D., 1980, Gravity investigation of the northern Triassic-Jurassic, New York [M.S. thesis]: Newark, New Jersey, Rutgers University, 103 p.
Manspeizer, W., 1980, Rift tectonics inferred from volcanic and clastic structures, *in* Manspeizer, W., ed., Field studies of New Jersey geology and guide to field trips: Annual meeting of the New York State Geological Association, no. 52, p. 314–350.
Manspeizer, W., Costain, J. K., deBoer, J. Z., Froelich, A. J., McHone, J. G., Olsen, P. E., Prowell, D. C., and Puffer, J. H., 1987, Post-Paleozoic activity in the Appalachians, *in* Hatcher, R. D., Jr., Thomas, W. A., and Viele, G. W., eds., The Appalachian/Ouachita regions; U.S.: Boulder, Colorado, Geological Society of America, The Geology of North America, v. F-2 (in press).
McHone, J. G., and Butler, J. R., 1984, Mesozoic igneous provinces of New England and the opening of the North Atlantic Ocean: Geological Society of America Bulletin, v. 94, p. 757–765.
Miller, B. B., and Puffer, J. H., 1972, The cordierite zone of hornfels near the base of the Palisades Sill at Weehawken, New Jersey [abs.]: New Jersey Academy of Science Bulletin, v. 17, p. 46.
Olsen, P. E., 1980, Triassic and Jurassic formations of the Newark Basin, *in* Manspeizer, W., ed., Field studies of New Jersey geology and guide to field trips: Annual Meeting of the New York State Geological Association, no. 52, p. 2–39.
——, 1984, Periodicity of lake-level cycles in the Late Triassic Lockatong Formation of the Newark Basin (Newark Supergroup, New Jersey and Pennsylvania), *in* Berger and others, eds., Milankovitch and climate, N.A.T.O. Symposium: Dordrecht, Netherlands, D. Reidel, pt. 1, p. 129–146.
Peters, J. J., 1984, Triassic taprock minerals of New Jersey: Rocks and Minerals, v. 59, p. 157–183.
Philpotts, A. R., 1979, Silicate liquid immiscibility in tholeiites: Journal of Petrology, v. 20, p. 99–118.
Philpotts, A. R., and Doyle, C. D., 1983, Effect of magma oxidation state on the extent of silicate liquid immiscibility in a tholeiitic basalt: American Journal of Science, v. 283, p. 967–986.
Philpotts, A. R., and Martello, A., 1986, Diabase feeder dikes for the Mesozoic basalts in southern New England: American Journal of Science, v. 286, p. 105–126.
Philpotts, A. R., and Reichenbach, I., 1985, Differentiation of Mesozoic basalts of the Hartford basin, Connecticut: Geological Society of America Bulletin, v. 96, p. 1131–1139.
Puffer, J. H., 1984, Copper mineralization of the Newark Basin, *in* Puffer, J. H., ed., Igneous rocks of the Newark Basin; Petrology, mineralogy, ore deposits, and guide to field trip: Geological Association of New Jersey, First Annual Field Conference, p. 127–136.
Puffer, J. H., and Laskowich, C., 1984, Secondary mineralization of Paterson area taprock quarries, *in* Puffer, J. H., ed., Igneous rocks of the Newark Basin; Petrology, mineralogy, ore deposits, and guide to field trip: Geological Association of New Jersey, First Annual Field Conference, p. 103–126.
Puffer, J. H., and Lechler, P., 1980, Geochemical cross sections through the Watchung basalts of New Jersey: Geological Society of America Bulletin, v. 91, pt. 1, p. 7–10; pt. 2, p. 156–191.
Puffer, J. H., and Philpotts, 1986, Eastern North American quartz tholeiites, *in* Manspeizer, W., ed., Triassic-Jurassic rifting; North America and North Africa: American Association of Petroleum Geologists Memoir (in press).
Puffer, J. H., Hurtubise, D., Geiger, F., and Lechler, P., 1981, Chemical composition of Mesozoic basalt units of the Newark Basin, New Jersey, and the Hartford Basin, Connecticut: Geological Society of America Bulletin, v. 92, pt. 1, p. 155–159; pt. 2, p. 515–553.
Puffer, J. H., Geiger, F. J., and Camanno, E. J., 1982, Mesozoic igneous rocks of Rockland County, New York: Northeastern Geology, v. 4, p. 121–130.
Ragland, P. C., and Whittington, D., 1983, Early Mesozoic diabase dikes of eastern North America; Magma types: Geological Society of America Abstracts with Programs, v. 15, p. 666.
Sanders, J. E., 1974, Guidebook to field trip to Rockland County, New York: Petroleum Exploration Society of New York, 87 p.
Shirley, D. N., and Warren, P. H., 1985, Differentiation of the Palisades Sill: Geological Society of America Abstracts with Program, v. 17, p. 716.
Tomkeieff, S. E., 1940, The basalt lavas of the Giant's Causeway, District of northern Ireland: Bulletin of Volcanology, ser. 11, v. 61, p. 89–143.
Van Houten, F. B., 1969, Late Triassic Newark Group, north central New Jersey and adjacent Pennsylvania and New York, *in* Subitsky, S. S., ed., Geology of selected areas in New Jersey and Pennsylvania, and Guidebook: New Brunswick, New Jersey, Rutgers University Press, p. 314–347.
Walker, K. R., 1969, The Palisades Sill, New Jersey; A reinvestigation: Geological Society of America Special Paper 111, 178 p.
Weigand, P. W., and Ragland, P. C., 1970, Geochemistry of Mesozoic dolerite dikes from eastern North America: Contributions to Mineralogy and Petrology, v. 29, p. 194–214.
Whittington, D., and Cummins, L. E., 1986, Diabase of Virginia and South Carolina-II—Plumes, perches, and the calc-alkaline connection: Geological Society of America Abstracts with Programs, v. 18, p. 272.

Niagara Falls and Gorge, New York-Ontario

Carlton E. Brett, Department of Geological Sciences, University of Rochester, Rochester, New York 14627
Parker E. Calkin, Department of Geological Sciences, State University of New York, Buffalo, New York 14226

LOCATION

The gorge of the Niagara River extends northward along the U.S.-Canada boundary for 7.1 mi (11.4 km) from Niagara Falls (Niagara County, New York–Welland County, Ontario), to the Niagara Escarpment at Lewiston, New York–Queenston, Ontario (Fig. 1). Sites, described below, are located on the U.S.G.S. Niagara Falls and Lewiston 7½-minute quadrangles.

For purposes of this guide, stops have been selected along Niagara Gorge, from Lewiston southward to Niagara Falls. All locations, except for Stop 1, are readily accessible from either the Robert Moses Parkway (New York) or the Niagara Parkway (Ontario), and are well-known tourist areas with well-marked parking lots. For detailed road logs of these and other sites on both U.S. and Canadian sides of the Niagara Gorge, see Krajewski and Terasmae (1981) and Friedman and others (1982).

The following sections of this guide provide an overview of the two major geologic aspects of Niagara Gorge: (1) Pleistocene-Holocene geomorphology and geology, and (2) Paleozoic stratigraphy. Description of stops and their locations then follows.

SIGNIFICANCE

Niagara Falls is the world's most accessible and visited waterfall; the Niagara Gorge displays one of the best exposures of Paleozoic rocks in the Northeast. Two distinct aspects make the falls and gorge also a classic geologic locality: (a) glacial, geomorphic, and hydrologic phenomena that resulted from development of the Niagara River with its falls and gorge from a multi-outlet and partly glacier-fed river-lake system; and (b) exposure in the gorge walls of an internationally recognized stratigraphic sequence of uppermost Ordovician through middle Silurian bedrock.

SITE INFORMATION

History of Studies. Charles Lyell (1837, 1845) gave some of the first authoritative geological accounts on the evolution of the Niagara drainage, but the major early studies took place near the turn of the century (e.g., Gilbert, 1891, 1907; Grabau, 1901; Spencer, 1894, 1907; Kindle and Taylor, 1913). Recent examination of the Pleistocene-Holocene sediments along the Niagara River clarify the late Quaternary history of the gorge and surrounding Great Lakes (see Calkin and Brett, 1978; Calkin and Feenstra, 1985).

Paleontologic and stratigraphic studies of Silurian rocks in the Niagara vicinity stem from James Hall (1838, 1843, 1852), who also coined the name "Niagara Group." This has been applied in modified form (Niagaran Series) in North America to Lower and Middle Silurian rocks.

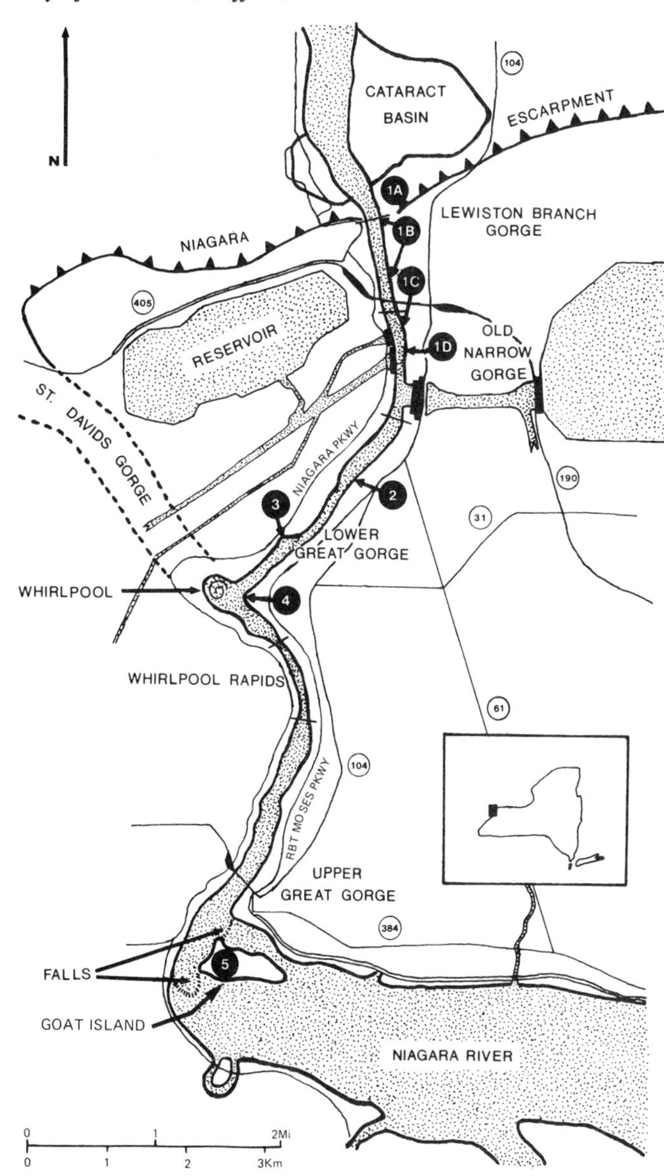

Figure 1. Location map of the Niagara Falls and gorge showing gorge segments and numbered stops described in the text. Data from Kindle and Taylor (1913), Calkin and Feenstra (1985), and Flint and Lalcama (1986).

Site 23-Geomorphologic Setting and Evolution of the Gorge. The Niagara River, unusual for its short length (32 mi; 51 km) and the immense storage capacity of its drainage basin, connects Lake Erie to Lake Ontario. From its head at the northeastern end of Lake Erie, at 571 ft (174 m) above mean sea level, the Niagara descends 75 ft (23 m) in 18.3 mi (31 km), in a shallow, broad span to the brink of the falls; from here, it drops vertically

174 ft (53 m) and then descends another 72 ft (22 m) over 6.8 mi (11 km) in the gorge to the Niagara Escarpment (Fisher, 1981). At the base of the Niagara Escarpment, the river broadens again and falls less than 3 ft (1 m) as it flows 5.6 mi (9 km) to Lake Ontario across the lake plain (Stop 1A). There it is bordered by lower bluffs of red Ordovician shale and late Pleistocene sediments. The mean natural flow of the Niagara, about 204,285 ft^3/sec (5,720 m^3/sec), is supplied largely by the excess discharge brought to Lake Erie from Lakes Superior, Michigan, and Huron (upper Great Lakes). The flow over the falls itself has been limited since 1905 by diversion for hydroelectric power generation; presently the river carries more than 50% of the natural flow during tourist hours and about 25% at other times (Calkin and Wilkinson, 1982).

The difference in elevation that caused the falls is related to the presence of the Niagara Escarpment (Stop 1A) with its cap of resistant Lockport dolostone; this divides the Lake Erie and Lake Ontario basins. The gorge of the Niagara has been cut southward from this escarpment into a section of jointed, gently south-dipping (¼° to ½°), Upper Ordovician through Middle Silurian rocks (Fig. 2).

Initiation of the Niagara River occurred at about 12,400 B.P., following the late Wisconsin, Port Huron stage of the Pleistocene, as the ice margin retreated northward from the Niagara Escarpment (Fig. 1) for the last time. Former channels through the Niagara Escarpment west of the falls, including the St. Davids Gorge (Flint and Lalcama, 1986; Fig. 1), were filled with glacial drift. Therefore, the Early Lake Erie outflow past Buffalo flooded the lowland between the Devonian Onondaga, and the Silurian Niagara Escarpments, forming a shallow river-lake called Lake Tonawanda. Multiple outlets initially carried waters across the Niagara Escarpment to glacial Lake Iroquois in the Ontario basin. Glacial rebound caused all outlets except the present to be abandoned by 10,900 B.P.

Gorge formation probably involved headward recession of one major falls line from the escarpment at Lewiston through most of the gorge length except through portions resurrected from the buried St. Davids Gorge (Stop 4). However the rate of back-cutting was erratic (Gilbert 1891). One major cause was the by-passing of Lake Erie and the Niagara River by drainage from the upper Great Lakes between about 11,800 B.P. and 5,000 B.P. This outflow passed directly to the Ontario basin and St. Lawrence valley via the Trent lowland and subsequently the Ottawa valley.

During retreat, the river, more than a few hundred meters above the respective cataract positions, must have resembled its present aspect above Goat Island (Stop 5), being broader and at a higher level. The former positions of this ancestral Niagara are recorded as terraces cut into the gravel, drift, and bedrock occurring along both margins of the Niagara Gorge. They are particularly well displayed in the general area of the Whirlpool Rapids and along the Niagara River Parkway north of the Whirlpool (Stop 4). Radiocarbon ages averaging 9,800 B.P. from mollusk shells within gravels from the lowermost of a set of these ancient river terraces at Whirlpool State Park (Calkin and Brett, 1978), indicate that the cataract itself may have been as little as 2,600 ft (800 m) downstream from here at that time. However, correlation with modern rates of falls recession suggests that the falls were farther north (perhaps just south of Devil's Hole) at that time. The rate of recession from Lewiston to the Devil's Hole area would have been on the order of about 4.3 ft/yr (1.3 m/yr) if recession started at 12,400 B.P.

Recession through the Whirlpool to the head of the Whirlpool Rapids section must have been very rapid; the Whirlpool formed some time after 9,800 B.P. when the southwestward-retreating falls intersected the buried St. Davids Gorge at nearly right angles (Fig. 1). Following breakthrough, the path of recession (probably rapids rather than a fall) turned abruptly south-southeast, reexcavating what is now the Whirlpool and an ancient plunge pool, called the "Eddy Basin," upstream. Retreat and excavation then continued through the drift fill of the Whirlpool Rapids Gorge section to the former head of the St. Davids Gorge. The broad and deep Upper Great Gorge (Fig. 1), extending 2.3 mi (3.7 km) to the falls from above the Whirlpool Rapids section, appears to have been cut by the full Niagara River discharge with drainage from the upper Great Lakes similar to that of the present. Thus, it seems to have been excavated during the past 3,800 years (Calkin and Feenstra, 1985). This correlation yields a somewhat lower rate of recession than historic data suggest. A rate of 3.6 ft/yr (1.1 m/yr) has been determined for the total years of record from 1842 through 1969 (American Falls International Board, 1974). Recent studies, considered below, suggest that recession rates must have been very irregular, despite the rather uniform discharge of this phase.

The present Niagara Falls comprise the Horseshoe (Canadian) Falls with 92% of the flow, which is separated by Goat Island (Stop 5) from the American Falls, with 8% of the discharge (American Falls International Board, 1974). Analysis of historic rates of recession of the Horseshoe Falls by the International Joint Commission (1953) suggested a rather uniform decrease in the rate from 4.2 ft/yr (1.28 m/yr) between 1842 and 1906 before diversion, to less than 2.2 ft/yr (0.67 m/yr) for the 1927 to 1950 period. The reduction was considered to be due to the southerly dip and thickening of the Lockport dolostone caprock and the increased diversion of water for hydroelectric power. However, more detailed analysis of historic recession by Philbrick (1970, 1974) indicates that the historic recession rate has decreased much more irregularly and largely has been controlled by the planimetric configuration of the falls rim. Since separation of the American and Horseshoe falls, perhaps about 600 years ago, the American Falls have retreated on the order of 200 ft (61 m) for a very low mean rate of 0.3 ft/yr (0.09 m/yr). Historic data, available since 1842, indicate an extremely sporadic recession (American Falls International Board, 1974).

Geological and engineering studies at the contemporary falls (Calkin and Wilkinson, 1982) throw light on past mechanisms of retreat (see also Gilbert, 1907). Erosion of the falls frequently involves sapping of Rochester Shale along numerous joints,

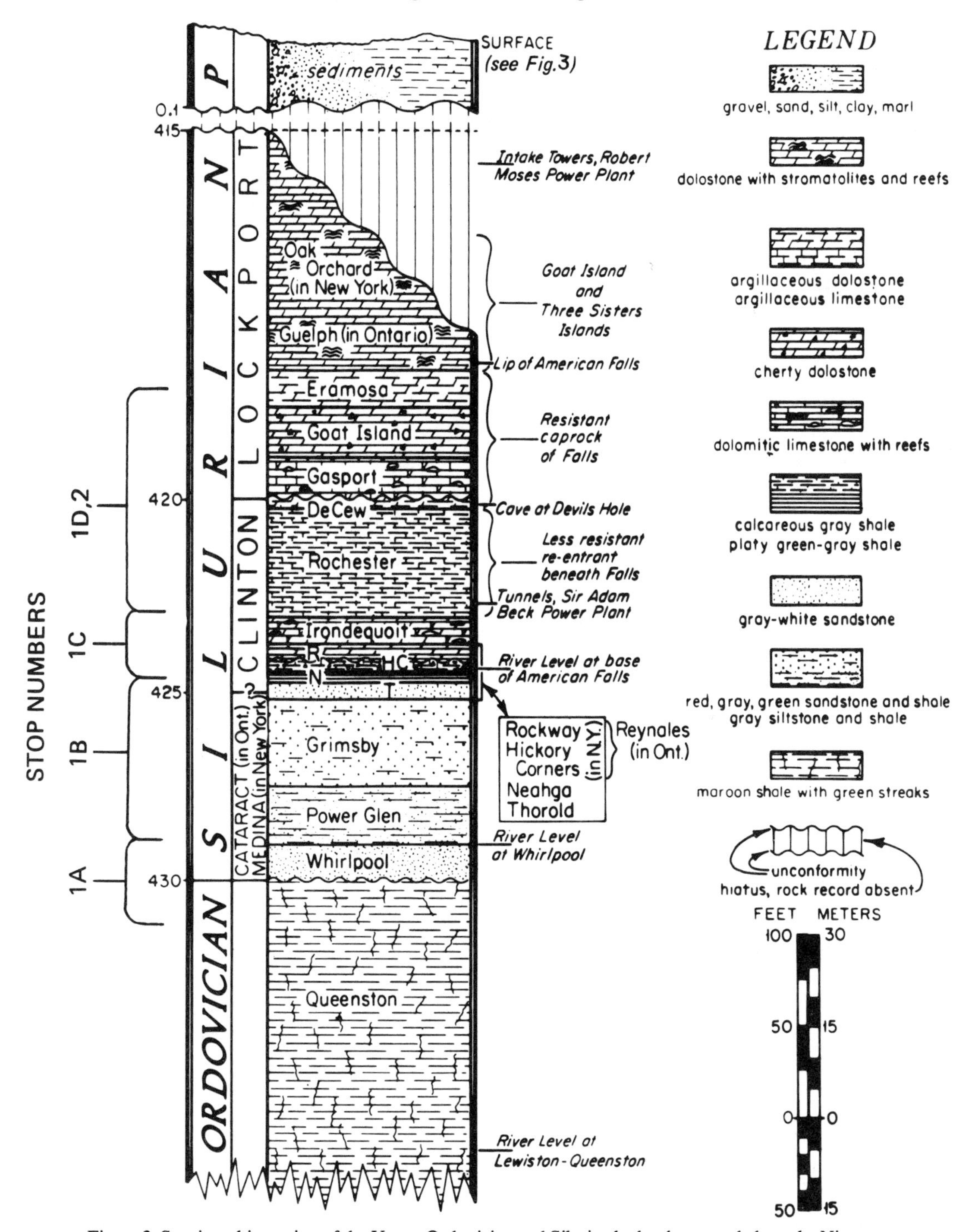

Figure 2. Stratigraphic section of the Upper Ordovician and Silurian bedrock exposed along the Niagara Gorge. Modified after Fisher and Brett (1981).

which ultimately undermines large blocks of the more resistant Lockport dolostone. Flowing and falling water, solution, seasonal ice buildup, wetting and drying, and release of residual stresses are major contributing factors. However, hydrostatic pressure acting in joints developed behind and parallel to gorge walls is also an important mechanism of rockfalls that occur without ordinary gravity collapse of the caprock (American Falls International Board, 1974; Calkin and Wilkinson, 1982).

The future of Niagara, as well as its past, has been the subject of speculation since the beginning of the century. Spencer (1894), in a much-quoted statement, suggested that the falls could go dry in about 5,000 years. This was incorrectly based on the

Pleistocene isostatic rebound rather than the barely detectable rates of contemporary vertical movement in this area (Calkin and Feenstra, 1985). In his *Principles of Geology,* Lyell (1837) suggested that on the basis of "contemporary" recession, the falls would eventually reach Lake Erie in about 30,000 years. Projection of the twentieth century rate of about 2 ft/yr (0.6 m/yr) via the shortest channel yields about 48,000 years. Such projections are based on fallacious assumptions, but the results may be as good as any number that can be generated considering the complicated scenario of irregular retreat.

Site 24-Paleozoic Stratigraphy. The classic stratigraphic section exposed along Niagara Gorge encompasses about 230 to 260 ft (70 to 80 m) of Late Ordovician, and Silurian strata with a veneer of surficial sediments (Fig. 2; see Kilgour and Liberty, 1981, for detailed stratigraphic descriptions). The well-exposed Silurian sequence forms an appropriate type section for the Niagaran Series. Rocks dip gently southward at ¼° to ½°, but are otherwise undeformed.

The lowest exposed rock unit consists of about 120 ft (36 m) of maroon shales with greenish bands and streaks, the upper portion of the Upper Ordovician (Ashgillian) Queenston Shale (Stop 1B). These sediments represent molasse—the "Queenston fan delta"—derived from erosion of tectonic highlands in the New England area, formed during the Ordovician Taconic orogeny. The Queenston beds in the Niagara region were deposited on a very broad, muddy alluvial-coastal plain and represent nonmarine to marginal marine environments (Friedman and others, 1982). No fossils have been found in the Queenston at Niagara, but exposures near Hamilton, Ontario, 43 mi (70 km) to the west, have yielded marine fossils (bryozoans, brachiopods).

Queenston beds are unconformably overlain by Lower Silurian (Llandoverian, A_2) Whirlpool Sandstone (Middleton and Rutka, 1986). Excellent exposures of the Ordovician/Silurian unconformity are accessible at Stop 1B. The base of the sandstone is sharp and contains abundant shale rip-up clasts, as well as impressions of desiccation cracks in the top of the Queenston.

Silurian rock units exposed in the gorge include, in ascending order, the: (1) Medina, (2) Clinton, and (3) Lockport groups. This is a highly condensed section containing rocks representative of the Llandoverian, Wenlockian, and lower Ludlovian series. See Rickard (1975) for detailed accounts of litho- and biostratigraphic relations. The lower Llandoverian Medina Group comprises three distinct siliciclastic facies: (a) basal, white cross-stratified, quartz arenite (Whirlpool Sandstone, 15 to 25 ft; 4.6 to 7.6 m), (b) gray, marine shales, with minor dolomitic siltstone and quartz arenite (Power Glen Formation, 33 to 36 ft; 10 to 11 m), and (c) red shales and red and white mottled sandstone (Grimsby Formation, 42 to 52 ft [12.8 to 15.8 m]; for details see Martini, 1971; Fisher, 1954; Bolton, 1957). Medina sediments are among the most economically important rock units in western New York; red and white sandstones have long been used as building- and curbstones, and the Grimsby Sandstone is an important natural gas reservoir rock.

The lower two-thirds of the Whirlpool Sandstone consists of medium- to fine-grained quartzose sandstone, with channels up to 300 ft (100 m) across, and large-scale tabular and festoon cross-bedding (Stops 1B, 4); this unit has recently been reinterpreted as the deposits of northwestward-flowing, braided rivers (Middleton and Rutka, 1986). In contrast, the upper 3 to 6 ft (1 to 2 m) of the Whirlpool has yielded a sparse fauna of ostracodes, lingulids, crinoids, and trilobite fragments, which is probably the oldest Silurian assemblage in North America (Kilgour and Liberty, 1981). The presence of hummocky cross-stratification (HCS), oscillation ripples, and wave marks suggests redeposition of this sand blanket in a transgressive, nearshore environment, probably a beach-shoreface environment (Seyler, 1982).

Power Glen shales are generally barren of megafossils, but Niagara Gorge exposures have yielded some of the oldest spore tetrads, as well as diverse acritarch assemblages (Gray and Boucot, 1971; Miller and Eames, 1982). Thin lenses of sandstone contain fossil assemblages of moderate diversity, including bryozoans, gastropods, bivalves, and crinoids (see Fisher, 1954), indicating a shallow, but fully marine, prodeltaic origin for these beds. A thin, sandy dolostone unit some 36 ft (11 m) above the base contains abundant bryozoans, mollusks, and pelmatozoan fragments with phosphatic pebbles.

Highly varied Grimsby beds include both marine and, in the upper third, nonmarine, deltaic facies. The base of the unit is demarcated from the underlying Power Glen Formation by an abrupt color change from green to red shales, but is otherwise gradational. Lower beds consist of red shales, planar to hummocky cross-laminated sandstones, and intraformational conglomerates, some of which contain abundant lingulids, bivalves, and bryozoans, indicating a shallow, storm-influenced marine origin. The upper Grimsby strata consist of lenticular to channeled, red-and-white mottled intraformational conglomerates and sandstones, interbedded with red silty shales. These units are generally devoid of body fossils, but some layers contain an abundance of the annulated burrow *Arthrophycus.* Ripple marks, small-scale cross stratification, soft sediment deformation, and intraformational conglomerates are abundant (Stop 1C). Collectively, upper Grimsby facies have been interpreted as representing a complex of shallow subtidal, tidal flat, channel, interdistributary bay, and overbank sediments formed on a prograding delta (Martini, 1971; Friedman and others, 1982).

The Whirlpool–Power Glen–Grimsby sequence records an Early Silurian marine transgression over the eroded Queenston deposits, followed by regression, resulting from active progradation of the Medina fringe delta. It is notable that the lower Whirlpool and Power Glen Formations are absent in the Genesee Gorge at Rochester about 72 mi (116 km) to the east; fully marine conditions apparently existed only as far east as Lockport.

The Clinton Group (mid-Llandoverian to Wenlockian) is a heterogeneous and highly incomplete sedimentary package that commences with siliciclastic facies similar to those of the lower Medina (for details see Gillette, 1947; Bolton, 1957; Kilgour, 1963). A widespread, thin, white quartzose sandstone, which yields rare ostracodes and *Arthrophycus* burrows (Thorold Sandstone, 5 to 10 ft; 1.5 to 3.0 m), is overlain by soft, greenish gray clay-shales (Neahga Shale, 6 to 7 ft; 1.8 to 2.1 m); the latter has produced a diverse palynomorph flora (Fisher, 1953). These siliciclastic units may represent reworking of upper Medina sediments in a shallow, transgressing sea. Neahga shales are overlain by a thin, argillaceous, phosphatic packstone and grainstone unit (Reynales or Hickory Corners Limestone, 2 to 4.3 ft; 0.6 to 1.3 m). Together, these three units apparently record a mid-Llandoverian (B_2–C_1) transgressive sequence.

The Hickory Corners Limestone is separated from the overlying Rockway Dolostone by an unconformity of considerable magnitude (Stop 1D), reflecting non-deposition or erosion of middle Clinton sediments. In west-central New York this interval is represented by more than 300 ft (100 m) of shales, carbonates, and the well-known hematite beds (Gillette, 1947; Kilgour, 1963).

Upper Clinton units (Llandoverian C-6 to middle Wenlockian age) are represented by (a) pale buff, argillaceous, sparsely fossiliferous Rockway Dolostone (11 ft; 3.4 m); (b) pinkish crinoidal pack- and grainstones of the Irondequoit Limestone (6 to 9.5 ft; 1.8 to 2.9 m); (c) thicker, calcareous gray shale and thin limestone, Rochester Shale (56 to 59 ft; 17 to 18 m); and (d) an upper argillaceous DeCew Dolostone (5.6 to 6.6 ft; 1.7 to 2 m); see Stop 1C, 1D. The abundant invertebrate faunas of the Irondequoit and lower Rochester, including more than 200 species of bryozoans, brachiopods, trilobites, ostracodes, mollusks, and pelmatozoan echinoderms, were among the earliest described in North America (Hall, 1838, 1852). Small bioherms of bryozoan- and algally(?)-bound micrites, occurring in the upper Irondequoit, were described from Niagara Gorge by Sarle (1901); see Stop 1D.

Upper Clinton sediments evidently accumulated on a shallow storm-influenced shelf. Irondequoit crinoidal sands were deposited in shallow water near wave base, whereas Rochester-DeCew muds accumulated in deeper muddy areas, within to slightly below storm wave base. Abrupt facies changes in the Rochester along the 7-mi gorge (11-km) section suggest the presence of a southward dipping paleoslope (Brett, 1982, 1983). The Irondequoit through DeCew sequence has been interpreted as recording a transgressive/regressive cycle with a subcycle in the Rochester Shale (Brett, 1983).

The Lockport Formation (late Wenlockian to Ludlovian), only partially exposed in bluffs along the top of the gorge (Stop 2), consists almost entirely of carbonates. This sequence has been described in detail by Zenger (1965). Lockport carbonates are quarried locally for crushed rock and building stone. The basal Gasport Member (15 to 45 ft; 4.6 to 13.6 m) consists of dolomitic, crinoidal grainstones, which are commonly cross-stratified and represent facies very similar to the older Irondequoit. However, the Gasport contains considerably larger micritic bioherms (reefs?), up to 160 ft (50 m) across and 30 ft (10 m) high, containing tabulate and rugose corals and domal to lamellate stromatoporoids (Crowley, 1973). Several excellent examples are exposed along Niagara Gorge (see Stop 5). The Goat Island Member (16.4 to 26 ft; 5 to 8 m) is dark brownish, saccharoidal, cherty dolostone with abundant mineral-filled vugs, some of which represent coral and stromatoporoid heads. Argillaceous, rarely oolitic, dolomicrite of the Eramosa Member (13.4 to 15 ft; 4.1 to 4.5 m) and brown stromatolitic dolostones of the Oak Orchard Member (incomplete and about 20 ft [6 m] thick) form the highest exposed units in the gorge and the cap rock of the falls. Together these represent a shallowing-upward sequence immediately prior to deposition of the evaporitic Salina Group.

DETAILED STOP DESCRIPTIONS

Stop 1. Niagara Escarpment and the mouth of Niagara Gorge. From Center Street (U.S. 104) in the town of Lewiston, New York, turn left (south) onto Fourth Street, which leads to the Earl W. Brydges Artpark (paid parking is available); by walking uphill to the level of the Visitor Center (above the theater), one has an excellent view of the escarpment and the mouth of the gorge.

Stop 1A. The north-facing cuesta (Niagara Escarpment; Fig. 1), which is overlain here by the Barre Moraine, forms an abrupt terminus of the gorge and stands 250 ft (76 m) above the adjacent Lake Ontario plain. The southern shore of glacial Lake Iroquois stood just north of the escarpment and is represented on the New York side by Ridge Road (U.S. 104) at 312 ft (95 m) above sea level. The town of Lewiston lies partly within the Cataract Basin (Fig. 1), a plunge pool formed by the first falls (or rapids) into Lake Iroquois. The Lewiston Branch Gorge and the similar Old Narrow Gorge form a straight section initiated during the Lake Tonawanda phase; this is now heavily modified by the power plants and the Lewiston-Queenston bridge.

Exposures of the Upper Ordovician upper Queenston Shale and its unconformable contact with basal Silurian Whirlpool Sandstone are visible along a short path, adjacent to the river, immediately south of the Artpark theater; pillars for an old suspension bridge rest on top of a small plateau on the Whirlpool Sandstone. Note here the sharply undercut erosional contact and local channeling and medium-scale cross stratification within the massive, whitish Whirlpool Sandstone; green shale clasts eroded from the Queenston occur near the base of the sandstone.

Stop 1B. Outcrops of the Silurian stratigraphic units above the Whirlpool Sandstone are accessible along an old haul road (old railroad grade modified during construction of the power plants) that leads southward from the Artpark Visitor Center into the gorge; the path continues south for about 0.9 mi (1.5 km) to the footings of the Lewiston-Queenston Bridge.

Additional excellent stratigraphic sections in the northern portion of Niagara Gorge include: (a) a large roadcut along the access road (South Haul Road) for the Robert Moses Power Plant (this is only accessed by obtaining permission from the Robert Moses Power Vista, Niagara Falls, New York, and taking an elevator from the Power Vista down to the road level; at the time of this writing, this section is closed to visitors); (b) a roadcut along the access road for the Sir Adam Beck Power Plant in

Queenston, Ontario; permission to travel this road must be obtained by contacting Ontario Hydro Limited, Niagara Falls, Ontario.

The complete Medina to lower Lockport sequence is continuously exposed (although largely inaccessible) in 150-ft-high (45-m) gorge walls from Lewiston southward for nearly 1.5 mi (2.5 km); this sequence is particularly well displayed in a small tributary gully (Fish Creek) about 600 ft (175 m) south of the gorge entrance. At the entrance to the gorge (edge of Niagara Escarpment) an isolated "butte" of Lower Silurian (Power Glen–lower Grimsby) bedrock between the path and the river, represents a remnant of a promontory in the gorge wall that was breached during excavation for the power plant (remains of an old railroad tunnel through the promontory are still visible in the gorge wall opposite the butte). Both the butte and the main gorge wall south to the bridge provide access to the Power Glen–Grimsby sequence. Abundant fallen debris also provides an excellent look at varied lithologies of the upper Medina, as well as Clinton and Lockport units. Caution is required in this exposure, as rock falls are common.

The Power Glen Formation at 1B consists of pyritic, greenish-gray clay shale (locally stained red from the overlying Grimsby), with a few thin, pale gray quartz arenite beds; shales are generally barren of macrofossils, but sandstones contain nautiloids, bivalves, crinoids, and bryozoans. A thin interval (5 ft; 1.5 m) of phosphatic, sandy dolostone, 25 to 30 ft (7.6 to 9 m) above the base, records maximum marine transgression within the Medina Group.

The Grimsby Formation consists of red shales and siltstones, with mottled pale red and white blocky sandstones; greenish-white reduction spots and fine liesegang banding are common in the sandstones. Exposures in the gorge walls display cyclic alterations of red shales and fining-upward, lenticular, cross-laminated sandstones and siltstones. Note abundant shale pebble conglomerates at bases of lenticular (channelized?) sandstone units, and thin, sandy coquinites containing *Lingula* fragments, bivalves, and rare bryozoans. Fallen blocks of Grimsby sandstone display a variety of sedimentary structures including oscillation and interference ripples, cross lamination (including HCS), rill and swash marks, load casts, desiccation cracks, and burrows (e.g., *Arthrophycus*). A 1.5- to 2-ft-thick (0.4 to 0.6 m) layer of mottled sandstone with prominent ball-and-pillow structures, about 18 ft (5.5 m) above the base of the Grimsby, is well exposed in the cliffs about 600 to 2,000 ft (200 to 700 m) north of the Lewiston-Queenston bridge.

Outcrops of Thorold Sandstone are inaccessible, but large fallen blocks can be examined in a landslide area just south of the Lewiston-Queenston Bridge. These consist of pale gray to nearly white quartz arenite with green shale pebble conglomerates; *Arthrophycus* burrows are prominent at some levels. The basal contact of the sandstone, well displayed in large, inverted blocks, shows spectacular load casts infilled with phosphate pebble-bearing sandstone.

Stop 1C. Units of the Silurian sequence above the Thorold Sandstone are inaccessible in the gorge cliffs north of the Lewiston-Queenston Bridge, but may be examined in fallen debris; for example, very large blocks of the Irondequoit crinoidal limestone occur in a large rock slide about 850 ft (260 m) south of the gorge entrance. The upper units are only accessible in outcrop by hiking along a very rough trail that ascends a slope immediately south of the footings for the Lewiston-Queenston Bridge. (Note: extreme caution is required in this section; permission for access should be sought in advance from the Niagara Parks commission or the Robert Moses Power Authority). At the top of the slope the upper Clinton, Irondequoit Limestone forms a prominent overhanging ledge, beneath which softer, buff gray Rockway Dolostone is recessed. About 1,300 ft (400 m) south of the Queenston Bridge, in and near a small stream (caution: contaminated water), are exposures of the upper Thorold, and overlying soft, greenish gray Neahga Shale (about 8 ft: 2.5 m thick); this shale is overlain, in turn, by about a meter of rubbly-weathering, bryozoan-rich, wacke- and packstone of the Hickory Corners (Reynales) Formation. The top of this thin unit is demarcated by a 4-in-thick (10-cm) greenish shale bed, containing abundant black phosphatic nodules; this unit marks the major mid-Clinton (Llandovery C-6) unconformity, and underlies the Rockway Dolostone.

Stop 1D. Approximately 1,100 ft (350 m) south of the Lewiston-Queenston Bridge and 1,800 ft (550 m) north of the Robert Moses Power Plant (immediately beneath electric power lines that cross the gorge) is a large talus cone composed of Rochester Shale slumped from the overlying cliffs (this is a good area to collect Rochester fossils, although it is dangerous due to the sharp dropoff into the gorge below). About 250 to 300 ft (80 to 100 m) south of this talus pile a section of Rockway and Irondequoit carbonates is exposed in a bluff above the path; notable here is the exposure of a small bioherm at the level of the Irondequoit-Rochester contact; now inaccessible, the bioherm was formerly at the level of a railroad cut and was extensively studied by Sarle (1901). South of this point are 80-ft-high (25-m), nearly vertical cliffs of Rochester Shale, about 60 ft (18 m) thick, and the overlying lower Lockport dolostone. The lower half of the Rochester Formation (Lewiston Member of Brett, 1982, 1983) is highly fossiliferous with lenticular bryozoan-, brachiopod-, and pelmatozoan-rich limestones near the base and top. The overlying Burleigh Hill Member is dark gray, sparsely fossiliferous, calcareous shale with thin calcisiltites; a few levels yield brachiopods and trilobites. These beds grade upward into thicker bedded DeCew Dolostone near the top of the cliffs. Note convoluted (enterolithic), soft sediment deformation in the DeCew.

The farthest south exposures in this section of the gorge, about 150 ft (50 m) north of the Robert Moses Power Plant, are of buff-weathering Lockport Formation; the lower, Gasport dolomitic limestone, displays spectacular cross stratification of crinoidal sands. The overlying Goat Island Member is light gray, vuggy dolostone (see further discussion under Stop 2).

Total walking distance in the gorge from Artpark parking area to this site is approximately 1.5 mi (2.5 km).

Stop 2. Devil's Hole State Park (New York). An overlook on the gorge, as well as a stone stairway and trail leading to the river level, are accessible from a free parking lot at Devil's Hole State Park (clearly marked) along the Robert Moses Parkway, 3.7 mi (5.2 km) south of the Lewiston exit and 4.3 mi (7 km) north of Niagara Falls.

Devil's Hole lies within the Lower Great Gorge and is a small reentrant in the east wall, marking the mouth of a former channel of the Niagara (Fig. 1). This channel, which is now occupied by Bloody Run Creek, existed for a short time after the falls had receded south of this point and before discharge was terminated by a drop in the level of Lake Tonawanda. The stairway and path leading into the gorge provide access to weathered exposures of the Lockport Formation and the DeCew Dolostone. At the head of the stairs, the major shear joint system of the area, with directions of N70° to 80°E and N to N38°E, is particularly well expressed in an exposed Lockport dolostone surface. Solution, localized along the main set (N70° to 80°E) has produced a small cave (4 ft high and 30 ft deep; 1.2 m by 9 m) in the DeCew Dolostone about 60 ft (18 m) below this point.

The DeCew displays spectacular soft-sediment deformation. Layers of intraformational conglomerate that are contoured into overturned isoclinal folds here may have been produced by slumping of fluidized sediment on a gentle paleoslope (Friedman and others, 1982). The overlying Gasport Member of the Lockport Formation (15 ft to 45 ft; 4.6 to 13.6 m) consists of fining upward dolomitic, cross-stratified crinoidal grainstones, interpreted as a channel-fill sequence by Friedman and others (1982, p. III-24). The basal contact of the Gasport displays rip-up clasts of dolomicrite, derived from the underlying DeCew. Overlying Lockport units (Goat Island and Eramosa members) are inaccessible.

Stop 3. Niagara Glen (Ontario) Provincial Park. Free parking for this provincial park is situated along the Niagara Parkway that runs along the western (Ontario) edge of the gorge; it is approximately 0.5 mi (0.8 km) north of the Whirlpool and 2 mi (3.2 km) south of the Lewiston-Queenston Bridge; a flat area immediately north of the park restaurant (Wintergreen Flats) represents the former Niagara River bed. A stairway near the north end of the restaurant leads down to river level, where one may observe large talus blocks and potholes.

A most striking feature of the Lower Great Gorge, Niagara Glen consists of a small promontory on the western wall of the gorge that bears an abandoned portion of the ancient Niagara River and its falls. When the falls retreated through this area, the flow was very wide but perhaps broken by a low island. The western side fall must have carried such a thin flow that it was eventually pirated of its water entirely, leaving the present dry bed, called "Winter Green Flats" (Kindle and Taylor, 1913). A talus accumulation of huge angular blocks, which are predominantly of the dolostone caprock, occurs below the crest and along a lower portion of the Glen near river level. Many of the boulders bear large potholes formed by turbulence at the base of the waterfalls.

Stop 4. Whirlpool State Park (New York). Free parking is available on the west (river) side of the Robert Moses Parkway, 2.2 mi (3.5 km) north of the entrance to the highway in Niagara Falls, New York, and 1.2 mi (1.9 km) south of Devil's Hole State Park. From the lot, walk through the park shelter and follow the sidewalk to the overlook above Whirlpool Point.

Whirlpool State Park provides an overview of the Eddy Basin, Whirlpool, and exposure of the drift-filled, interglacial St. Davids Gorge. Niagara River water rushing northward from the Eddy Basin immediately to the south, is funnelled through a narrow constriction before entering the broad Whirlpool basin. Sudden change in velocity results in a large gyre that normally rotates in a counterclockwise direction. However, during times of low flow, the rotation in the Whirlpool reverses (see Krajewski and Terasmae, 1981).

The Whirlpool basin resulted from a chance intersection of the modern Niagara River with, and erosion of, the drift from the buried St. Davids Gorge, visible directly opposite the Whirlpool overlook. The buried portion of the St. Davids Gorge in Ontario has been recognized as an ancient path of the Niagara since the time of James Hall and Charles Lyell (1845). From the escarpment, it extends north to Lake Ontario with distributaries to the lower Niagara River (Hobson and Terasmae, 1969; Flint and Lalcama, 1986). The St. Davids Gorge is nearly as deep and wide as the Upper Great Gorge, and its bottom may be marked by deep plunge pools and shallows like those of the re-excavated section, and of the Upper Great Gorge (Philbrick, 1970). A radiocarbon age of 22,800 ± 450 B.P. (GSC-816) was obtained on wood of interstadial swamp deposits occurring between drift sequences within this buried gorge north of the Whirlpool (Karrow and Terasmae, 1970). This, coupled with other studies, suggests that cutting occurred during, or more probably, before Middle Wisconsin time under nonglacial conditions when base levels were similar to those of the present (e.g., Sangamon interglaciation; see also Flint and Lalcama, 1986).

This locality is the type section of the Lower Silurian Whirlpool Sandstone, which forms the rapids and constriction just above the Whirlpool and is exposed as ledge or platform along its eastern margin. The overlying bedrock section (Fig. 2), particularly the Irondequoit through the Lockport interval, is clearly visible but inaccessible beneath the landings for the aerocar on either side of the St. Davids Gorge in Canada opposite Whirlpool Point.

Glacial sediments that fill the ancient Niagara River channel, including red till and overlying stratified drift (see Fig. 3), are exposed but not accessible above the Lockport Formation at the gorge margin just below the park overlook. Channel sand and gravel of the high, early stages of the Niagara River overlie this drift. Gastropods and bivalves from this deposit yielded the 9,800 B.P. radiocarbon age for gorge recession (Calkin and Brett, 1978).

Stop 5. Niagara Falls from Goat Island, Niagara Falls, New York. From New York 384 in the city of Niagara Falls a road (clearly marked) leads onto Goat Island; a perimeter drive

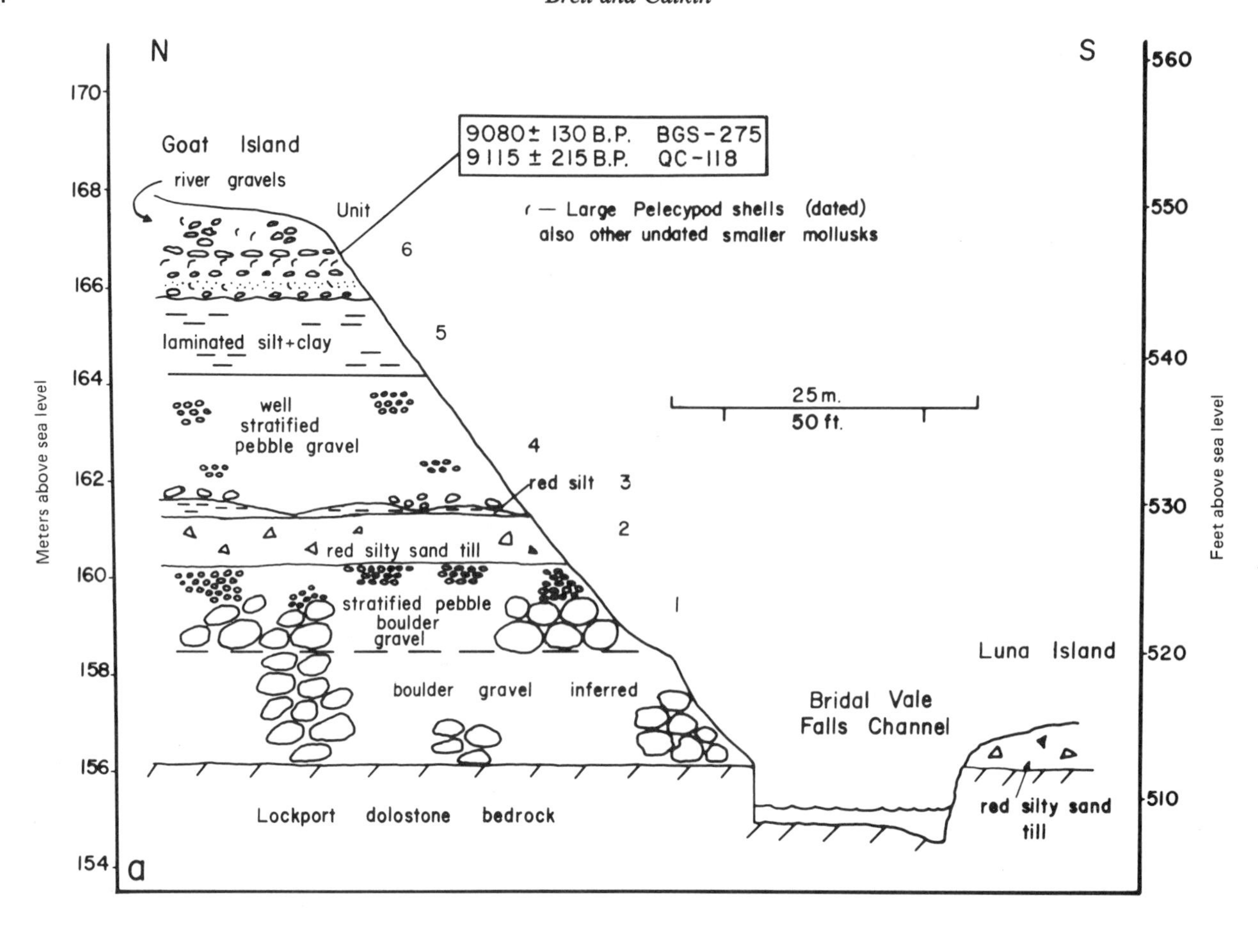

Figure 3. Stratigraphic section through the surficial deposits of Goat Island, Niagara Falls. The American Falls is just to the right. See Figure 1 for location. The strata exposed above the Lockport dolostone at Whirlpool Park are similar to this from the red till upward. After Calkin and Brett (1978).

leads to a paid parking lot. This is perhaps the most widely known tourist overlook and it provides access to both Horseshoe and American falls. Other views of the falls area are available from the Canadian side.

Bedrock exposures in this portion of the Niagara Gorge are generally poor and largely inaccessible. However, good exposures of the Rochester and Lockport units occur along the east side of the gorge at the Niagara Sewage Pumping station and along the access road for the Toronto Power facility in Niagara Falls, Ontario (see Brett, 1982, for detailed logs and descriptions of these sections). Large bioherms can be observed in the Gasport Member (Lockport Formation) in the cliff against the Horseshoe Falls, immediately above the Toronto Power Station and along the access road for the Niagara Sewage pumping station.

Detailed engineering geological data on the American Falls area are available in American Falls International Board (1974; see also Calkin and Wilkinson, 1982). Views westward across the gorge to Canada from between the two falls reveal the steep, cut-slope of the Niagara Falls Moraine. This moraine reaches 164 ft (50 m) above a narrow bedrock shelf along the gorge margin that is occupied by Queen Victoria Provincial Park in Ontario. The drift was swept from this area by the Niagara as it was deflected northward by the moraine at an earlier, high phase.

Goat Island itself consists of ancient channel fill of latest Wisconsin drift, which is topped by postglacial Niagara River gravels (Fig. 3); sections of these sediments are only accessible during excavations on the island, most recently in 1974. The underlying, northwest-sloping Lockport bedrock surface may be the eastern bank of an ancient south-trending Falls-Chippawa channel of Spencer (1907), or part of the north-trending St. Davids Gorge system (see Flint and Lalcama, 1986). Mollusks within the river gravels have yielded radiocarbon ages of 9,100 B.P. This suggests that Goat Island was abandoned as part of the Niagara River bottom about that time, or that more recent river scouring has removed any younger gravel material. A mastodon tooth was removed by James Hall from these river gravels at a locality near Prospect Point, across the American Falls from Goat Island (Calkin and Brett, 1978).

REFERENCES CITED

American Falls International Board, 1974, Preservation and enhancement of the American Falls at Niagara: Final report to the International Joint Commission; Appendix C, Geology and Rock Mechanics, 71 p. and Appendix D, Hydraulics, 41 p. (Out of print, but available for study at the Buffalo District Office of U.S. Army Corps of Engineers).

Bolton, T. E., 1957, Silurian stratigraphy and palaeontology of the Niagara Escarpment in Ontario: Canada Geological Survey Memoir 289, 145 p.

Brett, C. E., 1982, Stratigraphy and facies variation of the Rochester Shale (Silurian: Clinton Group) along Niagara Gorge, *in* Buehler, E. J., and Calkin, P. E., eds., Field trip guidebook: Amherst, New York State Geological Association Guidebook 54th Annual Meeting, p. 217–230.

—— , 1983, Sedimentology, facies, and depositional environments of the Rochester Shale (Silurian; Wenlockian) in western New York and Ontario: Journal of Sedimentary Petrology, v. 53, no. 3, p. 947–971.

Calkin, P. E., and Brett, C. E., 1978, Ancestral Niagara River drainage; Stratigraphic and paleontologic setting: Geological Society of America Bulletin, v. 89, p. 1140–1154.

Calkin, P. E., and Feenstra, B. H., 1985, Evolution of the Erie-Basin Great Lakes, *in* Karrow, P. F., and Calkin, P. E., eds., Quaternary evolution of the Great Lakes: Geological Association of Canada Special Paper 30, p. 149–170.

Calkin, P. E., and Wilkinson, T. A., 1982, Glacial and engineering geology aspects of the Niagara Falls and Gorge, *in* Buehler, E. J., and Calkin, P. E., eds., Field trip guidebook: Amherst, New York State Geological Association, 54th Annual Meeting, p. 247–283.

Crowley, D. J., 1973, Middle Silurian patch reefs in the Gasport Member (Lockport Formation), New York: American Association of Petroleum Geologists Bulletin, v. 57, p. 283–300.

Fisher, D. W., 1953, A microflora in the Maplewood and Neahga shales: Buffalo Society of Natural Sciences Bulletin, v. 21, no. 2, p. 13–18.

—— , 1954, Stratigraphy of the Medina Group, New York and Ontario: American Assocaition of Petroleum Geologists Bulletin, v. 38, p. 1979–1996.

—— , 1981, Introduction, *in* Tesmer, I. H., ed., Colossal cataract; The geologic history of Niagara Falls: Albany, State University of New York Press, p. 1–15.

Fisher, D. W., and Brett, C. E., 1981, The geologic past, *in* Tesmer, I. H., ed., Colossal cataract; The geologic history of Niagara Falls: Albany, State University of New York Press, p. 16–42.

Flint, J. J., and Lalcama, J. L., 1986, Buried ancestral drainage between Lakes Erie and Ontario: Geological Society of America Bulletin, v. 97, p. 75–84.

Friedman, G. M., Sanders, J. E., and Martini, I. P., 1982, Excursion 17A; Sedimentary facies; Products of sedimentary environments in a cross section of the classic Appalachian Mountains and adjoining Appalachian Basin in New York and Ontario: Field Excursion Guidebook, 11th International Congress on Sedimentology, McMaster University, p. M1–M30.

Gilbert, G. K., 1891, The history of the Niagara River: Annual Report of the Smithsonian Institution, v. 6, p. 231–257.

—— , 1907, Rate of recession of Niagara Falls: U.S. Geological Survey Bulletin 306, p. 5–25.

Gillette, T., 1947, The Clinton of western and central New York: New York State Museum Bulletin 341, 191 p.

Grabau, A. W., 1901, Guide to the geology and paleontology of Niagara Falls and vicinity: New York State Museum Bulletin 45, v. 9, 284 p.

Gray, J., and Boucot, A. J., 1971, Early Silurian spore tetrads from New York; Earliest new world evidence for vascular plants?: Science, v. 173, p. 918–921.

Hall, J., 1838, Third annual report of the Fourth Geological District: New York Geological Survey Annual Report 3, p. 287–339.

—— , 1843, Geology of New York, Part IV, comprising the survey of the Fourth Geological District: Albany, New York, Carrol and Cook, 683 p.

—— , 1852, Description of the organic remains of the lower middle division of the New York System, *in* Paleontology of New York, Volume 2: Albany, New York, C. Van Benthuysen, 362 p.

Hobson, G. D., and Terasmae, J., 1969, Pleistocene geology of the buried St. Davids Gorge, Niagara Falls, Ontario; Geophysical and palynological studies: Geological Survey of Canada Paper 68-67, 16 p.

International Joint Commission, 1953, Preservation and enhancement of Niagara: Washington, D.C., 354 p.

Karrow, P. F., and Terasmae, J., 1970, Pollen-bearing sediments of the St. Davids buried valley fill at the Whirlpool, Niagara River Gorge, Ontario: Canadian Jounral of Earth Science, v. 7, p. 539–542.

Kilgour, W. J., 1963, Lower Clinton (Silurian) relationships in western New York and Ontario: Geological Society of America Bulletin, v. 74, p. 1127–1141.

Kilgour, W. J., and Liberty, B. A., 1981, Detailed stratigraphy, *in* Tesmer, I. H., ed., Colossal cataract; The geologic history of Niagara Falls: Albany, State University of New York Press, p. 94–122.

Kindle, E. M., and Taylor, F. B., 1913, Description of the Niagara Quadrangle: U.S. Geological Survey, Geological Atlas, Folio 190, 26 p.

Krajewski, J., and Terasmae, J., 1981, Appendix A, Road guide for points of interest, *in* Tesmer, I. H., ed., Colossal cataract; The geologic history of Niagara Falls: Albany, State University of New York Press, p. 173–196.

Lyell, C., 1837, Principles of geology, Volume 1: London, John Murray (1st American Edition), 546 p.

—— , 1845, Travels in North America in the years 1841–42, with geological observations on the United States, Canada, and Nova Scotia: New York, Wiley and Putnam, 472 p.

Martini, I. P., 1971, Regional analysis of sedimentology of the Medina Formation (Silurian), Ontario and New York: American Association of Petroleum Geologists Bulletin, v. 55, p. 1249–1261.

Middleton, G. V., and Rutka, M., 1986, The nature of the marine transgression in the Lower Silurian of Ontario: Geological Society of America Abstracts with Programs, v. 18, p. 55.

Miller, M. A., and Eames, L. E., 1982, Palynomorphs from the Silurian Medina Group (Lower Llandovery) of the Niagara Gorge, Lewiston, New York, U.S.A.: Palynology, v. 6, p. 221–254.

Philbrick, S. S., 1970, Horizontal configuration and the rate of erosion of Niagara Falls: Geological Society of America Bulletin, v. 81, p. 3723–3732.

—— , 1974, What future for Niagara Falls?: Geological Society of America Bulletin, v. 85, p. 91–98.

Rickard, L. V., 1975, Correlation of the Silurian and Devonian rocks in New York State: Albany, New York State Museum and Science Service Map and Chart Series 24, 16 p.

Sarle, C. J., 1901, Reef structures in Clinton and Niagara strata of western New York: American Geologist, v. 28, p. 282–299.

Seyler, B., 1982, Depositional environments of Silurian Whirlpool Sandstone (Medina) in western New York: American Association of Petroleum Geologists Bulletin, v. 66, p. 1174.

Spencer, J. W., 1894, The duration of Niagara Falls: American Journal of Science, 3rd Series, v. 48, p. 455–472.

—— , 1907, Falls of the Niagara; Their evolution and varying relations to the Great Lakes; Characteristics of the power and the effects of its diversion: Geological Survey of Canada Publication 970, 490 p.

Zenger, D. H., 1965, Stratigraphy of the Lockport Formation (Middle Silurian) in New York State: New York State Museum and Science Service Bulletin 404, 210 p.

The Gowanda Hospital late Pleistocene site, western New York

Parker E. Calkin, *Department of Geological Sciences, State University of New York, Buffalo, New York 14226*

LOCATION

This site occurs 28 mi (45 km) south of Buffalo and 2.5 mi (4 km) north of the Village of Gowanda, Erie County, in west-central New York; it is within the Gowanda 7½-minute Quadrangle. The site encompasses a 62-ft (19-m)-high exposure of drift that extends 650 ft (200 m), mostly along the left bank of the South Branch, Clear Creek, near the edge of the Gowanda State Hospital property (Fig. 1). Gowanda is easily reached via Silver Creek or Angola exits off the New York State Thruway. The cut-bank exposure is on private property; its owner asks that scientists make their access along the creek edge from the New York 62 bridge over the creek.

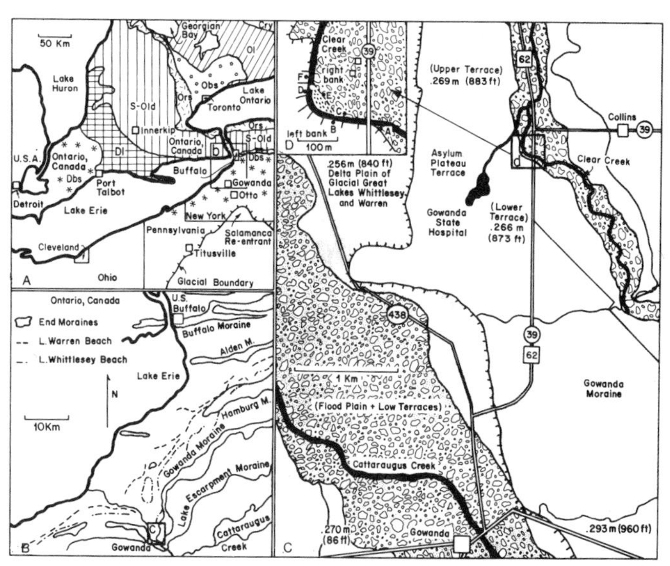

Figure 1. Location maps of the Gowanda Hospital Pleistocene site. A: Bedrock geology north of the site with symbols: Cry, Precambrian crystalline; Ol, Ordovician limestone; Obs, Ordovician black shale; Ors, Ordovician red shale; S-Old, Silurian and Ordovician limestone and dolomite; Dl, Devonian limestone; Obs, Devonian black shale. Letters A through F on insert D show locations in Figure 2. (After Calkin and others, 1982.)

SIGNIFICANCE

The Gowanda Hospital Pleistocene site is significant for its yield of stratigraphic, paleontologic, and radiometric information, which helps document the late Pleistocene history of western New York, as well as patterns and timing of glacial flow in the eastern Great Lakes area. Its stratigraphic importance derives from fossil-bearing silts and gravels, as well as dated wood of probable middle to early Wisconsin interstadial, or Sangamon interglacial age. These deposits overlie a thick weathered zone developed in the lowest of three tills. Only one other major Pleistocene organic site has been described from western New York, which is at Otto, 10 mi (16 km) southeast of Gowanda (Fig. 1A; Muller, 1964); however, the paleosol at the Gowanda Hospital site is unique in western New York. It is comparable to the late Pleistocene paleosols of Pennsylvania, Ohio, and nearby Ontario, as well as Indiana and Illinois that testify to long nonglacial intervals. Most of this summary is taken freely from a detailed analysis of the site as presented by Calkin and others (1982).

SITE INFORMATION

Geologic Setting

The South Branch, Clear Creek, has cut into a thick section of drift capped by terrace gravels at the Gowanda Hospital site; the drift was deposited in a deep embayment of the Lake Erie lowland, which extends southeastward into the margin of the glaciated, Allegheny Plateau (Fig. 1B). The embayment was in turn formed by the preglacial Allegheny River before the earliest ice advances caused its southward diversion to the Ohio River (Carll, 1880; Leverett, 1902; Fairchild, 1932). The embayment is now partly occupied by the much smaller, west-flowing Cattaraugus Creek, which is south and east of the site.

The South Branch of Clear Creek heads 5.6 mi (9 km) east of the site in the Lake Escarpment moraine (Muller, 1977) and descends through the younger Gowanda moraine 650 ft (200 m) east of the cut-bank off New York 62 (Fig. 1). The initial incision of this exposure by Clear Creek may have occurred as early as 13 to 14 ka, or soon after ice marginal retreat from the Lake Escarpment and closely associated Gowanda moraines, during the late Wisconsin substage (Calkin and McAndrews, 1980). As this recession occurred, waters of Cattaraugus Creek were imponded in front of the ice margin, forming Gowanda Lake (Fairchild, 1907). Alluvial deposits graded to this lake form the surface (Asylum Terrace) of the Gowanda Hospital site.

Stratigraphy

The units identified and their stratigraphic relations as expressed over a 25-year period of changing exposure, are given below (Table 1, Fig. 2). The gross stratigraphy may be traced across the exposure with the aid of seeps and marker beds. Minor digging is sometimes necessary to remove a thin clay colluvium.

Three distinct glacial diamictons have been distinguished: a red till (unit 1) named after exposures in the town of Collins, the "Collins Till"; a brown till (unit 3); and a gray till with interbedded silt beds (unit 6), named the "Thatcher Till," with a type section at Thatcher Brook, south of Gowanda (Calkin and others, 1982). The Collins Till, recognized in a few other localities at the margin of the Cattaraugus embayment, forms the lowest unit at

TABLE 1. STRATIGRAPHY OF THE GOWANDA HOSPITAL LATE PLEISTOCENE SITE

Exposure Unit	Section Description	Thickness (m)
	Terrace surface at about 266 m (873 ft).	
8	Gravel, oxidized, pebble-cobble; locally cross-bedded, moderately to deeply weathered; stones rounded with exotic lithologies numerous.	1.0
7c	Silt and clay, gray, interbedded; upper 0.3 m leached; 2 m of rhythmically bedded clay and silt or fine sand with couplets 1.3 to 0.6 cm.	4.4
7b	Till, gray silty clay; moderately calcareous; 5 percent pebbles mostly of shale with red granules of clay; occurs as lensoid masses within silt; fabric suggests source from north.	0.5
7a	Silt and interlaminated fine sand, gray; pebbly near base.	1.2
6e	Till, gray, clayey silt; weakly calcareous, 14 percent stones largely shale; includes sand silt clay lenses but more compact than laminated units above; gradational lateral and vertical contacts to stony silts; fabrics suggest source from north.	0.3
6d	Silt, gray, stratified, moderately calcareous.	0.6
6c	Till, gray, clayey silt; moderately calcareous, 2 percent stones largely shale; upper contact at seep contains contorted silt partings; lateral and vertical contacts gradational; fabrics indicate source from north.	2.8
6b	Sand and gravel, stratified, gray; some silt partings, compact; sharp lower contact.	0.6
6a	Till, gray, silty sand and stony (33 percent); mnoderately calcareous, compact; gradational to silt till to north; fabrics suggest source from north.	0.8
5	Gravel, pebble, well-sorted shaley; locally oxidized, crude imbrication suggests deposition from north.	<1.9
4b	Gravel, cobble-pebble, scattered lenses of gray silt; sparse-scattered wood fragments; pebbles of mixed lithology rounded, some striated, crude imbrication suggests flow from southeast; grades upstream to unit 4a silts and interfingered with brown till (unit 3) downstream.	6.0
4a	Silt, dark gray and brown sand, calcareous; contains terrestrial gastropod assemblage, sparse wood fragments, black pyritic granules in lenses a few cm long, and crossing shear fractures; deformed to 25° southwest dip; grades laterally to unit 4b but also sharply truncated above by younger horizontal beds of this unit.	4.0
3	Till, brown, silty clay; weakly calcareous; 6 percent stones, largely siltstone and carbonate residue or other oxidized rock fragments; lower half displays banded structure and shear surfaces dipping 10° north-northeast which incorporate deformed materials of underlying unit 2 dark silts and wood dated at 47,200±800 B.P. (QL-134); sharp contact to unit 4; fabrics suggest source to northwest.	6.5
2	Silt and sand, carbonaceous, often pebbly; contains abundant spruce twigs, roots, and logs to 15 cm diameter; many coalified, dated at >38,000 (W-866), > 48,400 B.P. (GrN-5486), 51,600 [+1900, -1500] B.P. (GrN-133).	1.7
1c	Buried paleosol developed in 1b; morphology described in separate section of paper.	2.7
1b	Till, red, silty clay; 6 percent stones, largely carbonate; very compact, strongly calcareous; fabrics suggest source from northwest.	5.0
1a	Till, red-gray, clayey silt; 6 percent stones, largely shale; moderately calcareous; locally separated from 1b by stone lag but appears to represent red till with assimilated local bedrock.	0.5
0	Bedrock, greenish-gray shale, Middle Devonian.	
Borehole Subsection*		
	Point bar surface at about 247 m (810 ft).	
3b	Gravel, cobble; point bar of modern Clear Creek.	0.6
2Bc	Silt, stony, laminated, black; markedly carbonaceous in upper 0.7 m.	1.7
2Bb	Silt, reddish-gray, stratified; calcareous with scattered pebbles; contains redeposited pre-Pleistocene spores and some small fragments of lignified plant tissue.	3.8
2Ba	Till-like stony silt and interbedded calcareous silt.	1.8
1B	Till, gray, silty clay, stony (16 percent); calcareous with some red sections; very compact (limit of auger).	0.2+

*Borehole units 1B and 2Ba-c are correlated with exposure units 1a-b and 2, respectively.

Table from Calkin and others (1982).

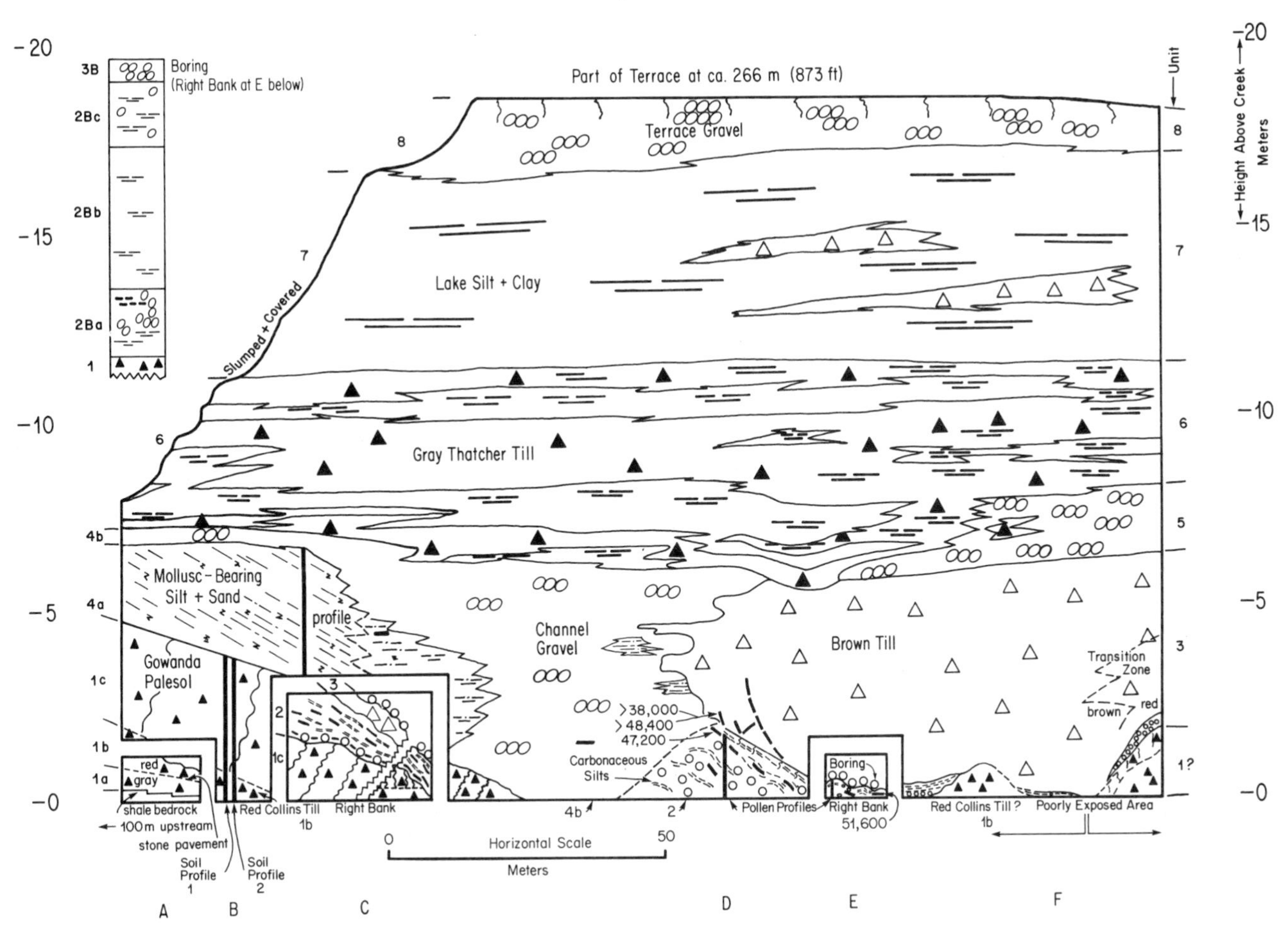

Figure 2. Stratigraphic sketch showing locations of right and left bank exposures and boring at Gowanda Hospital Pleistocene site along Clear Creek. Locations of letters A through F at base of figure are located in Figure 1D. (After Calkin and others, 1982.)

the site, resting on bedrock just east of the New York 62 bridge but buried beneath the gravelly silts of unit 2 in the middle point bar area (see boring log, Fig. 2). It is distinctly reddish-gray in color, except in areas where an apparently darker gray facies may result from reduction, as in the boring (Fig. 2), or from incorporation of gray shale (A, Fig. 2). The red color, together with texture, high carbonate content, and sand and clay mineralogy, clearly distinguish the Collins Till from the tills above, although visual separation from the overlying brown till is locally unclear (E and F, Fig. 2). Furthermore, these properties, with macro- and microfabric data, suggest lodgement occurred from a glacier that had moved southeastward across the area from the carbonate terrain of the Niagara Peninsula, Ontario (Fig. 1A). The paleosol developed in this till, is discussed separately below.

The brown till (unit 3) is very resistant and forms a small protrusion along the creek (D, Fig. 2). It resembles the Collins Till in texture and coherence but differs in properties that relate to degree of weathering. However, there is clearly no soil profile development as its properties are uniform throughout. Furthermore, shear surfaces suggestive of glacial lodgement contain woody detritus derived from the underlying silts of unit 2. Therefore the brown till appears to be glacially derived from weathered portions of Collins Till.

The diamicton units of the Thatcher Till (Table 1) display properties typical of surface tills of western New York that suggest incorporation of shale entrained as the ice flowed southward to this area (Muller, 1960; Calkin and McAndrews, 1980; Fig. 1A). Till fabrics support this flow path. The inclusion of discontinuous lacustrine interbeds indicate that deposition probably involved a thin, oscillating ice margin with alternating basal meltout and lacustrine deposition.

The dark, often pebbly organic silts of unit 2 (Table 2, Fig. 2) are often exposed as slightly deformed units beneath the brown till at river level; they were also once observed over the weathered Collins Till on the right bank (C, Fig. 2). The silt contains exceptionally well-preserved twigs and branches of spruce wood, sometimes with bark and needles. Two of four radiocarbon dates that have been obtained from this deposit (Table 2) are expressed as finite, but, like the other two, are near the limit of the conventional radiocarbon dating technique used. All suggest rapid, local burial, perhaps by glacial advance, more than 49,000 B.P. Amino acid ratios from wood (N. Rutter, personal communication,

TABLE 2. FIELD DESCRIPTION OF THE GOWANDA PALEOSOL

Horizon	Depth	Description
Locality C (Fig. 2)--Right bank exposure		
Bgb	1.6	Dark gray (5YR 2/1) gravelly loam; massive; very firm; scattered yellowish brown (10YR 5/6) stone ghosts among 40-50 percent stones; very weakly effervescent (pH ~7.5); upper boundary wavy.
Locality B (Fig. 2)*--Left bank exposure		
B1b	0-1.75 m	Weak red (10R 4/3) gravelly clay loam; massive; very firm; common fine pores with thin patchy clay films lining some of the pores; strong brown (7.5YR 5/8) and yellowish brown (10YR 5/6) stone ghosts; 20 percent coarse fragments increasing to 30 percent with depth, 40 percent stone ghosts decreasing to 25 percent with depth; neutral (pH ~7.2); clear wavy boundary.
B2tb	1.75-2.35 m	Reddish brown (5YR 4/3) silty clay; strong coarse angular blocky structure oriented 135° to 75° southwest and 220° to 75° southeast; very firm; common fine and very fine pores with clay linings; dark reddish gray (10YR 4/2) continuous clay films on red faces; 15 percent coarse fragments, 5 to 10 percent stone ghosts; moderately effervescent increasing to strongly effervescent with depth; clear wavy boundary.
B3b	2.35-2.7 m	Dark brown (7.5YR 4/2) silty clay; weak angular blocky structure oriented 135° to 75° southwest and 220° to 75° southeast; very firm; common fine pores; 15 percent coarse fragments; strongly effervescent; gradual wavy boundary.
C	2.7	Dark reddish gray (5YR 4/2) silty clay; massive; very firm; few fine pores; 15 percent coarse fragments; strongly effervescent.

*Description by D. Owens and P. Calkin. After Calkin and others (1982).

1979) may complement these ages (Calkin and others, 1982). Pollen recovered from unit 2 is not well preserved, but samples display a very strong spruce assemblage suggesting a closed boreal spruce forest environment (C. Winn and J. Terasmae, personal communication, 1979).

Fossiliferous silt (subunit 4a) overlies the paleosol at the upstream edge of the exposure, and the bedding dips about 25° west with the underlying Collins Till surface. The mollusk assemblage of largely terrestrial pulmonate gastropods has been examined by Leonard Kalas (personal communication, 1979) and subsequently by Barry Miller (personal communication, 1983). Kalas suggested that the assemblage may reflect initial milder conditions but that most of the sampled profile denoted harsher, tundra-like conditions. Amino acid ratios have been obtained from a few mollusk species (G. Miller, personal communication, 1979; W. McCoy, personal communication, 1983).

Most of unit 4 consists of gravel (subunit 4b), which interfingers with the silt upstream and with the brown till (unit 3) downstream. This suggests simultaneous deposition of units 3 and 4; however, an ice margin at this site would probably have been accompanied by high-level proglacial lakes, an environment apparently denied by the mollusk assemblage of the incorporated silt facies. Because of the gravel fabric and the mollusk assemblage, the gravel was considered by Calkin and others (1982) to be a channel fill cut through the brown till after considerable ice marginal retreat to allow lowering of any lake. The relationship of the gravel (unit 4b) and the brown till (unit 3) awaits further study.

Unit 7 (Table 1) is typical of regional glaciolacustrine deposition and includes a section of rhythmites suggesting 100 to 200 years of continuous deposition. However, there are included till lenses that appear to increase northward. These indicate settling or flow from floating or barely grounded ice (e.g., Evenson and others, 1977) that was nearby throughout the early period of unit 7 deposition. The upper surface of the lake clays is above 866 ft (264 m)—considerably above the projected levels of the glacial Great Lake phases recognized in this area (Calkin, 1970, Calkin and Feenstra, 1985). The clays may have been deposited in a local, proglacial lake, perhaps an early stage of the Gowanda Lake of Fairchild (1907).

Gravel (unit 8), forming the uppermost unit at the site, displays even bedding and sorting in its lower layers typical of deposition in standing water, whereas upper layers of the unit show imbrication suggesting westerly streamflow. The terrace developed in these gravels is at 873 ft (266 m) and is about 5,600 ft (1,700 m) wide. It terminates against the Gowanda moraine, 1.3 mi (2 km) to the south and east (Fig. 1). A similar distance to the north (off Taylor Hollow Road), the gravel thickens to 98 ft (30 m) and displays large-scale, truncated forset bedding that dips northwest. Deposition of unit 8 was into Gowanda Lake.

Gowanda Paleosol

The weathered zone in the Collins Till, named the "Gowanda paleosol" (Calkin and others, 1982), can usually be observed at the site and has been substantiated by detailed laboratory analyses. The soil (Table 2) occurs as a westwardly inclined, strikingly rusty brown-red (oxidized) zone of about 5 to 6 ft (1.5 to 2.0 m) sharply below calcareous sand and clay of unit 4a on the upstream edge of the cut-bank (B, Fig. 2). This brownish horizon contrasts with the uniform reddish gray of the unweathered Collins Till below. The soil was once observed as a dark gray (reduced?) till beneath black carbonaceous wood bearing silt a few ft (m) away on the opposite (point bar) bank.

The following are among the most important characteristics of the paleosol (Calkin and others, 1982). Oxidation has penetrated 6 to 10 ft (2 to 3 m) and leaching to 5 to 6 ft (1.5 to 2 m), leaving rusty ghosts of most carbonate clasts (Table 2). Shales and siltstones, red and gray quartzose sandstone, and granitic pebbles are also partly broken down throughout the oxidized zone. In general, the upper 3 to 6 ft (1 to 2 m) of the horizon is depleted in clay and fine silt, which increases downward, while the coarse silt and sand decrease downward. Much of the translocated clay is apaprently derived from the carbonates; no significant accumulation is detectable compared with the C horizon. The heavy mineral associations reflect the in situ weathered nature of the soil, as do heavy mineral weathering indicies (Calkin and others, 1982), which show the strongest weathering upward. Depletions of clinopyroxene and hornblende (Frye and others, 1960) relative to tourmaline and zircon in the weathered profile are about 80 and 65 percent, respectively. Etching of pyroxene grains is marked toward the surface. Gradual degradation of chlorite and illite clay minerals in the upper 3 to 4 ft (1 to 1.2 m) is evident by x-ray analyses of profile specimens with chlorite altering to vermiculite, and illite to a smectite species. These changes in clay mineralogy are similar to the alteration through both middle and early Wisconsin, and Sangamon in situ paleosols described from Pennsylvania and midwestern states (e.g., Droste and Tharin, 1958; Forsyth, 1965; Willman and others, 1966).

These characteristics indicate that the Gowanda paleosol is a well-developed, truncated profile that may be missing up to 10 to 25 percent of its solum. It is much better developed than surface soils now forming in the surrounding late Wisconsin tills. Profile development in tills near Buffalo that nearly duplicate the texture and composition of the Collins Till, and that have been exposed since retreat of the Port Huron ice margin 13 to 12 ka, reaches a depth of somewhat less than 3 ft (1 m). Soils are only slightly thicker in calcareous tills of the Valley Heads (Lake Escarpment) drift or on the middle to early Wisconsin Olean Till of the adjoining Allegheny Plateau of southern New York. Comparison of the Gowanda paleosol with surface and buried paleosols from Pennsylvania, Ohio, Indiana, and Illinois that were considered to be middle Wisconsin interstadial through Sangamon interglacial in age, suggested to Calkin and others that the weathering was somewhat more compatible to that in early Wisconsin deposits than in the Sangamon soils of these areas.

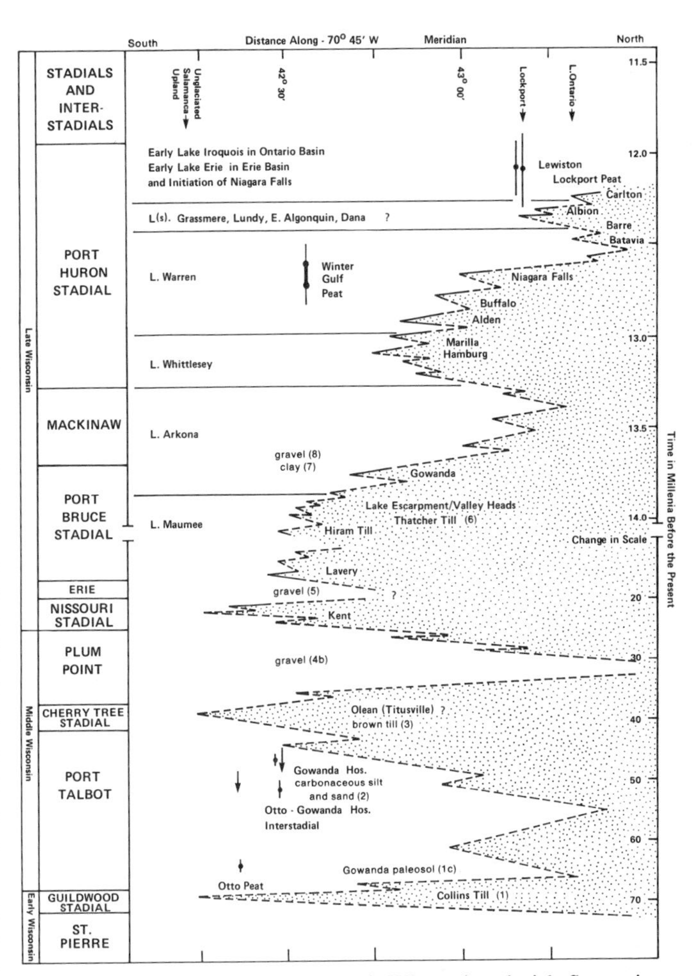

Figure 3. Suggested chronology of Wisconsin glacial fluctuations (stippled) through western New York with selected end moraines, till units, C-14 control ages, as well as stratigraphic units of the Gowanda Hospital site (numbered). Stades and interstades, and time scale for the eastern Great Lakes area after Dreimanis and Karrow (1972), Dreimanis and Raukas (1975), and Karrow (1984). The earliest Wisconsin unit (following the Sangamon interglaciation) is the Nicolet stade, not shown. (Modified after Muller, 1977, and Calkin, 1982.)

Chronology and Correlation

The following chronology appears to be recorded by the stratigraphy interpreted (Table 1) at the Gowanda Hospital Pleistocene site:

1. Deposition of the red Collins Till (unit 1) by ice that flowed over the Ontario Niagara Peninsula and southeastward across the Lake Erie basin.

2. Marked ice marginal retreat, almost certainly north of the Lake Erie basin, with subsequent development of the Gowanda paleosol in the Collins Till (unit 1a), over a period lasting considerably more than 15,000 years.

3. Deposition of a fluvial marsh deposit (unit 2) in association with a closed boreal spruce forest. This accompanied or followed soon after the Gowanda paleosol development.

4. Ice marginal readvance across the area 48 to 54 ka, or perhaps earlier. During glacier overriding of unit 2, organic detritus was incorporated and redeposited with the brown till (unit 3).

5. Stream erosion with subsequent gravel deposition (unit 4) by northwestward flow. This accompanied, or more likely, followed, deposition of the brown till and some ice marginal retreat. Moist, cold conditions persisted through most of the deposition, as indicated by the mollusk assemblage in the associated silt facies (unit 4a).

6. Deposition of the gray Thatcher Till sequence (unit 6)

occurred during an oscillating movement of the ice margin, which may have correlated with formation of the late Wisconsin Lake Escarpment moraine (Fig. 1B).

7. Thinning of the ice margin and final retreat northward across the site, probably from a stand at the Gowanda moraine position, with deposition of proglacial lake clays and waterlaid till lenses (unit 7) in proglacial Gowanda Lake.

8. Shoaling of Gowanda Lake with deposition of lacustrine and then deltaic gravel (unit 8).

A correlation by Calkin and others (1982) of the Gowanda units with some of those at the Otto, New York, site and in turn with the glacial oscillations in western New York is plotted in Figure 3 against a Wisconsin classification for the eastern Great Lakes area. This correlation suggests a very long ice-free middle Wisconsin (Port Talbot) interval prior to about 50 ka. Evidence is missing at Gowanda for glacial advances and retreats of the late Wisconsin glaciation preceding the Lake Escarpment glaciation (Muller, 1975) and coincident deposition of the Thatcher Till about 14 ka.

Reinterpretation of the Gowanda Hospital data in terms of subsequent studies of Pleistocene organic sites in the Great Lakes area led Fullerton (1986) to suggest a correlation of the Collins Till with Illinoian glaciation, the Gowanda paleosol with Sangamon interglacial weathering, and the brown till with early Wisconsin glaciation. Neither the correlation of Calkin and others (1982) nor of Fullerton (1986) are well confirmed at this time. Resolution is difficult because of uncertainties as to the reliability of radiocarbon ages that fall near the limit of the laboratory technique. However, there is now less evidence that an ice margin advanced through the Erie basin during middle Wisconsin time, which was when Calkin and others (1982) had suggested the brown till was deposited (Karrow, 1984; Fullerton, 1986).

REFERENCES CITED

Calkin, P. E., 1970, Strand lines and chronology of the glacial Great Lakes in northwestern New York: Ohio Journal of Science, v. 70, p. 78–96.

——, 1982, Glacial geology of the Erie lowland and adjoining Allegheny Plateau, western New York: Guidebook for New York State Geological Association, 59th Annual Meeting, Amherst, New York, p. 121–148.

Calkin, P. E., and Feenstra, B. H., 1985, Evolution of the Erie-Basin Great Lakes, *in* Karrow, P. F., and Calkin, P. E., eds., Quaternary Evolution of the Great Lakes: Geological Association of Canada Special Paper 30, p. 149–170.

Calkin, P. E., and McAndrews, J. H., 1980, Geology and paleontology of two late Wisconsin sites in western New York State: Geological Society of America Bulletin, v. 91, p. 295–306.

Calkin, P. E., Muller, E. H., and Barnes, J. H., 1982, The Gowanda Hospital Interstadial Site, New York: American Journal of Science, v. 282, p. 1110–1142.

Carll, J. F., 1880, A discussion of the preglacial and postglacial drainage in northwestern Pennsylvania and southwestern New York: Pennsylvania Geological Survey, 2nd, Report III, p. 1–10 and 330–397.

Dreimanis, A., and Karrow, P. F., 1972, Glacial history of the Great Lakes–St. Lawrence region, the classification of the Wisconsin(an) Stage, and its correlatives: International Geological Congress, 24th, Montreal, 1972, Quaternary Geology, sec. 12, p. 5–15.

Dreimanis, A., and Raukas, A., 1975, Did Middle Wisconsin, middle Weichselian, and their equivalents represent an interglacial or an interstadial complex in the Northern Hemisphere?, *in* Suggate, R. P., and Cresswell, M. M., eds., Quaternary studies: Wellington, Royal Society of New Zealand, p. 109–120.

Droste, J. B., and Tharin, J. C., 1958, Alteration of clay minerals in Illinoian till by weathering: Geological Society of America Bulletin, v. 69, p. 61–68.

Evenson, E. B., Dreimanis, A., and Newsome, J. W., 1977, Subaquatic flow tills; A new interpretation for the genesis of some laminated till deposits: Boreas, v. 6, p. 115–133.

Fairchild, H. L., 1907, Glacial waters in the Lake Erie basin: New York State Museum Bulletin 106, 86 p.

——, 1932, New York physiography and glaciology west of the Genesee Valley: Rochester Academy of Science Proceedings, v. 7, p. 97–135.

Forsyth, J. L., 1965, Age of the buried soil in the Sidney, Ohio, area: American Journal of Science, v. 263, p. 571–597.

Frye, J. C., Willman, H. B., and Glass, H. D., 1960, Gumbotil, accretion gley, and the weathering profile: Illinois State Geological Survey Circular 295, p. 1–39.

Fullerton, D. S., 1986, Stratigraphy and correlation of glacial deposits from Indiana to New York and New Jersey (IGCP volume) (in press).

Karrow, P. F., 1984, Quaternary stratigraphy and history, Great Lakes–St. Lawrence region, *in* Fulton, R. J., ed., Quaternary Stratigraphy of Canada—A Canadian Contribution to IGCP Project 24: Geological Survey of Canada Paper 84-10, p. 138–153.

Leverett, F., 1902, Glacial formations and drainage features of Erie and Ohio basins: U.S. Geological Survey Monograph 41, 802 p.

Muller, E. H., 1960, Glacial geology of the Cattaraugus County, New York: Friends of the Pleistocene Guidebook, Eastern Section, 23rd Reunion, Dunkirk, 1960, Syracuse, New York, Syracuse University, 33 p.

——, 1964, Quaternary section at Otto, New York: American Journal of Science, v. 262, p. 461–478.

——, 1975, Physiography and Pleistocene geology, *in* Tesmer, I. H., ed., Geology of Cattaraugus County, New York: Buffalo Society of Natural Sciences Bulletin 27, p. 10–20.

——, 1977, Quaternary geology of New York, Niagara Sheet: New York State Museum and Science Service, Map and Chart Series no. 28, scale 1:250,000.

Willman, H. B., Glass, H. D., and Frye, J. C., 1966, Mineralogy of glacial tills and their weathering profiles, Part II. Weathering profiles: Illinois State Geological Survey Circular 400, 76 p.

Alleghanian deformation within shales and siltstones of the Upper Devonian Appalachian Basin, Finger Lakes District, New York

Terry Engelder, Department of Geosciences, Pennsylvania State University, University Park, Pennsylvania 16802
Peter Geiser, Geology Department, University of Connecticut, Storrs, Connecticut 06268
Dov Bahat, Department of Geology and Mineralogy, Ben-Gurion University of the Negev, Beer Sheva, Israel

LOCATION

The effects of the Alleghanian Orogeny on marine sediments of the Catskill Delta may be examined at two locations in the Finger Lakes District of New York State (Fig. 1). The first location lies at the junction of New York 14 and 414 south of Watkins Glen, New York. Here, a 100-ft (30-m)-thick section of the upper Genesee Group may be followed for 1,300 ft (400 m) along the New York 414 roadcut southwest of the intersection. Park in town near the intersection and walk uphill from the intersection. Many benches may be reached by climbing over the guard rail of New York 414.

The second location is at Taughannock Falls State Park on New York 89 6.6 mi (11 km) north of Ithaca, New York, where a 200-ft (60-m)-deep gorge contains a continuous outcrop of the Tully Limestone and the overlying Genesee Group. Park at the entrance along New York 89. Outcrops may be viewed by walking westward from the entrance along the trail on the south side of Taughannock Creek. The pavement of the creek bed may be examined by climbing down into the creek bed at many points along the trail. The trail also permits excellent views of the rock walls of the gorge.

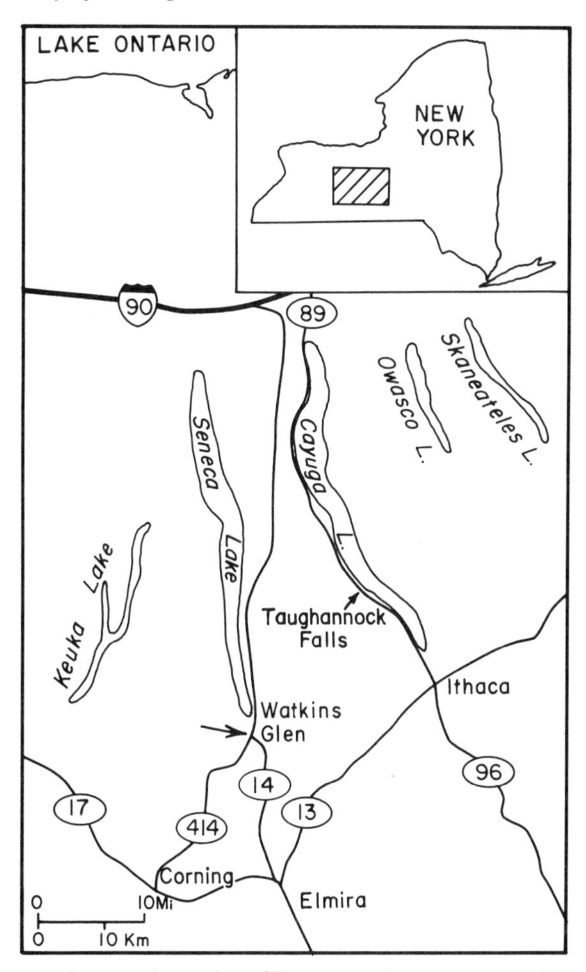

Figure 1. Geographic location of Taughannock Falls and Watkins Glen.

SIGNIFICANCE

The discovery of abundant evidence (deformed fossils) for layer parallel shortening in western New York (Engelder and Engelder, 1977) led us to recognize the extent to which the Alleghanian Orogeny affected the Appalachian Plateau. In addition to low-amplitude (<100 m), long-wave-length (≈15 km) folds, the Upper Devonian sediments of western New York contain many mesoscopic-scale structures that can be systematically related to the Alleghanian Orogeny. The Watkins Glen and Taughannock Falls locations are two of the best for introducing these structures.

Based on the nonorthogonality of cleavage and joints, the Alleghanian Orogeny in the Finger Lakes District of New York consists of at least two phases that Geiser and Engelder (1984) correlate with folding and cross-cutting cleavages in the Appalachian Valley and Ridge. To the southeast of the Finger Lakes District, the earlier Lackawanna Phase is manifested by formation of the Lackawanna syncline and Green Pond outlier and the development of a northeast-striking disjunctive cleavage within the Appalachian Valley and Ridge, mainly from the Kingston Arch of the Hudson Valley southwestward beyond Port Jervis, Pennsylvania (Fig. 2). Within the Finger Lakes District, New York, a Lackawanna Phase cleavage is absent; one finds instead a cross-fold joint set that is consistent in orientation with a Lackawanna Phase compression. In the Genesee Group this cross-fold joint set favors siltstones. In the valley and ridge, the Main Phase is seen as the refolding of the Lackawanna syncline and Green Pond outlier, as well as the development of the major folds in central Pennsylvania. Main Phase structures within the Finger Lakes District include an east-west striking disjunctive cleavage in the Tully Limestone, a pencil cleavage in the Geneseo shales (Engelder and Geiser, 1979), deformed fossils in the Genesee Group (Engelder and Engelder, 1977), and cross-fold joints that are orthogonal with the cleavage and deformed fossils in shales (Engelder and Geiser, 1980). Outcrops of Tully Limestone at Ludlowville (2.4 mi; 4 km east of Taughannock Falls) contain a

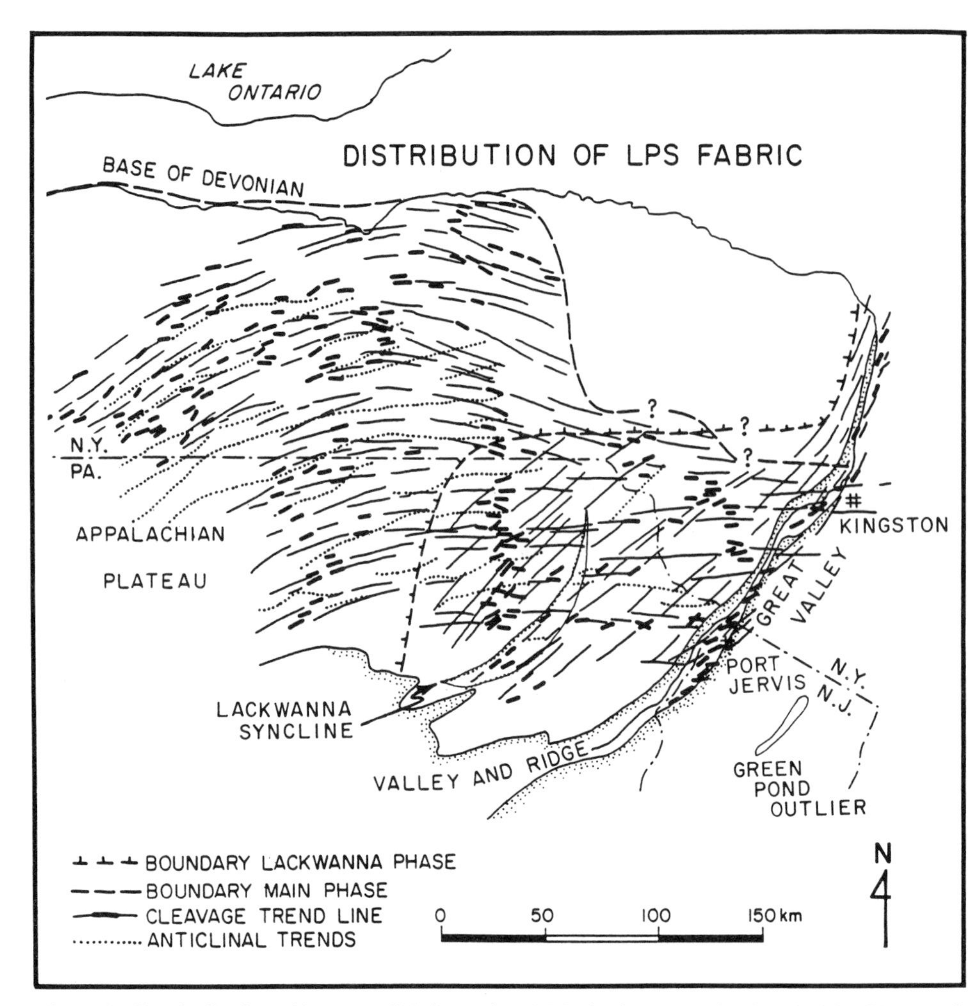

Figure 2. The distribution of layer-parallel-shortening (LPS) fabrics across the Appalachian Plateau of New York. The trend-line map was prepared by connecting data points (thick lines) with nearly parallel cleavage planes. The orientation of the cleavage planes is shown by a plot of the strike of cleavage planes (after Geiser and Engelder, 1983).

Main Phase disjunctive cleavage which truncates nonorthogonal Lackawanna Phase cross-fold joints, showing the age relationship between the Lackawanna and Main Phases (Engelder, 1985).

Many outcrops of the Appalachian Plateau contain more than one cross-fold joint set (Fig. 3). Those cross-fold joints attributed to the Lackawanna Phase strike counterclockwise from those attributed to the Main Phase. Cross-cutting cleavages in northeastern Pennsylvania have the same relationship (Fig. 2).

SITE INFORMATION

Stop 1: Watkins Glen. This roadcut is best viewed in the late morning when the sun strikes the joint surfaces at a high angle. After about 11:30 A.M. when the joint surfaces no longer receive direct sunlight, the surface morphology is far more difficult to see.

As this roadcut at Watkins Glen was excavated, benches were carved out by taking advantage of the jointed rock. The base of the roadcut is dominated by Upper Genesee Group shales. About mid-level in the exposed section, siltstone stringers are intercalated with the shale. At the top of the roadcut, siltstones dominate. In walking uphill along New York 414 from the town of Watkins Glen, find at the base of the roadcut a 6-in-thick (15-cm) siltstone with plume structures nicely developed on a joint face. Stratigraphic levels within the Genesee Group of this roadcut are referenced from the bottom of this 6-in-thick (15-cm) bed. Ten siltstone beds or groups of beds may be used as markers in describing the 112-ft-thick (34-m) roadcut. Major siltstone beds appear above bed #1 as follows: bed #2 ~13 ft (4 m); #3 ~40 ft (12 m); #4 ~53 ft (16 m); #5 ~69 ft (21 m); #6 ~73 ft (22 m); #7 ~83 ft (25 m); #8 and #9 between 89 and 99 ft (27 and 30 m); and #10 ~109 ft (33 m). Beds #7 through #10 are multiple beds. Key siltstone beds are a 7.6-in-thick (19 cm) siltstone stringer (#7) within a shale at the 84.8-ft (25.7-m) level and a 17.6-in-thick (44 cm) siltstone bed (#6) at the 73.9-ft level (22.4 m). Bed #6 stands out as an isolated bench just over half

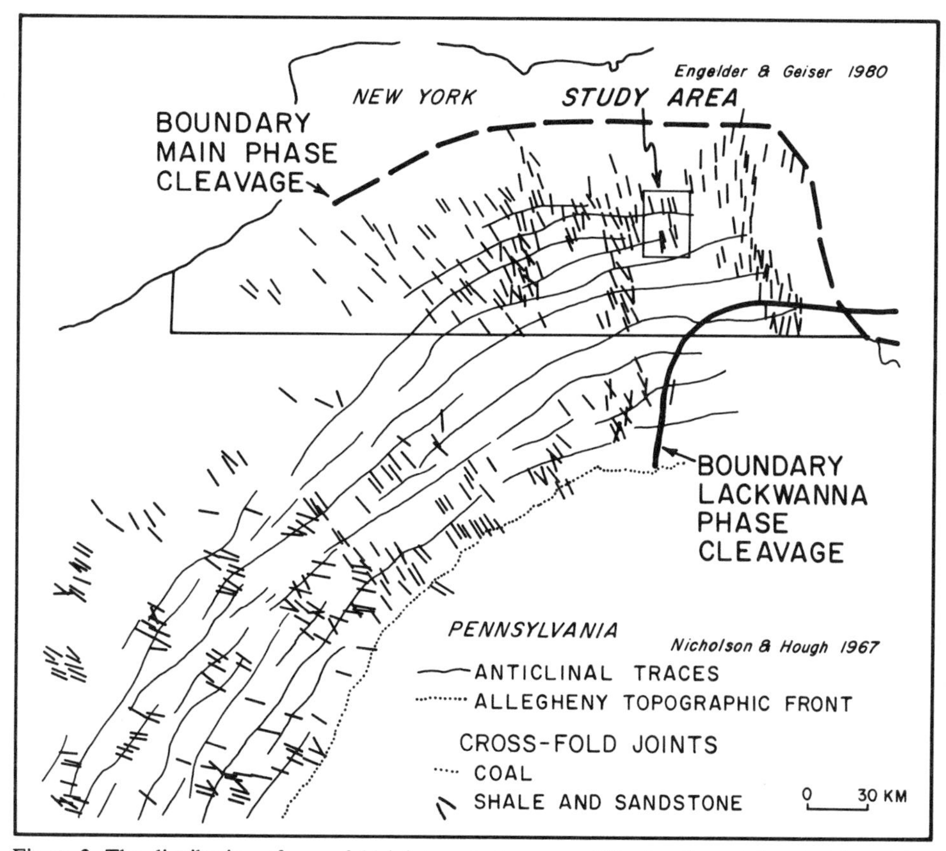

Figure 3. The distribution of cross-fold joints and cleavage within the Central Appalachians (after Engelder, 1985). Cleavage imprinted during the Lackawanna Phase of the Alleghanian orogeny affected the area south of the solid line whereas cleavage imprinted during the Main Phase of the Alleghanian orogeny affected the area south of the dashed line.

way up the roadcut. The key shale bed is the 13-ft-thick (4 m) bed found just above siltstone bed #6 and located between the 7.39-ft and 86.5-ft (22.4-m and 26.2-m) level.

One unusual aspect of this roadcut is the lithological control of the jointing. Vertical joints within the shales strike at 341°–343°, whereas vertical joints within the siltstone beds strike at 331°–334°. Another unusual aspect is the variety of well-developed markings on the surfaces of joints in both the siltstones and shales. Joint surface markings of particular interest include barbs (very fine hackles) and arrest lines. A composite of these barbs and arrest lines gives a delicate plumose marking on the surface of joints in siltstone (Bahat and Engelder, 1984).

The barbs consist of a fine roughness (low-relief elements) on the joint surface and were caused by local out-of-plane crack propagation. This roughness forms ridges parallel to the direction of rupture propagation. Out-of-plane propagation is believed to be caused by microscopic inhomogeneities, such as grain boundaries in the siltstone. Because the shale is more homogeneous on a microscopic scale, there is less tendency for out-of-plane crack propagation. Hence, the shales show no surface morphology equivalent to the barb and ancillary plumose structures on the siltstones.

Joint faces within siltstones contain three varieties of plumose patterns: the straight or s-type plumose marking, which is displayed in the 8-in-thick (19 cm) siltstone bed (#7) at the 84.8-ft (25.7-m) level (Fig. 4A); the curving or c-type plumose marking, which is best displayed in siltstone (bed #8) at the 95.7-ft (29-m) level (Fig. 4B); and the rhythmic c-type plumose marking, which is displayed on the 18-in-thick (44 cm) siltstone bed (#6) at the 73.9-ft (22.4-m) level (Fig. 4C). The straight plume has a linear axis parallel to bedding, whereas the curving plume commonly has an axis that divides into several branches that in turn may themselves divide. Barbs radiate from the plume axes of both the s-type and c-type plume patterns. The barbs form a fine surface morphology, that indicates the direction of rupture propagation, with the rupture moving from the plume axis outward toward the edge of the joint.

The plume structures may be traced backward to their initiation point. Cracks initiate at inclusions within the rock such as fossils, concretions, ripple marks, or microcracks. These inclusions are stress risers that permit the magnification of a far-field stress to overcome the local tensile strength of the rock. At the New York 414 roadcut, most initiation points are bedding plane boundaries. A good example of this effect is found on the joint

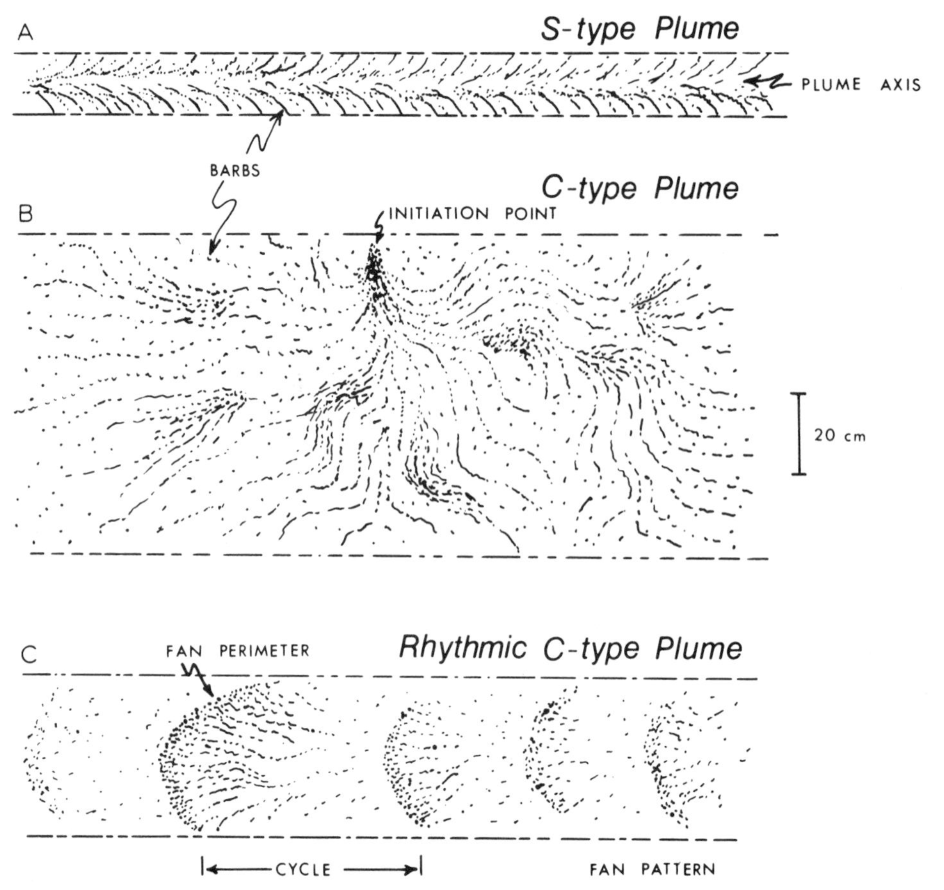

Figure 4. Various plume patterns observed on the surfaces of vertical joints in siltstones of the Finger Lakes District are shown: straight plume (bed #7), curving plume (bed #8), and rhythmic plume (bed #6) (after Bahat and Engelder, 1984).

surface of the 8-in-thick (19 cm) siltstone stringer (#7) at the 84.8-ft (25.7-m) level. The joint within this layer was forced by jointing in the adjacent shales to propagate in the direction of jointing in shale (i.e., 341°, which is highly unusual for jointing within siltstone beds of this roadcut). Four or more initiation points can be found along the 330 ft (100 m) of exposure of this bed. All initiation points are at the top of the bed. In contrast, the 18-in-thick (44 cm) stringer (bed #6) below the 13-ft-thick (4 m) shale bed is cut by a joint striking at 332° and showing initiation points on the bottom of the bed. Higher in the section (approximately 96-ft; 29-m level) siltstone beds are cut by 332° joints with initiation points within the bed.

A feature found on both shale and siltstone joints is arrest lines. These features mark the termination of propagation of individual cracks. The 13-ft-thick (4-m) shale bed, at the 79.2-ft (24-m) level, shows a large arrest line curving on the joint face with the convex side of the line facing in the direction of joint propagation (NNW). This scale shale bed contains the 8-in-thick (19 cm) siltstone stringer (bed #7) displaying an s-type plume pattern. Barbs of this s-type plume diverge in the direction of propagation, which is toward the NNW and compatible with the large arrest line within the shale. Within the same shale bed another joint terminates against the arrest line after propagating in the SSE direction, as indicated by the barbs on the s-type plume within the siltstone stringer (bed #7). Arrest lines can be observed on the 18-in-thick (44 cm) siltstone bed (#6) at the 74-ft (22.4-m) level. These arrest lines are part of the rhythmic c-type plume pattern found on joint faces cutting siltstone beds. There the arrest lines are spaced less than 3 ft (1 m) apart, in contrast to those on the thick shales which are separated by more than 165 ft (50 m).

The closely spaced arrest lines in bed #6 may be interpreted in terms of jointing mechanism. The deeper portion of the Appalachian Basin is undercompacted (Engelder and Oertel, 1985), which is taken to indicate that abnormal pore pressures once prevented normal pore collapse. Pore pressures approaching the weight of the overburden may have caused natural hydraulic fracturing. The closely spaced arrest lines within the siltstone beds suggest the type of cyclic rupture expected of a slow build-up of pore pressure followed by a fast decrease accompanying the incremental propagation of a joint. This process can repeat many times to leave a set of closely spaced arrest lines. Engelder (1985) distinguishes tectonic joints as those caused by abnormal pore pressures during tectonic compression. At Watkins Glen both cross-fold joints are tectonic joints.

The c-type plumes are found on joints striking at 331°–333° (this is considered the normal orientation for joints in siltstone layers in this roadcut), whereas small siltstone stringers (i.e., bed #7) in thick shales show the s-type plumes with joints striking at 341°. The s-type plume in bed #7 is believed to indicate a rapid rupture that extended more than 165 ft (50 m) in a horizontal direction. The length of the rupture is indicated by the distance between the initiation point 165 ft (50 m) to the SSE and the large arrest lines within the 13-ft-thick (4 m) shale layers. In contrast, the c-type plumes give the impression of a slower, less decisive rupture. Arrest lines spaced at less than a meter on the 332° joints confirm this notion.

The difference between joint propagation in the shales and joint propagation in the siltstones is further understood by placing the timing of their propagation in a regional context. On upper benches of the outcrop (at the 112-ft; 34-m level), deformed crinoid columnals show that the layer-parallel shortening (LPS) during the Alleghanian Orogeny was oriented at 341°. Geiser and Engelder (1983) interpret this LPS direction as a principal compression direction during the Main Phase of the Alleghanian Orogeny. At Watkins Glen the othogonality of shale joints and LPS indicated by deformed fossils suggests that the shale joints propagated during the Main Phase. If this is so, then when did the joints within the siltstone beds propagate?

The joints in the siltstone are believed to precede those in shales. First, early joints at Ludlowville, New York are cut by Main Phase cleavage. These joints strike a few degrees counterclockwise from the Main Phase LPS. Second, in a deeply buried siltstone-shale sequence, the siltstones are known to show a lower least principal stress compared to shales. If joints are hydraulic fractures propagating under high fluid pressures, the joints in beds with the lower least principal stress will propagate first. Third, the preferred orientation of chlorite within the siltstone beds is compatible with a LPS counterclockwise from the LPS affecting the shales (Oetel and Engelder, 1986). Elsewhere in the central Appalachians, the Lackawanna Phase LPS precedes the Main Phase LPS with a counterclockwise compression (Fig. 2).

Stop 2: Taughannock Falls State Park. Taughannock State Park features a U-shaped hanging valley and 165 ft (50 m) waterfall at the head of a 0.9-mi-long (1.5 km) gorge cut to the level of Cayuga Lake. Outcrops in the park consist of the Tully Limestone and Genesee shales within the stream bed of Taughannock Creek and the lower portion of the Genesee Group (the Geneseo shales) exposed on the walls of the gorge. About 660 ft (200 m) upstream from the park entrance, bedding surfaces of the Tully Limestone may be examined within the stream bed, whereas 3,300 ft (1,000 m) upstream from the park entrance the stream bed consists of the Geneseo shales.

On beds of the Tully Limestone 660 ft (200 m) from the park entrance, a disjunctive cleavage is well developed. The cleavage gives a faint herringbone pattern on the gray pavement of the Tully Limestone. Cleavage domains appear as a wavy trace of a dark selvage against the light gray background of Tully Limestone. Individual selvages extend for tens of cm before ending in many fine branches. The microlithons of Tully Limestone are 2 to 6 in (5 to 15 cm) thick. This spacing of cleavage domains constitutes a weak cleavage according to the classification of Alvarez and others (1978). The general trend of the cleavage is 077°, which is normal to the compression direction of the Main Phase of the Alleghanian orogeny in the vicinity of Ithaca, New York. Because the cleavage is wavy, any one cm-length of selvage might be misoriented from the 077° trend by as much as 15°. Close examination of the selvages will reveal short stylolites pointing in the direction of the Main Phase compression at about 347°. Further west at the Watkins Glen outcrop this compression direction is 341°.

The contact between the Tully Limestone and the Geneseo shales is a fine example of the relationship between disjunctive cleavage in the limestone and the development of pencil cleavage in the shales. Best examples are found on the north side of the creek about 1,320 ft (400 m) from the park entrance. The long axes of the pencils within the Genesee shales trend at 077°, which parallels the strike of disjunctive cleavage within the Tully Limestone. Here the pencils take the shape of blocky rectangular solids rather than being long and skinny. In other outcrops of the Appalachian Plateau the pencils are well developed enough to be like a pencil in shape. The two short dimensions of a pencil cleavage consist of bedding and a disjunctive cleavage normal to bedding.

At Taughannock Falls, the Tully Limestone contains neither of the cross-fold joints described at Watkins Glen. The best developed joints in the Tully Limestone are several sets of en echelon cracks found on the second bench of Tully Limestone about 1,000 ft (300 m) from the park entrance. Individual cracks within the en echelon set strike at 316°, whereas the shear couple indicated by the en echelon zone strikes at 324°. There seems to be no clear relationship between these en echelon cracks and the Alleghanian Orogeny.

Cross-fold joints start to appear within the Genesee shales about 3,300 ft (1,000 m) from the park entrance. On walking upstream, Taughannock Creek makes a righthand turn at this point. Here it is common to see later subparallel cross-fold joints (~330°) curving into and abutting earlier cross-fold joints (~340°). These latter joints are the same orientation as the cross-fold joints in shale at Watkins Glen and indicate a compression direction 7° counterclockwise from that indicated by the LPS in the Tully Limestone. However, the abutting of cross-fold sets was not found at Watkins Glen.

The rocks in the walls of Taughannock Creek gorge provide an example of the behavior of cross-fold joints within thick (>165 ft; 50 m) sequences of homogeneous shale. This is best seen at the point where the gorge makes a right turn 3,300 ft (1,000 m) from the park entrance. Looking up on the southeast side of the gorge, cross-fold joints are displayed on the southern wall; strike joints are displayed on the eastern wall. The cross-fold joints are better developed and more closely spaced. Engelder

(1985) distinguishes tectonic joints (those caused by abnormal pore pressures during tectonic compression) from release joints (those caused by erosion and controlled in orientation by a pervasive fabric such as disjunctive cleavage). At the right turn in Taughannock Creek, the cross-fold joints are tectonic joints and the strike joints are release joints. Evidence for the development of abnormal pressures during tectonic (cross-fold) joint propagation include the undercompaction of the Geneseo shales (Engelder and Oertel, 1985). Note that good examples of release joints are not common in the creek bed. In general the release (strike) joints tend to be less regular in profile than tectonic (cross-fold) joints.

Within the gorge of Taughannock Creek the tectonic joints can be traced continuously up the valley wall for a large fraction of the exposure of the Geneseo shales. The joints propagate so that their vertical dimension is as large or larger than their horizontal (parallel-to-strike) dimension. This is an example of vertical joint growth that is not impeded by bedding interfaces (Engelder, 1985). The Watkins Glen outcrop presents an example where vertical joint growth is impeded by bedding interfaces. At Watkins Glen the growth in the horizontal direction is extreme (>165 ft; 50 m), and vertical growth is limited (<13 ft; 4 m). This latter phenomenon can also be observed at the Falls Road Bridge, which crosses Taughannock Creek at the upper entrance to the park. Take Gorge Road which turns west off of New York 89 just south of the park. Proceed 1.2 mi (2 km) and then make a right turn at the T-intersection with Falls Road.

REFERENCES

Alvarez, W., Engelder, T., and Geiser, P., 1978, Classification of solution cleavage in pelagic limestones: Geology, v. 4, p. 698–701.

Bahat, D., and Engelder, T., 1984, Surface morphology on cross-fold joints of the Appalachian Plateau, New York and Pennsylvania: Tectonophysics, v. 104, p. 299–313.

Engelder, T., 1985, Loading paths to joint propagation during a tectonic cycle; An example from the Appalachian Plateau, U.S.A.: Journal of Structural Geology, v. 7, p. 459–476.

Engelder, T., and Engelder, R., 1977, Fossil distortion and décollement tectonics on the Appalachian Plateau, New York: Geology, v. 5, p. 457–460.

Engelder, T., and Geiser, P., 1979, The relationship between pencil cleavage and lateral shortening within the Devonian section of the Appalachian Plateau, New York: Geology, v. 7, p. 460–464.

—— , 1980, On the use of regional joint sets as trajectories of paleostress fields during the development of the Appalachian Plateau, New York: Journal of Geophysical Research, v. 85, p. 6319–6341.

Engelder, T., and Oertel, G., 1985, The correlation between undercompaction and tectonic jointing within the Devonian Catskill Delta: Geology, v. 13, p. 863–866.

Geiser, P., and Engelder, T., 1983, The distribution of layer parallel shortening fabrics in the Appalachian foreland of New York and Pennsylvania; Evidence for two non-coaxial phases of the Alleghanian orogeny: Geological Society of America Memoir 158, p. 161–175.

Nickelsen, R. P., and Hough, V. D., 1967, Jointing in the Appalachian Plateau of Pennsylvania: Geological Society of America Bulletin, v. 78, p. 609–630.

ACKNOWLEDGMENTS

This work was supported by NSF grants EAR-79-13000 and EAR-83-08650 and a contract with the Nuclear Regulatory Commission (T.E.).

Lower Devonian limestones, Helderberg Escarpment, New York

Donald W. Fisher, State Paleontologist (retired), New York State Geological Survey, State Museum, Albany, New York 12230

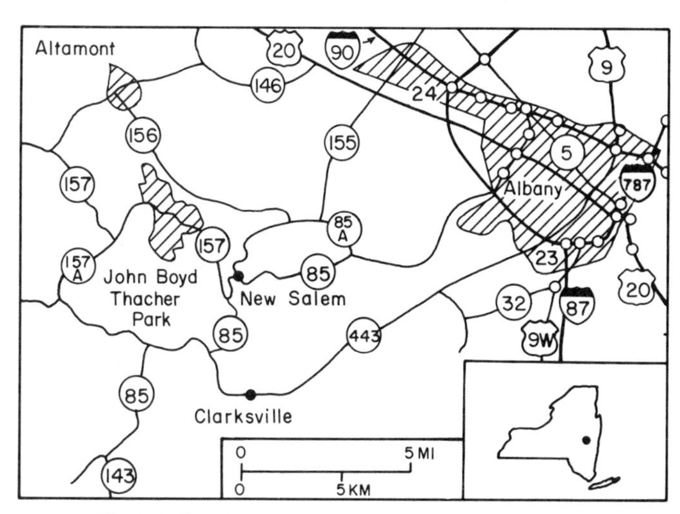

Figure 1. Road map to John Boyd Thacher State Park.

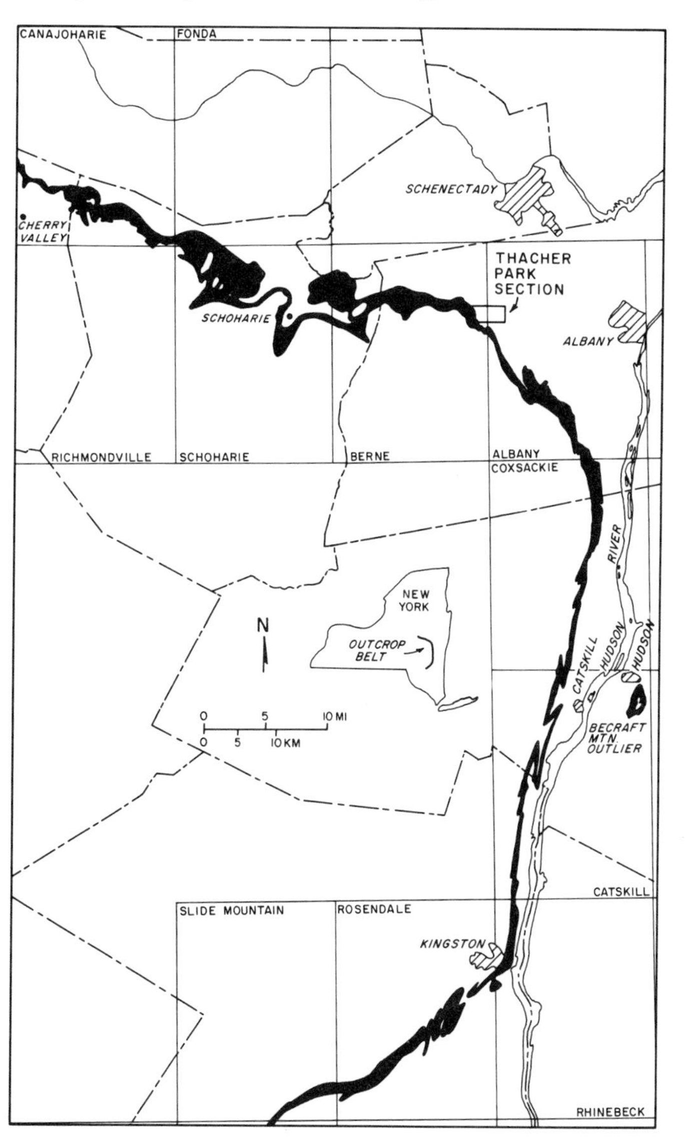

Figure 2. Helderberg outcrop belt. Modified from Rickard 1962.

LOCATION

The site discussed here is in Thacher State Park, which extends for 2 mi (3.2 km) along New York 157 in the extreme northwestern corner of Albany County (Fig. 1). The Park is 13 mi (20.8 km) due west of the State Capitol in downtown Albany at 42°38′N and 74°0′W. The limestone belt is accessible by bus, passenger car, and foot, with rock exposures on both public and private lands; hammering and collecting is prohibited within the state park. The Helderberg Escarpment extends from Kingston (Rosendale Quadrangle, Ulster County) on the south through Catskill (Catskill Quadrangle, Greene County) through Thacher State Park (Berne Quadrangle, Albany County) through Schoharie (Schoharie Quadrangle, Schoharie County) to Cherry Valley (Canajoharie Quadrangle, Otsego County) (Fig. 2).

SIGNIFICANCE

Since the early nineteenth century the Helderberg Escarpment has been much investigated stratigraphically, sedimentologically, and paleontologically (Chadwick, 1944; Fisher, 1979, 1980; Goldring, 1935, 1943; Grabau, 1906; Rickard, 1962, 1975). Because of its extensive exposure as shown in Figure 2, the Helderberg section has provided geologists with vital stratigraphic, sedimentologic, and paleontologic data, both along the regional sedimentary strike (north-south) and at right angles (west-east) to sedimentary strike. This section has made unusually reliable paleogeographic and paleoenvironmental interpretations possible over large portions of New York and Pennsylvania. Accordingly, the Helderberg Escarpment displays the finest Lower Devonian sequence in North America. This classic terrane demonstrates several Early Devonian (Helderbergian = Gedinnian) limestone facies, each with their diagnostic physical and organic characters.

Important unconformities occur at both the base and summit of the Helderberg Group, although neither erosion surface is observable within Thacher State Park. Both are nearly approachable, the lower one especially where the Indian Ladder Trail passes under Minelot Falls. The Late Ordovician (Edenian) Indian Ladder Shale is subjacent; the late Early Devonian (Ulsterian = Siegenian) Oriskany Sandstone is superjacent to the Helderberg Group. Proven Silurian strata are absent between south of Catskill and northwest of Thacher Park. Nonfossiliferous Rondout Dolostone at the base of the Thacher Park section has variously been classed as latest Silurian or earliest Devonian in

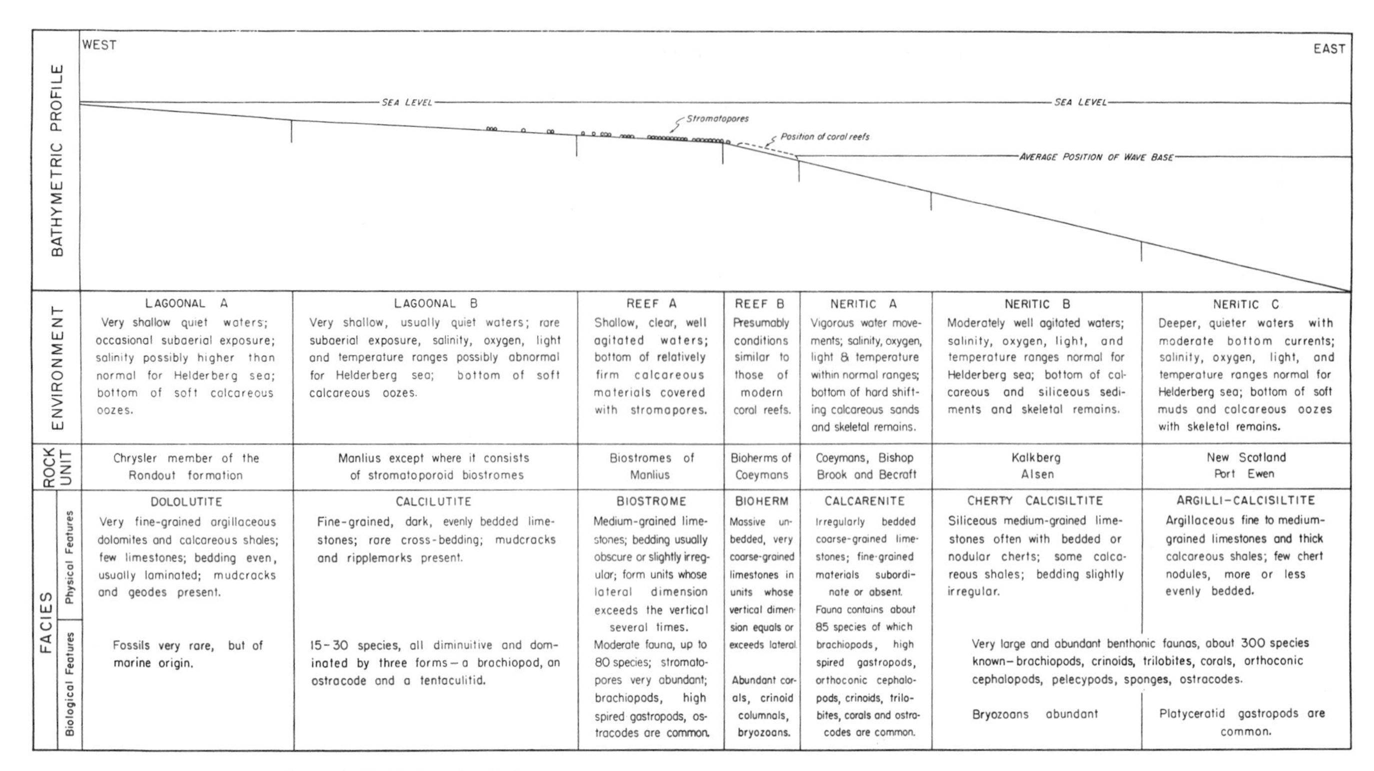

Figure 3. Helderbergian facies and their inferred environments. Modified from Rickard, 1962.

the eastern Berne, Albany, Coxsackie, and northern Catskill quadrangles. Probably no Silurian strata are present within Thacher Park (see Figs. 3 and 4 for stratigraphic and paleoenvironmental relationships).

SITE INFORMATION

Within the park, at the "Overlook" at the edge of the escarpment, one may view the Taconic and Berkshire mountains to the east and, in the foreground, the intervening Hudson River valley—once occupied by glacial (Wisconsinan) Lake Albany. The cliff's caprock consists of the Coeymans Limestone, whereas the bulk of the cliff face consists of the underlying Manlius Limestone. A prominent reentrant marks the position of the relatively thin basal Rondout Dolostone. Close examination of these three units is possible by hiking the "Indian Ladder Trail," consisting of stairs at both north and south ends descending to an unpaved soil and rock trail that follows the base of the carbonate sequence.

The Rondout Dolostone, a fine-textured, light gray "waterlime," is best seen at the point where Minelot Falls showers over the trail at the large overhang. No more than 3 ft (1 m) of disturbed Rondout is visible. The deformation has been attributed to collapse of strata owing to chemical solution. Many cavities, once occupied by saline minerals, are visible.

The overlying Thacher Member of the Manlius Limestone (58 ft; 17.4 m) includes three different lithologies that replace one another laterally (Fig. 4). The first type, the "ribbon" limestones or "Tentaculite" limestones, is more typical of the lower Thacher. It is a fine-grained, very dark gray to black, very thin-bedded, exceedingly pure limestone (95-98 percent calcium carbonate). The rock weathers very light gray to white. Bedding planes are smooth. Loose pieces often ring or clink when struck with a hammer. Fossils are relatively common but only a few species are present. A fauna of abundant individuals but few species is suggestive of abnormal environmental conditions; in this instance lagoonal conditions are inferred. A cricoconarid (*Tentaculites gyracanthus*), an ostracode (*Leperditia alta*), and a brachiopod (*Howellella vanuxemi*) are most prevalent.

The second rock type is similar but with thicker bedding and occasional shale interbeds. Grain-size is somewhat coarser and rare cross-bedding and ripple marks may be present.

The third rock type records a significantly different environment. It is a stromatoporoid biostrome that acted as a baffle to breaking waves at the outer edge of the lagoon (Fig. 3). This facies has indistinct-to-irregular bedding and resembles cabbage heads on weathering. Where present, it occurs at the top of the Manlius Limestone in Thacher Park. The biostrome is present at the southern access to the Indian Ladder Trail but is absent at the northern access.

At the top of the escarpment the overlying Coeymans Limestone (35 ft; 10.5 m) is a calcarenite facies, in contrast to the calcilutite facies of the Manlius (Fig. 3). The Coeymans is a medium- to coarse-grained, light gray to bluish gray limestone with irregular to indistinct bedding. Weathering is light gray. Fossils are abundant but frequently broken. Crinoid columnals are very abundant as are robust brachiopods, chief of which is the pentamerid *Gypidula coeymanensis*. Tabulate and rugose corals

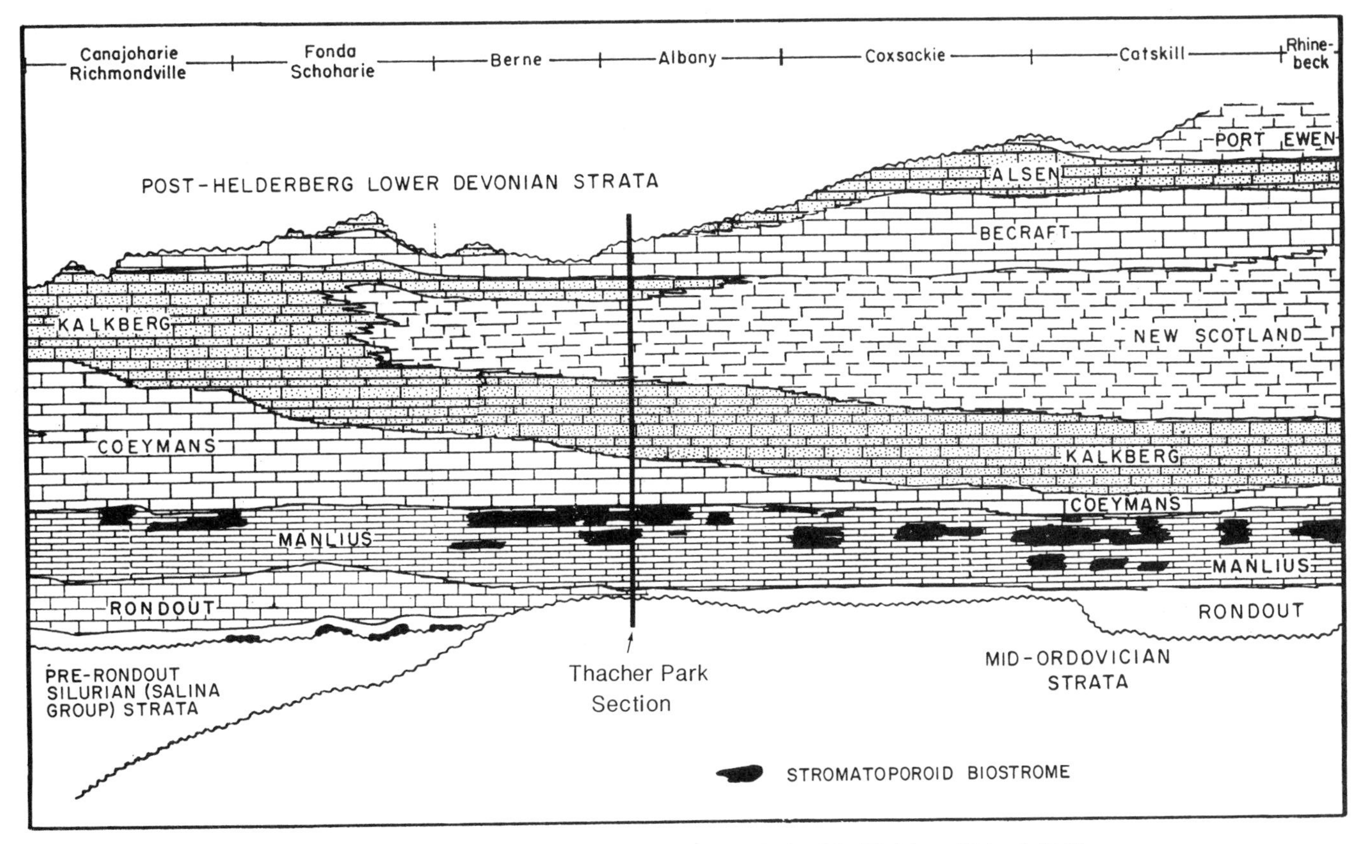

Figure 4. Stratigraphic relationships of Helderbergian facies. Modified from Rickard, 1962.

are common; locally, west of the area discussed, corals are profuse enough to construct reefs. A few species of trilobites, straight-shelled nautiloid cephalopods, gastropods, bryozoans, cystoids, and ostracodes round out the fauna.

From 50 to 250 ft (15 to 75 m) back from the lip of the escarpment is the overlying Kalkberg Limestone (49 ft; 14.7 m) well shown along Minelot Brook. This is a thin- to medium-bedded, fine- to medium-grained, medium to dark gray, siliceous limestone with nodules and beds of black and bluish-black chert and interbedded calcareous and argillaceous shales. Brachiopods are the predominant faunal element, closely followed by a great diversity of bryozoans, corals, ostracodes, and subordinate trilobites. In 1955, I was fortunate to discover a bentonite (volcanic ash) bed in the Kalkberg near Cherry Valley. This bed was subsequently traced eastward. Radiometric analysis on zircons in the ash by Donald Miller, Rensselaer Polytechnic Institute (written communication, 1957) gave an age of 395 ± 5 Ma for the deposition of the ash, making this the first radiometric age for the Early Devonian in North America.

Overlying the Kalkberg is the New Scotland Limestone (66 ft; 20 m), the most fossiliferous of the Helderberg units (more than 300 species have been reported). Several road cuts occur along New York 157 within the park. The New Scotland is an argillaceous limestone to calcareous shale, often quite siliceous and with some black chert beds and nodules. It weathers a distinct rusty-brown, denoting its relatively high iron oxide content. Brachiopods and bryozoans predominate.

Forming a second small terrace above the main Helderberg Escarpment is the Becraft Limestone (12 ft; 4 m), which directly overlies the New Scotland Limestone. The most accessible exposures are along and near New York 157 within the park. The Becraft is a very coarse-grained, massive, medium gray to pink limestone with such an abundance of crinoidal debris and robust brachiopods that it may be termed a shellrock or coquinite. Bedding is normally indistinct although some green shale partings do occur. The crinoid base *Aspidocrinus scutelliformis* is ubiquitous. Likewise, atrypid and uncinulid brachiopods are exceedingly abundant. Like the Coeymans, the Becraft Limestone was formed in a high-energy environment as evidenced by its fragmented and robust fossils.

Atop the Helderberg Group of limestones and west of New York 157 may be found the Oriskany Sandstone, Esopus Shale, Carlisle Center calcareous siltstone and shale, Schoharie siliceous limestone, and Onondaga Limestone. The last-named unit forms the third and highest terrace within the park limits. The higher hills of the Catskill Mountains proper consist of black, gray, and green shales and brownish green sandstones of the Middle Devonian Hamilton Group.

REFERENCES CITED

Chadwick, G. H., 1944, Geology of the Catskill and Kaaterskill quadrangles; Part II, Silurian and Devonian geology: New York State Museum Bulletin 336, 251 p., map scale 1:62,500.

Fisher, D. W., 1979, Devonian stratigraphy and paleoecology in the Cherry Valley, New York, region: New York State Geological Association Guidebook, 51st Meeting, Troy, Rensselaer Polytechnic Institute, 26 p.

—— , 1980, Bedrock geology of the central Mohawk Valley: New York State Museum Map and Chart Series 33, 44 p., scale 1:48,000.

Goldring, W., 1935, Geology of the Berne Quadrangle: New York State Museum Bulletin 303, 238 p., map scale 1:62,500.

—— , 1943, Geology of the Coxsackie Quadrangle: New York State Museum Bulletin 332, 374 p., map scale 1:62,500.

Grabau, A., 1906, Guide to the geology and paleontology of the Schoharie Valley in eastern New York: New York State Museum Bulletin 92, 386 p., map scale 1:62,500

Rickard, L. V., 1962, Late Cayugan (Upper Silurian) and Helderbergian (Lower Devonian) stratigraphy in New York: New York State Museum Bulletin 386, 157 p.

—— , 1975, Correlation of the Silurian and Devonian rocks in New York State: New York State Museum Map and Chart Series 24, 16 p.

Exposures of the Hudson Valley Fold-Thrust Belt, west of Catskill, New York

Stephen Marshak, *Department of Geology, University of Illinois, Urbana, Illinois 61801*
Terry Engelder, *Department of Geosciences, Pennsylvania State University, University Park, Pennsylvania 16802*

LOCATION

Outcrops visible along New York 23 and along Catskill Creek, about 1.2 mi (2 km) northwest of the town of Catskill, New York (about 33 mi [55 km] south of Albany), provide a nearly complete cross-sectional display of the Hudson Valley Fold-Thrust Belt (Marshak, 1983; 1986a). These exposures provide an excellent opportunity both to examine thin-skinned structural styles and to study facies relationships among shallow marine carbonate rocks. The roadcuts are accessible directly from the highway, and it is legal to park for short periods anywhere along the highway shoulder. Creek exposures are accessible from paths that run north and south from the southeast side of the New York 23 bridge over Catskill Creek, but it is necessary to obtain permission of landowners (see posted signs) to view these exposures.

The New York 23/Catskill Creek site can be reached from the New York State Thruway Exit 21 (Fig. 1). Leave the Thruway at Exit 21 and pass through the tollgate. At the end of the tollgate access road, turn left (southeast) onto New York 23B heading toward Jefferson Heights and Catskill. Continue on New York 23B for 0.2 mi (0.3 km) to the junction with New York 23. The outcrops along the west shoulder of New York 23B at this parking spot are composed of Kalkberg Formation and display several small thrust faults with adjacent "drag" folds. To reach outcrops N5 and N4 (labeled in Figs. 1, 2), park on the west shoulder of 23B just north of the entrance ramp that leads onto New York 23 heading northwest. Outcrop N5 is along the exit ramp from New York 23 northwest leading to 23B, and outcrop N4 is along the entrance ramp from 23B onto 23 northwest. To reach outcrops N3-N1 and S3-S1, drive onto New York 23 heading northwest toward Cairo. Outcrops N3 and S3 lie southeast of the Thruway, N2 and S2 lie between the Thruway and Catskill Creek, and N1 and S1 lie northwest of Catskill Creek.

Our discussion will consider only the New York 23 and Catskill Creek outcrops. Figure 2 (see also Babcock, 1966) illustrates that sufficient outcrops occur away from these cuts to permit mapping of the area. In addition, excellent exposures of the fold-thrust belt are available elsewhere along the Hudson Valley fold thrust belt (Fig. 3). Notable examples include (1) quarry exposures at Fuerra Bush (Marshak, 1986a), Becraft Mountain (Chapple and Spang, 1974), Mount Ida (Ratcliffe and others, 1975), Quarry Hill, West Camp (Leftwich, 1973; Zadins, 1983), and Kingston (McEachran, 1985; Tabor, 1985; Marshak, in preparation); (2) roadcuts along New York 23A (Marshak and Geiser, 1980), New York 199 and 32 in Kingston (Waines and Hoar, 1967), U.S. 9W in Kingston (Marshak, 1986b), and the New York State Thruway (written permission must be obtained from the Thruway Authority in Albany to view Thruway cuts); and (3) in Hasbrouck Park in the City of Kingston (Marshak, in preparation).

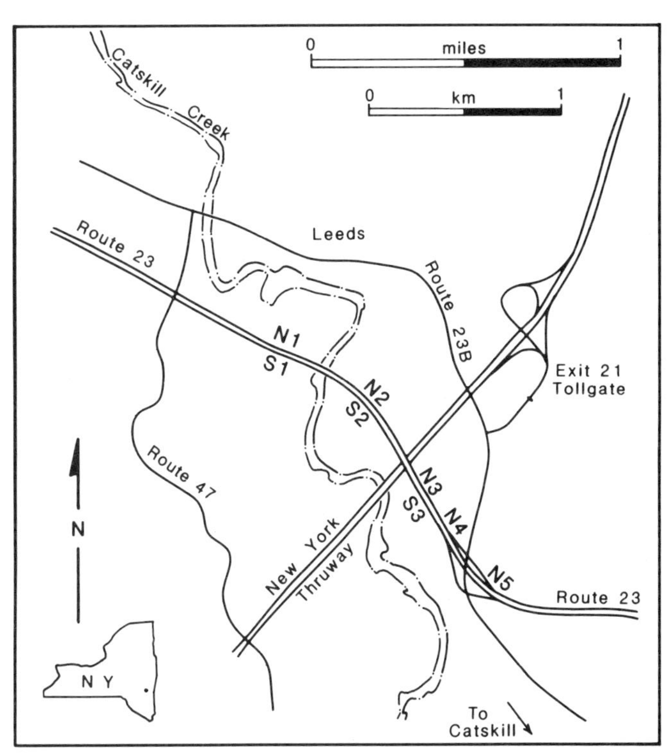

Figure 1. Sketch map of the Catskill area showing the location of the New York 23 and Catskill Creek outcrops discussed in the text.

SIGNIFICANCE

The Hudson Valley Fold-Thrust Belt (HVB) involves Upper Silurian through lower Middle Devonian strata, and lies along the west edge of the Hudson Valley between Kingston and Albany (Fig. 3). Structural features within the HVB are similar in style to those of most fold-thrust belts except that the dimensions of structures in the HVB are so small that most structures can be seen in their entirety in a single outcrop (Davis, 1882; 1883; Sanders, 1969). The HVB, in effect, provides scale models of range-scale structures that occur in large fold-thrust belts, such as the Canadian Rockies. In particular, the exposures near Catskill provide (1) numerous examples of thin-skinned structural geometries, (2) exposures of the Taconic angular unconformity (Rodgers, 1971), (3) examples of mesoscopic structures (e.g., tectonic cleavage and

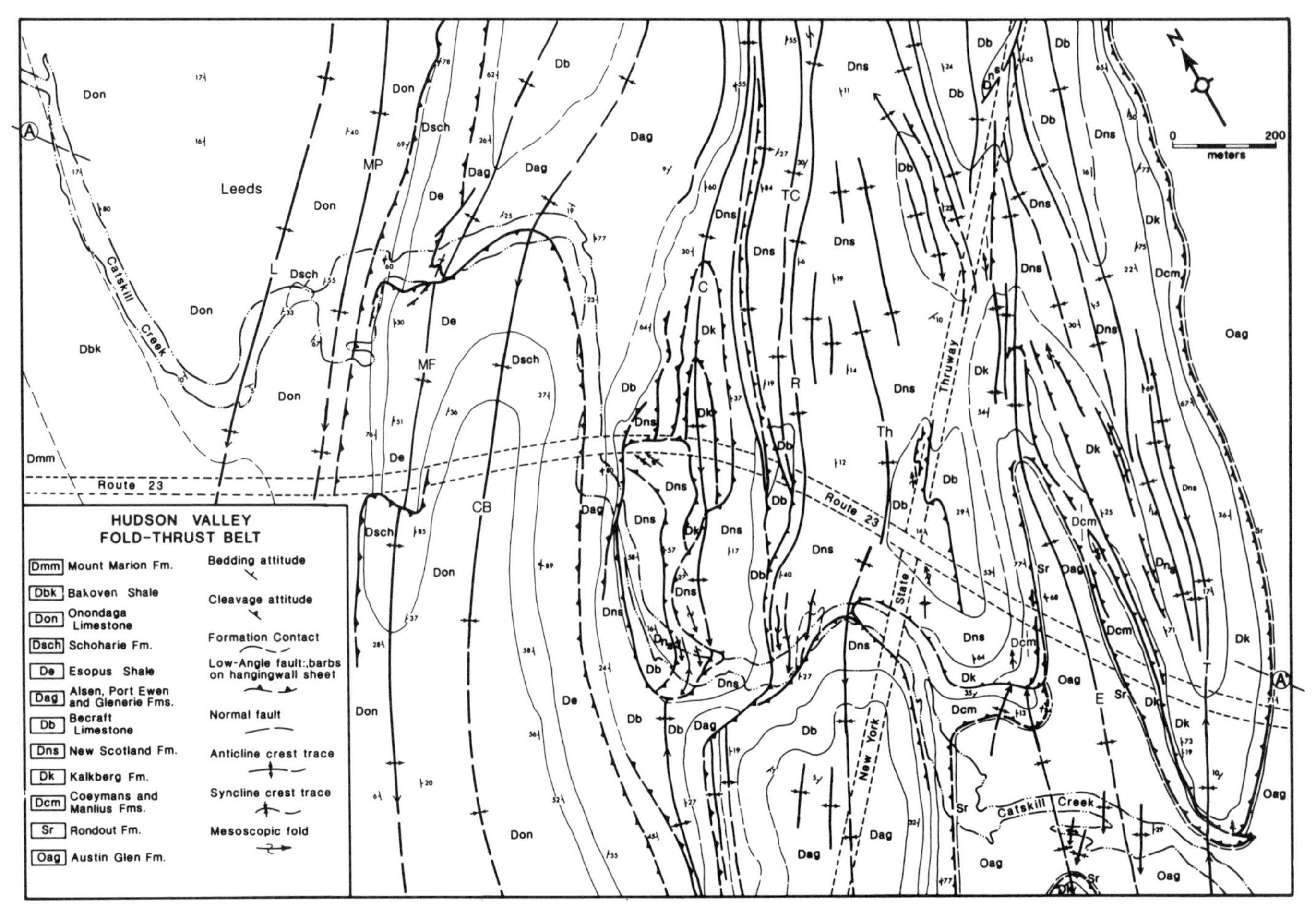

Figure 2. Geologic map of the HVB along New York 23 and Catskill Creek, northwest of Catskill, New York. T = Tollgate; E = Eastern; Th = Thruway; R = Rip van Winkle; TC = Town & Country; C = Central; CB = Creek Bend; MF = Mill Falls; MP = Mill Pond; L = Leeds.

slip fibers) that characteristically form in sedimentary rocks deformed under relatively low pressure and temperature conditions (Marshak and Engelder, 1985), (4) classic Lower Devonian North American faunas (see Chadwick, 1944; Goldring, 1943), and (5) examples of shallow-marine carbonate facies (Rickard, 1962; LaPorte, 1969). Studies of structures in the HVB also provide useful information concerning tectonics of the New England Appalachians (Marshak, 1986a).

SITE INFORMATION

The HVB involves units that lie above the Taconic unconformity (Fig. 4). Near Catskill, the subunconformity sequence is composed of the Austin Glen Formation, which consists of interbedded greywacke and shale. Locally, this shale contains well-developed pencil cleavage. The basal unit above the unconformity is the Rondout Formation, represented by 3 to 6 ft (1 to 2) m of sandy dolomitic limestone. (This unit thickens significantly to the south.) Above the Rondout Formation are the Helderberg and Tristates Groups, which include a range of units indicative of successive transgressions of a shallow sea (see Sanders, 1969). The only noncarbonate unit in this sequence is the Esopus Formation. Above the Tristates Group is the Onondaga Limestone, which is the youngest carbonate unit to be deposited prior to the deposition of the Catskill clastic wedge. Deformation features characteristic of the HVB are visible in the Bakoven Shale and Mount Marion Formation but cannot be found in younger units (Murphy and others, 1980). Below, we discuss structural features exposed along New York 23 and along Catskill Creek. Only brief descriptions are possible here; for further descriptions, see Marshak, 1986b.

Across the width of the HVB near Catskill, there are 10 major folds (named in Fig. 2) with amplitudes in the range of 164 to 394 ft (50 to 120 m). These folds generally trend north-south to N.15°E., and are gently plunging. Within these folds, there are many significant faults and, locally, zones of mesoscopic folding.

Outcrop N5 (Fig. 5) is the easternmost outcrop the HVB along New York 23. At the east end of this outcrop is an exposure of the Taconic unconformity, which here is a pronounced angular unconformity. The upper meter of the overlying Rondout Formation is intensely deformed and may be the location of a detachment fault, called the Rondout detachment, which acts as

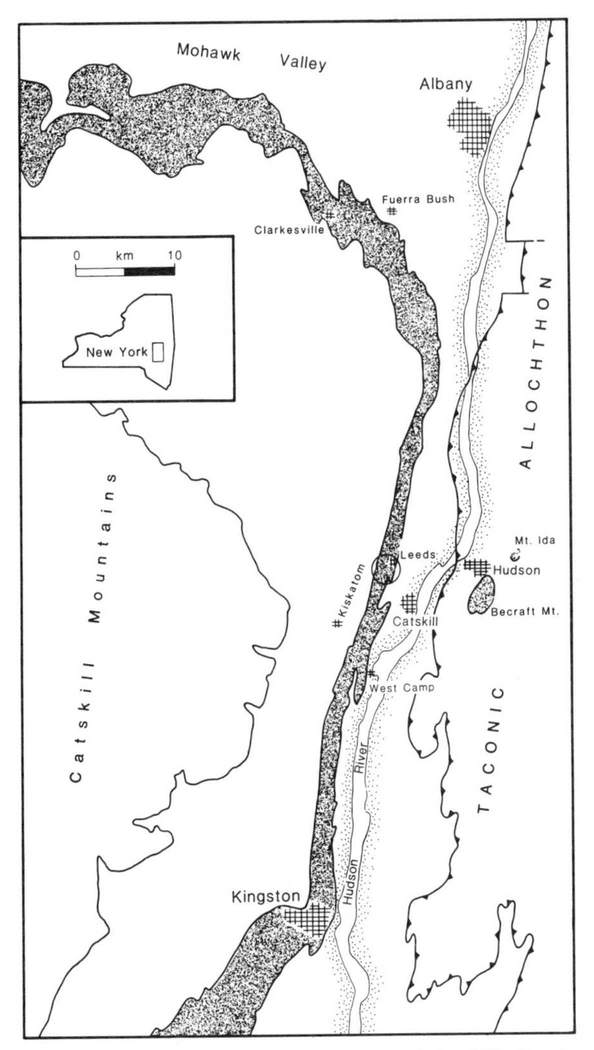

Figure 3. Location map of the HVB. The outcrop belt of Silurian through lower Middle Devonian strata is stippled.

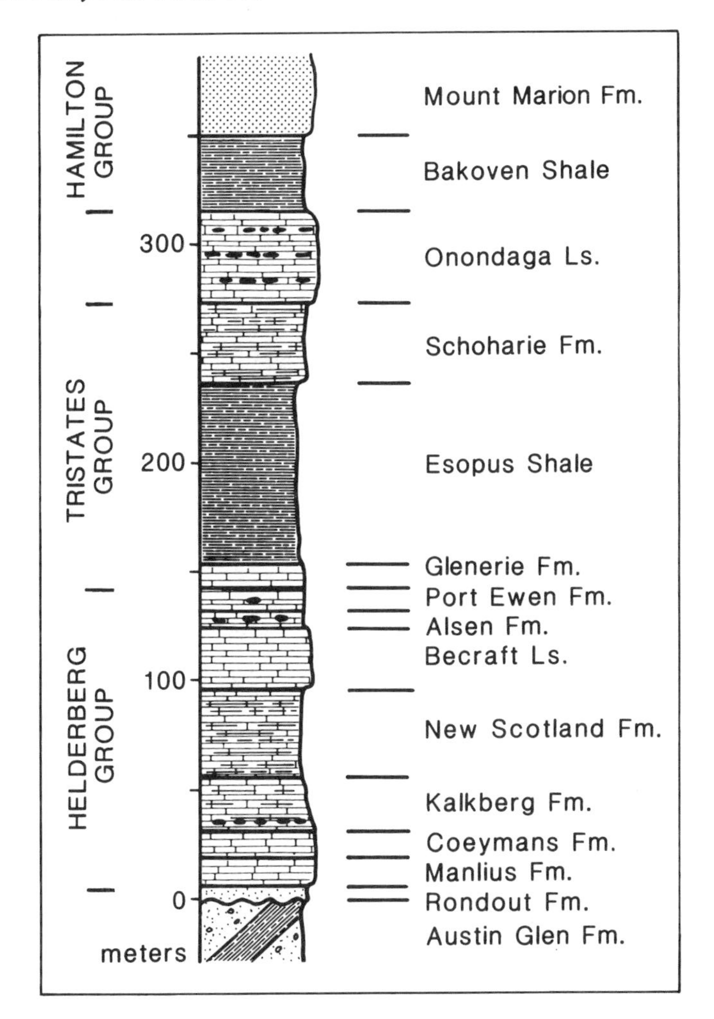

Figure 4. Stratigraphic column of units exposed in the HVB near Catskill.

the floor thrust with regard to faults exposed within Lower Devonian units (Marshak, 1986a, b). Over the Rondout Formation, in this outcrop, is a homoclinally dipping sequence of the lower Helderberg Group (Manlius-Kalkberg Formations). Note that here, as throughout the HVB, development of cleavage is lithologically controlled; cleavage occurs primarily in rocks containing greater than 10 percent clay.

Outcrop N4, on the western limb of the Tollgate Syncline (Fig. 2) provides another exposure of the lower Helderberg Group. In this outcrop, the section has been thickened as a result of movement on two well-exposed thrust faults. The lower fault has significant stratigraphic throw, for it brings the Manlius Formation over the Kalkberg Formation. These faults may be out-of-the-syncline faults (see Dahlstrom, 1970).

Outcrops N3 and S3 provide additional exposures of the Taconic Unconformity and of the Helderberg Group (through the Becraft Formation). Of particular note in these outcrops are the complex faults and folds in the Rondout and lower Manlius Formations; these structures are probably manifestations of movement on the Rondout detachment. Many bedding-plane slip surfaces, which developed during flexural-slip folding and are coated with sheets of calcite slip fibers, occur in outcrops N3 and S3. In the Kalkberg Formation, some of these slip surfaces are bounded by zones of nearly slaty cleavage. At the northwest end of outcrop N2, numerous mesoscopic folds, as well as two backthrusts, occur within the Becraft Limestone.

Outcrops N2 and S2 are the most spectacular of the New York 23 roadcuts. These exposures (which include rocks of the Manlius Formation through Becraft Limestone) display, from southeast to northwest, ramp faults with hanging wall anticlines (Rip van Winkle anticline), out-of-the-syncline forethrusts and backthrusts, folded ramps and flats (in the Central anticline), and zones of tectonic cleavage intensification (on the northwest limb of the Central anticline). Of particular note are the examples of fault bends (Suppe, 1983) at which bedding-parallel flats join cross-strata ramps. The Central anticline appears to be composed

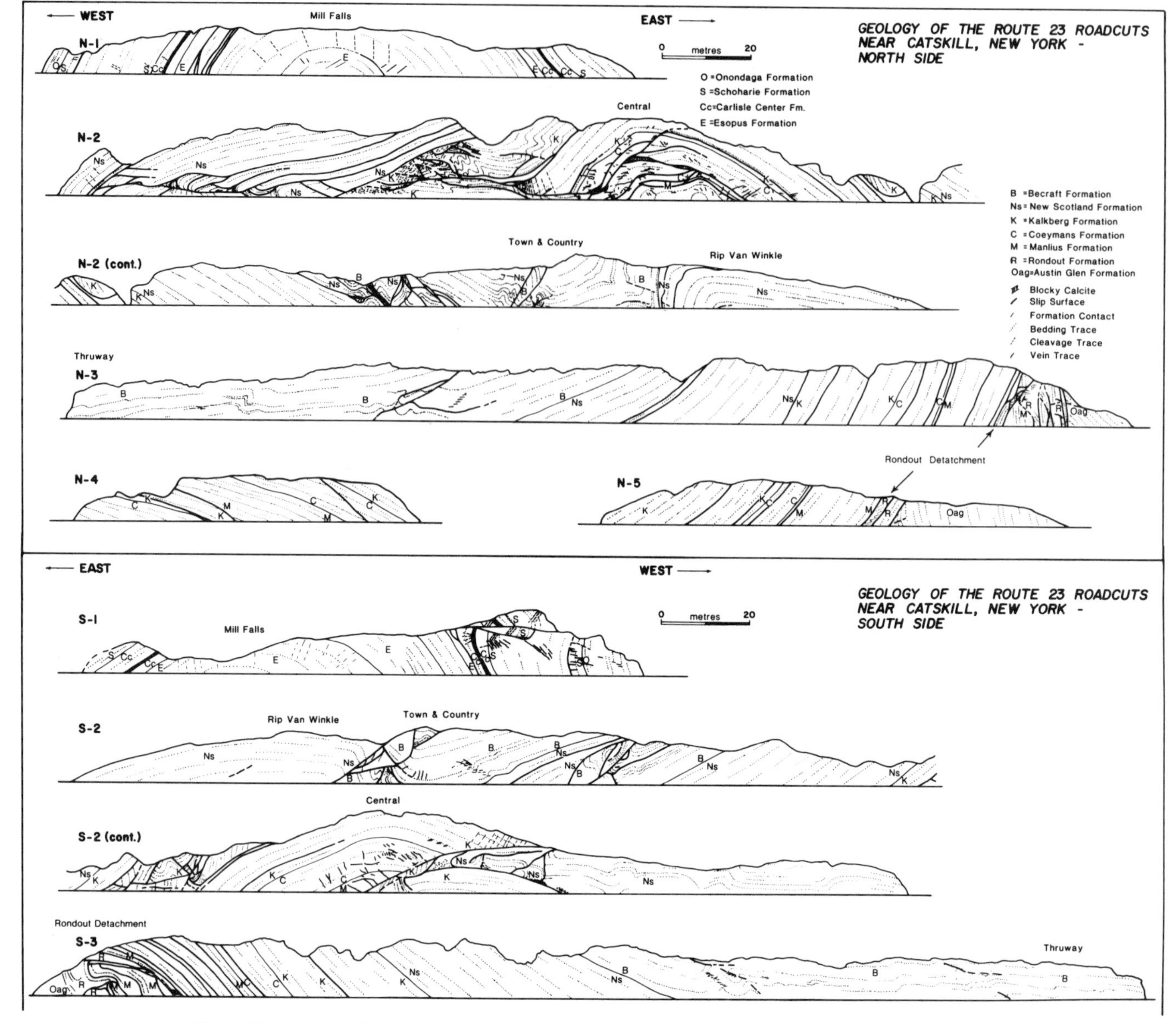

Figure 5. Structural features of the New York 23 roadcuts in the HVB. These cross-sections were constructed on a photomosaic base.

of a stack of fault-bounded horses (see Boyer and Elliott, 1984), one of which is internally deformed throughout by mesoscopic folds. Structures of outcrops N2 and S2 do not directly correlate across the highway, illustrating how rapidly structural geometry can change along strike in a fold-thrust belt. The contrast in structural geometry between outcrops on opposite sides of the road may reflect the occurrence of a lateral ramp in the interval that was excavated during construction of the highway.

Outcrops N1 and S1 expose the Esopus and Schoharie Formations, and the base of the Onondaga Limestone. These units are arched around the Mill Falls anticline. Nearly slaty cleavage occurs within the lower Esopus Formation. The upper few meters of the Esopus Formation are composed of beds of finely laminated mudstone and siltstone that have been crinkled into tiny folds. A subhorizontal fault is present at the top of outcrop S2.

Creek exposures provide both map-view and cross-section views of principal folds and faults in the HVB. North of the New York 23 bridge (locality C1), the Mills Falls anticline and the Creek Bend syncline are visible. South of the bridge (locality C2), the creek cuts through imbricate thrust sheets involving the Becraft and New Scotland Formations (Fig. 2).

The depth of exposure available at Catskill does not permit direct construction of a cross section down to the basal detachment of the belt. Figure 6 presents a reasonable cross-sectional model of the belt. In this model, the thrust system visible at the

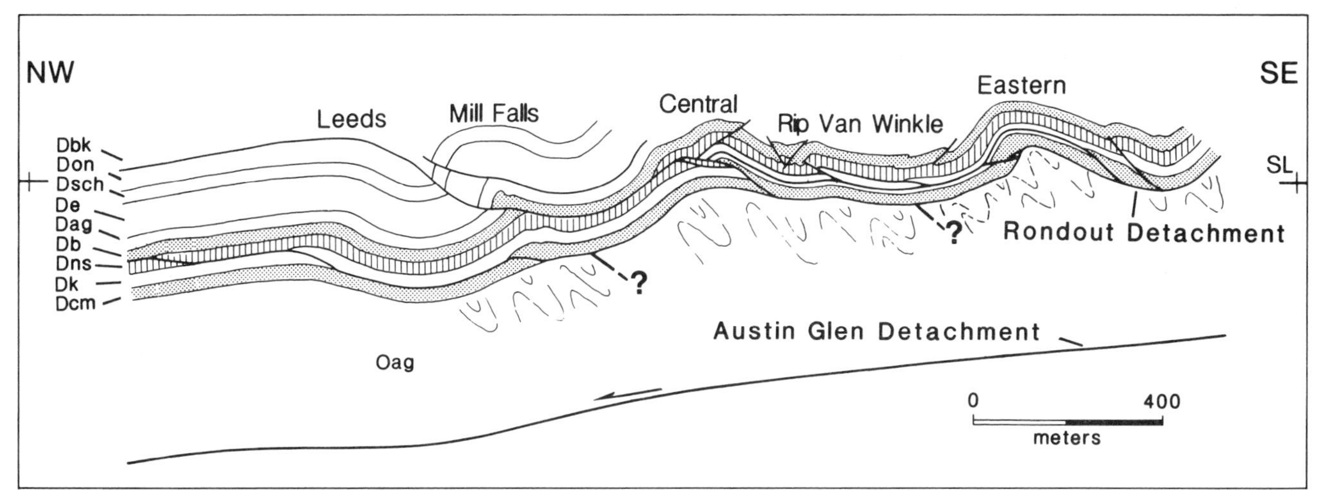

Figure 6. Interpretive cross section of the HVB at Catskill. Abbreviations refer to fold names identified in Figure 2. RD = Rondout detachment; AD = Austin Glen detachment.

field site lies entirely above the Rondout detachment. Some of the faults describe an upward-imbricate fan and others a duplex (see Boyer and Elliott, 1984). The Rondout detachment in this model is itself folded; this feature requires that there was shortening of the subunconformity Austin Glen flysch during the development of the HVB. A lower blind thrust, called the Austin Glen detachment (Marshak, 1986), may have developed at depth to accommodate this shortening. Deformation of the HVB dies out westward; west of Leeds, there appears to have been movement on blind thrusts, for the Bakoven Shale and the Mount Marion Formation contain tectonic cleavage. Present exposures of the HVB are probably only the western edge of what was once a much wider Acadian (?) fold-thrust belt that extended farther to the east, perhaps across the Taconic Mountains.

REFERENCES CITED

Babcock, E. A., 5th, 1966, Structural aspects of the folded belt near Leeds, New York [M.S. thesis]: Syracuse, New York, Syracuse University, 62 p.

Boyer, S. E., and Elliott, D., 1982, Thrust systems: American Association of Petroleum Geologists Bulletin, v. 66, p. 1196–1230.

Chadwick, G. H., 1944, Geology of the Catskill and Kaaterskill Quadrangles, Part II. Silurian and Devonian geology, with a chapter on glacial geology: New York State Museum Bulletin 336, 251 p.

Chapple, W. M., and Spang, J. H., 1974, Significance of layer-parallel slip during folding of layered sedimentary rocks: Geological Society of America Bulletin, v. 85, p. 1523–1534.

Dahlstrom, C.D.A., 1970, Structural geology in the eastern margin of the Canadian Rocky Mountains: Bulletin of Canadian Petroleum Geology, v. 18, p. 332–406.

Davis, W. M., 1882, The Little Mountains east of the Catskills: Appalachia, v. 3, p. 20–33.

—— , 1883, The folded Helderberg limestones east of the Catskills: Harvard College Museum Comprehensive Zoology Bulletin, v. 7, p. 311–329.

Goldring, W., 1943, Geology of the Coxsackie Quadrangle, New York: New York State Museum Bulletin 332, 374 p.

Laporte, L. F., 1969, Recognition of a transgressive carbonate sequence within an eperic sea, Helderberg Group (Lower Devonian) of New York State, *in* Friedmann, G. M., ed., Depositional Environments in Carbonate Rocks: A Symposium: Society of Economic Paleontologists and Mineralogists Special Publication 14, p. 98–119.

Leftwich, J. T., Jr., 1973, Structural geology of the West Camp area, Green and Ulster Counties, New York [M.A. thesis]: Amherst, Massachusetts, University of Massachusetts, 88 p.

Marshak, S., 1983, Aspects of deformation in carbonate rocks of fold-thrust belts of central Italy and eastern New York State [Ph.D. thesis]: New York, New York, Columbia University, 223 p.

—— , 1986a, Structure and tectonics of the Hudson Valley fold-thrust belt, New York: Geological Society of America Bulletin, v. 97, p. 354–368.

—— , 1986b, Guidebook to the Hudson Valley fold-thrust belt between Catskill and Kingston, New York: Field guide prepared to accompany the 1986 meeting of the Geological Society of America, Northeast section.

Marshak, S., and Engelder, T., 1985, Development of cleavage in limestones of a fold-thrust belt in eastern New York: Journal of Structural Geology, v. 7, p. 345–360.

Marshak, S., and Geiser, P., 1980, Guidebook to pressure-solution phenomena in the Hudson Valley: Field guide prepared to accompany the Penrose Conference on Pressure Solution and Dissolution, 49 p.

Marshak, S., Kwiecinski, P., McEachran, D., and Tabor, J., 1985, Structural geometry of the "orocline" in the Appalachian foreland, near Kingston, New York: Geological Society of America Abstracts with Programs, v. 17, p. 53.

McEachran, D. B., 1985, Structural geometry and evolution of the basal detachment in the Hudson Valley fold-thrust belt north of Kingston, New York [M.S. thesis]: Urbana, Illinois, University of Illinois, 97 p.

Murphy, P. J., Bruno, T. L., and Lanney, N. A., 1980, Décollement in the Hudson River Valley: Geological Society of America Bulletin, v. 91, pt. I, p. 258–262; pt. II, p. 1394–1415.

Ratcliffe, N. M., Bird, J. M., and Baharami, B., 1975, Structural and stratigraphic chronology of the Taconide and Acadian polydeformational belt of the central Taconics of New York State and Massachusetts: New England Intercollegiate Geology Conference Guidebook, 67th meeting, New York, p. 55–86.

Rickard, L. V., 1962, Late Cayugan (Upper Silurian) and Helderbergian (Lower Devonian) stratigraphy in New York: New York State Museum and Science Service Bulletin 386, 157 p.

Rodgers, J., 1971, The Taconic Orogeny: Geological Society of America Bulletin, v. 82, p. 1141–1178.

Sanders, J. E., 1969, Bedding thrusts and other structural features in cross-section through "Little Mountains" along Catskill Creek, (Austin Glen and Leeds Gorge), west of Catskill, New York, trip 19 *in* Guidebook for field trips in

New York, Massachusetts, and Vermont; New England Intercollegiate Geological Conference, 61st Annual Meeting, Albany, New York: SUNY-A Bookstore, p. 1–38.

Suppe, J., 1983, Geometry and kinematics of fault-bend folding: American Journal of Science, v. 283, p. 684–721.

Tabor, J. R., 1985, Nature and sequence of deformation in the southwestern limb of the Kingston orocline [M.S. thesis]: Urbana, Illinois, University of Illinois, 87 p.

Waines, R. H., and Hoar, F. G., 1967, Upper Silurian–Lower Devonian stratigraphic sequence, western mid-Hudson Valley region, Kingston to Accord, Ulster County, New York: New York State Geological Association Guidebook, 39th meeting, New Paltz, p. D1–D28.

Zadins, Z. Z., 1983, Structure of the northern Appalachian thrust belt at Cementon, New York [M.S. thesis]: Rochester, New York, University of Rochester, 137 p.

Strand-line facies of the Catskill Tectonic Delta Complex (Middle and Upper Devonian) in eastern New York state

Kenneth G. Johnson, Department of Geology, Skidmore College, Saratoga Springs, New York 12866

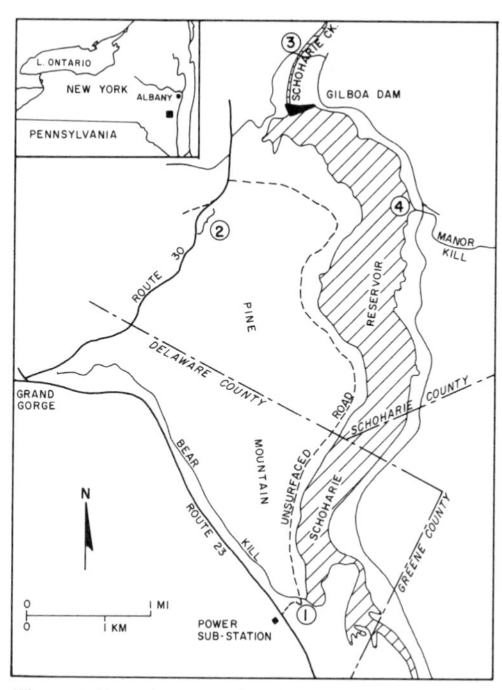

Figure 1. Location map showing Stops 1, 2, 3, and 4.

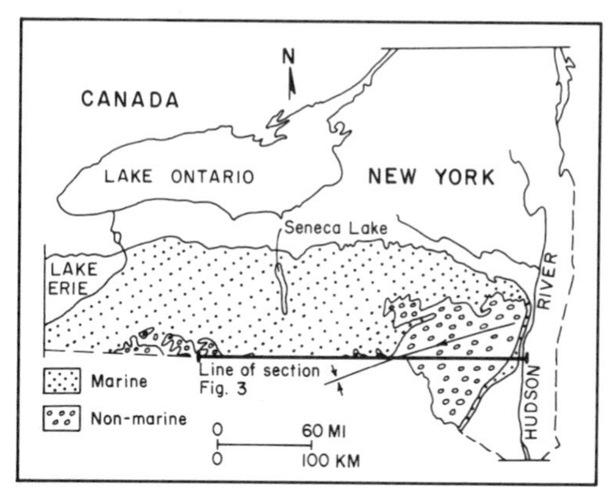

Figure 2. Generalized map of the Upper Middle and Upper Devonian bedrock of New York State (after Rickard, 1964) showing Figure 3 cross-section location.

LOCATION

Strand-line facies which evolved during a short-lived transgressive pulse that interrupted westward progradation of the Catskill delta complex are well exposed in four sections of the Schoharie Valley. These sections are approximately 40 mi (64 km) southwest of Albany on or very near the shore of Schoharie Reservoir of the New York City water supply system (Fig. 1).

Stop 1 (Prattsville Quadrangle) is 2.8 mi (4.5 km) southeast of Grand Gorge in Hardenburgh Falls and the hillsides above the falls. Park at the power substation on the southwestern side of New York 23 and walk northeast on the unsurfaced road to the bridge across Bear Kill. The falls are beneath the bridge. **Stop 2** (Gilboa Quadrangle) is 1.9 mi (3.0 km) northeast of Grand Gorge in the lower part of the roadcut where New York 30 extends down the western side of Pine Mountain. Park at the foot of the hill and proceed back up to the base of the section. Be very careful! This is a busy road and the visibility for motorists coming down the hill and around the curve is very poor. **Stop 3** (Gilboa Quadrangle) is 0.4 mi (0.6 km) north of Gilboa Dam at the western end of the bridge across Schoharie Creek. Examine the exhibit of fossil seed-fern stumps just south of the road and then walk beyond the exhibit to the outcrop on the west bank of the creek. An overgrown, abandoned quarry is located immediately west of the river-bank outcrop. **Stop 4** (Gilboa Quadrangle) is 1.2 mi (1.9 km) southeast of Gilboa Dam where the Manor Kill enters the Schoharie Reservoir. Park at the south end of the bridge across the Manor Kill and walk west down the path on the south side of the stream to the mouth of the Manor Kill. It is necessary to obtain permission to visit Stops 1, 3, and 4 at the New York City Board of Water Supply office, Tannersville, New York 12485; telephone (518) 589-5020.

SIGNIFICANCE

The sections at these stops exhibit a combination of sedimentary structures, lithology, geometric relationship with adjacent units, and biogenic structures, which indicate that they are in part of tidal origin, a facies not common in the clastic geologic record. These rocks are part of a sequence of Devonian strata that is some 10,000 ft (3,030 m) thick in the Catskill Mountains of eastern New York, thins to 2,500 ft (760 m) in the westernmost part of the state, and similarly thins significantly toward the southwest. To the north and northwest in New York, Devonian rocks have been completely removed by erosion. In New York State, the Devonian sequence was only slightly affected by later orogeny; for the most part it was deformed into gently undulating folds.

The Lower Devonian and Lower Middle Devonian sequence of eastern New York and Pennsylvania consists of marine and transitional carbonate rocks. These are overlain by Upper Middle and Upper Devonian strata that form a thick wedge of nonmarine rocks that have a progradational relationship to marine formations farther west. The distribution and spatial relationship of these rocks, which are a part of the Catskill tectonic delta complex, are shown in Figure 2.

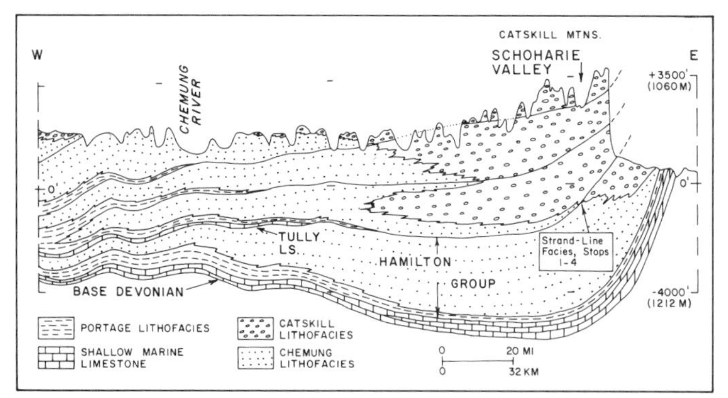

Figure 3. Cross-section of the Devonian system along the New York-Pennsylvania border (generalized from Fig. 17, Broughton and others, 1962).

The deltaic wedge is composed of detritus derived from a source area east of the present-day Catskill Mountains that was being elevated during the Acadian Orogeny. At the base of the clastic sequence is an interval of some 2,500 ft (760 m) of fossiliferous sandstones and shales (Hamilton Group) which thins toward the west and southwest (Fig. 3). It was deposited when the source terrain was in the beginning stages of uplift. The Tully Limestone, a transgressive carbonate tongue at the base of the Upper Devonian, represents the last significant limestone deposition in the New York Devonian prior to the overwhelming of the marine basin by clastic sediment influx.

Uplift in the east persisted after deposition of the Tully and continued on a large scale into the Mississippian. A thick wedge of clastic continental sediment (Catskill lithofacies) was deposited at the margin of the basin (Fig. 3). The red and green-gray sandstones, shales, and conglomerates of this wedge interfinger westward with littoral and shallow-marine (Chemung lithofacies) sandstones and shales. These grade into dark-colored shales and siltstones (Portage lithofacies) farther west that are of deeper marine origin. The irregular, interfingering contact between the continental beds and the marine formations, which rises stratigraphically toward the west, is the result of continental beds off-lapping marine strata. This prograding relationship, the result of displacement of the late Devonian sea by the expanding clastic wedge, is shown in Figure 3. A correlation chart by Rickard (1975) shows the relationship of the depositional phases of the Devonian rocks in New York State. Woodrow and Sevon (1985) provide an up-to-date summary of recent work on the Catskill delta complex and Sevon (1981) presents a concise account of the origin of the delta system within the context of plate tectonic theory.

The expression "delta system," as used by Sevon (1981) and herein, refers to multiple contiguous deltas operating in the same sedimentary basin at approximately the same time. That portion of the Catskill delta system located in eastern New York State and briefly described above has been characterized as a tectonic delta complex (Friedman and Johnson, 1966) in recognition of the fact that it represents "a deltaic complex built into a marine basin contiguous to an active mountain front and dominated by orogenic sandstone derived from the nearby tectonic highland." Sedimentary facies in the New York State delta complex range from braided stream in the eastern part of the Catskill lithofacies to distal offshore in the Portage lithofacies to the west. A reasonable modern analogue for the Devonian Acadian source terrain and the Catskill epeiric sea appears to be the Island of New Guinea and the Arafura Sea, immediately north of Australia.

SITE INFORMATION

Virtually all of the rocks exposed at the four stops described herein are shoreline facies of the easternmost Chemung lithofacies. An exception is approximately 20 ft (6 m) of the lower part of the section at Stop 4, which is Catskill lithofacies. The following summary of the lithologic, sedimentologic, and biogenic characteristics of the Chemung beds in the vicinity of Schoharie Reservoir applies, in general, for rocks exposed at all four stops. Not all of these characteristics can be observed at a single stop.

Lithology. Lithologies within the Chemung lithofacies are interbedded medium gray to dark gray, micaceous siltstone and shale and medium gray, very fine-grained sandstone with subordinate, medium gray, shell lag lenses. All of the sandstones of the lithofacies are submature or immature graywackes. They contain sporadic accumulations of shale pebbles as well as moderately common, small pyrite nodules. A few polymictic pebble conglomerates containing pebbles of light gray and greenish gray quartzite, medium gray slate, red and olive siltstone, and subordinate medium gray limestone are present (Stop 1). Siltstones and shales of the lithofacies are dark gray in color due to a high content of fine organic material. They are very micaceous and variably thinly cross-laminated to occasionally fissile. Flaser bedding is present in the Stop 1 section.

Elongate lenses of very fossiliferous sandstone range in thickness from a few inches to 18 in (45 cm) and in length from 3 ft (1 m) to 50 ft (15 m) and rest in channeled contact on underlying beds. Shell material in the lenses consists mostly of large spiriferid brachiopods, which in most fresh exposures are composed of calcium carbonate. Some of the lenses also contain pelecypod fragments as well as red siltstone pebbles. The valves have no preferred orientation, although some seem to be crudely imbricated. These lenses are exposed best at Stop 2.

Sedimentary Structures. Bedding thickness of sandstone ranges from medium to thick and very thick (terminology after Ingram, 1954). Virtually all of the strata in the lithofacies, with the exception of the fissile shales, are cross-bedded or cross-laminated. Even the very thick-bedded sandstones, which in some cases appear homogeneous, are well cross-laminated. Interference, oscillation, and current ripple marks are common (Stops 1 and 4). The current and oscillation ripple marks, which have wave lengths of several centimeters and amplitudes of only 1 in (2.5 cm) or less, provide reliable and plentiful evidence of variability of sedimentary strike and direction of transport. Cross-bedding is of both planar and trough types, and in many cases

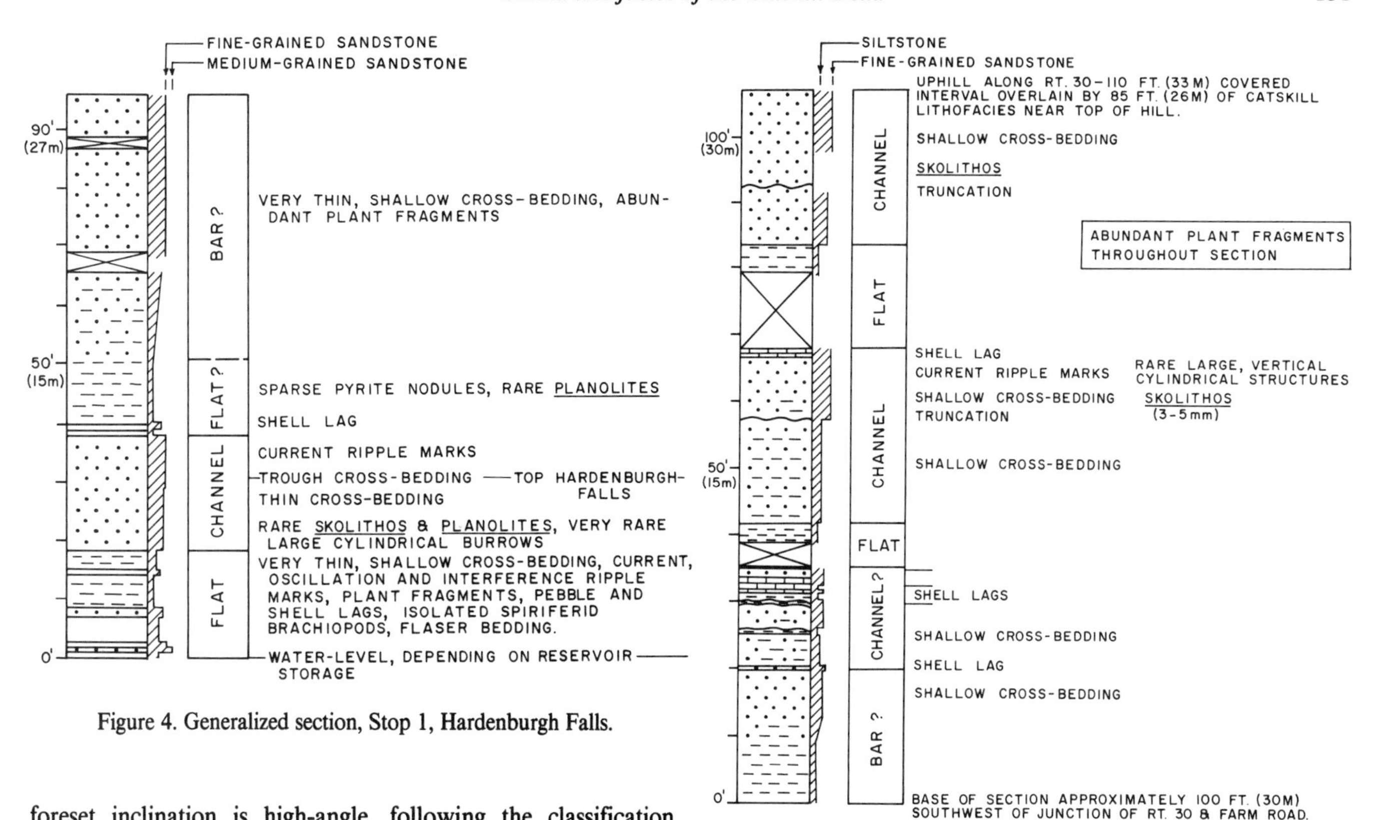

Figure 4. Generalized section, Stop 1, Hardenburgh Falls.

Figure 5. Generalized section, Stop 2, lower New York 30 roadcut, west side of Pine Mountain.

foreset inclination is high-angle, following the classification scheme of Pettijohn (1962).

Dessication cracks are well developed in the uppermost part of the Hamilton Group in the Schoharie Valley (Stop 4). Most of these occur as polygonal patterns of medium gray, very fine-grained sandstone infill on bedding surfaces of dark gray shaly siltstone. In one instance (Stop 4), numerous sandstone infills extend some 6 in (15 cm) perpendicular to bedding into a shale ledge.

Biogenic Structures. Biogenic structures in the Chemung lithofacies of the Schoharie Reservoir area are of three general types: 1) brachiopod and pelecypod body fossils, 2) ichnofossils, and 3) fossil seed ferns.

Brachiopods and pelecypods occur in both sandstones and shales as isolated specimens and as concentrations that appear to be allochthonous. Those found in allochthonous arrangements were considered, for purposes of this study, as biological sedimentary particles occurring in lithified sediment not necessarily that of their life environments. Ichnofossils in the Chemung lithofacies of the Schoharie Valley occur on bedding planes of sandstone as shallow, generally circular and ovoid depressions, which are slightly darker in color than the enclosing lithology. These occur in two sizes; those only about 0.5 in (1 cm) in diameter, and those 1 in (2.5 cm) or more in diameter. The smaller ones extend downward, perpendicular to bedding, for a distance of up to 6 in (15 cm). At one locality, abundant vertical burrows are 12 in (30 cm) in length. A few of the burrows have a Y or U pattern. Ichnogeneric names are indicated below in the description of each stop.

Fossil seed-fern stumps are present at three stratigraphic levels in the upper Hamilton beds near Gilboa Dam. More than 200 stumps were taken from the lowest of these levels (Stop 3) during quarrying operations just north of the dam (Goldring, 1924, 1927). They occur in light olive-gray, tabular and trough cross-bedded, fine-grained sandstones, some of which contain abundant *Skolithos* up to 12 in (30 cm) long. The beds are thick and very thick bedded, are in part slightly calcareous, and contain abundant casts of large spiriferid brachiopods at certain levels.

Stop 1. (Prattsville 7½-minute Quadrangle, Section 48 of Johnson, 1968.) At this point Bear Kill drops over Hardenburgh Falls and flows into Schoharie Reservoir. Beds here are assigned to the tidal channel, tidal flat, and possibly shallow bar facies (Fig. 4). The tidal channel facies is represented by gray, cross-bedded, fossiliferous, fine-grained sandstones and the tidal flat facies by very dark gray, very thin-bedded, in part conglomeratic, shales. Lag-concentrates in both facies are rich in shallow-marine brachiopod shells and pebbles. Plant fragments are abundant.

Stop 2. (Gilboa 7½-minute Quadrangle, Section 43 of Johnson, 1968.) Be very careful! This stretch of road is a narrow speedway with poor visibility. This section consists of cross-bedded sandstone of tidal channel, tidal flat, and possibly shallow bar origin with lag concentrations of shallow-marine spiriferid brachiopods (Fig. 5). Trace fossils include *Skolithos* and rare, large, vertical cylindrical structures.

Stop 3. (Gilboa 7½-minute Quadrangle, Section 44 of

Johnson, 1968.) The exhibit of seed-fern stumps at the western end of Gilboa bridge and the adjacent quarry from which they were taken represents the lowest of three locations where *Aneurophyton* was found (Goldring, 1924, 1927).

The other two sites, now flooded by the waters of Schoharie Reservoir, were 60 ft (18 m) and 160 ft (48 m) above Stop 3. A total of several hundred trees were discovered at these three sites, many of them in growth position. Those stumps in growth position had strap-like roots extending into dark gray shale and trunks surrounded by gray, fossiliferous, cross-bedded, burrowed sandstone. In some cases, mud cracks have been found on shale bedding surfaces and as sole markings on sandstone beds. These seed ferns, which are thought to have grown to heights of at least 30 to 40 ft (9 to 12 m), constituted a distal alluvial plain or tidal swamp that was buried by shifting tidal channel and nearshore bar sands during oscillations of the late Middle Devonian delta shoreline.

Stop 4. (Gilboa 7½-minute Quadrangle, Section 46 of Johnson, 1968.) The beds exposed in the Manor Kill Gorge are in the upper part of the Hamilton Group (Fig. 6). Those in the lower part of the section, adjacent to the Schoharie Reservoir, are trough cross-bedded, burrowed, medium-grained sandstone of the Chemung lithofacies assigned to the tidal channel facies. Some of these sandstones are rich in plant material and in a few places, during fall and winter low water stages of the reservoir, fossil seed fern stumps (*Aneurophyton*) in growth position may be seen. The remainder of the section upstream consists of interbedded red and green shales, dark gray shales, and medium gray, shallowly cross-bedded, fine-grained sandstone; these are interpreted, respectively, as distal alluvial plain, tidal flat, and tidal channel facies.

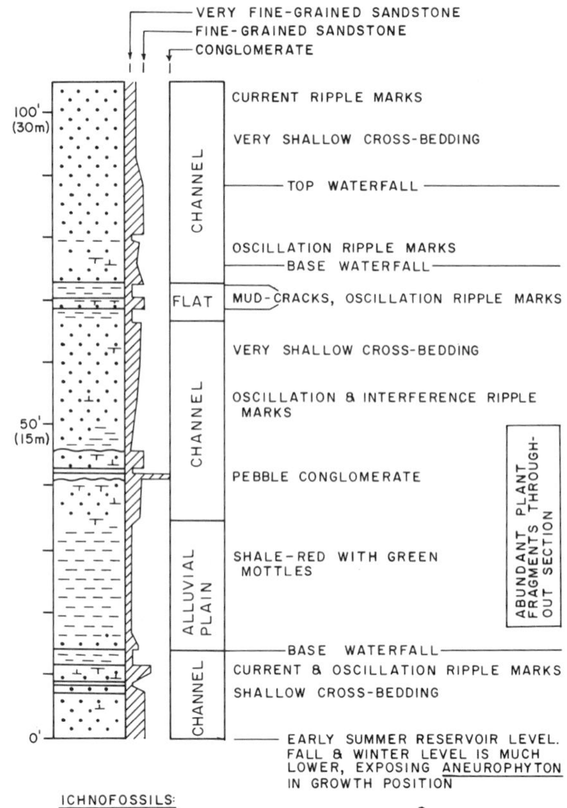

Figure 6. Generalized section, Stop 4, Manor Kill Gorge.

REFERENCES

Broughton, J. G., Fisher, D. W., Isachsen, Y. R., and Rickard, L. V., 1962, The geology of New York State (text): New York State Museum and Science Service Map and Chart Series 5.

Cooper, G. A., and Williams, J. S., 1935, Tully Formation of New York: Geological Society of American Bulletin, v. 46, p. 781–868.

Folk, R. L., 1965, Petrology of sedimentary rocks: Austin, Texas, Hemphill's Bookstore, 159 p.

Friedman, G. M., and Johnson, K. G., 1966, The Devonian Catskill deltaic complex of New York, type example of a "tectonic delta complex": *in* Shirley, M. L., ed., Deltas in their geologic framework: Houston Geological Society, p. 171–188.

Goldring, W., 1924, The Upper Devonian forest of seed ferns in eastern New York: New York State Museum Bulletin 251, p. 50–92.

——, 1927, The oldest known petrified forest: Scientific Monthly, v. 24, p. 515–529.

Harms, J. C., and Fahnestock, R. K., 1965, Stratification, bed forms, and flow phenomena (with example from the Rio Grande), *in* Middleton, G. V., ed., Primary sedimentary structures and their hydrodynamic interpretation: Society of Economic Paleontologists and Mineralogists Special Publication No. 12, p. 84–115.

Ingram, R. L., 1954, Terminology for the thickness of stratification and parting units in sedimentary rocks: Geological Society of America Bulletin, v. 65, p. 937–938.

Johnson, K. G., 1968, The Tully clastic correlatives (Upper Devonian) of New York State; Model for recognition of alluvial, dune (?), tidal, nearshore (bar and lagoon), and offshore sedimentary environments in a tectonic delta complex [Ph.D. thesis]: Troy, New York, Rensselaer Polytechnic Institute, 122 p. + appendices and plates.

——, 1972, Evidence for tidal origin of Late Devonian clastics in eastern New York State, U.S.A., *in* Proceedings, Section 6, Stratigraphy and Sedimentology, 24th International Geological Congress, Montreal, p. 285–293.

Johnson, K. G., and Friedman, G. M., 1969, The Tully clastic correlatives (Upper Devonian) of New York State; A model for recognition of alluvial, dune (?), tidal, nearshore (bar and lagoon), and offshore sedimentary environments in a tectonic delta complex: Journal of Sedimentary Petrology, v. 39, no. 2, p. 451–485.

Miller, M. F., and Johnson, K. G., 1981, *Spirophyton* in alluvial-tidal facies of the Catskill deltaic complex; Possible biological control of ichno-fossil distribution: Journal of Paleontology, v. 55, p. 1016–1027.

Pettijohn, F. H., 1962, Paleocurrents and paleogeography: American Association of Petroleum Geologists Bulletin, v. 48, p. 1468–1493.

Rickard, L. V., 1975, Correlation of the Silurian and Devonian rocks in New York State: New York State Museum and Science Service Map and Chart Series 24, 16 p. and 4 plates.

Sevon, W. D., 1981, The Middle and Upper Devonian clastic wedge in northeastern Pennsylvania, *in* Enos, P., ed., Guidebook for field trips in south-central New York: Binghamton, New York State Geological Association 53rd Annual Meeting, p. 31–63.

Woodrow, D. L., and Sevon, W. D., eds., 1985, The Catskill Delta: Geological Society of American Special Paper 201, 254 p.

Igneous and contact metamorphic rocks of the Cortlandt Complex, Westchester County, New York

***Robert J. Tracy,** Department of Geological Sciences, Virginia Polytechnic Institute and State University, Blacksburg, Virginia 24061*
***Nicholas M. Ratcliffe,** U.S. Geological Survey, M.S. 925, Reston, Virginia 22092*
***John F. Bender,** Department of Geography and Earth Science, University of North Carolina at Charlotte, Charlotte, North Carolina 28223*

LOCATION

The Cortlandt Complex is located in northern Westchester County, New York, just south and east of the city of Peekskill on the east bank of the Hudson (Fig. 1). The site is about a one-hour drive north of New York City; principal access routes from the south are U.S. 9 and the Taconic State Parkway. The complex comprises about 25 mi^2 (65 km^2) of exposure (Fig. 2) and has a roughly elliptical shape, elongated east-west. Individual localities within the overall site are indicated in Figures 1 and 2, and will be described in detail below.

Stop 1 is in the Popolopen Lake 7½-minute Quadrangle and is located some miles west of Peekstill. It may be reached by taking U.S. 6 west from Peekskill to Cranberry Hill, about 2 mi (3.2 km) east of the intersection of U.S. 6 and I-84. As a starting point of the trip, it may also be reached by taking the U.S. 6 (east) exit on I-84.

Stop 2, in the Peekskill Quadrangle, is reached by taking Welcher Avenue (U.S. 9A) south from Peekskill to the town of Buchanan. Turn right onto Tate Street, then bear right at a small traffic circle onto Westchester Avenue. Turn right at Buchanan Town Hall and park.

Stop 3 is in the Ossining Quadrangle; take New York 129 east from U.S. 9 in Croton-on-Hudson for about 2.5 mi (4 km), then turn north (left) on Croton Avenue before crossing the bridge over the Croton Reservoir. About 0.3 mi (0.5 km) north on Croton Avenue, stop at an abandoned quarry on the left.

Stop 4 is in the Mohegan Lake Quadrangle and is 1 mi (1.6 km) north of Stop 3 on Croton Avenue. Stop 5, in both Mohegan Lake and Peekskill quadrangles, is reached by going 3.4 mi (5.5 km) north from Stop 4 on Croton Avenue. Where Croton Avenue intersects Jacob Street, turn left on Jacob, then right at the next major intersection to stay on Croton Avenue. Stop is on small road to left with the sign "EMRI."

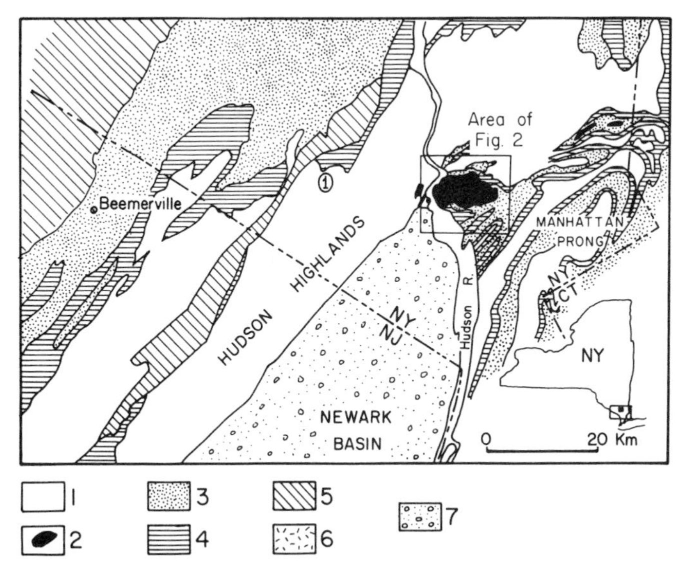

Figure 1. Generalized regional geologic map of southeastern New York and adjacent New Jersey illustrating the Cortlandt-Beemerville magmatic belt (after Ratcliffe and others, 1983). Patterns are as follows: 1, Proterozoic Y gneisses; 2, Cortlandt Complex and related smaller intrusives; 3, Cambro-Ordovician schists; 4, Cambro-Ordovician carbonates and quartzites; 5, Siluro-Devonian sediments; 6, Peeksill Granite; 7, Triassic-Jurassic rift sediments and igneous rocks. The circled 1 is the approximate location of suggested Stop 1 near Monroe, New York.

SIGNIFICANCE

The Cortlandt Complex is a Late Ordovician magmatic complex of alkaline affinity, which syntectonically or posttectonically intruded the deformed and metamorphosed Cambro-Ordovician rocks of the northern Manhattan Prong. Composed of a variety of igneous rock types ranging from peridotite to monzodiorite, the complex was intruded as six separate plutons whose intrusive sequence has been established (Fig. 2). Intrusion occurred into previously regionally metamorphosed country rocks that remained hot, giving rise to significant assimilation and magma contamination. REE and major-element chemistry suggests that the mantle-derived parental magma was alkali-basaltic and that the more typically calc-alkaline rocks, such as diorites and norites, were derived from this magma through assimilation of crustal material. Despite their alkaline nature, the Cortlandt plutons are clearly syntectonic rocks similar to other amphibole-bearing mafic and ultramafic igneous rocks found in orogenic belts throughout the world, such as the Appinite Suite of the Caledonide Belt in Great Britain. The Cortlandt Complex is part of a long linear belt of similar plutons that stretches from the Beemerville Dike Swarm in New Jersey to the Prospect Complex in Connecticut. The intrusion of these rocks is clearly part of the Taconian orogenic event, but their exact relationship to other orogenic features is currently unknown.

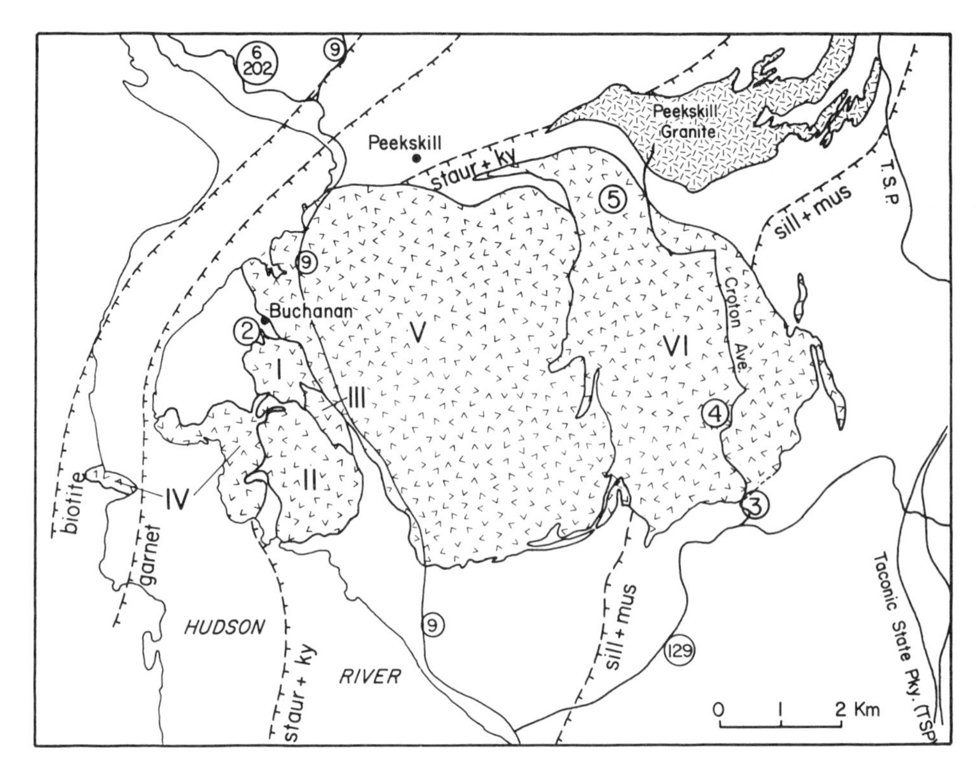

Figure 2. Map of the Cortlandt Complex showing the six plutons (I through VI in decreasing age) that make up the complex. Also shown are the Taconian regional metamorphic isograds, with tick marks pointing to higher metamorphic grade. Circled numbers (2 through 5) indicate approximate locations of suggested stops. Major roads are also indicated.

SITE INFORMATION

The Cortlandt Complex has long been recognized as an unusual geologic feature. The earliest serious geologic investigation was that of J. D. Dana (1881) who first throught the layered rocks were part of a metamorphic complex, but later realized their igneous origin. Petrographic studies by Williams (1886) accurately identified the major igneous rock types and their interrelationships. Perhaps the most important early study was that of Rogers (1911), which produced the first geologic map of the complex and emphasized the gradational nature of many of the internal intrusive contacts in an apparently composite intrusion. In a classic work of igneous petrology, Balk (1927) described the igneous flow layering and deduced the existence of several large basinal structures delineated by layering and foliation. Shand (1942) reexamined Rogers' petrology in light of Balk's structures and proposed a model of concentric lithologic zonation that discarded Rogers' model of composite intrusion. A gravity and magnetic study (Steenland and Woolard, 1952) supported the notion of eastern and western feeder funnels, but showed that the largest concentric structure, the central basin, did not extend to any great depth.

Recent work on the igneous rocks includes remapping by Ratcliffe (Ratcliffe and others, 1982; Ratcliffe and others, 1983), which shows a correlation between Balk's data and Rogers' mapping, and casts significant doubt on the Shand generalization. Other recent studies are geochemical and petrologic investigations by Bender (Bender and others, 1983a, b) and Tracy (Tracy and Kenzie, 1987).

Studies of contact metamorphism at the Cortlandt Complex have concentrated on the emeries that represent ultrametamorphism and metasomatism of aluminous country rock xenoliths completely engulfed by mafic magma. Friedman (1956) and Barker (1964) examined the petrography of the emery xenoliths and provided data documenting the metasomatic changes accompanying emery formation. More modern studies (Tracy and McLellan, 1985; Waldron, 1985, 1986) have provided compositional data to better illustrate the processes occurring during contact metamorphism. The contact metamorphic rocks here are particularly interesting to study because the contact metamorphism was superimposed on relatively high-grade regional metamorphic rocks, and therefore the contact rocks are not fine-grained hornfelses but are relatively coarse grained.

The most recent studies of the contact rocks (Waldron, 1985, 1986; Waldron and Tracy, 1986) have produced estimates of the physical conditions under which the contact assemblages formed. These data show that the pressures of regional metamorphism (5 to 6 kilobars) were about 2 kilobars less than those of contact metamorphism (7 to 8 kilobars). Waldron and Tracy (1986) have explained this data by proposing an abrupt episode of crustal thickening between the times of Taconian regional metamorphism and intrusion of the Cortlandt Complex; in this model, thickening was caused by late Taconian east-over-west thrusting that has left no geologic evidence, other than metamorphic mineral chemistry, because the allochthonous rocks have been completely eroded.

Documentation of the originally great crustal depth of the present erosional level of the Cortlandt Complex may have considerable significance for understanding the abundance of unusually mafic cumulate igneous rocks in the complex. High pressure has been shown to expand the temperature range of fractional crystallization of olivine and clinopyroxene from basaltic magma and to suppress the crystallization of plagioclase. The abundance of cumulate clinopyroxenites and wehrlites in Pluton VI at the eastern end of the complex is therefore entirely consistent with extensive fractional crystallization of alkali basaltic parental magma at about 15 mi (25 km) depth in the crust; if the plagioclase-enriched residual magma moved farther up in the crust, its crystallization products have subsequently been eroded away.

To summarize, it appears that the Cortlandt Complex (along with other bodies that comprise the Cortlandt-Beemerville Suite) developed from mantle-derived alkali basaltic parental magma generated in the late stages of the Taconic orogeny during the initiation of a strike-slip or transpressional episode of tectonics. This tectonic stage followed an earlier episode of collisional tectonics that created tectonic loading adjacent to the suture zone and permitted development of a Barrovian-type regional metamorphism. Along the trend of the igneous rocks from the foreland to the internal zones, the form of the plutonic bodies (diatremes in the west at Beemerville to deep-level plutons in the east at Cortlandt) and their chemistry (from strongly alkaline and relatively uncontaminated in the west to noritic and less alkaline in the east) changed principally as a function of the thickening tectonic cover toward the east. This may have affected the nature of the igneous suites in at least two ways: (1) greater depression of the mantle under thicker tectonic cover allowed a greater degree of partial melting and therefore less alkaline parental magma; and (2) greater opportunity for deep crustal modification of magmas because longer intrusive paths led to a reversed sequence at the Cortlandt of early noritic-dioritic contaminated magmas, followed by more primitive ones after conduits had been established. If this deep-seated intrusive event was accompanied by any concurrent volcanic activity at the surface, no evidence of it has yet been identified.

Stop 1. (Popolopen Lake Quadrangle). The purpose of this stop is to examine some of the mafic alkaline dikes presumed to be representative of the Cortlandt-Beemerville magmatic event. Several dark-colored kaersutite-bearing lamprophyre dikes with excellent ocellar structure and feldspar-rich leucophyre are exposed in the roadcut along Route 6 near the crest of Cranberry Hill where they cut Grenville age gneisses of the Hudson Highlands. Post-tectonic lamprophyric dikes crop out in a belt extending from the Cortlandt Complex west to Beemerville, New Jersey (Fig. 1) where the dikes are strongly alkalic. These dikes (camptonite, spessartite, andesite, and leucophyre) are associated with both the Rosetown and Cortlandt plutons. Along the Cortlandt-Beemerville trend, the dikes never cut rocks younger than Middle Ordovician and yield K-Ar dates consistent with Late Ordovician age. Many dikes contain excellent ocellar structures suggestive of liquid immiscibility (Bender and others, 1982). The ocelli commonly contain higher orthoclase/plagioclase ratios, more abundant quartz and calcite, and less (or no) mafic mineral than the groundmass between ocelli. For details of the chemistry of the dikes, see Bender and others (1982) and Ratcliffe and others (1983).

Stop 2. (Peekskill Quadrangle). The hillside across from Buchanan Town Hall has excellent outcrops of pegmatitic kaersutite gabbro (some with flow-oriented kaersutite crystals) that form the bulk of this pluton, which is believed to be the oldest and most-deformed pluton of the Cortlandt Complex. The contact zone with the Manhattan Schist is exposed to the northwest across the playing field. Large poikiloblastic muscovite in the outermost part of the aureole includes plicated sillimanite of contact origin; regional grade outside the aureole is garnet zone in this area.

Geochemical study of the igneous rocks at this locality indicates that rocks of Pluton I are relatively primitive magmas related to an alkali basalt parent by fractionation of kaersutite and plagioclase. The hornblendites here, as well as some of the more mafic gabbros, appear to have formed as concentrations of cumulate kaersutite. Some gabbroic rocks also appear to result from the mixing of basic melt with a pelitic component derived from the country rocks either by melting or in aqueous fluid; both major and trace element compositions reflect this mixing. Several inclusions of Manhattan Schist country rock can be seen here.

Stop 3. (Ossining Quadrangle). Proceeding north on Croton Avenue from New York 129, one crosses the southern contact of the complex where pyroxenite of Pluton VI is separated from country rock schist by a thick lens of leuconorite to monzonorite. In the abandoned small quarry at this locality, the monzonorite is a two-pyroxene rock consisting of hypersthene-augite-biotite-plagioclase-orthoclase-apatite-ilmenite. There are excellent exposures by the road where monzonorite may be traced outward into contact breccia and sillimanite schist with typical poikiloblastic muscovite. Regional metamorphic grade in the Manhattan Schist outside the aureole is sillimanite+muscovite here. The monzonorite has clearly resulted from the partial fusion of country-rock schist and the mixing of this granitic fraction with the partially to mostly crystalline ultramafic rock. Textures in thin section show both pyroxenes breaking down to form both hornblende and biotite.

Stop 4. (Mohegan Lake Quadrangle). Excellent typical exposures of cumulate-textured, well-layered ultramafic rocks of Pluton VI may also be seen in outcrops beside the road, down toward the reservoir shoreline to the east, and along the trail into the telephone pipeline to the west. Rock types found here are clinopyroxenite, olivine clinopyroxenite, wehrlite, and dunite, which occur as thin layers (0.8 to 3.3 ft; 0.25 to 1.0 m) in coarse websterite. The obvious layering dips northwest toward the center of this concentrically layered pluton.

The cumulate rocks consist almost entirely of augite, hypersthene, and magnesian olivine in quite variable proportions, although augite is typically dominant. The layering is best developed in the lower parts of this pluton and becomes more vague in the upper clinopyroxenite unit which crops out near the top of Dickerson Mountain. Cryptic compositional layering occurs in these rocks and is reflected by abrupt shifts in the Mg/Fe ratios of pyroxenes and olivine, in Al and Ti contents of pyroxenes, and in the Al/Cr ratios in green spinels.

Stop 5. (Mohegan Lake and Peekskill quadrangles). This stop is at the EMRI-CRETE emery quarry, a working quarry on private property; you must stop at the office to ask for permission (at present, the operators are John Leardi and Alan and John Lucas). The petrology of the extensive Emery Hill emery deposits has been studied in detail by Butler (1936), Friedman (1956), and Barker (1964). The emeries are highly aluminous hornfelses that were rapidly heated, recrystallized, and metasomatically altered during the intrusive event. Mineral associations in the emeries consist of various combinations of the phases staurolite, kyanite, sillimanite, garnet, cordierite, hornblende, gedrite, corundum, quartz, plagioclase, biotite, spinel, magnetite, and ilmenite. Superficially similar, but higher temperature, emeries occur also at Salt Hill on the southern contact of the complex. Both emery localities involve the interaction of mafic magma and Manhattan Schist; at Emery Hill, staurolite and kyanite, and perhaps other minerals, appear to be relict from the preintrusive regional metamorphism, whereas at Salt Hill no relict phases have been identified and the high-temperature assemblages that include aluminous hypersthene and sapphirine appear to represent complete recrystallization of Manhattan Schist. Both emery localities are still actively quarried; the principal uses for the crushed and sized emery are as abrasive and as non-slip coatings for concrete.

REFERENCES CITED

Balk, R., 1927, Die primare struktur des Noritmassivs von Peekskill am Hudson, Nordlish New York: Neues Jahrbuch Für Mineralogie Beilageband, v. 57, p. 249–303.

Barker, F., 1964, Reaction between mafic magmas and pelitic schists, Cortlandt, New York: American Journal of Science, v. 262, p. 614–634.

Bender, J. F., Hanson, G. N., and Bence, A. E., 1982, The Cortlandt Complex; Evidence for large scale liquid immiscibility involving granodiorite and diorite magmas: Earth and Planetary Science Letters, v. 58, p. 330–344.

—— , 1984, Cortlandt Complex; Differentiation and contamination in plutons of alkali basalt affinity: American Journal of Science, v. 284, p. 1–57.

Butler, J. W., 1936, Origin of the emery deposits near Peekskill, New York: American Mineralogist, v. 21, p. 537–574.

Dana, J. D., 1881, Origin of the rocks of the Cortlandt Series: American Journal of Science, v. 22, p. 103–112.

Friedman, G. M., 1956, The origin of the spinel-emery deposits with particular reference to those of the Cortlandt Complex, New York: New York State Museum Bulletin 351, 68 p.

Ratcliffe, N. M., Armstrong, R. L., Mose, D. G., Seneschal, R., Williams, N., and Baramonte, M. J., 1982, Emplacement history and tectonic significance of the Cortlandt Complex and related plutons and dike swarms in the Taconide zone of southeastern New York based on K-Ar and Rb-Sr investigations: American Journal of Science, v. 282, p. 358–390.

Ratcliffe, N. M., Bender, J. F., and Tracy, R. J., 1983, Tectonic setting, chemical petrology, and petrogenesis of the Cortlandt Complex and related igneous rocks of southeastern New York State: Field Trip Guide, Geological Society of America Northeastern Section, 93 p.

Rogers, G. S., 1911, Geology of the Cortlandt Series and its emery deposits: New York Academy of Science, v. 21, p. 11–86.

Shand, S. J., 1942, Phase petrology in the Cortlandt Complex, New York: Geological Society of America Bulletin, v. 53, p. 409–428.

Steenland, N. C., and Woolard, G. P., 1952, Gravity and magnetic investigation of the structure of the Cortlandt Complex, New York: Geological Society of America Bulletin, v. 63, p. 1074–1104.

Tracy, R. J., and Kenzie, M., 1987, High pressure fractional crystallization in the Cortlandt Complex, New York [abs.]: American Geophysical Union, Transasctions, in press.

Tracy, R. J., and McLellan, E. L., 1985, A natural example of the kinetic controls of compositional and textural equilibration, *in* Thompson, A. B., and Rubie, D. C., eds., Physical geochemistry: New York, Springer-Verlag, v. 4, p. 118–137.

Waldron, K. A., 1985, Muscovite breakdown above the second sillimanite isograd, Cortlandt Complex, New York: Geological Society of America Abstracts with Programs, v. 17, p. 741.

Waldron, K. A., 1986, High pressure contact metamorphism in the aureole of the Cortlandt Complex, southeastern New York [Ph.D. thesis]: New Haven, Yale University, 92 p.

Waldron, K. A., and Tracy, R. J., 1986, P-T relationships in Taconian regional metamorphism and post-orogenic contact metamorphism, northern Manhattan Prong, New York: Geological Society of America Abstracts with Programs, v. 18, p. 74.

Williams, G. H., 1886, The peridotites of the Cortlandt series on the Hudson River near Peekskill: American Journal of Science, v. 31, p. 26–31.

Geology of Manhattan Island and the Bronx, New York City, New York

Charles Merguerian, *Geology Department, Hofstra University, Hempstead, New York 11550*
Charles A. Baskerville, *U.S. Geological Survey, 922 National Center, Reston, Virginia 22092*

LOCATION

Four easily accessible localities in Manhattan and the Bronx (Fig. 1), have been chosen to illustrate the metamorphic geology of New York City. Streets surrounding the localities are shown in Figure 2. All outcrops are in public parks or in roadcuts that may be reached by car, bus, or subway.

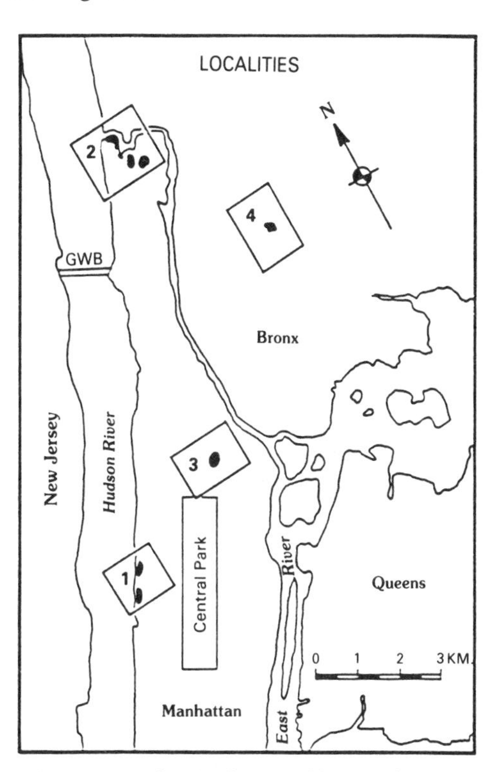

Figure 1. Index map showing the field localities mentioned in Manhattan and the Bronx. GWB, George Washington Bridge.

SIGNIFICANCE OF SITE

The four locations consist of exemplary exposures of high-grade metamorphic rocks that have challenged geologists since Merrill (1890). They are named the Fordham Gneiss, the Yonkers Gneiss, the Inwood Marble (called limestone by Merrill, 1890), and the Manhattan Schist. These names are well known in the geologic literature, although our understanding of them has changed greatly through time. Owing to the lack of fossils in these sillimanite/kyanite–grade rocks, advances in our understanding of their age and sequence have been made largely through detailed geologic mapping studies and comparisons of lithofacies with better known fossiliferous sections of Middle Proterozoic to Middle Ordovician rocks in eastern New York State and Vermont. Since 1964, Leo M. Hall has done more than any other worker to aid present understanding of this complex sequence of rocks through detailed stratigraphic and structural mapping. The interpretations presented here depend heavily on his major contributions, but include modifications for which we must accept responsibility.

The series of localities described here allows an overview of the complex high-grade metamorphic terrane that underlies Manhattan Island and the Bronx, as well as an appreciation of the problems geologists have faced in attempting to disentangle the structural and stratigraphic history of the region.

Two stratigraphic sequences can be recognized in these complexly faulted and folded metamorphic terranes. On the east, metamorphosed Cambrian and Ordovician eugeoclinal rocks (Hartland Formation) lie on a presumed oceanic basement. On the west, metamorphosed Cambrian and Ordovician miogeoclinal rocks (Lowerre Quartzite, Inwood Marble) overlie a Proterozoic (Fordham and Yonkers gneisses) continental basement. The miogeoclinal and eugeoclinal rocks are juxtaposed along a Taconian crustal suture zone, Cameron's Line (Hall, 1976; Merguerian, 1983a), which appears to be a complexly folded and faulted east-over-west thrust fault. In New York City, the "classic" Manhattan Schist of Merrill (1890) contains at least three mappable units, of which only the lower unit, correlative with member A of Hall (1968), clearly belongs to the autochthonous miogeoclinal sequence. The middle part of the Manhattan Schist is correlative with members B and C of Hall (1976) and is allochthonous (a thrust fault separates it from the underlying miogeoclinal rocks); however, its stratigraphic association (whether miogeoclinal or eugeoclinal) is not yet fully understood. The uppermost unit of the Manhattan Schist is correlative with the allochthonous, eugeoclinal sequence (Hartland Formation) in the upper plate of the Cameron's Line thrust.

SITE INFORMATION

In New York City, the northeast-trending, deeply eroded Middle Proterozoic to lower Paleozoic metamorphic rocks plunge southward beneath unmetamorphosed Mesozoic, Pleistocene, and Holocene sedimentary rocks (Fig. 3). The exposed bedrock includes (1) the autochthonous sequence: Middle Proterozoic Fordham Gneiss, Late Proterozoic Yonkers Gneiss, Cambrian and Ordovician Inwood Marble, and the lower part of the Ordovician Manhattan Schist, correlative with Manhattan Schist member A of Hall (1968); and (2) allochthonous rocks

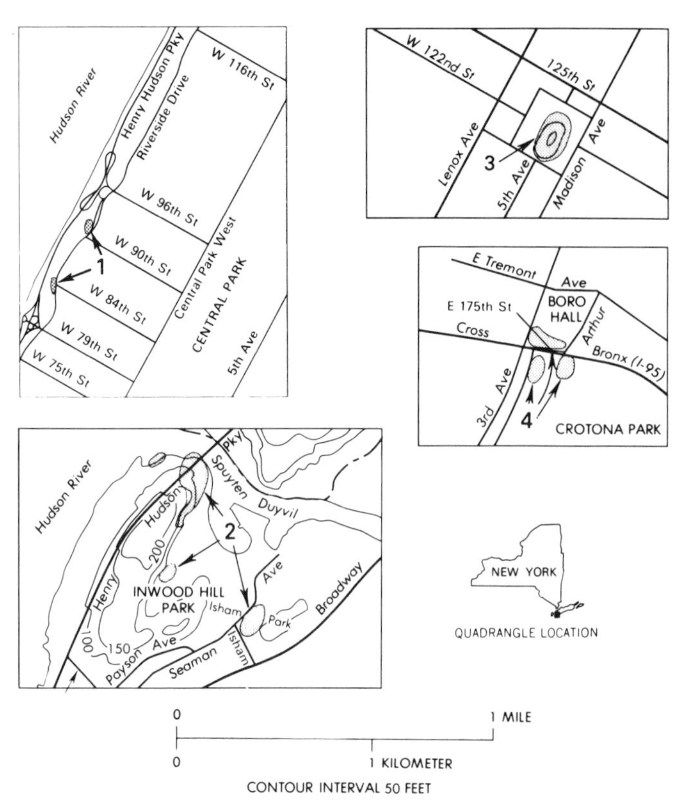

Figure 2. Cross-street locality maps showing outcrops discussed in text.

including: the middle and upper parts of the Manhattan Schist, which are correlative partly with Manhattan Schist members B and C of Hall (1976) and partly with the Hartland Formation.

STRATIGRAPHY

The Fordham Gneiss, oldest rock unit in the region, (Y in Fig. 3), is correlative with Middle Proterozoic Hudson Highlands gneisses (Hall and others, 1975; Hall, 1976). The Fordham Gneiss contains intrusive Late Proterozoic granitoid rocks such as the Yonkers gneiss (Zy in Fig. 3); it is overlain unconformably by the discontinuous Lower Cambrian Lowerre Quartzite (Hall, 1968). In White Plains, New York, the Fordham Gneiss is overlain unconformably by the Cambrian and Ordovician Inwood Marble (O-Ꞓi in Fig. 3). The Middle Ordovician Manhattan Schist, correlative with member A of Hall (1968; Oml in Fig. 3), containing calcite marble, unconformably overlies the Inwood Marble.

The above units constitute the autochthonous miogeoclinal basement-cover sequence of the Manhattan Prong (Hall, 1968), which includes the Bronx and Manhattan Island. Units overlying the Fordham may represent metamorphosed sedimentary rocks deposited on continental crust, now exposed only to the west of Cameron's Line. East of Cameron's Line in western Connecticut and southeastern New York, the rocks of Cambrian and Ordovician age belong to the Hartland Formation of Cameron (1951), Gates (1951), Rodgers and others (1959), or the Hutchinson River Group of Baskerville (1982b; see also Seyfert and Leveson, 1969, and Baskerville, 1982a). In contrast to the basement-cover sequence west of Cameron's Line, the Hartland Formation and correlatives (O-Ꞓh in Fig. 3) are a sequence of metasedimentary and metavolcanic rocks presumably deposited on oceanic crust accreted to North America during the Middle Ordovician Taconic orogeny (Hall, 1976; Merguerian, 1979, 1983a; Merguerian and others, 1984; Robinson and Hall, 1979).

On the basis of lithostratigraphic, structural, and geochronologic evidence, the schist on Manhattan Island is divided into three distinct sillimanite-grade lithostratigraphic units (Merguerian 1983b; Mose and Merguerian, 1985; Merguerian, unpublished data). All three, lower, middle, and upper parts of Merrill's Manhattan Schist of this report, are separated by thrust surfaces.

The lower part of the Manhattan Schist of Merrill (1890) (Oml in Fig. 3) crops out in northern Manhattan and the west Bronx (Fig. 3) where it lies unconformably above the Inwood Marble (O-Ꞓi in Fig. 3). It is composed of brown- to rusty-weathering, fine- to medium-grained, muscovite-biotite-quartz-plagioclase-sillimanite-garnet schist containing inch- to foot-scale (cm- to m-scale) calcite ±diopside marble interlayers (Locality 2). The unit is correlative with the Middle Ordovician Manhattan Schist member A of Hall (1968).

The lower part of Merrill's Manhattan Schist and the Inwood Marble are structurally overlain by the middle schist unit (Omm), which constitutes the bulk of the exposed schist of Manhattan (Localities 2–4; Fig. 3). This schist unit consists of rusty- to maroon-weathering, medium- to coarse-grained, biotite-muscovite-plagioclase-quartz-garnet-sillimanite-kyanite gneiss and schist. This unit is also characterized by sillimanite ± kyanite + quartz ± magnetite layers and lenses as much as 4 in (10 cm) thick, and inch to several feet thick (cm to m) layers of blackish amphibolite and quartzose granofels. The middle part of Merrill's Manhattan Schist, composed of schist and amphibolite, is lithologically similar to and correlative with the Manhattan Schist members B and C of Hall (1976) and, in part, the Waramaug and Hoosac formations in New England.

The structurally highest or upper part of Merrill's Manhattan Schist, correlative with the Hartland Formation (O-Ꞓh), is dominantly gray-weathering, fine- to coarse-grained, well-layered muscovite-quartz-biotite-plagioclase-garnet schist, gneiss, and granofels that contain inch- to several feet-thick scale (cm- to m-scale) layers of greenish amphibolite ± garnet (Locality 1).

STRUCTURAL GEOLOGY

The base of the middle part of Merrill's Manhattan Schist is truncated by a shear zone, the St. Nicholas thrust (Locality 3; open symbol in Fig. 3). Grain-size reduction occurs at the thrust zone with the formation of mylonitic layering, ribboned quartz, lit-par-lit granitization, and quartz veins parallel to the axial surfaces of F_2 folds (Merguerian, 1983b). During the deformation that produced the St. Nicholas thrust, a penetrative foliation (S_2) and metamorphic growth of lenses and layers of quartz ± sillimanite ± kyanite ± magnetite as much as 4 in (10 cm) thick, formed axial planar to F_2 folds; this produced large-scale recumbent folds that strike N50°W, and whose axial surfaces dip 25° SW.

Although the older metamorphic grain of bedrock in the city trends N50°W, geometry of map contacts is strongly affected by younger F_3 isoclinal to tight folds overturned toward the west (Fig. 3). The S_3 axial surface is oriented N30°E and dips 75° SE. It varies from a spaced schistosity to a transposition foliation in many places with shearing near F_3 hinges. At least three phases of crenulate to open folds and numerous brittle faults and joints are superimposed on the older ductile fabrics, but they have little effect on contacts at map scale.

Locality 1. Riverside Park, Hartland Formation. In Riverside Park, from West 116 Street southward to West 75 Street, the upper part of Merrill's Manhattan Schist, which is equivalent to the Hartland Formation, crops out in exposures on Riverside Drive near West 90–91 Streets and West 82–85 Streets (Figs. 1, 2, and 3).

The northernmost outcrops near West 90 Street consist of gray-weathering, well-layered, and slabby to laminated, lustrous muscovitic schist, containing interlayers of quartz-muscovite ± biotite granofels, 1-cm-thick glassy quartzose layers, and elliptical pods of recrystallized dark quartz.

In the outcrop, 2–3-cm metamorphic layering can be observed parallel to the axial surfaces of southeast-plunging long-limbed isoclinal and intrafolial F_2 folds. Tight southwest-plunging F_3 "s" folds are superimposed on the older folds and related metamorphic fabrics.

Outcrops to the south near West 82 Street are lithologically identical to the outcrops near West 90 Street, except the layering is thicker (6–8 cm). A 3 ft (1 m) thick laminated black-weathering, greenish-black biotite-quartz amphibolite layer is evident in the outcrop.

Locality 2. Inwood Hill and Isham Parks, Inwood Marble and Allochthonous Manhattan Schist. Inwood Hill Park, on the northern tip of Manhattan Island, is bordered by Dyckman Street on the south, the Hudson River on the west, Spuyten Duyvil (Harlem Ship Canal) on the north, and Payson and Seaman avenues on the east. Isham Park occupies the flat area east of Inwood Hill Park extending eastward to Broadway between Isham Street and West 214 Street from Seaman Avenue.

Isham Park contains extensive exposures of the Inwood Marble and the middle part of the Manhattan Schist of Merrill (1890). The thrust surface separating the autochthonous marble from the allochthonous schist is not exposed. Several lithologies of marble appear: coarse-grained dolomitic marble, fine-grained calcite marble, and foliated calc-schist—all containing differentially eroded diopside, tremolite, and quartzose layers.

When entering Inwood Hill Park at Seaman Avenue and Isham Street, follow the path past the playground. The first prominent ridge contains the middle part of Merrill's Manhattan Schist. Follow the path to its curve to the west side of the ridge where the valley is underlain by a south-plunging antiform; the exposed west limb shows a tan weathering gray-white Inwood Marble just off the left side of the path.

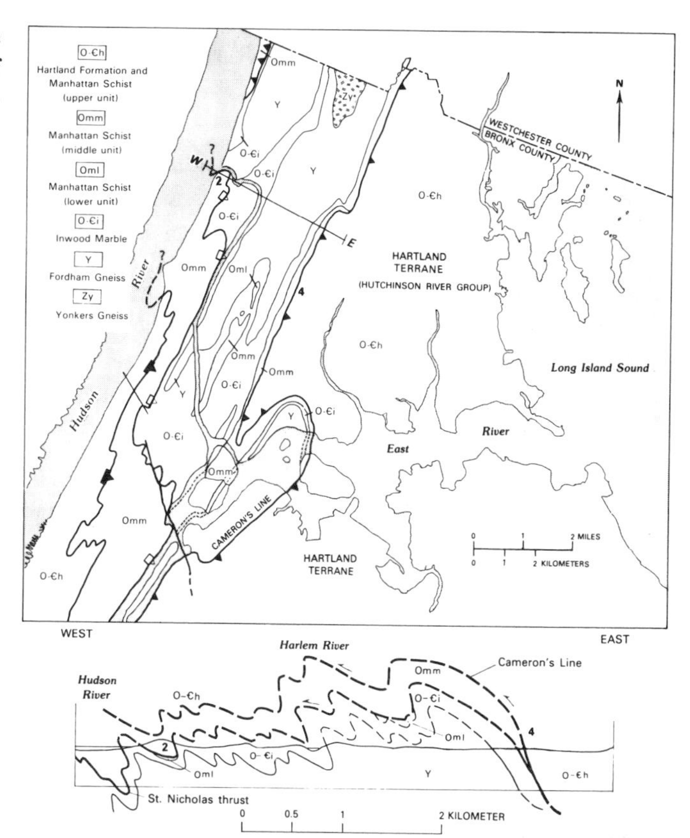

Figure 3. Simplified geologic map of the Bronx and Manhattan based on mapping by the authors. In the type locality of the Manhattan Schist in New York City, schistose rocks have been divided into a lower unit (Oml), a middle unit (Omm), and an upper unit (O-Ꞓh) (Merguerian, unpublished data). The open thrust symbol in Manhattan is the St. Nicholas thrust; the closed symbol is Cameron's Line—barbs show dip direction of foliation. The interpretive W-E section is from the Hudson River to the central Bronx. Numbers refer to localities described in text. Some contacts are projected above the line of section.

Along the path that extends northward (up-slope) on the side of the west ridge, massive brown-weathering blackish amphibolite of the middle part of Merrill's Manhattan Schist crops out. The rocks on top of the ridge above the amphibolite are massive muscovite-biotite-plagioclase-quartz-garnet-sillimanite-kyanite gneiss and schist containing weathered aluminosilicate nodules. The ridge is a south-plunging F_3 synform overturned toward the northwest.

At the top of the ridge beneath the Henry Hudson Bridge, a steep dirt trail leads down to the river, where a coarse-grained gray-white calcitic marble is exposed at low tide. Above the marble, the schist unit consists of biotite-quartz-plagioclase-sillimanite rock containing abundant garnet porphyroblasts. Here, the schist contains an S_2 mylonitic foliation composed of millimeter-scale ribboned quartz striking N45°E, and dipping 55°SE with a strong down-dip lineation that plunges 50° in a

S34°E direction. The zone is structurally complex, consisting of what may be, according to our interpretation, a thrust consisting of intercalated lithologies of the lower and middle schist units together with mylonitic amphibolite.

Locality 3. Mount Morris Park, Inwood Marble and Allochthonous Manhattan Schist. Mount Morris Park, centered at West 122 Street and Fifth Avenue, contains an erosional remnant of the middle part of Merrill's Manhattan Schist, which forms a klippe protruding above the Harlem Valley. The Inwood Marble crops out on the Madison Avenue side of the park. The marble is gray to tan weathering and contains schistose zones with layers and nodules that consist of diopside, tremolite, and quartz. The klippe is terminated to the south (West 120 Street) by a normal fault exposed at street level trending N75°W, dipping 72°SW.

The contact between the middle schist unit and the Inwood Marble is exposed in the outcrop, where it is marked by a shear surface. The shear surface truncates lithologic layering in at an angle of 10° and is accompanied by shearing and imbrication of lithologies, as well as 0.8 to 1.2 in (2 to 3 cm) S_2 mylonitic layering parallel to the axial surfaces of reclined F_2 folds. Disharmonic southwest-plunging F_3 folds are exposed on the thrust surface at the northern edge of the klippe.

Locality 4. Cross Bronx Expressway, Cameron's Line, Allochthonous Manhattan Schist and Hartland Formation. Boro Hall Park in the Bronx is surrounded by East Tremont Avenue on the north, Third Avenue on the west, Arthur Avenue on the east, and East 175 Street on the south, which also serves as the westbound service road for the Cross Bronx Expressway, (I-95).

Along the north side of East 175 Street, the outcrop nearest the intersection with Third Avenue is a brown- to rusty-weathering, medium-grained, gray, biotite-muscovite schist of the middle part of Merrill's Manhattan Schist (B and C units of Hall) containing several pegmatite dikes. Test borings for the I-95 overpass near this locality indicates that marble occupies the Third Avenue valley to the west.

East of the outcrop is a soil-covered, 100 to 130 ft (30 to 40 m) wide, N40°E trending shallow swale. The Hartland Formation crops out to the east of the swale and also south of I-95 along strike with this northern outcrop (Fig. 2). The Hartland rocks consist of brown- to tan-weathering gray muscovite-biotite-quartz-plagioclase-garnet schist and gneiss with granitoid sills and layers of greenish amphibolite. Cameron's Line, the contact between the middle part of Merrill's Manhattan Schist on the west and the Hartland Formation is not exposed, but presumably occupies the swale between the outcrops.

The simplified W–E cross section (Fig. 3) shows the structure of New York City and how the St. Nicholas thrust and Cameron's Line place the middle part of Merrill's Manhattan Schist and the Hartland Formation, respectively, above the Fordham, Inwood, and lower part of Merrill's Manhattan Schist basement-cover sequence. The major F_3 folds produce digitations of the structural and lithologic contacts, which dip gently south (downward from the cross section shown in Fig. 3 toward the viewer).

REFERENCES CITED

Baskerville, C. A., 1982a, The foundation geology of New York City, *in* Legget, R. F., ed., Geology under cities: Geological Society of America Reviews in Engineering Geology, v. 5, p. 95–117.

—— , 1982b, Adoption of the name Hutchinson River Group and its subdivisions in Bronx and Westchester counties, southeastern New York: U.S. Geological Survey Bulletin 1529-H, Stratigraphic Notes, 1980–1982, Contributions to Stratigraphy, p. H1–H10.

Cameron, E. N., 1951, Preliminary report on the geology of the Mount Prospect Complex: Connecticut Geological and Natural History Survey Bulletin 76, 44 p.

Fisher, D. W., Isachsen, Y. W., and Rickard, L. V., 1970, Geologic map of New York: The University of the State of New York, The State Education Department, State Museum and Science Service, Map and Chart Series No. 15, Lower Hudson Sheet, 1:250,000.

Gates, R. M., 1951, The bedrock geology of the Litchfield Quadrangle, Connecticut: Connecticut Geologic and Natural History Survey Miscellaneous Series 3 (Quadrangle Report No. 1), 13 p.

Hall, L. M., 1968, Times of origin and deformation of bedrock in the Manhattan Prong, Chapter 8, *in* Zen, E-an, White, W. S., Hadley, J. B., and Thompson, J. B., Jr., eds., Studies of Appalachian geology, Northern and maritime: New York and London, Interscience Publishers, p. 117–127.

—— , 1976, Preliminary correlation of rocks in southwestern Connecticut, *in* Page, L. R., ed., Contributions to the stratigraphy of New England: Geological Society of America Memoir 148, p. 337–349.

Hall, L. M., Heleneck, H. L., Jackson, R. A., Caldwell, K. G., Mose, D. G., and Murray, D. P., 1975, Some basement rocks from Bear Mountain to the Housatonic Highlands, *in* Ratcliffe, N. M., ed., New England Intercollegiate Geological Conference, 67th Annual Meeting Guidebook: City College of City University of New York, p. 1–29.

Merguerian, C., 1979, Dismembered ophiolite along Cameron's Line–West Torrington, Connecticut: Geological Society of America, Abstracts with Programs, v. 11, no. 1, p. 45.

—— , 1983a, Tectonic significance of Cameron's Line in the vicinity of Hodges complex; An imbricate thrust model for western Connecticut: American Journal of Science, v. 283, p. 341–368.

—— , 1983b, The structural geology of Manhattan Island, New York City (NYC), New York: Geological Society of America Abstracts with Programs, v. 15, p. 169.

Merguerian, C., Mose, D. G., and Nagel, S., 1984, Late syn-orogenic Taconian plutonism along Cameron's Line, West Torrington, Connecticut: Geological Society of America Abstracts with Programs, v. 16, p. 50.

Merrill, F.J.H., 1890, On the metamorphic strata of southeastern New York: American Journal of Science, v. 39, p. 383–392.

Mose, D. G., and Merguerian, C., 1985, Rb-Sr whole-rock age determination on parts of the Manhattan Schist and its bearing on allochthony in the Manhattan Prong, southeastern New York: Northeastern Geology, v. 7, p. 20–27.

Robinson, P., and Hall, L. M., 1979, Tectonic synthesis of southern New England, *in* Wones, D. R., ed., Proceedings, "The Caledonides in the U.S.A." International Geological Correlation Project 17–Caledonide orogen, 1979 meeting: Blacksburg, Virginia, Virginia Polytechnic Institute and State University Memoir 2, p. 73–82.

Rodgers, J., Gates, R. M., and Rosenfeld, J. L., 1959, Explanatory text for preliminary geological map of Connecticut, 1956: Connecticut Geological and Natural History Survey Bulletin 84, 64 p.

Seyfert, C. K., and Leveson, D. J., 1969, Speculations on the Hutchinson River Group and the New York City Group, *in* Symposium on the New York City Group, 40th Annual Meeting: Queens College Press, New York State Geological Association, Geological Bulletin 3, p. 33–36.

The Taconic Allochthon Frontal Thrust: Bald Mountain and Schaghticoke Gorge, eastern New York

William Bosworth, *Marathon International Oil Company, P.O. Box 3128, Houston, Texas 77253*
Steven Chisick, *Department of Geology, Rensselaer Polytechnic Institute, Troy, New York 12180*

LOCATION

Bald Mountain is located in the southeast corner of the Fort Miller 7½-minute Quadrangle, Washington County, eastern New York. Access is gained via New York 40 north from Greenwich, New York (Fig. 1). A paved secondary road (Bald Mountain Road) leads past the quarries on the west side of the mountain, and the quarry faces can be reached by walking along an overgrown gravel track (entered near the intersection of Bald Mountain and Lick Springs roads). The quarries are abandoned, but adjacent landowners should be notified of your intentions. Schaghticoke Gorge is located south of Bald Mountain in the Schaghticoke 7½-minute Quadrangle at Schaghticoke, New York. Access is also from New York 40 at Chestnut Street on the south side of the village. A dirt road is followed from the village street to the lower gorge of the Hoosick River (Fig. 1). The main exposures here lie in the river bottom, and are therefore best seen in the late summer and fall. The river bottom is public domain, but cars should not be driven over the old concrete bridge at the river (see Fig. 4), as this is owned by the power company.

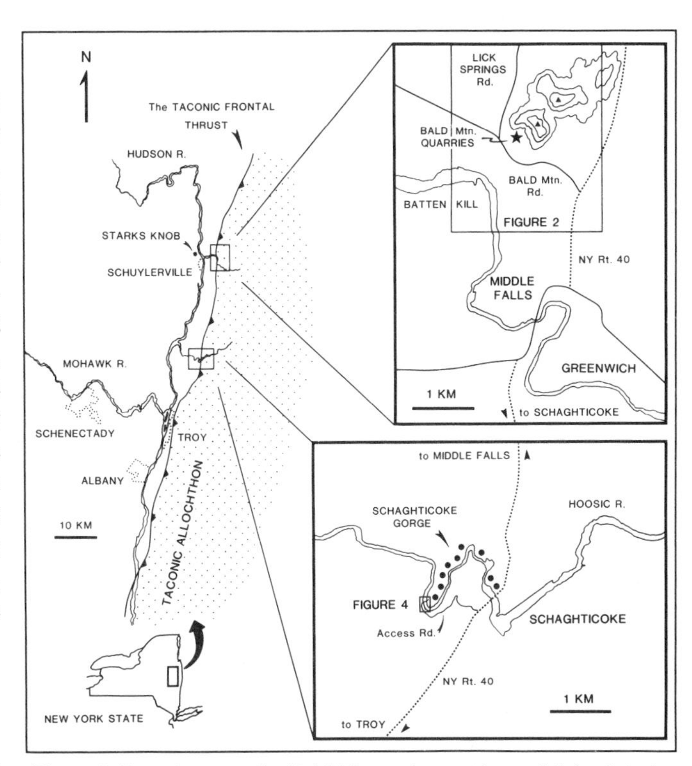

Figure 1. Location map for Bald Mountain quarries and Schaghticoke Gorge.

SIGNIFICANCE

The quarries at Bald Mountain provide exposures of the western edge of the Taconic Allochthon, the first recognized allochthonous terrane of the Appalachian orogen. A continuous section from far traveled pelitic rocks of continental rise affinity ("Taconic Sequence"), through a narrow fault zone and into carbonate shelf rocks ("New York Standard Sequence") is present at this historic locality. The carbonate rocks are themselves to some extent transported (parallochthonous), having been derived from sub-cropping shelf rocks a short distance to the present east. These Bald Mountain Carbonates have played a crucial role in all models proposed for the emplacement history of the Taconic Allochthon. They have been envisioned to represent: (1) fault slivers attached to the base of a large thrust sheet, (2) olistoliths within a melange zone at the base of either a large gravity slide or thrust sheet, or (3) a composite feature, the large carbonate blocks having been imbricated during hard-rock thrusting, followed by cannibalism, erosion, and subsequent underthrusting at emergent, subaqueous fault scarps. The overall structural geometry is here interpreted to comprise a duplex form, with the roof fault forming the Taconic Frontal Thrust. The floor fault separates the carbonate rock from underlying medial Ordovician synorogenic flysch and melange, and is exposed nearby in the gorge of the Batten Kill. An excellent exposure of this Taconic flysch is found a few mi (km) south at Schaghticoke, where imbrication of undisrupted flysch and melange can be observed at a structural level beneath the frontal thrust. The structural and stratigraphic relationships present at Bald Mountain and Schaghticoke Gorge provide important constraints on the environmental conditions prevailing during emplacement of the Taconic Allochthon, and on the deformational history of the Taconic orogen in general.

SITE INFORMATION

The Taconic Ranges of eastern New York and adjacent Vermont and Massachusetts are now recognized as one of the principal external allochthonous terranes of the Appalachian orogen (Zen, 1967; Rodgers, 1970; Bird and Dewey, 1970). Composed of a dominantly argillaceous rock sequence, the Taconic Allochthon is inferred to have been derived from a continental rise position in the lower Paleozoic paleogeography of North America. Its emplacement on the coeval carbonate shelf sequence of eastern New York occurred during the attempted subduction of North America beneath one or more west-facing (present coordinates) island arcs in medial Ordovician times (Chapple, 1973,

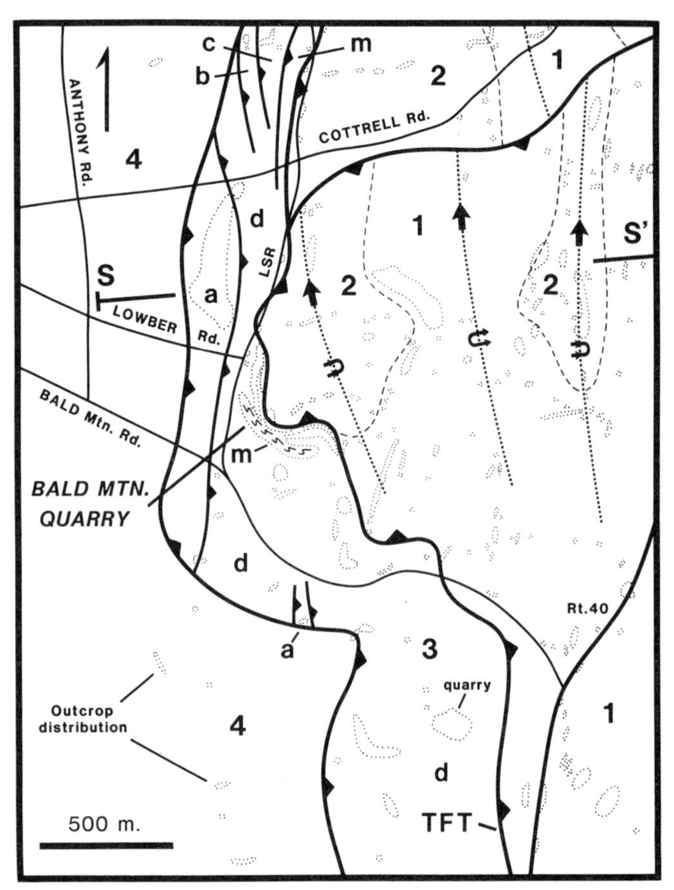

Figure 2. Geologic map of the Bald Mountain quarry area (reinterpreted from Platt, 1960, and Bosworth, 1980). LSR, Lick Springs Road; TFT, Taconic Frontal Thrust. Lithologic units are as follows: 1, Lower Cambrian Bomoseen olive-green graywackes and slates (also includes Truthville green slates); 2, Lower Cambrian Browns Pond black slates and limestones and Upper Cambrian Hatch Hill black slates and dolomitic sandstones; 3, Bald Mountain carbonate terrane, including: a, hard quartz arenites, micaceous in part; b, limestone pebble and cobble in sandy matrix conglomerate ("Rysedorph Conglomerate" of Ruedemann, 1901); c, thin bedded limestones and slates, extensively boudinaged in part; d, undifferentiated limestones and dolostones, largely massive bedded; m, large outcrop of melange at quarry; 4, Middle Ordovician Taconic flysch and melange.

1979; Rowley and others, 1979; Rowley and Kidd, 1981; Stanley and Ratcliffe, 1985). A minimum amount of transport for the Taconic Allochthon is established by the present, unrestored position of the Cambrian-Ordovician shelf edge, some 43 mi (70 km) to the present east (Rodgers, 1970).

Bald Mountain provides the best available exposure of the western edge of the Taconic Allochthon (Fig. 1). The massive carbonate rocks present on the west side of the mountain were quarried for limestone during the nineteenth and early twentieth centuries. Ebenezer Emmons published the first cross section through Bald Mountain (1844; 1847, p. 89), in which he inferred the carbonates to be resting unconformably on the Taconic slates. On the basis of lithologic similarities, Emmons correlated the Bald Mountain limestones and dolostones with the "Lower Silurian Calciferous sandrock" of the 'standard' New York sequence (the "Lower Silurian" now being referred to as Ordovician), supporting his contention that the underlying Taconic slates belong to a much older sequence of rocks, and that the term "Taconic System" was a necessary addition to the lower Paleozoic stratigraphic nomenclature of North America. Later geologists would show that Emmons had erred in many of his stratigraphic correlations at Bald Mountain, and "Taconic" would never succeed in replacing Sedgwick's "Cambrian" as the accepted base of the Paleozoic Era.

The structural and stratigraphic relationships at Bald Mountain were first correctly assessed by C. D. Walcott (1888). On the basis of faunal content, Walcott demonstrated that there exist slate and carbonate rock of several ages at Bald Mountain, and that major structural boundaries are also present. The peaks of the mountain are underlain by Lower Cambrian Bomoseen olive-green graywacke and slate, Lower Cambrian Browns Pond black slate and limestone, and Upper Cambrian Hatch Hill dolomitic sandstone and slate (Fig. 2; present-day Cambrian faunal assignments). These allochthonous Taconic units are separated from large blocks of Lower Ordovician (Ruedemann, 1914, p. 77) and Middle Ordovician (Platt, 1960, p. 50) shelf limestone and dolostone by a zone of phacoidally cleaved, dark gray shale containing small slivers and clasts of carbonate, graywacke, and broken quartz-calcite vein material. Walcott inferred a "Great Fault" to pass above this "contorted" zone (1888, p. 317), which he equated with the regional overthrust boundary of W. E. Logan (1861).

Shelf carbonates make up the bulk of the exposure in the Bald Mountain quarries (Fig. 2). The carbonate occurs as fault bounded blocks, enclosed laterally, in part, by phacoidally cleaved shale (Rodgers and Fisher, 1969). The largest blocks at the quarry are a few tens of ft (m) in exposed dimensions, but along strike continuous outcrops of several hundred ft (m) are found (Fig. 2; Bosworth, 1980, Plates 1 and 2; Bosworth and Kidd, 1985). The block-in-shale geometry has led to the interpretation that the Bald Mountain carbonates occur as olistoliths within melange (Rodgers, 1952, 1982; Rodgers and Fisher, 1969), perhaps at the base of a soft-sediment Taconic gravity slide (Rodgers, 1951). Ruedemann (1914), following Walcott, interpreted the Bald Mountain carbonates as fault slivers caught at the base of the Taconic (hard-rock) overthrust. Zen (1964, p. 15) similarly inferred that the entire carbonate terrane must have been emplaced as a tectonic sliver beneath the Taconic Allochthon, but with J. M. Bird (in Zen, 1967, p. 36–37) allowed that the quarry carbonate blocks were part of a polymict conglomerate deposited in front of the advancing allochthon, and hence structurally correlative with the Forbes Hill Conglomerate (Zen, 1961).

The base of the carbonate rocks is not exposed at Bald Mountain. It can be examined a short distance to the south in the gorge of the Batten Kill at Middle Falls, below the bridge for New York 40 (Fig. 1). There massive carbonate is in contact with

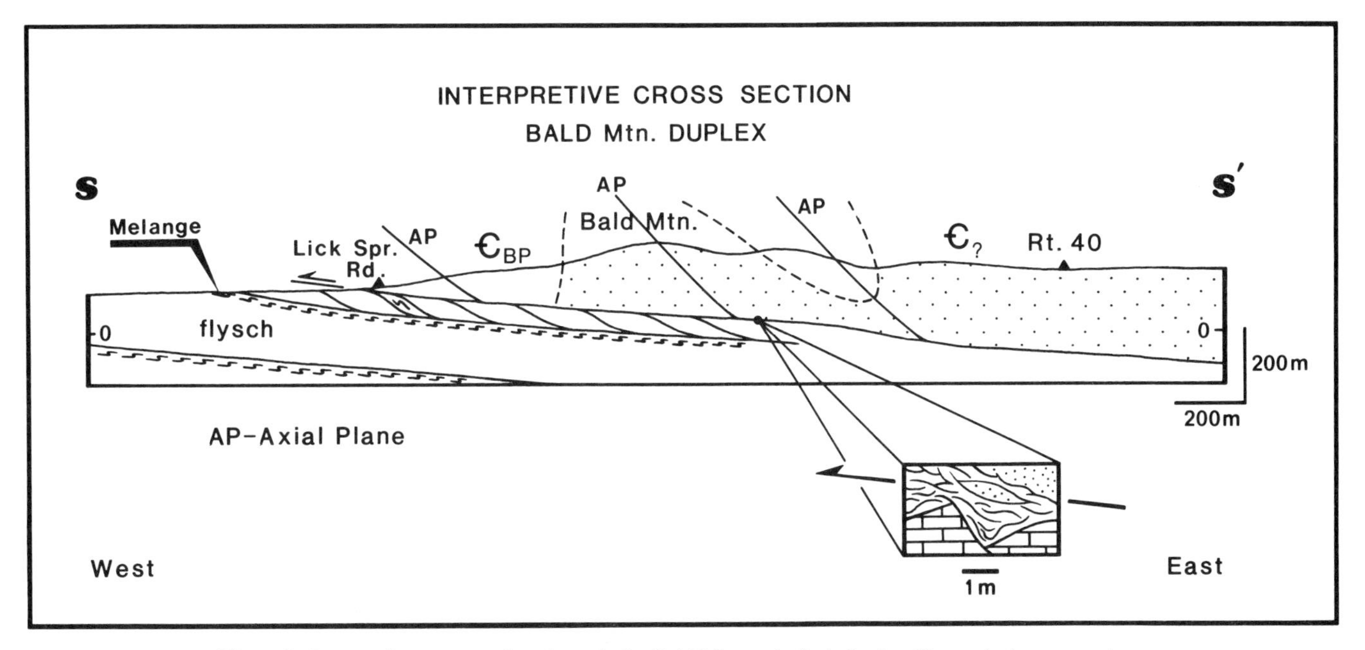

Figure 3. Interpretive cross section through the Bald Mountain fault duplex. No vertical exaggeration. Inset illustrates the complex nature of the duplex roof fault zone, as exposed at the quarries. €?, Cambrian?, Bomoseen graywacke and Truthville slate; $€_{BP}$, Cambrian BP, Browns Pond and Hatch Hill slate, limestone, and dolomitic sandstone.

parallochthonous medial Ordovician shale. This shale and associated graywacke and siltstone were deposited in a westerly migrating foreland basin at the time of the Taconic arc–continent collision. Similar, slightly older graywacke and shale is found at the top of the allochthonous Taconic sequence of rocks. The entire assemblage of synorogenic deposits, including both allochthonous and autochthonous units, has been informally referred to by Rowley and Kidd (1981) and Bosworth and Vollmer (1981) as the "Taconic flysch." The contact between flysch and shelf carbonate at the Batten Kill is a thrust fault and much of the shale has been disrupted to a melange fabric.

Although much of the Bald Mountain carbonate terrane demonstrably consists of small blocks in a fine-grained matrix, as originally recognized by Rodgers, and hence can be interpreted either as a sedimentary or tectonic breccia, there is evidence that suggests this is not the case for the main carbonate bodies (Bosworth, 1980; Rowley and Kidd, 1982; Bosworth and Kidd, 1985). At the Bald Mountain quarries, several of the massive carbonate blocks are cut by bands of strongly foliated carbonate a few centimeters wide that contain mylonitic fabrics. This suggests that the blocks were disrupted at least in part in a relatively high P/T environment, such as at the base of a large nappe complex. Fault breccias found along strike between blocks also point to a tectonic origin for the present large-scale block geometry, and eliminate any need to invoke sedimentary processes for the formation of the Bald Mountain carbonate bodies. The large size of the coherent carbonate masses, particularly north-south parallel regional strike, and the alignment of outcrops of similar lithologies as north-south bands within the carbonate terrane (Fig. 2), argue against a wholesale breccia origin for the carbonate blocks. In analogy with similar structural settings in other orogens, we interpret the Bald Mountain carbonates as a series of horses within a duplex structure, the roof fault being poorly exposed in the upper part of the quarry faces, and the floor fault exposed at the base of the carbonate in Middle Falls (Fig. 3; Bosworth and Kidd, 1985). Ruedemann similarly recognized a series of overthrust planes in the carbonate terrane and referred to it as "Schuppenstruktur" (1933, p. 134).

The contacts between individual horses in the carbonate terrane are locally extremely complex, and parts of the duplex appear to have been further disrupted subsequent to its initial imbrication, resulting in the sawtooth geometry visible at the southern Bald Mountain quarry face. The many years that have passed since the quarries at Bald Mountain were last active have taken their toll on the quality of the exposures in the upper parts of the quarry faces, and sketches made by Ruedemann (1914) early in this century are invaluable for interpreting the details of the frontal thrust structure. The contact between the large, massive carbonate blocks and the overlying block-in-shale has previously been interpreted as a karst surface (Sanders and others, 1961) and a regional unconformity (Zen and Bird *in* Zen, 1967). Zen has discussed why the karst interpretation is untenable (Zen, 1967). There do exist other features that are likely to be bonafide karsts entirely within single blocks of the massive limestones of the terrane, and these may prove useful in deciphering the detailed stratigraphic correlations for these units. A well-exposed karst surface can be seen at the south end of the large abandoned quarry, now a private park, just south of Bald Mountain Road

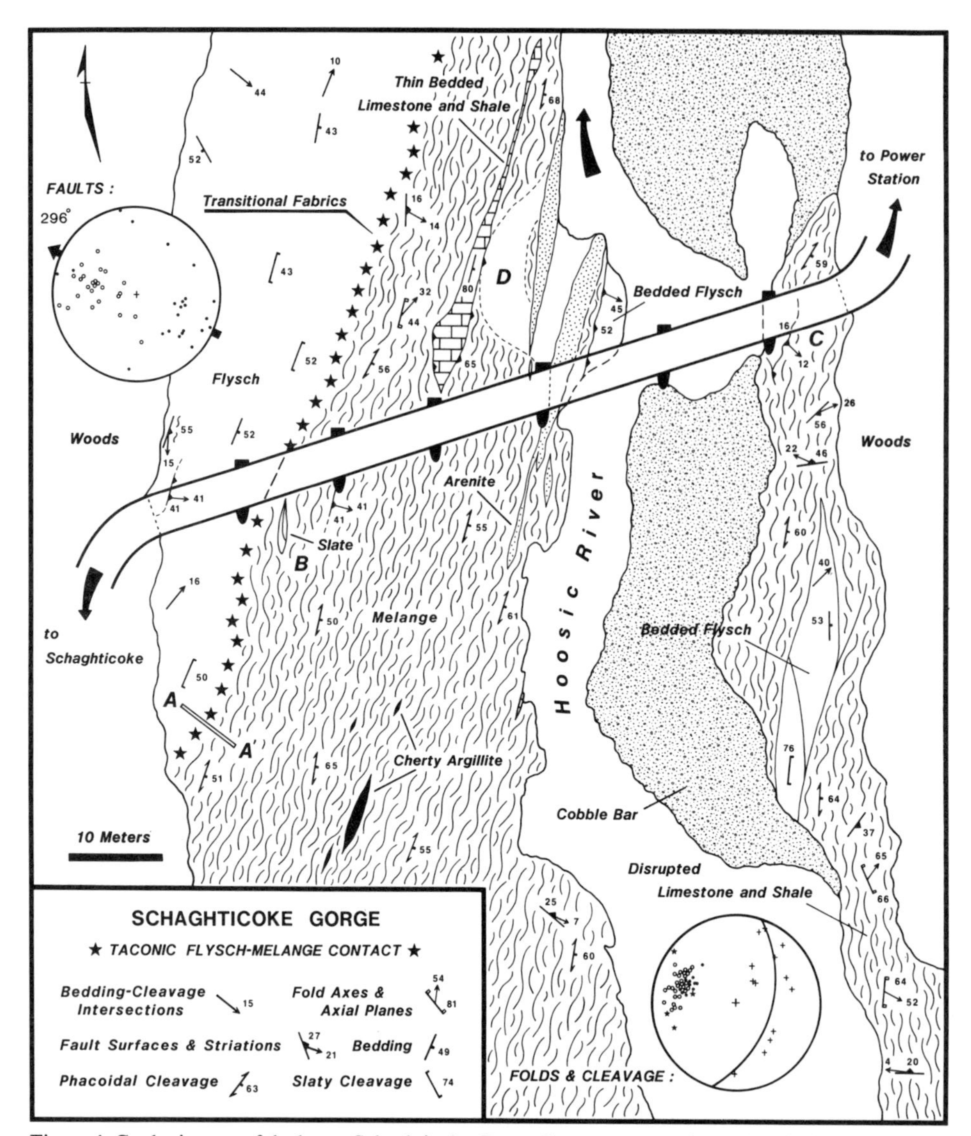

Figure 4. Geologic map of the lower Schaghticoke Gorge. Stereograms are lower hemisphere, equal area projections (Faults: open circles, poles to small-scale faults; solid circles, striations on faults. Folds and Cleavage: open circles, poles to phacoidal cleavage; solid circles, poles to slaty cleavage; stars, poles to small-scale axial planes; crosses, hinge lines.) Section A–A′ at the flysch/melange fabric transition has been studied in thin-section and demonstrates the overprinting of the slaty cleavage by an anastomosing phacoidal fabric. The block at "B" is a remnant of the slate unit with its borders being progressively cannibalized. Pebbly mudstones are particularly well exposed at "C," and the area around "D" illustrates the progressive disruption of the bedded flysch sequence through extensive boudinage and structural slicing. An overthrust direction of 296° (WNW) is inferred from the outcrop structural measurements, in agreement with numerous other transport determinations.

(labeled simply "quarry" in Fig. 2). This is also an excellent site to view the internal continuity of the larger carbonate "slivers."

The total outcrop length of the Bald Mountain carbonate terrane along the western edge of the allochthon is approximately 6 mi (10 km). The melange that exists at the top of the Bald Mountain quarries, between many of the carbonate horses and beneath the carbonates at Middle Falls also continues along strike and demarcates the boundary of the Taconic Allochthon in all localities where good exposures are present (Berry, 1962; Zen, 1967; Bird and Dewey, 1975). One of the best exposed sections of this melange occurs south of the Bald Mountain area in the lower gorge of the Hoosic River (Fig. 4). The upper gorge contains magnificent outcrops of allochthonous Taconic rocks (type locality of Ruedemann's [1903] "Schaghticoke Shale"), but the frontal thrust itself is not exposed. O'Brien (1960) mapped the general bedrock geology of the Schaghticoke area, but unfortunately, outside the gorge, exposures are very sparse.

A most interesting aspect of the Schaghticoke Gorge locality is that the transition from undisrupted Taconic flysch to melange is well exposed. The phacoidal melange fabrics can be seen to progressively overprint the slaty cleavage of the flysch, demonstrating that at least here, the main deformation within the melange occurred in well-lithified rocks (Bosworth and Vollmer, 1981). All regional studies support the association of the formation of this melange with the final emplacement of the Taconic Allochthon on the Cambrian-Ordovician shelf, hence establishing the physical properties of the underthrust material to be that of hard rocks, not soft sediments (further discussion in Rowley and Kidd, 1981; Bosworth and Vollmer, 1981; Bosworth and Rowley, 1984).

Schaghticoke Gorge also provides an opportunity to view the various lithologies that have been incorporated into the Taconic melange outside the Bald Mountain duplex zone. Figure 4 illustrates the position of several large blocks of cherty argillite, interbedded slate and siltstone, thin bedded limestone and shale, and massive siltstone and arenite. These are all rock types normally occurring within the undisrupted flysch (Ruedemann, 1914). Smaller clasts of dolomitic sandstone, red slate, "black quartz" graywacke, and green graywacke can be found in pebbly mudstone horizons as at point "C" in Figure 4. These are Taconic lithologies, and have been interpreted to represent small-scale olistostromes, deposited at subaqueous fault scarps and subsequently incorporated into the tectonically derived melange during Taconic overthrusting (Bosworth and Vollmer, 1981; Rowley and Kidd, 1982). The deformational history of the Taconic melange was exceedingly complex, and it is important to be very specific as to which parts of that history one correlates with regional structural events. Comparison of melange fabrics seen at Schaghticoke Gorge with those described for melanges associated with the regionally analogous Hamburg klippe of Pennsylvania (Lash, 1985) and Humber allochthon of western Newfoundland (Bosworth, 1985) reveals some of the similarities and dissimilarities present along the length of the Appalachian orogen.

Truly "exotic" blocks are actually uncommon in the Taconic melange, if one accepts the hypothesis that the main Bald Mountain carbonate masses form a duplex structure beneath the allochthon. Bird and Dewey (1975) describe a variety of small exotic blocks farther south along the frontal thrust, but the most unusual exotic occurs within the Hudson River melange zone at Northumberland, New York (located just north of Schuylerville on U.S. 4). The block forms "Starks Knob" (an historical park, ensuring easy access), and is composed of beautifully preserved pillow basalts (Woodworth, 1901; Ruedemann, 1914). This slab was probably detached from a position in the sub-cropping shelf sequence during imbrication of the Taconic flysch (Vollmer and Bosworth, 1984), a view that is supported by unpublished industry reflection seismic profiles. Please do not damage the rare internal pillow structures when visiting this locality.

The localities described above provide an opportunity to view a nearly complete section through the western edge of the Taconic Allochthon. The frontal thrust, a possible duplex complex or series of fault slivers, frontal melange, and underlying synorogenic flysch can all be examined during a single day in the field. These same outcrops have all played critical roles in the development of eastern American geologic thought. Fascinating reviews of the various "Taconic Controversies" can be found in Walcott (1888), Merrill (1924), and Zen (1967).

REFERENCES CITED

Berry, W.B.N., 1962, Stratigraphy, zonation and age of the Schaghticoke, Deepkill, and Normanskill shales, eastern New York: Geological Society of America Bulletin, v. 73, p. 695–718.

Bird, J. M., 1963, Sedimentary structures in the Taconic sequence rocks of the southern Taconic region, *in* Bird, J. M., ed., Stratigraphy, structure, sedimentation, and paleontology of the southern Taconic region, eastern New York: Geological Society of America Guidebook for Field Trip Three, p. 5–21.

—— , 1969, Middle Ordovician gravity sliding, Taconic region, *in* Kay, G. M., ed., North Atlantic geology and continental drift: American Association of Petroleum Geologists Memoir 12, p. 670–686.

Bird, J. M., and Dewey, J. F., 1970, Lithosphere plate-continental margin tectonics and the evolution of the Appalachian orogen: Geological Society of America Bulletin, v. 81, p. 1031–1060.

—— , 1975, Selected localities in the Taconics and their implications for the plate tectonic origin of the Taconic region, *in* Ratcliffe, N. M., ed., New England Intercollegiate Geological Conference 67th Annual Meeting, Guidebook for field trips in western Massachusetts, northern Connecticut and adjacent areas of New York, p. 87–121.

Bosworth, W., 1980, Structural geology of the Fort Miller, Schuylerville and portions of the Schaghticoke 7½-minute Quadrangles, eastern New York, and its implications in Taconic geology [Ph.D. thesis, pt. 1]: Albany, State University of New York, 237 p.

—— , 1985, East-directed imbrication and oblique-slip faulting in the Humber Arm Allochthon of western Newfoundland: structural and tectonic significance: Canadian Journal of Earth Sciences, v. 22, p. 1351–1360.

Bosworth, W., and Kidd, W.S.F., 1985, Thrusts, melanges, folded thrusts, and duplexes in the Taconic foreland, *in* Lindemann, R. H., ed., New York State Geological Association 57th Annual Meeting Guidebook, p. 117–147.

Bosworth, W., and Rowley, D. G., 1984, Early obduction-related deformation features of the Taconic Allochthon: analogy with structures observed in modern trench environments: Geological Society of America Bulletin, v. 95, p. 559–567.

Bosworth, W., and Vollmer, F. W., 1981, Structures of the medial Ordovician flysch of eastern New York; Deformation of synorogenic deposits in an overthrust environment: Journal of Geology, v. 89, p. 551–568.

Chapple, W. M., 1973, Taconic orogeny; Abortive subduction of the North American continental plate?: Geological Society of America Abstracts with Programs, v. 5, p. 573.

—— , 1979, Mechanics of emplacement of the Taconic Allochthon during a continental margin-trench collision: Geological Society of America Abstracts with Programs, v. 11, p. 7.

Emmons, E., 1844, The Taconic system: Albany, New York, 65 p. (reprinted in Agriculture of New York, v. 1, 1847, Albany, New York, p. 45–108).

—— , 1856, American Geology, v. 1, pt. 2: Albany, New York, Sprague and Company, 251 p.

Fisher, D. W., Isachsen, Y. W., and Richard, L. V., 1970, Geologic map of New York State: New York State Museum and Science Service Map and Chart Series 15.

Kay, G. M., 1935, Taconic thrusting and paleogeographic maps: Science, v. 82, p. 616–617.

Keith, A., 1912, New evidence on the Taconic question: Geological Society of America Bulletin, v. 23, p. 720–721.

Lash, G. G., 1985, Accretion-related deformation of an ancient (early Paleozoic) trench-fill deposit, central Appalachian orogen: Geological Society of America Bulletin, v. 96, p. 1167–1178.

Logan, W. E., 1861, Remarks on the fauna of the Quebec group of rocks, an the Primordial zone of Canada, *in* Hitchcock, E., and others, eds., Geology of Vermont, Part 1: Claremont, New Hampshire, p. 379–382.

Merrill, G. P., 1924, The first one hundred years of American geology: New Haven, Yale University Press, 126 p.

O'Brien, P. J., 1960, The geology of Schaghticoke and environs, New York [M.S. thesis]: Troy, New York, Rennselaer Polytechnic Institute, 58 p.

Platt, L. B., 1960, Structure and stratigraphy of the Cossayuna area, New York [Ph.D. thesis]: New Haven, Yale University, 126 p.

Potter, D. B., 1972, Stratigraphy and structure of the Hoosick Falls area, New York–Vermont, east-central Taconics: New York State Museum and Science Service Map and Chart Series 19, 71 p.

Rickard, L. V., and Fisher, D. W., 1973, Middle Ordovician Normanskill Formation, eastern New York; Age, stratigraphic and structural position: American Journal of Science, v. 273, p. 580–590.

Rodgers, J., 1951, La tectonique decoulement par gravite, gravity sliding tectonics (essay review): American Journal of Science, v. 249, p. 539–540.

—— 1952, *in* Billings, M. P., Rodgers, J., and Thompson, J. B., Jr., Geology of the Appalachian highlands of east-central New York, southern Vermont and southern New Hampshire: Geological Society of America 65th Annual Meeting, Boston, Guidebook, p. 1–71.

—— , 1970, The tectonics of the Appalachians: New York, Wiley-Interscience, 271 p.

—— , 1982, Discussion *of* "Stratigraphic relationships and detrital composition of the medial Ordovician flysch of western New England; Implications for the tectonic evolution of the Taconic orogeny": Journal of Geology, v. 90, p. 219–222.

Rodgers, J., and Fisher, D. W., 1969, Paleozoic rocks in Washington County, New York, west of the Taconic klippe, *in* Bird, J. M., ed., New England Intercollegiate Geological Conference, Guidebook, p. 6-1 to 6-12.

Rowley, D. B., Kidd, W.S.F., and Delano, L. L., 1979, Detailed stratigraphic and structural features of the Giddings Brook slice of the Taconic Allochthon in the Granville area: New York State Geologists Association Annual Meeting Field Trip Guidebook, no. 51, p. 186–242.

Rowley, D. B., and Kidd, W.S.F., 1981, Stratigraphic relationships and detrital composition of the medial Ordovician flysch of western New England; Implications for the tectonic evolution of the Taconic orogeny: Journal of Geology, v. 89, p. 199–218.

—— , 1982, Reply *to* Discussion *of* "Stratigraphic relationships and detrital composition of the medial Ordovician flysch of western New England; Implications for the tectonic evolution of the Taconic orogeny": Journal of Geology, v. 90, p. 223–226.

Ruedemann, R., 1901, Trenton conglomerate of Rysedorph Hill, Rensselaer County, New York: New York State Museum Bulletin 49, p. 3–14.

—— , 1903, The Cambric *Dictyonema* fauna in the slate belt of eastern New York: New York State Museum Bulletin 69, p. 934–958.

—— , 1914, *in* Cushing, H. P., and Ruedemann, R., Geology of Saratoga Springs and vicinity: New York State Museum Bulletin 169, 177 p.

—— , 1933, *in* Newland, D. H., and others, eds., The Paleozoic stratigraphy of New York: 16th International Geological Congress, Guidebook 4, Excursion A-4, p. 121–136.

Sanders, J. E., Platt, L. B., and Powers, R. W., 1961, Bald Mountain Limestone, New York; New facts and interpretations relative to Taconic geology: Geological Society of America Bulletin, v. 72, p. 485–487.

Stanley, R. S., and Ratcliffe, N. M., 1985, Tectonic synthesis of the Taconian orogeny in western New England: Geological Society of America Bulletin, v, 96, p. 1227–1250.

Vollmer, F. W., and Bosworth, W., 1984, Formation of melange in a foreland basin overthrust setting; Example from the Taconic orogen, *in* Raymond, L. A., ed., Melanges: Their nature, origin and significance: Geological Society of America Special Paper 198, p. 53–70.

Walcott, C. D., 1888, The Taconic system of Emmons and the use of the name Taconic in geologic nomenclature: American Journal of Science, v. 35, p. 229–242, 307–327, and 394–401.

Woodworth, J. B., 1901, New York State Geologist 21st Annual Report, p. r17–r29.

Zen, E-an, 1961, Stratigraphy and structure at the north end of the Taconic Range in west-central Vermont: Geological Society of America Bulletin, v. 72, p. 293–338.

—— , 1964, Taconic stratigraphic names; Definitions and synonymies: United States Geological Survey Bulletin 1174, 95 p.

—— , 1967, Time and space relationships of the Taconic Allochthon and Autochthon: Geological Society of America Special Paper 97, 107 p.

The mid-Ordovician section at Crown Point, New York

Brewster Baldwin, Department of Geology, Middlebury College, Middlebury, Vermont 05753

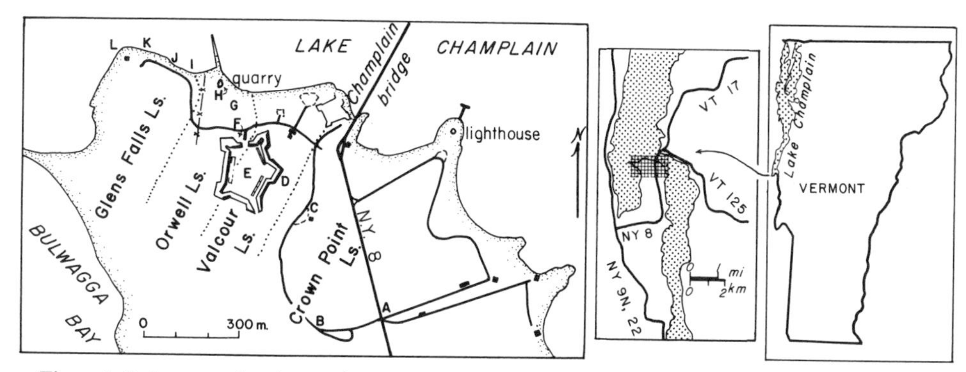

Figure 1. Index map showing stations (after Baldwin, 1980; permission from Northeastern Geology).

LOCATION AND ACCESSIBILITY

The site is in New York on Crown Point peninsula, at the south end of the wide part of Lake Champlain (Fig. 1). From Vermont, follow Vermont 125 or Vermont 17 west to the Champlain bridge; from New York, turn from New York 9N (New York 22) east onto New York 8 (Fig. 1, inset). The Crown Point section is exposed at the Crown Point State Historic Site; the entrance is 0.4 mi (0.7 km) south of the southwest end of Champlain bridge.

SIGNIFICANCE

At Crown Point, fossils and sedimentary features of the 400-ft (120-m) section of mid-Ordovician carbonates record the onset of the continent-arc collision known as the Taconic Orogeny. The lower half of the section shows strata deposited in the shore zone of a slowly subsiding, passive, cooling continental margin. The upper half shows the transition toward deeper water environments, preceding the collapse as the margin entered the subduction zone.

SITE INFORMATION

The site is scenic and historic. Fort St. Frederic, by the Champlain bridge, was built by the French in 1734. In 1759 the British built Fort Crown Point (E in Fig. 1). This is a State Historic Site; if the gate is closed, park out of the way and walk in. *Use of hammers and the collecting of samples are prohibited.* It is helpful and evidently not objectionable to circle fossils with chalk for the aid of others. Selleck (Selleck and Baldwin, 1985) identified some of the fossils; Mehrtens (Baldwin and Mehrtens, 1986) and Selleck have described aspects of the petrography of the rocks.

Beds in the section at Crown Point strike northeast and dip about 8° northwest. The places to see the several parts of the section are indicated by lettered stations on the map (Fig. 1) and columnar section (Fig. 2). For historical purposes, Figure 2 also shows Raymond's (1902) sections B and C; he listed a large number of fossils from those sections.

Abundant sedimentary features and fossils indicate a sequence of three environments of deposition. The Chazy Group (and the older sediments of west-central Vermont) accumulated at about sea level on the passive margin of proto-North America (Fig. 3). The onset of the Taconic Orogeny changed the nature of sedimentation, because subsidence began to exceed the accumulation rate, and the water deepened. The Orwell Limestone (Black River and lower Trenton groups) was deposited in the photic zone; the Glens Falls Limestone (Trenton Group) then accumulated in open shelf, sub-photic conditions. This transition was followed by rapid subsidence of the platform, when nearly 1 mi (1.5 km) of shale (exposed in northern Vermont and southern Quebec) was deposited on the Glens Falls Limestone (Baldwin, 1980).

Figure 3 uses the regional stratigraphy to illustrate this history. The time scale is from Harland and others (1964) and Churkin and others (1977). The deposited thickness (T_d) of the proto-American platform section and overlying deeper water shale was calculated from present thickness of strata in west-central Vermont (Baldwin, 1980), using compaction data of Baldwin and Butler (1985).

Following principles of McKenzie (1978), space for T_d was provided by crustal subsidence due to tectonism (C_t), plus crustal subsidence due to sedimentary loading (C_s). The amount of C_t reflects how much the continental crust was stretched and thinned during rifting. The rate of C_t was determined by thermal cooling.

Values of T_d and C_s were calculated for the end of the Cambrian, Lower Ordovician, and Chazy Group; T_d values were also calculated for the end of the Glens Falls Limestone and Trenton shale. To do this, average solidities were calculated for the accumulating section, using solidity values and burial depths of Baldwin and Butler (1985). The overall average solidity was 75%, indicating a sediment density of 2.3 g/cm^3. This value, compared with an assumed mantle density of 3.3 g/cm^3, indicates that C_s accounted for 70% of the total subsidence, and the remaining 30% was due to C_t.

The tectonic control of T_d is shown in Figure 3 which compares the west-central Vermont section with model curves of McKenzie (1978) for sediment thickness T_d. The McKenzie curves reflect thermal cooling accompanying crustal extension (β) of 1.25 and 1.5 times. The T_d curve for west-central Vermont suggests β of about 1.4 times.

Thus, the Cambro-Ordovician section of the ancient continental platform, up through deposition of the Chazy Group, fits McKenzie's data for a rifted and then thermally cooling, passive continental margin. Incidentally, Figure 3 also indicates that rifting ended about 550 Ma; this is in general agreement with more elaborate studies of Bond and others (1984). Strata younger than the Chazy Group were deposited on a rapidly subsiding continental margin that was being forced to enter the subduction zone.

THE CHAZY GROUP

Beds of the Chazy Group are seen at Stations A to E. Much of the interval is covered (Fig. 2). Here, the Chazy is characterized by the presence of beds of dolostone, which is absent in younger beds.

Dolostone beds can be recognized in several ways in the Crown Point section. Their weathering color ranges from cream to tan to brown, whereas the limestones weather light to medium gray. Dolostones weather less readily than limestone and stand in bolder relief. A few of these beds are more than 3 ft (1 m) thick and appear massive and structureless. In some beds, the hand lens reveals dolomite rhombs; in others, burrows are dolomitized.

The limestones are typically calcilutites and calcisiltites, with pellets and fossils, though there are a few beds with abundant cm-sized limestone clasts. Bimodally cross-bedded calcarenite occurs at Stations B and E; such beds at E contain a few clasts of fine-textured limestone.

Parallel bedding is typical, but there are some exceptions. At Station A, uneven laminae of brown-weathering dolostone separate beds every inch (2 cm) or so. At Station D, there is a scour channel about 6 in (15 cm) deep. Southwest of D, there are thin limestone and dolostone interbeds with low-angle cross-bedding. An erosional unconformity with 6 in (15 cm) of relief can be traced from Station D southwest to the south corner of the fort.

Most of the fossils in the Chazy Group are the algae *Girvanella,* burrows, fairly abundant *Maclurites* and other plano-spiral gastropods, a few straight nautiloids, trilobite fragments, and a few brachiopods. *Girvanella* is an oncolite resembling a cocktail onion; it commonly has a core of a rounded grain of crystalline calcite—possibly from *Maclurites,* or from a pelmatozoan. *Girvanella* is abundant at and near Station A. *Maclurites,* which is more than 4 in (10 cm) across, probably grazed on algae; it is found at Station A, and in beds from C to E, as well as in the Orwell Limestone.

The Chazy Group was deposited in environments ranging

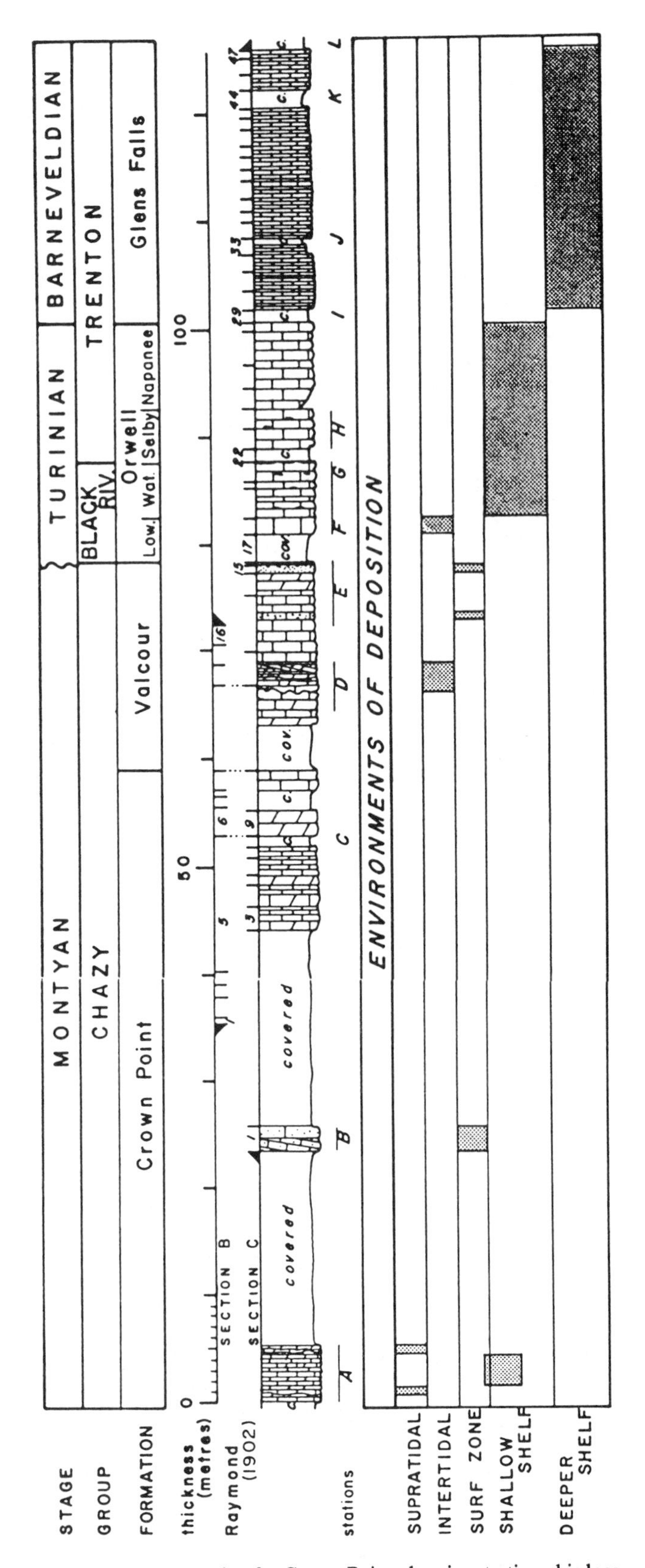

Figure 2. Columnar section for Crown Point, showing stratigraphic location of stations and environments of deposition of the formations (after Baldwin, 1980; permission from Northeastern Geology).

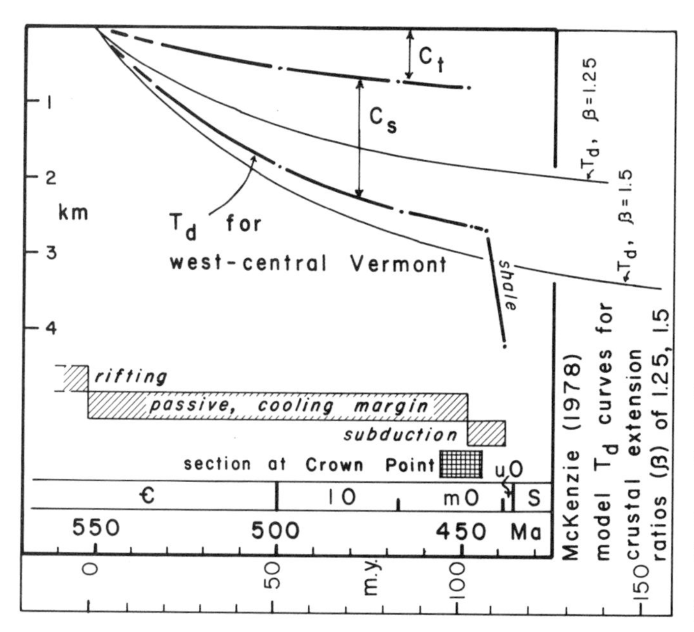

Figure 3. Cumulative thickness (T_d) of the west-central Vermont section, compared with model curves of McKenzie (1978): T_d = deposited thickness; C_s = crustal subsidence, sedimentary loading; C_t = crustal subsidence, tectonism; β = crustal extension ratio.

from tidal flat to shallow subtidal, as suggested by the presence of *Girvanella* and *Maclurites,* the burrowing, the scour channel, the calcarenite beds, and the beds of dolostone.

Station A. Nearly 20 ft (6 m) of limetone with dolomitic laminae are exposed in this outpost fort. The tan surface at about 5 ft (1.5 m) shows probable horizontal burrows; beneath this, the mottled limestone-dolostone is a burrowed unit. On the main vertical face, the limestone is poorly sorted; 2-mm rounded to elongate grains of crystalline calcite and fossil fragments occur in the calcilutite and calcisiltite matrix. Fossils include *Girvanella,* trilobite fragments, several brachiopods and nautiloids, and *Maclurites.* The dolostone laminae, algae, and *Maclurites* indicate a tidal flat to very shallow subtidal environment; the nautiloids could have been washed in.

Station B. The 6-ft (2-m) ledge that extends northeast from the road to the main highway is a calcarenite with bimodal cross-bedding. Many of the carbonate grains are calcilutite. Scattered grains of rounded quartz occur throughout and also are concentrated in laminae. This unit represents a sandy shoal or surf zone.

Station C. Poorly studied limestones and dolostones form the steep slope southwest of the Pavilion. *Maclurites* and some trilobite fragments are present. Similar beds crop out between the Pavilion and Station D.

Station D. A variety of features are displayed on the 10-ft (3-m) face outside the east corner ("Flag Bastion") of the British fort. A concrete sidewalk connects this place with the Visitor Center at the bottom of the hill. Almost at the base of this outcrop is an erosional unconformity with about 6 in (15 cm) of relief. The lower beds are limestone with fossil "hash" and dolomitized burrows. The upper half is interbedded limestone (light gray) and dolostone (darker; dolomite rhombs). The beds are cut by a scour channel about 6 in (15 cm) deep. Near the top of the face, the limestone beds are disrupted, with limestone clasts "floating" in the dolostone. In thin section, fecal pellets are abundant in the dolomitized beds. The scour channel, burrowing, dolostone, and disrupted limestone are features consistent with a tidal flat environment.

Follow these beds southwest in the moat, using the erosional surface as a marker. Study the 12-ft (4-m) cliff for low-angle cross-bedding, short vertical burrows, *Maclurites,* and "reefs" of snail shells.

Station E. In 1916, gunite was sprayed on the interior walls of the two barracks. Starting in 1976, the New York State Division for Historic Preservation began extensive renovation.

The prominent flat outcrop in front of the southern barrack is a bimodally cross-bedded calcarenite. Like the limestone at Station B, there are scattered rounded quartz grains. The highest beds in this outcrop have small plano-spiral snails and limestone clasts. The calcarenite represents a sandy shoal or surf zone; the quartz grains were probably wind-blown.

Above the calcarenite is a thick brown dolostone, which is better exposed in the moat on the southwest side of the fort. The western outcrop in the parade grounds is a 2-ft (0.5-m) quartzite with some 3 mm grains, burrows, and a *Maclurites* mold. The bed has not been traced past the southwest moat. A covered interval at least 10 ft (3 m) thick separates the quartzite from the overlying Orwell Limestone that is exposed just north of the fort.

THE ORWELL LIMESTONE

The Orwell Limestone, from Stations F to I, consists of Black River and lower Trenton strata. Bedding is massive and somewhat uneven; whole fossils are abundant. Dolostone beds are absent, though one or two beds have dolomitized burrows. Some horizons have nodules of black chert and one horizon is an essentially continuous chert bed up to 2 in (5 cm) in thickness. The Orwell Limestone has essentially no sign of deposition from, or agitation by, moving water; it is mostly massive calcilutite and calcisiltite. However, at Station F, there is a bed with cm-sized limestone clasts. Also, two beds of calcisiltite to fine calcarenite, at Stations G and H, show faint lamination and cross-bedding.

The Lowville beds, seen at Station F (and in building blocks in the southern barrack), are at the base of the Orwell. The Lowville beds are massive and weather very light gray, and, except for *Phytopsis,* there are very few fossils. Above the Lowville, whole fossils are abundant. *Lambeophyllum,* the oldest genus of solitary coral, has a deep calyx. *Foerstephyllum,* a honeycomb coral, forms isolated masses a few in to 3 ft (5 cm to 1 m) across. *Stromatocerium,* a sponge, looks like a cow-pie. Gastropods include the high-spired *Hormotoma,* the sharp-shouldered *Lophospira,* and opercula (lids) of *Maclurites.* Stick bryozoans are locally abundant, and straight nautiloids are not uncommon. Pelmatozoan stems are found locally, as is the double-ended trilobite, *Isotelus.* Burrows are not as evident as in the pre-Orwell strata.

The Orwell Limestone must have been deposited in low-energy shallow subtidal waters in the photic zone. Absence of dolostone seems to rule out intertidal environments, and the presence of corals and grazing snails rules out sub-photic conditions.

Station F. Just outside the British fort is a 15-ft (5-m) face. At the base, the Lowville beds, consisting of light-colored calcilutite, have vertical tubes of *Phytopsis.* Traditionally, the Lowville has been assigned to a tidal flat origin, though the environment may have been a zero-energy lagoon. Immediately above the Lowville beds are the more typical gray limestone beds with abundant fossils. Near the top of the face is a line of black chert nodules.

Station G. Go north across the road onto the extensive ledge-forming outcrop of the Orwell Limestone that can be followed to the shore of Lake Champlain. A cluster of *Lambeophyllum* occurs on one projection of the main ledge. Nautiloids, *Stromatocerium,* and gastropods can be seen on the several beds. Chert nodules tend to rest on horizontal burrows that are an inch or so (2 to 5 cm) across. A large *Foerstephyllum,* 3 ft (1 m) across, occurs in the northern end of one of the ledges. Where the main surface ends in brush, the 3-ft (1-m) ledge just above and to the west can be seen to be capped with a chert bed. Walk east into the grass, and then shoreward to get back onto the outcrop; the highest bed here is the same black chert with molds of costate brachiopods. Along the lake shore, the Orwell sequence from Lowville to the black chert horizon is present, but glacial erosion, including rounding and striations, obscures details.

Station H. From the black chert bed near the shore, walk west just south of the large quarry blocks. The water-filled part of the quarry is perhaps 6 ft (2 m) deep. The quarry blocks show pelmatozoan stems, stick bryozoans, brachiopods, and an occasional *Isotelus.* Look at the northwest wall of the quarry for vertical burrows, a layer of concave-down shells, chert nodules, and calcilutite like the Lowville. Cross-bedded limestone, uncommon in the Orwell, is exposed a few ft above the bench west of the quarry.

Between Stations H and I. Stations I, J, and K are covered intervals, where Lake Champlain has formed gravel beaches. At Station I, the boundary between the Orwell and Glens Falls limestones is hidden.

From the quarry at Station H, go north onto the sand beach next to the tree-lined artificial spit, and turn left (west) onto the glacially smoothed Orwell. Opercula (lids) of *Maclurites* are abundant just above low lake level; they resemble thick-walled and curved cones. In the upper beds of the Orwell Limestone, there are *Lambeophyllum, Foerstephyllum, Stromatocerium,* and a clam (*Ambonychia*).

THE GLENS FALLS LIMESTONE

The Glens Falls Limestone, of the Trenton Group, is exposed up-section from Station I. **Do not go onto private property (Stations K to L).** Return to your vehicle by retracing the route or walk up the road that is just inland.

The Glens Fall Limestone is dramatically different from the Orwell, because its beds are thin and distinct, and because it has no photic-zone features. It lacks the corals, *Stromatocerium,* and *Maclurites.* Brachiopods (*Rafinesquina, Dinorthis, Dalmanella,* and especially the smooth wide-hinged *Sowerbyella*), are very common and moderately diverse. There are abundant fragments of trilobites (*Flexicalymene* in the lower beds and *Cryptolithus* between Stations J and K). Pelmatozoans are fairly common, though quite small. Stick bryozoans (*Eridotrypa, Stictopora*) are also fairly common, and in beds up-section from Station J there are "derby-hat" *Prasopora.* One large nautiloid was found on the shore southwest of the private property.

The beds of the Glens Falls Limestone are 2 to 5 in (5 to 10 cm) thick. Many are graded, with most of the fossils in the lower half and a fossil-poor calcilutite in the upper half. Some bedding surfaces clearly were populated by attached benthonic animals. Argillaceous partings are present between many of the limestone beds; these were evidently the first signs of the influx of mud that formed the thick overlying Trenton shale.

The Glens Falls was deposited below the photic zone. The graded beds are probably turbidites transported from shallower and nearer shore environments; they also have been called "tempestites"—storm-reworked bottom sediments.

REFERENCES CITED

Baldwin, B., 1980, Tectonic significance of mid-Ordovician section at Crown Point, New York: Northeastern Geology, v. 2, p. 2–6.

—— , 1982, The Taconic orogeny of Rodgers, seen from Vermont a decade later: Vermont Geology, v. 2, p. 20–25.

Baldwin, B., and Butler, C. O., 1985, Compaction curves: American Association of Petroleum Geologists Bulletin, v. 69, p. 622–626.

Baldwin, B., and Mehrtens, C. J., 1986, The Crown Point section, New York: Vermont Geological Society Guidebook 1, Field Trip Guide D, Vermont Geology, v. 4, p. D1–D14.

Bond, G. C., Nickeson, P. A., and Kominz, M. A., 1984, Breakup of a supercontinent between 625 Ma and 555 Ma; New evidence and implications for continental histories: Earth and Planetary Science Letters, v. 70, p. 325–345.

Churkin, M., Jr., Carter, C., and Johnson, B. R., 1977, Subdivision of Ordovician and Silurian time scale using accumulated rates of graptolitic shale: Geology, v. 5, p. 452–456.

Harland, W. B., Smith, A. G., and Wilcox, B., eds., 1964, The Phanerozoic time scale: Geological Society of London, Quarterly Journal, v. 120s, 458 p.

McKenzie, D., 1978, Some remarks on the development of sedimentary basins: Earth and Planetary Science Letters, v. 40, p. 25–32.

Raymond, P. E., 1902, The Crown Point section: Bulletin of American Paleontology, v. 3, no. 14, 44 p.

Selleck, B., and Baldwin, B., 1985, Cambrian and Ordovician platform sedimentation—southern Lake Champlain valley: New York State Geological Association Field Trip Guidebook, 57th Annual Meeting, p. 148–168.

Geology of the Adirondack-Champlain Valley boundary at the Craig Harbor faultline scarp, Port Henry, New York

J. Gregory McHone, Department of Geological Sciences, University of Kentucky, Lexington, Kentucky 40506-0059

LOCATION

Port Henry is a small incorporated village on Lake Champlain, within the Town of Moriah in Essex County of northeastern New York. The Craig Harbor site is at approximately 44°03′30″N and 73°27′08″W, within the Port Henry 15-minute Quadrangle (Fig. 1). The major state highway through Port Henry is New York 9N/22, but the village can also be reached from I-87 (the Northway) off Exit 31 onto 9N south from Westport, or from Exit 30 east on County Road 6 to Mineville, south on County Road 7 to Moriah Center, and southeast on County Road 4 to Port Henry.

The Beach Road turns eastward off Main Street at the base of the fault-line bluff on which the village is situated, and parking is available at a public boat launch and picnic area to the north of the road (Fig. 1). Walk north through the Villez Marina (a local landmark facility!) and up onto the adjacent Delaware and Hudson railway. Exposures of the site extend north about 1,650 ft (500 m) to Craig Harbor, but excellent exposures also continue for many kilometers along the railway and lakeshore. Stay off the tracks themselves, as approaching trains are difficult to hear around bends in the railroad cuts and may appear suddenly.

SIGNIFICANCE

Deep railroad cuts, fault-line bluffs, and wave-cut exposures combine to illustrate the structures, Proterozoic basement rocks, and Lower Paleozoic stratigraphy that characterize the abrupt boundary between the eastern Adirondack Mountains and the western Lake Champlain Valley (e.g., Walton and deWaard, 1963; Isachsen and Fisher, 1971). This outstanding section along the Delaware and Hudson Railroad near Craig Harbor has stimulated interest in Adirondack geology for more than 150 years. The site shows field relations of Proterozoic "Grenville" marbles and calc-silicate rocks, older (?) basement granitic gneisses, lenses of metagabbro, a magnetite ore body, part of the shelf sequence of Paleozoic carbonate rocks, faults that juxtapose the rock units, and resultant structurally controlled landforms.

SITE INFORMATION

Partly because of important magnetite ores mined since the 1780s, the Delaware and Hudson Railroad was pushed through Port Henry with great effort in the mid-19th century. Many deep cuts and several tunnels were necessary to provide room for the railbed along the precipitous cliffs that border the Adirondack dome to the west and the Lake Champlain Valley to the east. This site provides a traverse across the northern edge of the Port Henry fault block on which the village is constructed (Fig. 1), and illustrates some of the Lower Paleozoic sedimentary units that overlie Proterozoic basement rocks along much of the eastern border of the Adirondacks.

Paleozoic Rocks. Although basal conglomerates of the Cambrian Potsdam Sandstone are well exposed in southern areas of Port Henry (Kemp and Ruedemann, 1910), rotation of the fault block down to the northeast has preserved the Cambrian Ticonderoga and Ordovician Whitehall carbonate units that overlie the Potsdam in this area (Fig. 1). The Paleozoic section is thickest (an estimated 800 ft; 250 m) and best exposed immediately south of Craig Harbor. Good summaries of the Paleozoic stratigraphy of the western Lake Champlain Valley are provided by Welby (1961) and Fisher (1968).

Along the railway above Villez Marina, dark sandy dolostone of the Cambrian Ticonderoga Formation is characterized by crossbedding, channel depressions, and blue-black chert layers and nodules. Some of the contortions and broken appearance of the dolostone mask the generally low, easterly dip angles of the layers. High-angle faults and attendant breccias are visible at several places in the cliff walls, with offsets of a few cm to a meter or more (Fig. 2A). The Ticonderoga Dolostone is measured by Welby (1961, p. 99) as 300 ft (90 m) thick in the Port Henry area.

Within a few hundred meters north of the marina, the cliff walls angle westward to open a swale that once contained a crushing mill. Although hidden behind woods now, a quarry near to the west once produced carbonate flux material for local iron furnaces and for road building. North of the swale there appears a lighter, more uniform "dove-colored" carbonate unit mapped as Ordovician Whitewall Dolostone, a lower unit of the Beekmantown Group that is also 300 ft (90 m) thick, according to Welby (1961). Beds of this dolostone unit dip to the northeast more steeply than do the Ticonderoga layers, and Kemp and Ruedemann (1910) inferred the presence of a fault through the swale that separates what was then known as Beekmantown divisions A and B (Fig. 2). Another possibility is that the northern outcrop of dolostone is part of the overlying Ordovician Cutting Formation, and the Whitewall Dolostone is not exposed because of downfaulting.

Proterozoic Rocks. The railroad bed crosses the mouth of Craig Harbor (known as Crag Harbor before being misnamed on the USGS maps of the 1890s) and should be traveled quickly so as to avoid conflict with trains. The northern wall of the embayment is a 165-ft-high (50-m)-high fault scarp representing the northernmost end of the Port Henry fault block where it bounds the Proterozoic with the Paleozoic sections.

In the upper west cliff face, a white "Grenville" marble

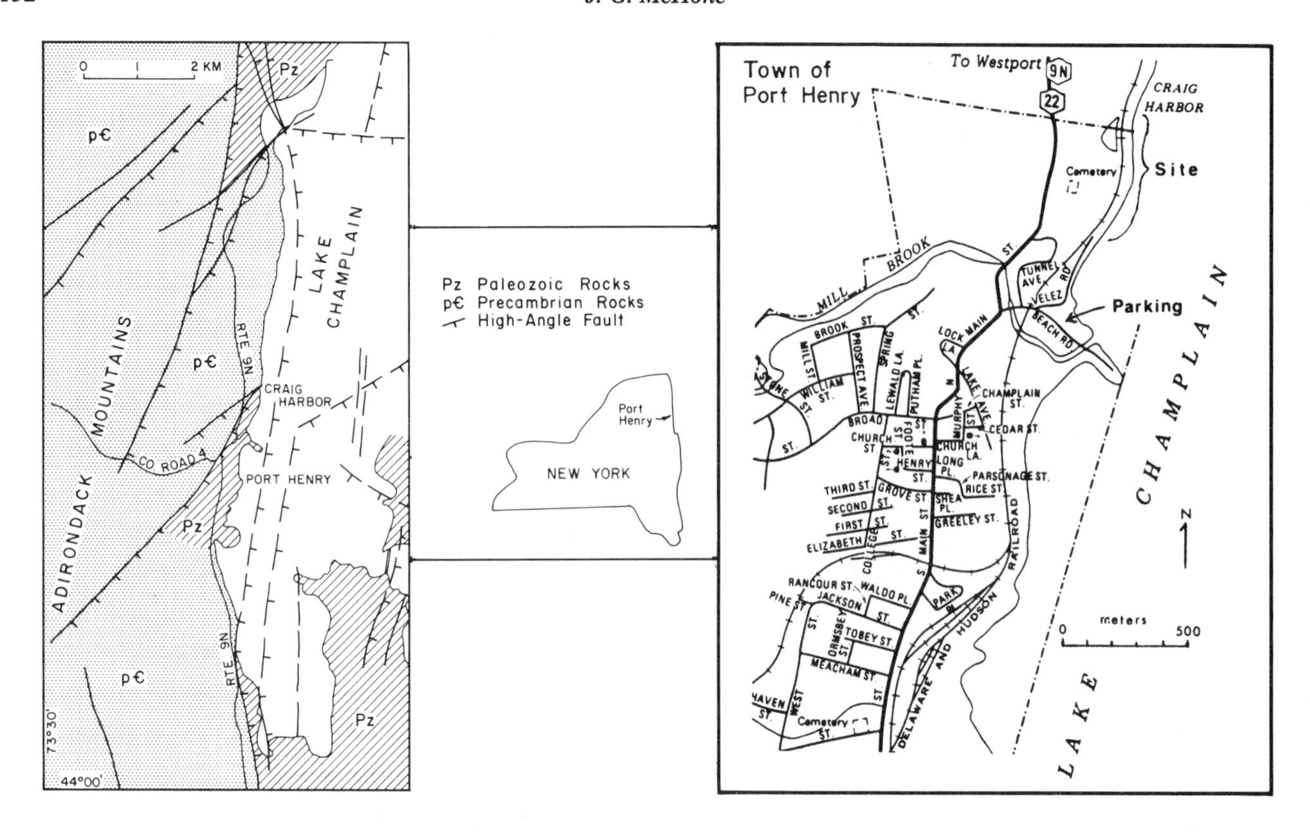

Figure 1. The location of Port Henry, New York and the Craig Harbor site is shown. Geology is generalized from Welby (1961) and Isachsen and Fisher (1971).

exposure contains folded and disrupted lenses of calc-silicate rocks and gneisses (Fig. 2B). Emmons (1842) apparently observed this and other such outcrops in his pioneering 1830s work on the geology of New York. Emmons cited the intrusive form of the marble, its inclusions, and its high-temperature minerals as evidence for an igneous origin of "primitive limestone" (1842, p. 37–59).

The coarsely crystalline, graphite-bearing marbles characterize the Paradox Lake Formation of Walton and DeWaard (1963), suggested as the basal member of a Grenvillian metasedimentary supergroup that overlies older basement paragneisses. Wiener and others (1984) place the marbles above the biotite-quartz-plagioclase Eagle Lake Gneiss, which they assign as the base of the Lake George Group. Charnockitic and granitic gneisses of the Piseco Group make up the older basement rocks that unconformably underlie the Lake George Group (Wiener and others, 1984, p. 22–25). Charnockitic gneiss is the first gneiss unit encountered in the cut north of Craig Harbor.

The granitic gneiss in the cliff face below the marble shows westward-dipping layers of orange-weathering hornblende-rich gneiss and a horizon of magnetite (Fig. 2B), noted in descriptions by Emmons (1842, p. 236–237) and by Kemp and Ruedemann (1910, p. 100–101). Polygonal sheets of gneiss are exfoliating from the cliff wall, but have not yet produced the large talus pile that might be expected from a vertical cliff.

Along the track a few tens of meters north of the harbor, the western wall of the railroad cut exposes the westward-dipping upper contact of metagabbro underlying the hypersthene-bearing, augite, hornblende, microcline, plagioclase gneiss (Fig. 2C). Within a few meters of the contact, the metagabbro is dark, fine-grained granular, and slightly gneissic, with altered plagioclase, brown hornblende, and augite in decreasing order of abundance. Kemp (1894) reported olivine in small amounts in his description of this exposure, but McHone (1971) did not observe olivine in 15 samples taken along this cut. Garnet, apatite, magnetite, and orthopyroxene are present as minor minerals.

To the north and farther below the contact, a coarse ophitic facies of the metagabbro appears. The widespread "Adirondack orange" weathering of the rock surfaces hides the texture except where rock falls and hammer blows have produced fresher, medium gray surfaces. This coarse facies is dominated by coronas of garnet, brown hornblende, biotite, and clear plagioclase surrounding cores of magnetite, augite, or hypersthene. Kemp (1894, 1921) and Gillson and others (1928) carefully documented the corona sequences and spinel clouding of the primary plagioclase, and understood them as metamorphic products. The textures are like those described from other Adirondack metagabbros by Whitney and McLelland (1975) and McLelland and Whitney (1980), who used them to estimate P-T conditions near 8 kb and 800° C (granulite facies). Many small high-angle faults are pres-

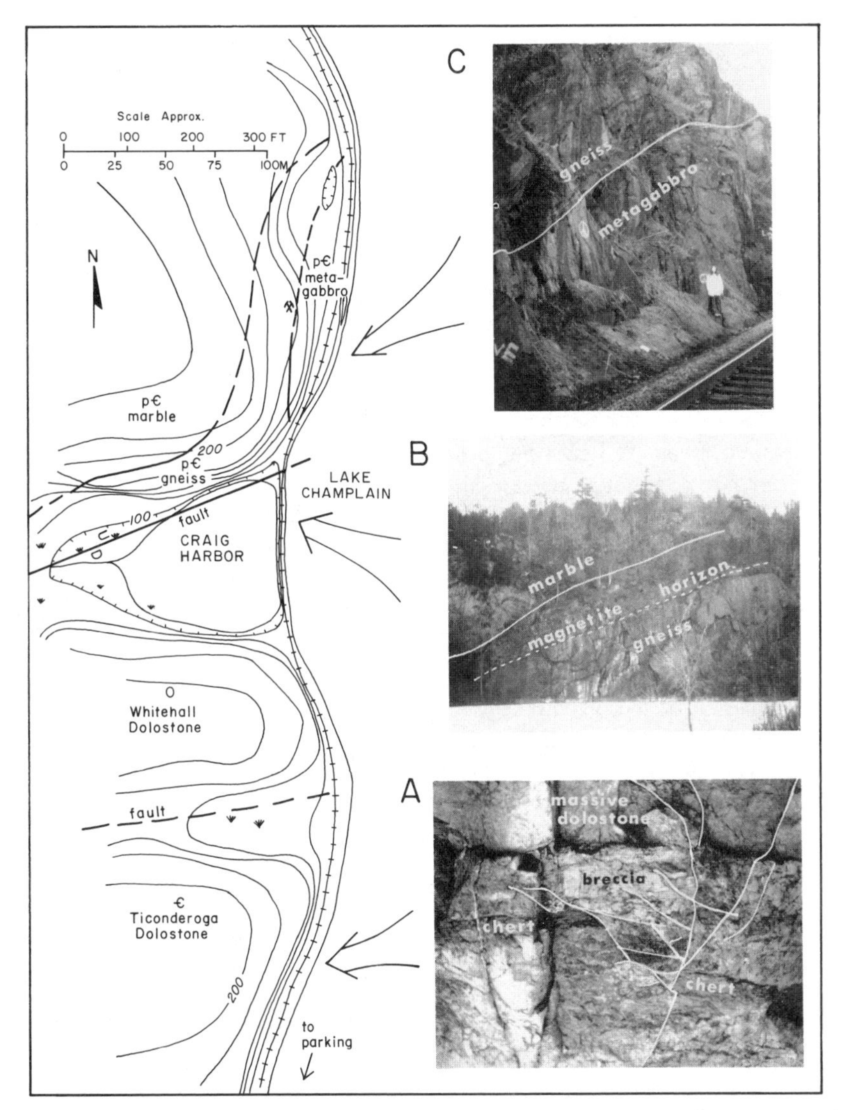

Figure 2. In the geologic sketch map of the Craig Harbor area, topography is only approximate. Photographs have faults and unit contacts emphasized by white lines: A) small fault zone (about 4 × 6 ft; 1.5 × 2 m) in Ticonderoga Dolostone; B) fault scarp on the north side of Craig Harbor, about 165 ft (50 m) high; C) contact of charnockitic gneiss and fine-grained metagabbro, western side of railroad cut about 130 ft (40 m) north of Craig Harbor.

ent in the metagabbro, showing glassy slickensides and zones of breccia with pseudotachylite veins.

In the gneiss above the gabbro, a small tunnel has been excavated into the magnetite horizon (Fig. 2). Although known before the 1840s, this deposit was not worked because the ore is unusually high in sulfur and especially titanium, making it difficult for early forges to handle. Lamellae of ilmenite are exsolved along octahedral crystal planes in polished specimens of magnetite from this prospect (McHone, 1971). Kemp (1898, 1899) and Kemp and Ruedemann (1910) noted the association of magnetite with metagabbro at this site and elsewhere in the region, and advanced a magmatic model for the ore that is similar in some respects to recent ideas for Feiss and others (1983) for the origin of Precambrian magnetite deposits in North Carolina.

Flow-folded marble, calc-silicate skarns, and anorthosite crop out along the railway cuts north of the gabbro. Several high-angle faults with slickensided surfaces are well exposed in the calc-silicate rocks. Garnet, diopside, and wollastonite are present in specimens similar to wollastonite ores mined about 24 mi (40 km) to the north in Willsboro, New York. The Grenville marble has plastically flowed to enclose "xenoliths" of adjacent rocks.

Age of Faulting. The timing of the Adirondack border fault activity is not well established. Because it is likely that much of the border faulting was associated with uplift of the Adirondack dome, the age of fault movement at Craig Harbor has regional significance. Similar high-angle faults are also mapped throughout the Lake Champlain Valley of Vermont and New

York, and extend northward into the St. Lawrence River Valley of Quebec (Quinn, 1933; Welby, 1961; Stanley, 1980).

Quinn (1933) believed that most Lake Champlain Valley faulting preceded Taconic overthrusting, but Welby (1961) showed that many high-angle faults offset the Champlain Thrust of the central Lake Champlain Valley. Snake Mountain, visible to the east of Craig Harbor in Vermont, is an example of a quartzite-capped segment of the Champlain Thrust that is bounded by high-angle faults (Welby, 1961).

Studies of the MacGregor border fault system about 80 km to the south have revealed a history of reactivation that may include movements in Late Precambrian, Paleozoic, and possibly Holocene times (Willems and others, 1983; Isachsen and others, 1983). At least minor Late Mesozoic or younger Lake Champlain Valley fault activity is shown by rare crosscutting relationships with Early Cretaceous alkalic intrusions in Vermont and New York (McHone, 1978; Stanley, 1980; Yngvar Isachsen, personal communication, 1981).

More than 600 ft (200 m) of stratigraphic offset exists on the Craig Harbor fault that postdates Lower Ordovician dolostone. The excellent preservation of the fault scarp and general freshness of slickensides and pseudotachylite in the metagabbro suggests youthful tectonism, an idea also advanced by releveling studies along the Adirondack-Champlain border by Isachsen (1975) and Barnett and Isachsen (1980). As part of this study and to assess the potential for direct radiometric dating, chloritic pseudotachylite collected from a fault breccia in the metagabbro at Craig Harbor was submitted for analysis. The sample yielded a K-Ar date of 665 ± 26 Ma (avg. ^{40}Ar = 0.02661 ppm; avg. %K = 0.478; ^{40}K = 0.570 ppm; analysis by Geochron Laboratories, 1984). Because this date precedes the considerable Cambro-Ordovician offset, an unknown amount of argon contamination from the approximately 1000 Ma metagabbro is inferred. Success in future efforts to date the fault movement may require purer rock or mineral material that formed in the fault zone during a single tectonic event.

CONCLUSION

Finally, to quote Ebenezer Emmons (1842, p. 87): "I would take this occasion to recommend to students in geology, to visit the shores of Lake Champlain. It is a field full of interesting and instructive phenomena; one in which the dynamics of geology may be studied to the best advantage. Moreover, the field is quite accessible, and every part may be visited at an expense not disproportionate to the advantages which may be obtained."

REFERENCES

Barnett, S. G., and Isachsen, Y. W., 1980, The application of Champlain water level studies to the investigation of Adirondack and Lake Champlain crustal movements: Vermont Geology, v. 1, p. 5–11.

Emmons, E., 1842, Geology of New York, Part II; Survey of the Second Geological District: Albany, 437 p.

Feiss, P. G., Goldberg, S., Ussler, W., Bailar, E., and Myers, L., 1983, The Cranberry magnetite deposit, Avery County, North Carolina, *in* Lewis, S. E., ed., Geologic investigations in the Blue Ridge of northwestern North Carolina: Carolina Geological Society Field Trip Guidebook, Boone, North Carolina, p. III1–III34.

Fisher, D. W., 1968, Geology of the Plattsburgh and Rouses Point, New York–Vermont, Quadrangles: Vermont Geological Survey Special Bulletin 1, 51 p.

Gillson, J. L., Callahan, W. H., and Miller, W. B., 1928, Adirondack studies; The age of certain of the Adirondack gabbros, and the origin of the reaction rims and peculiar border phases found in them: Journal of Geology, v. 36, p. 149–163.

Isachsen, Y. W., and Fisher, D. W., 1971, Geologic map of New York, Adirondack sheet: New York State Museum and Science Service Map and Chart Series 15, scale 1:250,000.

Isachsen, Y. W., 1975, Possible evidence for contemporary doming of the Adirondack Mountains, New York, and suggested implications for regional tectonics and seismicity: Tectonophysics, v. 29, p. 169–181.

Isachsen, Y. W., Geraghty, E. P., and Wiener, R. W., 1983, Fracture domains associated with a neotectonic, basement-cored dome; The Adirondack Mountains, New York: Salt Lake City, International Basement Tectonics Association Publication 4, p. 287–305.

Kemp, J. F., 1894, Gabbros on the western shore of Lake Champlain: Geological Society of America Bulletin, v. 5, p. 213–224.

——, 1898, Geology of the magnetites near Port Henry, N.Y., especially those of Mineville: American Institute of Mining and Engineering Transactions, v. 27, p. 146–203.

——, 1899, The titaniferous iron ores of the Adirondacks: U.S. Geological Survey Annual Report, pt. 3, p. 377–422.

——, 1921, Geology of the Mt. Marcy Quadrangle, Essex County, New York: New York State Museum Bulletin 119, 126 p.

Kemp, J. F., and Ruedemann, R., 1910, Geology of the Elizabethtown and Port Henry Quadrangles: New York State Museum Bulletin 138, 178 p.

McHone, J. G., 1971, Petrography of some highly metamorphosed Precambrian rocks near Port Henry, New York [abs.]: Vermont Academy of Science, Student Symposium, p. 3.

——, 1978, Distribution, orientations, and ages of mafic dikes in central New England: Geological Society of America Bulletin, v. 89, p. 1645–1655.

McLelland, J. M., and Whitney, P. R., 1980, Compositional controls on spinel clouding and garnet formation in plagioclase of olivine metagabbros, Adirondack Mountains, New York: Contributions to Mineralogy and Petrology, v. 73, p. 243–251.

Quinn, A., 1933, Normal faults of the Lake Champlain region: Journal of Geology, v. 41, p. 113–143.

Rodgers, J., 1937, Stratigraphy and structure in the upper Champlain Valley: Geological Society of America Bulletin, v. 48, p. 1573–1588.

Stanley, R. S., 1980, Mesozoic faults and their environmental significance in western Vermont: Vermont Geology, Bulletin 1, p. 22–32.

Walton, M. S., and deWaard, D., 1963, Orogenic evolution of the Precambrian in the Adirondack highlands, a new synthesis: Koninklijke Nederlandse Akademie Van Wetenschappen Proceedings, ser. B, v. 66, no. 3, p. 98–106.

Welby, C. W., 1961, Bedrock geology of the central Champlain Valley of Vermont: Vermont Geological Survey, Bulletin 14, 296 p.

Wheeler, R., 1942, Cambrian–Ordovician boundary in the Adirondack border region: American Journal of Science, v. 240, p. 518–524.

Whitney, P. R., and McLelland, J. M., 1973, Origin of coronas in metagabbros of the Adirondack Mountains, New York: Contributions to Mineralogy and Petrology, v. 39, p. 81–98.

Wiener, R. W., McLelland, J. M., Isachsen, Y. W., and Hall, L. M., 1983, Stratigraphy and structural geology of the Adirondack Mountains, New York; Review and synthesis, *in* Bartholomew, M. J., The Grenville event in the Appalachians and related topics: Geological Society of America Special Paper 194, p. 1–56.

Willems, H. X., Bosworth, W., and Putman, G. W., 1983, Evidence for complex strain history of the MacGregor fault, eastern Adirondacks; A possible reactivated Late Precambrian rift structure: Geological Society of America Abstracts with Programs, v. 15, p. 124.

Leucogranitic gneiss bodies and associated metasedimentary rocks, Adirondack lowlands, near Canton, New York

William D. Romey and William T. Elberty, Jr., Department of Geography, and Russell S. Jacoby, Department of Geology, St. Lawrence University, Canton, New York 13167

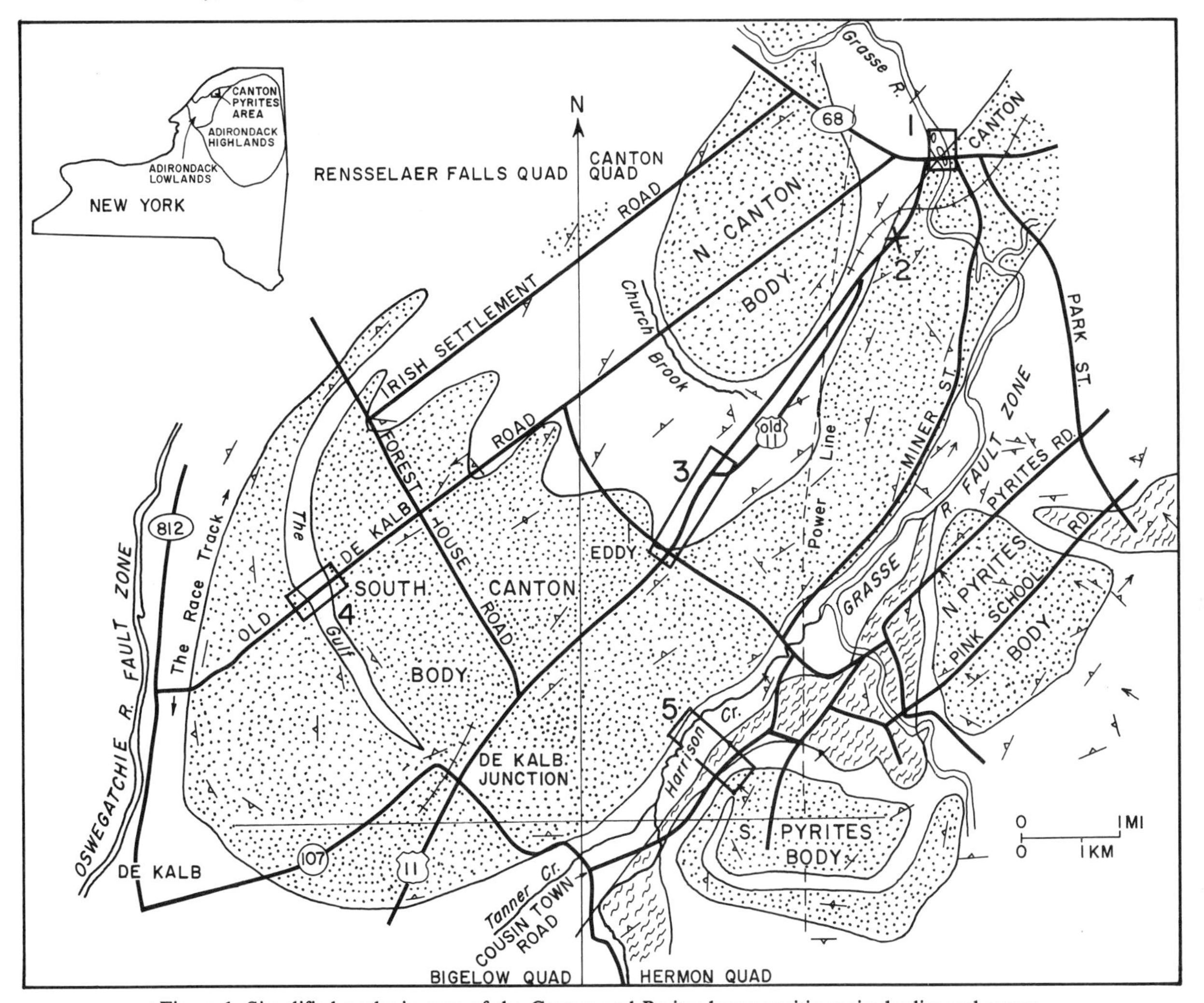

Figure 1. Simplified geologic map of the Canton and Pyrites leucogranitic gneiss bodies and recommended field-guide stops (based on map by Romey and others, 1980). Other regional details and sources are available in Fisher and others (1970). Key: dot pattern, leucogranitic gneiss; wiggly dashes, amphibolitic gneiss; unpatterned areas, undifferentiated marble, metasedimentary gneiss, and metaquartzite (with further details available in Romey and others, 1980). Generalized foliation and lineation directions are indicated.

LOCATION

Four large bodies of Precambrian leucogranitic gneiss occupy parts of the Canton, Rensselaer Falls, Hermon, Bigelow, 7½-minute Quadrangles in the northwest Adirondack lowlands of New York (Fig. 1). Recommended stops are numbered and keyed to locations described below. U.S. 11 provides access and runs parallel to the general N50°E structural grain of the Proterozoic Adirondack lowland fold belt and the Grenville Province, of which the Adirondacks are part. The Canton area can also be reached by New York 68 from Ogdensburg, New York or from the central Adirondacks (Tupper Lake) via New York 56 and 68.

SIGNIFICANCE

The origin and structural relationships of several large leucogranitic gneiss bodies (Fig. 2) in the Adirondack lowlands have long been subjects of controversy. Buddington (1929, 1939) called them phacoliths and interpreted them as igneous alaskite bodies intruded into the hinge areas of folds and subsequently metamorphosed during the Grenville orogeny. Engel and Engel (1963) proposed that they represent stratigraphic units, perhaps arkose that subsequently underwent metamorphism and anatexis. Silver (1965) suggested rhyolite as a possible parent rock, and Carl and Van Diver (1975) favored an ash-flow tuff. Foose and Carl (1977) and Romey and others (1980) accepted a metavolcanic origin and supported the idea that these gneisses form a single extensive, folded sheet that forms part of a metasedimentary-metavolcanic stratigraphic sequence, consisting, from bottom to top, of leucogranitic gneiss, marble, and pelitic gneisses. There is further controversy as to whether the leucogranitic gneiss forms the basal unit or whether a stratigraphically lower marble may underlie the gneiss (Weiner and others, 1984; de Lorraine and others, 1985). Recent work in dating the northern Adirondack leucogranitic gneiss bodies (at 1200–1300 Ma) and geochemical work promise to help resolve this question (Grant and others, 1984; Foose and others, 1981; Carl and others, 1986). The minimum melt composition of these rocks and the common presence of migmatite suggest at least local anatexis (Romey and others, 1980). A number of different types of intrusive granite have also been identified, including most recently the Antwerp and Rossie granitoids of Buddington's Antwerp type, now dated by Carl and others (1986) by Rb-Sr methods at 1131 Ma and 1151 Ma, respectively.

The South Canton, North Canton, South Pyrites, and North Pyrites bodies together cover more than 48 mi^2 (124 km^2), forming the largest well-exposed grouping in the Adirondack lowlands of what is generally interpreted, along with other fine-grained leucogranitic gneiss bodies in the northwest Adirondacks, to be part of a single sheet (although ductile flow of associated marble produced disruption and break-up). Structural relationships, as shown in the map pattern, Figure 1, indicate that the sheet is involved in a refolded isoclinal fold in the Canton area. Northwest of the Canton area the sheet may provide a stratigraphic marker unit that reveals the upper limb of a large-scale nappe with its hinge located near Parishville (Romey and others, 1980). The present isolated, domal outcrops of granitic gneiss in the Adirondack lowlands may reflect an "egg-carton" structure produced by multiple folding of this rock sequence (Romey and others, 1980; Wiener and others, 1984).

SITE DESCRIPTIONS

A Circle Route Across the North and South Canton Granitic Gneiss Bodies

The route suggested here (Fig. 1) leads from an initial point in Canton, at the contact between the South Canton leucogranitic gneiss body and overlying metasedimentary rocks, to the town of Eddy through a valley underlain by marble and flanked on the northwest and southeast by parts of the Canton leucogranitic gneiss sheet. It then passes onto the hook-shaped southwest tip of the South Canton body where evidence for the refolded nature of the sheet can be seen. The last part of the trip, to the South Pyrites body, provides further evidence for the refolded nature of the leucogranitic body and its relationship to enveloping marble and amphibolite.

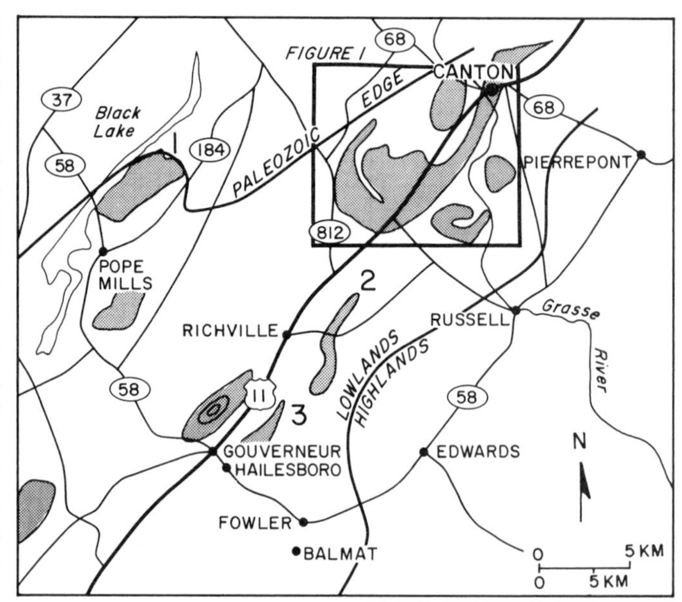

Figure 2. Location of several granitic gneiss bodies (shaded) in the Adirondack lowlands southwest of Canton that are possibly related to the Canton and Pyrites occurrences, including Fish Creek (1), Moss Ridge (2) and Reservoir Hill (3) mentioned in the text. For source data on these and other similar bodies in the lowlands, see Fisher and others (1970).

The leucogranitic gneiss throughout the area covered by this guide is fine-grained (0.5 to 2 mm) with an average grain size less than 1 mm. It comprises mainly microcline perthite (17% to 48%), plagioclase (10% to 30%), flat, ovoid quartz "leaves" (25% to 40%; which may reach 2 cm in largest dimension) and hornblende, biotite, opaque minerals and other dark accessories (2% to 10%). Aplitic textures occur in places and complex mortar (mylonitic) textures characterize some porphyroblastic varieties. A well-developed metamorphic foliation produced by alignment of dark minerals, mylonitic zones, and quartz leaves characterizes much of the rock. Granitic pegmatite and migmatite zones, crosscutting or parallel, disrupt the foliation in many areas. The overlying metasedimentary rocks include (in structurally ascending order) sillimanite-garnet gneiss, medium- to coarse-grained, white to gray or buff calcareous and dolomitic marble with thin quartzite units, and medium-grained amphibolite.

Stop 1. (0.0 mi; km) Park at the Island Park between the two Canton bridges on U.S. 11. Exposures at the southern tip of the island show typical pink, fine-grained, well-foliated, leucogranitic gneiss characteristic of the Canton and Pyrites bodies. Associated pegmatite bodies contain K-feldspar megacrysts up to 6 to 8 in (15 to 20 cm) long.

Walk back across the southern island and U.S. 11 and cross a small footbridge to the northern island. Extensive exposures of quartz-feldspar-diopside paragneiss and thin lenses of calc-silicate–bearing marble typical of metasedimentary rocks structurally overlying the gneiss are visible along the river on both sides of the island and underlying the rapids. These exhibit small, recumbent, isoclinal folds with axes plunging 15°–30°NE, reflecting the nature of folding in the northwestern Adirondacks. A prominent lineation, somewhat steeper than these fold axes (30°–45°NE), occurs on many foliation surfaces.

Stop 2. At a point 0.8 mi (1.3 km) southwest of Canton along U.S. 11, stop near the entry to the drive-in theater. Glacially rounded roches moutonnées of leucogranitic gneiss occur on the hillside leading northward, downhill, to the drive-in theater. The valley north of the road is underlain by marble and other metasedimentary rocks correlative to those at Stop 1. A shallow synform in the folded upper surface of the leucogranitic gneiss sheet demonstrates the regional "egg-carton" structure. The fold axis here trends NE-SW, paralleling the regional structural trend, and is roughly horizontal (Fig. 1).

Several roadcuts along U.S. 11 between Stops 2 and 3 expose, from northeast to southwest, contorted, impure marble (3.1 mi; 5 km), well-layered pelitic gneiss with layers of impure marble (3.6 mi; 5.8 km), and a deep cut into resistant diopside-microcline-dolomitic marble (4.3 mi; 6.9 km). This last roadcut contains a 4- to 8-in-thick (10 to 20 cm), recumbent, isoclinally folded sheet of resistant microcline gneiss.

Stop 3. (4.9 mi [7.9 km] southwest of Stop 1, on northwest side of road, 250 ft [75 m] from the road). A large exposure of medium-to-coarse–grained biotite-hornblende-garnet-plagioclase gneiss is cut in its central section by a fault along which a coarse-grained granite pegmatite was intruded. A small, flat, glacially-striated exposure opposite, on the southeast side of the road, exposes more of this gneiss and also a 1- to 3-ft-thick (0.3 to 1.0 m), unmetamorphosed basaltic dike of unknown age. Further southwest, at the junction of the Eddy-Pyrites road, a 10-ft-deep (3 m) drainage cut exposes coarse-grained quartz-diopside-garnet gneiss that commonly occurs at the contact between the enveloping metasedimentary sequence and the Canton and Pyrites leucogranitic gneiss bodies. This rock can be correlated with the rocks seen at Stop 1. The foliation of the pelitic gneiss here is subvertical and swings from N50°E to N80°E and then to N-S along the Eddy Road. Abundant small folds can be seen in the gneiss. Across U.S. 11 and across the Eddy-Pyrites road at the intersection beside the railroad tracks, leucogranitic gneiss of the South Canton body crops out. This is the contact zone between the South Canton body and the overlying metasedimentary sequence. In this area, the shallow, narrow axial trough between the North and South Canton bodies, with its infolded, deformed metasedimentary sequence, swings to the northwest from the NE-SW trend it has followed along U.S. 11. The granitic gneiss sheet is refolded around an axis that also trends NE-SW. The two fold generations are coaxial. The Eddy-Pyrites road to the northwest passes through an interfingering sequence of leucogranitic and metasedimentary gneiss and marble in the crumpled hinge zone of the large fold represented by the "hook" in the South Canton leucogranitic gneiss.

Continue southwest on U.S. 11 to the crossroad in the village of DeKalb Junction. Turn northwest across the railroad tracks on county road 107 and then southwest to DeKalb. The SW-NE trend of the road is nearly parallel to the strike of foliation in the underlying granitic gneiss for 2.0 mi (3.2 km) to the contact of the granitic gneiss with quartzite and marble. Quartzite and marble occur in small roadcuts near the intersection just beyond the contact.

At the village of DeKalb, turn north on New York 812, along the Oswegatchie River, on the Oswegatchie Fault, which disrupts both stratigraphy and structure in this area (Brown, 1983; Karboski and others, 1983; Romey and Elberty, 1983). At 1.4 mi (2.2 km), turn northeast along the Old DeKalb road. Just before ascending the road that climbs steeply up the hill, pause to view the distinct, flat-floored valley that extends north and south of the road. Mapped as marble (but unexposed), the valley is typical of the "race tracks" surrounding most of the northwest Adirondack leucogranitic gneiss bodies. Exposures, when found in these valleys, are marble, commonly with calc silicates. The existence of these valleys and occasional outcrops supports the idea of a uniform stratigraphic position for the granitic gneiss bodies (Romey and others, 1980; Weiner and others, 1984).

Continue northeastward 0.2 mi (0.3 km) across the contact of the South Canton leucogranitic gneiss body with its typical rocky roches moutonnées. Well-developed foliation strikes nearly north-south and dips 20°–60° west. The swing of foliation parallel to the contact confirms that the South Canton body is folded. Note the folding of a foliation, which may represent the axial surface of earlier isoclinal folds.

Stop 4. Continue across the upland underlain by leucogranitic gneiss to a long, narrow valley 1.8 mi (2.9 km) northeast of the Coopers Falls Junction at New York 812. This valley along Gulf Creek ("the gulf") is underlain by marble. The hillside on the northwest side of the road exposes leucogranitic gneiss, and, at its bottom, a garnet gneiss unit commonly associated with the contact between the leucogranitic gneiss and overlying marble. The foliation strikes N40°W (at right angles to the regional structural trend) and dips 25°–30°SW. The contact is overturned. In the stream valley 300 to 600 ft (100 to 200 m) southeast of the road, marble crops out. It extends southeastward along the gulf nearly to DeKalb Junction and northward in an arc extending into the Indian Creek Wetlands (the edge of the Paleozoic overlap at the north edge of the NW Adirondack lowlands). The northeast side of the gulf is leucogranitic gneiss with foliation parallel to that on the west side. The eastern contact is right-side up. As shown by the map pattern (Fig. 1), the gulf represents a refolded isoclinal infold into the top surface of the granitic gneiss.

From Stop 4, return northeast to Canton along this road, passing numerous outcrops of leucogranitic gneiss and crossing the refolded synclinal trough, floored by marble units, between Forest House Road and Church Brook. Forest House Road and Irish Settlement Road provide further opportunities to explore the relationships between the leucogranite and the enveloping rocks.

The Pyrites Leucogranitic Gneiss Bodies

Southeast of the main Canton leucogranitic gneiss bodies are two smaller, dome-shaped masses of leucogranitic gneiss labeled the North and South Pyrites bodies by Romey and others (1980).

Stop 5. The Arthur E. Green farm on Cousintown road provides a section across the contact of the South Pyrites body and demonstrates its stratigraphic and structural equivalency to the Canton body. This stop may be reached via the Canton-Pyrites Road (also called Barnes Road), which turns southwest from Canton's Park Street 2 mi (3.2 km) south of the village limit. The Canton-Pyrites road crosses the north tip of the North Pyrites body, passing outcrops of marble and leucogranitic gneiss for a distance of 3.3 mi (5.3 km), to a right-angle turn southward. Immediately after this turn, turn again to the southwest across the Grasse River bridge. At the "Y" take the left (more southerly) road toward Hermon, climbing up onto a substantial glaciofluvial delta. Continue for 1.3 mi (2 km) to Cousintown Road. Turn northwest (right), travel 0.2 mi (0.3 km) and turn southwest (left). The Green farm is on the southeast side 0.6 mi (1 km) southwest along Cousintown Road (Fig. 1). Ask permission at the farm to obtain access to the area.

Walk southeast across the meadow behind the house across a section of marble containing forsteritic olivine and serpentinite, across a small stream, uphill across the marble-leucogranitic gneiss contact onto the South Pyrites leucogranitic gneiss body. Stratigraphic units and the contact between the leucogranitic gneiss and the metasedimentary sequence extend southwestward for several hundred yards (meters) and then southward around the west end of this ovoid-shaped dome of leucogranitic gneiss. Next, walk back northwestward across this structural and stratigraphic section to the area of the farmhouse. Continue northwestward across the road and meadows to massive outcrops of well-foliated, medium-grained amphibolitic gneiss with foliation striking N45°E and dipping 30°NW. A prominent lineation plunges 20°NW.

Continue northwestward across this hill of amphibolite into the valley, which lies on the contact between the amphibolite and structurally overlying knobby serpentinite marble. From the north edge of this hill of knobby marble, view the Harrison Creek valley. The creek flows along a prominent NE lineament occupied farther northeast by the Grasse River. Disruption of stratigraphy and structure along this lineament supports the view that this is a major fault across the area. To the north of Harrison Creek, the southeastern contact of the South Canton leucogranitic gneiss body rises steeply from the valley floor on the northwest.

Cousintown Road continues for almost 2 mi (3.2 km) roughly parallel to the stratigraphic section, near the contact between marble and amphibolite, to the Hermon-DeKalb Junction road. Turn right to cross the contact onto the South Canton leucogranitic gneiss body. The road south to Hermon passes excellent outcrops of the marble, garnet amphibolite, and tailings piles of the now-defunct Stellaville pyrite mines (Prucha, 1957). Figure 2 shows several additional road sections that expose structural and petrologic relationships in the Proterozoic (Grenville) rocks of the northwest Adirondacks.

REFERENCES CITED

Brown, C. E., 1983, Mineralization, mining, and mineral resources in the Beaver Creek area of the Grenville lowlands in St. Lawrence County, New York: U.S. Geological Survey Professional Paper 1279, 21 p.

Buddington, A. F., 1929, Granite phacoliths and their contact zones in the northwest Adirondacks: New York State Museum Bulletin 281, p. 51–107.

—— , 1939, Adirondack igneous rocks and their metamorphism: Geological Society of America Memoir 7, 354 p.

Carl, J., and Van Diver, B., 1975, Precambrian Grenville alaskite bodies as ash-flow tuffs, northwest Adirondacks, New York: Geological Society of America Bulletin, v. 86, p. 1691–1707.

Carl, J., de Lorraine, W., and Mose, D., 1986, Northwest Adirondack stratigraphy; Intrusive origin and Rb-Sr age of Rossie and Antwerp granitoids: Geological Society of America Abstracts with Programs, v. 18, p. 8.

de Lorraine, W., Carl, J., Brown, C., and Foster, D., 1985, Reconsideration of stratigraphy in the northwest Adirondacks, New York: Geological Society of America Abstracts with Programs, v. 17, p. 15.

Engel, A.E.J., and Engel, C., 1963, Metasomatic origin of large parts of the Adirondack phacoliths: Geological Society of America Bulletin, v. 74, p. 349–352.

Fisher, D. W., Isachsen, Y., and Rickard, L., 1970, Geologic map of the State of New York: Albany, New York, New York State Museum and Science Service Map and Chart Series, no. 15, scale 1:250,000.

Foose, M. P., and Carl, J. D., 1977, Setting of alaskite bodies in the northwest Adirondacks, New York: Geology, v. 5, p. 77–80.

Foose, M. P., Mose, D., Nagel, M., and Tunsoy, A., 1981, The Rb-Sr ages and structural and stratigraphic relationships of Precambrian granitic rocks in the northwest Adirondacks, New York, Geological Society of America Abstracts with Programs, v. 13, p. 133.

Grant, N. K., Carl, J., Lepak, R., and Hickman, M., 1984, A 350-Ma crustal history in the Adirondack Lowlands: Geological Society of America Abstracts with Programs, v. 16, p. 19.

Karboski, F., Elberty, W. T., Jr., Romey, W. D., and Doyle, J., 1983, Sandstone and conglomerate enclaves in Grenville rocks, St. Lawrence County, New York; Evidence of basement remobilization: Geological Society of America Abstracts with Programs, v. 15, p. 673.

Prucha, J. J., 1957, Nature and origin of the Pyrite deposits of St. Lawrence and Jefferson counties, New York: Albany, New York, New York State Museum and Science Service Bulletin 357, 87 p.

Romey, W. D., and Elberty, W. T., Jr., 1983, Brittle deformation and metamorphism of post-Grenvillian sandstone, Adirondack Lowlands: Geological Society of America Abstracts with Programs, v. 15, p. 124.

Romey, W. D., Elberty, W. T., Jr., Jacoby, R. S., Christoffersen, R., Shrier, T., and Tietbohl, D., 1980, A structural model for the northwestern Adirondacks based on leucogranitic gneisses near Canton, New York: Geological Society of America Bulletin, Part II, v. 91, p. 505–588.

Silver, L. T., 1965, U-Pb isotopic data on zircon of the Grenville Series of the Adirondack Mountains [abs.]: American Geophysical Union Transactions, v. 46, p. 164.

Wiener, R. W., McLelland, J. M., Isachsen, Y. W., and Hall, L. M., 1984, Stratigraphy and structural geology of the Adirondack Mountains, New York; Review and synthesis, *in* Bartholomew, M. J., ed., 1984, The Grenville event in the Appalachians and related topics: Geological Society of America Special Paper 194, p. 1–55.

The geology of Cameron's Line, West Torrington, Connecticut

Charles Merguerian, Geology Department, Hofstra University, Hempstead, New York 11550

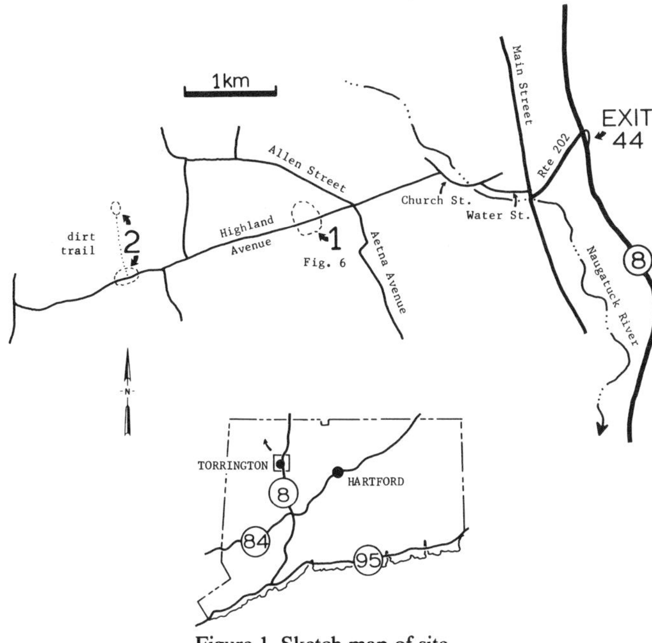

Figure 1. Sketch map of site.

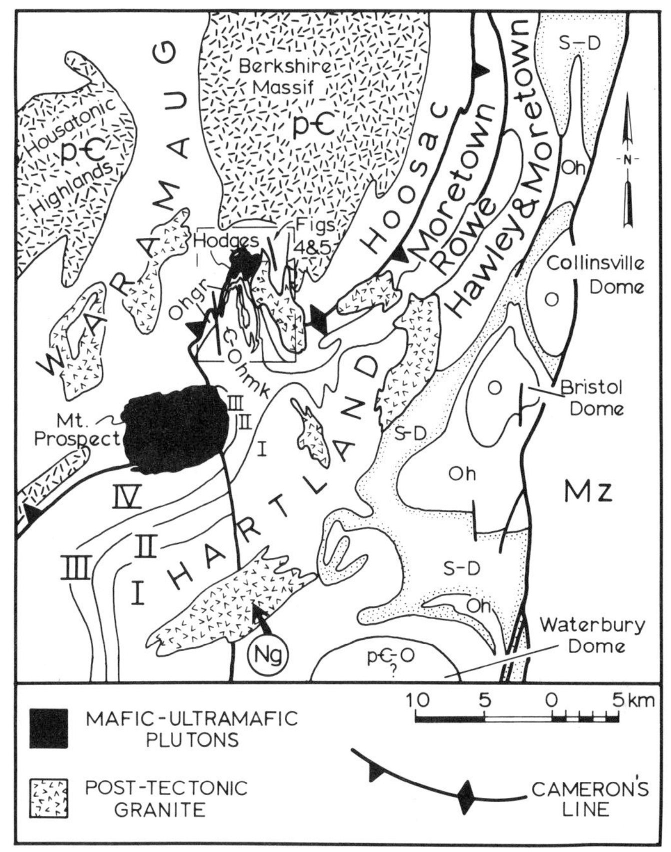

Figure 2. Regional geologic sketch map showing the regional geology around the West Torrington area (outlined). Same notation as in Figures 3 and 4 except pЄ, Proterozoic Y rocks; O, Ordovician rocks; S-D, Silurian and Devonian rocks; Mz, Mesozoic rocks of the Hartford Basin; Oh, Hawley Formation; Ng, Nonewaug Granite; and Roman numerals refer to Hartland stratigraphic units proposed by Gates (1967). Heavy lines are faults.

LOCATION

The site is situated in West Torrington, Connecticut, and consists of two stops in the West Torrington 7½-minute Quadrangle (Fig. 1). They can be reached from Exit 44 of Connecticut 8 by traveling southwestward on Connecticut 202 (East Main Street). To reach Stop 1, travel westward from the center of Torrington on Connecticut 202 to Water Street, then make a left onto Church Street, which ultimately leads to Highland Avenue (Fig. 1), and park 2,000 ft (600 m) west of the intersection with Allen Street near Patterson Pond. Stop 2 is located on Highland Avenue 1.2 mi (1.9 km) west of Stop 1. Some outcrops in this site are roadcuts or outcrops in stream beds and no special permission is required. Others, however, are on private property and permission *must* be sought from landowners.

SIGNIFICANCE

Modern interpretations (Robinson and Hall, 1980; Stanley and Ratcliffe, 1985) view the Middle Ordovician Taconic orogeny of the northern Appalachian mountain belt as the result of a collision between the lower Paleozoic continental margin of North America and a fringing volcanic arc. During the collision, the arc and intervening eugeosynclinal basin deposits were accreted to North America with imbrication of these disparate sequences at depth. The deformed strata now form a deeply-eroded tract of complexly-deformed high- to medium-grade metamorphic rocks. This site examines the geologic relationships exposed along Cameron's Line, a synmetamorphic ductile shear zone separating metamorphosed continental slope and rise(?) deposits on the west from eugeosynclinal deposits on the east.

The tectonic significance of Cameron's Line, long recognized as an important tectonic boundary in western Connecticut (Agar, 1927; Cameron, 1951; Rodgers and others, 1959; Gates and Christensen, 1965; Rodgers, 1985) and mapped as the Whitcomb Summit thrust in western Massachusetts (Stanley and Ratcliffe, 1985), can now be interpreted in the light of plate tectonics. The results of detailed mapping presented elsewhere (Merguerian, 1977, 1983, 1985a) suggest that in western Connecticut it marks a major deep-seated ductile fault related to the medial Ordovician Taconian imbrication of parts of the lower Paleozoic North American passive margin.

SITE INFORMATION

The lower Paleozoic metamorphic rocks of the New Eng-

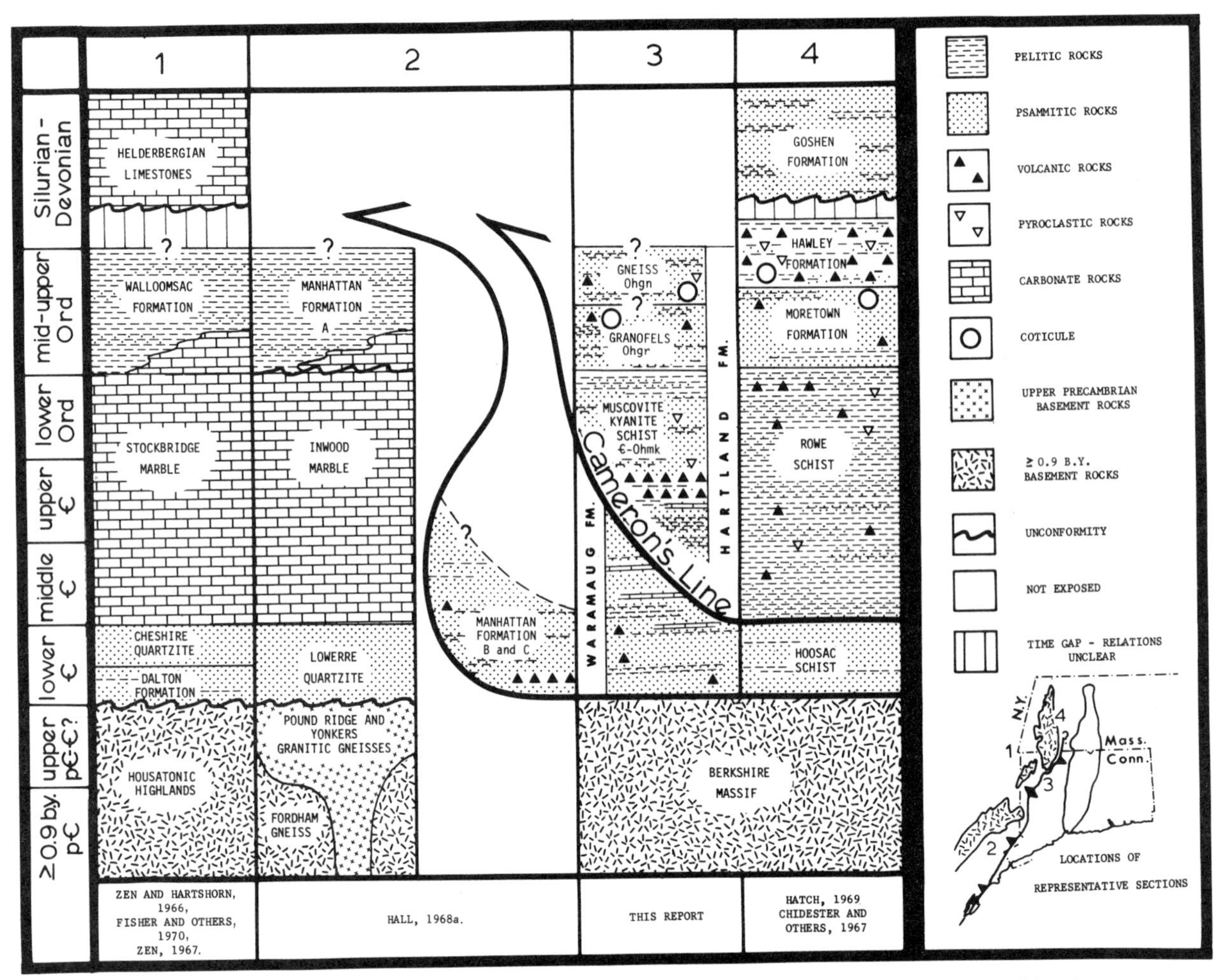

Figure 3. Stratigraphic correlation chart for southern New England showing the interpreted protoliths of the formations shown. Mixed symbols indicate relative, non-quantitative abundances of pre-metamorphic lithologies. Note the dominantly eugeosynclinal lithologic character of the lower Paleozoic rocks occurring east of Cameron's Line.

land Appalachians form a continuous northeast-trending crystalline belt across New England (Cady, 1969; Williams, 1978; Rodgers, 1985). They outcrop to the east of 900–1,100 Ma Proterozoic gneisses comprising, from south to north, the Hudson and Housatonic Highlands and the Berkshire and Green Mountains massifs (Fig. 2). The crystalline terrane of western Connecticut is bounded on the east by Mesozoic rocks of the Hartford Basin. To the west they decrease in metamorphic grade and are partly age and lithically correlative with weakly metamorphosed allochthonous strata forming the Taconic Mountains.

In northwestern Connecticut, Cameron's Line delimits the easternmost exposures of cratonal Proterozoic gneiss, massive Paleozoic carbonate shelf, and continental slope and rise (?) deposits. It separates a lower Paleozoic eugeosynclinal assemblage to the east (Hartland Formation) from partly coeval lower Paleozoic gneiss and schist with lithologic characteristics transitional between mio- and eugeosynclinal rocks (Figs. 2, 3). This western assemblage is known as the Waramaug or Hoosac Formation in Connecticut, as the Hoosac in Massachusetts, and as part of the Manhattan Formation in southeastern New York (Clarke, 1958; Gates, 1952; Hall, 1968a, 1968b; Hatch and Stanley, 1973; Merguerian, 1983; Merguerian and Baskerville, this volume). To the east, pre-Silurian rocks in the cores of the Bristol and Collinsville domes (Fig. 2) are unconformably mantled by Siluro-Devonian metamorphic rocks (the Straits Schist of Stanley, 1968; and Hatch and Stanley, 1973), which are correlative with rocks of the Connecticut Valley–Gaspé Synclinorium to the north (Goshen Formation in Fig. 3).

In West Torrington, there is a dramatic difference between the metamorphic rocks on either side of Cameron's Line (Figs. 2, 3). The Waramaug (p€-OWg) consists of massive quartzofeldspathic gneiss and schist with subordinate amphibolite and calc-silicate marble layers and lenses. The Hartland consists of dominantly well-layered micaceous schist, gneiss, and granofels (€-Ohmk, Ohgr, Ohgn) together with thick layers of complexly folded amphibolite (€-Oha, Ohau), and subordinate discontinuous layers and lenses of serpentinite and coticule (garnetiferous quartzite; Figs. 3, 4). Due to this profound difference in lithology and the lack of massive carbonate rocks and Proterozoic crust east of Cameron's Line, it seems probable that the Hartland protoliths were deposited on oceanic crust. The massive character and presence of calc-silicate and amphibolite in the Waramaug Formation suggests that their protoliths probably represent continental slope and rise (?) deposits. Thus, Cameron's Line separates sequences originally deposited adjacent to the lower Paleozoic shelf-edge of North America.

Structural Geology

Bedrock in the West Torrington Quadrangle provides evidence of at least five deformations that can be distinguished in the field by tracing axial surfaces of folds. The first two of these (D_1 + D_2) are probably continuous, developing two phases of isoclinal folds (F_1 + F_2) that culminate in the penetrative regional metamorphic fabric (S_1 + S_2) developed in both the Waramaug and Hartland formations. F_1 folds are rare but a local pre-existing S_1 foliation is folded by F_2 folds and S_1 is regionally truncated along with sub-units of the Hartland Formation along Cameron's Line. The D_2 structures are intensely developed in the vicinity of Cameron's Line. Furthermore, the regional parallelism and closer spacing of S_2 axial surfaces near Cameron's Line and the development of mylonite support the conclusion that Cameron's Line marks a synmetamorphic D_2 ductile fault between the Waramaug and Hartland formations.

The S_1 + S_2 regional foliation is locally warped by open F_3 folds that formed during intrusion of the Hodges Complex and the Tyler Lake Granite, but they have little effect on map patterns (Fig. 4). The S_1 + S_2 foliation and Cameron's Line is folded by dextral F_4 folds with steep southwest-plunging axes and northwest-dipping axial surfaces. A spaced biotite schistosity or crenulation cleavage (S_4) cuts all pre-existing structures as well as intrusive rocks of the Hodges Complex and Tyler Lake Granite. Finally, a weakly developed fifth deformation forms west- to northwest-trending spaced cleavage that cuts all previous structures and lithologic units, but has little effect on the map patterns. The superposition of these events has resulted in the complex map patterns shown in Figure 4 and in the interpretive cross sections of Figure 5.

Cameron's Line was a locus for lower Paleozoic plutonic activity. It acted as a weak zone in the crust enabling late synorogenic mafic-ultramafic magmas of the Hodges Complex and the younger Tyler Lake Granite to intrude along it (Merguerian, 1977). The Hodges Complex, engulfing Cameron's Line in West Torrington (Fig. 4), is a small composite mass of pyroxenite, hornblendite, gabbro, and diorite that is not sheared or offset by the fault. Rather, a narrow contact aureole that developed in the bounding Waramaug and Hartland formations overgrows the S_2 fold-fault fabric formed during the development of Cameron's Line.

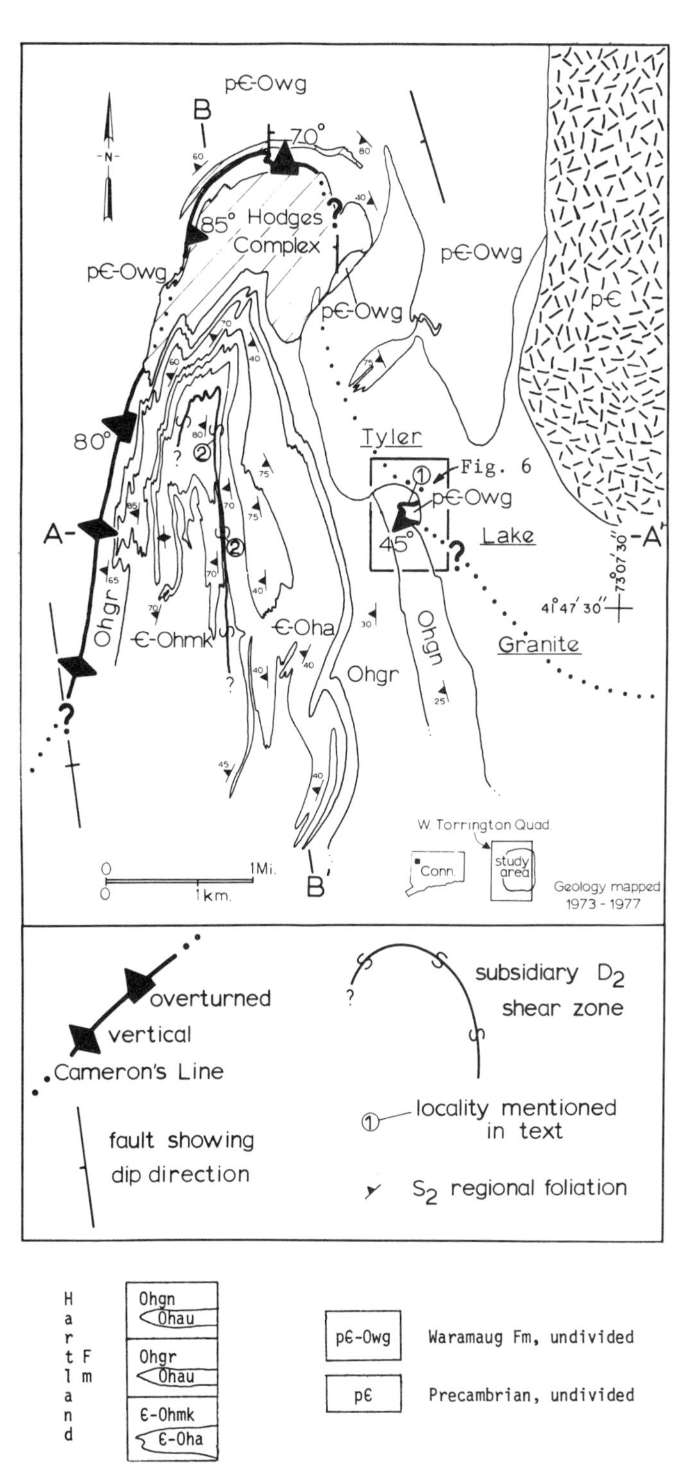

Figure 4. Simplified geologic map of part of the West Torrington Quadrangle, Connecticut. The complex map pattern of Hartland amphibolite (Є-Oha) is due to superposition of F_1 and F_2 isoclinal folds produced during the formation of Cameron's Line and major dextral folding by southwest-plunging folds with northwest-dipping axial surfaces (F_4). The approximate trace of Camron's Line is dotted through intrusive rocks.

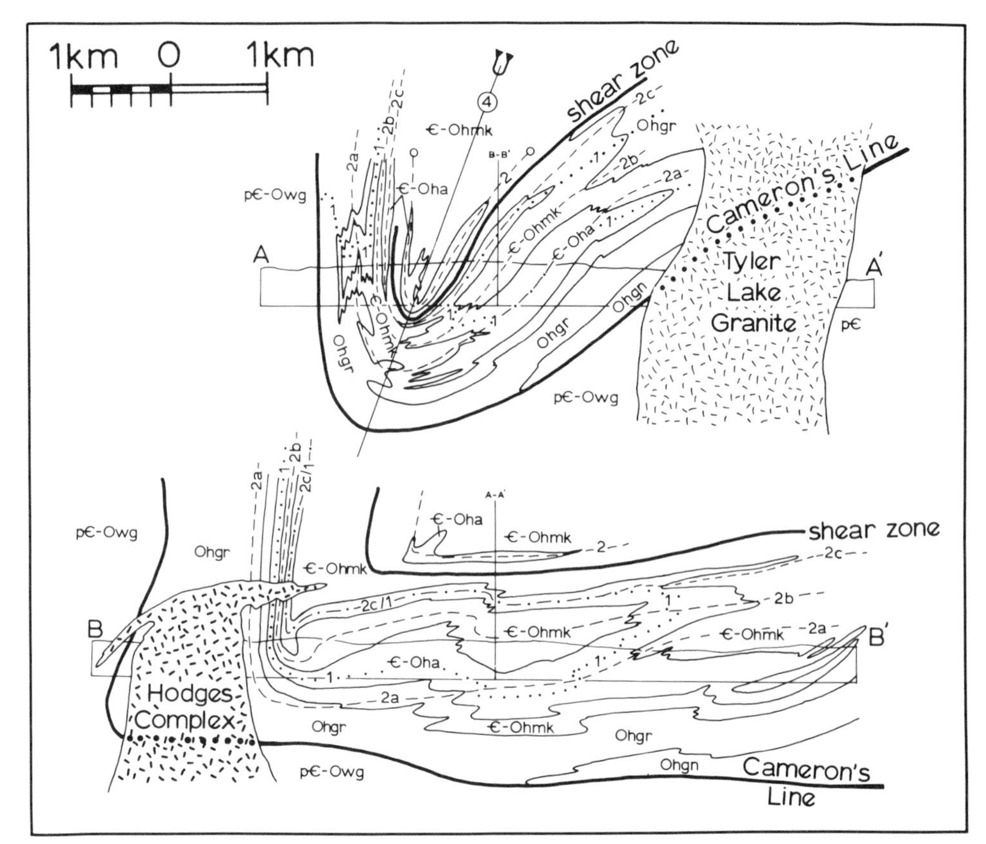

Figure 5. Geologic structure sections. Section lines shown in Figure 4. Numbered lines (1, 2a, 2b, 2c, 4) mark the projected axial surface traces of various fold generations. Note the regional parallelism of S_2 axial surface traces with Cameron's Line and the effects of F_4, which in section A–A′ plunges toward the reader. Also note the discordant contact relationship of both the Hodges Complex and the Tyler Lake Granite and truncation of Hartland gneiss sub-unit (Ohgn) against Cameron's Line. No vertical exaggeration.

The Hodges Complex is cut on the east by the Tyler Lake Granite (Fig. 4). The Tyler Lake yielded a middle Ordovician Rb/Sr age (Merguerian and others, 1984), which thus establishes a minimum age for the faulting on Cameron's Line. Therefore, the two stops described below offer a view of an ancient deep-seated collisional boundary that probably formed during or diachronously before the medial Ordovician Taconic orogeny (Merguerian, 1985b).

SITE DESCRIPTION

Stop 1. Cameron's Line Exposure. Biotite gneiss and schist of the Waramaug Formation crop out in the woods on both sides of Highland Avenue immediately to the west of the stream (Fig. 6). The rocks strike northwest and dip southwest. They are locally folded by F_3 folds with sub-horizontal axial surfaces and are locally injected by granitoid of the Tyler Lake Granite.

On the south side of Highland Avenue roughly 600 ft (200 m) west of the stream, mylonitic Harland amphibolite separates the Waramaug rocks from muscovitic schist, gneiss, and granofels of the Hartland Formation (Figs. 4, 6). Sub-units of the Hartland (Ohgn and Ohau) are regionally truncated along Cameron's Line (Figs. 3–6).

The rarely exposed contact, Cameron's Line, between the Hartland and Waramaug formations typically contains tectonically interleaved rocks from both formations. The contact, which has been mapped through the Hodges Complex by examining xenoliths and screens and identified elsewhere in the West Torrington quadrangle, is actually a zone 50 to 300 ft (15 to 90 m) thick that typically incorporates mylonitic amphibolite layers up to 10 ft (3 m) thick. Here, the S_1 metamorphic layering is locally preserved within F_2 fold hinges, but strong shearing parallel to S_2 axial surfaces forms a penetrative D_2 fold-fault fabric that marks Cameron's Line. Amphibolites away from Cameron's Line also show the effects of F_2 folds, but the high degree of shearing and disarticulation of the pre-D_2 fabrics is unique to the fault zone.

Trace the streambed 300 ft (100 m) northwestward from where it crosses Highland Avenue to where a 30-ft-thick (10 m) isoclinally folded serpentinite body occupies the contact zone between Waramaug rocks to the southeast and Hartland rocks to the west and northwest. Folded by F_2 folds (Fig. 7), the serpen-

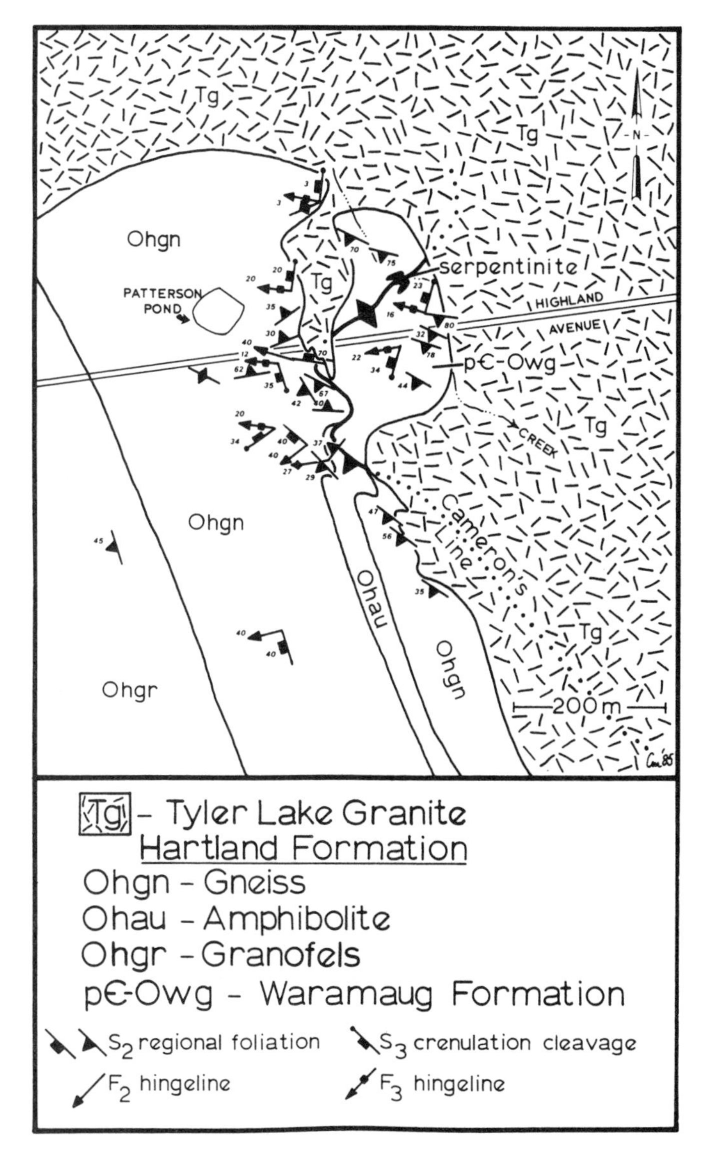

Figure 6. Sketch map showing geologic relations of Stop 1 (details in text). The warping of Cameron's Line and the S_2 regional foliation is due to shouldering near the Tyler Lake Granite and the effects of syn-intrusive F_3 folds. Same notation as in Figure 4.

Figure 7. Photograph (facing east) of an isoclinally folded serpentinite occurring at Cameron's Line with a steep S_2 axial surface trace (dashed line). The serpentinite is zoned and highly altered containing relict olivine and orthopyroxene with intergrown amphiboles. The compositional zoning (thin white line folded by F_2) is due to relative enrichment of matted cummingtonite and tremolite with minor serpentine in the upper part of the body in comparison to the dense black serpentine- and anthophyllite-enriched lower part.

tine body is compositionally zoned and strongly altered. It contains relict olivine and orthopyroxene with recrystallized cummingtonite, anthophyllite, and tremolite. It is distinct in mineralogy and texture from ultramafic rocks of the Hodges Complex and other nearby mafic-ultramafic plutons that post-date Cameron's Line.

Hartland rocks to the southwest (immediately southeast of Patterson Pond) contain layers of amphibolite and pink-orange 1 mm laminae of garnetiferous quartzite (coticule; Merguerian, 1979). The coticule rock, the overall eugeosynclinal nature of the Hartland rocks, and the structural position of the serpentinite together suggest that the serpentinite body represents metamorphosed dismembered oceanic lithosphere (ophiolite).

Stop 2. Subsidiary Shear Zone. Massive Hartland amphibolite occurs 1.2 mi (1.9 km) west on Highland Avenue (known here as Soapstone Hill Road) in excellent roadside outcrops. Here, the typical foliated but non-mylonitic character of the amphibolite is obvious in contrast to mylonitic amphibolite of Stop 1. F_2 isoclinal folds of the S_1 metamorphic foliation are exposed on the roadcut north of Soapstone Hill Road. Steep northwest-dipping S_4 slip cleavage related to southwest-plunging crenulations, cuts the $S_1 + S_2$ composite foliation and forms wedge-like angular pieces of amphibolite due to the intersection of structural fabrics.

At the sharp bend in Soapstone Hill Road at the western end of the roadcut, phyllonitic muscovite schist of the Hartland (Є-Ohmk) marks a subsidiary D_2 shear zone that imbricates the schist and amphibolite in the area (Figs. 4, 5). The shear zone is marked by F_2 isoclinal folds with sheared out limbs and a deeply-weathered zone to the south. About 2,300 ft (700 m) northward from Soapstone Hill Road along a dirt trail, it is marked by a soapstone-talc-chlorite schist quarry that may represent another ophiolite body. Away from the shear zone the Hartland schists and amphibolites are medium- to coarse-grained and do not show the abundant shearing parallel to F_2 axial surfaces that characterizes the ductile faults.

SUMMARY

Cameron's Line in West Torrington is a zone of ductile deformation that separates metamorphosed Upper Proterozoic (?) to lower Paleozoic continental slope and rise (?) rocks of the

Waramaug Formation from dominantly eugeosynclinal rocks of the Hartland Formation. Subunits of the Hartland are truncated along strike against Cameron's Line where early metamorphic fabrics (S_1) are obliterated by the intense fold-fault fabric (S_2). The S_2 fabric is the dominant regional foliation found on either side of the fault.

The S_2 axial surface traces are regionally parallel to the mapped trace of Cameron's Line. The spatial coincidence of intense localized isoclinal F_2 folds with sheared-out limbs and imbricated lithologies, including mylonitic amphibolite and rare serpentinite, suggests that Cameron's Line and its subsidiary shear zone mark deep-seated ductile faults. Indeed, many of the stratigraphic complications of the Hartland Formation in western Connecticut may result from other, as yet unrecognized ductile faults.

Available data suggest that Cameron's Line may be a thrust within the deep levels of a west-facing Taconian accretionary prism formed when the North American shelf and transitional sequence (Waramaug Formation) became juxtaposed with oceanic rocks of the Hartland Formation during closure of a fringing marginal basin. Today, the Hartland Formation represents a deeply-eroded portion of the accretionary complex of oceanic basin rocks formerly positioned between the lower Paleozoic passive margin of eastern North America and a volcanic archipeligo to the east.

REFERENCES CITED

Agar, W. M., 1927, The geology of the Shepaug Aqueduct Tunnel, Litchfield County, Connecticut: Connecticut Geological and Natural History Survey Bulletin 40, 46 p.

Cady, W. M., 1969, Regional tectonic synthesis of northwestern New England and adjacent Quebec: Geological Society of America Memoir 120, 181 p.

Cameron, E. N., 1951, Preliminary report on the geology of the Mount Prospect Complex: Connecticut Geological and Natural History Survey Bulletin 76, 44 p.

Chidester, A. H., Hatch, N. L., Jr., Osberg, P. H., Norton, S. A., and Hartshorn, J. H., 1967, Geologic Map of the Rowe Quadrangle, Massachusetts and Vermont: U.S. Geological Survey Geologic Quadrangle Map GQ-642, scale 1:24,000.

Clarke, J. W., 1958, The bedrock geology of the Danbury Quadrangle: Connecticut Geological and Natural History Survey Quadrangle Report 7, 47 p.

Fisher, D. W., Isachsen, Y. W., and Rickard, L. V., 1970, Geologic map of New York State, Lower Hudson Sheet: New York State Museum and Science Service, scale 1:250,000.

Gates, R. M., 1952, The geology of the New Preston Quadrangle, Connecticut, Part 1, The bedrock geology: Connecticut Geological and Natural History Survey Miscellaneous Series 5 (Quadrangle Report No. 2), p. 5–34.

——, 1967, Amphibolites—syntectonic intrusives: American Journal of Science, v. 265, p. 118–131.

Gates, R. M., and Christensen, N. I., 1965, The bedrock geology of the West Torrington Quadrangle: Connecticut Geological and Natural History Survey Quadrangle Report 17, 38 p.

Hall, L. M., 1968a, Geology in the Glenville area, southwesternmost Connecticut and southeastern New York, *in* Orville, P. M., ed., Guidebook for fieldtrips in Connecticut, New England Intercollegiate Geologic Conference, 60th annual meeting: Connecticut Geological and Natural History Survey Guidebook 2, Trip D-6, p. 1–12.

——, 1968b, Times of origin and deformation of bedrock in the Manhattan Prong, *in* Zen, E-an, and White, W. S., eds., Studies of Appalachian Geology; Northern and Maritime: New York, Wiley, p. 117–127.

Hatch, N. L., Jr., 1969, Geologic map of the Worthington Quadrangle, Hampshire and Berkshire counties, Massachusetts: U.S. Geological Survey Geological Quadrangle Map GQ-857, scale 1:24,000.

Hatch, N. L., Jr., and Stanley, R. S., 1973, Some suggested stratigraphic relations in part of southwestern New England: U.S. Geological Survey Bulletin 1380, 83 p.

Merguerian, C., 1977, Contact metamorphism and intrusive relations of the Hodges Complex along Cameron's Line, West Torrington, Connecticut [M.A. thesis]: The City College of New York, Department of Earth and Planetary Sciences, 89 p., with maps (also on open-file Connecticut Geological Survey, Hartford, Connecticut).

——, 1979, Dismembered ophiolite along Cameron's Line, West Torrington, Connecticut: Geological Society of America Abstracts with Programs, v. 11, no. 1, p. 45.

——, 1983, Tectonic significance of Cameron's Line in the vicinity of the Hodges Complex; An imbricate thrust model for western Connecticut: American Journal of Science, v. 283, p. 341–368.

——, 1985a, Geology in the vicinity of the Hodges Complex and the Tyler Lake granite, West Torrington, Connecticut, *in* Tracy, R. J., ed., Guidebook for fieldtrips in Connecticut and adjacent areas of New York and Rhode Island, New England Intercollegiate Geological Conference, 77th annual meeting: New Haven, Connecticut, Yale University, p. 411–442.

——, 1985b, Diachroneity of deep-seated versus supracrustal deformation in both the Appalachian Taconic and Cordilleran Antler orogenic belts: Geological Society of America Abstracts with Programs, v. 17, no. 7, p. 661.

Merguerian, D., Mose, D. G., and Nagel, S., 1984, Late syn-orogenic Taconian plutonism along Cameron's Line, West Torrington, Connecticut: Geological Society of America Abstracts with Programs, v. 16, no. 1, p. 50.

Robinson, P. and Hall, L. M., 1980, Tectonic synthesis of southern New England, *in* Wones, D. R., ed., Proceedings of the IGCP Project 27; The Caledonides in the U.S.A.; Blacksburg, Department of Geological Sciences, Virginia Polytechnic Institute and State University Memoir 2, p. 73–82.

Rodgers, J., 1985, Bedrock geological map of Connecticut: Connecticut Geological and Natural History Survey, scale 1:125,000.

Rodgers, J., Gates, R. M., and Rosenfeld, J. L., 1959, Explanatory text for the preliminary geological map of Connecticut, 1956: Connecticut Geological and Natural History Survey Bulletin 84, 64 p.

Stanley, R. S. 1968, Metamorphic geology of the Collinsville area, *in* Guidebook for fieldtrips in Connecticut, New England intercollegiate geologic conference, 60th annual meeting, Connecticut Geological and Natural History Survey Guidebook No. 2, 17 p.

Stanley, R. S., and Ratcliffe, N. M., 1985, Tectonic synthesis of the Taconian orogeny in western New England: Geological Society of America Bulletin, v. 96, p. 1227–1250.

Williams, H., 1978, Tectonic lithofacies map of the Appalachian orogen: Memorial University of Newfoundland Map No. 1, scale 1:2,000,000.

Zen, E-an, 1967, Time and space relationships of the Taconic allochthon and autochthon: Geological Society of America Special Paper 97, 107 p.

Zen, E-an, and Hartshorn, J. H., 1966, Geologic map of the Bashbish Falls Quadrangle, Massachusetts, Connecticut and New York: U.S. Geological Survey Geologic Quadrangle Map GQ-507, scale 1:24,000, 7 p., explanatory text.

Mesozoic sedimentary and volcanic rocks in the Farmington River Gorge, Tariffville, Connecticut

Norman H. Gray, Department of Geology and Geophysics, University of Connecticut, Storrs, Connecticut 06268

LOCATION

The basalt flows of the Hartford Basin crop out as prominent topographic ridges extending from Amherst in central Massachusetts, to New Haven in Connecticut. Only two streams traverse the ridge: the Westfield and the Farmington Rivers. The Farmington enters the basin near Bristol but makes a 9-mi (15-km) detour north to the Tariffville Gorge before crossing the basalt ridges. Tariffville is located in north-central Connecticut at the junctions of Connecticut Routes 189, 187, and 315. The sites discussed below are located by number in Figure 1. Outcrops along the roads are easily accessible. Access to the exposures on the banks of the Farmington River is more difficult but well worth the visit.

SIGNIFICANCE OF THE SITE

The Hartford Basin is an extensional basin related to, but predating, the Mesozoic opening of the modern Atlantic Ocean. The basin was filled by redbed-type arkosic sediments beginning first in mid to late Triassic time. The uniform pattern of the sedimentation was interrupted only briefly in the early Jurassic with the eruption of three basin-filling basalt flows. The vulcanism and associated tectonic activity profoundly disrupted the drainage in the basin. Playa and shallow lake sediments are well represented in the sedimentary rocks separating the flows. Outcrops in the Farmington River Gorge at Tariffville provide an unusually complete section through the three Mesozoic lava flows and intervening sediments. The site is particularly noteworthy for exposures of the upper contacts of the flows. These were cited by Rice (1886) and Davis (1898) to settle the controversy over the extrusive versus the intrusive nature of the trap sheets in the Connecticut River Valley.

SITE INFORMATION

A generalized stratigraphic column of the rocks that fill the Hartford Basin in northern Connecticut is illustrated in Figure 2. The division between the two main arkosic units, the Triassic New Haven and the Jurassic Portland, is marked by three major basalt flows. From oldest to youngest, these are today referred to as the Talcott, Holyoke, and Hampden (Lehmann, 1959). Although all are basalts, they are individually distinct and differ from one another chemically, mineralogically, and structurally. The Talcott and Hampden are each made up of at least two separate but closely related flows. Sediments of the Shuttle Meadow and East Berlin members of the Meriden formation separate the main flows.

The New Haven, Talcott, Shuttle Meadow, Holyoke, East Berlin, and Hampden rocks are exposed in the Tariffville Gorge (Schnabel and Eric, 1965). The more important outcrops and significant features of each unit are described below, keyed by number to the specific locations indicated on the map in Figure 1.

New Haven Arkose (Stops 1 and 9). A few meters of the red-colored New Haven arkoses and shales are exposed just beneath the Talcott flow along the Farmington River (Stop 9) and in a roadcut on Connecticut 189.

Talcott Basalt (Stops 1, 2, and 3). In central Connecticut the Talcott is a complex multiple sheet composed of many small flows piled on top of one another over a period of several years. Its total thickness averages 165 ft (50 m). At Tariffville the Talcott consists of a 50-ft (15-m)-thick single lower flow and a complex sequence of thinner upper flows for a total thickness exceeding 100 ft (30 m). Less than 1.2 mi (2 km) north of the gorge, the Talcott is completely absent. The erosional unconformity responsible is one of the interesting Talcott features exposed at Tariffville.

Regional mapping of flow indicators in the Talcott suggests a western source north of Meriden and a separate eastern source farther south (Gray, 1982; Gray and Simmons, 1985).

Chilled samples of the Talcott contain olivine, plagioclase, clinopyroxene, and scattered orthopyroxene phenocrysts. The original olivine is pseudomorphed by serpentine, quartz, calcite, or chalcedony, but the crystal form is unmistakable. The phenocryst assemblage is one of the most characteristic petrographic features of the flow (Philpotts and Reichenbach, 1985). Its chemical composition is somewhat more ambiguous, complicated as it is by widespread secondary alteration. Analyses of exceptionally fresh material suggest the Talcott is compositionally a typical Eastern North American high Ti-quartz tholeiite (Puffer and others, 1981).

At Tariffville the Talcott consists of two major flow units. The lower flow is about 50 ft (15 m) thick and massive, except for a pillowed base and vesicular upper surface. The upper flow is much more complex and consists of multiple thin, interfingering flow lobes, many of which are pillowed. An agglomerate of fragments mixed with red sediments marks the upper boundary of the lower flow. A complete secton through the lower flow, including both its upper and lower contacts, is well exposed in roadcuts along Connecticut 189 (Stops 1 and 2). The upper flow outcrops along the banks of the Farmington River but may be inaccessible at times of high water.

The pillows at the base of the lower flow are intimately intermixed with the underlying sediment. Spaces between and

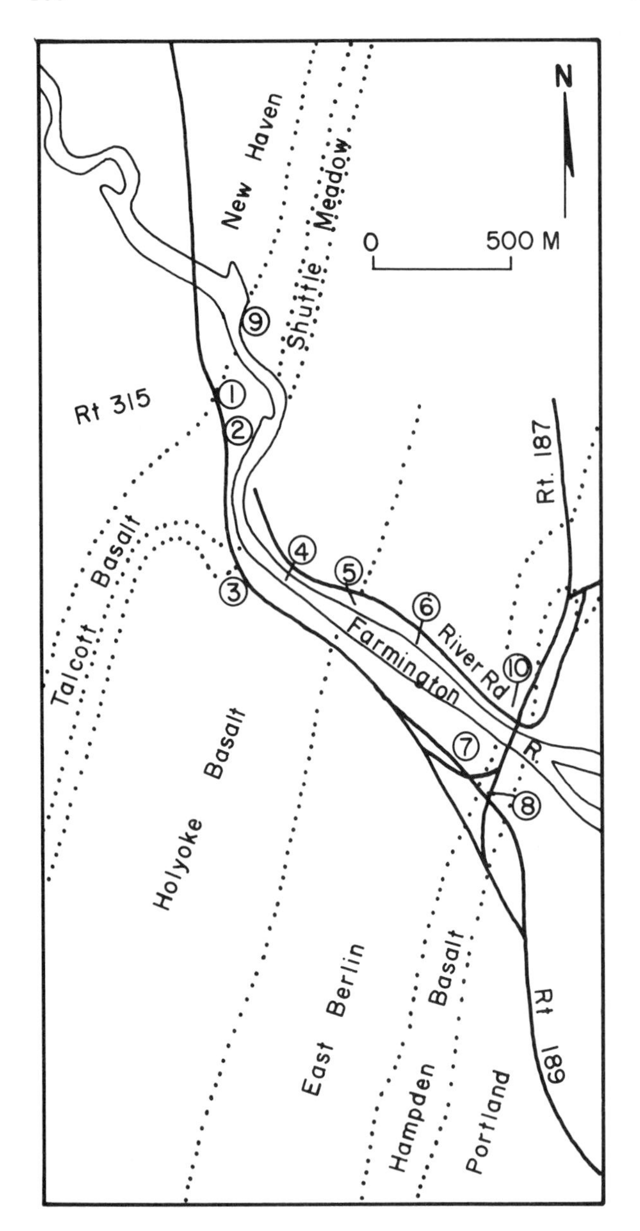

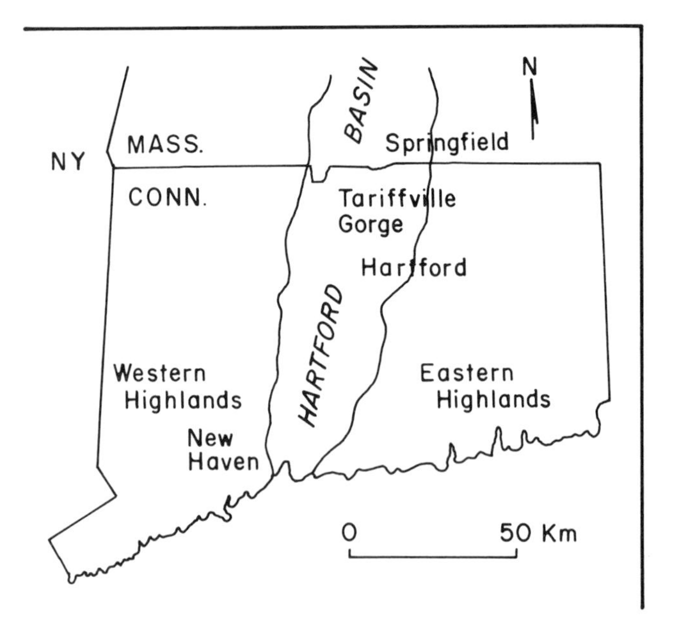

Figure 1. Location and geology of Tariffville Gorge, Connecticut. Locations referenced in the text are numbered 1 through 10.

fractures within the pillows are filled with a mixture of the red sediment and palagonitized fragments of pillow rinds. In outcrops on the north side of the Farmington River, pillows can be seen completely buried in red sediment 3 ft (1 m) below the main contact (Stop 9).

Less than 0.6 mi (1 km) to the south, on Lucy Hill, the entire thickness of the lower flow is pillowed. Presumably the Talcott at that site entered a shallow lake or pond. The pillows are elongated, lobate, tubular, and tongue-shaped bodies, 1.5 to 6 ft (0.5 to 2 m) in diameter, that are interconnected in overlapping distributary-like networks. The archtypical pillow, a separate completely detached lava sack, is rare.

As is characteristic of the Talcott regionally, contact metamorphic effects are conspicuous by their absence. The underlying New Haven sediments are unbleached and fissile right up to the base of the flow. The Holyoke and Hampden flows show more pronounced contact effects.

Reasonably well-developed cooling-related joints outlining columns 1.5 ft (0.5 m) in diameter originate in the massive basalt just above the basal pillows and extend upward 20 ft (6 m), where they appear to be "bent."

The upper half of the lower flow is characterized by the presence of flat-bottomed "half-moon" vesicles 2 to 8 in (5 to 20 cm) in diameter. These cavities probably originated near the base of the flow as large gas bubbles attached to slowly rising diapirs of vesicular lower crust. A variety of secondary minerals, principally calcite, quartz, prehnite, and datolite now fill the half-moon vesicles. Abundant cm-sized rectangular gash-like crystal cavities in these fillings record the former presence of anhydrite.

Half the way along the Talcott exposure on Connecticut

189, a peculiar 1.5-ft (0.5-m)-wide breccia dike cuts the flow. Fragments of basalt are set in a datolite-cemented arkosic matrix. Both Davis (1898) and Rice (1886) observed this dike in a railway cut along the route of the present road and suggested it represented the filling of an open fissure in the lower flow.

Shuttle Meadow Sedimentary Rocks (Stop 3). The Shuttle Meadow sediments are poorly exposed in northern Connecticut. The Tariffville section is one of the few localities where it does outcrop.

A complete 56-ft (17-m) section is exposed along the banks of the Farmington River below the Connecticut 189 roadcut. The site is best reached by descending a small ravine from Connecticut 189 directly above the river outcrops. The lower half of the section consists of clastic red sandstones and shales. Rounded pebbles and cobble-sized fragments of basalt are found in the sediments immediately above the Talcott. Calcite-filled evaporite crystal casts up to 2 in (5 cm) in size occur in abundance in some layers. The largest casts are found on the north bank of the river 300 ft (100 m) west of the old bridge pier. The crystal outline of the large cavities suggests the original mineral was glauberite. The smaller casts, which become more abundant higher in the section, are either collapsed or so poorly developed that their original identity cannot be established with certainty.

Finely laminated gray soda-rich mudstones form the upper half of the section. Microcrystalline albite and chlorite are the principal constituents. The origin of this rock is obscure, but it is likely that the present mineralogy is the consequence of reaction between an original clay-rich sediment and hot soda-rich pore waters during the cooling of the overlying Holyoke flow. A systematic decrease in fissility and an increase in the grain size of the authigenic albite and chlorite is noticeable near the Holyoke contact.

Holyoke Basalt (Stops 3, 4, and 5). Throughout central Connecticut the Holyoke sheet is stratigraphically relatively simple. South of Farmington, it consists of two massive flows 165 to 330 ft (50 to 100 m) thick. Only the top few ft (m) of each are vesicular. The lower flow thins just north of Farmington, and only the upper flow is present in northern Connecticut. Secondary alteration is extensive but the chemical data available suggest the Holyoke is a typical eastern North America iron-rich quartz tholeiite (Puffer and others, 1981).

Rare pipe vesicles at the lower contacts suggest a south-to-southwesterly source for both the upper and lower Holyoke flows in central Connecticut (Manspiezer, 1969).

Holyoke Basalt outcrops on Connecticut 189 just east of the Talcott roadcuts but the best exposures are found along the Farmington River (Stops 3 through 5). The base of the flow is visible at the Shuttle Meadow locality described above and on the north side of the river farther upstream (Stop 10). Its vesicular upper crust outcrops at river level just below a power line crossing (Stop 6).

Except for its vesicular upper surface the 300-ft (100-m) thick Holyoke flow is massive. The conspicuous splinter-like jointing of its lower third is typical of the Holyoke regionally. In contrast to the Talcott the lower contact is sharp and relatively featureless. Scattered tilted vesicles at the base indicate flow from the south. En echelon sills 1.5 ft (0.5 m) thick and dikes of pegmatitic diabase containing centimeter-size curved pigeonite crystals are common in the central section of the flow. Differential weathering of the pegmatitic diabase makes the sills most conspicous in the natural outcrops at river level (Stop 4), but they can also be found along Connecticut 189.

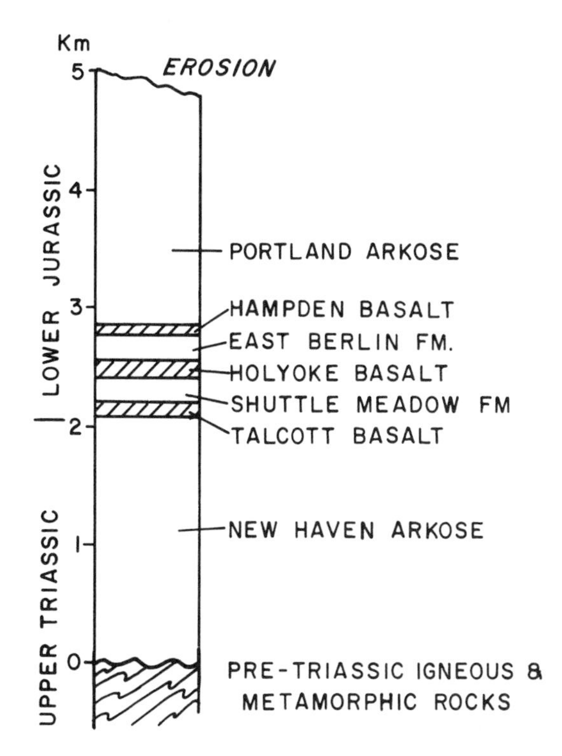

Figure 2. General stratigraphy of the Hartford Basin (Hubert and others, 1978).

East Berlin Sedimentary Rocks (Stops 6 and 7). The East Berlin lying between the Holyoke and Hampden flows outcrops both along Connecticut 189 and in the River. Neither section is complete and the roadcuts are much more accessible. Extremely low water or a boat is needed to examine the river exposures.

The East Berlin in the Farmington Gorge is almost identical petrographically and stratigraphically to the sections in central Connecticut described by Hubert and others (1978). The formation is dominantly red mudstone and arkose. Two separate 3.3 ft-thick (1-m) cycles of gray and black lacustrine shales are partially exposed in the roadcuts. Calcite-filled vertical tubes attributable to root rhizome are abundant in the red arkoses lying between the lake sediments.

Hampden Basalt (Stops 8 and 10). In central Connecticut the Hampden Basalt is typically massive and ranges in thickness from about 60 to 100 ft (20 to 30 m). Its base is commonly more vesicular than either the Holyoke or Talcott. Tilted pipestem vesicles are present at almost every exposure of its lower contact and generally indicate a southwesterly source (Ellefsen and Rydel, 1985). At Hartford and in southern Massachusetts the Hampden sheet appears to be made up of two separate flows (Colton and Hartshorn, 1966). However, because cooling joints cross this interflow contact, their regional significance is difficult to assess. Only one major flow seems to be present in the Hampden at Tariffville.

The Hampden is chemically unlike either the Talcott or Holyoke. Indeed, together with its correlative, the Hook Mountain Basalt of New Jersey, it forms a compositionally distinct group of eastern North American basalts (Puffer and others, 1981).

Unlike the other two flows, the Hampden is poorly exposed on the banks of the Farmington. It is best examined in roadcuts at the junction of Connecticut 187 and 189 (Stop 8).

The most conspicuous feature of the flow is the presence of large half-moon vesicles in its upper third. Like similar vesicles in the Talcott, these are tilted toward the northeast, matching the asymmetry of the basal pipestem vesicles exposed on the river road. The cavities are filled with the usual suite of secondary minerals—quartz, anhydrite (casts), prehnite, and datolite.

REFERENCES CITED

Colton, R. B., and Hartshorn, J. H., 1966, Bedrock geologic map of the West Springfield Quadrangle, Massachusetts and Connecticut: U.S. Geological Survey Quadrangle Map GQ-537.

Davis, W. M., 1898, The Triassic formation of Connecticut: U.S. Geological Survey, 18th Annual Report, pt. 2, p. 1–192.

Ellefsen, K. J., and Rydel, P. L., 1985, Flow direction of the Hampden basalt in the Hartford Basin, Connecticut and southern Massachusetts: Northeastern Geology, v. 7, p. 33–36.

Gray, N. H., 1982, Mesozoic volcanism in north-central Connecticut: 74th New England Intercollegiate Geological Conference Guidebook, p. 173–193.

Gray, N. H., and Simmons, R., 1985, Half moon vesicles in the Mesozoic basalt of the Hartford Basin; Their origin and potential use as paleoflow indicators: Geological Society of America Abstracts with Program, v. 17, p. 21.

Hubert, J. F., Reed, A. A., Dowdall, W. L., and Gilchrist, J. M., 1978, Guide to the Mesozoic Redbeds of Central Connecticut: State Geological and Natural History Survey of Connecticut, Guidebook No. 4, 129 p.

Lehmann, E., 1959, The Bedrock Geology of the Middletown Quadrangle: State Geological and Natural History Survey of Connecticut, Quadrangle Report 8, 40 p.

Manspeizer, W., 1969, Paleoflow directions in Late Triassic basaltic lava of the Newark Basin and their regional implication: Geological Society of America Abstracts with Programs for 1969, pt. 7, p. 142.

Philpotts, A. R., and Reichenbach, I., 1985, Differentiation of Mesozoic basalts of the Hartford Basin, Connecticut: Geological Society of America Bulletin, v. 96, p. 1131–1139.

Puffer, J. H., Hurtubise, D., Geiger, F., and Lechler, P., 1981, Chemical composition and stratigraphic correlation of Mesozoic basalt units of the Newark Basin, New Jersey, and Hartford Basin, Connecticut: Geological Society of America Bulletin, v. 92, pt. I, p. 155–159, pt. II, p. 515–553.

Rice, W. N., 1886, On the trap and sandstone in the gorge of the Farmington River at Tariffville, Connecticut: American Journal of Science, v. 132.

Schnabel, R. W., and Eric, J. H., 1965, Bedrock geological map of the Tariffville Quadrangle, Hartford County, Connecticut and Hampden County, Massachusetts: U.S. Geological Survey Quadrangle Map GQ-370.

The Willimantic fault and other ductile faults, eastern Connecticut

R. P. Wintsch, Department of Geology, Indiana University, Bloomington, Indiana 47405

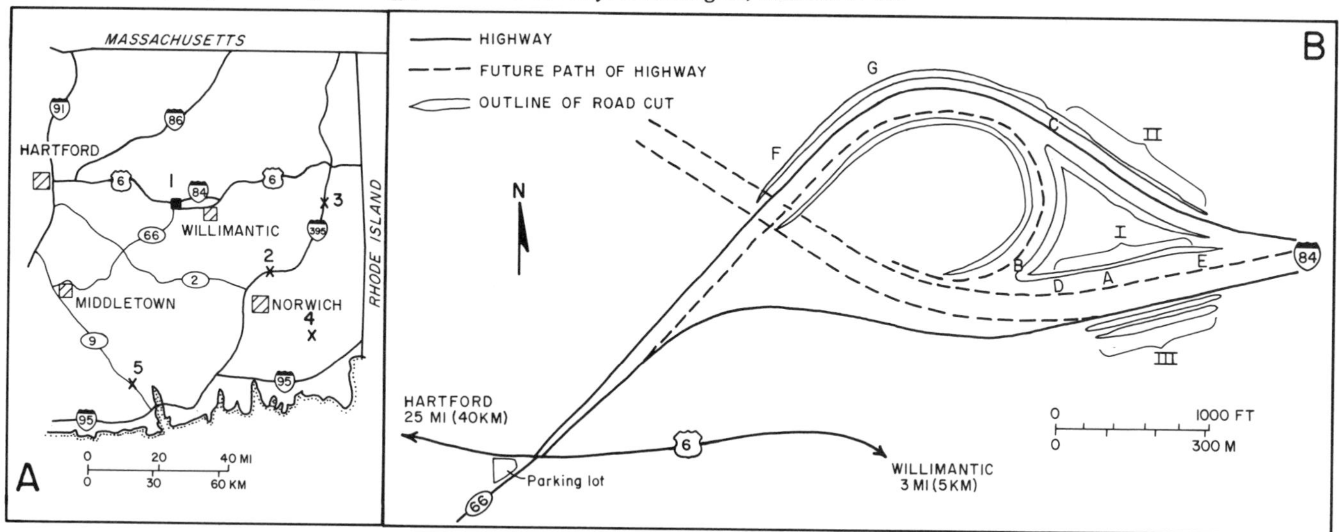

Figure 1. A, Regional location map; B, Map of the west end of I-84 and its junction with U.S. 6 and Connecticut 66, near Willimantic, Connecticut. Letters A through G and Roman numerals I and II refer to the locations of features described in the text.

LOCATION

Stop 1. The most spectacular exposures of any ductile fault in southeastern New England are those of the Willimantic fault in road cuts along the westernmost interchange of the 5 mi (8 km) segment of unfinished I-84 at its junction with U.S. 6 and Connecticut 66, 3.5 mi (5.7 km) west of Willimantic, Connecticut (Fig. 1). The outcrops at 41°43′24″ north latitude and 72°16′ west longitude are in the town of Coventry, Tolland County, east-central Connecticut (Fig. 2). They lie in the northwest quadrant of the Columbia 7½-minute Quadrangle, but are not shown on Snyder's (1967) bedrock map because the cuts were first opened in 1972. This interchange may be reached from Hartford, Connecticut by following I-84 east to U.S. 6, then east to the interchange; from Middletown, Connecticut by following Connecticut 66 east to the interchange; from Providence, Rhode Island, by following U.S. 6 west to the interchange; or from Norwich, Connecticut by following Connecticut 32 north to U.S. 6, then west to the interchange. Once at the interchange, park in the commuter parking lot between U.S. 6 and Connecticut 66, and walk up the median of I-84 to the cuts (see Fig. 1). These cuts are in part of a federal highway system! Stay well off the pavement and away from traffic! Please do nothing to jeopardize the use of these exposures by future geologists.

Stop 2. Tectonic blocks associated with the Tatnic fault are exposed in the north-bound exit ramp of the interchange of I-395 (formerly Connecticut 52, interchange 83), with Connecticut 97, 3.4 mi (5.5 km) northeast of Norwich in the Norwich Quadrangle (see Snyder, 1961).

Stop 3. Mylonites associated with the Lake Char fault are exposed in the road cuts along the south-bound ramp of Exit 88 off I-395 at its intersection with Connecticut 14A, 0.6 mi (1 km) north of Plainfield in the Plainfield Quadrangle (Dixon, 1965).

Stop 4. Mylonites derived from gabbro are exposed on the north side of old Connecticut 2, 2.8 mi (4.5 km) east of its junction with Connecticut 164, and are shown on the Old Mystic Quadrangle map of Goldsmith (1985a).

Stop 5. Rocks strongly deformed at very high grade are very well exposed along the north-bound entrance ramp to Connecticut 9 at Interchange 3 where 9 and 154 meet 1.2 mi (2 km) west of Essex, Connecticut (see Wintsch, 1985, Stop 2).

SIGNIFICANCE

Some of the most important structures in southeastern New England are numerous north- and east-striking ductile faults. These faults are important because they reveal much about the structure and rheology of the crust, and in New England they also mark the boundaries of exotic or suspect terranes. The rocks described in this guide expose structures of the ductile, Willimantic fault, which marks the boundary between the Putnam-Nashoba and the underlying Avalon terranes (Fig. 2). Most of the structures present in this fault were formed in the lower crust at depths of 6 to 12 mi (10 to 20 km), and thus these fault rocks share many properties of rocks regionally metamorphosed at amphibolite facies conditions. However, the Willimantic fault, like many other ductile faults, has been reactivated in the upper crust, at depths less than 6 mi (10 km). Consequently, granoblastic textures and high-grade mineral assemblages reflecting the early ductile deformation are partially overprinted by cataclastic tex-

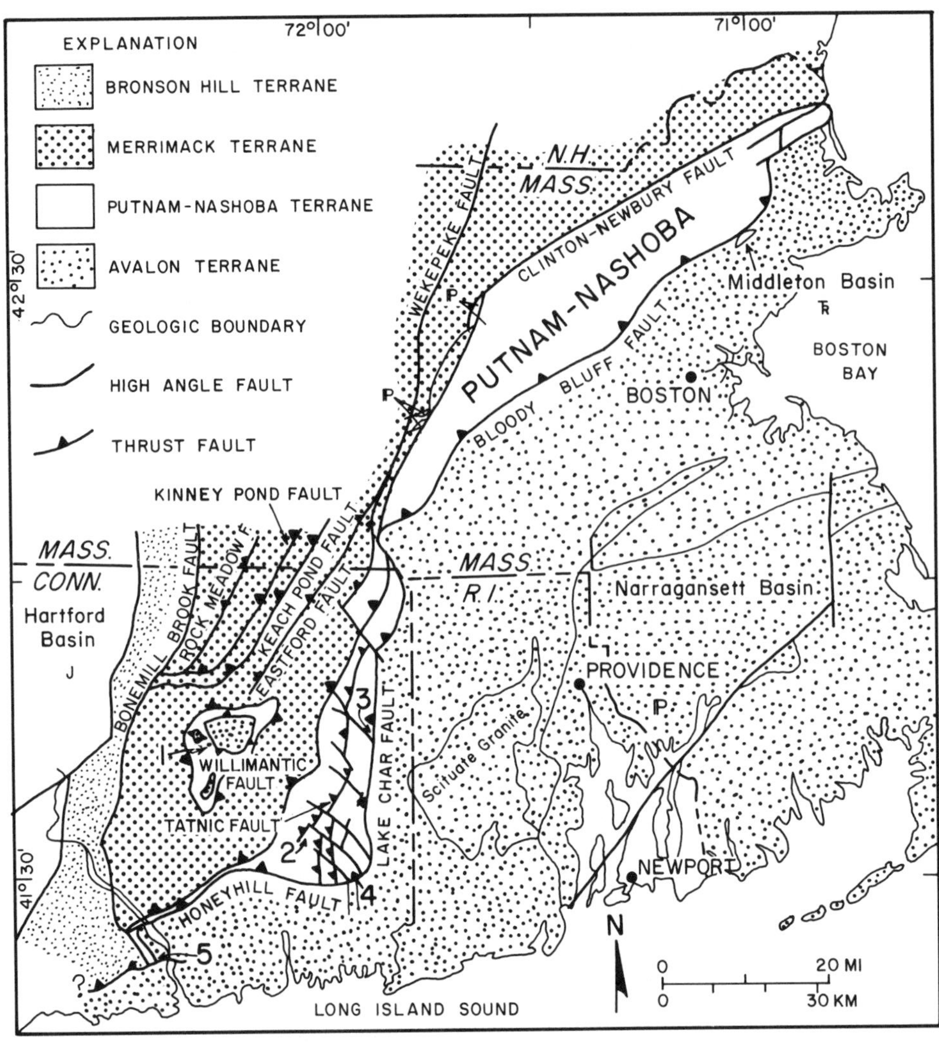

Figure 2. Map showing the distribution of major terranes in southeast New England, as well as several important faults. The map is simplified from that compiled by Rodgers (1985) by leaving all the units within terranes undivided, and by not including units of the Merrimack and Putnam-Nashoba terranes complexly infolded with units of the Avalon terrane. Stops included in this guide are indicated by the boxed numbers.

tures and mineral assemblages characteristic of greenschist or zeolite facies conditions. Most stops are very well suited to the study of the structural and petrologic complexities of ductile faults, in general, and some show segments of the Honey Hill fault system where it acts as a terrane boundary.

SITE INFORMATION

The identification of ductile faults in the field has been hampered in the past by the strong similarity of high-grade fault rocks and regional metamorphic rocks. However, the recognition of exotic terranes in southern New England (e.g., Williams and Hatcher, 1982; Zartman and Naylor, 1984) has refocused attention on terrane boundaries, and much field work on ductile faulting and ductile fault rocks in terrane boundaries has recently been completed (Castle and others, 1976; Goldstein, 1982; Gromet and O'Hara, 1985; Mergurian, 1985; Wintsch, 1979, 1985).

At least four terranes have been recognized in southeastern New England: Avalon, Putnam-Nashoba, Merrimack, and Bronson Hill (Fig. 2), and most of the boundaries between these terranes are now recognized as ductile faults. This guide focuses on faults along the boundary between the late Proterozoic Avalon terrane and the late Proterozoic(?) Putnam-Nashoba terrane. This is a complicated boundary with the different segments—Honey Hill, Lake Char, Bloody Bluff, and Willimantic faults (Fig. 2)—all recording unique aspects of the faulting. The stratigraphy of the rocks of the Putnam-Nashoba and Avalon terranes on either side of these faults is not well understood. However, along the Honey Hill and Lake Char faults, parts of the Rope Ferry Gneiss, Hope Valley Alaskite Gneiss, and Plainfield Formation all thin to

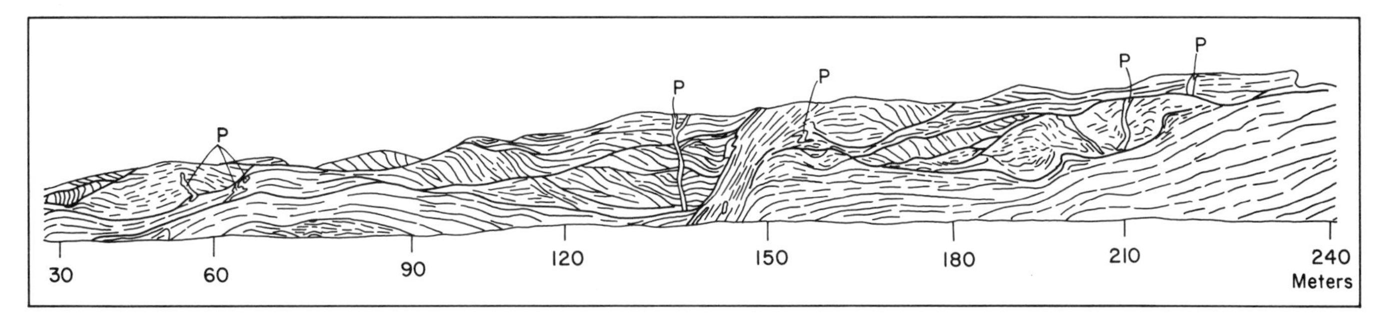

Figure 3. A structural profile of cut I, Figure 1 (from Wintsch, 1979, Plate I). All the rocks exposed here are part of the peltic lower member of the Tatnic Hill Formation. The rock is locally intruded by narrow (0.7 to 3.3 ft; 0.2 to 1 m) thick pegmatite dikes (P) and quartz veins (not shown). Scale in meters.

the east and north, and are probably cut out by these faults (Dixon and Lundgren, 1968). In the Putnam-Nashoba terrane, metavolcanic rocks of the Marlboro (Massachusetts) and the Quinebaug (Connecticut) formations are present above the Lake Char and Bloody Bluff faults respectively, but they are absent along the Honey Hill and Willimantic segments where the pelitic, Tatnic Hill Formation rests directly on Avalonian rocks. Again, this may be because of lack of deposition, but it is probably because they are cut out to the west (Wintsch and Fout, 1982). The circular outcrop pattern of the Willimantic dome is probably caused by the truncation of the top of a doubly plunging anticline in the Avalonian rocks below, and their exposure is thus in a tectonic window. Like many faults between the Bloody Bluff and Clinton-Newbury faults (not shown on Fig. 2), the Tatnic fault at the base of the Tatnic Hill Formation is internal to the Putman-Nashoba terrane. It is marked by a zone of intense imbrication (e.g., Dixon, 1982a), and because the Quinebaug Formation thins to the north, this fault may also cut down section.

Retrograde reactivation of these faults is variably developed. The most detailed information of this retrogression comes from the Honey Hill fault (Snyder, 1964b; Dixon, 1982b; Wintsch, 1984; Goldsmith, 1985b), where a large spectrum of structures and mineral assemblages from upper amphibolite to zeolite facies is present. Retrogression along the Tatnic fault is not strongly developed, while along parts of the Lake Char and the Bloody Bluff faults it dominates (see Nelson, this volume). Other evidence for this reactivation comes from crosscutting relationships. At its east end the Honey Hill fault cuts the Tatnic fault, and at its west end near the Bronson Hill terrane, it is cut by several later faults (Wintsch, 1985).

In general, the motion on the fault system has been thrusting to the southeast, but Goldstein (1982) has found evidence that at least the latest motion on the Lake Char was oblique slip with a component of normal motion. The amount of displacement on these faults is nowhere well constrained (Dixon and Lundgren, 1968; Castle and others, 1976), but must be at least 2.2 mi (3.5 km) locally, and could be much more (Wintsch, 1979, p. 390). The history of motion on this fault system is also complex. The retrograde reactivation of the faults in Connecticut must have occurred during the Permo-Triassic cooling of the Avalon terrane (Wintsch and Lefort, 1984). Wintsch and Sutter (1986) argue that earlier activity of these faults created a Carboniferous fold-thrust belt, and that the Pennsylvanian Narragansett basin was the foreland basin to these thrusts. If this is true, then a displacement of tens of miles (kilometers) is likely. Although late Paleozoic activity seems quite certain at this point, earlier Acadian, or even Taconian movements cannot be ruled out. However, the strong late Paleozoic fabrics on these faults will make recognition of any earlier structures difficult.

Stop 1. The most stunning features exposed in the road cuts near Willimantic are the large and very large tectonic blocks separated by anastomosing shear zones. The dome shape of these blocks leads to quaquaversal foliation patterns in natural exposures, and the Willimantic and Columbia Quadrangle maps of Snyder (1964a, 1967)—including the southern part of the Willimantic dome—show that this unusual foliation pattern is diagnostic of the basal Tatnic Hill Formation. Because they are usually 100 ft (30 m) or more long in the E-W direction, they are best viewed from a distance of 100 ft (30 m) or more. Face I (Figs. 1 and 3) is particularly well suited for viewing from a distance because I-84 is not finished; but for those especially interested in these structures, a walk through all the cuts is imperative. The lack of continuity of the blocks or the shear zones on either side of the road along the north side of the interchange (at II, Fig. 1) indicates that these structures are not longer than 100 ft (30 m) in the N-S direction, and thus the blocks must be lens- or turtle-shaped. Some of these blocks may have developed as large drag folds (e.g., 100 m, Fig. 3) with axes striking N30°E, evolving into these discrete blocks as strain was concentrated on the long limbs (Wintsch, 1979, Fig. 5; see also Soula and Bessiere, 1980). However, many blocks do not appear to be rotated, and some degree of mega-boudinage (e.g. 100 m, Fig. 3) was probably also involved. These blocks and most small-scale structures were probably produced during southeast motion of the Putnam-Nashoba terrane over the Avalon terrane (Wintsch, 1979), and at least the latter stages of this deformation occurred in the late Paleozoic.

Augen gneiss, blastomylonitic gneiss, mylonitic schist, and

mylonite are present in order of decreasing abundance in these rocks. The highest grade assemblages are best preserved in the augen gneisses inside and away from tectonic blocks (e.g., east end, 150 to 250 m, Fig. 3). These quartz-plagioclase-biotite-K-feldspar-garnet-sillimanite bearing gneisses locally contain K-feldspar and plagioclase augen up to 5.5 in (14 cm) in diameter. These augen probably grew as porphyroblasts, and did not crystallize from a melt. Evidence for this comes from rotated biotite inclusion trails in some crystals. The incorporation of these rotated inclusions demands growth of a rigid crystal in a solid matrix. Moreover, the bulk composition of the augen gneisses is strongly syenitic or monzonitic, and does not reflect the minimum melt composition of these pelitic gneisses, which should be close to a granitic eutectic. Sillimanite is ubiquitous in these gneisses, and commonly occurs as randomly oriented needles in fibrolite mats. However, some sillimanite does occur with a strong preferred orientation trending approximately N60°W, parallel to locally developed biotite streaks and quartz-feldspar rods. According to Wintsch (1979), this direction is parallel to the overall southeast transport direction.

The shear zones surrounding and cutting these blocks of high-grade gneiss contain blastomylonitic schist and gneiss with lower grade kyanite-bearing assemblages (e.g., localities A [140 m, Fig. 3], B, C if not collected out). Still lower grade, slabby, strongly lineated and layered mylonitic schists are well exposed on the natural cliff face III, south of the road. This schist appears to underlie all rocks exposed in the road cuts, and its strong N-S trending lineation parallel to tight isoclinal folds suggests a similar change in fault motion direction. Another less slabby, even textured, and finer grained mylonitic schist forms a gently folded 10 to 12 ft (3 to 4 m) thick layer at D (Fig. 1) between 60 and 120 m; Fig. 3). This later foliation is itself folded into small, isoclinal folds, but they are difficult to find because the rock almost totally lacks compositional layering. True mylonites are rare, but late, fine-grained, middle to lower greenschist facies mylonite up to 1 in (2 cm) thick cuts the other gneisses in the steeply west-dipping shear zone which cuts the entire exposure at A (Fig. 1; 140 m, Fig. 3). A brittle fracture zone occurs at E (Fig. 1) where the K-feldspar-chlorite-epidote bearing assemblage reflects very shallow alteration of this zone. Together these shear zones and their associated mineral assemblages demonstrate repeated deformation in this fault zone during shallower and lower metamorphic grade conditions.

Several other rock types are present in these cuts. Interlayered in the pelitic blastomylonitic schists and gneisses are at least 28 thin (12 in; 30 cm) amphibolite layers, all boudinaged (between A and D, Fig. 1). Successive boudinage of a larger (3 ft; 1 m) amphibolite boudin at its tapering neck can be seen at 215 m (Fig. 3). Several 12- to 20-in-thick (30 to 50 cm) layers of diopside-bearing marble are present at F (Fig. 1). The margins of one of these contains hornblende porphyroblasts up to 4 in (10 cm) in diameter. A 6 ft (2 m) diameter pod of ultramafic rock, now chlorite-talc schist is present at G (Fig. 1).

The exposures described above reveal the features of deep, ductile faults better than any other exposure in eastern Connecticut. However, other exposures of fault rocks are well worth examining. Some show that the large-scale tectonic blocks and boudins are widespread near the base of the Tatnic Hill Formation (Stop 2), others show the effect of deformation under lower amphibolite or greenschist facies conditions (Stops 3 and 4), and the last (Stop 5) shows the mineralogical and petrological consequences of ductile deformation under near anatectic conditions.

Stop 2. Tectonic blocks are characteristic of the base of the Tatnic Hill Formation. They are not well exposed along the Honey Hill fault (but see Lundgren, 1968, Stop 3; Wintsch, 1985, Stop 7), but excellent examples of tectonic blocks similar to those at locality 1 (Fig. 3) are well exposed along the Tatnic fault. The shear zones separating the blocks generally dip steeply; the blocks look as if they were 'standing on end.' The shear zones do not contain kyanite or other retrograde minerals, and thus developed under upper amphibolite facies conditions. They probably first formed with a low westward dip and were later folded or thrusted to their present steeply-dipping position.

Stop 3. Most of the deformation in the Willimantic and Tatnic faults occurred under upper amphibolite facies conditions, whereas many fault rocks along the Lake Char fault developed under lower amphibolite and greenschist facies conditions. An example of the latter is found in these cuts (see also Lundgren, 1968, Stop 12). Here strongly-layered, and well-foliated and folded mylonites and blastomylonites are well exposed. The common $FeO/TiO_2/MnO$ ratios of these mylonites with an andesite porphyry in Dixon's (1965) qlc unit in the Quinebaug Formation (Wintsch, unpublished data) support her suggestion that these mylonites were derived from this porphyry.

Stop 4. Even lower grade, finer grained mylonites are exposed at the junction of the Honey Hill and Lake Char faults (see Lundgren, 1968; Lundgren and Ebblin, 1972). These road cuts described by Sclar (1958) and Dixon (1982b) expose deformed Preston Gabbro. The deformation has caused a progressive grain size reduction of coarse-grained gabbro to fine-grained mylonite and phyllonite. Feldspars are more sodic in the mylonites than in the gabbro, and chlorite and epidote are surprisingly rare in the fine-grained rocks. Apparently H_2O-rich fluid did not have access to the mylonite during its development.

Stop 5. In contrast to the first four stops, Stop 5 is composed of rocks that show the response of metaigneous rocks to ductile deformation under near anatectic conditions. The rocks lie in the Essex Quadrangle, but the cuts were not present when Lundgren (1964) mapped these rocks as the metasedimentary Putnam Gneiss. Wintsch (1985) interpreted these rocks to be part of the Falls River fault zone that extend at least 3 mi (5 km) to the west. These cuts expose a strongly layered and folded sequence of blasto-mylonitic biotite-sillimanite schists and biotite-plagioclase gneisses. Their southern contact is not exposed, but their northern contact with a late Proterozoic intrusive orthogneiss of the Rope Ferry Gneiss (see Wintsch, 1985) is gradational. It is characterized by an increase in the quality of layering defined by the concentration of biotite into discrete foliation planes, and by an increase in the amplitude and a decrease in the wavelength of folding in these layered rocks. The presence of these sillimanite-bearing schists caused Lundgren (1964) to map these rocks as part of the pelitic Tatnic Hill Formation (Stops 1 and 2), but the gradational contact between these schists and the orthogneiss makes this interpretation difficult to defend.

To test whether these schists could have been derived from the enclosing orthogneiss of Essex, Wintsch and Dipple (1986) analyzed the rocks and minerals of the blastomylonites and the enclosing orthogneiss. They used the constant FeO:MgO:TiO_2 ratios of these schists and the orthogneiss to argue that these blastomylonitic rocks were derived from the orthogneiss. They proposed that the extraction of quartz and plagioclase from the orthogneiss: (1) concentrated biotite into discrete foliation planes, and (2) enriched the gneissic layers in quartz and plagioclase. This 'unmixing' occurred on the scale of centimeters to feet (meters) to create the strongly layered rock present in these cuts. Some of the quartz and plagioclase removed from the schists are also concentrated in vein and migmatitic structures. Supporting evidence for this hypothesis comes from several sources. The constant FeO: MgO:TiO_2 ratios in bulk-rock samples is consistent with this because these elements are housed primarily in biotite. The nearly constant compositions of plagioclase from both the orthogneiss and the schists and gneisses of the fault zone is further evidence for this parental relationship. Evidence that quartz and plagioclase were added to the gneissic layers comes from the large poikioblastic nature of the quartz and plagioclase grains that include an earlier biotite foliation. Thus all the evidence supports the hypothesis that the blastomylonitic rocks in these cuts were derived from the host orthogneiss through some structurally-induced metamorphic differentiation. Apparently this has not occurred in the other stops included here, and the very high-grade conditions of the deformation in this fault were probably important in allowing this differentiation to take place.

REFERENCES CITED

Castle, R. O., Dixon, H. R., Grew, E. S., Griscom, A., and Zietz, I., 1976, Structural dislocations in eastern Massachusetts: U.S. Geological Survey Bulletin 1410, 39 p.

Dixon, H. R., 1965, Bedrock geologic map of the Plainfield Quadrangle, Windham and New London counties, Connecticut: U.S. Geological Survey, Geological Quadrangle Map GQ-481, scale 1:24,000.

—— , 1982a, Bedrock geologic map of the Putnam Quadrangle, Windom County, Connecticut: U.S. Geological Survey Map GQ-156Z, scale 1:24,000.

—— , 1982b, Multistage deformation of the Preston gabbro, eastern Connecticut, *in* Joesten, R., and Quarrier, S. S., eds., Guidebook for fieldtrips in Connecticut and south-central Massachusetts: State Geological Natural History Survey of Connecticut Guidebook no. 5, p. 453–463.

Dixon, H. R., and Lundgren, L. W., 1968, Structure of eastern Connecticut, *in* Zen, E-an, White, W. S., Hadley, J. B., and Thompson, J. B., Jr., eds., Studies of Appalachian geology; Northern and maritime: New York, Wiley Interscience Publishers, p. 219–229.

Goldsmith, R., 1985a, Bedrock geologic map of the Old Mystic and part of the Mystic quadrangles, Connecticut, New York, and Rhode Island: U.S. Geological Survey Miscellaneous Investigations Series Map I-1524, scale 1:24,000.

—— , 1985b, Honey Hill fault and Hunts Brook syncline, *in* Tracy, R. J., ed., Guidebook for fieldtrips in Connecticut: State Geological and Natural History Survey of Connecticut Guidebook no. 6, p. 491–507.

Goldstein, A. G., 1982, Geometry and kinematics of ductile faulting in a portion of the Lake Char mylonite zone, Massachusetts and Connecticut: American Journal of Science, v. 282, p. 1378–1408.

Gromet, L. P., and O'Hara, K. D., 1985, The Hope Valley shear zone; A major late Paleozoic ductile shear zone in southeastern New England, *in* Tracy, ed., Guidebook for fieldtrips in Connecticut: State Geological Natural History Survey of Connecticut Guidebook no. 6, p. 277–295.

Lundgren, L. W., 1964, Bedrock geology of the Essex Quadrangle: Geological and Natural History Survey of Connecticut, Quadrangle Report no. 15, 36 p.

—— , 1968, Honey Hill and Lake Char faults, *in* Orville, P. M., ed., Guidebook for fieldtrips in Connecticut: State Geological and Natural History Survey of Connecticut Guidebook no. 2, p. F1-1–F1-8.

Lundgren, L. W., Jr., and Ebblin, C., 1972, Honey Hill fault in eastern Connecticut; Regional relations: Geological Society of America Bulletin, v. 83, p. 2773–2794.

Mergurian, C., 1985, Geology in the vicinity of the Hodges complex and the Tyler Lake granite, West Torrington, Connecticut, *in* Tracy, R. J., ed., Guidebook for fieldtrips in Connecticut: State Geological Natural History Survey of Connecticut Guidebook no. 6, p. 411–442.

Rodgers, J., 1985, Bedrock geological map of Connecticut: Connecticut Geological and Natural History Survey, scale 1:125,000.

Sclar, C. B., 1958, The Preston gabbro and the associated metamorphic gneisses New London County, Connecticut: Geological Natural History Survey of Connecticut Bulletin 88, 136 p.

Snyder, G. L., 1961, Bedrock geology of the Norwich Quadrangle, Connecticut: U.S. Geological Survey, Geological Quadrangle Map GQ-144, scale 1:24,000.

—— , 1964a, Bedrock geology of the Willimantic Quadrangle, Connecticut: U.S. Geological Survey, Geological Quadrangle Map GQ-335, scale 1:24,000.

—— , 1964b, Petrochemistry and bedrock geology of the Fitchville Quadrangle: Connecticut: U.S. Geological Survey Bulletin 1161-I, 63 p.

—— , 1967, Bedrock geologic map of the Columbia Quadrangle, east-central Connecticut: U.S. Geological Survey Geological Quadrangle Map GQ-592, scale 1:24,000.

Soula, J. C., and Bessiere, G., 1980, Sinistral horizontal shearing as a dominant process of deformation in the Alpine Pyrenees: Journal of Structural Geology, v. 2, p. 69–74.

Williams, H., and Hatcher, R. D., Jr., 1982, Suspect terranes and accretionary history of the Appalachian orogen: Geology, v. 10, p. 530–536.

Wintsch, R. P., 1979, The Willimantic fault; A ductile fault in eastern Connecticut: American Journal of Science, v. 279, p. 367–393.

—— , 1984, Multiple reactivation of the Honey Hill fault zone, southern New England: Geological Society of America Abstracts with Programs, v. 16, p. 71.

—— , 1985, Bedrock geology of the Deep River area, Connecticut, *in* Tracy, R. J., ed., Guidebook for fieldtrips in Connecticut: State Geological Natural History Survey of Connecticut Guidebook no. 6, p. 115–141.

Wintsch, R. P., and Dipple, G. M., 1986, Evidence for tectonic schists in the Falls River fault zone, Essex, Connecticut: Geological Society of America Abstracts with Program, v. 18, p. 76.

Wintsch, R. P., and Fout, J. S., 1982, Structure and petrology of the Willimantic dome and the Willimantic fault, eastern Connecticut, *in* Joesten, R., and Quarrier, S. S., eds., Guidebook for fieldtrips in Connecticut and south-central Massachusetts: State Geological Natural History Survey of Connecti-

cut Guidebook no. 5, p. 465–482.
Wintsch, R. P., and Lefort, J.-P., 1984, A clockwise rotation of variscan strain orientation in SE New England and regional implications, *in* Hutton, D.H.W., and Sanderson, D. J., eds., Variscan tectonics of the north Atlantic region: London, Blackwell, p. 245–251.
Wintsch, R. P., and Sutter, J. F., 1986, A tectonic model for the late Paleozoic of southern New England: Journal of Geology, v. 94 p. 459–472.
Zartman, R. E., and Naylor, R. S., 1984, Structural implications of some radiometric ages of igneous rocks in southeastern New England: Geological Society of America Bulletin, v. 95, p. 522–539.

ACKNOWLEDGMENTS

R. Goldsmith and A. Goldstein reviewed an earlier draft of the manuscript. This work was partially supported by NSF Grant EAR-83-13807.

Geological Society of America Centennial Field Guide—Northeastern Section, 1987

Ledyard recessional moraine, Glacial Park, Connecticut

***Richard Goldsmith**, U.S. Geological Survey, Reston, Virginia 22092*

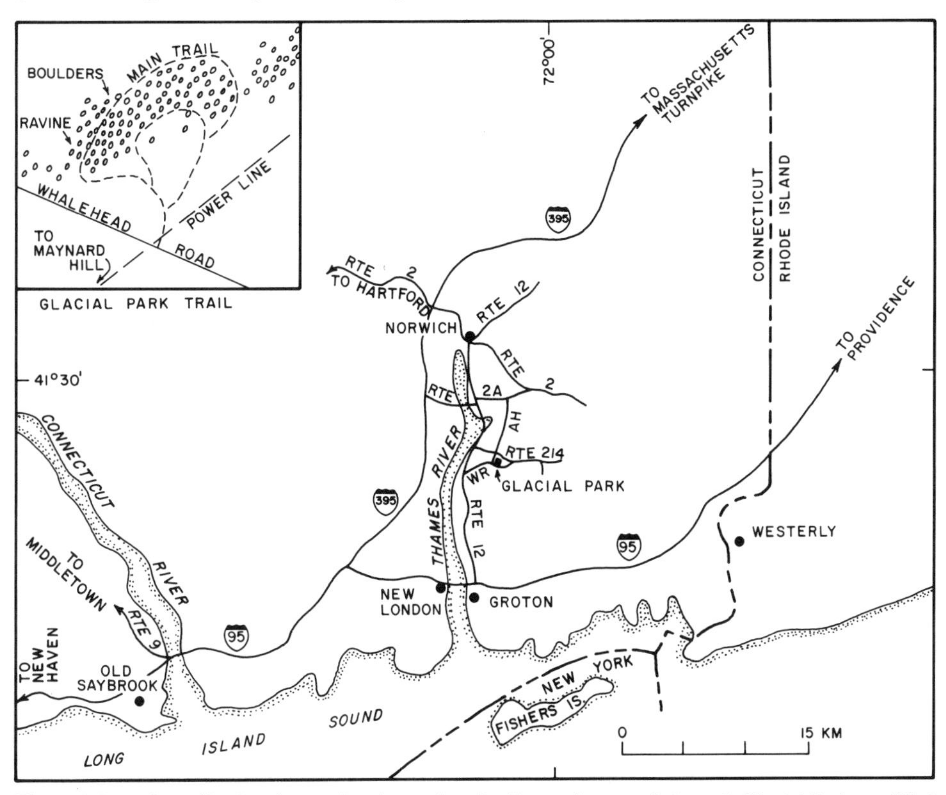

Figure 1. Location of Ledyard recessional moraine site. Insert shows trail through Glacial Park modified from Maire (1976). WR, Whalehead Road; AH, Avery Hill Road.

LOCATION

The exposure of the Ledyard recessional moraine is located on Whalehead Road off Connecticut 12 on the east side of the Thames River between Norwich and New London, Connecticut (Figs. 1, 5). From the south, turn northeast (right) off Connecticut 12 onto Whalehead Road (0.1 mi or 0.5 km north of the shopping center at Gales Ferry) and follow Whalehead Road 1.8 mi (3 km) to a power line and the sign on the left for "Glacial Park." From the north, turn east (left) off Connecticut 12 onto Stoddards Wharf Road (Connecticut 214), and travel 1.1 mi (1.8 km) to Avery Hill Road. Turn south (right) and travel 0.6 mi (1.0 km) to the intersection with Whalehead Road. Turn east (left) onto Whalehead Road and travel about 0.5 mi (0.75 km) to Glacial Park. The areas of interest are in three locations: Glacial Park, the northeast side of Maynard Hill, and the kettled area on either side of Avery Hill road near Stoddard Wharf Road. The site is located in the Uncasville 7½-minute Quadrangle.

SIGNIFICANCE

The site illustrates many of the features characteristic of small recessional moraines in southern New England that mark a succession of stillstands of the late Wisconsin Laurentide ice sheet as it receded from its terminal position on Long Island, Block Island, Martha's Vineyard, and Nantucket Island (Fig. 2; Schafer and Hartshorn, 1965). The identification of these recessional moraines in the early 1960s showed that, in the early stages of ice retreat, a systematic retreat of an active ice margin existed, and that R. F. Flint's early (1930) hypothesis of regional stagnation of ice in New England was not applicable, at least in the coastal area of southern New England. Identification of the moraines was aided by the high concentration of boulders in some segments. Although the mechanism for forming boulder concentrations, such as that at Glacial Park, is not well understood, a contributing factor is an abundant source of widely jointed granitic rock.

GENERAL SETTING

The recessional moraines of southeastern Connecticut and adjacent Rhode Island occupy a belt about 12 mi (20 km) wide north of the terminal moraines (Figs. 2 and 3; Flint and Gebert, 1976; Goldsmith, 1982). The recessional moraines are characterized by an alignment of morainic segments approximately perpendicular to the direction of ice movement as indicated by striations and grooves on bedrock surfaces (Fig. 3). Most of the

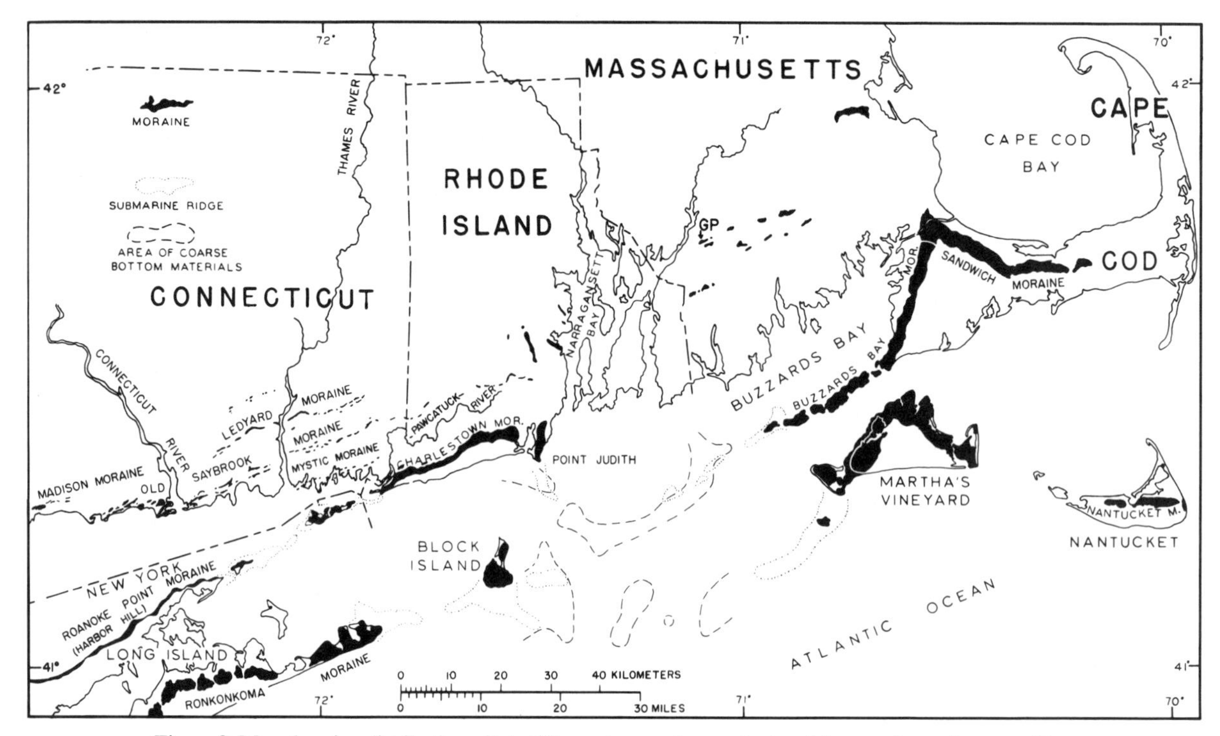

Figure 2. Map showing distribution of late Wisconsin moraines and related features in southeastern New England. GP, Glacial Park. (Modified from Goldsmith, 1982; copyright 1982 by Kendall/Hunt Publishing Company, reprinted by permission.)

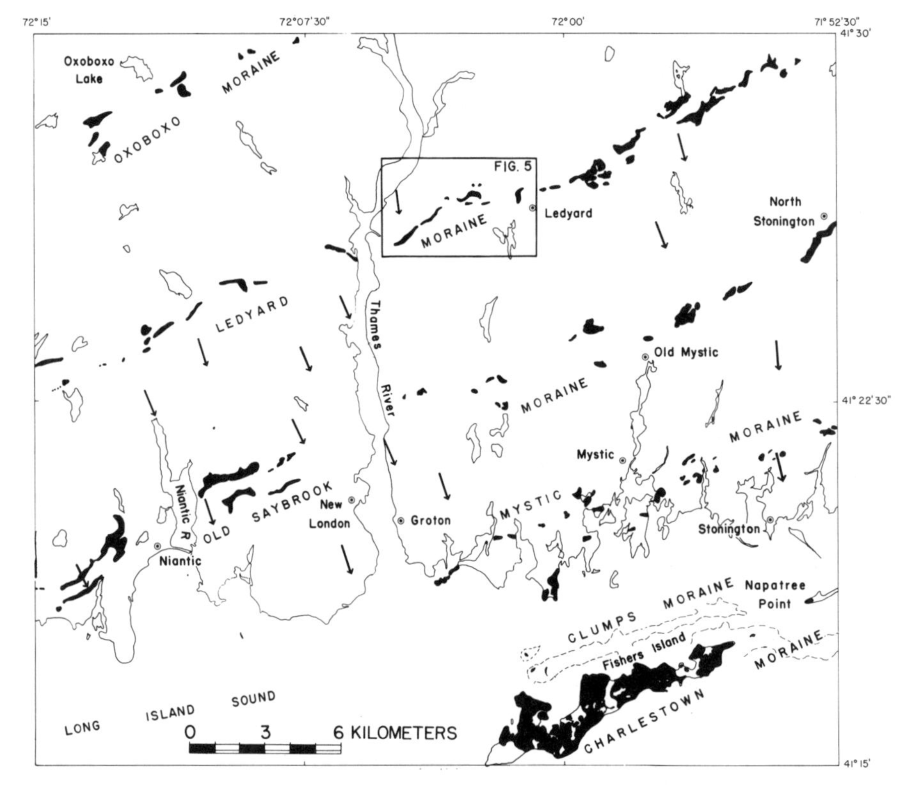

Figure 3. Distribution of morainal segments (black) in the New London area, Connecticut and New York, and direction of ice movement (arrows) as indicated by striations and grooves on bedrock surfaces. (Modified from Goldsmith, 1982; copyright 1982 by Kendall/Hunt Publishing Company, reprinted by permission.)

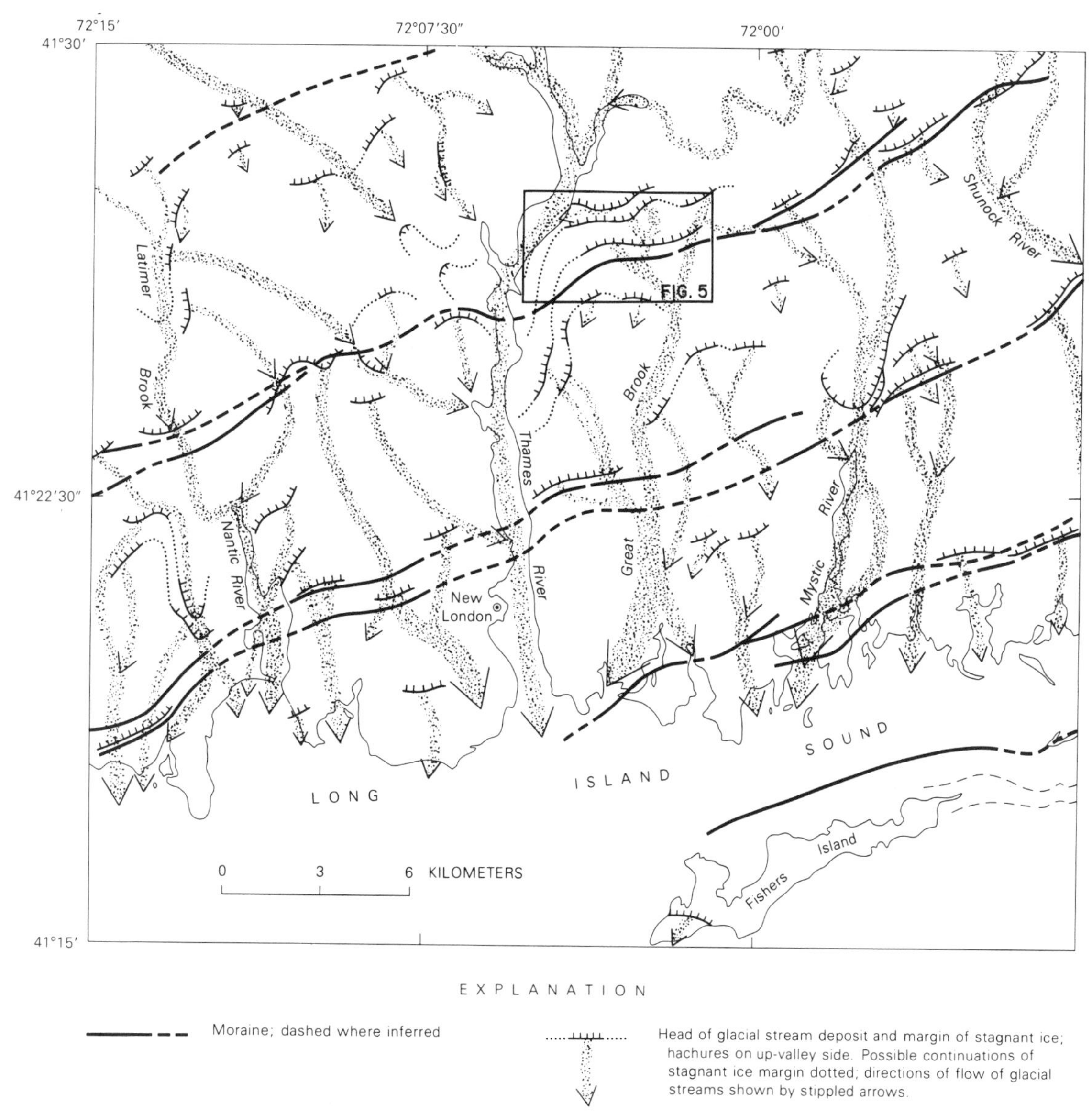

Figure 4. Distribution of heads of glacial stream deposits and their relation to recessional moraines in the New London area, Connecticut and New York. (Modified from Goldsmith, 1982; copyright 1982 by Kendall/Hunt Publishing Company, and reprinted by permission.)

segments are about 300 ft (100 m) wide and 6 to 20 ft (2 to 6 m) high and consist of sandy, bouldery, generally loose-textured glacial till, with or without beds and lenses of well to poorly sorted stratified drift. Exposures of the material are rare. Surface form may be smoothly rounded or undulating and more bouldery than adjacent smooth-surfaced till, or it may be irregular and hummocky and in places contain closed depressions. In some places, as at Glacial Park, the surface may consist almost entirely of boulders without interstitial fine material. The moraines are little affected by topography except for some inflection in large valleys such as that of the Thames River. This lack of influence of topography is attributed to the thickness of ice that must have exceeded the local topographic relief of 250 ft (75 m) when the moraines were formed over the tops of the hills at the active ice front. Contributory to the relationship between ice thickness, lines of parallel glacier flow, and a linear ice margin unaffected by topography is the fact that the seaward gradient of hilltops in eastern Connecticut north of the morainal belt is about 15 ft/mile (2.8 m/km). Within the morainal belt, the slope is about 25 ft/ mile (4.7 m/km). In places, the moraines are double or paired (Fig. 3). Deposits of glacial streams derived from melting ice overlap the moraines (Fig. 4). Ice-contact forms are found in glacial stream deposits located near the moraines. These mark the heads of streams that were derived from belts of ice lying at and north of the moraines after the active ice front had retreated northward to new stillstands or ephemeral positions. In eastern Connecticut,

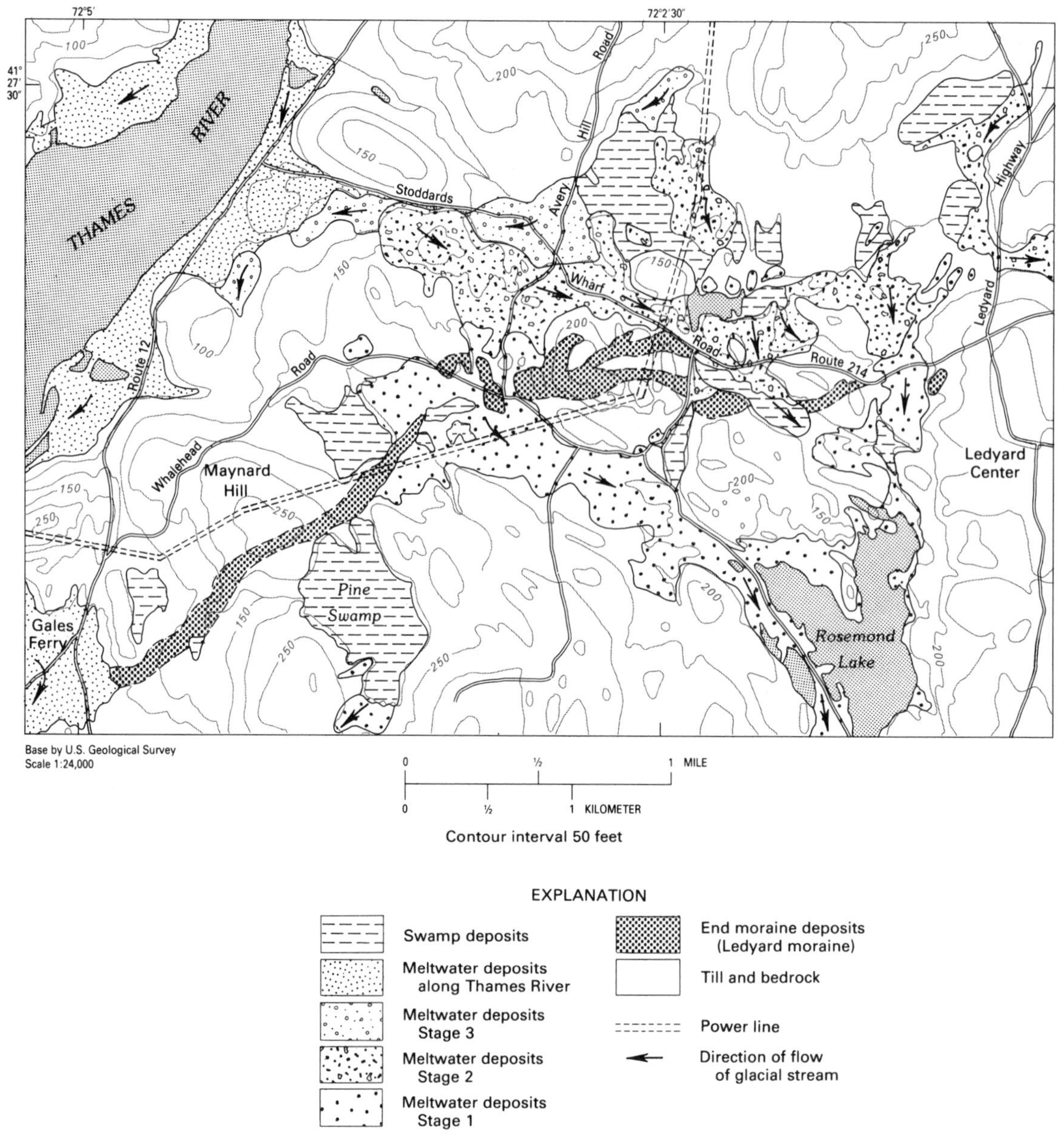

Figure 5. Ledyard moraine and glacial stream deposits in the Gales Ferry–Ledyard Center area. (Modified from Goldsmith, 1960.)

north of the morainal belt, recessional moraines are rare, probably because of an accelerating rate of ice retreat. The ice retreat in eastern Connecticut is recorded by a succession of deltaic deposits in valleys in which the drainage to the south was temporarily blocked by earlier deposits and isolated remnants of ice. These deposits formed near and in the temporary position of the fringe of stagnant ice marginal to the ice sheet as the ice retreated northward.

SITE INFORMATION

Stop 1, Glacial Park. A trail prepared by the Ledyard Conservation Commission (Maire, 1976) leads from Whalehead Road at the power line to an unusual concentration of boulders, first noted by Wells (1890), that forms this part of the Ledyard moraine (inset, Fig. 1; Fig. 5). The main trail proceeds clockwise from the south flank of the boulder concentration across a ravine, turns around a bedrock ledge and then back across the ravine, and traverses over the boulder concentration as it rises up the hill to the northeast. Near the top of the hill, the trail follows the north edge of the boulder concentration, recrosses the moraine (which here has a less bouldery aspect), and returns out along the south side of the moraine.

The unusual concentration of boulders without interstitial

COUNT OF BOULDERS IN MORAINE AT GLACIAL PARK, LEDYARD

EXPLANATION	Number of boulder counts	Percent	Long dimension in meters
Alaskite gneiss; Biotite granite gneiss	267[1]	90.5	1–6
Other rocks[2]	28	9.5	0.4–2
Sillimanite gneiss and schist	0[3]	0	—

[1]These rocks are distinguished from each other primarily by biotite content. An estimated 90% of this number is of alaskite gneiss

[2]Biotite gneiss, quartzite, calc-quartzite and schist, amphibolite and pegmatite

[3]One boulder was observed previously

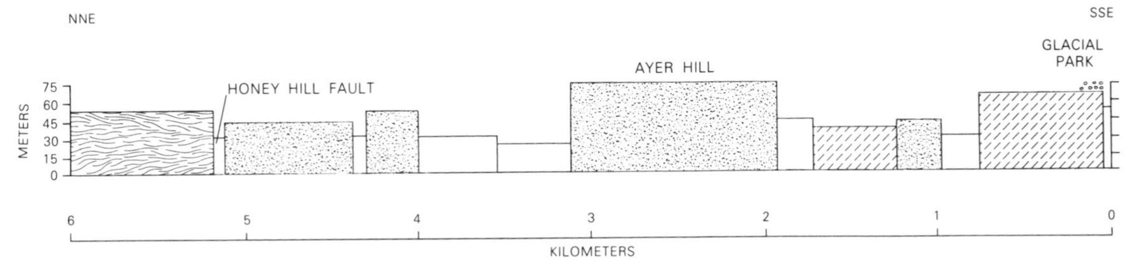

Figure 6. Source of boulders in boulder concentrations in the Ledyard moraine at Glacial Park, Ledyard, Connecticut. Section shows widths of outcrops of rock units along a line extending N10°W from Glacial Park, and generalized altitudes along and near the line. Rock units actually dip about 50°N. (Modified from Goldsmith, 1982; copyright 1982 by Kendall/Hunt Publishing Company, and reprinted by permission.)

fine material is attributed to an abundance of granitic source material accumulated by the "dirt machine" (Koteff, 1974) and to sorting of this material by slumping that was contemporaneous with winnowing by meltwater, possibly in a crevasse. The presence of ledges of bedrock beneath the moraine at the side of the ravine and to the east near the power line indicates that till is relatively thin in the area. The boulders at Glacial Park and elsewhere in the Ledyard moraine are derived primarily from rock types that have widely spaced joints and poor or no fissility. Most are derived from ledges within 0.6 to 1.2 mi (1 or 2 km) to the north and generally within less than 3 mi (5 km) (Fig. 6; Goldsmith, 1967), but a few have been transported greater distances. East-northeast of Glacial Park the moraine continues as local areas of boulder concentrations and as areas of bouldery till that appear to form a double moraine. The distribution of the moraine in this area has not been mapped in detail. Immediately west of Glacial Park, the moraine appears to be dispersed. Here, many large erratics are scattered on an irregular and largely bedrock-controlled surface. The moraine, however, emerges as a discrete feature from beneath stratified glacial deposits northeast of Maynard Hill (area 2, Fig. 5).

Stop 2, Maynard Hill. A walk of about 0.5 mi (0.8 km) west-southwest along the power line from the entrance to Glacial Park lies across the surface of the earliest and highest glacial stream deposit (stage 1, Fig. 5) at elevations of 190 to 180 ft (58 to 55 m) (see Uncasville Quadrangle). This deposit extends just north of the Ledyard moraine and was laid down soon after the ice front retreated north of the line of the moraine. Meltwater drainage was to the southeast and south through valleys in the upland because the Thames Valley at this time was filled with ice (Goldsmith, 1960). Beyond artificially ponded areas where the sand and gravel and swamp deposits have been removed, the power line crosses a ridge of bouldery till that is the Ledyard moraine. A ditch dug across the moraine (about 0.5 mi or 0.8 km from Glacial Park and to the north of the power line) affords an opportunity to examine the poorly sorted material in it. The moraine continues southwest as a characteristic ridge of bouldery till up the northeast side of Maynard Hill. This low ridge can be traced over the top of the hill just south of its crest down to and beneath one of the terraces along the Thames River. The moraine reappears west of the Thames River just south of Uncasville as a belt of bouldery hummocky till that trends west-southwest. The change in trend indicates a slight lobation of the active ice front down the Thames River valley (Figs. 3, 4). Exposures of the moraine also can be seen just south of the top of Maynard Hill by driving west from Glacial Park about 1.2 mi (2 km) on Whalehead Road to Friar Tuck Drive and then south up the hill. Cuts are located at the junction of Robin Hood and Maid Marian Drives, but permission should be obtained from property owners in the area before digging into the banks.

Stop 3, Kettled topography on Avery Hill Road. Drive or walk back to Avery Hill Road from Glacial Park and proceed north toward Stoddard Wharf Road (Connecticut 214). Very shortly the road drops from the 190- to 180-ft (58- to 55-m) surface (stage 1 glacial stream deposit) to the next younger surface (stage 2) at 180 to 170 (55 to 52 m). The glacial streams forming this surface were blocked by the earlier deposit from flowing directly southeast and at first flowed east before passing through gaps in the bedrock ridges to entrench themselves slightly downvalley into the deposits of stage 1. This surface is deeply kettled both east and west of Avery Hill Road, indicating deposition over masses of stagnant ice that remained just north of the moraine as the front of active ice receded to the north. Patches of still younger deposits (stage 3) that formed a surface at 130 ft (40 m) are present near the intersection of Avery Hill Road and Stoddard Wharf Road. The path for this meltwater drainage, however, was west to the Thames valley because the level of the ice in the Thames valley had lowered sufficiently to provide an outlet in this direction. Low terraces along the Thames River were deposited by a succession of still later glacial streams when ice remnants were at a low level in the Thames River valley.

REFERENCES CITED

Flint, R. F., 1930, The glacial geology of Connecticut: Connecticut Geological and Natural History Survey Bulletin, no. 47, 294 p.

Flint, R. F., and Gebert, J. A., 1976, Latest Laurentide ice sheet; New evidence from southern New England: Geological Society of America Bulletin, v. 87, p. 182–188.

Goldsmith, R., 1960, Surficial geology map of the Uncasville Quadrangle, Connecticut: U.S. Geological Survey Geologic Quadrangle Map GQ-138.

——, 1967, Bedrock geologic map of the Uncasville Quadrangle, New London County, Connecticut: U.S. Geological Survey Geologic Quadrangle Map GQ-576.

——, 1982, Recessional moraines and ice retreat in southeastern Connecticut, *in* Larson, G. J., and Stone, B. D., eds., Late Wisconsinan glaciation of New England: Dubuque, Iowa, Kendall/Hunt Publishing Company, p. 61–76.

Koteff, C., 1974, The morphologic sequence concept and deglaciation of southern New England, *in* Coates, D. R., ed., Glacial Geomorphology: Binghamton, State University of New York Publications in Geomorphology, p. 121–144.

Maire, B. L., 1976, Ledyard Glacial Park: Ledyard, Connecticut, Ledyard Conservation Commission, 19 p.

Schafer, J. P., and Hartshorn, J. H., 1965, The Quaternary of New England, *in* Wright, H. E., Jr., and Frey, D. G., eds., The Quaternary of the United States: Princeton, New Jersey, Princeton University Press, p. 113–128.

Wells, D. A., 1890, Evidences of glacial action in southeastern Connecticut: Popular Science Monthly, v. 37, p. 691–703.

Geological Society of America Centennial Field Guide—Northeastern Section, 1987

Geologic relationships of Permian Narragansett Pier and Westerly Granites and Jurassic lamprophyric dike rocks, Westerly, Rhode Island

O. Don Hermes, *Department of Geology, University of Rhode Island, Kingston, Rhode Island 02881*

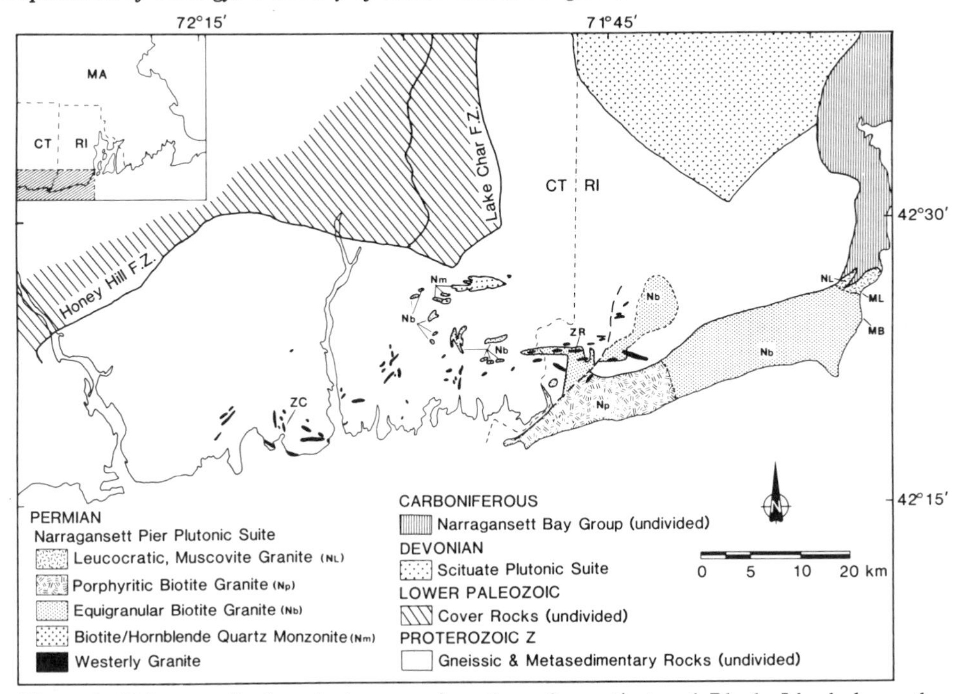

Figure 1. This generalized geologic map of southern Connecticut and Rhode Island shows the distribution of Westerly and Narragansett Pier Granites (after Zartman and Hermes, 1986). The Boston Platform is the terrane south and east of the Honey Hill-Lake Char fault system. ML and MB = monazite localities from leucocratic and biotite-bearing facies of Narragansett Pier Granite; ZC = Graniteville, CT zircon locality; ZR = Westerly, RI zircon locality. Geology after Quinn (1971), Goldsmith (1985), Hermes and others (1981), and Rodgers (1985).

LOCATION

The geologic features described at these sites are located in southwestern Rhode Island and adjacent Connecticut (Fig. 1) within the Ashaway Quadrangle (Feininger, 1965). All stops are along Rhode Island/Connecticut 78, or within short walking distances from the highway (Fig. 2). Vehicles may be parked at each stop by driving well off the shoulder of the road from either the east- or west-bound lanes. **Because of possible heavy traffic, cross the highway on foot with extreme caution.**

Stop 1. Outcrops on both the north and south sides of Rhode Island 78, 0.1 mi (0.16 km) east of the intersection with Rhode Island 3 (Fig. 2). Coarse-grained Narragansett Pier Granite is intrusive into Late Proterozoic amphibolitic gneisses and schists. The Narragansett Pier Granite is cut by dikes of fine-grained Westerly Granite, and both granites are intruded by dikes of lamprophyre that contain xenoliths and megacrysts of mantle derivation.

Stop 2. Abandoned granite quarries on the hills 0.1–0.4 mi (0.16-0.64 km) southwest of Stop 1 (Figs. 2 and 3). An abandoned dirt road at the northwest end of the roadcut at Stop 1 (just south of Rhode Island 3) provides access to most of these quarries. Sites at Stop 2 include typical occurrences of equigranular Narragansett Pier Granite cut by dikes of Westerly Granite, with locally abundant inclusions of xenolithic blocks and screens of amphibolite and granite gneiss contained in both granites.

Stop 3. Outcrop on the north side of Rhode Island 78, 0.3 mi (0.48 km) west of the intersection with Rhode Island 3 (Fig. 2). Major features to be observed include polydeformed amphibolite containing deformed aplites and pegmatites, which in turn are cut by dikes and veins of undeformed Narragansett Pier Granite, Westerly Granite, and related pegmatites.

Stop 4. Cloverleaf outcrops at the intersection of Rhode Island 78 and 2 in Connecticut, just west of the Rhode Island State line (Fig. 2). This site contains polydeformed amphibolite and granite gneiss, and massive undeformed granite dikes and

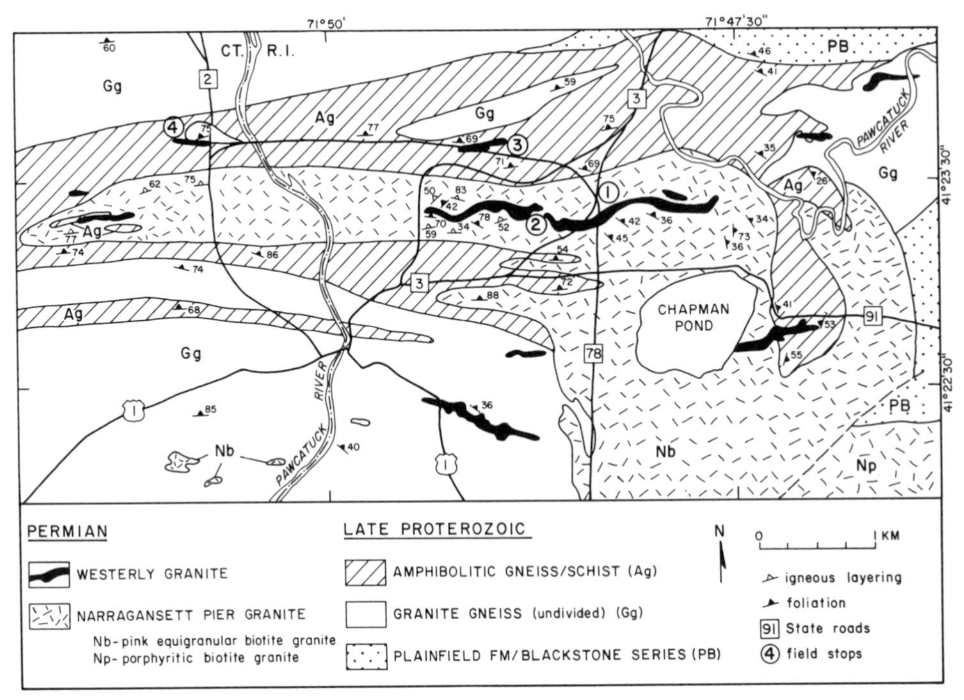

Figure 2. Geology of the Westerly, RI area showing field stop locations described in this paper (geology after Feininger, 1965; Kern, 1979).

veins. Also present are good exposures of Westerly Granite dikes that exhibit flow layering at high angles to fabric in the older rocks.

SIGNIFICANCE

The comagmatic Narragansett Pier and Westerly Granite plutonic suite, which is of Permian age, represent the youngest Paleozoic igneous activity in the New England Appalachians. Only the pegmatites at Middletown, Connecticut (Zartman and others, 1965) and the Milford Granite of southern New Hampshire and adjacent Massachusetts appear to be of comparable age (Aleinikoff and others, 1979). Intrusion of the Narragansett Pier Granite postdates D1 and D2 events that have deformed Carboniferous rocks of the Narragansett Basin and adjacent basement rocks. As such these rocks best exhibit the characteristics of Alleghanian igneous activity in southeastern New England.

Radiometric studies document a primary Permian igneous age of 275 Ma for Narragansett Pier Granite and Westerly Granite (Zartman and Hermes, 1984, 1986). An unexpected but significant finding of the isotopic study was the demonstration that magmas for both Narragansett Pier Granite and Westerly Granite were derived from crustal source material that contained an Archean (2,700 Ma) component. Such old crustal source material has not been recognized previously in the Appalachians. In particular, similar studies on the pre-Permian rocks in southeastern New England show no evidence of very old crust in the source region, and nowhere in New England do igneous plutons show components of inheritance older than the latest Proterozoic (Aleinikoff, 1979). The presence of such an old inherited Archean signature only in the post-Carboniferous igneous rocks of SE New England is interpreted to reflect underplating of this part of Avalon by material derived from northwest Africa during the closing of the Iapetus Ocean (Zartman and Hermes, 1986). As such this is the first *direct* evidence of an African connection with southeastern New England in the Late Paleozoic. The meta- to peraluminous nature of the Narragansett Plutonic Suite reflects a marked difference in composition compared to preceeding Paleozoic alkalic magmatism in this part of the Avalon Zone (Hermes and Zartman, 1985; Hermes and Murray, 1986), which also reflects a change in the crustal source material prior to the melting event that generated the magmas to produce these Late Paleozoic rocks (Zartman and Hermes, 1984, 1986).

Also unique to this area is the presence of monchiquite dikes (175 Ma) that were emplaced in the Jurassic. These dikes contain a diverse suite of chromium spinel-bearing lherzolites and other ultramafic rocks of mantle derivation. This locality was the first reported occurrence in New England, and remains unique.

SITE INFORMATION

Geologic Setting. Rocks of the Narragansett Pier Granite and Westerly Granite constitute a composite plutonic suite of batholithic size that extends eastward from southeastern Connecticut across southern Rhode Island to the eastern boundary of Narragansett Bay (Fig. 1). The batholith lies to the south and east of the Honey Hill-Lake Char fault system within the Boston Platform or Avalon Zone of southern New England. Rocks of the batholith are intrusive into a variety of Late Proterozoic gneisses, schists, and amphibolites, as well as into Carboniferous rocks of the Narragansett Basin (Fig. 1). Locally, the massive rocks of the pluton both truncate and parallel the fabrics in the older rocks.

Although the Narragansett Pier Granite pluton is mineralogically and chemically considerably homogeneous, subtle textural, mineralogical, and color variations have permitted four distinct and mappable facies to be recognized (Fig. 1): 1) pink, medium- to coarse-grained, equigranular, biotite granite, 2) pink, coarse-grained, porphyritic, biotite granite, 3) white, medium- to coarse-grained, equigranular to porphyritic, leucocratic, muscovite-garnet granite (Kern, 1979; Kocis, 1981), and 4) more mafic, hornblende, quartz monzonite. The pink granite contains perthitic microcline, plagioclase, quartz, and biotite with accessory muscovite, magnetite, ilmenite, apatite, monazite, sphene, zircon, allanite, pyrite, and chlorite. The pink color is due to oxidized iron along grain boundaries and fractures and included iron within feldspars. Grain size ranges from medium to coarse. The porphyritic facies generally is similar except that it contains phenocrysts of euhedral-subhedral feldspars up to 1.2 in (3 cm) in a medium-grained groundmass. The white granite, which occurs only in the eastern part of the pluton (Fig. 1), contains accessory euhedral primary muscovite and garnet. Some varieties of the white facies are extremely leucocratic and contain no biotite or opaque minerals. Sparse occurrences of the hornblende quartz monzonite, which contains a modestly developed foliation (but weaker fabric than the adjacent granite gneisses; e.g., Goldsmith, 1985), outcrop in southern Connecticut (Fig. 1).

Rocks of the pluton mainly are granite, with minor occurrences of granodiorite and quartz monzonite (Hermes and others, 1981). Geochemically, $K_2O > Na_2O$ and the rocks, which are meta- to peraluminous, contain 1–3% corundum in the norm. The overall range of chemistry exhibited is quite small and no significant difference between equigranular and porphpyritic pink varieties is apparent. The leucocratic granite is lower in CaO, MgO, TiO_2, and total Fe. The batholith is cut by simple to complex veins of aplite, pegmatite, and quartz. In addition, dikes of fine- to medium-grained, aplitic Westerly Granite (generally thought to be late-stage, but comagmatic) locally cut the granite and the older country rock, and are especially prominent in the western part of the pluton. Similarity of Narragansett Pier Granite and Westerly Granite in mineralogical and geochemical composition (Buma and others, 1971) is consistent with their probable comagmatic nature. Younger, and petrogenetically unrelated, lamprophyric dikes of monchiquite that contain mantle-derived lherzolite nodules and megacrysts locally cut both the Narragansett Pier Granite and Westerly Granite.

To the east the Narragansett Pier Granite clearly is intrusive into Carboniferous metasedimentary rocks of the Narragansett Basin (Hermes and others, 1981; Murray, this volume). In one case plant fossils of Stephanian B age have been recovered from a xenolithic carbonaceous-rich layer that is enclosed in the granite (Brown and others, 1978). U/Pb ages on monazite from two nearby samples of granite (Fig. 1) yield ages of 276 ±2 Ma (Kocis and others, 1978; Kocis, 1981). Zircon from Narragansett Pier Granite near Westerly exhibits mild inheritance, but is interpreted to yield an igneous age concordant with the monazite ages from the eastern part of the pluton. The inherited component in zircon from Westerly Granite is more prominent than in Narragansett Pier Granite and defines a chord on concordia that yields a lower intercept of ~275 Ma, but an upper intercept of 2,700 Ma, thus demonstrating an Archean crustal component in the magma source region (Zartman and Hermes, 1984, 1986).

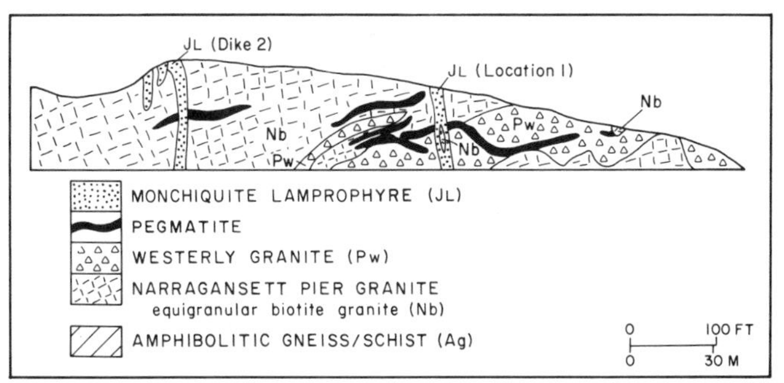

Figure 3. This generalized cross-section of the Stop 1 road cut (west side) shows Narragansett Pier Granite intrusive into older amphibolitic gneisses. Westerly Granite cuts the Narragansett Pier Granite and both granites are cut by lamprophyre dikes that contain lherzolite nodules.

Contacts between different facies of the Narragansett Pier Granite are nowhere exposed, and it is unclear whether they are gradational, intrusive, or faulted. On the other hand, contacts with country rock are exposed in a number of places; in nearly all cases they exhibit a complex lit-par-lit relationship, generally concordant with the foliation and/or bedding of the intruded rock. In many places cross-cutting dikes connect the lit-par-lit sills to form a patchwork of enclosed xenoliths. The simplest exposed contact is in Westerly where the Narragansett Pier Granite truncates the foliation of intruded amphibolite at a shallow angle, but even here, concordant veins of granite, aplite, and pegmatite intrude the amphibolite a few tens of meters to the north. Xenoliths in the granite are common near the contact zone and become less prolific southward within the main pluton. Xenolith type correlates with the adjacent country rock, and in no cases have restite-type xenoliths been observed. For example, xenoliths at Westerly are exclusively varieties of amphibolite and biotite schist that contain mineral assemblages and modal proportions quite similar to the adjacent country rock (Kern, 1979). A short distance to the east, granite gneisses form most of the xenoliths as well as the country rock. The long dimensions of xenoliths are parallel to foliation; moreover, the foliations exhibit similar orientation to foliations in the country rock. On a local scale, both Kern (1979) and Smith and Barosh (1981) have demonstrated that foliation in xenoliths wraps around, following the contact and the foliation in the country rock NE of Westerly (Fig. 2). Xenoliths in the leucocratic granite facies near Narragansett are schists, sandstones, and conglomerates identical to rocks of the intruded Carboniferous

Rhode Island Formation (Hermes and others, 1981; Kocis, 1981) and similarly exhibit foliation conformable to that in the country rock away from the immediate contact.

This consistent orientation of xenoliths and their structures, plus the remarkable alignment of the structural fabric between the blocks of country rock, testifies to the rather passive nature of the intrusion, in which only a few blocks were stoped and rotated. The country rock was separated along bedding and foliation planes to permit the lit-par-lit emplacement as a result of this gentle mode of emplacement. The resultant screens of included rock represent a crude ghost stratigraphy that locally reveals the pre-Narragansett Pier Granite structure.

All granite facies are structurally massive and nonfoliated, except for local areas where flow foliation has developed. Flow foliation and layering is locally prominent, especially in some of the quarries at Westerly. The layering is caused by concentration of biotite that exhibits crude cross-bedding and cut-and-fill structures.

In addition to containing key geologic relationships regarding Late Paleozoic igneous activity in New England, the presence of Narragansett Pier Granite and Westerly Granite permitted Westerly, Rhode Island, to become a major center for the granite industry from 1860–1915 (Macomber, 1958; Macaulay, 1985). Quarrying activity peaked after the War Between the States, when demand was high for Civil War monuments and elaborate tombstones. The attractive appearance and fine, even texture of Westerly Granite, as well as an available work force of skilled quarrymen, stone cutters, and carvers, made it one of the more popular building and monument stones. Also of historical interest is the fact that the relatively fine grain size and homogeneous character of Westerly Granite has resulted in its widespread use as a geochemical standard (Goldich and Oslund, 1956), as well as material for a variety of experimental studies, including statistical modal variation studies (Chayes, 1950, 1952) and rock deformational experiments (Brace and others, 1965, 1971; Stesky, 1978).

Stop 1. The highway outcrops show the intrusive contact of Narragansett Pier Granite into amphibolitic country rock. Also exhibited are two dikes of Westerly Granite that cut the Narragansett Pier Granite and two younger dikes of lamprophyre that cut both the Narragansett Pier Granite and Westerly Granite. General field relationships exposed in the outcrop are illustrated in Figure 3.

This is the only known place where Narragansett Pier Granite exhibits a simple intrusive relationship with the country rocks, as opposed to its more general lit-par-lit mode of emplacement elsewhere. The contact with biotite amphibolitic schist at the north end of the outcrop, which strikes N60°E and dips 60°N, is slightly discordant with the foliation of the schist (E-W, 65°N). The granite is equigranular and coarse grained next to the amphibolite. The amphibolite is rich in biotite and muscovite, but locally has hornblende-rich layers and less-common garnet. Small aplites and pegmatites (up to 3 ft; 1 m wide) extend from the granite into the amphibolite where they locally cut foliation.

The two dikes of fine-grained Westerly Granite that cut the Narragansett Pier Granite strike E–W. The smaller dike (approximately 3 ft; 1 m) exhibits a shallower dip upward in the outcrop, where it eventually terminates (Fig. 3). Upward terminating dikes of Westerly Granite also have been described in some of the nearby quarries at Stop 2 (Feininger, 1967). The lower contact of the larger dike of Westerly Granite (Fig. 3) is somewhat undulatory, but subparallel to the upper contact, which dips 15–20°S. The contacts against Narragansett Pier Granite lack chilled margins and commonly are characterized by discontinuous zones of pegmatite; in some cases these pegmatites follow the contact for a short distance and then abruptly cut into Narragansett Pier Granite or Westerly Granite. The coarse-grained pegmatites contain the same mineral assemblage as both Narragansett Pier Granite and Westerly Granite and appear to be an integral component of the petrogenetically related plutonic suite. The variable textures exhibited by the suite, as well as the absence of chilled contacts throughout the pluton, suggest that fluid pressures were highly variable during different stages of crystallization. Coarser textures formed during periods of high fluid pressure, whereas the finer-grained, aplitic rocks crystallized under relatively drier conditions. The general sequence of coarse granite to aplite (Westerly Granite) to pegmatite reflects oscillation in fluid pressure, which also is indicated by the smaller-scale complexly zoned pegmatites/aplites so common in the eastern part of the batholith near Narragansett (Kocis, 1981). Locations of Narragansett Pier Granite and Westerly Granite samples, which were collected for zircon isotopic analysis and yielded the Archean inherited signatures, are shown in Figures 1 and 3.

Two nearly vertical lamprophyre dikes (exposed widths up to 3 ft; 1 m) that strike N25°E cut both the Narragansett Pier Granite and Westerly Granite (Fig. 3). The dikes are monchiquite (Leavy and Hermes, 1977, 1978; Leavy, 1980) and are characterized by microphenocrysts of Ti-augite, kaersutite, olivine, phlogopite, titanomagnetite, and apatite enclosed in a matrix of analcite and calcite. K-Ar measurements on phlogopite from dike 2 (Fig. 3) has yielded a radiometric age of 175 Ma (Hermes and others, 1984). Primary flow fabric parallel to contacts is common in the dike rocks and ocellular textures are present that may indicate liquid immiscibility. Along with water, CO_2 was prominent in the fluid phase as indicated by the prominence of calcite in the matrix as well as its partial replacement of some of the primary silicate phases. At several places throughout the outcrop, closely spaced joints parallel to the lamprophyre dikes exhibit narrow stringers (several mm wide) of calcite. In addition, granite contacts within one meter of the lamprophyres are quite disaggregated and contain calcite along grain contacts that were derived during lamprophyre emplacement. At location 1 (Fig. 3), the dike bifurcates around an enclosed screen of Narragansett Pier Granite that has been severely disaggregated by the lamprophyre emplacement. These observations indicate that emplacement of the lamprophpyre and the accompanying H_2O and CO_2 fluids were rapid and forceful, quite different from the more passive emplacement exhibited by the distinct tholeiitic and mildly alkalic suites of Triassic-Jurassic dolerites so common in

New England (McHone, 1978; Ross, 1981, 1985; Hermes and others, 1984).

The monchiquite dikes contain a variety of megacrysts (olivine, varieties of low-Ti clinopyroxene, and ilmenites) and xenolithic nodules that include chrome spinel-bearing lherzolite, harzburgite, and wehrlite. Until the findings of McHone (1978) and Ross (1981, 1983), the dikes at Westerly were the only dikes known in New England to contain mantle-derived material. Based on the mineral assemblages and pyroxene chemistry, Leavy and Hermes (1977) and Leavy (1979) estimate that these nodules equilibrated at temperatures of 950–1200°C in a pressure regime between 9–25 Kb, clearly within the mantle. Although the dikes as now exposed contain only small nodules and abundant megacrysts, excellent quality material containing nodules up to 4 in (10 cm) was collected in 1975 during construction of the roadbed. The dikes were observed to widen downward, where they contained blocks with as much as 90% nodules and megacrysts immersed in a 10% matrix of monchiquite host. Apparently the upward narrower conduits permitted only smaller and sparser amounts of the xenolithic material to be incorporated in the presently exposed parts of the dikes. A substantial volume of nodules and dike rock was collected from the road base during construction and is stored in the URI collection. We will make samples available to anyone wishing to perform additional studies on high-quality material.

Stop 2. Inspection of the abandoned quarries in the hills to the west of Stop 1 shows a number of intrusive features characteristic of Narragansett Pier Granite and Westerly Granite. Clearly exhibited are: 1) apophyses of Westerly Granite in the Narragansett Pier Granite; 2) stoped blocks of Narragansett Pier Granite in the Westerly Granite; 3) stoped blocks of foliated amphibolite and granite gneiss in the Narragansett Pier Granite and Westerly Granite; 4) localized flow foliation in Westerly Granite and Narragansett Pier Granite; 5) upward terminating dikes of Westerly Granite; and 6) several sets of late-stage pegmatites.

Kern (1979) has shown that the fabric in many xenolithic blocks enclosed in Narragansett Pier Granite and Westerly Granite are conformable to those in the nearby country rock, indicating that most xenoliths represent unrotated screens that remained rather static while they were passively intruded by the younger granites. Although detailed studies of the quarries have not been published, a supplementary guide can be found in Feininger (1963).

Stop 3. Exposed in the long roadcut at Stop 3 are several varieties of amphibolitic gneiss and schist which are cut by foliated veins and dikes of granite, aplite, and pegmatite. Argon release patterns on hornblende and biotite from rocks similar to these amphibolites yield ages of 240–257 Ma (Dallmeyer, 1982), indicating that some of the fabric may be of Alleghanian age. However, relationships at Stop 4 indicate that the gneisses and amphibolites are polydeformed and the oldest fabric is likely to be of Late Proterozoic age. Clearly cutting these fabrics (but locally conformable) are unfoliated veins of massive granite, pegmatite, and aplite. These rocks are similar to those of the Narragansett Pier Granite and Westerly Granite seen at Stops 1 and 2, and demonstrate that their emplacement clearly occurred after peak deformational events during the Alleghanian orogeny. At the west end of the outcrop, a nearly horizontal sheet of Westerly Granite truncates fabrics in the country rock; this dike of Westerly Granite exhibits closely spaced, flat-lying joints quite anomalous compared to most occurrences of Westerly Granite.

Stop 4. These cloverleaf outcrops offer excellent three-dimensional exposures of the amphibolitic and granite gneiss units that are cut by dikes of Westerly Granite. The polydeformed schists, gneisses, and older granitic veins contrast with the more massive Westerly Granite. An older fabric in some of the amphibolitic units is clearly cut by some of the granite gneisses and a younger pervasive fabric is superimposed on both units. The granite gneiss probably is correlative with the nearby Potter Hill Granite Gneiss, whereas the amphibolite is likely to be a member of the Waterford Group so prominent in southern Connecticut (Rodgers, 1985). If both the granite gneiss and the amphibolite do correlate with such rocks of Late Proterozoic age, the structural features offer excellent evidence of a Late Proterozoic deformational event. The argon release data of Dallmeyer (1982) imply that these rocks underwent a subsequent Late Paleozoic heating and deformational event. These older fabrics are clearly cut by a number of massive, aplitic dikes of Westerly Granite. Flow foliation in some dikes of Westerly Granite (especially visible in the west walls of the roadcut) are locally conformable with the older fabrics, but can be observed to wrap around and cut these features where the dikes of Westerly Granite truncate the schistosity. The southern outcrop exposes a thick dike of Westerly Granite that contains xenoliths of amphibolite and clearly cuts the fabric of the country rock. Rocks of Stops 3 and 4 presently are being studied in detail by James Hutton as part of a master's thesis at the University of Rhode Island.

REFERENCES

Aleinikoff, J. N., Zartman, R. E., and Lyons, J. B., 1979, U-Th-Pb geochronology of the granite near Milford, south-central New Hampshire; New evidence for Avalonian basement and Taconic and Alleghanian disturbances in eastern New England: Contributions to Mineralogy and Petrology, v. 71, p. 1–11.

Brown, A., Murray, D. P., and Barghoon, E. E., 1978, Pennsylvanian fossils from metasediments within the Narragansett Pier Granite, Rhode Island: Geological Society America Abstracts with Program, v. 10, p. 35–36.

Buma, G., Frey, F. A., and Wones, D. R., 1971, New England granites; Trace element evidence regarding their origin and differentiation: Contributions to Mineralogy and Petrology, v. 31, p. 300–320.

Brace, W. F., and Jones, A. H., 1971, Comparison of uniaxial deformation in shock and static loading of three rocks: Journal of Geophysical Research, v. 76, p. 4913–4921.

Brace, W. F., Orange, A. S., and Madden, T. R., 1965, The effect of pressure on the electrical resistivity of water-saturated crystalline rocks: Journal of Geophysical Rsearch, v. 70, p. 5669–5678.

Chayes, F., 1950, Composition of the granites of Westerly and Bradford, Rhode Island: American Journal of Science, v. 248, p. 378–407.

——, 1952, The finer-grained calcalkaline granites of New England: Journal of Geology, v. 60, p. 207–254.

Dallmeyer, R. D., 1982, $^{40}Ar/^{39}Ar$ ages from the Narragansett basin and southern Rhode Island basement terrain; Their bearing on the extent and timing of Alleghanian tectonothermal events in New England: Geological Society of America Bulletin, v. 93, p. 1118–1130.

Feininger, T., 1963, Westerly Granite and related rocks of the Westerly–Bradford area: New England Intercollegiate Geological Conference, 55th Annual Meeting, Guidebook, p. 48–52/

——, 1965, Bedrock geologic map of the Ashaway Quadrangle, Connecticut–Rhode Island: U.S. Geological Survey Map GQ-403.

Goldich, S. S., and Oslund, E. H., 1956, Composition of Westerly Granite G-1 and Centerville diabase W-1: Geological Society of America Bulletin, v. 67, p. 811–815.

Goldsmith, R., 1985, Bedrock geologic map of the Old Mystic and part of the Mystic Quadrangles, Connecticut, New York, and Rhode Island: U.S. Geological Survey Map I-1524, scale 1:24,000.

Gromet, P. L., and O'Hara, K., 1984, Two distinct Late Precambrian terranes within the "Avalon Zone," southeastern New England, and their Late Paleozoic juxtaposition: Geological Society of America Abstracts with Programs, v. 16, p. 20.

Hermes, O. D., Barosh, P. J., and Smith, P. V., 1981, Contact relationships of the Late Paleozoic Narragansett Pier Granite and country rock, *in* Boothroyd, J. C., and Hermes, O. D., eds., Guidebook to field studies in Rhode Island and adjacent areas: New England Intercollegiate Geological Conference, 73rd Annual Meeting, University of Rhode Island, Kingston, Rhode Island, p. 125–152.

Hermes, O. D., and Murray, D. P., 1986, Middle Devonian through Permian plutonism and volcanism in the North American Appalachians: Geological Society of London Special Publication (in press).

Hermes, O. D., and Zartman, R. E., 1985, Late Precambrian and Devonian plutonic terrane within the Avalon Zone of Rhode Island: Geological Society of America Bulletin, v. 96, p. 272–282.

Hermes, O. D., Rao, J. M., Dickenson, M. P., and Pierce, T. A., 1984, A transitional alkalic dolerite dike suite of Mesozoic age in southeastern New England: Contributions to Mineralogy and Petrology, v. 86, p. 386–397.

Kern, C. A., 1979, Petrology of the Narragansett Pier Granite, Rhode Island [M.S. thesis]: Kingston, University of Rhode Island, 201 p.

Kocis, D. C., 1981, Petrology of Narragansett Pier Granite and associated country rocks near Narragansett Bay [M.S. thesis]: Kingston, University of Rhode Island, 132 p.

Kocis, D. E., Hermes, O. D., and Cain, J. A., 1978, Petrologic comparison of the pink and white facies of the Narragansett Pier Granites, Rhode Island: Geological Society of America Abstract with Program, v. 10, p. 71.

Leavy, B. D., 1979, Petrology of lamprophyre dikes and mantle-derived inclusions from Westerly, Rhode Island [M.S. thesis]: Kingston, University of Rhode Island, 114 p.

Leavy, B. D., and Hermes, O. D., 1977, Mantle xenoliths from southeastern New England, *in* The mantle sample inclusions in kimberlites and other volcanics: Proceedings of the 2nd International Kimberlite Conference, p. 374–381.

Macaulay, D., 1984, The strength of the stone: Westerly, Rhode Island, Westerly Historical Society, 14 p.

Macomber, S. W., 1958, The story of Westerly Granite: Westerly, Rhode Island, Westerly Historical Society, 38 p.

McHone, J. G., 1978, Distribution, orientations, and ages of mafic dikes in central New England: Geological Society of America Bulletin, v. 89, p. 1645–1655.

Quinn, A. W., 1971, Bedrock geology of Rhode Island: U.S. Geological Survey Bulletin 1265, 68 p.

Rodgers, J., 1985, Bedrock geological map of Connecticut: Connecticut Geological and Natural History Survey, scale 1:125,000.

Ross, M. E., 1981, Mafic dikes of northeastern Massachusetts, *in* Boothroyd, J. C., and Hermes, O. D., eds., Guidebook field studies in Rhode Island and adjacent areas: New England Intercollegiate Geological Conference, 73rd Annual Meeting, University of Rhode Island Kingston, Rhode Island, p. 285–302.

——, 1985, Mafic dike swarms of the Boston Platform, eastern Massachusetts [abs.]: International Conference on Mafic Dike Swarms, University of Toronto, p. 142–147.

Ross, M. E., Knowles, C. R., Chamress, J. S., 1983, Megacrysts, xenocrysts, and ultramafic xenoliths from a camptonite dike in Cambridge, Massachusetts: Geological Society of America Abstracts with Program, v. 15, p. 174.

Stesky, R. M., 1978, Mechanisms of high temperature frictional sliding in Westerly Granite: Canadian Journal of Earth Science, v. 15, p. 361–375.

Zartman, R. E., and Hermes, O. D., 1984, Evidence from inherited zircon for Archean basement under the southeastern New England Avalon Terrane: Geological Society of America Abstracts with Program, v. 16, p. 704.

——, 1986, Archean inheritance in zircons from Late Paleozoic granites from the Avalon Zone of southeastern New England: Earth and Planetary Science Letters (in press).

Zartman, R. E., Snyder, G., Stern, T. W., Marvin, R. F., and Buckman, R. C., 1965, Implications of new radiometric ages in eastern Connecticut and Massachusetts: U.S. Geological Survey Professional Paper 525-D, p. D1–D10.

The Alleghanian Orogeny in the Narragansett Basin area, southern Rhode Island

***Daniel P. Murray**, Department of Geology, University of Rhode Island, Kingston, Rhode Island 02881*

LOCATION

The locality consists of three sites along the western margin of Narragansett Bay, in southern Rhode Island (Figs. 1 and 2). Stop 1 is in the Wickford 7½-minute Quadrangle, while the other stops are in the Narragansett Pier 7½-minute Quadrangle.

Stop 1. Jamestown, Conanicut. Take Rhode Island 138 to the eastern end of the Jamestown Bridge, which connects the western side of Conanicut Island to the eastern shore of Narragansett Bay. Park near the Jamestown Shores Motel and take the path down to the shoreline exposures beneath the bridge. At the time of this writing the bridge is being replaced with a new one, approximately 330 ft (100 m) to the north; however, the directions given here should still apply in the future. There is free access and outcrops are best seen at low tide.

Stop 2. Stook Hill, Saunderstown. From Stop 1, proceed west on Rhode Island 138, over the Jamestown Bridge, to the intersection of Rhode Island 138 and 1A. Continue west on Rhode Island 138 for 1.35 mi (2.2 km) to the Stook Hill roadcuts (Fig. 3), where access is also free.

Stop 3. Cormorant Point, Narragansett. Return east to the intersection of Rhode Island 138 and 1A, and take 1A (Boston Neck Road) south 5.8 mi (9.3 km). Turn left (east) and take the secondary road 0.8 mi (1.3 km) (or seven "roadhumps") east, and park along the road. The shoreline exposures (Fig. 4) may be reached by walking approximately 0.25 mi (0.4 km) farther to the south, along a private road. The outcrops are shoreline exposures entirely on private land, and permission to visit them must be obtained from the people living in nearby houses. This is a "no hammer" stop (there is plenty of fresh, loose material available for collecting) and is unsuitable for large groups without prior permission.

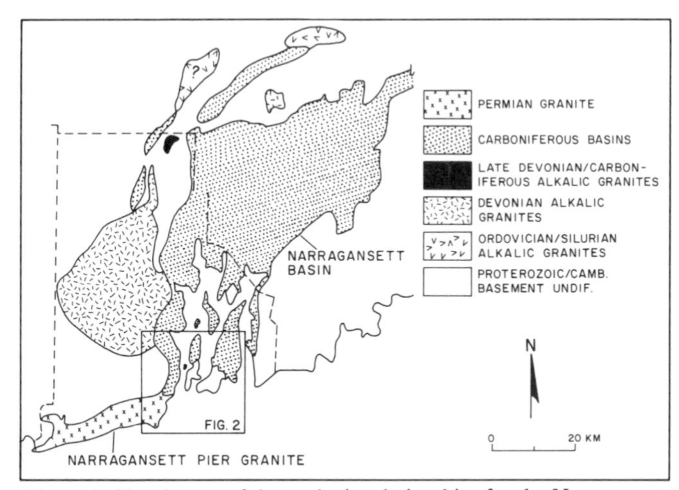

Figure 1. Sketch map of the geologic relationships for the Narragansett Basin area.

SIGNIFICANCE

Southeastern New England plays a key role in our understanding of the Appalachians, as it contains the most complete record of the Avalonian and Alleghanian Orogenies. The stops described here are chosen to illustrate some of the best evidence for the Alleghanian Orogeny in New England. This orogeny is a late Paleozoic event that probably is the cumulative result of not only the accretion of one or more microplates to eastern North America, but also the collision of North America with northwestern Africa. The relative timing and mechanics of these two independent(?) processes is incompletely understood, and is currently the subject of much controversy. Further discussions of the tectonic setting may be found in Hatcher and others (1987). The effects of the progressive regional metamorphism, complex deformation, and plutonism associated with the Alleghanian Orogeny are most clearly seen in the coal-bearing strata of the Narragansett Basin (Fig. 1), and the Permian granite that intrudes them. The first two stops examine exposures of amphibolite grade metasedimentary rocks, and the last stop considers the intrusive contact between the Permian Narragansett Pier Granite and the metasedimentary rocks of the basin. The latter contain graphite derived from the metamorphism of coal and dispersed organic matter, and as such have preserved one of the best records available anywhere of the response of organic matter to a wide range of conditions of metamorphism and deformation.

GEOLOGIC SETTING

The stops described here are located in the southwestern part of the Narragansett Basin, a 618 mi^2 (1,600 km^2) arcuate structural basin comprised of Mississippian to Late Pennsylvanian (and younger?) coal-bearing non-marine clastic sediments. These rocks have been variably deformed and metamorphosed, and intruded by the Permian Narragansett Pier Granite (Figs. 1 and 2). It is the largest and most informative of several Carboniferous basins in the area, of which the others are the Worcester Basin (Grew, 1973), Norfolk Basin, and arguably the Scituate and Woonsocket basins. The geology has been the subject of many studies over the last hundred years, and reviews of previous work as well as more recent studies can be found in Quinn (1971), Skehan and others (1979), Mosher (1983), and Hatcher and others (1987). The Carboniferous stratigraphy consists of the Narragansett Bay Group, which consists (from oldest to young-

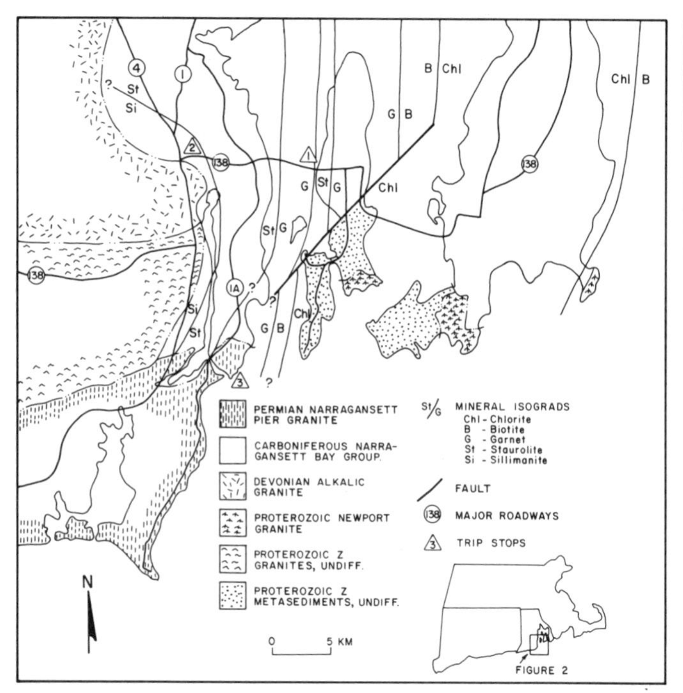

Figure 2. Geologic map showing the distribution of rock units, Alleghanian isograds, major roads, and stops.

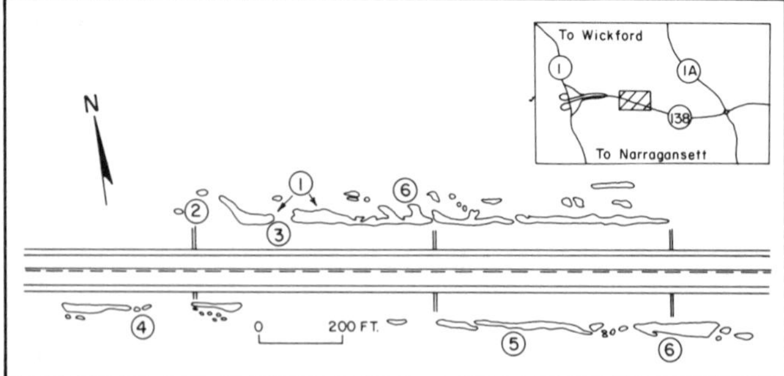

Figure 3. Sketch map of Stop 2, Stook Hill, showing the outcrop pattern and sites described in the text.

est) of the following five formations: (1) Pondville Formation, arkose and conglomerates; (2) Wamsutta Formation, redbeds and bimodal volcanic rocks; (3) Rhode Island Formation, sandstone, conglomerate, shale, and minor coal; (4) Purgatory Conglomerate, distinctive conglomerate comprised entirely of quartzite clasts; and (5) Dighton Conglomerate, another conglomerate. The Rhode Island Formation comprises 90% of the outcrop, and is the only formation of the Narragansett Bay Group that is seen in the stops.

Mineral assemblages (Grew and Day, 1972) and garnet-biotite–equilibria (Murray, unpublished data) imply peak metamorphic conditions of 600°C and 5 kb (Fig. 2). Retrograde metamorphism is ubiquitous in the southwestern portion of the Narragansett Basin, and is spatially associated with shear zones. The meta- to peraluminous Narragansett Pier Granite Plutonic Suite consists of five facies, the petrology of which is described by Hermes (this volume). One of those facies, the muscovite-garnet leucogranite, is restricted to the vicinity of contacts with the Rhode Island Formation, and in contrast to the more common pink varieties of the Narragansett Pier Granite, is white. Although in a gross sense metamorphic grade increases toward the contact with the Narragansett Pier Granite, the thermal maximum is located north of the contact, and that the isograds are truncated by the granite (Fig. 2). The absence of both a chill zone within the granite and thermal contact metamorphic effects within the metasedimentary rocks implies that the ambient temperatures of the host rocks were high at the time of intrusion (600°C and 4+ kb). Although some assimilation of metasediment by granite has taken place, petrologic arguments (mineral chemistry, mixing models, etc.) imply that the granite is *lit par lit* injected into metasedimentary rock, and not derived from them through anatexis. At Stop 3 the contact with Carboniferous metasediments is characterized by abundant roof pendants, and there the granite is bleached white through the reduction of ferric iron-bearing phases within the granite and the presence of muscovite + garnet in place of biotite. These changes in granite petrology are taken as evidence of widespread fluid interchange between carbonaceous metasediments and granite (Murray and Skehan, 1979; Murray, 1987).

Plant fossils within the metasedimentary rocks (Brown and others, 1978), U-Pb radiometric ages from the granite (Hermes and others, 1981; Zartman and Hermes, 1984), cooling ages from both the Narragansett Pier Granite and Narragansett Bay Group (Dallmeyer, 1982), and mesoscopic and microscopic fabric relationships (Grew and Day, 1972; Murray and Skehan, 1979; Burks and others, 1981; Murray and others, 1981; Mosher, 1983) can be used to construct the chronology shown in Figure 5. More detailed discussions of the geologic setting may be found in the aforementioned references, plus chapters in this volume by Hermes and by Mosher and Burks.

Stop 1. East end of the Jamestown Bridge. Here the Rhode Island Formation consists of interlayered carbonaceous schists, metasandstone, and metaconglomerate. Pennsylvanian plant fossils have been found in the schists 0.3 mi (0.5 km) to the south, while approximately 1.2 mi (2 km) to the north there are calc-silicate horizons and the remains (now nearly inaccessible) of a small graphite mine. The schist contains twinned staurolite, garnet, and biotite porphyroblasts within a matrix of biotite, muscovite, quartz, ilmenite, and graphite. Garnet-biotite mineral equilibria yield temperatures of 525°C (Murray, unpublished data). Porphyroblasts have overgrown the earliest fabric in the rock (S1), but have been deformed by a well-developed crenulation cleavage (S2). Chlorite grade retrograde metamorphism increases in intensity northward along the coast. S1 is approximately parallel to bedding, at N20°E 30°SE, and younging inferred from bedding/cleavage relationships suggests that the section is largely overturned. Local F1 fold hinges are also present, although the evidence is subtle and not entirely convincing. A NNE-striking crenulation cleavage (S2) plus minor folds (F2 or F3) are also present, and are progressively better developed to the

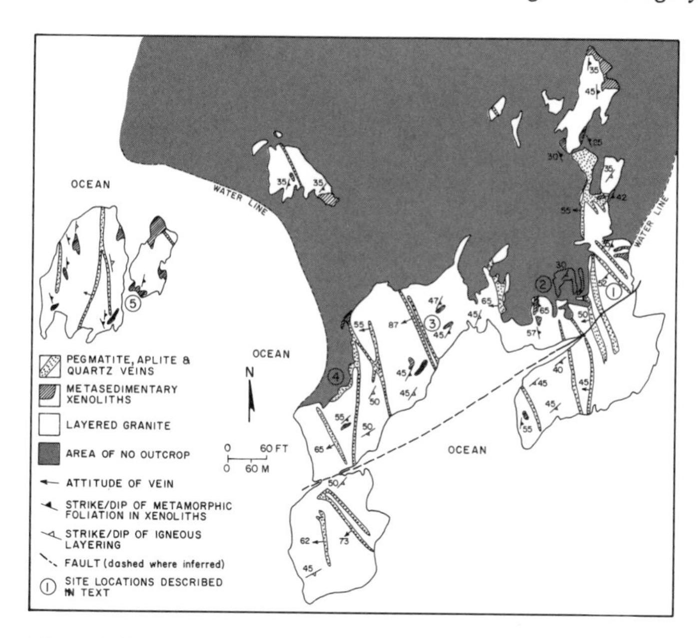

Figure 4. Sketch map of the field relationships at Stop 3, Cormorant Point, including location of sites described in the text.

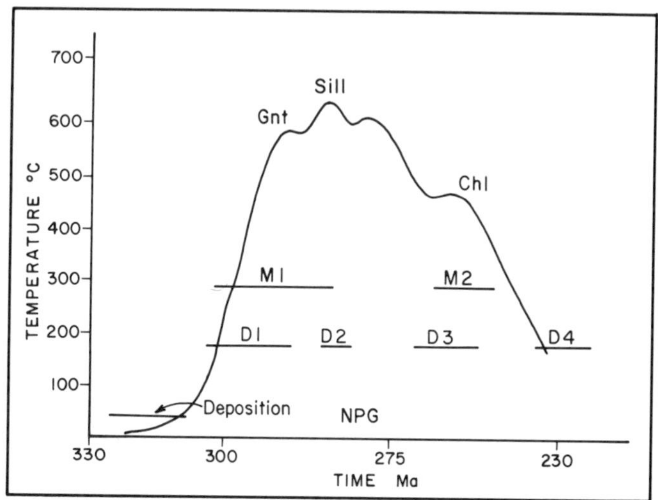

Figure 5. Timing of late Paleozoic metamorphism, plutonism and deformation in the Narragansett Basin area. References are given in the text. Legend: NPG (age of emplacement of the Narragansett Pier Granite); D-1 to D-4 (periods of deformation); M1 and M2 (periods of metamorphism); gnt-garnet, sill-sillimanite, and chl-chlorite (isograds).

north. Pebbles are oblate within the plane of S1. This stop is also described in Grew and Day (1972), Murray and Skehan (1979), Burks and others (1981), Hepburn and Rehmer (1981), and Murray and others (1981).

Stop 2. Stook Hill. These roadcuts are composed of upper amphibolite facies metasedimentary rocks with unusually well preserved primary features. In order of decreasing abundance, the lithologies present are metaconglomerate, metasandstone, pelitic schist, garnet amphibolite, and pegmatitic granite. The first four of these comprise the Rhode Island Formation, whereas the last is considered to be related to the leucocratic facies of the Narragansett Pier Granite. Figure 3 shows a map view of the roadcuts, together with key features described in the text. Primary structures are best seen on the northern side of the road, but diagnostic metamorphic assemblages are most abundant on the southern side. Parageneses within pelitic schists (various combinations of garnet, staurolite, biotite, kyanite, rare fibrolite, muscovite, quartz, graphite, and ilmenite) indicate metamorphic conditions of 600°C and 5 or more kb (Grew and Day, 1972; Murray, unpublished data). Similar or possibly higher temperatures and pressures are suggested by sillimanite-bearing assemblages (1) at Tower Hill, halfway between Stook Hill and Stop 3; and (2) in possible outliers of the Narragansett Basin Group that occur approximately 12 mi (20 km) to the southwest. In contrast to Stop 1, retrograde metamorphism is minor. Pegmatitic granite contains beryl, muscovite, and garnet, the latter of which contains up to 33 weight percent manganese (Murray, unpublished data). Although the shales have completely recrystallized to schists, primary fluviatile structures are well preserved in coarser grained rocks. Cross-bedding, erosional surfaces, and graded bedding all indicate that the section is predominantly overturned.

Bedding and cleavage (S1) are roughly parallel, and in thin section one can distinguish two schistosities. S1 strikes ENE (65°–70°), and dips to the north gently (25°–30° on the northern side of the road) to moderately (35° on the southern side). Tight macroscopic folds of S1 are limited to close proximity to small shear zones. Pegmatitic granite truncates S1 and occurs both as massive granite and as boudinaged dikes, with boudin axes roughly horizontal and trending E-W. An analysis of pebble geometry from this outcrop by several methods (D. Jones, personal communication, 1985) indicates that they have been deformed into "tongue" shapes, with their plane of flattening coincident with S1. Their direction of elongation is 30°NE, approximately coincident with the downdip direction of S1, fold axes, and pebble elongation directions seen elsewhere in the southern Narragansett Basin. The sequence of structural events seen here is considered to be: (1) Isoclinal folding about NE trending axes, associated with the development of S1 and regional metamorphism; (2) development of shear zones and local folding of S1; (3) emplacement of granite along fractures; and (4) boudinage of granitic dikes, possibly contemporaneous with open folding of earlier structures about horizontal E-W axes. This stop is also described in Grew and Day (1972), Murray and Skehan (1979), and Burks and others (1981).

Good examples of these important features may be seen at the following places shown on Figure 3: (1) primary structures indicating inversion of the stratigraphic section; (2) amphibolitic layers containing grossularitic garnet, epidote, and actinolitic hornblende; (3) post-S1 shear zone and tight folds; (4) kyanite (difficult to find)-staurolite schist; (5) pegmatitic granite and boudinaged dikes of pegmatitic granite; and (6) deformed pebbles.

Stop 3. Cormorant Point. The shoreline exposures seen

here straddle the contact between the white (i.e., leucocratic) facies of the Narragansett Pier Granite and the Rhode Island Formation (Fig. 4). In this outcrop the white facies may be subdivided into five field units—pegmatite, aplite, garnetiferous granite, massive granite, and quartz veins—that define a crude igneous layering. A map showing the distribution of all these varieties is included in Hermes and others (1981), and a simplified version of it is given in Figure 4. The mineralogy of all varieties is similar, consisting of quartz, perthitic microcline, oligoclase, primary muscovite, biotite, and accessory garnet, apatite, zircon, monazite, and allanite. Variations in granite fabric and mineralogy found here reflect variations in activity of water, and to a lesser extent assimilation of the surrounding metasediments.

Abundant lenses of metasedimentary rocks are prevalent throughout the area, consisting of pelitic and psammitic schist plus subordinate metaconglomerate. These lenses are characterized by a uniformly NE-trending schistosity (S1) and pebble elongation direction that roughly coincides with both the orientation of the layering within the granite and with the regional tectonic fabric in the country rock. This suggests that the metasedimentary rocks are roof pendants and that the current erosion surface is close to the top of the magma chamber. The granite truncates both the dominant schistosity (S1) as well as a crenulation cleavage (S2) in the schist, and appears to have intruded along fractures, suggesting that its emplacement was relatively passive and post-dated most of the ductile deformation observed in the host rock. The time of intrusion is well constrained by the following independent age determinations: (1) Stephanian A or younger (i.e., Late Pennsylvanian) plant fossils in one of the lenses (Brown and others, 1978); (2) U-Pb age of 275 Ma from igneous monazite (Hermes and others, 1981); and (3) Ar^{40}/Ar^{39} cooling age of 238 Ma (Dallmeyer, 1982).

The relationships described above can be best seen at the following places (labeled on Fig. 4). (1) Variations in granite texture. (2) Lens of carbonaceous schist from which plant fossils (none remaining) were collected. Nearby lenses contain folded quartz veins and pegmatites, and well-layered garnetiferous granite. (3) Several of the largest lenses of schist occur in this area. (4) Samples for the radiometric dates were collected from a small abandoned quarry, here. One can also clearly see the granite truncating a crenulated schistosity (S2) in the metasedimentary rocks. (5) This part of the outcrop (accessible only at low tide) contains the best examples of deformed metaconglomerate, and the deformed pebbles have roughly the same orientation as do the ones at Stop 2. Other descriptions of this stop may be found in Murray and Skehan (1979), Burks and others (1981), Hermes and others (1981), Kocis (1981), and Schwartz (1981).

REFERENCES CITED

Brown, A., Murray, D. P., and Barghoorn, E. E., 1978, Pennsylvanian fossils from metasediments within the Narragansett Pier Granite, Rhode Island: Geological Society of America Abstracts with Programs, v. 10, p. 35–36.

Burks, R. J., Mosher, S., and Murray, D. P., 1981, Alleghanian deformation and metamorphism of southern Narragansett Basin, *in* Boothroyd, J. C., and Hermes, O. D., eds., Guidebook for field studies in Rhode Island and adjacent areas: New England Intercollegiate Geological Conference, 73rd Annual Meeting, University of Rhode Island, Kingston, Rhode Island, p. 265–275.

Dallmeyer, R. D., 1982, $^{40}Ar/^{39}Ar$ ages from the Narragansett basin and southern Rhode Island basement terrain; Their bearing on the extent and timing of Alleghanian tectonothermal events in New England: Geological Society of America Bulletin, v. 93, p. 1118–1130.

Grew, E. S., 1973, Stratigraphy of the Pennsylvanian and pre-Pennsylvanian rocks of the Worcester area, Massachusetts: American Journal of Science, v. 273, p. 113–129.

Grew, E. S., and Day, H. W., 1972, Staurolite, kyanite, and sillimanite from the Narragansett Basin of Rhode Island: U.S. Geological Survey Professional Paper 800-D, p. 151–157.

Hatcher, R. D., Jr., Thomas, W. A., and Viele, G. W., eds., 1987, The Appalachian/Ouachita regions, U.S.: Boulder, Colorado, Geological Society of America, The Geology of North America, v. F-2 (in press).

Hepburn, J. C., and Rehmer, J., 1981, The diagenetic to metamorphic transition in the Narragansett and Norfolk basins, Massachusetts and Rhode Island, *in* Boothroyd, J. C., and Hermes, O. D., eds., Guidebook for field studies in Rhode Island and adjacent areas: New England Intercollegiate Geological Conference, 73rd Annual Meeting, University of Rhode Island, Kingston, Rhode Island, p. 47–67.

Hermes, O. D., Barosh, P. J., and Smith, P. V., 1981, Contact relationships of the late Paleozoic Narragansett Pier Granite and country rock, *in* Boothroyd, J. C., and Hermes, O. D., eds., Guidebook for field studies in Rhode Island and adjacent areas: New England Intercollegiate Geological Conference, 73rd Annual Meeting, University of Rhode Island, Kingston, Rhode Island, p. 47–67.

Kocis, D. E., 1981, Petrology of Narragansett Pier Granite and associated country rocks near Narragansett Bay [M.S. thesis]: Kingston, University of Rhode Island, 132 p.

Mosher, S., 1983, Kinematic history of the Narragansett Basin, Massachusetts and Rhode Island; Constraints on late Paleozoic plate reconstruction: Tectonics, v. 2, p. 327–344.

Murray, D. P., 1987, Post-Acadian metamorphism in the Appalachians, *in* Harris, A. L., ed., The Caledonide-Appalachian orogen: Geological Society of London Special Volume (in press).

Murray, D. P., and Skehan, J. W., 1979, A traverse across the eastern margin of the Appalachian-Caledonian orogen, southeastern New England, *in* Skehan, J. W., and Osberg, P. H., eds., The Caledonides in the U.S.A.; Geological excursions in the northeast Appalachians, International Geologic Correlation Program Project 27: Weston, Massachusetts, Boston College Department of Geology and Geophysics, p. 1–32.

Murray, D. P., Raben, J. D., Lyons, P. C., and Chase, H. B., Jr., 1981, The geological setting of coal and carbonaceous material, Narragansett Basin, southeastern New England, *in* Boothroyd, J. C., and Hermes, O. D., eds., Guidebook for field studies in Rhode Island and adjacent areas: New England Intercollegiate Geological Conference, 73rd Annual Meeting, University of Rhode Island, Kingston, Rhode Island, p. 175–195.

Quinn, A. W., 1971, Bedrock geology of Rhode Island: U.S. Geological Survey Bulletin 1295, 68 p.

Schwartz, S. Y., 1981, A study of the contact relations between the Narragansett Pier Granite and Pennsylvanian-aged metasediments at Cormorant Point, Rhode Island [Senior thesis]: Providence, Rhode Island, Brown University, 45 p.

Skehan, J. W., Murray, D. P., Hepburn, J. C., Billings, M. P., Lyons, P. C. and Doyle, R. G., 1979, The Mississippian and Pennsylvanian (Carboniferous) systems in the United States—Massachusetts, Rhode Island, and Maine: U.S. Geological Survey Professional Paper 1110-A, p. A1–A30.

Zartman, R. E. and Hermes, O. D., 1984, Evidence from inherited zircons for Archean basement under the southeastern New England Avalon Terrane: Geological Society of America Abstracts with Programs, v. 16, p. 704.

Alleghanian deformation in the southern Narragansett Basin, Rhode Island

S. Mosher, R. J. Burks, and B. H. Reck, Department of Geological Sciences, University of Texas, Austin, Texas 78713

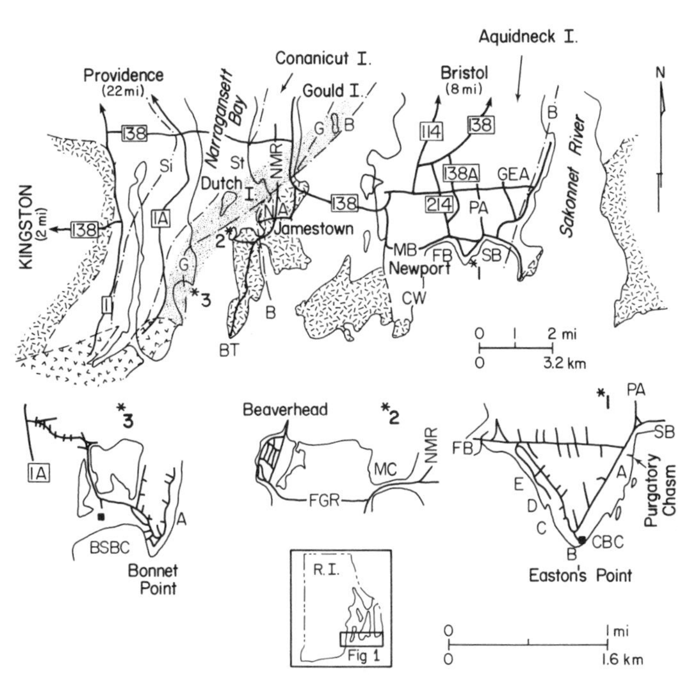

Figure 1. Location map for three stops with enlargements showing local roads for each stop. Geologic features shown: Beaverhead Shear Zone, shaded; basement, random dashes; Narragansett Pier Granite, v pattern, basin sediments, no pattern; isograds, dash-dot lines; Si, sillimanite; St, staurolite; G, garnet; B, biotite. Roads and places referred to in text: NA, Narragansett Avenue; NMR, North Main Road; MB, Memorial Boulevard; GEA, Green End Avenue; PA, Paradise Avenue; FGR, Fort Getty Road; BT, Beavertail; CW, Cliff Walk; FB, First Beach; SB, Second Beach; BSBC, Bonnet Shores Beach Club; MC, Mackeral Cove; CBC, Clambake Club. On maps for Stops 1 and 3, letters refer to localities mentioned in text.

LOCATION OF STOPS

Stop 1. Purgatory Chasm—Easton's Point, Newport, Rhode Island. Purgatory Chasm State Park is just west of Second Beach on the east side of Easton's Point (Fig. 1). Outcrops can be reached either from Second Beach parking lot at the southern end of Paradise Avenue or from Purgatory Chasm parking lot located on the road that parallels the coast west of Second Beach. Public access to continuous coastal outcrop along Easton's Point is provided by three public dirt paths, one on each side of the Clambake Club (at sharp curve on this road) and one at locality E. A complete or partial loop of the point can be made using the paths. Reconnaissance of the return path is advisable because it is often difficult to recognize from the outcrops. Most of this exposure is covered at high tide. Permission for small groups is unnecessary because all coastline between high and low tide is public domain. For large groups, permission from owners is advisable. Percival Quarry (east side of Paradise Avenue) is one of the few places where the conglomerate can be sampled. Permission can be obtained from quarry office, just north of entrance on Paradise Avenue.

Stop 2. Beaverhead, Jamestown, Rhode Island. Fort Getty Recreational Area is just west of Jamestown on southern Conanicut Island. Take North Main Road south from Rhode Island 138, cross Narragansett Avenue, and continue south past the beach at Mackeral Cove. Take the first right onto Fort Getty Road. Follow road past gatehouse and campground to northern pier, keeping right on unpaved roads. Exposures are located along the western shore and are accessible from numerous points along coastline. In summer, an entrance fee is charged, but the gatekeeper usually will admit small groups of geologists free. For larger groups, permission for free access is obtainable from Jamestown Town Office, located on the west side of North Main Road just south of Narragansett Avenue.

Stop 3. Bonnet Point, Bonnet Shores, Rhode Island. Bonnet Shores is a private subdivision located off Rhode Island. 1A (Boston Neck Road) approximately 3.5 mi (5.6 km) south of the intersection of Rhode Island 1A and 138. Turn east at sign for Bonnet Shores and take right fork which curves around past Bonnet Shores Beach Club. (This road makes a complete loop of the subdivision.) Take fourth right past Beach Club (at edge of large grassy field). The small road will curve left and rejoin previous road. Just east of southern terminus of this loop is a public dirt path to outcrops at southern end of point. No parking is allowed within subdivision without permission. Small groups can obtain permission from neighbors, larger groups should contact town administrator.

SIGNIFICANCE

Pennsylvanian age Narragansett Basin, Rhode Island and Massachusetts is a composite graben filled with wet alluvial fan sediments that were deposited during active faulting. The sediments have been affected by four distinct phases of Permian deformation. The first two phases (D1 and D2) are the result of nearly E-W compression and the later two (D3 and D4) are related to strike slip shearing along preexisting basement faults. A prograde Barrovian metamorphism was synchronous with D1 in the southwestern basin and with D2 in the southeastern and central basin. Intrusion of the Narragansett Pier Granite started during D3. Retrograde metamorphism was patchy in occurrence and locally pre- or post-dates D4. The first stop includes the classical outcrop of Purgatory Conglomerate at Purgatory Chasm and is significant, in itself, because the conglomerate has only been deformed by pressure solution and intercobble rotation.

SITE INFORMATION

Outcrops at the three stops demonstrate the nature of Alleghanian deformation in southeastern New England. The coarse-grained metasediments at Stops 1 and 3 are the best examples of alluvial fan sedimentation and D1 and D2 structures. D1 produced at least two stages of N-trending, W-verging, isoclinal folds with associated axial planar foliations. One of these foliations is pervasive and found throughout the basin. D2 produced regional N-trending, E-verging, open folds and an axial planar crenulation cleavage. The exposures of medium- to fine-grained metasedimentary rocks at Stop 2 provide the best easily accessible example of D3 and D4 effects. Only one other excellent exposure (Dutch Island; Berryhill, 1984) showing D3 and D4 structures is open to the public (accessible by private boat). D3 is expressed by a series of structures compatible with sinistral shearing on NNE- to NE—trending basement faults that include: ENE- to NE-trending folds that are often en echelon, multiple generations of noncoaxial, superimposed folds and crenulations that show a clockwise younging sense, sheath folds, brittle normal and sinistral strike slip faults, and tension gashes. D4 structures are compatible with dextral shearing on NW-, NE-, and ENE-trending basement faults and include: N-trending folds, multiple generations of noncoaxial, superposed folds and crenulations with an counter-clockwise younging sense, sheath folds, dextral strike and oblique slip faults, and boudinage with a N-S extension direction. D3 and D4 structures are found throughout the basin, but are best expressed within intense deformation zones that divide the southcentral basin into numerous rhomb-shaped domains, the most intense of which is the NE-trending Beaverhead Shear Zone. Information cited above comes from Mosher (1983, unpublished data), Thomas (1981), Farrens (1982), Berryhill (1984), Henderson (1985), and the following references. Information for Stop 1 is from Mosher (1976, 1978, 1980, 1981, 1987) and Mosher and Wood (1976); for Stop 2, Burks (1981, 1985); and for Stop 3, Reck (1985). Additional conglomerate stops are given in Mosher and Wood (1976) and other basin stops and references to older guidebooks are in Burks and others (1981).

Stop 1. At this outcrop the approximately 440-ft-thick (135-m), massive, clast-supported Purgatory Conglomerate units are interbedded with thin sandstones and magnetite-rich sandstone lenses. The conglomerate represents a proximal to very proximal facies of a wet alluvial fan that formed off the southeastern block-faulted margin of the basin. Clasts are generally prolate triaxial ellipsoids and range from pebbles to boulders in size (majority are cobbles). Clasts are predominantly quartzite, with rare granite and schist cobbles. The outcrop forms part of one of several elongate, N-trending ridges that mark the positions of major F1 and F2 fold limbs (Fig. 2A). Long cobble axes trend N10°E parallel to fold axes. Most deformation was caused by D1 and D2; only minor indications of D3 and D4 are present. Metamorphism to chlorite grade was syn- to post-D2.

Cobble deformation was achieved by pressure solution.

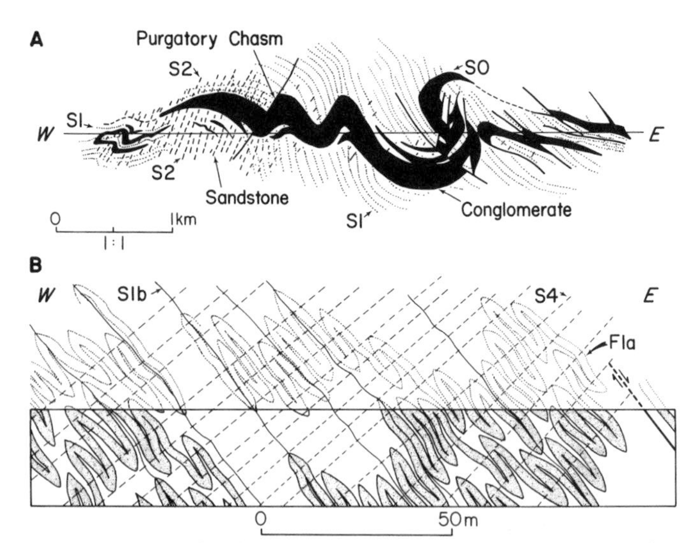

Figure 2. (A) Profile from eastern basin margin to First Beach after Mosher (1981) and Farrens (1982). (B) Cross section of Bonnet Point after Reck (1985); section line shown in Fig. 3.

Cobbles have tangential, almost planar, and deeply embayed contacts. Thin sections of cobble contacts show no evidence of quartz or mica deformation within cobbles nor any distortion of internal cobble bedding. Some cobbles contain numerous microstylolitic seams parallel to long axes, which give cobbles an internally deformed appearance in the field. Large, fibrous, quartz pressure shadows can be seen at long axis terminations of most cobbles. Where cobbles are in close contact, matrix is 1 to 3 mm thick, depleted in quartz (<3%) and enriched in residual material remaining after pressure solution. Shear fractures offset margins of some cobbles. Substantial redistribution of cobble volume (ΔV) has been measured for the conglomerate throughout the southeastern basin (23% hinges; 55% overturned limbs). Much of the apparent strain is caused by original cobble shapes; intercobble rotation during D1 and D2 realigned already ellipsoidal sedimentary cobbles into their present position. Real strains are constrictional (Cobble strains, Percival Quarry: $e_x = 0\%$, $e_y = -20\%$, $e_z = -11\%$, $\Delta V = -23\%$; west Easton's Point: $e_x = 0\%$, $e_y = -28\%$, $e_z = -26\%$, $\Delta V = -41\%$). Although pressure solution features can be observed everywhere on this outcrop, best localities are: Second Beach near large out-of-place conglomerate block, hilltop near Chasm, and just east of Chasm parking lot along N-trending gravel pathways. On hilltop near Chasm parking lot, conglomerate beds interfinger with sandstones (bedding: N10°E, 55°SE), showing right-side-up cross-bedding and both flat-lying (N4°E, 22°E) S1 and steep (N24°E, 70°NW) S2 foliations.

Purgatory Chasm is the result of weathering of closely spaced, quartz filled, 'joints' that may mark edges of large rectangular-shaped boudins. Similar features, in the same orientation, crop out along the northern Cliff Walk where conglomerate forms rectangular, sharp-ended boudins separated by quartz. (The most spectacular boudinage outcrops can be seen on Gould Island from a boat.) Exposed in the northern Chasm face is a

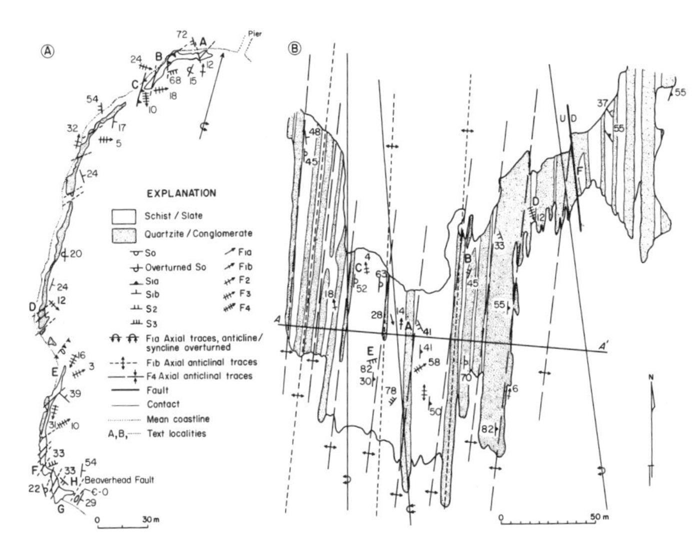

Figure 3. Geologic and structural maps. (A) Stop 2, Beaverhead after Burks (1981). (B) Stop 3, Bonnet Point after Reck (1985). Some thin schist units and major F1b, F4 synclinal hinges omitted for clarity. Each quartzite body represents a F1b axial trace (see cross section along line A–A′, Fig. 2B).

N-trending fault. Faults and the Chasm boudin are the only evidence of D3 and D4 events.

This exposure extends beyond the park boundary southward as continuous outcrop around Easton's Point to First Beach (for localities see Fig. 1, *1). Immediately south of the Chasm (on private property; locality A), numerous vertical E-trending rock faces beautifully display the interpenetrative nature of conglomerate clasts. At the point, units structurally, and presumably stratigraphically, below the Purgatory Conglomerate crop out and the massive conglomerate interfingers with Rhode Island Formation sandstones and siltstones that form the more distal portions of the alluvial fan complex. Cross-beds, pebble lag deposits, and fining upward sequences are abundant. At locality B, fine-grained units are folded into a well exposed, upward facing F2 antiform that is overturned to the east (Fig. 2A). On the west, the other limb is exposed and units again interfinger with Purgatory Conglomerate. Here, the conglomerate is thinner and shows evidence of at least one generation of F1 isoclinal folds (locality D). Exposures further north (locality E) have cross-beds and cleavage/bedding relations that indicate downward facing folds. At locality C, the pervasive S1 (N24°E, 30°W) is a crenulation of an earlier S1a. Between localities B and C are three dimensional exposures of boudins, many of which show chocolate tablet structures.

Stop 2. Outcrops at this stop lie within the Beaverhead Shear Zone. Evidence of D3 deformation is widespread, whereas D4 is much less pronounced. (At Gould Island, the reverse is true; and at Dutch Island, both are well preserved.) The main lithologies (Rhode Island Formation) are: graphitic biotite schist, chloritic quartzite, quartzite metaconglomerate, and graphitic slate. Units are generally overturned, forming the lower limb of a large, recumbent, upward facing F1 isocline. All outcrops show a pervasive, bedding-parallel, S1 schistosity, which generally strikes NE and dips 30°–40°SE. Metamorphism to biotite grade was syn-D1 and pre-D2.

Outcrops start 60 ft (20 m) west of the northern pier along the beach (for localities, see Fig. 3A). On west side of schist outcrop at locality A, several minor F1 isoclinal fold hinges in thin quartzite layers are exposed. A few show transposed bedding laminations. Small amplitude (0.4 to 4 in; 1 to 10 cm), NE-verging, F2 folds and an associated S2 crenulation cleavage (striking NNW and dipping steeply [>60°] SW) are also observable. Several generations of D3 and D4 crenulations and small folds are visible on S1. Between here and locality C, many units are juxtaposed across foliation-parallel thrust faults. At low tide, continue along coast to next set of slate outcrops, or return to top by retracing your steps or scaling broad rock face. Go down a few meters and take first steep path. At path base (locality B), large EW-trending, E-plunging, F3 chevron and box folds deform S1 and a rarely visible NNW-trending S2 crenulation lineation (best seen at low tide). Locally an axial planar S3 crenulation cleavage is well-developed. Several thin, pyrite-replaced layers are isoclinally folded by F1. Numerous low-angle faults cut the section. The structurally lowest fault-bounded slice (locality C) is a sheared quartzo-feldspathic schist similar to the basin's basal Pondville Formation. About 30 ft (10 m) south of this exposure, a ledge of quartzite and metaconglomerate continues southward along most of the coastline and can be reached at high tide from a path that parallels the coast. A few isoclinally folded metaconglomerate and quartzite contacts are visible (locality D); most of this section is overturned (cleavage/bedding relationships and scoured bedding features). Finite D1 pebble strain ratios (2.17/1/0.25) fall within the apparent flattening field, unlike the Purgatory Conglomerate. Minor EW-trending kink bands and broad warps are observed. Rocks south of thrust between localities D and E are generally right-side-up. Take path south of this fault and continue down the beach a few meters. Here interlayered schists, thin quartzites, and metaconglomerates are complexly deformed by D3 and D4 shearing events. A tight NE-trending F3a is folded by a later EW-trending F3b (locality E). The cliff exposure south of these folds contains complexly folded conglomerates and a complex pattern of low and high angle faults.

Several small promontories along the extreme southern coastline provide best viewing of the superposed shear-related structures because of their proximity to the Beaverhead Fault. Small, open, N-trending F4 folds reorient F2 and F3 folds and crenulations. Quartz veins are boudinaged in a N-S direction. At localities F and G, three sets of pervasive D3 crenulations young progressively clockwise from NNE to NE, indicating formation during sinistral shearing. In places, D3 crenulations can be seen to affect S2. More widely space D4 crenulations, younging from NNE to NNW, cut D3 crenulations. Under rock face at locality F are dextral en echelon vein arrays and NE- and NNE-trending superposed kink bands with a dextrally younging sense. A NW-oriented, dextral D4 zone contains ESE-trending crenulations cut by EW-trending ones (locality G). At locality H, ENE-trending F3 folds and S3 crenulations are visible deflected into the NE-trending Beaverhead Fault, and narrow, vertical, NE-trending dextral ductile shear zones clearly deflect earlier S3 crenulations that formed as the result of sinistral motion along the same faults. Southward verging, recumbent sheath folds deform D3 crenulations. The break in outcrop at the end of this sequence marks the trace of the Beaverhead Fault. Cambrian metasedimentary rocks are exposed approximately 50 ft (15 m) south of the Fault and continue south to Beavertail.

Stop 3. The most prominent features at Bonnet Point are parallel bodies of arkosic quartzite, interbedded with graphitic schist, that each form a mesoscopic F1b fold hinge. The interference pattern of NNW-trending F1a and NNE-trending F1b isoclinal folds (Figs. 2B, 3B) is determined by reversals in fold facing indicated by sedimentary structures. S1a is the prominent foliation in schist and quartzite (where foliated); S1b is developed locally. Metamorphism to biotite grade was syn-D1. A N-striking, W-dipping crenulation cleavage, S2, locally axial planar to small, closed F2 folds, occurs in the schist and thinly laminated quartzite. S2 is crosscut in a few places by a W-striking crenulation cleavage, S3, oriented appropriately to have formed during D3. All of these structures were then reoriented by NNE-trending, E-verging open folds, F4. These F4s fold pegmatite sheets associated with the Narragansett Pier Granite to the south.

At locality A (see Fig. 3B for localities), a small F1a fold can be seen refolded by a F1b hinge. Somewhat larger F1a folds (wavelengths of 2 m) are visible about 25 ft (7 m) west of this locality. S2 is visible in the schist at locality A, and F4 folding causes S1a and S1b to steepen and S2 to shallow from west to east. An excellent example of an F1b antiformal syncline hinge in quartzite is at locality B. At locality C, complex small scale folding results from interference of meso- and macroscopic F1a and F1b folds plus the F2 folding seen alone to the right. At locality D, an F4 fold deforms S1a, S1b, and S3 crenulations. At locality A of Figure 1, *3, F4 folds deform a pegmatite sheet. S3 crenulations can be seen crosscutting S2 at locality E. These rare, W-trending crenulations, plus some similarly oriented kinks (examples occur about 15 ft [5 m] SE), are the only clear evidence at Bonnet Point of N-oriented shortening associated with sinistral D3 shearing. A moderately E-dipping fault cuts quartzite and several earlier faults at locality F. These faults and others (dipping more steeply) farther north along Bonnet Shores postdate other structures and are probably Triassic normal faults. A large sheet of Narragansett Pier Granite is visible directly offshore, just south of Bonnet Point.

Bonnet Point is part of an almost continuous outcrop along Narragansett Bay's southwestern shore (southwesternmost exposures of Rhode Island Formation). At the southern end of the exposure, sediments are in intrusive contact with the Narragansett Pier Granite. The granite contains N-trending sheeted aplites and pegmatites and inclusions of schist that show S1a, S1b, S2, and small F2s. Between Bonnet Point and the granite, interleaved metasediments and N-trending pegmatites, typically parallel to S1, crop out. Above relationships indicate that the granite began to intrude sediments during D3. The granite crystallization age (272 Ma; Hermes and others, 1981) provides an approximate age for the transition from dominantly shortening deformation of the basin to shearing associated with strike slip motion.

REFERENCES CITED

Berryhill, A. W., 1984, Structural analysis of progressive deformation within a complex strike-slip fault system; Southern Narragansett Basin, Rhode Island [M.A. thesis]: University of Texas at Austin, 79 p.

Burks, R. J., 1981, Alleghanian deformation and metamorphism in southwestern Narragansett Basin, Rhode Island [M.A. thesis]: University of Texas at Austin, 93 p.

—— , 1985, Incremental and finite strains within ductile shear zones, Narragansett Basin, Rhode Island [Ph.D. thesis]: University of Texas at Austin, 138 p.

Burks, R. J., Mosher, S., and Murray, D. P., 1981, Alleghanian deformation and metamorphism of southern Narragansett Basin, *in* Boothroyd, J. C., and Hermes, O. D., eds., Guidebook to Geologic Field Studies in Rhode Island and adjacent areas, New England Intercollegiate Geologic Conference, 73rd annual meeting, p. 265–284.

Farrens, C. M., 1982, Styles of deformation in the southeastern Narragansett Basin, Rhode Island and Massachusetts [M.A. thesis]: University of Texas at Austin, 66 p.

Henderson, C. M., 1985, Narragansett Basin, Rhode Island; Role of preexisting intrabasinal horsts and grabens in Alleghanian deformation [M.A. thesis]: University of Texas at Austin, 72 p.

Hermes, O. D., Gromet, L. P., and Zartman, R. E., 1981, Zircon geochronology and petrology of plutonic rocks in Rhode Island, *in* Boothroyd, J. C., and Hermes, O. D., eds., Guidebook to geologic field studies in Rhode Island and adjacent areas, New England Intercollegiate Geologic Conference, 73rd annual meeting, p. 315–338.

Mosher, S., 1976, Pressure solution as a deformation mechanism in Pennsylvanian conglomerates from Rhode Island: Journal of Geology, v. 84, p. 355–364.

—— , 1978, Pressure solution as a deformation mechanism in the Purgatory Conglomerate [Ph.D. thesis]: University of Illinois at Urbana, 181 p.

—— , 1980, Pressure solution deformation of conglomerates in shear zones, Narragansett Basin, Rhode Island: Journal of Structural Geology, v. 2, p. 219–225.

—— , 1981, Pressure solution deformation of the Purgatory Conglomerate from Rhode Island: Journal of Geology, v. 89, p. 37–55.

—— , 1983, Kinematic history of the Narragansett Basin, Massachusetts and Rhode Island; Constraints on late Paleozoic plate reconstructions: Tectonics, v. 2, p. 327–344.

—— , 1987, Pressure solution deformation of the Purgatory Conglomerate, Rhode Island; Quantification of volume change, real strain, and sedimentary shape factor: Journal of Structural Geology, v. 9, (in press).

Mosher, S., and Wood, D. S., 1976, Mechanisms of Alleghanian deformation in the Pennsylvanian of Rhode Island, *in* Cameron, B., ed., Geology of southeastern New England: New England Intercollegiate Geologic Field Conference, 68th Annual Meeting, Boston, Massachusetts, p. 472–490.

Reck, B. H., 1985, Deformation and metamorphism in southwestern Narragansett Basin and their relationship to granite intrusion [M.A. thesis]: University of Texas at Austin, 77 p.

Thomas, K. J., 1981, Deformation and metamorphism in the central Narragansett Basin of Rhode Island [M.A. thesis]: University of Texas at Austin, 96 p.

Geological Society of America Centennial Field Guide—Northeastern Section, 1987

Cambrian stratigraphy and structural geology of southern Narragansett Bay, Rhode Island

James W. Skehan, S. J., and Michael J. Webster, Weston Observatory, Department of Geology and Geophysics, Boston College, Weston, Massachusetts 02193
Daniel F. Logue, Southeastern Division, EXXON Company, P.O. Box 61812, New Orleans, Louisiana 70161

LOCATION

This guide describes the rocks of southernmost Aquidneck and Conanicut Islands, Rhode Island. Follow Rhode Island 114 south into Newport, until it meets Thames Street. Follow Thames Street to Bellevue Avenue and turn right (south) on Bellevue Avenue at the Tennis Hall of Fame (0.0 mi; 0.0 km) toward the Newport Mansions. Follow Bellevue Avenue to Ocean Avenue; watch for the sharp left turn (Fig. 1; 2.2 mi; 3.5 km). Continue on Ocean Avenue to the northwest end of the Brenton Point State Park and stop in the parking lot (5.2 mi; 8.3 km). Walk a few hundred ft north along Ocean Avenue to Stop 1 (Fig. 2; 5.3 mi; 8.5 km).

To get to Stop 2 in Jamestown on Conanicut Island, retrace the route into Newport. Follow the signs for Rhode Island 138–Newport Bridge and Jamestown (Fig. 1), and take Rhode Island 138 west across Newport Bridge. From the toll booth (0.0 mi; 0.0 km), continue west for 1.6 mi (2.6 km) and turn south on North Main Road (Fig. 1). Cross Narragansett Avenue (Fig. 1) in Jamestown and proceed south on Southwest Avenue. At 2.6 mi (4.2 km), bear right and continue across the sandbar at Mackeral Cove. Follow Beavertail Road on the western island into Beavertail State Park. Park as close to the Beavertail Lighthouse as possible; coastal exposures form Stop 2 (Fig. 3; 5.3 mi; 8.5 km). Follow Beavertail Road eastward past the lighthouse and the original lighthouse monument, northward to the second parking lot on the eastern side to Stop 3, an extensive shoreline exposure (Fig. 4, 5.8 mi; 9.3 km). For further stops (8) within the Cambrian strata of Conanicut Island, the reader is referred to Skehan and others (1981); for additional stops in the Precambrian rocks of Newport, see Rast and Skehan (1981).

SIGNIFICANCE

The rocks of this site represent the most southerly outcrops of Avalonian Terrane that forms the eastern margin of the Northern Appalachians and one of the best exposures of Avalonian Terrane outside of its type area. The Cambrian fossils discovered at this site are, in part, similar to those found in the Boston area and at Hoppin Hill (near the Massachusetts–Rhode Island border) and form part of the basis for claiming southeastern Massachusetts and Rhode Island as part of the Avalonian Terrane (Skehan and others, 1978). This site is composed of a succession of rocks having features characteristic of Avalonian Terranes; it consists of: (1) Late Proterozoic metasedimentary strata and volcanic rocks, intruded by plutonic rocks dated at 595 ± 12 Ma (Smith and Giletti, 1978); (2) Lower and Middle Cambrian beds, in part abundantly fossiliferous; and (3) Pennsylvanian-age fluvial strata. At this site, only Lower and Middle Cambrian rocks, intensely and multiply deformed, will be investigated. The similar styles of deformation among the Cambrian and Pennsylvanian rocks suggest that the most significant orogenic episode(s) possibly occurred during Alleghanian collision(s) of the landmasses of Gondwana with Laurentia. Yet Taconian or Acadian ages, although less likely, cannot presently be ruled out. The older rocks have been lightly metamorphosed to low greenschist facies. The Pennsylvanian rocks have been variably metamorphosed, ranging up to sillimanite–K-spar zone in the southwesternmost exposures of the Narragansett Basin.

SITE INFORMATION

The Cambrian strata of this guide are well exposed along easily accessible seacoast exposures, mainly located within Beavertail State Park of the Narragansett Bay Park system. Nearly continuous exposures of these rocks also occur along most of the seacoast in southern Conanicut Island. Precambrian granite, cutting older hornfelsed mafic volcanic and metasedimentary rocks, is well exposed in Fort Wetherill and in coastal exposures along the east side of the island south of Jamestown (Fig. 1). Pennsylvanian strata are readily examined along the west coast of Conanicut and on southern Aquidneck Island, especially along northern Cliff Walk and in the seacoast exposures east of Newport. The later dominant deformational features of the Cambrian rocks may have formed contemporaneously with those produced in the Pennsylvanian strata of the adjacent overlying Narragansett Basin.

Lower Cambrian Rocks of Newport

Pirate Cave Formation. Graded metagraywacke, metasiltstone, and slate of the Precambrian Newport Neck Formation are overlain with angular unconformity by the Lower Cambrian rocks of the Pirate Cave Formation along the westernmost shore of Newport Neck (Fig. 1). In the south of the outcrop area (Fig. 2), the overturned green metasandstones and white tuffs of the Graves Point Member of the Newport Neck Formation are overlain by a basal pink limestone bed 10 ft (3 m) thick, of the Pirate Cave Formation. In the north part of the outcrop area (Fig. 2), this limestone bed overlies overturned, graded, green

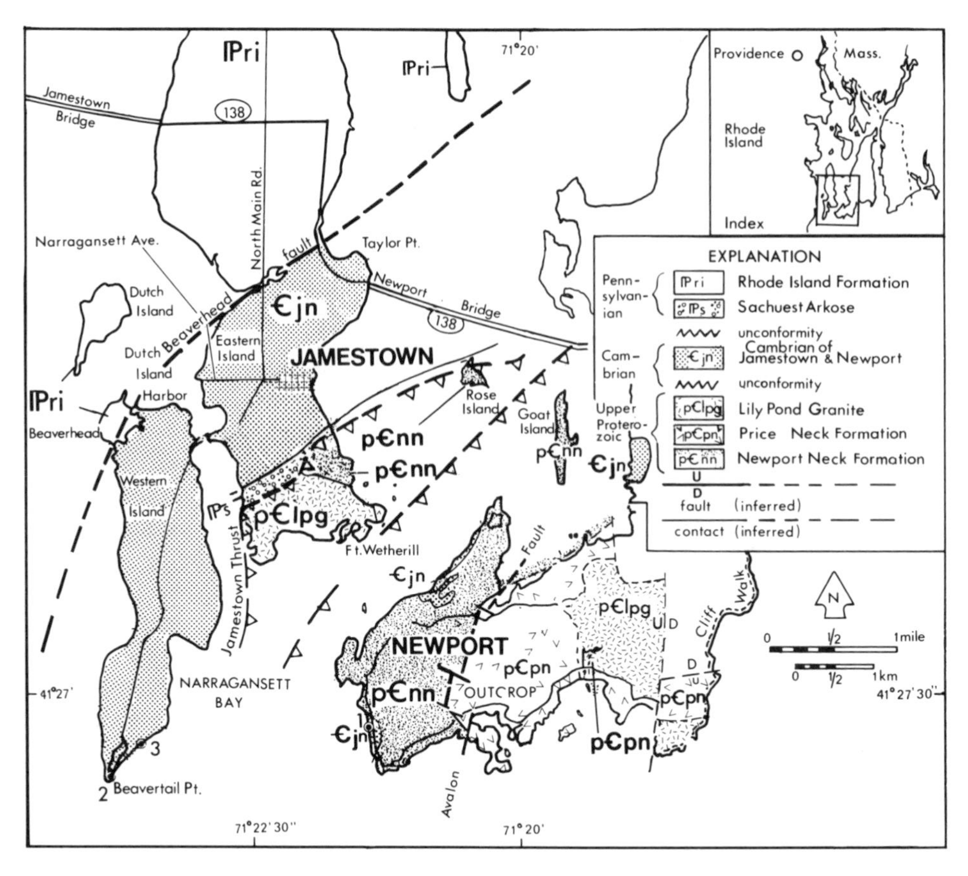

Figure 1. Generalized geologic map of the southern Narragansett Bay area showing place names and geologic stops mentioned in the text. (Modified from Skehan and others, 1981.)

metagraywacke, metasiltstone, and slate beds of the Castle Hill Member of the Newport Neck Formation. The Precambrian Graves Point Member grades upward into the Castle Hill Member; the contact is an angular unconformity (Webster, 1986). The rocks of the Pirate Cave Formation consist of two members: a basal limestone member and an overlying phyllite member.

Limestone Member. The base of the Pirate Cave Formation consists of several thin limestone layers interlayered with dark magenta slate. The strongly developed folding, foliation, boudinage, and hydrothermal activity imposed on this limestone and on several other limestones of the section have obscured the original texture, fabric, and thickness. However, a basal limestone bed (2 to 14 in; or 5 to 35 cm, thick), ranging from pink to red to orange, can be differentiated. In the northern part of the outcrop area (Fig. 2) it strikes east-northeast, and dips northwest at a shallow angle. This basal limestone is fine grained and thinly layered due to millimeter-scale foliation defined by chlorite and limonite. The intervening pinkish limestone is a microsparite with no evidence of such allochems as fossil fragments. Whitish areas of the rock are calcite spar-filled voids reflecting several stages of deformation. The basal and overlying thin limestone beds contain abundant shell debris. Ongoing micropaleontologic investigations have yielded diagnostic tube-like Early Cambrian hyoliths (Webster and others, 1986) that have been stretched and broken by tectonic deformation.

Phyllite Member. The basal limestone is overlain by a 50-ft (15-m)-thick succession of fine-grained varicolored phyllite that ranges from maroon and red to green, gray, and "steely silver." No sandstone and siltstone beds occur within this phyllite sequence; thus original bedding within the member is difficult to establish. Several orange lenses and brown, pyrite-rich phyllite layers may be indicative of original bedding. The mineralogy of the phyllites, in order of abundance, consists of muscovite, microcrystalline quartz, and minor chlorite.

Several horizons within the phyllite contain small, dark resistant patches or flattened nodules similar to phosphate nodules encountered in the Middle Cambrian of Jamestown. Several thin, greenish-white calcareous lenses are encountered throughout the section as well. As yet, no fossil debris or ichnofossils have been identified from the phyllite member.

East Passage Formation. The East Passage Formation is located along the northern coastline of Newport Neck and consists dominantly of red, brown, and gray sandstone and siltstone that is thinly bedded and nongraded. Rare phyllite and limestone and several horizons of volcanic rock and welded tuff also are

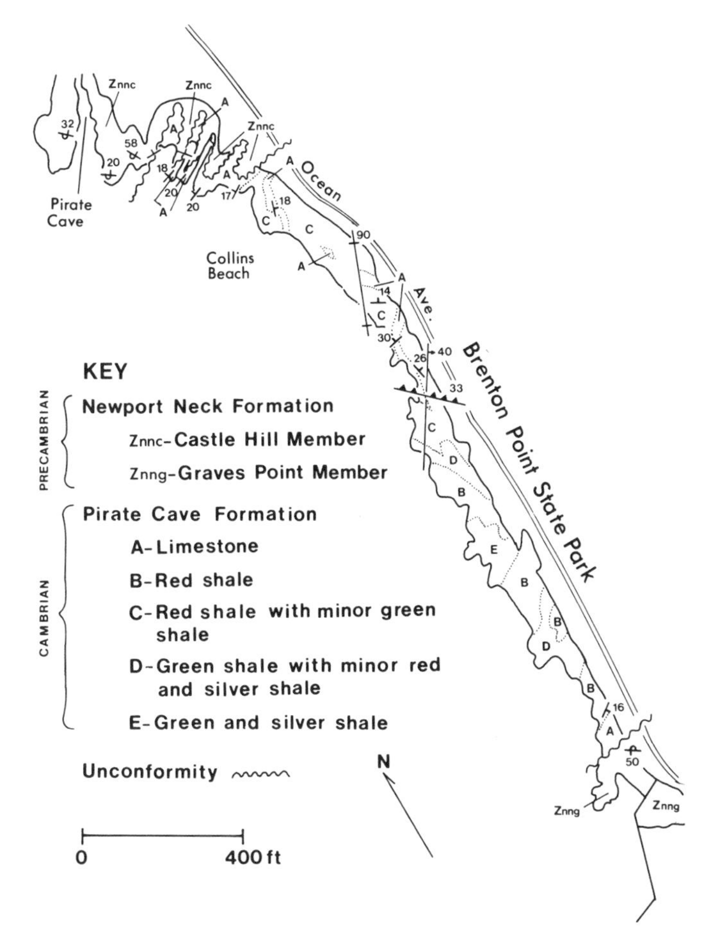

Figure 2. Geologic map of Stop 1 showing the outcrop of Lower Cambrian Pirate Cave Formation, the Upper Proterozoic Newport Neck Formation, and the angular unconformity between them, Brenton Point State Park (Webster, 1986). Location of map area is north of Brenton Point.

Figure 3. Location map for features near Beavertail Point and Lighthouse, Stop 2. (Modified from Skehan and others, 1978, 1981.)

present. The sandstones and siltstones vary from 3 ft to 12 in (1 m to 30 cm) in thickness and are composed of microcrystalline quartz, epidote, chlorite, sericite, detrital feldspar grains, and opaques. The tuffs are generally of uniform thickness (6 to 8 in, or 15 to 20 cm); contain dark flattened fragments and shards; and consist of large feldspar laths, microcrystalline quartz, epidote, opaques, and rare microcrystalline feldspar. The extremely fine-grained volcanic rocks have a mineralogy similar to that of the tuffs. The fault-bounded silvery phyllites are more foliated than the surrounding rocks and contain black porphyroblasts similar to the Middle Cambrian phyllites of Jamestown, Rhode Island. These rocks may be tectonically emplaced.

The lower contact of the East Passage Formation appears to be an angular unconformity, since it overlies both the Castle Hill Member and the Graves Point Member of the Newport Neck Formation in various locations. In several locations, shearing and intense foliation in the rocks at the base of the formation suggest that the contact may be a fault or tectonic slide. In either case, the contact is folded and further study is needed. No upper contact of the East Passage Formation is observed, but it is probable that those rocks underlie a good portion of the bay northwest of Newport Neck. Although a Precambrian age cannot be ruled out, it is highly probable that the East Passage Formation is of Cambrian age and possibly of Middle Cambrian, the faunally based age of the Beavertail Point Member.

Middle Cambrian Rocks of Jamestown

These Middle Cambrian rocks consist of three formations, with only the age of the Jamestown being constrained by fossils. From oldest to youngest, the Jamestown, Fort Burnside, and Dutch Island Harbor Formations all appear to be part of a coherent sedimentary package representing related environments of deposition. Thus the rocks of the Fort Burnside and of the Dutch Island Harbor formations may possibly range into Late Cambrian or Ordovician, although data are now lacking. The reader is referred to Skehan and others (1978, 1981) for a description of the historical development of the ages, including the paleontologically determined ages and stratigraphy of these rocks.

Jamestown Formation. The Jamestown formation forms

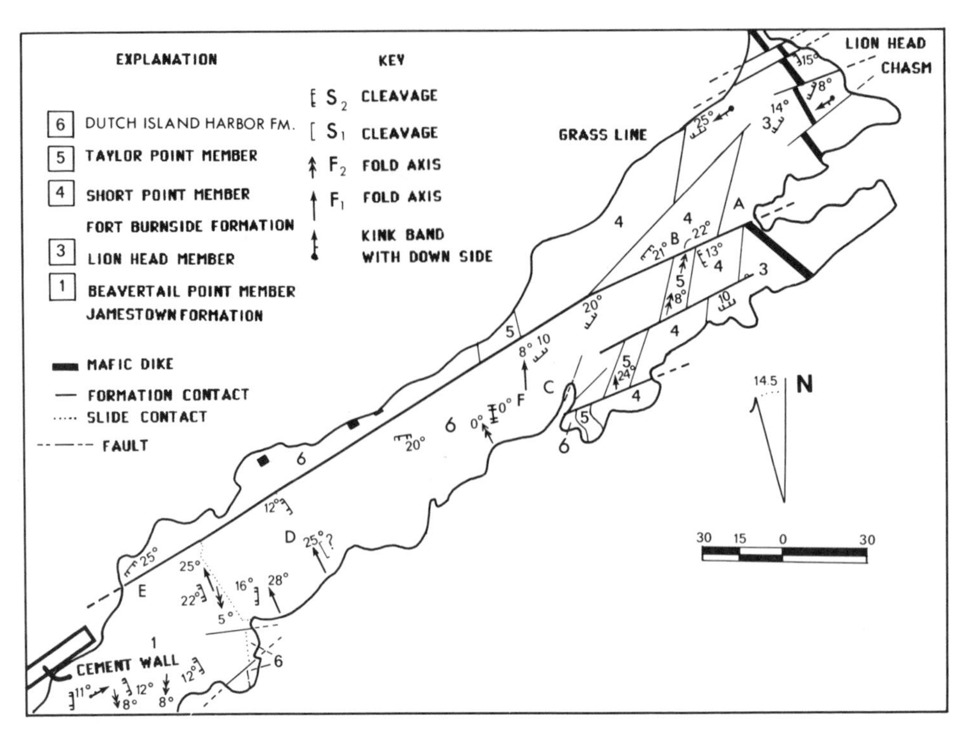

Figure 4. Geologic sketch map of the southeastern shore of Beavertail (Stop 3) showing a nearly complete succession of inverted Middle Cambrian strata; scale in meters. (Modified from Skehan and others, 1981.)

the basal part of the exposed Middle Cambrian succession and consists of fossiliferous green and gray phyllite, and rhythmically layered buff to pink siltstones with black and gray phyllite. This basal unit, approximately 600 ft (200 m) thick, consists of three members, but nowhere is the contact with older rocks seen, nor are all three of the members in contact with each other. The stratigraphic position of the faunally determined Lion Head Member, with respect to the other members of this formation, is inferred but not known with certainty. The Hull Cove Member is probably a facies equivalent of the Beavertail Point Member.

Beavertail Point Member. This unit consists of 80 to 90 percent green phyllite comprising the assemblage quartz + chlorite + muscovite + feldspar ± siderite ± paragonite. Buff and white siltstone makes up about 10 to 15 percent and is typically brown weathering; the buff siltstone in beds 2 mm to 30 cm thick is micaceous (approximately 10 percent muscovite) and may contain 40 percent dolomite or a ferroan carbonate. Where siltstone beds are present, the bedding is recognized as cyclical. The black phyllite (5 percent) has the same mineral assemblages as the green. Dolomite concretions (2.5 to 16 in; 6 to 40 cm) are less abundant than in the Dutch Island Harbor Formation. Ichnofossils may be examined at Stop 2 (Fig. 3).

Hull Cove Member. This member consists dominantly of green and gray phyllite. White siltstone beds are a distinctive component, are rarely 3.5 in (9 cm) thick, and consist of the assemblage quartz + chlorite + muscovite + carbonate. Rarer buff-colored siltstone is also present. The phyllite is cyclically bedded with cycles being up to 18 in (45 cm) thick and progressing upward from white siltstone through black to gray phyllite. Trilobite fragments, as well as minor ichnofossils and an infantile *Paradoxides,* have been observed in this member.

Lion Head Member. This massive gray phyllite without conspicuous bedding consists of the same mineral assemblages as the corresponding parts of the other two members of the Jamestown Formation, but it differs from them in the absence of coarser clastics and in the presence of fluorapatite nodules, abundant trilobites, and trilobite fragments. Bedding is recognized by a minor color variation and is on the order of 1 to 3 in (3 to 8 cm).

Fort Burnside Formation. Named for the U.S. Harbor Command Control Post on Beavertail, formerly occupying part of the land now forming the Bay Islands Park System, the Fort Burnside Formation consists of two members. This formation is sedimentologically similar in its finer grained components to the Jamestown Formation, but its age may range into Late Cambrian or Early Ordovician.

Short Point Member. The cyclical sedimentation units of this member consist of buff siltstone and black and gray phyllite. The cycles result from interlayered siltstone laminae and black phyllite that gradually give way to black phyllite without siltstone laminae. The siltstone beds are micaceous, calcareous, and have abundant ripples, and cross-lamination. Both phyllites appear to be structureless except that fluidization structures are comon in the lower 60 ft (20 m); soft sediment faults are common.

Taylor Point Member. This member is thin (30 ft, or 10 m) and differs from the Short Point in the absence of gray phyllite within sedimentary cycles. South of Lion Head chasm (Fig. 4), the siltstone tends to be micaceous, calcareous, and buff-weathering. Elsewhere, as north of Lion Head and at Taylor Point, the phyllites are whiter and cleaner, containing less than 5 percent carbonate and less than 10 percent mica.

Dutch Island Harbor Formation. This formation is a black rhythmically bedded phyllite whose beds are 1 to 4 cm thick, commonly featuring 1-cm-deep scour channels, and brown-weathering carbonate beds containing concretions 4 to 12 in (10 to 30 cm) long. The Dutch Island Harbor Formation is about 300 ft (100 m) thick (Skehan and others, 1981). Although this formation has certain sedimentologic similarities to the underlying formations, it may range upward in age beyond Middle Cambrian.

STRUCTURAL GEOLOGY

The following sequence in the structural evolution of the Lower Cambrian rocks of the Pirate Cave Formation of Newport is recognized: (1) D_2-regional deformation produced north-south–trending F_2-recumbent folds that are variable in size from millimeters to tens of meters. Precambrian turbidites of the Newport Neck Formation have been deformed regionally by F_1 and S_1; D_2-deformation is common to the Precambrian and Lower Cambrian rocks. An associated axial planar S_2-crenulation cleavage is the dominant fabric over the area, and is low-lying to horizontal with variable dip; (2) development of tectonic slides; (3) north-south–trending, broad, gentle F_3-folds that deform the S_2-cleavage domains as well as bedding; an S_3-vertical fracture cleavage is locally developed; (4) north-south–trending F_4-box folds and conjugate kink band formation; (5) development of variably striking, low-lying thrust faults; (6) late normal faulting and jointing.

Structures recognized in the Middle Cambrian strata of Jamestown include: (1) widespread tectonic slides over the outcrop area, resulting in the juxtaposition of several members and formations with each other, accompanied by the production of breccia along and near the slide boundary (Fig. 4); (2) local D_1-folds of small (Stops 2 and 3) and large size, (Fig. 4) with well-developed S_1-cleavage; (3) intrusion of metaminette dikes (Fig. 4); (4) the dominant S_2-cleavage, well represented throughout the area, associated folds and strongly aligned elongate ferroan carbonate porphyroblasts; the S_2-cleavage is responsible for the nearly flat, platform-like masses arranged like stairsteps over the eastern seashore; (5) the Jamestown thrust fault, whose branches are well displayed on Mackeral Cove, indicates that the Middle Cambrian and Pennsylvanian sequence of Jamestown has been overthrust by a succession of Precambrian rocks (Lily Pond Granite [previously and informally called Newport granite], Newport Neck and Price Neck formations). The Pennsylvanian is represented by the nonconformable, basal Sachuest Arkose (Skehan and others, 1981; Skehan and others, 1986); (6) the north-striking Jamestown thrust fault in Mackeral Cove changes strike at the northwest edge of the granite (new interpretation by Skehan and Rast, unpublished data, 1985); (7) the dominant S_2-cleavage is deformed by kink bands and is cut by quartz-filled and slickensided, normal and reverse faults (e.g., the Beavertail fault, Fig. 4) along which the "platformal stairsteps" are disrupted and rotated.

STOP DESCRIPTIONS

Stop 1: Near northwestern entrance to Brenton Point State Park on Ocean Avenue, just south of Pirate Cave and north of Collins Beach. Precambrian-Cambrian unconformity (Fig. 2). From the northwestern entrance of Brenton Point State Park, walk north along Ocean Drive approximately 600 ft (200 m). Enter the stone gateway on the left and walk 250 ft (75 m) west to the cliff exposures on the shoreline. On this view point and northward are well-graded beds of the Castle Hill Member of the Newport Neck Formation developed in a large overturned syncline. An F_2-fold on the southern face of a small island (250 ft; 75 m long) can be viewed to the north across Pirate Cave.

Southward along the coastline the Lower Cambrian limestones and phyllites are exposed. The unconformity, best observed at low tide, is exposed at the base of the nearby cliff that forms the south slope of this viewpoint, and is located south of Pirate Cave and north of Collins Beach (Fig. 2). The Precambrian rocks at this locality strike approximately 340° and dip variably northeast. The Lower Cambrian limestones in several places strike approximately 060° and dip at a shallow angle toward the northwest but are intensely and recumbently folded. Extensive exposures of the limestone beds are observable on the next point south, as is the unconformity with the coarse Precambrian turbidite sandstones. Although complexly folded by F_2-folds, two distinct limestone layers can be identified. Shell debris has been identified within both horizons, which on preliminary investigation appears to represent pretrilobite Lower Cambrian fossils (E. Landing, personal communication, 1986). Large, east-west–trending isoclinal recumbent folds repeat the limestone layers. A pervasive S_2-slaty, low-lying cleavage, the dominant fabric over the Lower Cambrian succession, is axial planar to such folds. This cleavage is locally and broadly warped by north-south–trending F_3-folds. Box folds and kink bands trending north-south to northeast-southwest deform the cleavage domains still further, and appear related to late-stage brittle faulting, probably associated with Alleghanian deformation.

Stop 2. Beavertail Point, Jamestown, Conanicut Island. The Beavertail Point Member of the Jamestown Formation and the Dutch Island Harbor Formation (Fig. 3) are well displayed here in a folded tectonic slide contact, in turn cut by the Beavertail fault zone. From Beavertail Lighthouse, walk to exposures along the southern point (below the old lighthouse foundation) and the western shoreline (Fig. 3). Geologic features of interest here include: (1) lithology and sedimentologic features of the Beavertail Point Member (type locality); (2) the tectonic slide at the contact of the Dutch Island Harbor Formation and the Beavertail Point Member (Fig. 3)—this is essentially the same type of slide contact on the north side of the Beavertail fault as will be observed at Stop 3, to the south of the fault; (3) ichnofossils or "worm trails" (Fig. 3); (4) an F_1-fold may be observed as well as F_2-folds, especially at the slide contact where the color contrast enhances the recognition of S_2-fold patterns.

A series of specific features may be examined here (Fig. 3): sedimentologic features of the Beavertail Point Member are such that similarities and differences may be noted between that Member and the Lion Head Member of Stop 3, the latter probably being a facies variant of the Beavertail Point Member. These features are exposed in cross sections of F_2-folds to which the dominant S_2-cleavage is axial planar. On the southeast side of the Beavertail fault, gray and green phyllite beds within the Beavertail Point Member are exposed. The tectonic slide contact between the Dutch Island Harbor Formation and the Beavertail Point Member is deformed by F_2-folds just northwest of the Beavertail fault. This contact and the units on either side are offset by the adjacent Beavertail fault, a late post-F_2-fault. A late quartz-breccia–filled fault (N87°E; 55°NW) offsets the tectonic slide contact about 23 ft (7 m) in a sinistral sense. The bedding within the Dutch Island Harbor Formation, about 23 ft (7 m) east of that contact, is intensely disrupted. Cutting through this point there is a mylonite zone 1 ft (0.3 m) thick; N15°W; vertical that truncates bedding. At this point the Dutch Island Harbor Formation is in contact with disrupted, brecciated beds that are deformed by upright F_1-folds plunging N25°W; 12° within this disrupted mass. Worm trails or ichnofossils, identified as *Palaeophycus* (= *Buthrotrephis*), *Planolites*, and *Helminthopsis*, are exposed on bedding planes where they are parallel to S_2-cleavage.

Stop 3. Southeastern shoreline of lower Conanicut Island, Beavertail State Park. Middle Cambrian succession, major structures and trilobite locality (Fig. 4); *no hammers and no trilobite collecting are allowed in the state park.* Walk northeasterly along the shore for about 1,600 ft (500 m) to Lion Head Chasm (Fig. 4). Stay near the upper shoreline exposures and walk carefully around the head of the chasm, along the only path there, to the beginning of the traverse immediately north of the chasm.

Along this traverse the entire inverted Middle Cambrian sequence of Jamestown may be seen, with the exception of one member. From the northeast to the southwest, examine the succession, which, from the base at Lion Head upward, consists of: the Lion Head Member of the Jamestown Formation, the Short Point and Taylor Point Members of the Fort Burnside Formation, and the Dutch Island Harbor Formation. At the southwestern end of the traverse, the Beavertail Point Member, in tectonic slide contact with the Dutch Island Harbor Formation, will be examined.

The following features may be examined: (1) The trilobite locality in the Lion Head Member, from which fossils diagnostic of a medial Middle Cambrian age were recovered (Skehan and others, 1977, 1978). (2) Structures associated with intraformational and interformational tectonic slides (Fig. 4, Locality F); a variety of soft-sediment deformation features possibly associated with slides and related instability of the sedimentational basin as between Localities A and B. (3) Folds and associated features; the dominant folds of this stop are F_2-folds and associated gently dipping S_2-cleavage. The F_1-folds of Localities C and D, with their associated S_1-cleavage, may be locally preserved features associated with intraformational slides. Conversely, F_1 folds at nearby locations appear to be part of larger F_1-structures of regional significance, subsequently deformed by the F_2-flattening event that produced the dominant S_2-cleavage. (3) Late brittle deformation features, such as faults and kink bands, which may be seen at Localities C and E, respectively, the latter being near the well-exposed trace of the Beavertail fault and its branches. Many of these brittle faults contain slickensided vein quartz in the plane of the fault, as at Locality F (Fig. 4), whereas others may be without quartz veins but truncate earlier veins. Throughout many parts of this Cambrian outcrop area on Conanicut Island kink bands are well developed, as north of Localities A and B (Fig. 4) along the higher parts of the outcrop. These kink bands appear to be closely associated with the Beavertail fault and some of its branches.

Note added in proof: Skehan and Rast have recently reinterpreted the deformation features at the Lower Cambrian–Precambrian contact as due to westward overthrusting of the Newport Neck Formation onto the Pirate Cave Formation.

SELECTED RECENT REFERENCES CITED*

Rast, N., and Skehan, J. W., 1981, The geology of Precambrian rocks of Newport and Middletown, Rhode Island, *in* Hermes, O. D., and Boothroyd, J. C., eds., Guidebook to the geology of Rhode Island and adjacent areas: New England Intercollegiate Geological Conference, 73rd Annual Meeting, Kingston, Rhode Island, p. 67–92.

Skehan, J. W., and Murray, D. P., 1979, Geology of the Narragansett Basin, southeastern Massachusetts and Rhode Island, *in* Cameron, B., ed., Carboniferous basins of southeastern New England: Field Guidebook for Trip No. 5, 9th International Congress of Carboniferous Stratigraphy and Geology, p. 7–35.

Skehan, J. W., Murray, D. P., Palmer, A. R., Smith, A. T., and Belt, E. S., 1978, Significance of fossiliferous Middle Cambrian rocks of Rhode Island to the history of the Avalonian microcontinent: Geology, v. 6, p. 694–698.

Skehan, J. W., Rast, N., and Logue, D. F., 1981, The geology of Cambrian rocks of Conanicut Island, Jamestown, Rhode Island, *in* Hermes, O. D., and Boothroyd, J. C., eds., Guidebook to the geology of Rhode Island and adjacent areas: New England Intercollegiate Geological Conference, 73rd Annual Meeting, Kingston, Rhode Island, p. 237–264.

Skehan, J. W., Rast, N., and Mosher, S., 1986, Paleoenvironmental and tectonic controls in coal-forming basins of the United States, *in* Lyons, P., and Rice, C., eds., Coal-forming basins in the United States: Geological Society of America Special Paper 210 (in press).

Webster, M. J., 1986, The structure of the Precambrian Newport Neck, the Lower Cambrian Pirate Cave, and the East Passage Formations, southeastern Rhode Island [M.S. thesis]: Chestnut Hill, Massachusetts, Boston College, 213 p.

Webster, M. J., Skehan, J. W., and Landing, E., 1986, Newly discovered fossiliferous Lower Cambrian rocks of the Newport Basin, southeastern Rhode Island: Geological Society of America Abstracts with Programs, v. 18, p. 75.

*In these recent references, other references cited in this paper may be found.

The Rowe Schist and associated ultramafic rocks along the Taconian suture in western Massachusetts

Norman L. Hatch, Jr., MS 926, U.S. Geological Survey, Reston, Virginia 22092
Rolfe S. Stanley, Department of Geology, University of Vermont, Burlington, Vermont 05405

LOCATION

Outcrops of ultramafic rock enclosed within Rowe Schist along the Taconian suture are just north of Cone Road in the township of Middlefield, western Massachusetts. The site is about 6.4 mi (10 km) north of the village of Chester, which, in turn, is located on U.S. 20 about 25 mi (40 km) west-northwest of the city of Springfield, Massachusetts. The site itself is in the northwest corner of the Chester 7½-minute Topographic Quadrangle (see also Hatch and others, 1970). Adjacent topographic quadrangles are Worthington (Hatch, 1969) to the north, Becket (Norton, 1974a) to the west, and Peru (Norton, 1974b) to the northwest. Access to the site is along country roads either northward from U.S. 20 or southward from Massachusetts 143 (Fig. 1).

From Chester village on U.S. 20, go north on Middlefield Road (paved). (The designation of "paved" or "gravel" for the roads in the area reflects their condition as of October 1985 and may or may not reflect their condition at future dates.) After about 1.4 mi (2.2 km) the road turns sharply right, crosses the West Branch of the Westfield River, and turns sharply left. About 0.5 mi (0.8 km) beyond the bridge you will encounter a series of roadside exposures of serpentinite. These exposures are in the largest of the 42 bodies of ultramafic rock enclosed within the Rowe Schist and the immediately adjoining Moretown Formation that are present in a narrow north-south belt across western Massachusetts. For those interested in the variations in lithology of the ultramafic rock in the Taconian suture belt, a stop at one of these exposures is appropriate. Exposures of ultramafic rock continue north to the Middlefield-Chester town line (also the Hampshire-Hampden County line) about 1.4 mi (2.2 km) past the bridge. Continue north on Middlefield Road (which changes in name to Chester Road at the town line) for about 2.5 mi (4.0 km) to the small village of Middlefield. In the village, turn right (east) opposite the church and just before the village store and Post Office onto Bell Road (gravel) and continue for 0.8 mi (1.3 km). Turn left (north) onto Arthur Pease Road (gravel) and proceed for 0.3 mi (0.5 km) before turning right (east) onto Root Road (gravel). After 0.6 mi (1 km) turn left (north) onto Chipman Road (gravel) and after another 0.7 mi (1.1 km) turn right (east) onto Cone Road (gravel). Access to the site is 0.8 mi (1.3 km) to the east, just before Cone Road begins a long steep downgrade.

To reach the site from the north, start from Massachusetts 143, which runs roughly east-west between Dalton and Williamsburg. From West Worthington on Massachusetts 143, go about 3.8 mi (6 km) east of the village of Peru or about 4.2 mi (6.8 km) west of the village of Worthington (named Worthington Corners on the Worthington 7½-minute Quadrangle) and then turn south onto River Road (paved) along the Middle Branch of the Westfield River. After 3.3 mi (5.3 km) the paved road swings left, following the river valley, and Cone Road (gravel) continues straight ahead (south) up a moderate to steep hill. Continue up Cone Road for about 0.9 mi (1.4 km) to the top of the first rise and the entrance to the site (Fig. 2). Although steep and somewhat rough, Cone Road should be passable in dry weather to a standard two-wheel–drive vehicle. The property at and immediately around the site is owned by Mr. Charles F. Siebold of Easton, Connecticut. Mr. Siebold has generously extended permission for interested geologists to examine the rocks at the site. However, all visitors should exercise great caution, particularly in and near the pits where the talc-bearing rocks are especially slippery. Mr. Siebold assumes no liability for any mishaps. Please respect his property and do not litter.

SIGNIFICANCE

The exposures at the site are part of the narrow belt of ultramafic and associated rocks that extends south across western New England from the Baie Verte–Brompton line of Quebec and Newfoundland (Williams and St. Julien, 1982). The ultramafic rocks are generally interpreted to be fragmented slivers of deep oceanic crust or mantle obducted onto the margin of North America during Taconian (Ordovician) closure of the Iapetus, or proto-Atlantic, Ocean (Stanley and others, 1984; Stanley and Ratcliffe, 1985). This closure resulted from collision between North America, now locally represented by the 1 b.y. old Berkshire massif to the west, and an island arc, now represented by the rocks of the Bronson Hill anticlinorium to the east. The ultramafic belt is thus thought to mark a Taconian suture. The rocks at the site include not only ultramafic rock, now consisting largely of talc, but also pelitic schist and amphibolite that are believed to be the slope-rise sediments and volcanic rocks onto or into which the ultramafic rock was emplaced. After deformation and metamorphism in the Taconian orogeny, all of the rocks of this belt also were extensively deformed and metamorphosed during the Devonian Acadian orogeny. The grade of this Acadian regional metamorphism along the ultramafic belt in Massachusetts ranges from garnet at the Vermont line to low sillimanite at the Connecticut line.

SITE INFORMATION

Initial mapping in the early 1960s of the rocks along the east

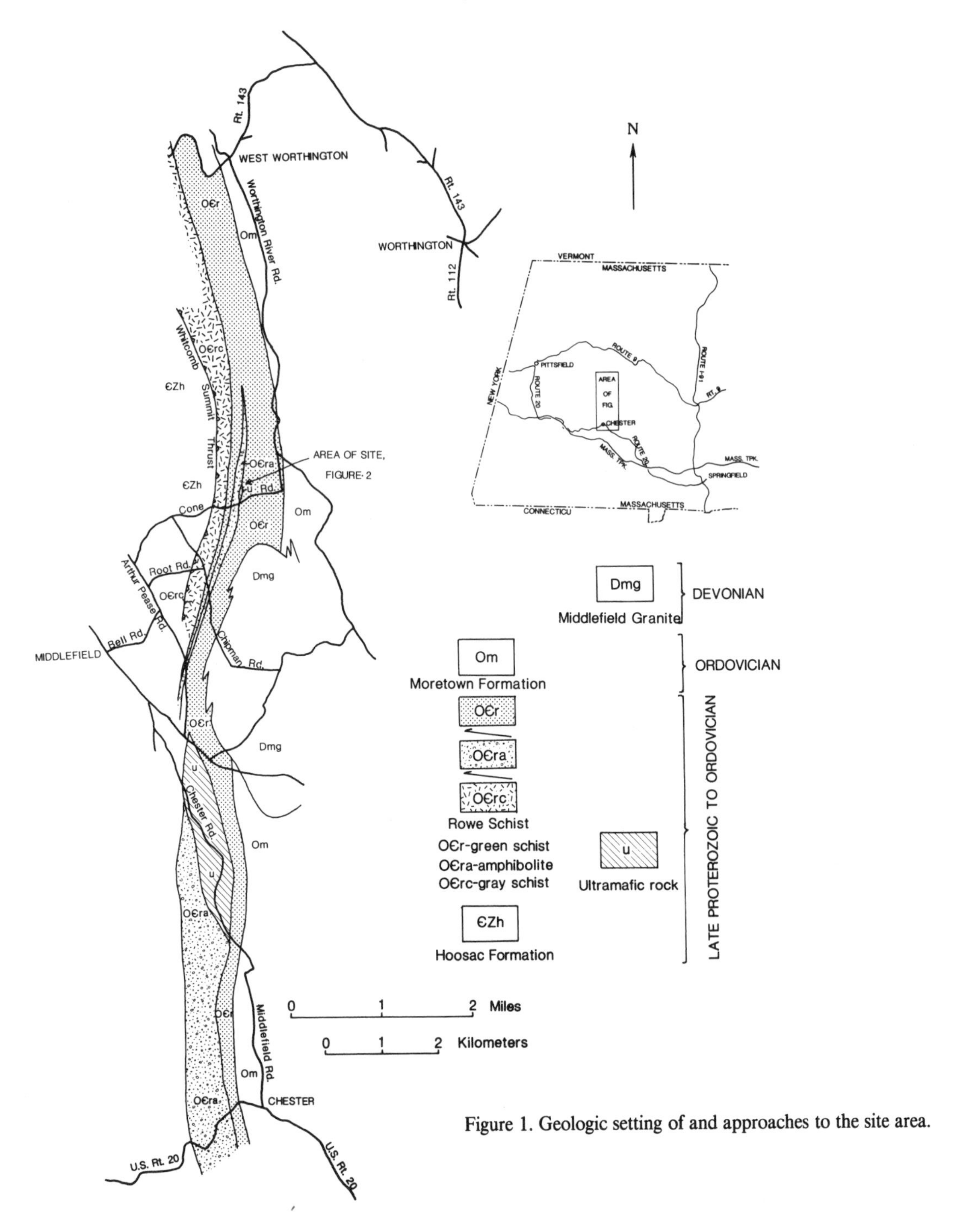

Figure 1. Geologic setting of and approaches to the site area.

flank of the Berkshire massif in Massachusetts convinced Hatch and his colleagues that the sequence of units, Pinney Hollow, Chester, Ottauquechee, and Stowe shown on the map by Doll and others (1961) in southern Vermont could not be mapped into Massachusetts. All of the appropriate lithologies of these units, principally light-green mica-quartz schist, gray carbonaceous mica schist, and amphibolite are present, but they are randomly repeated and interleaved in lenses rather than forming a consistent stratigraphic order. In order to resolve the problems of names for these rocks, while at the same time enabling mapping of the distinctive lithologies, the whole package or interval of lenses of these rocks, bounded by the Hoosac Formation on the west (below?) and the Moretown Formation on the east (above?), was named the Rowe Schist (Hatch and others, 1966), redefining an old term of Emerson (1898). Although Hatch and others (1966) originally interpreted these lenses to be sedimentary, we now believe that they are tectonic and represent slivers of carbonaceous and noncarbonaceous mud and basalt imbricately shuffled during Taconian collision between North America and the Bronson Hill island arc. We interpret the lenses of ultramafic rock to be tectonic slivers and/or olistoliths of oceanic crust or mantle emplaced into the rocks of the Rowe Schist during the subduction phase of this collisional process. Finally, we believe that this whole imbricated package of Rowe and ultramafic rocks was then thrust westward onto the Hoosac Formation on the Whitcomb Summit thrust (Stanley, 1978; Stanley and others, 1984).

Figure 2 is a detailed plane-table map of the site area from Hatch and Norton (1969). Ultramafic rock is best exposed in the two quarry pits, which are most safely approached from the narrow drainage channel connecting them. Within and immediately adjacent to the pits, the ultramafic rock is schistose to massive, light sea-green to white or pale-buff talc rock. Ferroan magnesite and lesser amounts of dolomite form discrete white crystals, coarse aggregates, and veinlets within the talc rock. Magnetite, the only other mineral recognized in hand specimen, is present in scattered grains. About 16.4 ft (5 m) west of the pits are two outcrops of serpentine rock, the northern of which is in exposed contact with the talc rock to its east. The serpentine rock is green and massive and consists predominantly of serpentine plus minor talc and magnesite. Talc rock is inferred to underlie the area without exposures west of the serpentine rock.

The rocks at and south of the site area are at staurolite grade of Acadian regional metamorphism. The large body of ultramafic rock exposed along the road north of Chester Village (Fig. 1) consists predominantly of serpentine plus local talc but also contains some relict olivine near the center of the body, suggesting that it was originally a dunite. Serpentinization, essentially hydration, probably took place during or soon after imbricate faulting and obduction, and this process went to completion in all but the cores of the larger ultramafic bodies. Subsequently, probably during Acadian regional metamorphism, the serpentine was partly dehydrated to talc that now forms an envelope around many of the ultramafic bodies and forms all or most of many of the smaller bodies. The quarries at the present site result from mining for talc, primarily during the 1850s and 1860s (oral communication, Chuck Vreeland of Middlefield, 1968).

West of the trail into the mine pits, outcrops of amphibolite have been traced as a narrow belt for a few miles (kilometers) to the north (Hatch, 1969) and about 2.5 mi (4 km) to the south (Hatch and others, 1970). A second body of similar amphibolite is exposed to the east at the intersection of Cone Road and the trail north to the mine pits. The amphibolites are schistose to massive, dark-green to black, medium- to coarse-grained hornblende-plagioclase rocks. The band west of the ultramafic rocks is at least 410 ft (125 m) wide whereas the band east of the ultramafic rocks is only about 33 ft (10 m) wide.

On its east side, the ultramafic rock is in contact with a light-green, fine-grained muscovite-quartz-chlorite schist in which bedding is faint to nondiscernible. Lenses of granular quartz 0.4 to 0.8 in (1 to 2 cm) thick and about 4 in (10 cm) long are abundant. Similar schist is exposed east of the eastern belt of amphibolite. It is characteristic of green mica-quartz schist of the Rowe mapped all across the state (unit O€r of Zen and others, 1983; Hatch, 1969; and Hatch and others, 1970).

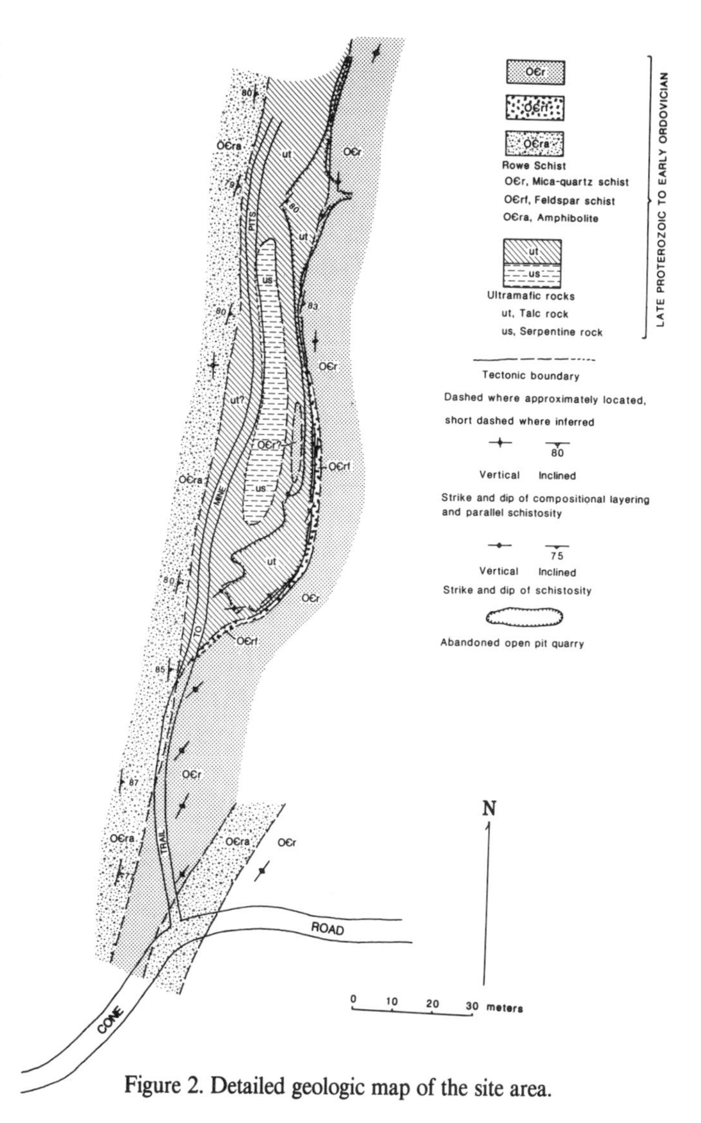

Figure 2. Detailed geologic map of the site area.

Light-green schistose plagioclase-rich rock with minor chlorite and epidote forms a band about 3 ft (1 m) wide at the east contact of the ultramafic rock with the muscovite-quartz-chlorite schist in and immediately north of the south mine pit (unit O€rf on Fig. 2). This band is not present to the north, has not been traced south of the south pit, and was not recognized elsewhere in the Rowe Schist belt. Its origin is uncertain.

The third member of the Rowe Schist in Massachusetts, the gray carbonaceous mica schist (designated O€rc by Zen and others, 1983, and on quadrangle maps) is not present in the immediate site area. Small exposures of it were mapped, however, about 1,000 ft (300 m) to the west and northwest (Fig. 1; Hatch and others, 1970).

The ages of the Rowe Schist and of the ultramafic rocks are very poorly constrained. The Massachusetts bedrock map (Zen and others, 1983) and the pertinent quadrangle maps give the

Rowe a letter symbol designation of O€, although the arrows on the correlation of map units on the state bedrock map indicate that the age could be as old as Late Proterozoic (Z). If we are correct in our model that the ultramafic rocks were tectonically emplaced into the Rowe, they, too, could be as old as the Late Proterozoic opening of Iapetus. The upper age limit for both the Rowe and the ultramafic rocks can only be given as (Early?) Ordovician, although lithic correlations with Taconic rocks to the west suggest that a Cambrian upper limit might be more probable. Recognizing these uncertainties and complexities, the letter symbol designations on the accompanying figures follow those of the Massachusetts bedrock map for compatibility, whereas the age brackets suggest a somewhat broader range.

The dominant fabric in all of the rocks at this site, and in most of the surrounding exposures, is schistosity that strikes north to northeast and dips vertically to very steeply east (Fig. 2). Examination of the green muscovite-quartz-chlorite schist of the Rowe Schist shows that this schistosity locally consists of at least two anastomosing schistosities at very low angles to each other. Where compositional layering is discernible, it is discontinuous and roughly parallel to the schistosity. This layering may represent relict fragments of highly disarticulated bedding. The schistosity is clearly composite. At least one of the schistosities can be traced westward into the fold-thrust fabric in the fault zones within and at the eastern margin of the Berkshire massif and is of probable Taconian age (Ratcliffe and Hatch, 1979; Norton, 1971, 1975). Another can be traced eastward into the axial plane cleavage of isoclinal folds that are extensively developed in the strata of the Lower Devonian Goshen Formation (Hatch, 1975) and therefore must be of Acadian age.

Our interpretation of the structural history of the rocks at this site is as follows. During closure of Iapetus and eastward subduction of oceanic crust beneath an approaching island arc (present Bronson Hill anticlinorium), imbricate westward-directed thrusting shuffled slices of the oceanic crust (or mantle) with rocks of the accretionary prism (the Rowe Schist), imparting a strong thrust-related fabric, now represented by part of the composite schistosity noted above, to all of the rocks. Subsequently, during a later stage of the Taconian orogeny, the intershuffled package was thrust west up onto the continental margin (Hoosac Formation) along the Whitcomb Summit thrust (Fig. 1), further accentuating the earlier fabric. During early stages of the Devonian Acadian orogeny, a strong schistosity that is axial planar to isoclinal folds to the east was formed subparallel to the earlier Taconian fabrics. This composite Taconian-Acadian schistosity was then deformed into its present sub-vertical orientation during subsequent stages of the Acadian orogeny.

REFERENCES CITED

Doll, C. G., Cady, W. M., Thompson, J. B., Jr., and Billings, M. P., 1961, Centennial geologic map of Vermont: Montpelier, Vermont Geological Survey, scale 1:250,000.

Emerson, B. K., 1898, Geology of old Hampshire County, Massachusetts, comprising Franklin, Hampshire, and Hampden counties: U.S. Geological Survey Monograph 29, 790 p.

Hatch, N. L., Jr., 1969, Geologic map of the Worthington Quadrangle, Hampshire and Berkshire counties, Massachusetts: U.S. Geological Survey Geologic Quadrangle Map GQ-857, scale 1:24,000.

—— , 1975, Tectonic, metamorphic, and intrusive history of part of the east side of the Berkshire massif, Massachusetts, [Chapter D], *in* Harwood, D. S., and others, Tectonic studies of the Berkshire massif, western Massachusetts, Connecticut, and Vermont: U.S. Geological Survey Professional Paper 888, p. 51–62.

Hatch, N. L., Jr., and Norton, S. A., 1969, A talc-serpentine body near Middlefield, Massachusetts: U.S. Geological Survey Open-File Report, 7 p., 1 map.

Hatch, N. L., Jr., Chidester, A. H., Osberg, P. H., and Norton, S. A., 1966, Redefinition of the Rowe Schist in northwestern Massachusetts, *in* Cohee, G. V., and West, W. S., eds., Changes in stratigraphic nomenclature by the U.S. Geological Survey, 1965: U.S. Geological Survey Bulletin 1244-A, p. A33–A35.

Hatch, N. L., Jr., Norton, S. A., and Clark, R. G., Jr., 1970, Geologic map of the Chester Quadrangle, Hampden and Hampshire counties, Massachusetts: U.S. Geological Survey Geologic Quadrangle Map GQ-858, scale 1:24,000.

Norton, S. A., 1971, Possible thrust faults between Lower Cambrian and Precambrian rocks, east edge of the Berkshire Highlands, western Massachusetts: Geological Society of America Abstracts with Programs, v. 3, no. 1, p. 46.

—— , 1974a, Preliminary geologic map of the Becket Quadrangle, Berkshire, Hampshire, and Hampden counties, Massachusetts: U.S. Geological Survey Open-File Report 74-92, 17 p. 1 pl., scale 1:24,000.

—— , 1974b, Preliminary geologic map of the Peru Quadrangle, Berkshire and Hampshire counties, Massachusetts: U.S. Geological Survey Open-File Report 74-93, 17 p., 1 pl., scale 1:24,000.

—— , 1975, Chronology of Paleozoic tectonic and thermal metamorphic events in Ordovician, Cambrian, and Precambrian rocks at the north end of the Berkshire massif, Massachusetts [Chapter B], *in* Harwood, D. S., and others, Tectonic studies of the Berkshire massif, western Massachusetts, Connecticut, and Vermont: U.S. Geological Survey Professional Paper 888, p. 21–31.

Ratcliffe, N. M., and Hatch, N. L., Jr., 1979, A traverse across the Taconide zone in the area of the Berkshire massif, western Massachusetts, *in* Skehan, J. W., and Osberg, P. H., eds., The Caledonides in the U.S.A., Geological excursions in the northeast Appalachians; Contributions to the International Geological Correlation Program (IGCP) Project 27; Caledonide orogen: Weston, Massachusetts, Weston Observatory, Department of Geology and Geophysics, Boston College, p. 175–224.

Stanley, R. S., 1978, Bedrock geology between the Triassic and Jurassic basin and east flank of the Berkshire massif, Massachusetts: Geological Society of America Abstracts with Programs, v. 10, no. 2, p. 87.

Stanley, R. S., and Ratcliffe, N. M., 1985, Tectonic synthesis of the Taconian orogeny in western New England: Geological Society of America Bulletin, v. 96, p. 1227–1250.

Stanley, R. S., Roy, D. L., Hatch, N. L., Jr., and Knapp, D. A., 1984, Evidence for tectonic emplacement of ultramafic and associated rocks in the pre-Silurian eugeoclinal belt of western New England; Vestiges of an ancient accretionary wedge: American Journal of Science, v. 284, p. 559–595.

Williams, H., and St-Julien, P., 1982, The Baie Verte—Brompton line; Early Paleozoic continent-ocean interface in the Canadian Appalachians, *in* St. Julien, P., and Beland, J., eds., Major structural zones and faults of the northern Appalachians: Geological Association of Canada Special Paper 24, p. 177–207.

Zen, E-an, ed., and Goldsmith, R., Ratcliffe, N. M., Robinson, P., and Stanley, R. S., compilers, 1983, Bedrock geologic map of Massachusetts: U.S. Geological Survey, scale 1:250,000, 3 sheets.

The Bloody Bluff fault zone near Lexington, Massachusetts

A. E. Nelson, U.S. Geological Survey, Reston, Virginia 22092

LOCATION

Three exposure areas displaying various ductile and brittle deformation fabrics in the Bloody Bluff fault zone are located within the Concord 7½-minute Quadrangle of eastern Massachusetts (Figs. 1 and 2). Stops 1 and 2 are in Lexington, and the rocks underlying Stop 1 form a prominent exposure on the west side of I-95, 0.76 mi (1.2 km) north of its intersection with Massachusetts 2 A (interchange 45). Stop 2 includes two rock exposures west of I-95 near interchange 45; the principal exposure is along Massachusetts Avenue on Fisk Hill, and the other is along an access road that leads to a highway maintenance storage area. It is just west of and parallels the southbound entrance ramp to I-95. The exposures at Stop 2 are just east of Bloody Bluff (Fig. 2) where the fault was first recognized (Cupples, 1961). Bloody bluff is also the site of one of the early battles of the revolutionary war. Stop 3 is in Lincoln and comprises two exposures along the north side of Massachusetts 117 near its intersection with Tower Road; one exposure is about 460 ft (140 m) northwest of Tower Road and the other is about 660 ft (201 m) southeast of the intersection. All exposures are adjacent to public roads, and space to park several vehicles is either adjacent or close to the rock exposures.

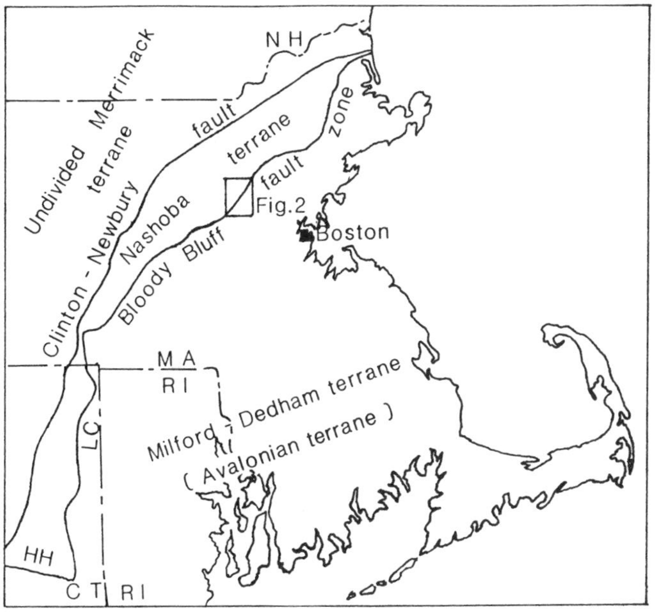

Figure 1. Generalized tectonic map of eastern Massachusetts and adjoining areas showing major terranes and location of Figure 2. HH, Honey Hill fault; LC, Lake Char fault.

SIGNIFICANCE

The Bloody Bluff fault zone (Fig. 1) is a major fracture-suture zone in eastern New England (Nelson, 1976; Skehan, 1968) extending from Connecticut to the Gulf of Maine and possibly to New Brunswick. The fault zone separates the Nashoba lithotectonic zone on the west from the Milford-Dedham lithotectonic zone to the east (Zen, 1983). These zones represent two distinct terranes that were "exotic" to North America prior to mid to late Paleozoic (Hepburn and Munn, 1984). Each of these terranes has a distinctive stratigraphic and plutonic rock assemblage and differing tectonic and metamorphic histories (Hatch and others, 1984; Hepburn and Munn, 1984). Exposures in the Bloody Bluff fault zone were chosen because fabrics in the rocks are suggestive of a long and complex tectonic history that is associated with the accretion of terranes in southeastern New England. Deformed rocks in the fault zone range from slightly to poly-deformed, and their textures include those formed during both brittle and ductile regimes of deformation.

BLOODY BLUFF FAULT ZONE

The Bloody Bluff fault zone forms a long topographic lineament that is defined by aligned but discontinuous streams and valleys. A prominent aeromagnetic lineament coincides with the fault (Zietz and others, 1972). The Nashoba terrane borders the west side of the zone and is composed of highly metamorphosed sedimentary rocks, volcanic rocks, and lesser marble of Proterozoic and/or Ordovician age that was intruded by Ordovician-Silurian Andover Granite (Zen, 1983) and by Silurian granodiorite and tonalite (Zartman and Naylor, 1984). Metamorphism in the Nashoba terrane is in the sillimanite zone and locally, in the sillimanite-K-feldspar zone. This metamorphism probably took place between the Taconian and Acadian orogenies (Hepburn and Munn, 1984). In contrast, the Milford-Dedham terrane, on the east side of the fault zone, is dominated by the 600–630 Ma calc-alkaline Dedham Granite (Nelson, 1975) that intrudes metamorphosed volcanic and sedimentary rocks. The Late Proterozoic rocks of the Milford-Dedham terrane in eastern Massachusetts were metamorphosed twice (Dawse, 1948; Nelson, 1975). The first metamorphism was mostly in the greenschist facies, but locally some rocks have amphibolite facies mineral assemblages. The age of this metamorphism is uncertain but could range from the Late Proterozoic to mid Paleozoic. A later lower grade greenschist facies metamorphism, which is probably Alleghanian in age (Gromet and O'Hara, 1984), retrograded much of the previously metamorphosed rock. The Late Proterozoic rocks are overlain by faintly metamorphosed Late Proterozoic to Cambrian volcanic and clastic rocks, by two small fault slices containing Upper Silurian to Devonian volcanic rocks (Newbury Basin), by a large basin of Pennsylvanian clastic and volcanic rocks (Naragansett Basin), and by several small basins of Triassic-Jurassic rocks off shore. Ordovician to Silurian and Devonian alkalic intrusive rocks were emplaced into the rocks of the terrane. The ages and faunal assemblages of the rocks constituting the Milford-Dedham terrane are similar to those of the Avalon zone of south-eastern Newfoundland (King, 1980). It has been suggested that much of the area from Newfoundland to Georgia was once part of a discontinuous Avalonian volcanic island arc or arcs (Rodgers, 1972), or microcontinents (Rast and Skehan, 1981).

Regional considerations suggest that the Bloody Bluff fault zone is a series of steeply dipping, southeast-directed thrust faults with pre-Mesozoic right lateral movement, and with probable later left lateral and nearly vertical movements during the Mesozoic (Nelson, 1976; Barosh, 1984). Although the Bloody Bluff fault zone is a major structural discontinuity, it is difficult to demonstrate large displacements of discontinuous rock lenses within the fault zone. Nevertheless the displacement is large as rock units can not be correlated across the fault zone. The zone consists mostly of Dedham Granite, metamorphosed volcanic and sedimentary rocks of the Milford-Dedham terrane, gabbro and mafic dikes of Ordovician(?) age, and some serpentine and unnamed granodiorite (Nelson, 1975). Rocks of the Nashoba terrane have not been identified in the fault zone.

The Bloody Bluff fault zone is locally as much as 2 mi (3.2 km) wide. Rocks in the zone range from undisturbed to polydeformed, and they were deformed under both ductile and brittle conditions. The terms used herein to describe the textures of the faulted rocks are adapted from Higgins (1971) and Sibson (1977). Cataclastic rocks refers to cohesive rocks that developed by brittle fragmentation and rotation of mineral grains in zones of dislocation. These rocks have a random fabric. Mylonitic rocks are cohesive and formed during intense intracrystalline plastic deformation, and their matrices commonly are mixtures of deformed and new unstrained grains (Fig. 3A and B). Under these conditions quartz grains in quartzofeldspathic rocks are reduced in grain size by dynamic recovery and recrystallization of highly strained grains (Bell and Etheridge, 1973; White, 1977). Mylonitic rocks have a mylonitic foliation. Rocks with cataclastic textures formed in lower temperature environments, closer to the surface, than rocks with mylonitic textures. The mylonitic rocks developed in a ductile environment whereas the cataclastic rocks deformed under brittle conditions along with fault breccias, gouge zones, and several generations of faults and fractures. Rock in some exposures formed in the transition zone where the deformation mode changed from brittle to ductile.

Sibson (1977) reported on the physical conditions for the formation of various rock fabrics within major fault zones. He showed that mylonite fabrics begin to form near the lower temperature boundary of greenschist facies conditions and that cataclastic fabrics begin to form at slightly lower temperatures. In places, both mylonitic and cataclastic fabrics are in close proximity within the Bloody Bluff zone. The mylonites probably developed as a result of shearing and flowage associated with faulting under greenschist or higher temperature metamorphic environments. The earlier deformation appears to have been mostly ductile, but locally there was concomitant brittle deformation. In places, mylonitic foliation is folded into small isoclinal folds. Later brittle deformation is recorded by gouge zones as well as cross-cutting fractures.

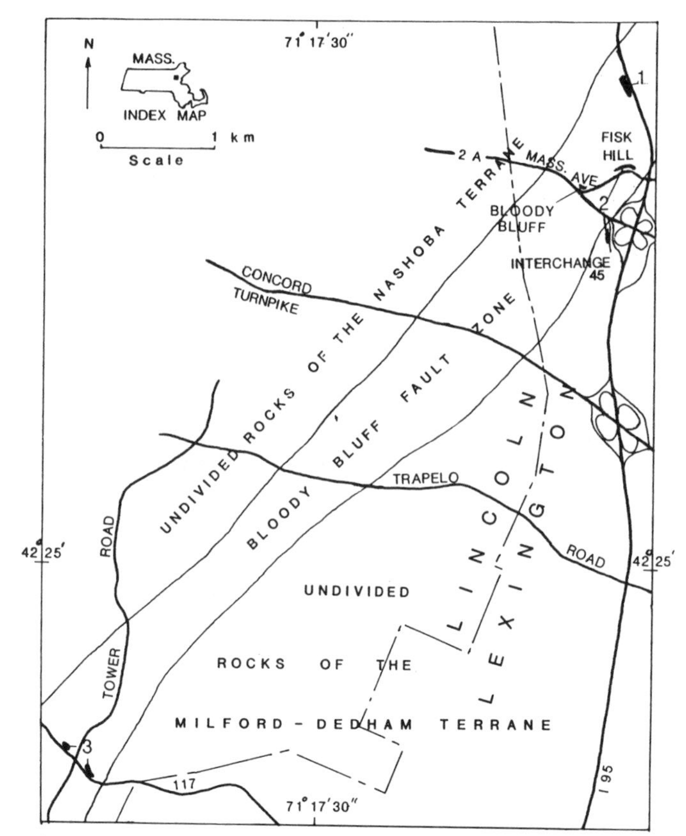

Figure 2. Showing location of stops 1, 2, and 3 and the Bloody Bluff fault zone in the east-central part of the Concord 7½-minute Quadrangle, Massachusetts.

The highly deformed rocks in the fault zone contrast markedly with the mostly undeformed to slightly deformed rocks of the adjacent Milford-Dedham terrane. With few exceptions, the mylonitic foliation seems to parallel the regional foliation. Locally, dissimilar rock types are deformed and juxtaposed. Irregularly interspersed with the mylonitic and cataclastic rocks are discontinuous lenses, more than several feet (meters) thick, of undeformed rock. The close spatial relationship between mylonites and cataclastic rocks and later faulting suggests that deformation took place intermittently along a series of parallel to subparallel faults over a long time interval.

Inconclusive data make it difficult to determine when mylonitic deformation was initiated and how long it lasted. Castle (1964, Fig. 22) showed mylonites of the Bloody Bluff zone to be intruded by the Middle Devonian Peabody Granite (Zartman, 1977). This suggests that ductile deformation in the fault zone began prior to the Devonian intrusive event. The Andover Granite, which has ages of 408–450 Ma and is of probable Ordovician or Silurian age (Zartman and Naylor, 1984), is locally mylonitized adjacent to the Bloody Bluff zone. Therefore, some of the

A

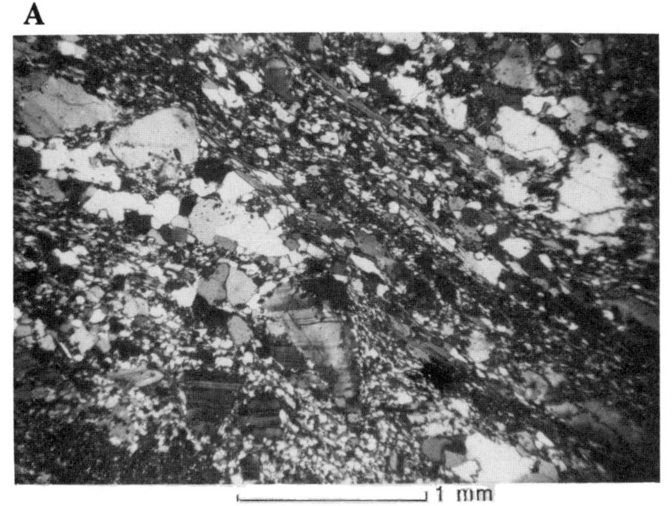

B

Figure 3. Photomicrographs of fault rocks from the Bloody Bluff fault zone near Lexington, Massachusetts, crossed polarizers. A, mylonitic foliation defined by thin directed bands of finely comminuted dark minerals between larger fragmented grains; some post mylonite microfaulting present. B, blastomylonite with well-developed mylonitic foliation, some fragmented feldspar grains in fine textured matrix where neomineralization-recrystallization is dominant over cataclasis.

mylonitized rocks in the fault zone post-date emplacement of the Andover. Zartman and Naylor (1984) suggest that the Nashoba and the Milford-Dedham terranes were not joined in their present relative positions until after the Middle Devonian. The minimum age for brittle deformation in the Bloody Bluff zone is Jurassic since Triassic-Jurassic fossiliferous rocks are present along a fault slice in the zone (Kaye, 1983).

Stop 1. At Stop 1 (Fig. 2) on I-95 approximately 660 ft (201 m) of the Bloody Bluff fault zone is exposed. This location is almost 660 ft (201 m) southeast of the western border of the fault zone. Most of the rocks are gray to grayish pink mylonitized granite, but some lensoidal masses of apparently undisturbed granite are also present. Massive, fine-grained dark-gray gabbro and some thin fine-grained mafic dikes 2 to 3 in (5 cm to 8 cm) thick, crosscut the mylonitized granite. Also present within the mylonitized granite are lenses of variably streaked, fine-grained, dark-gray to black metavolcanic rock(?). Quartz lenses and pegmatitic rocks fill steeply-inclined gash fractures.

Well-developed mylonitic foliation strikes north-northeast and mostly dips steeply west-northwest. In places it is planar but elsewhere it is anastomosing. All the rocks at stop 1 are highly fractured, with many fracture surfaces slickensided. Faults with small displacements, measured in centimeters, are present and one fault with a gouge zone clearly displaces the mylonite.

The mylonitic rocks at stop 1 are blastomylonite (Fig. 3B), mylonite gneiss, protomylonite, and mylonite. Most of the deformed rocks are mylonite gneiss, in which lenses or ovoids of coarse fragments of microcline and sericitized plagioclase lie in a matrix of recrystallized grains. Between the large plagioclase and microcline fragments, smaller fragments of these minerals lie in a mylonite matrix. Tails extending from the larger mineral fragments taper off into recrystallized blastomylonite bands.

Stop 2. Stop 2 consists of two exposures on opposite sides of the southeastern boundary of the Bloody Bluff fault zone. A little more than 660 ft (201 m) of grayish pink mylonitized granite, that has been locally intruded by nondeformed fine-grained gabbro, forms the northern exposure at Stop 2. This outcrop is located on Fisk Hill (Fig. 2) and is only about 330 ft (101 m) west of the eastern border of the Bloody Bluff fault zone. The mylonitized granite, which is locally faulted with displacements measured in centimeters, is commonly fractured and in places contains veins of quartz and veinlets of epidote. The mylonite has a well-developed mylonitic foliation that mostly strikes northeast and dips gently northwest. The mylonitic foliation, which ranges from planar to gently waving to strongly undulating, is folded and contains a few tiny rootless isoclines. The texture of the mylonite varies and in places lenses of coarser mylonite are juxtaposed against very fine grained mylonite (Fig. 3A). The gabbro is chilled at its contact with the granite mylonite. Small variably oriented inclusions of mylonite are present in the gabbro. Some mafic metavolcanic rocks also display a faint mylonitic foliation.

At this outcrop mylonite and protomylonite appear to be more abundant than blastomylonite and mylonite gneiss. The mylonite fabrics in thin section show that mechanical grain size reduction of porphyroclasts dominates over grain-size reduction of porphpyroclasts by recovery and recrystallization. The blastomylonites and mylonite gneisses have textures in which recrystallization and grain growth dominates over grain-size reduction. Generally, the mylonitic foliation in the blastomylonite contains fewer wavy or anastomosing segments than do similar structures in mylonite. Unlike plagioclase and microcline, quartz is not present as large grains. Instead, formerly large quartz grains now form ribbon-like zones that have been polygonized into strain-free areas of subgrains and aggregates of small recrystallized new grains (Bell and Etheridge, 1973). These linear zones alternate with finer grained mylonite zones containing numerous variably sized feldspar fragments and tiny micaceous grains. A rare rock type transitional between mylonite and cataclasite, has poorly developed mylonite foliation with many randomly oriented feldspar fragments. However, polygonized subgrains in a few ribbon-like elongate quartz grains are aligned in a crude mylonitic foliation. One thin section from another part of the exposure shows the mylonitic fabric to be cohesively brecciated and broken into numerous fragments.

The second, or southern, exposure at stop 2 (Fig. 2) along the highway maintenance road consists of variably layered amphibolite and a few very thin layers of siliceous rock. This exposure, which forms part of the Milford-Dedham terrane, is about 165 ft (50 m) from the eastern border of the Bloody Bluff fault zone. Compositional layers range from a few centimeters to as much as several feet (a meter) in thickness. Some of the very thinly layered rock may represent mafic tuffs, but most of the thicker layers probably represent basaltic lava flows. These flows contain many mineralized circular blebs that are probably amygdules. Small veins of epidote are widespread throughout the amphibolite. In places some of the layers are fragmented. Parts of some of these layers have been mylonitized, but the development of mylonite is not as widespread as in rocks of the Bloody bluff fault zone.

Stop 3. The eastern part of the exposure of stop 3, (Fig. 2) east of Tower Road (Fig. 2), is grayish pink mylonitic granite or felsic volcanic rock that has been intruded by massive fine-grained gabbro. These rocks contain inclusions of very fine-grained, altered mafic volcanic rocks. The mylonitic foliation at Stop 3 strikes almost east and has a moderate north dip, but the strike and dip varies due to local folding. The texture of the mylonitized rock is variably fine grained. Some rocks from this site were described as protomylonite to mylonite gneiss by Nelson (1976, see Fig. 4B), but Barosh (1984, Stop 10) considers them to be felsic tuffs and not mylonite. West of Tower Road the outcrop consists principally of medium- to coarse-grained granodiorite containing quartz, plagioclase, and hornblende. This rock grades into a protomylonite containing many ovoids of feldspar.

Rocks from these exposures show a variety of deformed fabrics, including mylonite, protomylonite, rare cataclasite, and blastomylonite. The cataclasite does not have a fluxion structure; however, in one thin section very faint lamellae suggest the beginning of a mylonitic foliation.

REFERENCES CITED

Barosh, P. J., 1984, The Bloody Bluff fault system, *in* Geology of the coastal lowland Boston, Massachusetts, to Kennebunk, Maine: Salem, Massachusetts, Department of Geological Sciences, Salem State College, p. 312–318.

Bell, T. H., and Etheridge, M. A., 1973, Microstructures of mylonites and their descriptive terminology: Lithos, v. 6, p. 337–348.

Castle, R. O., 1964, Geology of the Andover Granite and surrounding rocks, Massachusetts: U.S. Geological Survey Open-FIle Report, 550 p.

Cupples, N. P., 1961, Post-Carboniferous deformation of metamorphic and igneous rocks near the northern boundary fault, Boston basin, Massachusetts, *in* Short papers in the geologic and hydrologic sciences—Geological Survey research 1961: U.S. Geological Survey Professional Paper 424D, p. D46–D48.

Dawse, A. M., 1948, Geology of the Medfield-Holliston area, Massachusetts [Ph.D thesis]: Cambridge, Massachusetts, Radcliffe College, 133 p.

Gromet, P. L., and O'Hara, K. D., 1984, The Hope Valley shear zone; A major late Paleozoic ductile shear zone in southeastern New England: Salem, Massachusetts, Department of Geological Sciences, Salem State College, p. 103–114.

Hatch, N. L., Jr., Zen, E-an, Goldsmith, R., Ratcliff, N. M., Robinson, P., Stanley, R. S., and Wones, D. R., 1984, Lithologic assemblages as portrayed on the new bedrock map of Massachusetts: American Journal of Science, v. 284, p. 1026–1034.

Hepburn, C. J., and Munn, B., 1984, A geologic traverse across the Nashoba block, Eastern Massachusetts, *in* Geology of the coastal lowland Boston, Massachusetts, to Kennebunk, Maine: Salem, Massachusetts, Department of Geological Sciences, Salem State College, p. 103–114.

Higgins, M. W., 1971, Cataclastic rocks: U.S. Geological Survey Professional Paper 687, 97 p.

Kaye, C. A., 1983, Discovery of a Late Triassic basin north of Boston and some implications as to post-Paleozoic tectonics in northeastern Massachusetts: American Journal of Science, v. 283, no. 10, p. 1060–1079.

King, A. F., 1980, The birth of the Caledonides; Late Precambrian rocks of the Avalon Peninsula, Newfoundland, and their correlations in the Appalachian orogen, *in* Wones, D. R., ed., The Caledonides in the USA, I.G.C.P. Project 27, Caledonide orogen: Blacksburg, Virginia Polytechnic Institute and State University, Department of Geological Sciences Memoir 2, p. 3–8.

Nelson, A. E., 1975, Bedrock geologic map of the Framingham Quadrangle, Middlesex and Worcester counties, Massachusetts: U.S. Geological Survey Geologic Quadrangle Map GQ-1274, scale 1:24,000.

—— , 1976, Structural elements and deformation history of rocks in eastern Massachusetts: Geological Society of America Bulletin, v. 87, no. 10, p. 1377–1383.

Rast, N., and Skehan, J. W., 1981, Possible correlation of Precambrian rocks of Newport, Rhode Island, with those of Anglesey, Wales: Geology, v. 9, no. 12, p. 596–601.

Rodgers, J., 1972, Latest Precambrian (post-Grenville) rocks of the Appalachian region: American Journal Science, v. 272, no. 6, p. 507–520.

Sibson, R. H., 1977, Fault rocks and fault mechanisms: Journal of the Geological Society of London, v. 133, p. 191–213.

Skehan, J. W., 1968, Fracture tectonics of southeastern New England as illustrated by the Wachusett-Marlborough Tunnel, east-central Massachusetts, *in* Zen, E-an, White, W. S., Hadley, J. B., and Thompson, J. B. Jr., eds., Studies of Appalachian geology, Northern and maritime: New York, Interscience, p. 281–290.

White, S. H., 1977, Significance of recovery and recrystallization processes in quartz: Tectonophysics, v. 39, p. 143–170.

Zartman, R. E., 1977, Geochronology of some alkalic rock provinces in eastern and central United States: Annual Review of Earth and Planetary Science, v. 5, p. 257–286.

Zartman, R. E., and Naylor, R. S., 1984, Structural implications of some radiometric ages of igneous rocks in southeastern New England: Geological Society America Bulletin, v. 95, no. 5, p. 522–539.

Zen, E-an, ed., Goldsmith, R., Ratcliff, N. M., Robinson, P., and Stanley, R. S., compilers, 1983, Bedrock geologic map of Massachusetts: Reston, Virginia, U.S. Geological Survey, 3 sheets, scale 1:250,000.

Zietz, I., Gilbert, F., and Kirby, J. R., 1972, Aeromagnetic map of the Boston 1°×2° Quadrangle, Massachusetts and New Hampshire: U.S. Geological Survey Open-File Maps, 13 sheets, scale 1:250,000.

Geological Society of America Centennial Field Guide—Northeastern Section, 1987

Stratigraphy of the Boston Bay Group, Boston area, Massachusetts

Richard H. Bailey, Department of Geology, Northeastern University, Boston, Massachusetts 02115

LOCATION

The Boston Bay Group is preserved in the Boston Basin of eastern Massachusetts. The instructions below allow the reader to locate four different stops within the basin. For driving convenience it is easiest to visit the areas in either regular or reverse numerical sequence (Fig. 1). Persons not familiar with the greater Boston metropolitan area should have a detailed road map or road atlas. All of the sites are open to the public without permission during daylight hours and ample parking is available.

Stop 1. Old Mystic River Quarry, Sommerville (Fig. 2). From I-93 (traveling *north*) exit onto Massachusetts 28; turn left under interstate and immediately right onto Mystic Avenue (Massachusetts 38); after 0.3 mi (0.5 km) turn left onto Butler Drive; enter public housing project (that now occupies quarry) and turn left into parking area. Park next to basketball courts and walk about 290 ft (100 m) southwest across the field to the quarry face. The best exposures are to the northwest along the rock face. If heading to Stop 2, take Massachusetts 38 onramp to I-93 south; exit onto Storrow Drive (U.S. Route 1); exit onto Beacon Street at Kenmore Square. Continue west on Beacon Street.

Stop 2. Webster Conservation Area, Newton (Fig. 3). Outcrops in the vicinity of the intersection of Beacon Street and Hammond Pond Parkway may be reached by traveling west on Beacon Street (about 1.2 mi [2 km] beyond Boston College) or north on Hammond Pond Parkway from Massachusetts 9. Park on north side of Beacon Street (Fig. 3) and examine outcrops along and near sidewalk; then walk south down the west side of Hammond Pond Parkway about 500 ft (150 m) to the path that descends into the woods on right. Follow path about 200 ft 60 m) and take trail to right to a series of high ledges.

Stop 3. Squaw Rock Park, Quincy (Fig. 4). Take Exit 20 off Massachusetts 3 south (Southeast Expressway); follow Gallivan Boulevard east under Massachusetts 3 and bear right following signs for Massachusetts 3A to Quincy. Stay in left lane and bear left onto Quincy Shore Drive after crossing Neponset River Bridge. After about 0.9 mi (1.5 km) turn left at stop light onto Squantum Street and bear left (following shoreline) onto Dorchester Street. Just before causeway turn left through gate into the Veterans of Foreign Wars post parking area. Follow trails north and then west along shoreline to outcrop 3A (Fig. 4).

Stop 4. Rocky Neck at Worlds End Reservation, Hingham (Fig. 5). Follow Massachusetts 3A south through Quincy and Weymouth to Hingham. About 3 mi (5 km) east of the Weymouth/Hingham town line, enter traffic rotary and take second right onto Rockland Street; go about 0.6 mi (1 km) and turn left onto Martin's Lane. Follow Martin's Lane to gate at Worlds End Reservation. After paying a small admission fee, drive to unpaved lot at end of road. Follow dirt road on foot north about 1,640 ft (500 m) to intersection with road to right. Turn right and walk about 490 ft (150 m) north to where trail branches left (to west) and skirts pond. Outcrop 4A is about 16 ft (5 m) north of this intersection and 33 ft (10 m) east (to right) into woods. Outcrop 4B is immediately north of trail along steep bedrock scarp (Fig. 5). Outcrops 4C–4F begin at northern tip of point and extend south along the east side of the point.

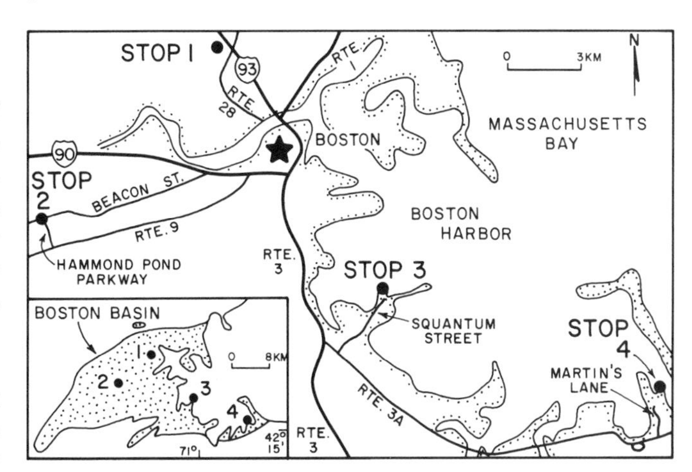

Figure 1. Location map for Boston Basin field stops in Boston metropolitan area. Inset map shows area of outcrop of Boston Bay Group with field stops.

SIGNIFICANCE

The Boston Bay Group is important in characterizing the tectonic/stratigraphic region of eastern Massachusetts known as the Boston terrane. The strata within the Boston Basin and the underlying volcanic and plutonic rocks record a significant Late Proterozoic tectonic event: continental margin rifting or back arc spreading, apparently related to the Avalonian orogeny. These strata may be correlated with late Precambrian rocks deposited in a similar paleotectonic setting within the Avalon terrane of southeastern Newfoundland (McCartney, 1969; King, 1980; Skehan and Rast, 1983).

SITE INFORMATION

The Boston Bay Group consists of about 3 mi (5 km) of clastic sediments and interbedded mafic volcanics. For many years it was thought that these rocks were of Carboniferous age and were related to the well dated non-marine strata of the nearby Narragansett and Norfolk basins (Rahm, 1962; Billings, 1976). Recent research (Bailey and Newman, 1978; Lenk and others, 1982) indicates that a late Precambrian age is more likely.

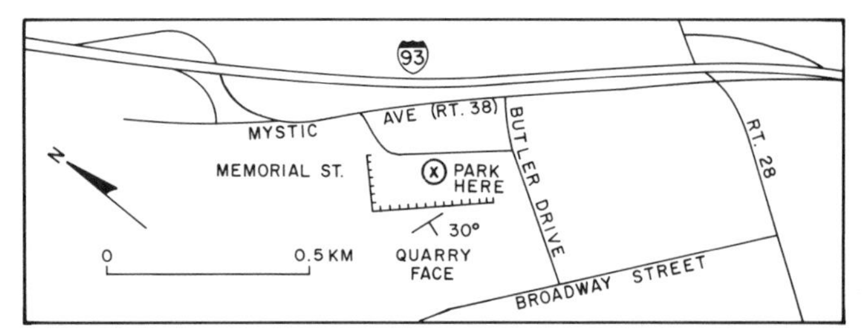

Figure 2. Location of Stop 1, former Mystic River Quarry, Sommerville, Massachusetts.

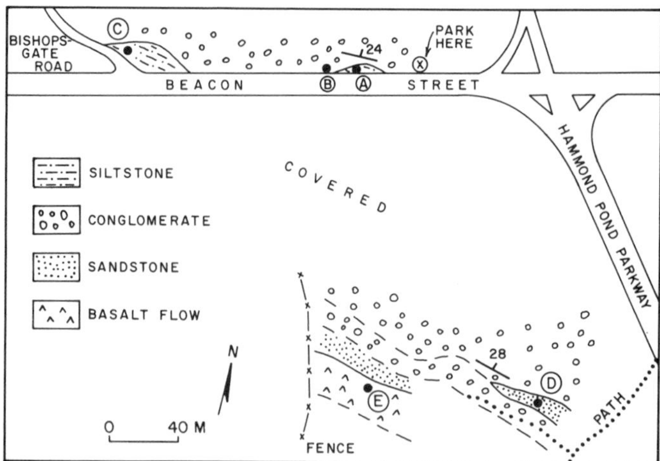

Figure 3. Generalized geologic map at Stop 2, Newton, Massachusetts (from Rehmer and Roy, 1976).

The Boston Bay Group unconformably overlies a 580 to 650 Ma volcanic-plutonic terrane (Kaye and Zartman, 1980; Zartman and Naylor, 1984) containing minor amounts of older metasedimentary rocks.

Fine clastics of the Cambridge Formation dominate the northern part of the basin; whereas, interstratified conglomerate, sandstone, and siltstone comprise the three members of the Roxbury Formation in the southern part (Fig. 6); Billings, 1976). Facies relationships and sedimentary structures indicate that sediments were eroded from a nearby rugged source area to the south and west. Very coarse clastics were carried by rivers through alluvial fan, gravel braid plain, and fan delta environments to a narrow gravelly marine shelf or ramp. Finer clastics and occasionally conglomerates (diamictites) were moved down the slope and into basinal environments by gravity mass transport processes. Deep erosion of the source area and concomitant subsidence of the basin caused fine clastic facies to onlap southward over coarser facies of the basin margin.

Stop 1. The Cambridge Formation at this stop is a black argillite with thin, gray, fine sandstone interbeds 0.1 to 4 in (0.2 to 10 cm) thick. Of particular interest are the deformed and scoured lower contacts of poorly graded sandstone beds. Much of the irregularity of bedding surfaces has been produced by loading and it is not uncommon to find structures created by injection of sand downward into underlying muds. Abundant platy intraclasts, some only 0.1 in (0.3 cm) thick and 1 to 12 in (2 to 3 cm) long, are good indicators of erosion of a cohesive bottom by turbidity currents. Soft sediment structures such as small slump folds, pinch and swell bedding, and micro-faults that affect only several laminae are indicative of syndepositional plastic flow. The outcrop, although possessing a rhythmic flysch-like character, lacks sedimentary structures usually associated with classic turbidites. The observed structures and the abundance of larger scale slump structures associated with intraclastic conglomerates at other localities suggest deposition by low energy muddy turbidity currents on a submarine prodeltaic slope.

Stop 2. At this stop the Brookline Member is a polymictic clast-supported conglomerate interbedded with sandstone and siltstone. Rehmer and Roy (1976) describe this stop and give measured stratigraphic sections. Outcrops 2A, B, and C can be examined by walking along the north side of Beacon Street (Fig. 3). The conglomerate forming the ledge north of Beacon Street fills a channel about 33 ft (10 m) deep eroded into a very thinly laminated siltstone and fine sandstone. At outcrop 2A the siltstone rises about 10 ft (3 m) above the road but quickly descends so that the conglomerate/siltstone contact is at road level at 2B. Careful study of the contact at 2B reveals probable load structures as well as commingling of the conglomerate and siltstone. After following the sidewalk west, turn about 33 ft (10 m) up Bishopsgate Road and examine the outcrop face behind the bushes (2C). The large penecontemporaneous intrastratal folds in the siltstone are all overturned to the northwest. Many of the folds are recumbent and in one case their tops have been truncated by erosion and are overlain by undeformed siltstone.

Outcrops in the woods to the south of Beacon Street give more evidence of conglomerate filled channels. At outcrop 2D, a 15 ft (4.6 m) thick feldspathic litharenite bed is exposed in a high ledge. Following the ledge a short distance to the west the sandstone is channeled down to a thickness of about 1.5 ft (0.5 m). The sandstone bed consists of a series of amalgamated fining upward sequences, each from 1 to 2 ft (0.3 to 0.6 m) thick. The lower part of a fining-upwards cycle begins with a coarse gravel lag followed by planar cross-bed sets 4 to 12 in (10 to 30 cm) high. The cycle often terminates with a rippled and shaly, fine reddish sandstone. Ripples are also exposed just below the sandstone bed. Similar sandstone beds are exposed at other outcrops in the Hammond Pond area. A short distance to the west and near the base of the slope, a portion of a basalt flow is exposed (2E). The flow is amygdaloidal near the top and large vesicular fragments have been eroded from the flow and incorporated into the overlying sandstone. A fan-delta depositional model may be utilized to explain the stratigraphy at this stop. Lateral migration of braided channels on the delta top or gravel braid plain allowed accumulation of upward fining sandstone sequences. High velocity flow during tectonic pulses or flood

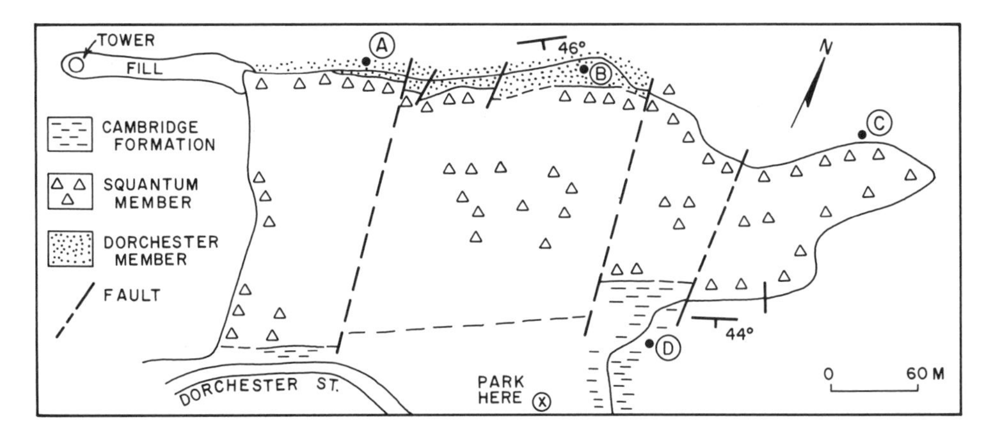

Figure 4. Geologic map of Stop 3, Squaw Rock Park, Quincy, Massachusetts.

events incised the conglomerate filled channels. Progradation of distributary channel facies over estuarine or proximal prodeltaic silts superposed coarse clastics over finely laminated slump folded siltstones as at outcrops 2A, B, and C (Fig. 3). Episodic uplift, changes in flow velocity, and base level shift due to sea level fluctuation may explain similar abrupt transitions of lithosomes frequently observed in the Brookline Member.

Stop 3. The full thickness of the Squantum Member is exposed on the small rocky peninsula known as Squaw Head. At outcrop 3A (Fig. 4), a gradational contact between the Squantum Member and the underlying Dorchester Member is represented by layers of pebbly mudstone interbedded with laminated mudstones and fine sandstones. Outsized clasts in thinly bedded sandstones and diamictites, and intraclasts up to 6 ft (2 m) in length are present in this interval. Here the Dorchester Member (3B) is a gray, thinly bedded sandstone containing graded beds and slump folds with amplitudes up to 6 ft (2 m). Many small scale folds are truncated by overlying graded beds. The Squantum Member (outcrop 3C) is a polymictic diamictite. Subrounded to subangular felsite clasts dominate and a number of granite and quartzite boulders are 2 to 3 ft (0.5 to 1.0 m) in diameter. The upward gradational transition to the overlying Cambridge Formation (outcrop 3D) is represented by an interval of reddish, thinly laminated mudstone and interbedded pebbly sandstones and sandy conglomerates. The origin of the Squantum Member has been debated for many years with some authors favoring glacial deposition (Wolfe, 1976) and others preferring a submarine mass flow or debris flow origin (Dott, 1961; Bailey and others, 1976). The field evidence is best explained by gravity mass flow on or near a submarine slope. In such an environment the diamictite represents a debris flow or olistostrome. Large distorted intraclasts, soft sediment folds, and thinly bedded turbidites are also typical of gravitational disruption and resedimentation in slope and basinal environments.

Stop 4. At Rocky Neck it is possible to study the basal part of the Brookline Member and to see the nonconformity between the Boston Bay Group and the Dedham Granite. Outcrop 4A is a granite knob with a thin veneer of reddish matrix-rich arkose

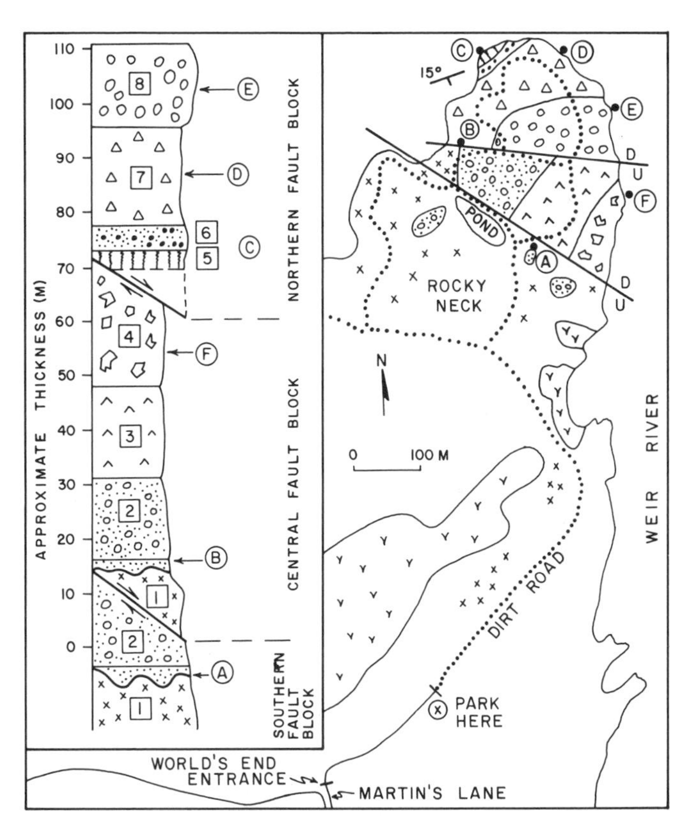

Figure 5. Geologic map and generalized composite stratigraphic section at Stop 4, Rocky Neck, World's End Reservation, Hingham, Massachusetts. Units exposed at outcrops 4A–F indicated along right side of section. Lithologic descriptions of rock units 1–8 (labeled with same symbols on map and section) are as follows: Unit 1, medium to coarse-grained granite; Unit 2, polymictic conglomerate and reddish arkose; Unit 3, massive mafic volcanic; Unit 4, polymictic sharpstone conglomerate containing mafic volcanic and granite boulders; Unit 5, mafic block flow or lahar; Unit 6, conglomerate and feldspathic litharenite; Unit 7, amygdaloidal mafic flow with pillow-like structures; Unit 8, polymictic roundstone conglomerate.

(Fig. 6). This basal arkose overlies the irregular basement surface and fills joint sets that were open prior to deposition. A conglomerate composed of almost all felsite clasts, not derived from the interbedded flows at this stop, overlies the granite at outcrop 4B. There is a paleogrus zone 1 to 2 ft (0.2 to 0.6 m) thick between the granite and the conglomerate. Starting at outcrop 4C and walking south along the eastern shore it is possible to go through a section of about 300 ft (100 m) of interbedded mafic flows and conglomerates (Fig. 5). Interpretation of stratigraphy is complicated by two easterly and southeasterly trending faults (which produce the northerly facing scarps) and by considerable relief on the nonconformity (Billings and others, 1939). Rock units on the northernmost fault block (Fig. 5) are difficult to correlate with those on the southern blocks, suggesting that the displacement of the northernmost fault may be significant. Rocks at 4C are brecciated mafic volcanics, perhaps emplaced as lahars, interbedded with lithic sandstones and conglomerates. This sequence is overlain by an amygdaloidal mafic flow that has pillow-like structures throughout. Between and overlying the "pillows" are highly distorted and irregular pods of thinly laminated silicified tuff. The conglomerates at outcrop 4F contain angular to subangular blocks, boulders, and cobbles of granite, assorted felsites, and extremely angular mafic volcanic fragments in a lithic sand matrix. A red chert, possibly representing a silicified tuffaceous matrix, cements and binds portions of the sandstone and conglomerate. The Rocky Neck sequence records tectonic instability during the early history of the Boston Basin. Rifting and thermal arching produced rugged exposures of granitic and felsic volcanic basement. As fault blocks of basement were uplifted they were rapidly eroded and buried in their own detritus. Mafic flows, probably both subaerial and subaqueous, were cannibalized by erosion to produce clasts, and in some cases volcanogenic materials were introduced directly as tuffs or lahars. Bell (1964) describes related localities in the nearby Nantasket area that are worth seeing.

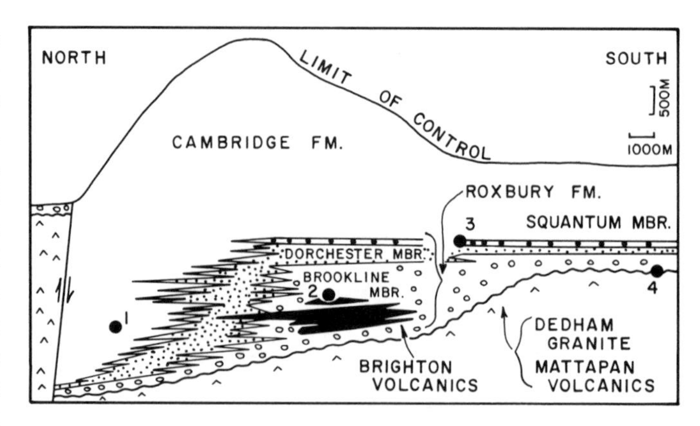

Figure 6. Schematic stratigraphic cross section of the Boston Basin showing facies relationships. Approximate locations of field stops indicated with numbers (modified from Billings, 1976).

REFERENCES CITED

Bailey, R. H., and Newman, W. A., 1978, Origin and significance of cylindrical sedimentary structures from the Boston Bay Group, Massachusetts: American Journal of Science, v. 278, p. 703–711.

Bailey, R. H., Newman, W. A., and Genes, A., 1976, Geology of the Squantum "Tillite," *in* Cameron, B., ed., Geology of southeastern New England: New England Intercollegiate Geologic Conference, 68th Annual Meeting, p. 92–106.

Bell, K. G., 1964, Structure and stratigraphy of the Nantasket locality, *in* Skehan, J. W., ed.: New England Intercollegiate Geological Conference, 56th Annual Meeting, p. 115–120.

Billings, M. P., 1976, Geology of the Boston Basin: Geological Society of America Memoir 146, p. 5–30.

Billings, M. P., Loomis, F. B., Jr., and Stewart, G. W., 1939, Carboniferous topography in the vicinity of Boston, Massachusetts: Geological Society of America Bulletin, v. 50, p. 1867–1884.

Dott, R. H., 1961, Squantum "tillite," Massachusetts; Evidence of glaciation or subaqueous mass movement?: Geological Society of America Bulletin, v. 72, p. 1289–1305.

Kaye, C. A., and Zartman, R. E., 1980, A Late Proterozoic Z to Cambrian age for the stratified rocks of the Boston Basin, Massachusetts, U.S.A., *in* Wones, D. R., ed., Proceedings of the Caledonides in the U.S.A., IGCP Project 27; Caledonide orogen: Blacksburg, Virginia Polytechnic Institute and State University Memoir 2, p. 257–262.

King, A. F., 1980, The birth of the Caledonides; Late Precambrian rocks of the Avalon Peninsula, Newfoundland, and their correlatives in the Appalachian orogen, *in* Wones, D. R., ed., Proceedings of the Caledonides in the U.S.A., IGCP Project 27; Caledonide orogen: Blacksburg, Virginia Polytechnic Institute and State University Memoir 2, p. 3–8.

Lenk, C., Strother, P. K., Kaye, C. A., and Barghoorn, E. S., 1982, Precambrian age of the Boston Basin; New evidence from microfossils: Science, v. 217, p. 619–620.

MCartney, W. D., 1969, Geology of Avalon Peninsula, southeast Newfoundland, *in* Kay, M., ed., North Atlantic; Geology and continental drift: American Association of Petroleum Geologists Memoir 12, p. 115–129.

Rahm, D. A., 1962, Geology of the Main Drainage Tunnel, Boston, Massachusetts: Journal of the Boston Society of Civil Engineers: v. 49, p. 319–368.

Rehmer, J. A., and Roy, D. C., 1976, The Boston Bay Group; The boulder bed problem, *in* Cameron, B., ed., Geology of southeastern New England: Intercollegiate Geologic Conference, 68th Annual Meeting, p. 71–91.

Skehan, J. W., and Rast, N., 1983, Relationship between Precambrian and lower Paleozoic rocks of southeastern New England and other North Atlantic Avalonian terrains, *in* Schenk, P. E., ed., Regional trends in the geology of the Appalachian-Caledonian-Hercynian-Mauritanide orogen, Dordrecht, Netherlands, D. Riedel, p. 131–162.

Wolfe, C. W., 1976, Geology of Squaw Head, Squantum, Massachusetts, *in* Cameron, B., ed., Geology of southeastern New England: New England Intercollegiate Geologic Conference, 68th Annual Meeting, p. 107–116.

Zartman, R. E., and Naylor, R. S., 1984, Structural implications of some radiometric ages of igneous rocks in southeastern New England: Geological Society of America Bulletin, v. 95, p. 522–539.

Glacial and coastal geology, Cape Cod National Seashore, Massachusetts

John J. Fisher, *Department of Geology, University of Rhode Island, Kingston, Rhode Island 02881*
Stephen P. Leatherman, *Laboratory for Coastal Research and Department of Geography, University of Maryland, College Park, Maryland 20742*

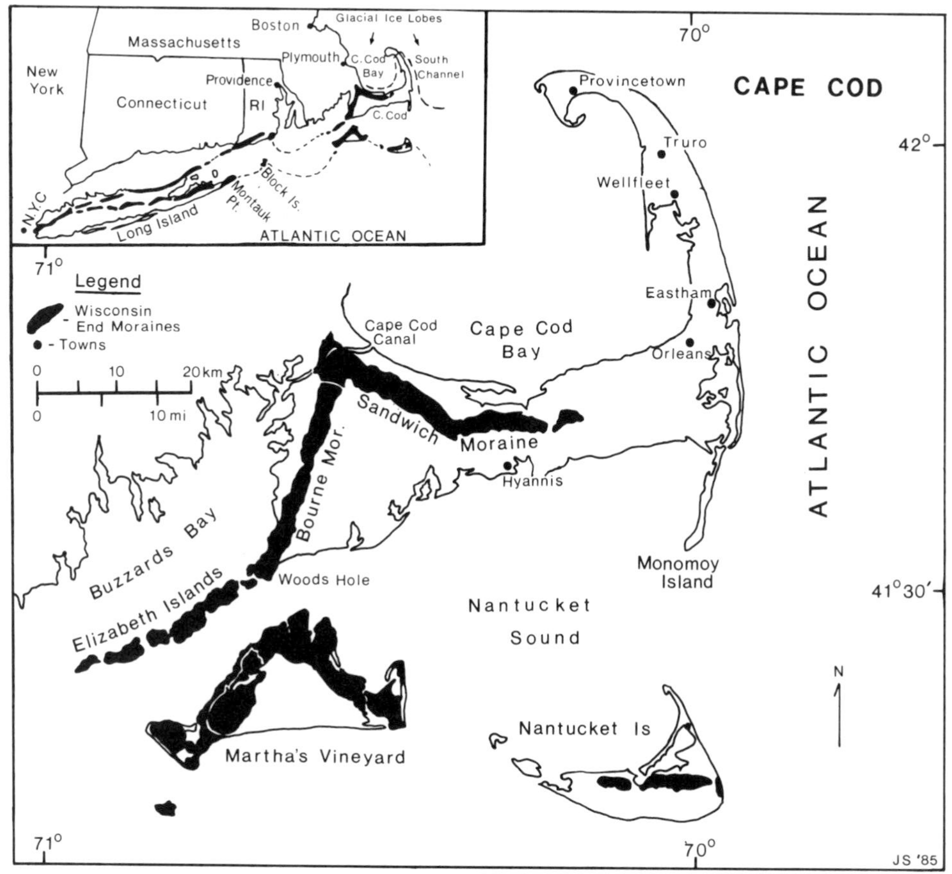

Figure 1. Map of southeastern New England, showing relationship of Cape Cod's glacial geology to regional glacial geology. Cape Cod is composed primarily of a late Wisconsian recessional moraine with extensive associated outwash plains. Martha's Vineyard and Nantucket Islands are composed of the same late Wisconsian terminal moraine. The same terminal and recessional moraines are also responsible for the development of Long Island.

LOCATION

Cape Cod National Seashore locality is located on a peninsula in southeastern Massachusetts (Fig. 1). It is 90 mi (145 km) from both Boston, Massachusetts, and Providence, Rhode Island. Driving time is 2 to 3 hours by divided highways (I-195 from Rhode Island, Massachusetts 3 from Boston) to the Cape Cod Canal and then U.S. 6 to the Orleans Visitors Center. Airports are at Hyannis and Provincetown. Because this Cape Cod locality is so far from major through routes, several stops have been described that are suitable for either a half or full day trip. Suggested field stops occur off U.S. 6, at easily accessible sites primarily on public lands and maintained trails in order to preserve private property and protect the fragile environment (Fig. 2).

Topographic quadrangles are: Orleans, Eastham, Wellfleet, North Truro, and Provincetown at 1:24,000.

Stop 1. Visitors Center. Cape Cod National Seashore Park Eastham is 2.8 mi (4.5 km) north of the Orleans traffic rotary on U.S. 6, then 0.1 mi (0.2 km) east of U.S. 6.

Stop 2. Doane's Rock. Leave Center, go east on Nauset Road, at 0.5 mi (0.8 km) continue east on Doane Road, 0.3 mi (0.5 km) to Doane's Rock parking lot on Pinecrest Road.

Stop 3. Coast Guard Beach. Leave Doane's Rock, go east on Doane Road, past intersection, 0.5 mi (0.8 km) to parking lot. During peak summer months only shuttle buses from the visitor's center is allowed in this lot.

Stop 4. Nauset Lighthouse. Leave Coast Guard Beach, go north 0.2 mi (0.3 km) to Ocean View Drive, then north 1.9 mi

(3.0 km) and east 0.1 mi (0.2 km) on Cable Road to parking lot.

Stop 5. Marconi Wireless Station. Leave the Nauset Lighthouse lot, head west on Cable Road 0.9 mi (1.4 km) to intersection and north on Nauset Road 1.0 mi (1.6 km) to U.S. 6. Go north on U.S. 6 2.5 mi (4.0 km) to Marconi Wireless Station-Beach Road, then east 0.1 mi (0.2 km) to intersection and northeast 1.0 mi (1.6 km) to Marconi Station parking (not Marconi Beach to southeast).

Motor Route—Ocean View Drive. Between Stops 5 and 6, an optional winding sea cliff drive, 12 mi (19.0 km) in length, is described under the Stops section.

Stop 6. Pamet River. Retrace the route from Marconi Station to U.S. 6, then go north 8.0 mi (12.8 km) to Pamet Roads where a 3.1 mi (5.0 km) round trip on south and north Pamet Roads leads to parking at Ballston Beach.

Stop 7. Highland "Cape Cod" Light. Leave Pamet Roads, go northwest on U.S. 6, 2.9 mi (4.6 km) to south Highland Road, then northeast 1.3 mi (2.1 km) to a short spur road and east 0.3 mi (0.5 km) to the lighthouse parking lot.

Stop 8. Pilgrim Heights. Leave Highland Light by retracing the spur road and turn north 0.1 mi (0.2 km) to Highland Road, then west 0.8 mi (1.3 km) to U.S. 6 and northwest 1.6 mi (2.6 km) to Pilgrim Heights parking lot.

Stop 9. Province Lands Visitor Center. Leave Pilgrim Heights lot and go to U.S. 6, continue 4.5 mi (7.2 km) northwest to Race Pt. Road, then northwest 1.2 mi (1.9 km) to the visitors center on the spur road 0.4 mi (0.6 km) to northeast.

Stop 10. Race Point Beach. Leave the visitors center on the spur road, go back to Race Point Road and then northwest 0.8 mi (1.3 km) to Race Point Beach parking lot.

(S. P. Leatherman was responsible for descriptions of Stops 3, 8, and 9; J. J. Fisher for the remaining stops and general material.)

SIGNIFICANCE

Outer Cape Cod is significant in that it has been the site of some of the earliest basic coastal (Davis, 1896) and glacial (Woodworth and Wigglesworth, 1934) geologic studies in the New England area. It continues to be a site for similar studies because it is the longest continuous unaltered coastline along the eastern United States. There are no jetties, groin, or seawalls along the entire outer cape shoreline. Unlike most of the New England coast, which is influenced by bedrock topography, this coast has been completely formed by marine agencies. It was and continues to be significant in glacial studies because it is the first east coast mainland site of the Wisconsinan recessional moraine and outwash system. In addition, the extensive sea cliffs produced by wave action provide excellent exposures of the Cape Cod interlobate outwash plain system. This outer Cape Cod coastline is also the northernmost coastal compartmented barrier shoreline as found along the middle Atlantic barrier coast from North Carolina to Cape Cod, consisting of a spit, headland, barrier island, and segmented barrier island segments (Fisher, 1982a).

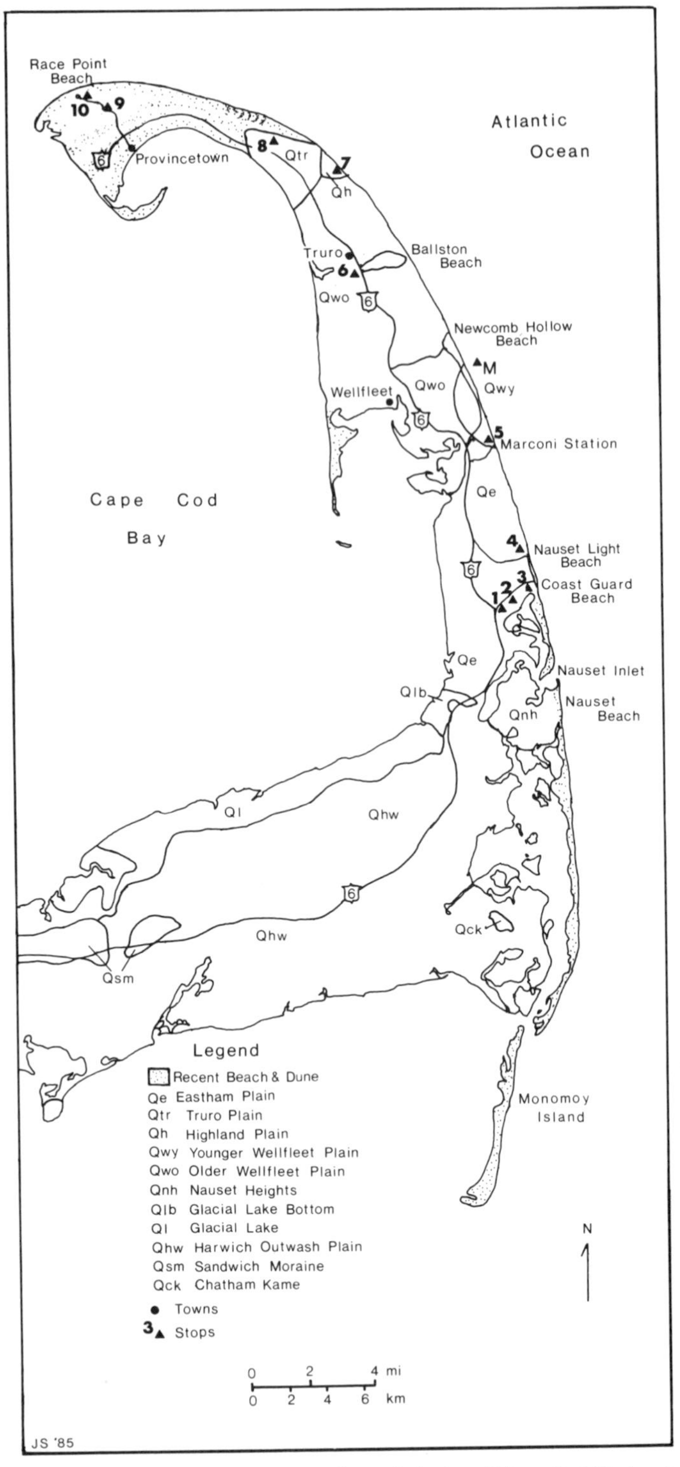

Figure 2. Map of eastern "outer" Cape Cod, site of Cape Cod National Seashore, indicating main highways, towns, field trip stops, and glacial geology. Outer Cape Cod is composed primarily of interlobate glacial outwash deposits developed between the Cape Cod Bay glacial ice lobe and the South Channel ice lobe offshore, in what is now the Atlantic Ocean. Post-glacial wave action during sea level rise has developed extensive spits such as at Provincetown and along the "outer" beach of Cape Cod.

Cape Cod's Outer Shoreline, first called "the Great Beach" by the well-known author Henry David Thoreau (1895), has changed little since the Pilgrims first landfall on its shores some 350 years ago. The National Park Service has attempted to preserve the original character of this shoreline as well as its unique coastal, glacial, and dune landscape by incorporating it into the National Park system.

SITE INFORMATION

Site 47. The southernmost extent of the late Wisconsin-age ice advance in southern New England is responsible for both the unconsolidated material that forms Cape Cod as well as all the major offshore islands (Fig. 1). The Wisconsin terminal moraine extends along the length of Long Island to Montauk Point, then Block Island, Martha's Vineyard, and Nantucket Island (Fig. 1, insert). Cape Cod contains the easternmost extent of the prominent late Wisconsin recessional moraine that also extends from New York City, along the north shore of Long Island, the southern coast of Rhode Island, the Elizabeth Islands, and finally to Cape Cod itself. From Woods Hole north to the Cape Cod Canal the system of recessional moraines is divided into the Bourne and Sandwich recessional moraines. South of this recessional moraine system on Cape Cod is an extensive sandy outwash plain (Hyanis Outwash Plain) pitted with kettle hole lakes. An interlobate outwash plain between the Cape Cod Bay lobe and South Channel ice lobe, in what is now the Atlantic Ocean, makes up the easternmost section of Cape Cod, that area set aside as the Cape Cod National Seashore.

Site 48. Holocene wave action eroding Pleistocene glacial deposits, primarily outwash sands, has developed depositional spits, baymouth beaches, barrier beaches, and barrier spits (Fisher, 1985). Wind action has developed extensive dunes in the Provincetown section of Cape Cod.

SITE DESCRIPTIONS

By following the stops in the indicated manner, one will encounter upland glacial features at Stops 1 and 2, then development of the southern depositional barrier spit shoreline at Stop 3; Stops 4, 5, 6, and 7 indicate the glacial and coastal erosional features along the sea cliffs. Finally, Stops 8, 9, and 10 show the depositional development of the northern spit coastline and the attendant dune systems. Regular changes in beach sediments in each of these coastal compartments can be observed at all adjoining beaches from Stops 3 to 10.

Stop 1. Visitor's Center, Cape Cod National Seashore Park, Eastham. This center is recommended for publications, maps, geologic and historic displays, and the overlook it provides of Salt Pond, a kettle hole drowned by rising sea level. In the distance, beyond Salt Pond Bay to the southeast, is Nauset Beach Spit, a southerly prograding barrier spit beach enclosing Salt Pond Bay. Nauset Inlet is at the barrier spit's southern end while Coast Guard Beach is at the spit's mainland northern end.

Stop 2. Doane's Rock. Hidden in a grove of trees to the south of the road is Doane's Rock (Enos Rock), the largest glacial erratic on Cape Cod. It is 45 ft (13.5 m) long, 25 ft (7.5 m) wide, and 18 ft (5.5 m) high but continues below ground to a depth of 12 ft (3.5 m). Doane's Rock is basaltic in composition; there are no similar bedrock materials beneath the Cape, but similar basaltic material north of Boston has been suggested as the source.

Stop 3. Coast Guard Beach Station. The Nauset Spit system has developed from the deposition of material eroded from the glacial cliffs of Eastham and Wellfleet and transported southward by littoral currents. This barrier beach system, consisting of three spit segments and two islands, formed by spit elongation within the last 5 to 6 thousand years. Since this time, the spit system has continued to evolve and has migrated landward, constantly reducing the overall dimensions of the enclosed bays. Nauset Spit is dependent upon the eroding headland (glacial) section of the outer Cape for its continued sediment supply. The shoreline position of the Spit is generally controlled by the erosion rate of these glacial bluffs (averaging 2.6 ft [0.8 m] per year; Leatherman and others, 1981), since a smooth contour of the outer Cape has been maintained through time by wave and longshore current processes.

Nuaset Spit–Eastham and Nauset Spit–Orleans (Fig. 3A) protect Nauset marsh and serve as the outer barrier for Nauset Harbor. This barrier consists of a double spit separated by Nauset Inlet. Prior to 1946, there was no south spit since Nauset Inlet was located against the glacial headlands at Nauset Heights. Since 1946, Nauset Spit–Orleans has rapidly grown northward with lateral inlet migration at the expense of Nauset Spit–Eastham. This change is occurring even today, despite the long-term southward direction of net littoral drift along this section of the Atlantic coast of Cape Cod.

Correlation of a series of cores into a stratigraphic section illustrates the long-term landward migration of the spit. Figure 3B is a cross-section constructed from cores taken at Site 1 on Nauset Spit-Eastham. Core C-1 was taken in the middle of a narrow washover throat in the dune line, and two cores (C-2 and C-3) were taken in the recently overwashed deposit in the salt marsh. An outcrop of peat (*Spartina alterniflora*) was present on the beach during the survey period. A second peat layer 24 in (60 cm) thick, dated at 815 ± 95 B.P. and typed as *S. patens* with lenses of *S. alterniflora,* was found at the bottom of a core C-1, more than 13 ft (4 m) below the present surface. Above this organic layer was an orange/white sandy section, characterized by coarse sandy zones and heavy mineral laminations. This material was subaerially deposited by overwash, with evidence of some deposition of layers by wind. The surface layers in cores C-1 and C-2 are recent washover deposits from the 1972 "northeaster" storm (Fig. 2; Leatherman, 1979). From these cores it is evident that a well-developed salt marsh existed behind the barrier dune on Nauset Spit–Eastham as early as 815 years ago. Washover deposits buried the salt marsh sometime after this time, and dunes subsequently formed on them. The presence of peat also indicates that an inlet has not existed in area of location during the last 800 years and that the dunes must have formed atop washovers.

Transgressive sandy barrier beaches along the northeastern coast of the United States are dominated by dune ridges and overall dune-type topography. Nauset Spit falls within this category and is retreating landward by inlet dynamics and overwash processes. Salt marshes have developed principally upon flood tidal deposits, related to the original growth and development of the Spit or subsequent cyclic breaching and downdrift inlet migration. Thus the lateral growth of this transgressive barrier is largely controlled by inlet dynamics, but overwash during major storms has resulted in substantial widening in some areas (Leatherman and Zaremba, 1986). Overwash tends to be most effective in barrier migration through the interaction with aeolian transport and dune-building processes in the subaerial zone.

Stop 4. Nauset Lighthouse Parking Lot. This lot is located on the Eastham outwash plain formed by deposition from the South Channel ice lobe that was offshore in what is now the Atlantic Ocean to the east. In 1839, three brick lighthouses occupied this site, but cliff erosion toppled all three into the ocean between 1892 and 1923. Geomorphologic landform analysis indicates that the outer Cape is a "mega-winged headland" with spits growing north and barrier spits and barrier islands growing south from the eroding outwash plain headlands. The identification of these Cape Cod regional coastal landforms by shoreline sedimentology has been investigated in a synoptic high tide sampling program after summer deposition (Fisher, 1972). The sediments show more clearly an interesting median grain size variation along the beach (Fig. 4), earlier suggested by Schalk (1938). Median grain size southward of Coast Guard Beach, (Stop 3) along the Nauset Spit-Eastham beach shows anomalous increase in grain size in the direction of long-shore drift. South of Nauset Inlet, along Nauset Spit-Orleans, grain size varies (although usually of medium sand size) but with no downdrift decrease in grain size. To the north of Coast Guard Beach, along the sea cliff section (Stops 4, 5, and 6), grain size increases to medium and then becomes coarse sand with very coarse sand on the Provincelands spit (Stop 10). Thus the coastal geomorphologic landforms (spit, sea cliffs, and barrier islands) can be further distinguished by the beach sediment patterns. Along the barrier spit section, the grain size varies around one value; along the sea cliff section it increases to the north at a moderate rate, but along the spit section it increases at a rate twice that of the previous sea cliff section. This indicates that longshore transport processes are most active along the spit section, less active along the sea cliff section and much less important along the barrier spit section. These differences in sediment size patterns can often be observed at the beaches at Stops 3, 4, 5, 6, 7, and 10.

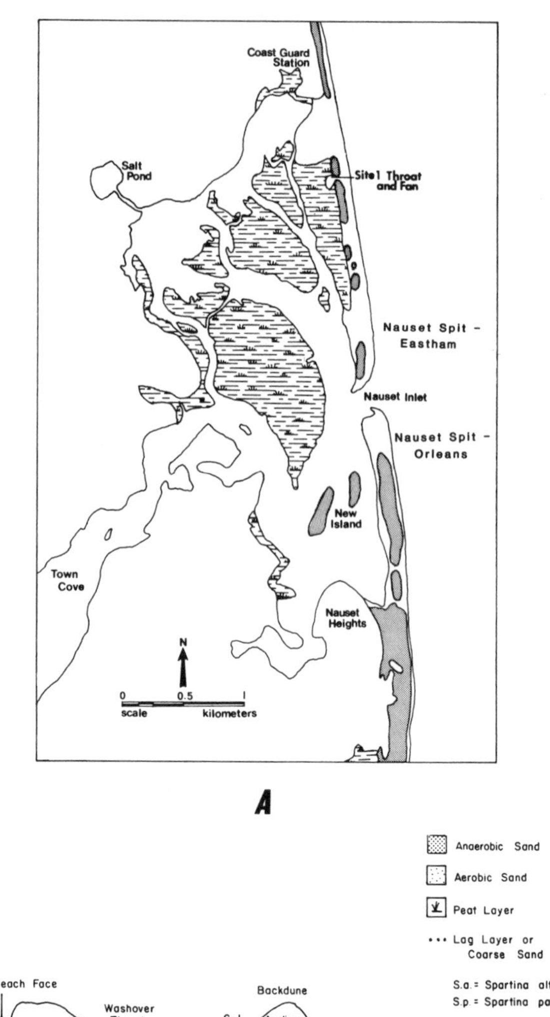

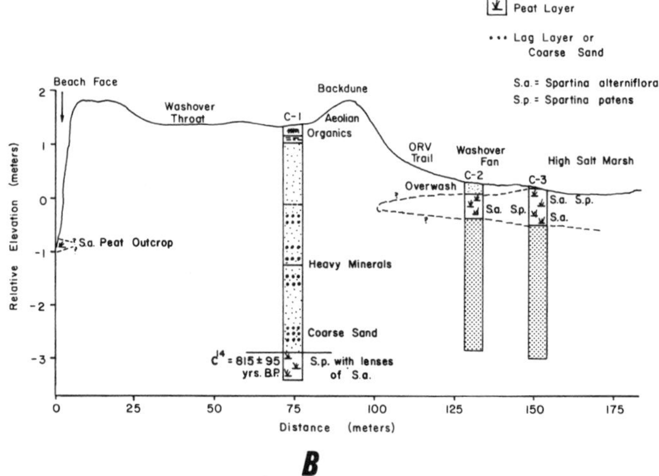

Figure 3. Coast Guard Beach Stop. A. Map of Nauset Spit-Eastham (post-1978 Northeaster). B. Stratigraphic cross-section, showing overwash deposition above salt marsh sediments.

Stop 5. Marconi Wireless Station. One approaches Marconi Station across the Wellfleet plain, as first described by Grabau (1897). A glacial outwash plain with a slight southwesterly slope, large entrenched valleys and numerous kettle holes ponds, the Wellfleet Plain is the oldest, highest and most extensive of the glacial interlobate outwash plains of the outer Cape. About two-thirds of the way down the 100-ft (30-m) cliff face at Marconi Station the contact between the upper and lower, Wellfleet deposits (Oldale, 1968) can be observed as a small bench. Extensive cliff erosion occurs along this coast. Marconi's four antenna towers were erected here and began transmitting in January, 1903. Since that time, the cliff has eroded 170 ft (52 m) and only the concrete bases of the two western towers remain. Average rates of cliff erosion along these cliffs were measured as 3 ft (0.9 m) per

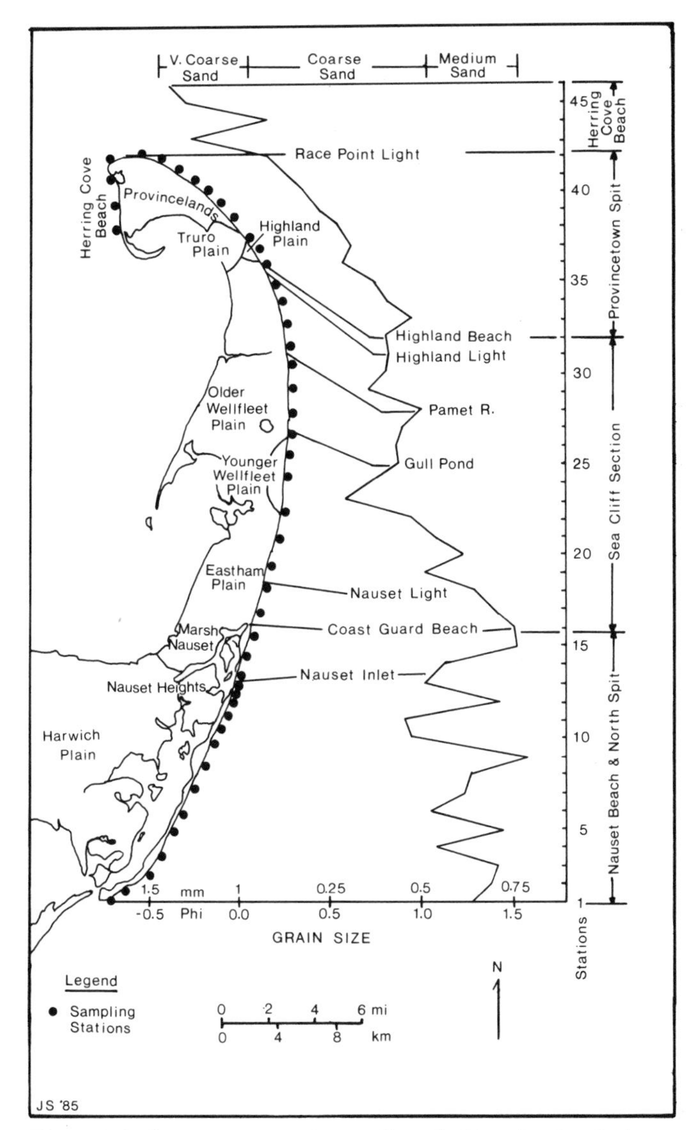

Figure 4. Sediment patterns of outer Cape Cod beaches in relation to coastal geomorphologic landforms. Mean grain size of high tide line fall season beaches indicate increasing from medium sand along the southern barrier spit beaches, through medium to coarse sand along the sea cliffs, with coarse and very coarse sand sizes along the Provincetown spit.

year in the late 1800s, decreasing to 2.5 ft (0.8 m) per year in the early 1900s and finally to 1.8 ft (0.6 m) in the 1970s. From this wireless station, there is a boardwalk trail 1.0 mi (1.6 km) long to the west down through White Cedar Swamp, which occupies a large, 0.2 mi (0.3 km) water-filled kettle hole. To reach the beach below the cliff, go back almost to U.S. 6 and take the road southeast to Marconi Beach where a maintained stair leads safely down.

Motor Route—Ocean View Drive. Taking the ocean route along the eastern cliff, continue from Marconi Station north 0.9 mi (1.4 km) to LeCount Hollow Road. Turn east onto LeCount Hollow Road, which follows a former proglacial outwash meltwater valley toward the sea cliffs. At 0.8 mi (1.3 km) is LeCount Hollow parking lot. The "hollow" or gap in the sea cliff represents the headward extension of the Blackfish Creek "pamet" which has been eroded by wave action from the east. Farther ahead, at Stop 6, is the type locality for this "pamet" landform. Turning north onto Ocean View Drive, one follows the sea cliff on the younger Wellfleet outwash plain, which is pitted with kettle holes. A road to the east at 1.7 mi (2.7 km) leads down to Cahoon Hollow, another lesser gap in the cliff edge. At 1.1 mi (1.8 km) Ocean View Drive bears east to Newcomb Hollow where one can park and reach the beach. Continue west on Gross Hill Road turning northwest at 0.7 mi (1.1 km) onto Gull Pond Road. Gull Pond, to the north, 0.5 mi (0.8 km) in diameter and more than 60 ft (18 m) deep, is the largest kettle hole pond on the outer Cape. The water level in the pond is governed by the regional water table since there are no true surface streams on the Cape. Upon reaching U.S. 6 at 1.4 mi (2.2 km), turn north to Stop 6, 3.3 mi (5.3 km) distant.

Stop 6. Pamet River, Truro. East is Pamet River, the largest "pamet" on the Cape and the type locality of such a feature (Fisher, 1982b). A pamet is defined as a glacial outwash stream channel on an outwash plain that no longer contains the stream that formed it; it is therefore a relict channel containing perhaps an "unfit" stream. In some cases, as on Cape Cod, along Cape Cod Bay, a tidal stream or estuary is present in the pamet, due to drowning of the relict channel. Where the head of this relict channel reaches the eroding sea cliffs and the head of the channel has been eroded away, the channel itself creates a "sag" or "gap" in the cliff edge, referred to locally as a "hollow." To the east, at 1.5 mi (2.4 km), is Ballston Beach, which can be reached by a 3.1-mi (5.0-km) round trip on South and North Pamet Roads. Ballston Beach forms a low "hollow" that is actually formed between dunes developed from longshore drifted beach sands across the pamet head.

Stop 7. Highland "Cape Cod" Light. Follow the trail at the southeast end of the parking lot, south to the cliff overlook. (Please do not enter the lighthouse property). The wave-cut cliffs at Highland Light are remnants of a once more extensive land mass to the east. Davis (1896) suggested that the "greatest retreat of the original shore" was about 2.5 mi (4.0 km), based on average rates of erosion during the past 3,000 years. Erosion still continues along these cliffs, from wave erosion below where, at high tide, the water reaches the cliff base, and from landslides at the top. Highland Light, 120 ft (36 m) above the beach, is situated on the Highland Plain, a small triangular area bordered on the south by the 50 ft (15 m) higher Wellfleet plain. These Highland Light glacial deposits are among the most well-known on Cape Cod, although there are differences in interpretations as to their age and depositional environments. In the past, the three major units here were correlated with the glacial Jameco Gravel, interglacial Gardiners Clay, and glacial Jacob Sand on Long Island (Woodworth and Wigglesworth, 1934). Other workers have since shown that these deposits were all late Wisconsian and laid down in a body of marine(?) water, dammed by the Wellfleet plain, the Cape Cod Bay ice, and the South Channel ice.

Stop 8. Pilgrim Heights and Province Land Dunes. Leaving U.S. 6, the road to Pilgrim Heights passes over the Truro plain, the most northerly of the glacial outwash plains of the outer Cape and about 50 ft (15 m) lower than the Highland plain. Truro plain deposits overlie the Highland plain deposits with an unconformable contact, indicating that it is younger than both the Highland and Wellfleet deposits. From the east end of the parking lot, a short walk along the Pilgrim Spring trail leads to a relict sea cliff above the "Head of the Meadow." The salt meadow separates the Truro plain from the Province Land spit. Davis (1896) pointed out that as the Province Lands spit grows by accretion and as the cliff retreats by erosion, there is a neutral point, or "fulcrum," of no change along the shoreline. With time, however, this fulcrum shifts toward the spit as the cliff erodes.

Looking to the northwest, Mt. Ararat is visible at the western end of Pilgrim Lake. The crest of this large dune is more than 80 ft (25 m) above mean sea level; the highest dunes in the Province Lands have elevations of approximately 100 ft (30 m). The migrating sand dunes to the north are parabolic or "U-shaped, with the open end of the dune facing into the prevailing northwest wind. From this vantage point it is easy to visualize High Head as an ancestral marine cliff, with the Province Lands spit growing from the fulcrum position along the shoreline. The first spit may have extended from High Head on the east, passed through the Mt. Ararat area, and continued west to form the land on which Provincetown is now situated. Pilgrim Monument is located on a dune ridge (Town Hill), which is nearly 100 ft (30 m) high. The Province Lands have been described historically as an extensive wasteland of undulating sand. The "walking dunes" threatened Provincetown and its harbor as early as the 1700s. By 1725 much of the valuable top soil had blown away, some houses in the town were actually buried, and thousands of tons of sand had to be removed from the streets each year. At the beginning of the 19th century the dunes were migrating into the town harbor at rates up to 90 ft (27 m) per year (Chamberlain, 1964). The sandy area in the Province Lands were estimated to exceed 6,000 acres (2,400 ha) in the early 1900s, and the dune migration rate was recorded at 15 ft (4.5 m) per year through the forest, toward the city and harbor.

At present, the dunes are continuing to march across the landscape, burying everything in their path. Zak and Bredakis (1963) reported that 10,000 yd^3 (7,600 m^3) of sand move onto U.S. 6 annually, which amounts to a dune migration rate of 18 ft (5.5 m) per year (Madore and Leatherman, 1981). The sand blasting effect to automobile windshields can be severe during the winter winds. In order to bring these dunes under control, it will be necessary to stabilize the sand as near the source area as possible, and to establish and maintain permanent vegetation on the sand dunes.

Stop 9. Province Lands Visitor Center. The Province Lands is an excellent example of a large coastal landmass created by marine and aeolian processes. Although most of Cape Cod is composed of glacial sediments in the form of recessional moraines and outwash plains, the Province Lands peninsula was formed by deposition of sediments eroded from these glacial deposits (Fig. 5A). This complex recurved spit, or peninsula, was created by a series of prograding sandspits. However, the location, shape, and orientation of the original sandspits, as well as their relationship to the extensive present-day system of parallel and curving dune ridges, is still debated.

The Province Lands Visitor Center was constructed on what appears to be a relict spit ridge. It is easy to visualize each of the dune lines as relict beach ridges, representing "lines of growth." Undulations in Race Point Road, toward the northwest, tend to accentuate ridge-and-trough topography of these supposedly relict spits. Beach ridges, which are the "ribs" of the Province Lands, developed within the last five or six thousand years (Zeigler and others, 1965), forming a large recurved spit of more than 7,500 acres (3,000 ha) today.

Currently, there is a controversy over the exact mechanism for the evolution of the Province Lands. Two theories have been proposed (Davis, 1896; Zeigler and others, 1965). Davis (1896) hypothesized that the peninsula was formed by a number of prograding sand spits built from sediment eroded from the glacial deposits of Truro and Wellfleet and transported to the northwest by longshore currents. Formation of the peninsula began about 6,000 years ago when the first sand spit was built northwest of the glacial headland (Fig. 5B). As these spits built northwest, they cut off areas of water in between. Davis (1896) also suggested that, as the glacial sediments of the outwash plain eroded toward the southeast, the change in sea cliff position shifted the fulcrum of the accreting sand spit system (Fig. 5C). This shift in the fulcrum was believed by Davis to result in the building of a new tangential spit to the north and seaward of the previous spit. Therefore, each of the sandspits that formed the peninsula built out from east to west over the entire length of the peninsula before being subsequently cut off by newly forming sandspits to the north (Fig. 5B). In this way the peninsula's length was established as early as the completion of the first sandspit. Subsequent spits forming to the north would increase the width of the Province Lands.

Zeigler and others (1965) presented a different interpretation of the origin of the peninsula, disputing the assumption by Davis (1896) that the present-day parallel dune ridge systems correspond to the location of the original prograding sandspits. According to Zeigler and others (1965), each sequential spit built a short distance to the west and then hooked to the south (Fig. 5D). In this way the present-day width of the Province Lands was established, while the peninsula as a whole slowly built farther to the west with the addition of each new spit. Thus, the westernmost end of the peninsula was only recently formed. Zeigler and others (1965) theorized that there was a change in deposition patterns approximately 2,000 years ago. Instead of the series of tight hooks that originally formed the peninsula, deposition tended to extend the whole northerly coast seaward to the north. This shift, which was believed to correspond to an abrupt reduction in the rate of sea-level rise, resulting in the widening of the end of the Province Lands hook. In the model by Zeigler and others (1965), it follows that there is no relationship between the

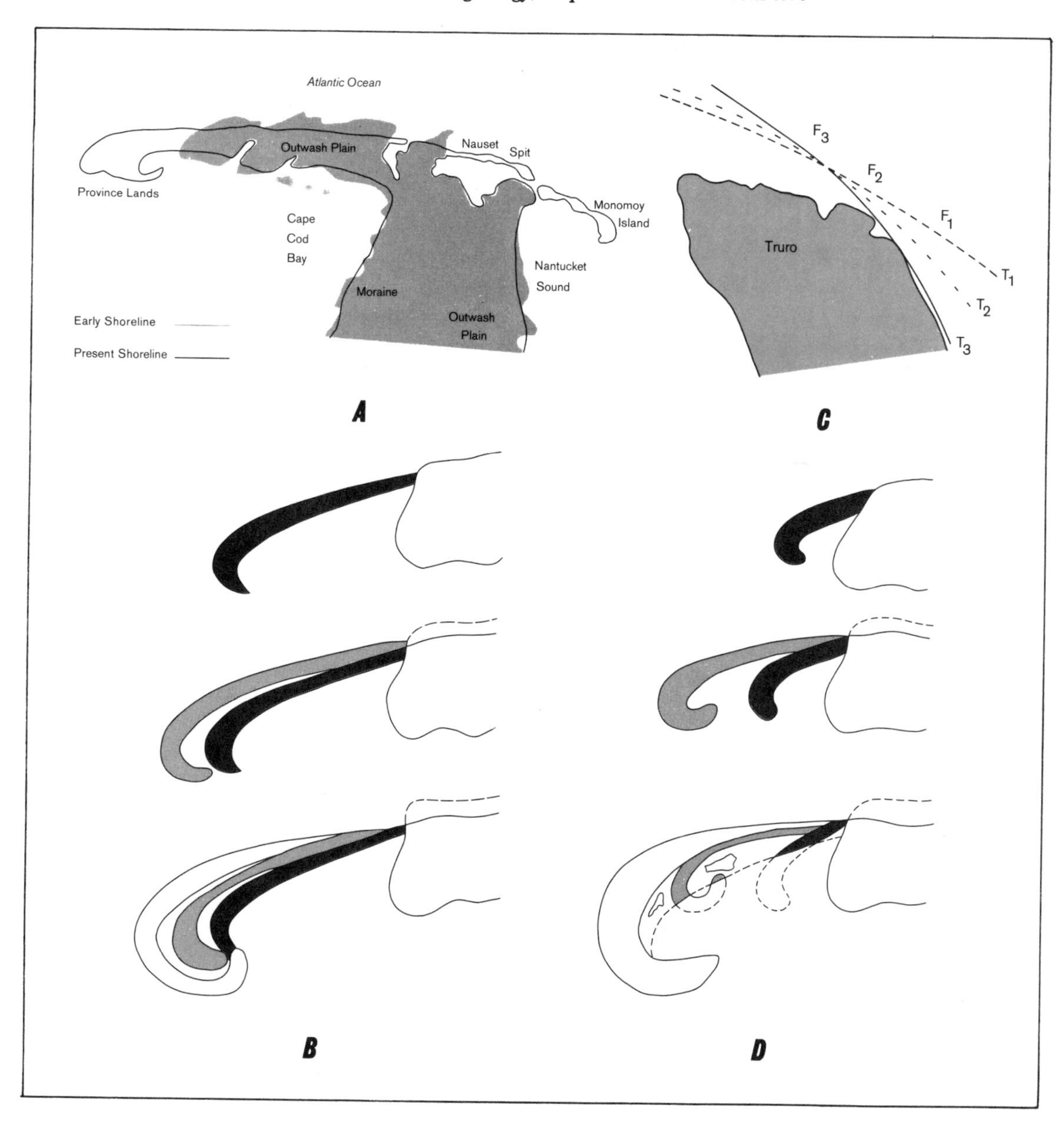

Figure 5. Province lands Stop. A. Sediments eroded from glacial deposits transported by longshore currents to form coastal features (after Davis, 1896). B. Conceptual evolution of a spit, according to Davis (1896). C. Movement of fulcrum point (F_1–F_3) northward with time (T_1–T_3) as Truro erodes and sediment accretes to form Province Lands (from Davis, 1896). D. Conceptual evolution of a spit according to Zeigler and others (1965).

present-day parallel dune ridges and the location of the previous shorelines (sandspits), except perhaps for the seaward-most of the sand ridges.

Shallow cores (10 ft; 3 m) and radiocarbon dating of recovered organic materials indicate that there is no direct correlation between the present sand ridges and the original sandspits (Leatherman, 1985). Most of the 41 cores taken in the Province Lands reveal thin surface organic material layers underlain by wind-deposited sediments. Many cores reveal vegetative and sedimentary information, which contradicts Davis's (1896) interpretation, but which could be adequately explained considering the spit orientation and age sequence hypothesized by Zeigler and others (1965). However, deeper cores are necessary to completely unravel the evolution of the spit during the last 6,000 years. This core evidence also confirms the conclusions drawn by McCaffrey and Leatherman (1979) regarding the massive disruption of the sand dune ecosystems by the Pilgrims.

The core data show quite clearly that dune migration was rampant throughout the Province Lands during and following the time of the Pilgrims. Many freshwater bogs and swamps were destroyed by sand encroachment and eventual burial. It appears that there was at least one period of extreme aeolian sand movement through the area. This time probably corresponded with the arrival and settlement of the area by the Pilgrims, but additional radiocarbon dates are necessary to precisely date this period. Cranberry bogs and ericaceous swamps (quaking bogs and freshwater marshes) have been established in the low-lying areas since this time, reflecting somewhat greater stability in recent years.

Stop 10. Race Point Beach. Along Race Point Road, at 0.6 mi (1.0 km), the Provincetown Airport occupies a wide trough. This trough probably represents a wide relict beach formed in front of the foredune ridge, much like the wide beach in front of present-day Race Point Beach. Undulations in Race Point Road as it heads northwest towards the beach reflect the ridge-and-trough topography of these relict spits. The most recent of these growing spit beaches is, of course, present-day Race Point Beach.

Offshore from Race Point Beach is a fairly permanent longshore bar, Peaked Hill Bar, that can sometimes be observed by the waves breaking on it offshore. Peaked Hill Bar actually begins tangent to the beach at the Highland Sea cliffs to the east (Stop 7) and extends westward, and offshore about 2,000 ft (320 m) to Race Point. This bar may thus be the beginning of a new offshore spit ridge as suggested by the Davis (1896) model for the development of the now relict sand ridges or sandspits. The coarsest sediment size on the Cape Cod outer beaches is found here on Race Point Beach. The average sediment size is almost very coarse sand (Stations 40 to 42, Fig. 4); patches of gravel are common. The source area for these coarse beach sands is along the sea cliff to south (Stops 5 and 7) where the adjacent beach sands are a finer, medium sand (Fig. 4). Thus, the beach sediment size along the Province Lands spit is increasing in a regular fashion, rather than decreasing in size in the direction of longshore transport away from the source area, the sea cliffs. A similar well-known situation, where grain size increases downdrift, has been reported from Chesil Beach, a shingle (pebble) beach in Dorset, southern England.

An explanation for this apparent paradox is that during the winter months, the dominant northeasterly winds, less common but more intense, selectively winnow out the finer sediments, then transport them back to the sea cliff area, leaving behind the coarser sediments. An offshore source of the coarser sands along the spit is also possible. Ending at Race Point Beach, one can continue back along Race Point Road and across U.S. 6, 0.3 mi (0.5 km) into the well-known resort and fishing town of Provincetown.

REFERENCES

Chamberlain, B., 1964, These fragile outposts: Garden City, New Jersey, Natural History Press, 327 p.

Davis, W. M., 1896, The outline of Cape Cod: American Academy of Arts and Science Proceedings, v. 31, p. 303–332.

Fisher, J. J., 1972, Field guide to geology of Cape Cod National Seashore, 5th Annual Meeting, American Association of Stratigraphic Palynologists: Department of Geology, University of Rhode Island, 53 p.

—— , 1982a, Barrier islands, *in* Schwartz, M. L., ed., Encyclopedia of Beaches and Coastal Environments: Stroudsburg, Pennsylvania, Hutchinson Ross Publishing Co., p. 124–133.

—— , 1982b, Pamet, *in* Schwartz, M. L., ed., Encyclopedia of Beaches and Coastal Environments: Stroudsburg, Pennsylvania, Hutchinson Ross Publishing Co., p. 630–631.

—— , 1985, Atlantic USA-North, *in* Bird, E.C.F., and Schwartz, M. L., eds., The World's coastline: New York, Van Nostrand Reinhold, p. 223–234.

Grabau, A. W., 1897, The sand plains of Truro, Wellfleet, and Eastham: Science, n.s., v. 5, p. 334–335.

Leatherman, S. P., 1979, Barrier Dynamics; Nauset Spit, Massachusetts, *in* Leatherman, S. P., ed., Environmental geologic guide to Cape Cod National Seashore: Society of Economic Paleontologists and Mineralogists, Eastern Section, p. 144–169.

Leatherman, S. P., and Zaremba, R. E., 1986, Dynamics of a northern barrier beach; Nauset Spit, Cape Cod, Massachusetts: Geological Society of America Bulletin, v. 97, p. 116–124.

Leatherman, S. P., Giese, G., and O'Donnell, P., 1981, Historical cliff erosion of outer Cape Cod: University of Massachusetts, NPS Research Unit Report 53, 59 p.

Madore, C., and Leatherman, S. P., 1981, Dune stabilization of the Province Lands, Cape Cod National Seashore, Massachusetts: University of Massachusetts, NPS Research Unit Report 52, 59 p.

McCaffrey, C., and Leatherman, S. P., 1979, Historical land use practices and dune instability in the Province Lands, *in* Leatherman, S. P., ed., Environmental geologic guide to Cape Cod National Seashore: Society of Economic Paleontologists and Mineralogists, Eastern Section, p. 207–222.

Oldale, R. N., 1968, Geologic map of the Wellfleet Quadrangle: U.S. Geological Survey Geologic Quadrangle Map GQ-750.

Schalk, M., 1938, A textural study of the Outer Beach of Cape Cod, Massachusetts: Journal of Sedimentary Petrology, v. 8, p. 41–54.

Strahler, A. N., 1966, A geologists view of Cape Cod: Garden City, New Jersey, Natural History Press, 150 p.

Thoreau, H. D., 1895, Cape Cod: W. W. Northon and Co., 300 p.

U.S. Army Corps of Engineers, 1971, Report on the National Shoreline Study: Washington, D.C., U.S. Government Printing Office, p. 18–19.

Woodworth, J. B., and Wigglesworth, E., 1934, Geography and geology of the region including Cape Cod, Elizabeth Island, Nantucket, Martha's Vineyard, No Mans Land, and Block Island: Harvard Collection, Museum of Comparative Zoology Memoirs, v. 52, 328 p.

Zak, J. M., and Bredakis, E., 1963, Dune stabilization at Provincetown, Massachusetts: Shore and Beach, v. 31, p. 19–24.

Zeigler, J. M., Tuttle, D. S., Tasha, H. J., and Giese, G. S., 1965, The age and development of the Provincelands, Outer Cape Cod, Massachusetts: Limnology and Oceanography, v. 10, p. R298–R311.

Wisconsinan and pre-Wisconsinan drift and Sangamonian marine deposits, Sankaty Head, Nantucket, Massachusetts

Robert N. Oldale, *Box 361, North Falmouth, Massachusetts 02556*

LOCATION

Sankaty Head is located on the eastern shore of Nantucket Island near its southeast corner (Fig. 1). Nantucket lies about 90 mi (150 km) southeast of Boston and about 30 mi (45 km) from Hyannis on Cape Cod. Access to the island is by sea, a 2.5-hour car ferry ride across Nantucket Sound from Hyannis, or by air from Boston and Hyannis. Once on the island, the U.S. Geological Survey 7½-minute Siasconset topographic map (1:25,000) and the geologic map of Nantucket (1:48,000; Oldale, 1985) will be useful guides for the trip to Sankaty Head. Whether arriving by sea or air, the route to Sankaty Head cliff begins with Milestone Road that starts near the village of Nantucket and passes close to the airport. The road traverses a broad outwash plain, bordered to the north by the Nantucket Moraine. Just east of Tom Nevers Road, Milestone Road drops into Phillips Run, an outflow route for a glacial lake that was dammed north of the moraine. From Siasconset, a colonial fishing village, the route turns north, along Atlantic Street, lined with seventeenth and early eighteenth century houses, and then along the Polpis Road. About 1.9 mi (3 km) from Siasconset center, the Hoicks Hollow Road turns northeast toward the beach. From the end of the road, the Sankaty Head section is a 0.7 mi (1.2 km) walk southward down the beach (Fig. 2). The route to the beach can also be traveled by bicycle. Milestone Road is paralleled by a bike path that starts in Nantucket village and ends in Siasconset. Traveling the beach by 4-wheel drive vehicle is *not* recommended. Soft deep sand, steep beach faces, and high tides that lap the dunes and cliff face make vehicular access dangerous. Access to the Sankaty section from the cliff top is also not recommended. Although the cliff face is public property, abutting owners are very sensitive to cliff erosion and are much concerned by people climbing down the cliff. For the same reason, geologists should keep digging at the exposure to a minimum. Storm-eroded cliff face may collapse and should be approached with caution.

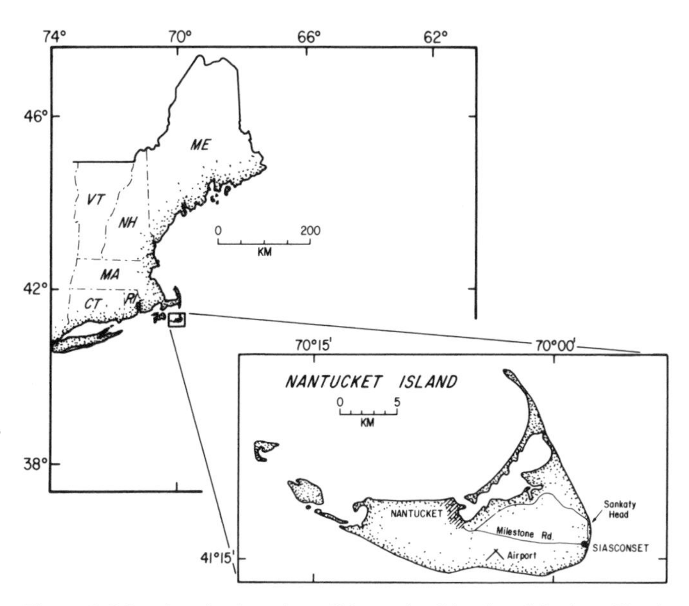

Figure 1. Map showing location of Nantucket Island and Sankaty Head.

SIGNIFICANCE

The fossiliferous Sankaty Sand has been dated using amino-acid–racemization and uranium-thorium analyses (Oldale and others, 1982). The results showed (1) the Sankaty Sand is Sangamonian in age and correlates with oxygen-isotope stage 5, (2) the underlying drift is Illinoian or older (pre-stage 5), and (3) the overlying drift is Wisconsinan in age. Fossils from the lower part of the Sankaty Sand indicate a marine climate somewhat warmer than present; those from the upper part imply a marine climate as cold or somewhat colder than present (Oldale and others, 1982). Involutions in the Sankaty Sand and atop the upper Wisconsin drift, and a deep channel in the lower part of the upper Wisconsin drift, denote episodes of severely cold subaerial climate.

The lower drift indicates a pre-Sangamonian glaciation whose extent was equal to or greater than that of the late Wisconsinan glaciation. Correlation of these deposits to older drift on Long Island and in New England suggests that the Montauk till and the "lower till" of New England are older than Wisconsinan (Oldale and others, 1982; Oldale and Eskenasy, 1983). Correlation of the Sankaty Sand faunas with the pre-Wisconsinan faunas of Long Island (Gustavson, 1976) implies similar Sangamonian or oxygen-isotope stage 5 ages for the latter.

SITE INFORMATION

The Sankaty Head section has attracted the attention of investigators for more than 130 years. The earliest description of the section was by Desor and Cabot (1849). Verrill (1875) published an extensive faunal list. He concluded that fossils in the lower part of the Sankaty Sand represent a warm marine climate and that fossils in the upper part represent a cold marine climate. Cushman (1904) divided the Sankaty Sand into four beds and compiled a faunal list for each. He proposed that the fossils in the lower part of the Sankaty Sand represent a marine climate similar to that off Nantucket today while those from the upper part record a colder marine climate. Wilson (1905) studied the Sankaty section at about the same time. He recognized that the

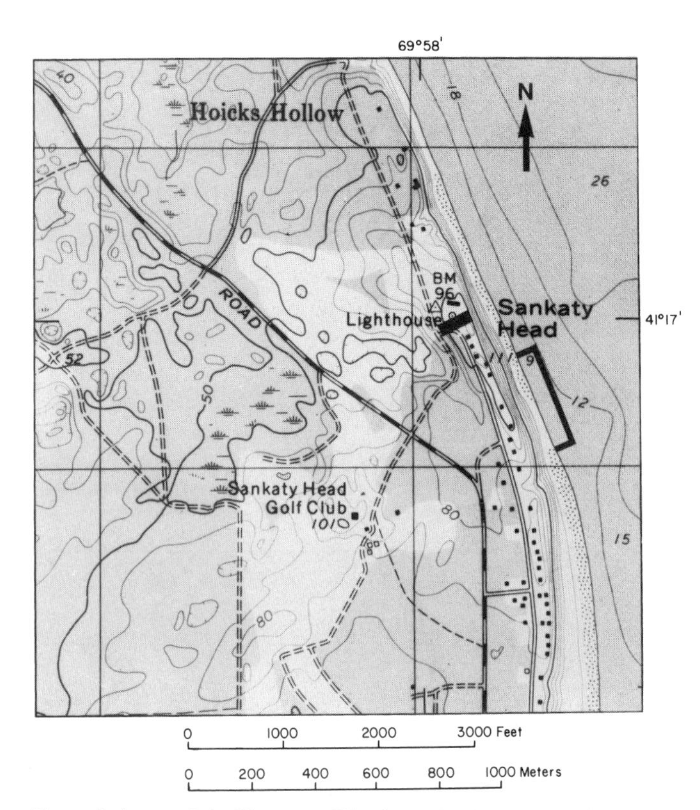

Figure 2. A part of the Siasconset 7½-minute Quadrangle showing beach route from Hoicks Hollow to the Sankaty Head exposure (bracket) and the location of the south fence (heavy bar) at Sankaty Head Light.

Figure 3. Southern end of Sankaty Head exposure showing: a, lower drift; b, basal marine gravel; c, bioturbated bed; d, shelly lower part of Sankaty Sand; e, serpulid bed of lower part of Sankaty Sand; f, upper part of Sankaty Sand; and g, medium to fine sand of upper drift. Note shovel immediately below b for scale. Photograph located at about station 10 + 50.

marine beds overlie glacial drift and agreed with Cushman's paleoenvironmental conclusions. The Sankaty Sand was formally named by Woodworth (Woodworth and Wigglesworth, 1934); however, he misinterpreted its stratigraphic position. Gustavson (1976) agreed with previous workers as to the marine climates represented by the upper and lower parts of the Sankaty Sand and suggested that the marine deposit is late Sangamonian to early Wisconsinan in age. This field guide is a result of the most recent geological study of Nantucket and the adjacent islands. Field work started in the spring following the great February blizzard of 1978. Waves from this powerful nor'easter eroded the base of the Sankaty Head cliff, providing an excellent exposure of the section.

The Sankaty Head section is portrayed in detail in two U.S. Geological Survey reports, a geologic profile based on 34 measured sections (Oldale and others, 1981) reproduced in part as Figure 4, and a series of 32 photographs keyed to the geologic section (Oldale, 1982). These publications are useful when studying the section and can be obtained from the Map Distribution Section, U.S. Geological Survey, Federal Center Building 41, Box 25286, Denver, CO 80255.

The condition of the Sankaty exposure varies greatly. Some of the observations that are reported here were made in the spring of 1978 and fall of 1980, following major storms that exposed the section down to beach level. Most of the section was measured in the fall of 1979 when the lower part was obscured by slump and small fans (Fig. 3). Horizontal measurements used here represent distances south of the south fence bounding the lighthouse property (Fig. 2), the only landmark visible from the beach. Elevations are measured from beach level, inferred to represent sea level. This imprecise datum is the only one readily available. Field measurements were made in feet and are so presented here along with metric values where possible. The most important part of the section, including the upper part of the lower drift and the marine beds of Sangamonian age, occurs south of Sankaty Head Light between station 7+00 (700 ft; 213 m south of the south fence) and station 11+00 (Fig. 4).

The lower stratigraphic units, including the older drift and marine deposits, dip southerly and hence become younger to the south. The oldest unit shown in Figure 4 is a silty medium to very course sand (Qps) that includes some silt and till beds. These older deposits are overlain by a dark gray to brown, massive, compact, stone-poor, silty to clayey till (Qt and Qto) that crops out at beach level between stations 7+75 and 8+65. The till in turn is overlain by oxidized beds of deformed sand, till, and silt (Qst), a water-laid subglacial facies of the massive till. The till is overlain by steeply dipping, foreset-bedded sand and gravel (Qfs). Near beach level in the vicinity of station 9+40, the sand and gravel interfingers with a silt and clay (Qlc) that appears to be varved in places. The interfingering and varves were exposed in 1978 and 1980, but were obscured by slumping at other times. South of station 9+40, the deformed sand, till, and silt crops out at beach level; south of station 10+00, it is capped by the unconformity that separates the older drift from the marine deposits of Sangamonian age.

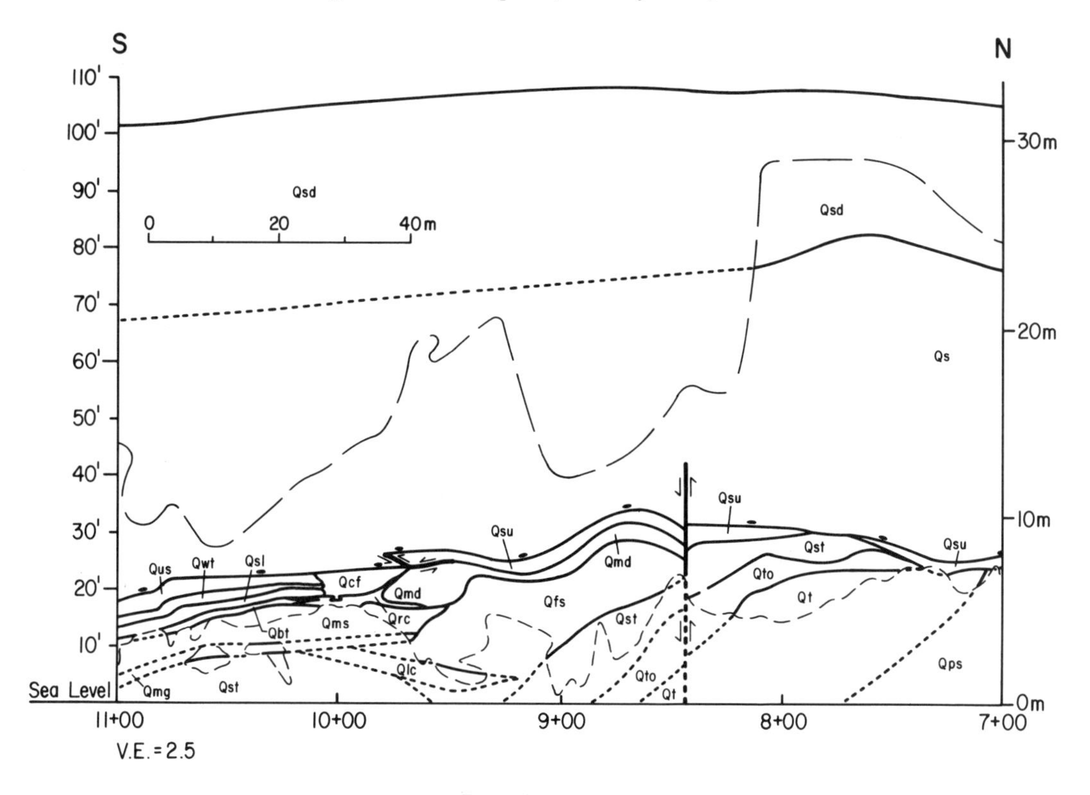

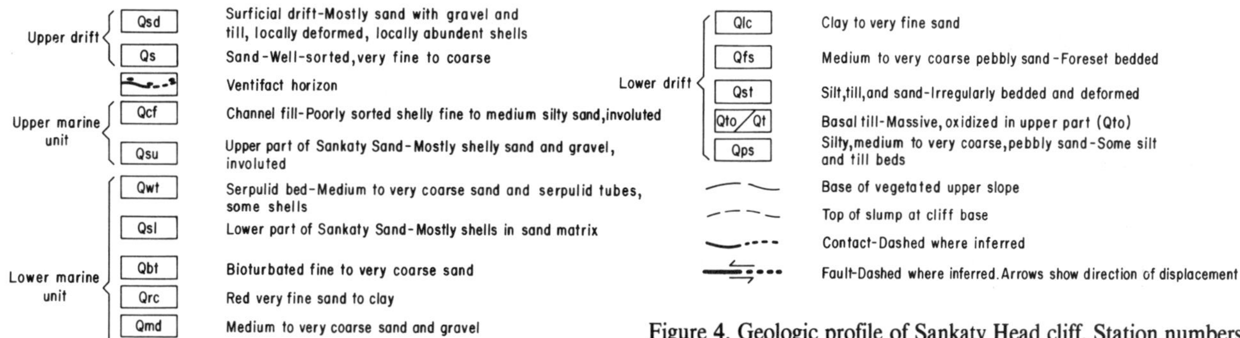

Figure 4. Geologic profile of Sankaty Head cliff. Station numbers indicate distance south of the south fence, Sankaty Head Light. Modified from Oldale and others, 1981).

Marine deposits at Sankaty Head, including the Sankaty Sand, are best displayed between stations 10 and 11. A well sorted ferruginous gravel (Qmg) marks the base of the marine section. Overlying the gravel is a well sorted, cross-bedded and planar-bedded sand (Qms) that becomes finer grained near the top. A bioturbated fine to very coarse sand (Qbt) separates the underlying deposits from fossiliferous Sankaty Sand. Near station 10+20, the bioturbated bed interfingers with a bright red ferruginous sandy clay (Qrc) that near station 9+80 interfingers with a foreset bedded sand and gravel (Qmd). The much studied Sankaty Sand caps the bioturbated sand and in the lower part (Qsl) consists of a bed containing abundant oyster (*Crassostrea virginica*) and clam (*Mercenaria mercenaria*) shells and a bed (Qwt) composed mostly of serpulid worm tubes and scattered articulated *Mercenaria.* A well-defined unconformity separates the lower and upper parts of the Sankaty Sand. The upper part (Qsu) is characterized, from bottom to top, by a fossiliferous gravelly sand that contains abundant shells and shell fragments, mostly *Mercenaria campechiensis*; a well-sorted sand; and a poorly sorted fine to medium sand containing finely divided fragments of mussel shell (*Mytilus edulis*). To the north, in the vicinity of station 9+90, the bioturbated sand and Sankaty Sand are cut out and replaced by a channel fill (Qcf) composed of poorly sorted fine to medium sand containing finely divided shell fragments. North of the channel, the upper part of the Sankaty Sand continues and can be traced to about station 4+40 where it is obscured by vegetation. The upper part of the marine deposits is involuted, indicating subaerial exposure to a severe cold climate.

Capping the marine deposits is an unconformity distinguished by a horizon of wind polished stones (ventifacts) that may also indicate cold subaerial conditions.

Shell and coral (*Astrangia* sp.) from the Sankaty Sand were used to obtain amino-acid–racemization age estimates and a uranium-thorium date (Oldale and others, 1982). Amino-acid analysis on *Mercenaria* from the Sankaty Sand suggest an age of about 120,000 to 140,000 yr. Uranium-series–disequilibrium analysis on coral from the upper part of the Sankaty Sand indicates an age of 133,000 ± 7,000 yr for that deposit. Radiocarbon ages ranging from 26,000 to 40,000 yr were obtained on shell and wood from the marine deposits and from the upper and lower drifts. However, the radiocarbon dates are internally inconsistent, and because of the amino-acid and uranium-thorium data, they were not used to estimate the ages of the deposits at Sankaty Head (Oldale and others, 1982).

The lower part of the Wisconsinan drift consists of a thick deposit of yellowish-white, well-sorted, very fine- to medium-grained, planar- and cross-bedded sand (Qs). The sand was deformed penecontemporaneously and contains widely scattered fragments of poorly preserved wood and shell. An unconformity marks the top of the fine to medium sand. The upper Wisconsin drift (Qsd) atop the unconformity consists mostly of gravel and gravelly sand discontinuously capped by till and is the surficial drift that occurs elsewhere on Nantucket.

Several other features of the Sankaty Head exposure are of interest. A thin bed of gravelly muddy sand occurs within the gravelly sand of the lower drift between stations 4+40 and 6+00. The deposit, inferred to be a flowtill, is compact, unbedded, and similar to the lower till at station 8+00 and to till from the upper drift (Oldale and Eskenasy, 1983). Significant faults are present in two places. A vertical normal? fault down thrown to the south, cuts the pre-Wisconsinan and lower part of the Wisconsinan deposits at station 8+45, and a small thrust fault repeats the upper part of the Sankaty Sand and the ventifact unconformity near station 9+75. Finally, an unusual feature occurs at about station 4+50. At this location a narrow channel with nearly vertical walls cut through the Wisconsinan medium to very fine sand, the upper part of the Sankaty Sand, and into the lower drift. The channel is part of the unconformity in the upper Wisconsinan deposits and is filled with sand and gravel of the surficial drift. The lack of mixing of sediment along the near vertical walls suggest that the channel, if erosional, formed in frozen sand or alternatively, the channel may be a rift formed when a block of frozen sand slid down-dip (southward) along a bedding plane.

The Sankaty Head exposure constitutes a unique and important upper Pleistocene stratigraphic section. The basal deposits represent a pre-Sangamonian glaciation equal to or greater in extent than the Laurentide continental ice sheet of Wisconsinan age. Correlation of the lower drift of Sankaty Head to pre-late Wisconsinan drift elsewhere in New England and Long Island can be argued (Oldale and others, 1982; Oldale and Eskenasy, 1983) and would indicate that the Manhassett Formation (Gustavson, 1976), including the Montauk till and its equivalents on Block Island and Martha's Vineyard, and the "lower till" in New England are pre-Sangamonian in age. The argument would also indicate that marine deposits below the Manhassett Formation, including the Gardners clay, are older than Illinoian or oxygen isotope stage 6. The Sangamonian marine beds at Sankaty Head represent an interglacial highstand of sea level, possibly the oxygen-isotope stage 5e highstand. However, glaciotectonic deformation precludes using the marine deposits to estimate past sea level. Molluscs in the lower part of the Sankaty Sand indicate a marine climate somewhat warmer than present and those in the upper part indicate a marine climate slightly cooler than present but not glacial. Features of possible periglacial origin formed in the Sankaty Sand, the lower part of the Wisconsinan deposits, and near the surface of the surficial drift indicate three episodes of severely cold subaerial climate. Sankaty Head is included in the glaciotectonically formed Nantucket Moraine (Oldale and O'Hara, 1984) of late Wisconsinan age (Oldale, 1985).

REFERENCES CITED

Cushman, J. A., 1904, Notes on the Pleistocene fauna of Sankaty Head, Nantucket, ssachusetts: American Geologist, v. 34, p. 169–174.

Desor, M. E., and Cabot, E. C., 1849, On the Tertiary and more recent deposits in the Island of Nantucket: Geological Society of London Quarterly Journal, v. 5, p. 340–344.

Gustavson, T. C., 1976, Paleotemperature analysis of the marine Pleistocene of Long Island, New York, and Nantucket Island, Massachusetts: Geological Society of America Bulletin, v. 87, p. 1–8.

Oldale, R. N., 1982, Photographs of the upper Pleistocene section at Sankaty Head cliff, Nantucket Island, Massachusetts: U.S. Geological Survey Miscellaneous Field Studies Map MF-1397.

—— , 1985, Geologic map of Nantucket and nearby islands, Massachusetts: U.S. Geological Survey Miscellaneous Geologic Investigations Map I-1580, scale 1:48,000.

Oldale, R. N., and Eskenasy, D. M., 1983, Regional significance of pre-Wisconsinan till from Nantucket Island, Massachusetts: Quaternary Research, v. 19, p. 302–311.

Oldale, R. N., and O'Hara, C. J., 1984, Glaciotectonic origin of the Massachusetts coastal end moraines and a fluctuating late Wisconsinan ice margin: Geological Society of America Bulletin, v. 95, p. 61–74.

Oldale, R. N., Eskensy, D. M., and Lian, N. C., 1981, A geologic profile of the Sankaty Head cliff, Nantucket Island, Massachusetts: U.S. Geological Survey Miscellaneous Field Studies Map MF-1304.

Oldale, R. N., Valentine, P. C., Cronin, T. M., Spiker, E. C., Blackwelder, B. W., Belknap, D. F., Wehmiller, J. F., and Szabo, B. S., 1982, Stratigraphy, structure, absolute age, and paleontology of the upper Pleistocene deposits at Sankaty Head, Nantucket Island, Massachusetts: Geology, v. 10, p. 246–252.

Verrill, A. E., 1875, On the post-Pliocene fossils of Sankaty Head, Nantucket Island, with a note on the geology by S. H. Scudder: American Journal of Science, v. 10, p. 364–375.

Wilson, J. H., 1905, The Pleistocene beds of Sankaty Head, Nantucket: Science, v. 21, p. 989–990.

—— , 1906, The glacial history of Nantucket and Cape Cod with an argument for a fourth center of glacial dispersion in North America: New York, Columbia University Press, 90 p.

Woodworth, J. B., and Wigglesworth, E., 1934, Geography and geology of the region including Cape Cod, the Elizabeth Islands, Nantucket, Marthas Vineyard, No Mans Land, and Block Island: Harvard College Museum of Comparative Zoology Memoir 52, 322 p.

The Champlain thrust fault, Lone Rock Point, Burlington, Vermont

Rolfe S. Stanley, Department of Geology, University of Vermont, Burlington, Vermont 05405

LOCATION

The 0.6 mi (1 km) exposure of the Champlain thrust fault is located on the eastern shore of Lake Champlain at the north end of Burlington Harbor. The property is owned by the Episcopal Diocesan Center. Drive several miles (km) north along North Avenue (Vermont 127) from the center of Burlington until you reach the traffic light at Institute Road, which leads to Burlington High School, The Episcopal Diocesan Center, and North Beach. Turn west toward the lake and take the first right (north) beyond Burlington High School. The road is marked by a stone archway. Stop at the second building on the west side of the road, which is the Administration Building (low rectangular building), for written permission to visit the field site.

Continue north from the Administration Building, cross the bridge over the old railroad bed, and keep to the left as you drive over a small rise beyond the bridge. Go to the end of this lower road. Park your vehicle so that it does not interfere with the people living at the end of the road (Fig. 1). Walk west from the parking area to the iron fence at the edge of the cliff past the outdoor altar where you will see a fine view of Lake Champlain and the Adirondack Mountains. From here walk south along a footpath for about 600 ft (200 m) until you reach a depression in the cliff that leads to the shore (Fig. 1).

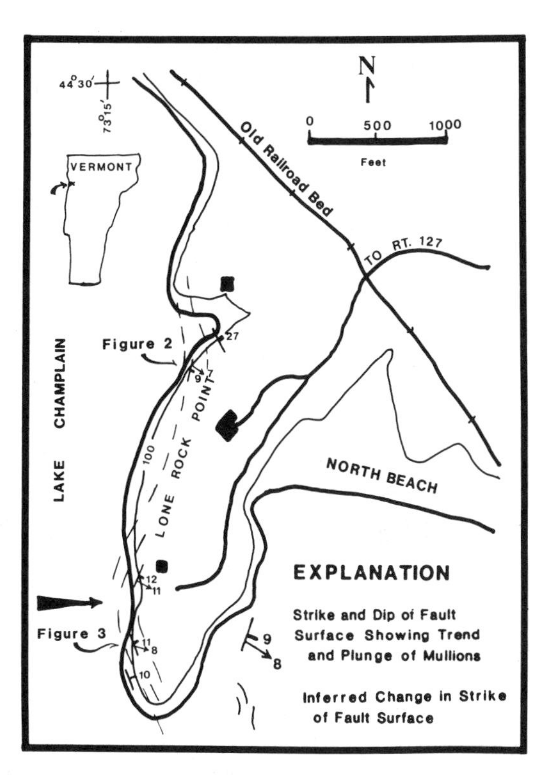

Figure 1. Location map of the Champlain thrust fault at Lone Rock Point, Burlington, Vermont. The buildings belong to the Episcopal Diocesan Center. The road leads to Institute Road and Vermont 127 (North Avenue). The inferred change in orientation of the fault surface is based on measured orientations shown by the dip and strike symbols. The large eastward-directed arrow marks the axis of a broad, late syncline in the fault zone. The location of Figures 2 and 3 are shown to the left of "Lone Rock Point." The large arrow points to the depression referred to in the text.

SIGNIFICANCE

This locality is one of the finest exposures of a thrust fault in the Appalachians because it shows many of the fault zone features characteristic of thrust faults throughout the world. Early studies considered the fault to be an unconformity between the strongly-tilted Ordovician shales of the "Hudson River Group" and the overlying, gently-inclined dolostones and sandstones of the "Red Sandrock Formation" (Dunham, Monkton, and Winooski formations of Cady, 1945), which was thought to be Silurian because it was lithically similar to the Medina Sandstone of New York. Between 1847 and 1861, fossils of pre-Medina age were found in the "Red Sandrock Formation" and its equivalent "Quebec Group" in Canada. Based on this information, Hitchcock and others (1861, p. 340) concluded that the contact was a major fault of regional extent. We now know that it is one of several very important faults that floor major slices of Middle Proterozoic continental crust exposed in western New England.

Our current understanding of the Champlain thrust fault and its associated faults (Champlain thrust zone) is primarily the result of field studies by Keith (1923, 1932), Clark (1934), Cady (1945), Welby (1961), Doll and others (1961), Coney and others (1972), Stanley and Sarkisian (1972), Dorsey and others (1983), and Leonard (1985). Recent seismic reflection studies by Ando and others (1983, 1984) and private industry have shown that the Champlain thrust fault dips eastward beneath the metamorphosed rocks of the Green Mountains. This geometry agrees with earlier interpretations shown in cross sections across central and northern Vermont (Doll and others, 1961; Coney and others, 1972). Leonard's work has shown that the earliest folds and faults in the Ordovician sequence to the west in the Champlain Islands are genetically related to the development of the Champlain thrust fault.

In southern Vermont and eastern New York, Rowley and others (1979), Bosworth (1980), Bosworth and Vollmer (1981), and Bosworth and Rowley (1984), have recognized a zone of late post-cleavage faults (Taconic Frontal Thrust of Bosworth and Rowley, 1984) along the western side of the Taconic Mountains. Rowley (1983), Stanley and Ratcliffe (1983, 1985), and Ratcliffe (in Zen and others, 1983) have correlated this zone with the Champlain thrust fault. If this correlation is correct then the Champlain thrust zone would extend from Rosenberg, Canada, to the Catskill Plateau in east-central New York, a distance of 199 mi (320 km), where it appears to be overlain by Silurian and Devonian rocks. The COCORP line through southern Vermont shows an east-dipping reflection that roots within Middle Proterozoic rocks of the Green Mountains and intersects the earth's surface along the western side of the Taconic Mountains (Ando and others, 1983, 1984).

The relations described in the foregoing paragraphs suggest that the Champlain thrust fault developed during the later part of the Taconian orogeny of Middle to Late Ordovician age. Subsequent movement, however, during the middle Paleozoic Acadian orogeny and the late Paleozoic Alleghenian orogeny can not be ruled out. The importance of the Champlain thrust in the plate tectonic evolution of western New England has been discussed by Stanley and Ratcliffe (1983, 1985). Earlier discussions can be found in Cady (1969), Rodgers (1970), and Zen (1972).

REGIONAL GEOLOGY

In Vermont the Champlain thrust fault places Lower Cambrian rocks on highly-deformed Middle Ordovician shale. North of Burlington the thrust surface is confined to the lower part of the Dunham Dolomite. At Burlington, the thrust surface cuts upward through 2,275 ft (700 m) of the Dunham into the thick-bedded quartzites and dolostones in the very lower part of the Monkton Quartzite. Throughout its extent, the thrust fault is located within the lowest, thick dolostone of the carbonate-siliciclastic platform sequence that was deposited upon Late Proterozoic rift-clastic rocks and Middle Proterozoic, continental crust of ancient North America.

At Lone Rock Point in Burlington the stratigraphic throw is about 8,850 ft (2,700 m), which represents the thickness of rock cut by the thrust surface. To the north the throw decreases as the thrust surface is lost in the shale terrain north of Rosenberg, Canada. Part, if not all, of this displacement is taken up by the Highgate Springs and Philipsburg thrust faults that continue northward and become the "Logan's Line" thrust of Cady (1969). South of Burlington the stratigraphic throw is in the order of 6,000 ft (1,800 m). As the throw decreases on the Champlain thrust fault in central Vermont the displacement is again taken up by movement on the Orwell, Shoreham, and Pinnacle thrust faults.

Younger open folds and arches that deform the Champlain slice may be due either duplexes or ramps along or beneath the Champlain thrust fault. To the west, numerous thrust faults are exposed in the Ordovician section along the shores of Lake Champlain (Hawley, 1957; Fisher, 1968; Leonard, 1985). One of these broad folds is exposed along the north part of Lone Rock Point (Fig. 2). Based on seismic reflection studies in Vermont, duplex formation as described by Suppe (1982) and Boyer and Elliot (1982) indeed appears to be the mechanism by which major folds have developed in the Champlain slice.

North of Burlington the trace of the Champlain thrust fault is relatively straight and the surface strikes north and dips at about 15° to the east. South of Burlington the trace is irregular because the thrust has been more deformed by high-angle faults and broad folds. Slivers of dolostone (Lower Cambrian Dunham Dolomite) and limestone (Lower Ordovician Beekmantown Group) can be found all along the trace of the thrust. The limestone represents fragments from the Highgate Springs slice exposed directly west and beneath the Champlain thrust fault north of Burlington (Doll and others, 1961). In a 3.3 to 10 ft (1 to 3 m) zone along the thrust surface, fractured clasts of these slivers are found in a matrix of ground and rewelded shale.

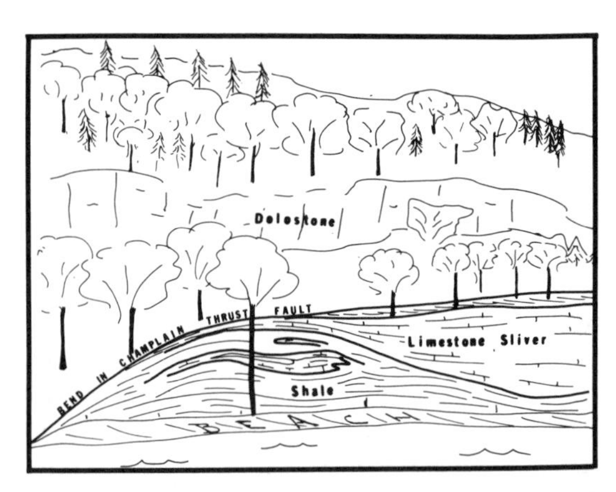

Figure 2. A sketch of the Champlain thrust fault at the north end of Lone Rock Point showing the large bend in the fault zone and the slivers of Lower Ordovician limestone. The layering in the shale is the S1 cleavage. It is folded by small folds and cut by many generations of calcite veins and faults. The sketch is located in Figure 1.

Estimates of displacement along the Champlain thrust fault have increased substantially as a result of regional considerations (Palmer, 1969; Zen and others, 1983; Stanley and Ratcliffe, 1983, 1985) and seismic reflection studies (Ando and others, 1983, 1984). The earlier estimates were less than 9 mi (15 km) and were either based on cross sections accompanying the Geologic Map of Vermont (Doll and others, 1961) or simply trigonometric calculations using the average dip of the fault and its stratigraphic throw. Current estimates are in the order of 35 to 50 mi (60 to 80 km). Using plate tectonic considerations, Rowley (1982) has suggested an even higher value of 62 mi (100 km). These larger estimates are more realistic than earlier ones considering the regional extent of the Champlain thrust fault.

Lone Rock Point

At Lone Rock Point the basal part of the Lower Cambrian Dunham Dolostone overlies the Middle Ordovician Iberville Formation. Because the upper plate dolostone is more resistent than the lower plate shale, the fault zone is well exposed from the northern part of Burlington Bay northward for approximately 0.9 mi (1.5 km; Fig. 1). The features are typical of the Champlain thrust fault where it has been observed elsewhere.

The Champlain fault zone can be divided into an inner and outer part. The inner zone is 1.6 to 20 ft (0.5 to 6 m) thick and consists of dolostone and limestone breccia encased in welded, but highly contorted shale (Fig. 3). Calcite veins are abundant. One of the most prominent and important features of the inner fault zone is the slip surface, which is very planar and continuous throughout the exposed fault zone (Fig. 3). This surface is marked by very fine-grained gouge and, in some places, calcite slickenlines. Where the inner fault zone is thin, the slip surface is located

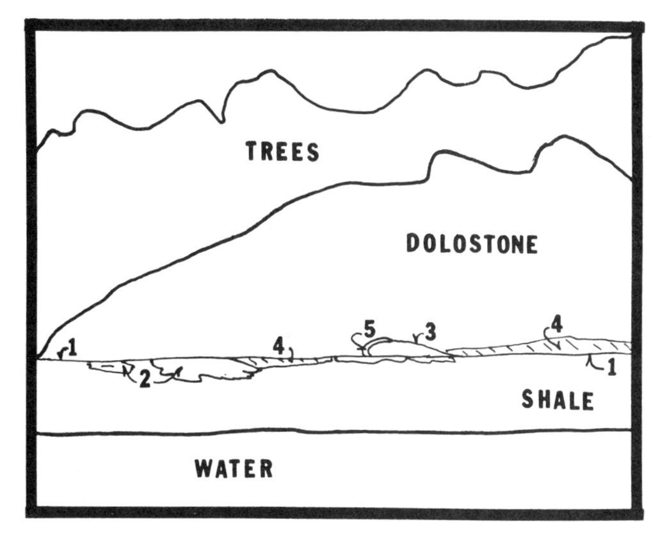

Figure 3. View of the Champlain thrust fault looking east at the southern end of Lone Rock Point (Fig. 1). The accompanying line drawing locates by number the important features discussed in the text: 1, the continuous planar slip surface; 2, limestone slivers; 3, A hollow in the base of the dolostone is filled in with limestone and dolostone breccia; 4, Fault mullions decorate the slip surface at the base of the dolostone; 5, a small dike of shale has been injected between the breccia and the dolostone.

along the interface between the Dunham Dolomite and the Iberville Shale. Where the inner fault is wider by virtue of slivers and irregularities along the basal surface of the Dunham Dolomite, the slip surface is located in the shale, where it forms the chord between these irregularities (Fig. 3). The slip surface represents the surface along which most of the recent motion in the fault zone has occurred. As a consequence, it cuts across all the irregularities in the harder dolostone of the upper plate with the exception of long wave-length corrugations (fault mullions) that parallel the transport direction. As a result, irregular hollows along the base of the Dunham Dolomite are filled in by highly contorted shales and welded breccia (Fig. 3).

The deformation in the shale beneath the fault provides a basis for interpreting the movement and evolution along the Champlain thrust fault. The compositional layering in the shale of the lower plate represents the well-developed S1 pressure-solution cleavage that is essentially parallel to the axial planes of the first-generation of folds in the Ordovician shale exposed below and to the west of the Champlain thrust fault (Fig. 4). As the trace of the thrust fault is approached from the west this cleavage is rotated eastward to shallow dips as a result of westward movement of the upper plate (Fig. 4). Slickenlines, grooves, and prominent fault mullions on the lower surface of the dolostone and in the adjacent shales, where they are not badly deformed by younger events, indicate displacement was along an azimuth of approximately N60°W (Fig. 4; Hawley, 1957; Stanley and Sarkesian, 1972; Leonard, 1985). The S1 cleavage at Lone Rock Point is so well developed in the fault zone that folds in the original bedding are largely destroyed. In a few places, however, isolated hinges are preserved and are seen to plunge eastward or southeastward at low angles (Fig. 4). As these F1 folds are traced westward from the fault zone, their hinges change orientation to the northeast. A similar geometric pattern is seen along smaller faults, which deform S1 cleavage in the Ordovician rocks west of the Champlain thrust fault. These relations suggest that F1 hinges are rotated towards the transport direction as the Champlain thrust fault is approached. The process involved fragmentation of

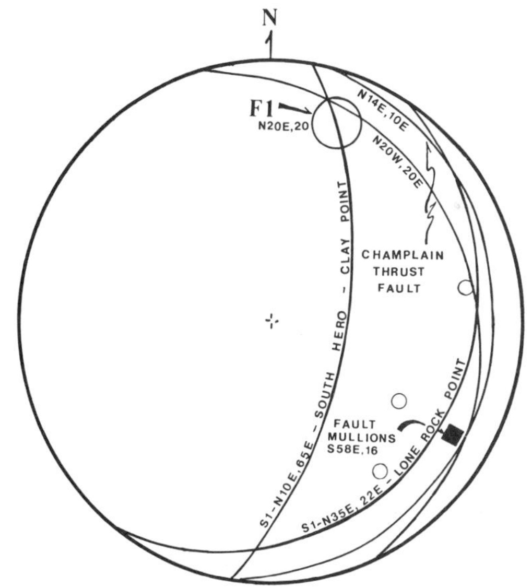

Figure 4. Lower hemisphere equal-area net showing structural elements associated with the Champlain thrust fault. The change in orientation of the thrust surface varies from approximately N20°W to N14°E at Lone Rock Point. The orientation of S1 cleavage directly below the thrust is the average of 40 measurements collected along the length of the exposure. S1, however, dips steeply eastward in the Ordovician rocks to the west of the Champlain thrust fault as seen at South Hero and Clay Point where F1 hinges plunge gently to the northeast. Near the Champlain thrust fault F1 hinges (small circles) plunge to the east. Most slickenlines in the adjacent shale are approximately parallel to the fault mullions shown in the figure.

the FI folds since continuous fold trains are absent near the thrust. Much of this deformation and rotation occurs, however, within 300 ft (100 m) of the thrust surface. Within this same zone the S1 cleavage is folded by a second generation of folds that rarely developed a new cleavage. These hinges also plunge to the east or southeast like the earlier F1 hinges. The direction of transport inferred from the analysis of F2 data is parallel or nearly parallel to the fault mullions along the Champlain thrust fault. Stanley and Sarkesian (1972) suggested that these folds developed during late translation on the thrust with major displacement during and after the development of generation 1 folds. New information, however, suggests that the F2 folds are simply the result of internal adjustment in the shale as the fault zone is deformed by lower duplexes and frontal or lateral ramps (Figs. 1, 2). The critical evidence for this new interpretation is the sense of shear inferred from F2 folds and their relation to the broad undulations mapped in the fault zone as it is traced northward along Lone Rock Point (Fig. 1). South of the position of the thick arrow in Figure 1, the inferred shear is west-over-east whereas north of the arrow it is east-over-west. The shear direction therefore changes across the axis of the undulation (marked by the arrow) as it should for a synclinal fold.

REFERENCES CITED

Ando, C. J., Cook, F. A., Oliver, J. E., Brown, L. D., and Kaufman, S., 1983, Crustal geometry of the Appalachian orogen from seismic reflection studies, *in* Hatcher, R. D., Jr., Williams, H., and Zietz, I., eds., Contributions to the tectonics and geophysics of mountain chains: Geological Society of America Memoir 158, p. 83–101.

Ando, C. J., Czuchra, B. L., Klemperer, S. L., Brown, L. D., Cheadle, M. J., Cook, F. A., Oliver, J. E., Kaufman, S., Walsh, T., Thompson, J. B., Jr., Lyons, J. B., and Rosenfeld, J. L., 1984, Crustal profile of mountain belt; COCORP Deep Seismic reflection profiling in New England Appalachians and implications for architecture of convergent mountain chains: American Association of Petroleum Geologists Bulletin, v. 68, p. 819–837.

Bosworth, W., 1980, Structural geology of the Fort Miller, Schuylerville and portions of the Schagticoke 7½-minute Quadrangles, eastern New York and its implications in Taconic geology [Ph.D. thesis]: Part 1, State University of New York at Albany, 237 p.

Bosworth, W., and Rowley, D. B., 1984, Early obduction-related deformation features of the Taconic allochthon; Analogy with structures observed in modern trench environments: Geological Society of American Bulletin, v. 95, p. 559–567.

Bosworth, W., and Vollmer, F. W., 1981, Structures of the medial Ordovician flysch of eastern New York; Deformation of synorogenic deposits in an overthrust environment: Journal of Geology, v. 89, p. 551–568.

Boyer, S. E., and Elliot, D., 1982, Thrust systems: American Association of Petroleum Geologists Bulletin, v. 66, p. 1196–1230.

Cady, W. M., 1945, Stratigraphy and structure of west-central Vermont: Geological Society of America Bulletin, v. 56, p. 515–587.

——, 1969, Regional tectonic synthesis of northwestern New England and adjacent Quebec: Geological Society of America Memoir 120, 181 p.

Clark, T. H., 1934, Structure and stratigraphy of southern Quebec: Geological Society of America Bulletin, v. 45, p. 1–20.

Coney, P. J., Powell, R. E., Tennyson, M. E., and Baldwin, B., 1972, The Champlain thrust and related features near Middlebury, Vermont, *in* Doolan, B. D., and Stanley, R. S., eds., Guidebook to field trips in Vermont, New England Intercollegiate Geological Conference 64th annual meeting: Burlington, University of Vermont, p. 97–116.

Doll, C. G., Cady, W. M., Thompson, J. B., Jr., and Billings, M. P., 1961, Centennial geologic map of Vermont: Montpelier, Vermont Geological Survey, scale 1:250,000.

Dorsey, R. L., Agnew, P. C., Carter, C. M., Rosencrantz, E. J., and Stanley, R. S., 1983, Bedrock geology of the Milton Quadrangle, northwestern Vermont: Vermont Geological Survey Special Bulletin No. 3, 14 p.

Fisher, D. W., 1968, Geology of the Plattsburgh and Rouses Point Quadrangles, New York and Vermont: Vermont Geological Survey Special Publications No. 1, 51 p.

Hawley, D., 1957, Ordovician shales and submarine slide breccias of northern Champlain Valley in Vermont: Geological Society of America Bulletin, v. 68, p. 155–194.

Hitchcock, E., Hitchcock, E., Jr., Hager, A. D., and Hitchcock, C., 1861, Report on the geology of Vermont: Clairmont, Vermont, v. 1, 558 p.; v. 2, p. 559–988.

Keith, A., 1923, Outline of Appalachian structures: Geological Society of America Bulletin, v. 34, p. 309–380.

——, 1932, Stratigraphy and structure of northwestern Vermont: American Journal of Science, v. 22, p. 357–379, 393–406.

Leonard, K. E., 1985, Foreland fold and thrust belt deformation chronology, Ordovician limestone and shale, northwestern Vermont [M.S. thesis]: Burlington, University of Vermont, 138 p.

Palmer, A. R., 1969, Stratigraphic evidence for magnitude of movement on the Champlain thrust: Geological Society of America, Abstract with Programs, Part 1, p. 47–48.

Rodgers, J., 1970, The tectonics of the Appalachians: New York, Wiley Interscience, 271 p.

Rowley, D. B., 1982, New methods for estimating displacements of thrust faults affecting Atlantic-type shelf sequences; With applications to the Champlain thrust, Vermont: Tectonics, v. 1, no. 4, p. 369–388.

——, 1983, Contrasting fold-thrust relationships along northern and western edges of the Taconic allochthons; Implications for a two-stage emplacement history: Geological Society of America Abstracts with Programs, v. 15, p. 174.

Rowley, D. B., Kidd, W.S.F., and Delano, L. L., 1979, Detailed stratigraphic and structural features of the Giddings Brook slice of the Taconic Allochthon in the Granville area, *in* Friedman, G. M., ed., Guidebook to field trips for the New York State Geological Association and the New England Intercollegiate Geological Conference, 71st annual meeting: Troy, New York, Rensselaer Polytechnical Institute, 186–242.

Stanley, R. S., and Ratcliffe, N. M., 1983, Simplified lithotectonic synthesis of the pre-Silurian eugeoclinal rocks of western New England: Vermont Geological Survey Special Bulletin No. 5.

——, 1985, Tectonic synthesis of the Taconic orogeny in western New England: Geological Society of America Bulletin, v. 96, p. 1227–1250.

Stanley, R. S., and Sarkesian, A., 1972, Analysis and chronology of structures along the Champlain thrust west of the Hinesburg synclinorium, *in* Doolan, D. L., and Stanley, R. S., eds., Guidebook for field trips in Vermont, 64th annual meeting of the New England Intercollegiate Geological Conference, Burlington, Vermont, p. 117–149.

Suppe, J., 1982, Geometry and kinematics of fault-related parallel folding: American Journal of Science; v. 282, p. 684–721.

Welby, C. W., 1961, Bedrock geology of the Champlain Valley of Vermont: Vermont Geological Survey Bulletin No. 14, 296 p.

Zen, E-an, 1972, The Taconide zone and the Taconic orogeny in the western part of the northern Appalachian orogen: Geological Society of America Special Paper 135, 72 p.

Zen, E-an, ed., Goldsmith, R., Ratcliffe, N. M., Robinson, P., and Stanley, R. S., compilers, 1983, Bedrock geologic map of Massachusetts: U.S. Geological Survey; the Commonwealth of Massachusetts, Department of Public Works; and Sinnot, J. A., State Geologist, scale 1:250,000.

Stratigraphy of the Cambrian platform in northwestern Vermont

Charlotte J. Mehrtens, Department of Geology, University of Vermont, Burlington, Vermont 05405

LOCATION

The Cambrian to Lower Ordovician stratigraphic sequence in northwestern Vermont outcrops in a north-south trending belt bordered on the east by the Green Mountain anticlinorium and on the west by deformed Middle Ordovician shales. Significant facies changes, both parallel and perpendicular to depositional strike, can be observed between Burlington and U.S. 2, 8 mi (13 km) to the north (Fig. 1). Stops 1 to 5 in the Lower Cambrian Dunham Dolomite, Middle Cambrian Monkton Quartzite, and Winooski Dolomite are located on a 3 mi (5 km) west-to-east series of outcrops along U.S. 2 near Milton, starting in an abandoned quarry on the north side of the road, 5 mi (8 km) east of the Sand Bar State Park. The section continues 0.9 mi (1.5 km) farther east along U.S. 2 at a large roadcut on the north side of U.S. 2 (Stop 2), and again another 0.8 mi (1.3 km) farther along Route 2 at a roadcut on the south side of the road (Stop 3). The final two stops along U.S. 2 are another 0.8 mi (1.3 km) (Stop 4) and 0.8 mi (1.3 km) (Stop 5, Chimney Corners). Total mileage along U.S. 2 is 3.3 mi (5.3 km). Stop 6 is located along the Winooski River in downtown Winooski, Vermont, where the supra-to-intertidal and shallow subtidal facies of the Monkton, Winooski, and Danby Formations are exposed. All outcrops are accessible by car; access does not require permission.

SIGNIFICANCE

Cambro-Ordovician siliciclastic and carbonate sediments in western Vermont were deposited on a tectonically stable shelf following late Precambrian rifting of the Iapetus Ocean (Rodgers, 1968). The alternating siliciclastic and carbonate units record sedimentation in supra-to shallow subtidal platform environments, which pass laterally into platform margin and basinal deposits. The basal Cheshire Quartzite, not included in this site, represents a shallow siliciclastic "blanket" over the Eocambrian rift topography (Myrow, 1983). Subsequent Cambrian units record the vertical upbuilding of the carbonate platform characteristic of the early Paleozoic continental margin in the Appalachians.

SITE INFORMATION

Background. The Lower Cambrian Dunham Dolomite lies in gradational contact with the Cheshire Quartzite (Fig. 2) and represents the initial carbonate deposition on the newly formed shelf (Gregory, 1982). The Dunham Dolomite is important because it records the initial carbonate facies development and the establishment of the platform geometry that are continued in subsequent units in northwestern Vermont. Upward building of the carbonate platform during Dunham time resulted

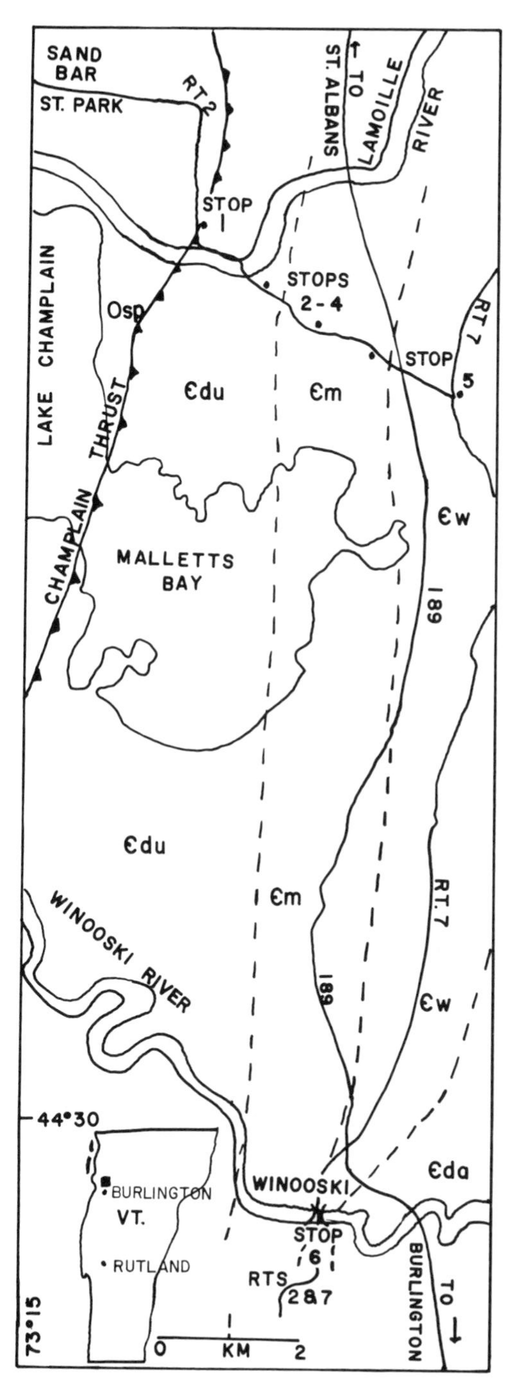

Figure 1. Locality map for stops in northwestern Vermont. Stops 1–5 occur along U.S. 2 between Sand Bar State Park and the intersection of U.S. 2 and 7 (Chimney Corners). Stop 6 occurs along the banks of the Winooski River in downtown Winooski, starting at the first ledges downstream of the bridge and extending upstream past the mill to the last ledges. Lithic designators: Cdu-Dunham Dolomite, Cm-Monkton Quartzite, Cw-Winooski Dolomite, Cda-Danby Quartzite, Osp-Stony Point Shale.

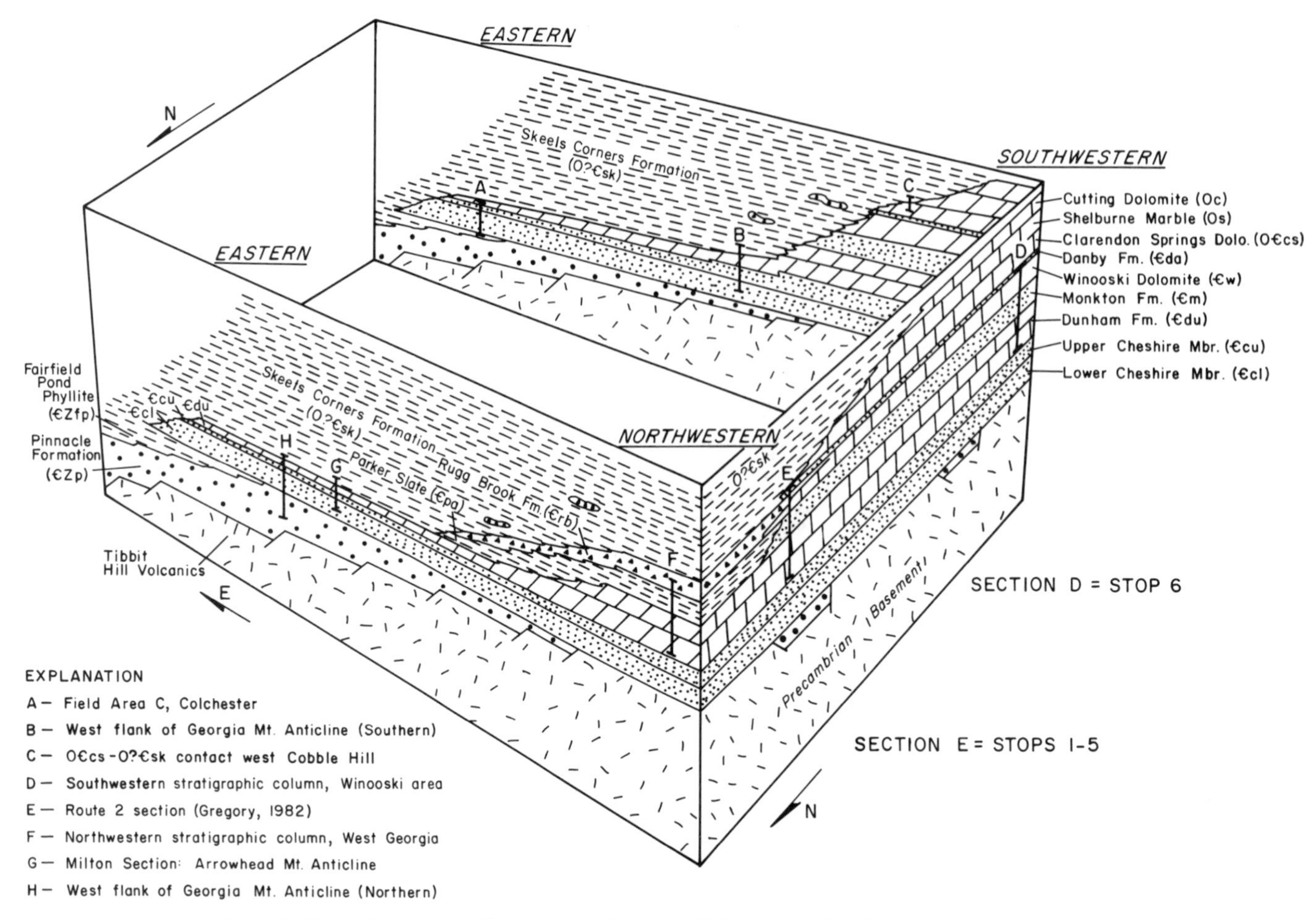

Figure 2. Block diagram from Dorsey, and others (1983) illustrating the platform stratigraphy and the eastward to northward facies changes associated with the platform margin transition. Sections represented by Stops 1–5 and 6 are sections E and D, respectively.

in the development of a platform-to-platform margin transition which was characterized by the abrupt pinch-out of shallow water facies into basinal shales and proximal talus slope breccias. This platform-to-basin transition remained localized in roughly the same paleogeographic position throughout the remainder of the Cambrian and Early Ordovician, probably as a result of movement on underlying rift-related faults. The Cambro-Ordovician platform was characterized by vertical upbuilding and little-to-no progradation into the adjacent basin.

The nature and distribution of lithofacies in the Monkton Quartzite (lower Middle Cambrian) can be compared to the underlying Dunham Dolomite; these also serve as models for interpreting the youngest siliciclastic unit, the Danby Quartzite. As seen in the Dunham Dolomite, the Monkton also records deposition in tidal, shallow subtidal, and platform margin environments. The distribution of lithofacies on the platform is also similar, recording both east-to-west and north-to-south facies changes into the adjacent shale basin. The lithofacies and environments of deposition of the Monkton Quartzite were studied and summarized by Rahmanian (1981). Rahmanian recognized seven lithofacies: three comprise mixed siliciclastic and carbonate sediments associated with shallowing-up cycles, three are pure siliciclastic deposits, and one is a pure carbonate (oolitic dolomite) facies. The 1,000-ft (300-m)-thick Monkton is composed of cyclic shallowing-up cycles characterized by repetitive packages of: (1) basal subtidal siliciclastic sand shoals and channels (Stop 4) overlain by, (2) interbedded siliciclastic sand and silt and carbonate intertidal flat sediments (Stop 6), capped by (3) carbonate muds of the high intertidal and supratidal flat (Stop 6). These shallowing-up cycles are interpreted to represent prograding tidal flat deposits, structurally similar to those in the underlying Dunham Dolomite (Mehrtens, 1986).

The Winooski Dolomite (Middle Cambrian) is a structureless dolomite with disseminated quartz sand and cryptalgalaminites common throughout. Little detailed sedimentology has been done on the Winooski, but based on its stratigraphic relationship with over- and underlying units as well as lithofacies analysis, it is interpreted to represent shallow subtidal (Stop 6) to platform margin (Stop 5) environments.

The Upper Cambrian Danby Quartzite is a mixed siliciclastic and carbonate unit (Butler, 1986) that represents deposition in intertidal to shallow subtidal as well as platform margin settings. Butler (1986) has interpreted much of the Danby as recording shallow subtidal, storm-influenced sedimentation.

The Cambro-Ordovician sequence seen in Vermont records alternating carbonate and siliciclastic sedimentation not seen elsewhere in the Appalachian-Caledonide Orogen. The environments of deposition represented by each formation are similar, regardless of composition, and contacts between units are gradational. These three characteristics: regionally localized cyclic carbonate or siliciclastic sedimentation, similarity of platform paleogeography, and consistently gradational contacts, suggests that the cyclicity is not a result of large-scale regressive events (Rowley, 1979), but rather local variations in siliciclastic sand supply and distribution.

Stop 1. Stop 1, an abandoned quarry, lies immediately above the Champlain Thrust, as the floor of the quarry is Middle Ordovician Stony Point Shale overlain by Lower Cambrian Dunham Dolomite. The Dunham Dolomite is approximately 1,300 ft (400 m) thick and is exposed along U.S. 2, near Milton (Stops 1–3), where rocks characteristic of the peritidal, subtidal/open shelf, and platform margin can be seen. Stop 1 exhibits fresh exposures of the peritidal facies, characterized by the rhythmic interbedding of white dolomite and red dolomitic siltstone in a "sedimentary boudinage" bedding style. Intraformational conglomerate, imbricated clasts, and burrows are locally common. It is strongly recommended that visitors to this stop avoid climbing on the quarry walls but confine themselves to large blocks lying around the quarry floor.

Stop 2. The overlying subtidal/open shelf facies of the Dunham Dolomite is characterized by shallowing-up cycles up to 30 ft (10 m) thick that have at their base massive beds of bioturbated and mottled dolomite with disseminated quartz and feldspar sand throughout. This lithology is capped by packages of the rhythmically interbedded dolomite and dolomitic siltstone of the peritidal facies. These shallowing-up cycles are interpreted to represent tongues of tidal flat sediments that prograded into the adjacent subtidal shelf. Shallowing-up cycles make up the bulk of the 1,000-ft (300-m)-thick subtidal facies, and these pass upsection into structureless, subtidal, burrowed muds before passing into the platform margin facies.

Stop 3. Stop 3 exhibits massive beds of polymictic breccia interpreted as proximal debris flows interbedded with graded dolomitic sandstone beds interpreted as turbidites. Clasts within the breccia are poorly sorted and angular. Beds are poorly developed and structureless. At other localities the Dunham platform margin lithofacies can be seen to pass laterally into basinal shales of the Parker Slate.

Stop 4. Stop 4, along U.S. 2, exhibits the subtidal and platform margin facies of the Monkton Quartzite. At this stop, subtidal/tidal channel, crossbedded sands pass up section into thickly bedded, crossbedded platform margin sand bodies. A smaller outcrop immediately to the east of Stop 4 exposes horizons of the platform margin polymictic breccia in a matrix of coarse-grained quartz sand. Note the variable clast composition and the angularity and poor sorting of the clasts. Compare the sedimentary structures and bed thickness at this exposure to the inter- and supratidal Monkton seen at Stop 6.

Stop 5. The Winooski Dolomite exposed at Stop 5 (Chimney Corners) consists of structureless dolomite which is environmentally nondiagnostic, but it is capped by horizons of polymictic breccia in a sand-rich dolomite matrix which has been interpreted as a platform margin breccia. The breccia can be compared to those seen in the underlying Monkton and Dunham Formations.

Stop 6. Stop 6, downstream of the bridge (southeast bank) over the Winooski River in Winooski, exhibits shallow-water facies of the Monkton, Winooski, and Danby Formations. The base of the section is in the Monkton Quartzite, where inter- and supratidal facies are exposed in shallowing-up cycles. Bedforms diagnostic of supra-, inter- and shallow subtidal environments are exposed on broad bedding planes. Several different ripple morphologies can be identified, along with mudcracks and vertical and horizontal burrows. Structureless beds of buff dolomite are interpreted as supratidal deposits.

The Winooski Dolomite is also exposed on the north side of the Winooski River (Stop 6) from below the bridge to 300 ft (100 m) upstream. The basal contact with the Monkton is underwater here, but in a quarry a few miles away it can be seen to be gradational over a 30-ft (10-m) interval. The upper contact of the Winooski with the Danby Quartzite (Stop 6, upstream on ledges above the mill) is also gradational, characterized by increasing sand content until the first quartzite bed of the Danby is reached. The Winooski here consists of structureless dolomite, which, with the exception of horizons of cryptalgalaminites, is environmentally nondiagnostic. The Winooski is interpreted as shallow subtidal in origin, based on the abundance of stromatolites(?) and its stratigraphic position relative to the underlying inter- and supratidal Monkton facies.

The Danby Quartzite (Stop 6) is exposed in a series of ledges upstream of the bridge over the Winooski River in Winooski. The outcrop at Stop 6 exhibits beds with a diverse assemblage of ripple morphologies, laterally discontinuous bedding, hummocky cross-stratification, stromatolites, and oncolites. The platform margin facies of the Danby is not exposed along Route 2, but occurs in the woods nearby. It is also characterized by polymictic breccia clasts in a crossbedded sand matrix, similar to those described for the Monkton Quartzite.

SUMMARY

The distribution of facies on the Cambrian platform records deposition on a flat-topped, low-gradient platform bordered on the east and north by a deep basin in which shale was deposited (Fig. 2). Regardless of siliciclastic or carbonate composition, all platform deposits appear to record similar facies: supratidal to shallow subtidal in the platform interior and platform margins characterized by carbonate or siliciclastic shoal deposits and talus slope breccias. Significant lateral migration of facies between units is not seen, suggesting that sedimentation on the platform was continuous and able to keep pace with subsidence.

REFERENCES

Butler, R. J., 1986, Sedimentology of the Upper Cambrian Danby Formation of northwestern Vermont; An example of mixed siliciclastic and carbonate platform sedimentation [M.S. thesis]: Burlington, University of Vermont, 137 p.

Dorsey, R., Agnew, P., Carter, C., Rosencrantz, E., and Stanley, R., 1983, Bedrock geology of the Milton Quadrangle, northwestern Vermont: Vermont Geologic Survey Special Bulletin 3.

Gregory, G., 1982, Paleoenvironments of the Dunham Dolomite (Lower Cambrian) in northwestern Vermont [M.S. thesis]: Burlington, University of Vermont, 91 p.

Mehrtens, C. J., 1986, Stratigraphic significance of shallowing-up cycles in the Lower Cambrian Dunham Dolomite and Middle Cambrian Monkton Quartzite: Geological Society of America Abstracts with Programs, v. 18, p. 54.

Myrow, P., 1983, Sedimentology of the Cheshire Formation in west-central Vermont [M.S. thesis]: Burlington, University of Vermont, 177 p.

Rahmanian, V., 1981, Transition from carbonate to siliciclastic tidal flat sedimentation in the Lower Cambrian Monkton Formation, west-central Vermont: Geological Society of America Abstracts with Programs, v. 13, p. 20–21.

Rodgers, J., 1968, The eastern edge of the North American continent during the Cambrian and Early Ordovician, *in* E-an Zen, White, W., Hadley, J., and Thompson, J., eds., Studies in Appalachian geology; Northern and maritime: New York, John Wiley and Sons, p. 141–149.

Rowley, D., 1979, Ancient analogues for the evolution of sedimentation at modern Atlantic-type margins; Example from eastern North America: Geological Society of America Abstracts with Programs, v. 11, p. 507.

Taconic sedimentary rocks near Fair Haven, Vermont

Brewster Baldwin, *Department of Geology, Middlebury College, Middlebury, Vermont 05753*
Andrew V. Raiford, *Department of Geology, Castleton State College, Castleton, Vermont 05735*

LOCATION

Fair Haven is west of Rutland, Vermont, a few miles east of the New York state line. It is served north-south by Vermont 22A and east-west by U.S. 4 (Fig. 1). Most of the outcrops are road cuts or are within a 5-minute walk of the road.

Stop 1: Eureka Quarry. Leave the southeast corner of the village green and drive south; go up the hill 0.2 mi (0.3 km) on Vermont 22A; U.S. 4A bears right. At 1.3 mi (2.1 km), turn left (east) onto Bolger Road. At 2.9 mi (4.6 km), take the left fork. At 3.1 mi (5.0 km), park near the entrance to the Eureka quarry of Vermont Structural Slate. Return to the village green.

Stop 2: Limestone "Conglomerate." Leave the northeast corner of the village green; continue north on U.S. 4A. At the blinker light, 0.5 mi (0.8 km), bear left (north) onto Scotch Hill Road. Cross the overpass above U.S. 4 and at 1.1 mi (1.7 km) park on the east shoulder, across from the long low outcrop. Return to the village green.

Stop 3: Browns Pond Formation. From the northwest corner of the green, drive northwest on Vermont 22A, past the U.S. 4 overpass. At 1.7 mi (2.7 km), park on the right shoulder just past Sheldon Road.

Stop 4: Deep-Sea Fan. From Stop 3, continue another 0.25 mi (0.4 km) northwest on Vermont 22A to the next right-hand curve. Park by the Edward J. Morris farmhouse, and ask permission to visit the site.

Stop 5: Wildflysch-Type Conglomerate. From Stop 4, continue northwest and north on Vermont 22A. At 2.6 mi (4.2 km) north of the village green, on a right curve, the base of the West Castleton Formation is exposed in the roadcut just southeast of the filled quarry hole. At 3.6 mi (5.8 km), the West Haven road joins from the left. At 4.0 mi (6.5 km), where the road widens to three lanes, park on the left (west side) in the parking area.

SIGNIFICANCE

The Taconic sequence consists of folded deep-water argillites and associated interbeds of carbonate and terrigenous material. The origin of these sediments can be studied in the Fair Haven area, which is at the west edge of the Taconic allochthon. Here, the strata are scarcely metamorphosed. As a result, the beds retain sedimentary features that shed light on the source areas and environments of deposition of the Taconic sequence, and even the nature of the Taconic Orogeny. Also, the Fair Haven area includes active commercial quarries of green and purple roofing slate.

SITE INFORMATION

Most of the Taconic strata were deposited on the continental rise of ancient proto-North America, offshore of a platform sequence of Cambrian and Ordovician carbonate and quartz sandstone beds. The age of the Taconic sequence is from approximately 600 Ma, when the proto-Atlantic ocean began to open, to about 450 Ma, with the emplacement of the Taconic allochthon onto the platform sequence during the Taconic Orogeny. The orogeny was due to collision of the proto-North American continental margin with a volcanic arc (Chapple, 1973; Rowley and Kidd, 1981).

The Taconic sequence now occurs as a set of thrust slices, with the lowermost exposed in the west and the uppermost in the east. Rowley and Kidd (1981) inferred that the slices were stacked westward; the easternmost slice is the highest and oldest, and the westernmost slice is the lowest and youngest. Stanley and Ratcliffe (1985) inferred that the slices were stacked eastward, with the youngest being the easternmost and highest. Whichever the order, the tectonic loading evidently drove out most of the pore water and the sediments lithified.

Perhaps the slices were pre-assembled on the ocean floor (Rowley and Kidd, 1981) and lost the pore water there. Then the Taconic allochthon was emplaced onto the platform sequence; this occurred when the continental margin subsided as it entered the east-dipping subduction zone, underthrusting the deep-water sediments. The platform sequence itself became folded and thrust faulted; some beds were dragged westward as parautochthonous slices beneath the Taconic allochthon, and this suggests that the allochthon was acting in a rigid manner. The allochthon itself was further deformed and imbricated into slices. Still later, the continental margin rose isostatically (Rowley, 1980; Baldwin, 1982), generating a source area for a westward-spreading terrigenous wedge.

Table 1 summarizes the stratigraphy for the western Taconics. It reflects work of W.S.F. Kidd and his students and is taken largely from Rowley (1983). However, the Hatch Hill Formation is used here in a narrow sense, as noted below. See Zen (1964) for a more complete record of stratigraphic names. Figure 1 (inset) shows the formations exposed in the Fair Haven area and their approximate thicknesses. The Taconic sequence in the Giddings Brook Slice, measured at right angles to bedding planes, is about 3,300 ft (1,000 m) thick.

The Fair Haven Area

The geology shown in Figure 1 is compiled mostly from Rowley (1983). The part southeast of Fair Haven is from Wright (1970), and the distribution of map units east of Lake Bomoseen is inferred from Zen (1961). In Figure 1 the oldest black slate

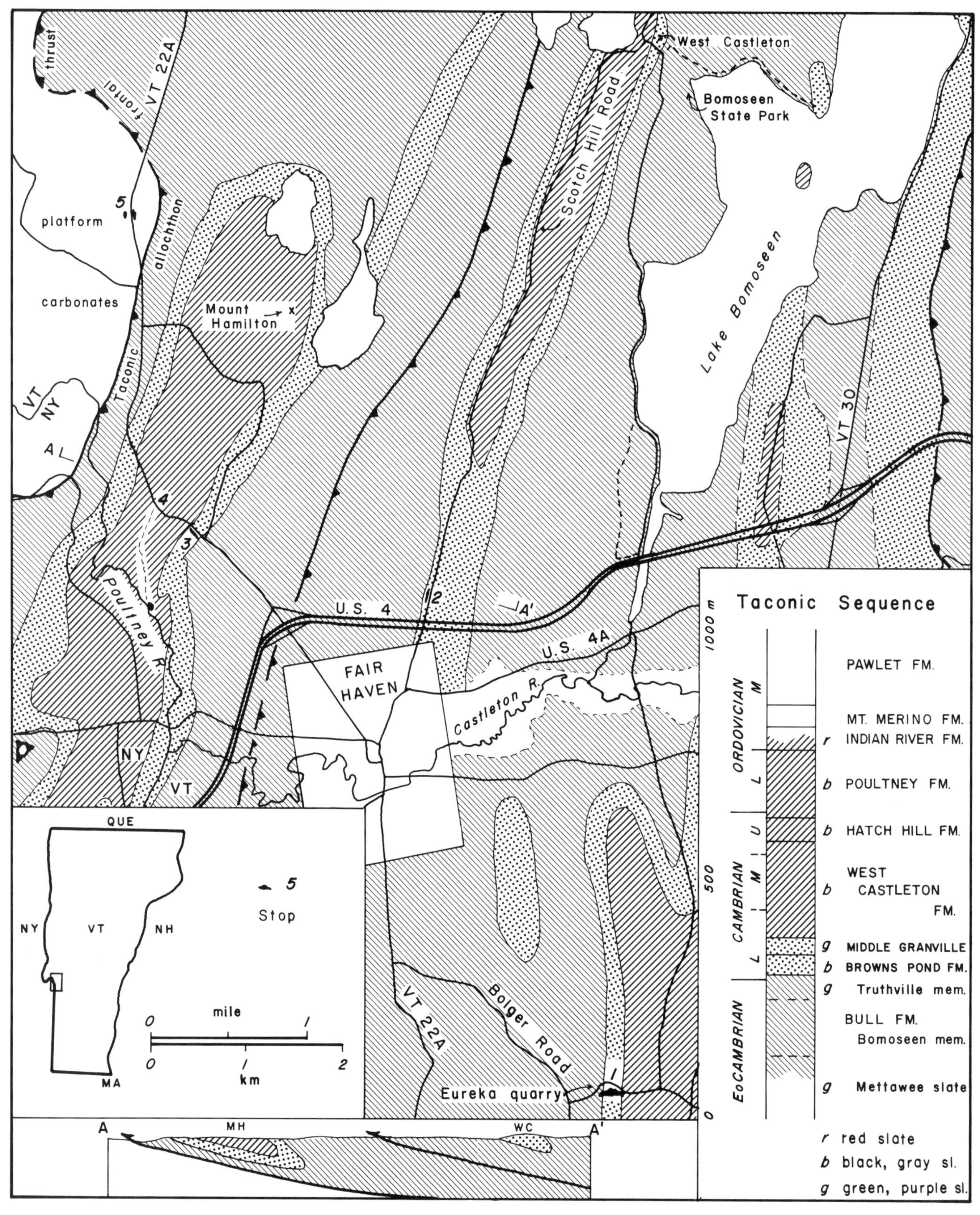

Figure 1. Geologic map of the Fair Haven area (modified from Rowley, 1983; Zen, 1961; Wright, 1970). In cross-section A-A′, MH, Mount Hamilton syncline; WC, West Castleton syncline.

TABLE 1. THE TACONIC SEQUENCE
(Modified from Rowley, 1983)

Formation	Characteristics	Age
WILLARD MOUNTAIN GROUP		
Pawlet Formation	graywacke and black slate	Middle Ordovician
Mt. Merino Formation	black slate, some dark chert	
MOUNT HAMILTON GROUP*		
Indian River Slate	orange-red and green slate	Middle Ordovician (Chazy)
Poultney Slate	black slate, with some interbeds of dolostone, dolomitic quartzite	Upper Cambrian to lower Middle Ordovician
Hatch Hill Formation*	quartzite and black slate	Upper Cambrian
West Castleton Formation*	black slate with some thin limestone beds	Lower to Middle Cambrian
Middle Granville Slate*	purple and green slate; roofing slate quarries	Lower Cambrian
Brown's Pond Formation*	black slate, siltstone, thin limestone beds (North Brittain "conglomerate" of Zen); includes the Mudd Pond quartzite (offshelf equivalent of the Cheshire Quartzite of the platform sequence)	lowermost Cambrian
NASSAU GROUP* (also includes Biddie Knob Formation and Rensselear Grit)		
Bull Formation*	green and purple slate (= Mettawee facies) with sandstone and limestone	Eocambrian to lowest Cambrian
Truthville Slate Member*	green, some purple slate; roofing slate quarries; this is the Mettawee facies above the Bomoseen Member	
Bomoseen Member	graywacke, grit	
Zion Hill Member	quartzite and graywacke	

*Formation can be seen at this site.

(Browns Pond Formation) and the overlying green slate (Middle Granville Formation) are shown as a single map unit that separates the main mass of green and purple slates from the overlying black and gray slates. Most of the stops are within the Mount Hamilton syncline and the West Castleton syncline (Fig. 1, Cross-section A–A′). Rocks in the Fair Haven area are in the Giddings Brook Slice. They are the least metamorphosed and must not have been buried tectonically by the highest slices.

The deep-water sediments of the Taconic sequence are varied in color, method of deposition, composition, texture, and structure. Green and purple slates of the Bull Formation, which characterize the lower half of the Taconic sequence, represent non-reducing conditions on the sea floor. Soon after the start of the Cambrian, however, conditions became reducing. The evidence is the black and gray slates of the West Castleton through the Poultney formations, which were deposited through the Early Ordovician. Moreover, the change from green-purple to black reversed briefly, because there is a transition zone in which the first black slates (Browns Pond Formation) are overlain by more green-purple slates (Middle Granville Formation).

Some sediment settled through quiet water, and other sediment was transported by density currents down the continental slope. Much or most of the slate was deposited by slow settling. However, one slate bed (not visited in these stops) is identified as a turbidite by a cluster of chert pebbles at the base; the pebbles were derived from a 1-cm chert bed within the depositional sequence. Whatever the sedimentation rate of a single bed, the net accumulation rate of Taconic muds was most probably controlled by the supply of particles to the ocean floor and continental slope (Baldwin, 1983). The Taconic slate accumulated at the imperceptible average rate of about 30 ft (10 m) per m.y., or 10 microns per year. Baldwin (1983) estimated the rate to be 13 ft (4 m) per m.y., but, as Rowley (1983) noted, outcrop width of vertical beds is a structural thickenss that may be less than half of the stratigraphic thickness.

The origin of the limestone beds in the Browns Pond and West Castleton formations is a question. The Browns Pond Formation is 60 ft (20 m) thick in some places and only 16 ft (5 m) in others (Campbell, 1983); Wright (1970) showed it to be absent in places north of the Eureka quarry. Graded bedding is rarely if ever seen, but if indeed the limestone beds are turbidites, why are they most commonly interbedded with black slates and less commonly with green-purple slates?

Interbeds of dolostone and quartz sandstone (and some associated limestone) flowed down from the continental platform. Twice, when the shoreline regressed toward the edge of the platform, shallow-water grains of carbonate and quartz were resuspended and easily transported into deeper water (Rowley, 1979) to form submarine fans. Sandstone turbidites of the West Castleton Formation can be correlated with distal platform quartz sands of the early Middle Cambrian Monkton Quartzite, and turbidites (Stop 4, Fig. 2) of the Hatch Hill Formation are correlated with platform quartz sands of the Upper Cambrian Danby Formation (Baldwin, 1983). A 13-ft (4-m) bed of conglomerate in the Hatch Hill Formation has a matrix of dolomite-cemented quartz sandstone, with sandstone clasts to 8 in (20 cm). The transition from the turbidite-rich Hatch Hill Formation to the turbidite-poor Poultney Formation probably marks the next

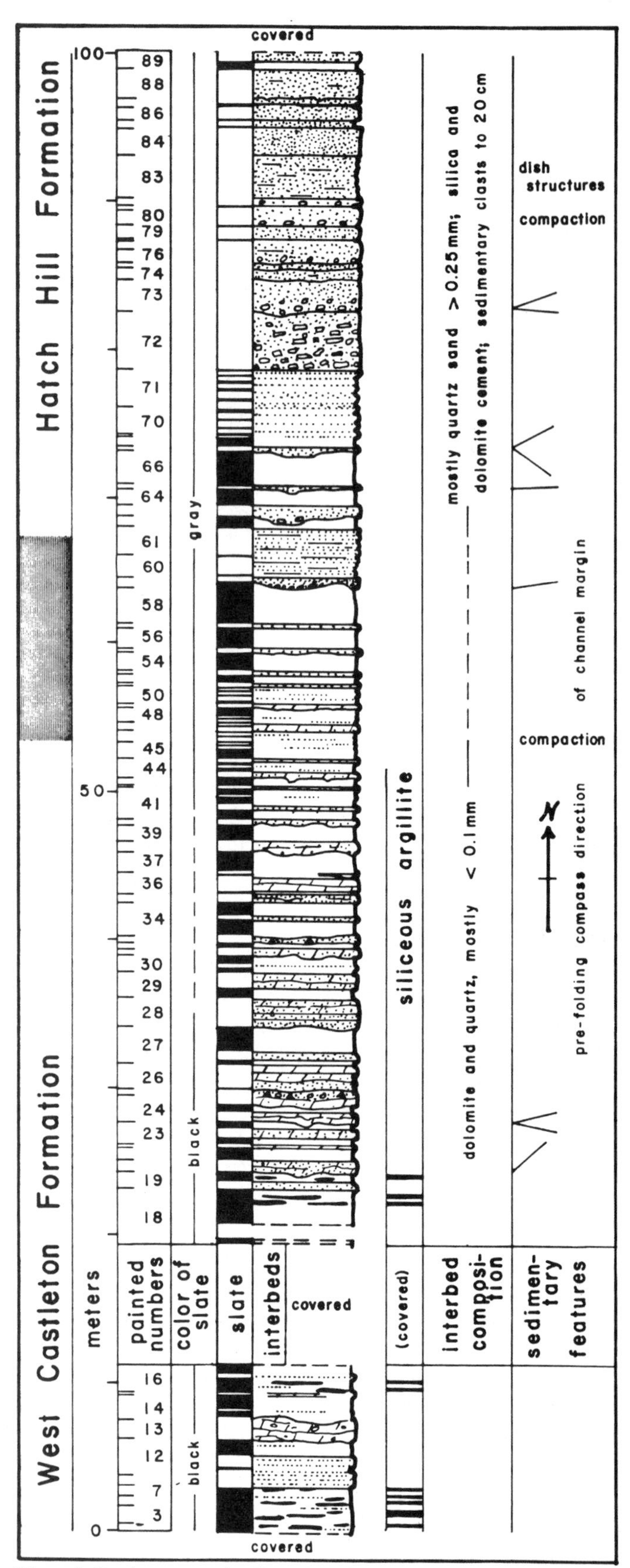

Figure 2. Section exposed at Stop 4, on Poultney River. (Modified from Baldwin, 1983).

transgression of the shoreline, shutting off the supply of turbidites. It should be noted that Rowley (1983) and Fisher (1984) define the Hatch Hill in a lithologically less restricted sense, as a Lower to Upper Cambrian formation that includes the West Castleton as a member.

Soft-sediment deformation occurs in the Browns Pond Formation (Stops 2, 3). One example is the large clasts of folded black slate suspended in massive, lighter colored slate. Another example is the round, unstrained quartz sand grains that are suspended in a massive gray muddy siltstone, which also has suspended clasts of fine-textured black slate. This is the "calcareous quartz wacke" of Rowley (1983); the grains and clasts may have become mixed by flowage. A third example is the limestone "conglomerate"; the clasts are the result of disrupted beds of limestone that are interbedded with the black slate. Because some of these clasts are bent, the limestone was not yet lithified at that stage.

The soft-sediment deformation occurred during sedimentation or during the stacking of the thrust slices, but before emplacement of the allochthon on the platform sequence. The rock-unit thickness of the Taconic sequence is only 3,000 to 6,000 ft (1,000 to 2,000 m), so that the solidity (complement of porosity) was only about 80% (Baldwin, 1971) before tectonic burial. That leaves enough pore water to permit soft-sediment deformation during stacking of the thrust slices, but not later, as noted above. The limestone "conglomerate" is commonly related to west-verging folds and thrusts (Campbell, 1983), and those folds were thus probably formed during the stacking of thrust slices.

The Indian River Formation, exposed in this area on Mt. Hamilton, is a Chazy-age orange-red slate that probably was derived from terra rossa soils formed during uplift of the continental platform (Bird and Dewey, 1970), just before the margin entered the subduction zone. The Pawlet Formation, seen by the bridge just upstream from Poultney, on Vermont 30, about 7 mi (10 km) south of U.S. 4, consists of slate and graywacke turbidites; the latter are composed of immature sands derived from a voclanic arc to the east (Rowley and Kidd, 1981).

STOP DESCRIPTIONS

Stop 1: Eureka Quarry. At the splitting shed (nearest the road), ask for permission to visit the inactive part of the upper (eastern) quarry. The orange-red slate on pallets near the splitting shed is from the Indian River Formation and has been brought in from the Granville area in New York. Some loose blocks of purple slate show green ellipsoids, which represent reduction spheres that were later deformed.

The main quarry is in the Truthville Slate Member of the Bull Formation. The Browns Pond Formation constitutes the top 16 ft (5 m) of the main quarry wall. Above the Browns Pond Formation, in the upper quarry, is 30 ft (10 m) of the Middle Granville Formation. The upper half of the upper quarry face is black slate and thin limestone of the West Castleton Formation.

Stop 2: Limestone "Conglomerate." Interbedded black slate and limestone form the north end of the outcrop. In most of the road cut, however, the limestone is disrupted into a "conglomerate," with some bent clasts of limestone. This is Zen's (1961) "North Brittain Conglomerate" and is part of the Browns Pond Formation.

Stop 3: Browns Pond Formation. The 300-ft (90-m) outcrop at Stop 3 is the Browns Pond Formation on the overturned east limb of the Mount Hamilton syncline. Pacing from the southeast end, several important rock types and features can be seen. From 0 to 86 ft (0 to 26 m), clasts of folded black slate, which are suspended in lighter colored structureless slate, record soft-sediment deformation; near 60 ft (20 m), clear black grains of round unstrained quartz are surrounded by gray muddy siltstone; at 86 ft (26 m), there is a disrupted sandstone; from 89 to 238 ft (27 to 72 m), there is well-bedded black slate with a sandstone bed; from 238 to 271 ft (72 to 82 m), the limestone "conglomerate" includes a few bent clasts; from 271 to 297 ft (82 to 90 m), there is monotonous green-gray slate of the Middle Granville Formation.

Stop 4: Deep-Sea Fan. The outcrop is next to Poultney River, a 10-minute walk (0.5 mi; 0.8 km) from the highway. (Leave the gates the way you find them—open or closed.) Walk south on the farm road, through the gate to the bottom of the hill. Continue south in the pasture, go up the hill on a wagon road between two fences. Go through a gate and up to the knoll where a transmission line crosses. Leave the wagon road and continue due south, obliquely to the left down the slope to the north end of several evergreens. From there, scramble down the open 50-ft (15-m) slope eroded in glacial deposits, to the extensive outcrops by Poultney River.

This is Section 3 of Baldwin (1983); the painted numbers identify arbitrarily selected units. **Please do not sample near the painted numbers.** Figure 2 illustrates important features of this glacially polished 300-ft (100-m) section. Above unit 19, the slate and interbeds represent part of a submarine fan (Baldwin, 1971); unit 72 is a channel deposit. The West Castleton/Hatch Hill boundary is here put somewhere between units 45 and 61, using the abundance of medium quartz sandstone as a criterion. The youngest quartz sandstone exposed is taken as the top of the Hatch Hill Formation; overlying beds in outcrops downstream are of the Poultney Slate. The decompacted geometry of units 46 and 80 has been illustrated and discussed by Baldwin (1971).

The source of the carbonate and quartzose interbeds lay to the west. The clasts of quartz sandstone and dolostone in unit 72 and in other channel breccias must have come from the stable continental platform. The strata are vertical and strike N10E, and several channels are exposed in cross-section. Margins of these channels are within 40° of vertical; if the vertical beds are unfolded about a horizontal cylindrical axis, the channel margins are rotated into an angle within 40° of S80E (Fig. 2).

Stop 5: Wildflysch-Type Conglomerate. This locality is just beneath the base of the Giddings Brook Slice. At the "No Littering" sign at the south end of the parking area, step over the fence; walk west 100 ft (30 m) down the faint trail, and then northwest through brush 80 ft (25 m) to the outcrop in the clearing. This is a "wildflysch-type conglomerate" in the autochthonous Hortonville Slate. Zen (1961) named this the Forbes Hill Formation and inferred that the lithified clasts of foreign rocks slid off the Giddings Brook Slice as the Taconic allochthon moved west across the muds of the former platform.

The rocks in the road-cut across from the parking area are fault slivers in the upper part of the platform-sequence limestones and shales. They are formed by the overriding Taconic allochthon (Rowley and Kidd, 1982, p. 225; citation included in citation for Rowley and Kidd, 1981). The limestones and slates are fossiliferous. Toward the north end, some slates contain quartzite clasts, and other slates have limestone clasts.

REFERENCES

Baldwin, B., 1971, Ways of deciphering compacted sediments: Journal of Sedimentary Petrology, v. 41, p. 293–301.

——, 1982, The Taconic orogeny of Rodgers, seen from Vermont a decade later: Vermont Geology, v. 2, p. 20–25.

——, 1983, Sedimentation rates of the Taconic sequence and the Martinsburg Formation: American Journal of Science, v. 283, p. 178–191.

Bird, J. M., and Dewey, J. F., 1970, Lithosphere plate-continental margin tectonics and the evolution of the Appalachian orogen: Geological Society of America Bulletin, v. 81, p. 1031–1060.

Campbell, H. K., 1983, Structure and stratigraphy of the North Brittain conglomerate [abs.]: Green Mountain Geologist, v. 10, n. 1, p. 6–7. (Also, unpublished senior thesis, Middlebury College, 57 p.)

Chapple, W. M., 1973, Taconic orogeny; Abortive subduction of the North American continental plate?: Geological Society of America Abstracts with Programs, v. 5, p. 573.

Fisher, D. W., 1984, Bedrock geology of the Glens Falls—Whitehall region, New York: New York State Museum Map and Chart Series No. 35, 58 p.

Rowley, D. B., 1979, Ancient analogs for the evolution of sedimentation at modern Atlantic-type margins; Example from eastern North America: Geological Society of America Abstracts with Programs, v. 11, p. 507.

——, 1980, Timor—Australian collision; Analog for Taconic allochthon emplacement: Geological Society of America Abstracts with Programs, v. 12, p. 79.

——, 1983, Operation of the Wilson cycle in western New England during the early Paleozoic, with emphasis on the stratigraphy, structure, and emplacement of the Taconic allochthon: [Ph.D. thesis]: State University of New York at Albany, Department of Geological Sciences; University Microfilm International, Ann Arbor, Michigan, 602 p.

Rowley, D. B., and Kidd, W.S.F., 1981, Stratigraphic relationships and detrital composition of the Medial Ordovician flysch of western New England; Implications for the tectonic evolution of the Taconic orogeny: Journal of Geology, v. 89, p. 199–218. Also: Discussion by J. Rodgers (and reply), 1982, v. 90, p. 219–226; discussion by P. Geiser (and reply), 1982, v. 90, p. 227–233.

Stanley, R. S., and Ratcliffe, N. M., 1985, Tectonic synthesis of the Taconian orogeny in western New England: Geological Society of America Bulletin, v. 96, p. 1227–1250.

Wright, W. H., III, 1970, Rock deformation in the slate belt of west-central

Vermont [Ph.D. thesis]: Urbana, University of Illinois, 91 p.
Zen, E-an, 1961, Stratigraphy and structure at the north end of the Taconic Range in west-central Vermont: Geological Society of America Bulletin, v. 72, p. 293–338.
——, 1964, Taconic stratigraphic names; Definitions and synonymies: U.S. Geological Survey Bulletin 1174, 94 p.

Barre granite quarries, Barre, Vermont

Dorothy A. Richter, Hager-Richter Geoscience, Inc., P.O. Box 572, Windham, New Hampshire 03087

LOCATION

The Barre granite quarry district is located in the southeastern section of Barre Town, Washington County, Vermont, near the villages of Graniteville and Websterville. The quarried area extends slightly into the northeastern corner of Williamstown, Orange County. Figure 1 shows the general location of the quarry district and the various village entities within Barre Town.

From I-89 the area is reached by taking Exit 6 (Barre, South Barre) and following the access road 5 mi (8 km) to the intersection with Vermont 14 (Fig. 1). Cross Vermont 14 and follow Middle Road to Graniteville.

Figure 2 shows the locations of the quarries discussed in this field guide. Public viewing platforms are available at the edge of the Rock of Ages quarry (behind the Rock of Ages Visitors Center, 0.9 mi [1.5 km] east of Lower Graniteville) and the Wells-Lamson quarry (along the Websterville Road at the north end of the quarry). During the summer months, Rock of Ages operates a short train ride for tourists that gives an excellent view of the active Smith Quarry and the abandoned Duffee Quarry.

Access to the several other active and abandoned granite quarries in the area may be obtained only by making prior arrangements with the Rock of Ages Corporation and the Wells-Lamson Quarry Company. Both companies are accommodating to serious visitors and researchers who wish to do detailed sampling or mapping. The Pirie/Adam, Smith, and Wetmore & Morse quarries are especially interesting to visit. There are more than 70 inactive quarries in the Barre area, but most are now filled with water; in fact, many are a part of the local public water supply. Dale (1923) provided a map showing the locations and names of many of the quarries active at that time.

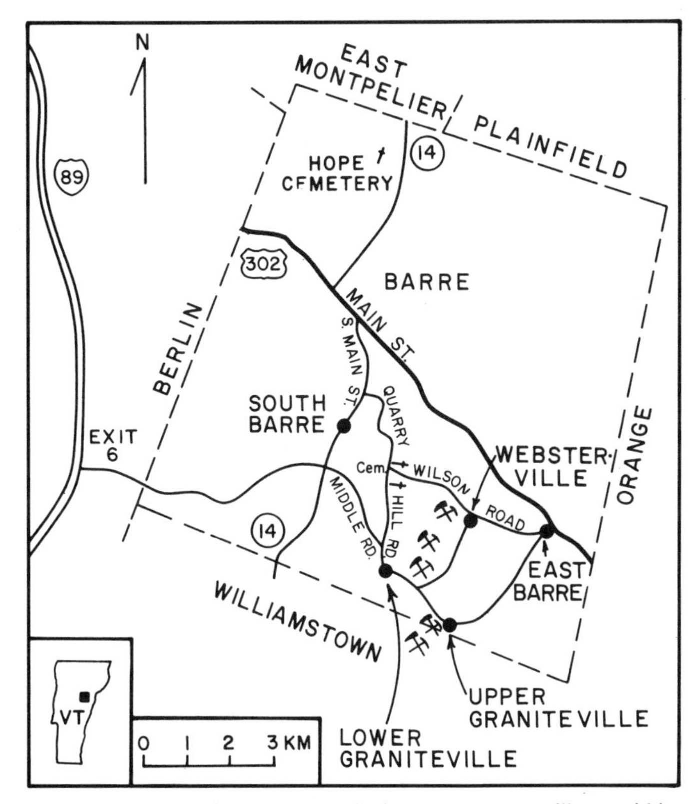

Figure 1. Location of Barre quarry district and numerous villages within the boundaries of Barre Town, Vermont.

SIGNIFICANCE

The Barre granite quarries expose a three-dimensional view of a small, well-studied pluton of the New Hampshire Plutonic Series. It is a Devonian calc-alkaline pluton formed by partial melting of Siluro-Devonian sedimentary rocks that had been folded and regionally metamorphosed during the Acadian orogeny and thus illustrates that the orogeny occurred rather suddenly in the region.

Contact relations with the Siluro-Devonian Gile Mountain Formation, internal igneous flow features, and various types of postconsolidation fracturing are very well displayed in the quarries. In addition, the Barre quarry district is one of the few localities in the Northeast U.S. where geologists and the public can observe the ancient and honorable tradition of quarrying large-dimension stone blocks.

REGIONAL SETTING

The Connecticut Valley–Gaspé synclinorium in eastern Vermont is characterized by tightly folded phyllitic and calcareous metasediments of the Gile Mountain and Waits River formations. The metasediments are generally interpreted to have been deposited in the remnants of the Iapetus ocean. A series of more than 20 two-feldspar, two-mica granitoids, ranging in size from small dikes <0.6 mi (<1 km) long to plutons >37 mi (>60 km) in diameter, intrude the metasediments in eastern Vermont. In general, the plutons are unzoned, internally uniform in composition, and not associated with igneous rocks more mafic in composition than diorite.

Contacts with the metasediments are mildly to strongly discordant. Metamorphic aureoles are associated with most of the intrusions (including the Barre pluton), indicating that the plutons were intruded after the peak of regional metamorphism. The regional metamorphic grade of the host metasediments ranges from garnet to kyanite grade in central Vermont. The regional metamorphic grade around the Barre pluton is staurolite grade.

Naylor (1971) obtained K-Ar dates on primary micas in

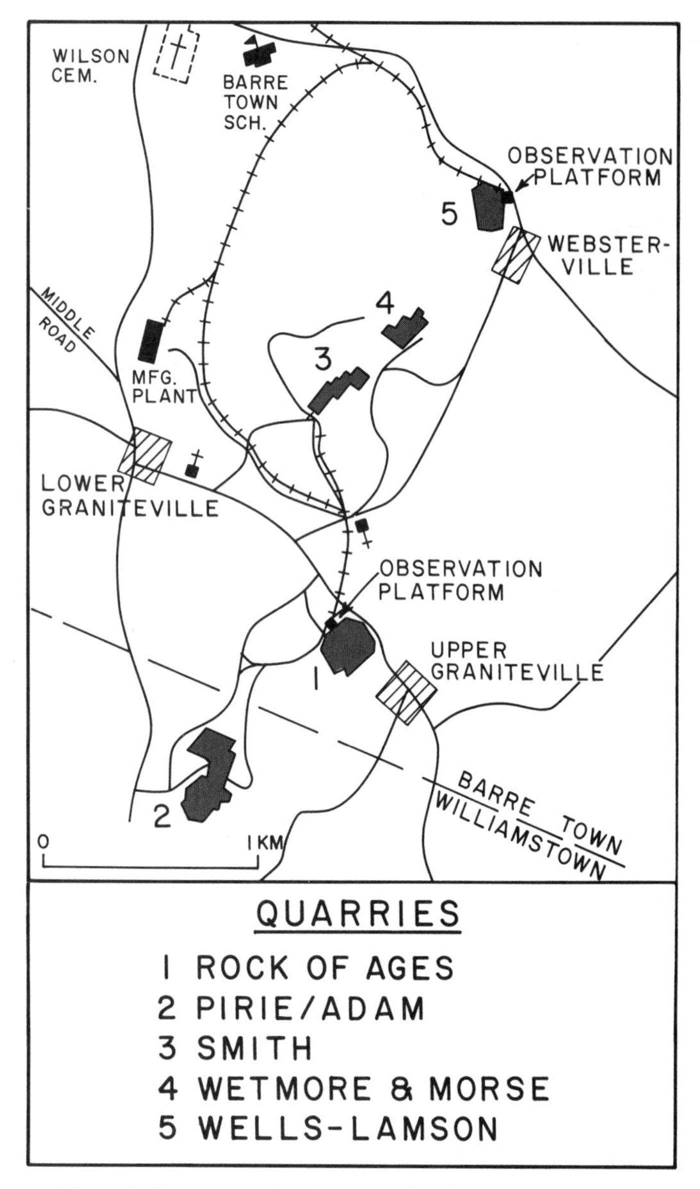

Figure 2. Sketch map showing quarry locations discussed in text.

several of the Vermont granitoids (including the Barre pluton) that clustered around 380 Ma. He concluded that the Acadian orogenic event in that part of the northern Appalachians must have been a brief but intense event, since the Devonian granitoids crosscut Devonian sediments that had been folded and regionally metamorphosed before intrusion.

BARRE AREA

Figure 3 is a generalized geologic map of the Barre area modified from Murthy (1957) and Doll and others (1961). The Barre pluton is contorted in outline but is approximately 1.9 mi (3 km) wide and 5 mi (8 km) long. The pluton is elongated parallel to the regional strike of the enclosing metasediments, the Gile Mountain Formation for the most part, and the Waits River Formation for a segment of the eastern contact. A much larger igneous body of similar composition, the Knox Mountain pluton, occurs a few miles northeast of the Barre pluton.

The Gile Mountain Formation is composed of fine-grained, dark gray mica schists and micaceous quartzites. The Waits River Formation contains similar rocks plus abundant calcareous schists. The stratigraphic relationships between the two host formations are complex and contradictory in places. Woodland (1977) and Fisher and Karabinos (1980) determined from structural analysis and graded bedding in the Royalton area, approximately 31 mi (50 km) south of Barre, that the Gile Mountain Formation is younger than the Waits River Formation and that the belt of Gile Mountain Formation in which most of the Barre pluton resides is the core of a north-plunging synform. Woodland (1977) has further suggested that the synform is the downward-facing nose of a recumbent nappe with an eastward root zone.

Contacts of the Pluton

Foliation of the metasediments near the Barre pluton is nearly vertical, strikes north-northeast, and more or less bulges around the igneous rock. The contacts of the pluton are thus semiconcordant with the foliation in most places. It appears that the pluton gently "shouldered" itself into position along zones of weakness in the country rock. Large xenoliths (>6 ft or >2 m) that appear not to have been reoriented are common, and septa of the metasediments (possible roof pendants) are present within the pluton.

A good pavement exposure of the contact occurs behind the storage garages at the south end of the Pirie quarry. The contact is exposed in places along the east walls of the Rock of Ages and Pirie quarries where quarrying terminated. A large septum of the Gile Mountain Formation roughly follows the road between Lower Graniteville and Websterville. Small outcrops along the road contain the contact and small dikes of granitic material invading the country rock. Contact with the same septum is visible along the east walls of the Smith, Wetmore & Morse, and Wells-Lamson quarries.

Composition

The Barre pluton is strikingly uniform in composition. It is a gray, fine- to medium-grained (1 to 5 mm) granodiorite. Note, however, that even among geologists the rock is widely known and commonly referred to by its commercial name, Barre granite.

In thin section, the rock is characterized by sericitized oligoclase laths (35 volume percent), partially chloritized reddish-brown biotite (9 percent), moderately to highly strained quartz (27 percent), and interstitial fresh-appearing microcline (21 percent). Primary muscovite (6 percent) is not common but does occur as disseminated ragged flakes. Common accessories include minute (<0.1 mm) zircon, apatite, sphene, metamict allanite, and opaques. Chayes's (1952) modal analyses on 21 samples from different quarries in the Barre district show little variation in modal abundances despite apparent visual variations in grain size and gray shades.

Aplites do not occur in the Barre pluton and pegmatitic veins are not common. The few pegmatites that occur are simple unzoned pegmatites composed of 4-6 cm creamy plagioclase, microcline, thin mats of muscovite and biotite, and interstitial quartz. A pair of semihorizontal pegmatite dikes, 3 ft (1 m) thick, occurs in the south wall of the Pirie quarry. A zone of pegmatitic material also occurs along the upper east side of the excavation for the Adam quarry. (Note: the Pirie and Adam quarries are converging as quarrying proceeds and may be considered one quarry by the owners.) The quarrymen refer to the pegmatitic material at the Adam quarry as "Gazeley," after the name of a quarry started in similar rock some time ago. Pegmatite veins are more abundant in the Knox Mountain pluton a few mi (km) to the northeast of the Barre pluton.

Internal Fabrics

Primary igneous flow patterns are faintly visible at most locations in the quarries. Balk (1937) measured the attitude of the lineations and concluded that there are several domelike intrusions represented in the quarries. In general, the flow patterns in the currently active quarries appear to dip gently to the southwest.

From the observation platform at the Rock of Ages quarry, prominent flow lines can be observed at the northeast corner of the quarry, about 15 ft (5 m) below an unused red hoist building. Another area of pronounced flow features occurs near the top of the Smith quarry at the U-1 derrick location. Here, the flow lines are wavy and commonly cross-cutting.

Xenoliths are common near the contacts of the pluton. They are particularly common in the southeastern limb of the body and can be seen in all stages of assimilation by the magma in the Rock of Ages and Pirie quarries. From the observation platform at the Rock of Ages quarry, one can observe large (6 to 15 ft; 2 to 5 m) xenoliths and biotite schlieren in the east wall of the quarry, and toward the bottom of the south end of quarry. Xenoliths are much less common in the Smith, Wetmore & Morse, and Wells-Lamson quarries, which are located in the central portion of the pluton.

Structure

Two styles of jointing are present in the Barre quarries: high-angle joints with a northeast strike, and sheeting joints that follow the topography. One type of joint is generally more prominent in each quarry.

In the Rock of Ages and Pirie/Adam quarries, a prominent set of joints strikes northeast and dips steeply eastward. The dip is about 70° in the Rock of Ages quarry and about 45° in the Pirie/Adam quarry. Joint surfaces commonly contain slickensides and are marked by greenish chloritized alteration products. Sheeting and cross-jointing are poorly developed in the Rock of Ages and Pirie/Adam quarries.

In contrast, high-angle joints are much less common and more widely spaced in the Smith, Wetmore & Morse, and Wells-Lamson quarries. These quarries, located on the flanks of Millstone Hill, are characterized by well-developed sheeting joints that are subparallel to the topography of the hill. The spacing between sheeting joints gradually thickens from <4 in (<10 cm) near the surface to >15 ft (>5 m) in the quarries.

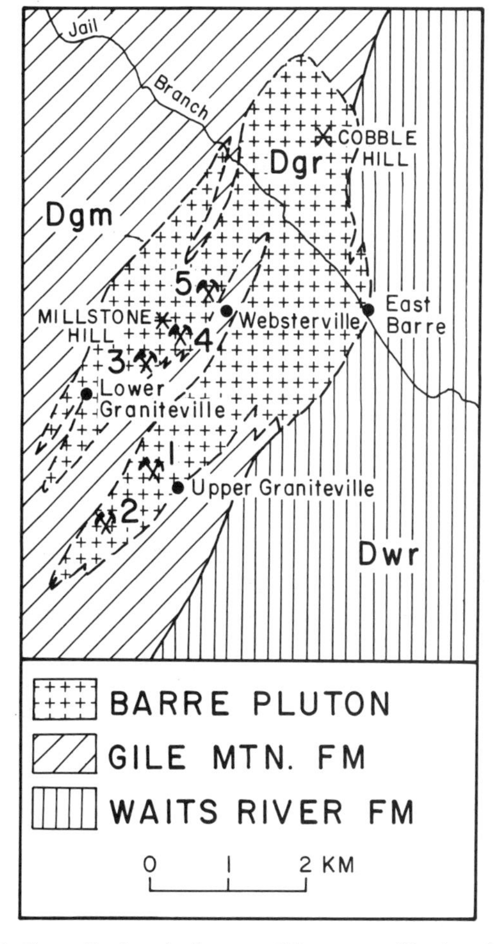

Figure 3. Generalized geologic map of Barre area. Numbered quarries are the same as shown in Figure 2. (Modified from Murthy, 1957, and Doll and others, 1961.)

Sheeting is generally thought to be formed during unloading as the rock is brought to the surface. It appears that because the southeast limb of the Barre pluton had a prominent set of high-angle joints along which the unloaded stresses could be relieved, the formation of sheeting was not favored. In the central part of the pluton, on the other hand, high-angle joints are not well developed, and in situ stresses during unloading could be relieved only by propagating sheeting joints.

Rock bursts, i.e., the spontaneous cracking of the granodiorite during quarrying, is a common phenomenon in the Barre quarries (White, 1946). The rock bursts are due to release of in situ stress and are more common in sheeted than in unsheeted quarries. Nichols (1975) found that the strain relaxation time for a quarried block of Barre granite was several days. It is interesting to note that quartz in thin sections from the central quarries appears less strained than quartz in thin sections from the southeastern quarries.

Several Mesozoic lamprophyre dikes intrude the Barre pluton. The dikes are generally 3 to 5 ft (1 to 1.5 m) wide and have a northeast strike similar to that of the high-angle joints. A pair of narrow dikes can be observed in the north wall of the Rock of Ages quarry on strike from the observation platform. One of the dikes is continuous under the face below the platform.

Economic Geology

The Barre pluton has been quarried for dimension stone continuously since the 1830s. The quarry industry became a major economic force in the Barre area after the Civil War, when the railroad was constructed and when, at about the same time, the market for cemetery monuments shifted from marble to the more durable granite. Most granite blocks produced in Barre today are manufactured into cemetery monuments. Other products include building facings, surface plates (precision ground flat tables for machine bases and inspection surfaces), and paper rolls. The paper rolls are cylinders up to 30 ft (10 m) long and 6 ft (2 m) in diameter, which are used to dewater crushed pulp in the manufacture of newsprint. The Barre district is one of the major producers of granite products in the United States.

Quarrymen are natural practitioners of rock mechanics in that they use jointing and the microstructure of the granite to split out regular rectangular blocks. In the dimension stone industry, the three directions in which a block of granite will split commonly are called "rift," "grain," and "headgrain" (or "hardway"), in order of decreasing ease of splitting. The three directions of splitting are termed "rift," "lift," and "head" in the Barre district only.

The rift is parallel to the sheeting in most quarry districts; in Barre, however, the rift is a nearly vertical plane that has a strike roughly parallel to the strike of the high-angle joints. It is independent of primary igneous flow features in the rock. The rift is defined by a set of open and partially healed intergranular microcracks that can be recognized in oriented thin sections (Douglass and Voight, 1969). Dale (1923) recognized that planes of secondary fluid inclusions in quartz are aligned with the rift of several commercially quarried granites. With practice, the rift in the Barre granite can be recognized in blocks and in outcrop by looking for minute cracks with a preferred orientation in quartz grains.

The optimum block size from the Barre quarries is 175 to 200 ft^3 (5 to 6 m^3), a size that permits the mass manufacture of cemetery monuments. Waste (called "grout" in the industry) is unusable stone that is undersized, irregularly shaped, fractured, or is "blemished" by xenoliths ("knots"), veins ("sand lines", "streaks"), or strongly pronounced flow features ("waves"). Saleable recovery from the Barre quarries varies from 25 to 40 percent. More than 700,000 ft^3 [20,000 m^3] of granite blocks are produced by the Barre district quarries annually. The quarry blocks are separated into five grades, depending on their color and texture. The darker gray, finer grained granite from the Rock of Ages and Pirie quarries commands the highest price. By the time the granite is processed into a finished product, the total recovery is said to be 15 percent.

The waste is discarded in huge piles since there is no economic use for it. The market for crushed stone is too small in the Barre area to make much use of the grout. The granite is too fine grained and the feldspars contain too much iron for separation into individual mineral products for the industrial minerals market.

The manufacturing process for the granite products is fascinating to most geologists, although it is too detailed to be described here. There is an observation platform at the Rock of Ages manufacturing plant in Lower Graniteville where one can see large automated polishing machines and also stone cutters hand-trimming granite monuments. Some of the smaller granite fabricators in the city of Barre also permit serious visitors. In addition, a visit to Hope Cemetery located on the west side of Vermont 14 just north of the city of Barre is highly recommended for those who are interested in seeing spectacular carved granite monuments.

REFERENCES CITED

Balk, R., 1937, Structural behavior of igneous rocks: Geological Society of America Memoir 5, 135 p.

Chayes, F., 1952, The finer grained calc-alkaline granites of New England: Journal of Geology, v. 60, p. 207–254.

Dale, T. N., 1923, The commercial granites of New England: U.S. Geological Survey Bulletin 738, 488 p.

Doll, C. G., Cady, W. M., Thompson, J. B., Jr., and Billings, M. P., 1961, Centennial geologic map of Vermont: Vermont Geological Survey.

Douglass, P. M., and Voight, B., 1969, Anisotropy of granites; A reflection of microscopic fabric: Geotechnique, v. 19, p. 376–398.

Fisher, G. W., and Karabinos, P., 1980, Stratigraphic sequence of the Gile Mountain and Waits River formations near Royalton, Vermont: Geological Society of America Bulletin, Part I, v. 91, p. 282–286.

Murthy, V. R., 1957, Bed rock geology of the East Barre area, Vermont: Vermont Geological Survey Bulletin 10, 121 p.

Naylor, R. S., 1971, Acadian orogeny; An abrupt and brief event: Science, v. 172, p. 558–560.

Nichols, T. C., Jr., 1975, Deformations associated with relaxation of residual stresses in a sample of Barre granite from Vermont: U.S. Geological Survey Professional Paper 875, 32 p.

White, W. S., 1946, Rock-bursts in the granite quarries at Barre, Vermont: U.S. Geological Survey Circular 13, 15 p.

Woodland, B. G., 1977, Structural analysis of the Silurian–Devonian rocks of the Royalton area, Vermont: Geological Society of America Bulletin, v. 88, p. 1111–1123.

Mascoma Dome, New Hampshire: An Oliverian gneiss dome

***Richard S. Naylor**, Earth Sciences Department, Northeastern University, Boston, Massachusetts 02115*

LOCATION

The Mascoma Dome is located in central western New Hampshire about 12 mi (20 km) east of the towns of Lebanon and Hanover (Fig. 1). Groups of outcrops at two localities are described. Most of the major rock units can be seen by visiting either locality separately, but some relationships between the units are displayed only at one locality or the other. Both localities can be reached by passenger car or by charter bus.

This section gives instructions for reaching each group separately. If both locality groups are to be visited on a single trip, the drive from one locality to the other can be shortened considerably by navigating from the Mt. Cube and Mascoma 15-minute topography maps. The starting point for reaching both localities is Exit 20 off I-89 about 1 mi (1.6 km) east of the junction of I-89 with I-91.

The first locality is at the Dartmouth College Skiway at Holts Ledge in Lyme Center, New Hampshire (Mt. Cube 15-minute Quadrangle). Leave I-89 at Exit 20 and travel north on New Hampshire 12A. In West Lebanon continue northward on New Hampshire 10. On the north side of the village of Lyme, turn right on the unnumbered paved road leading eastward through Lyme Center toward the Skiway. At 1.2 mi (1.9 km) east of Lyme Center, follow the paved road that forks right to the Skiway. Park in one of the parking lots and walk to the base of the beginner's slope (the southmost trail on the west side of the road). The trip continues on foot following instructions given later in the text. The property is owned by Dartmouth College. Permission is not required but please heed their signs.

The second locality is a group of exposures off Moose Mountain Road north of Enfield, New Hampshire. Proceed east on I-89 from Exit 20. Leave I-89 at Exit 17 and travel east 4 mi (6.4 km) on U.S. 4 to the crossroads in Enfield (at the bottom of a hill; presently marked by a park and a Post Office). Note mileage from this point and proceed directly to Stop 4, marking the locations of Stop 5 and Stop 6, which will be visited on the return trip. Mile 0.0, turn left (north) off U.S. 4. Mile 0.1 (0.16 km) fork left onto Moose Mountain Road. At mile 1.9 (3 km) as road enters woods and climbs at north end of farm, is the location of Stop 6. At mile 3.1 (5 km), where the road curves sharply right and enters woods, is the location of Stop 5. Continue to mile 3.5 (5.6 km) at abandoned road and cellar hole to right for Stop 4. Park on pavements of granite on left side of road and walk about 0.6 mi (1 km) east on the abandoned road (bear right wherever the trail forks) to the abandoned granite quarry at the top of the hill. Visit Stops 5 and 6 on the way back to Enfield.

SIGNIFICANCE

The Oliverian Domes lie in a belt that can be traced from central Connecticut through northern New Hampshire. The Ammonoosuc, Partridge, Clough, Fitch, and Littleton formations can be traced more or less continuously throughout this belt. Rocks structurally beneath the Middle(?) Ordovician Ammonoosuc Formation crop out discontinuously in the cores of discrete domes. To the extent that the domes are anticlinal, the core rocks would comprise the basement to one of the axial belts in the Appalachian orogen. Most of the core rocks are of igneous composition and many are massive in character. Some interpretations have suggested that the core rocks are younger plutons rather than basement (Billings, 1956; Chapman, 1939; Hadley, 1942).

The Mascoma Dome exposes a representative assortment of Oliverian rock types and displays critical contact relationships at several localities. The overprint of Acadian metamorphism is less than in the more southerly domes. Naylor (1969) demonstrated that in the Mascoma Dome some of the core rocks are stratified metavolcanic gneisses (Holts Ledge Gneiss) underlying the volcanic rocks of the Ammonoosuc Formation, whereas others are intrusives crosscutting the metavolcanic gneisses. At the Mascoma Dome it is unclear whether the intrusives cut the Ammonoosuc volcanic rocks, but they are demonstrably older than the Silurian Clough Formation.

The Dartmouth Skiway Localities, Stops 1–3. The Clough Formation, the Ammonoosuc Formation, and the stratified, metavolcanic core rocks (Holts Ledge Gneiss) are well exposed at the Dartmouth Skiway. A complete tour requires a walking traverse (about 2 mi or 3 km with a climb of about 1,000 ft or 300 m) lasting about 3 hours. Please heed the landowner's posted signs and make no attempt to use motor vehicles on the slopes. The unstratified, intrusive core rock (Mascoma Granite) can be seen a short drive farther along the road through the notch (this stop can be skipped by people planning to visit the Enfield localities, where the granite is better exposed).

The walking tour of Stop 1 starts from the base of the beginners slope (see Location). The slope closely follows a bedding surface near the top of the Holts Ledge Gneiss and the long axis of the dome passes very near this point. The traverse is thus parallel to the plunge at the northern end of the dome. About two-thirds of the way up the slope a faintly marked foot path leads down from the left margin of the ski trail. The cliff exposures near here comprise Stop 1A. The gray gneiss exposed in these cliffs is very typical of the upper part of the Holts Ledge Gneiss. The gneiss is massive but has a faint stratification marked partly by weathering surfaces and partly by subtle variations in lithology. Hadley (1942) interpreted this rock to be an intrusive orthogneiss and mapped it as part of the granodiorite unit of the Mascoma Group of the Oliverian Magma Series. Naylor (1969) reinterpreted the gray gneiss as metavolcanic. He formally re-

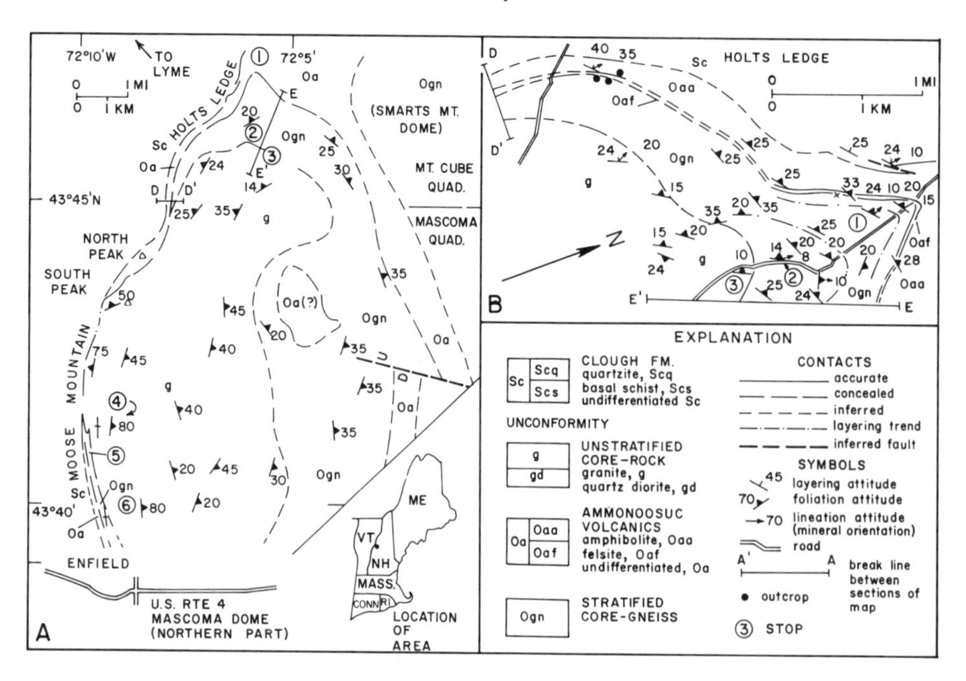

Figure 1. Location and bedrock geologic map of the northern part of the Mascoma Dome, New Hampshire (from Naylor, 1969; modified from Chapman, 1939; and Hadley, 1942). The crossroads in Enfield (Mile 0.0 for stops 4–6), the town of Lyme, and all of the stops are marked (Fig. 1A) but the starting point (Exit 20 off I-89) lies a few miles beyond the western edge of the figure. The lines DD′ and EE′ do not denote structure sections but rather show the limits of the enlarged detail map of Holts Ledge (Fig. 1B).

ferred to this rock as "stratified core rock." Here this rock is informally called "Holts Ledge Gneiss."

Especially note a layer of dark metabasalt about 30 ft (10 m) below the level of the ski trail. This layer can be traced about 300 ft (100 m) along the cliffs and appears to be entirely concordant to the stratification of the gray gneiss above and below. This layer is included with the Holts Ledge Gneiss but closely resembles rocks occurring in the overlying Ammonoosuc Formation. Below metabasalt layer and somewhat to the north is a cliff face in the Holts Ledge Gneiss exposing distinctive biotite-rich schlieren. These are possibly the deformed remains of flattened pumiceous bombs.

Return to the beginners slope and climb the rest of the way to its top. The outcrops near the lift head comprise Stop 1B. The lithology of most of the Holts Ledge Gneiss is plagioclase-quartz-biotite. Potassium feldspar is rare. A thin section from one of the outcrops near the lift head, however, displays about 20% potassium feldspar suggesting a layer of metarhyolite. These rocks bear no resemblance whatever to the unstratified core rock (Mascoma Granite), as they might be expected to if the granite had formed by Acadian remobilization of older metarhyolite. Also note the quartz-tourmaline veins that are fairly common in the core rock. The basal layers of the overlying Clough Formation are locally tourmaline-rich, suggesting that the veins relate to the intrusion of the Mascoma Granite.

Walk west more or less on the same level over to the Worden Schuss Trail. The contact between the Holts Ledge Gneiss and the Ammonoosuc Formation is nearby. All exposures on the ledges above and west of the Worden Trail are Ammonoosuc. The contact is exposed at several places near the top of the cliffs east and below the Schuss. Naylor (1969) defined the contact as a surface below which the coarse gray gneisses comprise more than 50% of the section. He argued this was consistent with the criteria used by contemporary geologists mapping other domes in the belt. M. P. Billings (oral communication, ca. 1970) suggested treating the gray gneisses as a lower member of the Ammonoosuc Formation. The author prefers the separate (albeit informal) designation, Holts Ledge Gneiss, for the following reasons: (1) the contact is readily mapped, (2) most workers in the belt have not included similar rocks in the Ammonoosuc, (3) the possibility remains that an unconformity occurs at or near the contact, and (4) the Holts Ledge Gneiss is conspicuously crosscut by the Mascoma Granite whereas relationships between the granite and the (upper) Ammonoosuc Formation have not been resolved in the Mascoma Dome.

Climb the middle third of the steep Worden Schuss Trail to a point about 150 ft (50 m) short of where the trail forks and climbs very steeply. The ledges above and west of the trail comprise Stop 1C. The first rocks are interbedded, finely crystalline, felsic and mafic gneisses of the Ammonoosuc Formation. Naylor (1969) found that the felsic rocks comprise a marker horizon that could be traced southward for almost 6 mi (10 km) along the western flank of the Mascoma Dome. Parallelism to this marker horizon is a major observation in the argument for the stratigraphic character of the contact between the Holts Ledge Gneiss and the Ammonoosuc Formation. A little higher in the section note the conspicuous epidote knots in the mafic volcanic rocks and the quartz-tourmaline veins. With a small group it is instructive to scramble up the cliff to the Poma Lift viewing the upper parts of the Ammonoosuc Formation. With more than six people in the group it is better to return to the Schuss.

Continue climbing the Schuss (or the Poma Lift) to the lift head. From here continue south about 150 ft (50 m) to the overlook at Holts Ledge. The fence is there with good reason. The rocks exposed at the top are in the lower part of the Ammonoosuc Formation. The mountain (with fire tower) nearby to the northeast is Smarts Mountain exposing the adjacent Smarts Mountain Dome. Mount Cardigan is conspicuous on the horizon slightly south of east at a distance of about 12 mi (20 km). On its summit Kinsman Granite is exposed within the nappe complex that overrides the Oliverian Domes. The rocks of Winslow's Ledge immediately to the northeast form the northeast flank of the Mascoma Dome. The notch below the two ledges displays a series of steps and terraces eroded into bedding surfaces in the Holts Ledge Gneiss. The elongate lake about 5 mi (8 km) to the south is Goose Pond in the granitic part of the dome core. There is another overlook about 150 ft (50 m) farther south along the rim trail. It is small (be careful) but offers a foreshortened view of the contact exposed in the cliff.

Return to the head of the Poma Lift and descend via the leftmost of the broad trails. This trail adheres to bedding surfaces near the contact between the Ammonoosuc Formation and the Clough Formation. Some of the Clough exposures show stretched pebbles. About halfway down is a spectacular glacial erratic of Bethlehem Granite. This granite appears to be an S-type granite and displays a conspicuous xenolith of Littleton Formation. The Bethlehem Granite is a slab-shaped intrusion associated with the nappes that override the domes. Thrust faults repeat the contact between the Ammonoosuc and Clough formations but the relationships are hard to see without very accurate base maps.

Return to cars. Continue southward (uphill) on the road through the notch. After about one mi (1.6 km), the road forks. Bear right. Past the fork there is a swampy pond to the east (left). Near the south end of the pond the road cuts through prominent ledges of Holts Ledge Gneiss. This is Stop 2. Stratigraphically this is one of the deepest exposures of the gneiss in the northern part of the dome. Sample MHL-1 (Naylor, 1969) was collected here and yielded a zircon age of 445 +/− 25 Ma.

Stop 3 is 0.5 mi (0.8 km) farther south where the road passes over a pavement exposure of granite. This is one of the northernmost exposures of the unstratified, intrusive core rocks of the dome. Similar rock is better exposed at Stop 4 and the reader is referred to the description for Stop 4. If Stop 4 is to be visited there is no need to go to Stop 3.

Enfield Localities, Stops 4–6. Stop 4 is a small granite quarry (shown with a quarry symbol on most editions of the Mascoma 15-minute Quadrangle). Except for a slightly greater proportion of microcline, the rock exposed here is representative of the unstratified core rocks of the Mascoma Dome. The author interprets it as a fairly shallow plutonic, intrusive rock contrasting with the metavolcanic rocks described at Stops 1, 2, 5, and 6 (Naylor, 1968, 1969, 1971).

The major minerals are plagioclase, quartz, microcline, and biotite. Except for possibly the weak foliation of the biotite, the author believes these are primary intrusive minerals retaining much of their original texture. Minor minerals visible in a hand sample include epidote, garnet, and magnetite. The epidote may have formed by reaction with plagioclase during Acadian metamorphism, but the garnet and magnetite are considered primary. D. R. Wones (oral communication) noted that the assemblage of microcline, biotite, and magnetite suggests crystallization at fairly low water pressure (because at higher water pressure all of the iron could be accommodated in biotite). Crosscutting aplite veins are common, especially at the north end of the quarry, and suggest that fluid pressure exceeded confining pressure during the late stages of intrusion. This in turn suggests that the intrusion was shallow, perhaps as shallow as 0.6 or 1.2 mi (1 or 2 km).

Naylor (1969, p. 414) reported Rb/Sr whole rock isochron ages of 431 +/− 45 Ma for granitic rocks of the Mascoma Dome and of the nearby Lebanon Dome. (All ages reported here have been recalculated to the decay constants of Steiger and Jager, 1977.) Two of the samples used in that study, MMQ-3 and MMQ-11, come from this quarry. It is noteworthy that aplite sample, MMQ-3, leftovers from which may still be seen at the north end of the quarry, lies on the 431 Ma isochron. This indicates that the aplite veins were intruded in the late stages of the primary intrusive episode and are not the result of Acadian remobilization. In the Mascoma area Acadian metamorphism reached kyanite-staurolite grade and may have been responsible for some of the scatter shown by the whole rock samples on the isochron plot. The author, however, could find no field, petrographic, or isotopic evidence for partial melting or remobilization of any of the dome rocks during the Acadian. This metamorphism was higher grade southward and some of the domes in Massachusetts and Connecticut may show such effects. Mineral separates from the Mascoma Dome yield whole rock–biotite isochrons of 245 Ma, possibly as a response to Permian effects that are more conspicuous to the south and east.

Naylor (1969, p. 416–417) reported an age of 445 +/− 25 Ma for three zircon separates from sample MMQ-11 from this locality. Given the scatter of the Rb/Sr whole rock samples, the zircon data probably give the better estimate of the time of intrusion. Zircons from sample MHL-1 (Stop 2) give the same age within experimental error, indicating the volcanism that produced the stratified core rocks (Holts Ledge Gneiss) immediately preceded the intrusion of the plutonic gneisses. Zartman and Leo (1985) report zircon ages from a number of additional localities that Naylor (1968, p. 238) suggested should correlate with the Mascoma Granite. By interpreting their new data as a suite of samples, rather than as individual points, Zartman and Leo

(1985) derive an age of 444 ± 8 Ma. They noted that the zircon data of Naylor (1969) plot within experimental error on the line defined by the newer and analytically more precise data.

A good view of Moose Mountain about 1.2 mi (2 km) to the west may be obtained from the highest points in the quarry. The white ledges near the top of the ridge are the Lower Silurian Clough Formation, which lies non-conformably on the granite.

Walk back to the cars and drive back towards Enfield. Stop 5 occurs after 0.4 mi (0.6 km), where the road curves to the left and leaves the woods. A small cut of mafic Holts Ledge Gneiss is visible at the edge of the road. The relationships among the units are described by Naylor (1969, 1971, p. 36). Here (at least before the brush grew so thick) and elsewhere it is clear that the Mascoma Granite cuts the Holts Ledge Gneiss. The Ammonoosuc Formation stratigraphically overlies the Holts Ledge Gneiss. The Clough Formation rests unconformably on the Ammonoosuc and non-conformably on the granite but has not been mapped in contact with the Holts Ledge Gneiss.

Stop 6 is 1.2 mi (1.9 km) closer to Enfield. The road descends, curves right, and enters the clearings of a small farm. The contact between the granite and the Holts Ledge Gneiss may be closely approximated between small roadcuts of both units (see Fig. 1). Small veins of plagioclase granite cut the Holts Ledge Gneiss. Possibly these originated as aplite and the potassium has subsequently reacted with the mafic gneiss.

REFERENCES CITED

Billings, M. P., 1956, The geology of New Hampshire, Part II, Bedrock geology: New Hampshire State Planning and Development Commission, 203 p.

Chapman, C. A., 1939, Geology of the Mascoma Quadrangle, New Hampshire: Geological Society of America Bulletin, v. 50, p. 127–180.

Hadley, J. B., 1942, Stratigraphy, structure, and petrology of the Mt. Cube area, New Hampshire: Geological Society of America Bulletin, v. 53, p. 113–176.

Naylor, R. S., 1968, Origin and regional relationships of the core-rocks of the Oliverian Domes, *in* Zen, E-an, White, W. S., Hadley, J. B., and Thompson, J. R., Jr., eds., Studies in Appalachian geology; Northern and maritime: New York, Interscience, p. 231–240.

—— , 1969, Age and origin of the Oliverian Domes, central-western New Hampshire: Geological Society of America Bulletin, v. 80, p. 405–428.

—— , 1971, Geology of the Mascoma mantled gneiss dome near Hanover, New Hampshire: *in* Lyons, J. B., and Stewart, G. W., eds., New England Intercollegiate Geological Conference, Guidebook, v. 63, p. 28–37.

Steiger, R. H., and Jager, E., 1977, Subcommission on geochronology; Convention on the use of decay constants in geo- and cosmochronology: Earth and Planetary Science Letters, v. 36, p. 359–362.

Zartman, R. E., and Leo, G. W., 1985, New radiometric ages on gneisses of the Oliverian Domes in New Hampshire and Massachusetts: American Journal of Science, v. 285, p. 267–280.

Metamorphic stratigraphy of the classic Littleton area, New Hampshire

Robert H. Moench, *U.S. Geological Survey, Box 25046, Denver Federal Center, Denver, Colorado 80225*
Katrin Hafner-Douglass, *Dartmouth College, Hanover, New Hampshire 03755*
Christian E. Jahrling II, *University of New Hampshire, Durham, New Hampshire 03824*
Anne R. Pyke, *University of Vermont, Burlington, Vermont 05401*

LOCATION

The Littleton area, in northwestern New Hampshire (Fig. 1, inset), is easily reached from Canada or southern New England via I-93 and a network of paved and graded roads. Directions to the numbered stops shown on Figure 1 are provided below. The following topographic quadrangle maps would be helpful: St. Johnsbury 15-minute (1:62,500), and the Lower Waterford, Littleton, Woodsville, Lisbon, Newbury, and East Haverill 7½-minute (1:24,000).

Site 55—Bronson Hill Sequence

Stop 1A. Roadcut in Dead River Formation (O€d) on the eastbound lane of I-93, 1.5 mi (2.4 km) east of interchange 42, and just east of a long roadcut through amphibolite of lower member of Ammonoosuc Volcanics (Oal).

Stop 1B. Outcrop of Dead River Formation (O€d) on the south bank of the Ammonoosuc River 0.3 mi (0.5 km) west of the intersection of U.S. 302 and New Hampshire 112.

Stop 1C. Stream-washed outcrops of volcanic-bearing member of Perry Mountain Formation (Spv), the Foster Hill fault, Dead River Formation (O€d), lower member of Ammonoosuc Volcanics (Oal), and Partridge Formation (Op) along Childs Brook, upstream and downstream from the Woodsville Road bridge, which is 0.7 mi (1.1 km) south of the west end of the covered bridge at Bath Village. The Foster Hill fault between O€d and Spv is exposed on the north bank of the brook about 500 ft (150 m) upstream from the bridge (Fig. 1).

Stop 2A. Wave-washed outcrops of mixed (Oaux) and mafic (Oaub) parts of upper member of Ammonoosuc Volcanics along the southeast shore of Moore Reservoir between New Hampshire 135/18 and I-93.

Stop 2B. Roadcuts in mixed part of upper member of Ammonoosuc Volcanics (Oaux) and Partridge Formation (Op) along New Hampshire 135/18, about 0.2 to 0.5 mi (0.3 to 0.8 km) west of the causeway at the south end of Moore Reservoir.

Stop 2C. Large roadcut in mixed part of upper member of Ammonoosuc Volcanics (Oaux) on Foster Hill Road, which joins New Hampshire 135/18 about 1 mi (1.6 km) west of the causeway across Moore Reservoir.

Stop 2D. Outcrops across contact between mixed (Oaux) and distal epiclastic (Oaue) parts of upper member of Ammonoosuc Volcanics at elevation 1,150 to 1,200 ft (350 to 366 m) at the southeast end of Foster Hill. Access from Foster Hill Road about 1 mi (1.6 km) southwest of Stop 2C. Visit in conjunction with Stop 8.

Stop 2E. Outcrops of Partridge Formation (Op), and mixed (Oaux) and distal epiclastic (Oaue) parts of upper member of Ammonoosuc Volcanics starting near Partridge Lake on Dodge Pond Road and extending over knolls to the west. The best starting point is 0.3 mi (0.5 km) southwest of the extreme northwest corner of Partridge Lake.

Stop 2F. Large stream-washed outcrop of Partridge Formation (Op) and overlying mixed part of upper member of Ammonoosuc Volcanics (Oaux) just downstream from covered bridge at Bath; obtain permission.

Stop 2G. Outcrops of lower member of Ammonoosuc Volcanics (Oal) and Partridge Formation (Op) on and near the west bank of the Ammonoosuc River 0.4 mi (0.6 km) southwest of the bridge at Lisbon village. Felsic rocks in the upper member of the Ammonoosuc (Oaux) are exposed under the bridge.

Stop 2H. Outcrops of lower member of Ammonoosuc Volcanics (Oau), Partridge Formation (Op), and mixed part of upper member of Ammonoosuc Volcanics (Oaux) along a north-trending power line southwest of Swiftwater. Access from Briar Hill Road along northeast-trending logging road; outcrops are plainly visible where the logging road crosses the power line, starting about 500 ft (150 m) to the south and ending about 2,500 ft (750 m) to the north.

Stop 2I. Large stream-washed outcrops of mixed part of upper member of Ammonoosuc Volcanics (Oaux) along the Wild Ammonoosuc River at Swiftwater.

Stop 2J. Stream-washed pavements of Partridge Formation (Op) overlain by mixed part of upper member of Ammonoosuc Volcanics (Oaux) along the Wild Ammonoosuc River 0.6 mi (1 km) upstream from Swiftwater.

Stop 2K. Large outcrops of Quimby Formation (Oq) along the Wild Ammonoosuc River 0.4 to 0.6 mi (0.6 to 1.0 km) downstream from Swiftwater.

Stop 2L. Outcrops of mafic facies of upper member of Ammonoosuc Volcanics (Oaub) along the Wild Ammonoosuc River 0.9 mi (1.5 km) downstream from Swiftwater.

Stop 2M. Outcrops of lower member of Ammonoosuc Volcanics (Oal), Partridge Formation (Op), and mixed part of upper member of Ammonoosuc Volcanics (Oaux) on the east bank of the Ammonoosuc River and along the lower reach of

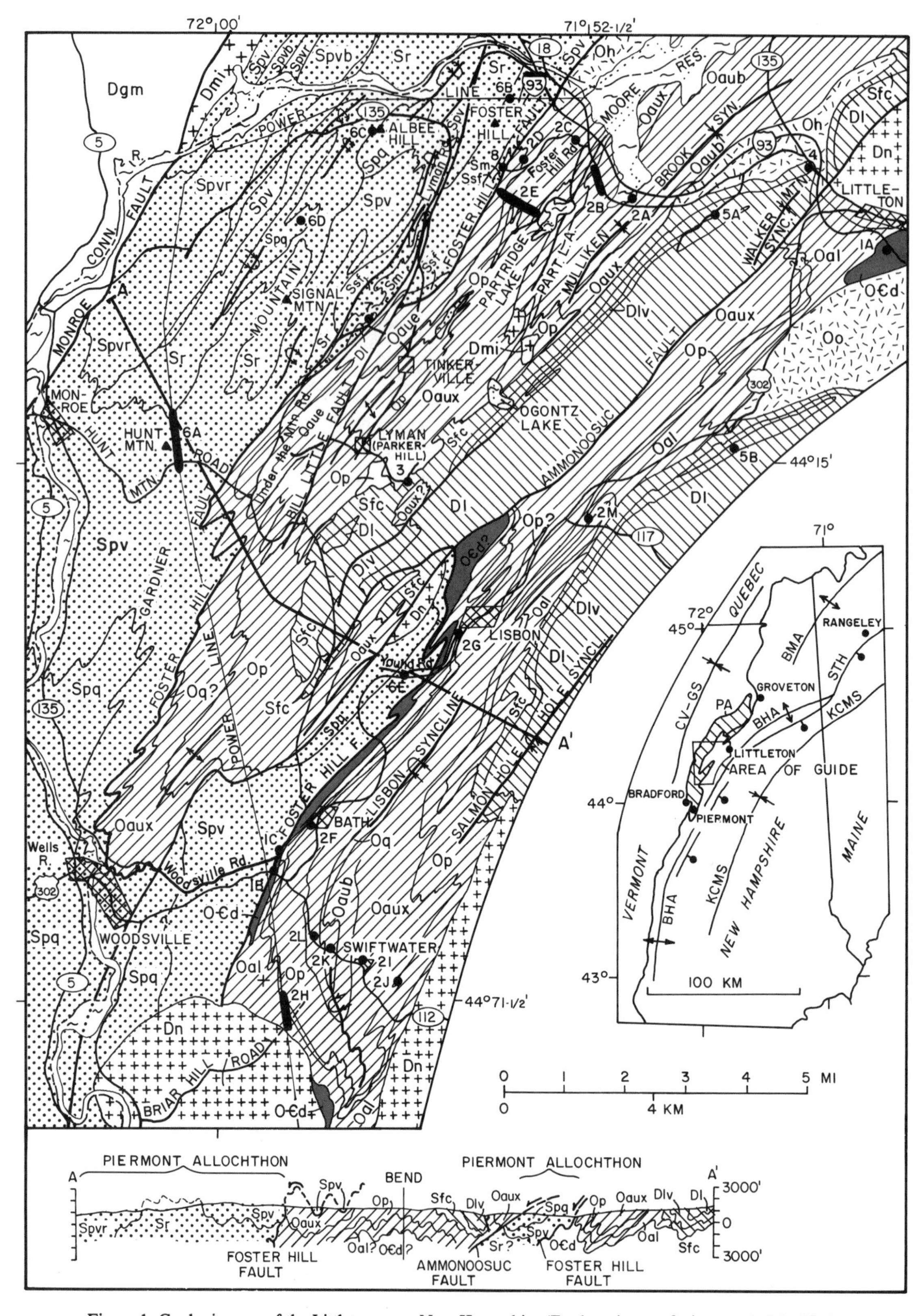

Figure 1. Geologic map of the Littleton area, New Hampshire (Explanation on facing page). Modified from Billings (1935), White and Billings (1951), Hall (1959), and Eric and Dennis (1958). Inset: Location of the Littleton area in relation to tectonic features: CV-GS, Connecticut Valley–Gaspe' synclinorium; PA, Piermont allochthon; BHA, Bronson Hill anticlinorium; BMA, Boundary Mountain anticlinorium; KCMS, Kearsarge–Central Maine synclinorium; STH, Silurian tectonic hinge.

EXPLANATION OF UNITS

	Dn Dmi	DEVONIAN PLUTONIC ROCKS—New Hampshire Plutonic Suite, Dn; Metagabbro and Metadiabase, Dmi
	Oo Oh	ORDOVICIAN PLUTONIC ROCKS—Oliverian Plutonic Suite, Oo; Highlandcroft Plutonic Suite, Oh
		STRATIFIED METAMORPHIC ROCKS—Greenschist facies northwest of Ammonoosuc fault; lower to middle amphibolite facies southeast of fault.
		PIERMONT ALLOCHTHON (Mainly Albee Formation of type area)
	D1	LITTLETON FORMATION (LOWER DEVONIAN)
	Sm	MADRID FORMATION (UPPER SILURIAN)
	Ssf	SMALLS FALLS FORMATION (UPPER SILURIAN
		PERRY MOUNTAIN FORMATION (SILURIAN)
	Spq	Schist and Quartzite Member
	Spv	Volcanic-bearing Member
	Spvb Spvr	Basalt-bearing, Spvb; Rhyolite-rich, Spvr
	Sr	RANGELEY FORMATION (LOWER SILURIAN)
		BRONSON HILL SEQUENCE (Autochthonous)
	Dl Dlv	LITTLETON FORMATION (LOWER DEVONIAN)—Main Body, Dl; Volcanic Member, Dlv
	Sfc	FITCH FORMATION (Sf) AND CLOUGH QUARTZITE (Sc) (SILURIAN)—Undivided
	Oq	QUIMBY FORMATION (UPPER ORDOVICIAN?)
		UPPER MEMBER OF AMMONOOSUC VOLCANICS (MIDDLE ORDOVICIAN)
	Oaue Oaub	Distal epiclastic, Oaue; Basaltic to Andesitic, Oaub
	Oaux	Mixed, mainly felsic
	Op	PARTRIDGE FORMATION (MIDDLE ORDOVICIAN)
	Oal	LOWER MEMBER OF AMMONOOSUC VOLCANICS (MIDDLE ORDOVICIAN)
	O€d	DEAD RIVER FORMATION (UPPER CAMBRIAN? AND LOWER ORDOVICIAN, Albee Formation outside the Piermont allochthon)

MAP SYMBOLS

CONTACT, Dotted under water; querried where conjectural

POSTMETAMORPHIC FAULT, Dotted under water

PREMETAMORPHIC FAULT, Ticks on inferred hangingwall side; dotted under water

UPRIGHT AND OVERTURNED SYNCLINE

UPRIGHT AND OVERTURNED ANTICLINE

●6B NUMBERED STOP DESCRIBED IN TEXT

Salmon Hole Brook, 0.25 mi (0.4 km) northeast of New Hampshire 117 near Salmon Hole.

Stop 3. Large outcrop of Clough Quartzite (Sc) at the intersection of Parker Hill Road and Stickney Road, about 1 mi (1.6 km) southeast of Lyman (Parker Hill); this outcrop is at the type locality of Clough Quartzite.

Stop 4. Roadcut in Fitch Formation (Sf) on the south side of New Hampshire 18, about 1 mi (1.6 km) west of Littleton and 0.25 mi (0.4 km) southeast of Highland Croft Farm (name is on barn).

Stop 5A. Slate quarry on the south side of Slate Ledge Road, 0.9 mi (1.5 km) south of junction with New Hampshire 135/18, exposing Littleton Formation (D1). Obtain permission at nearest house on north side. Volcanic member (D1v) is exposed along a power line 0.3 mi (0.5 km) southwest of the quarry.

Stop 5B. Stream-washed outcrops of volcanic member of Littleton Formation (D1v) along Gale River, about 0.25 to 0.4 mi (0.4 to 0.6 km) upstream from the Ammonoosuc River. Cross the Ammonoosuc River at Barrett (bridge) and walk south to the Gale River.

Site 56—Piermont Allochthon

Stop 6A. Outcrops of volcanic-bearing member of Perry Mountain Formation (Spv) and Rangeley Formation (Sr) along the north-trending power line where the Hunt Mountain Road passes over Gardner Mountain. The road is steep but graded on the west side, where it leads south from New Hampshire 135 at Monroe Village; it is steeper and deteriorating (as of 1985) on the east side.

Stop 6B. Outcrops of Rangeley Formation (Sr) and volcanic-bearing member of Perry Mountain Formation (Spv) along east-trending power line at north end of Foster Hill and knoll to north; access from west along power line, from New Hampshire 135.

Stop 6C. Outcrops of Rangeley Formation (Sr) overlain by schist and quartzite member of Perry Mountain Formation (Spq) at Albee Hill; access west side of hill via logging trail 1.1 mi (1.8 km) west of junction of New Hampshire 135 and Lyman Road; several outcrops of Sr occur about 300 to 1,000 ft (100 to 300 m) south along the trail; Spq is directly up the hill to the east.

Stop 6D. Outcrops of schist and quartzite member of Perry Mountain Formation (Spq) on the west side of Gardner Mountain, between elevations 1,500 and 1,700 ft (457 to 518 m) about 1.3 mi (2.1 km) north of Signal Mountain (a prominent peak on Gardner Mountain); access by foot traverse from New Hampshire 135.

Stop 6E. Large, stream-washed pavement of volcanic-bearing member of Perry Mountain Formation (Spv) on the north side of the Ammonoosuc River about 1.6 mi (2.6 km) southwest of the bridge at Lisbon Village via Young Road (beware of poison ivy).

Stop 7. Bench outcrops of Littleton Formation (D1) near logging trail about 300 ft (100 m) northwest of T-intersection with Under The Mountain Road, and 1.35 mi (2.2 km) south of Y-intersection of Under the Mountain Road with Lyman Road.

Stop 8. East-younging sequence of Perry Mountain (Spv), Smalls Falls (Ssf), and Madrid (Sm) Formations juxtaposed against Ammonoosuc Volcanics (Oaue) along Foster Hill fault. Access from Foster Hill Road about 1.4 mi (2 km) southwest of Stop 2C. Park near driveway to Wilbur Willey residence (obtain permission) at BM 1188. A roadside outcrop of Spv is on south side of Foster Hill Road about 800 ft (250 m) west of BM 1188; Spv and overlying Ssf are exposed in woods to south. Units Ssf and Sm are exposed on low knolls to the east on both sides of the road, 200 to 400 ft (60 to 120 m) southwest and northwest of BM 1188. The Foster Hill fault, between Ssf or Sm and basaltic greenstone of Ouae, can be bridged within 3 ft (1 m) in woods east of driveway to Willey residence, about 800 ft (250 m) north of BM 1188.

SIGNIFICANCE

The pioneering work of Billings (1935, 1937) in the Littleton-Moosilauke area laid a foundation that has guided geologic research in New England for the past five decades. He established the broad regional framework, and provided the basis, for example, for the definition of the nappes of central New England. This guide has been prepared as an update of the geologic framework of the region and highlights exposures that have been important in the present synthesis. The revisions are based on mapping by Moench and all co-authors from near Piermont to near Groveton, New Hampshire, in 1983–1986 (Fig. 1, inset), and by Hafner-Douglass (1986) in the Woodsville Quadrangle in 1985.

SITE INFORMATION

Because of space limitations and the need to describe the basis for major stratigraphic revisions, this guide emphasizes metamorphic stratigraphy. Stratigraphic features of the area are best exposed, or can be seen most conveniently, at the numbered stops on Figure 1. A few stops are roadcuts along I-93, and visitors are cautioned about possible restrictions against parking and walking along interstate highways.

TECTONIC SETTING AND SUMMARY OF STRATIGRAPHIC REVISIONS

The area of Figure 1 lies just northwest of the axial trace of the Bronson Hill anticlinorium and southeast of the Monroe fault, which marks the southeast margin of the Connecticut Valley–Gaspe synclinorium (Fig. 1, inset). The Kearsarge–Central Maine synclinorium is the western part of the Merrimack synclinorium of Billings (1956). A feature here named the Piermont allochthon and briefly described in this report extends across the Littleton area (Fig. 1). On Figure 1 and in the listing of numbered stops, units of the revised classic sequence of the Bronson Hill anticlinorium, here called the Bronson Hill sequence, are separated from those of the Piermont allochthon.

The metamorphic core of the region was generally assumed to be Precambrian until C. H. Hitchcock discovered fossils in an abandoned quarry 2 mi (3.2 km) northwest of the village of Littleton, in calcareous beds subsequently termed the Fitch Formation. Four decades later, F. H. Lahee discovered Early Devonian fossils in the same area west of Littleton, but in rocks later named the Littleton Formation. It was against this background (Billings and Cleaves, 1934) that Billings commenced his field studies.

Figure 2 illustrates the stratigraphic sequence Billings described, and the revisions that we propose on the basis of our mapping. We focused on the unfossiliferous pre-Clough formations; Billings' interpretation of the well-dated Clough, Fitch, and Littleton formations remains intact.

Problem of the Albee Formation and description of the Piermont allochthon

Billings (1935, p. 9) named the Albee Formation for "a group of slates and quartzites typically exposed on Gardner Mountain," in the area of low-grade rocks that lies between the Ammonoosuc and Monroe faults (Fig. 1). The type area of the Albee is a 6-mi (10-km) length of that mountain between Albee Hill, from which the name was obtained, and Hunt Mountain. Billings also mapped the Albee southeast of the Ammonoosuc fault in the area of Figure 1, and he and his associates subsequently mapped the Albee through a wide belt from Piermont, New Hampshire northeastward into Maine (e.g., Billings, 1956). According to published descriptions, the Albee is composed mainly of interbedded schist and quartzite; volcanic rocks are not reported as a major constituent. Billings assigned an Ordovician age to the Albee. He inferred that the Albee Formation lies conformably below the Ordovician Ammonoosuc Volcanics.

In this chapter, the Albee Formation in its type area is subdivided into a succession of units correlated with the Rangeley, Perry Mountain (two members), Smalls Falls, and Madrid Formations (all Silurian), and, at one locality, the Littleton Formation (Lower Devonian). Albee strata outside the allochthon are here considered to be the Cambrian(?) and Lower Ordovician Dead River Formation. The formation names according to this new interpretation of the rocks of the Albee Formation will be used in this chapter.

Our mapping indicates that rocks previously mapped as Albee Formation throughout northern New Hampshire are divisible into at least two rock assemblages. One is a pre-Middle Ordovician quartzite-schist turbidite assemblage that lies below the Middle Ordovician Ammonoosuc Volcanics. This assemblage closely conforms to Billings' definition of the Albee Formation, but it does not occur at the type area of the Albee, nor anywhere within the Piermont allochthon. In northeastern New Hampshire, these rocks are coextensive with and identical to the Dead River

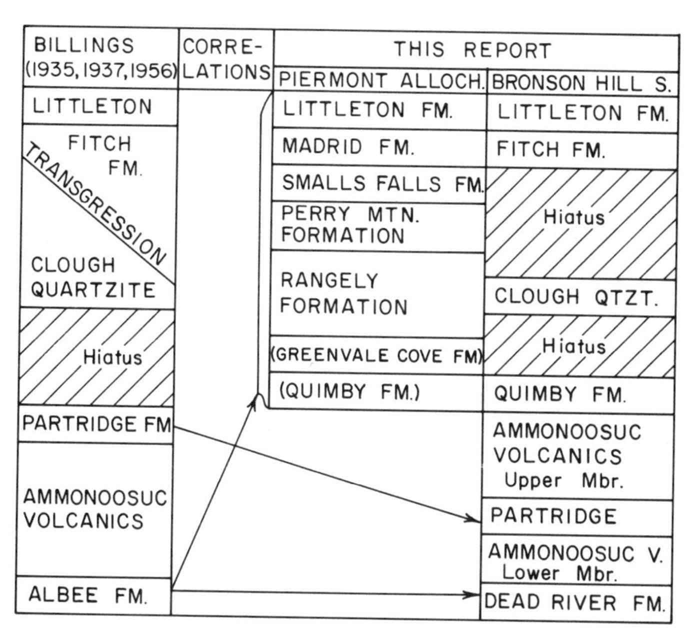

Figure 2. Proposed stratigraphic revisions in the Littleton area; allochthonous formations in parenthesis mapped only near Piermont (Fig. 1, inset).

Formation of Maine (Osberg and others, 1985). In the Littleton area, rocks that we assign to the Dead River Formation crop out in a narrow belt along the northwest limb of the Lisbon syncline, and at two other localities (Fig. 1).

The other assemblage, mainly Silurian in age (Moench and Aleinikoff, 1987), and including the type area of the Albee, characteristically contains volcanic and reworked metavolcanic rocks as well as schist and quartzite; it is divisible into several named formations, discussed below. These units occur within a fault-bounded tract at least 56 mi (90 km) long, shown as the Piermont allochthon on Figure 1 (inset). The name of the allochthon is taken from the village of Piermont, because both the age of the rocks and their tectonic setting were first recognized near the village as described briefly below. Prager (1980) was first to suggest that part of the Albee Formation might be Silurian; namely, massive quartzites at Gardner Mountain in rocks here tentatively assigned to a member of the Rangeley Formation.

We divided the rocks previously mapped as Albee Formation near Piermont and Bradford villages (Hadley, 1942; White and Billings, 1951) into several conformable units that can be correlated with the following formations of the Kearsarge–Central Maine synclinorium of western Maine (ascending order): Quimby (Upper Ordovician?), Greenvale Cove (Lower Silurian?), Rangeley, Perry Mountain, Smalls Falls, and Madrid (all Silurian), and Littleton (Lower Devonian). With the exceptions of carbonate-bearing rocks in the Greenvale Cove and Madrid Formations and conglomerates in the Rangeley, these are mainly siliciclastic turbidite formations. The type localities of all formations except the Littleton are near Rangeley, Maine (Moench and Boudette, this volume). Although the aggregate thickness of Rangeley equivalents of the Piermont area is much smaller than the type Rangeley (about 1600 versus 10,000 ft; 0.5 versus 3 km), the rocks can be correlated on a detailed member basis. Briefly, member A of the type Rangeley Formation is represented, near Piermont, by a few feet (meters) of basal polymictic debris-flow metaconglomerate; member B, by rusty-weathering, sharply interbedded dark pelitic schist, feldspathic metasandstone, and granule metaconglomerate; the lower part of member C, by interbedded quartz pebble metaconglomerate, feldspathic quartzite, and pelitic schist; and the upper part of member C is represented by rusty-weathering schist and feldspathic quartzite. Near Piermont, as well as Rangeley, the upper part of the Rangeley Formation grades upward to sharply interbedded light-colored, muscovite-rich schist and quartzite that typifies the Perry Mountain Formation. Unlike the Rangeley area, metavolcanic and reworked metavolcanic rocks, near Piermont and Bradford, occur in the upper parts of the Rangeley and Perry Mountain Formations, and locally within the Smalls Falls Formation. In the Littleton area farther north (Fig. 1) the metavolcanic rocks are contained mainly within the Perry Mountain Formation.

In the Littleton area (Fig. 1), rocks herein mapped within the Piermont allochthon are divided into five units (Stops 6A to 6E, 7, and 8), in ascending order:

(1) Rangeley Formation (Sr); at least 1500 ft (500 m) thick, but lower contact not exposed; correlated with the upper part of member C of the type Rangeley. Composed of interbedded rusty-weathering, gray to black slate or phyllite and feldspathic quartzite.

(2) Perry Mountain Formation; about 1500 ft (500 m) thick and divisible into a quartzite and schist member (Spq) similar to the type Perry Mountain, composed of sharply interbedded green or greenish-gray slate or schist and variably feldspathic quartzite, and a volcanic-bearing member (Spv) typically composed of green slate or schist and feldspathic quartzite, variable amounts of metamorphosed keratophyric ash- and lapilli-tuff, hypabyssal flows and shallow intrusives of quartz porphyry and granophyre, and sparse andesite and basalt. Locally, where Spv contains as much as 50 percent of basaltic greenstone or amphibolite, a basalt-bearing facies of Spv can be mapped, Spvb. In addition, poorly stratified pyritic quartz-sericite schist, typically with scattered quartz "eyes," mapped previously as Ammonoosuc Volcanics west of Gardner Mountain, is tentatively assigned to Spv as a rhyolite-rich facies, Spvr (Fig. 1).

(3) Smalls Falls Formation (Ssf); 30 to 60 ft (10 to 20 m) of rusty-weathering, sulfidic black slate or phyllite. Similar to but much thinner than type Smalls Falls (max. 60 ft [20 m] vs. max 2500 ft [750 m]).

(4) Madrid Formation (Sm); probably 30 to 60 ft (10 to 20 m) of gray, laminated, noncalcareous to weakly calcareous slate, punky brown-weathering calcareous phyllite, gray metasandstone, and local impure metalimestone. Somewhat similar to and considered coeval with Fitch Formation of the Bronson Hill sequence but assigned to the Madrid Formation to emphasize its conformable position above thick Silurian deposits that are

younger than the Clough Quartzite of the Bronson Hill sequence (Fig. 2).

(5) Littleton Formation (D1); one small remnant of gray slate thinly interbedded with graded feldspathic metasiltstone and metasandstone; may also contain felsic crystal metatuff.

The fault inferred to outline the Piermont allochthon was recognized first at Foster Hill, west of Littleton (Fig. 1), and is called the Foster Hill fault. It is a complexly-deformed, premetamorphic feature interpreted as a detachment fault of regional extent. Near Foster Hill (Stops 2D, 2E, 8), the fault is defined by the juxtaposition of west-younging rocks of the upper member of the Ammonoosuc Volcanics (Oaux and Oaue) against east-younging Albee rocks here mapped as the Rangeley (Sr), Perry Mountain (Spv, volcanic-bearing member), Smalls Falls (Ssf), and Madrid (Sm)) Formations, as shown on Figure 1. Although the fault was not seen in outcrop at Foster Hill, it was here that we first recognized that the type Albee cannot be demonstrated to lie stratigraphically below the Ammonoosuc Volcanics. At Stop 1C (Fig. 1) the fault is expressed by a few inches (centimeters) of metamorphosed mylonite. Near Piermont, the probable Quimby to Littleton sequence of the allochthon is deformed by tight, overturned folds that trend northwest and is sharply juxtaposed against less-deformed formations of the Bronson Hill sequence (Fig. 2). Here, the fault is a sinuous feature that clearly truncates mapped formations on both sides; at three outcrops, the fault surface is firm, knife-sharp, and shows no evidence of cataclasis.

On the basis of facies comparisons, the Rangeley Formation at the south end of the Piermont allochthon is interpreted to have accumulated near the Silurian tectonic hinge (Fig. 1, inset) because it represents a thin, transitional clastic shelf-to-basin facies relative to the much thicker type Rangeley, which accumulated in the basin southeast of the hinge. In contrast, the Clough Quartzite of the Bronson Hill sequence is interpreted as a shoreline facies of the Rangeley Formation, and is reasonably inferred to have accumulated in place. These relationships suggest that the Quimby to Littleton sequence of the allochthon accumulated some 9 to 15 mi (15 to 25 km; present distance) southeast of its present position and was transported westward during an early stage of the Acadian orogeny.

Modifications of the Middle to Upper Ordovician(?) succession

Billings (1935, p. 10) named the Ammonoosuc Volcanics for felsic-to-mafic metavolcanic rocks abundantly exposed along the valley of the Ammonoosuc River between Bath and Littleton, New Hampshire, southeast of the Ammonoosuc fault; U.S. 302 approximately follows the river (Fig. 1). The type area is northwest of the Ammonoosuc fault; it is a tract of about 2 mi^2 (5 km^2) that lies immediately southeast of Partridge Lake and northeast of Ogontz Lake. Billings further indicated that the Ammonoosuc is conformably overlain by the Partridge Formation, composed largely of black sulfidic slate or phyllite. The name comes from Partridge Lake, and the type area is the narrow belt of Partridge that extends from Ogontz Lake to Moore Reservoir (Fig. 1). Billings (1956, p. 19) suggested a Middle or Late Ordovician age for the Ammonoosuc Volcanics and Partridge Formation, in approximate accord with the presently accepted Middle Ordovician age of these formations.

Billings' interpretation that the Partridge Formation lies conformably above the Ammonoosuc Volcanics has long been a cornerstone of New England geology. According to his mapping northwest of the Ammonoosuc fault, the type Partridge lies along the troughlines of complex, doubly plunging synclines (Billings, 1937, Pl. 1, sections C–C′ to E–E′). On his map of the area just southeast of the fault, he depicts the Partridge as lying within an east-younging sequence above the Ammonoosuc and below the Clough Quartzite and Fitch Formation (Billings, 1937, Pl. 12, section C–C′).

In contrast, our mapping indicates that the Partridge Formation lies at a low stratigraphic level within the Ammonoosuc Volcanics (Fig. 2). Although the Partridge Formation might thus be considered to be a member of the Ammonoosuc, we prefer to retain formation status for the Partridge to allow for its use elsewhere in New England where Middle Ordovician black slate is greatly predominant over metavolcanic rocks.

The revised Bronson Hill sequence (Fig. 2) is complete southeast of the Ammonoosuc fault; the lower member of the Ammonoosuc has not been found northwest of the fault. Southeast of the fault, the lower member, a few feet (meters) to about 600 ft (200 m) thick, has the following composition: basaltic amphibolite, as flows, tuff, and agglomerate; thinly-bedded, fine-grained metamorphosed reworked volcanic and probable chemical sedimentary rocks; and local hypabyssal intrusions or domes of quartz porphyry. The lower and upper contacts are sharp and apparently conformable. The overlying Partridge Formation also is typically less than 600 ft (200 m) thick southeast of the Ammonoosuc fault. Northwest of the fault, it is exposed in wider belts and must be correspondingly thicker. The Partridge is composed largely of black sulfidic slate or phyllite; thin lenticular beds of white, sugary, probably exhalative quartzite occur near the lower contact, and felsic metatuff beds may be found near the upper contact. Although bedding features of the Partridge Formation are typically obscured by rust, repeated graded beds of mud-silt turbidite are displayed in one stream-washed pavement at the upper contact (Stop 2J), which is sharply gradational. Thick beds of felsic metatuff at the base of the overlying upper member of the Ammonoosuc commonly have rip-up clots and slices of black slate derived from the Partridge.

Where measurable southeast of the Ammonoosuc fault, the upper member of the Ammonoosuc Volcanics is about 3,200 ft (1 km) thick. Here it is divisible into a mixed but predominantly felsic lower part (Oaux) and a more mafic upper part (Oaub). The lower part contains metamorphosed thickly stratified felsic tuff, felsic lapilli tuff, matrix-supported felsite agglomerate and volcanic conglomerate, local reworked volcanic sediments, and minor basaltic flows or tuff. The upper part contains metamorphosed basaltic tuff and flows and intermediate pyroclastic rocks. Northwest of the Ammonoosuc fault, the lower part is lithologically similar to its equivalent southeast of the fault, except for containing more pyrite. East of Moore Reservoir, it is overlain by coarse mafic to intermediate agglomerate, pillowed basaltic greenstone, and thinly-bedded green phyllite and metatuff, shown as Oaub along the troughline of the Millikin Brook syncline (Fig. 1). West of Partridge Lake, the lower part is overlain by epiclastic rocks, Oaue, composed of interbedded green phyllite and feldspathic metatuff, basaltic greenstone (commonly magnetic), and minor felsic lapilli metatuff; it is interpreted to be a western distal facies of Oaub unit of the upper member.

West of Swiftwater (Fig. 1) the upper member of the Ammonoosuc is sharply succeeded by black sulfidic phyllite and dark graded metagraywacke (not previously mapped) assigned to the Quimby Formation. The Quimby occurs along the troughline of the Lisbon syncline, an early fold inferred to have been redeformed by the Salmon Hole Brook syncline (Fig. 1).

Clough, Fitch, and Littleton formations

Billings (1935, p. 13) named the Clough Quartzite for a group of interbedded quartzites and quartz conglomerates that lie below the Fitch Formation in the "Clough Hill district," southeast of Lyman Village (shown as Parker Hill on the Lower Waterford 7½-minute Quadrangle). Billings' comment that the Clough is the best key horizon in western New Hampshire remains true today. The Clough, zero to nearly 400 ft (120 m) thick in the Littleton area, forms bold, cliff-forming outcrops of massive or thickly bedded, vein quartz conglomerate and orthoquartzite, and more subdued outcrops of micaceous quartzite or thinly-bedded muscovite-rich schist and quartzite. Its lenticularity, style of bedding, and unconformable lower contact indicate that the Clough is a near-shore deposit, possibly, in part, the remnant of on-shore channel fillings. Early Silurian fossils are found in the Clough in southwestern New Hampshire and adjacent Vermont (Boucot and Thompson, 1963).

Billings (1935, p. 15) named the Fitch Formation for the site of a long-famous fossil locality in limestone and related sedimentary rocks on the farm (in 1934) of G. E. Fitch, 1.7 mi (2.8 km) west-northwest of Littleton. On the basis of conodonts found in the Fitch at two other localities in the Littleton area, and a review of the other fauna, Harris and others (1983) assigned a latest Ludlovian to at least early Pridolian age (latest Late Silurian) to the Fitch. In the low-grade zone, the Fitch Formation has been described variously as calcareous and noncalcareous slate, limestone, marble, metagrainstone, and metapackstone. In higher-grade zones, it becomes calcareous biotite or phlogopite schist, calc-silicate granofels or gneiss, and marble. Billings (1956, p. 26) reported a variable thickness to a maximum of about 800 ft (235 m) for the Fitch exposed within about 8 mi (13 km) of Littleton. Although the Fitch outcrops appear to lie conformably atop the Clough Quartzite (Billings, 1956, p. 24), the fossil data indicate that a considerable time gap separated deposition of the two formations.

Ross (1923) named the Littleton Formation for exposures of dark gray slate and sandstone on Walker Mountain, in the town of Littleton, 2.5 mi (4 km) southwest of Littleton, New Hampshire. Billings (1956, p. 27) described the type area as "a band 10 mi (16 km) long that lies northwest of the Ammonoosuc River between Lisbon and Littleton." Here the Littleton Formation is at least 4,600 ft (1,400 m) thick, but the top of the formation is not exposed. On the basis of fossils found in the Littleton area and elsewhere in northwestern New Hampshire, Boucot and Arndt (1960) and Boucot and Rumble (1980) assigned an Oriskany (Early Devonian) age to the formation. Geologic relationships suggest, however, that the lower part of the Littleton may be earlier Devonian in age (Helderberg), but not Silurian (see Harris and others, 1983, p. 736). The predominant lithology of the Littleton at low-metamorphic grade (D1) is dark gray slate or phyllite, but cyclically graded beds of metamorphosed wacke or siltstone to shale are the hallmark of the formation in New Hampshire. A bimodal volcanic member (Dlv), as much as 820 ft (250 m) thick, lies near the base of the formation in the Littleton area.

STOPS IN THE BRONSON HILL SEQUENCE AND THE FOSTER HILL FAULT

The Dead River Formation (Upper Cambrian? and Lower Ordovician, O€d) is best seen at Stops 1A, 1B, and 1C; the Foster Hill fault, inferred to mark the boundary of the Piermont allochthon, is exposed at Stop 1C.

The rocks at Stop 1A, on I-93, are greenish gray, magnetite-bearing pelitic schist of O€d, having conspicuous thin laminations and lenses of pink coticule; they are intruded by 3-ft-thick (1 m) dikes of metamorphosed felsite. This is an uncommon, quartzite-poor facies of O€d, but it is similar to the Aziscohos Formation of northeastern New Hampshire, which can be considered a pelitic facies of the Dead River Formation.

At Stop 1B, on the south bank of the Ammonoosuc River, thinly interbedded quartzite and dark pelitic schist of O€d is exposed at the upstream end of the outcrop; sharp, millimeter-thin laminations of pink coticule occur in the schist. Several feet (meters) downstream, O€d is faulted against thicker-bedded, more feldspathic rocks of the volcanic-bearing Perry Mountain Formation. The lithologic contrast between the formations is subtle, however, and the fault is more easily seen at Stop 1C.

At Stop 1C the fault is exposed on the north bank of Childs Brook, near the base of a steep cascade about 500 ft (150 m) upstream from the bridge; it is expressed by a thin zone of abundant quartz veins and a few centimeters of metamorphosed mylonite. The mylonitic fabric dips 50° NW, and mullions plunge 25°S75°W. The footwall rocks are thinly bedded pelitic schist and quartzite of O€d; the hangingwall rocks are thickly bedded rocks of the volcanic-bearing Perry Mountain Formation (Spv). Chalky-weathering, fine-grained feldspathic quartzite (or tuffaceous metasandstone) and green phyllite of Spv is well exposed where Woodsville Road crosses a power line 0.4 mi

(0.65 km) southwest of the bridge, particularly above cliffs overlooking the Ammonoosuc River. Just downstream from the Woodsville Road bridge, interbedded schist and quartzite of O€d are sharply overlain by basaltic amphibolite of the lower member of the Ammonoosuc Volcanics (here only a few feet [meters] thick), overlain in turn by and black sulfidic phyllite of the Partridge Formation.

The Partridge Formation, Op, and the upper member of the Ammonoosuc Formation, and its parts (all Middle Ordovician, Oaux, Oaub, Oaue), are exposed at low metamorphic grade at Stops 2A to 2E northwest of the Ammonoosuc fault.

At Stop 2A, along the shore of Moore Reservoir, layered white- to rusty-weathering, pyritic felsic metatuff at the top of Oaux is overlain to the east by mafic to intermediate greenstone agglomerate, tuff, flows (some pillowed), and interbedded green phyllite and chalky-weathering metatuff of Oaub.

Stop 2B is a series of roadcuts along New Hampshire 18/135. The eastern roadcuts expose massively-bedded, white-weathering felsic ash and lapilli metatuff of Oaux. At the west end is black sulfidic slate of Op, lying near the axial zone of the Partridge Lake anticline.

At Stop 2C, on Foster Hill Road, stratified agglomeratic pyroclastics of Oaux, possibly including lahar deposits, contain a wide variety of felsic clasts supported by a matrix of somewhat more mafic metatuff.

At Stop 2D, on the eastern end of Foster Hill, thickly bedded felsic to intermediate pyroclastic rocks of Oaux are gradationally overlain by interbedded green phyllite and white-weathering metatuff of Oaue. These rocks are exposed in a west-younging sequence on the west limb of Partridge Lake anticline; the sequence is demonstrated by several graded beds.

At Stop 2E, west of Partridge Lake, black sulfidic phyllite of Op (type area) is overlain to the east by mixed felsic pyroclastics of Oaux, which is gradationally overlain by interbedded green phyllite and white metatuff, basaltic amphibolite, and minor felsic lapilli metatuff of Oaue. Graded beds indicate that Oaue lies above Oaux on the west limb of the Partridge Lake anticline. Convincing graded beds of the Op-Oaux contacts were not found here, but basal Oaux metatuff in the area commonly has rip-up clots of black slate derived from Op.

Southeast of the Ammonoosuc fault, a complete post–Dead River, pre-Silurian sequence can be seen at medium grades of metamorphism at Stops 2F to 2M: the lower (Oal) and upper (Oaux and Oaub) members of the Ammonoosuc Volcanics (Middle Ordovician), the intervening Partridge Formation (Middle Ordovician, Op), and the overlying Quimby Formation (Upper Ordovician?, Oq).

At Stop 2F, at Bath, black sulfidic phyllite Op is sharply overlain by thickly bedded felsic metatuff of Oaux; thin rip-ups of black Op slate occur in basal tuff of Oaux.

At Stop 2G, at Lisbon, metamorphosed basaltic agglomerate and flows of Oal are sharply overlain by black sulfidic phyllite of Op, containing sugary, probably exhalative quartzite.

An east-younging sequence is exposed along the power line at Stop 2H, from south to north: thinly bedded feldspathic biotite schist, amphibolite, and minor quartz porphyry of Oal; black sulfidic phyllite with sugary quartzite of Op; and bedded felsic metatuff of Oaux.

The rocks at Stop 2I, at Swiftwater, are thickly bedded felsic tuff, lapilli tuff, and agglomeratic tuff of Oaux. Beds dip west and are upright, as shown by graded bedding and small channels.

A large, stream-washed pavement at Stop 2J exposes graded-bedded black sulfidic phyllite and metasiltstone of Op overlain by felsic metatuff of Oaux; these rocks are exposed across a small anticline on the east limb of the Lisbon syncline. Graded beds in Op and at the contact indicate that Oaux lies above Op.

Rocks assigned to the Quimby Formation, Oq, are exposed at Stop 2K, along the Wild Ammonoosuc River west of Swiftwater. The rocks are black sulfidic schist and dark, graded metagraywacke, best exposed on the north bank. The black schist is similar to typical Partridge, but Quimby lacks the basal sugary quartzite of Op, and Op lacks the metagraywacke of Oq.

At Stop 2L, farther west along the river, massive, matrix-supported, hornblende-bearing intermediate agglomerate is overlain by graded beds of basaltic metatuff, all within Oaub. Bedding is overturned to the east, on the west limb of the Lisbon syncline.

At Stop 2M, along the Ammonoosuc River and the lower reach of Salmon Hole Stream, thinly bedded volcaniclastic metasedimentary rocks, amphibolite, and some garnet- and anthophyllite-rich exhalative rocks of Oal are sharply overlain by black phyllite with sugary quartzite of Op, which is sharply overlain by felsic metatuff of Oaux.

Stops 3 and 4 are the most conveniently accessible outcrops of the Clough Quartzite (Lower Silurian, upper Llandoverian, Sc) and Fitch Formation (Upper Silurian, upper Ludlovian to Pridolian, Sf). At Stop 3, east of Lyman, is one of many lenses of white orthoquartzite and quartz conglomerate that are included in the type locality of the Clough. More extensive outcrops of Sc are on Black Mountain, in the East Haverill 7½-minute Quadrangle, just south of the area of Figure 1. Stop 4, west of Littleton, is one of two conodont localities of Harris and others (1983) in the Fitch. The rocks are thinly bedded, noncalcareous and calcareous gray slate and grainstone with scattered boulders of granodiorite. The Fitch is interpreted to lie unconformably on granodiorite, exposed nearby at the type locality of the Highlandcroft Plutonic Suite (Billings, 1935, p. 25).

Metasedimentary and metavolcanic rocks of the Littleton Formation (Lower Devonian, D1) can be seen at Stops 5A and 5B. Rocks at the quarry (Stop 5A) on Slate Ledge Road are repeated thin-graded beds of gray metasiltstone and darker gray slate, regionally characteristic of D1. Metamorphosed white-weathering felsic tuff and tuff breccia of Dlv can be seen in a field about 0.3 mi (0.5 km) southwest of the quarry. The stream-washed outcrops along the Gale River at Stop 5B display fine details of metamorphosed pillow basalt, felsic tuff, tuff breccia, schist-matrix volcanic mudflow deposits, and graded beds of gray pelitic schist and wacke of Dlv.

STOPS IN THE PIERMONT ALLOCHTHON

Stops 6A to 6E show characteristic features of rocks here assigned to the Rangeley (Lower Silurian, Llandoverian and Wenlockian?, Sr) and Perry Mountain (Silurian, Wenlockian?, Sp and Spv) Formations.

On the south side of Hunt Mountain Road at Stop 6A are large outcrops of massive, chalky-weathering, fine-grained quartz porphyry containing small euhedral crystals of quartz. Farther south are thick and thin beds of chalky-weathering, fine-grained, feldspathic metatuff and feldspathic green phyllite, and interbedded feldspathic quartzite and green phyllite; graded beds young southeast. These rocks are characteristic of the volcanic-bearing member of the Perry Mountain Formation, Spv; they are part of a sequence of interstratified, predominantly felsic metavolcanic and siliciclastic rocks mapped in detail by Margeson (1982) and Hafner-Douglass (1986) in the strike belt of the type area of the Albee Formation.

North of Hunt Mountain Road, younging directions in similar rocks alternate from northwest to southeast. Lowermost Spv near the contact with underlying gray phyllite and quartzite of the Rangeley, Sr, contains: interbedded green phyllite and white feldspathic metasandstone or felsic metatuff; a massive bed of rhyolite lapilli metatuff, 3 to 6 ft (1 to 2 m) thick (sampled for zircons and dated at 416 ± 3 Ma; Moench and Aleinikoff, 1987); and coarse-grained quartzite cut by irregular veins and masses of vein quartz. The contact with underlying gray to black phyllite and quartzite of Sr is isoclinally folded, but is exposed about 300 ft (100 m) north of the sampled metatuff; here graded bedding indicates that Spv lies above Sr.

Characteristic rocks of Sr and Spv are exposed in pavement outcrops at stop 6B, along the power line near the top of Foster Hill, here a north-trending ridge. West of the ridge crest is rusty-weathering, gray to black phyllite and quartzite of Sr; at the crest and to the east is nonrusty green phyllite, feldspathic quartzite, and lenticularly bedded ash metatuff of Spv. The contact zone is gradational by interbedding and sedimentary younging directions indicate that Spv lies above Sr.

Stop 6C is at Albee Hill, after which the Albee Formation was named. Here, interbedded feldspathic quartzite and black phyllite of Sr is conformably overlain by interbedded feldspathic, locally chalky-weathering quartzite and rusty-weathering but greenish phyllite of the schist and quartzite member of the Perry Mountain Formation, Spq. Rocks of Spq are exposed over the top of Albee Hill, where they are intruded by greenstone sills or dikes.

At Stop 6D, high on the west side of Gardner Mountain, are the best exposures of the schist and quartzite member of Spq in the Littleton area. The rocks display partial and complete Bouma turbidite sequences, and are identical to the type Perry Mountain Formation of western Maine.

Stop 6E is a large pavement of the volcanic-bearing member of the Perry Mountain Formation, Spv, washed by the Ammonoosuc River. Exposed here are punky-weathering biotite-feldspar schist, probably intermediate metatuff, succeeded by a thicker zone of lenticular-bedded, chalky-weathering metatuff and feldspathic green phyllite. The metatuff lenses are graded and some channel into the underlying phyllite; they range almost to lapilli metatuff in coarseness. These rocks are complexly, semi-concordantly injected by a fine-grained granitoid rock, possibly comagmatic with the metavolcanics; both are intruded by dikes of foliated greenstone.

Rocks here mapped as Littleton Formation (Lower Devonian, D1) are exposed at Stop 7, just west of Under The Mountain Road. This is the only known remnant of D1 north of the Piermont area. The rocks are "Littleton-gray" slate with graded laminations of feldspathic metasiltstone and metasandstone. The lower contact of D1 is not exposed here, but an east-younging sequence of the next-older Madrid, Smalls Falls, and Perry Mountain Formations is exposed about 2,000 ft (600 m) to the west along a small brook upstream from elevation 1150 ft (350 m).

Exposed at stop 8 are interbedded green phyllite and feldspathic quartzite of the Perry Mountain Formation, Spv, black slate of the Smalls Falls Formation, Ssf, and various gray- to brown-weathering, weakly to strongly calcareous slate and phyllite, metasandstone, and impure metalimestone of the Madrid Formations, Sm. The Perry Mountain Formation in this sequence contains abundant greenstone dikes. The Madrid Formation is exposed in a tight, south-plunging syncline whose east limb is truncated by the Foster Hill fault. Except for the syncline, this Silurian sequence youngs toward the Ordovician Ammonoosuc Volcanics. The closest Ammonoosuc to the eastern edge of the Silurian sequence (within 3 ft; 1 m) is greenstone of the epiclastic part of the upper member, Oaue, at the top of a generally west-younging sequence above Oaux (stop 2D).

REFERENCES CITED

Billings, M. P., 1935, Geology of the Littleton and Moosilauke quadrangles, New Hampshire: Concord, New Hampshire Planning and Development Commission, 51 p.

——, 1937, Regional metamorphism of the Littleton-Moosilauke area, New Hampshire: Geological Society of America Bulletin, v. 48, p. 463–566.

——, 1956, The geology of New Hampshire; Part II, Bedrock geology: Concord, New Hampshire Planning and Development Commission, 203 p.

Billings, M. P. and Cleaves, A. B., 1934, Paleontology of the Littleton area, New Hampshire: American Journal of Science, v. 28, p. 412–438.

Boucot, A. J., and Arndt, R., 1960, Fossils of the Littleton Formation (Lower Devonian) of New Hampshire: U.S. Geological Survey Professional Paper 334-B, p. 87–129.

Boucot, A. J., and Rumble, D., III, 1980, Regionally metamorphosed (high sillimanite zone, granulite facies) Early Devonian brachiopods from the Littleton Formation of New Hampshire: Journal of Paleontology, v. 54, no. 1, p. 188–195.

Boucot, A. J., and Thompson, J. B., Jr., 1963, Metamorphosed Silurian brachiopods from New Hampshire: Geological Society of America Bulletin, v. 74,

p. 1313–1334.
Eric, J. H., and Dennis, J. G., 1958, Geology of the Concord–Waterford area, Vermont: Vermont Geological Survey Bulletin 11, 66 p.
Hadley, J. B., 1942, Stratigraphy, structure, and petrology of Mt. Cube area, New Hampshire: Geological Society of America Bulletin, v. 53, p. 113–176.
Hafner-Douglass, K., 1986, Stratigraphic, structural, and geochemical analyses of bedrock geology, Woodsville Quadrangle, New Hampshire–Vermont [M.S. thesis]: Hanover, New Hampshire, Dartmouth College, 117 p.
Hall, L. M., 1959, Geology of the St. Johnsbury Quadrangle, Vermont and New Hampshire; Vermont Geological Survey Bulletin 13, 105 p.
Harris, A. G., Hatch, N. L., Jr., and Dutro, J. T., Jr., 1983, Late Silurian conodonts update the metamorphosed Silurian Fitch Formation, Littleton area, New Hampshire: American Journal of Science, v. 283, no. 7, p. 722–738.
Margeson, G. B., 1982, Iron-rich rocks of Gardner Mountain, New Hampshire, and their significance to base metal mineralization [M.S. thesis]: London, Ontario, The University of Western Ontario, 184 p.
Moench, R. H., and Aleinikoff, J. N., 1987, The Piermont allochthon of northwestern New Hampshire: Stratigraphic and isotopic evidence: Geological Society of America Abstracts with Programs, (in press).
Osberg, P. H., Hussey, A. M., II, and Boone, G. M., eds., 1985, Bedrock geologic map of Maine: Maine Geological Survey, scale 1:500,000.
Prager, G. D., 1980, Stratigraphy of the Gardner Mountain area, New Hampshire: Northeastern Geology, v. 2, p. 32–43.
Ross, C. P., 1923, The geology of part of the Ammonoosuc mining district, New Hampshire: American Journal of Science, 5th series, v. 5, p. 267–302.
White, W. S., and Billings, M. P., 1951, Geology of the Woodsville Quadrangle, Vermont–New Hampshire: Geological Society of America Bulletin, v. 62, p. 647–696.

ACKNOWLEDGMENT

Thanks to Douglass W. Rumble III for his guidance in the Piermont area.

Mount Washington–Crawford Notch area, New Hampshire

Richard P. Goldthwait, *P.O. Box 656, Anna Maria, Florida 33501*
Marland P. Billings, *RFD, West Side Road, North Conway, New Hampshire 03860*
John W. Creasy, *Department of Geology, Bates College, Lewiston, Maine 04240*

LOCATION

The Mount Washington–Crawford Notch area (Fig. 1) in the White Mountains of New Hampshire can be reached using U.S. 302 and New Hampshire 16. A network of hiking trails provides access to the off-road sites shown on Figures 2 and 3.

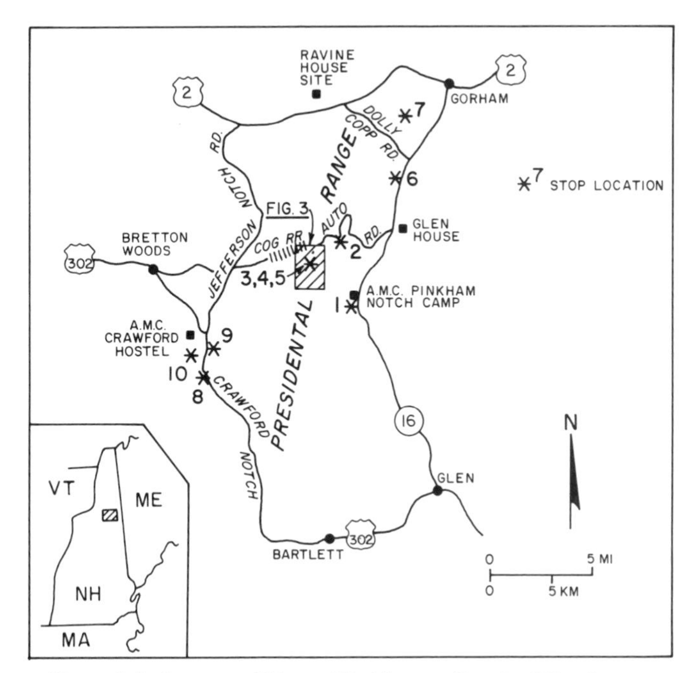

Figure 1. Index map of Mount Washington–Crawford Notch area.

SIGNIFICANCE

Mount Washington, altitude 6,288 ft (1,916 m), is the highest mountain in the northeastern United States. The well-developed Pleistocene glacial features, both alpine and continental, and the Arctic features, such as patterned ground, result from its continuously high altitude and make the area nearly unique in the Appalachians (Goldthwait, 1916; Goldthwait, 1939, 1944, 1979a, b; Johnson, 1933). The bedrock exposures are representative of the Paleozoic metamorphic and igneous rocks of the northern Appalachian orogen. Although paleontologically barren, these rocks are correlated with the fossiliferous rocks of western New Hampshire to the west and of western Maine to the east (Billings, 1941; Billings and others, 1979; Hatch and Moench, 1984). The area of this guide is just southeast of the hinge between a Silurian basin in the Merrimack synclinorium to the east and the Silurian shelf in the area of the Bronson Hill anticlinorium to the west. Mesozoic igneous rocks are representative of the extensional tectonic environment associated with the opening of the North Atlantic basin (Creasy and Eby, 1983).

Geologic maps (1:62,500) and accompanying reports for the Mount Washington (Billings and others, 1979), Crawford Notch (Henderson and others, 1977), and Gorham (Billings and Fowler-Billings, 1975) quadrangles are available. The map of Hatch and Moench (1984, 1:125,000) provides an excellent regional overview and discussion of the geology of northern New Hampshire. The Appalachian Mountain Guide (1983) provides detailed topographic and trail information.

SITE INFORMATION

Bedrock Geology. The area of this guide is underlain chiefly by high-grade metamorphic rocks and mesozonal granitic rocks. The Early Silurian Rangeley and Early Devonian Littleton Formations, composed largely of shales and sandstones metamorphosed to the sillimanite grade, form the resistant gneisses of the Presidential Range (Hatch and Moench, 1984). Both formations are widely exposed in central New Hampshire and western Maine and probably far exceed 3 mi (5 km) in thickness (Hatch and others, 1983). Two thin Late Silurian units occur between the Rangeley and Littleton Formations in this area. The Smalls Falls Formation, a rusty weathering sulfidic schist, and the Madrid Formation, a calc-silicate gneiss, are much thinner here than in the type localities in western Maine (Hatch and others, 1983). The middle (?) Silurian Perry Mountain Formation present between the Rangeley and the Smalls Falls in western Maine (Hatch and Moench, 1984; Moench and Boudette, this volume) is absent in this area. Although gneissification has destroyed the sedimentary features distinguishing these Silurian formations throughout much of this area (Hatch and others, 1983), distinctive lithologic features and stratigraphic sequence are retained. Previously, the stratified metamorphic rocks of this area were correlated (Billings, 1941) with the stratigraphy developed in western New Hampshire (Billings, 1937; see also Moench and Lyons, this volume), and more recently (Billings and others, 1979; Billings and Fowler-Billings, 1975) with the Littleton Formation. The Middle Ordovician Ammonoosuc Formation, composed of fine-grained biotite gneiss derived from rhyolites and amphibolite derived from andesites and basalts, is exposed to the northwest of the Presidential Range. In this area the Ammonoosuc is intruded by weakly foliated granitic rocks of the Late Ordovician Oliverian Plutonic Suite (Chapman and others, 1941). Devonian or Mississippian (?) two-mica granites (Lyons and Livingston, 1977) of the New Hampshire Plutonic Suite intrude all older rocks (Fig. 2). A strikingly different tectonic

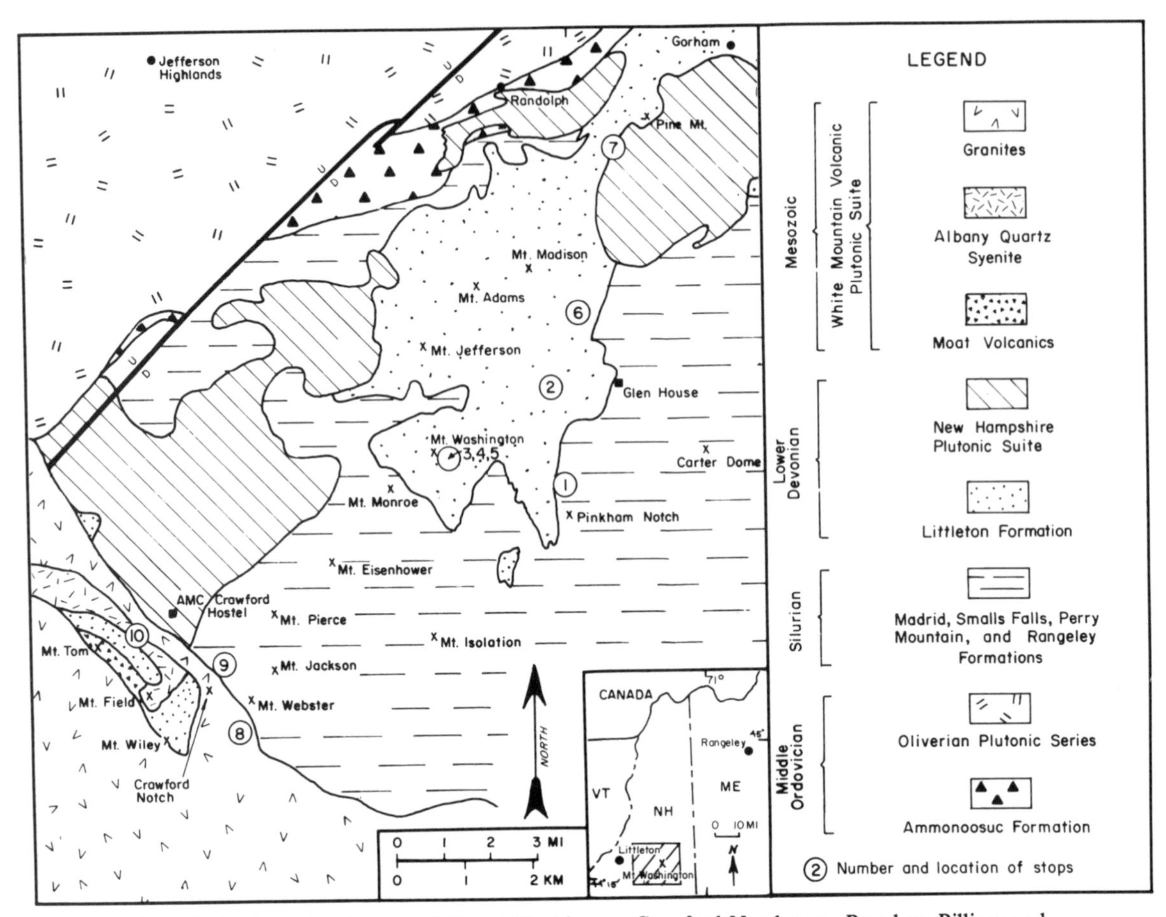

Figure 2. Geologic sketch map of Mount Washington–Crawford Notch area. Based on Billings and others (1979), Henderson and others (1977), Billings and Fowler-Billings (1975), and Hatch and Moench (1984).

environment is indicated by the Jurassic White Mountain batholith (Creasy and Eby, 1983; Eby and Creasy, 1983) present in the southwestern and southeastern parts of the area (Fig. 2). Three distinct structural levels of emplacement are represented by the silicic tuffs and breccias of the Moat Volcanics, the ring dikes of Albany prophyritic quartz syenite, and plutons of Conway and Mount Osceola granite (Creasy, 1986).

Geomorphology. The land surfaces of the Mount Washington–Crawford Notch area can be classified into four categories (Goldthwait, 1970a). The Presidential Upland is the highest and oldest (Tertiary ?) surface. It is a rolling graded surface at 4,500 to 5,500 ft (1,400 to 1,700 m) with slopes of 9° ± 9°, although it embraces the symmetrical summit cones like the cone of Mount Washington. The Alpine Garden and Bigelow Lawn (Fig. 3) are remnants of this surface. A second less-dissected surface is represented by the broad-facetted ends of forested ridges between 2,800 and 4,300 ft (850 to 1,300 m) that slope (15° ± 6°) with smooth concavity into the Ammonoosuc or Israel basins (west), the Dry basin (south), the Ellis-Saco basin (southeast), and the Peabody-Androscoggin basin (northeast). These "broad valley" surfaces are clearly relicts of former erosion surfaces, for they are sharply truncated and incised by a third group of surfaces defined by steep-walled valleys and ravines that encircle the Presidential Range at elevations of 3,000 to 5,000 ft (900 to 1,500 m). The V-shaped valleys (e.g., Ammonoosuc Ravine; Fig. 3) express stream erosion that is still active today. The cirques and broad-bottomed, U-shaped ravines (e.g., Tuckerman Ravine, Huntington Ravine, Great Gulf, and Oakes Gulf; Fig. 3) express the work of 9 to 12 mountain glaciers of Pleistocene age. The steep side slopes (36° ± 9°) of these glaciated valleys make remarkable contrast where they intersect the rolling Presidential Upland. A fourth group of surfaces represents rounding (0° to 25°) by the Wisconsin continental glacier. Rounding of peaks (e.g., the Southern Presidential Range) is clearly evident. Rounding of saddles as high as Lakes of the Clouds (Fig. 3) is evidenced by roches moutonées and grooves. In the lowest valleys, broad U-shaped troughs were begun. Crawford Notch is a spectacular U-shaped valley more than 1,000 ft (300 m) deep. Striae suggest a great funnelling of ice into these lower valleys; possibly a thousand times as much ice passed through Crawford Notch as over the high cols of the Presidential Range. Despite these modifications, Mount Washington still lies at the convergence of major east-

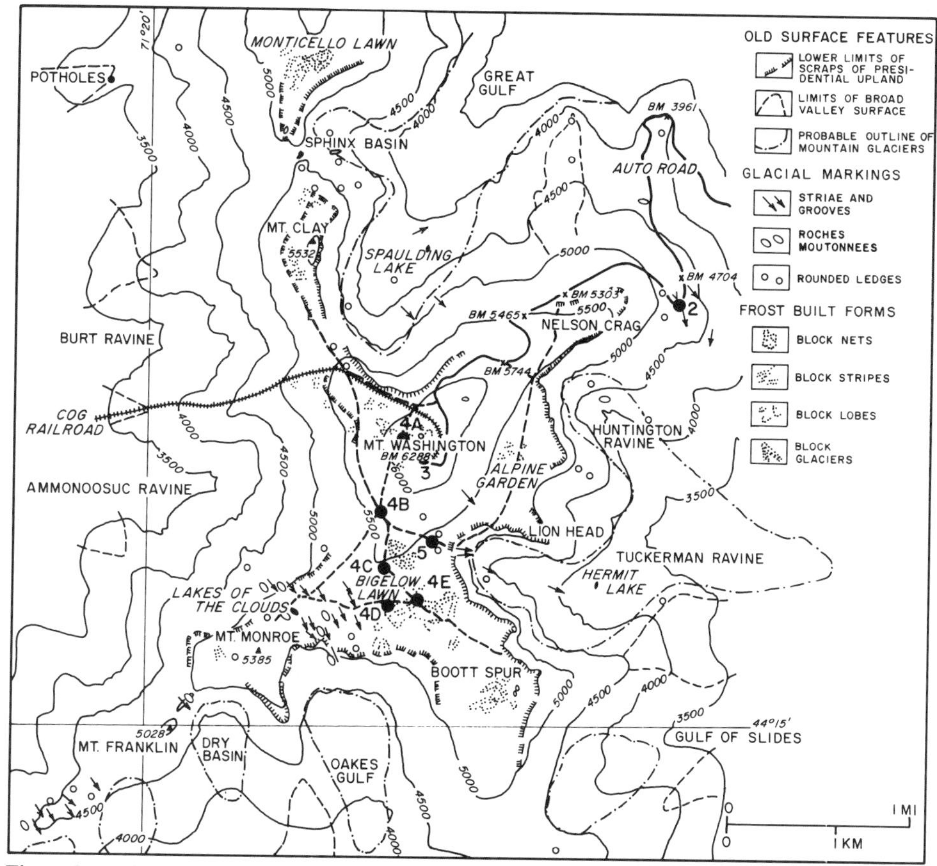

Figure 3. Surface features of the Presidential Range. Based on Goldthwait (1970b) and Billings and others (1979).

west watersheds established when the broad valley basins were completed.

Goldthwait (1970b) showed that the headwall of Tuckerman Ravine was striated by the southeasterly moving continental ice sheet, and concluded that the cirques were cut prior to the invasion by continental ice sheets. South of the site of the Ravine House (Fig. 1), are large erratics as much as 10 ft (3 m) in diameter, composed of rocks of the Littleton Formation, which crops out to the south. The country rock there is composed of granitic rocks of the Olivarian Plutonic Suite. Bradley (1981) concluded that these erratics were transported from King Ravine by a mountain glacier active in the waning stages of the melting of the continental ice sheet. Gerath and Fowler (1982) believe, however, that these boulders and the associated materials are debris-flow deposits unrelated to glaciation. For a discussion of the deglaciation of these mountains, see Goldthwait and Mickelson (1980) and Davis and others (1980).

Frost-built forms are among the most distinctive features of the Presidential Upland (Goldthwait, 1970a). The distribution of these forms is shown in Figure 3. After the continental ice sheet melted from the summits, intense frost action shattered the bedrock. Consequently the summit cones are covered by a "sea" of blocks (*felsenmeer*); photographs taken over the last century indicate that these blocks are fixed and not moving. Gentler slopes of the upland are till-covered, with boulders and locally derived schist blocks (2 to 4 ft; 0.6 to 1.2 m in size) sorted into polygonal nets, stripes, and lobes 10 to 20 ft (3 to 6 m) on centers. These characteristic features of Arctic permafrost also appear to be stabilized with no significant motion recorded by steel tape measurements during a period of 35 years (Goldthwait, 1970a). The shapes assumed by these sorted patterned soils on Mount Washington depend upon the slope of the Presidential Upland and to a lesser extent the amount of silt and clay in the soil (Goldthwait, 1970a). Subcircular block nets (Fig. 4) form on flat ground with a central grassy area surrounded by a 2 to 5 ft (0.6 to 1.5 m) rim of blocks. Block stripes (Fig. 5) are born on slopes of 5° to 10°; stripes are a few to 100 ft (30 m) long, as much as 10 ft (3 m) wide, and separated by grassy stripes 10 to 20 ft (3 to 6 m) wide. Block lobes (Fig. 6) form on slopes of 20° to 30°; a gently sloping grassy oval 50 to 100 ft (15 to 30 m) long and 20 ft (6 m) wide is enclosed, on the uphill side, by blocks as much as 3 ft (1 m) across. The front of the lobe slopes 30° to 40°.

Permafrost exists today only on the upper 200 ft (61 m) of Mount Washington, although sporadic frozen ground is encountered down to an elevation of 5,000 ft (1,524 m). Intense annual freezing under alpine tundra generates measurable annual activity

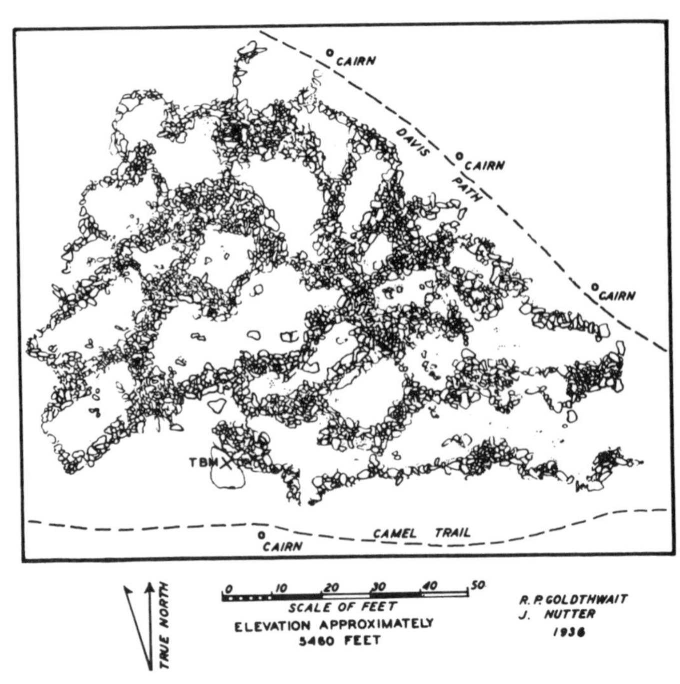

Figure 4. Block net on Bigelow Lawn (Goldthwait, 1944). Courtesy New Hampshire Academy of Science.

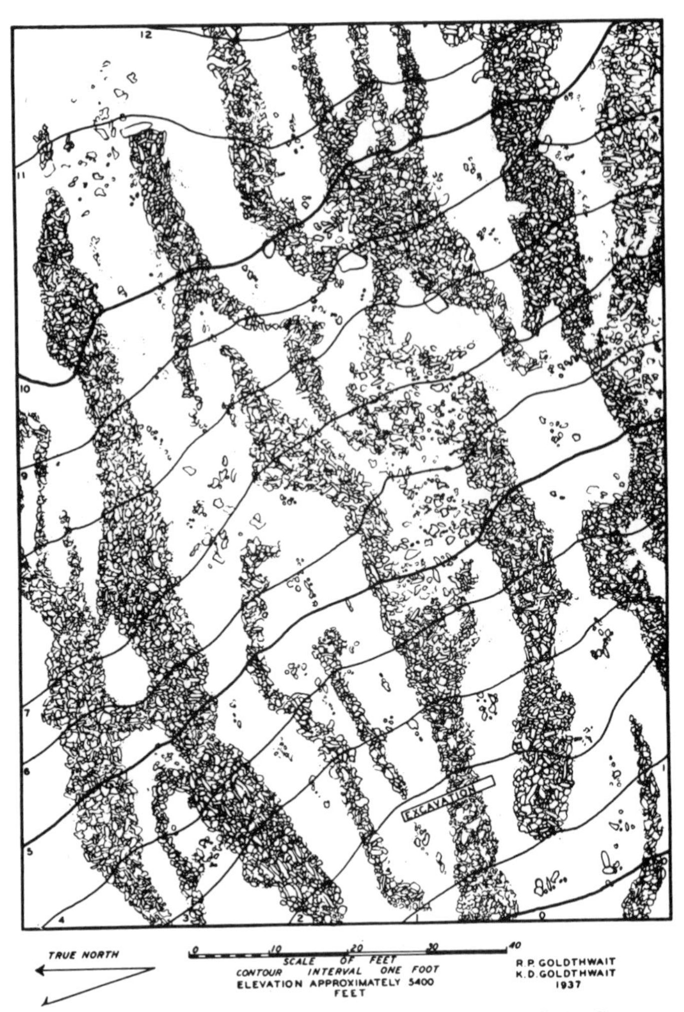

Figure 5. Block stripes on Bigelow Lawn (Goldthwait, 1944). Courtesy New Hampshire Academy of Science.

in small (1 to 2 ft; 0.3 to 0.6 m) stone polygons and stripes. At Lakes of the Clouds a small rock glacier comes down a narrow defile and fans out into the lowermost lake. A much larger rock glacier, 2,000 ft (610 m) long and 500 ft (152 m) wide, is present in Kings Ravine.

MOUNT WASHINGTON AREA

Stop 1. Appalachian Mountain Club (A.M.C.) Camp at Pinkham Notch. View to west of Mount Washington, Alpine Garden, and cirques. These are, from south to north, Gulf of Slides, Tuckerman Ravine, and Huntington Ravine.

Stop 2. Mount Washington Auto Road. Start at Glen House on New Hampshire 16, 2.5 mi (4 km) north of Pinkham Notch Camp (Fig. 1). From here to the summit of Mount Washington, the rocks are interbedded mica schist and micaceous quartzite of the Littleton Formation, except for a small Mesozoic volcanic vent 1.3 mi (2 km) from the entrance. Several folds in the Littleton Formation are well exposed at the big bend in the road 5.5 mi (8.8 km) from the entrance. The Alpine Garden is visible to the south over the top edge of Huntington Ravine.

Stop 3. Summit of Mount Washington (6,288 ft; 1,917 m). This stop (Fig. 3) may be reached using the auto road, the cog railroad (Fig. 1), or numerous foot trails. To identify the mountains within 6 mi (9.6 km), use the A.M.C. map of the Mount Washington Range. The Atlantic Ocean may be seen on clear days 80 mi (129 km) to the east. The Presidential Upland, a mature erosion surface of Tertiary(?) age (Goldthwait, 1970a), is preserved as an undulating shelf flanking this peak to the east (Alpine Garden) and south (Bigelow Lawn) at 5,000 to 5,500 ft (1,500 to 1,700 m) elevation (Fig. 3). The higher peaks rise above this surface as rounded hills. Below this surface on the ridge ends are broad Pliocene(?) valley slopes. These features are sharply incised by Pleistocene stream valleys to the southwest and cirques to the north, east, and south (Goldthwait, 1970a).

Stop 4. Bigelow Lawn. Several stations on Bigelow Lawn and the south flank of the summit cone of Mount Washington illustrate features produced by frost action (Fig. 3).

Station A. From the summit, descend about 0.2 mi (0.3 km) west on the Crawford Path to T-junction with the Gulfside Trail. A typical block lobe is located here at Station A. These are abundant, but generally restricted to northerly and westerly slopes of the summit cone. These tear-drop–shaped sorted solifluction terraces are the steep-slope end product of a continuum of patterned soil.

Station B. Continue descent of summit cone along the

Crawford Path about 0.3 mi (0.5 km) to junction with the Davis Path. Two hundred ft (60 m) east of this junction is one of a number of quartz boulder trains down the summit cone. This boulder train goes directly downslope perpendicular to the well-established direction of glacial motion. This demonstrates substantial post-glacial mass movement.

Station C. Proceed southeast on Davis Path to a point 200 ft (61 m) south of the junction with the Tuckerman Crossover. Southwest of the Davis Path are miniature patterned soils produced by present-day frost sorting. Stone nets are produced on crests where slopes are <5° and stone stripes where slopes are greater.

Station D. Continue southeast on Davis Path to junction with Camel Path; proceed 500 ft (150 m) west on Camel Path. Southwest of the Camel Path are block stripes consisting of soil centers 5 to 15 ft (1.5 to 5 m) apart, set between boulder gutters made of schist blocks up to 2 to 4 ft (0.6 to 1.2 m) long. These block stripes are on a 6° slope; forms on gentler slopes above and to the east are elliptical stretched polygons.

Station E. Return to junction of Camel and Davis paths. Large polygonal block nets are developed here on slopes of 0° to 3.5°. These forms and those at Station D have been stable over the last century (Goldthwait, 1970a).

Stop 5. Tuckerman Ravine from Tuckerman Junction. From Stop 4E, proceed 300 ft (91 m) southeast on Davis Path to junction with Lawn Cut-off; proceed north on Lawn Cut-off to Tuckerman Junction. Grooves and striae exposed under the cairn here on the edge of the Presidential Upland trend 113°, obliquely across the foliation of the Littleton Formation. These are clear evidence for the Wisconsin continental glaciation (Goldthwait, 1970a). The spectacular cirque below is Tuckerman Ravine. The upper headwall ranges from 600 to 1,100 ft (180 to 350 m) deep; the U-shaped floor hangs 300 ft (90 m) above a lower bowl (containing Hermit Lake) to form the Little Headwall.

Stop 6. West Branch of the Peabody River. Start in Gorham 15-minute Quadrangle at junction of West Branch with main stream, 100 ft (30 m) west of New Hampshire 16, 4.5 mi (7.2 km) north of Pinkham Notch Camp. A fine section of mica schist and micaceous quartzite of the Rangeley Formation is exposed for 1 mi (1.6 km) upstream. At 0.7 mi (1.1 km) from the stream junction a thin quartz conglomerate within the Rangeley is well exposed. The Smalls Falls and Madrid Formations are exposed at higher elevations in the West Branch and may be reached by continuing upstream or by using the Osgood Trail from New Hampshire 16. The Madrid Formation (130 ft [40 m] thick) is exposed 800 ft (250 m) downstream from the Osgood Trail crossing, at an elevation of 1,625 ft (495 m).

Stop 7. Pine Mountain (2,404 ft; 733 m). Follow Dolly Copp Road (Fig. 1) to the height of land and park near gravel road leading to the Horton Center. Hike up gravel road; at 0.8 mi (1.3 km) take the Ledge Trail to the right. For a detailed map of the geology see Plate 2 of Billings and Fowler-Billings (1975). The contact between the Littleton Formation and granite of the New Hampshire Plutonic Suite is exposed at an elevation of 2,000 ft (610 m). Fine exposures of interbedded mica schist and micaceous quartzite of the Littleton are present from here to the summit. At the South Cliffs, excellent graded beds top northwest. Around the summit are fine examples of roches moutonées made by Wisconsinan ice moving southward through Pinkham Notch. From the summit of Pine Mountain continue down the trail to the Horton Center and return to vehicles by the road.

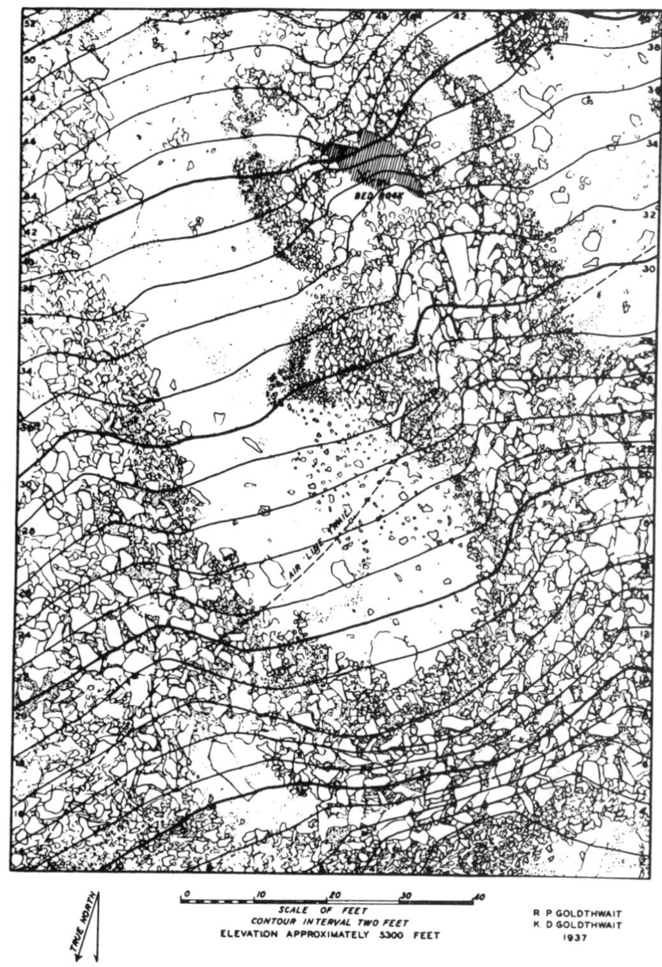

Figure 6. Block lobe northwest of Mount J. Q. Adams (Goldthwait, 1944). Courtesy New Hampshire Academy of Science.

CRAWFORD NOTCH AREA

Stop 8. Willey House. This site is in the south end of Crawford Notch, a U-shaped valley carved by the southeasterly flowing Wisconsin continental ice sheet. To the east is the smooth slope of Mount Webster rising 2,500 ft (762 m) above the valley floor. Most of this slope is Jurassic Conway Granite. The darker crest of the ridge is Silurian paragneiss and marks the roof of the Conway intrusion.

Stop 9. Conway Granite. Proceed 1.5 mi (2.4 km) north of the Willey House on U.S. 302 (Fig. 1). The contact of the

Conway Granite and the Silurian paragneiss viewed at Stop 9 is well exposed at road level.

Stop 10. Mount Willard (2,804 ft; 855 m). Park at A.M.C. Crawford Hostel (Fig. 1). Follow the Mount Willard Trail 1.4 mi (2.2 km) to the summit of Mount Willard, where Conway Granite is exposed. This point provides a magnificent vista of U-shaped Crawford Notch to the south.

REFERENCES CITED

Appalachian Mountain Club White Mountain Guide: 1983, Boston, Appalachian Mountain Club, 550 p.

Billings, M. P., 1937, Regional metamorphism of the Littleton-Moosilauke area: Geological Society of America Bulletin, v. 48, p. 463–566.

——, 1941, Structure and metamorphism in the Mount Washington area, New Hampshire: Geological Society of America Bulletin, v. 52, p. 863–936.

Billings, M. P., and Fowler-Billings, K., 1975, Geology of the Gorham Quadrangle, New Hampshire: Concord, New Hampshire, Department of Resources and Economic Development Bulletin no. 6, 120 p.

Billings, M. P., Fowler-Billings, K., Chapman, C. A., Chapman, R. W., and Goldthwait, R. P., 1979, The geology of Mount Washington Quadrangle, New Hampshire (second edition): Concord, New Hampshire, Department of Resources and Economic Development, 44 p.

Bradley, D. C., 1981, Late Wisconsinan mountain glaciation in the northern Presidential Range, New Hampshire: Arctic and Alpine Research, v. 13, p. 319–327.

Chapman, C. A., Billings, M. P., and Chapman, R. W., 1944, Petrology and structure of the Oliverian magma series in the Mount Washington quadrangle, New Hampshire: Geological Society of America Bulletin, v. 55, p. 497–516.

Creasy, J. W., 1986, Geology of the eastern portion of the White Mountain batholith, New Hampshire, *in* Newberg, D. W., ed., Guidebook for fieldtrips in southwestern Maine: New England Intercollegiate Geological Conference, 78th Annual Meeting, Lewiston, Maine, p. 124–137.

Creasy, J. W., and Eby, G. N., 1983, The White Mountain batholith as a model of Mesozoic felsic magmatism in New England: Geological Society of America Abstracts with Programs, v. 15, p. 549.

Davis, M. B., Spear, R. W., and Shane, W.C.K., 1980, Holocene climate of New England: Quaternary Research, v. 40, p. 240–250.

Eby, G. N., and Creasy, J. W., 1983, Strontium and lead isotope geology of the Jurassic White Mountain batholith, New Hampshire: Geological Society of America Abstracts with Programs, v. 15, p. 188.

Gerath, R. F., and Fowler, B. K., 1982, Discussion *of* 'Late Wisconsin mountain glaciation in the northern Presidential Range, New Hampshire': Arctic and Alpine Research, v. 14, p. 369–371.

Goldthwait, J. W., 1916, Glaciation in the White Mountains of New Hampshire: Geological Society of America Bulletin, v. 27, p. 263–294.

Goldthwait, R. P., 1939, Mount Washington in the Great Ice Age: New England Naturalist, no. 5, p. 12–17.

——, 1944, Glaciation in the Presidential Range: New Hampshire Academy of Sciences Bulletin no. 1, 41 p.

——, 1970a, Fieldguide for Friends of the Pleistocene, 33rd Annual Reunion, *in* Goldthwait, R. P., ed., Scientific aspects of Mount Washington in New Hampshire (unpublished manuscript): Columbus, The Ohio State University, 31 p.

——, 1970b, Mountain glaciers in the Presidential Range in New Hampshire: Arctic and Alpine Research, v. 2, no. 2, p. 85–102.

Goldthwait, R. P., and Mickelson, D. M., 1980, Glacier Bay; A model for the deglaciation of the White Mountains in New England, *in* Larson, G. J., and Stone, B. D., eds., Late Wisconsinan glaciation of New England: Dubuque, Kendall/Hunt, p. 167–182.

Hatch, N. L., Jr., and Moench, R. H., 1984, Bedrock geologic map of the wildernesses and roadless areas of the White Mountain National Forest, Coos, Carroll, and Grafton counties, New Hampshire: U.S. Geological Survey Map MF-1594-A, scale 1:125,000.

Hatch, N. L., Jr., Moench, R. H., and Lyons, J. B., 1983, Silurian–Lower Devonian stratigraphy of eastern and south-central New Hampshire; Extensions from western Maine: American Journal of Science, v. 283, p. 739–761.

Henderson, D. M., Billings, M. P., Creasy, J. W., and Wood, S. A., 1977, Geology of the Crawford Notch Quadrangle, New Hampshire: Concord, New Hampshire, Department of Resources and Economic Development, 29 p.

Johnson, D. W., 1933, Date of last glaciation in the White Mountains: American Journal of Science, 5th ser., v. 25, p. 399–405.

Lyons, J. B., and Livingston, D. E., 1977, Rb-Sr age of the New Hampshire Plutonic Series: Geological Society of America Bulletin, v. 88, p. 1808–1812.

Geological Society of America Centennial Field Guide—Northeastern Section, 1987

Geology of the Belknap Mountains Complex, White Mountain Series, central New Hampshire

Wallace A. Bothner, *Department of Earth Sciences, University of New Hampshire, Durham, New Hampshire 03824*
Marc C. Loiselle, *Maine Geological Survey, Augusta, Maine 04333*

LOCATION

The Belknap Mountains Complex is located in the Lakes Region of central New Hampshire. It is an easy two-hour drive from the Boston area via I-95 to Portsmouth, New Hampshire, and north on New Hampshire 16 and 11 (Fig. 1). The geologic map of the area, modified from Modell (1936) by Bothner and Gaudette (1971), occupies parts of the Winnipesaukee (Quinn, 1941), Gilmanton (Heald, 1955), Wolfeboro (Quinn, 1953), and Alton (Stewart, 1961) 15-minute Quadrangles. Most field localities are along main highways and secondary roads, foot trails, and ski slopes, as indicated on the topographic base. Permission should be requested from local landowners for examination of exposures away from main roads and on the shore and islands of Lake Winnipesaukee (e.g., Rattlesnake and Diamond islands, and the old Conway Granite Quarry).

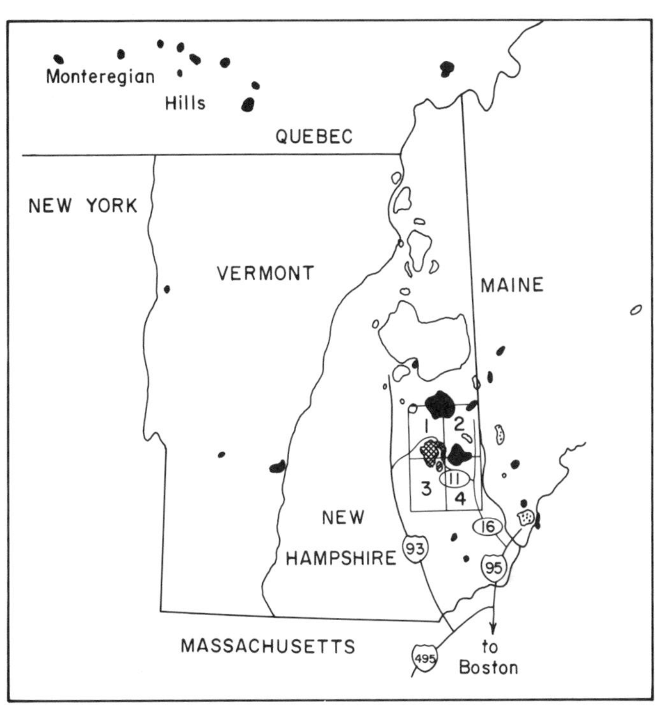

Figure 1. Distribution of White Mountain Plutonic-Volcanic Series bodies in New England and Canada (after McHone and Butler, 1984). Belknap Mountains Complex, crosshatched; stippled bodies, 240 to 200 Ma; unpatterned, 200 to 155 Ma; blackened, 135 to 100 Ma. Quadrangles: 1, Winnepesaukee; 2, Wolfeboro; 3, Gilmanton; and 4, Alton.

SIGNIFICANCE

The Belknap Mountains Complex is one of some 27 isolated and composite central intrusive complexes of the Mesozoic White Mountain Plutonic-Volcanic Series (Billings, 1956) in central New England. Abundant new outcrops afford an unusually good look at the internal magmatic "plumbing" and provide a basis for comparison with similar central complexes in New Hampshire (Kingsley, 1931) and Quebec, Scotland, and Africa. A well-developed, compositionally diverse, and genetically related suite of plutonic rock types, intrusive and vent breccias, and comagmatic volcanic rocks occur as small stocks, incomplete ring dikes, and down-dropped blocks and screens. Contact relations between rock types of the complex and country rocks (Devonian New Hampshire Plutonic Series [Lyons and Livingston, 1977] and older metasedimentary rocks) are well exposed. Structural relations displayed in outcrop support arguments for high-level intrusion by cauldron subsidence.

SITE INFORMATION

The Belknap Mountains Complex is about 7.5 mi long by 5 mi wide (12 by 8 km), with at least one ring dike beneath Lake Winnipesaukee (Fig. 2). Smaller, unmapped dikes are exposed south of the complex. Dated at 158 ± 3 Ma (K/Ar; Foland and Faul, 1977) and 168 ± 2 Ma (Rb/Sr; Loiselle, 1978), the complex represents the second of three pulses of the 140 Ma of igneous activity assigned to the Mesozoic White Mountain Plutonic-Volcanic Series (Foland and Faul, 1977; McHone and Butler, 1984).

Modell's (1936) detailed study examined the structure and petrology of the complex and is the standard reference for the area. Subsequent work by Gaudette and Bothner (1969) formed the base for a very complete geochemical study by Loiselle (1978). The Pine Mountain Complex, mapped also by Modell (1936), Heald (1955), and Dowse (1974) is a satellitic intrusion connected to the Belknaps at shallow depth (Joyner, 1963; Sharp and Simmons, 1978; Dowse and Bothner, 1980).

The distribution of rock types and successive truncation of ring dikes suggest migration of the magma chamber with time (Modell, 1936). The intrusive sequence was defined largely from crosscutting relations and by the "law of decreasing basicity" (Modell, 1936) and by geochemical character (Gaudette and Bothner, 1969; Loiselle, 1978). Igneous activity began with the emplacement of small stocks of gabbro and diorite, followed by ring fracturing and successive intrusion of monzodiorite, syenites, and quartz syenites. Porphyritic quartz syenite was emplaced during a second ring fracturing event, offset from the first. The Conway-type biotite granite was intruded into the diorite as a central stock and as thin marginal ring dikes. The volcanic rocks occupy an intermediate position in the sequence based on their chemical similarity to the syenites. Small vent agglomerates and crosscutting lamprophyre dikes represent the last episode of activity.

This field guide follows a general "differentiation sequence" from mafic to felsic rock type. Field stops (by number) are identified on Figure 2. Modal and chemical data for all rock types can be found in Modell (1936), Billings and Wilson (1964), and Loiselle (1978).

Gilford Gabbro (1) is poorly exposed on the west flank of Lockes Hill, Gilford. Reached by parking on New Hampshire 11 opposite the Smith Cove Marina and walking several hundred yards (meters) south to a topographic bench in a pine stand, it occurs in low ledges with no exposed contacts. The rock is a dark gray to black, knobby gabbro that is medium- to coarse-grained, poikilitic, and composed of labradorite, two amphiboles and pyroxene with accessory biotite, opaque oxides, pyrite, and apatite. Large (<1.2 in; 3 cm) poikilitic hornblende includes abundant plagioclase, clinopyroxene, and opaque, and produces the knots on weathered surfaces. Green hornblende replaces clinopyroxene (augite). Chlorite, white mica, epidote, and calcite are typical alteration minerals.

Endicott (Brecciated) Diorite (2) underlies the northern third of the complex. It is best exposed in recent highway cuts along New Hampshire 11 between Lake Shore Park and Glendale, where the contact with Kinsman Quartz Monzonite is knife-sharp and chilled. Abundant blocks of varying sizes of medium gray, medium-grained diorite include occasional blocks of Devonian quartz diorite, porphyritic quartz monzonite, and rare clasts of Littleton schist and are set in a matrix of Conway Granite. Large blocks are texturally uniform, but where smaller blocks have reacted with granite melt they are varied in color, irregularly shaped with cuspate borders, and commonly contain K-feldspar porphyroblasts (magma mixing?). "Unaltered" diorite contains sodic andesine, biotite, greenish hornblende, and minor clinopyroxene and quartz. "Altered," the mineralogy is highly varied with much of the mafic material chloritized. At Glendale (2A) biotite-rich clots, presumably partially digested metasedimentary xenoliths, are common and contain abundant green spinel euhedra in red-brown biotite, opaque, and probable serpentine.

The Ames Monzodiorite (3) is exposed on the north side of the complex, along New Hampshire 11 (0.2 mi; 0.3 km north of Ames) in new road cuts and north along an abandoned railroad bed. The monzodiorite is gray, medium-grained, and contains abundant andesine (sometimes mantled by K-feldspar), mottled hornblende, biotite, and very minor quartz. Chlorite-rich clots and small xenoliths of quartz diorite and metasedimentary rock are common. The unit is mapped to the shore of Lake Winnipesaukee. This ring was apparently truncated by the emplacement of the younger subporphyritic quartz syenite and by the central Conway stock.

The Gilmanton Monzodiorite (4) crops out along the southern margin of the complex between Piper Mountain and Goat Pasture Hill. Access is along the north shore of Manning Lake and on the hillside to the north; permission is needed from local property owners. The Gilmanton Monzodiorite is gray, medium-grained, and dominated by oligoclase, discrete microperthite, hornblende, biotite, clinopyroxene, and minor quartz. No mafic clots or xenoliths have been observed. Contact relations between the Gilmanton and metasedimentary host are sharp and steep; between the Belknap Syenite, gradational; and between the Albany Porphyritic Quartz Syenite, always interrupted by a thin band of granite (Modell, 1936, p. 1902).

Belknap and Cobble Hill Syenites were interpreted to be two stocks of the same magma (Modell, 1936; Bothner and Gaudette, 1971). Based on detailed geochemical work, Loiselle (1978) distinguished the two and revived Modell's earlier note (1936, p. 1904) that the syenite at Cobble Hill was more texturally varied than that exposed on Belknap and Gunstock mountains. The Belknap Syenite (5) is well exposed in pavement outcrops at the parking area at the head of the trail to the Belknap Mountain lookout tower. Exposures are typically light gray, coarse-grained, hornblende syenite. Microperthite, oligoclase with occasional included nepheline, minor biotite and clinopyroxene (as cores of hornblende), and minor quartz are present in thin section. A small iron deposit, last mined in the early nineteenth century, contains a group of epidote+magnetite+pyrrhotite veins on the western slope of Gunstock Mountain at an elevation of about 1,800 ft (550 m). The deposit is interpreted to have formed from late hydrothermal fluids emanating from the younger, nearby Conway Granite (Malkoski, 1976).

The Cobble Hill Syenite (6) crops out along the access roads throughout the Gunstock Acres development area (north of the entrance to the Mount Rowe Ski area on New Hampshire 11A) and spectacularly exposed beneath the Poorfarm Brook bridge (6A) at that entrance. Near the top of Cobble Hill, the light gray syenite occurs in pavements and in low road cuts; in Poorfarm Brook, downstream from the bridge, the syenite is in sharp contact with vent breccia choked with clasts of syenite. Syenite pegmatite, in which a few small grains of nepheline have been observed, occurs beside the Mount Rowe chair lift loading station. Petrographically, the Cobble Hill syenite is medium-grained, and dominated by perthite, hornblende, and minor biotite. Primary plagioclase and quartz are rare.

Sawyer Quartz Syenite (7) occurs as a 0.6 by 1.9 mi (1 by 3 km) NNW-trending lens from the north slope of Gunstock Mountain along the ridge toward Mount Rowe nearly to New Hampshire 11A. Accessible outcrops are along the ski trails from the top of the Gunstock chair lift where buff-pink, medium-grained, equigranular syenite is exposed. It consists of perthite, quartz, variable amounts of oligoclase, hornblende, and minor biotite. The rock contains two mappable pendants of Kinsman Quartz Monzonite, as well as numerous inclusions of Belknap Syenite (Modell, 1936).

Lake Quartz Syenite is exposed on Diamond and Rattle-

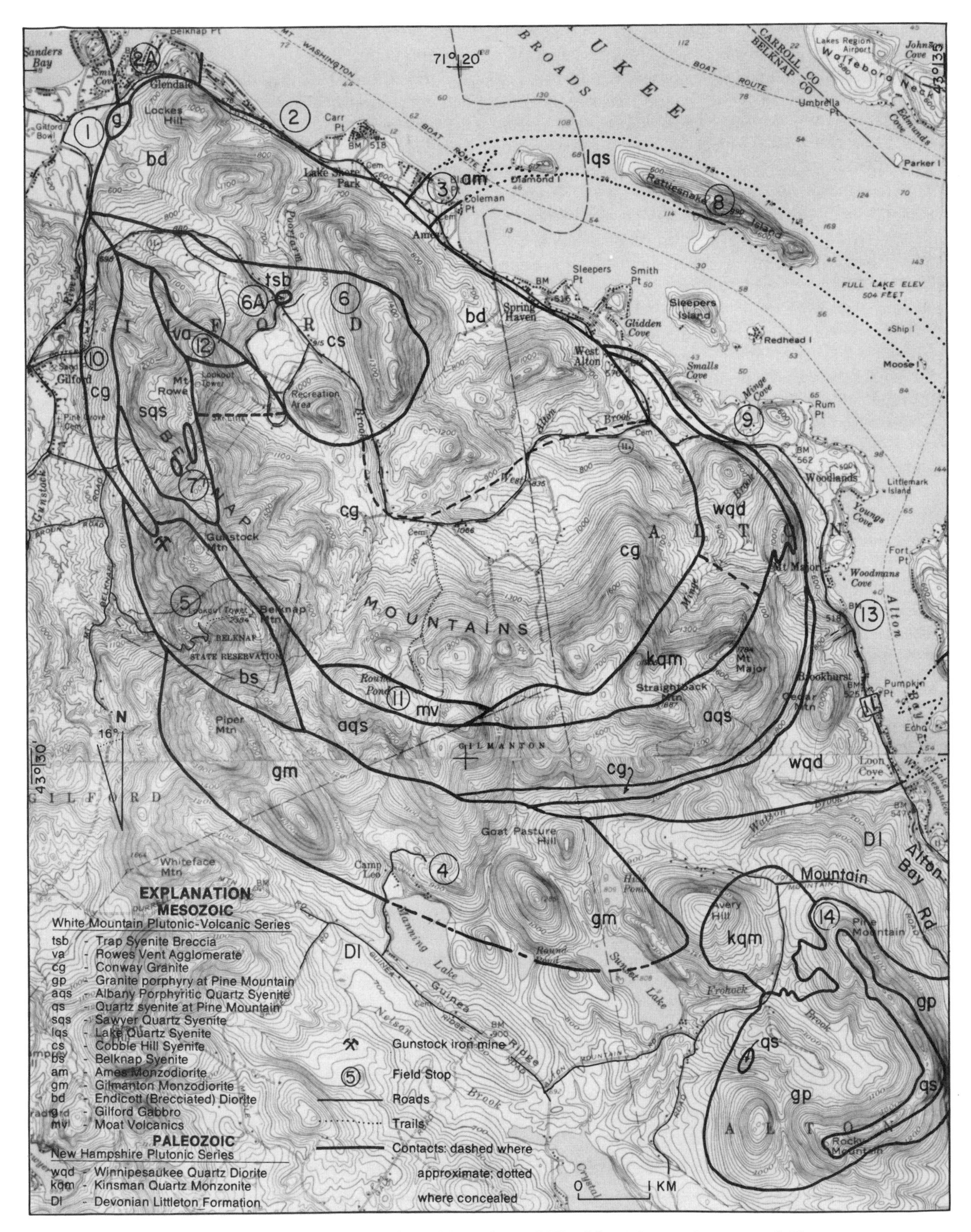

Figure 2. Geologic map of the Belknap Mountains and Pine Mountain complexes, central New Hampshire.

snake Islands (8). Access to the islands is by small boat; the best exposures occur along the northeast side of Rattlesnake Island. Request permission to land on the more sheltered southwestern shore of Rattlesnake Island to examine the contact between the quartz syenite and Winnipesaukee Quartz Diorite. The contact, near the break in slope, is marked by a fine-grained, banded (flow?) quartz syenite that contains abundant phenocrysts of alkali feldspar and mantled inclusions of diorite. Along the ridge of the island, the rock is a buff-pink, medium-grained, subporphyritic quartz syenite composed of microperthite, hornblende, quartz, and biotite. Spherical aggregates of ferro-magnesian minerals are clustered about small feldspar grains. Quartz is usually interstitial and nepheline was noted in several thin sections.

Albany Porphyritic Quartz Syenite rings nearly three-quarters of the Belknap Mountains Complex and ranges in width from a few tens of yards (meters) to more than 0.6 mi (1 km). The best and most accessible exposures occur on New Hampshire 11 at West Alton (9), the top of Mount Major, and below the Gunstock chair lift near the top of Gunstock Mountain. The New Hampshire 11 road cut exposes fresh, pink, coarsely porphyritic quartz syenite with a well-defined 4 to 6 in (10 to 15 cm) chill margin in contact with strongly sheared (mylonitized?) country rock. Occasional inclusions of dark gray to black aphanitic rock, perhaps an earlier chill phase, occur within the shallow dipping ring dike at this locality (elsewhere, the contact is nearly vertical). The quartz syenite consists of perthite phenocrysts, sometimes mantled by oligoclase (Rapakivi texture), isolated oligoclase, hornblende and minor biotite, and embayed euhedral quartz.

Conway Granite is exposed as the matrix material in the Endicott (brecciated) Diorite (2), in ledges on the trail to Round Pond, and as two incomplete ring dikes along the western and southern margin of the complex. A small, abandoned quarry (10) on the east side of Potter Hill Road, 0.4 mi (0.6 km) south of New Hampshire 11A toward Gilford, provides the freshest samples, but permission to cross private property is required. The granite is typically pink, medium- to coarse-grained, equigranular, and contains perthite, quartz, oligoclase, biotite, and minor hornblende. It differs little from biotite granite that occurs as the central stock of other central complexes and in the composite White Mountains batholith.

The Moat Volcanics and the Rowes Vent Agglomerate (and probably associated Trap Syenite Breccia at Poorfarm Brook) demonstrate the volcanic-plutonic relationships of the complex. The Moat Volcanics (11) are exposed 1.5 mi (2.5 km) south of New Hampshire 11A (following the trail) to the outlet brook from Round Pond and on the hillsides east and west of that drainage. They are in fault contact with the Albany Porphyritic Quartz Syenite. The volcanic rocks, downdropped during cauldron subsidence, are trachyte and rhyolite tuff. The trachyte is dark, dense, and porphyritic. Plagioclase phenocrysts are set in an orthoclase+oligoclase+biotite (now chlorite) matrix. The rhyolite tuff is light gray, fine-grained, and contains abundant altered plagioclase, quartz, and altered biotite. Gaudette and Bothner (1969) believe that the Moat Volcanics are coeval with the Belknap Syenite and therefore not the oldest unit of the complex (Modell, 1936).

The Rowes Vent Agglomerate (12) is the youngest rock unit in the Belknap Complex. It contains fragments of many of the older units including, significantly, the Conway Granite (Modell, 1936, p. 1910). It is exposed as pavement outcrops at the north end of Mount Rowe near the top of the Mount Rowe chair lift and is accessible by chair or by car, following Curtis Road from New Hampshire 11A up to the north side. Small outcrops of vent agglomerate and xenolithic dikes are also exposed in an old "rottenstone" quarry at the entrance gate to the Gunstock Ski area from the Mount Rowe area. Small clasts (<3 ft; 1 m) of Kinsman Quartz Monzonite, and rarer small fragments of Belknap and Sawyer syenites and Conway Granite, are set in a dark, fine-grained matrix of quartz and feldspar. Related dikes are dark gray, aphanitic, and contain fewer crystals and xenoliths.

Belknap Dike Rocks, shown as dashes on the geologic map (13), are spectacularly exposed in the Scenic View Area 3.5 mi (5.6 km) north of the town of Alton Bay on New Hampshire 11. Here the New Hampshire Series Winnipesaukee Quartz Diorite contains large inclusions of Kinsman Quartz Monzonite, both cut by diabase dikes and two White Mountain Series dikes. A magnificant view of the glacially sculpted Lake Winnipesaukee basin and the Ossipee complex to the northeast is an added attraction. The roadcut exposes medium gray, medium-grained, faintly foliated, quartz diorite consisting of andesine, quartz, and biotite. About 6 to 15 ft (2 to 5 m) disoriented inclusions of foliated porphyritic Kinsman Quartz Monzonite with 2 in ± (5 cm±) megacrysts of alkali feldspar, oligoclase, quartz, biotite+muscovite, and garnet are common. Coarse-grained pegmatites, containing microcline euhedra 4 in (10 cm) long, quartz, muscovite, apatite, magnetite, and sulfide crosscut these units.

A flow-banded, 10- to 13-ft-thick (3 to 4 m) rhyolite dike occurs near the southeast end of the outcrop; flow-banding follows a curviplanar surface, giving the dike a folded appearance. K-feldspar phenocrysts (some zoned) are rotated close to the contact and occur in a very-fine-grained, laminated, pinkish gray matrix. Embayed quartz euhedra are about 5 mm in diameter. The rock is dense and brittle near the contact, and in thin section contains small spherulites. The center of the dike is coarser grained, lighter colored, and more granular. Feldspars are hydrothermally altered, giving the dike a chalky appearance.

A second 15- to 20-ft-thick (5 to 6 m) dike toward the center of the outcrop is a green-gray quartz syenite containing small phenocrysts of alkali feldspar and quartz. Both dikes crosscut diorite, quartz monzonite, and several pegmatite dikes and have sharp, chilled contacts. Several Mesozoic diabase dikes are also exposed in the cut.

The Pine Mountain Complex is an isolated comagmatic stock separated from the Belknaps by a narrow (and probably thin) band of pelitic schist and Kinsman Quartz Monzonite (Modell, 1936; Heald, 1955; Dowse, 1974; Dowse and Bothner, 1980). Its age, 156±3 Ma (Foland and Faul, 1977), is comparable to the determination for the Belknap Complex. The Pine Mountain Complex consists of granite porphyry and an incomplete ring of quartz syenite forming the prominent ridge between Pine and Rocky mountains. Exposures are generally poor, with the most accessible outcrops (14) occurring along a southerly traverse from the crest of Mountain Road to the top of Pine Mountain (permission from the Dana Morse family on that road). Granite porphyry, exposed on the north flank of Pine Mountain, is a buff-to-pink rock characterized by quartz phenocrysts and microperthite glomerocrysts in a fine-grained matrix of quartz, feldspar, and minor hornblende and biotite. The quartz syenite has a similar mineralogy but contains less quartz and more feldspar. In both, quartz occurs as embayed euhedra. Several small masses of gray feldspar-hornblende porphyry and breccia occur near the top of the mountain. Field relations indicate that the quartz syenite was emplaced as a partial ring dike, followed by emplacement of granite porphyry engulfing the earlier syenite mass.

SYNTHESIS

The structural and petrologic evolution of the Belknap Complex generally followed the classic development of central intrusive complexes: early ring fracturing in the upper crust by ascending magma, eruption, and accompanying cauldron subsidence. Subsequent repetition of this process resulted in the emplacement of younger, compositionally distinct dikes and stocks. The central stock, commonly the most differentiated, was last to be emplaced. Late-stage doming resulted in "last-gasp" venting.

Petrologic models of Bothner and Gaudette (1971), Loiselle (1978), and Loiselle and Hart (1978), based on major, trace, and rare-earth element, and isotope geochemistry, refine Modell's (1936) gabbro-to-granite differentiation sequence. Importantly, a mantle-derived alkali basalt magma has been identified as the most probable parent for the sequence. Based largely on Loiselle's trace element and isotopic work, each unit of the Belknap Complex can be generated by: varying degrees of crystallization under varying pH_2O; assimilation, reaction, and partial melting of limited amounts of crustal materials; and cumulate processes.

The great similarity of the Belknap Complex with all or parts of other intrusives of the White Mountain Series in the 124-mi-long (200 km) north-northwest–trending zone from the New England coast to Canada still generates petrogenetic and structural questions (Fig. 1). The belt may reflect a deep-seated fracture system (Chapman, 1968), a Mesozoic "hotspot" trace (Coney, 1971), or the landward extension of a transform fault (McHone and Butler, 1984). The supporting arguments for these various hypotheses cannot be reviewed here, but what can be said with some certainty is: Whatever the structural control on magma generation and intrusion, the magmas evolved from a regionally similar upper mantle or lower crust to yield compositionally consistent products over a long time period and over a wide geographic area. In addition, the ascent of these magmas must have been facilitated by "easy" access to high crustal levels, as they often reached the surface as explosive volcanoes (Billings and Noble, 1967; Carr and others, 1980).

REFERENCES CITED

Billings, M. P., 1956, The geology of New Hampshire, Part II, Bedrock geology: Concord, New Hampshire State Planning and Development Commission, 203 p.

Billings, M. P., and Noble, D. C., 1967, Pyroclastic rocks of the White Mountain magma series: Nature, v. 216, p. 906–907.

Billings, M. P., and Wilson, J. R., 1964, Chemical analyses of rocks and minerals from New Hampshire; Part 19, Mineral resources survey: Concord, New Hampshire Division of Economic Development, 104 p.

Bothner, W. A., and Gaudette, H. E., 1971, Geologic review of the Belknap Mountain Complex: Guidebook for field trips in central New Hampshire and contiguous areas, New England Intercollegiate Geological Conference, Concord, New Hampshire, p. 88–99.

Carr, R. S., Lyons, J. B., Albaugh, D. S., Bothner, W. A., and Sharp, J. A., 1980, Ossipee restudied: Geological Society of America Abstracts with Programs, v. 12, p. 28.

Chapman, C. A., 1968, A comparison of the Maine coastal plutons and the magmatic central complexes of New Hampshire: *in* Zen, E-An, White, W. S. Hadley, J. B., and Thompson, J. B., Jr., eds., Studies of Appalachian geology; Northern and maritime: New York, Interscience, p. 385–396.

Coney, P. J., 1971, Cordilleran tectonic transitions and motion of the North American plate: Nature, v. 233, p. 462–465.

Dowse, M. E., 1974, The geology of the Pine Mountain Complex, Alton, New Hampshire [M.S. thesis]: University of New Hampshire, 67 p.

Dowse, M. E., and Bothner, W. A., 1980, Geology of the Pine Mountain Complex New Hampshire: Northeast Geology, v. 2, p. 74–80.

Foland, K. A., and Faul, H., 1977, Ages of the White Mountain intrusives; New Hampshire, Vermont, and Maine, U.S.A.: American Journal of Science, v. 277, p. 888–904.

Gaudette, H. E., and Bothner, W. A., 1969, Geochemistry of the Belknap Mountain Complex, New Hampshire: Geological Society of American Abstracts with Programs, Part I, p. 22.

Heald, M. T., 1955, Geology of the Gilmanton Quadrangle: Concord, New Hampshire State Planning and Development Commission, 29 p.

Joyner, W. B., 1963, Gravity in north-central New England: Geological Society of America Bulletin, v. 74, p. 831–858.

Kingsley, L., 1931, Cauldron subsidence of the Ossipee Mountains: American Journal of Science, 5th Series, v. 22, p. 134–167.

Loiselle, M. C., 1978, Geochemistry and petrogenesis of the Belknap Mountains Complex and Pliny Range, White Mountain Series, New Hampshire [Ph.D. thesis]: Massachusetts Institute of Technology, 302 p.

Loiselle, M. C., and Hart, S. R., 1978, Sr isotope systematics of the Belknap Mountains Complex: Geological Society of America Abstracts with Programs, v. 10, p. 73.

Lyons, J. B., and Livingston, D. E., 1977, Rb-Sr age of the New Hampshire plutonic series: Geological Society of America Bulletin, v. 88, p. 1808–1812.

Malkoski, M., 1976, Geology of the Mount Gunstock Iron Deposit, Gilford, Belknap County, New Hampshire [M.S. thesis]: University of New Hampshire, 103 p.

McHone, J. G., and Butler, J. R., 1984, Mesozoic igneous provinces of New England and the opening of the North Atlantic Ocean: Geological Society of America Bulletin, v. 95, p. 757–765.

Modell, D., 1936, Ring-dike complex of the Belknap Mountains, New Hamp-

shire: Geological Society of America Bulletin, v. 47, p. 1885–1932.
Quinn, A., 1941, Geology of the Winnipesaukee Quadrangle, New Hampshire: Concord, New Hampshire Department of Resources and Economic Development, 22 p.
——, 1953, Geology of the Wolfeboro Quadrangle, New Hampshire: Concord, New Hampshire Department of Resources and Economic Development, 24 p.
Sharp, J. A., and Simmons, G., 1978, Geologic/geophysical models of intrusions of the White Mountain Magma Series (WMMS): Geological Society of America Abstracts with Programs, v. 10, p. 85.
Stewart, G. W., 1961, Geology of the Alton Quadrangle, New Hampshire: Concord, New Hampshire Department of Resources and Economic Development, 33 p.

The Massabesic Gneiss Complex and Permian granite near Milford, south-central New Hampshire

John N. Aleinikoff, *U.S. Geological Survey, Box 25046, MS 963, Denver Federal Center, Denver, Colorado 80225*
Julian W. Green, *Department of Geology, Harvard University, Cambridge, Massachusetts 02138*

LOCATION

Road cuts are on the approach ramp to New Hampshire 101 from New Hampshire 101A (Fig. 1), 2.2 mi (3.5 km) east of intersection of New Hampshire 101A and 13 (center of Milford) and 0.5 mi (0.8 km) west of intersection of New Hampshire 101A and 122. Because the approach ramp is a busy thoroughfare, vehicles should be parked at the north end of the outcrop, either on the shoulder of New Hampshire 101A or in the parking lot of a nearby store about 300 ft (100 m) east on 101A. The three outcrops can be reached on foot; there is space of about 10 to 20 ft (3 to 6 m) between the edge of the road and the vertical faces.

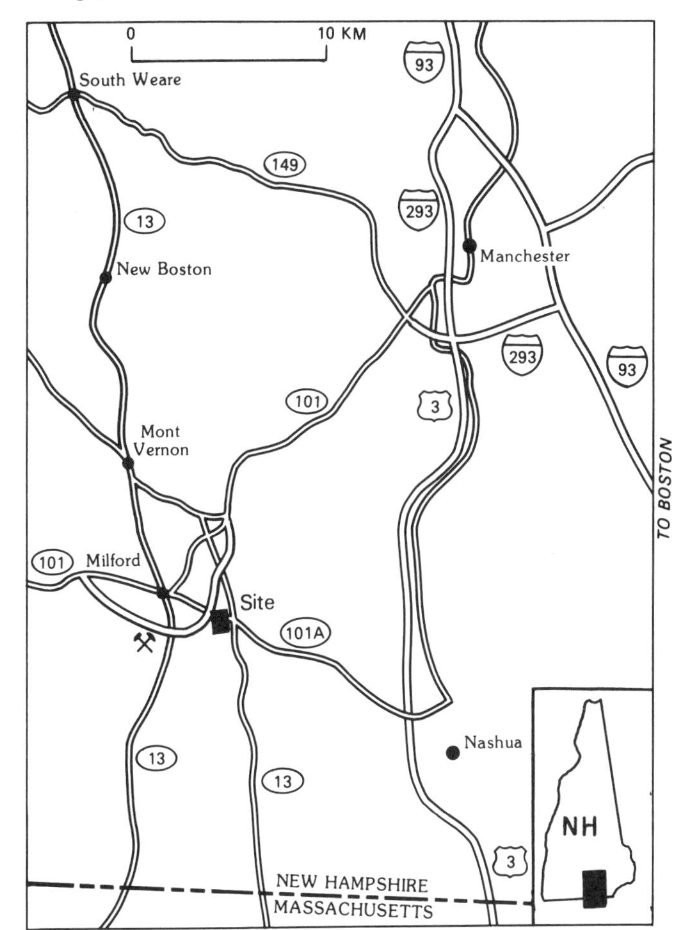

Figure 1. Location map of roadcuts of Massabesic Gneiss Complex and Permian granite.

SIGNIFICANCE

The Massabesic gneiss of Sriramadas (1962), now the Massabesic Gneiss Complex, is a sillimanite-zone migmatite that was included in the Fitchburg Granite (now the Fitchburg Complex) that forms the northernmost exposures of dated late Proterozoic (Avalonian) rocks in New England. The antiquity of this terrane and its importance to New England tectonics has been recognized for less than a decade (Lyons and others, 1982). The gneiss constitutes the core of the northeast-trending Massabesic anticlinorium (Lyons and others, 1982), and is flanked on the north by Silurian to Lower Devonian metasedimentary rocks of the Kearsarge–Central Maine synclinorium and on the south by pre-Silurian rocks of the Merrimack trough, which may be as old as Proterozoic. The Massabesic Gneiss Complex is composed of compositionally diverse Proterozoic paragneiss that is migmatitically intruded by Ordovician and late Proterozoic (foliated granite). Both of these rocks are intruded by the unfoliated Permian granite near Milford, New Hampshire, which is the northernmost known occurrence of 275 Ma granite. The extensive roadcuts at this site afford a rare view of the structural, petrologic, and age relationships among these rock assemblages.

SITE INFORMATION

The three roadcuts consist of the Massabesic Gneiss Complex, Permian granite, and two generations of pegmatites. The Massabesic Gneiss (named by Sriramadas, 1962, for excellent exposures around Lake Massabesic, east of Manchester, New Hampshire) is divided into two main units, "paragneiss" and "orthogneiss." The paragneiss is well-foliated biotite-quartz-plagioclase gneiss with minor amounts of pelitic schist, calc-silicate rock, amphibolite (as boudins), and sparse quartzite. The biotite-quartz-plagioclase gneiss is interpreted as metamorphosed intermediate-to-felsic volcanic rocks (Aleinikoff, 1978). The paragneiss contains at least two generations of folding; early isoclinal folds and later broad open folds. The biotite-quartz-plagioclase gneiss has been dated using the U-Pb zircon method at about 650 Ma (Aleinikoff and others, 1979).

The orthogneiss is pink to white, coarse-grained granitic gneiss. Although most of this rock approximately conforms to the foliation of the paragneiss, at many places it is sharply discordant. It often contains biotite-rich schlieren probably derived from partial assimilation of paragneiss. U-Pb dating of zircon from the orthogneiss in these outcrops indicates that it is about 475 Ma. Of very similar appearance is orthogneiss in outcrops east of Manchester, New Hampshire, which has been dated by Besancon and others (1977) and Kelly and others (1980) at about 650 Ma.

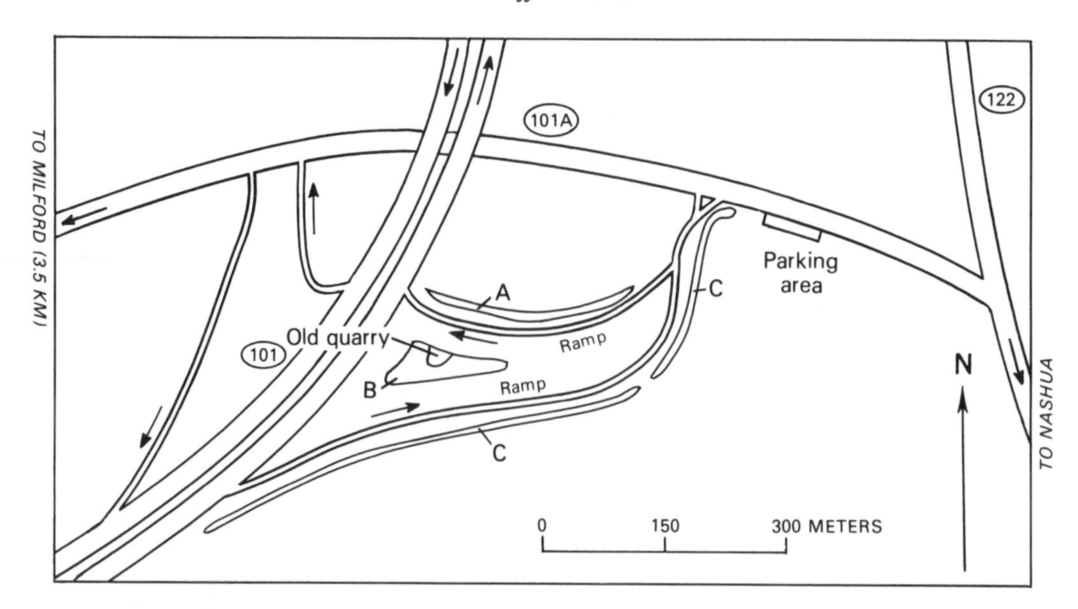

Figure 2. Sketch map of outcrops in relation to ramps between New Hampshire 101 and 101A. Note that outcrops A and C are primarily vertical walls (facing south and north, respectively); the apparent width as shown on the map is somewhat exaggerated. Outcrop B has exposures in both the south-facing wall and in the area shown in map plan.

Thus, there may be two generations of orthogneiss: (1) late Proterozoic, possibly related to the extrusion of the protolith of the paragneiss, and (2) Early Ordovician.

The terranes from which the metasedimentary rocks of the Massabesic Gneiss Complex were derived may be about 1500 Ma. This age was estimated by dating detrital zircons (Aleinikoff and others, 1979) from three localities: a layer of quartzite within the paragneiss near Bedford, New Hampshire; a bed of metasiltstone in the Oakdale Formation at Hickory Hills Lake in the Shirley (Massachusetts) Quadrangle; and a similar bed in the Oakdale Formation, also in the Shirley (Massachusetts) Quadrangle (Robinson, 1979). Although the Oakdale is assigned to the Silurian on the new Geologic Map of Massachusetts (Zen, 1983), this formation is coextensive with the Berwick Formation of southeastern New Hampshire, which is one of the group of formations that constitute the Merrimack trough and is probably no younger than Early Ordovician (Lyons and others, 1982).

Crosscutting all foliated rocks (paragneiss and orthogneiss) in many locations in these outcrops are dikes of the Permian granite. This granite is gray and composed of quartz, K-feldspar, biotite, plagioclase, and minor primary muscovite. It is massive, with only rare flow foliation indicated by weak alignment of biotite. The granite has yielded a U-Pb zircon age of about 275 Ma (Aleinikoff and others, 1979), making it among the youngest calcalkaline plutonic rocks in New England. It is the same age as some granites of southern New England such as the Westerly Granite (Zartman and Hermes, 1984) and the Narragansett Pier Granite (Kocis and others, 1978). The Permian granite at present is *not* referred to as the "Milford Granite," although this was the informal name used by Dale (1923) when many of the small pits in the Milford area were sites of active quarrying. The usage of Milford Granite today refers to a granite in southeastern Massachusetts, dated by Zartman and Naylor (1984) at about 625 Ma. Thus, to avoid confusion, the unfoliated granite in these outcrops near Milford, as mapped by Aleinikoff (1978), is referred to here as "Permian granite."

The 475-Ma orthogneiss and the 275-Ma granite are sometimes difficult to distinguish in outcrop, particularly when both contain xenoliths of paragneiss. However, the orthogneiss is generally white, coarse-grained, and foliated, whereas the granite is gray, medium-grained, and massive. As a guide in these outcrops, the Permian granite can be seen along the entire length of the roadcuts in the curbstone at the edge of the road. These rocks were quarried at the Kitledge operation and shaped into curbstone in the nearby Barreto Brothers shop.

At least two generations of pegmatites cut older rocks. They are distinguished primarily by color; pink pegmatites appear to be associated with intrusion of the orthogneiss and sometimes are obviously foliated. White pegmatites are related to the Permian granite and are never foliated. Both are coarse-grained and contain muscovite and biotite.

Figure 2 shows a map plan of the three outcrops exposed in roadcuts on the approach ramps between New Hampshire 101 and 101A in southern New Hampshire. Roadcut C, on the east and south sides of the road, is about 2,300 ft (720 m) of nearly continuous outcrop, ranging in height from slightly less than 10 ft (3 m) to about 50 ft (15 m). Layered paragneiss, in orientations ranging from subvertical to subhorizontal, occurs throughout this outcrop. About 250 ft (80 m) from the north end are excellent exposures of large (3 by 6 ft; 1 by 2 m) boudins of amphibolite. In the highest section, (about 1,500 ft or 475 m from the north end), paragneiss is seen cut by both orthogneiss and Permian granite.

Across the road at this location is outcrop B. This island of rock between the two roads contains the three principal rock units. This outcrop is particularly interesting because the rocks are exposed both in a vertical, south-facing wall and on a hillside sloping gently down to the west. Thus, three-dimensional views of the structural relationships are afforded by walking over this outcrop from northwest to southeast. On the hillside, dikes of the Permian granite cut the Massabesic Gneiss Complex. The gneiss at this location is migmatite composed of mixed orthogneiss and paragneiss. The granite extends southeastward across the road to the top of outcrop C where it is a sub-horizontal sheet.

Also present at outcrop B is an old, abandoned quarry of the Permian granite where rusted pins and old drill holes are relics of granite quarrying used at the turn of the century. Many small quarries of granite were in operation in the Milford area about 90 years ago. Today, only two remain open; a small quarry near Mason, New Hampshire, and a large commercial venture, the Kitledge quarry, due south of Milford, about 2 mi (3.2 km) southwest of these roadcuts. In contrast to the old, labor-intensive techniques of the past, modern granite quarrying using jet-fuel torches and continuous-wire cutting can be observed at the Kitledge quarry, where both massive granite and migmatite are cut. The migmatite (called "tapestry rock" by the quarry operators) is used for facades on buildings. The granite has many uses, including curb and paving stones, gravestones, and building stones.

Outcrop A is primarily an east-west exposure of the three principal rock types. It is much shorter than outcrop C (about 900 ft or 275 m) but at certain times of the day it may be better observed as its wall is south-facing, while outcrop C faces west and north.

Interpretation of the structural setting of the Massabesic Gneiss Complex is controversial. To the north across the Campbell Hill fault (Fig. 3) are the Silurian and Devonian metasedimentary rocks of Kearsarge–Central Maine synclinorium, as re-defined from the Merrimack synclinorium of Billings (1956) by Lyons and others (1982). To the south across the Silver Lake fault (Fig. 3) are the rocks of the Merrimack Group (Berwick, Eliot, and Kittery formations) that Hussey (1962) previously thought to be Silurian based on long range correlation with fossiliferous strata in Maine. The Merrimack Group is now known to be at least as old as Ordovician and possibly Proterozoic, since these formations are cut by the Exeter Diorite, dated at 473 ± 37 Ma using the Rb-Sr method by Gaudette and others (1984). In traverses from southeast to northwest through the Merrimack Group rocks into the Massabesic Gneiss Complex, one can observe a sharp increase in metamorphic grade but a similarity in structure and lithology. Xenoliths of presumed Berwick Formation can be seen in the Massabesic migmatite at localities east of Manchester (Bothner and others, 1984; Fagan, 1985). These relations have prompted recent workers to suggest that the Massabesic Gneiss Complex is simply a higher-grade variety of the Berwick Formation, indicating that the Berwick itself may be late

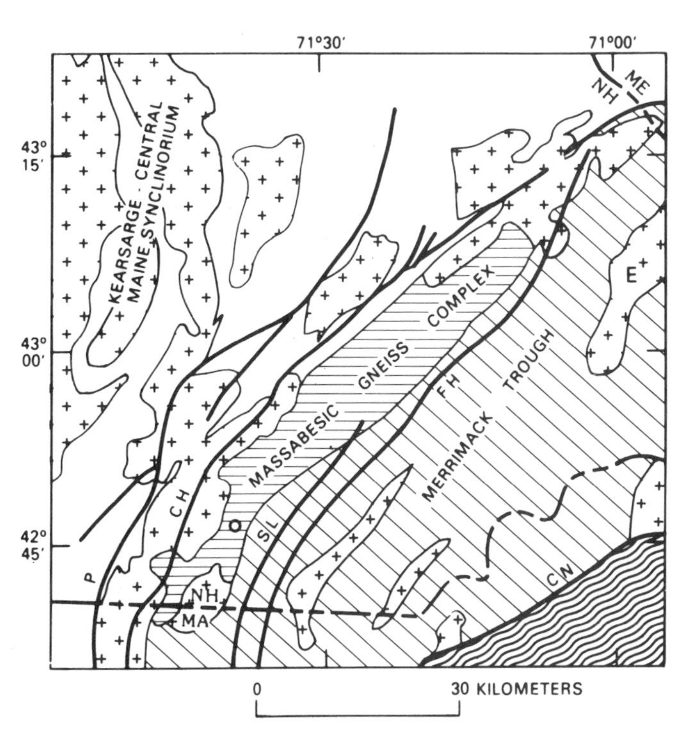

Figure 3. Simplified geologic map of south-eastern New Hampshire and north-eastern Massachusetts (modified from Lyons and others, 1982). Horizontal lines indicate area of the Massabesic Gneiss Complex; slanted lines indicate area of Ordovician or older Merrimack Group metasedimentary rocks; pluses are Paleozoic (Ordovician to Permian) plutonic rocks (E is Exeter Diorite); white areas are Paleozoic (Ordovician to Devonian) rocks of the Kearsarge–Central Maine synclinorium; wavey lines indicate area of Nashoba terrane. Dark lines are faults; where labeled they are: P, Pinnacle; C.H., Campbell Hill; S.L., Silver Lake; F.H., Flint Hill; C.N., Clinton-Newberry. Outcrops locality shown by circle in the south-central part of the Massabesic Gneiss Complex.

Proterozoic. The presence of detrital zircons of the same provenance age in both the Oakdale Formation (Berwick correlative in Massachusetts) and the Massabesic Gneiss Complex suggests that this explanation may be correct, but this interpretation is not unequivocal. The contact between the Massabesic Complex and Berwick Formation alternatively may be an unconformity or a fault.

Of similar uncertainty is the tectonic origin of the Massabesic Gneiss Complex itself. Lyons and others (1982) refer to its structure as the Massabesic anticlinorium, based on both its age in relation to younger rocks surrounding it and also because it appears to plunge under younger rocks at both ends. However, evidence for an actual structural arch is lacking. Because the gneiss is contained within the Campbell Hill and Silver Lake faults, it may also be either a horst or a thrust slice. Gravity studies (Aleinikoff, 1978; Anderson, 1978) indicate that the Massabesic is probably quite thin (up to 1.2 mi or 2 km), lending credence to the thrust sheet hypothesis. Future detailed mapping in southern New Hampshire, northern Massachusetts, and southern Maine may solve this problem.

REFERENCES CITED

Aleinikoff, J. N., 1978, Structure, petrology, and U-Th-Pb geochronology in the Milford (15′) quadrangle, New Hampshire [Ph.D. thesis]: Dartmouth College, 247 p.

Aleinikoff, J. N., Zartman, R. E., and Lyons, J. B., 1979, U-Th-Pb geochronology of the Massabesic Gneiss and the granite near Milford, south-central New Hampshire; New evidence for Avalonian basement and Taconic and Alleghenian disturbances in eastern New England: Contributions to Petrology and Mineralogy, v. 71, p. 1–11.

Anderson, R. C., 1978, The northern termination of the Massabesic Gneiss, New Hampshire [M.S. thesis]: Dartmouth College, 111 p.

Besancon, J. R., Gaudette, H. E., and Naylor, R. S., 1977, Age of the Massabesic Gneiss, southern New Hampshire: Geological Society of America Abstracts with Programs, v. 9, p. 242.

Billings, M. P., 1956, The Geology of New Hampshire; Part II Bedrock Geology: New Hampshire State Planning and Development Committee, 200 p.

Bothner, W. A., Boudette, E. L., Fagan, T. J., Gaudette, H. E., Laird, J., and Olszewski, W. J., 1984, Geologic framework of the Massabesic Anticlinorium and the Merrimack Trough, southwestern New Hampshire, *in* Hanson, L. S., ed., Geology of the coastal lowlands, Boston, Massachusetts, to Kennebunk, Maine: 76th Annual New England Intercollegiate Geologic Conference, p. 186–206.

Dale, T. N., 1923, The commercial granites of New England: U.S. Geological Survey Bulletin 738, 488 p.

Fagan, T. J., 1985, Does the Massabesic Gneiss Complex, southeastern New Hampshire, include a partial melt of Berwick Formation?: Geological Society of America Abstracts with Programs, v. 17, p. 18.

Gaudette, H. E., Bothner, W. A., Laird, J., Olszewski, W. J., and Cheatam, M. M., 1984, Late Precambrian/Early Paleozoic deformation and metamorphism in southeastern New Hampshire; Confirmation of an exotic terrane: Geological Society of America Abstracts with Programs, v. 16, p. 516.

Hussey, A. M., II, 1962, The geology of southern York county, Maine: Maine Geological Survey Special Geologic Studies Series, no. 4, 67 p.

Kelly, W. J., Olszewski, W. J., and Gaudette, H. E., 1980, The Messabesic orthogneiss, southern New Hampshire: Geological Society of America Abstracts with Programs, v. 12, p. 45.

Kocis, D. E., Hermes, O. D., and Cain, J. A., 1978, Petrological comparison of the pink and white facies of the Narragansett Pier granite, Rhode Island: Geological Society of America Abstracts with Programs, v. 10, p. 71.

Lyons, J. B., Boudette, E. L., and Aleinikoff, J. N., 1982, The Avalonian and Gander zones in central eastern New England, *in* St. Julien, P., and Béland, J., eds., Major structural zones and faults of the northern Appalachians: Geological Association of Canada Special Paper 24, p. 43–65.

Robinson, G. R., 1979, Bedrock geology of the Nashua River area, Massachusetts–New Hampshire [Ph.D. thesis]: Harvard University, 172 p.

Srirmadas, A., 1962, The geology of the Manchester Quadrangle, New Hampshire: New Hampshire Department of Resources and Economic Development Bulletin, v. 2, 78 p.

Zartman, R. E., and Hermes, O. D., 1984, Evidence from inherited zircon for Archean basement under the southeastern New England Avalon terrane: Geological Society of America Abstracts with Programs, v. 16, p. 704.

Zartman, R. E., and Naylor, R. S., 1984, Structural implications of some radiometric ages of igneous rocks in southeastern New England: Geological Society of America Bulletin, v. 95, p. 522–539.

Zen, E-An, ed., 1983, Bedrock geologic map of Massachusetts: Arlington, Virginia, U.S. Geological Survey, Scale 1:250,000.

Stratigraphy of the Rangeley area, western Maine

Robert H. Moench and Eugene L. Boudette, U.S. Geological Survey, MS 930, Box 25046, Denver Federal Center, Denver, Colorado 80225

LOCATION

This double site is in western Maine, about 5 hours drive north from Boston. Figure 1 shows the geology of the area and 16 numbered stops along and near Maine 4 between Oquossoc and Phillips. Most of the area is covered by geologic maps of the Kennebago Lake Quadrangle (Boudette, 1986) and the Phillips and Rangeley quadrangles (Moench, 1971); and a regional synthesis is provided by Moench and Pankiwskyj (1986). Rangeley village is the most convenient base for large groups. Directions to the numbered stops follow.

Stop 1. Large outcrop of Dead River Formation on the shore of Mooselookmeguntic Lake at the north end of the causeway to Spots Island. Obtain permission from residents on the island. Because the parking area at the causeway is limited, large groups should park in Oquossoc and walk (about 0.3 or 0.4 mi; 0.5 or 0.6 km).

Stop 2. Low road cut in black slate on the north side of Maine 4, 0.7 mi (1.1 km) east of its intersection with Maine 16 near Oquossoc.

Stop 3. Large road cut of Graywacke Member of Quimby Formation on Maine 4 near Dodge Pond, 3 mi (4.8 km) east of its intersection with Maine 16.

Stop 4. Road cut of shale member of Quimby Formation on Maine 4, on a hill at a T intersection, 1.2 mi (1.9 km) east of Stop 3.

Stop 5. Long, stream-washed outcrop of shale member of Quimby Formation in Nile Brook, 1.4 mi (2.3 km) south of Rangeley village, just upstream from Maine 4.

Stop 6. Road cuts in Greenvale Cove Formation along Maine 4, 2 mi (3.2 km) south of Rangeley village and extending, with small gaps, to Stop 7, in Rangeley Formation. For the best view of the sequences, visitors should walk from Stop 6 to Stop 7.

Stop 7. Natural outcrops and large road cuts in Member A of the Rangeley Formation along Maine 4, about 2.2 to 2.6 mi (3.5 to 4.2 km) south of Rangeley village.

Stop 8. Member B and type locality of Rangeley Formation. Includes (1) Riprap quarry and nearby outcrops along Cascade Stream, 3.3 mi (5.3 km) southeast of Rangeley village, just upstream from old Greenvale Cove schoolhouse; and (2) a belt of outcrops that extends sinuously northward from the quarry to the exposed upper contact of Member A, described under Stop 7. The line of outcrops is shown on the quadrangle map (Moench, 1971).

Stop 9. Hillside sequence of sparsely spruce-wooded pavements of Member C of the Rangeley Formation that extends nearly 1,000 ft (300 m) across strike, apparently an old burn that may become increasingly covered with vegetation. The pavements extend about N20°W. from the top of Hill 2248, which is about 0.75 mi (1.2 km) southeast of the outcrops in Cascade Stream at Stop 8 (see Moench, 1971). Climb to the top of Hill 2248, then descend along the pavements.

Stop 10. Road cut and nearby outcrops of Member C of the Rangeley Formation along State Highway 4, 3.9 to 4.2 mi (6.1 to 6.7 km) east of the eastern point of Long Pond (Fig. 1).

Stop 11. Large outcrops of Perry Mountain Formation and basal Smalls Falls Formation at a sharp bend in Maine 4, 5.0 to 5.3 mi (8.0 to 8.7 km) southeast of the eastern tip of Long Pond (Fig. 1). Several good exposures of the Perry Mountain also occur between Stops 10 and 11. Perry Mountain, which rises east of Maine 4 between Stops 10 and 11, is the type locality of the Perry Mountain Formation. Park on right, watch traffic, and walk south on the left side past large outcrops of light-colored schist and quartzite (uppermost Perry Mountain) to rust-encrusted outcrops (basal Smalls Falls).

Stop 12. Public picnic area at Smalls Falls, about 1 mi (1.6 km) south of Stop 11; type locality of Smalls Falls Formation. The picnic area includes a series of large falls and plunge pools in the Sandy River and Chandler Mill Stream, a major tributary from the west. The best exposures are at the upper part of the falls in Chandler Mill Stream, which are a short walk by a well-worn path from the picnic area.

Stop 13. Outcrops of the Smalls Falls and Madrid Formations (and their contact) along Sandy River on the north side of Maine 4, 2.5 mi (4 km) east of the entrance to the Smalls Falls picnic area. These outcrops are at the west end of about 0.9 mi (1.5 km) of continuous outcrop; the east end is at Madrid village. Descend the steep bank to the Sandy River.

Stop 14. Outcrops along the Sandy River and Saddleback Stream at Madrid village; type locality of Madrid Formation. The contact between the lower and upper members is exposed in Saddleback Stream just upstream from its confluence with the Sandy River.

Stop 15. Outcrops of the Carrabassett Formation along the Sandy River at the Maine 4 bridge 1 mi (1.6 km) south of Madrid.

Stop 16. Large road cut in Hildreths Formation along Maine 142, 1.5 mi (2.4 km) north of the intersection with State Highway 4.

SIGNIFICANCE

The Rangeley area contains the type localities of several regionally extensive formations that occur within one of the best exposed and most easily seen stratigraphic sequences of lower and middle Paleozoic rocks in the northern Appalachians. This sequence, moreover, has had a major impact on mapping and tectonic modelling from Maine to Connecticut. A composite thickness of approximately 6 mi (10 km) of weakly to strongly metamorphosed metasedimentary rocks, deposited apparently

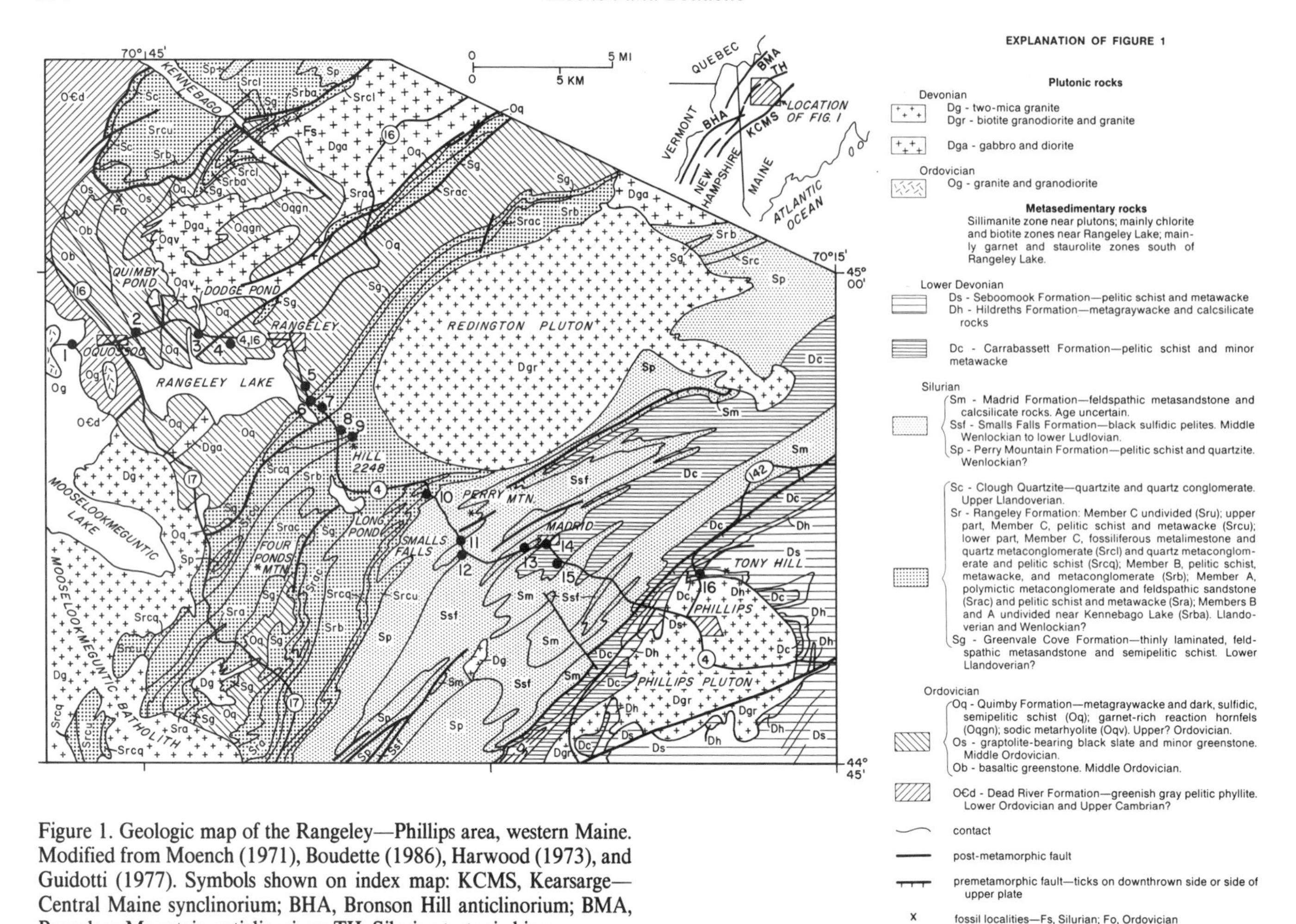

Figure 1. Geologic map of the Rangeley—Phillips area, western Maine. Modified from Moench (1971), Boudette (1986), Harwood (1973), and Guidotti (1977). Symbols shown on index map: KCMS, Kearsarge—Central Maine synclinorium; BHA, Bronson Hill anticlinorium; BMA, Boundary Mountain anticlinorium; TH, Silurian tectonic hinge.

without break from Middle Ordovician to Early Devonian time, can be seen in less than a day, ascending stratigraphically from Oquossoc to Phillips (Fig. 1). Original sedimentary features are generally well preserved. Although the finest sedimentary details are lost in the sillimanite zone, the major characteristics of all the formations are preserved at all metamorphic grades. The stratigraphy illustrates the control of sedimentation by a Silurian tectonic hinge between a source area to the northwest and a deep basin to the southeast. The source area probably was a mountain range that coincided with the present-day Bronson Hill and Boundary Mountain anticlinoria (Fig. 1, index). The ancestral basin coincided with the present Kearsarge–Central Maine synclinorium. The rocks of most of the guide area represent the deepest part of the northwestern side of the basin.

SITE INFORMATION

Formations and members are described in ascending order, according to the numbered stops (Fig. 1). Most of the field work on which this guide is based was done in the early and middle 1960s. That work was an outgrowth of largely unpublished, but pioneering geologic studies begun in 1948 and continued through the 1950s under the leadership of C. Wroe Wolfe of Boston University.

Stop 1. Dead River Formation (Upper Cambrian? or Lower Ordovician). The rocks here are strongly crenulated, greenish gray, pelitic phyllite that contains abundant irregular veinlets and pods of quartz. These rocks represent a pelitic facies of the Dead River Formation, which more typically contains subequal amounts of phyllite and quartzite. The type locality of the Dead River is in the Little Bigelow Mountain Quadrangle (Boone, 1973), and the name is now applied to equivalent rocks in western Maine originally assigned to the Albee Formation. The thickness of the Dead River Formation is unknown, but is certainly greater than 0.6 mi (1 km). The upper contact of the Dead River Formation is locally gradational, but regionally it appears to be an unconformity, for it truncates different facies of the formation. According to Boone and others (1984), sediments of the Dead River Formation were deposited as a carapace over inactive parts of an accretionary wedge composed of euxinic trench deposits and melange.

Stop 2. Black slate and basaltic greenstone (Middle Ordovician). These rocks were previously mapped as the Dixville Formation (Harwood, 1973), but they are left unnamed here until problems of regional correlation are resolved (see Moench and Pankiwskyj, 1986).

The rocks are rusty-weathering, black, sulfidic slate that locally contains Middle Ordovician graptolites (Fig. 1; Harwood and Berry, 1967). In the woods, about 200 ft (65 m) west of the road cut, strongly crenulated and brecciated phyllite of this unit is juxtaposed against foliated greenstone along a northwest-trending silicified fault zone. Massive basaltic greenstone also is exposed farther west, in road cuts near the intersection of Maine 4 and 16. To the east of Stop 2 are several small road cuts of thinly interbedded, rusty-weatheirng metagraywacke and black slate of the upper part of the black slate unit. The greenstone and black slate units comprise an intertonguing assemblage that is 7,000 to 10,000 ft (2,000 to 3,000 m) thick. The contact with the overlying Quimby Formation is gradational through several tens of ft.

The greenstone and black slate, and the felsic volcanic rocks at the base of the Quimby Formation, probably equivalent to the Ammonoosuc Volcanics and Partridge Formation (black sulfidic slate and schist of northern New Hampshire, represent an extensive Middle Ordovician assembalge of bimodal and mixed subaqueous volcanic rocks and euxinic basin sediments.

Stops 3, 4, and 5. Quimby Formation (Upper? Ordovician). These stops are in the type area of the Quimby Formation (Moench, 1969, Fig. 2). The rocks at Stop 3 are medium- to thinly-bedded volcaniclastic metagraywacke near the upper contact of the Graywacke Member of the Quimby. Stratigraphically lower rocks of the member are exposed along the highway a short distance to the west; they are more thickly bedded metagraywacke and conglomeratic metagraywacke. Rocks at Stops 4 and 5 represent the overlying shale member, composed of thinly-bedded, dark-gray to black, pelitic slate or phyllite and volcaniclastic, locally conglomeratic metagraywacke. Sodic metarhyolite, a facies of the lower member, can be seen about halfway between Stops 2 and 3, on a small knoll just northeast of the T intersection of Maine 4 and the road to Quimby Pond (permission needed from nearby farm house). The Quimby Formation is about 3,000 ft (900 m) thick in this area. Its upper contact is sharply gradational across a few centimeters.

The Quimby Formation accumulated in a euxinic basin that received abundant turbidity currents containing volcanic detritus, probably derived from dying Middle or Late Ordovician volcanoes of the anticlinoriums to the northwest.

Stop 6. Greenvale Cove Formation (Lower Silurian?, lower Llandoverian?). Stop 6 is the type locality of the Greenvale Cove Formation (Moench, 1969). The Greenvale Cove Formation, about 600 ft (200 m) thick, is composed of conspicuously interlaminated light- to medium-gray, semipelitic phyllite, and laminated, somewhat calcareous metasandstone. The upper contact is exposed on the hillside about 2,000 ft (600 m) to the northeast of Stop 6 (Moench, 1971). Here, the contact is sharp and conformable, but the uppermost 100 ft (30 m) of the

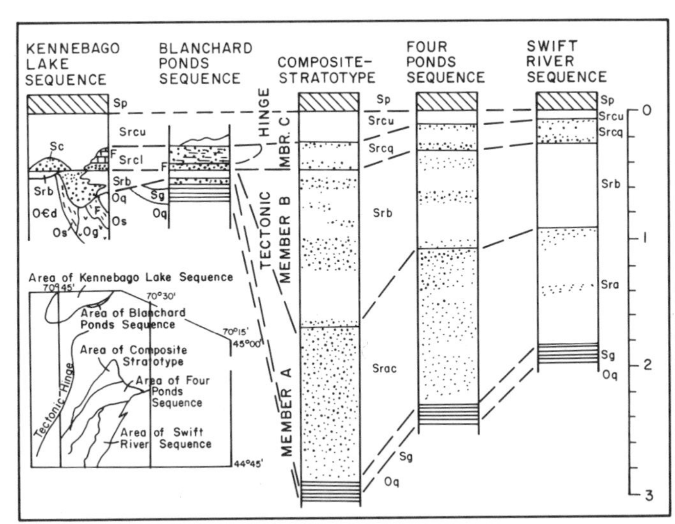

Figure 2. Stratigraphic columns showing internal facies of the Rangeley Formation and relationships to rocks here assigned to the Clough Quartzite. Unit symbols explained on Figure 1. Patterns of Rangeley Formation: coarse stipple, metaconglomerate; fine stipple, metasandstone; unpatterned, metashale; blocks, impure metalimestone.

Greenvale Cove contains scattered 3-ft-thick (meter-thick) lenses of Rangeley-like coarse-grained metasandstone.

The Greenvale Cove Formation is interpreted to represent a thin apron of possibly west-derived deltaic sediments; it is transitional to a new, Silurian sedimentary regime dominated by great differential uplift in the northwest and subsidence in the southeast.

Stops 7 to 10. Rangeley Formation (Lower Silurian, Llandoverian and Wenlockian?) and Clough Quartzite (Lower Silurian, upper Llandoverian). Figure 2 illustrates the complex facies variations within Members A, B, and C of the Rangeley Formation and the relationship of Member C to quartzite at Kennebago Lake assigned to the Clough Quartzite. The Clough of Figures 1 and 2 was previously mapped as a nearshore facies of the upper member of the Rangeley Formation (Moench and Pankiwskyj, 1986). Although the nearest known Clough is at its type locality near Littleton, New Hampshire, 75 mi (125 km) to the southwest, the quartzite at Kennebago Lake is similar to the type Clough in mode of occurrence, and age. The named sequence shown in each column on Figure 2 is a diagrammatic reconstruction of mapped relationships in the areas shown in the index. Stop 8 includes the type locality of the Rangeley Formation (Osberg and others, 1968, p. 251), at Cascade Stream. However, only part of one member (Member B) of the formation is represented along the stream. In order to provide a more complete description of the formation, three reference sections are established herein, as follows: reference section A, at Stop 7, to describe the lower part of the Rangeley Formation (Member A); reference section B at and near Stop 8, to describe the middle part of the formation (Member B); and reference section C, at Stop 9, to describe the upper part of the Rangeley (Member C).

Stop 7 (reference section A) is in the conglomerate and sandstone facies of Member A, which is 4,000 ft (1,200 m) thick south of Rangeley. Along Maine 4 south of Stop 6, beyond the southernmost outcrop of the Greenvale Cove Formation and a short covered interval, is a low but conspicuous, buff-colored outcrop of massive, coarse-grained, feldspathic metasandstone in Member A of the Rangeley Formation. This outcrop is succeeded to the south by extensive outcrops (mostly in woods) of exceptionally thickly bedded, massive metasandstone that becomes increasingly conglomeratic southeastward and stratigraphically upward; farther along is a long road cut of polymictic boulder to cobble metaconglomerate, about in the middle of Member A (Fig. 2). The clasts are of a wide variety of sedimentary, volcanic, and plutonic rocks, derived from the eroded region to the northwest along the Boundary Mountain anticlinorium. Most clasts are rounded but some are slabby. In the composite-stratotype (Fig. 2), Member A is about 4,000 ft (1,200 m) thick. The upper contact of Member A is exposed on a steep hillside about 3,000 ft (900 m) due north of Stop 8 (Moench, 1971). Here, polymictic boulder conglomerate is conformably and sharply succeeded by rusty-weathering, feldspathic metasandstone and gray pelitic schist of Member B.

Bedding styles at Stop 7 are characteristic of subaqueous conglomerates of the turbidite association (see Walker, 1975): particularly the great thickness of individual beds, internal massiveness and homogeneity, scattered out-sized cobbles in sandstone, and inverse-to-normal grading. These features and the relationships shown on Figure 2 indicate that Member A represents a large body of northwest-derived, subaqueous fanglomerate.

Outcrops at Stop 8 expose a large part of Member B of the Rangeley Formation. In the composite stratotype (Fig. 2), Member B is about 4,000 ft (1,200 m) thick and contains a lower part, about 1,500 ft (450 m) thick, composed of rusty-weathering, dark gray pelitic schist and thin to 3-ft-thick (meter-thick) beds of metasandstone, and an upper part in which similar rocks are interstratified with abundant quartz-rich, polymictic conglomerate. The lower part can be seen at the quarry and to the north, and characteristic rocks of the upper part are exposed along Cascade Stream. The schist at Stop 8 contains small porphyroblasts of staurolite, retrograded to chlorite and sericite. In the stream exposures, unsorted, schist-matrix, pebble-cobble conglomerate, interpreted as conglomeratic mudflow deposits, is interstratified with evenly-bedded pelitic schist and wacke. The conglomerate is polymictic, but it contains a much greater proportion of stable clasts (quartzite and vein quartz) than conglomerate of Member A. Also exposed in the stream pavement is a slump fold whose southeast limb is erosionally truncated by an intraformational unconformity. Metasandstone beds of Member B have characteristic features of intermediate to proximal turbidites: common poor grading with almost sharp or abruptly graded upper contacts (example in Hatch and others, 1983, Plate 1); ripup clots, load- and flute-casts, and sandstone dikes also occur. The conformable upper contact of Member B is marked by the first appearance of quartz conglomerate of the lower part of Member C.

Regionally, Member B, like A, thins abruptly and becomes more coarsely conglomeratic northwestward across the tectonic hinge; southeastward it thins gradually, probably by the loss of coarse clastic deposits (Fig. 2). The thick body of Member B conglomerate in the Kennebago Lake sequence contains large boulders; this body may have filled a deep channel on the shelf. The tectonic environment of Member B was similar to that of Member A, but the rocks express a somewhat reduced supply of gravel that was compositionally more mature than that of Member A, and a northwestward transgression of marine basin conditions and pelitic sedimentation.

In the composite stratotype, Member C is about 1,500 ft (450 m) thick. The rocks at Stop 9, mainly the lower part of the member, are interbedded dark gray, pelitic schist and thin to thick beds of feldspathic quartzite, and lenticular beds, as much as 10 ft (3 m) thick, of closely-packed quartz granule, pebble, and cobble conglomerate. The clasts are vein quartz, quartzite, and locally derived ripup mud clots that are now pelitic schist; rare cobbles of felsic volcanic rock may be found in the lowermost beds of the member. In the road cut at Stop 10, quartz pebble conglomerate, at the top of the lower part of Member C, is exposed at the nose of an anticline that plunges steeply northeast. The conglomerate is succeeded on both limbs of the anticline by characteristic rocks of the upper part of Member C. The upper part of Member C, composed of rusty-weatheirng, dark gray schist and typically massive, poorly graded feldspathic quartzite, grades by interbedding to the overlying Perry Mountain Formation, composed of light gray, muscovite-rich schist and nearly white, commonly cross-laminated quartzite. This transition can be seen in roadside outcrops within 0.2 mi (0.3 km) in both directions from the quartz conglomerate at Stop 10.

Quartz conglomerates at Stops 9 and 10, respectively, represent more distal marine basin facies. To the northwest across the tectonic hinge, equivalent quartz conglomerates of Member C pass into an inferred shelf facies (Blanchard Ponds sequence and southeast side of Kennebago Lake sequence), then to a shore-line facies (northwest side of the Kennebago Lake sequence) represented by the Clough Quartzite. Southwest of Kennebago Lake, the Clough rests unconformably on the northwest side of a body of coarse polymictic conglomerate in Member B, as mapped by Harwood (1973).

The shelf facies is composed of interbedded quartz conglomerate, feldspathic quartzite, and dark pelitic hornfels succeeded by an apparently local deposit of laminated impure limestone, now calc-silicate hornfels. Shelly fossils found by Robert J. Willard and Boudette occur in quartz conglomerate at the three indicates localities near the southeast end of Kennebago Lake (Fig. 1). Arthur J. Boucot (written communication, 1964) identified the fossils and assigned them a late Llandoverian age (brachiopod zones C4–C5). He assigned a less restricted span of brachiopod zones (C3–C5) to fossils that Boudette recovered from calc-silicate hornfels northeast of the area of this guide. These localities are extremely important, but they are not shown as a stop because they are very difficult to find.

The Rangeley–Perry Mountain transition expresses a gradual change from proximal to more distal sedimentation by turbidity flows that carried increasingly mature detritus. North of the hinge (Fig. 2), minor regression followed deposition of the pelitic and conglomeratic rocks of Member B; then, after stillstand when the shelf and shoreline deposits of the lower part of Member C and the Clough Quartzite accumulated, the sea margin transgressed northwestward.

Stops 11 and 12. Perry Mountain Formation (Silurian, Wenlockian?) and Lower Member of Smalls Falls Formation (Silurian, middle Wenlockian to lower Ludlovian).

Stop 11 displays typical rocks of the upper part of the Perry Mountain Formation and basal rocks of the Smalls Falls Formation. Stop 12 is the type locality of the Smalls Falls Formation. Although the Perry Mountain has not yielded fossils, it lies between the fossil-dated Lower Silurian (Llandoverian) Rangeley and Silurian Smalls Falls formations. The Smalls Falls is dated by graptolites found by Allan Ludman and J. R. Griffin in central Maine, in sulfidic slate originally mapped as the Parkman Hill Formation, but now assigned to the Smalls Falls (Pankiwskyj and others, 1976; Osberg and others, 1985; Moench and Pankiwskyj, 1986).

Rocks of the Perry Mountain Formation at Stop 11 are sharply interbedded white quartzite and light-colored, muscovite-rich pelitic schist that contains large retrograde pseudomorphs after staurolite and andalusite. The quartzites display graded bedding, and cross- and convolute-lamination in partial or complete Bouma turbidite sequences. Some beds are trough-cross laminated from top to bottom, which suggests reworking by bottom currents. The amount of quartzite in the Perry Mountain increases stratigraphically upward and northward in the region. The formation is about 2,000 ft (600 m) thick in this area. The upper contact of the Perry Mountain (covered at Stop 11) is sharp and conformable.

Outcrops of the Smalls Falls are characteristically rust encrusted; they produce an obnoxious sulfurous odor when broken. The lower member, about 2,200 ft (650 m) thick, is represented at Stops 11 and 12. The rocks are thinly interbedded sulfidic-graphitic schist and quartzite, and scattered 3-ft-thick (meter-thick) beds of massive quartz grit, which are more abundant farther north. The thin quartzite beds are graded and cross-laminated turbidites. Because of the abundant pyrrhotite (5%–10%), the Smalls Falls typically has iron-poor and magnesium-rich silicate mineral assemblages (Guidotti and Cheney, 1976). Accordingly, in the staurolite zone, as at Stops 11 and 12, staurolite and almandine are absent.

The abrupt contact between the Perry Mountain and Smalls Falls, both northwest-derived turbidite sequences, marks a change from well aerated to strongly reducing conditions, probably the result of abrupt subsidence that produced a closed, euxinic basin.

Stops 13 and 14. Upper Member of Smalls Falls Formation (Silurian, middle Wenlockian to lower Ludlovian) and Madrid Formation (Silurian?). The upper member of the Smalls Falls Formation is exposed at Stop 13. The member, about 500 ft (150 m) thick, is composed mainly of sulfidic, black to white calc-silicate rock. The original rocks were carbonate-cemented sandstone and siltstone that were deposited in a strongly reducing environment. Thinly laminated, garnet-rich ironstone, a metamorphosed chemical deposit, occurs near the upper contact. The upper contact of the member is sharp and conformable. It is marked by an upward decrease in the amount of pyrrhotite, and by the presence of lenticular beds of coarse-grained calcareous metasandstone, some with edgewise chip conglomerate at and near the base of the Madrid Formation. These features signal the arrival of energetic, oxygenated currents that ended the stagnant conditions of Smalls Falls deposition.

The Madrid Formation has not yielded fossils. It is assigned a Silurian(?) age on the basis of correlation with the well-dated Fitch Formation (Silurian, late Ludlow, and Pridoli) near Littleton, New Hampshire (Harris and others, 1983; Hatch and others, 1983).

The lower member of the Madrid, about 300 ft (100 m) thick, is exposed along the Sandy River between the basal contact at Stop 13 and the upper contact at Stop 14, in Saddleback Stream just upstream from the Sandy River. This member is composed mainly of thinly bedded, white, bluish, and violet calc-silicate rocks and semipelitic schist, which represent metamorphosed calcareous, siliciclastic sediments. Thick beds of metasandstone occur near the lower contact. The upper contact is marked by 7 ft (2 m) of black sulfidic schist, sharply succeeded by rocks of the upper member.

The upper member, about 650 ft (200 m) thick at the type locality, is composed of thickly bedded feldspathic metasandstone characterized by abundant lenses of calc-silicate rock, and about 20% of gray pelitic schist. The beds are graded; medium-scale cross lamination, too faint to see in shaded outcrops, can be seen in several places. The upper contact of the Madrid is covered at Stop 14. Elsewhere it is gradational by interbedding with gray pelitic rocks of the Carrabassett Formation.

The Madrid Formation represents a thick blanket of quartzo-feldspathic sandstone and subordinate shale known to extend from near the Maine–New Brunswick border to Massachusetts. Regional lithofacies suggest that provenance shifted near the beginning of Madrid time from the northwestern source to a volcanic tract in the east.

Stop 15. Carrabassett Formation (Lower Devonian). The type locality of the Carrabassett is in the Little Bigelow Mountain Quadrangle (Boone, 1973). Boone and Moench (in Moench and Pankiwskyj, 1986) have reassigned the upper calc-silicate member of Boone (1973) to the Hildreths Formation. Although no datable fossils have been found in the Carrabassett, the formation is correlated with the fossiliferous Littleton Formation, of Early Devonian age (Hatch and others, 1983; Harris and others, 1983).

The predominantly pelitic Carrabassett Formation is about 3,700 ft (1,300 m) thick in the Phillips Quadrangle. The rocks exposed directly under the bridge at Stop 15 are dark gray, staurolite-bearing pelitic schist that displays faint cyclic-graded bedding. Each bed, interpreted as a mud-silt turbidite, is several centimeters thick and is continuously graded from bottom to top. As shown by graded bedding, these rocks are underlain downstream by a more thickly bedded sequence of Madrid-like metasandstone and gray pelitic schist, but mapped within the Carrabassett Formation. Outcrops of massive, gray, pelitic schist and sparse thin- to thick-graded beds of wacke occur at a stratigraphically higher level, about 600 to 2,200 ft (200 to 700 m) upstream from Stop 15. Boone (1973) also described an impure metaquartzite member, as lenses that might be cross sections of shoestring sands; and an extensive, arenaceous, thinly layered semipelite member, which lies above the massive metapelite member and conformably below the Hildreths Formation.

Stop 16. Hildreths Formation (Lower Devonian). Stop 16 is in the sillimanite zone of metamorphism, near the contact of the Phillips pluton. The type locality of the Hildreths Formation is in the Dixfield Quadrangle, south of the Phillips Quadrangle (Osberg and others, 1968, p. 250).

The Hildreths Formation is typically only about 300 ft (100 m) thick and locally discontinuous. In the road cut, the rocks are dark metagraywacke (feldspathic biotite granofels), rusty biotite schist, two-mica schist, and coarsely crystallized, idocrase-bearing calc-silicate rock. White marble that was roasted for agricultural purposes at a small lime kiln south of the road cut was dug from a line of trenches up hill to the east. Regionally, the proportions of metagraywacke, calc-silicate rocks, and marble vary greatly, and calcareous rocks are locally absent. In the woods southeast of the road cut are several small outcrops of interbedded dark metagraywacke and schist of uppermost Carrabassett Formation.

The Hildreths represents an unusual mixed assemblage of probably volcaniclastic graywacke and carbonate rocks that occurs within an otherwise monotonous, thick sequence of regularly bedded mudstone, shale, siltstone, and wacke of the underlying Carrabassett and overlying Seboomook formations.

The overlying Seboomook Formation is widely exposed in the southeast corner of the Phillips Quadrangle, but no large pavement outcrops are conveniently accessible. Spectacular open pavements of the Seboomook occur, however, on Bald Mountain in the Dixfield Quadrangle. Follow Maine 142 south to Weld, in the Dixfield Quadrangle. The trail head to Bald Mountain is on Maine 156, about 5 mi (8 km) southeast of Weld village. The type locality of the Seboomook Formation is in northwestern Maine (Boucot, 1961), where it has been firmly dated by fossils at several localities. The Seboomook Formation of the Phillips and Dixfield quadrangles is similar to the Carrabassett Formation in style of bedding, but it is much more arenaceous than the Carrabassett. Pankiwskyj (1979) defined three members of the Seboomook, which for simplicity are not shown on Figure 1.

The Carrabassett-to-Seboomook sequence coarsens gradually upward, except for interruption by the thin Hildreths Formation, and it coarsens southeastward (Pankiwskyj, 1979; Moench and Pankiwskyj, 1986). Available data suggest that the bulk of these sediments came from the southeast, possibly from the newly emergent coastal volcanic belt.

REFERENCES CITED

Boone, G. M., 1973, Metamorphic stratigraphy, petrology, and structural geology of the Little Bigelow Mountain map area, western Maine: Maine Geological Survey Bulletin 24, 136 p.

Boone, G. M., Boudette, E. L., Hall, B. A., and Pollock, S. G., 1984, Correlation and tectonic significance of pre-Middle Ordovician rocks of northwestern Maine: Geological Society of America Abstracts with Programs, v. 16, p. 4.

Boucot, A. J., 1961, Stratigraphy of the Moose River synclinorium, Maine: U.S. Geological Survey Bulletin 1111-E, p. 153–188.

Boudette, E. L., 1986, Bedrock geology of the Kennebago Lake Quadrangle, Somerset County, Maine: U.S. Geological Survey Professional Paper (in press).

Guidotti, C. V., 1977, Geology of the Oquossoc 15-minute Quadrangle, west-central Maine: Maine Geological Survey Open-File Report 77-2, 26 p.

Guidotti, C. V., and Cheney, J. T., 1976, Margarite pseudomorphs after chaistolite in the Rangeley area, Maine: American Mineralogist, v. 61, p. 431–434.

Harris, A. G., Hatch, N. L., Jr., and Dutro, J. T., Jr., 1983, Late Silurian conodonts update the metamorphosed Fitch Formation, Littleton area, New Hampshire: American Journal of Science, v. 283, no. 7, p. 722–738.

Harwood, D. S., 1973, Bedrock geology of the Cupsuptic and Arnold Pond Quadrangles, west-central Maine: U.S. Geological Survey Bulletin 1346, 90 p.

Harwood, D. S., and Berry, W.B.N., 1967, Fossiliferous lower Paleozoic rocks in the Cupsuptic Quadrangle, west-central Maine: U.S. Geological Survey Professional Paper 575-D, p. D16–D23.

Hatch, N. L., Jr., Moench, R. H., and Lyons, J. B., 1983, Silurian–Lower Devonian stratigraphy of eastern and south-central New Hampshire; Extensions from western Maine: American Journal of Science, v. 283, no. 7, p. 739–761.

Moench, R. H., 1969, The Quimby and Greenvale Cove formations in western Maine: U.S. Geological Survey Bulletin 1274-L, p. L1–L17.

——, 1971, Geologic map of the Rangeley and Phillips Quadrangles, Franklin and Oxford counties, Maine: U.S. Geological Survey Miscellaneous Geologic Investigations Map I—605, scale 1:62,500.

Moench, R. H., and Pankiwskyj, K. A., 1986, Geologic map of western interior Maine; with contributions by G. M. Boone, E. L. Boudette, Allan Ludman, W. R. Newell, and T. I. Vehrs: U.S. Geological Survey Miscellaneous Investigations Map I-1692, scale 1:250,000 (in press).

Osberg, P. H., Moench, R. H., and Waner, J., 1968, Stratigraphy of the Merrimack synclinorium in west-central Maine, *in* Zen, E-an, White, W. S., Hadley, J. B., and Thompson, J. R., Jr., eds., Studies of Appalachian geology; Northern and maritime: New York, Interscience Publishers, p. 241–253.

Osberg, P. H., Hussey, A. M., II, and Boone, G. M., eds., 1985, Bedrock geologic map of Maine: Maine Geological Survey, Department of Conservation, scale 1:500,000.

Pankiwskyj, K. A., 1979, Geologic maps of the Kingfield and Anson quadrangles, Maine: Maine Geological Survey Map GM-7, scale 1:62,500.

Pankiwskyj, K. A., Ludman, A., Griffin, J. R., and Berry, W.B.N., 1976, Stratigraphic relations on the southeast limb of the Merrimack synclinorium in central and eastern Maine: Geological Society of America Memoir 146, p. 263–280.

Walker, R. G., 1975, Generalized facies models for resedimented conglomerates of turbidite association: Geological Society of America Bulletin, v. 86, p. 737–748.

Structure and stratigraphy of the Central Maine Turbidite Belt in the Skowhegan-Waterville region

Allan Ludman, Department of Geology, Queens College of the City University of New York, Flushing, New York 11367
Philip H. Osberg, Department of Geology, University of Maine, Orono, Maine 04469

LOCATION

The site consists of eight stops in the Kingsbury, Skowhegan, and Waterville Quadrangles of central Maine (Fig. 1). A current roadmap and geologic maps of these quadrangles (Osberg, 1968; Ludman, 1977, 1978) are recommended for navigation and for a more complete geologic description than can be given here. Stop locations follow, along with driving instructions for travel between stops.

Stop 1. NW Kingsbury Quadrangle at intersection of Maine 16 and 151 (45°06′12″N, 69°41′27″W). Drive south on Maine 151 to Athens and turn left (east) onto Maine 150. Drive 3.75 mi (6 km) to large roadcuts on south side of road at Lords Hill.

Stop 2. Skowhegan Quadrangle, series of outcrops on Maine 150 between Lords Hill and Harmony town line. Drive west on Maine 150 through Athens, park just west of Wesserunsett Stream, and walk north up long driveway to ask permission at the house. Stop 3 is in the stream east-northeast of the house.

Stop 3. Skowhegan Quadrangle in Wesserunsett Stream 800 to 1,000 ft (250 to 300 m) north of Maine 150 (44°55′28″N, 69°40′22″W. Follow Maine 150 south to Skowhegan and turn east onto U.S. 2. Park east of Coburn Park on south side of the Great Eddy of the Kennebec River.

Stop 4. SW Skowhegan Quadrangle at Great Eddy of Kennebec River and along U.S. 2 to the east (44°46′19″N, 69°42′30″W). Take U.S. 2 through Skowhegan and follow Maine 201 south to I-95, exiting at Exit 33 in Waterville. Turn left onto Kennedy Drive at stop sign and drive 0.9 mi (1.6 km) to outcrops on north side of road.

Stop 5. Waterville Quadrangle, road cut at intersection of Kennedy Drive and First Rangeway (44°32′N, 69°39′45″W). Continue east 0.6 mi (1 km). Pass light, cross bridge, and bear right onto Grove Street. After 0.5 mi (0.8 km) turn left onto Water Street. Travel 0.8 mi (1.4 km) and turn left onto Maine 201. Cross Kennebec River and turn left at traffic light. Drive 3.8 mi (6.8 km) north, cross Routes 100/139/11 at Benton Station, and continue north. Park at railroad crossing and walk tracks to the northeast.

Stop 6. Waterville Quadrangle, railway cut 1,200 ft (370 m) from River Road (44°35′47″N, 69°34′49″W). Retrace route to Maine 201 and take it south. After 0.2 mi (0.4 km) park in lot on right.

Stop 7. Waterville Quadrangle on Maine 201 1,000 ft (325 m) south of Waterville-Winslow bridge; exposure is over steep bank at river's edge (44°37′42″N, 69°37′32″W). Continue south on Maine 201 0.4 mi (0.75 km) and turn left onto Maine 137. Drive 2.5 mi (4.5 km) and park at junction of Patten Pond Road.

Stop 8. Waterville Quadrangle, roadcut at junction of Maine 137 and Patten Pond Road (44°31′24″N, 95°35″W).

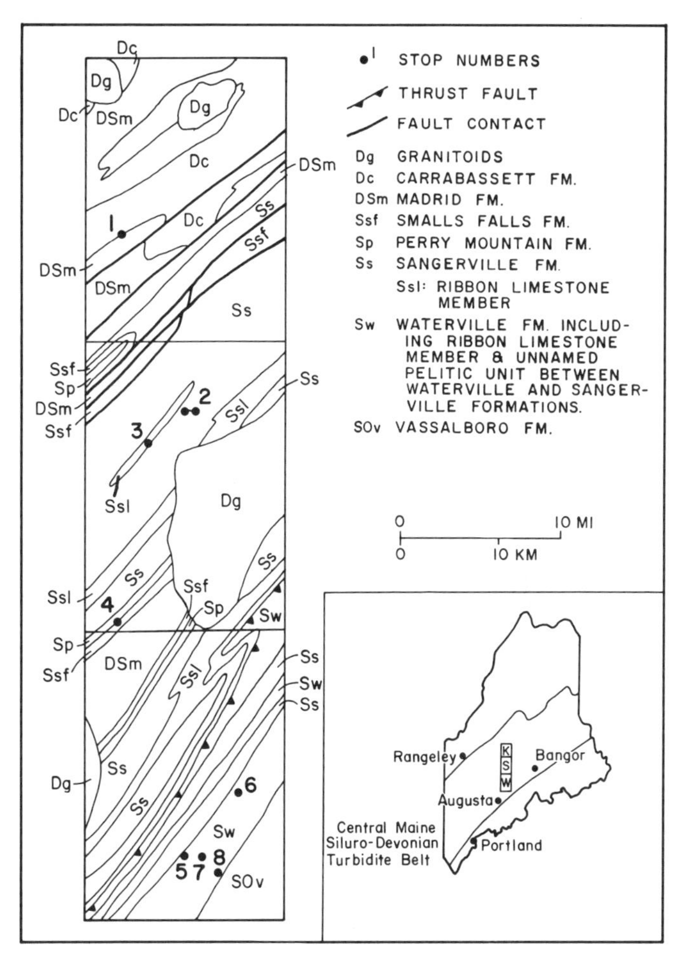

Figure 1. Simplified geologic map of the Kingsbury-Skowhegan-Waterville area. Inset shows location of Kingsbury (K), Skowhegan (S), and Waterville (W) Quadrangles.

SIGNIFICANCE

Siluro-Devonian turbidites underlie the Kearsarge–Central Maine Synclinorium (previously the Merrimack Synclinorium) in a belt 75 mi (120 km) wide across strike in Maine (inset, Fig. 1). This belt narrows rapidly to the south, but its rocks can be traced into the high-grade, complexly deformed terrain of northern Connecticut. In contrast, the sequence in central Maine is widely exposed at chlorite grade. As a result, primary sedimentary features are generally well preserved and graptolites have been collected from several localities. Two major contributions to the geology of the Northern Appalachians have come from the Central Maine Turbidite Belt, and particularly from the quadrangles covered in this site: (1) an understanding of a complex stratigraphy, complete with faunal age control, that is used to postulate post-Taconian facies relationships in eastern New England and to date rocks as far south as Connecticut; and (2) a multiphase deformation history that serves as a model for the Early Devonian Acadian orogeny. This site presents evidence for both structural and stratigraphic interpretations.

SITE DESCRIPTION

This site consists of eight outcrops that provide a traverse across the turbidite belt (Fig. 1) illustrating the lithologies and the styles and sequence of folding events. Several formation and intraformational member contacts are included so that detailed evidence is also presented for the interpreted stratigraphic sequence.

Stratigraphy. The stratigraphy of central Maine has been reconstructed from two continuous but geographically and structurally disjointed sections. The outcrops at Stops 1 through 4, in the Kingsbury and Skowhegan Quadrangles, define the upper part of this sequence whereas at Stops 5 through 8 in the Waterville area, they fix the lower part (Figs. 1, 2). A thrust fault probably separates the two parts of the section; thus they may have been more widely separated during deposition than present locations suggest. The upper part of the sequence is very similar to the section in the Rangeley area of western Maine (Moench and Boudette, 1970), and fossils from rocks in the Skowhegan-Kingsbury area have been used to pinpoint the ages of these western Maine correlatives. The lower part of the central Maine sequence, however, is less convincingly comparable to that of western Maine, and some stratigraphic problems remain. The central Maine Silurian rocks are interpreted as intermediate (and distal?) equivalents of the proximal rocks of western Maine (Pankiwskyj and others, 1978). The basin is thought to have been filled mostly from the west during Silurian times, but a shift to an eastern source during the Early Devonian has been suggested by Hall and others (1976).

Structure. Three episodes of folding are recognized in these Siluro-Devonian rocks: early recumbent folds (F_1), an intermediate set of upright folds (F_2), and late asymmetrical folds (F_3).

No large-scale folds attributable to the youngest event have been identified, although a well-developed S_3-cleavage is found throughout the region. Axial surfaces of F_3-folds strike within 10° of north and dip steeply. Bedding and earlier cleavages/schistosities are deformed, so that plunge and asymmetry of F_3-folds vary according to the attitude of the surface that has been deformed. These asymmetrical folds deform small dikes and aplites associated with the Hallowell Pluton (Osberg and others, 1985), and

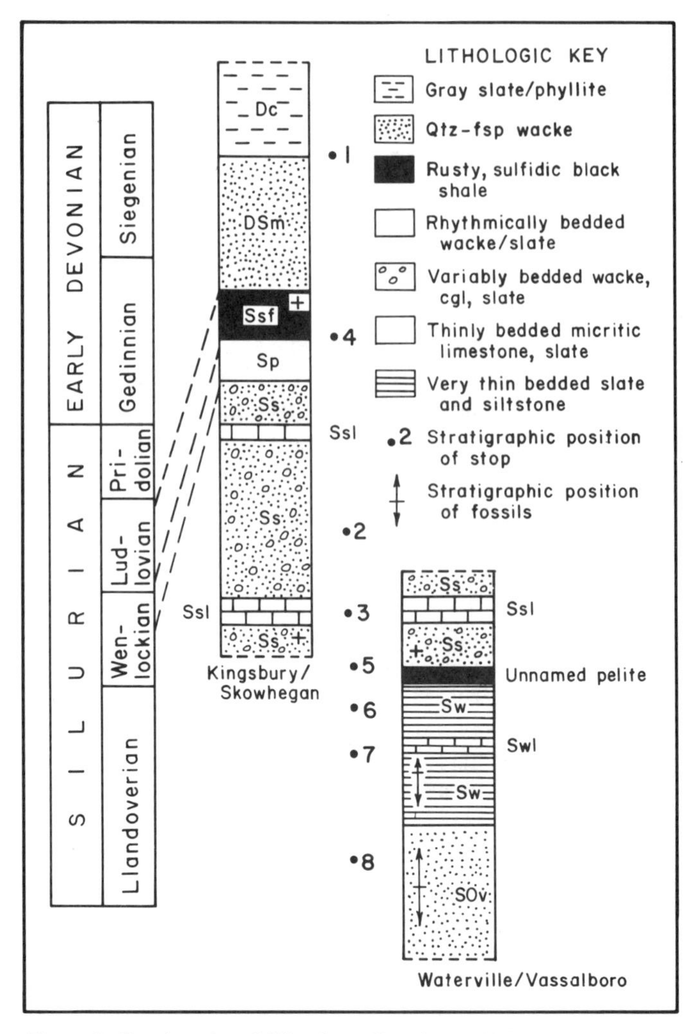

Figure 2. Stratigraphy of Kingsbury-Skowhegan-Waterville area. Formation symbols as in Figure 1.

that pluton has been dated as 387 ± 11 Ma by Dallmeyer and van Breemen (1978). Muscovite and sillimanite have grown in the axial plane cleavages of these folds. This metamorphic event (M_2 of Holdaway and others, 1982) has been dated by $^{40}Ar/^{39}Ar$ methods on hornblendes as ca. 360 Ma (Dallmeyer and van Breemen, 1978). The F_3-event is thus latest Devonian or Earliest Carboniferous in age.

The F_2-upright folds are recognized on both mesoscopic and regional scales, and control the regional map pattern (Osberg and others, 1985). The F_2-folds range from relatively open to highly flattened isoclinal structures, with form depending on the competence of the rocks affected. F_2-axial surface strike northeast and dip steeply. Pressure solution cleavage in low-grade rocks and a strong schistosity at higher grades parallel these axial surfaces. Plunges are generally less than 25° but some reach 45°. These folds face upward throughout some areas, but downward in others, indicating the presence of earlier folds. F_2-folds affect rocks as young as Siegenian, or possibly Emsian (Osberg and others,

1985), and are cut by plutons as old as 400 Ma (D. R. Lux, personal communication, 1984). F_2 is thus clearly associated with the Acadian orogeny. A weak Acadian metamorphism (M_1 of Holdaway and others, 1982) is synchronous with F_2-folding but is overprinted by the younger M_2-metamorphism.

Evidence for the F_1-recumbent folds is largely indirect. Local map patterns, downward-facing upright folds, and stratigraphic relationships have been used to infer early recumbent axial surfaces. Only at a single outcrop (see Stop 7) do minor folds display evidence of this episode of deformation. Little can be said about the form or vergence of F_1-folds, but the regional extent of downward-facing structures indicates that they are large-scale features. F_1-folds are probably of early Acadian origin but may represent large sedimentary slump structures.

STOP DESCRIPTIONS

Stop 1: Contact between the Carrabassett and Madrid Formations. Begin at the north side of Maine 16 at the culvert and walk west. Thick, massive beds of pale purplish gray, calcareous biotite-bearing wacke exposed at the culvert are typical of the Madrid Formation in this area. These pass westward into well-graded sandstones, siltstones, and very subordinate phyllites typical of the top of the formation. Calc-silicate pods, layers, and stringers are abundant here and throughout the entire Madrid. The beds thin and the amount of pelite gradually increases over about 250 ft (75 m) toward the west, until sandstone comprises only 60 percent of the outcrops. A consistent northwest facing is indicated here by graded beds, cross-bedding, flame, and cut-and-fill structures.

Cross to the south side of Maine 16 at the break in outcrop (the exposures are continuous in the woods to the north). The first exposures here are of the basal Carrabassett Formation: medium gray "clean" phyllite with very sparse, light gray siltstone laminae. A few believable graded beds indicate continued northwest facing, confirming that the Carrabassett is younger than the Madrid. Farther west, siltstone laminae become thicker and more abundant but remain subordinate to the phyllite.

Although no folds are visible here, S_2- and S_3-cleavages are well developed in the Carrabassett phyllites. They are characteristically only poorly exhibited by the Madrid sandstones. S_2 is dominant here and is nearly vertical. The phyllitic sheen parallel to S_2 is defined by aligned muscovite and biotite, and is attributed to the earliest metamorphic event (M_1). Small chloritoid and garnet porphyroblasts were also caused by M_1. A very steep north-trending cleavage (S_3) cuts S_2 in the pelites, and steep S_2/S_3-intersection lineations are prominent. Retrogradation of chloritoid to biotite and chlorite here is noted in thin section and is attributed to M_2.

Stop 2: Typical Sangerville Formation lithologies. Most of the Sangerville Formation consists of variably bedded calcareous, ankeritic quartzofeldspathic wackes interbedded with medium gray slates. Graded bedding is common, as are several other primary features associated with turbidites. The rocks here are typical of the finer grained, graded Sangerville sandstones and exhibit sole markings and cross-laminations. A gradual darkening of color related to original grain size and clay mineral content clearly denotes facing direction and is best seen on the glacially polished top of the exposure. S_2-cleavage is again dominant here, but the north-trending S_3 is also visible. A problematic subhorizontal cleavage is also present and has been assigned to a later deformational phase by Griffin (1973).

Drive east and park on the south shoulder at the power substation. Cross to the north side of the road and walk west to the outcrops. Several distinctive subordinate rock types occur within the Sangerville, two of which are well exposed here. Granule conglomerate containing abundant feldspar and lithic clasts is found locally in the lower part of the formation. The granule conglomerates occur either at the bases of Bouma sequences (as along Carson Hill just north of this outcrop), or as they do here—as massive beds with little sign of grading. Highly carbonaceous pyritiferous pelites commonly occur with these rocks, as is the case here. Two generations of pyrite are visible: early grains are flattened in S_2, but a younger set occurs as undeformed cubes. Graptolites from the black shales at this locality include two monograptid species that suggest a Wenlock age for this exposure.

Stop 3: Ribbon limestone member of the Sangerville Formation showing typical rock types and superb F_2 folds. These ribbon limestone exposures consist of gray micrite intercalated with noncalcareous medium gray slates and buff weathering calcareous sandstones. The micrite is dominant and occurs in beds 0.4 to 6 in (1 to 15 cm) thick. Sandstones range from 0.4 to 3 in (1 to 8 cm) thick and slates tend to be very thin. Sandstone and slate are less soluble than the micrite and stand out as ribs, whereas the limestone beds form narrow troughs. Crinoid columnal segments have been collected from this outcrop but have not provided useful age control.

These outcrops form a natural block diagram in which F_2-folds are magnificently displayed. They are very tight, with sharp hinges and small interlimb angles typical of F_2-folds in incompetent rocks. Plunges are gentle to the south. Cross-beds in some of the thicker sandstone beds indicate that the folds face upward, and are thus on the upright limb of an inferred F_1-fold.

Stop 4: Contact between the Perry Mountain and Smalls Falls Formations, with well-developed F_2-folds. Rocks exposed close to U.S. 2 are typical of the Perry Mountain Formation: rhythmically interbedded wackes and light gray phyllite in graded beds 1 to 4 in (2.5 to 10 cm) thick. Sandstone dominates over phyllite in all beds and cross-laminations are well developed. A few elliptical calcareous pods up to 12 in (30 cm) long are present in the wackes. Small, equant porphyroblasts best seen in the phyllite but also present in the wacke, are prograde pseudomorphs of biotite, muscovite, and quartz after ankerite. Nearly isoclinal F_2-folds are prominent, and primary facing features show that all face upward. A vertical north-trending S_3 cleavage is weakly developed in the phyllites.

Thin-bedded carbonaceous sulfidic slates of the Smalls Falls

Formation crop out at the river's edge and are in sharp contact with the Perry Mountain. A few thin sandstone beds separate the slates, but thicker bedded Smalls Falls rocks are well exposed along U.S. 2 just east of the Eddy and upsection from the river outcrops. These roadcuts are massive sulfidic quartzose sandstones in beds 4 in to 3 ft (10 cm to 1 m) thick, separated by rusty weathering sulfidic slates 1 to 12 in (2.5 to 30 cm) thick. The sequence here is almost identical to that described by Moench and Boudette (1970) at Smalls Falls in western Maine (and in this volume), and is critical in establishing regional correlations.

Stop 5: Contact between Sangerville Formation and an unnamed pelite unit, establishing relationships between Waterville area sequence and that of Skowhegan/Kingsbury. Rusty weathering, dark gray quartz-mica phyllite interbedded with dark gray quartzose layers typical of the unnamed unit is exposed at the east end of the outcrop. Pyrite is abundant. Crystals have a common orientation and do not have associated quartz-filled pressure shadows. Beds range from 0.5 mm to 1.5 cm in thickness.

The unnamed unit is in contact with the Sangerville Formation at the middle of the outcrop. The Sangerville here consists of light gray, slightly calcareous wacke and quartz-mica phyllite. Beds are 6 to 20 in (15 to 50 cm) thick and commonly grade from arenaceous bottoms to phyllite tops. Sandstone-phyllite ratios range from 1:1 to 8:1. Rip-up clasts and small feldspar grains are visible at the bases of some beds. Graded beds are best viewed at the top of the exposure and indicate that the Sangerville overlies the unnamed unit. Both units face west.

Because outcrops of the Waterville Formation can be seen in a drainage ditch that protects the airport runway just south of Kennedy Drive, the relationships here set the stratigraphic sequence: Sangerville Formation/unnamed phyllite/Waterville Formation.

Beds strike northeasterly and dip steeply. Schistosity cuts bedding at a low angle, and the traces of graded beds in schistosity indicate that the stratigraphic section is upright. Small F_2-isoclinal folds, some of which are sheared, can be seen to deform bedding on the vertical face. Quartz pods form boudin fillings, and other veins visible here are late features.

Stop 6: Evidence for F_1, F_2, and F_3 in Waterville Formation phyllites. This exposure is wholly within phyllites of the Waterville Formation. Although most of the formation is gray, the rocks here are light purplish to greenish gray quartz-mica phyllites interbedded with quartzite and ankeritic quartzite. Bed thickness ranges from 0.2 to 8 cm, but apparent thicknesses on cleavage faces are greater because of the small angle between bedding and cleavage. Quartzite beds have sharp boundaries, but color gradations in some pelitic beds reflect compositional differences thought to be due to original grading.

A prominent S_3-cleavage surface on the south wall of the cut strikes N12°E and is nearly vertical. The trace of bedding on this surface defines an open, somewhat symmetrical fold form. A joint nearly perpendicular to this cleavage displays a tight symmetrical fold form in the same bedding surface, so that the three-dimensional properties of the fold are readily observed. Its axial plane strikes N37°E and is nearly vertical; the fold plunges gently to the northeast. This structure is typical of F_2-folds in pelitic rocks, and the cleavage that cuts both of its limbs is S_3.

Directly across the cut on the north wall is a steeply dipping S_2-cleavage surface that strikes N35°E. The trace of bedding on this cleavage plunges gently to the northeast in harmony with the plunge of the upright fold. Color gradation in the beds suggests that they are upside down in the cleavage, and that these F_2-folds face **downward**. The local stratigraphy is thus upside down on the inverted limb of an F_1-recumbent fold.

About 165 ft (50 m) to the east, on the north wall of the cut, bedding and S_2-cleavage are essentially parallel. Dendroid graptolites and the trace fossils *Nereites* and *Phyllodacites* have been collected here. These are long-ranging forms, but are particularly common in the Silurian of Wales.

Stop 7: Relationships between all three fold episodes in and near the limestone member of the Waterville Formation. The rocks here belong to the limestone member and adjacent phyllites of the Waterville Formation. The pelitic rocks consist of gray quartz-mica phyllite intercalated with white to buff quartzite in graded beds 6 mm to 8 cm thick. The limestone member consists of gray, slightly micaceous micrite interbedded on a 6-mm to 12-cm scale with rusty weathering, buff-colored quartz-mica phyllite or micaceous quartzite. The contact between the two members here is sharp.

Four areas of this outcrop are particularly important (inset, Fig. 3). In Area 1, a northeast-trending F_2-isoclinal fold plunges gently to the north and displays a well-developed axial plane cleavage. Graded bedding near the hinge of this fold is tentatively interpreted to indicate that these folds face downward and that the stratigraphy here is inverted. This outcrop is thus on the inverted limb of a recumbent F_1-fold. Both limbs of the upright F_2-folds are cut by a steeply dipping S_3-cleavage. Right-handed asymmetric F_3-folds with axial surfaces parallel to this cleavage are found on both limbs of the upright folds.

In Area 2, an isoclinal F_2-synform deforms an earlier isoclinal fold (Fig. 3). The F_2-synform has an axial surface that strikes northeasterly and dips steeply; its axis plunges gently toward the southeast. Beds on the limbs of the earlier fold can be traced around the F_2-hinge, and the early cleavage can be seen only in the F_1-hinge. The plunge of this early fold has not been ascertained. Small-scale F_1-folds such as this are very rare.

At Area 3, beds of the limestone member are folded by gently plunging isoclinal F_2-folds. These have nearly plane flanks, sharp hinges, and northeast-striking axial surfaces. S_3 cleavage, best preserved in phyllite beds, cuts both limbs of F_2-folds.

In Area 4, a dike of light gray plagioclase-quartz-muscovite-chlorite-calcite-epidote granofels (meta-dacite?) strikes N33°E and cuts bedding at a low angle. Similar dikes elsewhere have been shown to cut F_2-folds (Osberg, 1968). This dike is cut by a faint S_3-cleavage, and is also broken into a "rotated" boudinage in which boudin lines plunge steeply. These relationships show

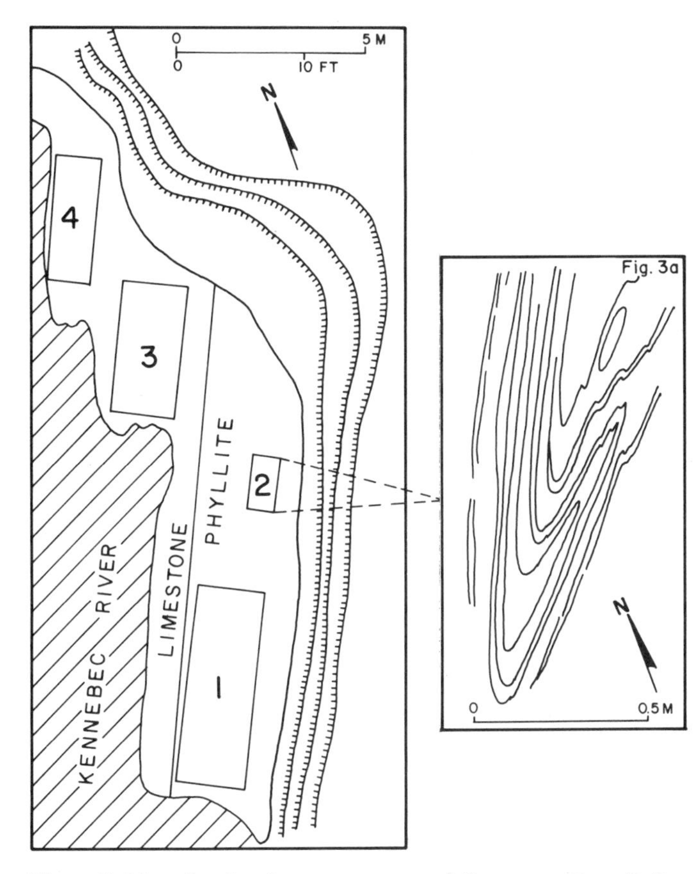

Figure 3. Map showing important structural features at Stop 7. 3a, refolded fold in Area 2.

that the dike intruded after the F_2-event but prior to F_3-folding and M_2-metamorphism.

Stop 8: F_2- and F_3-fold styles in sandstones of the Vassalboro Formation. (No Hammers, please!) Rocks here belong to the Vassalboro Formation and include light gray quartz-mica phyllite interbedded with blue-gray, slightly calcareous quartzwacke. Possible grading and cross-beds may be seen at the west end of the outcrop.

Open to isoclinal F_2 folds are visible at the west end of the outcrop. Their axial surfaces strike northeast and dip steeply, and their axes plunge to the southwest. The "openness" of these folds is controlled by the thick quartzose beds exposed on the south side of the road, and is typical of F_2-folds in the thicker bedded rocks of the region. In less competent rocks like those seen at Stops 3 through 7, F_2-folds are more nearly isoclinal. Primary sedimentary features here indicate that the folds face upward.

Cleavage parallels the axial surfaces of the upright folds. In competent beds it is a pressure solution cleavage, but in the more pelitic layers it is close-spaced with micas in the plane of cleavage. Cusp-shaped cross sections of thin quartzite beds interlayered with phyllite may be due to competency differences between the two rock types.

A set of north-trending F_3-folds deforms the F_2-structures, and S_3-cleavage is well developed. F_3-fold plunges vary, depending on the orientation of the surfaces that were folded. An interference pattern produced by F_2/F_3-interaction is beautifully displayed 100 ft (30 m) from the west end of the outcrop on the north side of the road.

REFERENCES CITED

Dallmeyer, R. B., and van Breemen, O., 1978, $^{40}Ar/^{39}Ar$ and Rb/Sr ages in west-central Maine; Their bearing on the chronology of tectonothermal events: Geological Society of America Abstracts with Programs, v. 10, p. 38.

Griffin, J. R., 1973, A Structural Study of the Silurian Metasediments of Central Maine [Ph.D. thesis]: Riverside, University of California, 157 p.

Hall, B. A., Pollock, S. G., and Dolan, K. M., 1976, Lower Devonian Seboomook Formation and Matagamon Sandstone, northern Maine; A flysch basin-margin delta complex, *in* Page, L., ed., Contributions to the Stratigraphy of New England: Geological Society of America Memoir 148, p. 57–63.

Holdaway, M. J., Guidotti, C. V., Novak, J. M., and Henry, W. E., 1982, Polymetamorphism in medium- to high-grade pelitic metamorphic rocks, west-central Maine: Geological Society of America Bulletin, v. 93, p. 572–584.

Ludman, A., 1977, Bedrock geology of the Skowhegan Quadrangle, Maine: Maine Geological Survey Geologic Map Series no. 5, 25 p.

—— , 1978, Bedrock geology of the Kingsbury Quadrangle, Maine: Maine Geological Survey Geologic Map Series no. 6, 31 p.

Moench, R. H., and Boudette, E. L., 1970, Stratigraphy of the northwest limb of the Merrimack Synclinorium in the Kennebago Lake, Rangeley, and Phillips Quadrangles, western Maine, *in* Boone, G., ed., Guidebook for trips in the Rangeley Lakes–Dead River area, western Maine: New England Intercollegiate Geological Conference, Syracuse, New York, p. 1–25.

Osberg, P. H., 1968, Stratigraphy, structural geology, and metamorphism of the Waterville–Vassalboro area, Maine: Maine Geological Survey Bulletin no. 20, 64 p.

—— , 1980, Stratigraphic and structural relationships in the turbidite sequence of south-central Maine, *in* Roy, D., and Naylor, R. S., eds., Guidebook to the Geology of Northeastern Maine and Neighboring New Brunswick: New England Intercollegiate Geological Conference, Presque Isle, Maine, p. 278–296.

Osberg, P. H., Hussey, A. M., III, and Boone, G. M., 1985, Bedrock Geologic Map of Maine: Augusta, Maine Geological Survey.

Pankiwskyj, K. A., Ludman, A., Griffin, J. R., and Berry, W.B.N., 1976, Stratigraphic relationships on the southeast limb of the Merrimack Synclinorium in central and west-central Maine, *in* Lyons, P. C., and Bronwlow, A. H., eds., Studies in New England Geology: Geological Society of America Memoir 146, p. 263–280.

Casco Bay Group, South Portland and Cape Elizabeth, Maine

Arthur M. Hussey II, Department of Geology, Bowdoin College, Brunswick, Maine 04011

LOCATION

This site consists of shoreline exposures of the Casco Bay Group between Spring Point, South Portland, and Two Lights State Park, Cape Elizabeth, and roadcuts along Maine 77 in Cape Elizabeth (Fig. 1). The following places are designated as stops where important aspects of the geology of the Casco Bay Group can be seen: (1) Southern Maine Vocational Technical Institute, South Portland; (2) Willard Beach, South Portland; (3) Danford Cove, South Portland; (4) Portland Head Light, Cape Elizabeth; (5) Chimney Rock, Cape Elizabeth; (6) road-cut exposures of the Scarboro Formation and Spurwink Metalimestone, Cape Elizabeth; and (7) Two Lights State Park, Cape Elizabeth. The public has access to the shoreline at Stops 1 through 4, 6, and 7, but at Stop 5 and other parts of the shoreline, permission to visit rock exposures must be obtained from individual landowners. The public does not have the right to walk along the intertidal zone without permission. In general, parking is very limited at private localities, and town ordinances generally prohibit parking along town roadways.

SIGNIFICANCE

The South Portland–Cape Elizabeth shoreline exposes a nearly complete section of the Casco Bay Group, a sequence of felsic (and lesser mafic) metavolcanics, metapsammites, and metapelites. Included in the group are the Cushing, Cape Elizabeth, Spring Point, Diamond Island, Scarboro, Jewell, and Macworth Formations, and the Spurwink Metalimestone. The Macworth and Jewell Formations are not exposed at this site.

The Casco Bay Group comprises an exotic terrane of island arc–back-arc basin affinities in the Coastal Lithotectonic Block of Maine. The age of the sequence, its place of origin, and time of accretion of the terrane to the northern Appalachian orogen are at present only poorly understood due to lack of age controls of rocks and structures. The rocks of the Casco Bay Group are interpreted to be in folded-thrust fault contact with Siluro-Ordovician rocks of the Kearsarge–Central Maine Synclinorium and Coastal Lithotectonic Block, establishing this thrusting as a structural feature of the Acadian orogeny. The Casco Bay rocks have been cut by splays of the post-metamorphic Norumbega Fault Zone, and injected by numerous basic dikes related to Juro-Triassic rifting associated with initial stages of the opening of the Atlantic Ocean.

SITE INFORMATION

The formations of the Casco Bay Group, with the exception of the Cushing, were named and defined by Katz (1917). He regarded the Cushing as a deformed intrusive granodiorite. Later, Bodine (1965) and Hussey (1968, 1971) recognized the volcanic character of these rocks and renamed the unit the Cushing Formation. The general lithologies of these formations within the South Portland–Cape Elizabeth area are summarized in the explanation for Figure 1.

The Cape Elizabeth Formation preserves evidence of two major deformations. F_1 folds are recumbent folds of unknown regional size, extent, vergence, and age. They are known only from minor recumbent parasitic folds seen in outcrop, and rare downward-facing F_2 folds. The F_2 folds are major north-northeast–trending upright folds of variable but generally gentle plunge, and they control the outcrop distribution of the formation. F_2 folds are regarded as structural effects of the Acadian orogeny. Because of lack of well-preserved bedding in the other formations, the multiple deformational geometry of these formations is not well known.

Regional metamorphism of the Casco Bay Group at this site ranges from chlorite to garnet grade in a low-pressure, intermediate-type facies series (andalusite is present in metapelites of the group just north of the site). The present prograde metamormphic mineral assemblage is regarded to be of Acadian age (Hussey, 1985). Retrograde metamorphism, particularly as manifested by the alteration of biotite and garnet to chlorite, is developed in proximity to faults of the Norumbega Fault Zone and is probably a result of that faulting.

Brookins and Hussey (1978) report Rb/Sr whole rock ages of 481 ± 40 Ma and 485 ± 30 Ma for the Cushing and Cape Elizabeth Formations respectively. These ages probably represent partial resetting of initial ages due to metamorphism. The Casco Bay Group is currently regarded to be of Precambrian to Ordovician age (Osberg and others, 1985) with a bias toward late Precambrian (Hussey, 1985). Correlations of the Cape Elizabeth are uncertain; possibilities are with the Rye Formation of southwestern Maine and southeastern New Hampshire, the Nashoba Formation of eastern Massachusetts, parts of the Massabesic Gneiss of southern New Hampshire, and part of the Ellsworth Formation of eastern Maine. The Cushing Formation may correlate in part with the Massabesic Gneiss and the Ellsworth Formation. Units of the Casco Bay Group above the Cape Elizabeth Formation have no known correlatives beyond their own outcrop belt.

Stop 1. Shoreline exposures on the east side of Spring Point at Southern Maine Vocational Technical Institute (SMVTI). To reach SMVTI from Exit 7, Maine Turnpike, turn left onto U.S. 1 at the end of the Turnpike connector. Turn right on Broadway, continuing on that road 3.6 mi (5.8 km) to Pickett Avenue, then turn right on Pickett. At the first stop sign, cross Fort Road and enter SMVTI campus. Outcrops to be examined are those in front of Hildreths Hall. The sea cliff here exposes the sequence from the

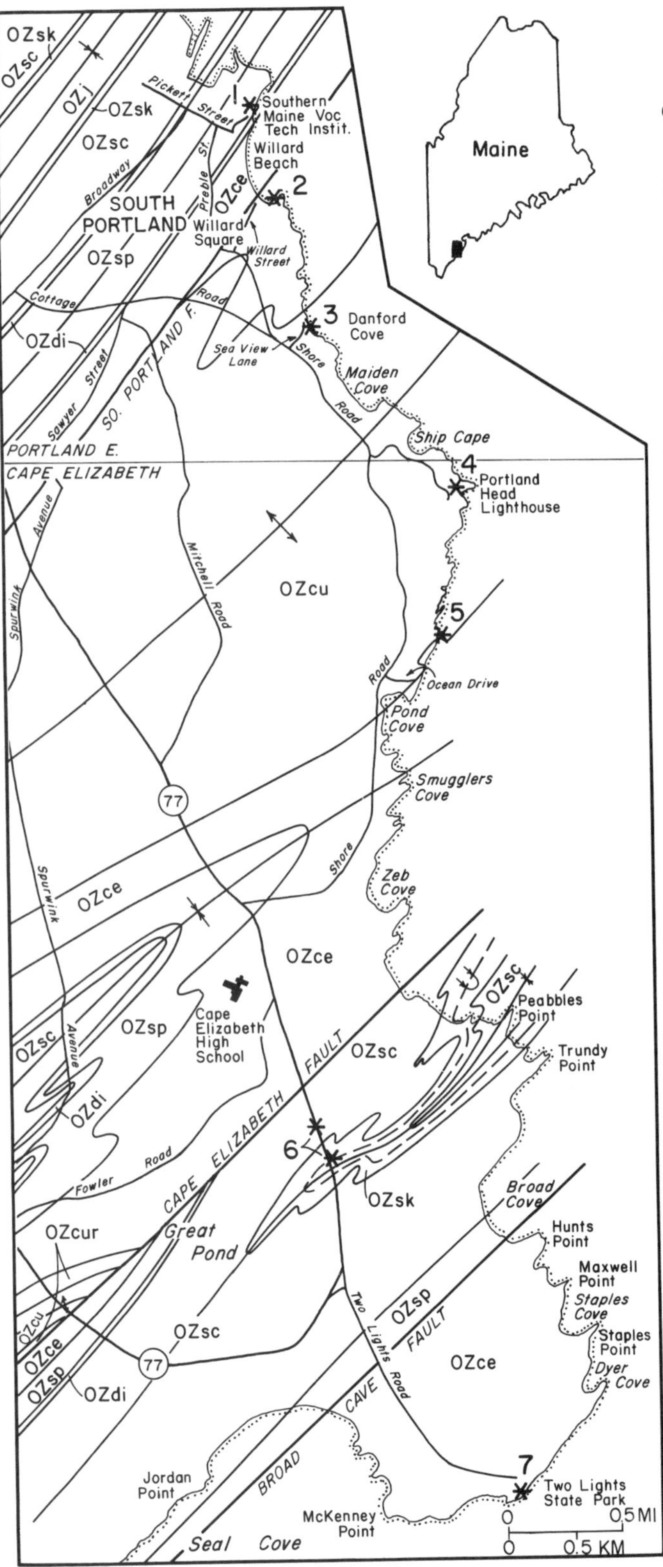

Figure 1. Geologic map of the South Portland–Cape Elizabeth shoreline area, southwestern Maine, showing stops of the field site.

EXPLANATION

OZj JEWELL FORMATION: Sulfidic and non-sulfidic muscovite-biotite-garnet schist; minor micaceous quartzite; rare 15 to 20 cm amphibolite beds.

OZsk SPURWINK METALIMESTONE: Medium gray, thinly interbedded fine-grained marble and biotite phyllite. Typically has ribbony-appearing weathered surface.

OZsc SCARBORO FORMATION: Same lithology as Jewell Formation.

OZdi DIAMOND ISLAND FORMATION: Sulfidic coal-black quartz-graphite-muscovite phyllite.

OZsp SPRING POINT FORMATION: Medium greenish-gray chlorite-garnet (manganiferous) phyllite; chlorite-actinolite-biotite phyllite; dark green blocky amphibolite; minor plagioclase-quartz-biotite ± garnet gneiss and granofels.

OZce CAPE ELIZABETH FORMATION: Medium-gray quartz-plagioclase-biotite-muscovite-garnet schist and dark gray biotite-chlorite-muscovite-garnet schist (in garnet zone metamorphism); buff-weathering ankeritic quartz-plagioclase-chlorite phyllite and dark gray phyllite (in chlorite zone).

OZcu CUSHING FORMATION: Quartz-plagioclase-biotite gneiss locally with relict fragmental volcanic structures; calc-silicate gneiss; and rusty-weathering feldspathic muscovite-biotite schist.
OZcur: rusty garnetiferous gneiss.

F_1 recumbent syncline
F_2 upright antiform
F_2 upright synform
High-angle post-Metamorphic fault

upper part of the Spring Point Formation (medium greenish gray actinolite-biotite-chlorite-plagioclase gneiss with numerous stretched felsic pyroclasts), through the Diamond Island Formation (100 ft-thick [30 m] sequence of black quartz-graphite-muscovite-pyrite phyllite with 1.6-ft-thick [0.5 m] ribbony metalimestone), to the lower part of the Scarboro Formation (nonrusty muscovite-biotite-garnet-quartz phyllite, very weakly bedded, with a 6- to 12-ft-thick (2 to 4 m) zone of metalimestone near the base). Rocks here are at garnet-grade regional metamorphism. This is the type locality for the Spring Point Formation.

Stop 2. Willard Beach. (From SMVTI campus, turn left onto Fort Road, and in 0.3 mi [0.5 km] turn left onto Prebble Street. Bear left at the stop sign at Willard Square, staying on Prebble Street. Turn left in 0.15 mi [0.24 km] onto Willard Street and proceed to the end of the street.) The sea cliff on the southeast side of Willard Beach opposite Stop 1 exposes garnet-grade Cape Elizabeth Formation. Between this exposure and those at Stop 1 is the South Portland Fault, one of the splays of the Norumbega Fault System. The fault is covered by sands of Willard Beach. Inland, about 1 mi (1.6 km) to the southwest, the position of the fault is more closely defined where outcrops of the Scarboro Formation lie within 250 ft (75 m) of the Cushing Formation.

Stop 3. Danford Cove. (Turn left from Willard Street onto Prebble Street. Turn left on Cottage Road at stop sign and in 0.1 mi [0.2 km] park on the right side of Cottage Road opposite Seaview Avenue. Walk down Seaview Avenue to the set of cement steps and descend to the base of the sea cliff.) Rocks of the Cushing Formation, here metamorphosed to garnet grade, retain their original pyroclastic structures, indicating that these rocks were felsic volcanic breccias and crystal tuffs. These rocks are massive, with only minor suggestion of current reworking. Feeble bedding is observed at very few areas in this outcrop belt of the Cushing Formation. A thin zone of feeble bedding, typical of, but rare in this part of the formation, can be seen just southeast of the steps to this stop. Grains of blue quartz and white plagioclase up to 3 mm in size are interpreted to be relict crystal fragments of an original crystal tuff. These crystal metatuffs have a very weak foliation, but a strong lineation (elongation of biotite clots parallel to the regional F_2 fold hinges). About 250 ft (75 m) south of the steps is a zone of coarse breccia, the fragments of which have been very strongly stretched parallel to regional F_2 fold hinges. Approximately 600 ft (200 m) north of the steps, the contact of the Cape Elizabeth with the Cushing is exposed on the wave-cut bench at the base of the sea cliff. The Cape Elizabeth Formation here consists mostly of biotite-chlorite-muscovite-garnet phyllite with thin interbeds of micaceous quartzite. Poorly developed graded beds near the contact indicate that the Cape Elizabeth Formation is on top of the Cushing, and that the sequence is upright.

Stop 4. Portland Head Lighthouse. (Continue 0.7 mi [1.1 km] on Cottage Road, which becomes Shore Road at the Cape Elizabeth Town line. Turn left into Fort Williams Park and drive to the Portland Head Lighthouse parking area. Walk along path to the north of the lighthouse until you can easily descend down over the sea cliff.) The Cushing Formation exposed here, close to the crest of the Cushing anticline, is a light gray quartz-plagioclase-biotite gneiss with minor muscovite and k-feldspar. Pyroclastic structures (relict crystal fragments and breccia blocks) are rare in the exposure, suggesting that these rocks represent fine-grained felsic tuffs. They preserve very little relict bedding. However, 300 ft (100 m) northwest of the lighthouse, close to the base of the sea cliff, light gray metatuff can be seen in contact with, and overlying, buff-colored metatuff. This is interpreted to be the contact between two separate pyroclastic units, perhaps ash flows. In general, the lack of interbedded volcanogenic metasediments and the lack of well-developed bedding indicative of current reworking or sorting of pyroclastic debris, are suggestive of a subaerial origin for this part of the Cushing. As at Danford Cove, a strong lineation parallel to F_2 fold hinges, and very weak foliation, is characteristic of the rock. Several unmetamorphosed basalt or diabase dikes of presumed Juro-Triassic age cut the Cushing in the general vicinity of the lighthouse. About 300 ft (100 m) north of the lighthouse a hackly fractured postmetamorphic felsic dike occupies a high-angle brittle fault zone, and it has been internally faulted parallel to its contacts with the Cushing.

Stop 5. Chimney Rock near Pond Cove. (From Fort Williams Park, turn left and proceed 0.9 mi [1.4 km] on Shore Road. Park on right side of road near, but not blocking, the private road marked by stone pillars and gate. Be careful not to park in areas marked by "No Parking" signs. They mean it! Walk northward along Ocean Drive and then down Lawson Road and ask permission of an owner of a shoreline home to pass through his/her property to shoreline ledges. This stop requires a time within 2 or 3 hours of low tide.) The exposures of the Cape Elizabeth Formation (thin-bedded dark gray phyllite and quartzose phyllite) and Cushing Formation (massive light gray felsic metavolcanics) at this stop demonstrate the presence of F_1 recumbent folds. The Cape Elizabeth Formation is structurally on top of the Cushing, but graded beds in the Cape Elizabeth close to the contact indicate that it is stratigraphically above the Cushing. This area is thus on the inverted limb of an F_1 fold of relatively local extent. Parasitic folds with well-developed axial plane cleavage exposed at this stop are upright F_2 folds. At the Cape Elizabeth Cushing contact is a 3-ft-thick (1 m) zone of dark gray fine granule-bearing phyllite suggesting that the contact may locally be an unconformity.

Stop 6. Maine 77 roadcuts. (From the parking area, continue on Shore Road to junction with Maine 77. Turn left on Maine 77, proceed 1.1 mi [1.8 km] and park beside the highway opposite the large, rusty-weathering roadcut. This stop consists of this roadcut and the next roadcut approximately 500 ft [150 m] southeast.) The rusty-weathering ledges at the first roadcut are typical of the sulfidic dark gray, slightly graphitic phyllite of the Scarboro Formation. Thin metapsammite beds show that the bedding here has a strike of N60°E and a dip of 80°NW, parallel to the phyllitic cleavage. A spaced cleavage dips moderately to the northwest and becomes more strongly developed toward the northwest edge of the road cut. Spaced cleavage of similar orientation has been developed in close proximity to a minor thrust fault close to the Cape Elizabeth Fault exposed along the shore approximately 3 mi (5 km) to the southwest, suggesting that the spaced cleavage at this stop may also be related to the same thrust.

The second road cut to the southeast exposes the ribbony-bedded Spurwink Metalimestone in the axis of the Peables Point Syncline, a refolded recumbent syncline.

Stop 7. Two Lights State Park. (Continue on Maine 77 for 0.6 mi [1 km] then turn left at sign for Two Lights State Park. Bear left again in 0.1 mi [0.2 km], and continue through the Park entrance to the parking lot near the shore. Walk to extensive shoreline exposures anywhere in the park. Since this is a state park, no collecting is permitted, and no hammers should be used.) Exposed throughout the park is chlorite-grade, thin- to thick-bedded, buff-weathering metasiltstones and dark gray phyllites of the Cape Elizabeth Formation. These rocks are separated from the biotite- or garnet-grade Spring Point Formation by Broad Cove Fault extending between Broad Cove and Seal Cove to the southwest. The metasiltstones are moderately calcareous and ankeritic, hence the buff-weathering color. These rocks have been affected by both F_1 and F_2 deformation. F_1 folds are east-verging parasitic recumbent folds. Upright graded bedding indicates that these are east-facing folds. F_2 folds are very gentle and open. Recumbent parasitic folds have an axial-plane spaced cleavage in the metasiltstone beds and a phyllitic cleavage in the dark phyllite beds. Both types of cleavage are parallel to axial planes of F_1 folds. The visitor to Two Lights State Park is referred to an illustrated popular account of the geology of this and nearby Crescent Beach State Park (Hussey, 1982; available for purchase at the park gatehouse or from the Maine Geological Survey, Augusta, Maine, 04333).

REFERENCES CITED

Bodine, M. W., Jr., 1965, Stratigraphy and metamorphism in southwestern Casco Bay, Maine: Guidebook, 57th New England Intercollegiate Geological Conference, Brunswick, Maine, p. 57–72.

Brookins, D. G., and Hussey, A. M., II, 1978, Rb-Sr age for the Casco Bay Group and other rocks from the Portland–Orrs Island area, Maine: Geological Society of America Abstracts with Programs, v. 10, p. 34.

Hussey, A. M., II, 1968, Stratigraphy and structure of southwestern Maine, *in* Zen, E-An, White, W. S., Hadley, J. B., eds., Studies of Appalachian geology; Northern and maritime: New York: Interscience Publishers, p. 291–301.

——, 1971, Geologic map of the Portland 15-minute Quadrangle: Maine Geological Survey, Geologic Map Series GM-1, scale 1:62,500.

——, 1982, Geology of the Two Lights and Crescent Beach state parks area, Cape Elizabeth, Maine: Maine Geological Survey Bulletin 26, 34 p.

——, 1985, Geology of the Bath and Portland 1° × 2° sheets: Maine Geological Survey Open-File Report 85-87, 82 p.

Katz, F. J., 1917, Stratigraphy of southwestern Maine and southeastern New Hampshire: U.S. Geological Survey Professional Paper 108, p. 165–177.

Osberg, P. H., Hussey, A. M., II, and Boone, G. M., 1985, Bedrock geologic map of Maine: Augusta, Maine, Maine Geological Survey, scale 1:500,000.

ACKNOWLEDGEMENTS

The geological interpretation of this site is based upon field work done for and supported by the Maine Geological Survey. The writer gratefully acknowledges the generous support by the survey and the encouragement and counsel given by the present State Geologist, Walter A. Anderson, and former State Geologists, Robert G. Doyle and John R. Rand.

Silurian–Lower Devonian volcanism and the Acadian orogeny, the Eastport area, Maine

Olcott Gates, *P.O. Box 234, Wiscasset, Maine 04578*

LOCATION

This geological traverse follows U.S. 1 for 47 mi (76 km) from Machias to Robbinston (Fig. 1). Applicable USGS 7½-minute Quadrangle maps are Machias, Machias Bay, Whiting, West Lubec, Lubec, Pembroke, Eastport, Red Beach, and Robbinston. Alternatively, *The Maine Atlas and Gazetteer,* DeLorme Publishing Company, Freeport, Maine, provides excellent regional geographical coverage and is sold in many stores. Geologic maps and field guides are Abbott (1978) and Gates (1975, 1978, 1983). Allow about six hours for the traverse, timed to arrive at the last few stops at low tide.

SIGNIFICANCE

This excursion demonstrates the classical criteria for deducing the occurrence of an orogeny: deformation, granitic plutonism, and an angular unconformity; all evidence for the Devonian Acadian orogeny in this area. It examines preorogenic folded volcanic rocks, the granitic rocks that intruded them, an erosion surface that bevelled both the volcanic rocks and the granites, and an overlying postorogenic alluvial conglomerate.

SITE INFORMATION

The preorogenic Silurian to Lower Devonian marine volcanic rocks (Fig. 1; Bastin and Williams, 1914; Gates, 1975; Gates and Moench, 1981) are bimodal, largely of rhyolite and basalt, and consist of flows, domes, ash flows, tuff-breccias, and bedded tuffs. Fossiliferous argillite, shale, siltstone, and minor limestone mark times of eruptive quiescence. During the Silurian, the sea became progressively shallower, leading to subaerial eruptions in the Early Devonian. Hypabyssal dikes, sills, and irregular small bodies of diabase and gabbro intrude the entire volcanic pile. Several linear belts of flowbanded rhyolite apparently are intrusions along contemporaneous faults.

The Cobscook anticline and the Lubec fault (Fig. 1) are the two principal structures in the Eastport area. The anticline plunges to the east-southeast and has a steeply inclined south limb against the Lubec fault zone and a more gently dipping east limb. It is an Acadian structure. Numerous faults, only a few of which are shown on Figure 1, reflect Acadian deformation, block faulting during the Late Devonian, and Carboniferous rifting best developed in New Brunswick. The Lubec fault of probable Carboniferous age separates the Cobscook anticline from a fault block containing the Lower Silurian Quoddy Formation engulfed in diabase and gabbro.

A variety of gabbros, diorites, and granites of Devonian age, only a small part of which are shown on Figure 1, intrude the folded volcanic rocks in a wide belt extending from the coast west of Machias into New Brunswick north of St. Andrews (Chapman, 1962; Amos, 1963; Pajari and others, 1974). The granitic rocks represent the plutonic phase of the Acadian orogeny.

The Upper Devonian Perry Formation (Schluger, 1973) of red terrestrial clastic rocks and two basalt flows was deposited in fault basins on an erosion surface that truncates both the folded volcanic rocks and the plutonic rocks that intruded them.

Rb-Sr whole rock dating of the volcanic rocks (Fullagar and Bottino, 1970) indicates that the folding phase of the Acadian orogeny in this area began after 408 Ma. Rb-Sr whole rock dates of 393–400 Ma on several of the granites (Spooner and Fairbairn, 1970; Pajari and others, 1974) place the granitic phase within the late Early Devonian. Uplift and consequent erosion that cut the unconformity surface is younger than the granites and older than the overlying Upper Devonian Perry Formation dated by plants (Schluger, 1973).

The volcanic rocks of the Eastport area are part of a coastal volcanic belt that includes Silurian–Lower Devonian marine volcanic formations in southern New Brunswick (Potter and others, 1979), in the Mount Desert and Penobscot Bay areas of Maine (Osberg and others, 1985), and in northeastern Massachusetts (Zen, 1983a). In Maine they are in fault contact or rest unconformably (Osberg and others, 1985) on Late Proterozoic to early Paleozoic slates and other metamorphic rocks of uncertain terrane affiliations within the northern Appalachians, but are commonly assigned to the Avalon terrane (Williams and Hatcher, 1982), which may itself be composite (Zen, 1983b).

Opinions differ as to whether the Acadian orogeny signals an Avalon–North America collision following progressive closing of the ancestral North Atlantic (Iapetus) during the Silurian, or whether Avalon originally docked near or alongside North America during the Ordovician Taconian orogeny and then was tectonically pushed farther westward against North America in the Acadian orogeny, and possibly moved transcurrently during the Carboniferous (e.g., Wilson, 1966; Dewey and Kidd, 1974; Osberg, 1978; Gates and Moench, 1981; Williams and Hatcher, 1982; Zen, 1983b; Ludman, 1986).

In the following traverse along U.S. 1 (Fig. 1), mileage (km) numbers begin at the railroad crossing just east of Machias and designate the approximate span of each formation exposed in outcrops along the highway. The traverse goes up-section stratigraphically. Bold mileages (km) within the text single out good exposures of a lithology within the formation.

The Quoddy Formation (Lower Silurian) does not crop out

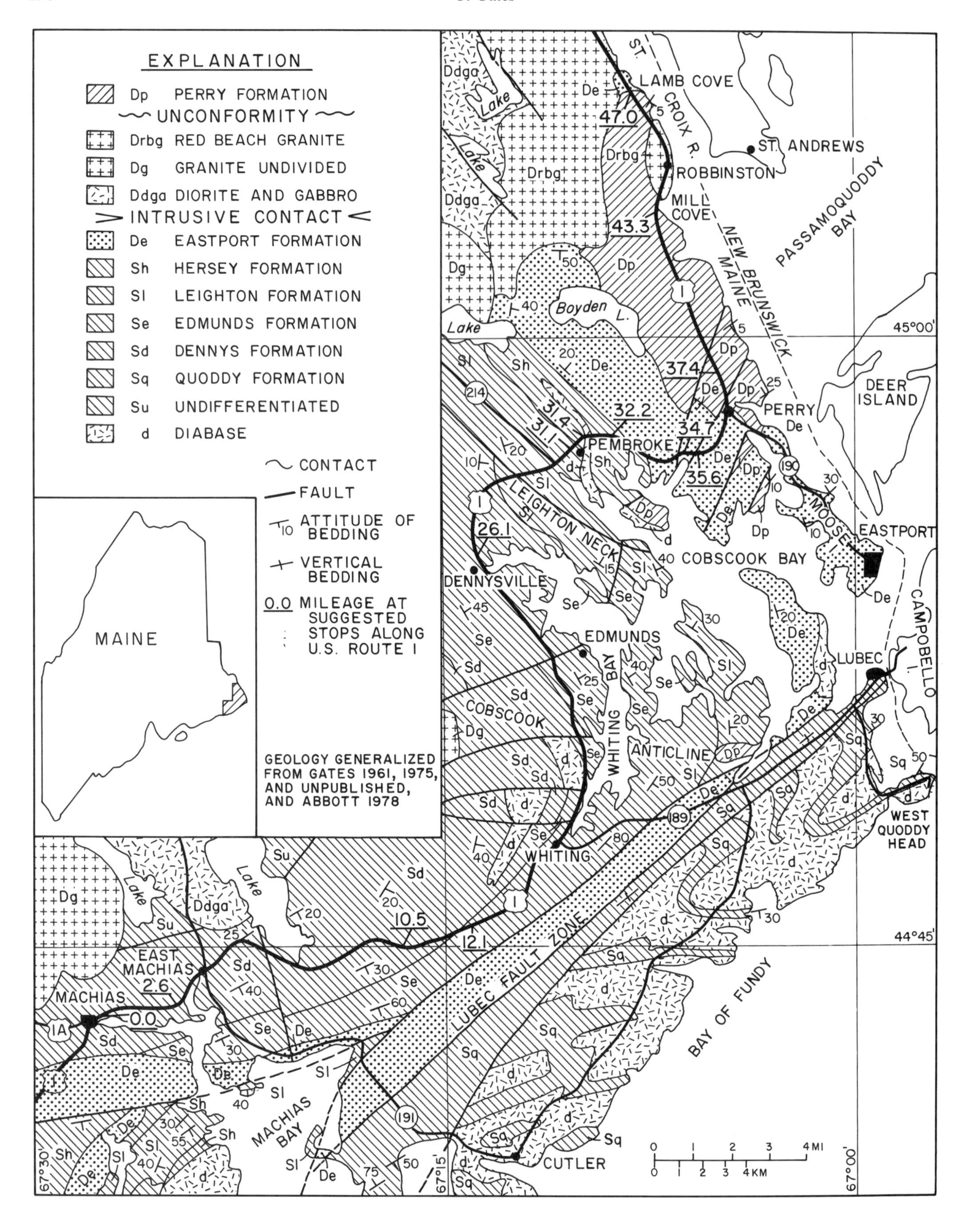
EXPLANATION
Dp PERRY FORMATION
UNCONFORMITY
Drbg RED BEACH GRANITE
Dg GRANITE UNDIVIDED
Ddga DIORITE AND GABBRO
INTRUSIVE CONTACT
De EASTPORT FORMATION
Sh HERSEY FORMATION
Sl LEIGHTON FORMATION
Se EDMUNDS FORMATION
Sd DENNYS FORMATION
Sq QUODDY FORMATION
Su UNDIFFERENTIATED
d DIABASE
CONTACT
FAULT
ATTITUDE OF BEDDING
VERTICAL BEDDING
0.0 MILEAGE AT SUGGESTED STOPS ALONG U.S. ROUTE 1
MAINE
GEOLOGY GENERALIZED FROM GATES 1961, 1975, AND UNPUBLISHED, AND ABBOTT 1978
LAMB COVE
ST. CROIX R.
ST. ANDREWS
ROBBINSTON
MILL COVE
NEW BRUNSWICK
MAINE
PASSAMAQUODDY BAY
Boyden L.
Lake
PERRY
DEER ISLAND
PEMBROKE
LEIGHTON NECK
COBSCOOK BAY
MOOSE I.
EASTPORT
DENNYSVILLE
EDMUNDS
WHITING BAY
COBSCOOK
ANTICLINE
CAMPOBELLO I.
LUBEC
WEST QUODDY HEAD
WHITING
LUBEC FAULT ZONE
EAST MACHIAS
MACHIAS
MACHIAS BAY
BAY OF FUNDY
CUTLER
45°00'
44°45'
67°30'
67°15'
67°00'
47.0
43.3
37.4
34.7
35.6
32.2
31.4
31.1
26.1
12.1
10.5
2.6
0.0
0 1 2 3 4 MI
0 1 2 3 4 KM

along the traverse. To see it, turn right at Whiting (mile 16.2; 26.1 km) onto Maine 189 and follow signs to West Quoddy Head, where it is well exposed along the north shore (Fig. 1).

0.0–11.0 mi (0.0–17.7 km). The Dennys Formation (Lower to Middle Silurian) consists of a lower section of keratophyric to rhyolitic flows, tuffs, and vent breccias commonly containing broken feldspar crystals and angular fragments of volcanic and sedimentary rocks and cut by diabase dikes, for example, at mile **2.6** (km **4.2**). Between 7.4 and 11.0 (11.9 and 17.7 km) there are outcrops of the overlying basaltic breccias, bedded tuffs, and flows (mile **10.5**; km **16.9**). As is common throughout the volcanic sequence, hydrothermal alteration has devitrified the once glassy rocks and changed original labradorite and augite to saussuritized plagioclase, chlorite, epidote, actinolite, and calcite. Original textures remain.

11.0–28.0 (17.7–45.0). The Edmunds Formation (Middle Silurian) contains pink, maroon, green, and purple dacite and rhyolite tuff-breccias together with rhyolite domes, a few basalt flows (mile **12.1**; km **19.5**), and interbedded siliceous tuffs and shales carrying a fauna of brachiopods, pelecypods, gastropods, trilobites, and corals. The fossiliferous rocks are best exposed on the shores of Whiting Bay. At **26.1** mi (**42.0** km) is a typical submarine avalanche or debris flow of tuff-breccia, an unsorted mixture of broken feldspar crystals, and angular to subangular fragments of volcanic and sedimentary rocks. Lenses and pods of siliceous-bedded tuff and shale within the tuff-breccia represent sediments swept upward into its base as the submarine avalanche moved across a muddy sea floor.

28.0–31.1 (45.0–50.0). The Leighton Formation (Upper Silurian) has a few basalt flows (**28.5** mi; **45.9** km) near its base, overlain by blue-gray shales and siltstones that underlie valleys and are best exposed along the northeast shore of Leighton Neck (Fig. 1). The fauna is a shallow-water one of pelecypods, gastropods, ostracodes, and a few brachiopods. At mi **31.1** (km **50**) a submarine avalanche deposit rests on shale swept upwards into its base and grades up into red beds of the Hersey Formation.

31.1–33.2 (50.0–53.4). The Hersey Formation (Upper Silurian to Lower Devonian) is a tidal flat or estuarian deposit of maroon mudstone and siltstone (mile **32.2**; km **51.8**) and nodular limestone with a fauna of pelecypods, gastropods, and ostracodes. At mi **31.4** (km **50.5**) a sill of ophitic gabbro intrudes the formation.

33.2–36.8 (53.4–59.2). The Eastport Formation (Lower Devonian) has a lower section of basaltic-andesite tuffs and flows, which at mi **34.7** (km **55.8**) have red-oxidized tops. Other lithologies include rhyolitic flow-banded flows (mi **35.6**; km **57.3**), ash flows, and tuff-breccias. Sedimentary rocks consisting of cross-bedded, ripple-marked, and mud-cracked varicolored shales and siltstones probably of tidal-flat origin and bearing a fauna of linguloids, pelecypods, gastropods, and ostracodes are well exposed on the north shore of Moose Island, along Maine 190 to Eastport (Fig. 1).

36.8–43.6 (59.2–70.1). The Perry Formation (Upper Devonian) is a post-orogenic terrestrial deposit of red coarse alluvial conglomerate and arkosic sandstone displaying cross-bedding, cut-and-fill, and a broad range of bedding thicknesses (Schluger, 1973). Lenses of red siltstone and shale may be overbank deposits. In the Eastport area there is one basalt flow (mi **37.4**; km **60.2**). At Mill Cove (mi **43.3**; km **69.7**; low tide required) cliffs along the southeast-trending shoreline expose typical Perry cross-bedded arkosic alluvial conglomerate. The rounded pebbles and cobbles are of basalts and rhyolites of the Eastport Formation and of hypabyssal diabase and gabbro, white vein quartz, and red Red Beach Granite, which is exposed along the highway on the hill north of the cove.

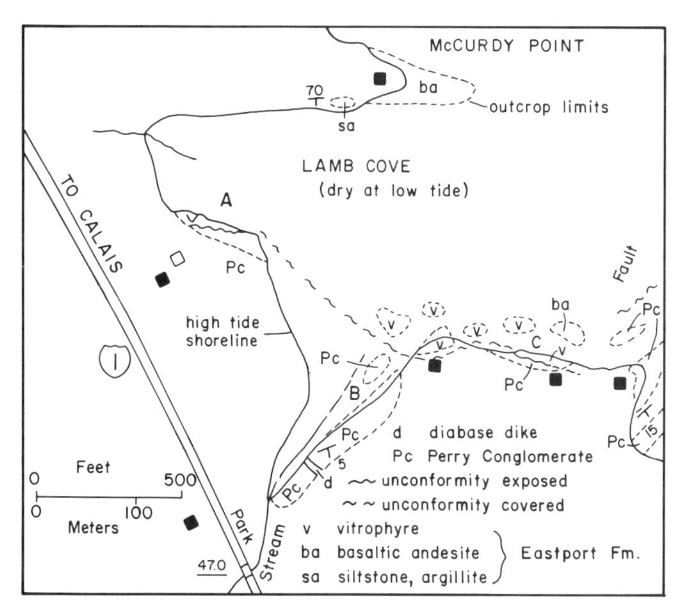

Figure 2. Sketch map of Lamb Cove (mile 47.0 [km 75.6], Figure 1) showing exposures of Acadian unconformity. Letters A, B, and C key specific outcrops discussed in text.

43.6–45.2 (70.2–72.3). The Red Beach Granite, dated at 400 Ma by Rb-Sr whole-rock analysis (Spooner and Fairbairn, 1970), intrudes folded basaltic-andesite of the Eastport Formation north of Boyden Lake 2 mi (3.2 km) west of Mill Cove (Fig. 1). Abbott (1978) distinguished five varieties of granite, one of which is the quartz-perthite-plagioclase-hornblende granophyre exposed here along Maine 1. The redness of the granite increases towards the unconformity, perhaps indicating Devonian weathering. Mariolitic cavities, granophyric texture, a weak metamorphic aureole, the generally small grain size, and electron microprobe analysis of mineral compositions led Abbott to conclude that the granite crystallized at about 700°C at a depth of 0.5 to 0.9 mi (0.8 to 1.5 km). The roadside park at mi **45.2** (km **72.7**) just north of Robbinston affords excellent exposures of the Red Beach granite and nice views across the St. Croix River to the red rocks of the Perry Formation north of St. Andrews, New Brunswick.

47.0 (75.6). The Acadian unconformity is exposed along the shore here at Lamb Cove (low tide essential). Gently dipping

Upper Devonian Perry Conglomerate overlies steeply dipping volcanic rocks of the Eastport Formation, part of a roof pendant in the Red Beach Granite, along an irregularly eroded surface. On Figure 2, letters A, B, and C key specific outcrops to the map. At McCurdy Point, porphyritic vesicular basaltic-andesite flows of the Eastport Formation underlie dark purple siltstone and argillite dipping steeply to the south. At A, unbedded Perry Conglomerate composed of rounded boulders and cobbles of Red Beach Granite and of gabbro together with pebbles of white vein quartz, common in the granites, and of aphanitic volcanic rocks rests on a jointed and fractured, purple and maroon-mottled devitrified vitrophyre. The surface of the contact dips about 45° to the south in contrast to the 5°–15° dips of the Perry Conglomerate in the Lamb Cove area. Perhaps the unconformity surface here marks the side of stream channel cut in the vitrophyre. At B, Perry cobble-and-pebble conglomerate, arkose, and red siltstone, cut by a diabase dike, form an overhanging cliff. At C, in the low cliff, the vitrophyre is much redder than in the beach area, perhaps due to the circulation of groundwater through the overlying Perry Formation and along the unconformity. The basal Perry here consists of a breccia of angular fragments of the underlying vitrophyre and rests on a sharp and irregular erosion surface. The breccia is overlain by interbedded lenses of arkose and grit made up largely of chips of vitrophyre. These are probably slope-wash deposits.

The outcrops along Maine 1 north to Calais are of gabbro, diorite, Red Beach Granite, and other granitic rocks of the plutonic belt.

REFERENCES CITED

Abbott, R. N., Jr., 1978, Geology of the Red Beach Granite, *in* Ludman, A., ed., Guidebook for field trips in southeastern Maine and southwestern New Brunswick (New England Intercollegiate Geological Conference 70th Annual Meeting): Flushing, New York, Queens College Press, p. 17–37.

Amos, D. H., 1963, Petrology and age of rocks, extreme southeastern Maine: Geological Society of America Bulletin, v. 74, p. 169–194.

Bastin, E. S., and Williams, H. S., 1914, Eastport folio, Maine: U.S. Geological Survey Folio 192, 15 p.

Chapman, C. A., 1962, Bays-of-Maine igneous complex: Geological Society of America Bulletin, v. 73, p. 883–888.

Dewey, J. F., and Kidd, W.S.F., 1974, Continental collisions in the Appalachian-Caledonian orogenic belt; Variations related to complete and incomplete suturing: Geology, v. 2, p. 543–546.

Fullagar, P.D., and Bottino, M. L., 1970, Rb-Sr whole rock ages of Silurian–Devonian volcanic rocks from eastern Maine; Shorter contributions to Maine geology: Maine Geological Survey Bulletin 23, p. 49–52.

Gates, O., 1961, The geology of the Cutler and Moose River quadrangles, Washington County, Maine: Maine Geological Survey, Quadrangle Mapping Series no. 1, 67 p.

—— , 1975, Geologic map and cross-sections of the Eastport Quadrangle, Maine: Augusta, Maine Geological Survey Geologic Map Series GM-3.

—— , 1978, The Silurian–Lower Devonian marine volcanic rocks of the Eastport Quadrangle, Maine, *in* Ludman, A., ed., Guidebook for field trips in southeastern Maine and southwestern New Brunswick (New England Intercollegiate Geological Conference 70th Annual Meeting): Flushing, New York, Queens College Press.

—— , 1983, The Silurian–Lower Devonian volcanic rocks of the Machias-Eastport area, Maine, *in* Hussey, A. M., II, and Westerman, D. S., eds., Field trips of the Geological Society of Maine 1978–1983: Geological Society of Maine, Maine Geology Bulletin no. 3, p.1–7.

Gates, O., and Moench, R. H., 1981, Bimodal Silurian and Lower Devonian volcanic rock assemblages in the Machias-Eastport area, Maine: U.S. Geological Survey Professional Paper 1184, 32 p.

Ludman, A., 1986, Timing of terrane accretion in eastern and east central Maine: Geology, v. 14, p. 411–414.

Osberg, P. H., 1978, Synthesis of the geology of the northern Appalachians; IGCP Project 27, Caledonian–Appalachian orogen of the North Atlantic region: Geological Survey of Canada Paper 78-13, p. 137–147.

Osberg, P. H., Hussey, A. M., II, Boone, G. M., 1985, Bedrock geologic map of Maine: Augusta, Maine Geological Survey, 1 sheet, scale 1:500,000.

Pajari, G. E., Jr., Trembath, L. T., Cormier, R. F., and Fyffe, L. R., 1974, The age of Acadian deformation in southwestern New Brunswick: Canadian Journal of Earth Sciences, v. 11, p. 1309–1313.

Potter, R. R., Hamilton, J. B., and Davies, J. L., 1979, Geologic map of New Brunswick: Fredericton, New Brunswick Department of Natural Resources, 1 sheet, scale 1:500,000.

Schluger, P. R., 1973, Stratigraphy and sedimentary environments of the Devonian Perry Formation, New Brunswick, Canada, and Maine, U.S.A.: Geological Society of America Bulletin, v. 84, p. 2533–2548.

Spooner, C. W., and Fairbairn, H. W., 1970, Relation of radioactive age of granitic rocks near Calais, Maine, to time of the Acadian orogeny: Geological Society of America Bulletin, v. 81, p. 3663–3670.

Williams, H., and Hatcher, R. D., Jr., 1982, Suspect terranes and accretionary history of the Appalachian orogen: Geology, v. 10, p. 530–536.

Wilson, J. T., 1966, Did the Atlantic close and then reopen?: Nature, v. 211, p. 676–681.

Zen, E-an, 1983a, Bedrock geologic map of Massachusetts, U.S. Geological Survey, 2 sheets, scale 1:250,000.

—— , 1983b, Exotic terranes in the New England Appalachians; Limits, candidates, and ages; a speculative essay, *in* Hatcher, R. D., Williams, H., and Zietz, I., eds., Contributions to the tectonics and geophysics of mountain chains: Geological Society of America Memoir 158, p. 55–81.

ACKNOWLEDGMENTS

The Maine Geological Survey has supported my field work in the Eastport and adjacent quadrangles.

Traveler Rhyolite and overlying Trout Valley Formation and the Katahdin Pluton; A record of basin sedimentation and Acadian magmatism, north-central Maine

Douglas W. Rankin, *U.S. Geological Survey, Reston, Virginia 22092 (Site 64)*
Rudolph Hon, *Boston College, Chestnut Hill, Massachusetts 02167 (Site 65)*

LOCATION

These sites cover approximately 15 by 31 mi (25 by 50 km), most of it wilderness, in and adjacent to Baxter State Park in northern Maine (Fig. 1). Important exposures are widely separated; much of the area is roadless and trailess. One should not deviate from the described routes without provisions, good maps, and adequate gear. In fact, this description assumes that the visitor has an adequate local map. It is not possible to visit all of the stops in one day. Stops 1 to 4 are within Baxter State Park, open May 15 to October 15. There is an entrance fee and camping is controlled and by reservation (Reservation Clerk, Millinocket, Maine 04462); no pets are allowed. Collecting permits are required to remove material from the park, and current rules disallow the use of hammers. A number of the stops include walks over difficult terrain. The hazards are stated under the discussion of each stop.

The 1:62,500 topographic quadrangle maps covering the sites are the Katahdin and Harrington Lake Quadrangles in the south and the Traveler Mountain Quadrangle in the north. Stops 1 to 3 may be reached from I-95 via Patten and Shin Pond. From I-95, either take Exit 58 at Sherman and follow Maine 11 north to Patten, or take Exit 59 at Island Falls and follow Maine 159 east to Patten. Stop 1 is near the Matagamon Gatehouse of Baxter State Park along the Grand Lake Road. For Stops 2 and 3, follow the Grand Lake Road for another 7.3 mi (11.7 km) beyond the Gatehouse to Trout Brook Crossing and from there take the side road 2.3 mi (3.7 km) to South Branch Ponds Campground. Stops 4 to 7 are within the Mount Katahdin region, accessible from I-95 via Exit 56 at Medway and Maine 11/157 to Millinocket. An alternate approach is from the southwest from Greenville. Stop 4 is within Baxter State Park, and its starting point at Roaring Brook Campground (Fig. 1) is reached from the Millinocket–Ripogenus Dam road by taking a side road north 2 mi (3.2 km) to the Togue Pond Gatehouse at the southern entrance to the Park and continuing another 8 mi (12.9 km) to the campground.

Stops 5 to 7 are along an unnumbered State Road between Millinocket and Abol Bridge and a private road open to the public between Abol Bridge and Ripogenus Dam. Land owners may operate entrance gates at various locations. Stop 5 is at Ripogenus Dam, about 31 mi (50 km) northwest from Millinocket. Stop 6 is at Big Ambejackmockamus Falls, 4.8 mi (7.7 km) east of Ripogenus Dam, 2.8 mi (4.5 km) east of Telos road turnoff, and 7.3 mi (11.7 km) west of Abol Bridge. Stop 7 is at Pockwockamus Falls, reachable from Abol Bridge via a dirt road along the southern side of West Branch Penobscot River. The turn to the dirt road is immediately west of the bridge, and the falls (rapids) are 2.0 mi (3.2 km) down the dirt road.

SIGNIFICANCE

Geologic studies of the region covered in this report have contributed significantly to the understanding of the Acadian orogeny and its igneous activity, early land-plant evolution, and igneous processes. The close spacial association of a geochemically and petrologically unique extrusive-intrusive sequence, the fossil control of the underlying and overlying strata, and the low grade of regional metamorphism are among the important characteristics that make this region one of the most critical in elucidating the Acadian orogeny. The focus is on the northern part of the very large (521 mi^2; 1,350 km^2), postorogenic Katahdin pluton (text by Hon) which intrudes folded Lower Devonian strata as well as its own volcanic pile, the Traveler Rhyolite (text for this and for the Trout Valley Formation by Rankin). The Traveler is unconformably overlain by the little deformed (probably post-Acadian) Trout Valley Formation of late Early to Middle Devonian age. The Trout Valley contains abundant detritus of the Traveler Rhyolite and, according to Kasper (1980), one of the richest early-land-plant deposits in the world. Geologic relations of the site pose tantalizing questions concerning the age of the Acadian orogeny.

SITE INFORMATION

The Katahdin-Traveler Igneous Suite is an exhumed magma chamber and a preserved part of its volcanic carapace. That the Acadian orogeny was short lived in this area is suggested by observations that the crystallized magma chamber (the Katahdin Granite, formerly Katahdin Quartz Monzonite) crosscuts folded Lower Devonian strata yet appears to be undeformed, whereas the ancestral volcanic carapace (Traveler Rhyolite) was erupted onto those same Lower Devonian strata while they were still poorly consolidated and was folded along with the underlying strata (Neuman and Rankin, 1980).

The Traveler Rhyolite is mostly a series of welded ash-flow sheets preserved within a structural depression interpreted to be a caldera. Its 10,500 ft (3,200 m) thickness within the caldera, and its conservatively estimated volume of 190 mi^3 (800 km^3) based

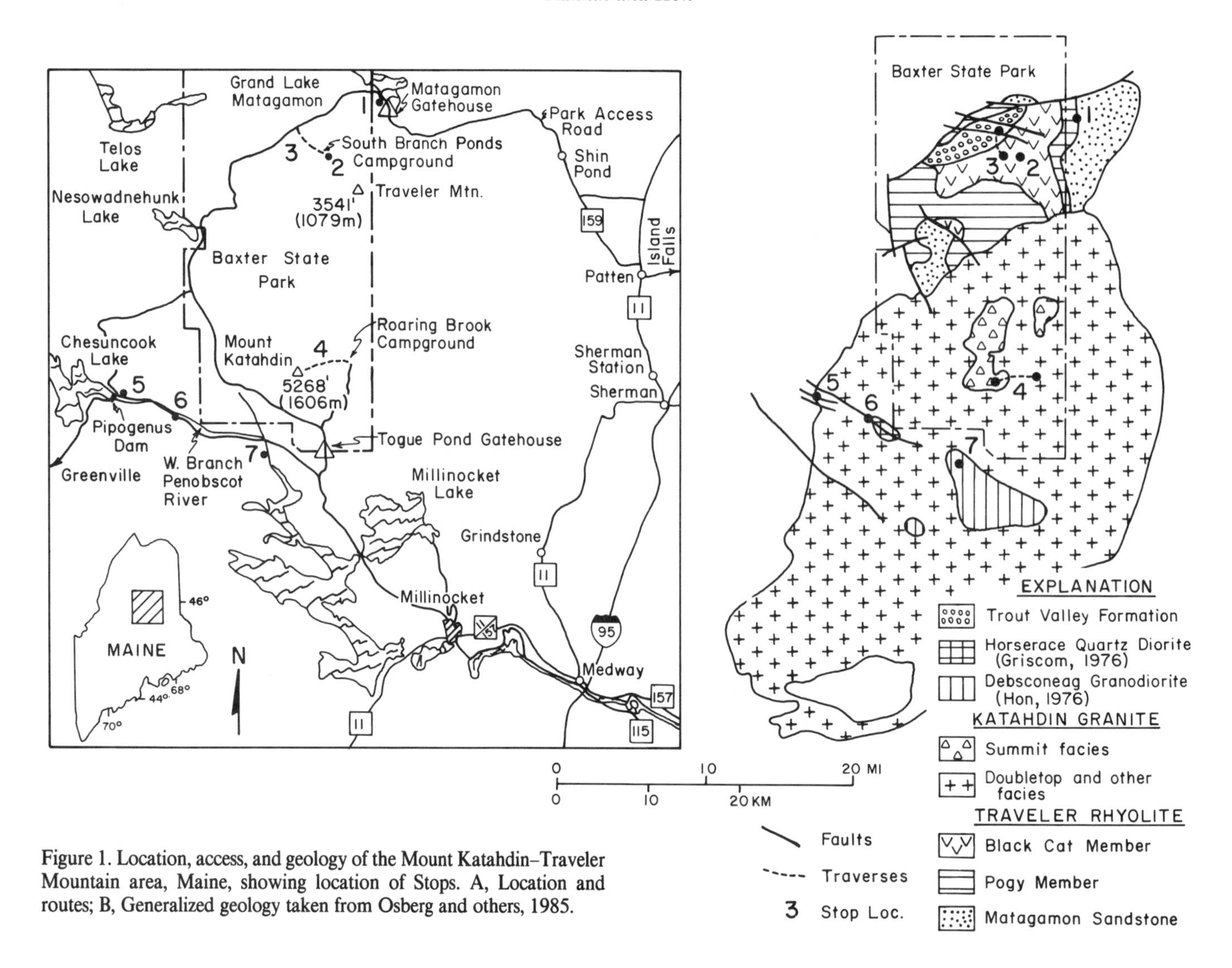

Figure 1. Location, access, and geology of the Mount Katahdin–Traveler Mountain area, Maine, showing location of Stops. A, Location and routes; B, Generalized geology taken from Osberg and others, 1985.

on the existing exposures, make it a significant world occurrence of ash-flow tuff. Hon argues that the volcanic carapace was far more extensive and that the entire Katahdin pluton was subvolcanic to an immense caldera that would qualify as among the largest volcanic features known in the world. The estimated dimensions near the base are, according to Hon, about 40 mi by 9 to 15 mi (65 km by 15 to 25 km); the estimated total volume of magmatic material is on the order of 1,200 mi^3 (5,000 km^3). The present erosion level permits inspection of 50% of the lower part of the complex.

Granites of the Katahdin pluton and rhyolites of the Traveler show remarkable chemical homogeneity given the large volume of both bodies. The granites are homogeneous within the limits of analytical precision; the rhyolites show a small degree of fractionation. Both are metaluminous to slightly peraluminous, and their unusually high scandium abundances suggest a distinctive petrogenetic province. The Katahdin Granite is mostly medium-grained biotite granite that shows large textural, though not mineralogical or geochemical, variations from a "normal" granitic texture in the interior of the pluton to a granophyric and miarolitic fine-grained granite near the summit of Mount Katahdin.

The **Traveler Rhyolite** is the youngest stratified unit intruded by the Katahdin pluton. It is the northernmost and by far the largest of a discontinuous belt of rhyolitic and shallow granophyric intrusive rocks of Early Devonian age that extends 100 mi (160 km) across north-central Maine in the Moose River synclinorium and its counterpart to the northeast, the Matagamon synclinorium. They are collectively called the Piscataquis volcanic belt (Rankin, 1968), which lies northwest of a nearly coeval belt of Acadian mafic and felsic plutons, the Greenville plutonic belt (Hon, 1980). The rhyolites were erupted on shallow-water marine fossiliferous sandstone of Early Devonian age (Becraft-Oriskany), the Matagamon Sandstone and its correlatives, in or at the margin of a deeper water marine basin represented by the flyshoid Seboomook Formation. Five major volcanic centers have been identified and the rhyolites are more peraluminous (higher percentage of normative corundum) toward the southwest. Some of the rhyolites contain prominent almandine garnet phenocrysts that are, however, extremely rare in the Traveler Rhyolite.

Little sedimentary rock and no paleosol horizons have been identified within the Traveler volcanic sequence, suggesting that the rhyolite was erupted in a relatively short time. The dominance of welded tuff argues for subaerial eruptions requiring a change from the marine paleoenvironment of the Seboomook and Matagamon (Rankin, 1960, 1965). Evidence that the Traveler Rhyolite is welded ash-flow tuff comes from both outcrop and thin section. The features preserved include flattened pumice lumps (fiamme; Fig. 4), deformed (flattened) shards, columnar jointing, and massive units without obvious bedding. No glass remains in these Devonian rocks, and welding per se cannot be demonstrated, but flattening of shards and pumice lumps certainly can.

The Traveler is a flinty aphanitic rhyolite that contains small (1 to 3 mm) phenocrysts. Color ranges from light gray through various shades of green, greenish gray, and bluish gray to nearly black. The darkest rocks contain the least altered phenocrysts and the best preserved primary textures.

The Traveler Rhyolite is composed of two named members and at least two additional members, which are at present poorly understood and are not exposed along the route of the field guide.

The basal Pogy Member, approximately 3,000 ft (900 m) thick, is characterized by moderately compacted welded ash-flow tuff containing about 15% phenocrysts, of which one-third are quartz and most of the rest are plagioclase (with oscillatory zoning, but generally progressing from cores of about An_{56} to rims of about An_{35}; Rankin, 1961). Sanidine (Or_{67}) occurs with plagioclase and quartz in a few samples. Mafic silicate phenocrysts may be absent or may constitute as much as 10% of the phenocryst population. They are generally altered beyond recognition, but clinopyroxene is present in one thin section, and the remnants of biotite phenocrysts were observed in a few. Garnet occurs in two samples. Lithic fragments, including rhyolite, diabase, sandstone, and shale, are present in most samples.

The overlying Black Cat Member, about 7,500 ft (2,300 m) thick, is characterized by highly compacted and welded ash-flow tuff containing 10% phenocrysts. Typically, 75% of these are plagioclase (with oscillatory zoning, but generally progressing from cores of An_{61} to rims of An_{33}), 20% are augite (optically determined to be about $Wo_{30}En_{28}Fs_{42}$), and 5% are magnetite (Rankin, 1961). Quartz phenocrysts are typically not present, but rarely may constitute as much as 10% of the phenocrysts. Biotite phenocrysts coexist with augite locally, and rarely, hornblende is the only mafic silicate present. Fayalite coexisting with augite and biotite is present in one thin section, and garnet phenocrysts occur in two samples. Glomeroporphyritic texture, rare in the Pogy Member, is common in the Black Cat.

To a first approximation, bulk compositions of both members fall within the synthetic, steam saturated, granite system close to the minimum melting compositions. The major mineralogic difference between the two members is that the Pogy Member contains quartz phenocrysts and the Black Cat Member typically does not. The Pogy Member contains a somewhat higher total percentage of phenocrysts, is more altered, and more commonly contains lithic clasts than the Black Cat. Pumice lumps in the Black Cat are characteristically more compacted than those of the Pogy. Length to thickness ratios of 10:1 to 20:1 are typical for the collapsed pumice lumps of the Black Cat. Ratios as high as 60:1 have been observed. Some Black Cat ash flows were apparently hot enough at the time of emplacement that they continued to flow during or after welding as evidenced by rotated phenocrysts and flow folds. In the Pogy, on the other hand, length to thickness ratios of the pumice lumps of 2:1 to 4:1 are more common.

These observations led to the following model, presented in more detail by Rankin (1968, 1980). The basal Pogy Member was erupted from the cooler and more volatile, H_2O-rich, upper part of the magma chamber and the overlying Black Cat Member followed, probably relatively quickly, from deeper, hotter, and drier parts of the magma chamber. Hon (1976) calculated that the phenocrysts of the Black Cat Member form an equilibrium assemblage at $T = 800°C$, $P(H_2O) = 1100$ bars, and $f(O_2)$ near the fayalite-magnetite-quartz buffer curve. The temperature for the Pogy Member is approximately 40° lower.

The **Katahdin Granite** is clearly intrusive into the Traveler Rhyolite as documented by numerous injections of granite into the rhyolite and by a narrow contact zone (300 ft; 100 m) of recrystallized rhyolite. Over 95% of the Katahdin pluton is composed of one of the several textural facies of Katahdin Granite. The remaining 5% is the Debsconeag Granodiorite of Hon (1976), in the center of the pluton, and the Horserace Quartz Diorite of Griscom (1976), along the West Branch Penobscot fault system. The pluton is considered to be a laccolith 25 mi in diameter and about 3 mi thick (40 km by ~5 km), not yet fully unroofed along its northern to northwestern contacts with Traveler Rhyolite. The longer axis of the pluton trends northeast, also the direction of decreasing regional metamorphic grade. By assigning reasonable lithostatic pressures to each of the metamorphic zones, Hon (1980) calculated that post-Acadian uplift tilted the crust about 5° so that the southwestern part of the pluton was originally 2 mi (3.5 km) deeper than the northeastern part. Thus the present land surface is an oblique section through the pluton.

The core of the pluton is formed by a massive, structureless, medium- to fine-grained biotite granite with remarkably constant mineralogy: 34% alkali feldspar (Or_{70}), 26% normally zoned plagioclase (An_{34} to An_{25}), 34% quartz, 6% biotite, and accessories of apatite, allanite, zircon, tourmaline, opaques, epidote, and rare fluorite and fayalite. Muscovite and titanite are notably absent. The Katahdin Granite is subdivided into distinct textural types delineated by Griscom (1976) and Hon (1976). The most common is the Doubletop facies of "normal" granitic texture, which forms most of the interior of the pluton. This facies grades upwardly into the Chimney facies characterized by finer-grained granophyric intergrowths that account for more than 60% of the rock. Toward the summit of Mount Katahdin the rocks become miarolitic, hence the Summit facies. At first appearance the miarolitic cavities are small and entirely filled with tourmaline, epidote, and rarely fluorite. Farther into the Summit facies the cavities are larger (about 1 up to several cm) and open.

The granites of the Summit facies, because of their finer grain size and "interlocking" granophyric texture, are more resistant to erosion than the more granular varieties. Once this erosional shield is broken, the more granular varieties are much more easily disintegrated and eroded, which explains the unusual geomorphology of Mount Katahdin, namely the existence of steep slopes surrounding the Tableland. This gently rolling high plateau (more than 4,400 ft; 1340 m in elevation) is underlain by a 300- to 1,600-ft-thick (100 to 500 m) layer of Summit facies, whereas the steep slopes are underlain by rocks of the Doubletop facies.

The best estimates of crystallization conditions for the Katahdin Granite are 710°C, $P_t = P(H_2O) = 1{,}200$ bars, and $f(O_2)$ near the fayalite-magnetite-quartz buffer curve (Hon, 1976). Determination of the age of crystallization for the Katahdin Granite poses a problem encountered whenever appreciable spread in Rb/Sr ratios is lacking. Naylor and others (1974) and later Loiselle and others (1983) suggested ages between 380 and 414 Ma. More recent $^{40}Ar/^{39}Ar$ data (Denning and Lux, 1985) and re-evaluation of the original Rb/Sr data (Hon and others, unpublished data) show that the granite crystallized between 395 and 400 Ma and that they postdate the rhyolite by about 5 to 10 m.y.

The **Debsconeag Granodiorite** is a hornblende-bearing biotite granodiorite, forming two stocks in the central and southern portion of the Katahdin pluton (Griscom, 1976). It is exposed along the rapids of the West Branch Penobscot River (Abol, Pockwockamus, and Debsconeag falls). Both stocks intrude the Doubletop facies and at least one of them is intruded by a dike swarm of Horserace Quartz Diorite. The Debsconeag Granodiorite is mineralogically and texturally quite similar to the Doubletop facies. The noticeable differences include a finer grain size of biotite and quartz (25% to 35% modally), a lower modal content of alkali feldspar (15% to 25%), higher proportion of biotite (8% to 15%) and plagioclase (35% to 40%), and ubiquitous, though subordinate presence of hornblende. Alkali feldspar tends to be more euhedral giving the rocks a subporphyritic appearance.

The **Horserace Quartz Diorite** forms a small elongated stock (1 by 3.5 mi; 1.6 by 5.6 km) between Big Ambejackmockamus Falls and Nesowadnehunk Deadwater of the West Branch Penobscot River (Griscom, 1966; Brown and Hon, 1985). Allingham (1960) used aeromagnetic data to model the shape of the stock to a depth of at least 2 mi (3 km) and suggested that the stock appears to be conical with outwardly dipping contacts at angles of about 55°. Its elongation along the West Branch Penobscot fault system and its injection as a dike swarm in the same direction suggest its emplacement was, at least in part, tectonically controlled.

The intrusion is concentrically zoned from hornblende quartz diorite at the margins grading inward to biotite granodiorite (Griscom, 1966; Brown and Hon, 1985). Modal content of hornblende varies from trace amounts to about 15%, that of biotite from a few to 12%, plagioclase from 55% to 70%, alkali feldspar from trace amounts to 15%, and quartz from 8% to 25%. Common accessories include titanite, apatite, zircon, oxides, and sulphides. $^{40}Ar/^{39}Ar$ age determinations on hornblendes and biotites show that the age of crystallization is 375 ±3 Ma and that it cooled to temperatures below 400°C very rapidly (Denning and Lux, 1985).

The following model is suggested for the origin of the Katahdin-Traveler Igneous Suite. At approximately 405 Ma, mafic magma of yet unknown type intruded the lower crust (at depth of about 30 km) causing a 30% partial fusion of geosynclinal sedimentary rocks. The newly formed magmas at temperatures of 750° to 800°C then moved slowly upward, and formed a magma chamber at a depth of 2.5 to 3 mi (4 to 5 km) below the surface. From this depth (water pressure compensation depth) periodic eruptions (and probably also periodic recharges) deposited numerous ash-flow sheets. When temperature in the magma chamber fell to about 700° to 720°C (about 5 to 10 m.y. later), the magma crystallized as granites, each of the textural varieties reflecting different cooling rates. Higher cooling rates led to development of granophyric and miarolitic textures, whereas slow cooling rates resulted in "normal" granitic textures. Sometime later the stocks of Debsconeag Granodiorite were intruded into the granites and thereafter, at 375 Ma, the Horserace stock was emplaced as the last igneous activity in this region. The Horserace Quartz Diorite (and possibly the Debsconeag Granodiorite) parental magma is clearly unrelated to the Katahdin-Traveler sequence and probably represents a calc-alkalic magma originally formed within the mantle but later modified in the crust.

STOP DESCRIPTIONS: SITE 64

Stop 1. Base of cliffs of Horse Mountain and shore of Grand Lake Matagamon (Traveler Mountain 15-minute Quadrangle). This stop offers exposures of the stratigraphic contact between the Traveler Rhyolite (Pogy Member) and the underlying Matagamon Sandstone (Fig. 2).

Station 1. Park along the Grand Lake Road 0.4 mi (0.6 km) north of the boundary of Baxter State Park (1.4 mi; 2.3 km northwest of the bridge over the East Branch of the Penobscot River), roughly opposite the south end of an island in Grand Lake Matagamon. At the park boundary, pause and note the 600 ft (180 m) cliffs of welded ash-flow tuff of the Pogy Member forming the east face of Horse Mountain above you. Columnar joints are prominent. Many columns are more than 3 ft (1 m) in diameter. Climb about 200 ft (60 m) up steep, partly wooded, hazardous scree slopes. The Matagamon Sandstone underlies the scree slope and is exposed in scattered outcrops toward the top. Note the detrital muscovite, mudstone chips, and rusty spots from weathered ankerite. The sandstone strikes approximately N15°E and locally dips as much as 40°W. Some outcrops display trough cross-bedding. Ash-flow tuffs of the Pogy Member

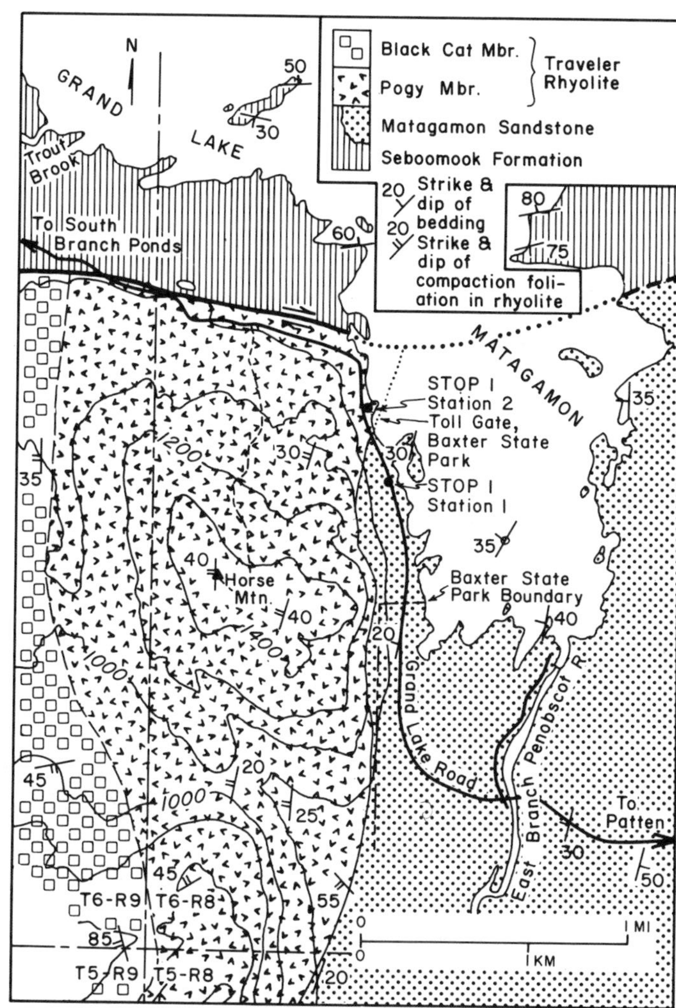

Figure 2. Geology of the Horse Mountain area, Stop 1. Topography from the U.S. Geological Survey Traveler Mountain 15-minute Quadrangle. Contour interval 200 ft (61 m). Geology by D. W. Rankin.

form the cliffs above. The contact, defined as the first appearance of recognizable layers of rhyolite, is exposed in places at the top of the scree slope and dips 20°W. The top 20 ft (6 m) or so of the Matagamon contains scattered pebbles of felsite and beds of tuffaceous sandstone, indicating that some volcanic activity preceded the main body of rhyolite. In the vicinity of Station 1, an irregular layer of sandstone up to 1.6 ft (0.5 m) thick was observed within the lower several feet (meters) of rhyolite roughly parallel to the compaction foliation in the ash-flow tuffs. Elsewhere sandstone dikes, some anastomosing and some as thick as 8 ft (2.5 m), have been observed in the base of the rhyolite indicating that the sand was unconsolidated at the time of the first major ash flows. The contact as seen along the base of the cliffs, is an irregular surface with relief of as much as 16 or 20 ft (5 or 6 m). This irregularity may be due to scouring by ash flows.

The basal 2 to 4 ft (0.6 to 1.2 m) of the massive-appearing rhyolite is composed of nonwelded tuff in which devitrified but nondeformed shards are clearly visible (Rankin, 1980, Fig. 1A). This nonwelded tuff grades up into welded ash-flow tuff that makes up most of the rhyolite of the Horse Mountain cliff. Fragments of collapsed pumice are visible in the rhyolite about 3 ft (1 m) above the base. Flattened and deformed shards are visible in a thin section collected from this station, 5 ft (1.5 m) above the base (Rankin, 1980, Fig. 1B).

Station 2. Gatehouse, Baxter State Park, about 0.3 mi (0.5 km) north of Station 1. The Pogy Member is well exposed along the lake shore from the gate house out to a small point. Crude columnar joints about 1 ft (0.3 m) in diameter plunge steeply southeast into the lake behind the gate house. Compaction foliation here strikes N15°E and dips 35°W. The ash-flow tuff is unusually highly compacted for the Pogy Member.

Stop 2. Center Ridge, Traveler Mountain (Traveler Mountain Quadrangle). This stop offers excellent exposures of welded ash-flow tuff of the Black Cat Member as well as near and distant perspectives of the ash-flow sheets and their structural setting. To reach the stop follow the Pogy Notch Trail south from the southeast corner of the South Branch Ponds Campground about 1.4 mi (2.3 km) around Lower South Branch Pond to the beginning of the Center Ridge Trail. Parts of this trail are steep and this ascent of Center Ridge should not be attempted in bad weather.

Before starting along the trail, walk down to the shore of the pond. Loose cobbles and pebbles of the Black Cat Member along the shore offer some of the best examples of the crude planar fabric (eutaxitic texture) typical of welded tuffs. This fabric is thought to be caused by the collapse of hot pumice lumps (fiamme) after the ash flow had largely come to rest. Notice the high length-to-thickness ratios of the fiamme in many of the samples indicating extreme compaction and implying a high emplacement temperature. You may be able to spot rotated phenocrysts and small folds of the fiamme in some samples.

At about 0.9 mi (1.5 km), the Howe Brook Trail forks to the east (left). The Pogy Notch Trail here crosses an old (Pleistocene?) debris flow that came out of the Howe Brook valley and caused the present separation of Upper and Lower South Branch Ponds.

The Center Ridge Trail leaves the Pogy Notch Trail on a bluff overlooking Upper South Branch Pond. The Center Ridge Trail terminates at Peak of the Ridges (3,220 ft; 980 m), but all of the features described may be seen by the time one has reached a prominent level stretch of the ridge at an elevation of about 2,000 ft (670 m; an approximately 0.5 mi; 0.8 km walk).

Columnar joints are obvious in many outcrops along Center Ridge. They range in diameter from a few in (cm) to 6 ft (2 m); the largest ones are as much as a few tens of yards (meters) long. Columns also may be discerned on distant cliffs. In ascending this ridge, one crosses a series of sloping terraces, which suggests the erosion of a layered sequence. Commonly, the rises of the terraces are zones of well-developed columnar joints, whereas the treads are zones of indistinct columns or no columns. In a few sequences, the size of the pumice lumps changes from one terrace rise to another, suggesting that the rises represent different flow units of a welded tuff sequence. Some outcrops along the lower part of Center Ridge are characterized by pumice lumps as large as 20 in (50 cm) across. In even fewer sequences it has been observed that the degree of flattening decreases upward in a terrace rise. Detailed mapping eventually may demonstrate that individual flow units are traceable.

The compaction foliation is commonly roughly perpendicular to the axes of the columns. Where this relationship holds true, the direction of compaction was roughly perpendicular to the cooling surface—by analogy with young welded tuffs, a quasi-horizontal surface that may be treated as bedding. Look across the South Branch Ponds at Black Cat Mountain. Note that the terraces or benches are tilted to the northwest. The benches on Center Ridge are tilted to the northeast. Both of these directions are roughly parallel to the attitude of the compaction foliation on the respective mountains. Hence the glacial valley of the South Branch Ponds is along the axis of an open, north-plunging anticline in the Black Cat Member. Dips on the limbs of 20° to 40° are typical. The northwest trend of the ridges of the Traveler (Pinnacle, Center, and North) and Big Peaked Mountain are controlled by the attitude of the compaction foliation on the east limb of the structure.

The intrusive contact of the Katahdin Granite with the Pogy Member crosses the south side of Pogy Notch. The north-plunging anticline of the Pogy Notch–South Branch Ponds valley (South Branch Ponds anticline) may be a resurgent dome within the Traveler caldera. From Center Ridge one can look north-northwest down South Branch Pond Brook, the locus of Stop 3, and into the broad flat valley of Trout Brook underlain by the Trout Valley Formation. The gentle dips of the Trout Valley Formation do not reflect the South Branch Ponds anticline.

Stop 3. Traverse from Lower South Branch Pond, down South Branch Ponds Brook to The Crossing on the Grand Lake Road (Traveler Mountain Quadrangle). The one-way distance is about 3 mi (4.8 km; 3 hours); there is no trail and it requires repeated crossing of the stream. This traverse offers excellent stream exposures of Black Cat ash-flow tuff, the basal rhyolite conglomerate lentil of the Trout Valley Formation, numerous exposures of the main body of the Trout Valley containing land plant fossils, and altered diabase of unknown age, intrusive into the Trout Valley and probably of intermediate composition.

Go north along the South Branch Ponds Road to the top of a long hill (ca. 0.6 mi; 1 km from Lower South Branch Pond) and a small parking area on west side of road, marked "X" on Figure 3. A marked trail leads a short distance east to The Ledges—exposures of welded tuff of the Black Cat characterized by large pumice lumps and columnar jointing. Mount Katahdin is visible over the South Branch Ponds. From parking area walk west down steep slope through open woods to South Branch Ponds Brook. Turn right and walk downstream.

Station 1. Exposures of Black Cat Member. Note the pattern of concentric joints in outcrop on corner at stream level. Actually the joints are concentric about more than one center, giving rise to a pattern of intersecting curving joints. Distinctive

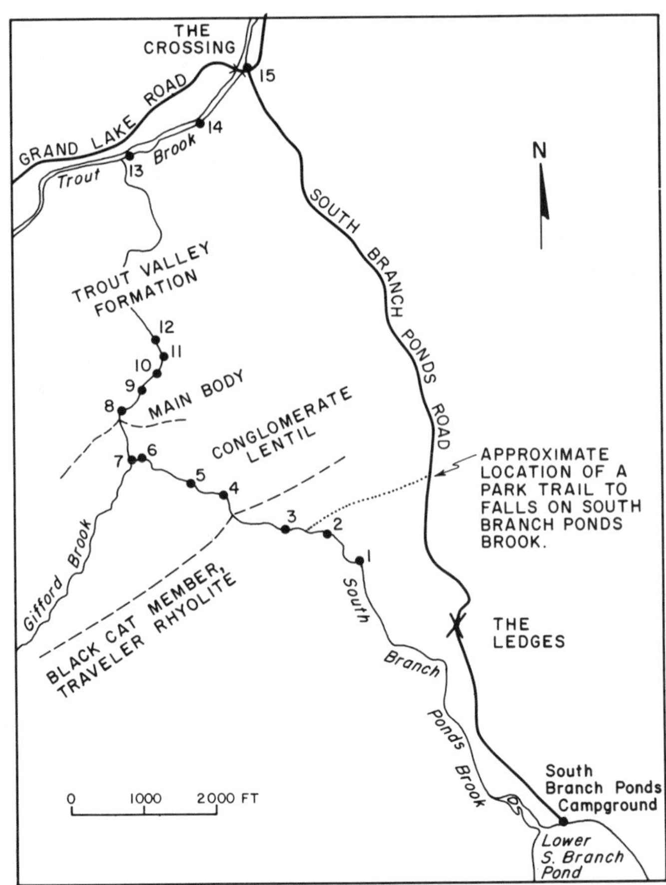

Figure 3. Sketch map of the South Branch Ponds Brook area, Stop 3. Base is an uncorrected aerial photograph. Station numbers on figure refer to localities described in text.

columnar joints of small diameter are present above the stream on the east bank. Higher and slightly downstream is another flow unit having columnar joints of larger diameter. Note the compaction foliation in the unit (eutaxitic texture) brought out by the presence of very thin collapsed pumice lumps.

Station 2. First of a series of sloping joint surfaces across which the stream flows. Close examination shows that these are dip surfaces of ash flows. Compaction foliation is roughly parallel to the surfaces, and, in places, crude columnar joints are roughly perpendicular to the surfaces. The ash flows strike about N70°E and dip 30°N. Local areas of crosscutting breccia are also present in this outcrop on the left bank of stream.

Station 3. Last of the series of dip slopes. Compaction foliation is visible on a number of steep joint surfaces (Fig. 4).

Station 4. In stream bed on east bank just upstream from steep gravel bank at corner. This is the lowest exposure of conglomerate of the Trout Valley Formation, which forms a lentil, probably a deltaic deposit at the base of the Trout Valley.

Station 5. Joints cut cobbles in conglomerate: first clue that this poorly consolidated conglomerate is not a Pleistocene till. Some cobbles are offset along the joints. Note that clasts (pebbles,

Figure 4. Photograph of outcrop of welded tuff of Black Cat Member, Traveler Rhyolite near Station 3, Stop 3 along South Branch Ponds Brook. Collapsed (flattened) pumice lumps (fiamme) give rise to a planar compaction foliation (eutaxitic structure) generally analogous to bedding in sedimentary rocks.

Figure 5. Photograph of outcrop of the basal conglomerate lentil of the Trout Valley Formation, Station 6, Stop 3 along South Branch Ponds Brook. Note joint surface on right that cuts cobbles in the conglomerate. A sandstone bed within the conglomerate is visible two-thirds of the way up the exposure. Pleistocene till overlies the conglomerate.

cobbles, and boulders) are well rounded and that *all* of them are rhyolite. The clasts are so weathered that many of them can be broken apart by hand. The weathering may date from the Devonian period. The conglomerate is cross-bedded and dips gently north, away from the volcanic rocks.

Station 6. About 33 ft (10 m) of Trout Valley conglomerate exposed in the canyon wall (Fig. 5). Where is the contact with the overlying Pleistocene till? Note the sandstone bed in the conglomerate near the top of the exposure and lenses of black sandy carbonaceous shale near the bottom.

Station 7. Junction with Gifford Brook.

Station 8. Large exposure at curve of stream on left bank. Coarse conglomerate is no longer dominant. You are now in the main body of the Trout Valley Formation above the basal conglomerate lentil. Numerous black chert lenses are visible and some have a vague internal structure. E. S. Barghoorn, in Dorf and Rankin (1962), identified one of these as *Prototaxites,* which is generally regarded as of algal affinities. Also present are siderite concretions and thin beds of sideritic ironstone.

Station 9. Upstream: sill of diabase and a 6-in-thick (15 cm) bed of ironstone. Downstream: lens of carbonaceous black shale, from which plant fossils were first collected in 1955 (Rankin, 1961).

Station 10. Light gray-green fine-grained diabase dike about 3 ft (1 m) thick. The dike trends N30°W and dips 40°N. Note the chilled dike margin against the sedimentary rocks.

Station 11. This is "Locality 4" of Dorf and Rankin (1962), from which the best specimens of flattened spiny stems of *Psilophyton* were collected.

Station 12. Gently dipping sill of diabase about 10 ft (3 m) thick. Note the net veining within the sill. Plant remains can be found in nearly every outcrop of sedimentary rock.

Station 13. Junction of South Branch Ponds Brook and Trout Brook. If time is short and water is not high, Trout Brook may be forded here. Grand Lake Road is a short distance north of the north (far) bank of Trout Brook. Turn right (east) on road for The Crossing. Alternatively, stay on the south bank of Trout Brook and turn right (east). Follow a faint fisherman's track along the right (south) bank of Trout Brook.

Station 14. This is "Locality 1" of Dorf and Rankin (1962) and Kasper (1980). Here a long outcrop of gently dipping sandstone and shale of the Trout Valley Formation is present. The sandstone is calcareous and current bedded. Kasper (1980) reports that this has been the most productive fossil site in the formation. The megafossils of land plants in the Trout Valley Formation include several species of naked plants (plants without leaves) that were formerly included under the name *Psilophyton,* now considered to be an artificial taxon (Kasper, 1980). On the basis of these, an age of late Early Devonian or earliest Middle Devonian was suggested by Andrews and others (1977). More recently, a lycopod, *Leclercqia complexa,* elsewhere known only from the Middle Devonian (Hamilton or Givetian of European terminology) has been described from the Trout Valley, but not from this station (Kasper and Forbes, 1979). Kasper (1980) reports the following assemblage from this site: *Psilophyton dapsile* (the best preserved plant at this location), ?*Psilophyton* sp., *Kaulangiophyton akantha, Taeniocrada* sp., *Thursophyton* sp., and *Sciadophyton* sp. Eurypterid scales were also found at this outcrop. A fault of unknown displacement cuts the southwest end of the exposure. Please remember that collecting permits are required to remove material from the park.

Station 15. The Crossing. Poorly preserved but large plant fossils occur in the outcrops beneath the bridge abutments.

STOP DESCRIPTIONS: SITE 65

Stop 4. Hike up the Helon Taylor Trail to Pamola, crossing the Knife Edge to Baxter Peak, the highest point in Maine. Katahdin Quadrangle map and altimeter are recommended. Return to Roaring Brook via the Saddle and Chimney Pond trails or by other trails, from either Pamola or Baxter Peak. The traverse covers nearly 10 mi (16 km) and 4,200 ft (1,280 m) of relief. You will need about 6 to 9 hours to complete the hike. Do not make this hike in inclement weather. Reserve several optional days and choose the clearest one.

Along the Helon Taylor Trail from Roaring Brook Campground (1,200 ft; 366 m) to Pamola Peak (4,902 ft; 1,494 m), you will observe the gradual textural transition in the Katahdin Granite from the Doubletop facies through Chimney facies to Summit facies (representing the roof of the magma chamber) as well as the porphyritic dike of Cathedral facies. The trail to an elevation of about 2,600 ft (792 m) passes through boulder fields and moraines, and passes outcrops of a finer-grained variety of Doubletop facies. From an elevation of 2,600 ft to about 3,800 ft (792 to 1,158 m) we suggest that you carry with you small pieces of different textural varieties to help you see the very gradual and often subtle changes. The granite at first becomes coarsely granophyric. As seen through a hand lens, smaller satellite quartz grains surround larger grains and are optically continuous with the larger grains. As you proceed upward, these satellite grains, intergrown with feldspar, become more abundant and finer grained. This is a characteristic texture of the Chimney facies. The biotite flakes also become somewhat thinner and larger. Near the transition into the Summit facies, more than 50% of the rock is composed of granophyric intergrowths that are extremely fine grained (you may need to moisten the rock to see the intergrowths with your hand lens). From here (3,800 ft; 1,158 m) to the summit of Pamola you will observe further advancement of the granophyric intergrowths and more importantly, the development of small and, at first, fully filled miarolitic cavities. Near 4,500 ft (1,372 m) you will see the first open cavities. The Summit facies is defined by the first occurrence of these miarolitic cavities. The cavities, which may be from 5 to 20 mm or more in diameter, were formed by trapped water-rich exsolved vapor bubbles. On emplacement, this magma contained about 4.5% water by weight (Hon, 1976), which was gradually purged out during crystallization, eventually escaping into the surrounding country rocks. On rapid cooling, however, these individual water-rich bubbles became trapped, forcing the magma to crystallize around them. The much faster diffusion rates in the vapor phase, as contrasted with the more sluggish rates in the melt, allowed the individual crystals at the perimeter of the bubbles to grow larger. Because of the absence of growth obstruction, the crystals projecting into the cavities commonly have euhedral terminations. Hydrothermal minerals inside the cavities include dark blue tourmaline, epidote, and rarely fluorite.

From Pamola the trail continues along The Knife Edge toward Baxter Peak. Return to Roaring Brook via a trail of your choice and observe the same textural changes over again but in reverse order.

Stop 5. Ripogenus Dam (Harrington Lake Quadrangle). Drive slowly over the dam and park on the northern side of the river. Walk 1 mi (1.6 km) east (down river), following a trail to its end along the northern side of Ripogenus Gorge; then very carefully descend the steep hillside toward the river bank just opposite an outlet from the power station. This stop illustrates the typical character of the contact between granite and country rocks on the northwest side of the Katahdin Pluton. At the contact, granite intrudes into the country rocks (Silurian calcareous sandstones of Ripogenus Formation; Griscom, 1966) a few tens of ft (m) as apophyses a few in (cm) to 1 ft (0.3 m) wide. A few large blocks of the country rocks occur as xenoliths close to the contact, but beyond 50 ft (15 m) from the contact, the granite is entirely free of any inclusions. This contact is near the tapering edge of the laccolith, but on its bottom side, because the contact dips inward. A short way north the contact dips outward. The thickness of the laccolith in this area was probably not more than a few hundred ft (m). Also, note the absence of a major thermal overprint in the country rocks. The siliceous magma has been emplaced along a major propagating subhorizontal fracture by pushing away the country rock and flowing into the resulting space. The magma flowed sideways under its own weight from the main reservoir at a depth of about 2 to 3 mi (3 to 5 km) near the center of the pluton. At this depth, the stoping magma reached equilibrium between the lithostatic pressure and the partial pressure of water.

Stop 6. Big Ambejackmockamus Falls where the river turns abruptly 90° to the north (Harrington Lake Quadrangle). The road is heavily traveled, particularly by pulp-hauling oversized trucks. Park your cars well off the main road and exercise caution while walking on the road. Intrusive contact of Horserace Quartz Diorite into Katahdin Granite is exposed along the river which in this stretch roughly follows the contact between the Horserace stock and the surrounding granite. The contact and some partial mixing of both magmas can be observed in the exposures on the east and south banks of the river. Note that the granite must have been in a partially molten state, probably as the result of reheating by the Horserace Quartz Diorite. This is documented by numerous embayments and undulations along the contact that is seen in the roadcut exposures opposite the river and along the eastern river bank. Follow the bank down the river to observe the mineralogical variation that is modeled by a concurrent fractional removal of plagioclase and hornblende (Brown and Hon, 1985). You may see some cumulate gabbro.

Stop 7. Pockwockamus Falls (Katahdin Quadrangle). On the drive downstream from Abol Bridge you may be able to enjoy one of the most spectacular views of Katahdin. Park your car in one of several turnoffs and walk toward the river. The exposures along the rapids are of Debsconeag Granodiorite. These rocks are the least understood of all the igneous units within the Katahdin pluton and are best explained by a mixing of two magmas and subsequent fractional crystallization of hornblende, biotite, and plagioclase. The two sources of magmas are the Katahdin Granite and Horserace Quartz Diorite. Mineralogical and textural variations are similar to the ones observed in the Horserace Quartz Diorite but are considerably less pronounced.

REFERENCES CITED

Allingham, J. W., 1960, Use of aeromagnetic data to determine geologic structure in northern Maine: U.S. Geological Survey Professional Paper 400-B, p. B117–B119.

Andrews, H. N., Kasper, A. E., Forbes, W. H., Gensel, P. G., and Chaloner, W. G., 1977, Early Devonian flora of the Trout Valley Formation of northern Maine: Review of Paleobotany and Palynology, v. 23, p. 255–285.

Brown, C., and Hon, R., 1985, A crystal fractionation model for the origin of a zoned calc-alkalic quartz diorite-granodiorite body; The Horserace Quartz Diorite of north-central Maine: Geological Society of America Abstracts with Programs, v. 17, p. 8.

Denning, A. S., and Lux, D. R., 1985, ^{40}Ar–^{39}Ar incremental release dating of two igneous bodies in the Katahdin batholith, north-central Maine: Geological Society of America Abstracts with Programs, v. 17, p. 15.

Dorf, E., and Rankin, D. W., 1962, Early Devonian plants from the Traveler Mountain area, Maine: Journal of Paleontology, v. 36, p. 999–1004.

Griscom, A., 1966, Geology of the Ripogenus Lake area, Maine, *in* D. W. Caldwell, ed., Guidebook for field trips in the Mount Katahdin region, Maine: New England Intercollegiate Geological Conference, 58th Annual Meeting, 1966, p. 36–41.

—— , 1976, Bedrock geology of the Harrington Lake area, Maine [Ph.D. thesis]: Cambridge, Massachusetts, Harvard University, 373 p.

Hon, R., 1976, Geology, petrology, and geochemistry of Traveler Rhyolite and Katahdin Pluton (north-central Maine) [Ph.D. thesis]: Cambridge, Massachusetts Institute of Technology, 239 p.

—— , 1980, Geology and petrology of igneous bodies within the Katahdin pluton, *in* Roy, D. C., and Naylor, R. S., eds., Guidebook to the geology of northeastern Maine and neighboring New Brunswick: New England Intercollegiate Geological Conference, 72nd Annual Meeting, 1980, Presque Isle, Maine, p. 65–79.

Kasper, A. E., Jr., 1980, Plant fossils (psilophytes) from the Devonian Trout Valley Formation of Baxter State Park, *in* Roy, D. C., and Naylor, R. S., eds., Guidebook to the geology of northeastern Maine and neighboring New Brunswick: New England Intercollegiate Geological Conference, 72nd Annual Meeting, 1980, Presque Isle, Maine, p. 127–141.

Kasper, A. E., and Forbes, W. H., 1979, The Devonian lycopod *Leclercqia* from the Trout Valley Formation of Maine: Maine Geology—Geological Society of Maine Bulletin no. 1, p. 49–59.

Loiselle, M., Hon, R., and Naylor, R. S., 1983, Age of the Katahdin Batholith, Maine: Geological Society of America Abstracts with Programs, v. 15, p. 146.

Naylor, R. S., Hon, R., and Fullagar, P. D., 1974, Age of the Katahdin Batholith, Maine: Geological Society of America Abstracts with Programs, v. 6, p. 59.

Neuman, R. B., and Rankin, D. W., 1980, Bedrock geology of the Shin Pond–Traveler Mountain region, *in* Roy, D. C., and Naylor, R. S., eds., Guidebook to the geology of northeastern Maine and neighboring New Brunswick: New England Intercollegiate Geological Conference, 72nd Annual Meeting, 1980, Presque Isle, Maine, p. 86–97.

Osberg, P. H., Hussey, A. M., II, and Boone, G. M., eds., 1985, Bedrock geologic map of Maine: Augusta, Maine Geological Survey, scale 1:500,000.

Rankin, D. W., 1960, Paleogeographic implications of deposits of hot ash flows: International Geological Congress, 21st, Copenhagen 1960, Report, pt. 12, p. 19–34.

—— , 1961, Bedrock geology of the Katahdin–Traveler area, Maine [Ph.D. thesis]: Cambridge, Massachusetts, Harvard University, 317 p.

—— , 1965, The Matagamon Sandstone, a new Devonian formation in north-central Maine: U.S. Geological Survey Bulletin 1194-F, 9 p.

—— , 1968, Volcanism related to tectonism in the Piscataquis volcanic belt, an island arc of Early Devonian age in north-central Maine, *in* Zen, E-an, White, W. S., Hadley, J. B., and Thompson, J. B., Jr., eds., Studies of Appalachian geology; Northern and maritime: New York, Interscience Publishers, p. 355–369.

—— , 1980, The Traveler Rhyolite and its Devonian setting, Traveler Mountain area, Maine, *in* Roy, D. C., and Naylor, R. S., eds., Guidebook to the geology of northeastern Maine and neighboring New Brunswick: New England Intercollegiate Geological Conference, 72nd Annual Meeting, 1980, Presque Isle, Maine, p. 98–113.

Geomorphology and glacial geology of the Mount Katahdin region, north-central Maine

D. W. Caldwell, *Department of Geology, Boston University, Boston, Massachusetts 02215*
Lindley S. Hanson, *Department of Geological Sciences, Salem State College, Salem, Massachusetts 01971*

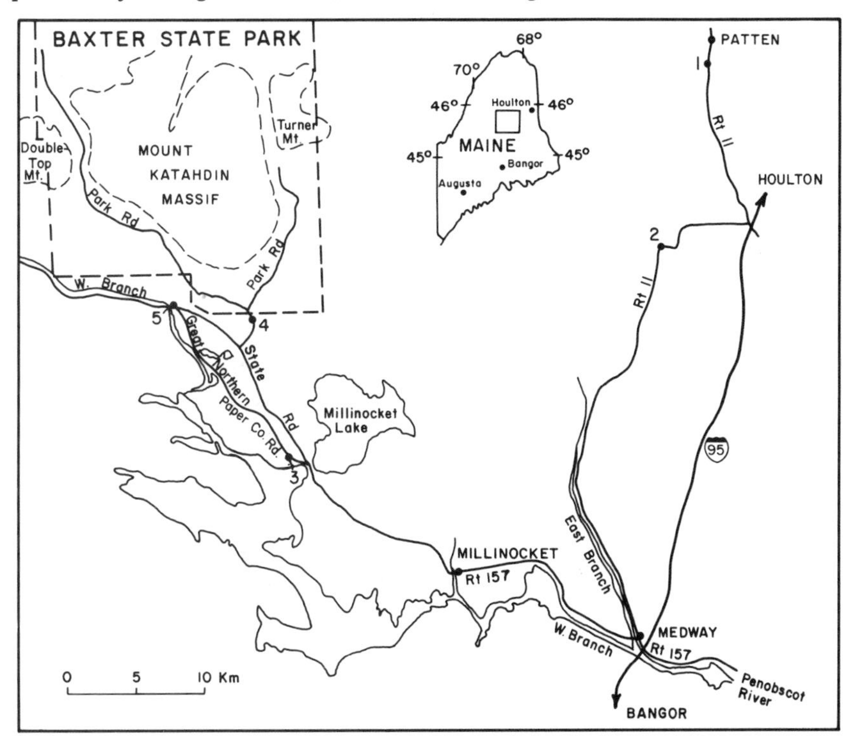

Figure 1. Location map of Mt. Katahdin and Baxter State Park. Numbers refer to locations of features or observation points discussed in text: 1, Top-of-the-World, view point near Patten; 2, Stacyville, east of Katahdin; 3, Rogen moraines near Millinocket Lake settlement; 4, Baxter Park gatehouse; 5, Abol bridge and view of south side of Mt. Katahdin.

LOCATION

Mount Katahdin is located west of Millinocket within Baxter State Park in north-central Maine. The Katahdin region (Fig. 1) is most easily reached from the east. To get to the park, leave I-95 at Medway (Exit 56) and take Maine 157 to Millinocket. From Millinocket, follow unmarked road about 16 mi (25 km) to the Toque Pond Road, which leads directly to the park gate. Although the roads themselves are not labeled, there are numerous signs indicating the direction to the park entrance. The park is open from May 15 to October 15, and there is a modest fee for vehicles with out-of-state registrations.

Beyond the gate, the park road follows the crest of a large esker. After 600 ft (200 m), take the right fork off the esker and follow this road for about 6 mi (10 km) to the Roaring Brook campground and park. Details of the mountain's geomorphology may be viewed from sites along the various trails that start here. Trails and site locations on the mountain are shown in Figure 2.

Many of the features discussed in the text can be viewed in panorama from the east shore of Sandy Stream Pond, about 0.3 mi (0.5 km) by trail from the Roaring Brook campground. For close observation of the geomorphic features, follow the Chimney Pond trail from Roaring Brook campground to the North and Great Basins. This 4.5-mi (7 km) hike may prove strenuous for those in poor condition. At about 1.9 mi (3 km), the trail crosses the Basin Pond moraine complex. At 2 mi (3.2 km), the North Basin cutoff trail leads to the right for about 0.6 mi (1 km), where it joins the Chimney Pond–North Basin trail. Turn right onto this trail for about 1,000 ft (300 m) to the crest of Blueberry Knoll. A well-marked trail leads along the floor of the North Basin cirque, to the headwall and a deposit that may be a rock glacier (Fig. 2). To reach Chimney Pond, return along the North Basin trail about 1 mi (1.6 km), where it joins the Roaring Brook trail. Turn right for 1,650 ft (500 m) to the Chimney Pond campground and an unobstructed view of the South Basin cirque and the Knife Edge arete. From Chimney Pond several trails lead to the uplands of Katahdin.

The south side of Katahdin is visible from the Abol bridge (Fig. 1, location 5). A good view of the geomorphology of the Katahdin massif and associated volcanic rocks can be seen from

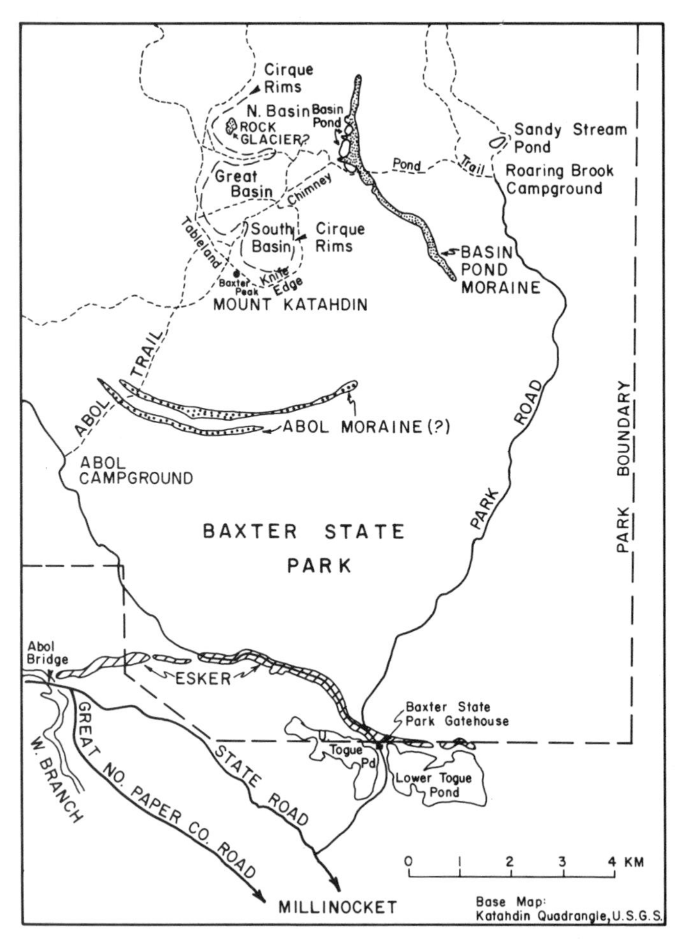

Figure 2. Map of Mount Katahdin and vicinity, showing roads, trails, and other points of interest discussed in text.

Top-Of-The-World, a roadside rest area about 1 mi (1.6 km) south of Patten and 5 mi (8 km) north of exit 58 from I-95 on Maine 11 (Fig. 1, location 1). There is a similar view in Stacyville, east of Katahdin.

The topography of Mt Katahdin is shown on the Mt. Katahdin and Harrington Lake Quadrangles, both at a scale of 1:62,500, as well as on the Mt. Katahdin and Doubletop Mountain orthophotoquads at a scale of 1:24,000. At this writing, the Katahdin and the Harrington Lake Quadrangles are being remapped at a scale of 1:24,000.

SIGNIFICANCE

Mount Katahdin is the only region in Maine with such well-developed features formed by alpine glaciation. Mt. Katahdin is also the highest mountain in Maine, and is one of the highest mountains in New England composed of plutonic rock, the Katahdin Granite (Griscom 1976; Hon, 1980). The high, steep-sided mountains north of Katahdin are composed of the Traveler rhyolite, the volcanic equivalent of the Katahdin granite.

Mount Katahdin is of particular geomorphic interest, as both continental and alpine glaciers had a profound influence on the geomorphology of the Mt. Katahdin area. Erratics on the summits of Katahdin and many adjacent mountains suggest that ice sheets were sufficiently thick to cover Mt. Katahdin (Caldwell, 1972; Davis and Davis, 1980), although there is no conclusive evidence to indicate when during the Pleistocene this occurred. Large cirques on the east side of Mt. Katahdin were thought by Caldwell and others (1985) to have been occupied by alpine glaciers during the waning stages of Wisconsinan glaciation. No other high mountains in Maine or in the rest of New England have features related to late-Wisconsinan alpine glaciation, with the possible exception of the Presidential Range in New Hampshire (Bradley, 1981). Davis and Waitt (1986), on the other hand, have concluded there is no evidence of post-Laurentide mountain glaciation in the White Mountains.

SITE INFORMATION

General Geomorphology. The Katahdin pluton lies at the northeastern end of a belt of mountainous terrain that is 150 mi (250 km) long by 30 mi (50 km) wide and is cored by Acadian plutons of the Greenville plutonic belt. Plutonic rocks in this belt have generally undergone granular disintegration and underlie basins. The surrounding mountains are resistant hornfels formed in contact aureoles around the plutons. The highest mountains in the Devonian terrane, other than the Katahdin and Traveler massifs, are those composed of this hornfels (Espenshade and Boudette, 1967).

Several theories have been proposed to explain the origin of the broad upland surface of Mt. Katahdin called the Tableland (Figs. 2, and 4). Goldthwait (1914) attributed the apparently flat upland to peneplanation; Thompson (1960, 1961) suggested leveling by cryoplanation; and Davis (1983) postulated that the surface was beveled by overriding ice.

Hon (1980) identified the Tableland as the roof of the Katahdin laccolith. The upper part of this pluton was chilled to a glass in contact with the country rock and has subsequently devitrified to form a granophyre called the Summit facies (Hon, 1980). The resistant granophyric caprock protects the easily disintegrated phaneritic rock of the underlying Doubletop facies, much like a resistant sandstone overlying shale. Removal of the granophyre is followed by rapid slope retreat and decline forming irregular hills and broad lowlands. Hence, the Katahdin pluton has the greatest relief of any in Maine, underlying both the highest peaks and the lowest basins of the region. This relationship is illustrated in Figure 3, a panoramic view as seen from Stacyville (Fig. 1, Location 2).

Glacial Geology

The glacial deposits in the lowlands near Katahdin are greatly affected by the course-grained Doubletop facies of the Katahdin pluton. Erosion of this rock has produced sandy, boulder-rich till that forms numerous moraine ridges (Hanson and Caldwell, 1983). These ridges have been identified as ribbed moraines by Thompson and Borns (1985) and as Rogen mo-

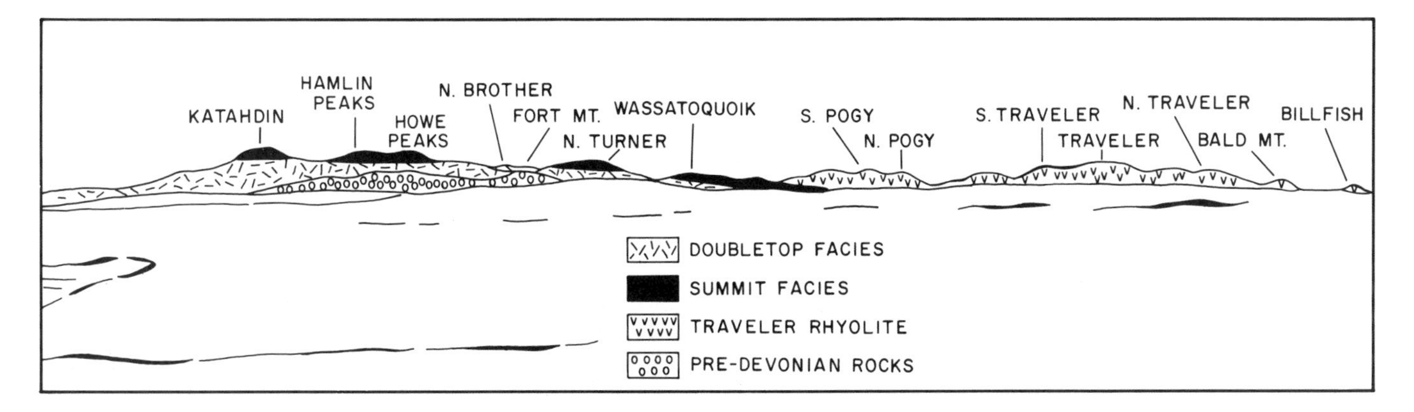

Figure 3. Panorama of the Katahdin and Traveler Ranges as seen from Stacyville, east of Katahdin. Facies within Katahdin pluton, as well as other nearby rock types, shown. Highest elevation are underlain by Summit facies of Katahdin pluton. South of Katahdin Range Summit, the low ground is underlain by Doubletop facies of Katahdin pluton. Sketch covers a distance of about 22 mi (35 km) from north to south. Vertical relief in Katahdin massif is about 4,000 ft (1,220 m) and about 2,000 ft (610 m) in Traveler area.

raines by Caldwell and others (1985). This type of moraine may be viewed at close hand on the Great Northern Paper Company road a few hundred meters west of the settlement at Millinocket Lake (Fig. 1, Location 3). Because of its broad areal extent, characteristic topographic expression, and texture, the Katahdin till forms a mappable unit (Thompson and Borns, 1985).

The meltwater deposits in the lowlands near Katahdin are dominated by long esker systems, which continue for more than 90 mi (150 km) to their terminus near the coast. Near the southern entrance to Baxter State Park (Fig. 2, Location 4), the park road follows one of these eskers for several kilometers. In the vicinity of Abol Bridge there are several pits in a large delta complex formed in a small glacial lake. Glacial-marine deposits consisting of deltas and silt and clay extend into inland Maine to near Medway (Fig. 1).

The Katahdin uplands contain remarkably well-preserved features formed by alpine glaciation. These features have recently been described by Caldwell and others (1985) and by Davis and Davis (1980). The landforms discussed below are shown in Figure 4.

Cirques. There are 10 or so cirques in this region, most of which are poorly developed, contain till deposited by continental ice, and have eroded headwalls. This evidence suggests that most of these cirques were not occupied by alpine ice during the late stages of deglaciation. The three large cirques on the east side of Katahdin, however, lack continental till, have steep rock headwalls ranging from 330 to more than 2,300 ft (100 to more than 700 m) high, and flat boulder-strewn floors. Thus the evidence from these eastern cirques suggests that they were more recently occupied by alpine glaciers. It may have been that snow blown from the Tableland increased the accumulation and thus the glacial activity in the east-facing cirques. Presently the greatest accumulation of snow on Katahdin is in these three cirques from blowing snow.

Aretes. Hamlin Ridge separates two cirque basins, North Basin and Great Basin. The Cathedral Ridge is apparently an eroded remnant of an arete developed between South Basin and Great Basin. The Knife Edge is an arete formed between continental ice flowing along the south flank of the mountain and alpine ice on the north side in South Basin (Fig. 2). Several workers have thought that these narrow serrated ridges could not have survived being overridden by continental ice; rather, they must have been reshaped by alpine glaciation following the continental glaciation (Antevs, 1932; Tarr, 1900; Caldwell and others, 1985). Caldwell and Hanson (1986) have suggested that Katahdin was not covered during the late Wisconsinan. If Katahdin was a nunatak throughout this period, it would explain why the serrate topography is so well developed and why the cirques appear to be so fresh when compared to others in the area that were covered by the last ice sheet.

Moraines. Caldwell (1972) identified moraines within each of the three large cirques, whereas Davis (1983, 1976) and Davis and others (1980) questioned the existence of these moraines. The small ridges of granite boulders within North Basin are believed to be end moraines of an alpine glacier. They can be studied at close range along the trail that leads from Blueberry Knoll to the headwall of the cirque. The floors of the other cirques are below timberline, making it difficult to study their topography.

The glacial feature on Mt. Katahdin that has been the most controversial has been the Basin Pond Moraine. Early workers (Tarr, 1900; Antevs, 1932; Caldwell, 1972) held that this feature was a medial moraine formed between the combined flow of ice from the cirques above and the waning Wisconsinan ice sheet. They pointed out that there is an abundance of sedimentary and volcanic rock clasts on the east or ice sheet side of the moraine, as compared with the west or alpine glacier side, which is composed of largely granite clasts. Other evidence that supports the two-glacier origin of the Basin Pond moraine, includes the three lobes in the mountain side of the moraine, which we believe were formed by the glaciers that emanated from the three cirques and merged with, or perhaps ramped onto, the continental ice. Davis and Davis (1980) and Davis and others (1980) have reported that this feature is a lateral moraine formed by the late Wisconsinan ice sheet, with no contribution from alpine glaciers.

About halfway up the great south wall of Katahdin there are other moraines or terrace-like features (Fig. 2) that have been the subject of much debate (Caldwell and Davis, 1983). This deposit is visible from Abol Bridge and is crossed by the Abol Trail. Davis believed that this is a moraine formed at the same time and in the same way as the Basin Pond moraine. Caldwell and others (1985) have interpreted this feature as a kame terrace because it consists of sand and gravel with scattered erratic boulders. From Abol Bridge, this deposit has an obvious slope from west to east, whereas the Basin Pond Moraine viewed from the east appears to be nearly horizontal. When viewed on vertical aerial photographs, it is also clear that the Basin Pond Moraine is much more prominent than the Abol deposit (Fig. 4). This difference in size may be explained by the effect of alpine glaciers transporting large quantities of drift to the site of the Basin Pond Moraine and the lack of such transport to the Abol features (Caldwell and others, 1985).

Figure 4. Vertical aerial photograph of Mt. Katahdin and vicinity; compare with Figure 2. Features discussed in text include: cirques, North Basin, 1; Great Basin, 2; South Basin, 3; aretes, Hamlin Ridge, 4; Knife Edge, 5; moraines, Basin Pond moraine, 6; Abol moraine(?), 7. Also shown are the Tableland, 8; and Baxter Peak, 9. Photograph courtesy of Baxter State Park, Millinocket, Maine.

REFERENCES CITED

Antevs, E., 1932, Alpine zone of Mt. Washington Range: Auburn, Maine, Merrill and Webber, 119 p.

Bradley, D. C., 1981, Late Wisconsinan mountain glaciation in the northern Presidential Range, New Hampshire: Arctic and Alpine Research, v. 13, p. 319–327.

Caldwell, D. W., 1972, The geology of Baxter State Park and Mt. Katahdin: Augusta, Maine Geological Survey Bulletin 12, 57 p.

—— , 1980, Alpine glaciation of Mt. Katahdin, *in* Roy, D., and Naylor, R., eds., A guidebook for field trips in northeastern Maine and Neighboring New Brunswick: Presque Isle, Maine, New England Intercollegiate Geological Conference, p. 80–85.

Caldwell, D. W., and Davis, P. T., 1983, The timing of the alpine glaciation of Mt. Katahdin, *in* Caldwell, D. W., and Hanson, L. S., eds., A guidebook for the Grenville–Millinocket regions; northcentral Maine: Greenville, Maine, New England Intercollegiate Geological Conference, p. 79–82.

Caldwell, D. W., and Hanson, L. S., 1986, The nunatak stage on Mt. Katahdin persisted through the late Wisconsinan: Geological Society of America Abstracts with Programs, v. 18, p. 8.

Caldwell, D. W., Hanson, L. S., and Thompson, W. B., 1985, Styles of deglaciation in central Maine, *in* Borns, H., LaSalle, P., and Thompson, W., eds., Late Pleistocene history of northeast New England and Adjacent Quebec: Geological Society of America Special Paper 195, p. 45–58.

Davis, P. T., 1976, Quaternary glacial history of Mt. Katahdin, Maine [M.S. thesis]: Orono, University of Maine, 155 p.

—— , 1983, Glacial sequence, Mt. Katahdin, north-central Maine: Geological Society of America Abstracts with Programs, v. 15, p. 124.

Davis, P. T., and Davis, R. B., 1980, Interpretation of minimum-limiting radiocarbon dates for deglaciation of Mount Katahdin area: Geology, v. 8, p. 396–400.

Davis, P. T., and Waitt, R. B., 1986, Cirques in the Presidential Range revisited; No evidence for post-Laurentide mountain glaciation: Geological Society of America Abstracts with Programs, v. 18, p. 11.

Davis, P. T., Jacobson, G., and Denton, G. H., 1980, Quaternary geology and paleoecology, Mt. Katahdin, Maine: Guidebook for American Quaternary Association biennial meeting, Orono, Maine.

Espenshade, G. H., and Boudette, E. L., 1967, Geology and petrology of the Greenville Quadrangle, Piscatquis and Somerset Counties, Maine: U.S. Geological Survey Bulletin 1241-F.

Goldthwait, J. W., 1914, Remnants of an old graded upland on the Presidential Range of New Hampshire: American Journal of Science, 4th ser., v. 37, p. 451–463.

Griscom, A., 1976, The geology of the Harrington Lake area, Maine [Ph.D. thesis]: Cambridge, Massachusetts, Harvard University.

Hanson, L. S., and Caldwell, D. W., 1983, The glacial geology and shoreline features in Maine lakes: Geological Society of America Abstracts with Programs, v. 15, p. 193.

Hon, R., 1980, Geology and petrology of igneous bodies within the Katahdin pluton, *in* Roy, D. and Naylor, R., eds., A guidebook for field trips in northeastern Maine and neighboring New Brunswick: Presque Isle, Maine, New England Intercollegiate Geological Conference, p. 65–79.

Tarr, R. S., 1900, Glaciation of Mt. Ktaadn, Maine: Geological Society of America Bulletin, v. 11, p. 433–438.

Thompson, W. F., 1960, The shape of New England mountains; Part 1: Appalachia, no. 131, p. 145–159.

—— , 1961, The shape of New England mountains; Part 2: Appalachia, no. 132, p. 316–335.

Thompson, W. B., and Borns, H. W., Jr., 1985, Surficial geologic map of Maine: Augusta, Maine Geological Survey.

Type section of the Early Ordovician Shin Brook Formation and evidence of the Penobscot orogeny, northern Penobscot County, Maine

Robert B. Neuman, *Department of Paleobiology, E-501 National Museum of Natural History, Smithsonian Institution, Washington, D.C. 20560*

LOCATION

The type section of the Early Ordovician Shin Brook Formation is along Shin Brook near Crommet Spring, about 11 mi (18 km) northwest of Patten and 100 mi (161 km) north of Bangor. It is accessible from the east via I-95 from the Island Falls or Sherman exists; through Patten and the Shin Ponds on Maine 159 (Fig. 1); or from the west about 12 mi (19 km) east of the Mattagamon gate of Baxter State Park.

From Crommet Spring walk approximately 2,300 ft (700 m) WNW (downhill) along the graveled hauling road to the bridge across Shin Brook; east of the bridge the Grand Pitch Formation is exposed in and adjacent to the roadbed. The section is measured northwestward (downstream) from the bridge.

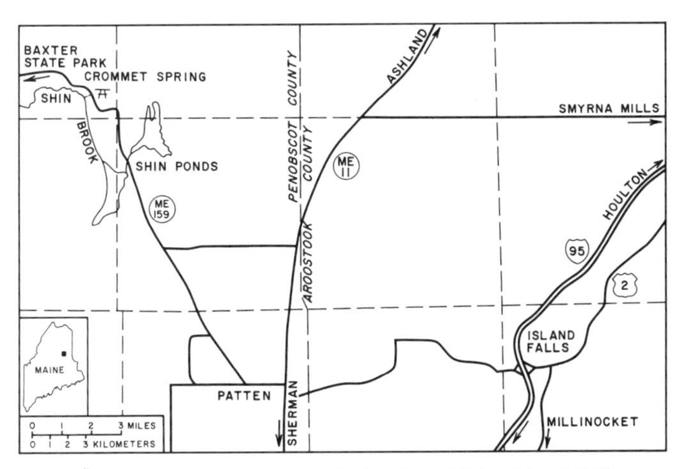

Figure 1. Access to Crommet Spring from I-95 at Island Falls.

SIGNIFICANCE

Tuffs and tuffaceous sedimentary rocks of the Early Ordovician Shin Brook Formation are underlain by pelite and quartzitic sandstone of the Cambrian(?) Grand Pitch Formation. The Shin Brook Formation in this area (Fig. 2) is unique in the State of Maine. Brachiopods, trilobites, and other fossils from several exposures are characteristic of assemblages that lived around islands in the Early Ordovician Iapetus Ocean. The Grand Pitch Formation, one of several largely unfossiliferous units of altered shale and quartzite that occupy the cores of anticlines in Maine and beyond (Boone and others, 1984), is assigned a Cambrian(?) age based on the occurrence of the trace fossil *Oldhamia smithi* Ruedemann (1942) in red shale 10 mi (16 km) SW of Crommet Spring (Smith, 1928).

The lithic contrasts between the Grand Pitch and Shin Brook formations are accompanied by contrasts in their structural styles. The former, here and elsewhere, contains evidence of multiple deformation, while the latter was deformed only once, presumably by Acadian events of Devonian age. Quartzite clasts in conglomerate at the base of the Shin Brook Formation confirm that this contact is an important unconformity, the result of the Penobscot orogeny, an event named from exposures in this vicinity where its age is constrained by these fossiliferous formations.

SITE INFORMATION

The route from Patten into Baxter State Park through Mattagamon gate traverses several of Maine's important geologic features (Neuman, 1967; Neuman and Rankin *in* Roy and Naylor, 1980). Silurian rocks of the northwestern margin of the Kearsarge–Central Maine synclinorium (Osberg and others, 1985) crop out between Patten and the Shin Ponds. Northwest of the Shin Ponds the route crosses the Weeksboro–Lunksoos Lake anticlinorium, one of several anticlinal structures of northern Maine in which pre-Silurian rocks are exposed. Silurian and Lower Devonian rocks on the northwest flank of this structure are encountered on this route north and west of the Seboeis River.

Preservation of the Shin Brook Formation in a shallow syncline in the central part of the Weeksboro–Lunksoos Lake anticlinorium permits insights into the geologic history of the region not available elsewhere. Evidence of multiple deformation of the rocks of the Grand Pitch Formation and of but one deformation of the rocks of the Shin Brook at this site and at other places in the region have led to the acceptance of the post-Cambrian–pre-Early Ordovician Penobscot orogeny here; and other criteria (Drake, 1987) have led to the identification of effects of this event over a much larger area.

The rocks and fossils of the Shin Brook Formation are of special interest. The Early Ordovician age of the formation, determined from its fossils, establishes it as the oldest volcanogenic unit in the region. The occurrence of welded tuff at one place on Sugarloaf Mountain, where it is overlain by richly fossiliferous tuff (Locality A of Fig. 2), suggests subaerial volcanism—a volcanic island. That this and other islands in the ancient Iapetus

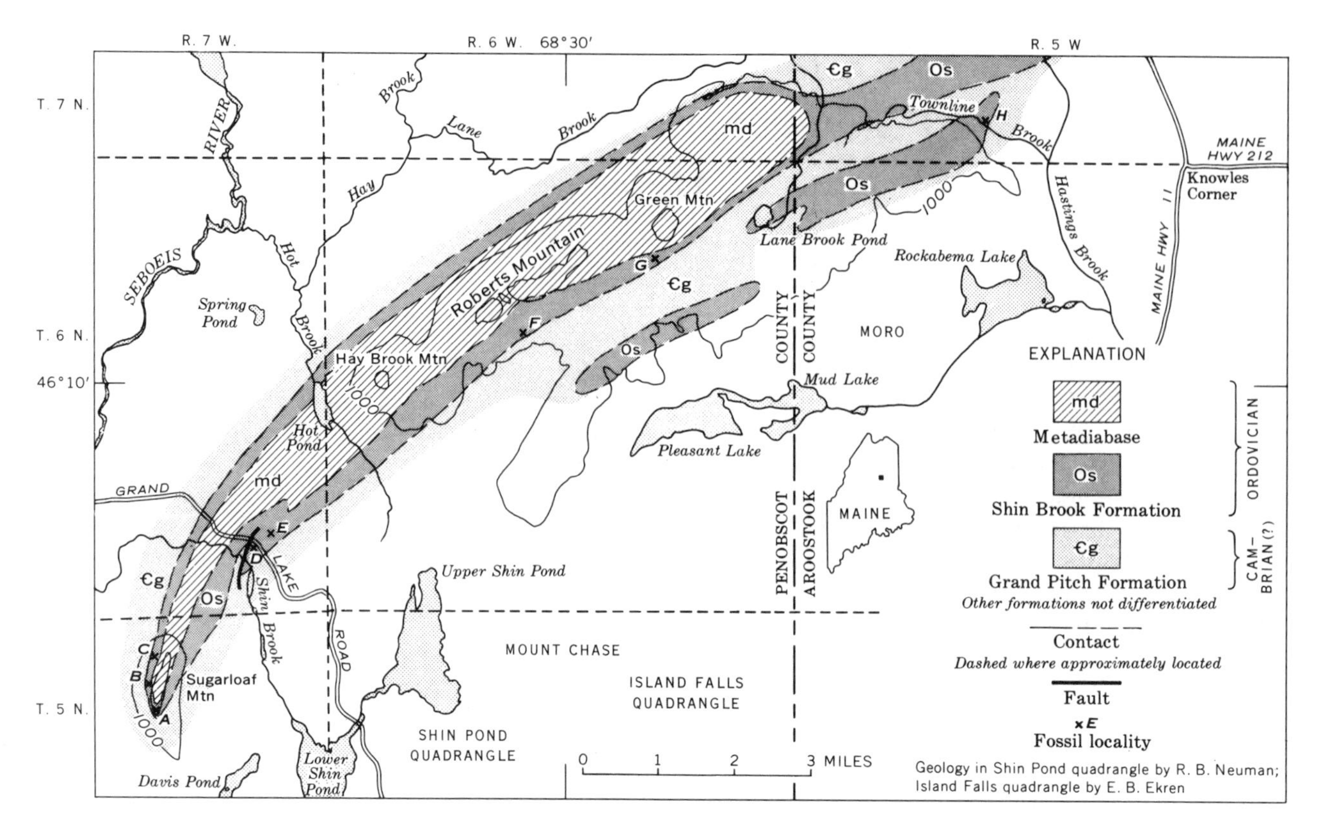

Figure 2. Map showing distribution of Shin Brook Formation (from Neuman, 1964). For more comprehensive geologic maps see the Shin Pond (Neuman, 1967) and Island Falls (Ekren and Frischknecht, 1967) quadrangles.

Ocean lay far seaward of the North American continental margin was inferred from these fossils and those from similar rocks in New Brunswick and Newfoundland (Neuman, 1984).

Identifiable fossils were found at several places in the outcrop belt of the Shin Brook Formation, but none have been found in the exposures along Shin Brook. Fossils nearest the brook were found in a small ledge SW of the road, about 0.75 mi (1.2 km) NW of Crommet Spring (Locality D of Fig. 2), but this ledge was covered by road construction shortly after it was discovered.

No dated rocks are known to overlie the Shin Brook Formation. The metadiabase that occupies the trough of the syncline (Figs. 2 and 3) is probably a sill, the roof of which has been eroded away.

MEASURED SECTION OF SHIN BROOK FORMATION, ADAPTED FROM NEUMAN (1964); SEE FIGURE 3

Shin Brook Formation: 902 ft (280 m) measured **Thickness**

19. Tuffaceous sandstone and siltstone in graded layers 3.2 to 12 in (8 to 30 cm) thick, with coarse-grained sandstone in their basal parts and finely laminated siltstone at their tops; siltstone more abundant than sandstone; unit includes two layers of crystal tuff, 6 and 12 in (15 and 30 cm) thick, respectively .... 35 ft (11 m)

18. Crystal tuff, greenish-gray; crystals are green altered plagioclase .... 10 ft (3 m)

17. Covered .... 3 ft (1 m)

16. Tuffaceous sandstone, grit, and conglomerate; finer grained part is well laminated, part not laminated includes fragments of porphyritic and nonporphyritic fine-grained igneous rock; ledge in streambed has distorted bedding structures that suggest deformation prior to lithification; base concealed .... 7 ft (2 m)

15. Tuffaceous sandstone, fine- to medium-grained, calcareous, strongly sheared with weathered pits that may have been concentrations of fragmentary fossils deformed beyond recognition .... 18 ft (5 m)

14. Covered .... 70 ft (21 m)

13. Crystal tuff, greenish-gray, with scattered angular cognate rock fragments 0.4 to 6 in (1 to 15 cm) average diameter; crystals of both matrix and fragments are green altered plagioclase; lacks primary layering; quartz veins abundant .... 20 ft (6 m)

12. Covered .... 375 ft (114 m)

11. Crystal tuff; light-green altered plagioclase crystals in a darker aphanitic matrix; fractured .... 45 ft (29 m)

10. Covered .... 95 ft (29 m)

9. Tuff, fine-grained, light greenish-gray; abundant, carbonate; strongly sheared, no bedding

preserved .... 30 ft (9 m)
8. Covered .... 30 ft (10 m)
7. Volcanic conglomerate; granules of aphanitic volcanic rock and dark slate, strongly sheared; bedding obliterated .... 50 ft (15 m)
6. Tuffaceous sandstone, gray, medium- and fine-grained; abundant carbonate; bedding obliterated by strong shearing .... 20 ft (6 m)
5. Covered .... 30 ft (9 m)
4. Volcanic granule conglomerate and coarse-grained sandstone; light-gray, strongly sheared .... 25 ft (8 m)
3. Sandstone and conglomerate; conglomerate beds 8 to 10 in (20 to 25 cm) thick, sandstone somewhat thinner; volcanic rocks and fine-grained quartzite clasts in conglomerate angular to subrounded, maximum 1.6 in (4 cm) average diameter .... 35 ft (11 m)
2. Phyllite, light-gray, probably tuffaceous ..... 2 ft (0.6 m)
1. Slate, dark-gray, with abundant small (0.25 to 0.5 mm) rhombic white grains (altered plagioclase?) ...... 2 ft (0.6 m)

Grand Pitch Formation: approximately 200 ft (61 m) exposed in streambank and roadbed. Gray slate and lesser amounts of gray quartzite and dark-gray feldspathic quartzite in thin, discontinuous beds, some tightly folded.

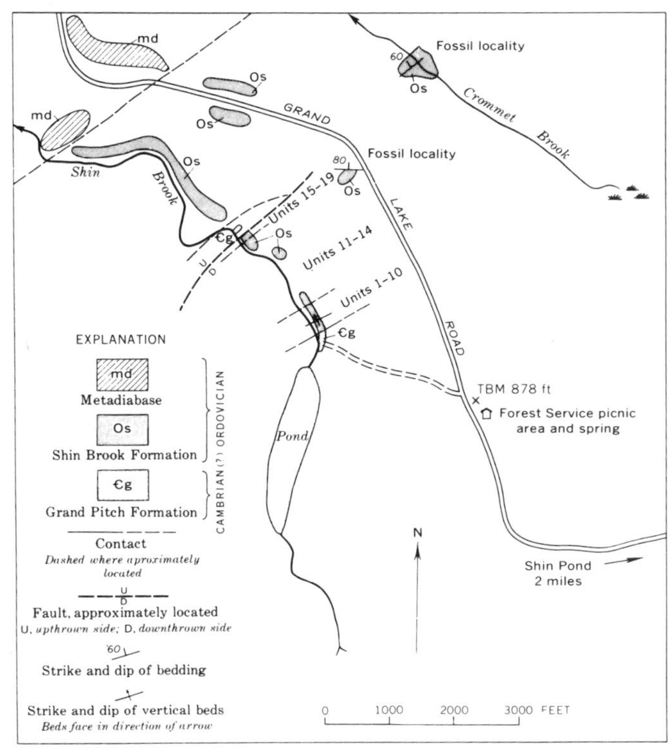

Figure 3. Geologic sketch map of the type section of the Shin Brook Formation; units 1–19 described in text. Outlined areas indicate outcrops (from Neuman, 1964).

REFERENCES CITED

Boone, G. M., Boudette, E. L., Hall, B. A., and Pollock, S. G., 1984, Correlation and tectonic significance of pre-Middle Ordovician rocks of northwestern Maine: Geological Society of America Abstracts with Programs, v. 16, p. 4.

Drake, A. A., Jr., 1987, The Taconic Orogen: *in* Hatcher, R. D., Jr., Viele, G. W. and Thomas, W. A., The Appalachian and Ouachita regions: Geological Society of America, The Geology of North America, v. F2 (in press).

Ekren, E. B., and Frischknecht, F. C., 1967, Geological-geophysical investigations of bedrock in the Island Falls Quadrangle, Aroostook and Penobscot counties, Maine: U.S. Geological Survey Professional Paper 527, 36 p.

Neuman, R. B., 1964, Fossils in Ordovician tuffs, northeastern Maine, with a section on Trilobita by H. B. Whittington: U.S. Geological Survey Bulletin 1181-E, p. E1–E38.

—— , 1967, Bedrock of the Shin Pond and Stacyville Quadrangles, Penobscot County, Maine: U.S. Geological Survey Professional Paper 524-1, 1-37.

—— , 1984, Geology and paleobiology of islands in the Ordovician Iapetus Ocean; Review and implications: Geological Society of America Bulletin, v. 95, p. 1188–1201.

Osberg, P. H., Hussey, A. M. II, and Boone, G. M., eds., 1985, Bedrock Geologic Map of Maine: Maine Geological Survey, scale 1:500,000.

Roy, D. C., and Naylor, R. S., 1980, The geology of northeastern Maine and neighboring New Brunswick: New England Intercollegiate Geological Conference, 72nd, Presque Isle, Maine, Guidebook, Chestnut Hill, Massachusetts, Boston College Press, 296 p.

Ruedemann, R., 1942, *Oldhamia* and the Rensselaer Grit problem: New York State Museum Bulletin 327, p. 5–18.

Smith, E.S.C., 1928, The Cambrian in northern Maine: American Journal of Science, 5th series, v. 15, p. 484–486.

The Mapleton Formation: A post-Acadian basin sequence, Presque Isle, Maine

David C. Roy and Kenneth J. White, Department of Geology and Geophysics, Boston College, Chestnut Hill, Massachusetts 02167*

LOCATION

The Mapleton Formation outcrops in a small area in the Presque Isle 15-minute Quadrangle west of Presque Isle, Maine, which lies along U.S. 1, 42 mi (67 km) north of the terminus of I-95 at Houlton. Stops are located along, and north of, Maine 163, an east-west highway from Presque Isle, through Mapleton, to Ashland. Stop 1 is a roadcut on the north side of Maine 163, 2.6 mi (4.2 km) west from the junction of U.S. 1 and Maine 163 in downtown Presque Isle. The exposures of Stop 1 are on a curve near the crest of a hill, so care should be exercised in parking and during study of the rocks. Stop 2 is reached by driving 1.0 mi (1.6 km) west on Maine 163 from Stop 1 to a gravel road to the north. Travel the gravel road 1.0 mi (1.6 km) to the second possible right turn to the east; drive east for 0.5 mi (0.7 km) and park. The outcrops of Stop 2 are located in a small flagstone quarry in a pasture on the south side of the road. Permission should be obtained from the owners.

SIGNIFICANCE

The Mapleton Formation is composed of nonmarine sandstone and conglomerate deposited in an intermontaine basin during the late Middle Devonian. The formation is the oldest clearly post-Acadian sedimentary sequence in the Northern Appalachians and rests with angular unconformity on deformed rocks of Silurian and Early Devonian age. Well-dated plant remains in the Upper Member of the Mapleton Formation and abundant shelly fauna in the pre-Mapleton formations indicate a stratigraphic age span of Acadian orogenesis in northern Maine from the Siegenian (mid-Early Devonian) to early Givetian (late Middle Devonian); a time interval of about 20 m.y., using the time scale of Palmer (1983).

SITE INFORMATION

Gregory (Williams and Gregory, 1900) described the largely reddish sandstones and conglomerates of what he called the "Mapleton Sandstone." Gregory recognized that the Mapleton Sandstone rests unconformably on the surrounding rocks and is the youngest sedimentary bedrock unit in the region. Many years later, Boucot and others (1964) provided the first modern description of the Mapleton and placed it in its proper tectonostratigraphic position. The Mapleton Sandstone has most recently been divided into two members and has been renamed the "Mapleton Formation" (White, 1975; White and Roy, 1975). The regional importance of the formation is that it provides a close stratigraphic dating of the Acadian Orogeny within the interior of the orogen and gives clues about deposition in an intermontane basin in the mountain system. Although the members of the Mapleton Formation are highlighted in this site, the maps in Figures 1 and 2, together with guidebooks of the New England Intercollegiate Geological Conference (Roy and Naylor, 1980; Osberg, 1974) and the regional summary by Roy and Mencher (1976), permit exploration of the pre-Mapleton rocks as well.

Pre-Mapleton Stratigraphy

The clastic rocks of the Mapleton Formation were derived from pre-Mapleton formations exposed in nearby uplands. These pre-Mapleton rocks in the Presque Isle area form a thick, fossiliferous succession of sedimentary and volcanic rocks that range in age from Late Ordovician to Early Devonian (Fig. 1; Roy and Mencher, 1976; Roy, 1980a, 1980b; Naylor, 1980). With the exception of a hiatus in the latest Silurian, the sequence represents virtually continuous sedimentation of deep-water shale/micritic limestone (Carys Mills Formation), bioturbated limestone and laminated calcareous mudstone (Spragueville Formation), and laminated graptolitic shale and siltstone (Jemtland Formation). To the west of Presque Isle, as seen in the Castle Hill Anticline (Fig. 1), the Ordovician sequence consists of mafic and felsic volcanic rocks, chert, black slate, and graywacke of the Winterville Formation, which is essentially coeval with slate and graywacke of the Madawaska Lake Formation. The Early Silurian sequences overlying these possibly arc-related Ordovician units (Hynes, 1976) consist of conglomerate and lithic sandstone (Frenchville Formation) interfingering eastward with calcareous slate, minor graywacke and micrite, and manganiferous ironstone (New Sweden Formation). The Frenchville Formation—the northernmost formation in a lithofacies belt that includes the Rangeley Formation in western Maine—contains coarse clastic material derived from a western upland ("Taconia") produced by uplift late in the Taconian Orogeny (Roy and Mencher, 1976; Roy, 1980a). The Jemtland Formation overlies the three Early Silurian formations conformably and was also derived from the west as the basin deepened and the western source became subdued and inundated (Roy and Mencher, 1976).

In the Presque Isle area the Silurian and older units were

*Present address: CH2M Hill, 3840 Rosin Court, Suite 110, Sacramento, California 95834.

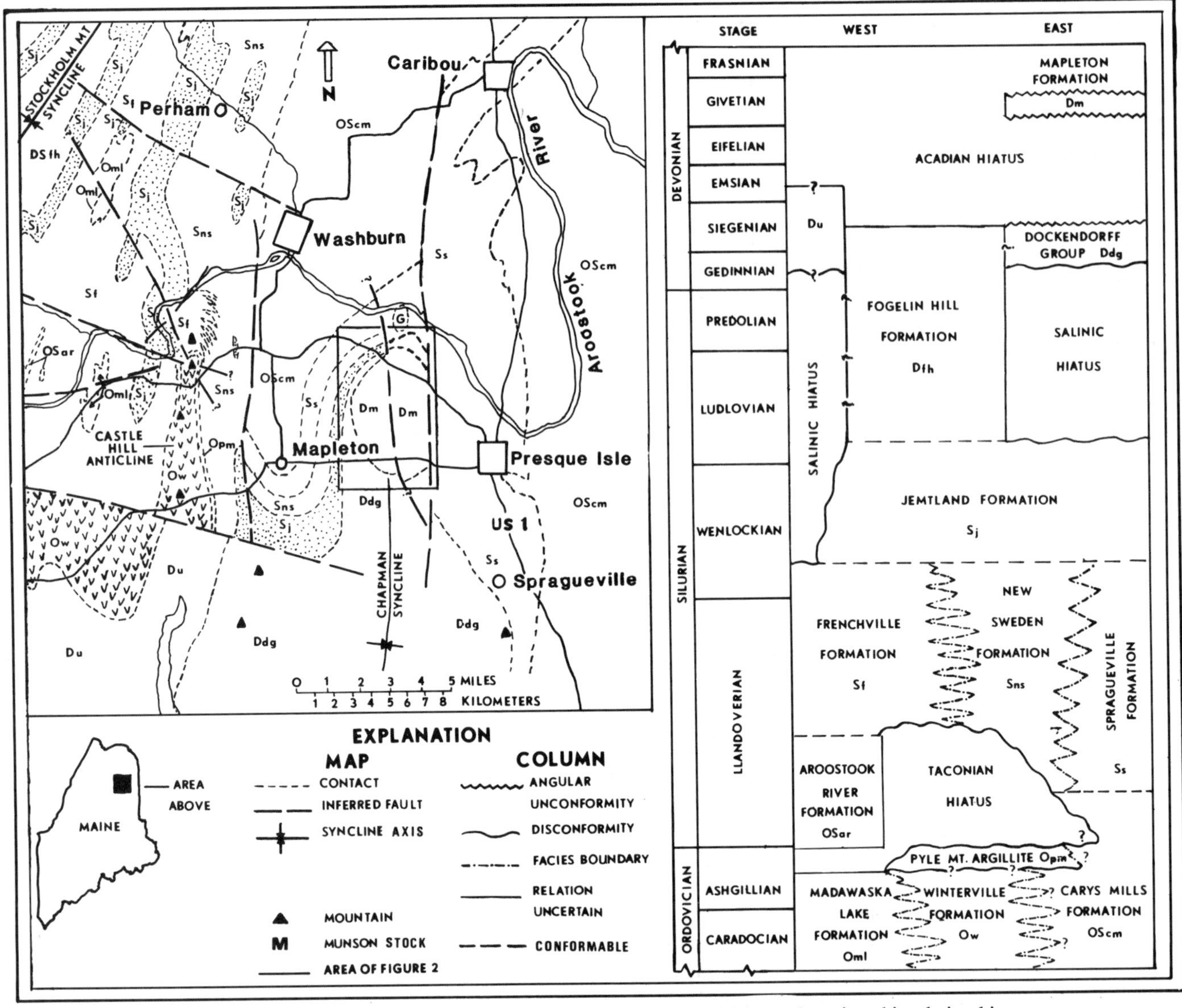

Figure 1. Geology of Presque Isle–Caribou region of northeastern Maine and stratigraphic relationships of the formations.

slightly deformed, uplifted, and eroded during the latest Silurian (Boucot and others, 1964; Pavlides, 1978; Roy, 1980a). The deformation and erosion are assigned to the "Salinic Disturbance" (Boucot, 1962; Boucot and others, 1964) and were succeeded by the eruption of 2,100 m of volcanic rocks of the Lower Devonian Dockendorff Group and the deposition of associated shallow-water sedimentary rocks (Boucot and others, 1964; Sargent, 1985). The volcanic rocks of the lower part of the group (Hedgehog Formation) are volumetrically mostly rhyolitic crystal and lithic tuffs with lesser basalt and andesite, whereas the volcanic rocks of the upper part of the group are andesitic (Edmunds Hill Andesite). The age of the Dockendorff Group is late Gedinnian, based largely on shelly fauna from the Chapman Sandstone with scattered localities in the sedimentary rocks of other formations within the group (Boucot and others, 1964). Nearby to the west and southwest, sedimentary rocks equivalent in part to the Dockendorff Group, and as young as Siegenian, are present and represent the youngest paleontologically dated pre-Acadian rocks in northeastern Maine (Roy, 1980a).

At Haines Hill, just north of the outcrop area of the Mapleton Formation, monzonite of the Munson Stock cuts both bedding and cleavage of Silurian units (Fig. 2; Boucot and others, 1964). The monzonite of the stock is poorly exposed but its presence is marked by a contact aureole of hornfels. The nearby Mapleton rocks are not affected by the contact metamorphism.

The Acadian Orogeny

During the span of late Siegenian-through-Eifelian of Early and Middle Devonian, the pre-Mapleton stratified rocks were folded and mildly metamorphosed during the compressional phase of the Acadian Orogeny (Roy, 1980a). This deformation accounts for the map pattern of the formational units in northeastern Maine and is responsible for the Chapman Syncline that underlies the structural basin of the Mapleton Formation (Mapleton Syncline). The single prominent cleavage that is generally north-south and steeply dipping, seen in the pre-Mapleton rocks of the Presque Isle area, is assigned to this deformation. During the deformation the rocks were subjected to pressures and temperatures no greater that those of the Prehnite-Pumpellyite grade of regional metamorphism (Richter and Roy, 1974, 1975). This low-grade metamorphism suggests that the rocks were buried only a few kilometers and remained structurally higher than their equivalents along strike in western and southwestern Maine (Roy, 1980a; Hon, 1980). The uplift stage of the Acadian caused substantial erosion before and during the deposition of the Mapleton Formation.

Deposition and Petrology of the Mapleton Formation

The Mapleton Formation lies in a structural basin formed by erosion of the doubly plunging Mapleton Syncline (Fig. 2). The original sedimentary basin was probably considerably larger than the limits of present outcrop. The Mapleton Syncline is bounded on the east by the Hanson Brook Fault that cuts the stratigraphy of the basin. A second subparallel fault just to the west cuts the syncline and passes through Haines Hill. The basal contact of the formation can be traced between outcrops of the formation and older rocks but is nowhere visible, although it may be closely approached on Griffin Ridge. An angular unconformity between the Mapleton Formation and older units is inferred, based on clasts of the older units seen in the Mapleton Formation; and generally much steeper dips of the older units along the flanks of the Chapman Syncline, as compared to bedding dips in the Mapleton within the almost coaxial Mapleton Syncline.

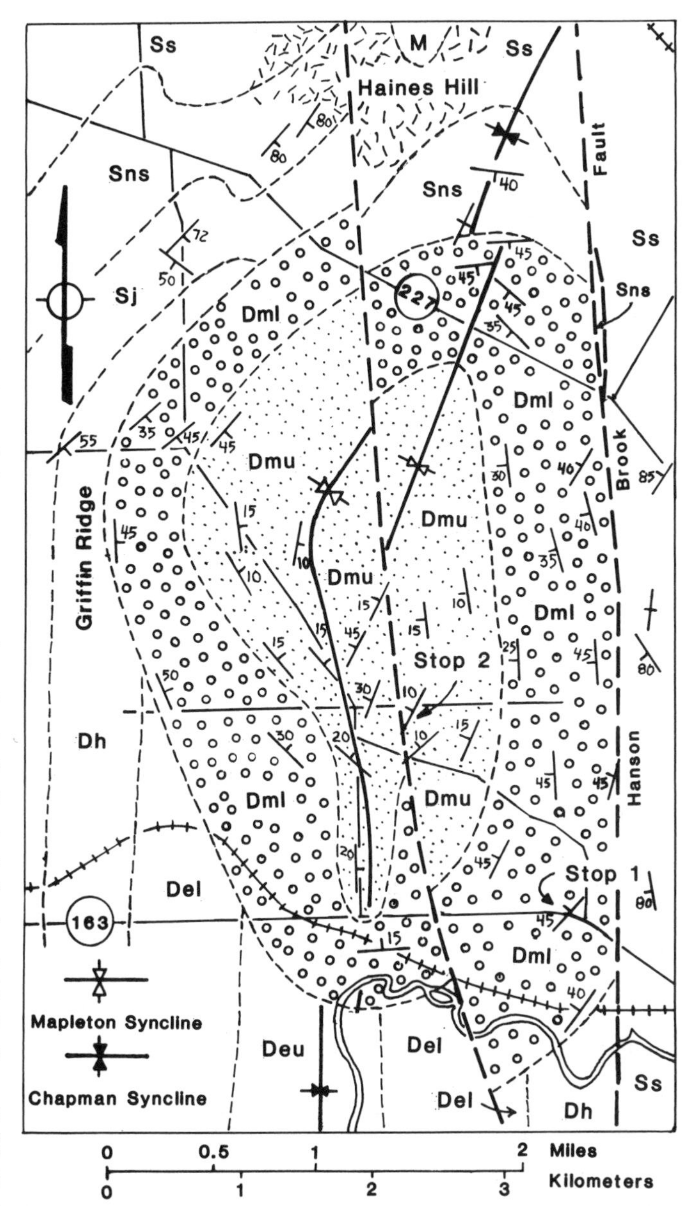

Figure 2. Geologic map of the Mapleton Syncline showing locations of Stops 1 and 2. Map notations for pre-Mapleton formations are given in the column of Figure 1. Mapleton Formation members are: Dml, Lower Member (circles); Dmu, Upper Member (stipling). Random dashes around Haines Hill indicate the extent of hornfels associated with the Munson Stock (M).

The Lower Member is composed of red pebble and cobble roundstone conglomerate with minor lenses, and beds of medium- to coarse-grained lithic sandstone. The member varies in thickness from 2,300 ft (700 m) in the east to about 1,000 ft (300 m) in the northwest. There is a general upward decrease in mean clast size within conglomerate beds of the member, from cobbles at the bottom to pebbles at the top with a concurrent slight increase in sandstone. The Lower Member grades upward into, and interfingers to the west, with the Upper Member. All clasts in the conglomerate beds are well rounded to subrounded and generally equant; limestone and slate fragments are, however, usually tabular and locally imbricated. In outcrop, clasts of vein quartz, gray- and brown-weathering argillaceous micrite, variably colored chert, felsic and mafic volcanic rocks, fine-grained sandstone, shale/slate, and coarsely crystalline calcite are common. The relative abundances of the rock types vary considerably, but usually volcanic and micritic limestone clasts predominate. No clasts of monzonite from the Munson stock have been identified. The interframework matrix in the conglomerate beds is usually sandstone, but in some beds calcite cement forms most of the interstital material. Sandstone beds of the Lower Member are typically lithic arenites with a small percentage of feldspar, markedly variable quartz:rock-fragment ratios, and small amounts of recrystallized argillaceous matrix and carbonate cement (Fig. 3).

The Upper Member consists almost exclusively of fine- to coarse-grained sandstone with much less siltstone and pebble conglomerate. There is a general decrease in mean grain size vertically within the member; this continues the trend seen in the

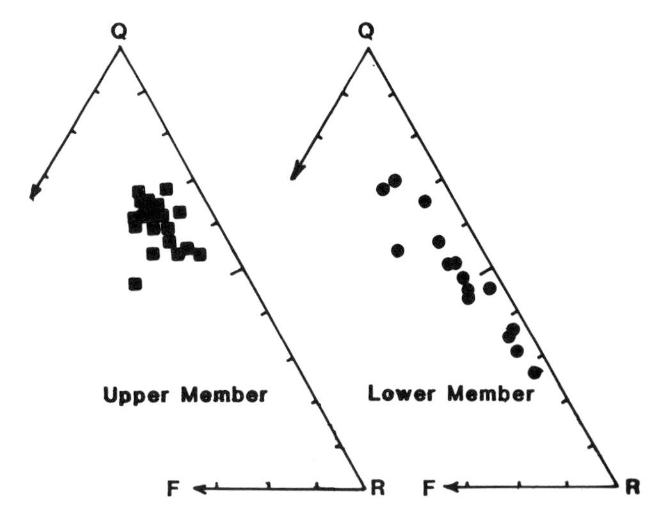

Figure 3. Compositions of medium-grained sandstones of Mapleton Formation members plotted on QFR diagram. Q-pole, monocrystalline and polycrystalline quartz grains and quartz crystals in phaneritic quartz-feldspar detrital grains; F-pole, monocrystalline and polycrystalline feldspar grains and feldspar crystals in phaneritic quartz-feldspar detrital grains; R-pole, fine-grained rock fragments of all types.

Lower Member. The color of the sandstones varies among red, green, and gray. The majority of the sandstone beds are composed of poorly sorted, compositionally immature lithic arenite, but some sandstones contain enough detrital matrix to be classified as lithic graywackes. On average, sandstones of the Upper Member have principal component compositions of about 60 percent quartz, 15 percent feldspar, and 25 percent fine-grained rock fragments (Fig. 3). The sandstones of the Upper Member are more uniform in composition than those of the Lower Member but otherwise quite similar. At some localities, and especially at Stop 2, there are laminated plant-rich and pyritiferous green-gray sandstone layers interbedded with laminated and relatively organic-poor sandstone. Although "mudchips" are common in some sandstone beds, shale beds are virtually absent from this generally finer grained of the two members. East of the axis of the Mapleton Syncline approximately 1,300 ft (400 m) of the Upper Member is preserved, whereas west of the axis 1,700 ft (520 m) of section is present.

The clasts in the conglomerates and sandstones of the Mapleton Formation can be accounted for in the pre-Acadian formations in the surrounding region. In the conglomerates, rhyolitic tuff and andesitic clasts are quite similar to rocks of the underlying Dockendorff Group, and the micritic limestone clasts are clearly derived from the nearby Carys Mills Formation to the east. Cobbles of fossiliferous sandstone similar in lithology and fauna to the pre-Acadian Chapman Sandstone (of the Dockendorff Group) have been reported by Boucot and others (1964) from conglomerates at unspecified localities. Basaltic fragments and chert are possibly derived from either the Winterville Formation to the west, or more likely from temporally and lithologically equivalent formations to the south (for example, the Dunn Brook Formation; Pavlides, 1965, 1973) or to the east in the Miramichi uplands of New Brunswick. As shown in the sandstone provenance diagrams in Figure 4, there is a variable ratio of volcanic fragments to fine-grained sedimentary rocks (mostly pelitic rocks, chert, and micrite) with the average near unity. Phaneritic composite grains of feldspar and quartz-feldspar are typically subordinate to fine-grained rock fragments (Fig. 4,A). In addition, felsic volcanic fragments are generally more abundant than more mafic fragments (Fig. 4,B), suggesting principal derivation from the local Dockendorff section and/or a greater loss of the mafic clasts during weathering and transportation. The increases in modal quartz and phaneritic rock fragments in the Upper Member are probably not due to a reduction in grain size since all modes were performed on medium-grained sandstones. The increases may be caused by the "unroofing" of hypabyssal intrusive rocks and shallow Acadian plutons seen at the surface today to the east and south.

The small size and scattered distribution of the exposures of the Mapleton Formation preclude detailed analysis of the depositional environment of the formation. Taken collectively, the sedimentary structures, bedding style, and rock types suggest fluviatile deposition. We interpret the preserved section to have been deposited in one or more alluvial fans, with the Lower Member forming the proximal fan and the Upper Member representing the more distal outer fan. Thickening of the Conglomerate Member to the east is taken to indicate derivation of the detritus from that general direction. A few paleocurrent measurements based on imbrication of clasts in the Lower Member show north-south directions with both north and south senses (White, 1975), suggesting that sediment transport directions in that member were complicated and not easily reconciled at present with a particular fan geometry. Measurements of parting lineations in sandstone beds of the Upper Member suggest more uniform, and generally northeast-southwest, flow directions.

Post-Mapleton Deformation

The Mapleton Syncline is nearly coaxial with, but clearly younger than, the Acadian Chapman Syncline. The age of the Mapleton Syncline is unknown, since rocks unquestionably younger than the Mapleton are not present. Nearby, to the northeast in New Brunswick, essentially nondeformed marine Mississippian strata are present (Rose, 1936), which might be argued to post-date Mapleton deformation (Boucot and others, 1964), but it is also possible that deformation as low in intensity as suggested by the Mapleton Syncline might not have been regionally uniform in either intensity or style. The faults that cut the Mapleton Syncline postdate the Acadian, but are otherwise also of unknown age.

Stop 1. The exposures at Stop 1 are representative of the conglomerates and sandstones of the Lower Member. These exposures are approximately in the middle of the member and the outcrop displays about 120 ft (37 m) of secton. The strata strike northeast, dip about 45° northwest, and consist of pebble and cobble conglomerate interlayered with lithic sandstone in three fining-upward sequences. The fining-upward sequences are 30 to 50 ft (10 to 15 m) thick, with a lower conglomerate interval overlain by a thinner bed of laminated sandstone. The conglomerate of the 51-ft (16 m) lowest sequence contains lenticular sandstone lenses that may be parts of eroded sequences that were less than 3 ft (1 m) thick. Abundant sandstone matrix and rounded pebbles and cobbles are typical of the conglomerates here, and in the Lower Member generally, as is the predominance of sedimentary clasts. A pebble count from conglomerate with sandstone matrix in the middle fining-upward sequence of the section shows a composition of approximately 23% micritic limestone, 13% vein quartz, 5% chert, 3% felsic and 2% mafic volcanic rocks, 2% slate/shale, 3% vein(?) carbonate, and 49% sandstone matrix. Micritic limestone clasts seem to be more abundant in conglomerates in the eastern than in the western part of member, a possible reflection of the prominence of the Carys Mills Formation in the eastern source area. An approximately 10-ft (3-m) interval of the conglomerate of the middle sequence is partially cemented with calcite that is probably the result of dissolution and precipitation of clast-carbonate.

Looking southeast from this stop, the west-dipping (about 70°) volcanic rocks of the lower Dockendorff Group can be seen in the "hogback" of Green Mountain in the distance, and in the nearer Quaggy Joe Mountain along the eastern flank of the Chapman Syncline. Quaggy Joe is located in Aroostook State Park, south of Presque Isle, and may be climbed, via a well-marked trail, to see exposures of the Hedgehog Formation, the more felsic part of the group. The upper volcanic rocks of the Dockendorff Group are present in Hobart Hill almost due south of Stop 1 and in Edmunds Hill just to the right of Hobart Hill, in the middle distance. Squa Pan Mountain, a long ridge in the distant southwest, is composed of the Hedgehog Formation along the western Flank of the Chapman Syncline.

Stop 2. The outcrops of this stop are reached by walking through the pasture gate and along a farm road for about 150 ft (45 m) and turning right into a flagstone "quarry" in an evergreen grove. Here sandstone beds of the Upper Member are nearly horizontal and only a few meters of section are available. Of importance here are the abundant and well-preserved plant remains, seen both in-place and in blocks that have been removed during quarrying. The plant assemblage has been described by Schopf (1964; Appendix I, Boucot and others, 1964), who found algae and both spiny and smooth psilophytes in abundance, as well as representatives of the genera *Barrandeina Stur* and *Calamophyton.* The fragmented plant remains are found in green-gray, variably calcareous, laminated, pyritiferous sandstone that may be storm deposits. The interbedded sandstone beds have much less preserved plant material, commonly contain mudchips, are parallel an cross-laminated, and show parting lineations, suggesting a spread in current directions about east-west.

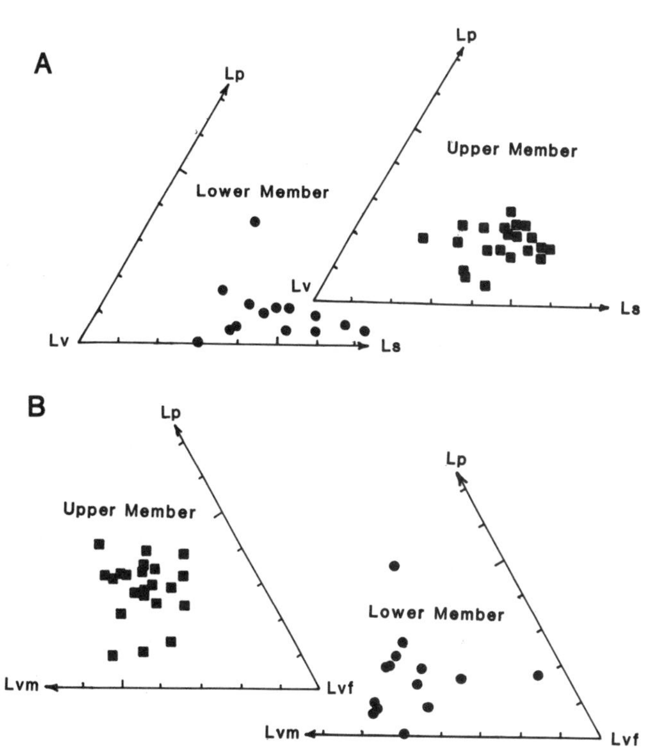

Figure 4. Sandstone provenance diagrams for Mapleton Formation. A, Relative proportions of grains interpreted to be derived from plutonic sources (Lp), volcanic rocks (Lv), and sedimentary rocks (Ls). Included in Lp are micrographic quartz-feldspar grains, perthite grains, and composite phaneritic grains of feldspar and quartz-feldspar. Chert is included in Ls. B, Proportions of igneous rock fragments. Lp pole is same as in A; Lvm, mafic/intermediate volcanic fragments; Lvf, felsic volcanic rock fragments.

REFERENCES

Boucot, A. J., 1962, Appalachian Siluro-Devonian, *in* Coe, K., ed., Some aspects of the Variscan fold belt: 9th Inter-University Geological Congress, Manchester University Press, p. 155–163.

Boucot, A. J., Field, M. T., Fletcher, R., Forbes, W. H., Naylor, R. S., and Pavlides, L., 1964, Reconnaissance bedrock geology of the Presque Isle Quadrangle, Maine: Maine Geological Survey Quadrangle Mapping Series, no. 2, 123 p.

Hon, R., 1980, Geology and petrology of igneous bodies within the Katahdin Pluton, *in* Roy, D. C., and Naylor, R. S., A guidebook to the geology of northeastern Maine and neighboring New Brunswick: New England Intercollegiate Geological Conference, 72nd Annual Meeting, Presque Isle, Maine, p. 65–79.

Hynes, A., 1976, Magmatic affinity of Ordovician volcanic rocks in northern Maine and their tectonic significance: American Journal of Science, v. 276, p. 1208–1224.

Naylor, R. S., 1980, Bedrock geology of the Presque Isle area, *in* Roy, D. C., and Naylor, R. S., eds., A guidebook to the geology of northeastern Maine and neighboring New Brunswick: New England Intercollegiate Geological Conference, 72nd Annual Meeting, Presque Isle, Maine, p. 169–178.

Osberg, P. H., 1974, ed., Guidebook for field trips in east-central and north-central Maine: New England Intercollegiate Geological Conference, 66th Annual Meeting, Orono, Maine, 240 p.

Palmer, A. R., 1983, The Decade of North American Geology, 1983 Geologic

Time Scale: Geology, v. 11, p. 503–504.
Pavlides, L., 1965, Geology of the Bridgewater Quadrangle, Aroostook County, Maine: U.S. Geological Survey Bulletin 1206, 72 p.
—— , 1968, Stratigraphic and facies relationships of the Carys Mills Formation, northeastern Maine and adjoining New Brunswick: U.S. Geological Survey Bulletin 1264, 44 p.
—— , 1973, Geologic map of the Howe Brook Quadrangle, Aroostook County, Maine: U.S. Geological Survey Quadrangle Map GQ-1094.
—— , 1978, Bedrock geologic map of the Mars Hill Quadrangle and vicinity, Aroostook County, Maine: U.S. Geological Survey Quadrangle Map I-1064.
Richter, D. A., and Roy, D. C., 1974, Subgreenschist metamorphic assemblages in northern Maine: Canadian Mineralogist, v. 12, p. 469–474.
—— , 1975, Prehnite-pumpellyite facies metamorphism in central Aroostook County, Maine, *in* Lyons, P. C., and Brownlow, A. H., eds., Studies in New England geology: Geological Society of America Memoir 146, p. 239–261.
Rose, B., 1936, Plaster Rock area, New Brunswick: Canadian Department of Mines Paper 36-19, 10 p.
Roy, D. C., 1980a, Tectonics and sedimentaton in northeastern Maine and adjacent New Brunswick, *in* Roy, D. C., and Naylor, R. S., A guidebook to the geology of northeastern Maine and neighboring New Brunswick: New England Intercollegiate Geological Conference, 72nd Annual Meeting, Presque Isle, Maine, p. 1–21.
—— , 1980b, Ordovician and Silurian stratigraphy of the Ashland Synclinorium and adjacent terrane, *in* Roy, D. C., and Naylor, R. S., A guidebook to the geology of northeastern Maine and neighboring New Brunswick: New England Intercollegiate Geological Conference, 72nd Annual Meeting, Presque Isle, Maine, p. 142–168.
Roy, D. C., and Mencher, E., 1976, Ordovician and Silurian stratigraphy of northeastern Aroostook County, Maine, *in* Page, L. R., ed., Contributions to the stratigraphy of New England: Geological Society of America Memoir 148, p. 25–52.
Roy, D. C., and Naylor, R. S., 1980, A guidebook to the geology of northeastern Maine and Neighboring New Brunswick: New England Intercollegiate Geological Conference, 72nd Annual Meeting, Presque Isle, Maine, 296 p.
Sargent, S. L., 1985, Petrology and geochemistry of the volcanic rocks of the Dockendorff Group and the Munson Pluton, northeastern Maine [M.S. thesis]: Chestnut Hill, Massachusetts, Boston College, 150 p.
Schopf, J. M., 1964, Middle Devonian plant fossils from northern Maine: U.S. Geological Survey Professional Paper 501D, p. D43–D49.
White, K. J., 1975, The Mapleton Formation, an immediately post-Acadian basin fill [M.S. thesis]: Chestnut Hill, Massachusetts, Boston College, 70 p.
White, K. J., and Roy, D. C., 1975, The Mapleton Formation; An immediately post-Acadian basin fill: Geological Society of America Abstracts with Programs, v. 7, p. 133.
Williams, H. S., and Gregory, H. E., 1900, Contributions to the geology of Maine: U.S. Geological Survey Bulletin 165, 212 p.

Komatiites in Munro Township, Ontario

N. T. ***Arndt,*** *Max Plank Institute, Mainz, West Germany*
A. J. ***Naldrett,*** *Department of Geology, University of Toronto, Toronto, Ontario M5S 1A7, Canada*

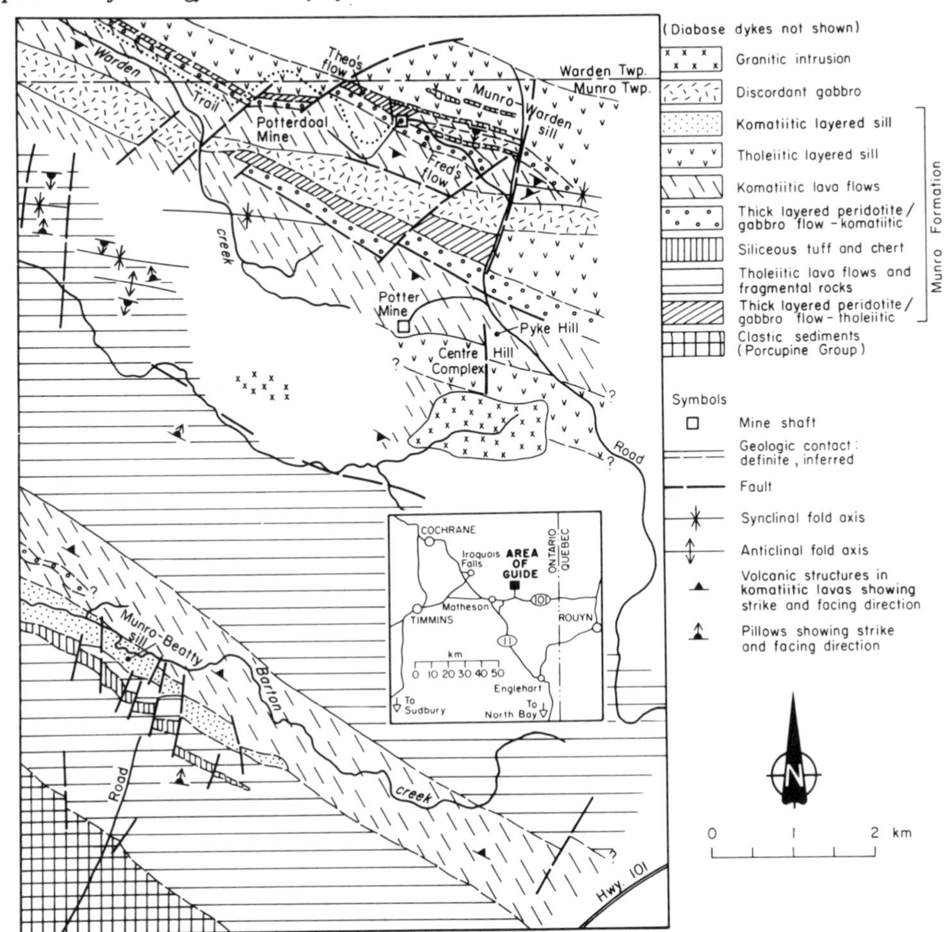

Figure 1. Geological map of Munro Township (modified after Arndt and others, 1977).

LOCATION

Munro Township is a 36-mi^2 (93.6 km^2 township in northwestern Ontario. It is reached by Ontario 101, which intersects Trans-Canada 11 at the town of Matheson, Ontario, and runs east to Dupaquet, Quebec (Fig. 1). The turn-off to the areas of geologic interest described here is at the top of a small hill (the Munro esker) about 17 mi (27 km) east of the bridge over the Black River in Matheson. It is presently marked by two signs, one labeled "Potter Mine" and the other "Hedman Mine." Turn north at this intersection and continue northward on a good gravel road along the esker for 5 mi (8 km) to Stop 1. Directions to Stops 2 to 5 are given in the stop descriptions.

SIGNIFICANCE

Munro Township contains one of the best exposed and least altered sequences of Archean komatiites in the world. Ultramafic magmas, which crystallized to give rise to large amounts of skeletal olivine, were first recognized 36 mi (60 km) west of Munro by Naldrett and Mason (1968). Viljoen and Viljoen (1969) identified similar rocks in the Barberton Mountain Land of South Africa as lavas, christening them "komatiites" after the Komati River area where they were best exposed. The same rocks also attracted interest in Western Australia because of their association with a new type of Ni-sulfide deposit (Woodall and Travis, 1969).

SITE INFORMATION

The geology of much of Munro township is shown in Figure 1. The stratigraphic succession of the komatiitic rocks is illustrated in Figure 2. Three cycles of volcanism are present; in general, successive flows within each cycle become progressively

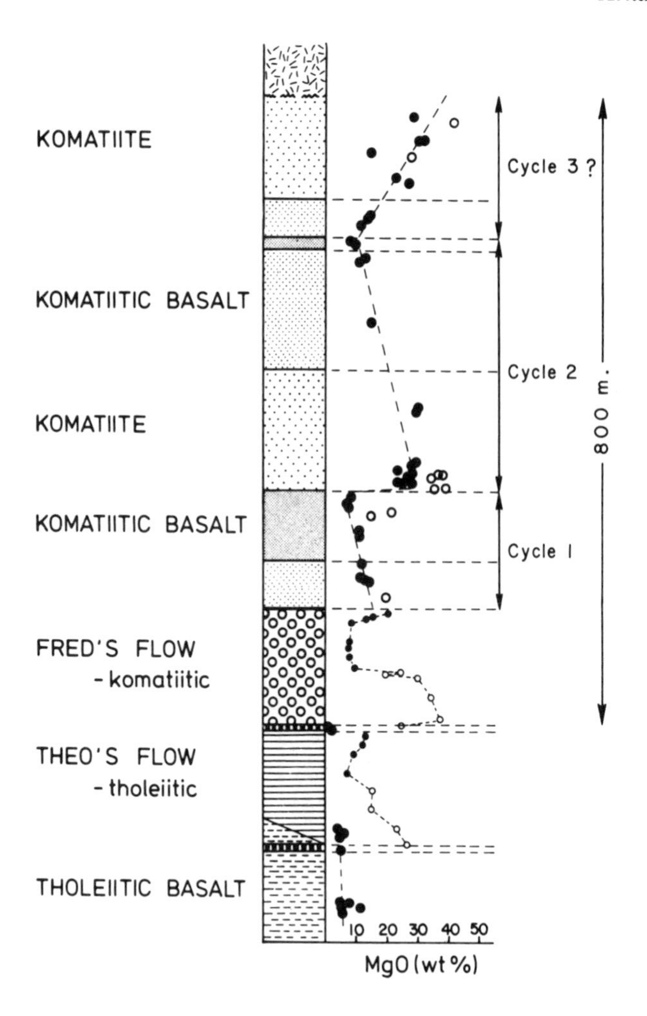

Figure 2. Stratigraphic succession of komatiitic rocks, Munro Township, showing the variation in MgO in cumulate (open circles) and non-cumulate (solid circles) rocks (modified from Arndt and others, 1977).

less magnesian, usually starting with komatiites and passing upward to basalt.

Nesbitt (1971) described textures within komatiites as resulting from the growth of skeletal olivine; he likened them to a local variety of grass and, drawing on an old miner's term, named the textures "spinifex." Pyke and others (1973) were the first to report komatiites from Munro and also, as a result of their work in the area designated as Stop 1 in this guide, were able to differentiate individual units making up the flows. Arndt and others (1977) documented a wide variety of flows in Munro Township in terms of both physical features and chemical composition. Chemical compositions of rocks from the township have been further documented by Arth and others (1977), Basaltic Volcanism Study Project (1981), and Arndt and Nesbitt (1982).

Stop 1, Pyke Hill. *Please do not use hammers or in other ways attempt to dislodge samples from areas A and C at this location. The quality of the exposure here is unique and should remain so.* This is a hill known in local geological circles as Pyke Hill. It is relatively bare, but has a sparse covering of fir trees. It lies to the left of the access road, about 1,000 ft (300 m) before the road divides, with the left branch going to the Potter Mine and the right branch leading to the Hedman asbestos mine and the northern part of the township. A short track leads to the northwestern end of the hill, by the tailings pond of the Potter Mine, and is a convenient place to park. Walk up onto the tailings and immediately climb the bare outcrop to your left. You are now in the area marked A in Figure 3.

The outcrop is made up of particularly well exposed and fresh komatiite lava flows which were first described by Pyke and others (1973). Most spectacular are flows with a spinifex-textured upper portion and smooth weathering cumulate lower portion. This type of flow is an end member of a continuum (Arndt and others, 1977), with uniform flows showing no spinifex texture at the other end (Fig. 4).

The spinifex-textured flows contain the following zones:

(1) A1 zone—This zone is made up of sparse olivine phenocrysts (<10%) set in a matrix of altered glass in which small randomly oriented blades of olivine have developed. It is interpreted to be the chilled top to the flow. Joints (some of them thought to be cooling cracks) break it up into polyhedra varying from 1 to 8 in (3 to 20 cm) in diameter.

(2) A2 zone—From the A1 zone downwards, the skeletal olivine blades become progressively coarser. The top of the A2 zone is defined as the horizon at which this fabric becomes easily visible to the eye. The blades are randomly oriented, up to 0.4 in (1 cm) long, and are set within a matrix of altered glass, within which skeletal needles of augite have developed.

(3) A3 zone—The olivine blades develop as a series of randomly oriented "books," each book being made up of small, parallel blades ranging from less than 1 cm to several cm in length. In the A3 zone the books lose their random orientation and develop sub-perpendicularly with respect to the plane of the flow. The books are larger than in the A2 zone and the constituent olivine blades are both thicker and longer.

(4) B1 zone—This zone is relatively thin, made up of tabular skeletal olivine grains oriented roughly parallel to the plane of the flow.

(5) B2 zone—In this zone the olivine is polyhedral and equant, and skeletal development is very rare. The proportion of olivine is greater than in the spinifex-textured zones, reaching 60 to 80 modal percent. Space between the olivine is occupied by skeletal augite needles and altered glass.

(6) B3 zone—This knobby-weathering zone contains equant accumulations of augite-rich matrix material. It is well developed

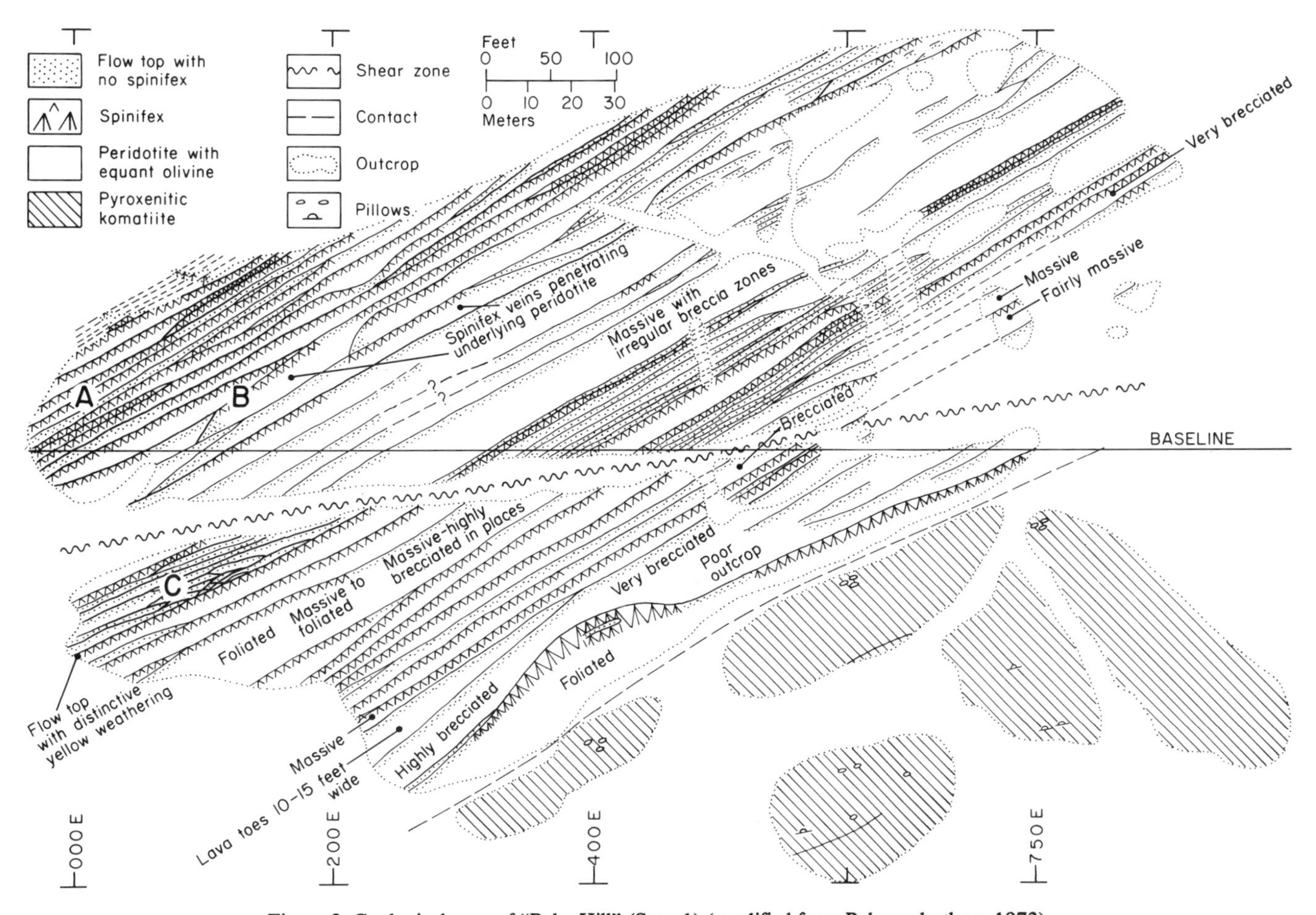

Figure 3. Geological map of "Pyke Hill" (Stop 1) (modified from Pyke and others, 1973).

in many of the Munro flows but is not common in many other areas of komatiites.

(7) B4 zone—Downwards across this zone olivine becomes less common and the proportion of altered glass, carrying rare olivine phenocrysts, increases. It is regarded as the basal chill zone of the flow.

More than half of the olivine porphyritic flows exposed at Pyke Hill have no spinifex texture and are of the type shown as type C in Figure 4. These are pervaded by the polyhedral joint systems seen in the A1 zone of the Type A flows. In the spinifex-free flows the joints are close-spaced and rectangular at the top and bottom of the flows, breaking the rock up into a series of rectangular or polyhedral masses. Near the center, they are more widely spaced and generally gently curving. The rock is composed of 45 to 55 modal percent olivine, set in altered glass containing skeletal, dendritic, radiating, and plumose sprays of pyroxene. The olivine is equant, with rare skeletal overgrowths.

A typical example of a flow with limited spinifex texture (Type B in Fig. 4) can be observed at Area B in Figure 3. In this flow, a polyhedrally-jointed A1 zone is underlain by a spinifex zone with the characteristics of the A2 zone of the type A flow. This A1 zone is underlain in turn by B2 type material in which the proportion of non-skeletal olivine increases downwards to a point three quarters down through the flow, below which it decreases to the base. Small schlieren with a fine-grained spinifex texture occur in a zone 3 to 6 ft (1 to 2 m) thick immediately below the spinifex zone of Type B flows. These schlieren are commonly 1 to 4 in (2 to 10 cm) thick, 20 in (50 cm) to several metres long, develop parallel to the plane of the flow, and have a crescent shape, concave upwards.

In the southwest portion of the outcrop (Area C in Fig. 5), 20 in to 10 ft (50 cm to 3 m) lava toes are well exposed (see Fig. 7). These have massive interiors but their margins are defined by zones of strongly-developed polyhedral jointing.

Stop 2—The Top of Fred's Flow. Continue along the main gravel road, taking the right fork to the Hedman Mine. In about 0.6 mi (1 km) you will see a cable across the road, obstructing the approach to the mine. At this point, there is a track heading off to the west (on the left). Follow this across a creek. (You may have to proceed on foot, depending on the level of the creek and the activity of the beavers!) Continue for 0.6 mi (1 km) from the gravel road to a point about 300 ft (100 m) before the track cuts through a grave pit. Turn left off the track along a blazed trail and walk 600 ft (200 m) south to the first outcrops.

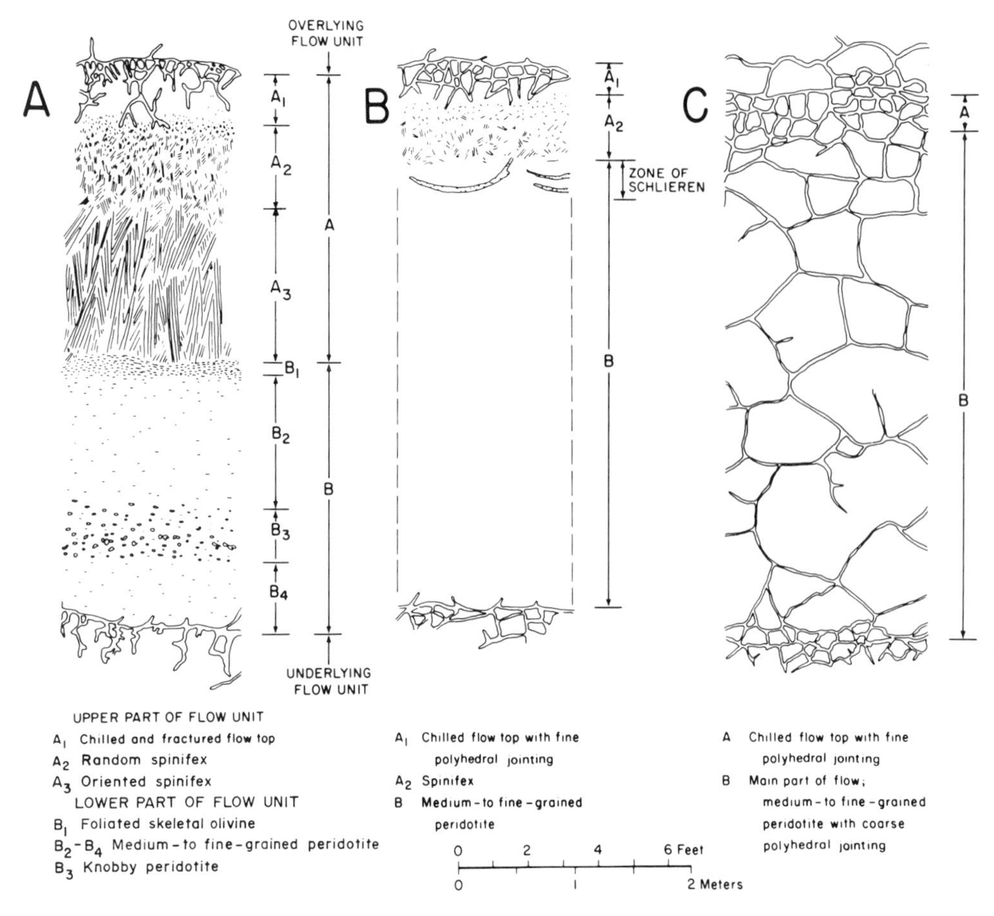

Figure 4. Profiles through three typical komatiitic flows: A) spinifex-rich flow, B) partially settled flow with minor spinifex texture, C) unsettled flow with polyhedral joint systems but no spinifex texture (modified after Arndt and others, 1977).

Follow these northwest until you cross a left-handed fault. You have now reached location 2 on Figure 6, shown enlarged in Figure 7.

Fred's Flow is a 400-ft thick (120-m) differentiated komatiitic flow (Arndt and others, 1977), in which the original magma appears to have contained 20 wt percent MgO. Only the upper layers of the flow and the underlying gabbroic differentiates are exposed at this locality. Go to the northeastern edge of the outcrop and examine the coarse gabbro exposed there. From here to the southwest, you will be walking stratigraphically up-section. Note the decrease in grain size of the gabbro, followed by the appearance of pyroxene needles, first with random orientation and then increasing in proportion and growing parallel to one another; the needles are orientated sub-perpendicular to the flow, giving rise to a texture known as pyroxene "string beef" spinifex. The needles have cores of altered pigeonite with augite mantles. The pyroxene spinifex grades upwards into an olivine-bearing type, which is best seen on the face of an abrupt cliff that marks the top of the spinifex zone. The top of the flow consists of 16.5 ft (5 m) of flowtop breccia, exposed below the cliff, and consists of cobbles of brownish komatiite in a matrix of similar composition.

Fred's Flow is overlain at this location by a komatiitic basalt flow, 33 ft (10 m) thick, with columnar jointing in places near the base and a pillowed top. Despite its light colour, the flow contains 9 wt percent MgO.

Stop 3—The Base of Fred's Flow. Return to the track and continue past the gravel pit to a drill hole on the left from which good, albeit rather rusty-tasting water is constantly flowing. Walk 60 ft (20 m) south from the road to where the basal

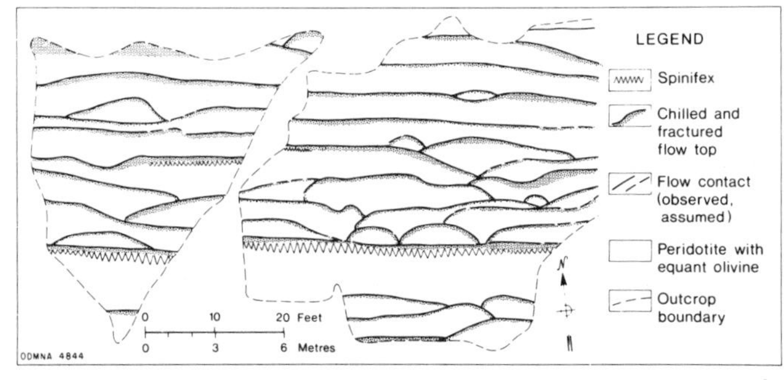

Figure 5. Plan showing 'bun-shaped' bodies in area C of Figure 4 (after Pyke and others, 1973).

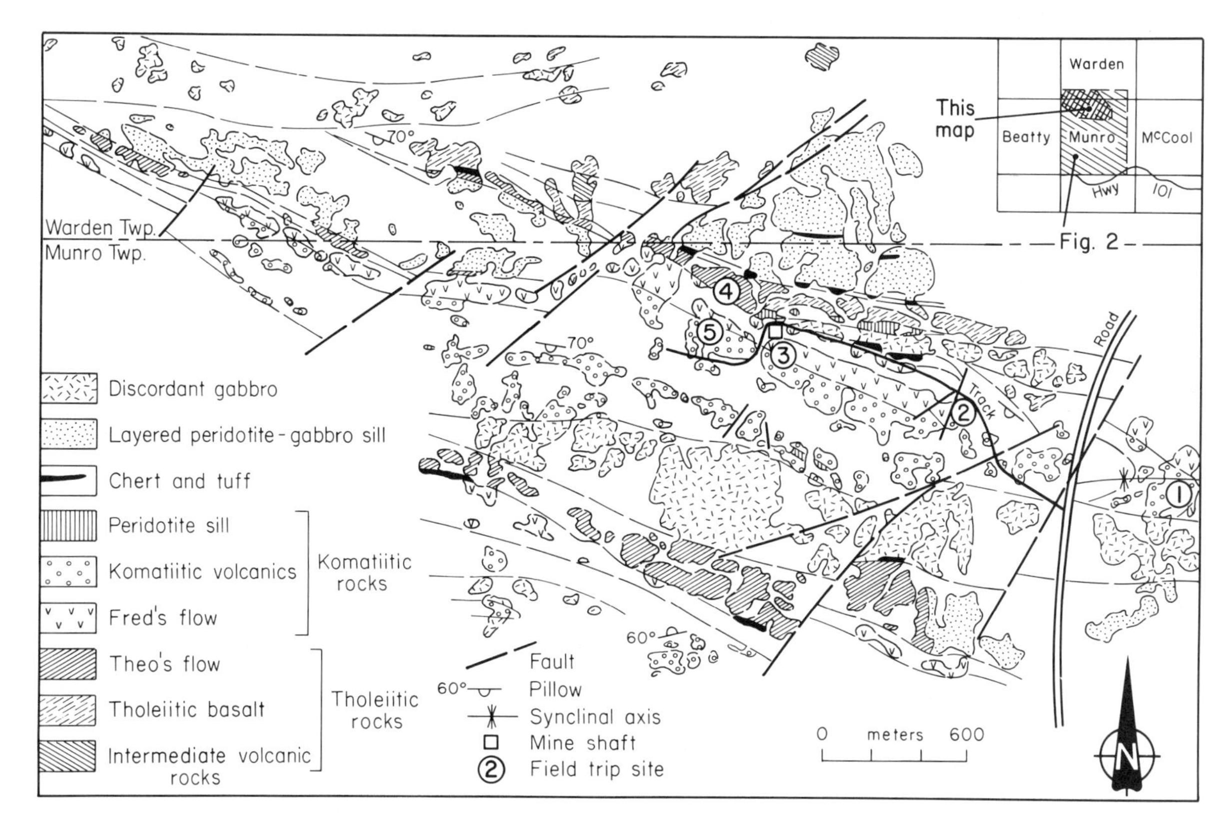

Figure 6. Geological map of northern Munro Township (modified after Arndt and others, 1977) showing the locations of Stops 2 to 5.

peridotite of Fred's Flow is well exposed (Stop 3 in Fig. 6). Lenses, 3 to 10 ft (1 to 3 m) in length, with a well-developed olivine spinifex texture, occur within the peridotite. If you cross the swamp ahead of you, you will cross (but won't see) the upper part of the peridotite, and a thin zone of two pyroxene pyroxenite (augite: bronzite ratio of approximately 4:1), arriving at the gabbro on the far side of the swamp.

Stop 4—Theo's Flow. Return to the track and continue west to a clearing that opens to the south, beyond which lies an old shaft. Walk 60 ft (20 m) north, through the bush, onto aphanitic pyroxenite and gabbro. These are at the eastern end of the area shown in Figure 8. Walk west along the outcrops for 300 ft (100 m) to good exposures of Theo's Flow, a differentiated tholeiitic flow.

Using Figure 8 you will be able to examine all of the rock units that make up this flow. Walking to the northwest, you will find, at the northern extremity of the outcrops, exposures of a thin veneer of peridotite that marks the base of the flow. This peridotite is overlain, in turn, to the south by an augite gabbro, an aphanitic pyroxene-rich rock (a chilled border phase), and hyaloclastite. The hyaloclastite may be seen on the steep face at the southern limit of the main outcrop, but is best exposed in fault-repeated slivers to the south of the main outcrop. Dikes of hyaloclastite penetrate downward through the zone of aphanitic pyroxenite to the gabbro.

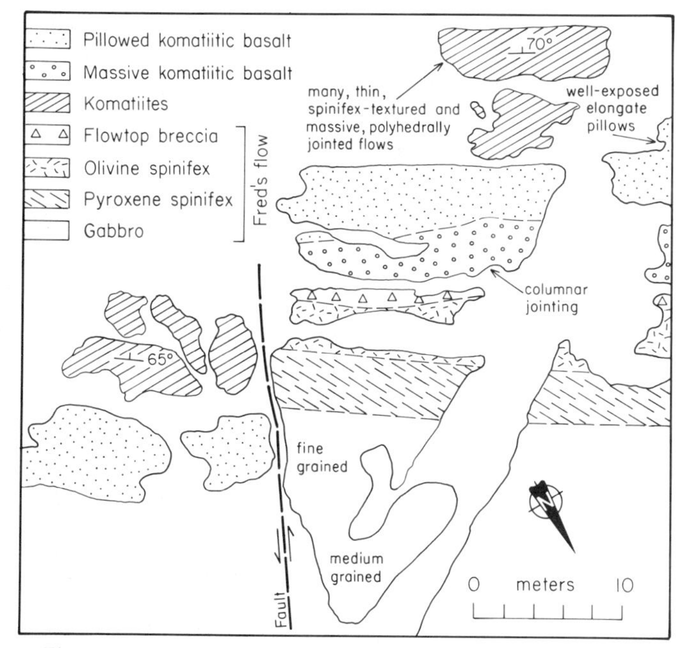

Figure 7. Geological map of Stop 2 showing the top of Fred's Flow.

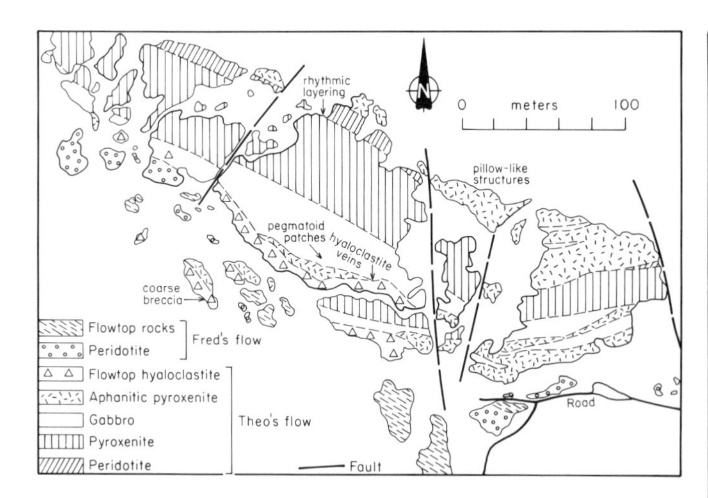

Figure 8. Geological map of part of Theo's Flow (Stop 4).

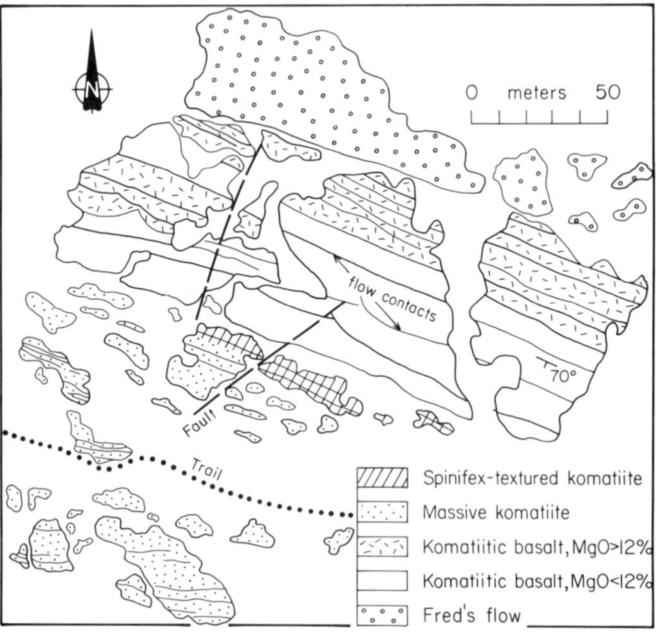

Figure 9. Geological map of komatiitic flows overlying Fred's Flow at Stop 5.

Note the disseminated Fe and Cu sulfides in the hyaloclastite. The Potter Mine, passed on the way in, consists of lenses of pyrrhotite, chalcopyrite, and sphalerite, interbanded with fine-grained ash. The lenses lie above 130 to 250 ft (40 to 80 m) of hyaloclastite identical in all respects to that seen here. Coad (1977) has concluded that a geothermal system, activated by the Centre Hill differentiated sill, concentrated sulfides such as those visible here to form the ore lenses at the Potter Mine.

Stop 5—Fred's Flow and Komatiites Above it. Walk south from Theo's Flow across the intervening 160 ft (50 m) of swamp onto more good outcrop. The swamp once again covers the ultramafic cumulates of Fred's Flow. The outcrop (Fig. 9) that you reach at the south of the swamp is gabbro. Continue across the gabbro to the spinifex zones and breccia that mark the top of the flow (similar to those at Stop 2).

A sequence of much thinner basaltic flows (3 to 30 ft; 1 to 10 m thick) overlies Fred's Flow. The lowermost of these flows contains about 14 wt percent MgO and is rich in pyroxene, but those higher in the sequence show a gradual decrease in MgO. These flows are mostly massive, although tops may have varioles, pillows, or, rarely, fine-grained pyroxene spinifex. Towards the base of some of them, cumulus enrichment in clinopyroxene has taken place. The uppermost basalt is overlain, with an abrupt change in composition, by komatiites forming the base of the third (second komatiitic) cycle. Some are spinifex-textured and others are massive. Upwards, the proportion of flows showing spinifex texture decreases. Stratigraphically, these komatiites are probably the equivalent of those seen at Stop 1.

Continue walking south across the komatiitic flows until you reach a track. Turn left (east) and continue along the trail until you reach the familiar territory of the turn to the old shaft (mentioned under Stop 4), and thence back to your transport.

REFERENCES

Arndt, N. T., and Nesbitt, R. W., 1982, Geochemistry of Munro Township basalts, *in* Arndt, N. T., and Nisbet, E. G., eds., Komatiites: London, Allen and Unwin, p. 309–329.

Arndt, N. T., and Nisbet, E. G., 1982, Komatiites: London, Allen and Unwin, 539 p.

Arndt, N. T., Naldrett, A. J., and Pyke, D. R., 1977, Komatiitic and iron-rich tholeiitic lavas of Munro Township, northeast Ontario: Journal of Petrology, v. 18, p. 319–369.

Arth, J. G., Arndt, N. T., and Naldrett, A. J., 1977, Genesis of Archean komatiites from Munro Township, Ontario; Trace element evidence: Geology, v. 5, p. 590–594.

Basaltic Volcanism Study Project, 1981, Mafic and ultramafic volcanism during the Archean, *in* Basaltic volcanism on the terrestrial planets: New York, Pergamon Press, Inc., p. 5–29.

Coad, P. R., 1977, The Potter Mine [M.S. thesis]: Toronto, Ontario, University of Toronto, 239 p.

Gresham, J., and Loftus-Hills, G., 1981, The geology of the Kambalda nickel field, Western Australia: Economic Geology, v. 76, p. 1373–1416.

Naldrett, A. J., and Mason, G. D., 1968, Contrasting Archean ultramafic igneous bodies in Dundonald and Clergue townships, Ontario: Canadian Journal of Earth Sciences, v. 5, p. 111–143.

Nesbitt, R. W., 1971, Skeletal crystal forms in the ultramafic rocks of the Yilgarn Block, Western Australia; evidence for an Archaean ultramafic liquid: *in* Symposium on Archaean Rocks, Geological Society of Australia Special Publication 3, p. 331–346.

Pyke, D. R., Naldrett, A. J., and Eckstrand, O. R., 1973, Archean ultramafic flows in Munro Township, Ontario: Geological Society of America Bulletin, v. 84, p. 955–978.

Viljoen, M. J., and Viljoen, R. P., 1969, The geology and geochemistry of the lower ultramafic unit of the Onverwacht Group and a proposed new class of igneous rock: Geological Society of South Africa, Upper Mantle Project Special Publication 2, p. 221–244.

Woodall, R., and Travis, G., 1969, The Kambalda nickel deposits, Western Australia: Publications of the Ninth Commonwealth Mining and Metallurgical Conference, Paper no. 26, 17 p.

The Sudbury Structure, Sudbury, Ontario

P. E. Giblin, Ontario Ministry of Northern Development and Mines, 199 Larch St., Sudbury, Ontario P3E 5P9, Canada

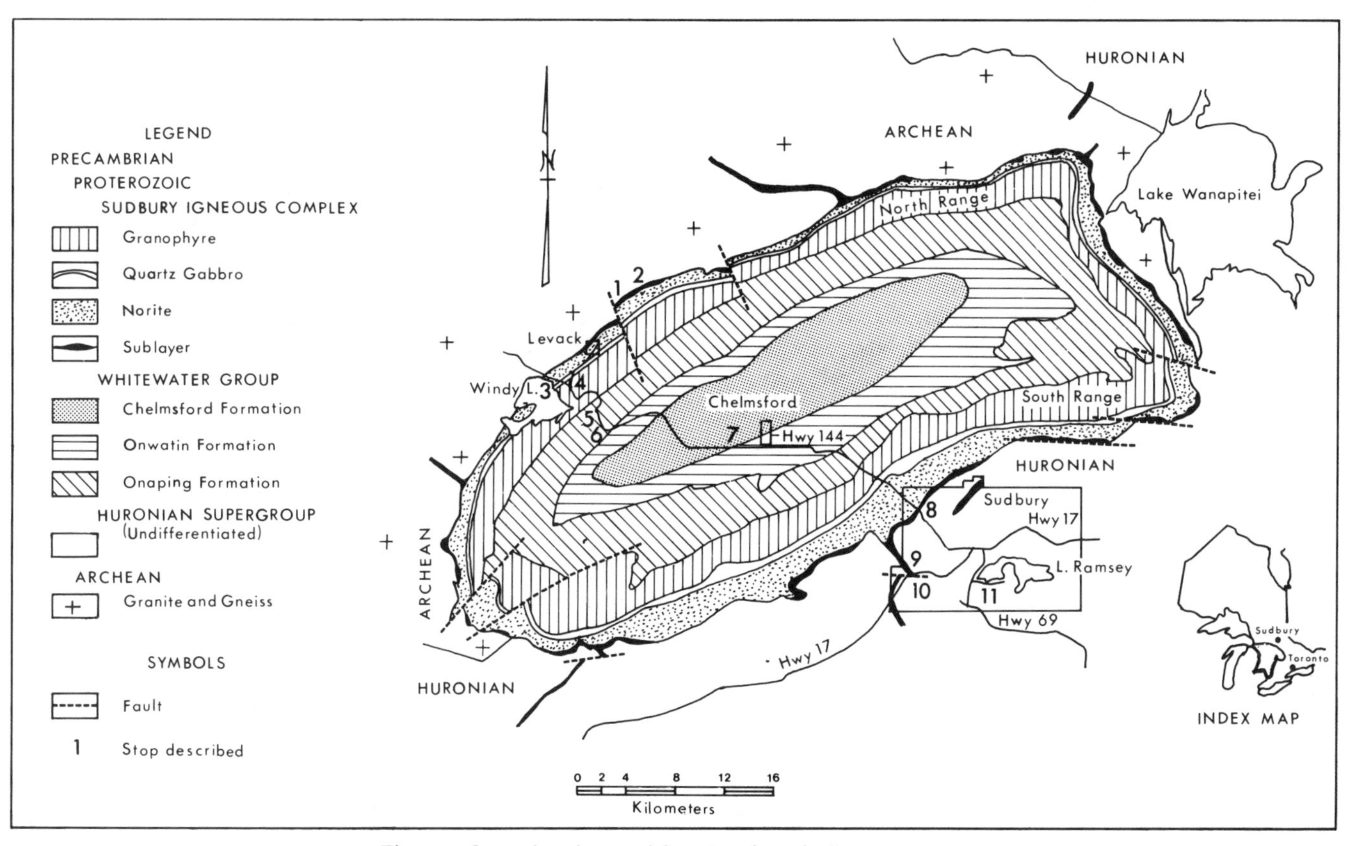

Figure 1. General geology and Stop locations, Sudbury Structure.

LOCATION

Sudbury is located in east-central Ontario, approximately 250 mi (400 km) north of Toronto. The 11 stops illustrating the Sudbury Structure are located, for the most part, on or near Ontario 144 northwest of the city of Sudbury and within the city (Fig. 1).

Stops 1 and 2 are on the north side of the Coleman Mine Road, a private road east of the town of Levack, and are located 1.6 mi (2.6 km) and 2.6 mi (4.2 km), respectively, east of the gate at the Levack Mine. Permission to enter must be obtained in advance from the Mines Exploration Department, Inco Ltd., Copper Cliff, Ontario P0M 1N0. Stop 3 is on the west side of Ontario 144, 0.9 mi (1.4 km) north of its intersection with Regional Road 8, on the northwest corner of the Highway's intersection with a secondary road. Stop 4 is on the south side of Regional road 8, 0.1 mi (160 m) east of its intersection with Ontario 144. Stop 5 is on the west side of Ontario 144, 2.2 mi (4.5 km) south of the intersection of Ontario 144 and Regional Road 8, while Stop 6 is at a small roadside picnic area 3.1 mi (6.0 km) south of the intersection. Stop 7 is on the north side of Ontario 144, 3.7 mi (5.9 km) east of the Vermillion River crossing, 1.2 mi (2.0 km) west of the Whitson River crossing. Stop 8 is at the developed historic site on the northeast side of Ontario 144, approximately 3.5 mi (5.6 km) east of the town of Azilda, and 1.2 mi (1.9 km) west of the intersection with Regional Road 30 (Godfrey Drive). Stop 9 is on the east side of Regional Road 30, 1.8 mi (2.9 km) south of its intersection with Ontario 144, and 1.4 mi (2.3 km) north of its intersection with Balsam Street in the town of Copper Cliff. Stop 10 is at the northwest corner of Trans-Canada 17 and Balsam Street, Copper Cliff. Stop 11 is located on the south side of Ramsey Lake Road (Regional Road 39) in Sudbury, 0.8 mi (1.3 km) east of its intersection with Paris Street (Regional Road 80).

SIGNIFICANCE

The rocks of the Sudbury area provide abundant evidence of a Middle Precambrian catastrophic explosive event, which led to formation of the Sudbury Structure. The nature of the event is uncertain. It has been variously attributed to unusually violent explosive volcanic activity and to the explosive effects of a meteorite impact.

The explosive event, known as the Sudbury Event, apparently triggered intrusion of the Sudbury Igneous Complex, a major mafic layered body from which segregated the world's largest known concentration of nickel-copper orebodies. The Complex differs from nearly all other layered complexes in that it contains no graded or fine scale igneous layering and is unusually rich in silica.

SITE INFORMATION

The Sudbury area lies in the Southern (structural) Province of the Canadian Shield, near the junction of the Southern, Superior, and Grenville Provinces. The general geology and the ore deposits have been described recently by Pye and others (1984).

The area is underlain predominantly by Middle Precambrian (Proterozoic) volcanic and clastic sedimentary rocks of the Huronian Supergroup, deposited between 2550 Ma and 2110 Ma, and deformed during the Middle Precambrian Penokean orogeny at about 1900 Ma. Near Sudbury, two granite plutons, dated at 2333 Ma, have intruded the lower part of the Huronian sequence. The Huronian rocks lie unconformably upon a basement of Early Precambrian (Archean) rocks, which consist predominantly of quartz monzonite plutons (2500–2700 Ma), minor metavolcanic and metasedimentary rocks, and hybrid migmatitic gneisses.

The Huronian and Archean rocks were affected about 1850 Ma by a catastrophic explosive event that resulted in formation of the Sudbury Structure, an elliptical northeast-trending feature underlying an area at least 60 mi (100 km) long by 30 mi (50 km) wide. It consists of 3 principal elements: (1) a funnel-shaped intrusive body, the Sudbury Igneous Complex, which outcrops as an elliptical ring 37 mi (59 km) long, 17 mi (27 km) across, with an average exposed width of 3 mi (5 km); (2) the Sudbury Basin, an area enclosed by the inner contact of the Complex, in which rocks of the Whitewater Group are preserved; and (3) a zone of intensely brecciated country rocks surrounding the outer ring of the Complex.

Brecciation of the Huronian and Archean country rocks surrounding the outer ring of the Sudbury Igneous Complex is characterized by the presence of a pseudotachylitic breccia known as Sudbury Breccia, by Footwall Breccia, shatter cones, and by microscopic shock-metamorphic features.

Sudbury Breccia (Stop 10) occurs as dikes and irregular bodies cutting all rocks older than the Sudbury Igneous Complex. Breccia bodies range in size from a few centimeters to nearly 6.6 mi (11 km) in length. Fragments are generally of the adjacent wallrock, but exotic fragments are common, and most commonly range in size from a few millimeters to a few meters in diameter. Larger fragments are usually rounded. The matrix consists of dark colored rock and mineral flour.

Footwall Breccia (Stop 2) occurs as discontinuous lenses along the outer, lower contact of the North and East Ranges of the Sudbury Igneous Complex. It is a heterolithic breccia, with angular to subrounded fragments generally of footwall rocks, but near the Complex it contains fragments of the Complex and occasional sulphide fragments. The matrix is primarily a granoblastic mosaic of plagioclase, quartz, pyroxene, and amphibole.

Shatter cones (Stop 11) occur in the country rocks surrounding the outer ring of the Sudbury Igneous Complex. They are distinctive conical fracture surfaces generated by high pressure shock waves and believed by some to be indicative of meteorite impact (e.g. Dietz, 1964).

Shock-metamorphic features in the footwall rocks include kink bands in biotite and planar features in quartz and feldspar (Dressler, 1984). Such features are known to form at very high pressures, above those generally considered achievable by volcanic processes, and are believed by some to be evidence of meteorite impact (French, 1968).

Rocks of the Whitewater Group are preserved only within the Sudbury Basin. The basal Onaping Formation consists largely of breccias of uncertain origin, and is overlain by siltstones of the Onwatin Formation and turbidites of the Chelmsford Formation.

The Onaping Formation is about 5,900 ft (1,800 m) thick and is divided into three members: the Basal Member, the Gray Member, and the uppermost Black Member, each consisting of breccias intruded by irregularly distributed igneous Melt Bodies.

The thin, discontinuous Basal Member consists of fragments of country rocks in a quartz-feldspar matrix having an igneous to sub-igneous texture.

The overlying Gray Member (Stop 5) is continuous around the Sudbury Basin, and is 650 to 2,300 ft (200 to 700 m) thick. It consists of a variety of heterolithic to monolithic breccias whose fragments consist of country rocks, devitrified glass fragments, fluidal-textured fragments, microbreccia fragments, composite fragments, and crystal and sulphide fragments (Muir and Peredery, 1984). Fluidal-textured material with conformable banding may coat country rock fragments, breccia fragments consisting of lithic and glassy fragments, and fluidal-textured material with contorted and truncated banding. Fragments range in size from less than 1 centimeter to many meters, but most range in size from 1 to 64 cm. The matrix consists of finely comminuted fragments of glass and other materials.

The Black Member (Stop 6) is also continuous around the Sudbury Basin; it ranges in thickness from 2,600 to 3,900 ft (800 to 1,200 m). It consists of breccias generally similar in fragment types to those of the Gray Member, but characterized by much smaller fragment size, a black carbonaceous matrix, and relatively common sulphide fragments.

Microscopic shock-metamorphic features occur in about 10 percent of the country rock fragments in the Gray and Black Members. These features include oriented planar features in quartz and feldspar, evidence of the former presence of maskelynite, and evidence of partial melting of zircon and titanite (French, 1972).

The Onaping Formation is considered to have been formed as a direct result of the explosive Sudbury Event; theories of the origin of the Sudbury Structure turn largely on interpretation of the origin of the Onaping Formation and on the significance of shock-metamorphic features. Proponents of an endogenic origin for the Sudbury Structure (e.g. Muir, 1984) consider the Onaping Formation breccias to be volcanic breccias deposited as a result of multiple explosive events, the Melt Bodies to be hypabyssal intrusive rocks with possible lava flows, and suggest the shock-metamorphic features formed endogenically. Proponents of an exogenic origin (e.g., Peredery and Morrison, 1984) consider the Onaping Formation breccias to be fallback debris and the Melt Bodies to be impact-generated melts. They point to the presence of shock-metamorphic features and the apparent inability of volcanic or tectonic processes to produce such features.

The Sudbury Igneous Complex intrudes the Onaping Formation, Huronian, and Archean rocks. It is a layered, mafic, essentially funnel-shaped body with generally inward-dipping contacts consisting of four members: (1) a Lower Zone comprised predominantly of norites (Stops 3 and 8); (2) a Middle Zone consisting of gabbros (Stop 4); (3) an Upper Zone consisting of granophyres (Stop 4); and (4) the Sublayer—a discontinuous, inclusion and ore-bearing, noritic to gabbroic rock that occurs around much of the outer margin of the Complex. Termed the Contact Sublayer (Stop 1), it also occurs as a filling for the so-called Offset Dikes (Stop 9). Norite of the Lower Zone has been dated at 1850 ±1 Ma (Krogh and others, 1984).

The nickel-copper ores occur in Contact Sublayer bodies and, on the North Range, in Footwall Breccia bodies (Stops 1 and 2, respectively); in Offset Dikes (Stop 9), and in fault-related deposits. The ores occur principally as massive to disseminated sulphide bodies in which pyrrhotite, pentlandite, and chalcopyrite predominate.

Stop 1: Contact Sublayer. The key outcrop is 500 ft (150 m) northwest of the road, on the northeast edge of a former gravel pit. Contact Sublayer, the principal host rock for ore at Sudbury, is a fine to medium-grained igneous rock consisting of plagioclase, hypersthene, and augite, with minor amounts of quartz, micrographic intergrowths of quartz and feldspar, and iron oxides. A wide variety of inclusions are generally present, consisting of fragments of recognizable wallrocks, and in places, xenoliths of mafic to ultramafic rocks not exposed at surface. The latter are believed to have been derived from one or more hidden layered complexes. Sulphide minerals are present in highly variable amounts, ranging from traces to more than 60 modal percent, and where sufficiently concentrated, form ore deposits. Mineralization is associated with Sublayer that carries mafic and ultramafic inclusions thought to be derived from one or more hidden layered intrusions.

At this stop, round to sub-round xenoliths of mafic igneous rocks, together with lesser amounts of quartzitic and granitic xenoliths, are set in a matrix of medium-grained, igneous-textured gabbroic rock. Patches of sulphide minerals occur in the matrix and in some of the mafic fragments. Please do not use hammers to collect samples at this stop.

Stop 2: Footwall Breccia. Footwall breccia is exposed on the north side of the road opposite the ore-loading tower of the Strathcona Mine. The rock is the principal host rock for ore deposits on the North Range. It is a heterolithic breccia, consisting largely of fragments of footwall rocks, together with mafic inclusions belonging to the suite of inclusions found in the Sublayer. Sulphides are common. The matrix is a granoblastic mosaic of plagioclase, quartz, pyroxene, and amphibole.

Stop 3: Felsic Norite. The Lower Zone of the Complex on the North Range consists of a mafic norite member and an overlying, predominant, felsic norite member approximately 1,500 ft (450 m) thick. Felsic norite, exposed at this stop, is a coarse-grained hypidiomorphic-granular rock consisting of plagioclase, hypersthene, augite, minor biotite, and interstitial micrographic intergrowths of quartz and feldspar.

Stop 4: Quartz Gabbro and Granophyre. The quartz gabbro member of the Middle Zone overlies felsic norite, is about 650 to 1,300 ft (200 to 400 m) thick, and consists principally of plagioclase, augite, amphibole, and intercumulus micrographic intergrowths of quartz and feldspar. Opaque oxides are present in amounts up to 8 modal percent, and apatite is common. Plagioclase exhibits a characteristic white color owing to sericitic alteration.

Walking southwesterly towards Ontario 144, one can observe in outcrops on the south side of the road a gradation of the gabbro into overlying granophyre of the Upper Zone, marked by a gradual increase in the number of small disseminated patches of pink micrographic intergrowth, until the rock becomes predominantly pink in color. The granophyre is about 3,000 ft (900 m) thick in this area, and consists of about 3 parts micrographic intergrowth to one part plagioclase. Quartz, biotite, amphibole, chlorite, epidote, and opaque minerals are also present.

Stop 5: Onaping Formation, Gray Member. Park in the small graveled area on the west side of Ontario 144 and walk south 1,000 ft (300 m) to the south end of the road-cut. The Gray Member here consists of a varied assemblage of angular to subround fragments, ranging in size from <2 mm to >64 mm, set in a fine to very fine-grained felsic matrix. The fragments consist of greenish-gray aphanitic fluidal-textured material and of country rocks, principally quartzite, granite, and gneiss, which are often coated by fluidal-textured material.

Stop 6: Onaping Formation, Black Member. Outcrops in the picnic area and on the west side of Ontario 144 display the small fragment size and black carbonaceous matrix typical of the Black Member. Some country rock fragments are enclosed by aphanitic flow-banded glassy material. Some structures (termed mud balls) are present; they consist of matrix material enveloped by aphanitic material with a high carbonaceous component, and are interpreted as sedimentary in origin (Peredery, 1972).

Stop 7: Chelmsford Formation. At this stop, several turbidite beds display well developed Bouma divisions; in particular, Divisions A, B, and C are well developed. Rusty-weathering concretions are common; they represent concentrations of ferruginous carbonate.

Stop 8: Quartz-Rich Norite and Contact Sublayer (Discovery Site). Park in the small roadside parking lot and walk about 100 ft (30 m) towards the railroad tracks. The weakly mineralized Contact Sublayer is poorly exposed on the west side of the railway. Quartz-rich norite is exposed on the east side of the railway. It comprises the basal unit of the Sudbury Igneous Complex on the South Range and has a thickness of 1,300 to 1,600 ft (400 to 500 m) (Naldrett and Hewins, 1984). The rock consists principally of plagioclase, hypersthene, and quartz, with minor clinopyroxene, biotite, and amphibole. Quartz content of the rock ranges from about 8 to 20 modal percent. Disseminated sulphides are common at this locality.

High-grade nickel-copper mineralization was discovered near this stop in 1883 during construction of the Canadian Pacific Railway.

Stop 9: Sublayer (Offset Dike Type). The Copper Cliff Offset is a dike of quartz diorite that extends southerly from the main mass of the Sudbury Igneous Complex. The Offset starts as a funnel-shaped embayment, then rapidly becomes a narrow, steeply dipping dike. It extends about 6 mi (10 km) into the footwall rocks and contains several orebodies (Cochrane, 1984; Grant and Bite, 1984).

At this stop, the rock is a medium to fine-grained quartz diorite containing abundant disseminated sulphides and small fragments of country rocks. The dike is less than 200 ft (60 m) thick at this point and can be seen in contact with the Creighton granite of Proterozoic age.

Stop 10: Sudbury Breccia. The outcrop is in a small grassed area, 100 ft (30 m) west of Balsam Street. Turbidites of the Huronian McKim Formation display well developed flame structures. The outcrop is cut by narrow zones of Sudbury Breccia, and the turbidite fragments locally exhibit wide divergences in strike. Please do not use hammers on this outcrop.

Stop 11: Shatter Cones. Park on the south shoulder of the road, walk east about 300 ft (100 m) to the road-cut on the south side of the road. Please do not use hammers or collect specimens from this stop.

Abundant shatter cones occur in cross-bedded feldspathic sandstone of the Huronian Mississagi Formation. This outcrop has some of the best-formed and most abundant shatter cones found to date in the Sudbury area. The cone surfaces are best seen when obliquely illuminated by the late afternoon sun.

The shatter cones are conical striated fracture surfaces whose striae fan from an apex. The striae are sharp grooves between intervening ridges and the cone surfaces are often micaceous and shiny. The cones range in length from a few centimeters to 3 meters. Large cones often have small cones developed on their flanks. Cone surfaces are exposed only where they parallel the outcrop surface: on other surfaces their intersecting crescentic fractures give a characteristic shattered appearance to the rock.

REFERENCES

Cochrane, L. B., 1984, Ore deposits of the Copper Cliff Offset, *in* Pye, E. G., Naldrett, A. J., and Giblin, P. E., eds., The geology and ore deposits of the Sudbury Structure: Ontario Geological Survey Special Volume 1, p. 347–359.

Dietz, R. S., 1964, Sudbury Structure as an astrobleme: Journal of Geology, v. 72, p. 412–434.

Dressler, B. O., 1984, The effects of the Sudbury Event and the intrusion of the Sudbury Igneous Complex on the footwall rocks of the Sudbury Structure, *in* Pye, E. G., Naldrett, A. J., and Giblin, P. E., eds., The geology and ore deposits of the Sudbury Structure: Ontario Geological Survey Special Volume 1, p. 97–136.

French, B. M., 1968, Shock metamorphism as a geological process, *in* French, B. M., and Short, N. M., eds., Shock metamorphism of natural materials: Baltimore, Mono Book Corporation, p. 1–17.

—— , 1972, Shock-metamorphic features in the Sudbury Structure, Ontario; A review, *in* Guy-Bray, J. V., ed., New developments in Sudbury geology: Geological Association of Canada Special Paper Number 10, p. 19–28.

Grant, R. W., and Bite, A., 1984, Sudbury quartz diorite offset dikes, *in* Pye, E. G., Naldrett, A. J., and Giblin, P. E., eds., The geology and ore deposits of the Sudbury Structure: Ontario Geological Survey Special Volume 1, p. 275–300.

Krogh, T. E., Davis, D. W., and Corfu, F., 1984, Precise U-Pb zircon and baddeleyite ages for the Sudbury area, *in* Pye, E. G., Naldrett, A. J., and Giblin, P. E., eds., The geology and ore deposits of the Sudbury Structure: Ontario Geological Survey Special Volume 1, p. 431–446.

Muir, T. L., 1984, The Sudbury Structure; Considerations and models for an endogenic origin, *in* Pye, E. G., Naldrett, A. J., and Giblin, P. E., eds., The geology and ore deposits of the Sudbury Structure: Ontario Geological Survey Special Volume 1, p. 449–489.

Muir, T. L., and Peredery, W. V., 1984, The Onaping Formation, *in* Pye, E. G., Naldrett, A. J., and Giblin, P. E., eds., The geology and ore deposits of the Sudbury Structure: Ontario Geological Survey Special Volume 1, p. 139–210.

Naldrett, A. J., 1984, Summary, discussion, and synthesis, *in* Pye, E. G., Naldrett, A. J., and Giblin, P. E., eds., The geology and ore deposits of the Sudbury Structure: Ontario Geological Survey Special Volume 1, p. 533–569.

Naldrett, A. J., and Hewins, R. G., 1984, The main mass of the Sudbury Igneous Complex, *in* Pye, E. G., Naldrett, A. J., and Giblin, P. E., eds., The geology and ore deposits of the Sudbury Structure: Ontario Geological Survey Special Volume 1, p. 235–251.

Naldrett, A. J., Hewins, R. H., Dressler, B. O., and Rao, B. V., 1984, The contact sublayer of the Sudbury Igneous Complex, *in* Pye, E. G., Naldrett, A. J., and Giblin, P. E., eds., The geology and ore deposits of the Sudbury Structure: Ontario Geological Survey Special Volume 1, p. 253–274.

Peredery, W. V., 1972, Chemistry of fluidal glasses and melt bodies in the Onaping Formation, *in* Guy-Bray, J. V., ed., New developments in Sudbury geology: Geological Association of Canada Special Paper Number 10, p. 49–59.

Peredery, W. V., and Morrison, G. G., 1984, Discussion of the origin of the Sudbury Structure, *in* Pye, E. G., Naldrett, A. J., and Giblin, P. E., eds., The geology and ore deposits of the Sudbury Structure: Ontario Geological Survey Special Volume 1, p. 491–511.

Pye, E. G., Naldrett, A. J., and Giblin, P. E., eds., 1984, The geology and ore deposits of the Sudbury Structure: Ontario Geological Survey Special Volume 1, 603 p.

Rousell, D. H., 1984, Onwatin and Chelmsford Formations, *in* Pye, E. G., Naldrett, A. J., and Giblin, P. E., eds., The geology and ore deposits of the Sudbury Structure: Ontario Geological Survey Special Volume 1, p. 211–218.

Transect along the Central Metasedimentary Belt/Central Gneiss Belt Boundary Zone, Grenville Province, Ontario Highway 35, Minden Area, Ontario

R. M. Easton, Precambrian Geology Section, Ontario Geological Survey, 77 Grenville Street, Toronto, Ontario M7A 1W4, Canada

LOCATION

This site displays deformed rocks in the boundary zone between the Central Metasedimentary Belt and the Central Gneiss Belt of the Grenville Province (Fig. 1) and consists of five stops (A–E) located along Ontario 35 between the villages of Norland and Dorset, Ontario (Fig. 1). Additional stops along Ontario 35 and elsewhere in the Minden area are shown in Figure 2 and are described in the field guides of Davidson and others (1984) and Easton and others (1984).

The traverse starts in Norland at the junction of Ontario 35 and 503. Norland is located about 90 mi (140 km) northeast of the city of Toronto and 40 mi (60 km) north of the town of Lindsay and can be reached from the south by Ontario 35. Norland is located about 95 mi (150 km) southwest of the town of Bancroft and can be reached from the east via Ontario 503 or Ontario 121 and 35.

Stop A (Stop 6, Easton and others, 1984) is located on Ontario 35 5.9 mi (9.5 km) north of Norland and consists of a series of outcrops (Fig. 4). Parking is available on the shoulder of the road. Stop B (Stop 7, Easton and others, 1984; Stop 27, Davidson and others, 1984) is located on Ontario 35, 8.6 mi (13.8 km) north of Norland at Miners Bay. Parking is available at Miners Bay Lodge located at the west end of the stop. Between Stops B and C important reference points are: Main Street Turnoff on Ontario 35 at Minden (16.6 mi, 26.7 km north of Norland); Ontario Ministry of Natural Resources Minden District Office (17.8 mi; 28.7 km); Ontario 35 and 121 to Bancroft junction (18.2 mi; 29.3 km). Stop C is located at the top of a hill at the south end of Mountain Lake 20.4 mi (32.8 km) north of Norland. Stop D (Stop 24, Davidson and others, 1984) is located 30.3 mi (48.8 km) north of Norland and is a road cut on the right hand (east) side of the highway. Parking is available on the left hand side of the highway 600 ft (200 m) past the outcrop. Stop E (Stop 21, Davidson and others, 1984) is a series of outcrops located on the left hand (west) side of the highway just after the bridge over Kushog Lake (37.5 mi; 60.3 km).

SIGNIFICANCE

This transect examines deformed rocks lying in the boundary zone between two major tectonic divisions of the Grenville Province: the Central Metasedimentary Belt consisting mainly of metamorphosed supracrustal rocks, including abundant marble; and the Central Gneiss Belt consisting mainly of metamorphosed quartzofeldspathic gneisses (Wynne-Edwards, 1972; Figs. 2, 3).

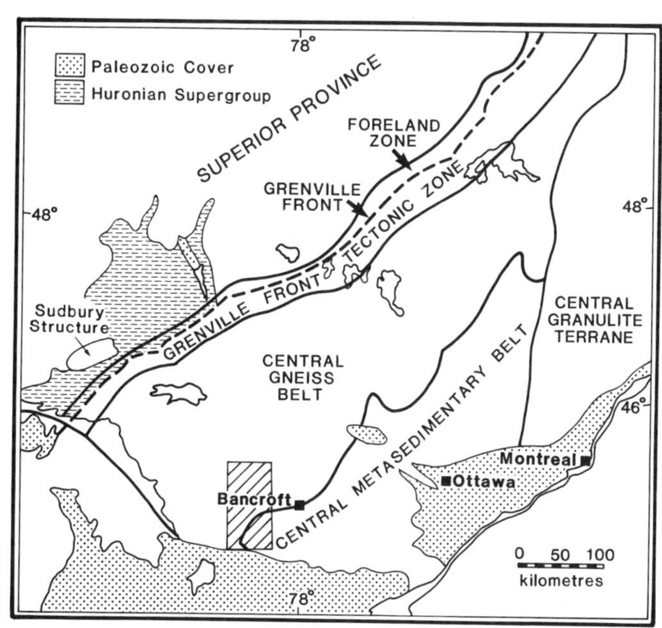

Figure 1. Location of site with respect to main tectonic divisions of the Grenville Province.

This transect emphasizes the deformed and disrupted character of the gneisses (tectonites) present within the boundary zone; as well as the general geology of the boundary zone.

SITE INFORMATION

The general geology of the area is shown in Figure 2 and relative ages of the main units are shown in Table 1. The area consists of four main tectonic elements. From east to west these are as follows.

1. The Glamorgan Gneiss Complex (Central Metasedimentary Belt), an igneous-metamorphic complex consisting mainly of tonalite gneiss, which may have been basement to the Grenville Supergroup. Grenville Supergroup carbonate and siliceous clastic metasedimentary rocks intruded by diorite, gabbro, nepheline-bearing syenite and monzodiorite, and granodiorite and monzogranite intrusions lie southeast of the Glamorgan Gneiss Complex. Stratigraphic analysis is possible in these metasedimentary rocks.

2. The Denna Lake Structural Complex is a zone of tectonically disrupted rocks consisting mainly of marble tectonic breccia, but also containing blocks, slices, and large zones of clastic siliceous metasedimentary rocks and disrupted granitoid rocks.

Most rock types observed in the Denna Lake Structural Complex have relatively undeformed counterparts in the Grenville Supergroup rocks southeast of the Glamorgan Gneiss Complex. It is possible to divide the Denna Lake Structural Complex into a number of zones based on dominant clast lithology and the presence of competent units such as quartzarenite horizons (Fig. 3). These zones preserve on a broad scale a ghost stratigraphy similar to that present in Grenville Supergroup rocks southeast of the Glamorgan Gneiss Complex (Easton, 1985; Easton and others, 1985), which may indicate that rocks in the Denna Lake Structural Complex have not been transported far.

3. The Central Metasedimentary Belt Boundary Zone (Stops A–D) consists of tectonically modified rocks that dip shallowly (10° to 40°) to the east and southeast. Within the boundary zone, different units can be mapped based on characteristic lithologies and textures (Figs. 2, 3). In the Digby area west of Stops A and B (Figs. 2, 3), rocks in the western part of the boundary zone may represent deformed equivalents of rocks within the Central Gneiss Belt (Easton and Van Kranendonk, 1984). Although the tectonites in the boundary zone are mainly quartzofeldspathic gneisses, small layers and lenses of marble, marble breccia, and calc-silicate gneiss are present locally. The eastern boundary of the zone with the Central Metasedimentary Belt is sharp, and is less than 3 ft (1 m) wide. The contact separates highly deformed granite gneisses from marble breccia of the Denna Lake Structural Complex. The western contact is a broad zone of deformation, marked in part by a shear zone along which brecciated quartzofeldspathic gneisses are present. Deformation related to this shear zone, and to the boundary zone, persists into the Central Gneiss Belt for several mi (km) (Easton and others, 1985; Fig. 3).

A cross section across the boundary zone north of Miners Bay on Gull Lake indicates that there is a three-fold division to the boundary zone (Fig. 3). From east to west these are as follows: (1) A zone of well-layered, highly-strained granite and granitoid gneisses of indeterminate protolith; including 'straight' and 'transposed' gneisses of Hanmer and Ciesielski (1984). (2) Within this zone are highly disrupted 'irregular-layered' quartzofeldspathic gneisses interlayered with calc-silicate gneisses and dolomitic marbles. Also present are the 'porphyroclastic gneisses' of Davidson and others (1982) and Hanmer and Ciesielski (1984), which consist of potassium feldspar augen in a fine-grained mylonitic matrix. Rocks within this part of the boundary zone are more disrupted, in a brittle sense, than rocks in the other two parts. (3) A zone of mafic gneisses in the north, and a zone of granitic gneisses to the south, which may be derived from similar rocks with lower strain in the adjacent Central Gneiss Belt. The thinness of the boundary zone in the southern part of the site area may be due to structural onlap of the Denna Lake Structural Complex.

The nature of the boundary zone has been controversial. Davidson and others (1982), Culshaw and others (1983), and Hanmer and Ciesielski (1984) regard the boundary zone as being a zone of major thrusting and deformation related to northwest-

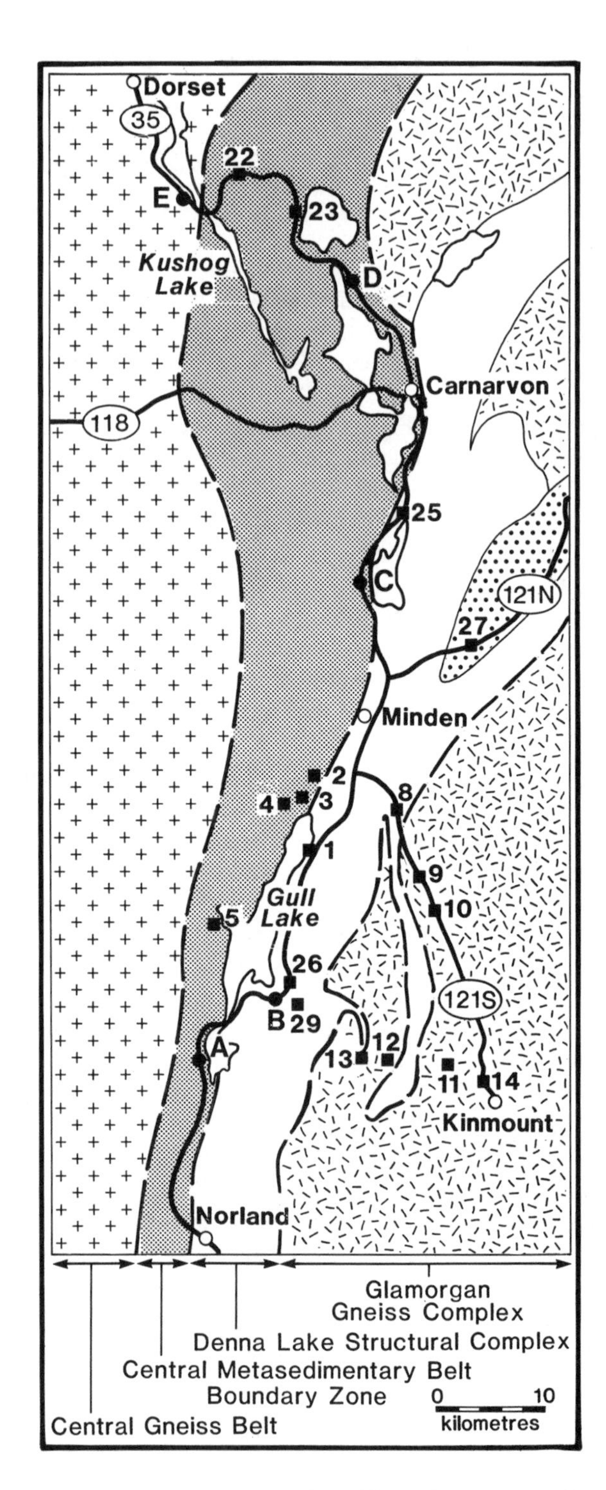

Figure 2. General geology and location of field trip stops along Ontario 35. Stops A–E, this guide. Stops 22, 23, 25, and 26, Davidson and others (1984); Stops 27 and 29, Bartlett and others (1984); Stops 1–5, Easton and others (1984, Day 1); Stops 8–14, Easton and others (1984, Day 2).

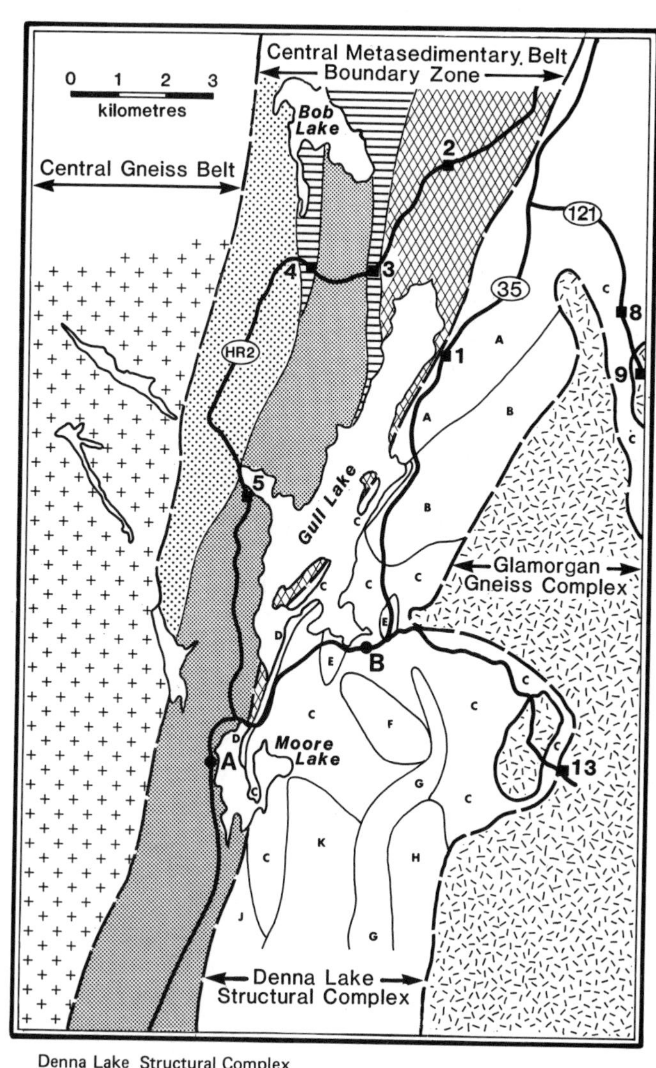

Denna Lake Structural Complex.

B marble breccia (subdivisions A-K based on clast size)

Central Metasedimentary Belt Boundary Zone:

- mafic gneisses
- irregularly layered gneisses
- marble, calc-silicate, and gneiss of possible metasedimentary origin
- "straight" gneiss
- anorthosite

Central Gneiss Belt:

- + + undivided granitic and diorite orthogneisses

Figure 3. Detailed geology of the Boundary Zone in the Digby area. Subdivisions of the Denna Lake Structural Complex breccias (A–K on figure) are based on clast size, abundance, and composition. Each letter represents a distinct breccia type. For details see Easton (1985) and Easton and others (1985).

directed thrusting of the Central Metasedimentary Belt over the Central Gneiss Belt, and that the boundary may approximate the boundary between the upper and the middle crust. Lumbers (1982) and Schwerdtner and Lumbers (1980) consider the zone, although locally deformed by diapiric uplift of the Algonquin Batholith that underlies the Central Gneiss Belt, as preserving an unconformable relationship between the Algonquin Batholith and the Central Metasedimentary Belt. They also consider that many of the quartzofeldspathic rocks in the boundary zone are metamorphosed clastic sedimentary rocks representing a basal clastic sequence developed along the unconformity. After examining the rocks of this site, the reasons for this controversy, as well as for its lack of resolution, will become evident.

4. Muskoka Domain–Fishog Subdomain, Central Gneiss Belt (Stop E, Fig. 2), consists of metamorphosed rocks of predominantly igneous origin. In the southern part of the site area, these rocks can be subdivided into five groups on the basis of lithology, structure, and cross-cutting relationships. In order of interpreted decreasing age, these are: an older diorite group; meta-anorthosite and related rocks; monzogranite and monzonite plutons; granodiorite plutons; and syenogranite sills and plutons.

TABLE 1. AGES OF MAIN ROCK UNITS IN SITE AREA

Unit	Age
1. Fishog Subdomain-- Central Gneiss Belt	ca. 1400-1500 Ma
2. Glamorgan Gneiss Complex	ca. 1300-1400 Ma
3. Grenville Supergroup	ca. 1250-1280 Ma
4. Denna Lake Structural Complex	formed ca. 1100-1050 Ma
5. Boundary Zone	formed ca. 1100-1025 Ma

Stop A: Granitoid Gneisses/Disrupted Pegmatite/Trachyandesite Dike. This stop shows several aspects of the geology of the Boundary Zone, but in particular addresses the question of genesis of the abundant pegmatite dikes and sills present within gneisses in the Boundary Zone. In particular, do feldspar augen in 'porphyroclastic gneisses' of the Boundary Zone represent crystals isolated by tectonic processes, or are they porphyroblasts? Evidence seen in outcrops at this stop is ambiguous, and favours both interpretations. At Stops C and D, there is good evidence for a tectonic origin. Specific areas of interest to be seen in this series of outcrops are shown in Figure 4 and include: (A) Fine- to medium-grained granitoid layers at this point on the outcrop may be derived from tectonic disruption of pegmatite; alternatively, they could be highly-strained, pre-existing granitic layers. Note the cross-cutting pegmatite dike at this site. (B) Outcrops here, and along Ontario 35 south to Norland, consist of mafic (dioritic) gneisses with varying proportions of granitic material, both fine-grained and pegmatitic. (C) At this point, a granitic dike is present that is relatively undeformed in its core, but along its margin has been disaggregated into isolated feldspar augen. (D) Thin, granitoid leucosome layers in this migmatite may be a relict of the

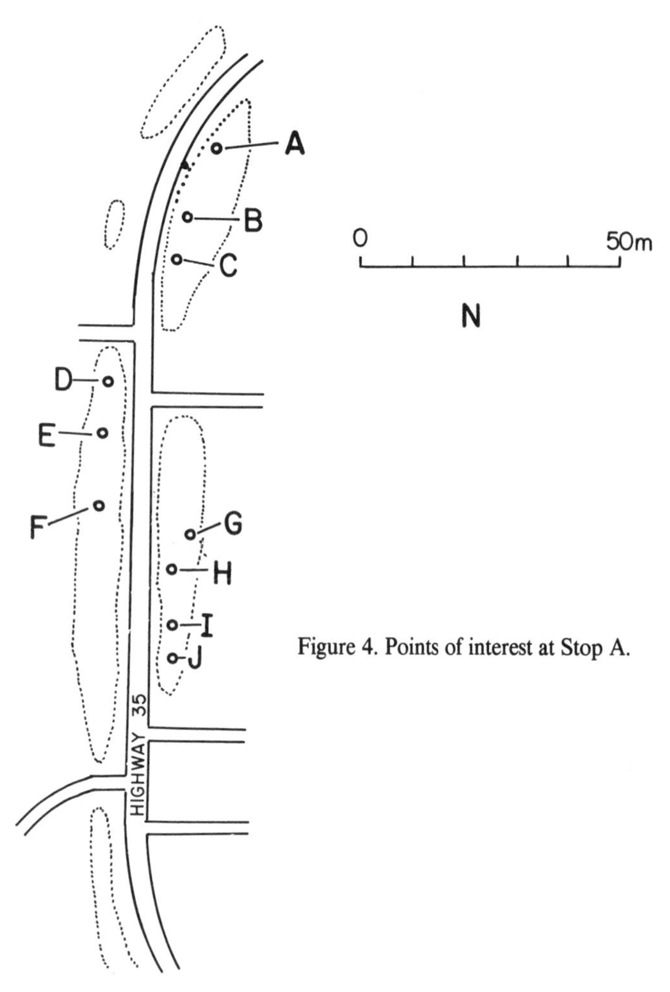

Figure 4. Points of interest at Stop A.

protolith, or they may be the product of mechanical crushing. Textures at this spot suggest both possibilities. (E) Large rafts of metagabbro within irregular layered gneisses. Note various ages of cross-cutting pegmatite. (F) Metamorphosed late-tectonic diabase dike cuts gneisses of the Boundary Zone at a low angle. Similar dikes cut strata in the Denna Lake Structural Complex. (G) Late shear zone cuts across the outcrop. (H) Trachyandesite dike intrudes gneisses of the Boundary Zone and is early Phanerozoic or Proterozoic in age. This is the most westerly occurrence of magmatic rocks of this age in southern Ontario; all other occurrences are related to the Ottawa-Bonnechere graben. Note stoping effects of the dike. (I) An example of a pegmatite dike showing varying degrees of disaggregation. (J) It is not clear if the abundance of pegmatite and granitic material in this outcrop is due to processes occurring during deformation, or whether the boundary zone has served as a locus for invasion by granitic material (i.e., Are the pegmatites close to the site of partial melting or have they been intruded far from the site of generation?). Textures at this spot suggest both alternatives.

Stop B: Marble Tectonic Breccia at Miners Bay. This spectacular roadcut exposes a thick unit of marble tectonic breccia of the Denna Lake Structural Complex. Much of the area of the Denna Lake Structural Complex (Figs. 2, 3) is underlain by similar rocks. The marble breccia has a marble matrix, is matrix supported, and contains clasts of different types of disrupted silicate rocks; some of which were once continuously interlayered calc-silicate gneiss, other perhaps intrusive phases. Metamorphic reaction selvages are well-developed around some blocks. Blocks of particular silicate rock types define large folds in the roadcut. Individual blocks are themselves folded and twisted in a recrystallized calcite matrix that shows a faint flow structure.

At the west end of the outcrop, a mass of pink granite intrudes the marble breccia at the base of the outcrop. Pieces of the same granite can be seen in the upper part of the outcrop, detached from the main granite body, to form blocks within the breccia. This granite was intruded after initial formation of the marble breccia, but prior to final movement.

Stop C: "Straight Gneiss" With Isolated Feldspar Augen. This stop is located in granitic gneisses near the eastern boundary of the Boundary Zone near the contact with the Denna Lake Structural Complex. The fine grain size of these rocks is probably due to mechanical deformation, as can be seen in the extremely attenuated quartz grains in the outcrop. Isolated feldspar augen in the outcrop show evidence of rotation and have well-developed 'tails.' Such augen have been used as an indicator of tectonic transport (Davidson and others, 1982, 1984; Culshaw and others, 1983; Hanmer and Ciesielski, 1984) and in this case are taken to indicate northwest-directed thrusting. Within the Boundary Zone and the Central Gneiss Belt, no augen showing an opposite sense of structural transport have been reported.

At the south end of the outcrop, broad, east-trending folds cause a gentle undulation in gneissic layering, and rusty-weathering dioritic gneisses of possible metasedimentary origin structurally overlie the granitic gneisses. The dioritic gneisses may in fact be part of the Denna Lake Structural Complex. Older maps of the area showed this contact as a possible unconformity. If it was an unconformity, tectonism has obliterated the original nature of the contact.

Stop D: Kinematic Indicators In Gneissic Tectonites. Pink and gray, layered, quartzofeldspathic gneisses in these road cuts dip gently to the east. Amphibolite layers are pulled apart, forming pods and lenses. The road cut is a section roughly perpendicular to gneissosity and parallel to lineation, and allows examination of several different structural elements from which a sense of displacement can be evaluated (Davidson and others, 1984). A number of the structural elements present in this outcrop have been used as indicators of transport direction during tectonism (Davidson and others, 1982; Culshaw and others, 1983). Points of interest at Stop D (Fig. 5) are as follows. (a) A narrow layer of calc-silicate gneiss containing a little calcite displays sheath folds; these appear as elliptical rings on outcrop surfaces that cut across the sheath axes, which are parallel to the southeast-plunging lineation in the outcrop. This lineation occurs throughout the Boundary Zone and the Denna Lake Structural Complex. (b) A good example of a rotated feldspar augen with 'tails' can be seen near the top of the outcrop below the power line. The sense of rotation gives displacement of the upper side to the north-northwest (up-lineation). Nearby, quartzofeldspathic lenses show quartz lenticle orientation in the flattening plane at an angle to the main gneissosity. The mutual relationship of these two planes indicate the same kinematic sense as that given by the rotated augen. (c) A hornblende-rich layer displays asymmetric boudinage, with back rotation of the boudins and offset on shear plane between boudins. The orientation of the boudins with respect to the gneissosity also indicates a north-northwest sense of displacement. (d) A pegmatite dike cuts up structural section to the northwest, and is boudined and offset in that direction. This feature is interpreted by Davidson and others (1984) as resulting from rotation of an originally steeper pegmatite toward the shear plane, which is consistent with a sense of displacement to the north-northwest. (e) An isoclinal fold is truncated by a disaggregated pegmatite, now represented by a string of feldspar augen.

Stop E: "Green Rock"—"Pink Rock" Transition, Muskoka Domain, Central Gneiss Belt. This stop lies on the west shore of Kushog Lake and is in the Muskoka Domain of the Central Gneiss Belt. Rocks of the Boundary Zone are located on the east side of the lake. This stop provides an opportunity to contrast rocks of the Muskoka Domain, in this case greenish 'charnockitic' orthogneiss, with the deformed and disrupted rocks seen at Stops A–D in the Boundary Zone.

For several mi (km) to the north, almost to the village of Dorset, Ontario 35 crosses numerous outcrops of assorted pink, grey, and greenish brown orthogneisses, of the eastern Muskoka Domain. In places, the orthogneisses retain retrograded granulite facies assemblages (two pyroxenes, locally with garnet), but elsewhere hornblende and biotite prevail in migmatitic rocks showing no evidence of retrogression. In a few places the charnockitic orthogneiss bodies appear to have intruded highly deformed pink and gray migmatitic orthogneiss, but in most cases a gradational relationship is most likely.

This stop illustrates gradational variation between another

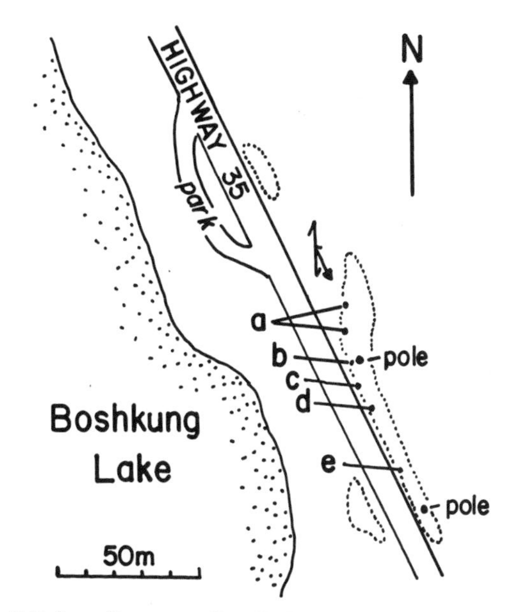

Figure 5. Points of interest at Stop D. From Davidson and others, 1984.

type of pink granite orthogneiss and yellow-green charnockitic rock. Here the retrogression is patchy and not apparently related to a deformational or intrusive event. Thin section work by A. Davidson indicates that the pink orthogneiss (metagranite) contains plenty of quartz and that the yellow-green rock contains very little and is syenitic (Davidson and others, 1984). Pyroxene occurs only in the syenitic rock, which also contains a mineral tentatively identified as fayalite (Davidson and others, 1984). The distribution of colour variation at this stop is composition dependent, and may be an original igneous feature. The patchy distribution of granulite facies minerals is common in orthogneisses in the Central Gneiss Belt.

ANNOTATED REFERENCES

Bartlett, J. R., Brock, B. S., Moore, J. M., Jr., and Thivierge, R. H., 1984, Grenville Traverse A, Cross-sections of parts of the Central Metasedimentary Belt: Geological Association of Canada, Joint Meeting, May 14–16, 1984, London, Ontario, Field Trip Guidebook 9A/10A, 64 p.

Culshaw, N. G., Davidson, A., Nadeau, L., 1983, Structural subdivisions of the Grenville Province in the Parry Sound–Algonquin Region, Ontario: Geological Survey of Canada Paper 83-1B, p. 243–252 (Local geology).

Davidson, A., Culshaw, N. G., and Nadeau, L., 1982, A Tectono-metamorphic framework for part of the Grenville Province, Parry Sound Region, Ontario: Geological Survey of Canada, Paper 82-1A, p. 175–190 (Local geology).

Davidson, A., Culshaw, N. G., and Nadeau, L., 1984, Grenville Traverse B, Cross-section of part of the Central Gneiss Belt: Geological Association of Canada, Joint Annual Meeting, May 14–16, 1984, London, Ontario, Field Trip Guidebook 9B/10B 79 p. (Related fieldtrip guide).

Easton, R. M., 1983, Howland Area; Haliburton, Peterborough, and Victoria counties, *in* Wood, J., White, O. L., Barlow, R. B., and Colvine, A. C., eds., Summary of field work, 1983, by the Ontario Geological Survey: Ontario Geological Survey Miscellaneous Paper 116, p. 74–79 (Local geology).

——, 1985, Stratigraphy along the Central Metasedimentary Belt boundary zone near Minden; Implications for mineral exploration [abs.]: Ontario Geological Survey, Geoscience Research Seminar, p. 15 (Local Geology).

Easton, R. M., and Van Kranendonk, M., 1984, Digby—Lutterworth Area; Haliburton and Victoria counties, *in* Wood, J., White, O. L., Barlow, R. B., and Colvine, A. C., eds., Summary of field work, 1984, by the Ontario Geological Survey: Ontario Geological Survey Miscellaneous Paper 119, p. 75–81 (Local geology).

Easton, R. M., Heaman, L. M., McNutt, R. H., Shaw, D. M., and Easton, M. G., 1984, Geology of the Minden and Chandos township areas, Central Metasedimentary Belt, Ontario: Friends of the Grenville 1984 Fieldtrip Guidebook, Brampton, 54 p. (Related fieldtrip guide).

Easton, R. M., Van Kranendonk, M., and Sanderson, D., 1985, Precambrian geology of the Digby-Lutterworth area: Ontario Geological Survey Map P.2951 and P.2952, Geology 1984, scale 1:15,840, color (Local geology).

Freeman, E. B., 1979, Geological Highway Map, Southern Ontario: Ontario Geological Survey Color Map 2441, scale 1:800,000 (Regional geology).

Hanmer, S. K., and Ciesielski, A., 1984, A structural reconnaissance of the

Central Metasedimentary Belt, Grenville Province, Ontario and Quebec: Geological Survey of Canada Paper 84-1B, p. 121–131 (Local geology).

Lumbers, S. B., 1982, Summary of metallogeny, Renfrew County Area: Ontario Geological Survey Report 212, 58 p.

Schwerdtner, W. M., and Lumbers, S. B., 1980, Major diapiric structures in the Superior and Grenville provinces of the Canadian Shield, *in* Strangway, D. W., ed., The continental crust and its mineral deposits: Geological Association of Canada Special Paper 20, p. 149–180.

Wynne-Edwards, H. R., 1972, The Grenville Province, *in* Price, R. A., and Douglas, R.J.W., eds., Variations in tectonic styles in Canada: Geological Association of Canada Special Paper 11, p. 263–334.

Supracrustal rocks of the Central Metasedimentary Belt of the Grenville Province, Madoc-Marmora-Havelock area, Ontario

R. M. Easton, Precambrian Geology Section, Ontario Geological Survey, 77 Grenville Street, Toronto, Ontario M7A 1W4, Canada
T. R. Carter, Petroleum Resources Laboratory, Ministry of Natural Resources, 458 Central Avenue, London, Ontario N6B 2E5, Canada
J. R. Bartlett, 84 Pembroke Street, Toronto, Ontario M5A 2N8, Canada

LOCATION

This site displays supracrustal rocks of the Grenville Supergroup as well as some intrusive rocks cutting the Grenville Supergroup within the Central Metasedimentary Belt of the Grenville Province (Fig. 1). The trip consists of five stops (1–5) located on or near Ontario 7 between the towns of Havelock and Madoc, Ontario (Fig. 2). One other stop (6) can be easily reached as part of a trip to this site and complements the main stops. Other stops in the area of Ontario 7 are described by Bartlett and others (1984) and Easton and others (1986). Easton and others (1984) describe stops in the Chandos area 30 mi (50 km) to the north-northwest in an area of similar geology.

The field trip starts in Havelock, at the junction of Ontario 7 and 30, about 80 mi (130 km) east-northeast of the City of Toronto and 25 mi (40 km) east of the city of Peterborough; this point can be reached from the east or west via Ontario 7. Proceed east from Havelock on Ontario 7 toward Marmora. Turn north onto Peterborough County Road 48 about 10 mi (15 km) east of Havelock. Proceed north on Peterborough County Road 48 about 5.6 mi (9 km), turn west onto a private gravel road that leads to Belmont Lake. Park in the grassy clearing where the road curves to the right, and walk back to the track on the left. Follow the track to the cottages overlooking Belmont Lake (about 1,200 ft; 350 m, Fig. 4). This is Stop 1 (Stop 4 of Bartlett and others, 1984). Please respect cottage owners' property and do not damage the shoreline outcrops. Retrace route back to County Road 48, proceed north on County Road 48 to a stop sign (2.0 mi; 3.2 km), turn west, proceed to the next stop sign (0.4 mi; 0.6 km), turn north, and proceed 0.3 mi (0.5 km) to where the powerline crosses the road. Park here and proceed east of the road to the second powerline tower, about 1,200 ft (350 m) to Stop 2 (Stop 3 of Bartlett and others, 1984). Proceed back to the stop sign (0.3 mi; 0.5 km), turn east and proceed to the next stop sign (0.7 mi; 1.1 km), continue east (straight) through the intersection 1.6 mi (2.6 km) to the town of Cordova Mines. Turn left at the T-junction in Cordova Mines and proceed 300 ft (100 m) north to Stop 3. Return to T-junction, turn east onto Hastings County Road 3 and proceed 7.8 mi (12.4 km) to Marmora. In Marmora, turn left at the crossroads and follow Ontario 7 1 mi (1.7 km) to the east. Turn right on Marmora Township 5th line road and proceed south for 0.5 mi (0.8 km). Park at Marmoraton Iron Mine (Stop 4; Stop 7 of Bartlett and others, 1984). Return to Ontario 7 and proceed east. Park by the roadside cut in massive pink granite located 3.5 mi (5.6 km) east of Marmora Township 5th Line Road (Stop 5; Stop 8 of Bartlett and others, 1984). Stop 6 (Stop 14 of Bartlett and others, 1984) can be reached by proceeding on Ontario 7 to Kaladar. Turn left at Kaladar and proceed north on Ontario 41 for 2.6 mi (4.1 km). Park by the road cut near the crest of the hill for Stop 6.

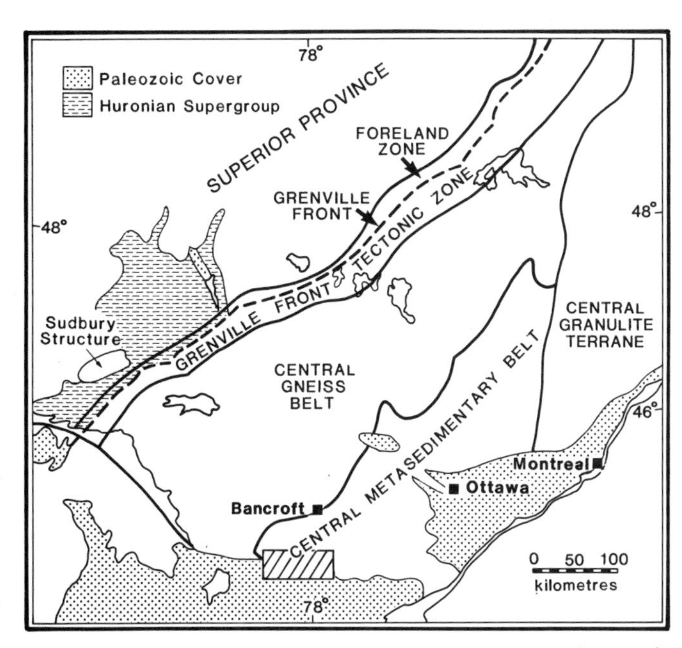

Figure 1. Location of site (slashed area) with respect to the main tectonic divisions of the Grenville Province.

SIGNIFICANCE

This site shows the variety of rock types found in the Central Metasedimentary Belt of the Grenville Province (Wynne-Edwards, 1972), with particular emphasis on the character of the metasedimentary and metavolcanic rocks present within the Grenville Supergroup in the Central Metasedimentary Belt, as well as displaying some of the igneous rocks that intrude this supracrustal sequence.

SITE INFORMATION

The exposures in the Madoc-Marmora-Havelock area are in the lowest metamorphic grade part of the Central Metasedimentary Belt (Elzevir Terrane). Protoliths of supracrustal units are

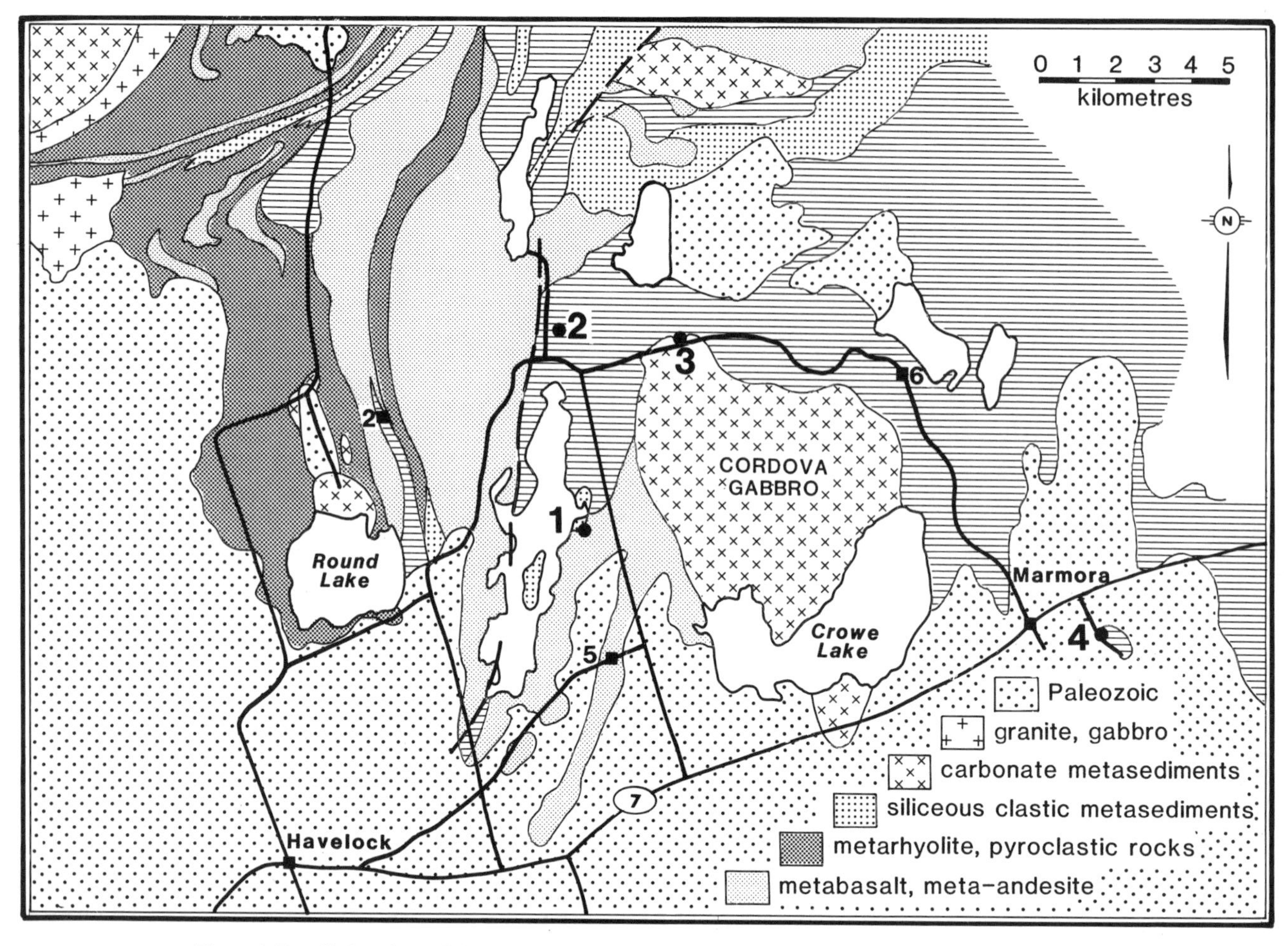

Figure 2. Detailed geology of the area and location of Stops 1–4. Small numbers refer to additional stops described in Bartlett and others (1984); geology after Bartlett and others (1984).

unequivocal and the principles of stratigraphy can be applied. Supracrustal rocks of the Central Metasedimentary Belt are generally part of the Grenville Supergroup, and individual groups and formations, mainly of informal status, have been identified in many places (Fig. 3). One group, the Flinton Group, has been identified as separate from the Grenville Supergroup by Moore and Thompson (1980). The Flinton Group lies unconformably on the Grenville Supergroup, although it is itself deformed and metamorphosed.

More specifically, the Havelock-Marmora area is underlain by a 40,000 ft (12-km)-thick metamorphosed supracrustal succession that has been intruded by a variety of plutons ranging in composition from pyroxenite to alkali feldspar granite and nepheline syenite (Fig. 2). Non-metamorphosed Paleozoic sedimentary rocks, mainly limestone, unconformably overlie the Precambrian rocks in the southern and western parts of the area and occur as numerous map-scale outliers throughout the remaining parts (Bartlett and Moore, 1985).

In the region between and north of Round and Belmont Lakes (Fig. 2), the rocks constitute an upright, moderately dipping, east-facing homocline that reflects the first phase folding (D_1). To the west, the homoclinal arrangement of strata is affected by open-to-closed macroscopic D_2 folds. Map-scale faults occur in and north of Belmont Lake.

All Precambrian rocks of the area have been regionally metamorphosed to upper greenschist facies (Bartlett and Moore, 1985). Grade rises northwesterly to middle amphibolite facies in eastern Belmont and northwestern Marmora Townships. Contact metamorphism accompanied the intrusion of the Cordova gabbro and the Deloro granite, and produced calc-silicate and magnetite-rich skarns in adjacent carbonate rocks (as seen at Stop 4). Volcanic rocks in the area are part of the Belmont Lake metavolcanic complex, which contains four mafic to felsic volcanic cycles (Bartlett and Moore, 1985).

The geologic history of the area can be summarized as follows: 1) Deposition of Grenville Supergroup sedimentary and volcanic rocks (Stops 1 and 2) at about 1250 Ma; 2) Intrusion of a variety of intrusions between 1220 and 1240 Ma, including a tonalite-trondjhemite suite of plutons, a gabbro-diorite suite (Cordova Gabbro, Stop 3), and nepheline-syenite and other al-

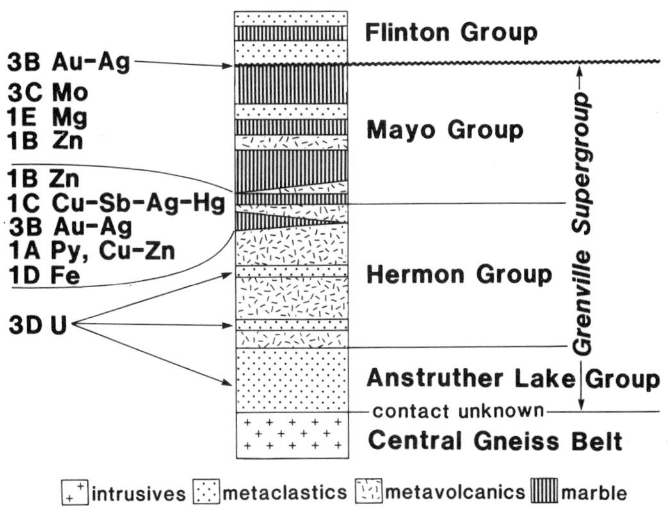

Figure 3. Generalized stratigraphic column of the Grenville Supergroup. Modified from Carter (1984).

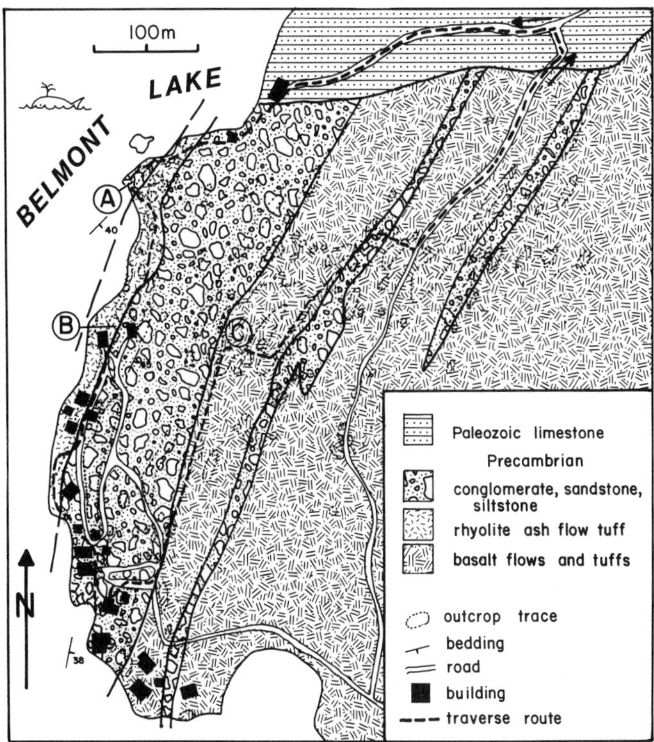

Figure 4. Detailed geology and traverse route at Stop 2. From Bartlett and others (1984).

kalic rocks. Skarns produced some mineral deposits (Stop 4, Marmorton Iron Mine); 3) Uplift and erosion, leading to deposition of the Flinton Group (Stop 6) at about 1080–1100 Ma; 4) Metamorphism and deformation at about 1070–1080 Ma. Intrusive activity (granodiorite, granite) also occurred during deformation; 5) Intrusion of late plutons such as the Deloro Granite (Stop 5) at about 1060 Ma; and 6) Intrusion of pegmatite dikes between 1020 and 1060 Ma.

STOP DESCRIPTIONS

Stop 1: Volcanic Rocks of the Grenville Supergroup. This stop displays volcanic rocks of the upper part of Cycle III (Belmont Lake formation) and the lower part of Cycle IV (Crowe River formation) of the Belmont Lake metavolcanic complex (Bartlett, 1983; Bartlett and Moore, 1985; Bartlett and others, 1984). This stop consists of a traverse that proceeds up-section through conglomerate, sandstone, siltstone, and ash flow tuff of Cycle III into volcanic breccias and flows of Cycle IV. The route is shown in Figure 4.

The lowest exposed rocks are cross-bedded feldspathic sandstones, best seen at station A. A sharp contact with overlying rhyolitic ash flow tuff can be seen at station B when lake level is low. At station B, the lower 13 ft (4 m) of the 60 ft (20-m)-thick ash flow tuff unit is coarsely vesicular, owing to original trapped gas. Mesoscopically, the unit is relatively uniform and massive, save for a weak foliation imparted by elongate lithic fragments and fiamme. Relict eutaxitic foliation is observed in thin section. Fifteen percent of each of plagioclase and blue-to-colourless quartz phenocrysts are present. The pink, beige-weathering ash flow tuff is readily distinguished from the gray, gray-to-brown-weathering sandstones adjacent to it. The ash flow tuff has a preliminary U-Pb zircon age of about 1250 Ma (D. W. Davis, personal communication, 1984), consistent with zircon and Rb-Sr ages from the Tudor Volcanics (Hermon Group) of the Grenville Supergroup in the Kaladar area to the east.

Overlying the ash flow tuff are feldspathic sandstones, conglomerates, and siltstones. The conglomerates are generally polymict and poorly sorted, with well-rounded clasts of felsic volcanic rocks (including flows and material similar to the underlying ash flow tuff), calcite marble, and silstone. Sandstone and siltstone, commonly interbedded, display cross-bedding, flame structures, ripple marks, and primary slump folds, all of which indicate that the succession is upright.

The sharp upper contact of the clastic rocks with basalt of the Crowe River formation (Cycle IV) is exposed at station C. Here the overlying rocks are mainly massive lavas that change eastward to volcanic breccias and reworked tephra. Some of the breccias contain bombs, and hence can be more properly termed agglomerates. The shapes of the bombs and the distribution of vesicles in the bombs indicate that some of the bombs were still plastic on impact, suggesting that eruption and deposition were subaerial.

Stop 2: Stromatolites in the Marmora Formation, Grenville Supergroup. Proceed along the powerline east of the road (up-section) to the second tower, about 1,200 ft (350 m); rocks along the way are mainly dark gray weathering calcite marbles and light gray to brown-weathering dolomite marbles, interlayered on the scale of meters to tens of meters. These rocks, belonging to the Marmora formation, have been intruded by small irregular bodies of gabbro and leucogabbro. Fossil stromatolites may be observed in dolomite marble west of the second tower. At the base of the tower, columnar stromatolites occur in locally derived blocks and, less clearly, in outcrop. By analogy with modern marine sediments, these stromatolites are indicative of an intertidal to supratidal environment of deposition. *Please do not hammer the stromatolites.*

Stop 3: Cordova Gabbro. This outcrop shows layering within the medium-grained Cordova gabbro pluton. The Cordova Pluton is one of several diorite to gabbro plutons present within the Central Metasedimentary Belt. A U-Pb zircon age of 1241 Ma from the Cordova Pluton (D. W. Davis, personal communication, 1985) is similar to the age of the volcanic rocks in the area, as well as the Tallan Lake Sill, a mafic intrusion located in Chandos Township 30 mi (50 km) to the northwest (Easton and others, 1984).

Stop 4: Marmoraton Iron Mine. One of the first Canadian Government aeromagnetic surveys outlined a strong positive anomaly centered in an outcrop area of Ordovician limestone southeast of Marmora. Drilling revealed a magnetite body within clinopyroxene-garnet skarn at the contact between a syenite-diorite mass (of undetermined dimensions) and calcitic marlbe of the Marmora formation. After stripping 130 ft (40 m) of limestone, Bethlehem Steel began mining the ore body in 1955. By 1978 the pit had reached a depth of 730 ft (220 m) and could be deepened no farther. Waste dumps are being exploited for crushed stone at present, and the pit is filling with water. This stop also provides a good section through the Gull River and Shadow Lake Formations of Middle Ordovician age, which unconformably overlie the Precambrian rocks here.

Stop 5: Deloro Granite. Most of this pluton is composed of coarse, massive perthitic leucogranite, in which the main mineral is altered riebeckite. Two-feldspar biotite granite is also present, and a fine-grained granophyre forms the core of the body. Although the plutonic rocks of this late-tectonic granite show little bulk strain, perthite is locally recrystallized; riebeckite is replaced by aggregates of hornblende, biotite, iron oxide and stipnomelane: Thus, part of the regional metamorphic history postdates emplacement of this pluton. The age of the granite is poorly known. An Rb-Sr age of 1094 ± 46 Ma was reported by Wanless and Loveridge (1972), but the Rb-Sr systematics are discordant (Bell and Blenkinsop, 1979).

Stop 6: Kaladar Conglomerate. This coarse, polymict metaconglomerate is correlated with the Lessard Formation of the Flinton Group, which unconformably overlies rocks of the Grenville Supergroup. The Flinton Group was deposited roughly between 1060 and 1090 Ma (Moore and Thompson, 1980). To the north, the Flinton Group rests unconformably on the Northbrook tonalite-granodiorite batholith. To the south, it is in tectonic contact with the Addington granite. The metaconglomerate contains clasts of similar lithology to these plutonic bodies, as well as quartzarenite, mafic and felsic volcanic rocks, marble, and calc-silicate rock. The matrix contains calc-silicate minerals, and exhibits a mixed sedimentary/volcanic provenance (Bartlett and others, 1984).

REFERENCES CITED

Bartlett, J. R., 1983, Stratigraphy, physical volcanology and geochemistry of the Belmont Lake metavolcanic complex, southeastern Ontario [M.S. Thesis]: Ottawa, Carleton University, 218 p.

Bartlett, J. R., and Moore, J. M., Jr., 1985, Geology of Belmont, Marmora, and Southern Methuen townships, Peterborough and Hastings Counties: Ontario Geological Survey Open-File Report 5537, 136 p.

Bartlett, J. R., Brock, B. S., Moore, J. M., Jr., and Thivierge, R. H., 1984, Grenville traverse A, cross-sections of parts of the Central Metasedimentary Belt: Geological Association of Canada, Joint Annual Meeting, May 14-16, 1984, London, Ontario, Field Trip Guidebook 9A/10A, 64 p.

Bell, K., and Blenkinsop, J., 1979, Regional synthesis of the Grenville Province of Ontario and western Quebec; Part III; Rubidium-Strontium isotopic studies in the Grenville Province of southeast Ontario: Current Research, Part B, Geological Survey of Canada Paper 79-1B, p. 167–170.

Carter, T. R., 1984, Metallogeny of the Grenville Province, southeastern Ontario: Ontario Geological Survey Open-FIle Report 5515, 422 p.

Easton, R. M., Heaman, L. M., McNutt, R. H., Shaw, D. M., and Easton, M. G., 1984, Geology of the Minden and Chandos township areas, Central Metasedimentary Belt, Ontario: Brampton, Friends of the Grenville Field Trip Guidebook, 54 p.

Easton, R. M., Carter, T. R., and Springer, J. S., 1986, Mineral resources of the Central Metasedimentary Belt, Grenville Province: Geological Association of Canada, Joint Annual Meeting, May 19-21, 1986, Ottawa, Ontario, Field Trip Guidebook 3, 60 p.

Hewitt, D. F., 1964, Geological notes for maps 2053 and 2054, Madoc-Gananoque Area: Ontario Department of Mines Geological Circular #12, 33 p.

Moore, J. M., Jr., and Thompson, P. H., 1980, The Flinton Group; A late Precambrian metasedimentary succession in the Grenville Province of eastern Ontario: Canadian Journal of Earth Sciences, v. 17, p. 1685–1707.

Wanless, R. K., and Loveridge, W. D., 1972, Rubidium-strontium isochron age studies—Report 1: Geological Survey of Canada Paper 72-23, 77 p.

Wynne-Edwards, H. R., 1972, The Grenville Province, *in* Price, R. A., and Douglas, R.J.W., eds., Variations in tectonic styles in Canada: Geological Association of Canada Special Paper 11, p. 263–334.

Paleozoic-Precambrian unconformity near Burleigh Falls, Ontario Highway 36, Ontario

R. M. Easton, Precambrian Geology Section, Ontario Geological Survey, 77 Grenville Street, Toronto, Ontario M7A 1W4, Canada

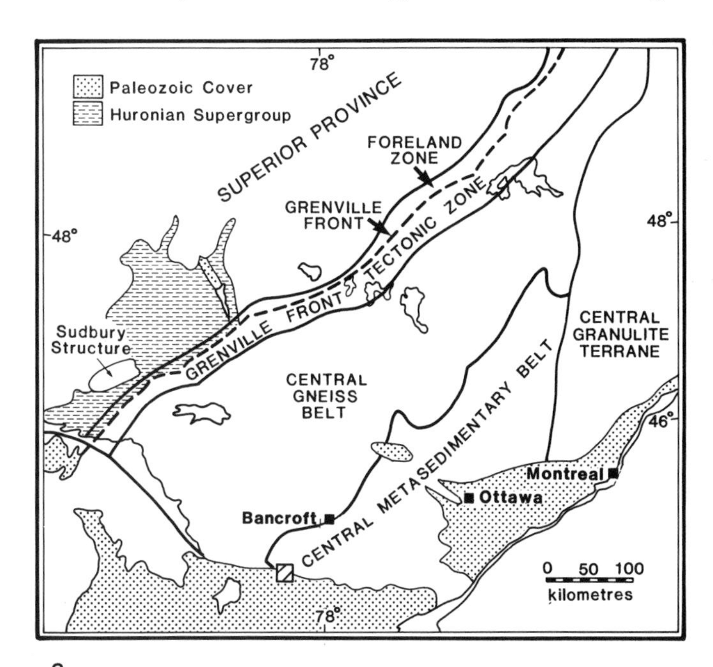

a

b

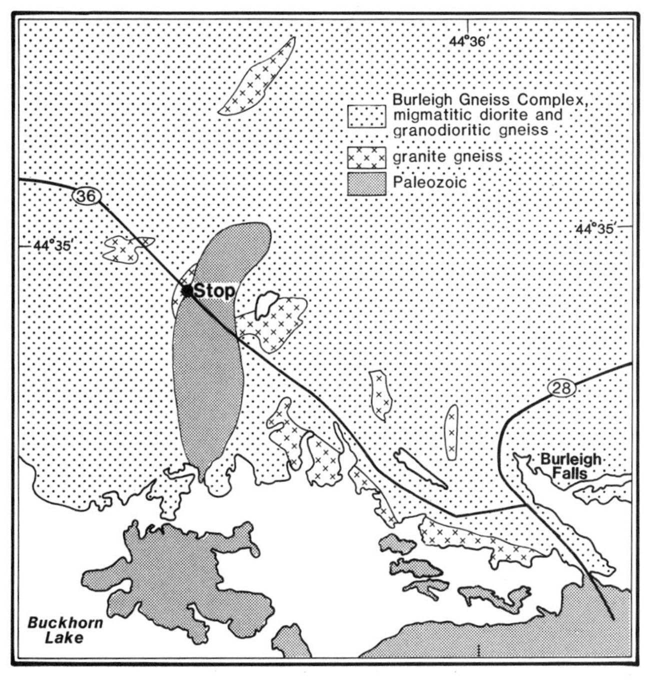

c

Figure 1. (a) Location of the site in central Ontario. (b) General geology showing the distribution of the Precambrian-Paleozoic unconformity in central Ontario, as well as main highways. (c) General geology in the vicinity of the site. Adapted from Morton (1983).

LOCATION

This site is a spectacular road-cut that exposes the unconformity between Middle Ordovician sedimentary rocks and Grenvillian gneisses in central Ontario. The stop is located on Ontario 36, 2.3 mi (3.7 km) west of the village of Burleigh Falls, Ontario (Fig. 1). Burleigh Falls is located on Ontario 28, 88 mi (140 km) northeast of the city of Toronto, and 22 mi (35 km) north of the city of Peterborough. It can be reached from the north (Bancroft) and south via Ontario 28, and from the west via Ontario 36.

SIGNIFICANCE

The site lies along the Precambrian-Paleozoic contact in central Ontario, and allows comparison of the lithologic character of the Precambrian terrane and the almost horizontal, overlying Paleozoic strata. This contact is not usually exposed, and the site is one of the best exposures of this contact in central Ontario.

SITE INFORMATION

The general geology of the site is shown in Figure 1c,

adapted from Morton (1983). From the junction of Ontario 36 and 28 in Burleigh Falls, Ontario 36-west cuts through migmatitic gneisses of the Burleigh Gneiss Complex for 1.8 mi (2.9 km). These cuts provide good exposures of the Precambrian rocks in the area. The gneisses dip to the east at roughly 45° and have an age of about 1250 Ma. Between 1.8 and 2.3 mi (2.9 and 3.7 km) west of Burleigh Falls, Ontario 36 rises up onto Paleozoic strata, and the site is located on the west side of the hill where the road descends to the Precambrian peneplain. The site is located along a 1,600-ft-long (500-m) road cut that exposes, from west to east, (up-section) sheared granitic gneisses of the Grenville Province, overlain by about 6 to 10 ft (2 to 3 m) of red and green arkose, conglomerate, and shale of the Middle Ordovician Shadow Lake Formation (Liberty, 1969). The Shadow Lake Formation is in turn overlain by limestones of the Gull River Formation (Liberty, 1969; Fig. 2). Figure 3 is a sketch showing the stratigraphy at the site.

The basement rocks are sheared, probably due to a fault that passes to the west of the Precambrian-Paleozoic contact. This faulting has contributed somewhat to the altered character of the basement rocks, although the hematite stain in the upper part of the gneisses may be related to a period of Cambrian weathering (Hewitt, 1964). The uneven nature of the Precambrian erosion surface can be seen by tracing the contact along the road cut.

Red weathering, arkosic and pebbly conglomerate of the Shadow Lake Formation immediately overlie the Precambrian rocks. These sedimentary units contain large, sub-angular to angular grains of quartz and feldspar and were probably derived in part from weathering of granitic pegmatite dikes in the area, as well as the Burleigh Gneiss Complex. The coarser sedimentary units grade upward into red and green weathering shales, with the occasional arkosic arenite interbed. The shales and arkoses are capped by a white-weathering, lithographic limestone of the Gull River Formation. The contact between the Shadow Lake and Gull River formations in this area is interdigitating. Further west, the contact is sharp.

REFERENCES CITED

Freeman, E. B., 1979, Geological Highway Map, southern Ontario: Ontario Geological Survey Color Map 2441, scale 1:800,000.

Hewitt, D. F., 1964, Precambrian–Paleozoic contact relationships in eastern Ontario: American Association of Petroleum Geologists, Guidebook, 1964 Annual Meeting, Toronto, Ontario, p. 9–13.

Liberty, B. A., 1969, Paleozoic geology of the Lake Simcoe area: Geological Survey of Canada Memoir 355, 201 p.

Morton, R. L., 1983, Geology of Harvey Township, Petersborough County: Ontario Geological Survey Report 230, 50 p., with Color Map 2475, scale 1:31,360.

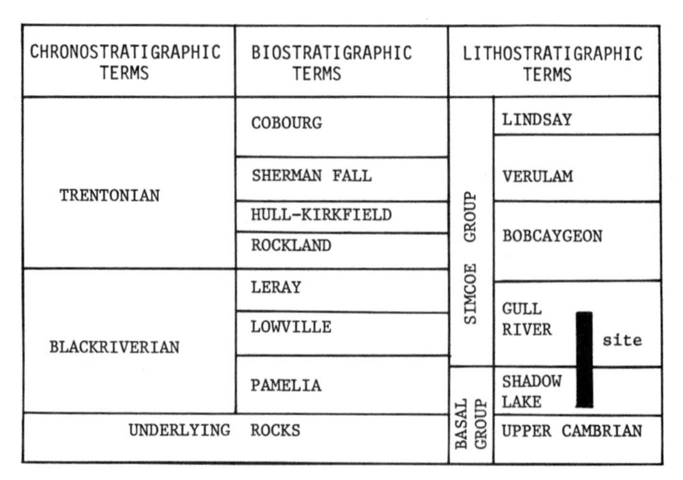

CHRONOSTRATIGRAPHIC TERMS	BIOSTRATIGRAPHIC TERMS		LITHOSTRATIGRAPHIC TERMS
TRENTONIAN	COBOURG	SIMCOE GROUP	LINDSAY
	SHERMAN FALL		VERULAM
	HULL-KIRKFIELD		BOBCAYGEON
	ROCKLAND		
BLACKRIVERIAN	LERAY		GULL RIVER (site)
	LOWVILLE		
	PAMELIA		SHADOW LAKE (site)
UNDERLYING ROCKS		BASAL GROUP	UPPER CAMBRIAN

Figure 2. Middle Ordovician stratigraphic nomenclature in central Ontario (after Liberty, 1969).

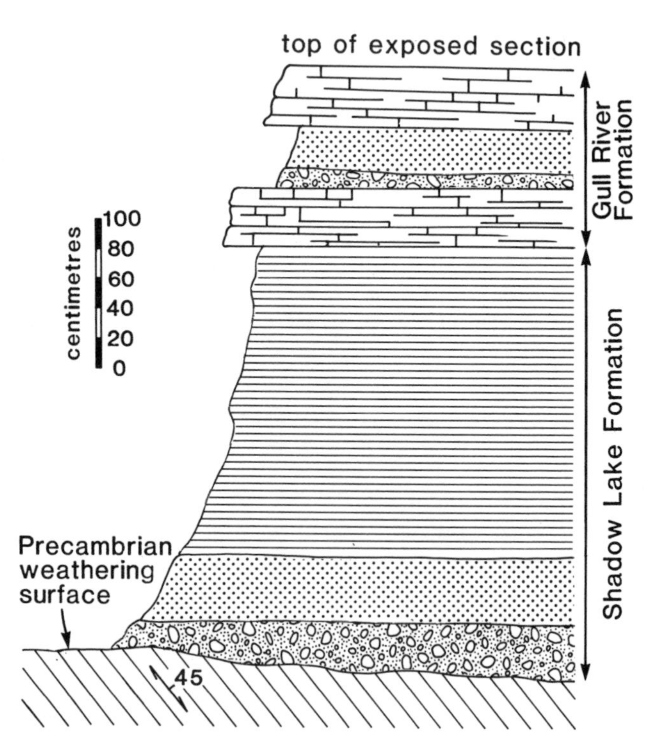

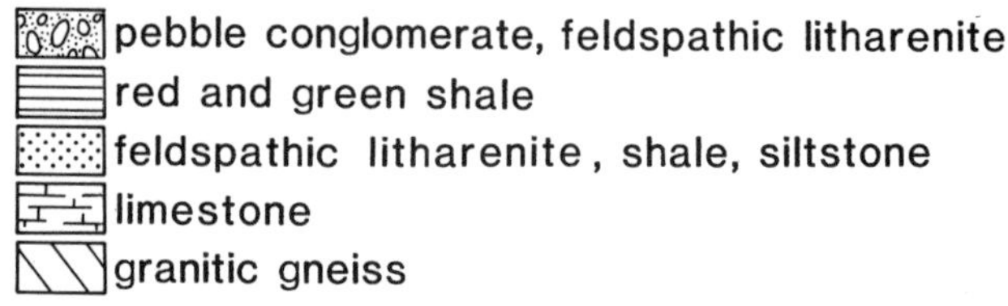

Figure 3. Stratigraphic sequence at the site.

Quaternary geology of Toronto area, Ontario

David R. Sharpe, *Geological Survey of Canada, 601 Booth Street, Ottawa, Ontario K1A OE8, Canada*

a

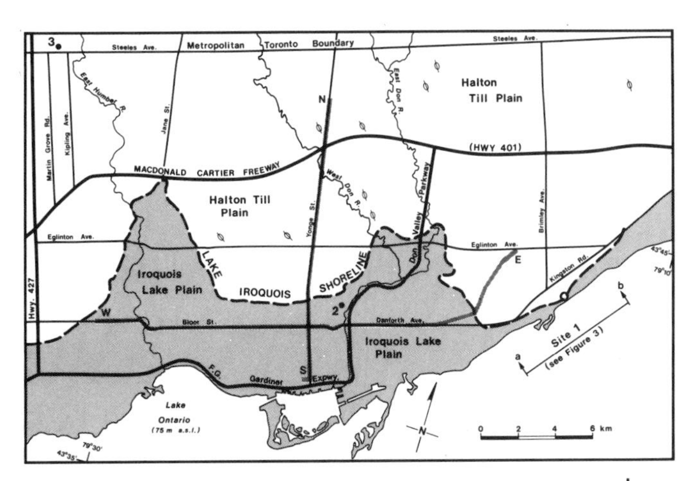

b

Figure 1. (a) Location of field site in Ontario. (b) Location map of field sites and areal geology of Metropolitan Toronto, Ontario: Site 1. Scarborough Bluffs (Sunnypoint), Site 2. Don Valley Brickyards, and Site 3. Woodbridge. Sites 2 and 3 are supplemental.

LOCATION

Three sites have been selected to demonstrate Toronto's Quaternary geology (Fig. 1); only one stop is described here in detail. The other two are supplementary and more detail about them is available in Barnett and others, 1987. All stops lie within NTS sheet 30 M 11 and 30 M 13 (1:50,000). A coloured map of the Quaternary geology of Toronto and Surrounding Area, Ontario Geological Survey Map P2204 (Sharpe, 1980) shows a road grid sufficient to obtain access to the three major sites of interest. The map is available from the Public Service Centre, Ministry of Natural Resources, Whitney Block—Room 1640, 99 Wellesley St. West, Toronto, Ontario M7A 1W3 (416 965-1348).

Stop 1: Sunnypoint Section, Scarborough Bluffs. To visit the Sunnypoint section, drive south on Brimley road and east via Barkdene Hills Road to Broadmead Avenue into Sunnypoint Park (Fig. 2). Alternatively, one may walk 0.6 mi (1 km) east from Bluffer's Park to the first large gully. Proceed north along a small creek until fresh exposure is observed along the east face of the gully (Fig. 2). Bluffer's Park also provides an excellent overall view of the Scarborough Bluffs, especially the stratigraphy and the main lithostratigraphic units. Stops 2 and 3 are supplemental sites that reveal the oldest units, the York Till and the Don interglacial beds (Table 1).

Stop 2: Don Valley Brickyards. To visit the Don Valley Brickyards, proceed on the Don Valley Parkway to the Bloor Street exit west. About 0.6 mi (1 km) west, take Bayview Avenue north and find the brickyards at about 0.3 mi (0.5 km) on the left side (north) of the road. Walk from the parking lot along the east side of the Quarry to the exposed sediments on the north face of the Quarry. (This classic site is being considered for parkland and preservation by the Ontario Government. For permission to visit, write to the Ontario Ministry of Natural Resources.)

Stop 3: Woodbridge Railroad Cut. To reach the Woodbridge site, drive west on Steeles Avenue to Kipling Avenue, turn right (north) and proceed several hundred metres to a large gate. Proceed on foot 500 ft (150 m) north, cross railway track and find the exposed bluffs on north side of tracks west of woodlot. Caution is needed crossing busy railway line. (Obtain key at Metro Toronto Regional Conservation Authority (416-661-6600), 5 Shoreham Drive, east of Jane, south of Steeles).

SIGNIFICANCE

The Quaternary geology of the Toronto area is significant for two reasons. First, a major portion of the last interglacial (Sangamonian) and glacial (Wisconsinan) stratigraphic record of the eastern Great Lakes region and Eastern Canada occurs here (Dreimanis and Karrow, 1972; Karrow, 1984b). The Toronto area strata include Illinoian till, interglacial deposits with warm-climate fauna, four Wisconsinan diamictons (probably tills) and interbedded proglacial lake deposits (Table 1). Well-preserved and fossil-rich deposits indicating separate warm-climate and cool-climate conditions have made the Toronto beds famous (Coleman, 1933). Secondly, a re-examination of the Toronto (Scarborough Bluffs) area has begun with application of facies model methods (Eyles and Eyles, 1983a; Eyles and others, 1983; Kelly and Martini, 1982, 1986; Karrow, 1984a; Sharpe, 1985; and Sharpe and Barnett, 1985). The latter sedimentological analysis questions the correlation of channel cutting events at Toronto with peat beds dated at 75,000 BP at St. Pierre, Québec, in the

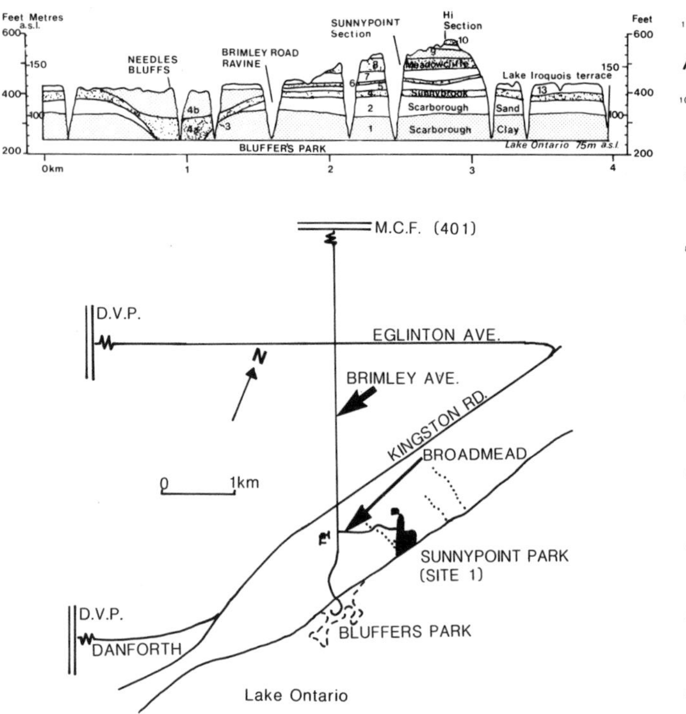

Figure 2. Cross-section and location map of the major site in the Toronto area: Site 1. Scarborough Bluffs (Sunnypoint); section shows the major strata exposed for 4 km of the Scarborough Bluffs near Bluffer's Park. Note large channel cut into the Scarborough sand and clay west of Bluffer's Park and west of Site 1 at Sunnypoint Section. Key: marble pattern is diamicton; dot pattern is silt and clay; blank is sand. 1 = Scarborough clay, 2 = Scarborough sand, 3 = Pottery Road Formation, 4a = Sunnybrook diamicton, 4b = Sunnybrook drift (Bloor member, rhythmites), 5 = Thorncliffe Formation, 6 = Seminary diamicton, 7 = Thorncliffe Formation, 8 = Meadowcliffe diamicton, 9 = Thorncliffe Formation, 10 = Halton drift, 13 = Iroquois sand. (See Table 1 also).

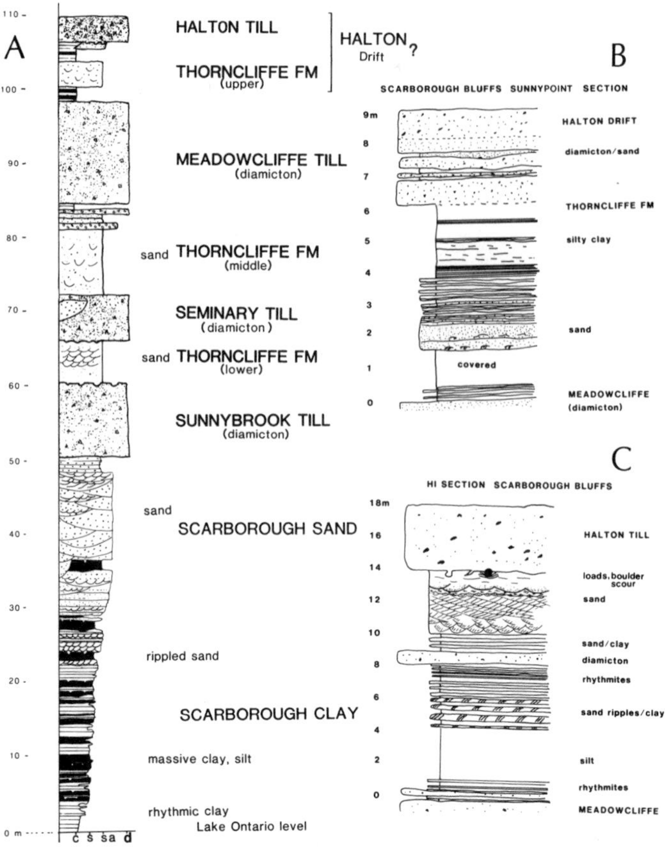

Figure 3 (a). Summary of lithofacies exposed at the Sunnypoint Park section (refer also to Karrow, 1967, section A1011, Kelly and Martini, 1986, Figure 8; and Eyles and Eyles, 1983, section 1281 and 1481). Halton drift refers to diamicton and stratified sediment deposited by a glacial episode. Textural key: C is clay, S is silt, Sa is sand and d is diamicton. (b, c) Sediment sequences associated with the Halton drift Scarborough Bluffs. (b) Sunnypoint park, showing interbedded character of Halton. (c) Hi Section, showing load structures and boulder scour at base of Halton diamicton.

lower St. Lawrence valley. The high pro-glacial lake sequence at Toronto (Thorncliffe Formation) may be in conflict with suggestions of multiple Wisconsinan deglaciation of Hudson Bay (Andrews and others, 1983). The spectacular bluff exposures at Scarborough will provide focus for the evolution of models of ice-marginal and basin sedimentation and for lithostratigraphic correlation to adjacent areas (Sharpe and Barnett, 1985).

SITE INFORMATION

Stop 1: Scarborough Bluffs—Sunnypoint Section

Background. The Scarborough Bluffs are spectacular lake bluffs up to 300 ft (90 m) high that extend for 9 mi (15 km) along Lake Ontario (Fig. 2). The Sunnypoint section reveals the most complete sequence of exposed Wisconsinan strata in the Toronto area and eastern Canada (outside Hudson Bay lowlands). The Sunnypoint section reveals a sequence of four diamictons and intercalated stratified sediment (sand, silt and clay) as shown in Figure 3a. This section includes the surface Halton drift (Fig. 3b,c) and subglacial landforms (Fig. 1b). The Pre-Wisconsinan drift (York Till) and the Sangamonian Interglacial beds (Don Formation) are present below lake level but can be seen at the Don Valley Brickyard (stop 2) and Woodbridge (stop 3). The regional chronostratigraphy of the Quaternary of Toronto can be seen in Figure 4.

Two radically different viewpoints as to the interpretation, origin, and significance of the basic units have arisen. Karrow (1967, 1969, 1974), following early studies by Hinde (1878) and Coleman (1895, 1913, 1933), interpreted the diamictons as tills deposited by glacier ice, and the stratified material as interstadial lake sediments deposited during ice retreats. Lithological studies on diamicton matrix and clasts and studies of organic remains in some of the stratified beds formed the basis of these interpretations. Recently, direct correlations have been attempted to the Scarborough Bluffs from 18 mi (30 km) east at the Bowmanville-

TABLE 1. QUATERNARY DEPOSITS IN THE TORONTO AREA

Formation or Member	Deposit	Environment of Deposition	Age	Location (Site)
Lake Iroquois (13)	sand, silt, gravel	littoral (45)	post-glacial	1, 2
Peel Ponds (12)	clay, diamicton	ice-marginal lake (110+)	L.W.	3
Wildfield drift (11)	silt, diamictons, rhythmites	ice-marginal	L.W.	3
Halton drift (10)	diamicton, sand, silt	subglacial	L.W.	1, 3
Upper Thorncliffe (6)	sand, silt, clay	subaqueous (60+)	L.W.?	1
Meadowcliffe (8)	silty diamicton	ice-proximal (subglacial?)	M.W.?	1
Middle Thorncliffe (7)	sand, silt, clay	subaqueous (high lake) (60+)	M.W.?	1
Seminary (6)	sandy silt diamicton interbedded with gravel	ice-proximal (subglacial?)	M.W.?	1
Lower Thorncliffe (5)	sand, stratified	subaqueous (high lake) (60+)	M.W.?	1
Sunnybrook Drift (4a)	clayey diamicton	ice-proximal (sub-glacial?)	E.W.?	1, 2, 3
(4b)	rhythmic beds			
Pottery Road (3)	sand and gravel	channel fill	E.W.?	1, 2
Scarborough (1,2)	sand over peaty clay	ice-marginal lake (50)	E.W.?	1, 2, 3
Don (II)	sand, silt, clay	fluvial/deltaic (20)	Interglacial	2, 3
York Till (I)	sandy silt diamicton with sand and gravel	subglacial	Illinoian	2, 3
Unconformity				
Georgian Bay Fm.	shale and carbonate	marine	Paleozoic	2

Notes: (1) L.W. = Late Wisconsinan; M.W. = Middle Wisconsinan; E.W. = Early Wisconsinan. The sequence from Sunnybrook to Halton could be M.W. to L.W.
(2) Numbers in column 1 refer to Figure 2.
(3) Elevations given in column 3 (Environment of Deposition) are meters above sea level, measured at the Brickyards (Stop 2). They indicate a raised lake environment supported by glaciers (except for the Don Formation).

Port Hope bluffs (Brookfield and others, 1982; Martini and others, 1984).

New sedimentological studies of the Scarborough Bluffs have concluded that stratified sediments and massive diamictons were deposited on the floor of a large lake, by lake basinal processes, without the direct presence of glacier ice (Eyles and Eyles, 1983a). As a result, Eyles and Eyles (1983a) have proposed the abandonment of the Scarborough Bluffs as a chronostratigraphic model. They have, however, used it as a depositional model for glaciated basins (Eyles and Eyles, 1983b; Eyles and others, 1983). Alternatively, it may be that glacier ice was present in the Lake Ontario basin during deposition of these thick sediments but not necessarily over the site (Sharpe, 1983; Sharpe and Barnett, 1985). Facies assemblages and associations, contact relationships between individual facies, overall geometry of the strata, and their position in the basin are important features needed to assess these views during field visits (Figs. 2, 3).

Major Units. The major strata exposed at the Scarborough Bluffs are described (and measured) from the base of the section up and listed, for convenience, following the terminology of Karrow (1969) with modification. See Figure 3a for strata and elevations.

The Scarborough Formation is exposed along the base of the bluffs at lake level (165 ft; 50 m; Fig. 3a): it contains a lower clay and an upper sand unit . Kelly and Martini (1982, 1986) defined 13 subfacies and 7 facies associations to describe the Scarborough delta. The Scarborough clay (0–92 ft; 0–28 m) is characterized by rhythmic clay, silt, and fine sand beds varying from 0.4 to 6 in (1 to 15 cm) thick. Thin clay beds are massive or normally graded. Small-scale cross-lamination, ripples, graded beds, and burrow-traces are evident in the beds of silt and fine sand. Thick beds of massive silty sand are a special feature of this unit. Minor deformation (load structures) occurs in silty laminae and paleo-current directions are generally to the south. Organic matter is present in cross-laminated forsets toward the top of the clay sequence. The Scarborough sand (92–165 ft; 28–50 m) unit is ripple cross-laminated, graded, trough cross-bedded, plane-bedded, and massive (Kelly and Martini, 1986). Beds vary from 4-in (10-cm) ripple sequences to 3-ft (1-m) thick troughs. Thin (1–2 cm) silt, clay layers may be interbedded. Paleoflow directions are south to southeast and organic matter is common throughout. Mud cracks were observed below the uppermost facies association. This facies may belong to the overlying Sunnybrook. Most contacts are sharp-based and gradational at the top, typical of underflow deposits.

Overlying the Scarborough Formation is the lowermost exposed diamicton unit (165–198 ft; 50–60 m) of the Sunnypoint section. Identified as the Sunnybrook Till (Karrow, 1969), it is a massive-to-stratified, thick and thin-bedded silty clay diamicton. The Sunnybrook diamicton is matrix-supported and low in clast content (1%). The diamicton is interbedded upwards with rhythmically laminated and varved clays, graded beds, and dropstone units defined as Bloor Member of the Sunnybrook Drift (Karrow, 1969). The diamicton beds are structureless-to-stratified and may show flow features. Bed contacts with the Scarborough sand may be sharp, loaded or gradational and they are further discussed below.

The Thorncliffe Formation (198–215 ft; 60–65 m, lower: 238–274 ft; 72–83 m, middle: 323–350 ft; 98–106 m, upper) comprises trough cross-stratified sand (lower), massive sand with loaded bed contacts (middle), and massive sand intercalated with rippled sand, rhythmitic silt, and clay (proximal varves). Leaves and seeds of subarctic plants occur locally in these beds and they have been dated at 28 ka and 32 ka (Berti, 1975).

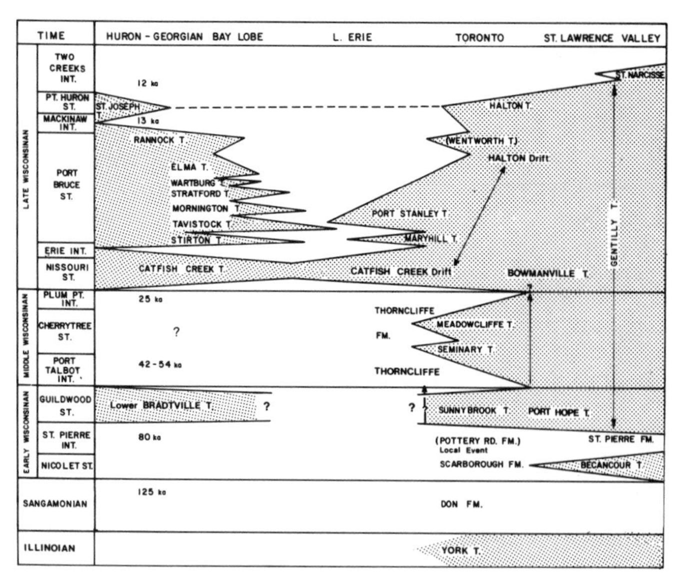

Figure 4. Time-distance diagram of the Wisconsinan stratigraphy of Southern Ontario and Quebec (after Karrow, 1984b) showing the regional significance of the Toronto area. The correlation of the St. Pierre peat beds (75,000 years BP) with the Pottery Road Formation is not valid. The whole Wisconsinan timing is uncertain, e.g. the sequence from Sunnybrook to Halton could be much younger.

The Thorncliffe Formation is separated by two diamicton units, Seminary and Meadowcliffe Tills, along the bluff (Fig. 3a) but these units have not been identified very far inland (Karrow, 1969). The Seminary member (215–238 ft; 65–72 m) consists of massive to stratified silty diamicton. Rapid lateral and vertical facies changes, as indicated by interbedding of sandy diamicton with massive and rippled sand, gravel, and breccia, occur within the Seminary unit. Most bed contacts are sharp or gradational.

The Meadowcliffe member (274–323 ft; 83–98 m) consists mainly of massive silty clay diamicton with few clasts. Interbeds of rhythmic silt and clay are common at the bottom and top of the thick diamicton sequence. Bed contacts may be sharp, loaded, or gradational (interbedded).

The Halton drift (350–363 ft; 106–110 m) consists of stratified sediment and massive silty diamicton (Halton Till), which has more clasts than lower diamicton units (Fig. 3b,c). Bed contacts are sharp, loaded, and gradational in places where interbedding with silt, clay, and sand occurs. The major lithologic characteristics of the above diamicton units are listed in Karrow (1967).

A level terrace of Glacial Lake Iroquois truncates the sequence west and east of Sunnypoint Park at Bluffer's Park at 429 ft (130 m) a.s.l. (Fig. 2). It was formed by a pro-glacial lake supported by ice retreating in the eastern Lake Ontario basin between 13,000 and 12,000 years B.P.

Discussion. The deposits at the Scarborough Bluffs basically represent a glacial and glaciolacustrine sequence in the Lake Ontario basin. Discussion is concerned with whether ice is over, adjacent to, or far away from the site of deposition.

The sediments of the Scarborough Formation are thought to have been deposited as a delta at the mouth of a large river, which flowed into Lake Scarborough (Coleman, 1941). Lake Scarborough had a level similar to that of Lake Iroquois, which is greater than 150 ft (45 m) above the level of Lake Ontario (Karrow, 1967); this lake may have drained via the Hudson Valley, as glacier ice occupied the lower St. Lawrence River valley.

The contact between the Scarborough sand and the underlying Sunnybrook diamicton, was considered to be a major unconformity having regional significance (Karrow, 1967, 1984b). In this view, subaerial channels, graded to a low base level, cut deeply into the Scarborough delta. Subaerial streams apparently filled these channels with sand and gravel, named the Pottery Road Formation. These inferred events were correlated to peat beds (low base level) at St. Pierre, Québec, dated at 75 ka.

The Scarborough-Sunnybrook contact, however, may only be a local unconformity, defined where channel forms occur. For example, Eyles and Eyles (1983a) concluded that the channel forms may be subaqueous delta lobe channels, while Sharpe and Barnett (1985) consider that the features may result from subaqueous, proglacial, or subglacial channel cut and fill. Thirdly, they may be subaqueous slump, scar, and fill features. The contact is interbedded and gradational, however, between the channels and may represent semi-continuous sedimentation.

The contact relationships of all strata are of importance in deciphering the depositional history of the Scarborough bluffs sequence. Eyles and Eyles (1983a) suggested that the Scarborough bluffs sequence lacks any substantial erosional breaks. This view considers that deposition occurred in a high pro-glacial lake, which may have existed in the Lake Ontario basin from the formation of the Scarborough delta until Lake Iroquois drained. Eyles and Eyles (1983a) maintained that debris rained-out from floating ice and was resedimented to form diamicton assemblages (Sunnybrook, Meadowcliffe and Seminary). Bed contacts showing gradation and interbedding and load, rip-up, and flame structures were cited as support for continuous lake sedimentation away from a glacier margin. Similar evidence of bed contacts and sedimentary structures was found by Kelly and Martini (1986) yet they suggested an ice-proximal source of sediment for construction of the Scarborough delta.

The relationships between other diamictons and adjacent stratified beds may also indicate ice proximal conditions. It is significant that the Meadowcliffe and the Sunnybrook diamictons occur as thin diamicton beds and stratified beds below and above thick massive diamicton. This facies arrangement is a common feature at the Scarborough Bluffs, as well as other lake bluffs, and has been used to suggest a proglacial/subglacial sediment model in the Lake Erie basin (Barnett, 1985).

The character of the Halton drift is also significant because it forms the surface landform at Scarborough. The Halton drift displays sedimentary structures (interbedding, load structures, sediment flows, see Fig. 3b,c) similar to those in the lower strata, (Sunnybrook, Seminary, Meadowcliffe) which are interpreted as lake-bottom deposits by Eyles and Eyles (1983a). The Halton strata, however, were formed in a subglacial depositional environment as shown by boulder scours and by surface forms, including drumlins, flutings and till plains (Sharpe, 1985; Fig. 1). The Halton drift may be part of a continuous depositional sequence with lower beds (Thorncliffe units) and hence it provides clues as to the environment and mode of deposition of the lower and older strata. Thus, the lower beds may also be subglacial or at least ice proximal deposits because of the similarity in sedimentological features. The Scarborough bluffs may be a useful predictive model for glaciolacustrine sedimentation if one emphasizes glacial processes in addition to lacustrine processes.

Lakes and Fossil Remains. The Quaternary stratigraphy exposed at Toronto comprises several lake phases (Table 1 and Sharpe, 1980, Table 1). The elevations of these former lakes indicates whether ice dammed drainage through the St. Lawrence Lowlands. For example the Don Formation occurs below present lake level at Scarborough bluffs and indicates ice free conditions. The Thorncliffe lake occurs about 200 ft (60 m) above Lake Ontario, implying glacial control on the outlet. These lakes were also important for preserving fossil remains.

The presence of fossil remains has been important to stratigraphic and paleoenvironmental reconstruction at the Scarborough Bluffs. The Scarborough Formation, comprising the Scarborough clay and sand (Fig. 2), contains plant fossils (Terasmae, 1960; Berti, 1975), beetles (Williams and others, 1981), molluscs, ostracods, (Poplowski and Karrow, 1981), and diatoms (Duthie and Mannada-Rani, 1967) used to interpret colder climates (9–11°F; 5–6°C) than today. Deltaic sediments and fossils were deposited up to 165 ft (50 m) above present day Lake Ontario in glacial Lake Scarborough (Table 1).

The lower Thorncliffe sandy sediments contain plant detritus dated at greater than 53 ka (Karrow, 1984b), while the upper Thorncliffe deposits have plant detritus dated at 28 and 32 ka (Mörner, 1971; Berti, 1975). The organic matter is not in situ and may have been carried in by large floods: it only dates periods of non-ice cover in adjacent highlands.

Trace fossils occur within rhythmic sediments and their regular arrangement suggests annual deposition. Varved clay within the upper Thorncliffe sediment also reflects the proximity of glacial ice in the Lake Ontario basin during this interval. Evidence for the ice-marginal Peel pond deposits (Table 1) may be represented by the interbedded character of the Halton drift at Sunnypoint (Fig. 3b). The youngest ice-supported lake in the Toronto area, Lake Iroquois, is represented by a sand-covered surface at the top of the section at Bluffer's Park. Freeman (1976) has produced a pamphlet highlighting other fossil remains of the Toronto area.

Don Valley Brickyards

Of prime interest at the Don Valley brickyards are the Illinoian York Till and the stratified sand and clay of the Don Formation interglacial beds. The York till occurs as a thin (3.3 ft; 1 m) clayey sand diamicton overlying Ordovician shale, with striae indicating ice flow northward from the Lake Ontario basin (Terasmae, 1960). The York Till is best exposed at Woodbridge (Site 3) and is further discussed there.

The Don beds represent fluvial to deltaic sand, gravel, and clay deposited up to 65 ft (20 m) above present Lake Ontario (Table 1). The plants and molluscs of the Don Formation were considered to indicate warmer climate (+3.6°F; +2°C) than present (Terasmae, 1960). Vertebrate remains for groundhog, deer, bison, bear, giant beaver, catfish, and perch have been reported (Karrow, 1969). Kerr-Lawson (1985) has reported on the gastropods and plant macrofossils of the Don beds. Amino acid dating suggests that Sangamonian age assignment is reasonable (Karrow, 1984).

The contact between the Scarborough sand and the overlying Sunnybrook drift and the occurrence of the channel fill deposits (Pottery Road Formation) is again of interest. Deep, trough cross-bedded sets suggest that subaqueous channel or slump formation may explain deposition of the Pottery Road Formation. The present view is that subaerial streams cut gullies below the current level of Lake Ontario. This apparent low water level for the Pottery Pond Formation was used to correlate it to peats at St. Pierre, Québec, dated about 75,000 years B.P. (Karrow, 1984b; LaSalle, 1984). Re-interpretation of this channel sequence alters the present chronology of the Wisconsinan events in the Toronto area (Fig. 4). This site is reported more fully in an INQUA fieldtrip, Quaternary History of Southern Ontario (Barnett and Kelly, 1987).

Woodbridge Railway Cut

The Woodbridge railway cut has the best exposure of the Illinoian York Till and the interglacial Don Formation. The York Till is up to 16 ft (5 m) thick, massive, oxidized yellowish brown, with inclusions of unoxidized grey pebbly, sandy silt till. Weathering of the York Till and associated stratified sediment indicates subaerial exposure possibly during the subsequent interglacial interval. Ice wedge casts have also been identified directly on top of the Don Formation interglacial beds: these beds contain large fresh water shells (Unios) as well as other fossils.

The Scarborough Formation consists of peaty silt and sand with wood and represents a local pond. Plant remains indicate a climate cooler than present. The overlying diamicton units were identified by White (1975) and Karrow and Morgan (1975) as separate depositional events: Sunnybrook, Wentworth, Halton, and Wildfield Tills. The latter three diamictons, however, are probably facies of the late Wisconsinan Halton Till and represent a single depositional succession. The sandy silt Wentworth Till of Karrow (1969, 1984b) is probably the sandy lower portion of the Halton drift. Above the massive silty Halton diamicton occurs a massive clayey diamicton with interbedded silt clay laminae representing Peel pond sediments deposited by retreating Halton ice (Sharpe, 1980). This site is discussed further in an INQUA field guide (Barnett and Kelly, 1987).

IMPORTANT REFERENCES

Andrews, J. T., Shilts, W. W., and Miller, G. H., 1983, Multiple deglaciations of the Hudson Bay Lowlands, Canada, since deposition of the Missinaibi (last interglacial?) formation: Quaternary Research, v. 19, p. 18–57.

Barnett, P. J., 1985, A model of Glacial-Glaciolacustrine sedimentation, Lake Erie Basin: Geological Society of America Abstracts with Programs, v. 17, p. 279.

Barnett, P. J., and Kelly, R. I., 1987, Quaternary history of Southern Ontario, Field Excurision A11, INQUA Congress XII, Ohara.

Berti, A., 1975, Paleobotany of Wisconsin interstadials, eastern Great Lakes region, North America: Quaternary Research, v. 5, p. 591–619.

Brookfield, M. E., Gwyn, Q.H.J., and Martini, I. P., 1982, Quaternary sequences along the north shore of Lake Ontario; Oshawa—Port Hope: Canadian Journal of Earth Sciences, v. 19, p. 1836–1850.

Coleman, A. P., 1895, Glacial and interglacial deposits near Toronto: Journal of Geology, v. 3, p. 622–645.

—— , 1913, Geology of the Toronto Region, *in* Fault, J. H., ed., The natural history of the Toronto region, Ontario, Canada: Canadian Institute, p. 51–81.

—— , 1933, The Pleistocene of the Toronto region: Ontario Department of Mines Annual Report 41, pt. 7, p. 1–55.

—— , 1941, Coleman, A. P., 1941, The last million years: University of Toronto Press, 216 p.

Dreimanis, A., 1984a, Sedimentation in a large lake; A reinterpretation of the late Pleistocene stratigraphy at Scarborough Bluffs, Ontario, Canada; Comments: Geology, v. 12, no. 3, p. 185–186.

—— , 1984b, Lithofacies types and vertical profile analysis; Comments on the paper by Eyles, N., Eyles, C. H., and Miall, A. D.: Sedimentology, v. 31, p. 885–886.

Dreimanis, A., and Karrow, P. F., 1972, Glacial history of the Great Lakes—St. Lawrence Region; The classification of the Wisconsin(an) Stage, and its correlatives: 24th International Geological Congress (Montreal), Section 12, p. 5–15.

Duthie, H. C., and Mannada-Rani, R. G., 1967, Diatom assemblages from Pleistocene interglacial beds at Toronto, Ontario: Canadian Journal of Botany, v. 45, p. 2249–2261.

Eyles, C. H., and Eyles, N., 1983a, Sedimentation in a large lake; A reinterpretation of the late Pleistocene stratigraphy at Scarborough Bluffs, Ontario, Canada: Geology, v. 11, p. 146–152.

—— , 1983b, A glaciomarine model for Late Precambrian diamictites of the Port Askaig Formation, Scotland: Geology, v. 11, p. 692–696.

Eyles, N., Eyles, C. H., and Miall, A. D., 1983, Lithofacies types and vertical profile analysis; An alternative approach to the description and environmental interpretation of glacial diamict and diamicite sequences: Sedimentology, v. 30, p. 393–410.

Freeman, E. B., 1976, Toronto's Geological Past; An Introduction: Ontario Division of Mines Miscellaneous Publication.

Hann, B. J., and Karrow, P. F., 1984, Pleistocene paleoecology of the Don and Scarborough Formations, Toronto, Canada, based on cladoceran microfossils at the Don Valley Brickyard: Boreas, v. 13, p. 377–391.

Hinde, G. J., 1878, The glacial and interglacial strata of Scarboro' Heights and other localities near Toronto, Ontario: Canadian Journal, n.s. 15, p. 388–413.

Karrow, P. F., 1967, Pleistocene geology of the Scarborough area: Ontario Department of Mines, Geological Report 46, 108 p.

—— , 1969, Stratigraphic studies in the Toronto Pleistocene: Geological Association of Canada Proceedings, v. 20, p. 4–16.

—— , 1974, Till stratigraphy in parts of southwestern Ontario: Geological Society of America Bulletin, v. 85, p. 761–768.

—— , 1984a, Sedimentation in a large lake; A reinterpretation of the late Pleistocene stratigraphy at Scarborough Bluffs, Ontario, Canada; Comments: Geology, v. 12, p. 185.

—— 1984b, Quaternary stratigraphy and history, Great Lakes—St. Lawrence region, *in* Fulton, R. J., ed., Quaternary Stratigraphy of Canada; A Canadian contribution to IGCP Project 24: Geological Survey of Canada Paper 84-10, p. 137–153.

Karrow, P. F., and Morgan, A. V., 1975, Quaternary Stratigraphy of the Toronto Area: Geological Association of Canada Fieldtrip no. 6, Annual Meeting, Waterloo, Ontario, p. 161–179.

Karrow, P. F., Cowan, W. R., Dreimanis, A., and Singer, S. N., 1978, Middle Wisconsinan stratigraphy in southern Ontario: Geological Society of America—Geological Association of Canada, Toronto '78 Field Trips Guidebook, p. 17–27.

Kelly, R., and Martini, I. P., 1982, Scarborough Formation, *in* Karrow, P. F., Jopling, A. V., and Martini, P., eds., Late Quaternary Sedimentary Environments of a Glaciated Area: International Association of Sedimentology, Guidebook for Excursion 11A, p. 74–80.

—— , 1986, Pleistocene Glacio—Lacustrine deltaic deposits of the Scarborough Formation, Ontario, Canada: Sedimentary Geology, v. 47, p. 27–52.

Kemmis, T., and Hallberg, G., 1985, Lithofacies types and vertical profile models; An alternative approach to the description and environmental interpretation of glacial diamict and diamictite sequences; Discussion: Sedimentology, v. 31, p. 886–890.

Kerr-Lawson, L. J., 1985, Gastropods and plant microfossils from the Quaternary Don Formation (Sangamonian Interglacial), Toronto, Ontario [M.S. thesis]: University of Waterloo, 193 p.

LaSalle, P., 1984, Quaternary stratigraphy of Quebec; A review, *in* Fulton, R. J., ed., Quaternary Stratigraphy of Canada; A Canadian contribution to IGCP Project 24: Geological Survey of Canada Paper 84-10, p. 155–171.

Martini, I. P., Brookfield, M. E., and Gwyn, Q.H.J., 1984, Quaternary stratigraphy of the coastal bluffs of Lake Ontario east of Oshawa, *in* Mahaney, W. C., (ed.), Quaternary Dating Methods; Developments in paleontology and stratigraphy: Amsterdam, Elsevier, v. 7, p. 417–427.

Mörner, N. A., 1971, The Plum Point Interstadial; Age, climate, and subdivision: Canadian Journal of Earth Sciences, v. 8, p. 1423–1431.

Poplowski, S., and Karrow, P. F., 1981, Ostracodes and paleoenvironments of the Late Quaternary Don and Scarborough Formations, Toronto, Ontario: Canadian Journal of Earth Sciences, v. 18, p. 1497–1505.

Sharpe, D. R., 1980, Quaternary Geology of Toronto and Surrounding Area: Ontario Geological Survey Preliminary Map P. 2204, Geological Series, scale 1:100,000, compiled 1980.

—— , 1984, Sedimentation in a large lake; A reinterpretation of the late Pleistocene stratigraphy at Scarborough Bluffs, Ontario, Canada; Comments: Geology, v. 12, p. 185.

—— , 1985, Landform—Sediment Relationships: Geological Society of America Abstracts with Programs, v. 17, p. 326.

Sharpe, D. R., and Barnett, P. J., 1985, Significance of sedimentological analysis to the Wisconsinan Stratigraphy of Southern Ontario: Géographie Physique et Quaternaire, XXXIX, no. 3, p. 255–273.

Terasmae, J., 1960, Contributions to Canadian palynology No. 2: Geological Survey of Canada, Bulletin 56, 41 p.

White, O. L., 1975, Quaternary geology of the Bolton area: Ontario Division of Mines, Geological Report 117, 119 p.

Williams, N. E., Westgate, J. A., Williams, P. D., Morgan, A., and Morgan, A. V., 1981, Invertebrate fossils (Insecta: Trichoptora, Diptera, Caleptera) from the Pleistocene Scarborough Formation at Toronto, Ontario, and their paleoenvironment significance: Quaternary Research, v. 16, p. 146–166.

Geological Survey of Canada Contribution 31686

The Port Talbot interstadial site, southwestern Ontario

Aleksis Dreimanis, Department of Geology, University of Western Ontario, London, Ontario N6A 5B7, Canada

LOCATION

The stratotype of the Tyrconnell Formation, also informally called the Port Talbot interstadial deposits, is visible along the base of Lake Erie bluffs below a cluster of cottages locally known as Bradtville, since they were built on the farm of Harry Bradt. Bradtville is in Elgin County, Dunwich Township, concession XIII, lot 18.

The site may be found with the help of the Port Stanley (40-I/11) 1:50,000 topographic map of the Canadian National Topographic Series A 751. Its grid location is 685184. Although the name "Bradtville" is not marked on the map, the cluster of cottages is shown about 1.2 mi (2 km) northeast of the nearest name on the map—Plum Point.

When using an Ontario road map, look for Port Talbot, which was located at the mouth of Talbot Creek at the beginning of the 19th century, but today is only a name on the map. If coming from the north, cross Talbot Creek at Port Talbot and travel 2.1 mi (3.5 km) southwestward along the Talbot Trail (Elgin County Road 16), until you reach the lot 18/19 road, which goes northwestward (Fig. 1). A few more metres past this road, turn left (south) on a private road on the opposite (Lake Erie) side that will lead you towards Bradtville.

Drive past a farmhouse for 0.6 mi (0.9 km) across a shallow gully, and park on the side of the cottagers' road, just before reaching the cottages. Walk along the road towards its termination marked by a low chain. Continue past the chain along a washed-out lane leading southwestward down to the mouth of Bradtville gully at the lake shore. An alternative, for descending into Bradtville gully, is a stairway at the southwest end of the cottages. At the beach, turn sharply left, toward the northeast, and cross a small brook coming out a culvert. From here on, for about 425 ft (130 m) northeastward, the lower 10 ft (3 m) of the bluffs consist of the Tyrconnell Formation (Figs. 2, 3), usually partly covered by slump.

The most commonly exposed subsection is at the 100 m mark of Figure 3, northeastward from the culvert, at a small gully. Here, a glaciotectonically folded and sheared anticline of the silt and gyttja of Member C, described below, is often visible.

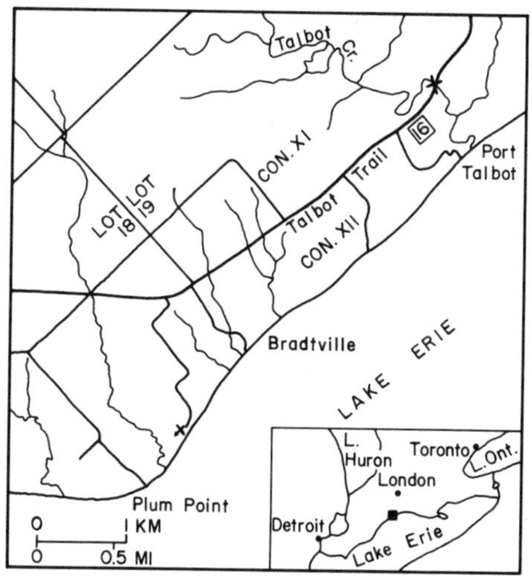

Figure 1. Location map (after Port Stanley 1:50,000 map 40-I/II).

SIGNIFICANCE

This was one of the initial key sections which led to the development during the late 1950s, of the concept of a threefold division of the last glaciation in the eastern Great Lakes region (Forsyth and La Rocque, 1956; Dreimanis, 1957 and 1958; Goldthwait, 1958). This threefold division, with the dominantly glacial Early and Late (called also "Main" or "classical") Wisconsinan substages, separated by an essentially nonglacial but interstadial Middle Wisconsinan substage in this region, was gradually recognized as a general stratigraphic pattern for the south-central and south-eastern part of Canada (Fulton, 1984; p. 4) and the adjoining part of the United States (Fullerton, 1986, Chart 1).

The Middle Wisconsinan Port Talbot interstadial (Dreimanis, 1958) sediments, formally named the Tyrconnell Formation by Dreimanis and Karrow (1972), were thoroughly investigated and dated, particularly during the 1960s and early 1970s (Dreimanis and others, 1966; Quigley and Dreimanis, 1972; Berti, 1975). Whenever the section is exposed by wave erosion, further investigations continue, and one of its latest descriptions is in Hicock and Dreimanis (1985).

SITE INFORMATION

The Bradtville site of the Tyrconnell Formation is in the core of a buried end moraine (Fig. 2) that was formed at the beginning of the Late Wisconsinan Nissouri Stadial around 25 ka. (For stratigraphic terms see Table 1.)

Although only the upper part of the Tyrconnell Formation is exposed above the lake level at the site, the entire formation has been encountered repeatedly in test drillings (see Fig. 2), and its stratigraphic relationship with the underlying Early Wisconsinan and the overlying Middle and Late Wisconsinan glacial deposits is well established. However, there is an erosional hiatus at Bradtville—the uppermost Middle Wisconsinan Wallacetown Formation, also called the Plum Point interstadial deposits, is missing. It is present at the Boy Scout Camp Gravel Pit to the southwest (Fig. 2).

Although the Tyrconnell Formation is the main object of interest at this site, other formations exposed along the bluffs

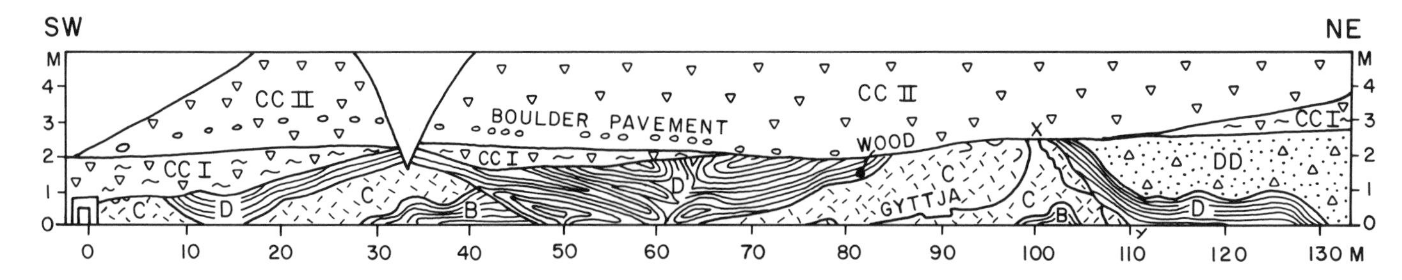

Figure 2. General profile section from Plum Point to Port Talbot along Lake Erie bluffs. A—section described in text and shown in Figure 3; Dunwich Drift; E—Erie Interstadial deposits (Malahide Formation); vertical dashed lines are test holes. After Dreimanis and Barnett (1985), updated.

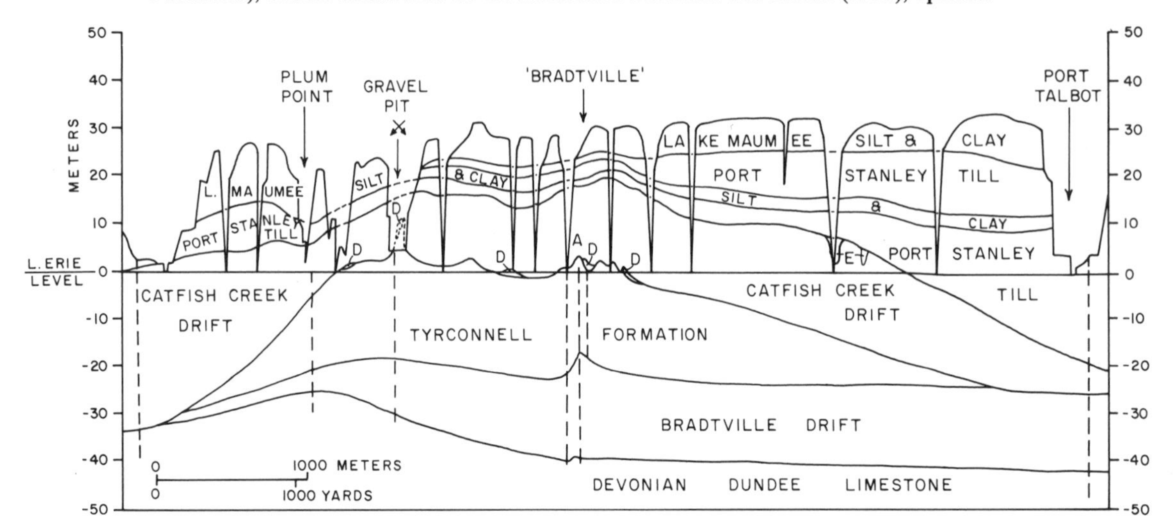

Figure 3. The basal 6 m of Lake Erie bluffs at Bradtville: the 35–125 m interval northeastward from the Bradtville gully culvert at high water level (May 1985); Members B, C, D are in the Tyrconnell Formation; DD are the Dunwich Drift sediments; (CC I) is the locally lowermost member of the Catfish Creek Drift, consisting of interbedded layers of Catfish Creek flow till and interbedded layers of Catfish Creek flow till and glaciolacustrine sediments; (CC II) is massive subglacial Catfish Creek till, with a boulder pavement 0.1–0.8 m above its base. Most of the Tyrconnell Formation is deformed by shearing and overthrusting, and some contacts have been turned into thrust planes, for instance X-Y. Symbols: triangles—till; short wavy lines and triangles—interbedded flow till and glaciolacustrine sediments; dots and triangles—interbedded sand and flow till; subparallel lines—stratified lacustrine sediments; short slanting lines—silt.

(Fig. 2) have been repeatedly investigated by applying multiple criteria, and some reference also will be made to them.

As already mentioned, the upper contact of the Tyrconnell Formation is erosional at the site (Fig. 3). It is overlain by the Dunwich and the Catfish Creek Drifts deposited at the beginning of the Nissouri Stadial.

The following members of the Tyrconnell Formation (Table 1), most of which are glaciotectonically deformed—particularly in the 70-130 m interval, appear in the section (Fig. 3), with Member D being the youngest; the range of their thickness exposed is given in ft (m).

Member D: 0–10 (0–3). Glaciolacustrine, carbonate-rich (30–45%), brownish buff, clayey silt; some layers are rhythmically laminated; others are massive and more clayey. Sediment flows are also present, for instance, at 45-75 m. The 0.4-1.6 in (1-4 cm) thick rhythmites (varves?) contain many thin laminae, particularly in their silty parts. This unit was called "glaciolacustrine II" in Dreimanis and others (1966) and "layer (e)" in Dreimanis (1958). Only an erosional remnant of Member D is exposed here. It is thicker in test drillings and other sections to the southwest.

Member C: 7–13 (2–4). Massive to laminated lacustrine silt, ranging from fine-sandy silt to clayey silt and containing 35–55% carbonate, mainly clastic dolomitic silt particles. The silt is gray when nonoxidized, but rapidly oxidizes yellow to buff when exposed to air. Organic remains are present in the silt: thin shells of molluscs and ostracods, pollen, seeds, and occasional pieces of wood (Dreimanis and others, 1966, p. 314–315; Berti, 1975, p. 598–602). A spruce log, found at the contact of Members C and D, at the 80-m mark (Fig. 3) was dated as 45,200 + 630 years B.P. (GSC-2352-2).

A 4–8-in (10–20-cm) thick, dark brown to black, gyttja

layer, a lacustrine organic mud deposit, is traceable through the middle of the silt layer (Fig. 3). It is rich in plant remains and shells, listed in the above references. The ^{14}C dates of the gyttja: 47,600 ± 400 (GrN-2601) and 46,700 ± 1400 (GrN-2570) years B.P.

The silt layer is up to 40 ft (13 m) thick in test drillings in the southwest and it contains abundant detrital wood at the Bradtville gully, about 25 ft (8 m) below lake level. A peat layer outcrops somewhere offshore, since pieces of woody peat of about the same age as the gyttja are thrown out by waves on the beach during times of low lake levels. Eight dates on wood and peat, from test drillings and peat balls washed ashore, range from 42,700 ± 1200 to 47,690 + 1190 years B.P. (Terasmae and others, 1972, Table 1, p. 57–58). This member was called "Port Talbot II" in Dreimanis and others (1966) and "layer (c)" in Dreimanis (1958).

Member B: 16–40 (5–12). Only the top 1 m is exposed, but the entire member is encountered in all test drillings in the area. Member B consists of rhythmically laminated, carbonate-rich (30–50%) varves, 0.8–6 in (2–15 cm) thick, each of which contains thin laminations, similar to the rhythmites in Member D. About 100 varves were estimated in the nearly continuous core at test drilling P.T.3 at the 80-m mark of Figure 3 (Dreimanis and others, 1966: Fig. 4). This member was called "glaciolacustrine I" in Dreimanis and others (1966) and "layer (b)" in Dreimanis (1958).

In all testholes drilled in the area between Plum Point and Bradtville (Fig. 2), Member B is underlain conformably by Member A, the lowest layer of the Tyrconnell Formation, the so-called Port Talbot I green clay—a diamictic, massive, silty clay and clayey silt, with some sand and grit. It contains only about 1% carbonate (more at its lower and upper transitional contacts; see Figure 4 in Dreimanis and others, 1966) and some primary vivianite; most feldspars in this layer are weathered; the principal clay minerals are illite, iron-chlorite and smectite; pollen is also abundant. The "green clay" has been interpreted by Quigley and Dreimanis (1972) as an accretion gley deposit accumulated in poorly drained depressions in a reducing environment. Since the base of the green clay is about 60 ft (18 m) below the present lake level, the lake level in the interstadial ancient-Erie depression must have been about as low as during the late-glacial Lake Erie, when the Niagara outlet area was isostatically depressed.

During the deposition of Tyrconnell Formation, the climate was cooler than it is now, judging from the dominantly boreal plant remains. The mean July temperatures were probably 15°–21°C during the Port Talbot I and 10°–15° during the Port Talbot II (Berti, 1975). Open landscapes with spruce and pine woodlands, also containing oak during the Port Talbot I, were probably present on the land adjacent to the ancient lake in the Erie basin. The lake level was below the present level during the Port Talbot I and Port Talbot II, but high proglacial lakes existed during the deposition of Member B (glaciolacustrine I) and Member D (glaciolacustrine II).

The Tyrconnell Formation is underlain by the Early Wisconsinan upper Bradtville till (Fig. 2), which may be recognized in test holes by its reddish colour and a higher content of limestone and calcite than in the tills immediately overlying the Tyrconnell Formation.

TABLE 1. STRATIGRAPHIC DIVISIONS OF MIDDLE WISCONSINAN AND ITS SEDIMENTS, AND THE OVERLYING LATE WISCONSINAN DEPOSITS IN THE PLUM POINT-PORT TALBOT AREA

Chronostratigraphic divisions	Lithostratigraphic divisions	
Stadials	Formations	Members
Late Wisconsinan		
Port Bruce Stadial	Port Stanley Drift	
Erie Interstadial	Malahide Formation	
Missouri Stadial	Catfish Creek Drift Dunwich Drift	
Middle Wisconsinan		
Plum Point Interstadial	Wallacetown Formation	
(Cherrytree Stadial?)→	Tyroconnell Formation	D. upper laminated and massive silty clay
		C. silt, gyttja, peat
Port Talbot Interstadial		B. lower laminated silty clay
		A. green diamictic clay

The Tyrconnell Formation is overlain by Late Wisconsinan Dunwich and Catfish Creek Drifts (Fig. 2). The Dunwich Drift consists of stratified to massive silty sand till interbedded with glaciolacustrine sand and silt. The till is rich in carbonates (42–45%) and contains more dolomite than any other tills in this area: dolomite is more abundant than limestone among pebbles, and the calcite/dolomite ratio in till matrix is 0.2–0.4 The Dunwich Drift was deposited by the Lake Huron–Georgia Bay lobe coming from the north, while the overlying Catfish Creek Drift was deposited by the Erie lobe coming from the east. The Catfish Creek Drift has been investigated in considerable detail at Bradtville (Stankowski, 1974; May and Dreimanis, 1976; May and others, 1980). Its stratified facies has been studied repeatedly northeast of Plum Point (Evenson and others, 1977; Gibbard, 1980; Dreimanis, 1982). Catfish Creek till is either massive or stratified, with a silty sand or sandy silt matrix, rich in clasts. Limestone, dolostone, and Precambrian rocks dominate the clasts. The presence of Precambrian coarse dolomitic marble, tillites and jasper conglomerate, as well as an abundance of red garnets in Catfish Creek till (Dreimanis and others, 1966), suggests the eastern Grenville and Superior provinces of the Canadian Shield as the distant provenance sources. Striae on intratill boulder pavements and the alignment of clasts indicate a local northwestward glacial movement during the deposition of Catfish Creek till (Hicock and Dreimanis, 1985). However, till fabrics are not strong. Although the till has been interpreted as a subglacial deposit, a considerable portion of it probably consists of subglacial flows (May and others, 1980) or diamictons deposited by undermelting under a partially floating glacier.

The overlying Port Stanley Drift (Fig. 2) consists of two clayey and silty till members interbedded with glaciolacustrine silts and clays deposited in Lake Maumee (Dreimanis and Barnett, 1985).

CHANGES IN PUBLISHED DATA AND THEIR INTERPRETATIONS

The Tyrconnell Formation has been studied, with some interruptions, since 1951; detailed reinvestigations of recent exposures and samples from test-holes are still going on in order to cross-check earlier conclusions and to test possible reinterpretations. Therefore, it is not surprising that some of the early conclusions had to be changed when new data become available.

Thus, the Dunwich Drift (or lowermost "till (a)" of Dreimanis 1958) was considered to be older than Port Talbot II, in Dreimanis (1960) and subsequent publications, but later it was found to be definitely younger (Dreimanis, 1980 and 1981). Now it is considered to be the lowermost deposit of the Late Wisconsinan Nissouri Stadial (Hicock and Dreimanis, in preparation).

The Southwold Drift has been interpreted either as the lowermost stratigraphic unit of the Late Wisconsinan substage (Dreimanis and others, 1966) or as a Middle Wisconsinan deposit belonging to the Cherrytree Stadial (in most references of 1957–1978). The latest investigations have revealed it to be part of the Late Wisconsinan Catfish Creek Drift (Dreimanis, 1980 and 1981; Dreimanis and Barnett, 1985).

The section shown in Figure 2 of Dreimanis (1958, Fig. 2) is essentially correct above the lake level for the Tyrconnell Formation, considering the minor changes resulting from erosion exposing new sections. However, the then proposed below-the-lake correlation is erroneous, because it was drawn before the test drillings. Both (a) and (d) of Dreimanis (1958, Fig. 2) belong to the Dunwich Drift.

The most up-to-date measurements and interpretations of glaciotectonic deformations in the Tyrconnell Formation are given in Hicock and Dreimanis (1985): the deformations indicate an earlier glacial thrust from the northeast, followed by a later one from the southeast, both caused by the Erie lobe.

OTHER NEARBY SECTIONS OF THE TYRCONNELL FORMATION

The silt unit (Member C) is occasionally exposed along the base of the bluff on the southwest side of a gully 0.2 mi (0.3 km) northeast of the Bradtville gully. On the northeast side of the same gully, intensely folded and faulted glaciolacustrine silty clays, probably of Member D, have been seen in the basal 10 ft (3 m) of the bluff.

The glaciotectonically deformed silty clay and silt of the Tyrconnell Formation are also exposed along the base of the bluff 0.2–0.3 mi (0.4–0.5 km) southwest of the Bradtville gully, and at several other sections as far as Plum Point.

REFERENCES

Berti, A. A., 1975, Paleobotany of Wisconsinan interstadials, eastern Great Lakes region, North America: Quaternary Research, v. 5, p. 591–619.

De Vries, H., and Dreimanis, A., 1960, Finate radiocarbon dates of the Port Talbot interstadial deposits in southern Ontario: Science, v. 131, p. 1738–1739.

Dreimanis, A., 1957, Stratigraphy of the Wisconsin glacial stage along the northwestern shore of Lake Erie: Science, v. 126, p. 166–168.

—— 1958, Wisconsin stratigraphy at Port Talbot on the north shore of Lake Erie, Ontario: Ohio Journal of Science, v. 58, p. 65–84.

—— , 1960, Pre-classical Wisconsin in the eastern portion of the Great Lakes region, North America: 21st International Geological Congress, Norden, Copenhagen, Reports, Part 4, p. 108–119.

—— , 1980, Field trip to the 'Bradtville'–Plum Point area, north shore of Lake Erie: London, Contribution of Department of Geology No. 533, University of Western Ontario, 12 p.

—— , 1981, Middle Wisconsin substage in its type region, the eastern Great Lakes, Ohio River basin, North America: Quaternary Studies in Poland, v. 3, p. 21–28.

—— , 1982, Two origins of the stratified Catfish Creek Till at Plum Point, Ontario, Canada: Boreas, v. 11, p. 173–180.

Dreimanis, A., and Barnett, P., 1985, Quaternary geology, Port Stanley area, Southern Ontario: Ontario Geological Survey Geological Series–Preliminary Map, Map P.2827, scale 1:50,000.

Dreimanis, A., and Karrow, P. F., 1972, Glacial history of the Great Lakes–St. Lawrence region, the classification of the Wisconsin(an) Stage, and its correlatives, *in* Proceedings, 24th International Geological Congress, Montreal, section 12, p. 5–15.

Dreimanis, A., Terasmae, J., and McKenzie, G. D., 1966, The Port Talbot Interstade of the Wisconsin Glaciation: Canadian Journal of Earth Sciences, v. 3, p. 305–325.

Evenson, E. B., Dreimanis, A., and Newsome, J. W., 1977, Subaquatic flow tills; A new interpretation for the genesis of some laminated till deposits: Boreas, v. 6, p. 115–133.

Forsyth, J. L., and LaRocque, J.A.A., 1956, Age of the buried soil at Sidney, Ohio: Geological Society of America Bulletin, v. 67, p. 1696.

Fullerton, D. S., 1986, Stratigraphy and correlaton of glacial deposits from Indiana to New York and New Jersey, *in* Richmond, G. M., and Fullerton, D. S., eds., Quaternary glaciations in the United States of America, Part 1 of Quaternary glaciations in the Northern Hemisphere: London, Pergamon Press (in press).

Fulton, R. J., 1984, Summary; Quaternary stratigraphy of Canada, *in* Fulton, J. R., ed., Quaternary stratigraphy of Canada; A Canadian contribution to IGCP project 24: Geological Survey of Canada Paper 84-10, p. 1–5.

Gibbard, P., 1980, The origin of the stratifed Catfish Creek Till by basal melting: Boreas, v. 9, p. 71–85.

Goldthwait, R. P., 1958, Wisconsin age forests in western Ohio; I=Age and glacial events: Ohio Journal of Science, v. 58, p. 209–219.

Hicock, S. R., and Dreimanis, A., 1985, Glaciotectonic structures as useful ice-movement indicators in glacial deposits; Four Canadian case studies: Canadian Journal of Earth Sciences, v. 22, p. 339–346.

May, R. W., and Dreimanis, A., 1976, Compositional variability in tills, *in* Legget, R. F., ed., Glacial till: Royal Society of Canada Special Publication 12, p. 99–120.

May, R. W., Dreimanis, A., and Stankowski, W., 1980, Quantitative evaluation of clast fabrics within the Catfish Creek Till, Bradtville, Ontario: Canadian Journal of Earth Sciences, v. 17, p. 1064–1074.

Quigley, R. M., and Dreimanis, A., 1972, Weathered interstadial green clay at Port Talbot, Ontario: Canadian Journal of Earth Sciences, v. 9, p. 991–1000.

Stankowski, W., 1974, Changeability of mechanical and petrographic composition of Catfish Creek drift sediments (in Polish): Geografia, v. 10, Poznań, p. 215–227.

Terasmae, J., Karrow, P. F., and Dreimanis, A., 1972, Quaternary stratigraphy and geomorphology of the eastern Great Lakes region of southern Ontario: 24th International Geological Congress, Montreal, Excursion A42 Guidebook, 75 p.

Structure and Ordovician stratigraphy of the Ottawa area, southern Ontario

D. A. Williams, Ontario Ministry of Northern Development and Mines, Tweed, Ontario KOK 3JO, Canada
P. G. Telford, Ontario Geological Survey, Toronto, Ontario M5S 1B3, Canada

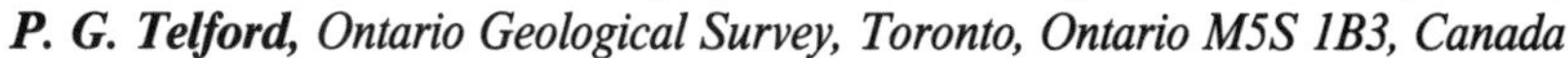

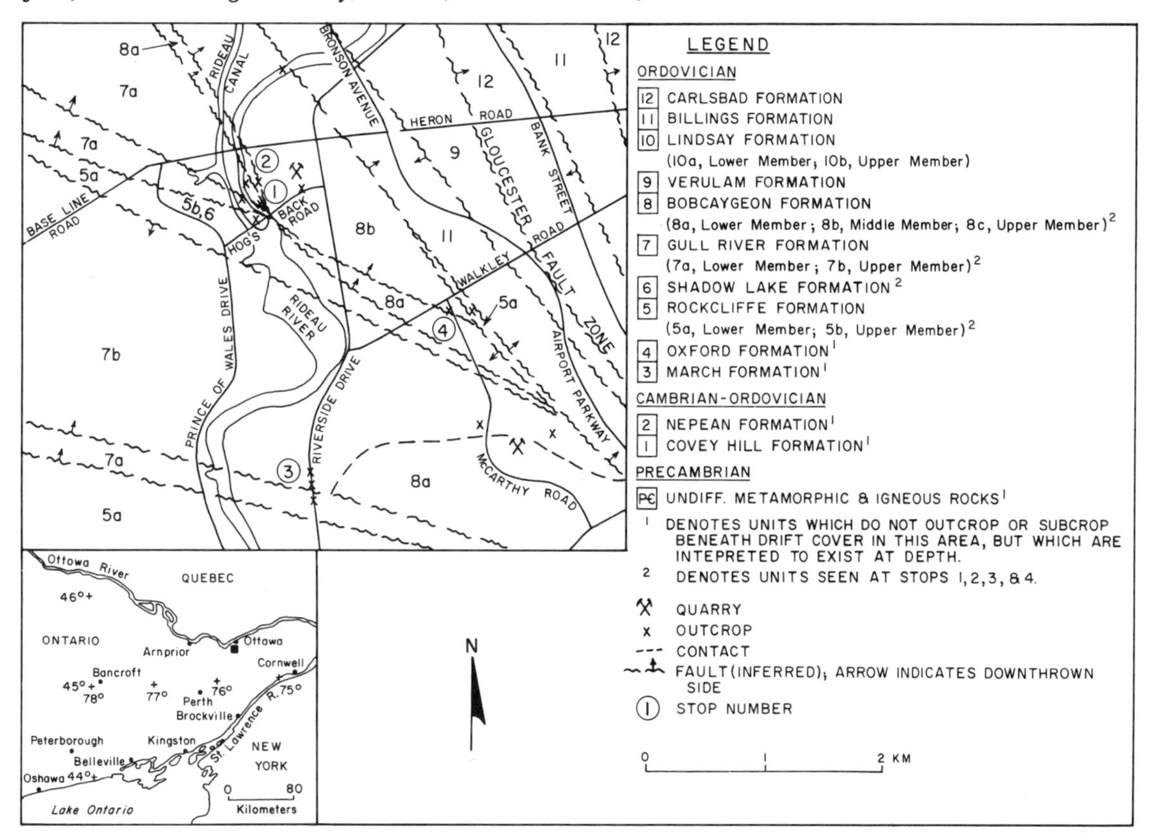

Figure 1. Bedrock geology of southeastern Ottawa (modified from Williams and others, 1984).

LOCATION

The Ottawa area is located in the eastern part of southern Ontario (Fig. 1). The structural and stratigraphic aspects of the bedrock geology are well illustrated by four localities situated in the southern part of the urban area, within 1.9 mi (3 km) of each other. Bronson Avenue provides access from downtown Ottawa. Stop 1 (Fig. 1) consists of exposures along both banks of the Rideau River at Prince of Wales Falls in Hog's Back Park, immediately west of the parking lot located to the north of Hog's Back Road. There is no charge for parking, and generally no difficulty in finding a parking space. Care must be taken when examining the outcrop, which is adjacent to the falls and only partially accessible. Stop 2 (Fig. 1) is also in Hog's Back Park and consists of an exposure along the east bank of the Rideau River, immediately south of the Heron Road bridge. A footpath joins Stops 1 and 2, which are about 1,310 ft (400 m) apart. Stop 3 (Fig. 1) is a roadcut on Riverside Drive, 1.9 mi (3 km) south of Heron Road. Vehicles may be parked along the turnoff to the west of Riverside Drive at the north end of the outcrop, and care must be taken when crossing the road. Stop 4 (Fig. 1) is a roadcut on McCarthy Road, 0.6 mi (0.9 km) east of Riverside Drive and immediately south of Walkley Road. McCarthy Road can be crossed at the traffic light located at the corner of Walkley Road. Vehicles may be parked along Provost Drive, a side road to the east of McCarthy Road, which is located one block south of Walkley Road.

SIGNIFICANCE

The middle part of the Cambrian-Ordovician sequence of the Ottawa–St. Lawrence Lowland is exposed at Stops 1, 2, 3, and 4. The lower and upper members of the Rockcliffe Formation, the Shadow Lake Formation, the lower and upper members of the Gull River Formation, and the lower and middle members of the Bobcaygeon Formation (Fig. 1) can be examined. Overlying stratigraphic units outcrop or subcrop beneath drift cover in the vicinity. Normal faults of the Gloucester fault zone (Fig. 1) are observable at all of the sites.

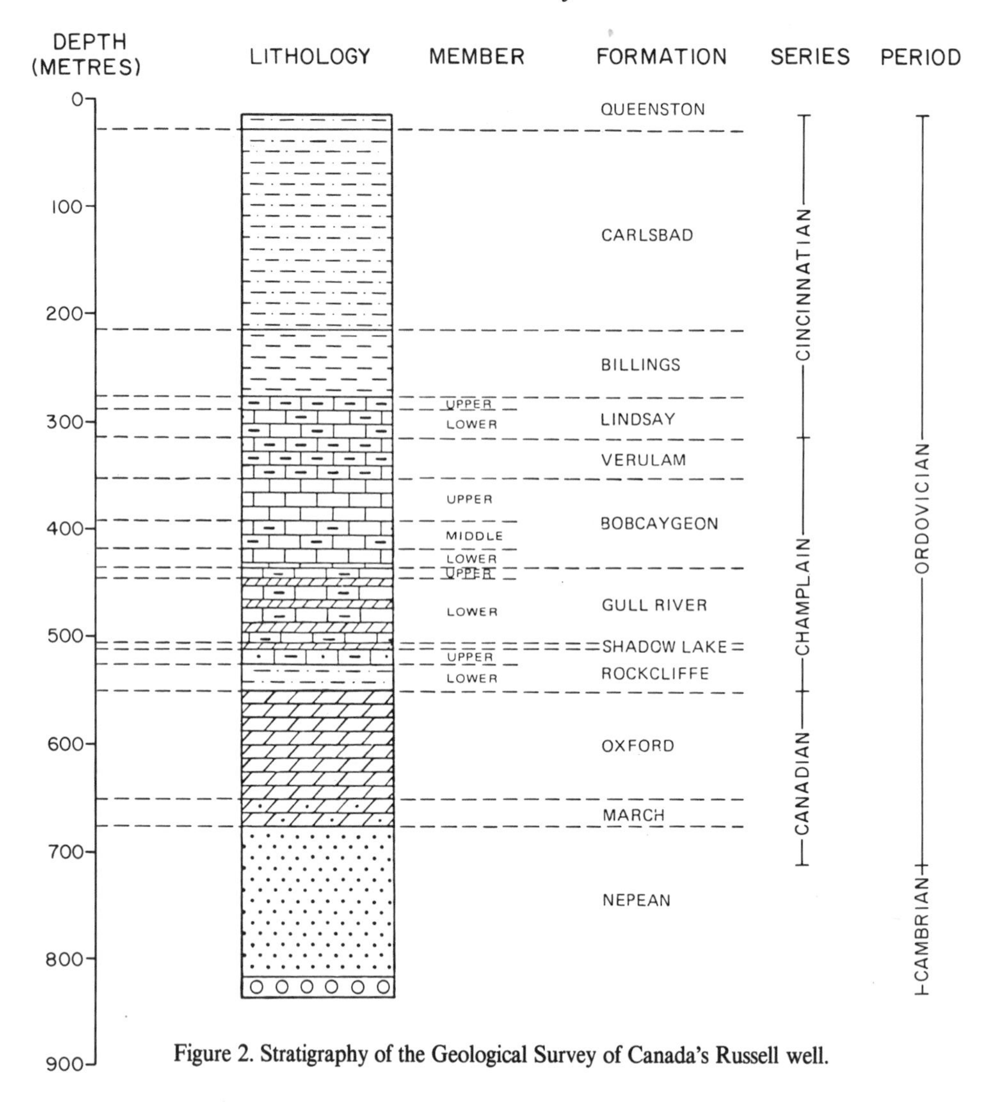

Figure 2. Stratigraphy of the Geological Survey of Canada's Russell well.

SITE INFORMATION

Stratigraphic control for the Ottawa area is provided by drillcore from the Russell well, drilled 15 mi (25 km) to the southeast of the city by the Geological Survey of Canada (GSC). The core is located at the GSC's core storage facility in Ottawa, and the section is shown in Figure 2. The stratigraphic nomenclature of Figure 2 was originally defined in southern Ontario. Usage of the terminology is based on studies by Wilson (1946), Liberty (1969), and Williams and Telford (1986). The Covey Hill Formation, originally defined in southern Quebec by Clark (1966) and constituting the lowermost unit of the Cambrian-Ordovician sequence, was not intersected by the Russell well. The feldspathic conglomerates and sandstones of the Covey Hill Formation were deposited in a terrestrial environment, and the overlying units in an intracontinental shelf environment. The red colour of the Queenston Formation, the uppermost unit, is probably due to post-depositional oxidation of the iron content. The terrigenous clastics of the lower part of the sequence (up to the Gull River Formation) were derived from Precambrian uplands, and those of the upper part of the sequence from the rising Appalachian Mountains to the east. The dolostones and lithographic to finely crystalline limestones of the lower part of the sequence were deposited in a supratidal-to-subtidal environment, the common occurrence of hypersalinity being indicated by the presence of algal lamination and sulphate nodules (generally calcitized). The limestones of the Bobcaygeon Formation and succeeding units were deposited in a low to high energy marine environment, as indicated by the alternation of beds varying from sublithographic to coarsely crystalline.

The Gloucester fault zone strikes generally northwestward across Ottawa (Fig. 1). Along the fault zone there is a vertical displacement of approximately 1,700 ft (520 m), with the downthrown block to the northeast (Williams and others, 1984). The fault zone consists of a series of steeply dipping individual faults, and is characterized by the branching of faults from fault junctions. Bedding in the formations is normally close to horizontal, but adjacent to faults it commonly dips steeply toward the downthrown block. The Gloucester fault zone is a component of the Ottawa Valley rift zone, and is considered to have resulted from

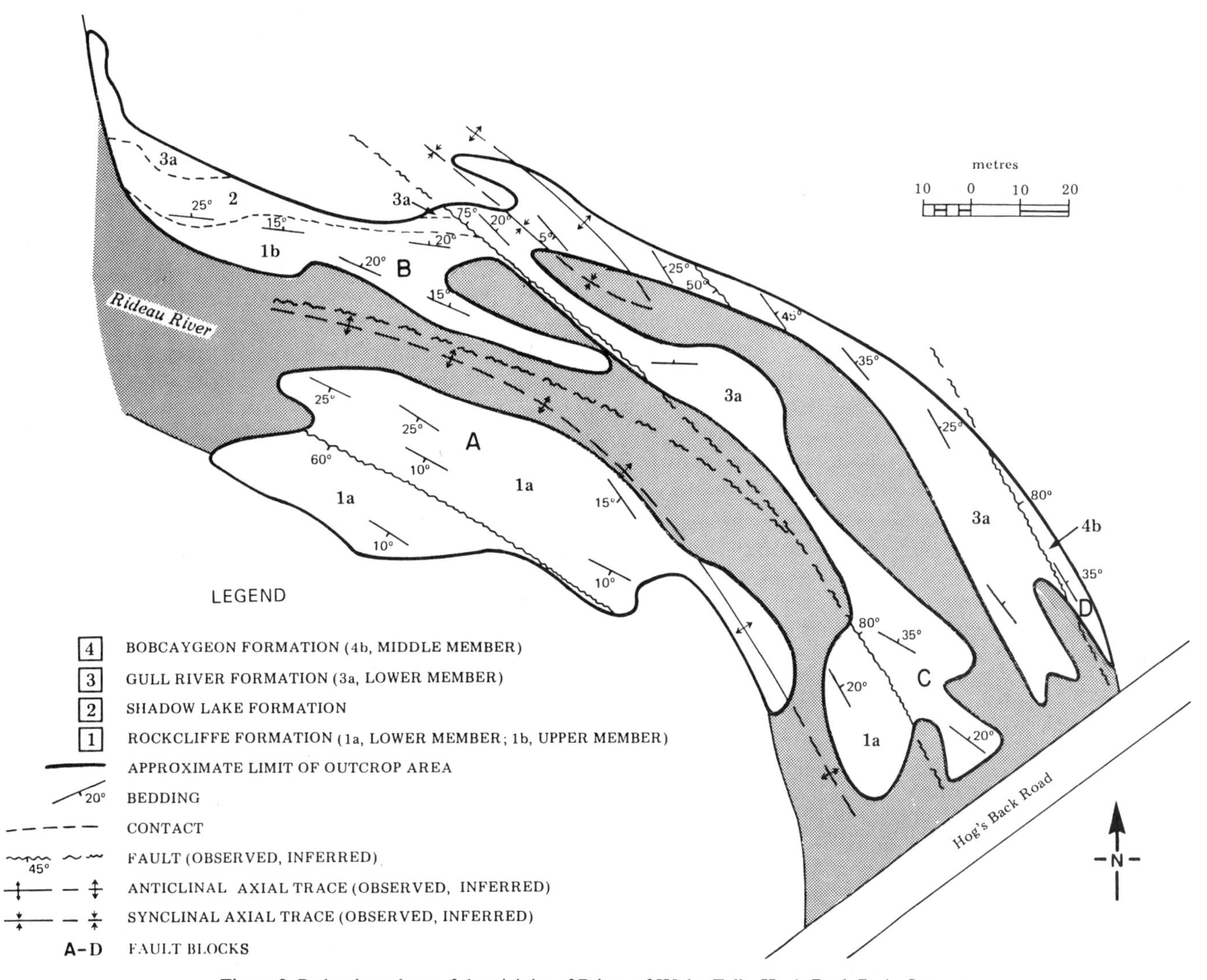

Figure 3. Bedrock geology of the vicinity of Prince of Wales Falls, Hog's Back Park, Ottawa.

reactivation of preexisting faults during the Mesozoic. The area was under extension during the Cretaceous, as indicated by the occurrence elsewhere in the Ottawa–St. Lawrence Lowland of carbonatite dikes of probable Cretaceous age.

Stop 1. As shown in Figure 3, three major faults occur at Stop 1 (Prince of Wales Falls). Two of the faults are exposed along the northeast bank of the Rideau River, and the third one is underwater. The relative movement along all three faults is northeast-side-down. Two faults of relatively minor displacement are also indicated on Figure 3.

The exposure along the southwest bank of the Rideau River (in fault block A of Fig. 3) consists of quartz sandstone with shaly partings of the lower member of the Rockcliffe Formation. A thickness of 19.5 ft (5.94 m) of this member was measured by Vollrath (1962, p. 8).

The exposure in the southwesternmost fault block along the northeast bank of the Rideau River (fault block B of Fig. 3) is of particular significance because it shows a part of the sedimentary sequence that rarely outcrops. About 8 ft (2.45 m) of the Shadow Lake Formation (silty dolostone, with thin interbeds of quartz sandstone) are underlain by 30.2 ft (9.2 m) of the upper member of the Rockcliffe Formation (interbedded quartz sandstone, shale, silty dolostone, and limestone) and overlain by 20 ft (6 m) of the lower member of the Gull River Formation (interbedded silty dolostone, limestone, and shale). Ripple marks and dolomitic mottling are observable in the lowermost limestone bed of the Gull River Formation. The base of the section is at the water level, in the extreme southeastern part of the fault block. The exposure was also described by Raymond (1911, p. 190–191) and by Vollrath (1962, p. 19–25) as section 3.

The exposure in the middle fault block along the northeast bank of the Rideau River (fault block C of Fig. 3) consists of interbedded silty dolostone, shale, fine-grained quartz sandstone, and limestone, of the lower member of the Gull River Formation.

The section, described by Raymond (1911, p. 192), measured 45 ft (13.5 m); Raymond's bed 7 is a stromatolitic limestone exposed along the upper part of the river bank in the northeastern part of the fault block. The lower part of the section is inaccessible.

The exposure in the northeasternmost fault block along the northeast bank of the Rideau River (fault block D of Fig. 3) consists of medium crystalline limestone of the middle member of the Bobcaygeon Formation.

Stop 2. A major fault is exposed at Stop 2 (Heron Road bridge). The relative movement is northeast-side-down. Southwest of the fault, the lower member of the Gull River Formation (which is also exposed at Stop 1 in fault block C of Fig. 3) can be examined. Burrowed silty dolostone is overlain successively by shale, fine-grained quartz sandstone, and stromatolitic limestone. The section, described by Vollrath (1962, p. 29–31) as section 4, is 10 ft (3 m) thick. To the northeast of the fault, the contact between the upper member of the Gull River Formation (lithographic to finely crystalline limestone with shaly partings) and the overlying lower member of the Bobcaygeon Formation (generally massive sublithographic to finely crystalline limestone) is exposed. The coral *Tetradium* occurs in abundance below the contact. This section was also described by Vollrath (1962, p. 32), who measured 5 ft (1.6 m).

Stop 3. Two major faults, 165 ft (50 m) apart, are exposed at Stop 3 (Riverside Drive). The relative movement along both faults is north-side-down. Bedding is close to horizontal except between the faults, where it dips northward at approximately 30°. Several other faults of relatively minor displacement also occur. The exposure in the southern fault block consists of interbedded silty to sandy dolostone and ostracod-bearing shale, with a few limestone interbeds up to 2 in (5 cm) thick, belonging to the lower member of the Gull River Formation. The exposure in the middle fault block consists of interbedded lithographic to finely crystalline limestone and silty dolostone, with a few shale interbeds up to 2 in (5 cm) thick, also of the lower member of the Gull River Formation. The lower part of the section exposed in the northern fault block consists of lithographic to finely crystalline limestone, with silty dolostone interbeds, belonging to the upper part of the lower member of the Gull River Formation. The upper part of the section exposed in the northern fault block consists of lithographic to finely crystalline limestone of the upper member of the Gull River Formation; some beds are intraclastic and oolitic.

Stop 4. Two major faults are exposed at Stop 4 (McCarthy Road). The relative movement along both faults is north-side-down. Bedding dips westward at approximately 25° in the northern part of the outcrop, and northward at approximately 65° in the southern part of the outcrop. Several other faults of relatively minor displacement also occur. In the southern fault block, the contact between the upper member of the Gull River Formation and the overlying lower member of the Bobcaygeon Formation is exposed; as at Stop 2, the coral *Tetradium* occurs in abundance below the contact. The middle member of the Bobcaygeon Formation is exposed in the middle and northern fault blocks; both sections contain sublithographic to finely crystalline limestone, the northern section also containing beds of medium crystalline limestone up to 6 ft (2 m) thick.

REFERENCES CITED

Clark, T. H., 1966, Chateauguay area: Quebec Department of Natural Resources Geological Report 122, 63 p.

Liberty, B. A., 1969, Paleozoic geology of the Lake Simcoe area, Ontario: Geological Survey of Canada Memoir 355, 201 p.

Raymond, P. E., 1911, Preliminary notes on the "Chazy" Formation in the vicinity of Ottawa: The Ottawa Naturalist, v. 24, p. 189–197.

Vollrath, J. D., 1962, Geology of the Hog's Back area from the White Bridge to Hog's Back [B.S. thesis]: Ottawa, Carleton University, 68 p.

Williams, D. A., and Telford, P. G., 1986, Paleozoic geology of the Ottawa area: Geological Association of Canada—Mineralogical Association of Canada—Canadian Geophysical Union, Joint Annual Meeting, Ottawa, Field Trip 8 Guidebook, 25 p.

Williams, D. A., Rae, A. M., and Wolf, R. R., 1984, Paleozoic geology of the Ottawa area, southern Ontario: Ontario Geological Survey Map P. 2716, Geological Series–Preliminary Map, scale 1:50,000.

Wilson, A. E., 1946, Geology of the Ottawa–St. Lawrence Lowland, Ontario and Quebec: Geological Survey of Canada Memoir 241, 66 p.

Glacial stratigraphy of the central St. Lawrence Lowland near Lac St. Pierre, Quebec

Nelson R. Gadd, Geological Survey of Canada, 601 Booth Street, Ottawa, Ontario K1A OE8, Canada

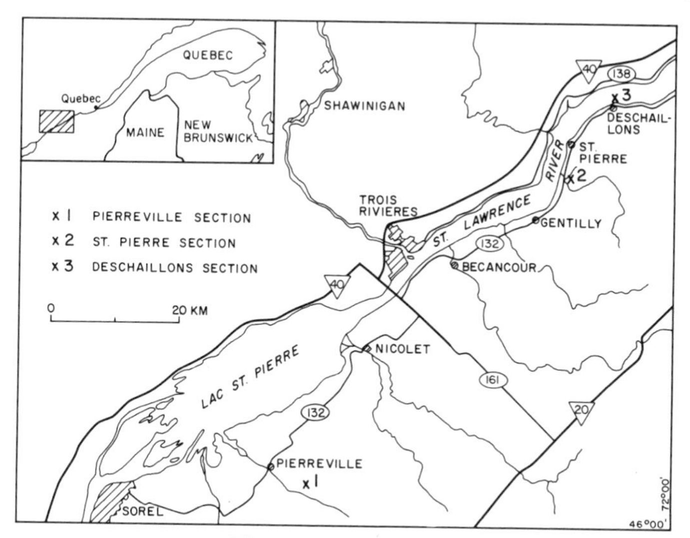

Figure 1. Location map.

LOCATION

Access to the three sections described here is relatively simple. The Lac St. Pierre region can be reached by way of either of two major highways that connect Montreal and Quebec city; Canadian Autoroute 40 follows the north shore of the St. Lawrence, and Canadian Autoroute 20 parallels the south shore at a distance of some 20 to 30 mi (35 to 40 km) southeast of the river.

Quebec 161 connects the two main arteries at Trois Rivières, where there is a high-level bridge over the St. Lawrence. The principal access road for the three stops in the site is Quebec 132, which runs generally east-west from Quebec 161, closely paralleling the south shore of the St. Lawrence River. Pierreville is on Quebec 132, 20 mi (35 km) southwest of the junction with Quebec 161 (Trois Rivières Bridge).

The Pierreville Section (Stop 1, Fig. 1) lies about 1.5 mi (2.6 km) southeast of the village on the right bank of St. François River; access is by a gravel road from the village. Permission from the farm proprietor on the east side of the road is required.

The St. Pierre Section (Stop 2, Fig. 1) is in a narrow ravine that crosses Quebec 132 about 1 mi (1.6 km) southwest of the village of St. Pierre (or Les Becquets on some road signs). Permission to cross private farm property may be obtained from the owner of the first farm house on the south side of the highway southwest of the ravine; very steep-sided ravine banks at this section are about 600 ft (200 m) southeast of the farm house and highway. The village of Deschaillons, on Quebec 132, is 6 mi (10 km) northeast of the village of St. Pierre (Les Becquets).

The Deschaillons section (Stop 3, Fig. 1) is in the clay-pit of the brickyard of Montreal Terra Cotta Company at the base of the St. Lawrence River escarpment near the eastern limit of the village. Permission may be obtained from the company office on the site.

TABLE 1. DESCRIPTION OF PIERREVILLE SECTION (STOP 1)*

Thickness (ft)	(m)	Lithology	Cumulative Thickness (ft)	(m)
		Gentilly Stade Units		
7.5	2.3	Brownish grey calcareous sandy till	7.5	2.3
		St. Pierre Interval Units		
17.5	5.3	Stratified, fine-grained buff to brown sand, becoming silty in lower 5 ft	25.0	7.5
1.0	0.3	Compressed peat with some wood	26.0	7.8
9.0	2.7	Stratified, fine-grained sand and silty sand grading downward into subjacent silt	35.0	10.5
		Bécancour Stade Units		
6.5	5.0	Grey to brownish grey varved silt (in lower part of section, beds are repeated by slumping: some contacts apparently are fault contacts)	51.5	15.5
5.5	7.7	Section covered by slump debris	77.0	23.1
		Gentilly Stade Units		
3.0	0.9	Brownish grey calcareous sandy till	80.0	24.0
		St. Pierre Interval Units		
2.0	0.6	Stratified fine-grained buff sand	82.0	24.6
1.0	0.3	Compressed peat with some wood (This peat was reported by Joseph Keele, 1916)	83.0	24.9
		Bécancour Stade Units		
7.0	2.1	Grey and brownish grey varved silt	90.0	27.0

*From Gadd, 1972, Appendix 1, p. 107-108; measured downward from surface (0 ft) as in Figure 2.

SIGNIFICANCE OF SITE

The sediments of the type sections described here represent two glaciations and one non-glacial interval. At the onset of each glaciation, the St. Lawrence valley was blocked in a narrow part of the Lowland somewhere near Quebec city. This resulted in the formation of a proglacial lake that was overridden by Laurentide ice as glaciation progressed.

The first glaciation of record eroded underlying bedrock, which locally is the brick-red Queenston shale (Bécancour River Formation). Thus, early varves and till have a characteristic

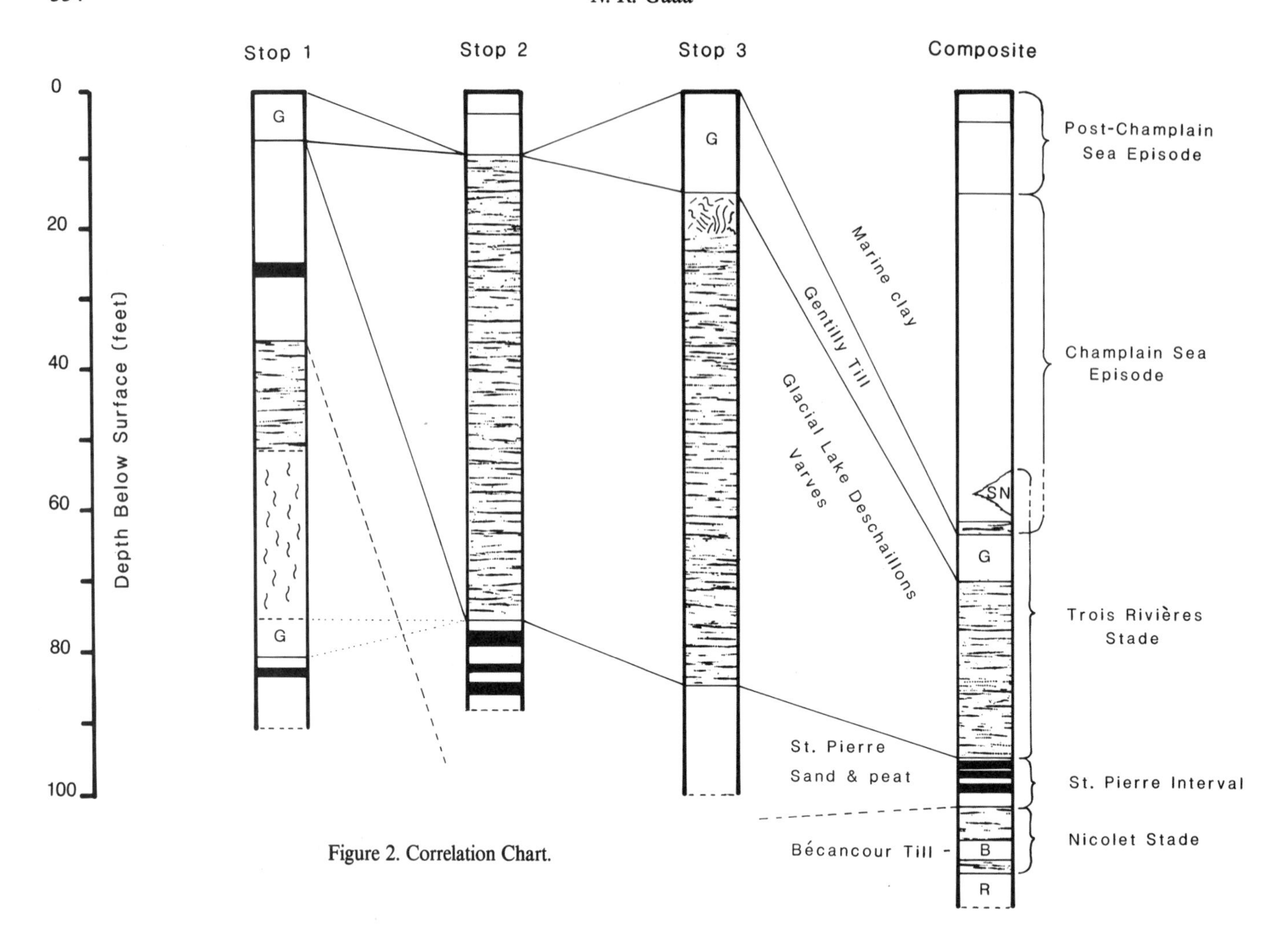

Figure 2. Correlation Chart.

brick-red colour in the sections described here. Retreat of ice at the end of this glaciation produced more varves and no record of marine submergence. This suite was initially designated as the Bécancour glaciation, represented by Bécancour Till (Gadd, 1972).

There followed a non-glacial fluvial cycle during which the St. Pierre sand and St. Pierre peat beds were laid down. The dating of peat and wood from the strata at about 70 ka allowed very tentative correlation of the St. Pierre Interval with either very early Wisconsin or late Sangamon; the correlation of Bécancour Till, therefore, has been given as 'early Wisconsin or older.'

The second glaciation of the area was preceded by the formation of Glacial Lake Deschaillons whose lowermost thin varves preserve the ripple-marked surface of underlying St. Pierre sands. Gadd (1972) has suggested that, therefore, Lake Deschaillons varves and St. Pierre sediments are very close in age. The Deschaillons varves are overridden and incorporated in the base of Gentilly Till, the surficial gray till of the region. In a few places within the area described here, and at a number of other localities within the St. Lawrence Lowland, the younger till is overlain by varves or varve-like sediments that grade upward into the more massive marine sediments of Champlain Sea that submerged most of the St. Lawrence Lowland at deglaciation around 12 ka. The apparent continuity of sedimentation, from ripple marks on the St. Pierre sediments through glaciation to Champlain Sea deposition, has permitted the hypothesis that Gentilly glaciation of the central St. Lawrence Lowland spanned all the time between St. Pierre Interval (ca. 70 ka) and Champlain Sea (ca. 12 ka) (Gadd, 1972).

Because of the redundancy of using the names Bécancour and Gentilly for both the main tills and the glacial stade in which they were deposited (Gadd, 1972), Dreimanis and Karrow (1972) proposed the Name 'Nicolet Stade' to replace "Bécancour Stade" and Occhietti (1978, 1982) proposed 'Trois Rivières Stade' to replace "Gentilly Stade." These new terms are in common usage in the stratigraphic literature.

A correlation chart (Fig. 2) shows the relationship of the three stops described to the regional stratigraphy. Varves of the Nicolet glaciation, older than the St. Pierre Interval, are exposed only at Stop 1; these overlie Bécancour Till (B) in other sections nearby. Marine clay of the Champlain Sea episode overlies varves of Glacial Lake Deschaillons at Stop 2, where Gentilly Till (G) is missing. In the composite section, there is an overlap of events between the Trois Rivières Stade and the Champlain Sea episode

TABLE 2. DESCRIPTION OF ST. PIERRE SECTION (STOP 2)*

Thickness (ft)	(m)	Lithology	Cumulative Thickness (ft)	(m)
		Post-Champlain Sea Units		
3.5	1.1		3.5	1.1
		Champlain Sea Units		
5.0	1.5	Massive, soft, light grey silt	8.5	2.6
		Gentilly Stade Units		
66.5	19.9	Approximately 500 varves of calcareous grey silt, thin-bedded at top and bottom of section	75.0	22.5
		St. Pierre Interval Units		
1.5	0.46	Fine- to medium-grained sand heavily charged with disseminated fragments of organic matter; brown	76.5	23.5
1.75	0.53	Compressed peat with abundant wood as flattened stems, branches, twigs	78.25	23.5
3.0	0.9	Stratified medium- to fine-grained sand, some silt; alluvial	81.25	24.4
0.5	0.2	Compressed peat, some wood	81.25	24.4
1.5	0.46	Medium and fine-grained sand, a few pebbles; alluvial	83.25	25.0
1.25	0.38	Compressed peat; beetle remains most common in this bed	84.25	25.4
3.0	0.9	Fine- to medium-grained sand and silty sand; alluvial	87.5	26.3

*From Gadd, 1972, Appendix 1, p. 195; measurements in ft below surface (0 ft).

TABLE 3. DESCRIPTION OF DESCHAILLONS SECTION (STOP 3)*

Thickness (ft)	(m)	Lithology	Cumulative Thickness (ft)	(m)
15.0	4.5	Buff till	15.0	4.5
70.0	21.2	Grey, thin-bedded, varved silts and clays	85.0	25.7
15.0	4.5	Stratified, yellow, cross-bedded sand; wood has been found in this layer	100.0	30.2

*After Karrow, 1957, Figure 11, locality 400, Appendix A, p. 92; measurements in ft below surface, as in Figure 2.

in order to accommodate the fact that St. Narcisse Moraine (SN) formed within Champlain Sea time—an event that is also considered to be a late phase of the Trois Rivières Stade.

SITE INFORMATION

Stop 1—Pierreville Section. Buried compressed peat beds, first reported by J. Keele (1915) in a slumped section were later shown (Gadd, 1972) to be associated with non-glacial fluvial sediments underlain by varved clay (in turn underlain by Bécancour Till in nearby sections) and overlain by Gentilly Till. Gadd correlated the peat and wood with similar beds at the St. Pierre Section (see below). Groeningen laboratory dating of wood collected at this site by H. De Vries gave the first finite date for the St. Pierre Sediments, 67,000 ± 2000 ^{14}C B.P. (Sample number GrN-1711, Vogel and Waterbolk, 1963) and provided the basis for interpreting an early Wisconsinan interstadial status for the non-glacial St. Pierre Interval. Later collections of wood from this site provided material for a dating of 74,000 $^{+2700}_{-2000}$ ^{14}C B.P. (Stuiver and others, 1978; sample QL-198) which is commonly rounded to ~75 ka in current literature. Table 1 describes the Pierreville Section (Fig. 1).

Stop 2—St. Pierre Section. This is the discovery and type section for the St. Pierre Sediments described by Gadd (1953, 1960, 1972) and botanically analyzed by Terasmae (1958). Three relatively thick beds of compressed peat are interstratified with sand and silty sand near the base of the section. The uppermost compressed peat bed provided wood, collected by H. De Vries and dated 65,300 ± 1400 B.P. (Vogel and Waterbolk, 1963; Sample GrN-1799). These St. Pierre beds are overlain by a thick deposit of varves that may be traced laterally and almost continuously from this section to the Deschaillons section (Stop 3). Pollen stratigraphy of the St. Pierre beds is continuous into the base of the varves (Terasmae, 1958). In other sections at the St. Lawrence shore nearby, the St. Pierre sediments are underlain unconformably by brick-red Bécancour Till and associated red and gray varves. Thus the St. Pierre sediments clearly separate two glacial events, and their age suggests an early Wisconsinan interstadial event. The St. Pierre Section is described in Table 2.

Stop 3—Deschaillons Section. St. Pierre sands at the base of the section contain undated, transported wood. The ripple-marked surface of the sand is preserved by overlying varves sediments in which the ripples are obvious for several centimeters above the contact. This combines with the pollen stratigraphy observation at the St. Pierre Section (Stop 2) as the main criteria used (Gadd, 1972) to indicate that Lake Deschaillons followed closely the end of the St. Pierre Interval. The top of the Deschaillons varve section (estimated to contain 4,000 to 5,000 varves (Hillaire-Marcel and Pagé, 1981) is disturbed and overlain by Gentilly Till, the younger of the two tills of the region. The upper till postdates the St. Pierre Interval and may be in part contemporaneous with early Champlain Sea; therefore the last glaciation may have occupied most of Wisconsinan time. The Bécancour Till, which predates the St. Pierre Interval, may prove to be pre-Wisconsinan. However, controversial datings on carbonate concretions in varves at both Pierreville and Deschaillons sections have suggested a maximum age of 35,000 B.P. for the last glaciation (Hillaire-Marcel and Pagé, 1981). Table 3 describes the Deschaillons Section (Fig. 1).

REFERENCES CITED

Dreimanis, A., and Karrow, P. F., 1972, Glacial history in the Great Lakes-St. Lawrence region, the classification of the Wisconsin (an) Stage, and its correlatives: *in* Quaternary Geology, 24th International Geological Congress Proceedings, Section 12, p. 5–15.

Gadd, N. R., 1953, Interglacial deposits at St. Pierre, Quebec [abs.]: Geological Society of America Bulletin, v. 64, p. 1426.

——, 1960, Surficial geology of the Bécancour map-area, Quebec: Geological Survey of Canada Paper 59-8.

——, 1972, Pleistocene geology of the central St. Lawrence Lowland: Geological Survey of Canada Memoir 359, 153 p.

Gadd, N. R., LaSalle, P., Dionne, J-C., Shilts, W. W., and McDonald, B. C., 1972, Quaternary geology and geomorphology, southern Quebec: XXIV International Geological Congress, Montreal, Quebec, Guidebook to Excursions A44-C44, 70 p.

Hillaire-Marcel, C., and Pagé, P., 1981, Paléotemperatures isotopiques du lac glaciaire de Deschaillons: *in* Mahaney, W. C., ed., Quaternary Paleoclimates: Norwich, U.K., Geobooks, University of East Anglia, p. 273–297.

Karrow, P. F., 1957, Pleistocene geology of the Grondines map-area, Quebec [Ph.D. thesis]: Urbana, University of Illinois.

Keele, J., 1915, Preliminary report on the clay and shale deposits of the Province of Quebec: Geological Survey of Canada Memoir 64, 280 p.

Occhietti, S., 1978, Le Quaternaire de la région de Trois-Rivières-Shawinigan, Québec; Contribution à la paléogéographie de la vallée moyenne du Saint-Laurent et corrélations stratigraphiques [Ph.D. thesis]: Université d'Ottawa.

——, 1982, Synthèse lithostratigraphique et paléoenvironnements du Quaternaire au Québec méridional; Hypothèse d'un centre d'englacement Wisconsinien au Nouveau-Québec: Géographie physique et Quaternaire, v. 36, No. 1-2, p. 15–49.

Stuiver, M., Heusser, C. J., and Yang, I. C., 1978, North American glacial history extended to 75,000 years ago: Science, v. 200, p. 16–21.

Terasmae, J., 1958, Contributions to Canadian palynology—Pt. 1, The use of palynological studies in Pleistocene stratigraphy; Pt. 2, Non-glacial deposits in the St. Lawrence Lowlands, Quebec; Pt. 3, Non-glacial deposits along Missinaibi River, Ontario: Geological Survey of Canada Bulletin 46, 35 p.

Vogel, J. C., and Waterbolk, H. T., 1963, Groningen Radiocarbon Dates IV: Radiocarbon, v. 5, p. 163.

The late mid-Ordovician transgressive sequence and the Montmorency Fault at the Montmorency Falls, Quebec

John Riva, *Laval University, Ste. Foy, Quebec G1K 7P4, Canada*
R. K. Pickerill, *University of New Brunswick, Fredericton, New Brunswick E3B 5A3, Canada*

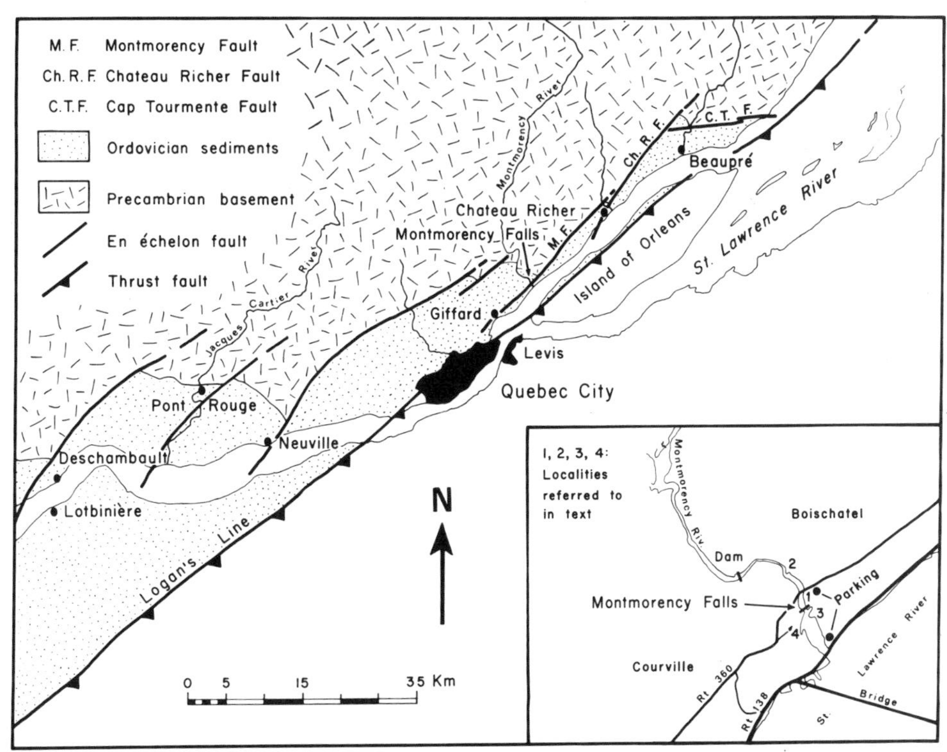

Figure 1. Generalized geologic map of area near Quebec City showing the location of the Montmorency Falls. The inset shows the approximate position of stops mentioned in text.

LOCATION

The Montmorency Falls is located about 6 mi (10 km) northeast of the city of Quebec, along the north shore of the St. Lawrence River (Fig. 1). From Quebec City the falls can be easily reached by first taking Route 40 and then 138, which passes below the falls, or Route 360 (or the parallel Boulevard des Chutes), which passes above it (Fig. 1, inset map). Ample parking is available both below and above the falls (Fig. 1, inset) as the site is partly a provincial and partly a municipal park that attracts a constant flow of visitors. No permission is needed to visit the falls or the park.

The falls are 277 ft (84 m) high, higher than the Niagara Falls. They were named by Samuel de Champlain in 1608 the day before he landed a few miles (kilometers) to the west to found the city of Quebec. The name Montmorency belongs to one of the most ancient and noble families of France, which had sponsored Champlain's ventures in the New World. The last Montmorency duke, Henry II, was the titular Vice-Roy of New France and was beheaded in 1632 for having led the last rebellion of the nobles against the absolutism of Louis XIII and Cardinal de Richelieu.

SIGNIFICANCE

The mid-Ordovician transgression of the Trenton Group onto the Precambrian basement is well exposed at Stops 1 and 2 (Fig. 1, inset map) on the left bank of the Montmorency River above the falls. This is one of the few localities in northeastern North America where this unconformable contact can be examined at ease. At the falls, the fault scarp of the Montmorency Fault is exposed. The fault belongs to a regional system of en echelon faults along which the autochthonous Ordovician stratigraphy of the St. Lawrence Lowlands has been downdropped with respect to the Precambrian basement. The left wall of the gorge, below the falls, shows a magnificent section of SE-tilted clastic sediments belonging to the flysch succession that transgressed on the collapsed or downfaulted Trenton platform at the onset of the

Figure 2. Schematic profile showing relationship of the Trenton Group to the underlying Black River Group and, or, Precambrian gneisses between Pont Rouge and Beaupré. No vertical scale is implied. For details see Harland and Pickerill, 1982, 1984. Modified after Harland and Pickerill, 1984.

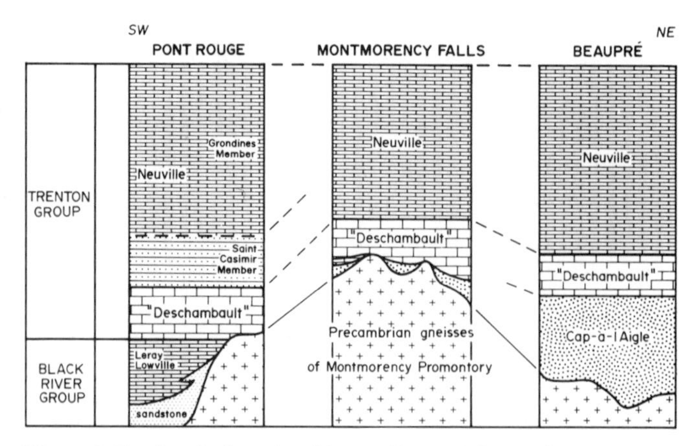

Figure 3. Erosional pinnacle of *in situ* Precambrian gneiss draped and unconformably overlain by thin-bedded, fine-grained limestones of the Trenton Group. Locality 1 of Figure 1 on the left bank of the Montmorency Falls.

Taconic orogeny. A thin layer of Utica Shale separates the Trenton limestone from the flysch beds. This section is best appreciated from the observation platform at the Montmorency House (Manoir Montmorency; Stop 4, Fig. 1, inset).

THE PRECAMBRIAN BASEMENT AND THE MID-ORDOVICIAN OVERLAP CARBONATE SEQUENCE ABOVE THE MONTMORENCY FALLS

In the St. Lawrence Lowlands of Quebec, upper Middle Ordovician strata of the Trenton Group are well exposed from Montreal in the southwest to St. Siméon in the northeast, a distance of approximately 300 mi (500 km). The Trenton Group is essentially a mixed carbonate sequence, and is relatively unfolded and flat-lying; it has been subdivided into a number of stratigraphic units whose nomenclature varies depending on geographic location, but many of which are candidates for synonymy (Harland and Pickerill, 1982). To the southwest of Quebec City, at Pont Rouge (Fig. 1), the Trenton is subdivided into a lower Deschambault Formation and an upper Neuville Formation. To the northeast of Quebec City, the group also includes a basal sequence of arkosic sandstones termed the Cap-à-l'Aigle Formation (Rondot, 1972), which also crop out at the Montmorency Falls (Fig. 2). These sandstones are followed by a heterogeneous assemblage of carbonate sediments that have previously been referred to as the Deschambault Formation, but which in no way resemble the more typical Deschambault strata of the Pont Rouge area (Harland and Pickerill, 1984). For this reason we prefer to refer to this assemblage of lithotopes as "Deschambault" in Figure 2. This assemblage is followed stratigraphically by the Neuville Formation, which is much like the type section of the formation at Neuville (Fig. 1).

North and northeast of Quebec City, the Trenton rests unconformably on a Precambrian (Grenvillian) gneiss basement and represents the deposits of a progressively deepening, marine shelf environment (Riva, 1972; Riva and others, 1977; Harland and Pickerill, 1982). The absence of Cambrian and older Ordovician strata in the area, which form an integral part of the total stratigraphic lower Paleozoic sequence in the Montreal area, suggests that in late Middle Ordovician time the sea onlapped a major topographic feature that has been termed the Montmorency Promontory by Belt and others (1979). At the time of deposition, the Montmorency Promontory probably stood out as a rugged archipelago (Harland and Pickerill, 1984). Within this archipelago, the lowest Trenton sediments (the Cap-à-l'Aigle and "Deschambault" formations) were deposited in relatively high-energy, rocky-shore environments (Harland and Pickerill, 1984) and formed a complicated spatial and temporal lithofacies mosaic. Such sediments, which are clearly exposed at and above the Montmorency Falls, are seldom preserved in the geological record because of their low preservation potential.

Above the Montmorency Falls, the basal sediments of the Trenton Group are seen in unconformable contact with underlying Precambrian gneisses at two locations: on a small ledge immediately alongside the eastern lip of the Falls (Stop 1 of Fig. 1), and a large domal island on the east bank of the Montmorency River, approximately 1,000 ft (300 m) upstream from the falls (Stop 2 of Fig. 1). At both locations (easily accessible through the park pathways), the unconformity is spectacularly exposed (Fig. 3). The Precambrian basement has an essentially eastwest vertical gneissosity and the Trenton strata are virtually horizontal, dipping only a few degrees downstream. At Stop 1, erosional pinnacles and boulders and cobbles of gneiss are surrounded and overlain by *Solenopora* gravels or fine-grained limestones of the "Deschambault" Formation; at Stop 2, the *Solenopora* gravels infill potholes, erosion-enlarged joints developed parallel to the

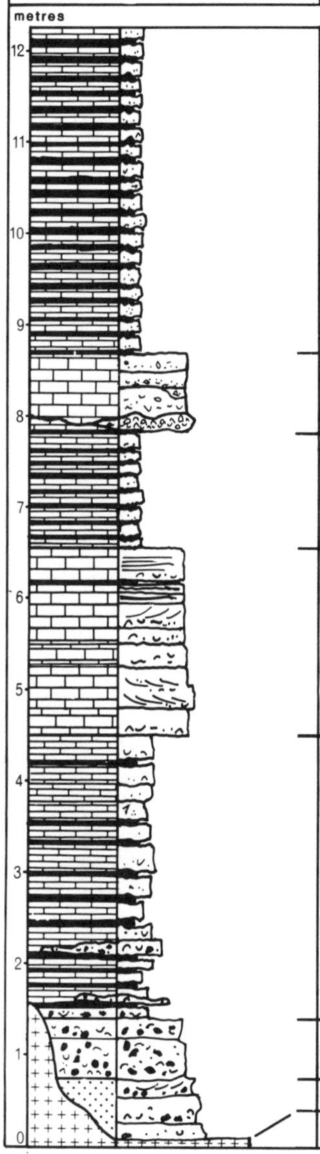

Lithology	Sedimentary facies	Ichnofauna	Biota	Interpretation
metres				
	transitional upward into evenly bedded lime mudstones and shales **Bryozoan limestones** irregularly bedded, with shale partings, abundant delicate bryozoans, "firm-ground" surfaces	intensely bioturbated including *Chondrites*	delicate ramose bryozoans with brachiopods, trilobites (including *Cryptolithus*) and crinoids	Offshore low energy shallow shelf muds
	Coquina limestones grainstones with brachiopod coquinas, scoured base and internal scours	unbioturbated	abundant inarticulate and articulate brachiopods	Subtidal channels
	Bryozoan limestones as above	as above	as above	Offshore low energy shallow shelf muds
	Skeletal limestones poorly cemented grainstones, part indistinctly cross bedded, top part flaggy, abundant reworked bryozoans and *in situ* mats	unbioturbated	*Hallopora* and other ramose bryozoans with brachiopods, crinoids, and gastropods	?ephemeral beach Nearshore high energy skeletal shoal and bar sands
	Fine grained limestones irregularly, partly flat, bedded, shale partings, intraclastic layers and sheets of bryozoans includes lenses of *Solenopora* gravel	*Chondrites, Planolites* & *Palaeophycus*	*Diplotrypa* sheets and other rare bryozoans, crinoids, brachiopods and trilobites (including *Cryptolithus*)	Nearshore mainly low energy sheltered lime muds
	***Solenopora* gravels** sandy matrix and boulders	unbioturbated	abundant, including *in situ Solenopora*	Algal maerls and rock pool clusters
	Arkosic sandstones	unbioturbated	rare, reworked	Marginal clastics
	Precambrian gneisses			

Figure 4. Sedimentological log of the sequence forming the basal part of the Trenton Group at Stop 2 above the Montmorency Falls, with environmental inferences.

gneissosity and shallow depressions on the Precambrian basement. Although most are clearly reworked, several residual hemispherical colonies of *Solenopora* still adhere to the gneiss surface in their position of growth. At low water levels, the flanks of the large domal island of Precambrian gneiss (now stripped of its Ordovician cover except for the *Solenopora* gravels) are onlapped by arkosic sandstones of the Cap-à-l'Aigle Formation, *Solenopora* gravels and thinly-bedded, fine-grained limestones of the "Deschambault" Formation.

The composite stratigraphic log of the sequence forming the basal part of the Trenton Group at Montmorency Falls, together with environmental interpretations, is given in Figure 4 (Harland and Pickerill, 1984). Although six lithofacies can be clearly identified, their environment of deposition in association with rocky shorelines makes them complexly related in space and time. Broadly, however, they form an upward-fining sequence.

Apart from the sparsely fossiliferous, arkosic sandstones, which contain only rare brachiopods such as *Lingula* and *Dinorthis,* bryozoans, and gastropods, the remaining five lithofacies (Fig. 4) are abundantly fossiliferous. Complete faunal lists of individual lithofacies, together with their relative abundance and taphonomy, are given in Harland and Pickerill (1984).

Upstream from the large domal island, for approximately 2,000 ft (600 m) up to the Hydro-Québec dam (Fig. 1), the

Figure 5. View of the Montmorency Falls and of the tilted strata constituting the downfaulted block below the falls, as seen from the observation platform at the Montmorency House (Manoir Montmorency).

Neuville Formation is extensively exposed on the eastern bank of the river and consists of variably interbedded light-, medium-, and dark-coloured, fine- to coarse-grained limestones with thin shale partings. Bedding is regularly developed in the lower part of the formation, but becomes irregular and even nodular in the upper part. The relative proportion of shale increases vertically within the succession, so that the contact with the overlying Utica Shale (seen below the falls) is gradational. Harland and Pickerill (1982) have interpreted the sequence as having been deposited initially in shallow subtidal environments and later in a progressively deepening shelf. Fossils are extremely abundant and well preserved and extensive faunal lists, in addition to those listed above, can be found in a number of publications, to which the reader is referred (Clark and Globensky, 1975, 1976a, 1976b, 1976c; Brun and Globensky, 1977).

THE MONTMORENCY FAULT

The Montmorency Fault is a large normal fault that belongs to a system of en échelon faults extending, along the northern limits of the St. Lawrence Lowlands, from north of Montreal to Quebec City and Montmorency Falls and from there to the Malbaie area, about 62 mi (100 km) to the northeast (Fig. 1). On the southeastern side of this fault system, the Precambrian basement and its cover of mid- to Upper Ordovician sediments have been downdropped and largely preserved to the present time, whereas on the northwest side only the Precambrian basement remains with, locally, a thin cover of mid-Ordovician carbonates. Downdragging accompanied this downfaulting with the result that strata nearest to the fault dip steeply away from it before flattening out. An excellent example of this is the downdropped section exposed on the left wall of the gorge below the Montmorency Falls (Fig. 5) where strata dip from 37° to 45° away from the Montmorency Fault.

The oldest strata in the downfaulted block outcrop all along the ravine to the left of the falls. They consist of about 33 ft (10 m) of thin-bedded, nodular limestone with shaly interbeds belonging to the upper Neuville Formation of the Trenton Group. The occurrence of these beds, coupled with the fact that the base of the Trenton is exposed just over the lip of the falls, allows us to arrive at a rough estimate of the net slip of the fault, which must be at least 820 ft (250 m) by taking into account the thickness of the Trenton Group (490 ft; 150 m), the height of the falls (276 ft; 84 m) and the general dip of the fault (60°). In addition, patches of oxidized fault breccia and slickenside marks on the gneiss surface are ample evidence of the type and direction of movement on the fault plane as are the secondary faults and fractures cutting the Trenton beds adjacent to the fault. The upper Trenton strata can be followed up the ravine to the upper terrace, suggesting that the downdrop along the Montmorency Fault was certainly not uniform.

The Montmorency Falls, the fault itself, and its geology attracted the attention of the first geologists to visit Quebec. In 1841, Ebenezer Emmons identified the limestone lying horizontally on the Precambrian gneiss above the falls with the Trenton Limestone of New York rather than as Carboniferous, as earlier geologists had done on the belief that the graptolites found a few miles (kilometers) away at Levis (Fig. 1) were plant remains. He also interpreted the Montmorency Fault as the probable extension of the big fault which, in New York, separated his Taconic System from the less disturbed strata of the Hudson River Valley. The next year, Charles Lyell visited the falls during his first travels in North America. Lyell (1845), though primarily interested in the post-glacial sediments lying unconformably on the Ordovician strata on the left wall of the gorge and their implication insofar as ancient sea levels were concerned, also studied the rock mass at the falls, which he identified as gneiss and qualified as "truly primary." An entirely different interpretation was reached by Jules Marcou (1891), a Swiss geologist and explorer of the Grand Canyon, who last visited the falls on August 30, 1863, at the end of a dry summer that had reduced the flow of water to a trickle. Marcou concluded, and steadfastly maintained to the end of his life, that the rock mass at the falls was a quartzite that was conformably followed downstream by Emmons' Taconic System. The downfaulted strata on the left side of the gorge he interpreted as a landslide mass of Trenton Limestone and Utica Shale related to the strata above the falls. These conflicting views were largely resolved in the 1850s by the work of W. E. Logan and his co-workers who made a point to study the Quebec area in detail. Logan (1863) traced the Montmorency Fault southwest along the present Boulevard des Chutes to Giffard and northwest towards Chateau Richer, parallel to the course of the St. Lawrence (Fig. 1). Emmons' great fault, or thrust, was recognized as passing along the northern edge of the Island of Orleans, about 2.5 mi (4 km) southeast of the falls (Figs. 1, 5), and Logan mapped its course southwesterly along the frontal edge of the Appalachians to the east side of Lake Champlain and Vermont. This thrust, or combination of thrusts, eventually came to be known in Canada and the United States as Logan's Line.

Recently Putman (1945) mapped the area in the proximity of the falls and published the first modern geologic map. Riva (1972) traced the Montmorency Fault to the north of Chateau Richer, where it dies out to be replaced by the Chateau Richer Fault (Fig. 1). This fault extends for about 12 mi (20 km) to the north of Beaupré where, however, the parallelism among en échelon faults is lost and the succeeding fault diverges at about 45° in the direction of the St. Lawrence (Fig. 1F).

FLYSCH SUCCESSION BELOW THE MONTMORENCY FALLS

The Trenton beds beyond the observation platform at the foot of the falls (which is perennially bathed by the spray from the falls) gradually pass up into about 105 ft (32 m) of brownish-black shale, which is identical to the Utica Shale of the Mohawk Valley of New York. These transitional Trenton-Utica beds contain graptolites of the *Orthograptus ruedemanni* zone of Riva (1969) and the Utica Shale itself yields abundant graptolites of the lower *Climacograptus spiniferus* zone. The transitional Trenton-Utica passage is poorly exposed here because of tectonic deformation; it is, however, well exposed on the left bank of the Jacques Cartier River, below the village of Pont Rouge (Riva, 1969, p. 527–29, Fig. 7a) about 28 (45 km) to the southwest (Fig. 1). The regional implication of this gradational passage and the unconformity that marks the Trenton-Utica contact elsewhere in NE North America have been discussed elsewhere (Riva, 1969; Clark and others, 1972, p. 51–52.

The Utica Shale is followed upwards abruptly by a thick clastic sequence consisting of thin- to medium-bedded graywacke, flaggy sandstone, siltstone, and shale with a few tan-weathering dolomitic beds near the top. This sequence occupies almost the entire cliff, from the platform at the foot of the falls to the highway to the south (Fig. 5). It is about 820 ft (250 m) thick. For a long time, namely since Logan (1863), it was believed to be the extension of the Utica-Lorraine formations in the Quebec area, but in 1969 Riva (p. 531) showed that this "Lorraine" yielded graptolites of the *C. spiniferus* zone and was, therefore, both older and distinct from the typical Lorraine, which yields graptolites of the much younger *Paraclimacograptus manitoulinensis* zone and a plethora of shelly fossils not found at Montmorency Falls (Walters, 1977). This led to a broad revision of the mid-Ordovician clastic stratigraphy of the Quebec area by Belt and Riva (Belt and others, 1979) and the recognition of two new stratigraphic units in the area between Montmorency Falls and Beaupré. The Utica Shale was restricted to the brownish-black shale at the foot of falls, and the thick graywacke beds overlying the Utica were assigned to the Beaupré Formation, a unit well developed at Beaupré. The bulk of the clastic section was assigned to the newly-redefined Lotbinière Formation, a unit first recognized by Clark and Globensky (1975) to the southwest of Quebec City at Lotbinière (Fig. 1), where it intervenes between the Utica and the Lorraine Group (Nicolet River Formation). The Lotbinière section at the Montmorency Falls, being much better exposed than the type section at Lotbinière, is now regarded as an excellent reference section for the formation.

Belt and others (1979) also proposed a new interpretation of the sedimentological history of the area. The disappearance of the carbonates of the Trenton Group and their replacement by the black Utica Shale was attributed to the downwarping or collapse of the outer margins of the Laurentian Platform caused by the approach of the Taconic allochthon from the southeast. This collapse was followed first by the deposition of the typical oceanic sediments of the Utica Shale and then by an influx of flysch sediments (Beaupré and Lotbinière Formations) derived from the erosion of the advancing Taconic masses. The Beaupré Formation itself was interpreted as having formed a large submarine fan with its center at Beaupré; only its western margins are represented in the section at the Montmorency Falls.

The flysch sedimentation ceased at the end of Middle Ordovician time with the final emplacement of the Taconic allochthon, and was followed by the prodeltaic regressive sediments of the Lorraine Group of Late Ordovician age, which are only preserved in the axial part of the St. Lawrence Lowland southwest of Quebec City. These prodeltaic sediments gradually filled the shallowing basin that had formed northwest of the newly-emplaced Taconic Mountains, and were later followed by the continental red shale and sandstone of the Queenston Group. None of these units are now preserved at the Montmorency Falls or anywhere northeast of the city of Quebec.

REFERENCES CITED

Belt, E. S., Riva, J., and Bussières, L., 1979, Revision and correlations of late Middle Ordovician stratigraphy northeast of Quebec City: Canadian Journal of Earth Sciences, v. 16, p. 1467–1483.

Brun, J., and Globensky, Y., 1977, Stratigraphy, petrography, and sedimentology of the Mohawkian of the Laurentian Platform (Joliette and Quebec areas): 69th Annual Meeting of the New England Intercollegiate Geological Conference, Laval University, Quebec, Excursion A7, 33 p.

Clark, T. H., and Globensky, Y., 1975, Grondines area: Quebec Department of Natural Resources Geological Report 154, 159 p.

——, 1976a, Laurentides area: Quebec Department of Natural Resources Geological Report 157, 112 p.

——, 1976b, Sorel area: Quebec Department of Natural Resources Geological Report 155, 151 p.

—— 1976c, Trois-Rivières area: Quebec Department of Natural Resources Geological Report 164, 87 p.

Clark, T. H., Globensky, Y., Riva, J., and Hofmann, H., 1972, Stratigraphy and structure of the St. Lawrence Lowland of Quebec: International Geological Congress, 24th Session, Guidebook for Excursion 52, 82 p.

Emmons, E., 1841, Geology of the Montmorenci: American Journal, November issue, p. 146–150. (Reprinted 1888, American Journal, v. 2, p. 94–100.)

Harland, T. L., and Pickerill, R. K., 1982, A review of Middle Ordovician sedimentation in the St. Lawrence Lowland, eastern Canada: Geological Journal, v. 17, p. 135–156.

——, 1984, Ordovician rocky shoreline deposits; The basal Trenton Group around Quebec City, Canada: Geological Journal, v. 19, p. 271–298.

Logan, W. E., 1863, Geology of Canada; Report of progress from its commencement to 1863: Geological Survey of Canada, Annual Report Series, 983 p.

Lyell, C., 1845, Travels in North America in 1841–1842: New York, Wiley and Putnam, 221 p.

Marcou, J., 1891, Geology of the environs of Quebec, with maps and sections: Proceedings of the Boston Society of Natural History, v. 25, p. 202–227, plates 7–9.

Putman, H. M., 1945, Discontinuité tectonique de Montmorency entre le ruisseau Lottainville (Petit Pré) et Loretteville: Naturaliste Canadien, v. 72, p. 289–308.

Riva, J., 1969, Middle and Upper Ordovician graptolite faunas of the St. Lawrence Lowlands of Quebec and of Anticosti Island, *in* Kay, M., ed., North Atlantic: Geology and continental drift: American Association of Petroleum Geologists Memoir 12, p. 513–556.

——, 1972, Geology of the environs of Quebec City: 24th International Geological Congress, Montreal, Guidebook for Excursion B19, 53 p.

Riva, J., Belt, E. S., and Mehrtens, C. J., 1977, The Trenton, Utica and flysch succession of the platform near Quebec City, Canada: 69th Annual Meeting of the New England Intercollegiate Geological Conference, Laval University, Quebec, Guidebook for Excursion A8, 37 p.

Rondot, J., 1972, La transgression ordovicienne dans le comté de Charlevoix, Quebec: Canadian Journal of Earth Sciences, v. 9, p. 1187–1203.

Walters, M., 1977, Middle and Upper Ordovician graptolites from the St. Lawrence Lowlands, Quebec, Canada: Canadian Journal of Earth Sciences, v. 14, p. 932–952.

Geological Society of America Centennial Field Guide—Northeastern Section, 1987

The Oak Hill Group, Richmond, Québec: Termination of the Green Mountains–Sutton Mountains Anticlinorium

Robert Marquis, Jacques Béland, and Walter E. Trzcienski, Jr., Département de Géologie, Université de Montréal, C.P. 6128, Succursale "A", Montréal, Québec H3C 3J7, Canada

LOCATION

The section of late Precambrian to early Paleozoic geology near Richmond, Québec, Canada, is well exposed in new road cuts created by recent construction of Québec 55 from Sherbrooke to Drummondville in the Eastern Townships of Québec (Fig. 1). All exposures of interest are to be found along the highway, to which one can gain access at Drummondville, Richmond, or Sherbrooke (topographic map sheet 31H/9, Richmond, Québec). Heading north from Vermont, Vermont I-91 becomes Québec 55 upon entering Québec. The road is a major two-lane highway with high-speed traffic, but outcrops are 45 to 90 ft (14 to 27 m) from the road surface. *Be careful,* nonetheless. Generally there is no problem in gaining access to the outcrops as they are on a provincial highway; however, if one has any questions there is a Québec Provincial Police detachment located just off Québec 116 in Richmond. As a number of different road cuts will be used to characterize the geology, the site description will start at the east end of the section and continue northwestward on Québec 55. The eastern-most point is an outcrop of serpentinite (Stop 1) that lies 5 mi (8 km) east of the intersection of Québec 55 and 116, the highway that leads into the village of Richmond off of Québec 55. This starting point, if one is driving from Sherbrooke, is 14.7 mi (23.6 km) northwest of the Québec 10 overpass of Québec 55.

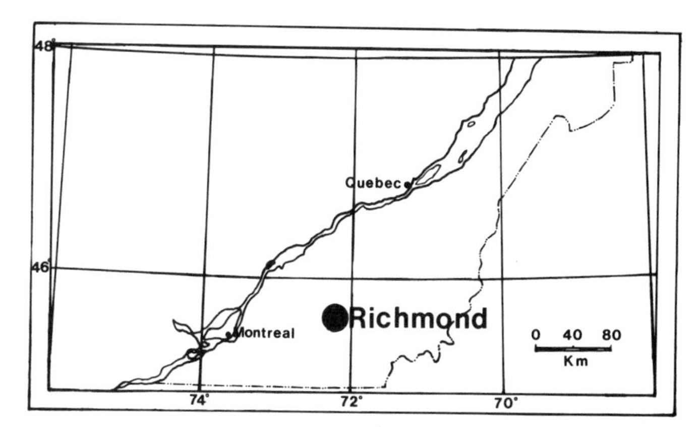

Figure 1. Regional location map of the Richmond area, Québec.

SIGNIFICANCE

The most recent synthesis of the southern Québec Appalachians, using a plate tectonic model, is that provided by St. Julien and Hubert (1975). According to the model, Grenville-like basement began rupture in late Precambrian time, giving rise to oceanic crust with its associated mafic and ultramafic rocks to the east and a continental shelf, slope, and rise sequence to the west. By late Early Ordovician time, extension and associated processes ceased. Deformation that followed closed the ocean created earlier and produced a sequence of imbricated nappes and obducted oceanic crust, all by the end of Middle Ordovician time. The remnant crust of this Ordovician ocean (Iapetus) is present along the Baie Verte–Brompton Line (BBL; Fig. 2), an alignment of ultramafic rocks that is interpreted (Williams and St. Julien, 1982) to represent an early Paleozoic continent-ocean interface. Rocks lying to the west of the BBL are those produced along the rifted margin, and then tectonically emplaced during ocean closure. These margin rocks are today exposed in a large antiformal structure (Fig. 3), the Green Mountains–Sutton Mountains Anticlinorium (Osberg, 1965), which plunges toward the northeast.

In an interval of 10 mi (16 km; 5 mi [8 km] on either side of Richmond), along Québec 55, one sees remnant oceanic crust and the metamorphosed equivalents of the volcanic and sedimentary rocks that bordered the ancient ocean to the west. The processes that led to the present configuration have left their signature on these rocks, and in the discussion below some of the observations and arguments used in interpreting the processes will be presented.

SITE INFORMATION

The first detailed geologic description and synthesis of southern Québec was that provided by Clark (1934, 1936) in which he recognized four major thrust slices, bordered to the east by the Memphremagog Complex. Osberg (1965) carried out a structural analysis of the Knowlton-Richmond Area in which he described the complexity of Clark's Oak Hill Slice (Fig. 2). The scenario leading to the assembling of these rocks in their present position has been discussed by St. Julien and Hubert (1975); a more detailed stratigraphic, structural, and metamorphic study of the Richmond area is currently being undertaken by the authors.

Stop 1. Oceanic Crust. The oceanic crust (5 mi [8 km] east of Richmond) is represented here by massive serpentinite. Originally an igneous rock composed primarily of olivine and pyroxene, the serpentinite is presently composed of serpentine minerals, mainly antigorite. Some disseminated chromite is also present. This serpentinization has resulted from the hydration of the original minerals at a relatively low temperature. Along with hydration, the original rock has been highly deformed, as seen in the outcrop where numerous slickensided surfaces are visible.

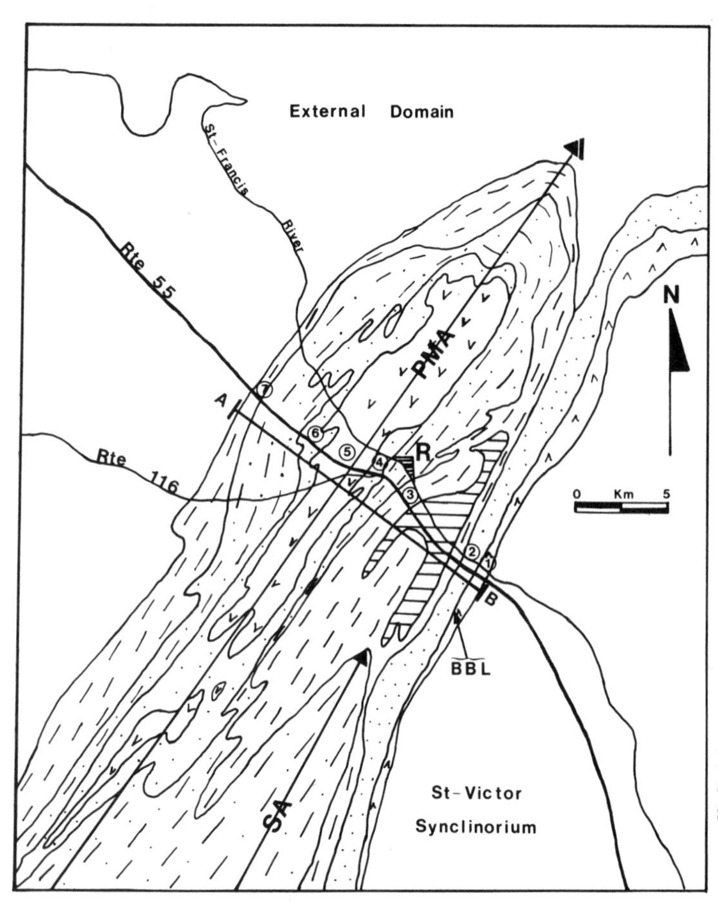

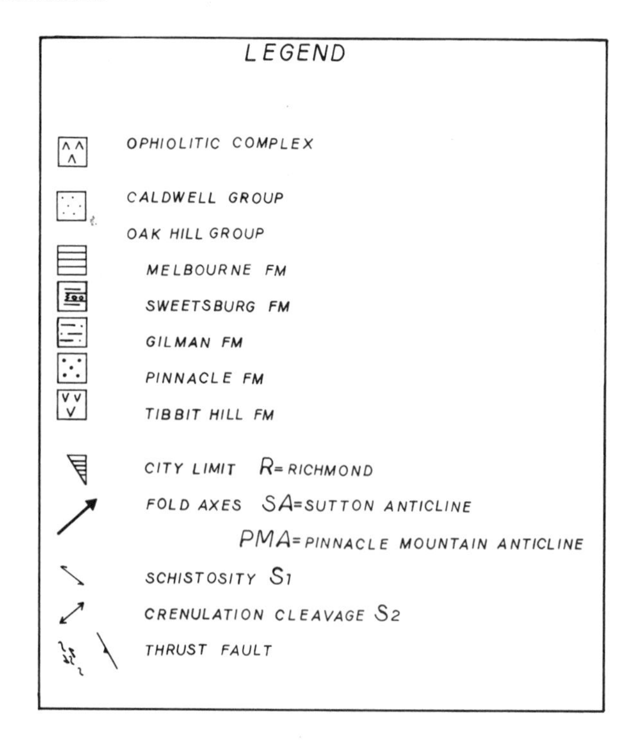

Figure 2. Geologic map of the Richmond area, Québec (modified after Osberg, 1965; St. Julien and Hubert, 1975) with designated stops described in the text.

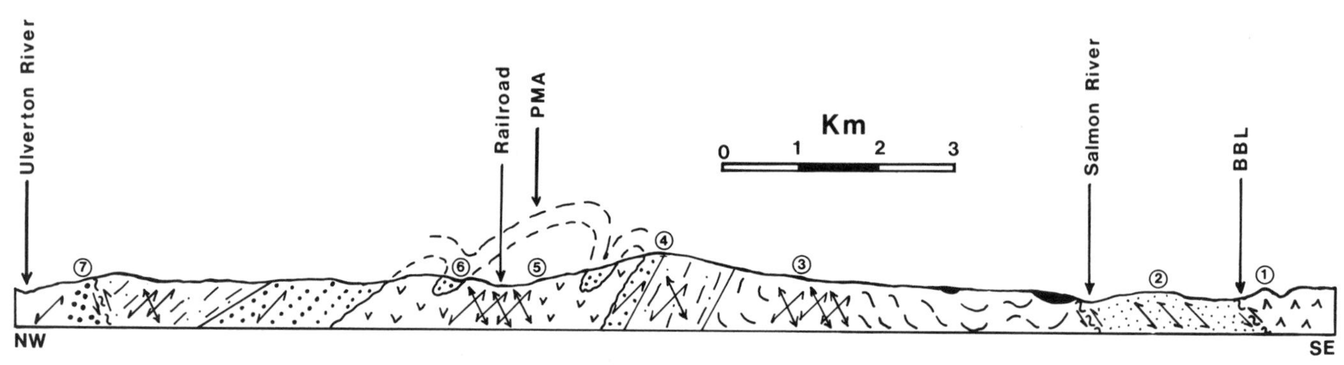

Figure 3. A geological cross section through the Richmond area, Québec (line AB, Fig. 2). Legend is the same as Figure 2 with addition of the Dunham Formation at the NW end of the section (large dots). Stops numbered as in Figure 2.

Because of this intense deformation, much of the original texture of the igneous rock has been lost. The major observable structure is a regional cleavage that is better seen in rocks farther to the west.

Stop 2. Caldwell Formation. Commonly, the serpentine rocks in the Eastern Townships have been interpreted (St. Julien and Hubert, 1975) to have been obducted onto the continental margin rocks by westward transport on an interbedded sequence of sandstones, shales, and volcanic rocks. This sequence, the Caldwell Formation, is the northward extension of the Stowe Formation in Vermont. Where seen (4.3 mi [6.9 km] east of Richmond), the Caldwell is a series of green, quartz-rich sandstones (some of the quartz grains are blue) that uncommonly show graded bedding with tops facing to the east. The volcanic rocks are low-grade greenstones containing albite, chlorite, and quartz as the most common metamorphic assemblage. In places the greenstones contain greenish spots that in thin section are seen to be rosettes of green stilpnomelane. Minor beds of carbonate in

Figure 4. Stop 3, Sweetsburg Formation with F_1 folding overturned toward the SE. Refraction of the S_1 axial planar schistosity is not evident at this scale. Bar scale is 6 ft (1.8 m) long.

the Caldwell weather to a buff color. Red and green slates, although not common here, elsewhere form a major part of the Caldwell Formation. The Caldwell probably represents sedimentary conditions varying from relative quiescence to high turbulence on a continental margin. A northeasterly trending regional cleavage is seen everywhere. In the slates especially, a late slip cleavage postdates the more prevalent regional cleavage. Several generations of quartz veinlets are observed, some of which are commonly folded.

Oak Hill Group (Stops 3 to 7). In 1934, Clark introduced the name Oak Hill Slice to describe a fault-bounded, lithostratigraphic sequence outcropping in the Sutton area, 10.5 mi (16.9 km) north of the Québec-Vermont border. Two years later he presented the type localities and gave a full description of each formation in the sequence (Clark, 1936). The northeastern extension of these rocks into Shefford and Brome counties of the Eastern Townships of Québec was assigned to the Oak Hill Group by Cooke and others (1962). Booth (1950) recognized what he called the "Oak Hill Succession" in Vermont, where it occupies the region between the eastern margin of the Champlain Lowland and the western limit of the Green Mountains. Some of the units were found to be traceable southward into west-central Vermont. In Québec the Oak Hill Group strikes northeasterly for 60 mi (100 km) from the Vermont border to the Richmond map area. The rocks outcrop mainly on the west flank of a major structural element, the Pinnacle Mountain Anticline (PMA, Fig. 2), which is equivalent to the Lincoln–Enosburg Falls Anticline in Vermont (Osberg, 1967). The outcrops described below are located near the northeastern extremity of the anticline where the formations can be seen on both flanks of the structure (Fig. 3).

Stop 3. Sweetsburg Formation. Spectacular outcrops (1.2 mi [1.9 km] east from Québec 55-116 intersection) display black phyllites interlayered with thin, buff-colored, dolomitic sandstone beds. This assemblage is characteristic of the Sweetsburg Formation in the Richmond area as well as in the Sutton area (Charbonneau, 1980). In Vermont, equivalent rocks have been called the Ottauquechee Formation (Christman, 1959). At this stop the bedding is tightly folded. The axial plane of the fold shown in Figure 4 strikes northeasterly and dips 60° northwest; it is parallel to the regional schistosity and overturned toward the southeast, as are the folds in the Pinnacle and Gilman formations described below. Thus, the entire anticlinal structure is overturned to the southeast, contrary to the assumed transport direction of the "Québec Allochthon" (St. Julien and Hubert, 1975). On the mesoscopic scale, one observes the refraction of the penetrative schistosity where it crosscuts centimeter-wide sandstone beds. This early schistosity also exhibits small corrugations. These minute crinkles are produced by the intersection of the schistosity with a late crenulation cleavage that strikes northeasterly and dips 80° southeast.

Melbourne Formation. The Melbourne Formation is a black, highly graphitic, crystalline limestone that overlies the Sweetsburg Formation. It may be equivalent to the thin-bedded

Figure 5. Stop 4, Pinnacle Formation showing an angular relationship between bedding, S_0, and schistosity, S_1, on the overturned limb of the PMA. Bar scale is 3 ft (0.9 m) long.

graphitic limestone of the Ottauquechee Formation (Christman, 1959). Although there is no good outcrop of the Melbourne along Québec 55, it may be seen in road cuts at the type locality (Cooke, 1952) in the village of Melbourne. The limestone unit overlies conformably the Sweetsburg Formation. Along a small brook south of the village, a gradational contact between Sweetsburg and Melbourne can be seen. From this relationship, the Melbourne Formation, not the Sweetsburg Formation, should be considered the uppermost unit of the Oak Hill Group (Cooke, 1952).

Stops 4 and 6. Pinnacle Formation. The Pinnacle Formation of the Sutton area was originally described by Clark (1936) as a lithic graywacke, commonly containing magnetite-rich beds varying in thickness from several millimeters to 4 in (10 cm). In the Richmond area this lithologic type represents less than 10% of the Pinnacle Formation. Instead, it consists mainly of a white and green sandstone with some arkosic beds and dark bluish phyllite. Near the base, a green schist is interstratified with the sandstone. Close to the contact with the underlying Tibbit Hill Formation, a fine-grained chloritoid schist also occurs in many places. This chloritoid schist could be the metamorphic equivalent of Clark's (1936) Call Mill Slate. However, chloritoid schist occurs also within the Tibbit Hill and the Pinnacle formations. At Stop 4 (interchange of Québec 55 and 116) the early schistosity is spectacularly refracted across the sandstone layers (Fig. 5). The bedding-schistosity relationship shows that the outcrop is on the overturned flank of the PMA (Fig. 3). Note also the presence of highly folded thin layers of a soft, yellowish sericite-chlorite schist (Fig. 6), which might represent horizons of felsic tuff.

The contact between the Pinnacle and the Gilman formations is exposed here as well, near the eastern end of the cliff on the southeast side of Québec 55. The rocks of the Gilman Formation, as seen here, are green chlorite-sericite phyllites with a highly penetrative schistosity. Cubic pyrite grains on the schistosity planes impart a spotted look to the rock. Interlayered with the phyllites are thin beds of dolomitic sandstone, now disrupted into centimeter-long lenses parallel to the regional schistosity. The folding related to the regional schistosity is not easily observed here because of the intense shearing and slippage along the schistosity planes. However, along Québec 116, 3.8 mi (6 km) southwest of the Québec 55-166 intersection, folds can be seen whose axial planes are parallel to the penetrative schistosity. Here again, folding is overturned towards the southeast. Axial planes dip 50° northwest and fold axes plunge slightly to the southwest.

At Stop 6 (1.8 mi [2.9 km] west of Québec 55-116 interchange) an anticlinal structure exhibits a strong contrast in the manner by which competent sandstone layers and incompetent dark bluish phyllite layers are folded (Fig. 7). It is also evident at this locality that the folds are overturned to the southeast, which is again contrary to the overall northwesterly transport direction assumed for this segment of the Appalachian fold belt.

Stop 5. Tibbit Hill Formation. The Tibbit Hill volcanic rocks constitute the basal unit of the Oak Hill Group. Although no good radiometric age data are available, the rocks are considered to be either late Precambrian or Early Cambrian. This age is

Figure 6. Stop 4, tight folding of a phyllitic unit between seemingly little-deformed sandstone strata in the Pinnacle Formation. Bar scale is 6 ft (1.8 m) long.

based on fossil evidence from overlying sedimentary rocks in which Cambrian fossils are known (Clark, 1936; Clark and McGerrigle, 1944; Shaw, 1954).

Interpreted as rift facies volcanics (0.6 to 0.8 mi [1 to 1.3 km] west of the Québec 55-116 interchange), the rocks vary from well-cleaved green phyllites to amygdaloidal metabasalt. Petrographically the rocks are most commonly a fine-grained chlorite-albite-amphibole-epidote-quartz assemblage. The amphibole varies from plain actinolite to a sodic-rich blue amphibole interpreted to indicate a moderately high pressure of metamorphism (Trzcienski, 1976). Some of these blue, sodic amphiboles contain cores of kaersutite indicating an alkali basalt parentage. A recent geochemical study of the Tibbit Hill metavolcanics in the Richmond area (Pintson and others, 1985) is consistent with this interpretation.

Epidote is ubiquitous in the rock matrix but is most spectacularly developed as large segregations in the greenstone (Fig. 8). In the most schistose parts of the Tibbit Hill Formation, the early regional schistosity is well displayed. This schistosity is commonly affected by a late crenulation cleavage, axial planar to minor folds that can be observed locally. Other than the uncommon kaersutite cores of some amphiboles, no other original igneous characteristics are presently found in the Tibbit Hill Formation.

Stop 7. Dunham Formation. About 4.5 mi (7.2 km) west of the Québec 55-116 interchange on the northwest flank of the anticline, a thin unit of quartz conglomerate is well exposed at the

Figure 7. Stop 6, a minor fault and fold on the upright flank of the PMA. Bar scale is 5 ft (1.5 m) long.

Figure 8. Stop 5, epidote nodules in a schistose part of the Tibbit Hill volcanics. The scale is 7 in (17.8 cm) long.

northwest extremity of the outcrop. This conglomerate and the sandstone that constitutes the majority of the outcrop belong to the Dunham Formation (Clark, 1936; Cooke and others, 1962). A dolomitic matrix imparts a distinctive brown coloration to the conglomerate.

REFERENCES CITED

Booth, V. H., 1950, Stratigraphy and structure of the Oak Hill Succession in Vermont: Geological Society of America Bulletin, v. 61, p. 1131–1168.

Charbonneau, J. M., 1980, Region de Sutton (W): Ministere de l'Energie et Ressources, Québec, DPV 681, 89 p.

Christman, R. A., 1959, Geology of the Mount Mansfield Quadrangle, Vermont: Vermont Development Commission, Bulletin 12, 75 p.

Clark, T. H., 1934, Structure and stratigraphy of southern Québec: Geological Society of America Bulletin, v. 45, p. 1–20.

——1936, A Lower Cambrian Series from southern Québec: Transactions of the Royal Society of Canada, v. 21, pt. 1, p. 135–151.

Clark, T. H., and McGerrigle, H. W., 1944, Oak Hill Series, Farnham Series and Phillipsburg Series, *in* Geology of Québec: Ministere des Richesses Naturelles du Québec, RG-20, v. 2, p. 386–407.

Cooke, H. C., 1952, Geology of parts of Richmond and Drummondville map areas, Eastern Townships, Québec: Ministere des Richesses Naturelles, DP. 467, 45 p.

Cooke, H. C., Eakins, P. R., and Tiphane, M., 1962, Shefford map area, Shefford and Brome counties, Eastern Townships, Québec: Ministere des Richesses Naturelles, DP 187, 145 p.

Osberg, P. H., 1965, Structural geology of the Knowlton–Richmond area, Québec: Geological Society of America Bulletin, v. 76, p. 223–250.

——1967, Lower Paleozoic stratigraphy and structural geology, Green Mountain-Sutton Mountain Anticlinorium, Vermont and Québec, American Association of Petroleum Geologists Symposium, Memoir 12, p. 687–700.

Pintson, H., Kumarapeli, P. S., and Morency, M., 1985, Tectonic significance of the Tibbit Hill volcanics, Geochemical evidence from Richmond, Québec, *in* Current Research, Part A: Geological Survey of Canada Paper 85-1A, p. 123–130.

St. Julien, P., and Hubert, C., 1975, Evolution of the Taconian orogen in the Québec Appalachians: American Journal of Science, v. 275A, p. 337–362.

Shaw, A. B., 1954, Lower and lower Middle Cambrian faunal succession in northwestern Vermont: Geological Society of America Bulletin, v. 65, p. 1033–1046.

Trzcienski, W. E., Jr., 1976, Crossitic amphibole and its possible tectonic significance in the Richmond area, Québec: Canadian Journal of Earth Sciences, v. 13, p. 711–714.

Williams, H., and St. Julien, P., 1982, The Baie Verte-Brompton Line; Early Paleozoic continent-ocean interface in the Canadian Appalachians, *in* St. Julien, P., and Beland, J., eds., Major structural zones and faults of the northern Appalachians: Geological Association of Canada Special Paper 24, p. 177–207.

ACKNOWLEDGMENTS

Comments by D. Roy on an earlier version of the manuscript were most helpful. Financial support from the Ministere de l'Energie et Ressources, Québec, to R. Marquis and J. Beland and from the Natural Sciences and Engineering Research Council and the Geological Survey of Canada to W. E. Trzcienski, Jr., are gratefully acknowledged.

Geological Society of America Centennial Field Guide—Northeastern Section, 1987

Thetford Mines ophiolite, Québec

Roger Laurent, *Department of Geology, Université Laval, Québec, Québec G1K 7P4, Canada*
Brewster Baldwin, *Department of Geology, Middlebury College, Middlebury, Vermont 05753*

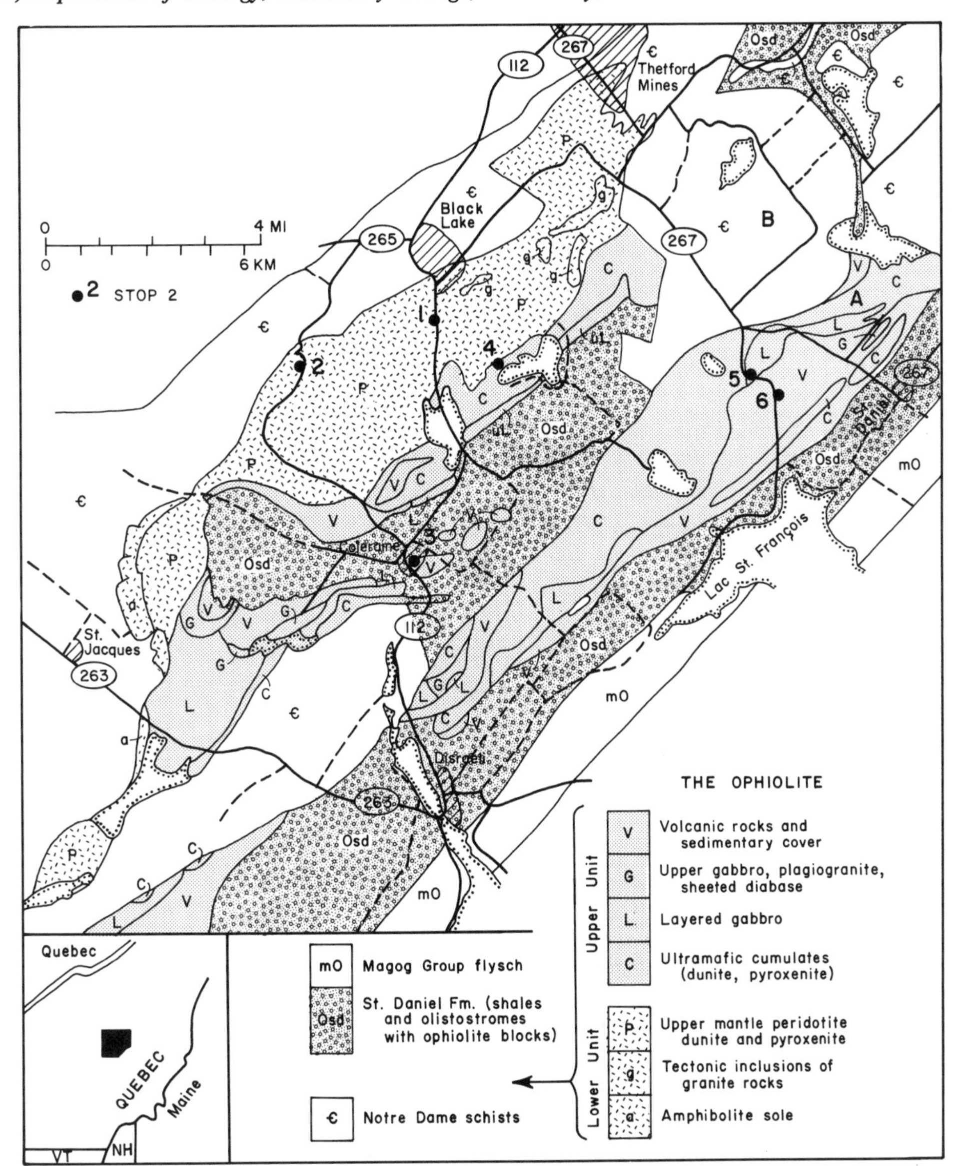

Figure 1. Roads and geology in the Thetford Mines area A, Mont Adstock; B, Bécancour Dome.

LOCATION AND ACCESSIBILITY

Thetford Mines is 50 mi (80 km) due south of Québec City (Fig. 1). The road log for each stop begins (0.0 mi; 0.0 km) at the turnoff from Québec 112, a given distance from the junction of Québec 112 and Québec 267 in Thetford Mines.

Stop 1. Scenic View—Mine Lac d'Amiante. Park on the shoulder of Québec 112 6.7 mi (10.7 km) south of the junction, headed south.

Stop 2: Lower Unit at Vimy Ridge Mine. The turnoff is 11.0 mi (17.7 km) south of the junction and 0.2 mi (0.3 km) south of the north edge of Coleraine, by a grocery store and hotel. 0.0, turn northwest toward the community of Vimy Ridge. At 4.9 mi (7.8 km), just beyond the hilltop, slow and park near the fire hydrant on right, opposite the highest of three houses.

Stop 3: Coleraine Breccia. The turnoff is 11 mi (18.4 km) south of the junction, opposite the Hotel de Ville (city hall) of Coleraine. Drive 0.09 mi (0.15) km northeast up Messel Avenue and turn right into the far end of the cemetery.

Stop 4: Chromite Mine, Base of Upper Unit. The turnoff is 8.8 mi (14.2 km) south of the junction (1 mi; 1.6 km north of a railroad crossing). 0.0, turn east, cross the railroad tracks, and

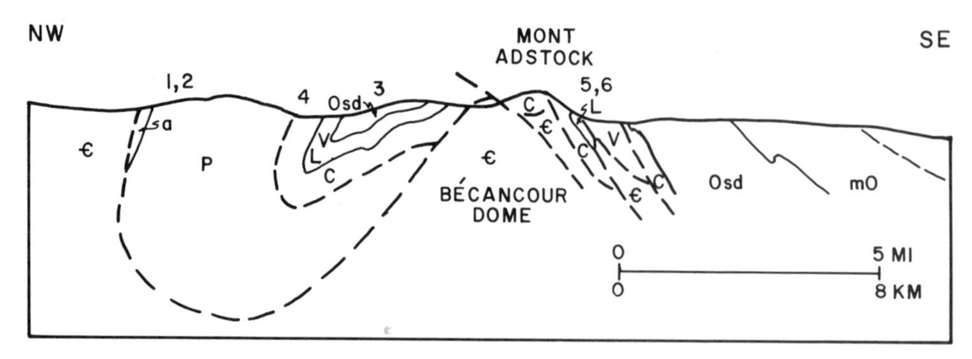

Figure 2. Diagrammatic cross section. The numbered stops indicate the represented parts of Figure 1. Key for cross section given on Figure 1. (Modified from Laurent, 1975.)

turn sharp right. At 0.8 mi (1.3 km), turn left and go northeast uphill. At 3.0 mi (4.8 km), take the left fork, heading northwest. At 4.6 mi (7.4 km), continue straight, past a road on the right. At the shore of Lake Caribou, turn sharp left. At 4.8 mi (7.7 km), turn sharp right, along the shore. At 5.8 mi (9.3 km), park near the entrance to a private drive on left. Walk in; ask permission to walk across small bridge.

Stop 5: Sheeted Intrusives in Gabbro. 0.0 is the Québec 112/267 junction. Drive southeast on Québec 267. At 6.9 mi (11.1 km), take the right fork, leaving Québec 267. At 7.9 mi (12.7 km), park on a left curve, opposite the cliff.

Stop 6: Boninitic Pillow Basalt. At 0.6 mi (1 km) past Stop 5, and after a sharp right curve park opposite the broad low outcrop.

SIGNIFICANCE OF LOCALITY

The Thetford Mines ophiolite and associated olistostromes mark the tectonic boundary between the eastern cratonic margin of proto–North America and accreted terranes of the Appalachian orogen. This boundary, the Baie Verte–Brompton Line (Williams and St. Julien, 1982) extends discontinuously from Baie Verte in northwestern Newfoundland to Brompton Lake near the international border in southern Québec. A string of slivers of ophiolites and olistrostromes, younging to the southeast, occurs along the boundary. The ophiolites are "flakes" of ancient upper mantle and ocean crust. Of these, the Thetford Mines ophiolite is especially interesting because the sequence is accessible and essentially complete; it is not severely dismembered (Fig. 2). Associated mineral deposits include asbestos, chromite (Kacira, 1982), copper sulfides, and noble metals (Oshin and Crockett, 1982). Baldwin and others (1985) discuss the history of understanding of the Thetford Mines area.

SITE INFORMATION

As a result of extensive work done in the Thetford Mines area (for example, St. Julien and Hubert, 1975; Laurent, 1975, 1980; Hebert, 1983), the geologic evolution can be outlined. The Notre Dame schists (Caldwell and Rosaire formations) occur northwest of the Baie Verte–Brompton Line. These had accumulated as muds and sands on the tectonically quiet continental margin of proto–North America, on the northwest side of the proto-Atlantic Ocean (Iapetus).

The Thetford Mines ophiolite is about 5 mi (8 km) thick (Fig. 3). The lower unit, 2.8 mi (4.5 km) thick, formed as a peridotite tectonite. The upper unit formed as a stratiform sequence of about 1.4 mi (2.2 km) of ultramafic cumulates and layered gabbros; these plutonic rocks are capped by 0.8 mi (1.3 km) of pillow lavas, volcaniclastic tuff, breccia, red chert,

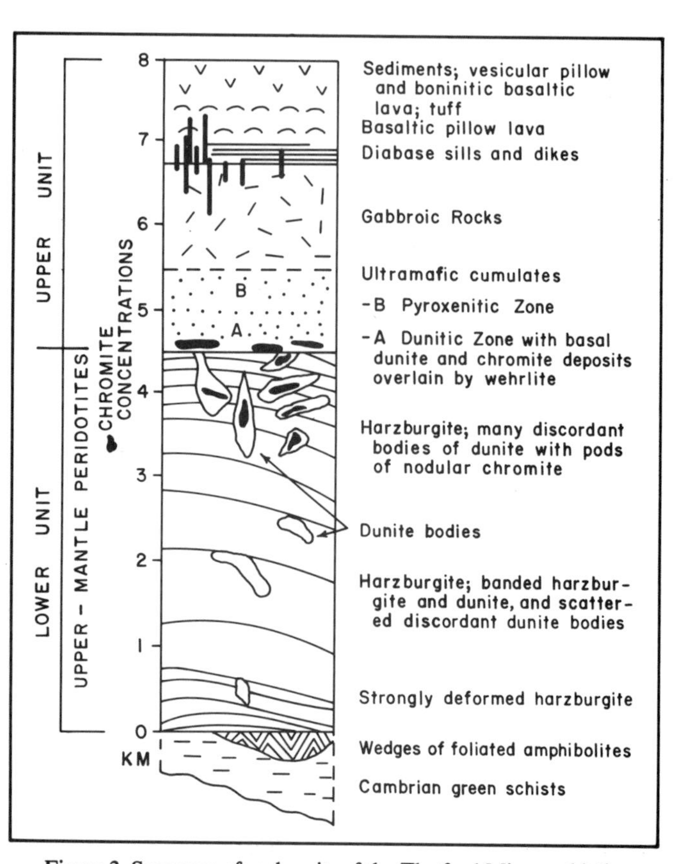

Figure 3. Sequence of rock units of the Thetford Mines ophiolite.

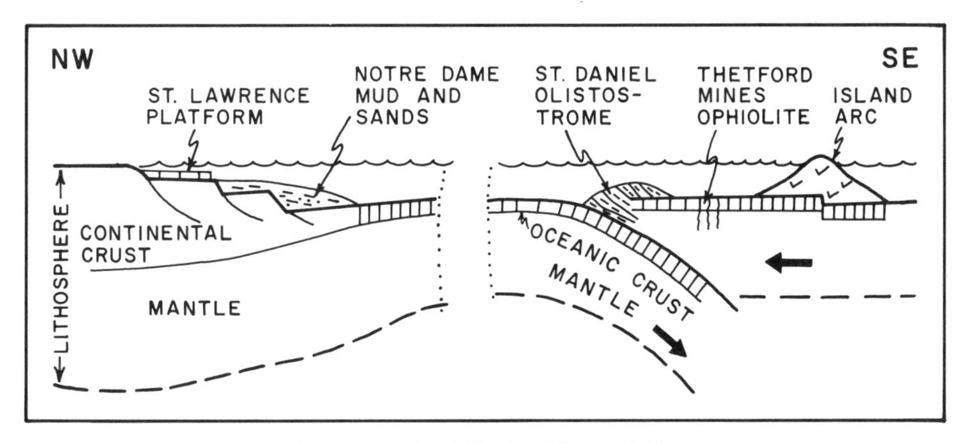

Figure 4. Origin of Thetford Mines ophiolite.

argillite, and siltstone. Most of the ophiolite was generated along a Cambrian ridge of the proto-Atlantic Ocean. When the ocean began to close, the proto–North American plate subducted southeastward, generating an island arc (Laurent, 1975). Lavas of the upper unit are boninitic and were extruded through the Thetford Mines ophiolite along fractures associated with block faulting of the forearc basement (Fig. 4).

When the ophiolite was sheared off the oceanic lithosphere, it generated an amphibolite sole (Feininger, 1981) that is dated at 491 Ma (see Clague and others, 1985). The lower unit sheared off at the base of the serpentinized peridotite. Late in the Ordovician, the arc-continent collision—the Taconic orogeny—obducted the Notre Dame schists and the ophiolite northwestward onto the craton as a series of nappes and thrusts (St. Julien and Hubert, 1975).

Peridotite sheets in the Notre Dame schists occur on thrust faults, and the ophiolite is separated from the underlying schists by a high-angle east-dipping fault. The schists and ophiolite are flanked on the southeast by the Ordovician St. Daniel Formation, an olistostrome that contains pieces of both sedimentary and ophiolitic rocks. The St. Daniel is overlapped on the southeast by the Middle Ordovician Magog flysch; in turn, the flysch is flanked by the volcanics of the Ascot-Weedon Formation (Figs. 1, 2).

STOP DESCRIPTIONS

Stop 1: Scenic View—Mine Lac d'Amiante. The openpit asbestos mine is in serpentinized peridotite of the lower unit. The contact with the underlying Cambrian metagraywackes (Notre Dame schists) is a vertically dipping shear zone on the north side of the pit. The peridotite is mostly harzburgite (65%–90% olivine, 10%–35% orthopyroxene) and dunite (olivine 90%–100%).

Across the pit are light-colored tectonic inclusions of foliated and partly rodingitized granodiorite and granite. These rootless rocks occur only in the lower unit (Laurent and others, 1984). The granite was deformed at low pressure and less than 500 °C. The $^{87}Sr/^{86}Sr$ ratios indicate a continental origin; whole-rock Sr ages indicate crystallization about 456 Ma, the probable age of the asbestos veins (Clague and others, 1985).

Stop 2. Lower Unit at Vimy Ridge Mine. Cross the road to the highest house and walk along the right (downhill) side of the house. Get on a level path that bears right; within 330 ft (100 m), bear right at the head of the abandoned open pit Vimy Ridge Mine, and go onto the weathered outcrop.

The outcrop shows features typical of altered upper mantle peridotites and may represent residual mantle material depleted by partial melting. Most of the outcrop is brownish-orange harzburgite tectonite with chrysotile asbestos veins. Bold-weathering orthopyroxenes form porphyroclasts up to 0.5 in ($\approx$1 cm) long. They are elongated at a high angle to lines of several tiny black chrome spinel grains that strike N20°W and dip steeply SW. The primary fabric evidently developed at a high temperature through solid flow, during formation of the ocean lithosphere. Forsteritic olivine grains, 0.5 to 5 mm in size, form a microblastic matrix.

Dunite, weathering yellowish orange, is exposed in the eastern part of the outcrop. It is chromite-bearing and its smooth surface is due to the absence of asbestos veins and orthopyroxene. The dunite lenses are concordant with the main foliation and (where present) metamorphic layering of the enclosing harzburgite. They represent either late-stage melting in the harzburgite, or early low-pressure fractional crystallization in magma channels.

The lower unit of the ophiolite is mostly a homogeneous harzburgite. There are, however, dunite pods near the base, and isoclinal folds of mineralogic layering. The lower unit is cut by dikes of orthopyroxenite and gabbro, and (as noted at Stop 1) contains fault slices of granitic rock. A 1.6 ft (0.5 m) quartz monzonite sheet-like body with talc-chlorite rims occurs in the re-entrant on the far side of the pit.

The first serpentinization is a pervasive but partial (about 50%) alteration of olivine to lizardite, and pyroxene to bastite (Laurent and Hebert, 1979). It post-dates the blasto-mylonitic fabric and formed at a temperature below 644°F (340°C; Cogulu and Laurent, 1984). The alteration occurred in the ocean basin before obduction.

A "spaced" cleavage, which strikes ENE and dips N at a moderate angle, is assigned to the initial stages of obduction. After obduction, asbestos veins with pure serpentine borders record the second serpentinization, probably from meteoric waters in brittle fractures in the basal and northwestern part of the harzburgite. Faults that trend N35°E cut the veins.

Stop 3: Coleraine Breccia. Study the glaciated knob on the far side of the cemetery (no hammers, please!). There are other polished outcrops upslope from the cemetery.

The breccia has a clayey to sandy silicified matrix and is chaotic. Clasts of all rock types of the ophiolite and of Cambrian metasediments are found. Many clasts indent the large green and red argillite clasts, attesting to a soft and soupy sediment.

The Coleraine Breccia occurs at the top of the ophiolite. The rock may actually be a big block within the surrounding St. Daniel olistostromes (Fig. 1; Hébert, 1983). The St. Daniel is exposed in a roadcut on Québec 267, 10.2 mi (16.3 km) southeast of Québec 112. In this cut, which is 0.2 mi (0.3 km) northwest of St. Daniel, only sedimentary rocks form the clasts.

Stop 4. Chromite Mine, Base of Upper Unit. Walk around the west end of Lake Caribou, on tailings outwash, and turn right on the north side of the lake. About 330 ft (100 m) in, walk up the mine dump to the west and along a little creek to the abandoned openpit mine. The walk takes ten minutes.

The mine shows layered dunite and chromitite. These cumulates had been deposited directly on the mantle peridotite that formed the floor of the magma chamber. In the center of the pit are four or five graded beds of chromitite, 4 to 24 in (10 to 60 cm) thick. The layering strikes N50°E and dips about 75 NW; chrysotile veins are uncommon.

The cumulate dunite forms a belt several hundred meters wide and has been traced along the southeast face of the adjacent ridge for 1.5 mi (2.5 km; Cooke, 1937). It is tightly and isoclinally folded and is in tectonic contact with lower-unit peridotite in the hills to the north. Most of the near-surface high-grade sheets and lenses of chromite have been mined, but much disseminated 10% ore remains.

Stop 5. Sheeted Intrusives in Gabbro. Walk across the road to the base of the cliff. Climb to the first good exposures.

This roof of the plutonic section of the ophiolite is a fine-grained uralitized gabbro that is invaded by many criss-crossing intrusive sheets, leaving few large remnants of gabbro. The nearly vertical sheets are mostly diabase but they range in composition from mafic to felsic, and textures range from aphanitic to medium grained.

Stop 6. Boninitic Pillow Lava. The outcrop that slopes west toward the road shows many features characteristic of pillow basalts (Seguin and Laurent, 1975). No hammers, please! Fresh samples can be collected across the road.

These lavas are faulted against the gabbros of Stop 5. The pillows face the road, indicating a northwest-verging fold in a southeast-facing sequence. Look for narrow feeder necks, the bulbous cylindrical shape of tubes, and the concentric zoning of pillows. Quench textures are preserved in the margins of the pillows. The matrix between the pillows is volcanic glass (devitrified) and some breccia from adjacent pillows. There are a few dikes. Southward, volcanic breccia forms a layer between these pillows and those in the fresh cut across the road. Compositionally, the lavas are olivine tholeiites and andesites. They are high in MgO, Ni, and Cr; they are low in TiO_2 and in incompatible elements such as Zr, Y, and REE (Seguin and Laurent, 1975). Unpublished new geochemical data support Shaw and Wasserburg (1984), whose isotopic data show that the lavas are similar to present-day boninites from the Marianas Trench (Shaw and Wasserburg, 1984).

REFERENCES CITED

Baldwin, B., Laurent, R., and Doolan, B. L., 1985, Thetford Mines area, P. Q.; Field trip guide A: Vermont Geological Society Guidebook 1, Vermont Geology, v. 4, p. A1–A31.

Clague, D. A., Frankel, C. S., and Eaby, J. S., 1985, The age and origin of felsic intrusions of the Thetford Mines ophiolite, Québec: Canadian Journal of Earth Sciences, v. 22, p. 1257–1261.

Cogulu, E., and Laurent, R., 1984, Mineralogical and chemical variations in chrysotile veins and peridotite host-rocks from the asbestos belt of southern Québec: Canadian Mineralogist, v. 22, p. 173–183.

Cooke, H. C., 1937, Thetford, Disraeli, and eastern half of Warwick map areas, Québec: Geological Survey of Canada Memoir 211, 159 p.

Feininger, T., 1981, Amphibolite associated with the Thetford Mines ophiolite complex at Belmina Ridge, Québec: Canadian Journal of Earth Sciences, v. 18, p. 1878–1892.

Hébert, Y., 1983, Etude petrologique du complexe ophiolitique de Thetford Mines, Québec [These de Ph.D.]: Québec, Québec, Université Laval, 426 p.

Kacira, N., 1982, Chromite occurrences of the Canadian Appalachians: Canadian Institute of Mines and Metallurgy Bulletin, v. 75, p. 73–82.

Laurent, R., 1975, Occurrences and origin of the ophiolites of southern Québec, northern Appalachians: Canadian Journal of Earth Sciences, v. 12, p. 443–455.

——, 1980, Environment of formation, evolution, and emplacement of the Appalachian ophiolites from Québec: Proceedings of the International Ophiolite Symposium, 1979, Geological Survey, Nicosia, Cyprus, p. 628–636.

Laurent, R., and Hébert, Y., 1979, Paragenesis of serpentine assemblages in harzburgite tectonite and dunite cumulate from the Québec Appalachians: Canadian Mineralogist, v. 17, p. 857–869.

Laurent, R., Taner, M. F., and Bertrand, J., 1984, Mise en place et pétrologie du granite associé au complexe ophiolitique de Thetford Mines, Québec: Canadian Journal of Earth Sciences, v. 21, p. 1114–1125.

Oshin, I. O., and Crockett, J. H., 1982, Noble metals in Thetford Mines ophiolites, Québec, Canada, part I; Distribution of gold, iridium, platinum, and palladium in the ultramafic and gabbroic rocks: Economic Geology, v. 77, p. 1556–1570.

St. Julien, P., and Hubert, C., 1975, Evolution of the Taconian orogen in the Québec Appalachians: American Journal of Science Special Publication, v. 275-A, p. 337–362.

Seguin, M. K., and Laurent, R., 1975, Petrological features and magnetic properties of pillow lavas from the Thetford Mines ophiolite (Québec): Canadian Journal of Earth Science, v. 284, p. 319–349.

Shaw, H. F., and Wasserburg, G. J., 1984, Isotopic constraints on the origin of Appalachian mafic complexes: American Journal of Science, v. 284, p. 319–349.

Williams, H., and St. Julien, P., 1982, The Baie Verte—Brompton Line; Early Paleozoic continent ocean interface in the Canadian Appalachians: Geological Association of Canada Special Paper 24, p. 177–207.

The Ordovician-Silurian boundary succession at Cap Henri, Anticosti Island, Québec

Allen Alexander Petryk, *Ministère de l'Énergie et des Ressources, Québec City, Québec*

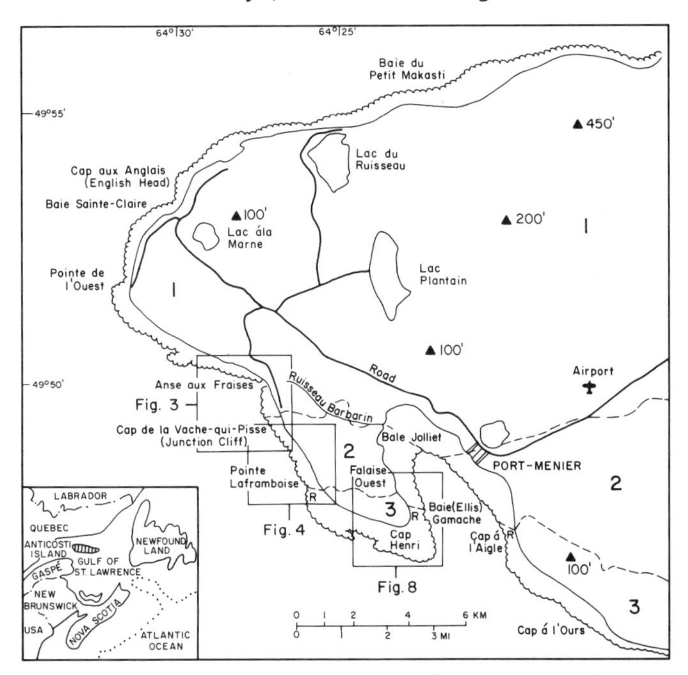

Figure 1. Geologic and location map of northwestern Anticosti Island, Québec, Canada. Numbered areas 1, 2, and 3 correspond to the Vauréal, Ellis Bay, and Becscie formations, respectively. Geological contacts are indicated in broken lines. R, reefal unit. The extensive wave-cut terrace is exposed during low tides. Insets locate Figures 3, 4, and 8.

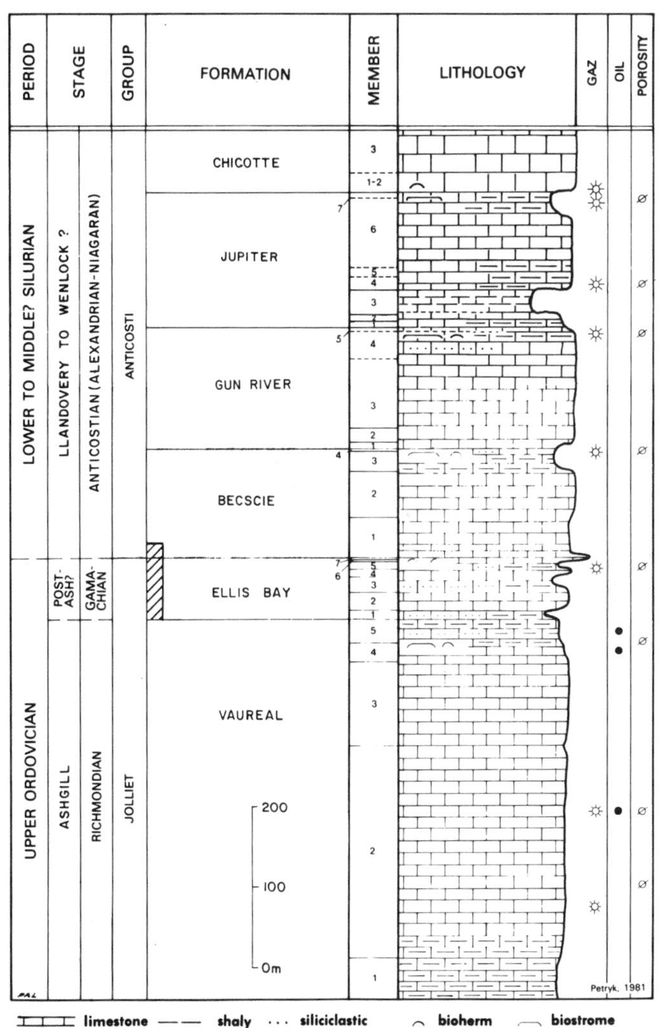

Figure 2. Composite stratigraphic section of Anticosti Island. The bar denotes the stratigraphic limits of the site area. Group and formational names are formal. Proposed formal and informal (numbered) member designations are (downsection): Becscie Formation: Cap Henri Member - 1; Ellis Bay Formation: Pointe Laframboise Reefal Member - 7; Pointe aux Ivrognes Member - 6; Ruisseau Delta Member - 5; Cap Blanc Member - 4; Falaise Ouest Member - 3; Junction Cliff Member - 2; Cap de la Vache-qui-Pisse Member - 1; Vauréal Formation: Anse aux Fraises Member - 5; Pointe aux Foins Member - 4. Note that the Ordovician-Silurian boundary is based on conodont microfossils.

LOCATION

Anticosti Island lies between the Estuary and the Gulf of St. Lawrence of eastern Canada, about 19 mi (30 km) south of the north shore, 46 mi (75 km) northeast of Gaspé Peninsula, and 108 mi (175 km) west-northwest of Newfoundland (Fig. 1). The municipality of Port-Menier, to the east of the site area, has regular air service links with Québec City, via Sept-Iles, and Gaspé. Accommodations are available at the hotel and in other facilities in Port-Menier. Access to the two site localities is readily available by vehicle (mini-bus or truck), boat, or by foot; a rental vehicle is preferred. Anticosti Island is a renowned hunting and fishing reserve. A permit (no cost) must be obtained from park authorities to circulate on the island; this guards against the marring of the island ecosystem and the nonscientific collection of fossil and rock material. To get to Stop 1, drive northwest 9 mi (14.5 km) from Port-Menier, then south to Anse aux Fraises until the road intersects the beach near a small delta, and then southeast along the beach to Cap de la Vache-qui-Pisse. The Stop 2 shoreline section occurs on the west side of Baie Ellis and can be reached by driving via Anse aux Fraises to Pointe aux Ivrognes and Cap Henri or by small boat from the quay across the bay to the wave-cut platform edge. For the former, travel 13.3 mi (22.2 km) from Port-Menier around Cap Henri to the bioherms, walk about 0.3 mi (0.5 km) to Falaise Ouest (West Cliff).

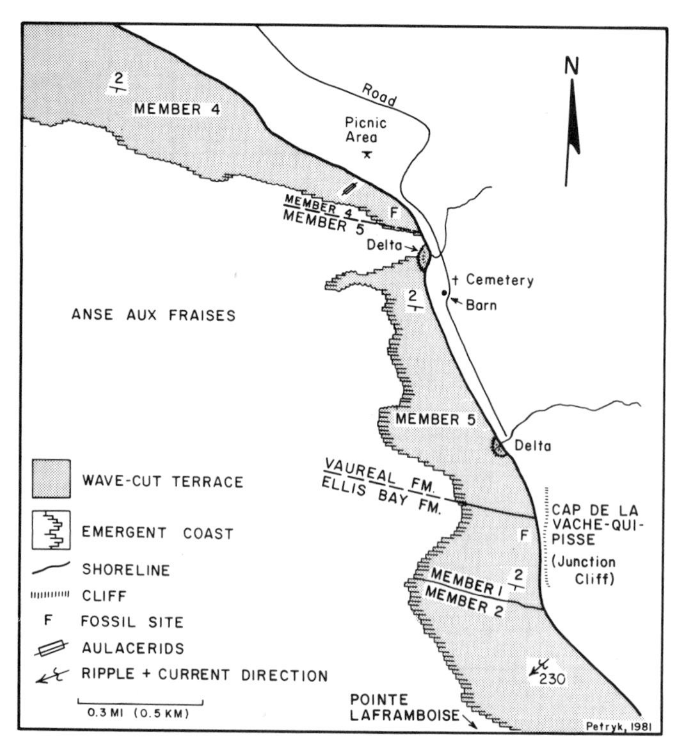

Figure 3. Geologic map of the Anse aux Fraises area showing the Vauréal–Ellis Bay interformational boundary and member locations (see Fig. 2 for geographic location and explanations).

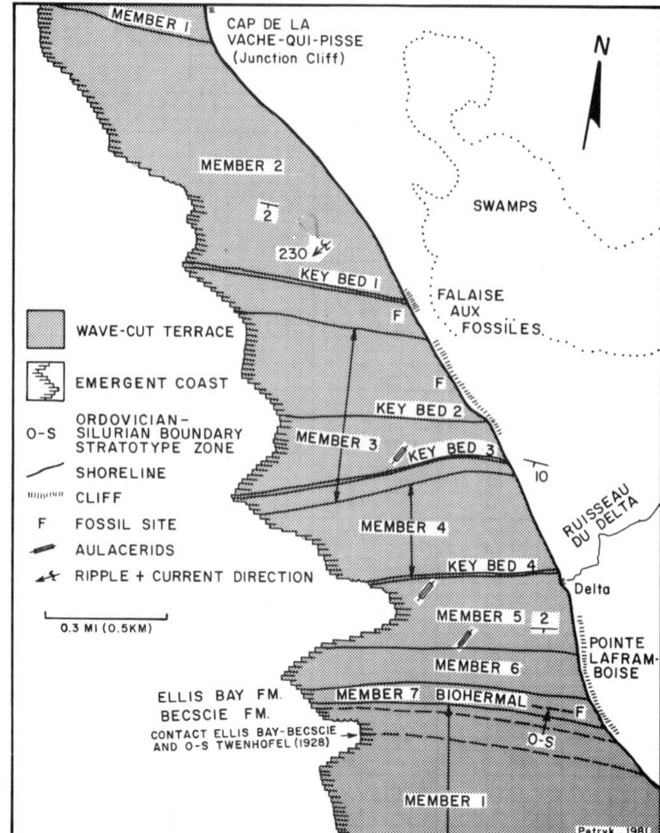

Figure 4. Geologic map of the Junction Cliff to Pointe Laframboise area showing the Ellis Bay–Becscie interformational boundary, member locations, and the Ordovician–Silurian systemic boundary (see Fig. 2 for geographic location and explanations).

SIGNIFICANCE

The Anticosti Island formations form the most complete and continuous sequence of Late Ordovician to possibly Middle Silurian strata in eastern North America, if not in the world, for the shelly facies. The island is largely underlain by nondeformed, rarely faulted, fossiliferous, carbonate shale and lesser transitional, carbonate-siliciclastic, shallow marine platform deposits. Because of its continuous, rich, and varied shelly macrofaunal and microfaunal succession, the Ordovician-Silurian boundary (O-S; ± 438 Ma) in the Cap Henri area of northeastern Anticosti, has been considered as a possible site for the international boundary stratotype (standard) for the base of the Silurian System.

SITE INFORMATION

Anticosti Island is part of the eastern region of the St. Lawrence Platform or Lowlands. The approximately 3,630-ft-thick (1,100 m) sequence on the island occurs as a shallow (about 2°), southeast-dipping, nondeformed monocline (Fig. 2). The area lies between the Canadian Shield to the north and the northern Appalachian orogen to the south and east. The ancient Anticosti platform was largely spared the stresses of the Taconic, Acadian, and subsequent orogenic pulses by its location to the north and west of the ancient geosuture known as Logan's Line. In the subsurface, however, Lower and Middle Ordovician Beekmantown, Chazy, Trenton, Black River, and Utica strata were apparently disrupted by tectonic forces. Two thick diabase dikes on the north shore of the island attest to the effect of Lower Cretaceous tensional forces on the paleoplatform.

The uppermost Ordovician, Ellis Bay Formation, is characterized by a cyclic alternation of (a) thin-bedded, relatively light "clean" limestone (represented by Members 2, 4, 6, and 7); and (b) generally poorly bedded, dark "dirty" argillaceous nodular limestone and shale members (represented by Members 1, 3, and 5), that represent marine sedimentation during inferred transgressive-regressive, interglacial-glacial, glacio-eustatic cyclic events, respectively. In contrast, the conformable, lowermost Silurian, Becscie Formation (except for a ±3-ft-thick [1 m] basal transitional zone), is comprised of remarkably clean thin-bedded limestones. The microfossil, conodont-based, cryptic, O-S chronostratigraphic boundary occurs in a ±10ft-thick (3 m), reefal complex of patch reefs (bioherms) and bedded, interreefal (biostromal) strata of the uppermost member of the Ellis Bay Formation. The Ellis Bay–Becscie interformational boundary zone (±3 ft; 1 m above the O-S chronostratigraphic boundary), marks the turning point in an inferred regressive-transgressive, eustatic, megacycle spanning Late Ordovician and Early Silurian time respectively.

Figure 5. Outcrop section of the Ellis Bay Formation, Members 5 to 7 at Pointe Laframboise. Recessive (regressional), subnodualr, argillo-calcareous wackestone of Member 5 is succeeded progressively by less argillaceous, calcisiltite and grainstone and relatively clean, homogeneous, resistant (transgressional), current-bedded, sparcely fossiliferous lime grainstone of Member 6. A mm-thick shale parting separates the principal key bed of Anticosti Island, the Oncolitic Platform Bed of Pointe Laframboise (OPB) and the basalmost Member 7, from Member 6 (note man pointing to this contact). The OPB marks the onset of the accretion of reefal boundstone during an inferred regressive–transgressive turning point of latest Gamachian or Llandovery Time.

Figure 6. Ellis Bay Formation, Member 7 (EB-7), patch-reef complex overlain and interdigitated with the basal Becscie Formation, Member 1 (B-1) at Pointe Laframboise. Note that the inferred, conodont based, Ordovician-Silurian (O-S) boundary is approximately where the 5 ft 10 in (1.8 m) man is pointing. Only the highest tides reach the outcrop (see Figs. 2 and 4 for location information).

Two separate sections make up the broad, site area: Stop 1, from Cap de la Vache-qui-Pisse (Junction Cliff) to Pointe Laframboise (Figs. 1 to 7); and Stop 2, from Ruisseau Barbarin to Cap Henri (Figs. 1, 8, and 9). Figure 2, a composite stratigraphic section of all the formations of Anticosti Island, shows the limits of the section of the site area. The thicknesses of Members 1 to 7 of the type section of the Ellis Bay Formation are 31.5, 72.5, 54.8, 32.1, 32.8, 8.2, 3.3 to 13.1 ft or 9.6, 22.1, 16.7, 9.8, 10.0, 2.5, 1.0 to 4.0 m, respectively.

Stop 1. The shoreline section from Cap de la Vache-qui-Pisse to Pointe Laframboise is regarded as the Ellis Bay type section or unit stratotype (Figs. 3 and 4; Petryk, 1981). The upper part of the recessive Member 1 (about 8 ft; 2.5 m) and the lower part of Member 2 (in excess of 30 ft; 10 m) are exposed in the undercut cliff at Cap de la Vache-qui-Pisse. For the next 2 mi (3.4 km) southeast along the beach, beds outcrop in the wave-cut platform and locally as low cliffs or outcrops above the beach. At about 0.7 mi (1.1 km) from Junction Cliff at Falaise aux Fossiles, the highly fossiliferous and shaley Member 2 outcrops. Farther, parts of Members 3 and 4 are exposed above the beach. Much of the upper part of Member 5 and all of Members 6 and 7 are exposed in low cliffs at Pointe Laframboise (Fig. 5). A distinctive, resistant, brown, skeletal (algal) lime grainstone (key bed 3: 1.3 ft; 0.4 m) occurs almost at the top of Member 3, and is well exposed at the shoreline and on the wave-cut platform about 0.3 mi (0.5 km) north-northwest of the small delta north of Pointe Laframboise (Fig. 4). Another distinctive, resistant, and key bed of Anticosti, the Oncolitic Platform Bed of Pointe Laframboise (5.1 to 11.8 in; 13 to 30 cm), occurs at the base of Member 7. The faintly greenish, ivory-colored bed is algal and fossil-rich in its upper part (Figs. 6 and 7). Delicately ribbed, round (about 0.7 in; 1.8 cm), thin-shelled brachiopods occur in its lower part, as well as in the whitish fine-grained, homogeneous, interference-rippled lime grainstone of the underlying Member 6.

Fossils are common throughout the Ellis Bay: pelmatozoans, crinoids, brachiopods, bryozoans, trilobites, gasteropods, pelecypods, ostracodes, nautoloid cephalopods, corals, aulacerid stromatoporoids and others, algae, scolecodonts, condonts, graptolites, sponges, annelids, trace fossils, and others (see Bolton, 1972; Lespérance, 1981). They are abundant in Members 1, 3, and 5, but are most abundant and varied in the uppermost

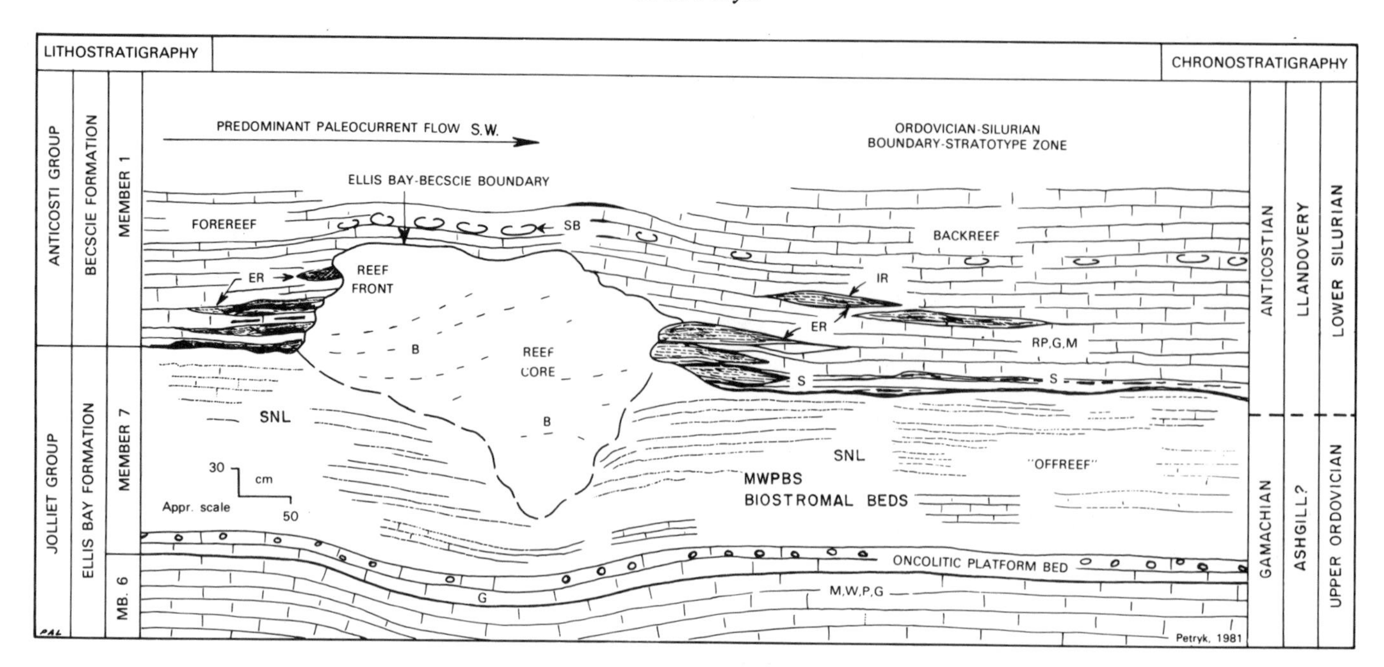

Figure 7. Lithostratigraphic, chronostratigraphic, lithofacies, and paleoenvironmental relationships of the Ordovician-Silurian boundary type-area of Anticosti Island, Québec, Canada. L, limestone; B, boundstone; ER, rudaceous skeletal encrinite; G, grainstone; IR, intrarudstone (intraformational conglomerate or breccia); M, mudstone; P, packstone; R, rudstone; W, wackestone; S, shale; SN, subnodular; SB, ball-and-pillow slump bed.

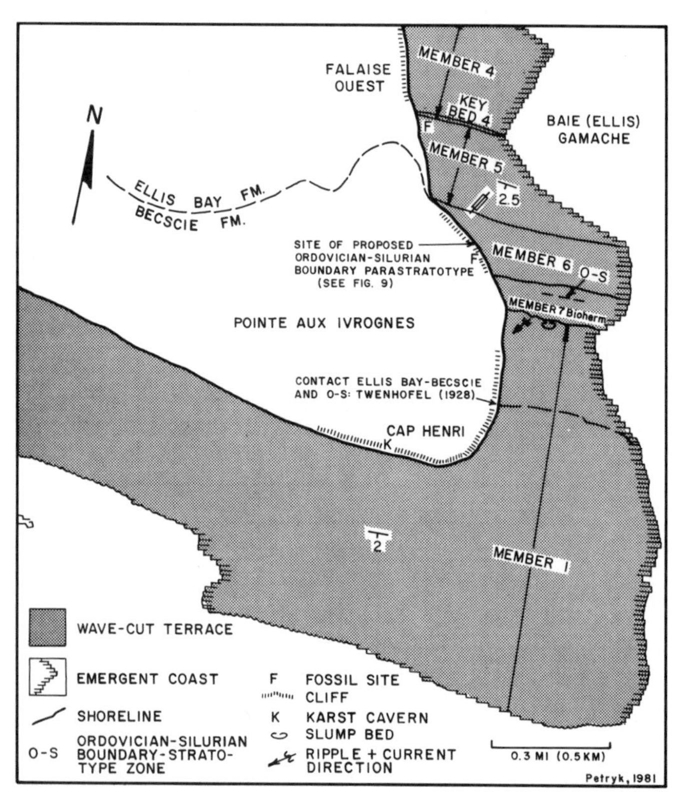

Figure 8. Geologic map of the Cap Henri area showing the Ellis Bay–Becscie interformational boundary, member locations, and the proposed Ordovician-Silurian stratotype boundary (see Fig. 2 for geographic location and explanations).

Members 2 and 4, and in the reefal, biohermal, Member 7. *Aulacera* spp., a cylindrical, log-like stromatopore (calcisponge?) of ±4 in (10 cm) diameter and ±12 in (30 cm) length fragments, is abundant just above the Members 3 to 4 contact, and below the Member 4 to 5 contact (Petryk, 1982). The next 1.4 mi (2.8 km) along the beach have exposures of Member 1, Becscie Formation, restricted largely to the wave-cut terrace before the cliff exposures are reached at Pointe aux Ivrognes and Cap Henri (Fig. 8). Except for the highly fossiliferous lensy encrinitic beds of the lowermost 3 ft (1 m) of Member 4, the unit is comprised of thin to medium, current-bedded, commonly bioturbated but poorly fossiliferous lime mudstone, calcisiltite, packstone, wackestone, grainstone, and pebble rudite.

Members 1, 3, 5, and basal 7 sequences of the Ellis Bay Formation are inferred to mainly represent regressive, marine sedimentation during latest Ordovician glacial epochs (Petryk, 1982). The intervening, Members 2, 4, 6, and upper 7 sequences would mainly represent transgressive episodes of marine sedimentation during interglacial epochs. The lack of intertidal or subaerial sedimentary features suggest that at least the Anticosti region of the paleo-platform remained submergent during the glacio-eustatic fluctuations in the uppermost Ordovician (Petryk, 1982).

Stop 2. Members 1 and 2 of the Ellis Bay Formation are poorly exposed here and are best seen at Stop 1. Most of Member 3 and the lower part of Member 4 are well exposed at Falaise Ouest (Figs. 1 and 8). Proceed back to Cap Henri along the beach; a short covered interval obscures the middle part of Member 4 but the upper part is exposed in low bluff exposures on

Figure 9. Proposed Ordovician-Silurian parastratotype boundary (O-S) section between Falaise Ouest and Cap Henri, western Anticosti Island (see Figs. 2 and 8 for location information and Figs. 5 and 6 for outcrop comparison). Note geological hammer and metre stick for scale. OPB = Oncolitic Platform Bed.

a small headland. The resistant marker or key bed 3 is present on the wave-cut terrace at the southeast tip of Falaise Ouest. The lower part of the dark gray-green argillaceous, subnodular, lime wackestone of Member 5 is not exposed at Falaise Ouest but is exposed at low tide in the wave-cut terrace. The upper part of the member outcrops in low cliffs below the more resistant whitish, lime grainstone beds of the lower Member 6. The contact between Members 5 and 6 can be traced around the next headland to the start of a long stretch of low cliffs in which the lowermost Becscie is exposed (at low tide) to the wave-cut terrace edge south of Cap Henri. The dark, lenticular rudaceous encrinitic bed of the lowermost Becscie seen in cliff sections occurs, locally in the wave-cut terrace, as a conspicuous, asymmetrical ripple bed (paleocurrent directed 235°) overlying the Member 7 reefal zone. The interval of Members 6 and 7 is well exposed in the cliff sections and on the adjacent wave-cut terrace about 0.3 mi (0.5 km) north of the tip of Cap Henri; the most detailed faunal and conodont collecting across the conodont-defined Ordovician-Silurian boundary has been made here (see Lespérance, 1981; Figs. 8 and 9). This section was proposed as the stratotype for the Ordovician-Silurian boundary (Barnes and McCracken, 1981; Barnes and others, 1981). (Note that the facies and sedimentary structures of the members at Stop 2 are identical with those of Stop 1.)

REFERENCES CITED

Barnes, C. R., and McCracken, A. D., 1981, Early Silurian chronostratigraphy and a proposed Ordovician-Silurian boundary stratotype, Anticosti Island, Québec, *in* Lespérance, P. J., ed., International Union of Geological Science Subcommission on Silurian Stratigraphy and Ordovician-Silurian Boundary Working Group, Field Meeting, Anticosti-Gaspé, Québec, 1981, Volume II, Stratigraphy and Paleontology: Québec, Université de Montréal, p. 71–79.

Barnes, C. R., Petryk, A. A., and Bolton, T. E., 1981, Anticosti Island, Québec, Field-Guidebook, Part I, *in* Lespérance, P. J., ed., International Union of Geological Science Subcommission on Silurian Stratigraphy, Ordovician-Silurian Boundary Working Group, Field Meeting, Anticosti-Gaspé, Québec, 1981, Volume I, Guidebook: Québec, Université de Montréal, p. 1–24.

Bolton, T. E., 1972, Geological map and notes on the Ordovician and Silurian litho- and biostratigraphy, Anticosti Island, Québec: Geological Survey of Canada Paper 71-19, p. 1–45.

Lespérance, P. J., ed., 1981, Field Meeting, Anticosti–Gaspé, Québec, 1981, Volume II, Stratigraphy and Paleontology, International Union of Geological Science Subcommission on Silurian Stratigraphy and Ordovician-Silurian Boundary Working Group: Québec, Université de Montréal, 321 p.

Petryk, A. A., 1981, Stratigraphy, sedimentology, and paleogeography of the Upper Ordovician–Lower Silurian of Anticosti Island, Québec, *in* Lespérance, P. J., ed., International Union of Geological Science Subcommission on Silurian Stratigraphy, Ordovician-Silurian Boundary Working Group Field Meeting, Anticosti-Gaspé, Québec, 1981, Volume I, Guidebook: Québec, Université de Montréal, p. 11–39.

——, 1982, Aulacerid ecostratigraphy of Anticosti Island, and its bearing on the Ordovician-Silurian boundary and the Upper Ordovician glacial episode, *in* Mamet, B., and Copeland, M. J., eds., Proceedings, Third North American Paleontological Convention, Volume 2, p. 393–399.

Twenhofel, W. H., 1928, Geology of Anticosti Island: Geological Survey of Canada Memoir 154, 481 p.

Silurian stratigraphy and facies, Port-Daniel/Gascons and Black-Cape areas, southern Gaspé Peninsula, Québec

Pierre-André Bourque, *Département de Géologie, Université Laval, Québec, Québec G1K 7P4, Canada*

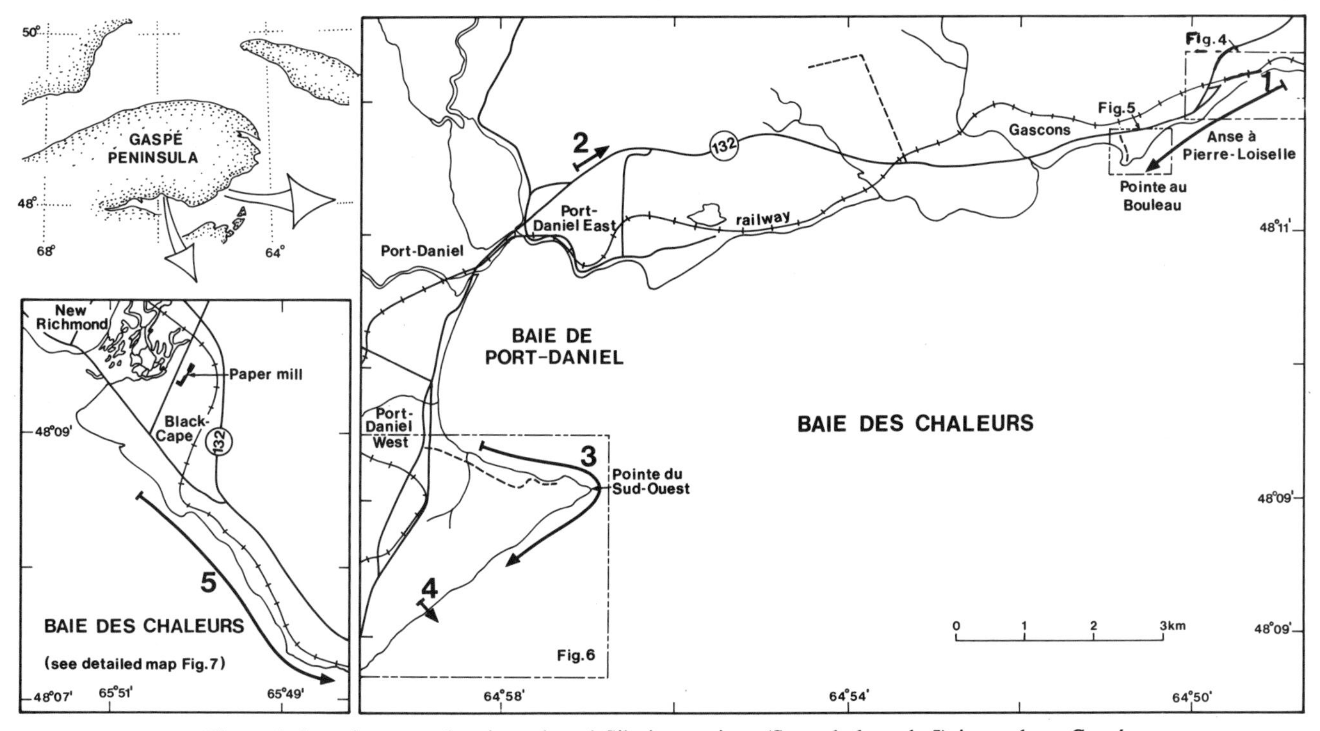

Figure 1. Location map showing selected Silurian sections (Stops 1 through 5) in southern Gaspé Peninsula (Port-Daniel/Gascons and Black-Cape areas). Arrows indicate stratigraphic top of sections. Base maps are 22 A/2 and 22 A/4 at the scale of 1:50 000. (From Surveys and Mapping Branch, Department of Energy, Mines and Resources of Canada.)

LOCATION

The five stops described herein illustrate the Silurian succession of the southern Gaspé Peninsula in Québec. They are located along the northern coast of the Baie des Chaleurs. Four of the maps are in the Port-Daniel/Gascons area; the other is in the Black-Cape vicinity (Fig. 1). Stop 1 consists of the coastal cliff section located in the Anse à Pierre-Loiselle, at 1.25 mi (2 km) east of the village of Gascons. Stop 2 is the second roadcut along Québec 132, at 0.75 mi (1.2 km) east of the Catholic Church of Port-Daniel East. Stop 3 is the coastal cliff section situated at Pointe du Sud-Ouest, south of the Baie de Port-Daniel, at the end of the Route de la Pointe in Port-Daniel West. Stop 4 consists of the shore outcrop at foot of Colline Daniel in Port-Daniel West. Stop 5 is the shore-cliff section south of the small village called Black-Cape, adjacent to the eastern limit of the Town of New Richmond, 50 mi (80 km) west of Port-Daniel.

SIGNIFICANCE

Coastal sections of Port-Daniel and Black-Cape areas have attracted geologists and paleontologists, as well as rock and fossil hunters, for more than a century. This interest was initiated by the pioneer work of Sir William Logan (1863). Subsequently, numerous field campaigns and writings of John M. Clarke (1912, 1913), Charles Schuchert (*in* Schuchert and Dart, 1926) and S. A. Northrop (1939) have contributed to the interest in these sections.

The Silurian succession of the southern Gaspé Peninsula is of particular geologic interest for the following reasons: (1) it is a complete Silurian succession without any significant breaks in deposition; (2) it is a very thick sequence, reaching thicknesses of as much as 16,400 ft (5 km), as compared to the relatively thin Silurian succession of the North American interior; (3) it displays a wide variety of sedimentary facies, many highly fossiliferous, from nearshore to offshore siliciclastic facies, peritidal algal and stromatolite limestone facies, reef and peri-reef facies, terrestrial red bed, and also subaerial volcanic facies; (4) it contains a unique, well-preserved reef and bank complex reaching thicknesses of 2,600 ft (800 m); and (5) all facies are particularly well preserved, not having been affected by metamorphism, as is the case in many other parts of the Appalachians. All together, these

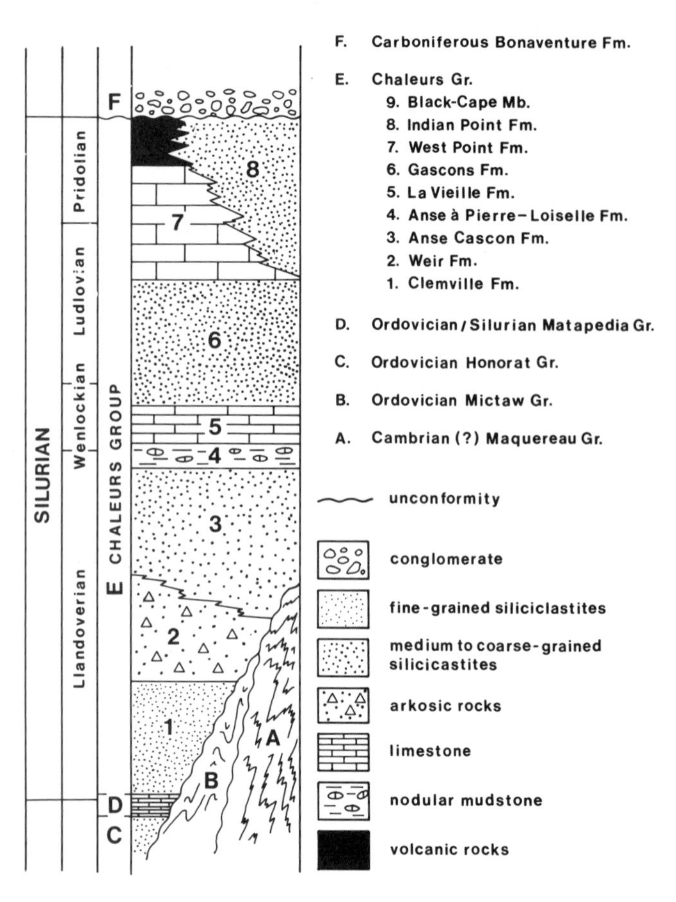

Figure 2. Silurian stratigraphy of southern Gaspé Peninsula (Chaleurs Baie Synclinorium). (Summarized from Bourque and Lachambre, 1980.)

features make the Silurian succession of the southern Gaspé Peninsula unique in the Appalachians orogen and a key sequence in understanding the evolution of the orogen.

SITE INFORMATION

The five sections described in the Port-Daniel/Gascons and Black-Cape areas of Baie des Chaleurs represent well the main facies of the Silurian succession (Fig. 2) and give the visitor an almost complete overview, with the exception of the two lowest formations (Clemville and Weir), which cannot be observed along the seashore, since they outcrop only inland (see Bourque, 1981, stop BC-1, for description of a section inland where both formations can be observed). Lithology and depositional facies are summarized for each formation in Table 1. The five stops showing significant sections chosen to represent the Silurian succession of the southern Gaspé Peninsula are shown in Figure 1. They are represented as columnar sections in Figure 3, where the relative stratigraphic position of each is indicated.

More recent stratigraphic work has been done by Burk (1964), Ayrton (1967), Bourque (1975), and Bourque and Lachambre (1980). Three field guides by Bourque (1979, 1981, and

TABLE 1. SUMMARY OF LITHOLOGY AND DEPOSITIONAL FACIES OF LITHOSTRATIGRAPHIC UNITS OF SILURIAN SUCCESSION OF SOUTHERN GASPE PENINSULA (CHALEURS BAY SYNCLINORIUM)

UNITS	LITHOLOGY	DEPOSITIONAL FACIES
Black-Cape Member	Basic lava flows and volcanic breccias and conglomerates; a few units of limestone and red mudstone.	Sub-aerial volcanism
Indian Point Formation	Thick-bedded siliciclastic mudstone, siltstone, and fine-grained sandstone.	Offshore-type fine-grained sedimentation
West Point Formation	Made up of three superposed complexes: 3. Upper reef complex: stromatoporoid-coral-bryozoan-algal boundstone, debris units, stromatoporoid rubble, cryptalgal laminites, nodular dolomite, laminated sandstones, mud-cracked red beds. 2. Middle bank complex: crinoid-stromatoporoid biostromes, gravity flow units. 1. Lower reef complex: interbedded coral and stromatoporoid-rich limestone and fine-grained siliciclastites, red stromatactis calcilutite, algal boundstone, gravity flow units.	Red bed coastal plain flanked by reef platform, with individualized intertidal, lagoonal reef margin and fore-reef zones. Crinoid bank and forebank complex Shallowing upward sequence of early-cemented sea floor facies, stromatactis mounds and culminating algal reefs.
Gascons Formation	Thick-bedded siliciclastic mudstone, siltstone, and fine-grained sandstone.	Offshore-type fine-grained sedimentation
La Vieille Formation	Made up of three superposed members: 3. Upper member: nodular calcilutite, shale interbeds. 2. Middle member: laminite-oncolite-stromatolite-fenestral limestones, oolite and peloid grainstones, coral-algal boundstones. 1. Lower member: nodular calcilutite, shale interbeds.	Offshore-type muds. Peritidal mud flat assemblage rimmed with coral-algal knobs. Offshore-type muds.
Anse à Pierre-Loiselle Formation	Nodular fine-grained siliciclastites, one sandstone and conglomerate unit in middle of formation.	Offshore-type muds, with an offshore bar.
Anse Cascon Formation	Fine to medium-grained quartz sandstone shale and mudstone interbeds. Laminated and cross-bedded. Highly bioturbated.	Beach to lower shoreface sand sheets.
Weir Formation	Arkosic sandstone and conglomerate, shale and mudstone interbeds.	Immature shoreline sands and gravels.
Clemville Formation	Siliciclastic mudstone and siltstone with laminated and cross-bedded fine-grained sandstone and calcilutite.	Outer shelf muds, storm layers.

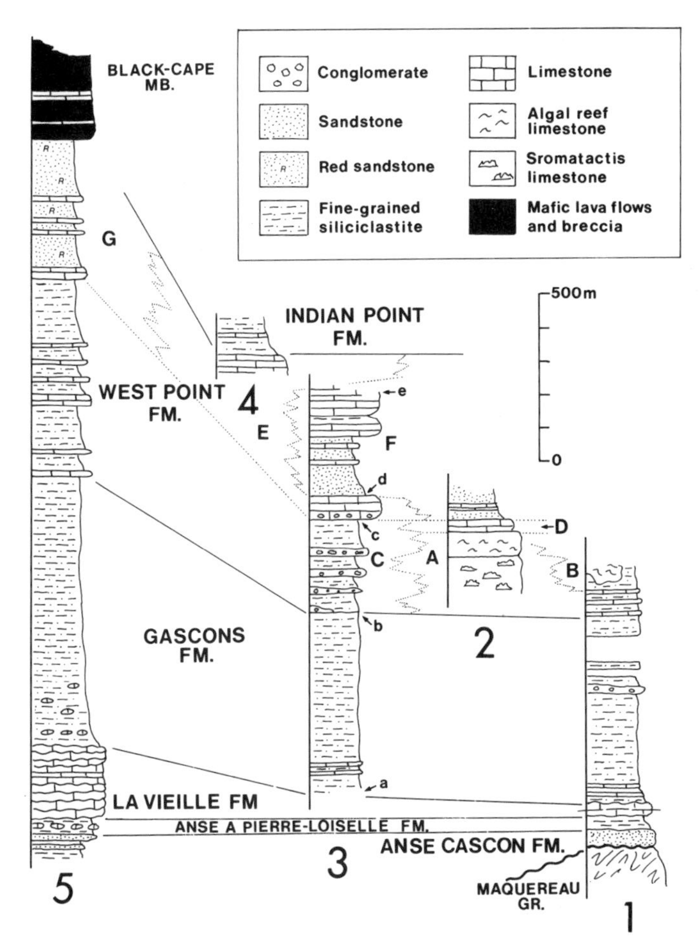

Figure 3. Stratigraphic correlation of the selected sections: 1, Anse à Pierre-Loiselle section; 2, Québec 132 section; 3, Pointe du Sud-Ouest section; 4, Colline Daniel section; 5, Black-Cape section. Facies of the West Point Formation are as follows: A, shallowing upward sequence of the lower reef complex; B, C, proximal fore-reef, distal fore-reef facies of the lower reef complex; D, crinoid bank complex facies; E, F, reef margin and back-reef to intertidal facies of the upper reef complex; G, supratidal red bed facies adjacent to upper reef complex platform. Lower case letters in section 3 refer to locations along the shore section at Stop 3. (From Bourque and Lachambre, 1980, Fig. 2.)

1982) deal with the sections here considered. Geologic maps relevant to the described stops are found in Bourque and Lachambre (1980, maps 1929, 1930, 1953). A summary paper on the La Vieille Formation and the West Point carbonate buildup can be found in Bourque and others (1986).

Stop 1—Coastal cliff section at Anse à Pierre-Loiselle. The base of this section is reached easily from a secondary paved road adjacent to Québec 132. From the intersection of that road with a railway, walk east along the railway for a distance of 1,000 ft (300 m) to a limestone outcrop (La Vieille Formation) on the north side of the railway, as indicated in Figure 4; and then take a small trail leading to the cliff at a point called Cap Rouge (previously La Vieille). Atop that cape, there is a superb view of the section to be visited toward the east. From the cape, the climb

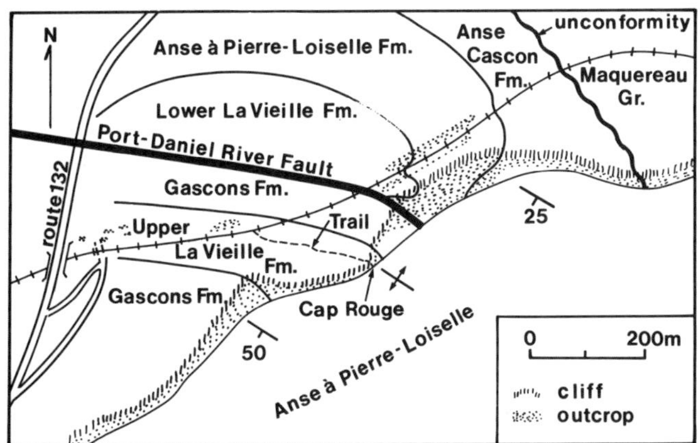

Figure 4. Geologic map of lower part of Anse à Pierre-Loiselle section (section 1). (From Bourque and Lachambre, 1980, Fig. 12.)

down to the beach is relatively easy, although care should be exercised. Walk along the beach about 1,600 ft (500 m) to the base of Silurian section, which can be reached *only at low tide.*

The Silurian section here dips southwesterly at about 25° (Fig. 4). Its base is seen to unconformably overlie the metamorphic rocks of the Cambrian(?) Maquereau Group at the eastern limit of the map in Figure 4. Between the unconformity and Cap Rouge, the Anse Cascon and Anse à Pierre-Loiselle Formations are exposed. The Anse Cascon contains abundant trace fossils, and affords study of the relationships between bioturbation and the rate of sedimentation. The overlying Anse à Pierre-Loiselle Formation is very fossiliferous (brachiopods, corals, and stromatoporoids) in places, particularly in the upper part of the formation, which is best observed atop the cliff along the railway. Immediately east of Cap Rouge, there is a highly sheared zone that corresponds to an important regional fault, the Port-Daniel River Fault, which, at beach level, brings the Anse à Pierre-Loiselle and the Gascons Formations in contact. It is unfortunate that this section constitutes the stratotype of the La Vieille Formation, since most of that formation is cut out by the fault. A complete sequence of the La Vieille Formation can be observed at Black-Cape (Stop 5). A faunal list and paleoenvironmental interpretation of the lower part of the section is available in Bourque (1981) and there is an account on trace fossils in Pickerill and others (1977).

The second part of the section, between Cap Rouge and Pointe au Bouleau (Fig. 1), is accessible by climbing back to the top of Cap Rouge (Fig. 4) and going down again to the beach on the west side of the cape. Strata here dip southwesterly an average of 50°. Walk west along the beach *at low tide* for a distance of about 1 mi (1.6 km) to the base of the West Point Formation at Pointe au Bouleau. Along that segment of the shore, only the Gascons Formation is observed. Except for a debris flow unit, a few brachiopod-rich beds and local trace fossils, it is a rather homogeneous and somewhat monotonous sequence of fine-grained siliciclastic rocks. A very interesting section of the Gas-

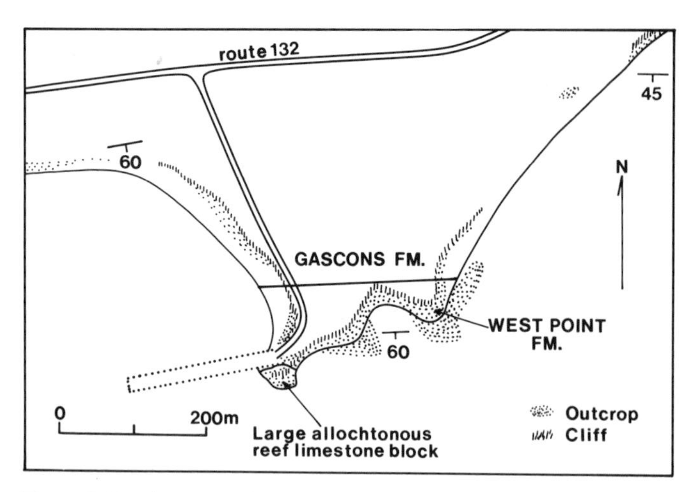

Figure 5. Detailed map of Pointe au Bouleau that constitutes upper part of Anse à Pierre-Loiselle section (Stop 1). (From Bourque, 1979, Fig. 12.)

cons Formation can be more easily observed in the Baie de Port-Daniel, west of Pointe du Sud-Ouest (Stop 3).

At Pointe au Bouleau (Fig. 5), the base of the West Point Formation has been drawn at the first occurrence of limestone beds. That unit of interbedded limestones and fine-grained siliciclastic rocks constitutes the foundation facies of the West Point reef and bank complex. The section is particularly fossiliferous, displaying beautiful large coral and stomatoporoid colonies. Be careful not to spoil the section by oversampling. The massive limestone unit that constitutes the small promontory at the topmost part of the section is a large allochthonous algal reef–derived block (130 by 330 ft; or 40 by 100 m in minimum size) floating in the fine-grained siliciclastic facies of the fore-reef (section 1 in Fig. 3). This block has been studied by Laliberté (1982).

Stop 2—Road cut section along Québec 132. The road-cut section of Stop 2 is best observed on the north side of the road. The section here is vertically dipping and tops toward the east. Several facies of the West Point reef and bank complex can be observed (Fig. 3, section 2). Proceeding upsection, one passes from the red stromatactis calcilutite facies and gray algal reef limestone facies of the lower reef complex, through the crinoid-rich biostromes of the middle bank complex, and into the laminated and rippled fine-grained sandstone and cryptocyanobacterial laminites of the intertidal facies of the upper reef complex. The red stromatactis calcilutite has been studied by Gignac (1980) and Bourque and Gignac (1983); the crinoid-rich biostromes, by Gosselin (1981).

Stop 3—Pointe du Sud—Ouest shore section. The section here referred to as the Pointe du Sud-Ouest section comprises that section on the north shore of Baie de Port-Daniel; it extends from a small brook northwest of Anse Elliot to Pointe du Sud-Ouest (previously called West Point) and the shore section from the point toward the southwest for a distance of approximately 1.4 mi (2.3 km) (Figs. 1, 6). The whole section can be walked quite easily *at low tide.* The public road ends at Chaleurs Chalet Resort in the Anse Beebe; permission to walk or drive to the point should be obtained from Alice or Rod Hayes, the owners of the resort.

Strata of this section dip 40° to 60° to the south, and a complete section from the lower part of the Gascons Formation to the upper part of the West Point Formation can be observed. The section can be divided into four segments representing four different depositional facies (Figs. 3, 6). Shore segment a-b displays a homogeneous sequence of fine-grained siliciclastic rocks of the Gascons Formation representing offshore mud-type facies. Shore segment b-c shows the same fine-grained siliciclastic facies as in the Gascons Formation, but with a few debris flow units derived from the lower reef and the middle bank complexes of the West Point Formation. In shore segment c-d, coral and stromatoporoid limestone of the reef margin facies of the upper reef complex of the West Point Formation is seen. Finally, shore segment d-e contains fine-grained siliciclastic rocks, cryptocyanobacterial laminites, amphiporid beds, stromatoporoid rubble units, nodular dolomites, stromatoporoid biostromes, all lithologies representing the intertidal, lagoonal, and back-reef facies of the upper reef complex of the West Point Formation. The two upper segments of the section are particularly fossiliferous (stromatoporoids and corals). Like the upper part of the section at Pointe au Bouleau (Stop 1), these beds can be easily harmed by oversampling. West of point e, walking along the shore becomes hazardous, even at low tide, so that it is advisable to go to the

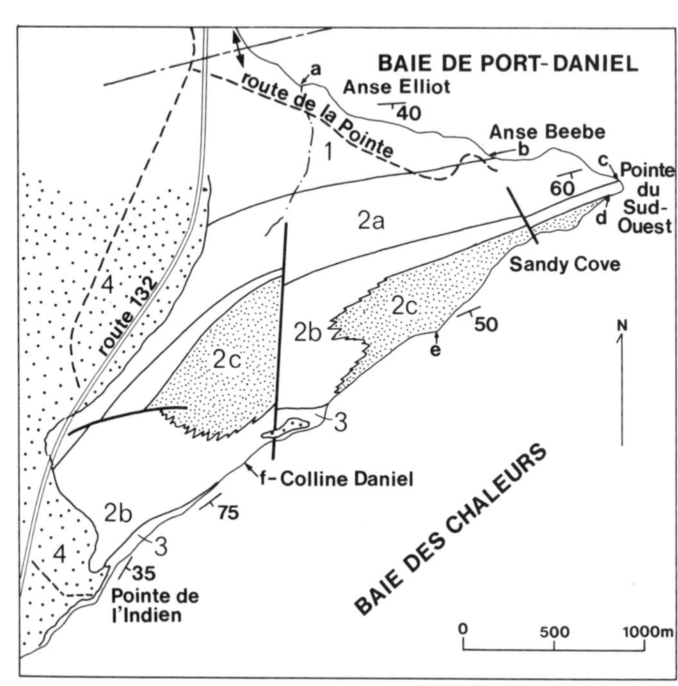

Figure 6. Geologic map of Pointe du Sud-Ouest–Pointe de l'Indien area in Port-Daniel West (Stops 3 and 4): 1, Gascons Formation; 2, West Point Formation; 2a, distal fore-reef facies of lower reef complex; 2b, 2c, reef margin and back-reef to intertidal facies of upper reef complex; 3, Indian Point Formation; 4, carboniferous cover. (From Bourque, 1979, Fig. 13.)

section at the foot of Colline Daniel, east of Pointe de l'Indien (Stop 4) to observe the upper part of the West Point Formation. A detailed description of the section has been written by Bourque (1979, Stops 9 and 10; 1982, sections 1-3).

Stop 4—Shore outcrop at foot of Colline Daniel. This outcrop is easily reached by a small farm road from Québec 132, leading to the beach at Pointe de l'Indien (Fig. 6). The foot of the Colline Daniel (point f in Fig. 6) is reached by walking northeast along the beach *at low tide,* for a distance of about 1,600 ft (500 m).

At the foot of the Colline Daniel, limestones of the reef margin and the shallow fore-reef facies of the upper reef complex of the West Point Formation can be observed. Large blocks that have fallen from the cliff are particularly worth examining because they offer a good sampling of the section and avoid the need to climb the cliff. Walking back to Pointe de l'Indien permits observation of the relatively deep-water fine-grained siliciclastic facies of the Indian Point Formation abruptly overlying the reef margin limestone facies.

Section 5—The Black-Cape section. The base of the section is reached by "rue du Port" in Black-Cape that leads to the beach (Fig. 7) where the section begins outcropping continuously along the shore toward the southeast, for a distance of more than 1.8 mi (3 km) without a break. The section can also be reached at two other points: (1) about midway, by a small road leading to Black-Cape public beach; and (2) at the southeastern end of the section, by a small road leading to Black-Cape Cemetery and from there to the beach at Les Caps Noirs (Black Capes). The section can be walked throughout its length without getting wet *only at low tide.*

The section at Black-Cape dips steeply and faces southeast. A complete succession, from the upper Anse Cascon Formation to the volcanic rocks of the Black-Cape Member that overlies the West Point Formation, can be observed. Various facies are worthy of examination, including the peritidal algal limestones (laminites, oncolites, and stromatolites) of the middle member of the La Vieille Formation, the reef-derived debris beds of the lower part of the West Point Formation, the red bed sequence of the upper member of the West Point Formation, and the volcanic lava flows and breccias of the Black-Cape Member. A detailed geologic map of Pointe Howatson and a log of the middle algal member of the La Vieille Formation has been provided by Bourque (1981, Stop BC-3), and an outcrop map of Plage Woodmans and a detailed columnar log of the red bed sequence of the West Point Formation is in Bourque (1982, Stop 2-1).

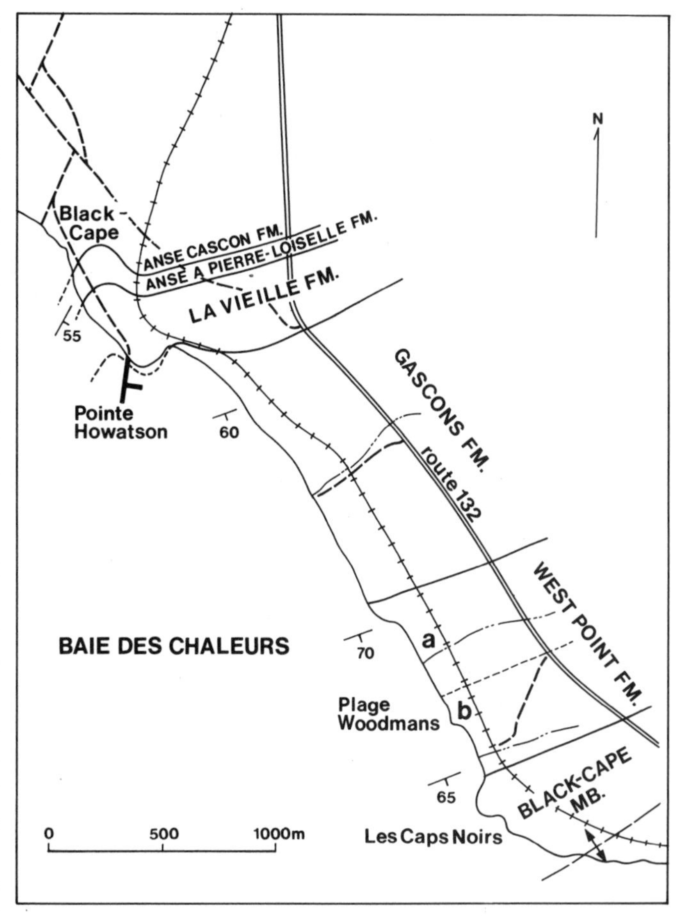

Figure 7. Geologic map of Black-Cape section (Stop 5): a, distal fore-reef facies of lower reef complex; b, red bed sequence of West Point Formation. (From Bourque and Lachambre, 1980, map 1930.)

REFERENCES CITED

Ayrton, W. G., 1967, Chandler–Port-Daniel area: Québec Department of Natural Resources Geological Report 120, 91 p.

Bourque, P. A., 1975, Lithostratigraphic framework and unified nomenclature for Silurian and basal Devonian rocks in eastern Gaspé Peninsula, Québec: Canadian Journal of Earth Sciences, v. 12, p. 858–872.

——, 1979, Facies of the Silurian West Point Reef Complex, Baie des Chaleurs, Gaspésie, Québec: Geological Association of Canada Annual Meeting, Québec City, 1979, Guidebook to Field Trip B-2, 29 p.

——, 1981, Baie des Chaleurs area, *in* Lespérance, P. J., ed., Field Meeting Anticosti-Gaspé, Québec, 1981; International Union of Geological Sciences, Subcommission on Silurian Stratigraphy, Ordovician-Silurian Boundary Working Group: Département de Géologie, Université de Montréal, v. 1, Guidebook, 56 p.

——, 1982, An Upper Silurian reefal limestone platform; From supratidal plain to marginal slope, *in* Hesse, R., Middleton, G. V., and Rust, B. R., eds., Paleozoic continental margin sedimentation in the Québec Appalachians: International Association of Sedimentologists, 11th International Congress, Hamilton, Canada, 1982, Guidebook to Excursion 7B.

Bourque, P. A., and Gignac, H., 1983, Sponge-constructed stromatactis mud mounds, Silurian of Gaspé Peninsula, Québec: Journal of Sedimentary Petrology, v. 53, p. 521–532.

Bourque, P. A., and Lachambre, G., 1980, Stratigraphie du Silurien et du Dévonien basal du Sud de la Gaspésie: Québec Department of Energy and Resources Special Paper ES-30, 123 p., 5 maps.

Bourque, P. A., Amyot, G., Desrochers, A., Gignac, H., Gosselin, C., Lachambre, G., and Laliberté, J. Y., 1986, Silurian and Lower Devonian reef and carbonate complexes of the Gaspé Basin, Québec; A summary: Canadian Petroleum Geology Bulletin (in press).

Burk, C. F., Jr., 1964, Silurian stratigraphy of Gaspé Peninsula, Québec: American Association of Petroleum Geologists Bulletin, v. 48, p. 437–464.

Clarke, J. M., 1912, A remarkable Siluric section of the Bay of Chaleur, *in* Notes on the Geology of the Gulf of St. Laurence: New York State Museum Bulletin 158, p. 120–126.

——, 1913, Excursion in eastern Québec and the Maritime Provinces; Dalhousie and the Gaspé Peninsula: 12th International Geological Congress, Canada, 1913, Guidebook 1, p. 85–108, 110–118.

Gignac, H., 1980, Etude du faciès de Gros Morbe du Complexe récifal inférieur de West Point, région de Port-Daniel, Gaspésie [M.S. thesis]: Québec, Université Laval, 49 p.

Gosselin, C., 1981, Etude paléo-environnementale d'un faciès à Crinoïdes silurien, Formation de West Point, Port-Daniel, Gaspésie, Québec [M.S. thesis]: Québec, Université Laval, 41 p.

Laliberté, J. Y., 1982, Etude des faciès de Bouleaux et d'Anse à la Barbe de la Formation de West Point à la Pointe au Bouleau, Baie des Chaleurs, Gaspésie, Québec [M.S. thesis]: Québec, Université Laval, 53 p.

Logan, W. E., 1863, Report on Geology of Canada: Geological Survey of Canada Report of Progress to 1863, 983 p.

Northrop, S. A., 1939, Paleontology and stratigraphy of the Silurian rocks of the Port Daniel-Black Cape region, Gaspé: Geological Society of America Special Paper 21, 302 p.

Pickerill, R. K., Roulston, B. V., and Noble, J.P.A., 1977, Trace fossils from the Silurian Chaleurs Group of southeastern Gaspé Peninsula, Québec: Canadian Journal of Earth Sciences, v. 14, p. 239–249.

Schuchert, C., and Dart, J. D., 1926, Stratigraphy of the Port-Daniel–Gascons area of southeastern Québec: Geological Survey of Canada Bulletin 44, p. 35–58, 116–121.

Late Ordovician sedimentary rocks and trace fossils of the Aroostook-Matapedia Carbonate Belt at Runnymede, Restigouche River, northern New Brunswick

R. K. Pickerill, Department of Geology, University of New Brunswick, Fredericton, New Brunswick E3B 5A3, Canada

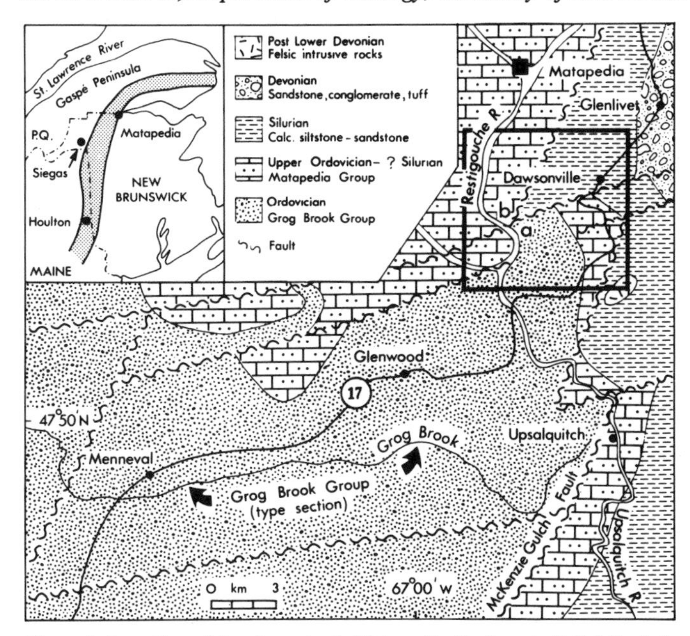

Figure 1. Location of the Aroostook-Matapedia Carbonate Belt (stippled area in inset) and simplified geological sketch map of the Matapedia district, northern New Brunswick (modified from St. Peter, 1977; Nowlan, 1983). Boxed area is illustrated in more detail in Figure 2, and a and b correspond respectively to outcrops of Grog Brook Group and Matapedia Group strata.

LOCATION

Runnymede is a small hamlet located 1,640 ft (500 m) north of the confluence of the Upsalquitch and Restigouche rivers, 4.3 mi (7 km) south of Matapedia, on the northwestern New Brunswick–eastern Quebec border. Like its more famous historical counterpart on the Thames River, southern England, where the Magna Carta was granted by King John in 1215, Runnymede, Quebec, is also located on the inside of a large meander but on the Restigouche River. The Runnymede site is situated on the outside, the New Brunswick side, of the same meander, the Quebec–New Brunswick border being situated along the centre of the Restigouche River. The site is subdivided into two major sections separated by the Rafting Ground Brook, a small tributary of the Restigouche River. Location and geological details are indicated in Figures 1 and 2.

The site is accessible by any sized vehicle at any time of the year. A car park with picnic tables is located just to the southeast of the confluence of the Rafting Ground Brook and Restigouche River (Fig. 2). Permission is not required to examine the site, which can be done thoroughly in approximately three hours.

SIGNIFICANCE

The Runnymede site exposes Late Ordovician sedimentary rocks of the Matapedia Group and partially coeval and underlying Grog Brook Group, the two major lithostratigraphic units of the New Brunswick part of the Aroostook-Matapedia Carbonate Belt, which extends from eastern Gaspé through New Brunswick and into Maine (Fig. 1). The contact between the two groups is faulted but both units are well- and continuously-exposed and exhibit a variety of well-preserved sedimentological and palaeontological features that cannot be more conveniently and accessibly observed elsewhere in the belt. Sedimentological features provide insight into sediment transport mechanisms and depositional environments of the two groups. Both groups at the Runnymede site have been precisely dated palaeontologically as Late Ordovician in age based on both graptolite and, more particularly, conodont faunas.

Of particular interest is the contained assemblage of trace fossils that constitutes at least 19 ichnogenera. The Aroostook-Matapedia Carbonate Belt contains the most diverse and abundant Ordovician flysch ichnofauna yet recorded. The Runnymede site enables examination of a wide variety of these trace fossils.

SITE INFORMATION

The site is located within the Aroostook-Matapedia Carbonate Belt of Ayrton and others (1969), which is equivalent to the Aroostook Anticlinorium of Pavlides (1968) and Rodgers (1970) or the Matapedia Basin of Fyffe and others (1981). The belt comprises a narrow tectonostratigraphic zone extending from eastern Gaspé through New Brunswick and into northern Maine (Fig. 1), where it merges with the Merrimack Trough (Bradley, 1983). Strata in this belt have been assigned various group and, or, formational names at different locations along its length. In Gaspé, strata are referred to the late Middle Ordovician–Early Silurian Honorat and Matapedia groups, the latter including the Late Ordovician–Early Silurian White Head Formation (Nowlan, 1981). In New Brunswick the strata include the late Middle Ordovician–Early Silurian Grog Brook and Matapedia groups (St. Peter, 1977; Nowlan, 1981), which farther south are referred to as the Carys Mills Formation of latest Ashgillian or earliest Silurian age (Rickards and Riva, 1981). Thickness of the Grog Brook Group has been estimated at 25,000 ft (7,600 m) and the laterally equivalent and overlying Matapedia Group between 4,000 and 13,000 ft (1,250 and 4,000 m; St. Peter, 1977).

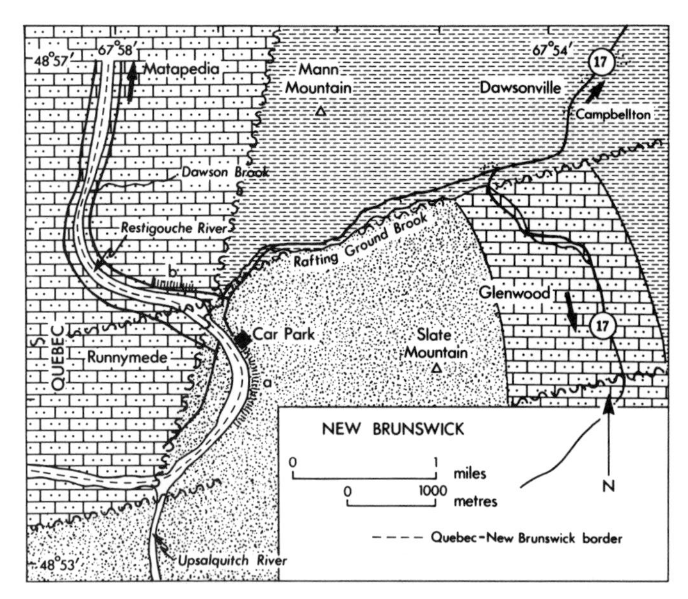

Figure 2. Detailed locality map of the Runnymede site on the east bank of the Restigouche River: a, Grog Brook Group strata; and b, Matapedia Group strata. Legend as in Figure 1 and geology after St. Peter (1977).

In New Brunswick these late Middle Ordovician–Early Silurian sequences are conformably overlain by Lower Silurian to Lower Devonian (Siegenian-Emsian) clastics in the northwest and by a suite of Late Silurian–Early Devonian (Gedinnian) volcanic units localized along the southern margin of the northeasterly trending belt (Fyffe and others, 1981). All strata within the belt are metamorphosed to sub-greenschist facies. Margins with adjacent tectonostratigraphic zones are commonly fault bounded.

In the Matapedia-Runnymede area of northwestern New Brunswick, contacts between the Grog Brook and Matapedia groups are commonly faulted (Fig. 1) and the strata are deformed by upright folds and possess a steeply dipping cleavage. The Runnymede site exposes sections within both groups (a and b of Fig. 2), but, as elsewhere, the contact between them is faulted.

Grog Brook Group. At Runnymede this group (Fig. 2, section a) comprises 738 ft (225 m) of fine- to medium-grained sandstones and interbedded argillites. Lenticular calcarenites are locally developed from which Nowlan (1983) has recorded diverse conodonts of Late Ordovician (probably Gamachian) age. The conodonts are of mixed provincial affinity including North Atlantic Province representatives (e.g., *Hamarodus, Icriodella, Periodon,* and *Protopanderodus*) and Midcontinent representatives (e.g., *Rhipidognathus* and *Phragmodus*). The strata are steeply but variably dipping, face northward and strike approximately normal to the Restigouche River, and are intruded by several mafic dykes up to 20 ft (6 m) thick.

The lower 295 ft (90 m), observed at the most southerly end of the exposure, are comprised of sandstone and argillite with sandstone beds typically 8 to 16 in (20 to 40 cm) but up to 6 ft (2 m) thick, the sandstone representing approximately 70% of the total thickness. Between 295 and 377 ft (90 and 115 m), sandstones decrease in thickness to an average of 2 to 4 in (5 to 10 cm) and argillite constitutes 50% of the section. From 377 to 492 ft (115 to 150 m) sandstones are reduced to 0.8 to 2 in (2 to 5 cm) thick, argillite constituting 70% of the section; above this the remainder of the section consists of 50% each of argillite and sandstone in beds typically 0.8 to 2 in (2 to 5 cm) thick (Nowlan, 1983).

Internally, the majority of sandstones are parallel and/or cross-laminated with many exhibiting spectacular examples of convolute lamination. Several exhibit parallel- passing into cross-laminated sets. Lower bedding surfaces may be planar and nonerosive or may exhibit morphologically variable flute and groove markings. Load and flame structures are also present. Upper surfaces are sharp and typically planar, though several, particularly in the upper part of the sequence, exhibit linguoid ripples. Both upper and, more typically, lower sandstone surfaces contain abundant trace fossils.

Sandstones of the Grog Brook Group are interpreted as turbidites (St. Peter, 1977; Pickerill, 1980), but, although St. Peter (1977) has suggested a bathyal or even abyssal environment of deposition, absolute depth *per se* is difficult to realistically assess. The trace fossils are, however, typically "deep water" ichnogenera (e.g., *Cosmorhaphe, Paleodictyon, Spirorhaphe,* etc.).

Matapedia Group. At Runnymede this group (Fig. 2, section b) is well exposed in a continuous section of approximately 165 ft (50 m) on both sides of the road leading to Matapedia. Detailed examination is best undertaken adjacent to the Restigouche River, that is, immediately below the western side of the road. The group consists of thinly interbedded (0.4 to 2 in; 1 to 5 cm) calcic and ankeritic limestones and calcareous argillites. Diffential solution has etched the more limy beds to a greater extent so that the rocks have a markedly ribbed appearance and may be referred to as 'ribbon limestones' (cf. Ayrton and others, 1969). The limy beds are generally massive but a few (<10%) exhibit parallel-and, or, cross-lamination. Most have planar bases and, rarely, some are erosive. Isolated small-scale channelized lenticular beds may also be observed, which are typically infilled with cross-lamination sets. Interbedded calcareous argillites are rarely finely laminated but more typically extremely bioturbated by the *Chondrites*-producing organism.

Rickards and Riva (1981) have recorded the graptolite *Dicellograptus* cf. *D. complanatus* from the location, thus attesting to its Late Ordovician age. Stringer and Pickerill (1980) have outlined criteria suggesting that the Matapedia Group represents slope deposits formed by hemipelagic and normal bottom-following contour currents with periodic introduction of calcisiltite by turbidity currents. While details of sediment transport directions remain to be resolved, these slope deposits are considered to have been deposited marginal to the northeast trending Miramichi Anticlinorium (see Rodgers, 1970; McKerrow and Ziegler, 1971; Fyffe and others, 1981), the tectonostratigraphic zone to the southeast of the Aroostook-Matapedia Carbonate Belt.

Ichnology. Ichnology is the study of trace fossils or biogenic sedimentary structures and although there is no formalized code, most researchers adopt a binomial nomenclatural system of ichnogenera and ichnospecies. These are named respectively on distinctive significant and accessory morphological characteristics. The actual producer of a trace fossil is invariably impossible to determine as the structures simply reflect behavioural activity, and unrelated organisms can behave in a similar fashion depending on physical or biological parameters. A peculiarity of ichnology is that the best collections are typically made in loose or float material. In this context the Runnymede site is no exception, but the abundant float material available on the river banks facilitates observation and collecting. Additionally, trace fossils can also readily be observed in situ. Although trace fossils commonly occur in strata of the Aroostook-Matapedia Carbonate Belt and lateral equivalents in south-central Maine (see for example, Emmons, 1844; Raymond, 1931; Roy and Naylor, 1980; Pickerill, 1980, 1981) the Runnymede site contains the most abundant, diverse, and well-preserved ichnoassemblage from a single location within the belt.

The following is a list of trace fossils identified at the ichnogeneric level at Runnymede, with the relative abundance indicated by A = Abundant, C = Common, and R = Rare. Figure 3 is a sketch of all these ichnogenera and is included to facilitate identification. A brief diagnosis of individual ichnogenera is included at the end of the section. Further details can be found in Pickerill (1980) and Pickerill and Forbes (1980).

Grog Brook Group. *Asteriacites* (A), *Buthotrephis* (A), *Cosmorphaphe* (R), *Glockerichnus* (R), *Gordia* (A), *Helminthopsis* (A), *Neonereites* (C), *Nereites* (R), *Palaeophycus* (A), *Paleodictyon* (R), *Planolites* (A), *Protopaleodictyon* (C), *Spirodesmos* (R), *Spirorhaphe* (R), *Stroliborhaphe* (R), *Taenidium* (A).

Matapedia Group. *Chondrites* (A), *Helminthopsis* (C), *Planolites* (A), *Scalarituba* (A), *Syncoprulus* (A).

Elsewhere in the Matapedia area Pickerill (1980) has also recorded the ichnogenera *Asterosoma, Belorhaphe, Cochlichnus, Diplichnites, Gyrochorte, Helminthoida,* and *Yakutatia* as well as several questionable forms but, to date, none of these have been recorded from Runnymede. Clearly, however, the total assemblage is consistent with the *Nereites* ichnofacies of Seilacher (1967), that is the ichnofacies characteristic of deep-sea environments.

At Runnymede, the Matapedia Group contains a trace fossil assemblage of great abundance but generally low diversity. All the traces represent activity by deposit-feeders. The generally low energy environment was not conducive to inhabitation by suspension feeders and related groups. In contrast, the Grog Brook Group contains a very diverse assemblage and includes traces reflecting deposit feeding, filter feeding, and resting activity. The increased diversity is related to increased energy conditions that permitted additional feeding and behavioural groups to become firmly established. Thus, the variation in type and diversity of trace fossils in the two groups is probably a reflection of variation in original energy levels, which in turn resulted in different nutrient levels, substrate types, and ecological interactions. Variation in water depth *per se* was probably unimportant.

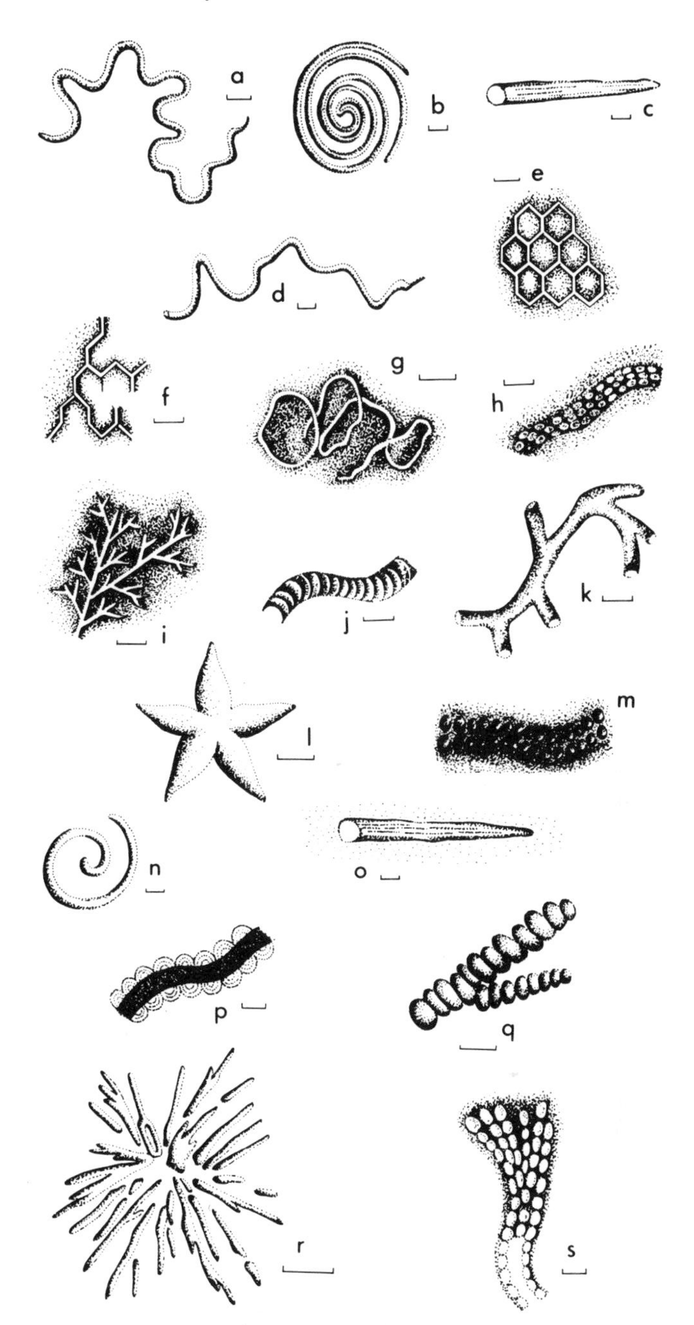

Figure 3. Artistic illustrations of trace fossils observed at the Runnymede site. Letters correspond to a, *Cosmorhaphe*; b, *Spirorhaphe*; c, *Palaeophycus*; d, *Helminthopsis*; e, *Paleodictyon*; f, *Protopaleodictyon*; g, *Gordia*; h, *Syncoprulus*; i, *Chondrites*; j, *Scalarituba*; k, *Buthotrephis*; 1, *Asteriacites*; m, *Neonereites*; n, *Spirodesmos*; o, *Planolites*; p, *Nereites*; q, *Taenidium*; r, *Glockerichnus*; s, *Stroliborhaphe.* Bar scale in each case is 0.4 in (1 cm).

Finally, it is noteworthy that strata of the Aroostook-Matapedia Carbonate Belt contain the most diverse assemblage of trace fossils yet recorded from an Ordovician flysch sequence (Pickerill, 1980). The Runnymede site is an excellent one to examine this apparently anomalous and diverse trace fossil assemblage.

Asteriacites. Small starfish-shaped trace fossil resting impressions preserved on lower sandstone surfaces in positive relief with five arms projecting from a centrak knob or subcircular area.

Buthotrephis. Irregularly branched cylindrical or subcylindrical burrows, parallel to stratification with a burrow fill that is coarser or finer-grained than surrounding material.

Chondrites. Systematically branching dendritic tunnel-like systems of uniform diameter; parallel, normal, or oblique to stratification. Burrow fill identical to or differs from enclosing sediment.

Cosmorhaphe. Meandering burrow preserved in positive relief on lower surfaces possessing two orders of meanders or undulations with small second order meanders superimposed on larger first order undulations.

Glockerichnus. Star-shaped radiating trace, parallel to stratification, with many branched or unbranched ribs radiating from a diffuse central zone.

Gordia. Slender non-branching burrows of uniform diameter that are parallel to stratification and exhibit non-systematic winding, looping and meandering patterns, and typically cross-cut.

Helminthopsis. Irregularly meandering unbranched cylindrical burrows parallel to stratification but never cross-cutting. *Helminthopsis* is typically of larger diameter in comparison to *Gordia.*

Neonereites. Curved or straight chain of biserially arranged sediment pods (= *N. biserialis*) preserved in positive relief on lower surfaces.

Nereites. Curved or straight horizontal burrow with arcuate leaf-like projections paralleling both sides of the burrow walls.

Palaeophycus. Curved or straight horizontal cylindrical or subcylindrical horizontal burrow with burrow fill of identical grain size to enclosing matrix.

Paleodictyon. Regular hexagonal honeycomb-like network of ridges preserved in positive relief on sandstone soles.

Planolites. Curved or straight horizontal cylindrical or subcylindrical horizontal burrow with burrow fill of different grain size to enclosing matrix.

Protopaleodictyon. Short and narrow zigzagging network of ridges preserved in positive relief on sandstone soles. Similar to *Paleodictyon* but networks are never complete.

Scalarituba. Straight or sinuous horizontal burrows characterized by closely spaced scalariform ridges.

Spirodesmos. Spirally coiled trace with individual coils widely spaced and preserved in positive relief on sandstone soles.

Spirorhaphe. Spirally coiled burrow in which a single tube coils inwards towards the centre of the structure then back outwards parallel to and between the initial coils. Preserved in positive relief on sandstone soles.

Stroliborhaphe. Horizontal straight to curved burrow consisting of a central stem that ramifies into knobs arranged in a cone-like form.

Syncoprulus. Horizontal straight to curved burrow stuffed with ovoid or elliptical faecal pellets.

Taenidum. Branched and unbranched cylindrical burrows preserved on sandstone upper surfaces and possessing annular constrictions or transverse segmentation.

REFERENCES CITED

Ayrton, W. G., and 6 others, 1969, Lower Llandovery of the northern Appalachians and adjacent regions: Geological Society of America Bulletin, v. 80, p. 459–484.

Bradley, D. C., 1983, Tectonics of the Acadian orogeny in New England and adjacent Canada: Journal of Geology, v. 91, p. 381–400.

Emmons, E., 1844, The Taconic System; Based on observations in New York, Massachusetts, Maine, Vermont, and Rhode Island: Albany, Caroll and Cook, 68 p.

Fyffe, L. R., Pajari, G. E., Jr., and Cherry, M. E., 1981, The Acadian plutonic rocks of New Brunswick: Maritime Sediments and Atlantic Geology, v. 17, p. 23–36.

McKerrow, W. S., and Ziegler, A. M., 1971, The Lower Silurian paleogeography of New Brunswick and adjacent areas: Journal of Geology, v. 79, p. 635–646.

Nowlan, G. S., 1981, Late Ordovician–Early Silurian conodont biostratigraphy of the Gaspé Peninsula; A preliminary report, *in* Lesperance, P. J., ed., Stratigraphy and paleontology Volume II: Subcommission on Silurian Stratigraphy, Ordovician–Silurian Boundary Working Group, field meeting, Anticosti-Gaspé, P.Q., p. 257–291.

——, 1983, Biostratigraphic, paleogeographic, and tectonic implications of Late Ordovician conodonts from the Grog Brook Group, northwestern New Brunswick: Canadian Journal of Earth Sciences, v. 20, p. 651–671.

Pavlides, L., 1968, Stratigraphic and facies relationships of the Carys Mills Formation of Ordovician and Silurian age, northeastern Maine: U.S. Geological Survey Bulletin 1264, 44 p.

Pickerill, R. K., 1980, Phanerozoic flysch trace fossil diversity; Observations based on an Ordovician flysch ichnofauna from the Aroostook-Matapedia Carbonate Belt of northern New Brunswick: Canadian Journal of Earth Sciences, v. 17, p. 1259–1270.

——, 1981, Trace fossils in a Lower Palaeozoic submarine canyon sequence; The Siegas Formation of northwestern New Brunswick, Canada: Maritime Sediments and Atlantic Geology, v. 17, p. 37–58.

Pickerill, R. K., and Forbes, W. H., 1980, Ordovician and Devonian sediments, fossils, and ichnofossils of northern New Brunswick: in Pickerill, R. K., ed., Canadian Paleontology and Biostratigraphy Seminar, Field trip guidebook, Fredericton, New Brunswick, p. 15–29.

Raymond, P. E., 1931, Notes on invertebrate fossils with descriptions of new species: Trails from the Silurian at Waterville, Maine, Bulletin of the Museum of Comparative Zoology, v. 55, no. 4, p. 184–194.

Rickards, R. B., and Riva, J., 1981, *Glyptograptus? persculptus* (Salter), its tectonic deformation, and its stratigraphic significance for the Carys Mills Formation of Northeast Maine, U.S.A.: Geological Journal, v. 16, p. 219–235.

Rodgers, J., 1970, Tectonics of the Appalachians: New York, Interscience, 271 p.

Roy, D. C., and Naylor, R. S., eds., 1980, A guidebook to the geology of northeastern Maine and neighbouring New Brunswick: 72nd annual meeting of the New England Intercollegiate Geological Conference, Presque Isle, Maine, 296 p.

Seilacher, A., 1967, Bathymetry of trace fossils: Marine Geology, v. 5, p. 413–428.

St. Peter, C., 1977, Geology of parts of Restigouche, Victoria, and Madawaska counties, northern New Brunswick: Mineral Resources Branch, Department of Natural Resources, New Brunswick, Report of Investigation 17, 69 p.

Stringer, P., and Pickerill, R. K., 1980, Structure and sedimentology of the Siluro-Devonian between Edmundston and Grand Falls, New Brunswick, *in* Roy, D. C., and Naylor, R. S., eds., The Geology of northeastern Maine and neighboring New Brunswick: New England Intercollegiate Geological Conference, p. 262–277.

Stratigraphy and tectonics of Miramichi and Elmtree terranes in the Bathurst area, northeastern New Brunswick

L. R. Fyffe, New Brunswick Department of Forest, Mines, and Energy, P.O. Box 6000, Fredericton, New Brunswick E3B 5H1, Canada

LOCATION

The city of Bathurst is located on Chaleur Bay in Gloucester County, northeastern New Brunswick (Fig. 1). Ordovician rocks of the Tetagouche Group that can be examined near the city are within the Bathurst Quadrangle (NTS 21 P/12), whereas those of the Fournier and Elmtree groups are seen to the north in the Pointe Verte Quadrangle (NTS 21 P/13). All described sites in the area are public and readily accessible by car; shore sections are best visited at low tide.

Stop 1 is located at the provincial picnic site at Tetagouche Falls on the road to South Tetagouche. It is reached by taking Exit 310 off the Bathurst bypass (New Brunswick 11) and proceeding 6 mi (10 km) west (Figs. 1 and 2). Stop 2 is located on the Bathurst bypass 2.2 mi (3.5 km) north of Exit 310 (Figs. 1 and 3). Stop 3 is located on New Brunswick 11 about 0.6 mi (1 km) north of the Elmtree River bridge (Fig. 1). Stop 4 is located on New Brunswick 11 at Exit 333 overpass (Fig. 1). Stop 5 is reached by taking the access road (Rue de la Gare) to Pointe Verte at Exit 333 on New Brunswick 11; proceed 3 mi (5 km) east along access road; pull off to the right just past railway crossing (Green Point Station) and walk south to quarry (Figs. 1 and 4). Stop 6 is reached from Rue Principale (New Brunswick 134); proceed 2.8 mi (4.5 km) north on New Brunswick 134 from Rue de la Gare in Pointe Verte; turn right to shore on paved side road just past the Belledune Village limit sign (Figs. 1 and 5). Stop 7 is accessed from Rue de la Bateau which leads to the shore off Rue Principale in Pointe Verte (Fig. 1). Stop 8 is accessed from the Doucet Road, which leads off Rue Principale to the northern end of Limestone Point on Chaleur Bay (Fig. 1).

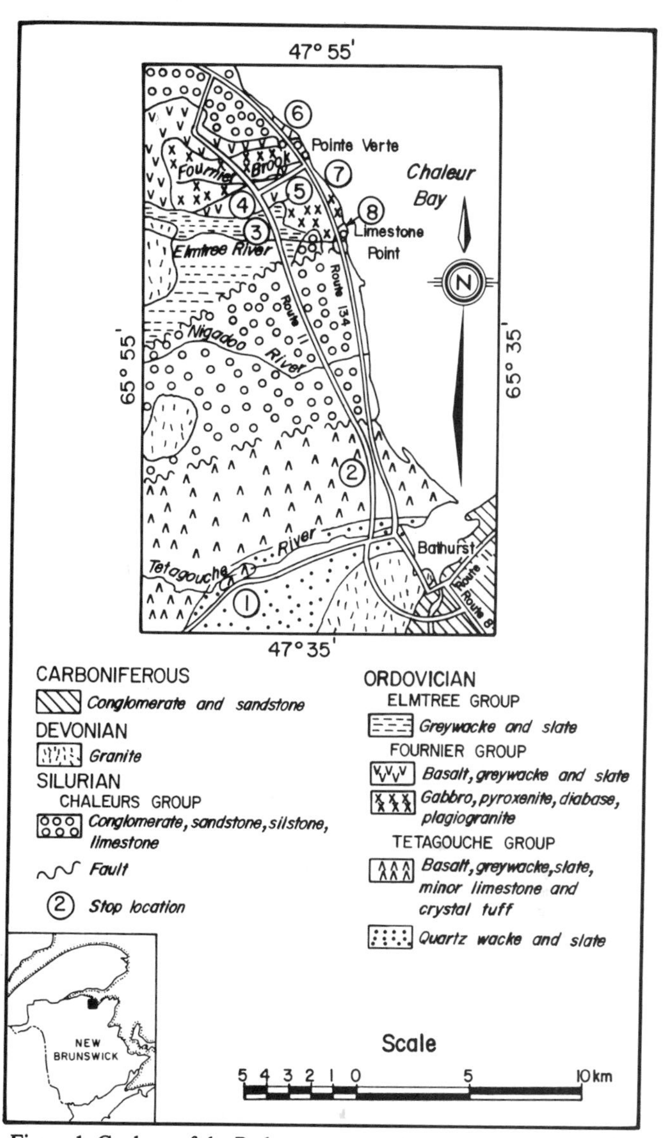

Figure 1. Geology of the Bathurst area, northeastern New Brunswick.

SIGNIFICANCE

Ordovician rocks of the Bathurst area comprise two markedly different tectonostratigraphic zones or terranes separated by a graben of Silurian sediments (Fig. 1). The Miramichi terrane to the south of the graben is underlain by the Tetagouche volcanic arc sequence (Stops 1 and 2), which extends southwestward from Bathurst to Woodstock and then across the Maine border. Twelve mi (20 km) southwest of Bathurst, iron formation interbedded with felsic volcanic rocks of the Tetagouche Group is host to one of the world's largest stratiform base metal ore deposits. The Elmtree terrane to the north of the Silurian graben contains the Fournier ophiolite complex (Stops 3 to 8) marking the site of a former ocean basin. The Miramichi and Elmtree terranes correspond respectively to the Gander and Dunnage terranes of Williams and Hatcher (1982).

Those wishing to visit the Brunswick Number 12 base metal mine should contact Bill Luff, mine geologist, Brunswick Mining and Smelting Corporation, Limited at (506) 546-6671. Additional geological sites of interest in northern New Brunswick are listed on the Geological Highway Map of New Brunswick and Prince Edward Island (Ferguson and Fyffe, 1985) obtainable from the New Brunswick Department of Forests, Mines, and Energy, Mineral Resources Division, P.O. Box 50, 495 Riverside

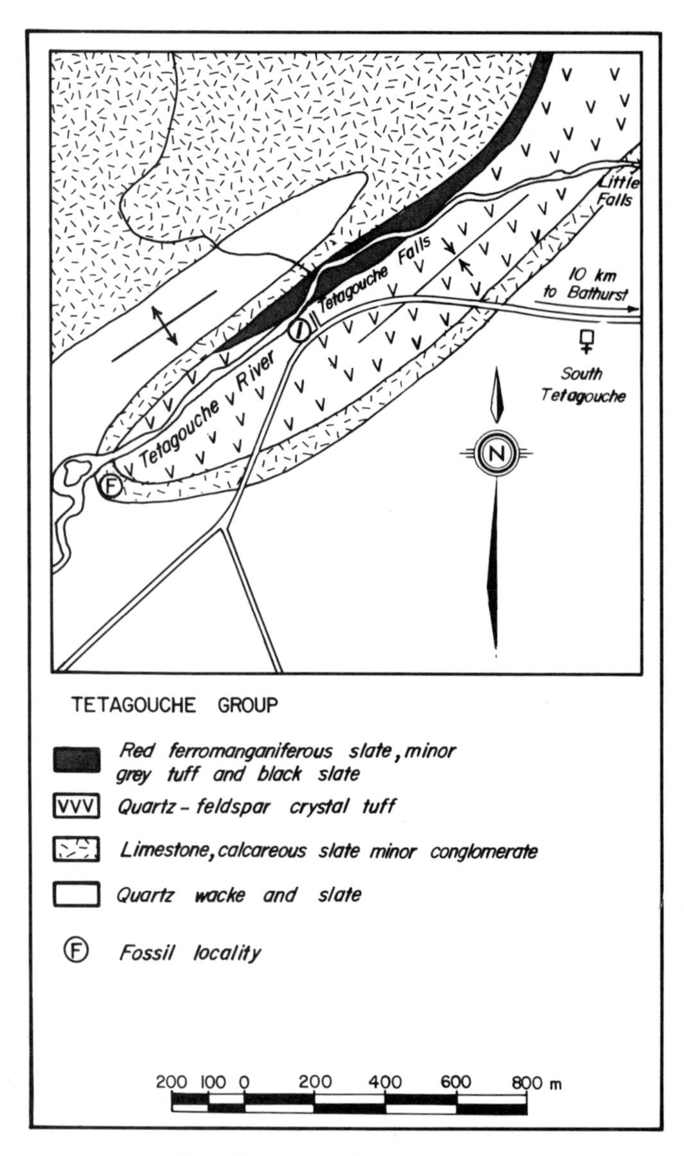

Figure 2. Geology of Tetagouche Falls.

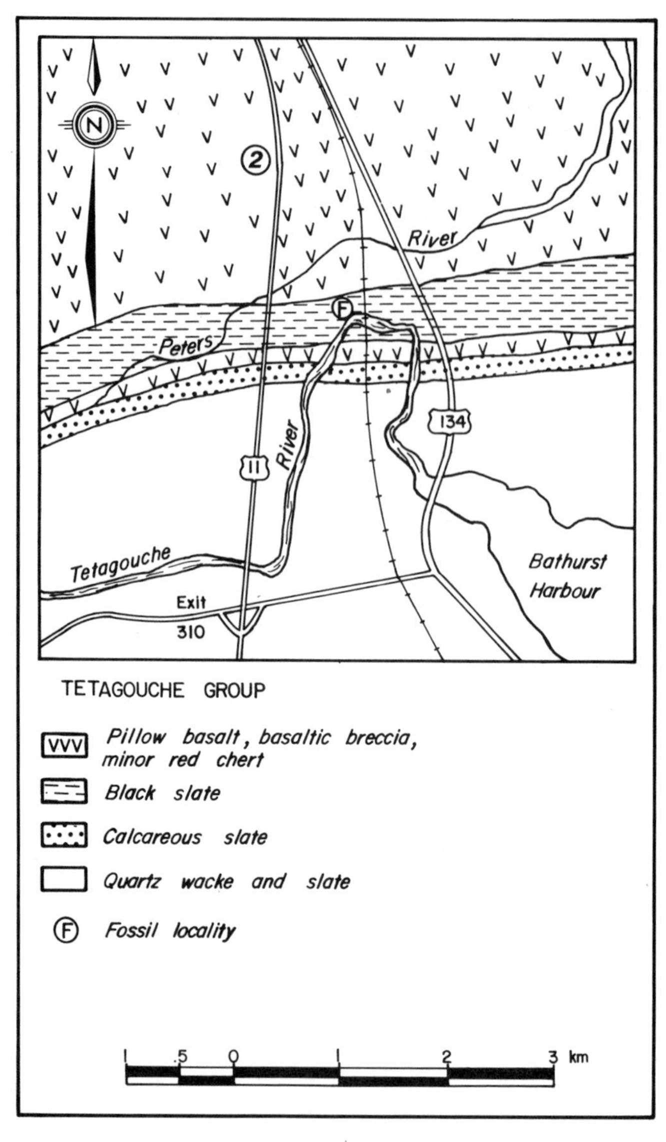

Figure 3. Geology of Bathurst bypass, showing location of Stop 2.

Drive, Bathurst, N.B., E2A 3Z1 or P.O. Box 6000, Fredericton, N.B., E3B 5H1.

SITE INFORMATION

Miramichi Terrane. The lower part of the Tetagouche Group contains a thick sequence of quartzose graywacke, quartzite, and slate which, in central and northern New Brunswick, is locally overlain by a thin calcareous siltstone containing an Early Ordovician (Arenig) brachiopod fauna (Neuman, 1968, 1984; Fyffe, 1976). Graptolites of about the same age (Tremadoc and Arenig) are found in black slate intercalated with quartzite in the southwestern part of the terrane (Fyffe and others, 1983). These quartzose rocks have been interpreted as a late Precambrian (Hadrynian) to Early Ordovician rift facies developed off the northern margin of the Avalonian Platform (Rast and others, 1976; Poole, 1976; Ruitenberg and others, 1977).

The upper part of the Tetagouche Group comprises rhyolite, quartz-feldspar crystal tuff, and pillow basalt intercalated with red and black slate and chert, iron formation, and minor limestone. Lithic graywacke containing abundant detritus eroded from the underlying volcanic rocks forms the youngest part of the group (Helmstaedt, 1971; Skinner, 1974). Graptolites in black slate and trilobites and conodonts from limestone indicate that the upper Tetagouche is Middle Ordovician (Caradoc) in age (Alcock, 1935; Kennedy and others, 1979; Nowlan, 1981).

The broadly circular distribution of the Ordovician felsic volcanic rocks in the Bathurst area suggests the presence of a large caldera complex (Davies, 1966; Harley, 1979). The abundance of these felsic volcanics and granites of both Ordovician and Devo-

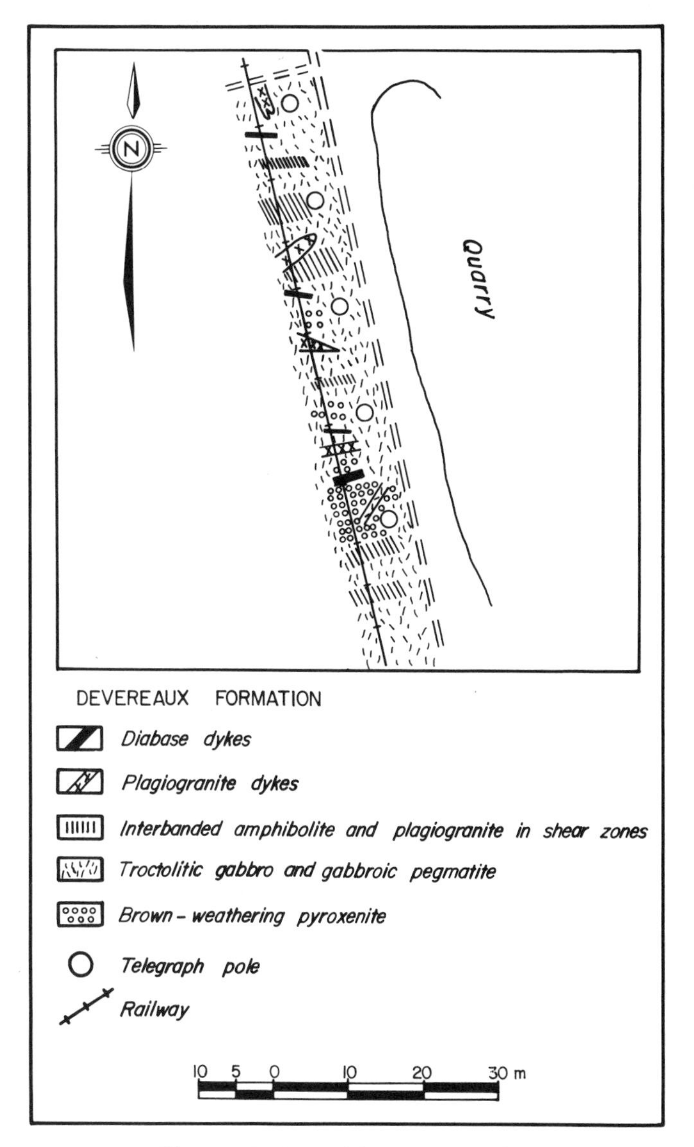

Figure 4. Geology of Green Point Station.

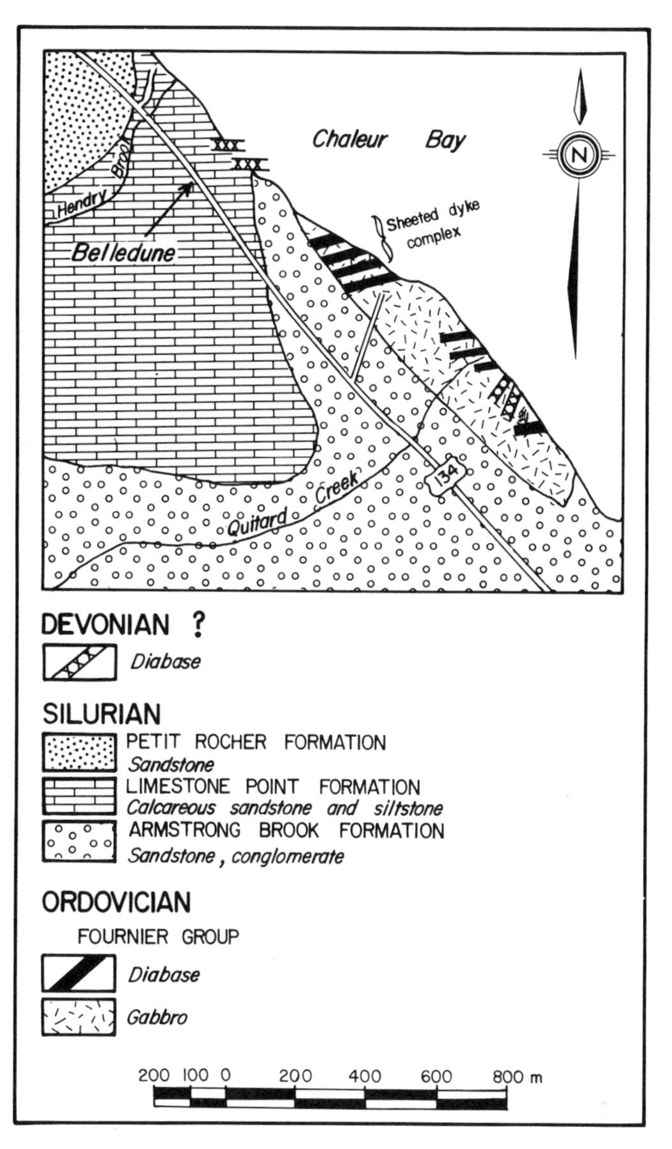

Figure 5. Geology of Belledune area.

nian ages implies that the terrane is a fragment of continental crust (Fyffe and others, 1981).

Chemical analyses of the Tetagouche volcanic rocks reveal a bimodal distribution with a conspicuous absence of andesitic compositions. The basaltic rocks are spilitic, which makes it impossible to use alkalis to classify them. However, titanium contents indicate that both alkaline and subalkaline basalts are present. A comparison of titanium, zirconium, and yttrium ratios further indicates that the tholeiitic basalts have ocean floor affinities, whereas the alkali basalts are intraplate. The felsic volcanics range from dacite to rhyolite in composition and have locally been affected by potassium metasomatism (Whitehead and Goodfellow, 1978).

The stratigraphic succession as well as the bimodal and trace element characteristics of the Tetagouche volcanic rocks are similar to the Carboniferous submarine volcanics of the Iberian Pyrite Belt of southern Spain, which have been interpreted to have formed in an extensional intracontinental basin (Munha, 1979). Such a tensional setting for the Ordovician volcanics of the Tetagouche Group could exist if the relative motion of the continental plate represented by the Miramichi terrane was directed away from the subducting oceanic plate.

Conglomerate of the Silurian Chaleurs Group, unconformably overlying the northern margin of the Miramichi terrane (Helmstaedt, 1971), brackets the period of Taconian deformation, metamorphism, and uplift between the Middle Ordovician (Caradoc) and Late Silurian (Ludlow).

Elmtree Terrane. An Ordovician inlier within Silurian rocks to the north of the Miramichi terrane contains an ophiolitic complex, termed the Fournier Group, that is conformably overlain by turbidites of the Elmtree Group. Deformed gabbro and peridotite intruded by dikes and veins of plagiogranite (Devereaux Formation) form the core of the complex. The core is enveloped by pillow basalt interbedded with dark gray slate, graywacke, and mélange (Pointe Verte Formation). Sheeted dikes are present locally (Pajari and others, 1977; Rast and Stringer, 1980). The Elmtree Group is composed of lithic and quartzose graywacke interbedded with slate and minor limestone, conglomerate, and mafic volcanic rocks. Lower Silurian conglomerate of the Chaleurs Group unconformably overlies rocks of the inlier.

A lack of an associated gravity anomaly over the Fournier Group and its intrusion by a stock of Devonian granite suggest that the complex is allochthonous and, therefore, represents an obducted remnant of Ordovician ocean crust. Volcanic rocks of the Tetagouche Group may have been produced by subduction of this ocean crust to the southeast under the continental margin of the Miramichi terrane.

The Taconic orogeny was caused by contraction of the Iapetus Ocean in the mid-Ordovician (Stevens, 1970; Bird and Dewey, 1970; Church and Stevens, 1971). In particular, Taconian deformation in the Maritime Appalachians was the result of collision between the Miramichi arc terrane and the North American foreland with consequent obduction of the Fournier ophiolite.

STOP DESCRIPTIONS

STOP 1. Felsic Tuffs, Tetagouche Group. Southward dipping, red ferromanganiferous slate interbedded with thin, gray tuff beds underlie Tetagouche Falls. Crystal tuff is exposed in a road-cut across from the entrance to the park. There has been interest in the red slates as a source of manganese and iron since 1842. This locality is estimated to contain 39 million tons, averaging 5.5% manganese (Skinner, 1974).

A lower part of the stratigraphic sequence is exposed 2,500 ft (750 m) upstream from Tetagouche Falls where boudinaged quartzose graywacke is overlain to the east with apparent conformity by dark gray calcareous slate containing very deformed brachiopods identified by Neuman (1984) as *Orthambonites, Rugostrophia, Tritoechia, Orthisocrania,* and *Multispinula* of Arenigian age. The lower succession is repeated 0.6 mi (1 km) downstream from Tetagouche Falls where quartzose graywacke and slate at the foot of Little Falls are disconformably overlain to the west by cross-bedded sandy limestone containing abundant pebbles of the graywacke at its base. The conodont species *Protopanderodus rectus* and *Drepanoistodus basiovalis,* which have been recovered from the limestone, range in age from middle Arenig to early Llanvirn (Nowlan, 1981). This recent age assignment strengthens the interpretation that the sandy limestone is a shallow-water facies equivalent of the calcareous slate found above Tetagouche Falls as suggested by Fyffe (1976) on the basis that both are underlain by quartzose graywacke and overlain by crystal tuff (Fig. 2).

STOP 2. Pillow Basalt, Tetagouche Group. Walk along gravel road to quarry on west side of highway (Location 2, Fig. 3). Large, well-preserved pillows, locally with red chert between them, show that these alkalic basalts young northward. Basaltic agglomerate underlies the pillow basalt on the Bathurst bypass, 650 ft (200 m) south of the quarry road. Another 0.6 mi (1 km) south, massive basalt contains thin interbeds of red chert.

The age of the basaltic rocks can be determined from graptolites found in underlying black slate just upstream from the railway bridge over the Tetagouche River. J. Riva (written communication, 1975) identified the following species that are indicative of a mid-Caradocian age: *Dicellograptus* cf. *forchammeri, Orthograptus* of the *quadrimucronatus* group, *Dicranograptus* sp., *Climacograptus* cf. *mohawkensis, Glyptograptus*?.

STOP 3. Graywacke and Slate, Elmtree Group. This road-cut exposes thick-bedded graywacke and interbedded dark gray slate typical of the Elmtree Group. Graded bedding and rip-up clasts demonstrate the turbiditic nature of these beds. The only fossil ever recovered from the Elmtree Group is a single stipe of *Orthograptus* sp. found about a 0.6 mi (1 km) downstream from the bridge (R. B. Rickards, written communication, 1975).

STOP 4. Contact between Fournier and Elmtree groups. This road-cut exposes the contact between the Pointe Verte Formation of the Fournier Group and the overlying Elmtree Group. Although the section is extensively sheared, there appears to be a gradational change between the two groups. Pillow basalts of the Pointe Verte Formation (upper part of Fournier Group) pass upsection to the south into a sequence of interbedded basalt, dark gray slate, and thick graywacke beds. The lower contact of the Elmtree Group is defined as the top of the uppermost lava flow.

Samples of interpillow limestone collected at this site by W. H. Poole of the Geological Survey of Canada contained a conodont assemblage indicative of the *Pygodus anserinus* zone of Llandeilo (Middle Ordovician) age (Nowlan, 1983).

STOP 5. Devereaux Formation, Fournier Group. The rocks seen in the railway cut to the west of the quarry are the most complexly deformed and highly metamorphosed part of the Devereaux Formation and presumably represent the deepest exposed portion of the ophiolitic sequence.

Dark green troctolitic gabbro locally varying to pegmatite intrudes brown-weathering pyroxenite toward the southern end of the exposure (Fig. 4). East-trending shear zones transecting igneous layering in the gabbro contain syntectonic, lit-par-lit injections of amphibolite and plagiogranite. Although the lit-par-lit structure is generally parallel to the shear foliation, plagiogranite bands locally exhibit folding. Emplacement of thick diabase and plagiogranite dikes postdates the shearing since their margins truncate the foliation in the shear zones.

STOP 6. Sheeted Dikes, Fournier Group. Sheeted dikes of the Fournier Group occur in a small inlier surrounded by Silurian rocks at Belledune (Fig. 5). The multiple diabase dikes, exposed a short distance northward along the shore, are subvertical and trend east-west. No septa of country rocks are present at this locality. Continue northwestward along shore to where Silurian conglomerate unconformably overlies the Fournier Group.

STOP 7. Pointe Verte Formation and Melangé, Fournier Group. Basaltic and sedimentary rocks of the Pointe Verte Formation are exposed a short distance northward along the shore. Pillows show that these spilitic basalts young to the north. The base of the basalt conformably overlies black shale that forms the top of a thick-bedded graywacke-shale sequence. Just to the north, the graywacke beds are faulted against the same basalt in an outcrop exposed only at low tide.

Other outcrops of basalt seen some distance to the south along the shore appear to be completely surrounded by boudin-aged beds of graywacke suggesting that the basalt may be present as olistoliths within a melangé rather than as continuous horizons (Rast and Stringer, 1980).

STOP 8. Contact between Fournier and Chaleurs groups. Northward, along the shore, graywacke of the Pointe Verte Formation is exposed. To the south, on the northern end of Limestone Point, red-stained slate of the Point Verte Formation is unconformably overlain by a thin unit of sandy limestone that forms the base of the Silurian Chaleurs Group in this area. Massive thick-bedded Lower Silurian red conglomerates conformably overlie the sandy limestone. The conglomerates are, in turn, overlain by a thick unit of Middle Silurian nodular limestone.

REFERENCES CITED

Alcock, F. J., 1935, Geology of Chaleur Bay region: Geological Survey of Canada Memoir 183, 146 p.

Bird, J. M., and Dewey, J. F., 1970, Lithosphere plate-continental margin tectonics and the evolution of the Appalachian orogen: Geological Society of America Bulletin, v. 81, p. 1031–1060.

Church, W. R., and Stevens, R. K., 1971, Early Paleozoic ophiolite complexes of the Newfoundland Appalachians as mantle-oceanic crust sequences: Journal of Geophysical Research, v. 76, no. 5, p. 1460–1466.

Davies, J. L., 1966, Geology of the Bathurst–Newcastle area, New Brunswick, *in* Poole, W. H., ed., Guidebook Geology of parts of Atlantic provinces: Geological Association of Canada, 155 p.

Ferguson, L., and Fyffe, L. R., 1985, Geological highway map of New Brunswick and Prince Edward Island: Atlantic Geoscience Society Special Publication, no. 2.

Fyffe, L. R., 1976, Correlation of geology in the southeastern and northern parts of the Miramichi Zone: New Brunswick Department of Natural Resources, 139th Annual Report, p. 137–141.

Fyffe, L. R., Pajari, G. E., and Cherry, M. E., 1981, The Acadian plutonic rocks of New Brunswick: Maritime Sediments and Atlantic Geology, v. 17, p. 23–36.

Fyffe, L. R., Forbes, W. H., and Riva, J., 1983, Graptolites from the Benton area of west-central New Brunswick and their regional significance: Maritime Sediments and Atlantic Geology, v. 19, p. 117–125.

Harley, D. N., 1979, A mineralized Ordovician resurgent caldera complex in the Bathurst–Newcastle mining district, New Brunswick, Canada: Economic Geology, v. 74, p. 786–796.

Helmstaedt, H., 1971, Structural geology of Portage Lakes area, Bathurst–Newcastle district, New Brunswick: Geological Survey of Canada Paper 70-28, 52 p.

Kennedy, D. J., Barnes, C. R., and Uyeno, T. T., 1979, A Middle Ordovician conodont faunule from the Tetagouche Group, Camel Back Mountain, New Brunswick: Canadian Journal of Earth Sciences, v. 16, p. 540–551.

Munha, J., 1979, Blue amphiboles, metamorphic regime, and plate tectonic modelling in the Iberian pyrite belt: Contributions to Mineralogy and Petrology, v. 69, p. 275–289.

Neuman, R. B., 1968, Paleogeographic implications of Ordovician shelly fossils in the Magog Belt of the northern Appalachians region, *in* Zen, E-An, White, W. S., Hadley, J. B., and Thompson, J. B., Jr., eds., Studies of Appalachian geology, northern and Maritime: Wiley Interscience Publishers, p. 35–48.

——, 1984, Geology and paleobiology of islands in the Ordovician Iapetus Ocean; Review and implications: Geological Society of America Bulletin, v. 95, p. 1188–1200.

Nowlan, G. S., 1981, Some Ordovician conodont faunules from the Miramichi Anticlinorium, New Brunswick: Geological Survey of Canada Memoir 345, 35 p.

——, 1983, Report on three samples from limestone in pillow basalts from the Pointe Verte Formation (northern New Brunswick): Geological Survey of New Brunswick, Report No. 002-GSN-1983, 2 p.

Pajari, G. E., Rast, N., and Stringer, P., 1977, Paleozoic volcanicity along the Bathurst–Dalhousie geotraverse, New Brunswick and its relations to structure, *in* Baragar, W.R.A., Coleman, L. C., and Hall, J. M., eds., Volcanic regimes in Canada: Geological Association of Canada, Special Paper 16, p. 111–124.

Poole, W. H., 1976, Plate tectonic evolution of the Canadian Appalachian region: Geological Survey of Canada Paper 76-1B, p. 113–126.

Rast, N., and Stringer, P., 1980, A geotraverse across a deformed Ordovician ophiolite and its Silurian cover, northern New Brunswick: Tectonophysics, v. 69, p. 221–245.

Rast, N., O'Brien, B. H., and Wardle, R. J., 1976, Relationships between Precambrian and lower Paleozoic rocks of the "Avalon Platform" in New Brunswick, the northeast Appalachians and the British Isles: Tectonophysics, v. 30, p. 315–338.

Ruitenberg, A. A., Fyffe, L. R., McCutcheon, S. R., St. Peter, C. J., Irrinki, R. R., and Venugopal, D. V., 1977, Evolution of pre-Carboniferous tectonostratigraphic zones in the New Brunswick Appalachians: Geoscience Canada, v. 4, p. 171–181.

Skinner, R., 1974, Geology of Tetagouche Falls, Bathurst and Nepiusiguit Falls map areas, New Brunswick: Geological Survey of Canada Memoir 371, 133 p.

Stevens, R. K., 1970, Cambro–Ordovician flysch sedimentation and tectonics in west Newfoundland and their possible bearing on a Proto-Atlantic Ocean, *in* Lajoie, J., ed., Flysch sedimentology in North America: Geological Association of Canada Special Paper 7, p. 165–177.

Whitehead, R.E.S., and Goodfellow, W. D., 1978, Geochemistry of volcanic rocks from the Tetagouche Group, Bathurst, New Brunswick, Canada: Canadian Journal of Earth Sciences, v. 15, p. 107–219.

Williams, H., and Hatcher, R. D., 1982, Suspect terranes and accretionary history of the Appalachian orogen: Geology, v. 10, p. 530–536.

Oil shales of the Albert Formation at Frederick Brook in southeastern New Brunswick

C. St. Peter, Mineral Development Branch, New Brunswick Department of Forests, Mines and Energy, P.O. Box 6000, Fredericton, New Brunswick E3B 5H1, Canada

LOCATION

Oil shales occur along Frederick Brook in Albert County in southeastern New Brunswick (Figs. 1, 2). The area is publicly accessible via automobile by driving 13 mi (21 km) south of Moncton along New Brunswick 114 to the town of Hillsborough. To reach the site, continue south along New Brunswick 114 2.5 mi (4 km) to an unnumbered highway on the right, designated the Albert Mines Road. Proceed west on this road for 2.2 mi (3.6 km) nearly to the bottom of a large hill where there is a small church and a secondary road leading to the right (northwest). Follow this unnumbered road for about 1.1 mi (1.8 km) to a shale pile on the right (or west) side of the road. The pile lies near the northeastern end of the underground altertite vein and is indicated as Stop 6 in Figure 2. Note that the final 2,300 ft (700 m) of road leading to the site is not a hard surface.

SIGNIFICANCE

The area at Frederick Brook, and Albert County in general, became famous in 1849 following the discovery of a unique natural substance that became known as albertite (Dawson, 1868). Albertite is a black, glassy, solid bitumen that occurs in a vein along Frederick Brook (Fig. 2). The discovery led to the opening of an underground mine that began production in 1860. By 1879, the mine was depleted: a total of 200,000 tons of albertite had been exported to the United States for the production of kerosene oil and gas.

The Albert Formation underlies the Moncton Subbasin (Fig. 1); it is composed of three members. In ascending order, they are: the Dawson Settlement Member, the Frederick Brook Member, and the Hiram Brook Member (Greiner, 1962; Pickerill and others, 1985). The oil shales at Frederick Brook constitute the type section of the medial Frederick Brook Member. The shales represent a spectacular example of thinly laminated strata deposited in a fresh-water lacustrine environment in Early Carboniferous (Tournaisian) time. Some shale beds contain well-preserved complete specimens of Palaeoniscid fishes and abundant fish scales.

The Albert shales are highly deformed and folded into upright, south-plunging mesoscopic folds. In addition to these larger scale folds, there are smaller scale recumbent folds present in exposures along Frederick Brook. The Frederick Brook rocks are bounded on the north and south by regional east-northeast–trending faults, and on the west and northeast by northwest-trending normal faults. To the southeast, the deformed Albert strata are unconformably overlain by the post-tectonic Late Mississippian or Pennsylvanian Enrage Formation.

SITE INFORMATION

The Albert Formation underlies an area of about 1,200 mi^2 (3,000 km^2) in the Moncton Subbasin of southeastern New Brunswick. The Albert strata constitute the medial formation of the Upper Devonian to Lower Mississippian Horton Group (Fig. 1). The Horton rocks represent the first sediments to accumulate in the Moncton Subbasin. The Subbasin formed in Late Devonian time as a result of normal faulting or strike-slip faulting of deformed Middle Devonian and older crystalline rocks (Poole, 1967; Bradley, 1982; Pickerill and others, 1985). The Albert Formation is conformably underlain and overlain by the Memramcook and Moncton Formations, respectively. Both the Memramcook and Moncton Formations are composed of alluvial fan and fluviatile red beds. The Horton Group rocks are exposed along the south flank of the Subbasin, where in most places the group is in fault contact with the crystalline basement of the Caledonia Uplift. On the north side of the Moncton Subbasin, the Horton rocks are exposed in a northeast-trending belt where they are typically unconformably overlain by Upper Mississippian– or Pennsylvanian-age strata.

The Horton rocks are overlain by Visean-age carbonates, sulfates, and evaporites of the Windsor Group. The Windsor represents the only known marine sequence in the Carboniferous of New Brunswick. Economically important potash beds occur in the upper evaporites of the Windsor in the Sussex area (Fig. 1).

The Windsor strata in the Moncton Subbasin are unconformably to conformably overlain by the Late Mississippian to Early Pennsylvanian Hopewell Group. Along the southern margin of the subbasin, the Hopewell typically comprises coarse-grained alluvial fan deposits lying with angular discordance on Windsor Group and older rocks. In the axial subsurface part of the subbasin, the Hopewell strata are mostly fine-grained red beds lying concordantly on Windsor Group evaporites.

Pennsylvanian-age, predominantly gray fluviatile beds of the Petitcodiac Group overlie the Hopewell Group within the Moncton Subbasin (Carr, 1968). In numerous places, the Petitcodiac strata overstep older Carboniferous rocks along the southern part of the Subbasin and lie atop the Caledonia basement. Similarly, the Petitcodiac beds overstep the basement north and east of the Moncton Subbasin where they underlie the entire eastern part of New Brunswick and constitute the New Brunswick Platform (Fig. 1).

The Carboniferous rocks of New Brunswick have received the attention of many geologists since the early writings of Dawson (1868). None of the rocks has been more inspected or has raised more interest than the Albert Formation; this is particularly

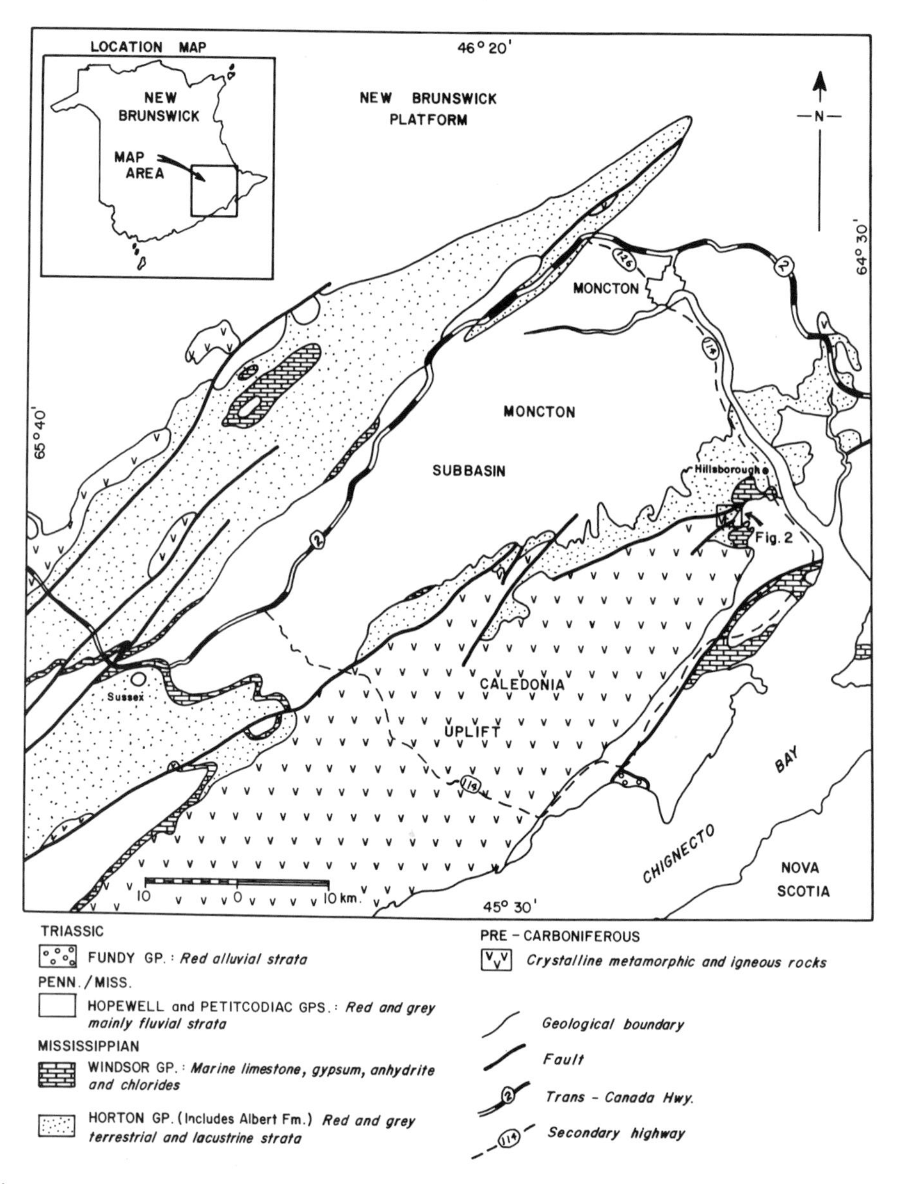

Figure 1. Regional geology map of southeastern New Brunswick, (index rectangle shows the location of the Frederick Brook–type section; see Fig. 2).

true of the oil shales at Frederick Brook. The medial Frederick Brook Member of the Albert Formation in the southeastern part of the Moncton Subbasin is underlain by the Dawson Settlement Member and overlain by the Hiram Brook Member (Greiner, 1962; Pickerill and others, 1985). The two latter members comprise an assemblage of alluvial and fluvial interbedded gray shale, sandstone, and conglomerate, with minor lacustrine oil shale, limestone, and dolostone.

The medial Frederick Brook Member is best exposed at the type section along Frederick Brook (Fig. 2). The type section comprises those rocks exposed along the brook on the north side of the road from Stop 2 down the brook to the northwest-trending fault about 660 ft (200 m) north of Stop 6. The rocks in the type section are brown-weathering, fissile, laminated, dolomitic, kerogenous siltstone, and rare beds of brown massive dolostone and fine-grained sandstone.

The characteristic oil shales of the formation are best seen at Stop 1 in a shale pit in an abandoned field on the east side of Frederick Brook. The tailings from this pit are stockpiled just north of the pit. The stockpile can be easily seen across the field for 660 ft (200 m) when approaching the site along the road immediately to the east. Parking is possible on the narrow access track that leads west from the road to the pit. The oil shales in the pit are among the best grade in Canada and may soon be exploited for their shale oil (Macauley and Ball, 1982). When split into layers, the shale can be ignited with a match. The beds average 100 litres of shale oil per ton of rock over a stratigraphic interval of 80 ft (25 m).

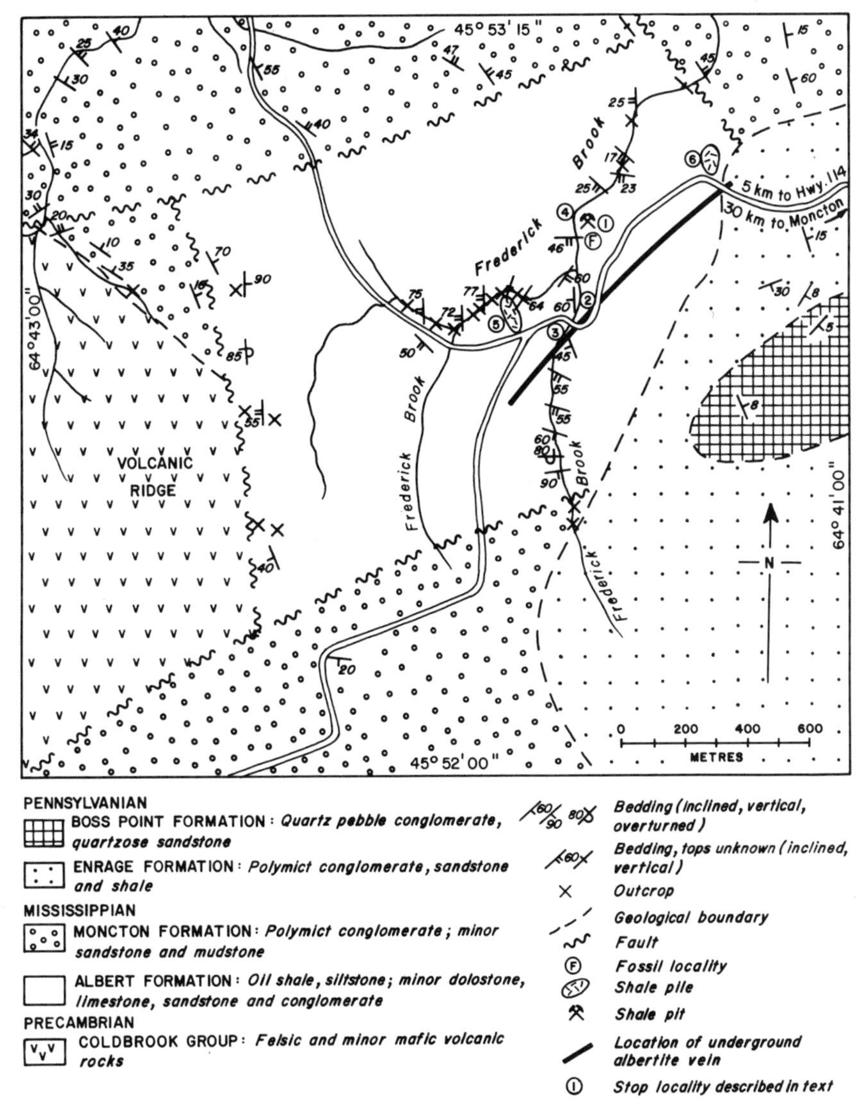

Figure 2. Geological map of the Frederick Brook area, New Brunswick.

The shales in the pit and those exposed at Stop 2, midway up the east bank of Frederick Brook, and immediately north of the road are known to contain complete fossil fish that are ascribed to *Rhadinichthys alberti* (Dawson, 1868; Lambe, 1909, 1910). *R. alberti* is a rather small fish averaging about 3.5 in (8.5 cm) in length (Fig. 3). It has well-developed fins and diamond-shaped scales. Clusters of black vitreous scales, ranging from 2 to 4 mm in length, are common on bedding planes.

The Frederick Brook section has been assigned an Early Carboniferous (Tournaisian) age based on miospores (Varma, 1969). The miospores imply a fresh-water origin for the oil shales (Varma, 1969). The kerogen laminations in the oil shales are probably of planktonic algal lacustrine origin (King, 1963; Macauley and others, 1984). The excellent preservation of *R. alberti* suggests that the lake was eutrophic, and had restricted circulation and few bottom scavengers.

The rocks along Frederick Brook reveal a complex folding history. Many of the more organic-rich oil shale beds display convoluted and chaotic synsedimentary slump folds that range from microscopic to a metre or more in wavelength. Superimposed on these sedimentary folds are two different types of tectonic folds. Tectonic folds of the first type are upright, open to tight, symmetrical to slightly asymmetrical, and south plunging, and range in wavelength from less than 3 ft (1 m) to more than 650 ft (200 m). Larger scale structures of this style were mapped by Wright (1922) in tunnels associated with the nineteenth-century Albertite mine. A small-scale fold of this type can be seen at water level in Frederick Brook about 80 ft (25 m) south of the road at Stop 3. This locality can be reached by walking south along the brook about 130 ft (40 m) to the outcrop exposed above the water line on the east bank.

The second style of tectonic fold is recumbent, tight, and asymmetrical, with wavelengths ranging up to 6 ft (2 m). A fold of this kind is exposed above the water level on the south side of a

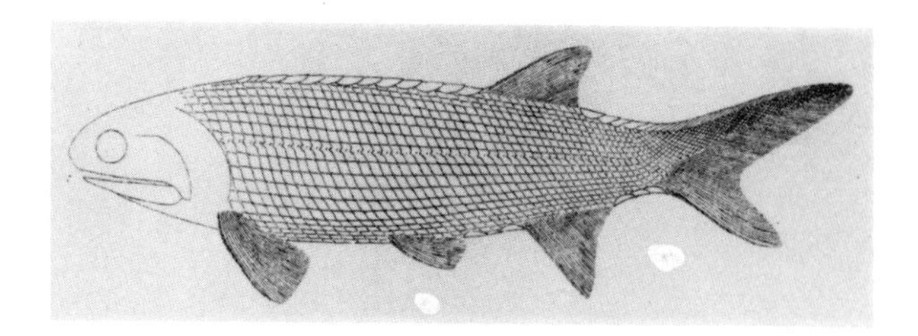

Figure 3. *Rhadinichthys alberti* (Jackson), restored outline (Plate III, Fig. 1 after L. M. Lambe, 1910; about actual size).

Figure 4. Recumbently folded oil shale beds (S_0) at Frederick Brook, with axial surface cleavage (S_1) developed in the hinge zone.

pool at the bottom of a small waterfall at Stop 4. This fold, depicted in Figure 4, has an axial surface cleavage (strike 085°, dip 30° S.) in the hinge that dies out toward the limbs of the structure. The relative ages of the upright and recumbent tectonic folds are unknown because interference fold structures have not been observed.

The folded shales at Frederick Brook are in fault-contact to the north and south with Moncton Formation conglomerate and mudstone (Fig. 2). Both bounding faults are regional structures that have been traced for at least 25 mi (40 km). Where the traces of these structures are seen at surface, the faults are marked by steeply dipping shear zones.

The Albert Formation is in fault-contact with a high-standing ridge of Precambrian volcanic rocks to the west. A borehole in the Albert rocks just east of the ridge reveals the presence of thick alluvial fan polymict conglomerates within the lacustrine oil shale sequence (Macauley and Ball, 1982). This stratigraphic interlaying implies that the basement-bounding fault is a normal structure with a pronounced offset. The rising volcanic ridge provided coarse alluvial fan detritus directly into the adjacent lake.

The tectonic events affecting the Carboniferous rocks in the Moncton Subbasin are ascribed to the post-Acadian Maritime Disturbance (Poole, 1967). The complex array of sedimentary and tectonic structures associated with this disturbance are well represented at Frederick Brook. The major deformational events were largely concluded by late Mississippian time. This is evidenced to the southeast, where the Albert rocks and their attendant structures are unconformably overlain by the subhorizontal Pennsylvanian Enrage and Boss Point formations.

Frederick Brook is most popularly known as the historical site of Albert Mines. A vein of solid bitumen—albertite—was mined from the area in the mid-nineteenth century. All that remains as evidence of the old mine are two shale piles at Stops 5 and Stop 6. Samples of black vitreous albertite are still numerous and easily collected from the shale piles.

REFERENCES CITED

Bradley, D. C., 1982, Subsidence in Late Palaeozoic basins in the Northern Appalachians: Tectonics, v. 1, no. 1, p. 107–123.

Carr, P. A., 1968, Stratigraphy and spore assemblages, Moncton map-area, New Brunswick: Geological Survey of Canada, Paper 67-29, 47 p.

Dawson, J. W., 1868, The geological structure, organic remains, and mineral resources of Nova Scotia, New Brunwick, and Prince Edward Island, Acadian Geology, second edition: London, MacMillan and Company, p. 231–248.

Greiner, H. R., 1962, Facies and sedimentary environment of the Albert Shale, New Brunswick: American Association of Petroleum Geologists Bulletin, v. 46, p. 219–234.

King, L. H., 1963, Origin of the Albert Mines oil shale (New Brunswick) and its associated albertite: Ottawa, Mines Branch Research Report R115, Department of Mines and Technical Surveys, 24 p.

Lambe, L. M., 1909, The fish fauna of the Albert shales of New Brunswick: American Journal of Science, v. 178, p. 165–174.

—— , 1910, Palaeoniscid fishes from the Albert shales of New Brunswick: Contributions to Canadian Paleontology, Geological Survey of Canada Memoir 3, v. III (quarto), pt. V, 69 p.

Macauley, G., and Ball, F. D., 1982, Oil shales of the Albert Formation, New Brunswick: Calgary, Institute of Sedimentary and Petroleum Geology, Geological Survey of Canada, Open File Report No. 82-12, 173 p.

Macauley, G., Ball, F. D., and Powell, T. G., 1984, A review of the Carboniferous Albert Formation oil shales, New Brunswick: Canadian Petroleum Geology Bulletin, v. 32, no. 1, p. 27–37.

Pickerill, R. K., Carter, D., and St. Peter, C., 1985, Albert Formation–oil shales, lakes, fans, and deltas: Geological Association of Canada, Mineralogical Association of Canada, Guidebook Excursion 6, 75 p.

Poole, W. H., 1967, Tectonic evolution of the Appalachian region of Canada, *in* Neale, E.R.W., and Williams, H., eds., Geology of the Atlantic Region: Geological Association of Canada Special Paper 4, p. 9–51.

Varma, C. P., 1969, Lower Carboniferous miospores from the Albert oil shales (Horton Group) of New Brunswick, Canada: Micropaleontology, v. 15, no. 3, p. 301–324.

Wright, W. J., 1922, Geology of the Moncton map area: Geological Survey of Canada, Canada Department of Mines, Memoir 129, 69 p.

Cambrian stratigraphy in the Hanford Brook area, southern New Brunswick, Canada

S. R. McCutcheon, New Brunswick Department of Natural Resources, P.O. Box 1519, Sussex, New Brunswick E0E 1P0, Canada

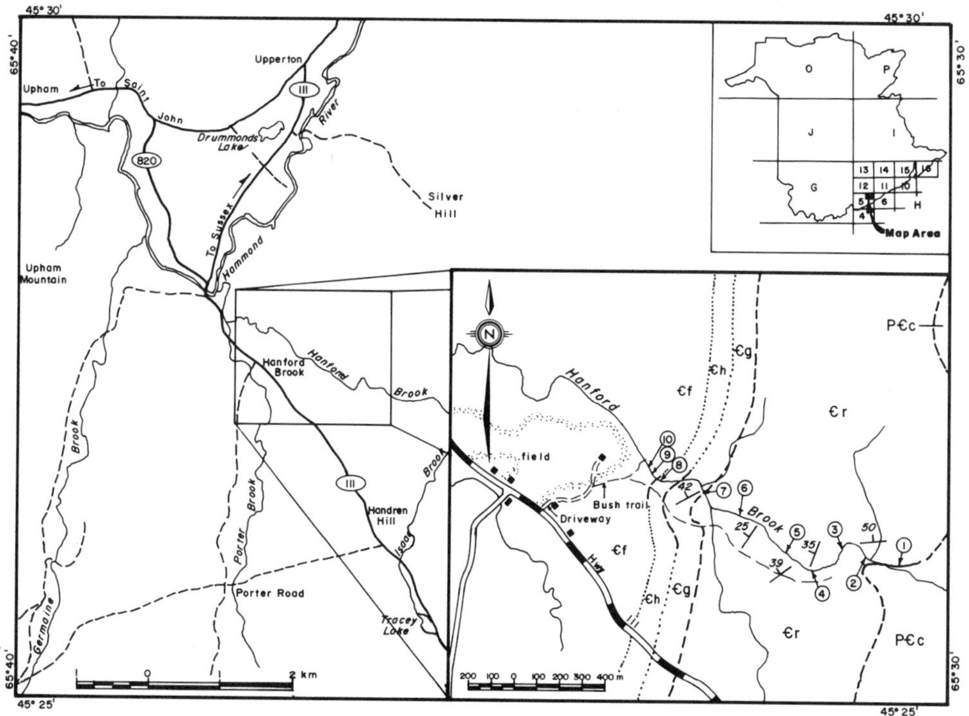

Figure 1. Location maps. The inset map of New Brunswick at upper right shows the National Topographic Series Grid (letters and numbers) and the position of the large map. The inset at lower right depicts the geology of the Hanford Brook section. P€, Late Precambrian Coldbrook Group; €r, Early Cambrian Ratcliffe Brook Formation; €g, Early Cambrian Glen Falls Formation; €h, late Early Cambrian Hanford Brook Formation; €f, Middle Cambrian Forest Hills Formation and younger rocks.

LOCATION

The Hanford Brook area is about 25 mi (40 km) northeast of Saint John, near the boundary between Saint John and Kings Counties. It can be reached by New Brunswick 111 and 820 from Saint John or by New Brunswick 111 from Sussex, which is about 22 mi (35 km) to the north.

The site (Fig. 1) is reached via a driveway, on the north side of New Brunswick 111, that is about 0.9 mi (1.4 km) south of the bridge across Hammond River. You can (1) drive to the house at the end of the lane and request permission to leave your vehicle in the landowner's yard; then proceed eastward across the field to Hanford Brook; *or* (2) park on the main road and walk along the lane and partly overgrown bush trail to the stream. In either case Stop 10 is the first low-relief outcrop to be encountered on the south bank, proceeding upstream and down stratigraphic section. *Note:* The water level in the stream is critical; if it is too high, much of the outcrop will be concealed. Therefore, the section is best viewed in mid- to late summer.

Other localities in Saint John where stratigraphically equivalent rocks can be seen are described in field trip guides by Patel (1972, 1973a).

SIGNIFICANCE

Good faunal and lithologic correlation exists between late Lower Cambrian to Tremadocian sequences of the Acado-Baltic faunal province on both sides of the North Atlantic (Henningsmoen, 1969), and it appears that the Cambrian–Ordovician boundary in the Acado-Baltic and North American faunal provinces is approximately equivalent (Landing and others, 1978). However, correlation of earliest Lower Cambrian rocks is less certain (Bengtson and Fletcher, 1983), and international agreement has not yet been reached on the position of the Cambrian–Precambrian boundary.

The Hanford Brook section is important for two reasons. First, the unconformable boundary between the late Precambrian Coldbrook Group and the Cambro-Ordovician Saint John Group is exposed. These two groups represent different depositional cycles (Ruitenberg and others, 1977; McCutcheon, 1981), herein referred to as the Avalonian and Iapetus cycles.

Second, there is a significant faunal gap within the Lower Cambrian part of the section that is apparent when it is compared with Lower Cambrian sections in other parts of the Acado-Baltic faunal province (see Anderson, 1981; Bengston and Fletcher,

1983). The oldest trilobites present are those of the late Lower Cambrian *Protolenus* Zone (Matthew, 1895). Older trilobite faunas are unrepresented and, therefore, a stratigraphic break exists between beds containing *Protolenus* (and other fossils), and underlying beds that lack trilobites and constitute the greater part of the section. Walcott (1900) reported finding hyolithids in the latter but their presence has not been confirmed. However, the pre-trilobite beds do contain Paleozoic-type trace fossils (Patel, 1976).

The significance of this faunal gap is open to interpretation. However, it could have regional importance if the Saint John Group represents a passive continental margin sequence as implied by Ruitenberg and others (1977). In such sequences a phase of erosion commonly occurs between the initial "rift stage" of development and the thermal subsidence or "drift stage," resulting in the "breakup unconformity" of Falvey (1974). If the Hanford Brook section is viewed in this way, then the red beds of the basal Ratcliffe Brook Formation probably represent the "rift stage" deposits, whereas the overlying Glen Falls (or perhaps the Hanford Brook) and younger formations are the "drift stage" deposits.

A corollary of Falvey's (1974) passive continental margin model is that the earliest oceanic crust in the opening ocean is about the same age as the oldest rocks in the "drift stage." This means that the oldest oceanic crust in the Iapetus cycle should be equivalent in age to the Glen Falls (or Hanford Brook) Formation.

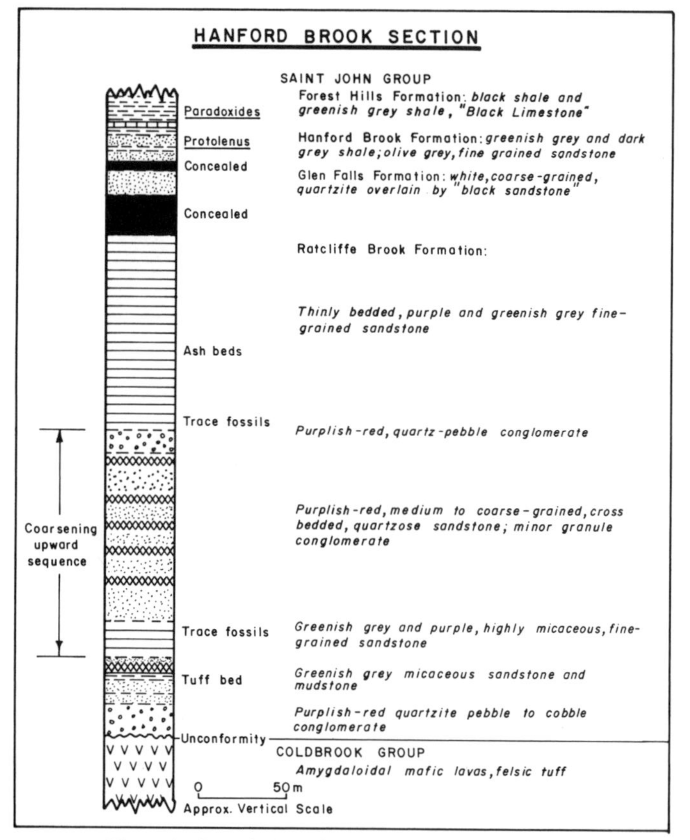

Figure 2. Graphic log of the Hanford Brook Section.

SITE INFORMATION

The Hanford Brook section, which is within the southern part of the Avalon Zone of Williams (1979) or the central part of the Caledonia Zone of Ruitenberg and McCutcheon (1982), comprises rocks of the Saint John and Coldbrook Groups. These two groups are separated by an angular unconformity (Matthew, 1890; Hayes and Howell, 1937; Patel, 1977; McLeod and McCutcheon, 1981).

The Coldbrook Group (Matthew, 1863; Alcock, 1938), Avalon Zone rocks *senso stricto,* has been divided into the northeast trending Eastern, Central, and Western Volcanic Belts (Giles and Ruitenberg, 1977; Ruitenberg and others, 1977, 1979), and each has its own characteristic rock assemblage. The Central Belt in which the Hanford Brook section lies, is typified by terrestrial felsic and mafic volcanics with minor intercalated sedimentary rocks of shallow subaqueous origin (Ruitenberg and others, 1979). Cleavage is generally absent and in many places prehnite-pumpellyite facies metamorphic assemblages are preserved. Amygdaloidal mafic lavas and red felsic tuff outcrop on Hanford Brook just below the Saint John Group.

The Saint John Group (Matthew, 1863, 1890; Hayes and Howell, 1937) includes seven formations (Pickerill and Tanoli, 1985). Only the lower four are present on Hanford Brook (Table 1).

The Ratcliffe Brook Formation is predominantly a redbed unit that Hayes and Howell (1937) considered to be terrestrial in spite of the fact that Matthew (1890) and Walcott (1900) reported marine fossils from it. It is variable both in thickness and lithology from locality to locality, but commonly there is a distinctive purplish-red conglomerate at the base.

On Hanford Brook, this formation is at least 820 ft (250 m) thick and divisible into six readily distinguishable units (Fig. 2). (1) A purplish red, basal conglomerate in which the pebble to cobble-size clasts are mainly of quartz and quartzite, with the latter predominant. (2) It is overlain by a unit characterized by greenish gray, fine- to coarse-grained, micaceous sandstone in parallel beds up to 3.3 ft (1 m) thick; minor mudstone and one 2 to 4 in (5 to 10 cm) thick, buff-weathering crystal tuff bed are also contained in this unit, which is exposed twice on Hanford Brook because of the combined effects of a change in strike and a U-shaped bend in the stream's course. (3) Thinly bedded, greenish gray and purple, highly micaceous, fine-grained sandstone and siltstone. (4) Purplish-red, medium to coarse-grained, cross-bedded, quartzose sandstone and granule conglomerate (a coarsening upward unit). (5) A purplish-red, quartz-pebble conglomerate. (6) Purple and greenish gray, thinly bedded, fine-grained sandstone and siltstone; the sandstone is locally calcareous toward the base and it includes, in its lower part, two thin (<2 in; <5 cm), highly siliceous layers (originally volcanic ash?).

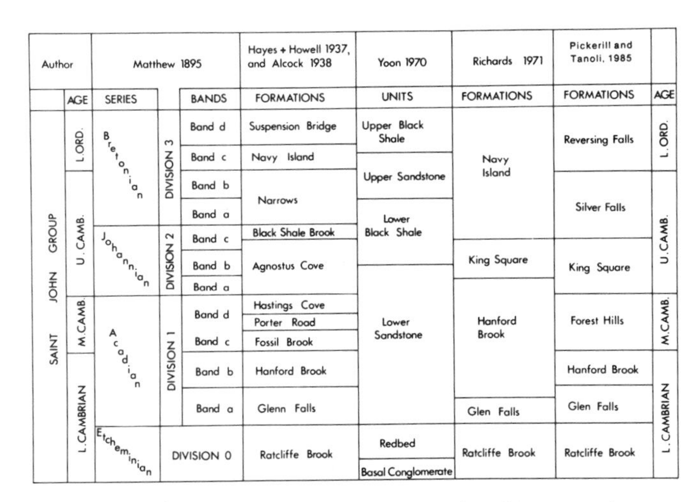

Author		Matthew 1895			Hayes + Howell 1937, and Alcock 1938	Yoon 1970	Richards 1971	Pickerill and Tanoli, 1985	
	AGE	SERIES		BANDS	FORMATIONS	UNITS	FORMATIONS	FORMATIONS	AGE
SAINT JOHN GROUP	L. ORD.	Bretonian	DIVISION 3	Band d	Suspension Bridge	Upper Black Shale	Navy Island	Reversing Falls	L. ORD.
				Band c	Navy Island	Upper Sandstone			
	U. CAMB.			Band b	Narrows			Silver Falls	U. CAMB.
				Band a		Lower Black Shale			
		Johannian	DIVISION 2	Band c	Black Shale Brook		King Square		
				Band b	Agnostus Cove			King Square	
				Band a		Lower Sandstone	Hanford Brook		
	M. CAMB.	Acadian	DIVISION 1	Band d	Hastings Cove			Forest Hills	M. CAMB.
					Porter Road				
				Band c	Fossil Brook				
	L. CAMBRIAN			Band b	Hanford Brook			Hanford Brook	L. CAMBRIAN
				Band a	Glenn Falls		Glen Falls	Glen Falls	
		Etcheminian	DIVISION 0		Ratcliffe Brook	Redbed	Ratcliffe Brook	Ratcliffe Brook	
						Basal Conglomerate			

Table 1. Evolution of Saint John Group stratigraphic nomenclature.

The Glen Falls Formation is a light-gray or greenish to pinkish white, coarse-grained quartzite from 16.4 to 49 ft (5 to 15 m) thick with a thin coarse-grained "black sandstone" at the top (Pickerill and Tanoli, 1985). Structurally, this formation is conformable with units above and below. On Hanford Brook, both contacts are concealed but the unit is at least 33 ft (10 m) thick.

The Hanford Brook Formation consists of olive gray to dark gray, fine-grained sandstone and shale; the latter is most abundant in the upper 16 ft (5 m) of the formation (Pickerill and Tanoli, 1985). This unit contains the late Lower Cambrian *Protolenus* fauna described by Matthew (1895). On Hanford Brook, it attains its maximum thickness of 75 ft (23 m).

The Forest Hills Formation is made up of three parts. At the base there is a fossiliferous argillaceous limestone or calcareous gray shale that is up to 5.7 ft (1.75 m) thick, the "Black Limestone" of Hayes and Howell (1937). This limestone marks the base of the Middle Cambrian. It is overlain by greenish gray shale and then by dark gray shale with a few thin, fine-grained sandstone interbeds and fossiliferous lime-rich lenses (Pickerill and Tanoli, 1985). This formation is up to 148 ft (45 m) thick and contains the *Paradoxides* fauna. On Hanford Brook, the limestone is about 3.3 ft (1 m) thick and only the lower few feet (meters) of the overlying section are exposed. However, an outcrop of dark gray shales typical of the upper part can be found along the bush trail near Hanford Brook (i.e., along the route leading to the section).

Descriptions of the remaining formations in the Saint John Group can be found in Pickerill and Tanoli (1985).

Correlation of the Hanford Brook section with the Lower Cambrian sequence in eastern Newfoundland is speculative, but worth attempting because it is one area where the stratigraphic relationships of the oldest Cambrian rocks are documented. Lithologically, the Glen Falls Formation is very similar to parts of the Random Formation (Anderson, oral communication, 1985), a unit that was deposited in a tidal to shallow subtidal environment during a time of global sea level rise (Hiscott, 1982). The Random, of early Lower Cambrian (Tommotian) age (Anderson, 1981) is almost everywhere bounded above and below by disconformities (Hiscott, 1982). However, in the Fortune–Burin region the Random is underlain conformably by the Chapel Island Formation, a unit composed of "green, red, and brown siltstones, argillites and sandstones with minor limestones" (Anderson, 1981). In the upper part of this formation, there are shelly fossils (*Aldanella attleborensis* assemblage), that establish its age as Tommotian, and in the lower part there are trace fossils (Bengtson and Fletcher, 1983). The diversity and abundance of the latter suggest that virtually the whole of the Chapel Island Formation is of earliest Cambrian (Tommatian) age (Crimes and Anderson, 1985). The Ratcliffe Brook Formation may be the time equivalent of some part of the Chapel Island Formation.

STOP LOCATIONS

Stop locations in the Hanford Brook section are numbered in stratigraphic order, 1 through 10 in Figure 1. However, the starting point is at the top of the section, and therefore, the stops are described below in reverse order.

Stop 10. Greenish gray shales of the Forest Hills Formation outcrop in low relief along the south bank above the bed of the stream. Walcott (1900) reported abundant remains of *Paradoxides* in these beds.

Stop 9. The "Black Limestone" of Hayes and Howell (1937) is easily located because it is resistant to erosion. A number of large blocks of it lie along the south bank, a short distance upstream from Stop 10. This unit contains abundant disarticulated trilobite detritus, shale clasts, and grains or nodules of phosphate, chamosite, and glauconite (Pickerill and Tanoli, 1985).

Stop 8. Immediately below the "Black Limestone," the Hanford Brook Formation is almost continuously exposed. Greenish gray shale with subordinate fine-grained sandstone interbeds up to 4 ft (1.2 m) thick, characterize the upper part. In the lower part, dark gray *Protolenus*-bearing shale with olive gray sandstone interbeds overlie a basal olive gray sandstone. The sandstone in this unit commonly contains glauconite, chamosite, and phosphate nodules (Pickerill and Tanoli, 1985).

Stop 7. Light gray to white, coarse-grained, crossbedded quartzite of the Glen Falls Formation forms prominent outcrops on both sides of the stream. This is the former location of the dam (mill-pond) shown in Matthew's (1890) sketch map. Along the south bank, about halfway through the concealed interval between this quartzite and the olive gray sandstone of the Hanford Brook Formation, there is a small, low-relief outcrop near the waterline. The greenish black sandstone of this outcrop is characteristic of the top of the Glen Falls Formation.

Stop 6 to Stop 5. Stop 6, on the north side of the stream near the waterline, is the first outcrop to be encountered after the quartzite. Between Stops 6 and 5, thinly bedded purple and greenish gray, fine-grained sandstone is exposed, but most of the outcrop occurs in the bed of the stream. Toward the top of this section the rocks are greener and contain glauconite (Matthew, 1890; Patel, 1973a). About halfway through this interval, there are two thin (<2 in; <5 cm) cherty layers in an outcrop on the south bank, which are probably ash beds. Toward the base of this section, a few of the beds are calcareous and trace fossils are common. It was in this part of the section that Walcott (1900) reported finding *Hyolithes* but it has never been relocated.

Stop 5. Purplish red rounded quartz pebble to cobble, matrix-supported conglomerate, which is about 33 ft (10 m) thick, is best exposed in the bed of the stream. This unit, abruptly overlain by the fine-grained sandstone cropping out between Stops 6 and 5, appears to be gradational below with coarse-grained, crossbedded quartzose sandstone with interbeds of granule conglomerate. It marks the top of a coarsening upward sequence, about 328 ft (100 m) thick, that begins at Stop 4.

Stop 4. Greenish gray and purple, highly micaceous, fine-grained sandstone and siltstone form low relief outcrops along the south bank of the stream. Trace fossils are present in these beds.

Stop 3. A thin (2 to 4 in; 5 to 10 cm), buff weathering, crystal tuff layer occurs in an interbedded greenish gray, micaceous sandstone and mudstone sequence on the north side of the stream.

Stop 2. Purplish red, clast-supported conglomerate forms prominent outcrops on either side of the stream; the subrounded pebbles and cobbles are mainly of quartzite and vein quartz. Clasts of granitic and volcanic rocks are also present. Maximum clast size is about 4 in (10 cm). This conglomerate marks the base of the Ratcliffe Brook Formation but the contact is not exposed at this locality. Beyond this point, the contact follows the stream for about 492 ft (150 m) with conglomerate on the north side and dark green to dark purple amygdaloidal lavas on the south.

Stop 1. On the north bank of the stream the conglomerate described above is in contact with reddish brown volcanic(?) rocks. Patel (1977) attributed this reddish brown color to paleo-weathering effects and interpreted these contact rocks as fossil soil.

REFERENCES CITED

Alcock, F. J., 1938, Geology of Saint John region, New Brunswick: Geological Survey of Canada Memoir 216, 65 p.

Anderson, M. M., 1981, The Random Formation of southeastern Newfoundland; A discussion aimed at establishing its age and relationship to bounding formations: American Journal of Science, v. 281, p. 807–830.

Bengtson, S., and Fletcher, T. P., 1983, The oldest sequence of skeletal fossils in the Lower Cambrian of southeastern Newfoundland: Canadian Journal of Earth Sciences, v. 20, p. 525–536.

Crimes, P. T., and Anderson, M. M., 1985, Trace fossils from Late Precambrian–Early Cambrian strata of southeastern Newfoundland (Canada); Temporal and environmental implications: Journal of Paleontology, v. 59, p. 310–343.

Falvey, D. A., 1974, The development of continental margins in plate tectonic theory: Australian Petroleum Exploration Association Journal, v. 14, p. 95–106.

Giles, P. S., and Ruitenberg, A. A., 1977, Stratigraphy, paleogeography and tectonic setting of the Coldbrook Group in the Caledonia Highlands of southern New Brunswick: Canadian Journal of Earth Sciences, v. 14, p. 1263–1275.

Hayes, A. O., and Howell, B. F., 1937, Geology of Saint John, New Brunswick, Geological Society of America Special Paper 5, 146 p.

Henningsmoen, G., 1969, Short account of Cambrian and Tremadocian of Acado-Baltic province, *in* Kay, M., ed., North Atlantic geology and continental drift: American Association of Petroleum Geologists Memoir 12, p. 110–114.

Hiscott, R. N., 1982, Tidal deposits of the Lower Cambrian random formation, eastern Newfoundland; Facies and paleoenvironments: Canadian Journal of Earth Sciences, v. 19, p. 2028–2042.

Landing, E., Taylor, M. E., and Erdtmann, B. D., 1978, Correlation of the Cambrian-Ordovician boundary between the Acado-Baltic and North American faunal provinces: Geology, v. 6, p. 75–78.

Matthew, G. F., 1863, Observations on the geology of Saint John County, New Brunswick: Canadian Naturalist and Geologist, v. 8, p. 241–260.

—— , 1890, On Cambrian organisms in Acadia: Transactions of the Royal Society of Canada, sec. 4, p. 135–143.

—— , 1895, The *Protolenus* fauna: Transactions of the New York Academy of Sciences, v. 14, p. 101–153.

McCutcheon, S. R., 1981, Revised stratigraphy of the Long Reach area, southern New Brunswick; Evidence for major northwestward directed Acadian thrusting: Canadian Journal of Earth Sciences, v. 18, p. 646–656.

McLeod, M. J., and McCutcheon, S. R., 1981, A newly recognized sequence of possible Early Cambrian age in southern New Brunswick; Evidence for major southward-directed thrusting: Canadian Journal of Earth Sciences, v. 18, p. 1012–1017.

Patel, I. M., 1972, Field Trips Stops 8-1 and 8-2, *in* Glass, D. J., ed., Appalachian geotectonic elements of the Atlantic Provinces and southern Quebec, International Geological Congress Field Excursion A63-C63, p. 102–106.

—— , 1973a, Saint John area, *in* Rast, N., ed., NEIGC Field Guide to Excursions, Trip A-13 and B-10, p. 115–118, University of New Brunswick, Fredericton.

—— , 1973b, Sedimentology of the Ratcliffe Brook Formation (Lower Cambrian?) in southeastern New Brunswick: Geological Society of America Abstracts with Programs, v. 5, no. 2, p. 206.

—— , 1976, Lower Cambrian of southern New Brunswick and its correlation with successions in Northeastern Appalachians and parts of Europe: Geological Society of America Abstracts with Programs, v. 8, no. 2, p. 243.

—— , 1977, Late Precambrian fossil soil horizon in southern New Brunswick, and its stratigraphic significance: Geological Society of America Abstracts with Programs, v. 9, no. 3, p. 308.

Pickerill, R. K., and Tanoli, S. K., 1985, Revised lithostratigraphy of the Cambro-Ordovician Saint John Group, southern New Brunswick; A preliminary report, *in* Current Research: Geological Survey of Canada Paper 85-1B, pt. B, p. 441–449.

Richards, N. A., 1971, Structure in the Precambrian and Paleozoic rocks at Saint John, New Brunswick [M.S. thesis]: Ottawa, Ontario, Carleton University, 73 p.

Ruitenberg, A. A., and McCutcheon, S. R., 1982, Acadian and Hercynian structural evolution of southern New Brunswick, *in* St. Julien, P., and Béland, J., eds., Major structural zones and faults of the Northern Appalachians: Geological Association of Canada Special Paper 24, p. 131–148.

Ruitenberg, A. A., and 5 others, 1977, Evolution of pre-Carboniferous tectonostratigraphic zones in the New Brunswick Appalachians: Geoscience Canada, v. 4, p. 171–181.

Ruitenberg, A. A., and 5 others, 1979, Geology and mineral deposits, Caledonia area: New Brunswick Department of Natural Resources, Mineral Resources Branch Memoir 1, 213 p.

Walcott, C. D., 1900, Lower Cambrian terrain in the Atlantic province *(sic)*: Proceedings of the Washington Academy of Sciences, v. 1, p. 301–339.

Williams, H., 1979, Appalachian orogen in Canada: Canadian Journal of Earth Sciences, v. 16, p. 792–807.

Yoon, T., 1970, The Cambrian and Lower Ordovician stratigraphy of the Saint John area, New Brunswick [M.S. thesis]: Frederickton, University of New Brunswick, 92 p.

The Avalonian terrane around Saint John, New Brunswick, and its deformed Carboniferous cover

K. L. Currie, Geological Survey of Canada, 601 Booth Street, Ottawa, Ontario K1A 0E8, Canada

LOCATION

The Saint John district of southern New Brunswick (Fig. 1), readily reached by major highways, forms an Avalonian terrane of which the essential elements can be examined in one day by means of a road traverse. A street map of the city would be helpful in following the instructions, but the provincial road map should be adequate. Commence at the interchange at Main Street and New Brunswick 1 in downtown Saint John. Procede east on New Brunswick 1 (toward Moncton) for 2.8 mi (4.5 km) to Stop 1, a large quarry on the south side of the highway. Procede another 300 yd (275 m) on New Brunswick 1 to Stop 2. Reverse directions on New Brunswick 1 at the Rothesay interchange, 0.3 mi (0.5 km) farther east, exit from New Brunswick 1 westbound at Thurston Drive (about 0.8 mi [1 km] west of the Rothesay interchange) and follow signs to Cherry Brook Zoo (Stop 3). Proceed southwest on Sandy Point Road 1.1 mi (1.8 km) to the corner of University Avenue (Stop 4). Continue southwest on Sandy Point Road 1.0 mi (1.6 km) to Hazen White School, a large yellow brick building on the west side of the road (Stop 5). Continue on Sandy Point Road south to Hawthorn Avenue (0.7 mi; 1.1 km), turn left on Hawthorn Avenue, cross the Crown Street overpass over New Brunswick 1, and follow signs to Rothesay Avenue. Continue on Rothesay Avenue for 2.2 mi (3.5 km) to the intersection with Golden Grove Road. Procede 1.0 mi (1.6 km) east on Golden Grove to the corner of Roxbury Avenue, turn left on Roxbury, cross a small stream, and continue to a sharp left turn at the top of a hill (Stop 6, about 300 yd [275 m] from Golden Grove Road). Return to Golden Grove Road, turn right (west), then left almost immediately on Westmorland Road. Follow Westmorland to its end at Loch Lomond Road (New Brunswick 111). Turn right (west) on Loch Lomond. Drive 200 yd (180 m) and turn left on Bayside Drive. Continue south to the Saint John Dry Dock (Stop 7). Parking is available on Willett Avenue east of the dry dock. Procede south about 0.5 mi (0.8 km) on Bayside and turn right on Red Head Road. Drive to the end of the road (about 10 mi; 16 km at Cape Spencer (Stop 8). Return via Red Head Road, Bayside Drive, and Rothesay Avenue to New Brunswick 1.

SIGNIFICANCE

The Saint John region demonstrates a small displaced terrane juxtaposed against terranes of different sedimentological, igneous, and structural histories. The terrane exhibits an unusually complete late Precambrian and Cambrian section typical of "Avalonian" terranes, as well as probable basement. The terrane was last deformed in Carboniferous time. Dextral transcurrent faulting produced spectacular small-scale thrust allochthons and zones of intense deformation and quartz veining due to curvature and splaying of the master fault. These phenomena form an interesting contrast to those observed along the metamorphic front of an orogenic belt.

SITE INFORMATION

Although the geology of the Saint John district has been studied for more than 150 years, many details of the stratigraphy and structure remain obscure and/or controversial. The extensive older literature is summarized by Hayes and Howell (1937) and Alcock (1938), whose work is now mainly of historical interest. Geologists wishing a detailed view of the geology can use the outcrop map of Currie (1983) or McCutcheon and Ruitenberg (1984).

Most rocks in the Saint John region can be assigned to one of seven major units, listed in order of decreasing age as follows: (1) Brookville Gneiss, a hornblende-dominant, migmatitic gneiss that yields Middle Proterozoic upper intercept zircon ages; (2) Green Head Group, a Precambrian platformal sedimentary assemblage of marble (locally stromatolitic), quartzite, and minor pelite; (3) Golden Grove Suite, Late Proterozoic, hornblende-dominant, I-type intrusions ranging from altered gabbro to leucogranite; (4) Coldbrook Group, Late Proterozoic volcanic rocks ranging from pillowed basalt to rhyolite with abundant lahars, tuffs, volcanic breccia, and minor sedimentary rocks; (5) an Eocambrian sequence of red volcaniclastic conglomerate, sandstone, and tuff with intercalated subaerial basalt; (6) Saint John Group, a Lower Cambrian to Lower Ordovician transgressive sequence of fine-grained sedimentary rocks; (7) a sequence of Carboniferous coarse clastic sedimentary rocks. Silurian and Devonian rocks do not occur in the Saint John terrane. Triassic conglomerate occurs in fault-bounded troughs. The Late Proterozoic volcanic sequence and Cambrian sedimentary sequence identify the Saint John region as an "Avalonian" terrane similar to several others in Canada and the northeastern United States.

The major units form a crudely antiformal sequence with a central core of Brookville Gneiss (Fig. 1) flanked by successive belts of Green Head, Coldbrook, Saint John, and Carboniferous rocks. The core of Brookville Gneiss was remobilized during the Golden Grove intrusive event, and locally intrudes the Green Head Group, which was also remobilized along the contact to form marble and quartzite dykes. Plutons within the Brookville Gneiss have migmatitic aureoles, whereas those in the Green Head Group exhibit contact metamorphic aureoles, suggesting

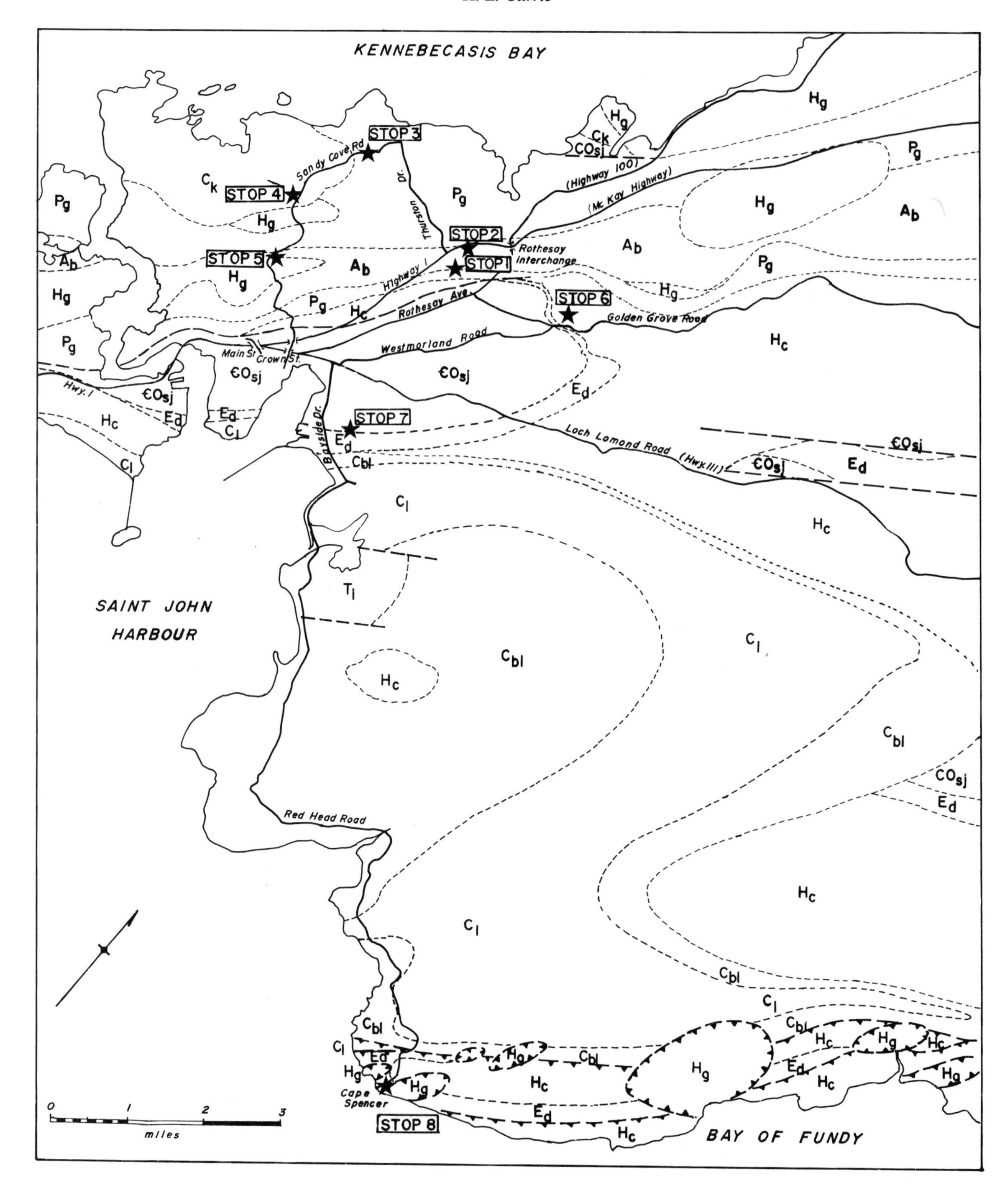

Figure 1. Geological sketch of the Saint John region, New Brunswick. Roads and highways referred to in the text are marked, and the described stops are marked by numbered stars. Geology simplified from Currie (1984). Explanation of symbols on facing page.

Triassic
T_1 Lepreau Formation; brown torrential conglomerate
-----**Unconformity**-----
Carboniferous
C_1 Lancaster Formation; gray lithic arenite, minor pebble conglomerate and black siltstone (Westphalian)
-----**Gradational Contact**-----
C_{bl} Balls Lake Formation; conglomerate, red shale with conglomerate lenses (Westphalian)
-----**Relations Unknown**-----
C_k Kennebecasis Formation; red conglomerate and sandstone (Tournaisian)
-----**Unconformity**-----
Cambro-Ordovician
$€O_{sj}$ Saint John Group; gray-green siltstone and sandstone quartzite, black shale
-----**Disconformity**-----
Eocambrian
E_d Red volcaniclastic sandstone and tuff, basalt
-----**Unconformity?**-----
Late Proterozoic
H_c Coldbrook Group; rhyolite, andesite, basalt, lahars, ignimbrites, minor cherty sediments
-----**Gradational Contact**-----
H_g Golden Grove Suite; granite, diorite, gabbro
-----**Intrusive Contact**-----
Middle Proterozoic
P_g Green Head Group; marble, quartzite, minor pelite
-----**Reactivated Unconformity**-----
Middle Proterozoic or older
A_b Brookville Gneiss; tonalitic to granodioritic gneiss and migmatite

Heavy broken line, fault; Heavy broken line with teeth, thrust fault with teeth on thrust block; dashed line, geological contact.

substantial relative (diapiric?) uplift of the Brookville Gneiss. The Coldbrook Group and disconformably overlying Eocambrian strata are down-faulted against the central crystalline core. Observations west of Saint John show that the Coldbrook Group is genetically related to the Golden Grove intrusions (Currie, 1986). The Saint John Group lies unconformably upon older rocks in tightly folded en echelon basins steeply overturned to the northwest. The Carboniferous section, divided into locally derived conglomerate (Kennebecasis Formation), and a southeasterly derived alluvial fan sequence (Balls Lake and Lancaster formations) lies unconformably on all older rocks, but exhibits structural style similar to that of the Saint John Group.

The Brookville Gneiss and Green Head Group were strongly deformed and metamorphosed during emplacement of the Golden Grove Suite. The Green Head Group, in contact with Brookville Gneiss, reaches sillimanite grade, but elsewhere is only at andalusite grade (Wardle, 1977). Late Proterozoic, Eocambrian, and Cambro-Ordovician rocks were folded and weakly metamorphosed (to epidote-chlorite-muscovite assemblages) in post-Ordovician time. The timing of this deformation and metamorphism is uncertain, but the similarity of structural style and metamorphic grade between Carboniferous and older rocks suggests the event was related to Fammenian to post-Westphalian D transcurrent faulting, which not only led to the final emplacement of the Saint John terrane, but created small pull-apart basins and thrust nappes within it.

The southeastern margin of the Saint John terrane appears to lie approximately along the shore of the Bay of Fundy where a steeply dipping cataclastic zone overprints earlier thrusts. Bouguer anomaly data show a sharp discontinuity along this zone from positive values of about 10 milligals on the landward side to −10 milligals beneath the bay (Chandra, 1982). The latter values occur over a thick Triassic sedimentary section revealed by oil-drilling operations. In the absence of outcrop, the identity of the terrane southeast of the Saint John terrane remains conjectural, but it probably is the Meguma terrane of southern Nova Scotia. The northwest margin of the Saint John terrane can be observed on the Long Reach of the Saint John River, 15 mi (24 km) north of Saint John, where fossiliferous, essentially unmetamorphosed rocks of the Saint John Group contact granite and hornfels of the Silurian Jones Creek Formation along a breccia zone some tens of yards (meters) wide. The Saint John terrane south of this fault (Wheaton Brook Fault) consists of Late Proterozoic sub-volcanic plutons, dykes, and volcanic rocks with infolds of Eocambrian and Cambrian sedimentary rocks. The terrane to the north (Mascarene-Nerepis belt) comprises a thick Silurian sedimentary section with Silurian and Devonian igneous rocks, and lacks recognizable Precambrian rocks.

The traverse described below takes the observer through the major units of the Saint John terrane approximately in order of decreasing age. New Brunswick 1 from Maine Street to Stop 1 passes first through gray-green cleaved sandstone of the Saint John Group, then through a large road cut in pale green epidotized rhyolite of the Coldbrook Group. Just before Stop 1 several small outcrops of Green Head Group occur on the roadside.

Stop 1. The Brookville Gneiss (Wardle, 1977) forms a generally tonalitic migmatite characterized by very regular compositional banding. In the quarry, biotite is the dominant mafic mineral, but elsewhere in this unit hornblende is dominant. Potash feldspar is absent from the finely banded gneiss, but occurs in migmatitic schlieren. Numerous randomly oriented rootless isoclinal folds are thought to indicate large-scale plastic deformation. The gneiss is cut by pale pink, near-massive tourmaline leucogranite of the Golden Grove Suite. Note the innumerable hybrids between gneiss and granite. Olszewski and Gaudette (1982) and Currie and others (1981) reported age determinations on zircons from this quarry that suggest a large inheritance from a 1.2 to 1.5 Ga source, emplacement at 0.6 to 0.8 Ga, and some recent lead loss. The age of the granite is not directly known, but it resembles phases of the Golden Grove Suite, dated by Rb-Sr at 615 Ma. Currie and others (1981) interpreted the Brookville Gneiss to be basement reactivated during emplacement of the Golden Grove Suite.

Proceding from Stop 1 to Stop 2, the road cuts are in Green Head Marble. Stop 2 lies at the Brookville–Green Head contact, about 300 yd (275 m) southwest of the exit ramp at the Rothesay interchange. At this point several dyes of Green Head marble cut the Brookville. The largest dyke follows a diabase dyke, and clearly crosscuts the gneissosity. Another 100 yd (90 m) to the southwest, past a screen of bushes, complex flow and brecciation can be observed in the marble that has completely broken up a diabase dyke. Note the presence of boudined Green Head quartzite and pelite at this locality.

New Brunswick 1 follows the Brokville–Green Head contact for 3 mi (5 km) northeast of this point. Geologists interested in the complexities of reactivated basement-cover relations may wish to examine the many road cuts in this stretch, which show abundant mutually intrusive relations of Green Head and Brookville (Currie and others, 1981).

Stop 3. The route from Stop 2 to Stop 3 passes across marble of the Green Head Group. Across the road from Cherry Brook Zoo, a large bulldozed exposure of marble displays strong gneissosity and numerous breccia zones. At the west end of this exposure, on the south side of Sandy Cove Road (the same side as Cherry Brook Zoo), a small outcrop of marble contains good, only slightly stretched, stromatolites. Deformation of the Green Head Group, and presumably of other units, appears very heterogeneous, ranging from plastic flow as seen at the last locality, through varying degrees of shear strain and brittle fracture seen here, to isolated regions with relatively little strain, such as the stromatolite locality.

Stop 4. The route from Stop 3 to Stop 4 passes roughly along the contact of the Green Head Group and Kennebacasis Formation. At Stop 4 the road cuts in the Kennebecasis Formation expose a boulder conglomerate with minor beds of red pebbly siltstone and sandstone. Clasts consist mainly of Green Head marble and quartzite with minor diorite. Outcrops of all these lithologies are visible from Stop 4. The Kennebecasis Formation accumulated in small fault-bounded troughs. The lowest, fluvial part of the formation contains Tournaisian fossils, but the conglomerate locally contains clasts with Visean faunas, indicating a considerable range in age of deposition.

About 400 yd (365 m) to the southwest of University Avenue, road cuts on Sandy Cove Road expose the Mayflower Lake diorite pluton, a pluton of the Golden Grove Suite emplaced in the Green Head Group. The pluton consists essentially of andesine and hornblende with minor biotite and quartz, and secondary epidote alteration on fractures. The hornfels aureole in pelites of the Green Head Group can be observed along the road to the dump just south of the road cuts.

Stop 5. At Hazen White School, cliffs of Green Head marble and pelite outcrop behind the playing field. Across the road, in the parking lot of Kingdom Hall, the rock faces expose the northern tip of the Fairville Pluton, a Golden Grove Suite pluton emplaced in Brookville Gneiss. Much of this pluton consists of megacrystic massive monzogranite, but this marginal exposure exhibits complex reactivation and metasomatic phenomena. Diabase dykes have been broken up, partially assimilated, and strewn through the granitoid rocks. The rock exhibits epidote-chlorite alteration, which has been overprinted by large, ovoid feldspars. The phenomena are thought to result from reactivation and assimilation of the Brookville Gneiss by the Fairville Pluton (Currie and others, 1981). The age of the Golden Grove Suite is not well known. The work of Olszewski and Gaudette (1982), as well as regional studies (Currie, 1986), suggest magmatism in at least two stages, with an early stage at about 770–800 Ma mobilizing the Brookville Gneiss and possibly emplacing early basalts of the Coldbrook Group, and a late stage at about 600 Ma emplacing granitic plutons like the Fairville, as well as the upper part of the Coldbrook Group.

Stop 6. Volcanic rocks of the Late Proterozoic Coldbrook Group outcrop here. The roadcuts consist of laharic material containing both acid and mafic volcanic fragments in a dark, volcanoclastic matrix. Note the strong cleavage and intense flattening of fragments, despite the good preservation of primary texture. The cleavage parallels that in the nearby Saint John Group, and may be of Carboniferous age. A small knoll 100 yd (91 m) east of the roadcuts, readily reached by an obvious footpath, consists of red ignimbrite. The small stream below the knoll is Cold Brook after which the group is named. These exposures typify the upper parts of the Coldbrook Group. The lower part of the group, exposed along New Brunswick 111 east of Saint John, consists mainly of basaltic rocks, commonly pillowed, with minor rhythmically banded cherty sediments. The age and chemistry of the Coldbrook Group are poorly known. The poorly defined Rb-Sr age of 750 Ma obtained by Cormier (1969) may refer to the lower, basaltic part of the group, but the upper, salic part is known to grade to plutons thought to be emplaced about 600 Ma (Currie, 1986). Preliminary chemical work by McCutcheon (in Ruitenberg and others, 1979) suggest calc-alkaline affinities for the Coldbrook Group but pervasive alteration makes this conclusion doubtful.

About 200 yd (180 m) west of the roadcuts, along Reading Crescent and Upland Avenue, Cambrian quartz-granule sandstone of the Glen Falls Formation outcrops on the lawn of Glenview United Church. This formation, thought to represent a beach deposit, is the most characteristic member of the lower Saint John Group. Glen Falls, after which the formation is named lies on Cold Brook just below the church.

Stop 7. The area around the Saint John Dry Dock (Fig. 2) displays the most complete single section of Eocambrian and Cambrian stratified rocks in the Saint John region. North of the railway line in the large quarry and various cuttings the gray-green, ripple-marked sandstone belongs to the Kings Square Formation of the Saint John Group (terminology of Pickerill and Tanoli, 1985). Behind Willetts Food Warehouse, black siltstone and shale of the underlying Forest Hills Formation (partly cut out by a minor fault) passes transitionally southward into Hanford Brook Formation, readily identified by black phosphatic nodules up to 5 mm in length. Near the stairway to the parking lot the rocks grade to reddish muscovitic sandstone of the Ratcliffe Brook Formation, the basal unit of the Saint John Group. The whole section faces west and dips steeply east, forming part of the overturned limb of the Saint John syncline.

Behind the old iron kiln at the east edge of the parking lot,

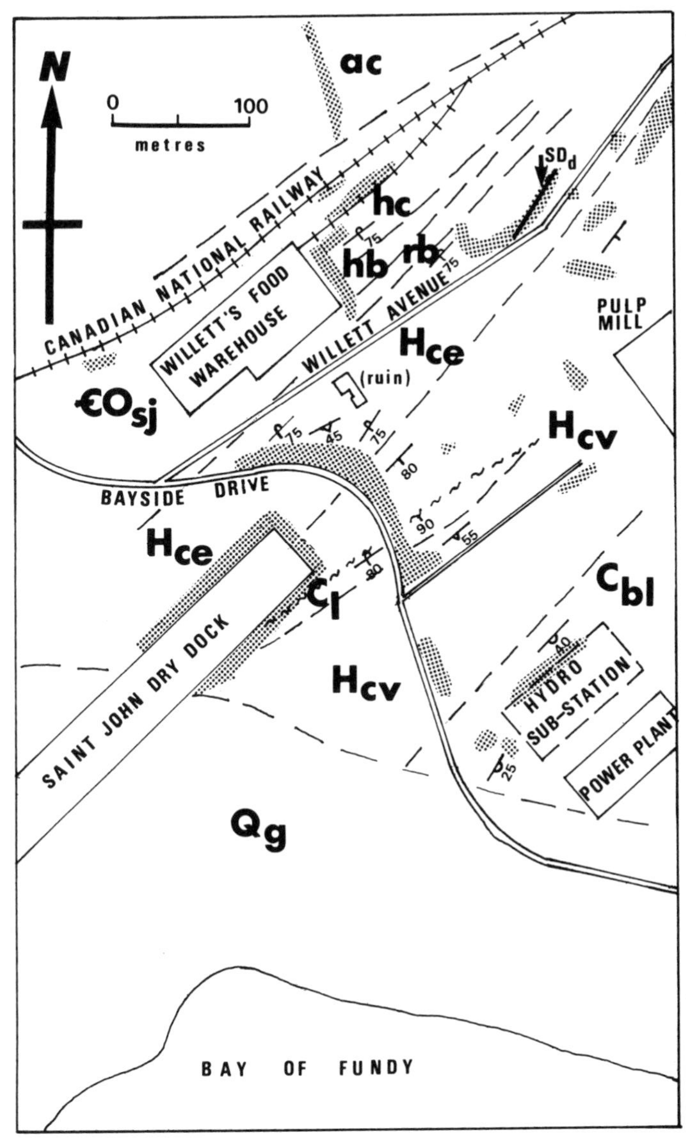

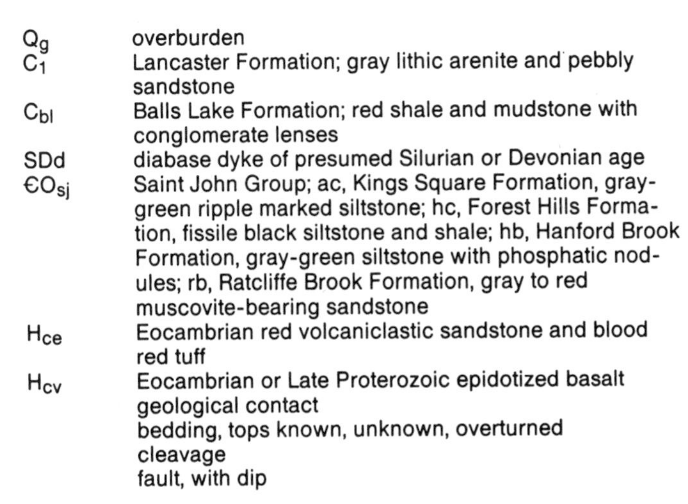

Figure 2. Geological sketch of the area around the Saint John Dry Dock. Outcrop areas are stippled.

pink sandstone, rich in feldspar crystals, likewise faces west and dips east, but there is about a 10° discordance in attitude with the Saint John Group. This sandstone, assigned to the Eocambrian by Currie (1984), is hornfelsed by a large diabase dyke visible on low knolls just north of Willett Avenue. This distinctive brown-weathering diabase occurs only within the Saint John Group and Eocambrian section. South of Willett Avenue basalt of the Eocambrian section outcrops under the power pylons.

The northern end of the roadcut on Bayshore Drive consists of west-facing, east-dipping red to pink feldspathic sandstone with 2 to 4 in (5 to 10 cm) blood red tuffaceous beds. These rocks rest approximately conformably on massive but chloritized and epidotized basalt. These rocks appear correlative to the section on strike on Willett Avenue. Near the sharp curve on Bayside Drive the basalt is truncated by a vertical fault against dark gray pebbly sandstone of the Carboniferous Lancaster Formation, which also faces west but is 40° discordant in strike with the sandstone. Near the southern margin of the Lancaster, a 12 in (30 cm) fault slice of basalt is present, but the most interesting contact is the southern contact of sandstone against basalt about 50 ft (15 m) north of the entry to the Rothesay Paper Mill. The sandstone contains recognizable pebbles of basalt, and despite the welded appearance of the contact, microscopic examination shows no contact metamorphism of the sandstone. The contact is interpreted to be an unconformity with Carboniferous sandstone resting on Eocambrian basalt. The basalt continues another 160 ft (50 m) beyond the entry to the Rothesay Paper Mill where it is overlain by red mudstone with conglomerate lenses that faces and dips east at a moderate angle. A large cutting in this Carboniferous Balls Lake Formation can be seen through the fence in the transformer station.

The dry dock area can be interpreted to display a continuous west-facing section through the overturned limb of the Saint John syncline, passing downward through the lower part of the Saint John Group and the underlying Eocambrian section. According to this interpretation, these older rocks are overlain with back-to-back unconformity by Carboniferous strata, and the slice of Lancaster Formation is part of the overturned limb of a small fold, similar to that exposed at the mouth of Little River, 0.5 mi (0.8 km) to the south. A very different interpretation, offered by Rast and others (1978), supposes that all of the Lancaster and Balls Lake formations—as well as the basalt, and possibly the red beds—form parts of far-travelled allochthons. This interpretation assumes that a "Variscan front" separates southern allochthonous rocks, significantly metamorphosed and intruded in Carboniferous time, from autochthonous rocks to the north that did not undergo such an event. Viewers of this section may wish to conduct their own search for the major thrust plane that must be present according to this interpretation. The alternatives were discussed by Currie (1984).

Stop 8. The route from the dry dock to Cape Spencer passes through abundant outcrop of Carboniferous clastic sedimentary rocks (Lancaster and Balls Lake formations), which form a southerly derived alluvial-fan complex (Currie and Nance, 1983).

The rocks are upright and gently dipping, but exhibit increasingly intense ("Variscan") deformation toward the south. The last 2 mi (3.2 km) to Cape Spencer lie in an intensely overthrust and cataclastically deformed zone.

Walking from the gate at Cape Spencer to the lighthouse, the first outcrops are highly deformed mafic volcanics, while those closer to the lighthouse are extremely sheared purplish sandstone and siltstone. Both rocks are interpreted to belong to the Eocambrian sequence, like those seen at the dry dock. The whole sequence is here strongly silicified and laced with quartz veins, some of which contain significant amounts of gold. From the lighthouse, skirt the low knoll of heavily sheared greenish granite to the west and follow the gravelled path to the contact on the cliff. *CAUTION: Steep slopes and loose rock!* Highly deformed granite can be seen thrust over sheared feldspathic sandstone. The youngest cleavage here strikes northeast and dips moderately northwest. The granite forms a keel-like mass plastered against the cliff and bounded by a folded thrust surface. The opposite (northwest-dipping) contact can be seen on the sea cliff about 0.5 mi (0.8 km) east of this point (accessible to the determined via a path from the lighthouse and a short scramble through the bushes). A similar but smaller keel of granite can be seen on the cliff to the north across the small cove.

The deformed belt, of which Cape Spencer forms a part, exhibits the classical tectonic reversal of stratigraphy, passing southward from Carboniferous through Eocambrian and Coldbrook Group strata to the tectonically highest slices of Precambrian plutons. Although this observation, and local movement indicators, clearly indicate transport to the northwest, all of the slices are locally derived, and none show metamorphism higher than biotite grade. The zone of allochthons is very narrow, generally less than 1 mi (1.6 km) wide. Several of lines of evidence suggest the deformation results from dextral movement on a major transcurrent fault (Currie, 1986), which drilling and geophysical evidence indicate must lie just offshore from Cape Spencer. The nature of the terrane on the other side of this fault is uncertain, but regional considerations suggest that it is probably the Meguma terrane of southern Nova Scotia, visible on the horizon from Cape Spencer on a clear day. The deformation in the Cape Spencer belt has been variously termed "Variscan" (Rast and others, 1978) and "Alleghanian" (Nance and others, 1985). Since both terms refer to orogenic episodes, and Carboniferous deformation in the Saint John region is not associated with significant metamorphism or plutonism, neither term is really appropriate.

REFERENCES CITED

Alcock, F. J., 1938, Geology of the Saint John region, New Brunswick: Geological Survey of Canada Memoir 216, 65 p.

Chandra, J., 1982, Bouguer anomalies, Loch Lomond map area: New Brunswick Department of Natural Resources, Mineral Development Division Plate 82-14, scale 1:50,000.

Cormier, R. F., 1969, Radiometric dating of the Coldbrook Group of southern New Brunswick: Canadian Journal of Earth Sciences, v. 6, p. 393–398.

Currie, K. L., 1983, The geology of the Saint John region, New Brunswick: Geological Survey of Canada Open-File 1027, scale 1:50,000.

—— , 1984, A reconsideration of some geological relationships near Saint John, New Brunswick: Geological Survey of Canada Paper 84-1A, p. 193–201.

—— , 1986, The boundaries of the Avalon tectonostratigraphic zone, Musquash Harbour–Loch Alva region, southern New Brunswick: Geological Survey of Canada Paper 86-1A, p. 333–341.

Currie, K. L., and Nance, R. D., 1983, A reconsideration of the Carboniferous rocks of Saint John, New Brunswick: Geological Survey of Canada Paper 83-1A, p. 29–36.

Currie, K. L., Nance, R. D., Pajari, G. E., and Pickerill, R. K., 1981, Some aspects of the pre-Carboniferous geology of Saint John, New Brunswick: Geological Survey of Canada Paper 81-1A, p. 23–30.

Hayes, A. O., and Howell, B. F., 1937, Geology of Saint John, New Brunswick: Geological Society of America Special Paper 5, 146 p.

McCutcheon, S. R., and Ruitenberg, A. A., 1984, Saint John, New Brunswick: New Brunswick Department of Natural Resources, Geological Surveys Branch Plate 84-1, scale 1:50,000.

Nance, R. D., Warner, J. B., and Caudill, M. R., 1985, The Mispec Group, Saint John; A tectonostratigraphic record of the Alleghanian (Variscan) event in southern New Brunswick: Geological Association of Canada Abstracts with Programs, v. 10, p. A43.

Olszewski, W. J., and Gaudette, H. E., 1982, Age of the Brookville Gneiss and associated rocks, southeastern New Brunswick: Canadian Journal of Earth Sciences, v. 19, p. 2158–2166.

Pickerill, R. K., and Tanoli, S. K., 1985, Revised lithostratigraphy of the Cambro-Ordovician Saint John Group, southern New Brunswick; A preliminary report: Geological Survey of Canada Paper 85-1B, p. 441–449.

Rast, N., Grant, R. H., Parker, J.S.D., and Teng, H. C., 1978, The Carboniferous deformed rocks west of Saint John, New Brunswick, *in* Ludman, A., ed., Guidebook for fieldtrips in southeastern Maine and southwestern New Brunswick: Queens College Geological Bulletin, v. 6, p. 162–173.

Wardle, R. J., 1977, The stratigraphy and tectonics of the Green Head Group; Its relations to Hadrynian and Paleozoic rocks, southern New Brunswick [Ph.D. thesis]: Fredericton, University of New Brunswick, 294 p.

A classic Carboniferous section; Joggins, Nova Scotia

Martin R. Gibling, Department of Geology, Dalhousie University, Halifax, Nova Scotia B3H 3J5, Canada

LOCATION

The Joggins section is located in the vicinity of Joggins Village on the Bay of Fundy in Cumberland County, Nova Scotia. Approximately 4,600 ft (1,400 m) of strata are exposed along 3.6 mi (6 km) of coast from Lower Cove to Ragged Reef Point (Fig. 1). These beds form part of an almost continuous 30 mi (50 km) coastal section from Minudie to Squally Point that, excluding the lowermost Horton Group, contains more than 14,500 ft (4,400 m) of Carboniferous strata. The section can be reached by leaving Trans-Canada 104 at Amherst (inset in Fig. 1), the closest community with readily available accommodation; garages and stores are located at Joggins and River Hebert.

The section can be divided into three geographic subsections, from stratigraphic base to top (Fig. 1): 1) Lower Cove to Joggins Village, 2) Joggins to MacCarrons River, and 3) MacCarrons River to Ragged Reef Point. For logistical ease, the visitor should begin at Lower Cove and continue up section, using the parking places shown on Figure 1. The first two together, and the third subsections each form a good day's study. Access from the parking places is by public rights of way, except at Ragged Reef Point, which is the property of Mr. Arnold Mills (see Fig. 1), from whom permission should be obtained for access by field parties. Public access to the beach is also possible at Two Rivers, farther south.

The Fundy tides are among the world's highest, with a mean tidal range of 34 ft (10 m) at Joggins (45 ft (13.7 m) for exceptional tides). Tide tables can be obtained from the Canadian Government Publishing Centre, Supply and Services Canada, Ottawa, Ontario K1A OS9, Canada, or designated distributors. Daily times are also included in most local papers; the tide at Joggins is 15 to 30 minutes later than at St. John, New Brunswick. The visitor should always be certain of the tidal stage before moving onto the beach, as the tide rises with great speed, especially under conditions of onshore winds and high waves. The section generally can be studied safely for three hours either side of low tide. Retreat to beach areas above tide level is not always possible. Hardhats are recommended, as the cliffs are unstable and overhang in places, and care should be taken to avoid soft, muddy parts of the foreshore. Nevertheless, with simple precautions, the sections can be studied safely.

The cliffs from Lower Cove to Hardscrabble Point form a protected locality. Special permission is required for sampling from the Curator of Special Places, Nova Scotia Museum, 1747 Summer St., Halifax, N.S., B3H 4A6, Canada. Abundant material for collection, however, is usually available on the beach.

SIGNIFICANCE OF SECTION

Sir Charles Lyell (1871, p. 409) described Joggins as "the finest example in the world of a natural (Carboniferous) exposure in a continuous section ten miles long. . . ." Local pride aside, it is certainly the finest Pennsylvanian section in eastern North America. Its historic significance stems from visits in the 1840s by Lyell, William Logan, and J. W. Dawson, who presented the first detailed description of a coal-bearing sequence and noted the cyclic nature of the sedimentation.

The section contains superb, three-dimensional exposures in sea cliffs and wave-cut platforms of alluvial channel and floodplain deposits that include coals and fossiliferous limestones. The preservation of many vertical trees is of great interest, especially as amphibian and reptilian remains have been found entombed within the trunks. The reptiles represent some of the earliest known. Trackways of giant arthropods are an additional paleontological feature.

SITE INFORMATION

Regional Setting

The strata belong to the Cumberland Group of late Westphalian A to early Westphalian C (Pennsylvanian) age (Bell, 1944; Hacquebard and Donaldson, 1964). No formations are presently recognized, although several divisional systems have been proposed (Logan, 1845; Bell, 1914; Shaw, 1951; Copeland, 1958). The section described belongs to Divisions 2 (basal part), 3, and 4 of Logan (numbered from the top down), and the Westphalian B Lower Fine Facies (Facies B) of Shaw (1951) and Copeland (1958).

The Cumberland Group was deposited in the Cumberland Basin, a deeply subsiding trough centered over Chignecto Bay and lying between the Cobequid Hills to the south and the Caledonia Hills of New Brunswick to the northwest (inset, Fig. 1). During the Pennsylvanian, the basin opened northeastward into the Gulf of St. Lawrence Basin, and formed part of the extensive Fundy Basin system of Atlantic Canada, in which short-lived, actively subsiding depocentres were associated with major strike-slip faulting in Late Devonian to Early Permian times, following the Acadian Orogeny (Belt, 1968; Bradley, 1982). Apart from marine incursions of the Visean Windsor Group, the strata are continental. Rapid subsidence of the alluvial plains and the periodically humid climate encouraged the growth of peat, now found as economic coals.

The Joggins section lies on the gently dipping (less than 25°) northern limb of the northeasterly-trending Athol Syncline. Faults are rarely observed, and the section is relatively undeformed. The predominantly fine-grained strata at Lower Cove and the overlying coal-bearing succession appear to pass south-

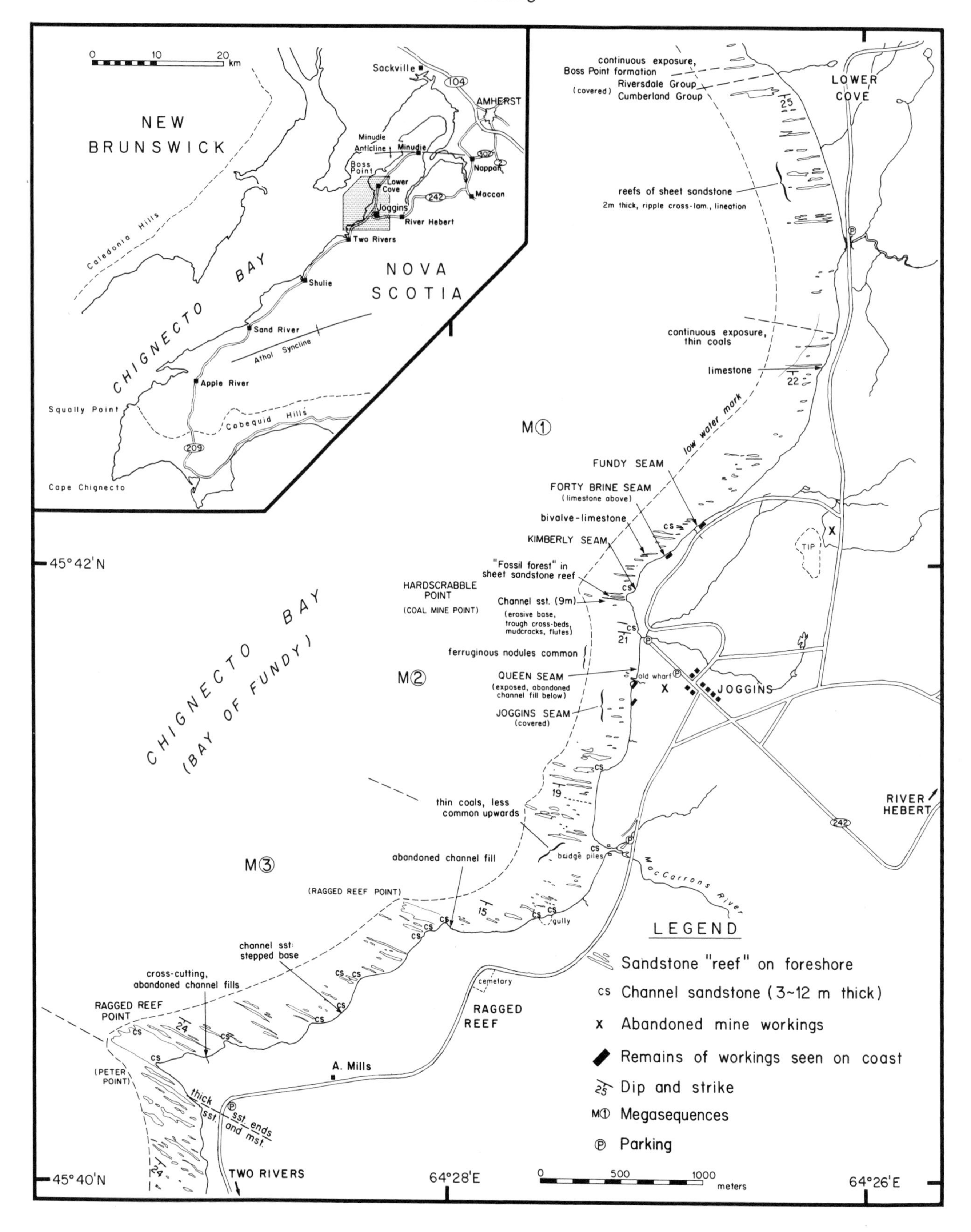

Figure 1. Location and geological features of the Joggins section, Nova Scotia.

ward across the synclinal axis into coarser facies that include pebbly sandstones and conglomerates towards the Cobequid Hills.

General Lithology and Stratigraphy

The 4,620 ft (1,400 m) section consists of red and gray mudstone and brown sandstone, with minor coal, dark limestone, and ferruginous carbonate nodules. The thicker sandstone units form resistant headlands that extend across the broad foreshore as "reefs." Perusal of Figure 1 shows that the proportion of sandstone reefs varies systematically along the section. The three geographic subsections correspond to three megasequences, each several hundred meters thick, in which the proportion and thickness of the sandstone units increase progressively upward until thick mudstone successions reappear. The lower two megasequences (Division 4 of Logan, 1845) can reasonably be ascribed to the progressive proximity of sandy channel belts to the depositional area. The third (Division 3 of Logan, 1845), which shows a concomitant upward change in fluvial style and the appearance of extrabasinal gravel at Ragged Reef Point, suggests that tectonic rejuvenation took place, accompanied by a slope increase.

The strata exhibit alternate gray and red intervals, tens to hundreds of meters thick. The gray intervals contain coals and limestones and appear poorly bedded due to root bioturbation, whereas the red intervals contain well-defined sandstone and mudstone units, thick channel sandstones, few coals, and show less bioturbation. The colour is clearly associated with lithology, especially with the abundance of vegetation, evidence that strongly supports a syndepositional or early diagenetic "fixing" of the color. The main gray, coal-bearing sequences occur south of Lower Cove, just north and south of Joggins Village, and south of MacCarrons River (Fig. 1).

Sedimentary Features

Channel Sandstones. Channel sandstones are relatively uncommon from Lower Cove to MacCarrons River, but become increasingly abundant towards Ragged Reef Point. They range from 10 to 41 ft (3.0 to 12.5 m) in thickness and are recognized by their thick-bedded nature and tendency to form headlands. The sandstones are remarkable for their narrow three-dimensional form, (width:thickness ratios commonly less than 10:1); indeed, in the cliffs, many can be seen to pass laterally into thin-bedded strata, or they form thick foreshore reefs that wedge out before reaching the cliffs.

The channel bases are erosional and show steps suggesting that lateral migration of the channels took place periodically and was accompanied by rapid filling. The basal beds contain mudstone fragments and logs up to 3 ft (1 m) long, with groove and tool casts visible on the undersurfaces. The overlying beds consist of medium-grained sandstone in the form of trough cross-beds up to 3 ft (1 m) thick that resemble the deposits of dune systems in modern channels. However, many fills appear structureless or display faint remnants of contorted troughs and show little upward change in grain size; inclined layers (lateral accretion sets) are present at the top of some fills. Many channels contain several subunits a few meters thick, separated by irregular erosional surfaces, which suggest that infilling took place in several stages.

Hardscrabble Point, north of Joggins village, is a good example of a thick channel sandstone (30 ft; 9 m) with mudstone and plant fragments at the base. Its lower part is thick bedded with trough cross-beds seen in full relief on the exposed bedding surfaces at the point. Its upper part is more thinly bedded. Flute casts, current lineation, and desiccation cracks can be seen on bed undersurfaces towards the top. An abandoned-channel fill of finer material is present as part of the channel body south of the point.

The channels formed networks of narrow, relatively deep (average 11.5 ft; 3.5 m) watercourses that, judging by the variable paleocurrent directions shown by the trough cross-beds, were highly sinuous. Their rapid vertical filling and frequent changes of course suggest that they were akin to modern anastomosed systems, in which the floodplain is traversed by a system of stably positioned channels (Rust and others, 1984). The stability of the channels in this case can be ascribed to the abundance of rooted vegetation that acted to bind the sediment and hinder erosion of the channel banks. Cross-beds in the channel sandstones south of MacCarrons River indicate paleoflow from the Caledonia Hills northwest of the section.

Sheet Sandstones. Sheet sandstones, present throughout the section, form units from a few centimetres to several metres thick that can be identified readily by their thin-bedded nature and continuity across the cliff faces and foreshore. Thin mudstone interbeds form an appreciable part of most sheets. The fine-grained sandstones contain a variety of sedimentary structures that include ripple cross-lamination, primary current lineation, groove casts, desiccation cracks, and rainprints (uncommon). However, many beds appear structureless due to the activities of roots, which can be seen as vertical, carbonized or clayey traces especially at the tops of the sheets. Many sheets show upward trends in bed thickness that include thinning upward, thickening upward, and thickening followed by thinning upward. Some sheets wedge out gradually across the cliff, and some adjacent sheets appear to wedge out in opposite directions. Sheet sandstones are the prime hosts of the vertical trees, described later.

Small channel fills, up to 6.6 ft (2 m) thick, are common in the sheets, and are narrow (width:thickness less than 15:1; Duff and Walton, 1973). They contain inclined layers of sandstone and mudstone that, in some suitable exposures, can be traced all across the outcrop. The sheet sandstones tend to be thicker and contain more beds in the vicinity of the small channels.

The sheet sandstones are interpreted as crevasse splay deposits formed by many individual floods that carried coarse sediment across the floodplain from the main channels (Way, 1968). The small channels probably acted as conduits for the floodwaters. Where the sheets are abundant, the depositional area probably lay close to a major channel (not necessarily seen in the section), as channel and sheet sandstones generally show a close associa-

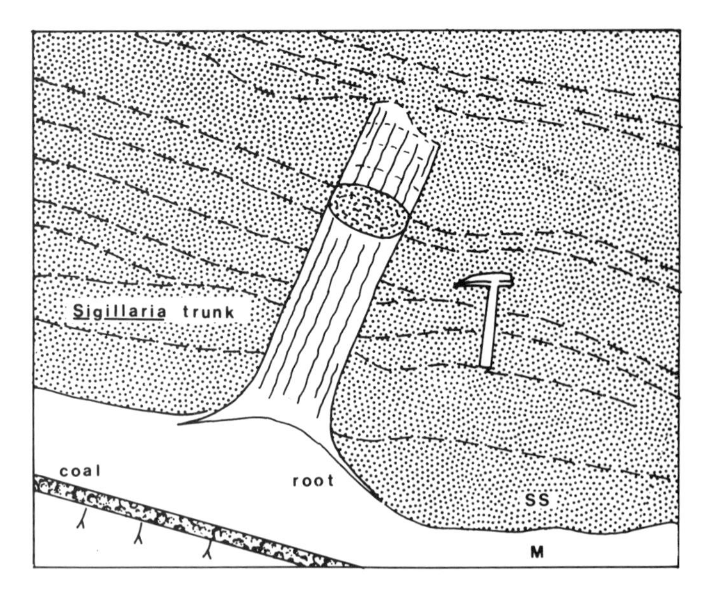

Figure 2. Sketch of *Sigillaria* trunk preserved in vertical position in a sheet sandstone (SS: stippled) above a coal and mudstone (M), 0.9 mi (1.5 km) south of MacCarrons River. The lower trunk is a mould, and the upper part forms a sandstone cast. Note the lenticular bedding close to the trunk, suggesting turbulent flow and/or compactional deformation. Hammer is 12 in (30 cm) high.

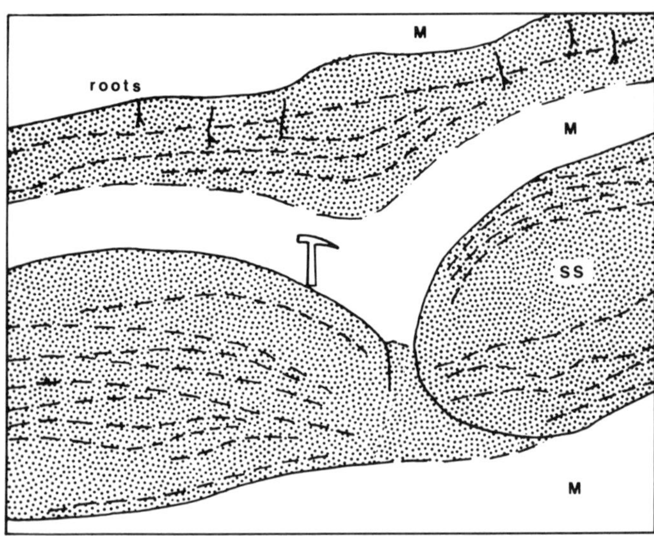

Figure 3. Sketch of deformational structure in sheet sandstone in cliffs on southern bank of MacCarrons River. Sandstone (SS) is stippled. Some sandy layers appear to have been deflected downwards, but overlying and underlying beds are undeformed. Roots are common in the mudstones (M) and in the uppermost sandstone unit; the structure probably represents syndepositional to early diagenetic subsidence associated with trees (not presently visible) and/or roots. Hammer is 12 in (30 cm) high.

tion in the cliffs. The vertical trends in bed thickness can best be explained by relative distance of the site from the main channel; thickening-upward sheets, for example, represent the advance of the main channel towards the site, depositing progressively thicker and coarser layers of sediment with each flood as the channel approached the site.

Trees were often buried where they stood by the sudden influxes of sediment. Turbulence around the projecting vegetation during floods produced undulating bedding surfaces and lenticular beds within the sheets (Fig. 2). Several unusual deformational structures (Fig. 3) appear to be due to localized subsidence associated with the trees, and in a sheet sandstone 0.6 mi (1 km) south of MacCarrons River, a "pot hole" 5 ft (1.5 m) deep and 2 ft (60 cm) wide filled with sandstone may have had a similar origin.

Bedding planes within the sheet sandstones show the trackways of a giant terrestrial arthropod, probably *Arthropleura* (Ferguson, 1975). These traces somewhat resemble paired motorcycle tracks, and a huge specimen—possibly Canada's largest known fossil—can be seen in the Geology Department at Mount Allison University, Sackville, New Brunswick (Ferguson, 1966).

Mudstones. The mudstones form units up to several meters thick and are red, gray, or mottled in colour. Roots are abundant but are often difficult to identify; clues to their presence are provided by the conchoidal weathering of the mudstone (laminae were destroyed during root penetration) and by red/gray mottling (diagenetic reduction around the roots). Thin, carbonaceous partings, generally underlain by rooted zones, represent vegetated soils, and can be considered "failed coals."

Brown, ferruginous nodules up to 8 in (20 cm) in diameter are present in many mudstone beds as scattered concretions or as vertically-oriented pipes that enclose roots. Such nodules in the Fundy Basin, generally composed of siderite with minor amounts of calcite and dolomite, display concentric structure and radially-arranged fissures of septarian type. The nodules are common in the strata just south of Joggins village. Siderite forms in modern swamp sediments under reducing conditions.

The mudstones represent floodplains that at times were thickly vegetated. The mud, the finest material transported by the rivers, was deposited in areas distant from the channels. Gray coloration suggests poor drainage and a high water table (reduced soils), whereas red coloration suggests good drainage and a low water table (oxidized soils).

Some thick mudstone beds contain concave-up surfaces with about 10 ft (3 m) of relief containing rubbly mudstone and thin beds of sandstone. They represent abandoned channels filled with blocks from collapsed muddy banks and with water-laid sediment. Several good examples can be seen towards Ragged Reef Point.

Coals. Logan and Dawson identified 71 coal seams in the present section, ranging from a few centimetres to 5 ft (1.5 m) in thickness. Most seams contain thick partings of dark carbonaceous mudstone so that the individual layers of bright-coal are less than 12 in (30 cm) thick and are not prominent in the cliffs. The coals rest on nonbedded mudstones that are generally pale and show abundant roots (seatearths). Five seams (Fundy, Forty

Brine, Kimberly, Queen, and Joggins) were mined in the area, from the early 1700s until 1979, and the positions of these seams in the cliffs (Fig. 1) can be identified by relics from the excavations. All except the Joggins Seam are visible in the cliffs.

The coals formed in peat swamps that were established locally on the floodplain surface during periods of reduced detrital supply. A 12-in- (30-cm)-thick coal layer represents about 10 ft (3 m) of original peat prior to compaction. The partings and high ash content reflect influxes of sediment into the swamps.

Limestones. Several dark limestone beds overlie coals and mudstones on the foreshore between Lower Cove and Hardscrabble Point (Fig. 1), where they form extensive, resistant scarps up to 28 in (70 cm) thick. The beds show wavy lamination due to thin layers of abundant compacted bivalve shells. The bivalve genera *Naiadites* and *Curvirimula* have been identified, along with ostracodes, conchostracans, annelid tests (*Spirorbis* sp.), mysidacid and diplopod arthropods, arachnids, fish fragments, and algal material (Copeland, 1957; Carroll and others, 1972; Duff and Walton, 1973).

The limestones formed in shallow lakes on the alluvial floodplain, most commonly where a rise in the groundwater table caused peat swamps to be drowned to a depth too great for rooted vegetation to persist, perhaps a few meters depth. The carbonate was mainly biogenic (shells), with a probable contribution from inorganic and/or biochemical precipitation. The faunal assemblage is of low diversity and the individual species tend to populate separate layers or units suggesting that the environment was stressful for most organisms. Although some of the biota suggest brackish conditions, no firm evidence attests to marine conditions.

Special Features

Trees. Trees preserved in upright position include the large lycopod tree *Sigillaria,* which can be seen as individual trunks up to one meter in diameter and several meters in height (Fig. 2). Lyell (1871) recorded a trunk 25 ft (7.6 m) high in the cliffs, and Dawson (1878, p. 188) mentioned a trunk, presumably horizontal, traced to the length of 40 ft (12.2 m) in the roof of the Joggins Seam. The trunks show patterns of leaf scars on the bark in well-preserved specimens and longitudinal corrugations in other examples. They can locally be observed to pass into the substrate as tuberculated roots known as *Stigmaria,* which are up to 4 in (10 cm) in diameter and can extend for 16 ft (5 m) away from the trunk. The horse-tail, *Calamites,* forms individual stems or clumps, each stem up to about 10 cm in diameter, characterized by closely spaced longitudinal striations and transverse joints; some preserved stems are more than 3 ft (1 m) long.

All the specimens are preserved as carbonized bark that encloses sandstone and mudstone, with little evidence of internal tissue. As in many modern species, the bark retained its coherence while the interior rotted to leave a hollow structure.

The trees are rooted in mudstone or sit on top of the coals, but are embedded in the sheet sandstones (Fig. 2). Their preservation was thus due to rapid burial by the crevasse splays, which, in modern alluvial plains, can deposit several feet of sediment during a flood. Because the hollow trunks were filled with sediment, they have suffered relatively little compaction during burial. The conditions surrounding their preservation were discussed with remarkable detail and lucidity by Dawson (1878).

Vertical trees are best seen in the Lower Cove to Joggins subsection but can be found throughout the section. They can often be observed high in the cliffs from a distance, but as they weather out rapidly, their locations cannot be guaranteed; however, casts of stems are common on the beach. The sheet sandstone that underlies the channel sandstone at Hardscrabble Point and extends as a reef across the foreshore (Fig. 1) was reported by Dawson to contain at least 24 upright lycopods (see Fig. 12 of Carroll and others, 1972). More than 30 *Calamites* and several *Sigillaria* trunks are still visible there.

Terrestrial vertebrates. Scattered vertebrate material has been found at several levels in the section, but complete skeletons and fragments from more than 200 individuals have been found within the lycopod stumps at many levels, a remarkable occurrence. The eleven genera identified include labyrinthodont amphibians (rhachitomes and embolomeres, which form the bulk of the assemblage), microsaurian amphibians, and reptiles (cotylosaurs and pelycosaurs, among the earliest known forms). Although such material is unlikely to be found by casual visitors, the perusal of fallen trunks could reveal specimens. Tetrapod tracks have also been found in the sheet sandstones.

The terrestrial vertebrates, most rather small forms, fell into the hollow trunks when the trunk rims were flush with the sediment surface. Their fragmented state suggests that later arrivals fed on the remains of their predecessors, while the abundance of coprolitic material suggests that they survived within the trunks for some time. The shells of land snails, and tests of millipedes, annelids, and ?eurypterid arthropods have also been found within the trunks. For a detailed description of the assemblage, see Carroll (1964, 1966, 1967) and Carroll and others (1972).

REFERENCES CITED

Bell, W. A., 1914, Joggins Carboniferous section, Nova Scotia: Geological Survey of Canada Summary Report 1912, p. 360–371.

——— , 1944, Carboniferous rocks and fossil floras of northern Nova Scotia: Geological Survey of Canada Memoir 238, 277 p.

Belt, E. S., 1968, Carboniferous continental sedimentation, Atlantic Provinces, Canada, *in* Kay, M., Late Paleozoic and Mesozoic continental sedimentation, northeastern North America: Geological Society of America Special Paper 106, p. 127–176.

Bradley, D. C., 1982, Subsidence in Late Paleozoic basins in the northern Appalachians: Tectonics, v. 1, p. 107–123.

Carroll, R. L., 1964, The earliest reptiles: Journal of the Linnean Society (Zoology), v. 45, p. 61–83.

——— , 1966, Microsaurs from the Westphalian B of Joggins, Nova Scotia: Proceedings of the Linnean Society, London, v. 177, p. 63–97.

——, 1967, Labyrinthodonts from the Joggins Formation: Journal of Paleontology, v. 41, p. 111–142.

Carroll, R. L., Belt, E. S., Dineley, D. L., Baird, D., and McGregor, D. C., 1972, Vertebrate paleontology of eastern Canada: 24th International Geological Congress, Montreal, Excursion A59, p. 64–80.

Copeland, M. J., 1957, The arthropod fauna of the Upper Carboniferous rocks of the Maritime Provinces: Geological Survey of Canada Memoir 286, 64 p.

——, 1958, Coalfields, west half Cumberland County, Nova Scotia: Geological Survey of Canada Memoir 298, 89 p.

Dawson, J. W., 1878, Acadian Geology: London, Macmillan, 694 p.

Duff, P. McL. D., and Walton, E. K., 1973, Carboniferous sediments at Joggins, Nova Scotia: 7th International Congress Carboniferous Stratigraphy and Geology, v. 2, p. 365–379.

Ferguson, L., 1966, The recovery of some large track-bearing slabs from Joggins, Nova Scotia: Maritime Sediments, v. 2, p. 128–130.

——, 1975, The Joggins section: Maritime Sediments, v. 11, p. 69–76.

Hacquebard, P. A., and Donaldson, J. R., 1964, Stratigraphy and palynology of the Upper Carboniferous coal measures in the Cumberland Basin of Nova Scotia, Canada: 5th International Congress Carboniferous Stratigraphy and Geology, v. 3, p. 1157–1169.

Logan, W. E., 1845, A section of the Nova Scotia coal measures as developed at the Joggins, on the Bay of Fundy, in descending order, from the neighborhood of the west Ragged Reef to Minudie, reduced to vertical thickness: Geological Survey of Canada Report of Progress 1843, Appendix, p. 92–153.

Lyell, C., 1871, Elements of Geology: New York, Harper, 640 p.

Rust, B. R., Gibling, M. R., and Legun, A. S., 1984, Coal deposition in an anastomosing-fluvial system; The Pennsylvanian Cumberland Group south of Joggins, Nova Scotia, Canada, *in* Rahmani, R. A., and Flores, R. M., Sedimentology of coal and coal-bearing sequences: International Association of Sedimentologists Special Publication 7, p. 105–120.

Shaw, W. S., 1951, Preliminary map, Springhill, Cumberland, and Colchester Counties, Nova Scotia (map and structural sections): Geological Survey of Canada, Paper 51-11.

Way, J. H., 1968, Bed thickness analysis of some Carboniferous fluvial rocks near Joggins, Nova Scotia: Journal of Sedimentary Petrology, v. 38, p. 424–433.

Jurassic basalts of northern Bay of Fundy region, Nova Scotia

George R. Stevens, Department of Geology, Acadia University, Wolfville, Nova Scotia B0P 1X0, Canada

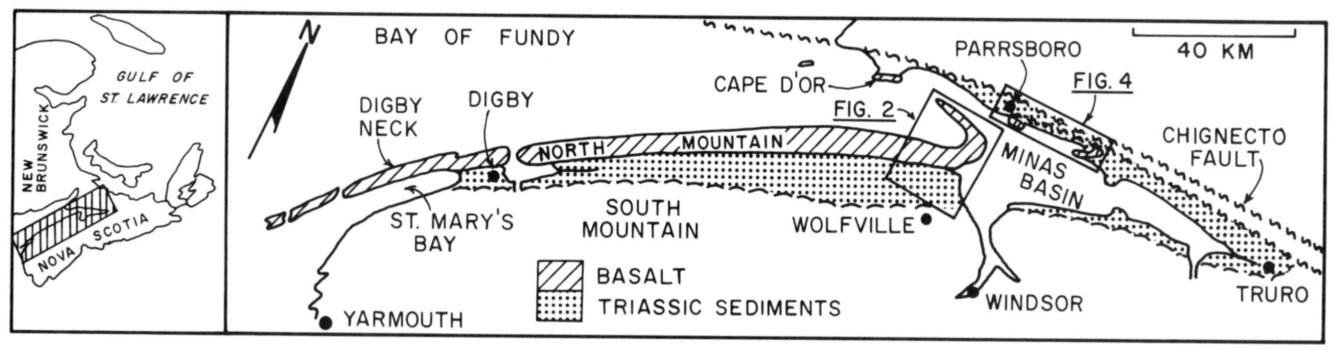

Figure 1. Index map to Bay of Fundy region showing regional extent of Jurassic basalt exposures and localities cited in this guide.

LOCATION

The upper flow units of the North Mountain Basalt Formation extend in continuous exposure for 124 mi (200 km) along the Bay of Fundy. The massive lower unit is exposed at several places along the escarpment face of the North Mountain cuesta, and continuously along Saint Mary's Bay (Fig. 1). The five stops described here center on the remarkable exposures near Scots Bay and along the North Shore of the Minas Basin, where structural and stratigraphic relations are well displayed. Stops are situated on the Parrsboro and Wolfville 1:50,000-scale map sheets 21-H8 and 21-H1, National Topographic System, available from the Canada Map Office, 615 Booth Street, Ontario K1A 0E9. The Geology Map of The Cobequid Highlands (sheet 2), and the Geological Highway Map of Nova Scotia (Bujak and Donahoe, 1980) both available from the Nova Scotia Department of Mines and Energy, Halifax, are useful for locating optional sites.

Caution. The tidal range of the Bay of Fundy region attains 50 ft (15 m). Shoreline field studies are severely constrained by the times of the tides, which vary daily. Consult the published current tide tables or daily Halifax newspaper to plan for a day having an early morning high tide. The shore should be vacated well before high tide to pass promontories that flood early.

North Mountain and Scots Bay Syncline

Stop 1 is a road cut and quarry at the base of the ridge, 2.5 mi (4 km) north of Sheffield Mills on the road to Glenmont (Fig. 2). Stop 2 is on the intertidal shelf at Ross Creek and Bennett's Bay, reached by a road leading north 3.7 mi (6 km) from its intersection with the road between Glenmount and "The Lookoff" (Fig. 2). A country church marks the intersection. Stop 3 on the Cape Split peninsula is reached by a 1,100 ft (350 m) walk on the intertidal shelf, westward from the parking site at the end of the road from Scots Bay Village (Fig. 2). The entire shelf, extending for several miles (kilometers) eastward and westward from the parking site, displays features of major interest (Fig. 3).

North Shore Minas Basin

The cliffs and spur at Old Wife Point in the Five Islands Provincial Park comprise Stop 4 (Fig. 4). The park is reached by a 1.9 mi-long (3 km) paved access road leading south off Nova Scotia 2, 1.2 mi (2 km) east of the center of Five Islands Village. A short walking trail leads from the camping grounds to the top of a high spur (a basaltic dike) from which geologic relations can

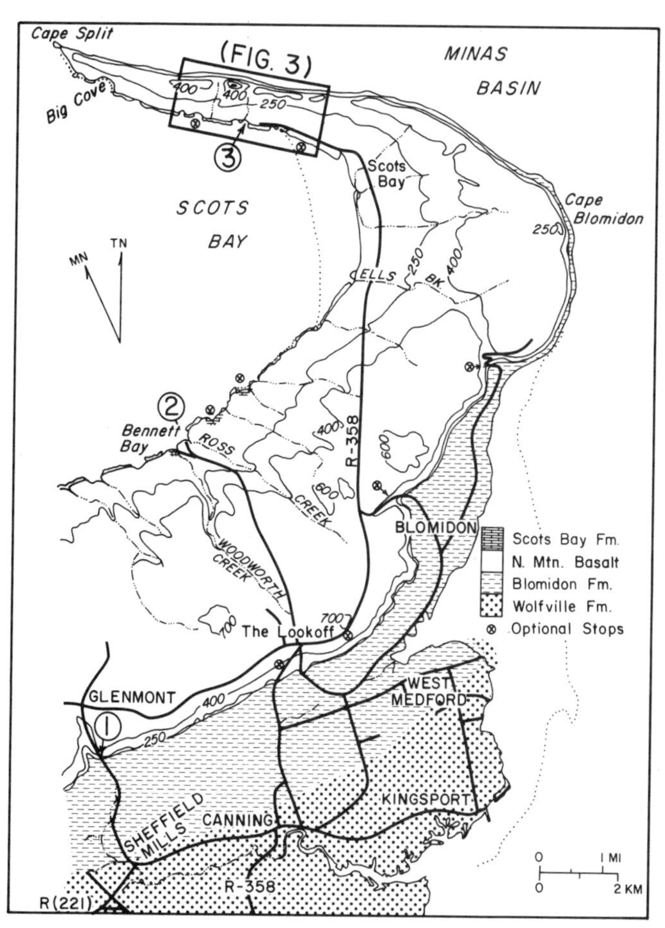

Figure 2. Geologic map of Scots Bay region showing the general geology and location of Stops 1, 2, and 3.

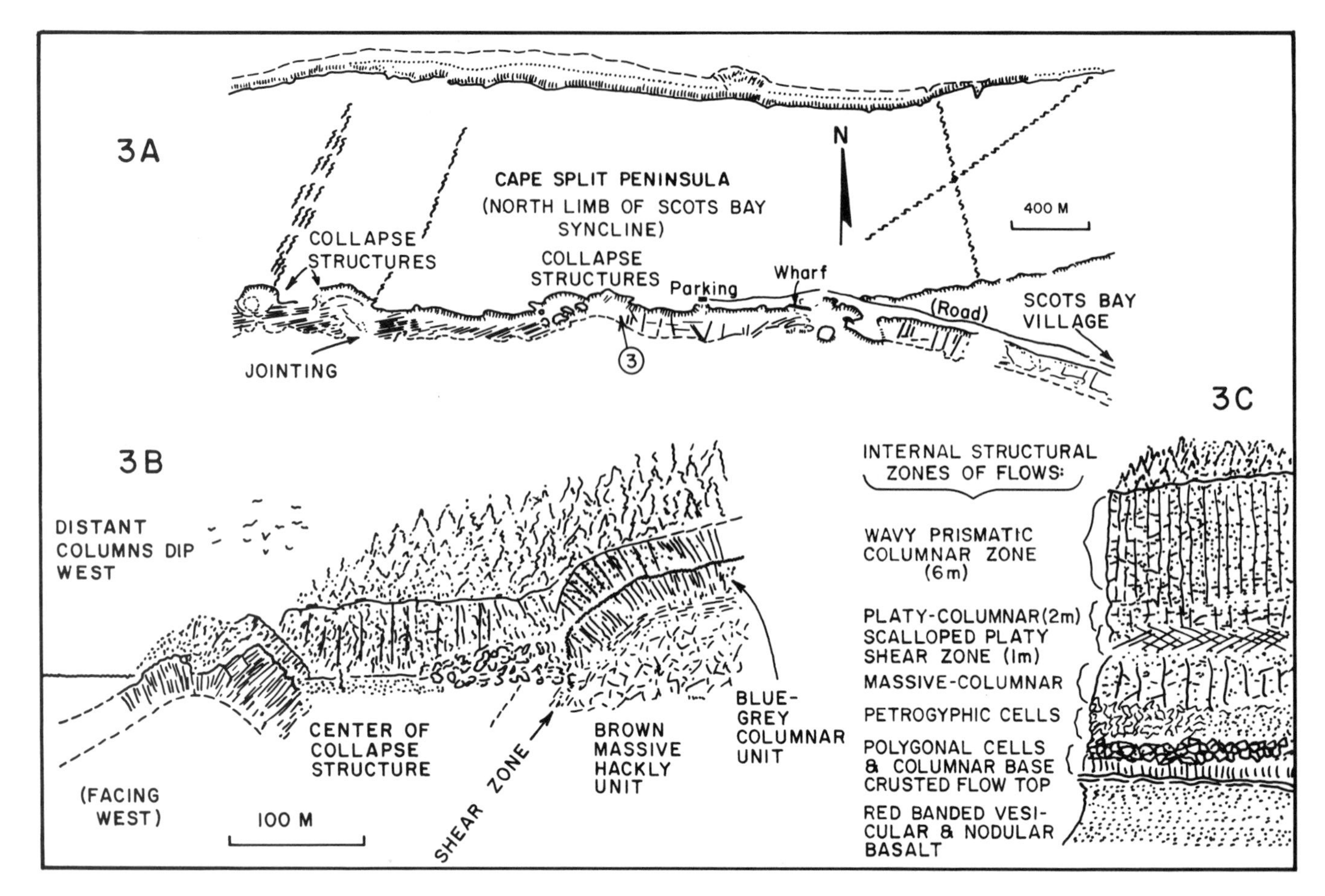

Figure 3. Collapse features and internal zones within upper flow, as seen at Stop 3, Cape Split Peninsula. 3A, Location map; 3B, North-south cross section of collapse feature; 3C, Schematic section showing internal structural zones of upper flow unit.

be examined from above. Beach-level access is gained from a picnic shelter in the lower park grounds. Walk west, then south along the beach for 0.3 mi (0.5 km) to reach the prominent spur of dark basalt. Stop 5 consists of the entire 5,000-ft (1,500-m) cliff-face at Wasson Bluff. The access trail to the shore begins at a parking site 5.6 mi (9 km) east from the causeway in Parrsboro, on the Greenhill Road. Station 5A is on the west side of the mouth of Wasson Brook, reached by a short walk to the shoreline along the woods trail from the parking site.

Optional stops of considerable geological interest are at Economy Mountain on Nova Scota 2 (basal flow unit), Blue Sack (agglomerate, lahars), MacKay Head (comformable contact of flows with Scots Bay Formation), and at Clarke Head (baked, faulted contact of basalt with the Blomidon Formation; Fig. 4). In addition, major exposures of the North Mountain basalt occur on the headlands west of Parrsboro at Partridge Island, and at Cape D'Or (Fig. 1).

SIGNIFICANCE

During Triassic and Jurassic time, opening of the Atlantic produced a swarm of large graben along the eastern margin of North America. These graben received up to 23,000 ft (7,000 m) of subaerial clastic sediments during subsidence, plus tholeiitic basalt flows, dikes, and sills. These basalt magmas represent the igneous signature of Mesozoic continental separation; thus their compositions, differentiation trends, and structures are of considerable tectonic significance (Philpotts, 1978).

The Bay of Fundy graben is one of the largest and the northernmost (Stevens, 1977). The coastal exposures along the dip slope of the North Mountain cuesta provide nearly continuous sea-cliff outcropping for 125 mi (200 km), and both the upper and basal units are exposed and accessible. Coastal erosion has also produced many good exposures of syntectonic volcaniclastic facies along the fault-controlled north shoreline of the Minas Basin. Structural and stratigraphic relationships are well displayed there (Liew, 1976).

The North Mountain Basalts are typical subaerial tholeiites with labradorite, augite, pigeonite (less), and opaques, commonly in a glassy groundmass. They are quartz-normative with felsic and mafic indices coincident with those of the Palisades sill. Though many individual flow units can be seen at sites along the

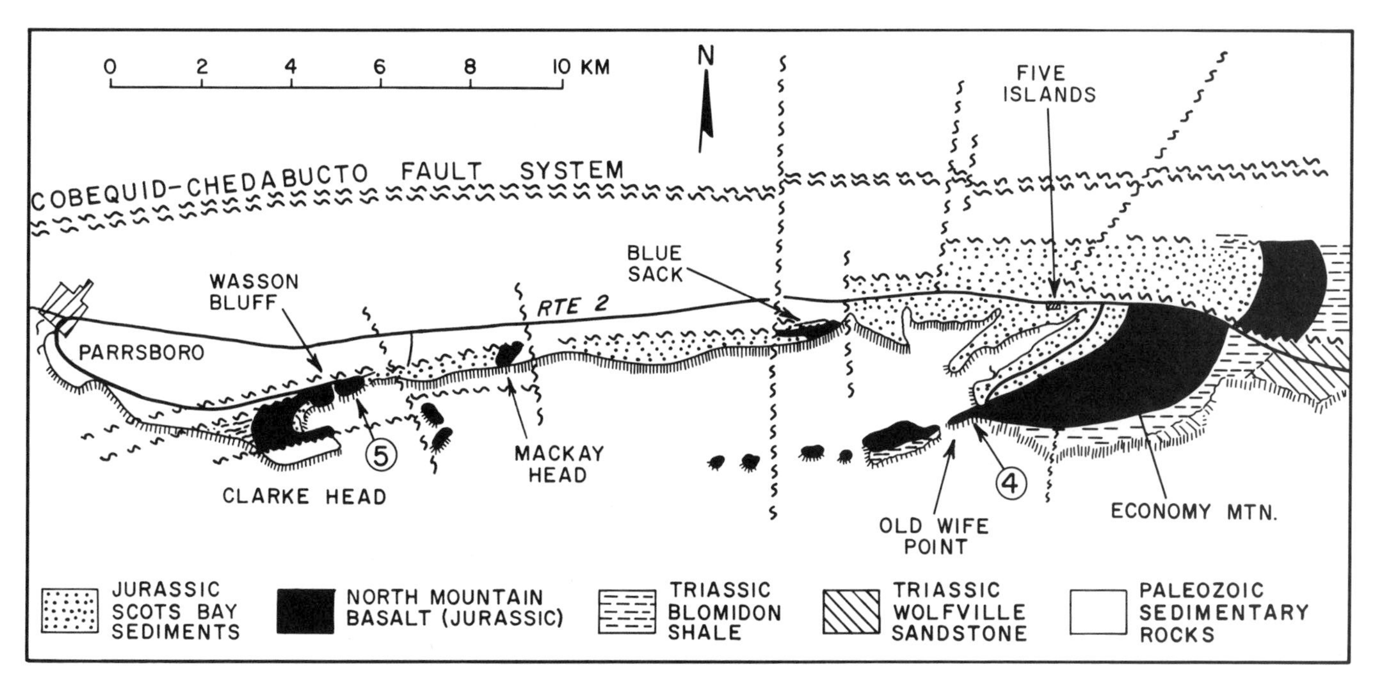

Figure 4. General geology of the North Shore of the Minas Basin showing the locations of Stops 4 and 5, and optional exposures of the North Mountain Formation.

shore, the entire sequence can be subdivided into three general units: a thick lower massive flow (up to 650 ft; 190 m thick), an intermediate zone of reddish gray amygdaloidal flows (up to 400 ft; 90 m thick), and an upper group of interlayered massive and amygdaloidal flows, generally blue-gray in color (up to 600 ft; 150 m thick). Individual flows range in thickness from 15 to 50 ft (5 to 15 m) each.

Columnar structure is common, plus an extensive catalogue of other primary structures related to flow and cooling. Two sets of early joints are filled with either zeolite or chalcedony, concomitant with subsidence and warping of the basin. Coarsely crystalline quartz (rarely amythystine) commonly fills the central plane of thicker agate and jasper veins and the larger gas cavities. Vesicles are zeolite filled only. Stilbite, heulandite, and chabazite are most common, though natrolite, apophyllite, laumonitite, mordenite, mesolite, and analcite also occur. Pillows have been reported at only two places, possibly fluviatile rather than marine. Polyhedral cell structures related to internal cooling within flows are common, and can be misleading because of their pseudopilloidal form. Large circular collapse features on the upper flow occur near Scots Bay (Stop 3). Some of them have been infilled with limy lacustrine sediments, with associated siliceous hot-spring deposits. On Digby Neck (Fig. 1), large circular bedrock hills, each with concentric depression bands, indicate erosional remnants of volcanic necks or plugs.

SITE INFORMATION

North Mountain, which forms the Fundy coast of Nova Scotia, is a continuous cuesta of tholeiitic basalt flows on a floor of red shales and sandstones (Stevens, 1985). Jurassic continental clastics overlie the basalts and compose the floor of the Bay of Fundy.

Prebasalt Triassic sedimentary rocks (Wolfville and Blomidon formations) thicken westward from 3,300 ft (1,000 m) in the Annapolis Valley of Nova Scotia to 13,000 ft (4,000 m) under the Bay of Fundy near Grand Manaan Island, as recorded from the Mobil-Chinampas borehole. Post-basalt sediments (Scots Bay Formation) attain a thickness of 2,300 ft (700 m) in the borehole. The basalt flows vary in total thickness between 1,000 and 1,500 ft (300 and 430 m; Lollis, 1959).

The Fundy graben is internally faulted and folded, and is marked by a major synclinal warp along its eastern edge. Scots Bay and Cape Blomidon are on the axis of this southwest-plunging fold, whose exposed limbs are the North Mountain cuesta and the Cape Split Peninsula.

The entire North Shore of the Minas Basin is part of the Chignecto-Glooscap fault system, or Minas Geofracture. It extends westward under the Bay of Fundy to the Gulf of Maine and eastward across Nova Scotia to Chedabucto Bay, where it joins the Orpheus Graben offshore. The entire system is therefore about 620 mi (1,000 km) long. Faulting was continuous through Triassic and Jurassic time, culminating well after deposition of the Scots Bay Formation. Vulcanism, sedimentation, and faulting were synchronous (Stevens, 1980). The coastal segment between Clarke Head and Economy Mountain is a severely cross-faulted synclinal graben on the flank of the larger Minas structure.

Paleozoic and Mesozoic sediments along this zone are faulted against Jurassic basalts and agglomerates (Liew, 1976). Basalts of North Mountain age (191 Ma; Hyatsu, 1979) are ex-

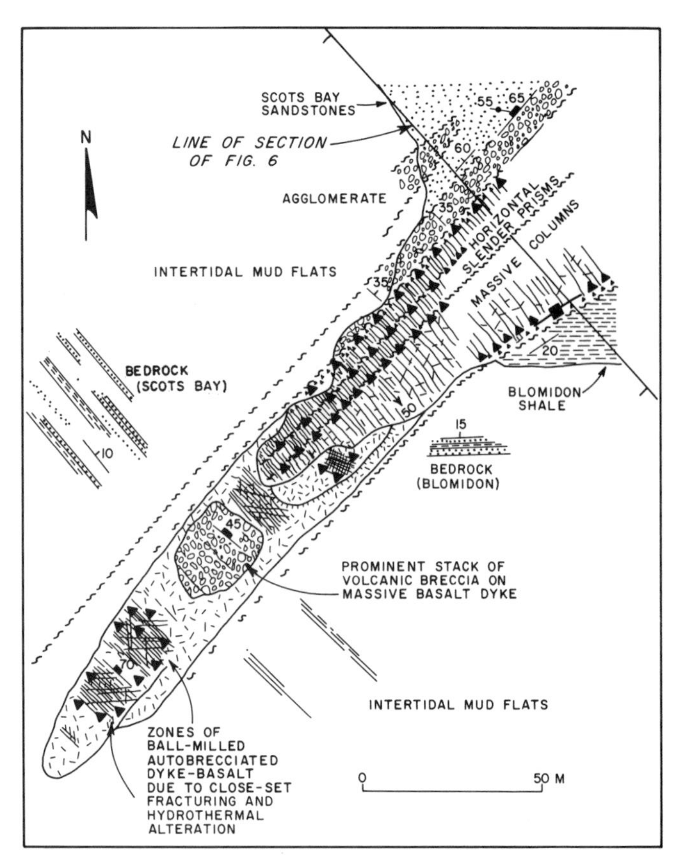

Figure 5. Sketch map of geologic relations at Stop 4, Old Wife Point, Five Islands Provincial Park. Triangle symbol indicates zones of autobrecciation.

posed at several places: most notably at Economy (Gerrish) Mountain in the east; at the Five Islands, McKay Head, Wasson Bluff, and Clarke Head; and at Cape D'Or in the west (Figs. 1, 4). Red clastics of pre- and postbasalt ages are also well exposed. Volcanic facies in this area are indicative of more violent activity.

Two tectonically contrasting exposures of basalt can be examined at the site. The first, in uppermost flows near Scots Bay (Stops 1 to 3), formed under relatively static conditions. The second group includes flows, dikes, pyroclastic rocks, and lahar agglomerates emplaced during active tectonism along a major regional fault zone (Stops 4, 5). Both tectonic and primary internal structures will be seen, including faults, mineralized joints, ball-milled hydrothermal breccia, and several types of gas structures. Many types of primary geometric structures related to cooling and flow occur (columns, cells, nodules, glyphs), which produce a gross internal zonation and layering of flows.

There are many excellent optional sites, including the coastal villages along the Bay of Fundy (upper flows) and at several places along the escarpment face (lower flow). Possible intrusive plugs can be seen on Digby Neck near Sandy Cove. A standard road map will suffice to locate such places.

Stop 1: Glenmount Notch. The massive lower basalt unit is well exposed both in the roadcut and in the adjacent quarry. Five gently dipping internal zones can be discriminated here by variation in styles of columnar jointing, over a road distance of 1,000 ft (300 m). A thickness of at least 200 ft (60 m) is exposed. Columns are generally inclined, leaning northwest. This may indicate flow direction at this stop. At the south (downhill) end of the roadcut, the basalt paraconformable contact with the red Blomidon shale may be seen dipping 2° to 4° northwest. The contact is usually wet with spring water, and may be obscured by local muck. Except for quartz veins and rare specularite in the upper zones, mineralization is lacking.

Stop 2: Ross Creek and Bennett Bay. The uppermost flow unit is exposed in the cliff and underfoot on the intertidal shelf. Walk westward (left) from Ross Creek. Zeolite-filled amygdules and veins (heulandite, stilbite) are general in the zone near the flow top (Colwell, 1980). Erosion has emphasized primary jointing and cooling structures, best seen in the wave-scoured shelf. Pseudo-pilloidal and "glyphic" forms (like hieroglyphs) result from differential weathering of cooling cells and joint-bounded blocks. These concentrically nested polyhedral shells of basalt are textural and mineralogic indicators that they are early features, produced within the basalt during cooling, and controlled by accellerated heat removal along primary joints. Nodular, botryoid, and cellular ("honeycomb") structures can be seen in the basalts here and elsewhere, emphasized by differential weathering of the less resistant hydrous groundmass between the anhydrous nodules. Gradual passive heat loss results in early hardening of the pasty lava (and dehydration) at randomly distributed centers. These centers enlarge with time to form nodules and botryoid forms. The planar interfaces of "packed" nodules produce polygonal cells ("honeycomb"), which approximate dodecahedra in form. Some workers have confused these forms with pillows.

Continue walking along cliffs westward to Bennett Bay (600 ft; 200 m). A circular arc of white platy Scots Bay limestone is seen in the intertidal zone, and the contact with the basalt is exposed on the two spurs that flank Bennett Bay. The inward dip and circular array indicate deposition within a collapse funnel in the basalt. Large masses of jasper and siliceous tubular encrustations occur in the bluffs at the head of the bay, the result of hot-spring activity during deposition of the limestone. Similar collapse features (devoid of limestone) are also seen at Stop 3. Exposures showing two more filled collapse structures may be reached by a short walk (0.6 mi; 1 km eastward) along the beach from Ross Creek.

Stop 3: Cape Split Peninsula. A collapse embayment exposed here is 600 ft (200 m) in diameter. Dark columnar basalts encircle the collapse, and all column axes lean inward (Fig. 3). The steeply dipping upper columnar zone of a prominent flow lobe seen on the north wall rests on a thick internal shear zone, formed as the upper zone slid on the pasty upper part of the massive basalt underneath. Some nearly horizontal columns can also be seen on the north wall. The zone of internal shear can be traced across the north wall into a silica-mineralized band with

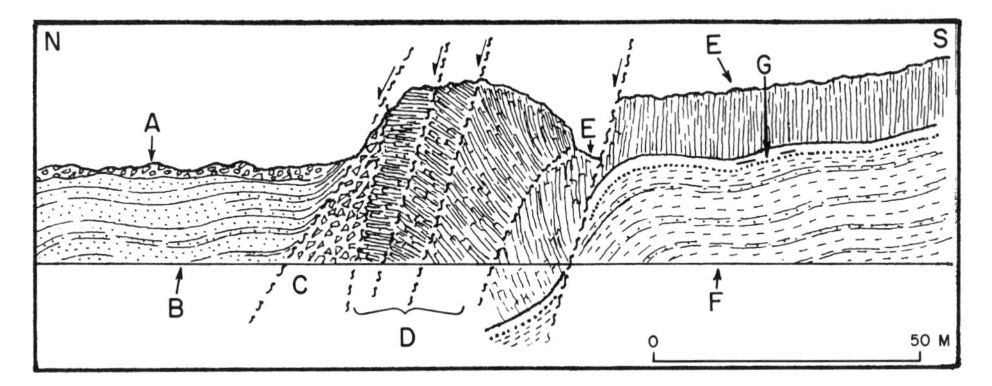

Figure 6. North-south structure cross secton at Stop 4. See Figure 5 for location of section. A, glacial till cover: B, Scots Bay Sandstone; C, volcanic agglomerate; D, dike basalt with breccia bands along faults; E, North Mountain Basalt; F, Gypsiferous red shale of Blomidon Formation; G, baked, bleached contact zone.

slickensides indicating slip direction. This collapse is adjacent to another, which has the same characteristics. The north wall of the second collapse exposes a red-oxidized flow top at the base of the cliff. Internal shear planes are visible within the massive upper flow there. Another pair of collapse embayments occurs 0.6 mi (1 km) west of Stop 3. During the return walk to the car park, observe subhorizontal slickensided sheets of quartz in the shear level within the flow, visible on the shelf underfoot.

About 0.3 mi (0.5 km) east of the parking site on the intertidal shelf are numerous large, bell-shaped gas cavities (spiracles?) filled with jasper, quartz, and vesicular basalt, often with rinds of black glass. They range between 8 in and 3 ft (20 cm and 1 m) in width and height and are typically either elliptical or D-shaped ("bell") in cross section. Long axes and symmetry axes are preferentially aligned, showing the east-west flow directions of the basalt. The "bells" occupy a particular layer within the host flow, between the upper columnar zone and a lower zone of polyhedral cellular basalt. The "bells" are not observed in other levels within the flow. Note that silica veins occupy north-trending joints, and zeolite veins (younger) follow east-trending joints.

Stop 4: Five Islands–Old Wife Point. The spur consists of volcanic agglomerate (north face) and a dike basalt with subhorizontal columns (Figs. 5, 6). They are in normal-fault contact with the Scots Bay strata, clearly exposed on the north side of the bluff. This fault is one of several parallel normal faults at Old Wife Point. Structurally and morphologically, the rocks of the spur form a dike.

The prominent erosional stack (Fig. 5) rising from the wave-cut bench of the spur consists of volcaniclastic breccia; it shows a crude internal banding, which dips about 45° northeast. It rests sharply on massive columnar basalt of the dike, which shows autobrecciation in places. Baked, jasperized veins of mobilized red sediments occur within the stack. The stack might be an erosional remnant of vent breccia, but this is uncertain..

Ball-milled pseudo-agglomerate. Conjugate extensional joints in northwest-trending zones cause rhomboidal fracture at several places on the wave-cut shelf of the dike, and in the longitudinal fault zones (Figs. 5, 6). Movement along these zones and hydrothermal activity have mobilized these rhomboidal clasts and partially rounded them by chemical and mechanical abrasion, resulting in pseudo-breccias, which superficially resemble volcanic agglomerates (exposed nearby). The black waxy matrix is shown by x-ray diffraction to be a hydrobiotite-vermiculite paste.

The view to the northeast along trend of the spur from its end provides a good cross-sectional view of contact relations and internal structures (Fig. 6). Steeply dipping normal faults are seen to have dropped the (north) hanging wall, bringing the Scots Bay sandstones into contact with the agglomerate, which itself is in fault contact with dike basalt. The well-developed prismatic structure is nearly horizontal in the north fault slice. Other faults cut the dike internally, with production of breccia bands. In the center the basalt is massive, and the large columns are moderately inclined, splaying upward into steeper, more splintery prisms. Nearly identical columnar structure and format occur in the dike at Wasson Bluff (Stop 5), including similar steeply dipping autobrecciated pseudo-agglomerate bands normal to column axes. Ball-milled breccia and hydrothermal paste (matrix) occur at several places throughout Old Wife Point. To the south a strongly curved internal fault can be seen bounding a block with nearly vertical massive columns (North Mountain Basalt). This block is downfaulted against the Blomidon Formation, visible to the south. Milled breccia also occurs in this fault zone. Fibrous gypsum veins parallel to bedding are common in the shales. Note the prominent baked and bleached contact zone at the top of the Blomidon Formation just under the North Mountain Basalt.

Stop 5: Wasson Bluffs. *Station 5-A.* Best cross-sectional view of the dike is westward, from the east end of the outcrop (Fig. 7). East of the fault and against it is an erosional remnant of North Mountain Basalt resting conformably on west-dipping Blomidon shales. Their contact is baked. The mass of Wasson Bluff consists of dike basalt faulted into contact with sedimentary rocks. Alternate subvertical sheets of prismatic-columnar basalt and breccia (pseudo-agglomerate) extend parallel to the shore.

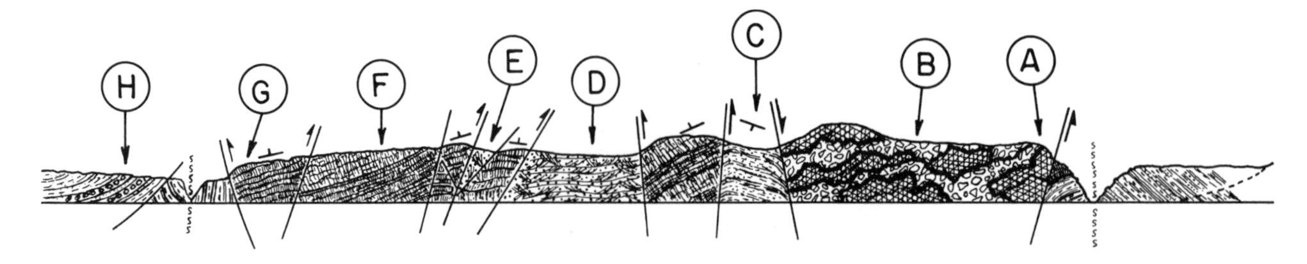

Figure 7. Geologic cross section at Wasson Bluffs (Stop 5), near Parrsboro, Nova Scotia. Stations A to H are described in text. Approximate width of section is 4,920 ft (1,500 m). Facing north.

Columns and prisms are inclined or subhorizontal, making this exposure structurally identical to the dike seen at Old Wife Point (Stop 4). Binoculars greatly enhance study of the internal structural zones of the dike. Brecciation due to movement on parallel internal faults or to sequential injection of dike sheets, creates thick breccia bands (pseudo-agglomerate), which separate bands of prismatic basalt.

Station 5-B. Walk west along shore from Station A. The cliffs at Station B are vertical intercalated sheets of prismatic basalt and breccia, striking east-west. A prominent sea stack here consists of pseudo-agglomerate (dike breccia). The vertical basalt sheets exhibit the end-on aspect of prismatic structure.

Station 5-C. This is an upthrown block of north-dipping gypsiferous sedimentary rocks, either the Blomidon Formation or possibly Windsorian strata (Mississippian). Gypsum veins parallel to bedding are common in the upper Blomidon shales, as at Old Wife Point (Stop 4) and at Station 5A. Possible thermal effects are to be seen in the sedimentary rocks adjacent to the faults. The presence of "clean" sandstone (similar to that at Station 5-D) at the base of slope next to the west fault suggests that the shales above could even be of Scots Bay age.

Station 5-D. Thick cross-bedded, aeolian, weakly consolidated sandstones of the Scots Bay Formation are in fault contact with basalt flows. Weathered boulders of basalt are scattered throughout these sandstones, with greatest abundance near the contacts. This distribution of basalt is an indication that faults were active during sand deposition, with the rising blocks of basalt acting as the source of the boulders. Alternatively, some of the boulders might be ballistic ejecta. Centimeter-sized and calcite-cemented concretionary nodules are abundant. Native copper occurs as the nucleii of small malachite-stained nodules in the sandstone adjacent to the faults, indicating thermal activity there after deposition of the sandstones. These sandstones are lithologically similar to those of the Wolfville Formation (pre-Blomidon) and reflect a similar desert-aeolian origin, but at a time of active rifting and vulcanism.

Station 5-E. The outcrop here is a narrow fault-bounded wedge between blocks of flow-basalt. A sandy talus cone has blanketed much of the exposure. The talus contains loose slabs of zeolite vein material weathered from the adjacent basalt and commonly in clusters of large perfect crystals (stilbite, chabazite, heulandite). This place is a famous locality for zeolite collection. The adjacent block of basalt is seen to have many anastomosing zeolite veins and cavities, but these are inaccessible.

Station 5-F. A thick sequence of west-dipping flows of the North Mountain Basalt may be seen here. Euhedral crystals of analcite up to 1.2 in (3 cm) in size are common in this sector.

Station 5-G. Basalt flows in reverse fault contact with post-basalt (= Scots Bay) sedimentary rocks are present here. Vertical and subhorizontal strata are faulted together near the contact.

Station 5-H. Postbasalt red and purple shales, containing several interbedded lobes of basalt conglomerate, or lahars, are seen here. These indicate proximal vulcanism, and erosion of the rising volcanic blocks.

REFERENCES CITED

Bujak, J., and Donohoe, H., 1980, Geological highway map of Nova Scotia: Halifax, Atlantic Geoscience Society Special Publication 1, scale 1:638,000.

Colwell, J. A., 1980, Zeolites in the North Mountain Basalt, Nova Scotia: Field Trip Guidebook 18, Geological Association of Canada Annual Meeting, Halifax, Nova Scotia, 16 p.

Donohoe, H. V., and Wallace, P. I., 1982, Geology map of the Cobequid Highlands: Halifax, Nova Scotia Department of Mines and Energy, sheet 2, scale 1:50,000.

Hyatsu, A., 1979, K-Ar isochron age of the North Mountain Basalt, Nova Scotia: Canadian Journal of Earth Science, v. 16-4, p. 973–975.

Liew, M.Y.C., 1976, Structure, geochemistry, and stratigraphy of Triassic rocks, North Shore of Minas Basin, Nova Scotia [M.S. thesis]: Wolfville, Nova Scotia, Acadia University, 181 p.

Lollis, E. W., 1959, Geology of Digby Neck, Long and Brier islands [Thesis]: New Haven, Connecticut, Yale University, 120 p.

Philpotts, A. R., 1978, Rift-associated igneous activity in eastern North America, *in* Petrology and geochemistry of continental rifts: Olso, North Atlantic Treaty Organization Advanced Study Institute, Series C, v. 36, p. 133–154.

Stevens, G. R., 1977, Geology and tectonic framework of the Bay of Fundy–Gulf of Maine region, *in* Fundy tidal power and the environment: Wolfville, Nova Scotia, Acadia University Institute Publication 28, p. 82–100.

—— , 1980, Mesozoic vulcanism and structure; Northern Bay of Fundy region, Nova Scotia: Field Trip Guidebook 8, Geological Association of Canada Annual Meeting, Halifax, Nova Scotia, 40 p.

—— , 1985, *Entries for* North Mountain Basalt and Scots Bay Formation, *in* Lexicon of Canadian stratigraphy, Volume VI; Atlantic region: Calgary, Canadian Society of Petroleum Geologists, p. 273–274, 334–335.

The Late Precambrian Antigonish Terrane (a periarc basin) unconformably overlain by Paleozoic overstep rocks typical of the Avalon Composite Terrane, near Antigonish, Nova Scotia

J. B. Murphy, Department of Geology, St. Francis Xavier University, Antigonish, Nova Scotia B2G 1CO, Canada

J. D. Keppie, Department of Mines and Energy, P.O. Box 1087, Halifax, Nova Scotia B3J 2X1, Canada

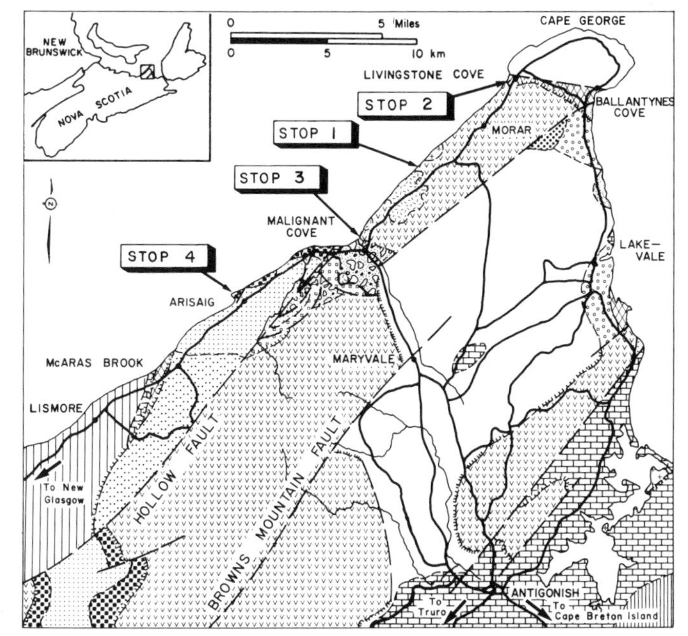

Figure 1. Geological map of the northern Antigonish Highlands with legend showing locations of field trip stops.

AGE	GROUP		CODE	FORMATION		LITHOLOGY	
CARBONIFEROUS LATE	Riversdale			Port Hood		Sandstone, shale, conglomerate, mudstone	
	Canso			Lismore		Wacke, mudstone, siltstone, conglomerate	
CARBONIFEROUS EARLY	Windsor			Ardness		Sandstone, shale limestone	
				Martin Road		Sandstone, shale, conglomerate	
	Horton			Wilkie Brook		Sandstone, siltstone, conglomerate, shale, limestone	
				(undivided)		Conglomerate, wacke, shale, siltstone, sandstone	
DEVONIAN LATE				(unnamed)		Conglomerate, sandstone, siltstone, shale	
DEVONIAN MIDDLE				McAras Brook		Basalt, sandstone, conglomerate	
DEVONIAN EARLY				Knoydart		Siltstone, wacke, shale, mudstone	
SILURIAN	Arisaig			Stonehouse		Wacke, siltstone	
				Moydart		Mudstone, wacke siltstone, limestone	
				McAdam		Mudstone, wacke, shale	
				French River		Wacke, mudstone	
				Ross Brook		Mudstone, shale, wacke	
				Beechhill Cove		Wacke, siltstone, conglomerate	
(?) LATE ORDOVICIAN				McGillivary Brook		Rhyolite, ignimbrite, laterite	
				Dunn Point		Basalt, laterite, rhyolite	
EARLY ORDOVICIAN CAMBRIAN	Iron Brook	McDonalds Brook		Ferrona	Arbuckle Brook	Quartzite, ironstone, calcareous tuff	Mafic volcanic rocks, limestone rhyolite, tuff, slate, conglomerate
				Little Hollow		Slate, limestone, quartzite	
				Black John	Malignant Cove	Conglomerate, sandstone	Conglomerate, sandstone, slate
LATE PRECAMBRIAN	Georgeville				South Rights	Slate, minor basalt, greywacke	
					Clydesdale	Basalt, greywacke, slate	
				Livingstone Cove	James River	Conglomerate, greywacke, mudstone, siltstone, chert, limestone	
				Morar Brook	Maple Ridge	Mudstone, siltstone, chert, limestone	
				Chisholm Brook		Mafic flows, tuff, marble, mudstone	

LOCATION

These outcrops are located along the shore of the Northumberland Strait, north of Antigonish (Fig. 1). They may be reached by driving north along Nova Scotia 245 from Antigonish to Malignant Cove, which is located at the Junction with Nova Scotia 337. To reach Stop 1, proceed 3.3 mi (5 km) east on Nova Scotia 337 to a road on the north side of the road leading to the Georgeville Quarry (Fig. 2). This road has a metal bar-type gate at its entrance. Park by the gate and walk down the road to the shore. Proceed east along the shore about 150 ft (50 m) to the first outcrops—this is Stop 1A (Fig. 3). Stops 1A-1P are located northeast of here, along the shore, and may be completely traversed only at low tide. At the end of the traverse, walk up the track along the brook between Stops 1N and 1P back to Nova Scotia 337 and then back to the starting point. Resume driving northeast along Nova Scotia 337 to Livingstone Cove (Fig. 2). Park near the Livingstone Cove Pier and walk 1,500 ft (500 m) southwest along the shore to reach Stop 2. To reach Stop 3, return to Malignant Cove along Nova Scotia 337 (Fig. 2). Stop 3 is located at Malignant Cove, about 300 ft (100 m) south of the junction of Nova Scotia 345 and 337. Park in the parking spaces by the side of the road and follow the short footpath to the outcrops in Malignant Brook, which runs along the eastern side of the road. From here, drive southeast along Nova Scotia 245 to Arisaig (Fig. 1), where you take the turn down to Arisaig Pier for Stop 4 (Fig. 6).

SIGNIFICANCE

The Late Precambrian Antigonish Terrane is an oceanic, deep marine sequence of rocks interpreted to have been deposited in a periarc basin. This terrane contrasts with other Late Precambrian terranes in the Avalon Composite Terrane, which are generally composed of either bimodal continental volcanic rocks or cratonic volcanic arc sequences (Keppie, 1982a, b, 1985). On the other hand, the Cambrian and Silurian rocks in the Antigonish Highlands are very similar to correlative units occurring elsewhere in the Avalon Composite Terrane, suggesting that the distinct Late Precambrian terranes were amalgamated by the end

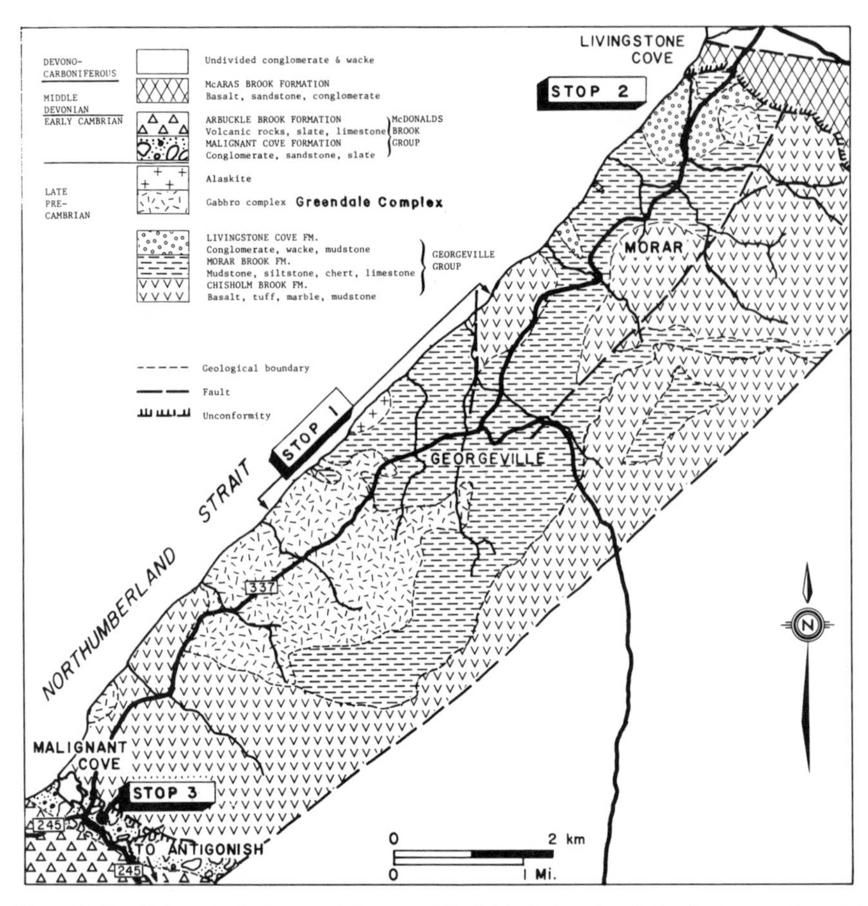

Figure 2. Detailed geological map of the coastal fault block showing field trip Stops 1, 2, and 3.

of the Precambrian. The stops described here allow one to examine these rocks for comparison with equivalent units elsewhere in the Avalon Composite Terrane. Also exposed in the Late Precambrian section are some interesting relationships between a volatile-rich gabbroic intrusion, an alaskite plug, and the country rocks.

SITE INFORMATION

Geological Setting

The Late Precambrian rocks of the northern Antigonish Highlands have been mapped as the Georgeville Group (Murphy and others, 1982, 1986), which has been subdivided into several formations (Fig. 1). Of these, three occur in the coastal fault block (Fig. 2), where the lowest unit exposed is the Chisholm Brook Formation. The base of this formation is not exposed. It consists of interlayered, ocean-floor, tholeiitic mafic flows (Fig. 4), tuff, marble, and mudstone in the west and mafic flows and tuff pervasively intruded by dikes in the east (Fig. 3). The overlying Morar Brook Formation is mainly composed of finely laminated, black porcellaineous mudstone with thin siltstone and chert horizons, and rare thin limestone and conglomerate beds. Cross bedding, graded bedding, and slump structures are present in these beds. The contact between the Morar Brook and Chisholm Brook Formations is gradational, with the boundary placed at the top of the highest volcanic unit. These rocks are overlain conformably by the Livingstone Cove Formation, which contains conglomerates interbedded with black mudstones and siltstones. The conglomerate horizons commonly have channeled bases and the clasts are generally supported by a sandy volcaniclastic matrix exhibiting both normal and reverse grading. Pebbles in the conglomerates are predominantly of volcanic origin with some granite; others are of locally derived black mudstone. The base of the Livingstone Cove Formation is placed where conglomerate enters the section. The top of the Livingstone Cove Formation is truncated by the angular unconformity at the base of the Middle Devonian McAras Brook Formation.

The environment of deposition of the Georgeville Group is inferred to have been a quiet deep water basinal area into which volcaniclastic deposits entered in turbidity currents with detritus supplied by erosion of rapidly rising volcanic highlands (Fig. 5). The marble horizons within the group are thought to have been deposited in shallow seas fringing volcanic islands.

During the Late Precambrian Cadomian Orogeny, the Georgeville Group was deformed first by thrust faults and

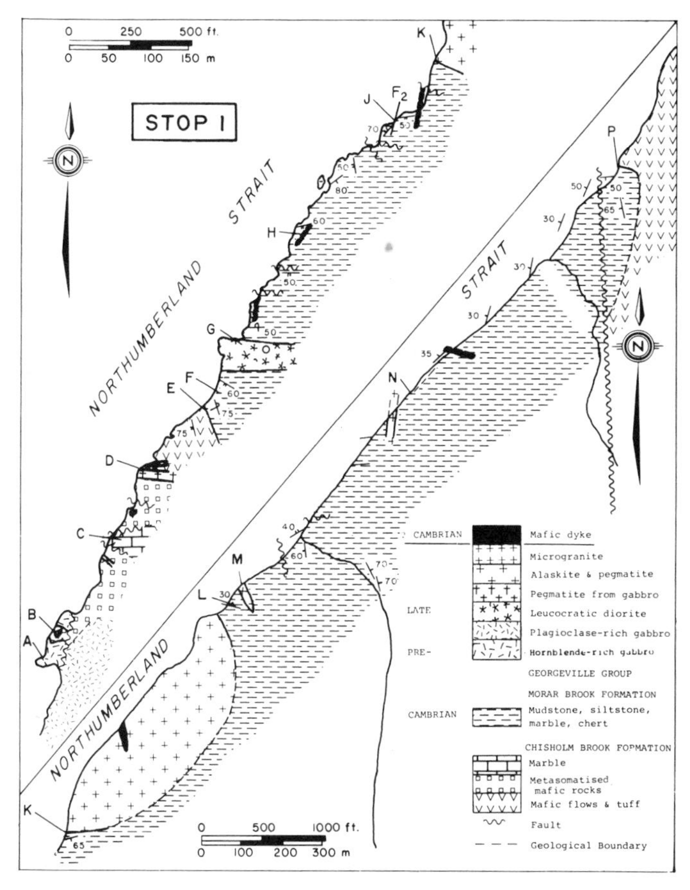

Figure 3. Detailed geological map of the Georgeville coast showing the localities A–P of Stop 1.

NNW–SSE trending, tight to isoclinal, recumbent, westerly vergent folds with a weak axial planar cleavage defined by chlorite, sericite, and minor biotite. This was followed by refolding about NE–SW and E–W upright folds. F_1 was accompanied by greenschist facies metamorphism. The deformation was followed by sporadic plutonism, exemplified in the coastal fault block by the post-tectonic gabbroic Greendale Complex and alaskite. $^{40}Ar/^{39}Ar$ incremental release data on large metasomatic hornblende crystals from this gabbro display a disturbed spectrum with ages ranging from 653 ± 8 to 599 ± 9 Ma. This age suggests that the gabbro is Late Precambrian and the disturbance noted in the spectrum is inferred to be due to the subsolidus metasomatism recorded by Murphy (1982). The alaskite has been dated by Cormier (personal communication, 1982) using Rb-Sr on whole rocks at 535 ± 13 Ma (initial $^{87}Sr/^{86}Sr$ ratio = 0.765 ± 0.03). However, a K-Ar age on muscovite (%K = 7.83) from the alaskite yielded an age of 604 ± 14 Ma (Wanless, personal communication, 1982). The older age is inferred to be closer to the time of intrusion and extensive deuteric alteration evident in the alaskite is inferred to have redistributed the Rb and Sr isotopes.

The Cambrian rocks in the northern Antigonish Highlands occur in a small, fault-bounded pull-apart basin produced by dextral movements on the NE–SW bounding faults (Murphy and others, 1986). They have been divided into two groups: the mainly sedimentary, 675-ft (225-m) thick, Iron Brook Group and the predominantly volcanic, 1,500-ft (500-m) thick, McDonalds Brook Group (Murphy and others, 1982, 1986) (Fig. 1). At the base of each group, red, nonmarine conglomerate, sandstone, and slate (Black John and Malignant Cove Formations) unconformably overlie rocks of the Georgeville Group. In the Iron Brook Group, these redbeds are conformably overlain by interbedded red, green, and black slate, thin limestone, and quartzite. The limestones have yielded a late Early Cambrian fauna of trilobites and paraconodonts (Landing and others, 1980). Equivalent rocks in the McDonalds Brook Group are the bimodal, strongly alkalic, within-plate rift volcanic rocks (Keppie and Dostal, 1980; Murphy and others, 1985) and thin interbedded units of red slate, siltstone, and fossiliferous limestone of the Arbuckle Brook Formation. The Little Hollow Formation is concordantly overlain by interbedded quartzite, and ironstone, of the Ferrona Formation, which has yielded some inarticulate brachiopods: *Obolus (Lingulobus) spissa* and *Lingulella*(?) of late Cambrian to early Ordovician age (Williams, 1914). This suggests that a hiatus corresponding to the Middle Cambrian is present in the succession.

During the Ordovician, the Cambro-Ordovician rocks in the central part of the basin were affected by polyphase deforma-

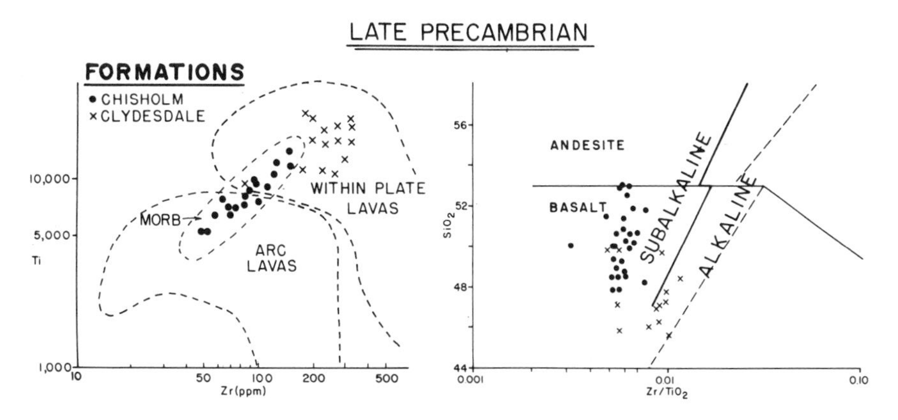

Figure 4. Geochemistry of the volcanic rocks in the Georgeville Group.

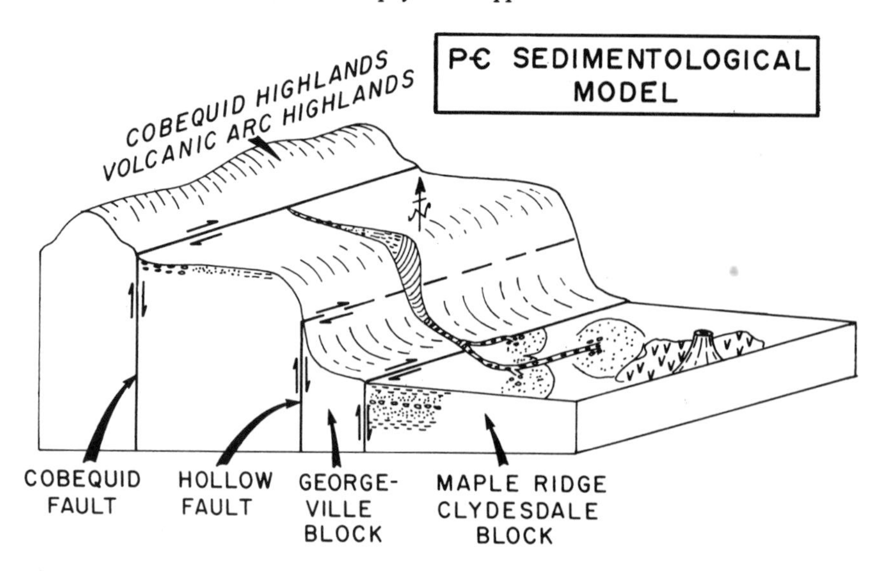

Figure 5. A schematic block diagram of the depositional environment of the Georgeville Group.

tion, which dies out toward the basin margins. This deformation consists of recumbent folds with a penetrative axial planar cleavage (defined by chlorite, muscovite and biotite) and accompanying thrusts, all of which were refolded by upright–inclined NW–SE trending folds. The deformation is inferred to be the result of sinistral motions along the NE–SW faults bounding the Cambrian pull-apart basin.

These deformed Cambrian rocks are unconformably overlain by the Arisaig Group. In the Arisaig area (Fig. 6), the Late Ordovician–Silurian succession starts with 270–360 ft (90–120 m) of subaerial, bimodal, tholeiitic (-alkalic), within-plate rift basalts, rhyolitic flows, and ignimbrites with interbedded laterites of the Dunn Point and McGillivray Brook Formations (Boucot and others, 1974; Keppie and others, 1979). The rhyolites have yielded a whole rock Rb-Sr isochron of 434 ± 15 Ma (recalculated from Fullager and Bottino, 1968). These rocks are disconformably to unconformably overlain by shallow marine conglomerate, sandstone, and siltstone of the Early Llandoverian Beechhill Cove Formation. The Beechhill Cove Formation lies unconformably upon the Iron Brook Group in Iron Brook. The Beechhill Cove Formation grades upwards into the Mid-Late Llandoverian black shales, muddy siltstone, tuff, and areneaceous limestone of the Ross Brook Formation. These formations are followed by shallow marine siltstone, shale, and limestone of the rest of the Arisaig Group (Fig. 1). The total thickness of the Arisaig Group is about 3,800 ft (1,245 m). The Arisaig Group passes upward into >3,000 ft (>1,000 m) of generally nonmarine silty mudstones, marls, sandstones, and conglomerates of the Early Devonian Knoydart Group. These late Ordovician–Early Devonian rocks were deformed during the late Early–Middle Devonian by NE–SW, upright to asymmetric, gently to steeply plunging folds. The accompanying metamorphism is subgreenschist facies. Westward along the shore, Devono-Carboniferous rocks unconformably overlie all of the older units.

Stop 1. At this stop, Late Precambrian Chisholm Brook and Morar Brook Formations (part of the Georgeville Group) are intruded by the gabbroic, appinitic Greendale Complex and alaskite on the shore northeast of the Georgeville Quarry, along the shore of the Northumberland Strait (Fig. 2).

Stop 1A. Two phases of the Late Precambrian, post-tectonic Greendale gabbro may be observed here—plagioclase-rich and hornblende-rich gabbros—with a sharp contact between them. The plagioclase crystals enclose enclaves of the matrix, indicating that it is a late crystallizing mineral, probably of metasomatic origin.

Stop 1B. Here, a basalt xenolith in the gabbro shows evidence of increasing contamination towards its margin. The basalt consists of albite, epidote, prehnite, and actinolite. The amphiboles in the basalt increase in size toward the contact with the gabbro. Coarse pegmatites with sinuous contacts and a microgranite dyke represent a late felsic stage.

Stop 1C. A fault slice containing metasomatized marble and interbedded impure serpentinized marble and basalt occurs here.

Stop 1D. The intrusive contact between a late pegmatitic felsic phase of the Greendale Complex and the country rocks (marble and tuff) is exposed here. The pegmatite contains large crystals of amphibole (up to 8 mm) locally enclosing plagioclase cores, in a matrix of orthoclase, quartz, and plagioclase. The pegmatite is cut by aplite dykes.

Stop 1E. At this locality, the contact between the Chisholm Brook and Morar Brook Formations is exposed. Interbedded tuff and agglomerate of the Chisholm Brook Formation grade conformably upward into black, thinly laminated mudstone of the Morar Brook Formation. The beds here are overturned.

Stop 1F. Syndepositional slump folds in the Morar Brook Formation may be observed here, especially if the rocks are wet.

Stop 1G. An intrusive leucocratic diorite sheet containing xenoliths of country rock occurs here. Both contain late stage

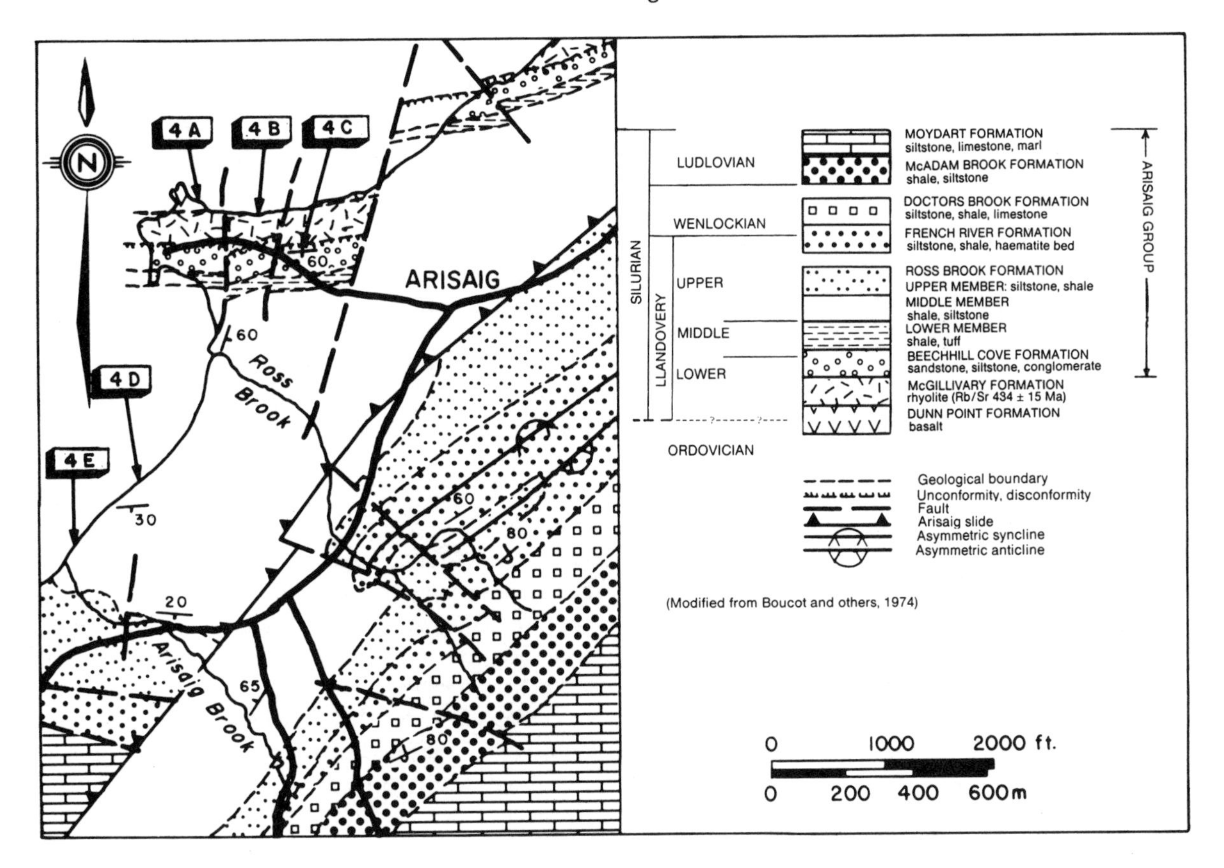

Figure 6. Geological map of the Arisaig area, northern Antigonish Highlands, showing the locations of field Stops 4A–4E.

metasomatic plagioclase porphyroblasts. The eastern contact of the sheet is broadly concordant with the bedding in the Morar Brook Formation. Load casts in the mudstone/siltstone here indicate overturned bedding.

Stop 1H. A diabase, which varies from a sill to a dike, intrudes interbedded mudstones, siltstones, marble, and chert typical of the Morar Brook Formation at this locality.

Stop 1J. An example of a NE–SW, upright F2 fold in the Morar Brook Formation occurs here.

Stop 1K. The intrusive contact between the post-tectonic alaskite stock and mudstones of the Morar Brook Formations is exposed at this location. Note the contact metamorphic effects. The alaskite consists of quartz, plagioclase, and microcline with minor chlorite and green amazonite. Note the extensive deuteric alteration. The alaskite is inferred to be derived from the gabbro by volatile extraction. Alkalic, mafic dikes intrude the alaskite, and are offset by dextral faults parallel to the Hollow Fault.

Stop 1L. An example of an easterly facing, isoclinal F_1 fold with a poorly developed fracture cleavage occurs in the Morar Brook Formation here.

Stop 1M. A coarse alaskitic pegmatite intrudes the Morar Brook Formation at this locality.

Stop 1N. Thin veins of alaskite intrude the mudstones here. They are undeformed, clearly documenting their post-tectonic age relationship.

Stop 1P. Tuff, basalt flows, and mudstone of the Chisholm Brook Formation are intruded by numerous, comagmatic mafic dikes at this location. Typical mineralogy in the basalts is albite, epidote, actinolite, and chlorite.

Stop 2: Livingstone Cove (Fig. 2). Overturned, gently dipping mudstones, siltstones, and volcaniclastic conglomerates of the Livingstone Cove Formation (part of the Georgeville Group) are here cut by numerous mafic dikes. Along its northern margin, the Livingstone Cove Formation is unconformably overlain by the red conglomerates and sandstones of the Middle Devonian McAras Brook Formation. This unconformity is well exposed on the shore; in places the unconformity is cut by minor transverse faults. Farther northeastward along the shore, an amygdaloidal basalt within the McAras Brook Formation crops out and is truncated by a fault.

Stop 3: Malignant Cove (Fig. 2). At this locality, the basal unit of the Cambrian (Malignant Cove Formation) is exposed. It consists of about 800 ft (250 m) of red, fluviatile conglomerate, breccia, sandstone, and slate. The conglomerates are generally well sorted and contain pebbles of mafic and felsic volcanic rocks, sericitic shale, quartzite, red siltstone, jasper, and perthitic feldspars. The matrix consists of quartz, sericite, chlorite, and minor biotite. The mafic dikes, which cut the formation, are chemically similar to the overlying basalts and probably acted as feeders.

Stop 4: Arisaig (Fig. 6). Stop 4A. At this locality, the contact between the Beechhill Cove and McGillivray Brook Formations is exposed. Here, conglomerate of the Beechhill Cove Formation lies disconformably upon rhyolite capped by lenses of laterite of the Dunn Point Formation. Most of the pebbles in the conglomerate are rhyolite and ignimbrite, presumably derived from the McGillivray Brook Formation. Some jasper pebbles are also present. Rapid changes in thickness of the conglomerate suggest deposition in channels. The rhyolite is made up of quartz and feldspar phenocrysts set in a devitrified glass matrix, with variable degrees of alteration to albite, sericite, and haematite.

Stop 4B. Here, basalt flows of the Dunn Point Formation are interbedded with laterite horizons and overlain by rhyolite flows of the McGillivray Brook Formation. The basalts are generally vesicular, particularly towards their tops, and grade upward from spheroidally weathered basalt through a rubbly laterite-basalt mixture into a blocky indurated red laterite capped by small irregular lenses of bedded laterite. This sequence is interpreted as an undisturbed red lateritic soil profile, with the bedded laterite being deposited by wind. In places, the basalt flows have incorporated pieces of the underlying laterite in their basal portions. The intimate association of vesicles with laterite trains suggests that the laterites were soft and wet at the time of extrusion of the basalt. The basalts are generally aphyric and consist of plagioclase, epidote, chlorite, calcite, opaques, and rare clinopyroxene relics. Vesicles are filled with quartz, chlorite, and prehnite. The basalts are tholeiitic with some alkaline tendencies, and were extruded during rifting within the continental plate (Keppie and others, 1979).

Stop 4C. The rhyolite flows of the McGillivray Formation display spectacular primary features here, such as flow banding, flow folds, and auto-brecciation.

Stop 4D. Here, the Middle Member of the Ross Brook Formation crops out as gray shales with some thin argillaceous siltstones. The shales are generally parallel laminated and in places bioturbated. The siltstones display parallel lamination, cross-bedding, and load casts. Fossils include brachiopods (including the diagnostic Early Silurian, shallow marine Eocelia hemisphaerica), pelecypods, worm burrows, and trails (Boucot and others, 1974). Suspension-feeding bivalves are often preserved in their life positions. These rocks were deposited in a shallow marine shelf environment, which was generally quiet, with clear water and only a few storm-generated currents. These shales have been deformed by variably-oriented late folds.

Stop 4E. At the abandoned mouth of Arisaig Brook, similar rock types of the Middle Member of the Ross Brook Formation have been deformed by a NE–SW, upright fold with an axial planar slaty cleavage, which is inferred to be Early Devonian in age.

REFERENCES CITED

Boucot, A. J., Dewey, J. F., Dineley, D. L., Fyson, W. K., Hickox, C. F., McKerrow, W. S., and Ziegler, A. M., 1974, Geology of the Arisaig area: Geological Society of America Special Paper 139, 191 p.

Fullager, P. D., and Bottino, M. L., 1968, Radiometric age of the volcanics at Arisaig, Nova Scotia, and the Ordovician—Silurian Boundary: Canadian Journal of Earth Sciences, v. 5, p. 311–317.

Keppie, J. D., 1982a, The Minas Geofracture: Geological Association of Canada Special Paper 24, p. 263–280.

——, 1982b, Tectonic Map of Nova Scotia: Nova Scotia Department of Mines and Energy. Scale 1:500,000.

——, 1985, The Appalachian Collage, *in* Gee, D. G. and Sturt, B., eds., The Caledonide Orogen, Scandinavia, and related areas: New York, J. Wiley and Sons, p. 1217–1226.

Keppie, J. D., and Dostal, J., 1980, Paleozoic volcanic rocks of Nova Scotia; International Geological Correlation Programme Project 27; Caledonide Orogen Proceedings: Blacksburg, Virginia Polytechnical Institute and State University Memoir 2, p. 249–256.

Keppie, J. D., Dostal, J., and Zentilli, M., 1979, Early Silurian volcanic rocks at Arisaig, Nova Scotia: Canadian Journal of Earth Sciences, v. 16, p. 1635–1639.

Landing, E., Nowlan, G. S., and Fletcher, T. P., 1980, A microfauna associated with Early Cambrian trilobites of the Callavia Zone, northern Antigonish Highlands, Nova Scotia: Canadian Journal of Earth Sciences, v. 17, p. 400–418.

Murphy, J. B., 1982, Tectonics and magmatism in the northern Antigonish Highlands [Ph.D. thesis]: Montreal, McGill University, 243 p.

Murphy, J. B., Keppie, J. D., and Hynes, A., 1982, Geological map of the northern Antigonish Highlands: Nova Scotia Department of Mines and Energy, Map 82-5, scale 1:50,000.

Murphy, J. B., Cameron, K., Dostal, J., Keppie, J. D., and Hynes, A., 1985, Cambrian volcanism in Nova Scotia, Canada: Canadian Journal of Earth Sciences, v. 22, p. 599–606.

Murphy, J. B., Keppie, J. D. and Hynes, A., 1986, Geology of the northern Antigonish Highlands: Nova Scotia Department of Mines and Energy (in press).

Williams, M. Y., 1914, Arisaig-Antigonish district, N.S.: Geological Survey of Canada Memoir 60, 173 p.

Glacial advances and sea-level changes, southwestern Nova Scotia, Canada

D. R. Grant, Geological Survey of Canada, 601 Booth St., Ottawa, Ontario K1A 0E8, Canada

LOCATION

Coast of Clare, between Yarmouth and Digby; centered on 44°05′N Lat.; 66°10′W Long. (Fig. 1). A nearly continuous 48-mi (80-km) exposure reveals a Quaternary sequence that records the major geological events of the last 130,000 years. Yarmouth can be reached by air from Boston and Halifax. Ferries connect Yarmouth and Digby with Bar Harbour and Portland, Maine, and with Saint John, New Brunswich, respectively. Rail service and Nova Scotia 101 and 103 link Yarmouth and Digby with Halifax. A network of roads gives access to each site. Accommodation is plentiful en route; car and bus rentals are available in Yarmouth and Digby. All sites are reached by public roads and by short walks along public beaches. No permission is required, except at Pembroke Cove. Additional maps: Nova Scotia Highway Map—free at entry points; Geological Highway Map and Pleistocene Geology Maps 7 and 8 may be ordered from the Nova Scotia Department of Mines, Box 1087, Halifax, B3J 2X1, Canada; and 1:50,000 topographic maps (21 A5 and 12; 21 B 1 and 8) may be ordered from Canada Map Office, Ottawa, K1A 0E4. Taylor (1969) treats the bedrock geology; Goldthwait (1924) and Roland (1982) the physiography.

SIGNIFICANCE

Thirteen lithostratigraphic units document four eustatic and isostatic marine transgressions and four glacial advances by both regional ice sheets and local ice caps. Weathered horizons demarcate intervals of ice retreat. Richly fossiliferous beds contain a molluscan fauna of warmer water association, partly in place and partly reworked from offshore sources, some possibly Pliocene. Glaciotectonic deformation of bedrock and transport of sediment masses are local features. Neotectonic faulting can be demonstrated. This area presents the most complete Quaternary sequence in Nova Scotia. Together with type areas in central and northern parts of the province, it forms the basis of a regional stratigraphic framework.

ADVANCE SUMMARY

Six sites, plus other features noted on Figure 1, form the basis for the following sequence of events since 130,000 B.P.

1. During the interval of higher warmer seas of the last interglacial period, an intertidal platform was cut in rock, while beach sediment (Sandford Gravel), fossiliferous nearshore sand (Salmon River Sand), and offshore mud (Barton Silt) were deposited.

2. At the onset of Wisconsin glaciation, cooling climate caused rock debris to shed from slopes onto the old beach.

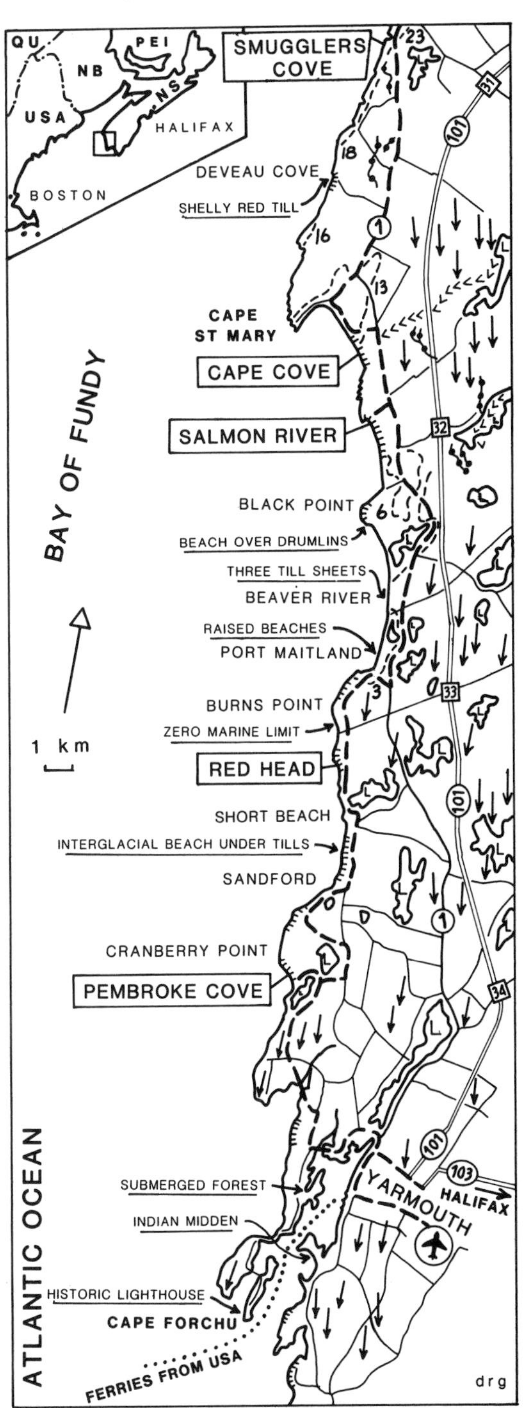

Figure 1. Detailed access map for sites with miscellaneous annotations. (Symbols: arrows = drumlins and glacial striations; beaded line = major end moraine; chevrons = esker; dashed line = marine limit with elevation (m); pecked line = till bluff; bold line = excursion route; double line = limited access highway; fine line = other roads; L = lake).

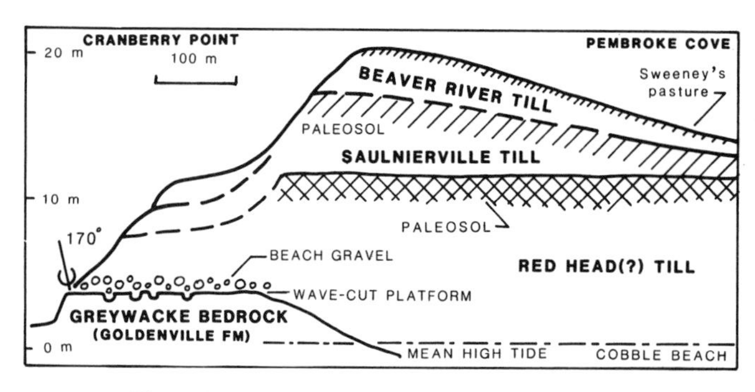

Figure 2. Pembroke Cove, geological cross section.

3. An ice cap spreading from the Nova Scotia upland deposited a locally derived drift (Little Brook Till).

4. During the last glacial maximum, a southward-flowing regional ice sheet crossed the Bay of Fundy from New Brunswick and laid down a reddish drift (Red Head Till) containing northern erratics and shell fragments from the sea floor.

5. Glacial shrinkage relocated the ice centre onto Nova Scotia, produced a hybrid drift (Saulnierville Till) by mixing local debris with the red till. Further retreat left outwash (Cape Cove Gravel), which became weathered during a nonglacial interval perhaps 30,000–40,000 years ago.

6. Late Wisconsinan re-expansion of Nova Scotian ice deposited a thin rubbly surface drift (Beaver River Till).

7. During deglaciation the DeGeer Sea inundated the depressed coast, ca. 14,000 B.P., reworked tills, and deposited beach gravel (Port Maitland Gravel) while sublittoral mud (Gilbert Cove Clay) was laid down locally in lower areas.

8. Continued rebound lowered relative sea level below its present level about 10,000 years ago. Thereafter, eustatic recovery coupled with crustal subsidence and increase of tidal range caused a renewed transgression to attack the coast and lay down tidal muds (Amherst Silt) in estuaries.

INTRODUCTION

Early geological mapping in the area led to hypotheses of regional and local glaciation by Dawson (1893, p. 168), and Bailey (1898, p. 26M). Flint (1951) and Hickox (1962) reaffirmed the local ice-cap concept. The first stratigraphic observations noted the shell-bearing sediments and the contrasting till provenance (Grant, 1968, 1971, 1976). This account is based on a more detailed field trip guide (Grant, 1980b) for the Geological Association of Canada. U/Th and amino-acid chronometry is currently in progress.

Of the many exposures available, the six field trip sites were selected to cover the entire sequence. The coast faces west, so afternoon illumination greatly enhances the colours and contacts. Tides range from 6.7 to 10 ft (2 to 3 m) in height, but pose no risk or impediment. If time is short, the Cape Cove site will suffice. Figure 1 is annotated with additional features between stops.

BEDROCK GEOLOGY

The distribution of rock types explains the composition of tills as a result of glacier movements. The coast and hinterland are underlain by gray sedimentary and volcanic rocks; the interior is a white biotite granite. These rocks have been dispersed radially by a Nova Scotian ice cap and form gray, locally-derived tills. To the north, Digby Neck is a cuesta of amygdaloidal Triassic basalt, famous for amethyst and zeolites. On the floor of the Bay of Fundy are red Triassic mudstones that impart a distinctive colour to the till deposited by ice sheets crossing the bay. The red till also contains shells dredged from the sea floor, as well as other indicator erratics such as the basalts and a suite of red granite, conglomerate, and pyroclastics from New Brunswick.

DESCRIPTION OF SITES

1. Pembroke Cove—Cranberry Point. Leaving Nova Scotia 1 north of Yarmouth (Fig. 1) the site is reached via the shore road to the farm of V. Sweeney, from whom permission should be obtained to reach the beach. The site is a 2,600-ft long (800-m) bluff exposing an ancient beach buried by three tills (Fig. 2). On Cranberry Point, southward-tilted graywacke is truncated by a fossil intertidal abrasion platform, 13 ft (4 m) above present high tide. Like the modern one, the emerged beach has potholes, runnels, and a veneer of rounded cobbles. This beach is assigned to the higher sea level of the warmest part of the Sangamonian Interglaciation (Stage 5a) preceding the deposition of the tills of the last glaciation (Wisconsin) (Grant, 1980a).

Of the overlying tills, the lowest is brownish-gray, contains shell fragments, and is weathered to a darker brown colour in its upper 10 ft (3 m) (paleosol). The middle till is light gray and also has a brownish paleosol. Both contain rounded erratics from New Brunswick and were deposited by a glacier flowing southward onto Nova Scotia. In contrast, the surface till is locally derived and the clasts are angular, suggesting weak and/or brief transport by ice flowing westward to the coast.

2. Red (High) Head. Four mi (7 km) north, divert at Short Beach and stop 0.6 mi (1 km) farther, opposite Allen Lake (Fig. 1). Walk along the beach 300 ft (100 m) to High Head (formerly Red Head), which forms a bold bluff of bright red shell-bearing drift resting on iron-cemented beach gravel over an emerged Sangamonian wavecut platform on schist (Fig. 3). Besides being the type locality of Red Head Till, this site features the juxtaposition of modern and emerged shore platforms and glacial folding of bedrock. At half tide or lower, one can wander over the truncated rock surface and observe how bedrock is disaggregated in the intertidal zone by wetting, drying, and seasonal freezing so that waves may further smooth the surface by the grinding of boulders. In this way potholes and runnels are fashioned, while sea stacks remain. The emerged interglacial platform has most of the same features yet stands 20 ft (6 m) above present low tide (= modern platform level). Sea level in the present interglacial period has not yet attained its previous position.

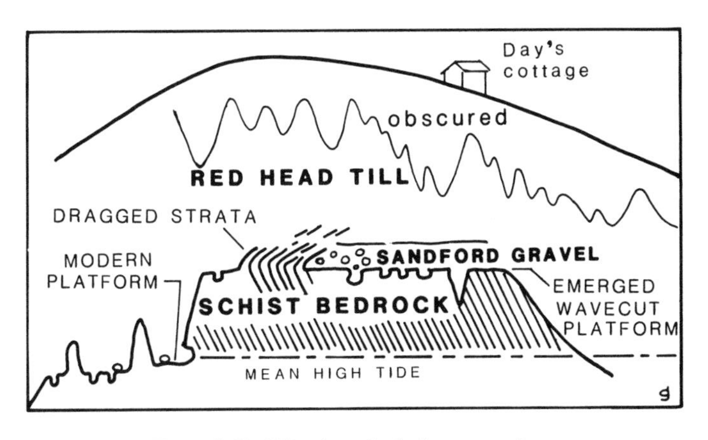

Figure 3. Red Head, geological cross section.

TABLE 1. SPECIES OF MOLLUSCS IN THE SALMON RIVER SAND (from Wagner, 1977)

Thracia conradi	Clinocardium ciliatum
Cyclocardia borealis	*Mercenaria mercenaria
*Atractodon stonei	Mya arenaria
Colus pygmaea	Mya truncata
Crepidula plana	Nucula delphinodonta
Nassarius trivittatus	Nucula proxima
Neptunea decemcostata	Phacoides filosus
Lunatia heros	*Placopecten magellanicus
Buccinum undatum	Yoldia limatula
*Astarte subequilatera	Litorina littorea
Astarte undata	Lunatia triseriata
?Skeneopsis planorbis	

*indicate the most abundant species.

The rock platform is locally deformed by glacial shear. The upper 6.6 ft (2 m) of strata are bent over southward in smooth curves that trail off as wisps into the overlying till and enclose pods of old beach gravel. The shear direction matches that of striations trending 120°–170°, indicating ice advance onshore from Bay of Fundy.

3. Salmon River. The site is reached either from Exit 32 of Nova Scotia 101, or by continuing northward on Nova Scotia 1 (Fig. 1). On this route, the southern limit of deglacial raised pebble beaches of the DeGeer Sea (ca. 14,000 years ago) is near Port Maitland. At a church 0.7 mi (1.2 km) north of the bridge over Salmon River, a lane leads to the shore exposure (Fig. 4). This is the type locality of the Salmon River Sand, a 10-ft (3-m) thick bed of shell-bearing, gray, stony mud sandwiched between a lower red drift with New Brunswick erratics (Red Head Till) and two overlying gray drifts (Saulnierville Till and Beaver River Till) derived from inland Nova Scotia. From the sand a variety of clam and snail shells (Table 1) can be collected. One fossil, the gastropod *Atractodon stonei,* became extinct during the last interglacial, about 100,000 years ago (Clarke and others, 1972). The assemblage is typical of depths of about 60 ft (20 m) and water temperatures not presently found north of Long Island (Gustavson, 1976). The deposit is assigned to the warmest part of the last interglaciation when temperatures are known from fossil vegetation to have exceeded present values (Mott and Grant, 1985).

Contacts with the Red Head Till show glaciotectonic features. The lower is jagged and interpenetrative. Lenses of red till occur in the shell sand; the sand bed pinches out southward as a parting plane within the till, suggesting that the Salmon River Sand bed was dragged from the floor of the Bay of Fundy while frozen to the base of the glacier, then released as the red till was deposited. Thus it is an older displaced block within and overlying a younger till.

The middle light gray drift (Saulnierville Till) has a rusty iron-manganese zone at its top, which is a Middle Wisconsinan paleosol. The surface drift (Beaver River Till) forms a moraine ridge that marks the terminus of the last advance of Nova Scotian ice. Inland, small recessional moraines and eskers leading coastward show the radial flow pattern (Fig. 1). Well sorted gravel spreads seaward from the moraine at 36 ft (11 m) elevation. It was deposited during a high-level stage of Bay of Fundy, called the DeGeer Sea, which began about 14,000 years ago.

4. Cape Cove. About 1.2 mi (2 km) north, Cape Cove fronts a 0.6-mi (1-km) stretch of bluffs cut in a complex Quaternary succession. The prime exposure in the area, it shows three marine episodes and three glacial advances, plus glacio-tectonics and postglacial faulting. It is reached by side roads from the north and south off Nova Scotia 1 (Fig. 1).

Beginning in the sand dunes of Mavillette Beach Provincial Park, nine stratigraphic units onlap southward in order of increasing age (Fig. 5). In the lagoon behind the beach, Holocene mud (Amherst Silt) is accumulating in salt marshes as sea level continues to rise since 7,000–10,000 years ago, due to eustatic recovery, tidal amplification, and crustal subsidence (Grant, 1970). The next oldest sediment is a 6.6 ft (2 m) thick mantle of loose light-brown gravel (Port Maitland Gravel) which pinches out at 44.5 ft (13.5 m) elevation. It represents the action of a high level deglacial marine stage called DeGeer Sea, which overlapped the depressed coast and reworked gravels from glacial deposits. The paleoshore has been upwarped by differential crustal rebound from present tide level near Red Head (Fig. 1) to 150 ft (45 m) near Digby.

The beach veneer has been reconstituted from a 50-ft (15-m) thick underlying unit (Cape Cove Gravel), a fluvial conglomerate whose long inclined beds with scour-and-fill structures project as ledges cemented by iron and manganese hydroxides. Perhaps it is a glacial outwash deposit laid down partly as a delta when sea level was lower than at present. Local distortion of the strata (isoclinal folds, vertical bedding, thrust faults) indicate deformation, probably caused by the glacial readvance, which deposited the till overlying it nearby. The gravel was evidently cemented during a lengthy period of subaerial exposure, presumably during a Middle Wisconsinan period of ice recession, which is dated by elephant fossils in other areas of Nova Scotia to the period 30,000–40,000 years ago.

The conglomerate rests conformably on reddish-brown drift

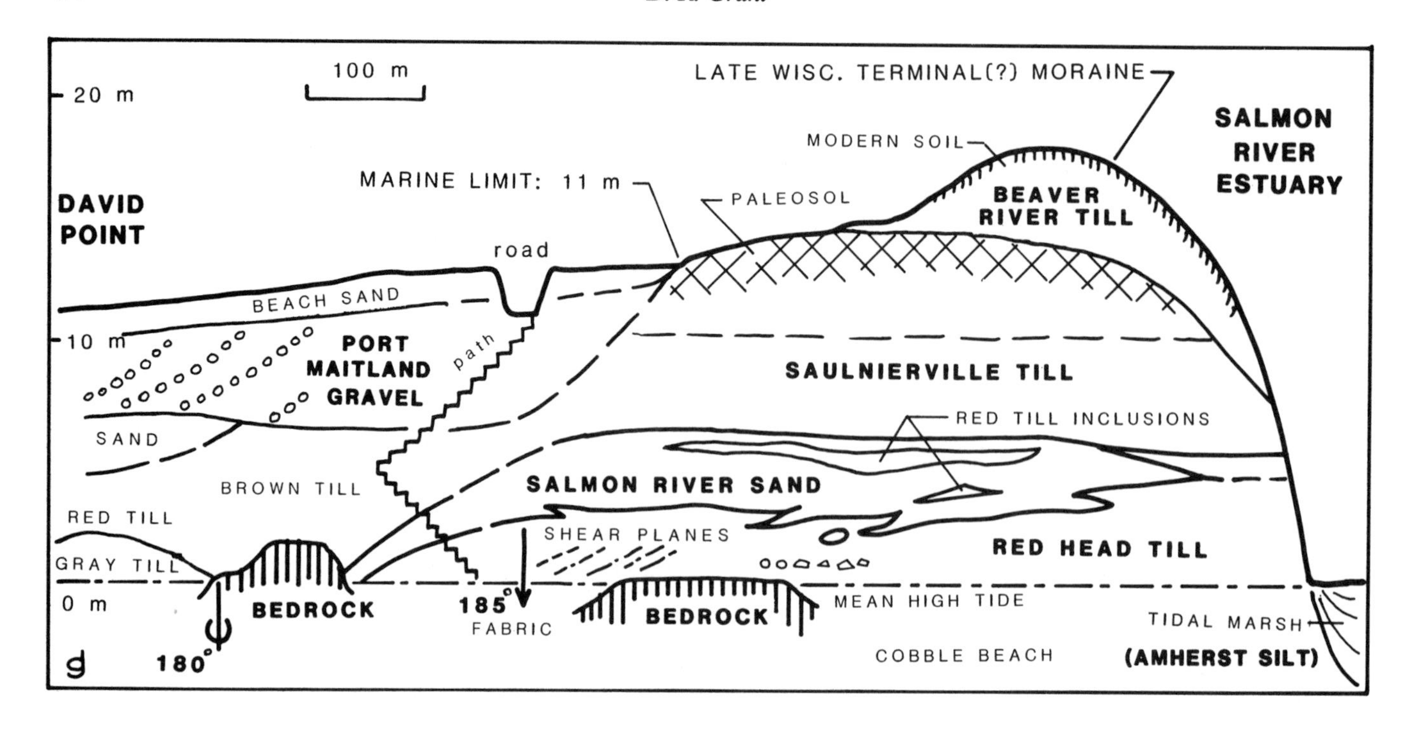

Figure 4. Salmon River, geological cross section.

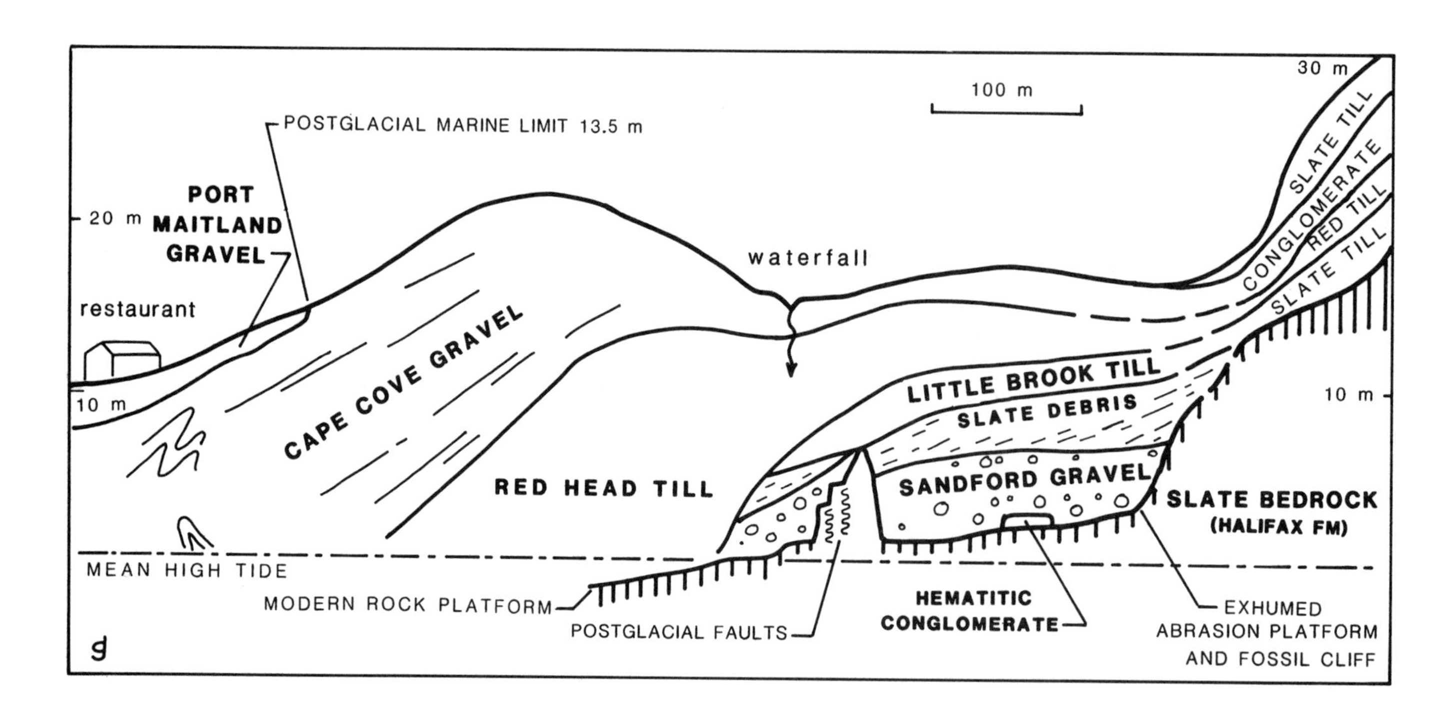

Figure 5. Cape Cove, geological cross section.

(Red Head Till) containing fragments and whole valves of the molluscs that occur in the Salmon River Sand (Table 1). This till represents the regional ice flood that crossed the Bay of Fundy from New Brunswick early in the last (Wisconsin) glaciation. It rests on a light gray till (Little Brook Till?) derived from the local slate bedrock, presumably during initial ice buildup on the Nova Scotia upland. Underneath is grayish-brown slate debris, interpreted as colluvium, that was produced in the periglacial climate before the onset of glaciation. Near the south end of the beach, slate bedrock emerges from the surf zone as a modern intertidal platform. A slightly higher rock bench forms the base of the drift sequence; it is overlain by a few metres of rusty cobble gravel (Sandford Gravel) thought to be a beach deposit corresponding to the period of wave planation. At one place the platform has a carapace of hematitic shingle conglomerate, which may be an older beach deposit from an earlier transgression.

A fossil sea stack projects from the platform. Formed of vertically cleaved slate, the top of the stack shows southward-pointing glacial striations (related to the red till phase), which are offset in a series of small steps along vertical faults with a total displacement of at least one metre. The direction of movement is up to the south, as are most of the other postglacial faults of the Maritimes region (e.g., Matthew, 1894; Goldthwait, 1924; Oliver and others, 1970). The movement is tentatively ascribed to deglacial elastic recovery and/or isostatic rebound, which might be localized along weaker bedrock zones, such as slate cleavage.

*5. **Smugglers Cove.*** The finest example in Canada of the last interglacial shoreline, as represented by a fossil rock platform, dead cliff, and weathered beach gravel is found near Smugglers Cove Provincial Picnic Park ("Comeau Cove" on topographic maps) (Fig. 1). From the parking lot, a shore path leads south for 1,900 ft (580 m) where it ends at a recess in the rock cliff. The spot is known locally as L'Anse aux Bôys ("cauldron cove") in reference to the potholes, which indent an abrasion platform standing 18 ft (5.5 m) above its modern counterpart. The bench is being exhumed by storm waves from its cover of shelly Red Head Till, and the surface has a mature patina of secondary iron precipitate. Remnants of hematite-cemented beach gravel cling to the inner reaches where a polished notch marks the upper limit of the last interglacial sea.

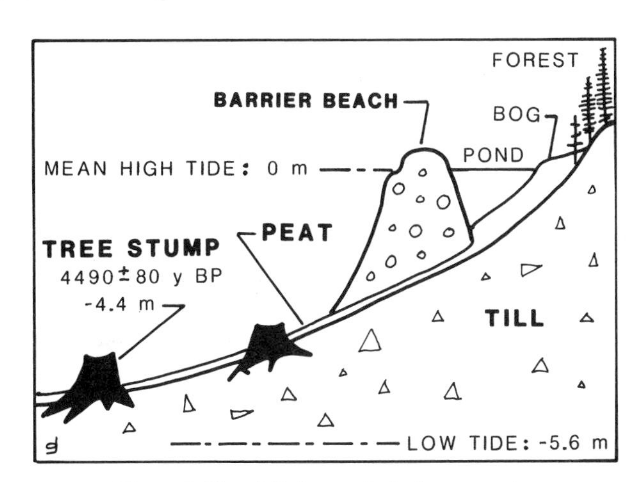

Figure 6. Church Point, geological cross section.

*6. **Church Point.*** About 12 mi (20 km) north of the area outlined in Figure 1, Church Point illustrates the rapid sea-level rise, which constitutes the final phase of postglacial crustal adjustment. An important process of modern environmental change, the rising sea erodes the open shore and deposits tidal mud in estuaries. The site is reached from Nova Scotia 1 in the vicinity of Ste. Anne College, where Lighthouse Road ends on a knoll overlooking a lagoon and barrier.

The beach face is a "forest" of tree slumps 6.7 to 20 ft (2 to 6 m) below high tide level; the stumps date about 4,000 years old (Fig. 6). Partly buried by freshwater bog peat, which grades upward to salt marsh peat, the trees show that rising sea level raised the groundwater table and caused a sphagnum bog to kill the forest before invasion by salt water. The same process continues—the pond is at high tide level and is drowning the modern forest. The submergence is due to eustatic rise of sea level and to regional subsidence of the crust. In the Bay of Fundy, an additional factor has been the increase in tidal range as basin geometry gradually changed with rising sea level (Grant, 1970).

REFERENCES

Bailey, L. W., 1898, Report on the geology of south-west Nova Scotia: Geological Survey of Canada Annual Report, v. IX, pt. M, 154 p.

Clarke, A. H., Jr., Grant, D. R., and Macpherson, E., 1972, The relationship of *Atractodon stonei* to the Pleistocene stratigraphy and paleoecology of southwestern Nova Scotia: Canadian Journal of Earth Sciences, v. 9, p. 1030–1038.

Dawson, J. W., 1893, The Canadian Ice Age, Montreal, 301 p.

Flint, R. F., 1951, Highland centres of former glacial outflow in northeastern North America: Geological Society of America Bulletin, v. 62, p. 21–38.

Goldthwait, J. W., 1924, Physiography of Nova Scotia: Geological Survey of Canada Memoir 140, 179 p.

Grant, D. R., 1968, Recent submergence in Nova Scotia and Price Edward Island: Geological Survey of Canada Paper 68-1A, p. 163–164.

—— , 1970, Recent coastal submergence of the Maritime Provinces, Canada: Canadian Journal of Earth Sciences, v. 7, p. 676–689.

—— , 1971, Glacial deposits, sea-level changes, and pre-Wisconsin deposits in southwest Nova Scotia, *in* Report of Activities, Part B: Geological Survey of Canada Paper 71-1B, p. 110–113.

—— , 1976, Reconnaissance of Early and Middle Wisconsinan deposits along the Yarmouth-Digby coast of Nova Scotia, *in* Report of Activities, Part B: Geological Survey of Canada Paper 76-1B, p. 363–369.

—— , 1977, Glacial style and ice limits, the Quaternary stratigraphic record, and changes of land and ocean level in the Atlantic Provinces, Canada: Géographie physique et Quaternaire, v. 31, p. 247–260.

—— , 1980a, Quaternary sea-level change in Atlantic Canada as an indication of crustal delevelling, *in* Mörner, N-A., ed., Earth Rheology, Isostasy, and Eustasy: New York, John Wiley and Sons Ltd., p. 201–214.

—— , 1980b, Quaternary stratigraphy of southwestern Nova Scotia; Glacial events and sea-level changes: Geological Association of Canada, Field Trip Guidebook, Trip 9, 63 p.

Grant, D. R., and King, L. H., 1984, A Quaternary stratigraphic framework for the Atlantic Provinces region, *in* Fulton, R. J., ed., Quaternary Stratigraphy of Canada; A Canadian contribution to IGCP Project 24: Geological Survey of Canada Paper 84-10, p. 173–191.
Gustavson, T. C., 1976, Paleotemperature analysis of the marine Pleistocene of Long Island, New York, and Nantucket Island, Massachusetts: Geological Society of America Bulletin, v. 87, p. 1–8.
Hickox, C. F., Jr., 1962, Late Pleistocene ice cap centered on Nova Scotia: Geological Society of America Bulletin, v. 73, p. 505–510.
Matthew, G. F., 1894, Postglacial faults at St. John, New Brunswick: American Journal of Science, v. 148, p. 501–503.
Mott, R. J., and Grant, D. R., 1985, Pre-Late Wisconsinan paleoenvironments in Atlantic Canada: Géographie Physique et Quaternaire, v. 39, p. 239–254.
Oliver, J., Johnson, J., Dorman, J., 1970, Postglacial faulting and seismicity in New York and Quebec: Canadian Journal of Earth Sciences, v. 7, p. 579–590.
Roland, A. E., 1982, Geological background and physiography of Nova Scotia: Halifax, The Nova Scotian Institute of Science, 311 p.
Stea, R. R., and Grant D. R., 1982, Pleistocene geology and till geochemistry of southwestern Nova Scotia: Nova Scotia Department of Mines and Energy, Map 82-10, sheets 7 and 8.
Taylor, F. C., 1969, Geology of the Annapolis-St. Mary's Bay map area, Nova Scotia: Geological Survey of Canada Memoir 358.
Wagner, F.J.E., 1977, Paleoecology of marine Pleistocene Mollusca, Nova Scotia: Canadian Journal of Earth Sciences, v. 14, p. 1305–1323.

Sedimentology, structure, and metallogeny of the Goldenville Formation, Meguma Terrane, Isaacs Harbour, Nova Scotia

***J. D. Keppie**, Department of Mines and Energy, P.O. Box 1087, Halifax, Nova Scotia B3J 2X1, Canada*

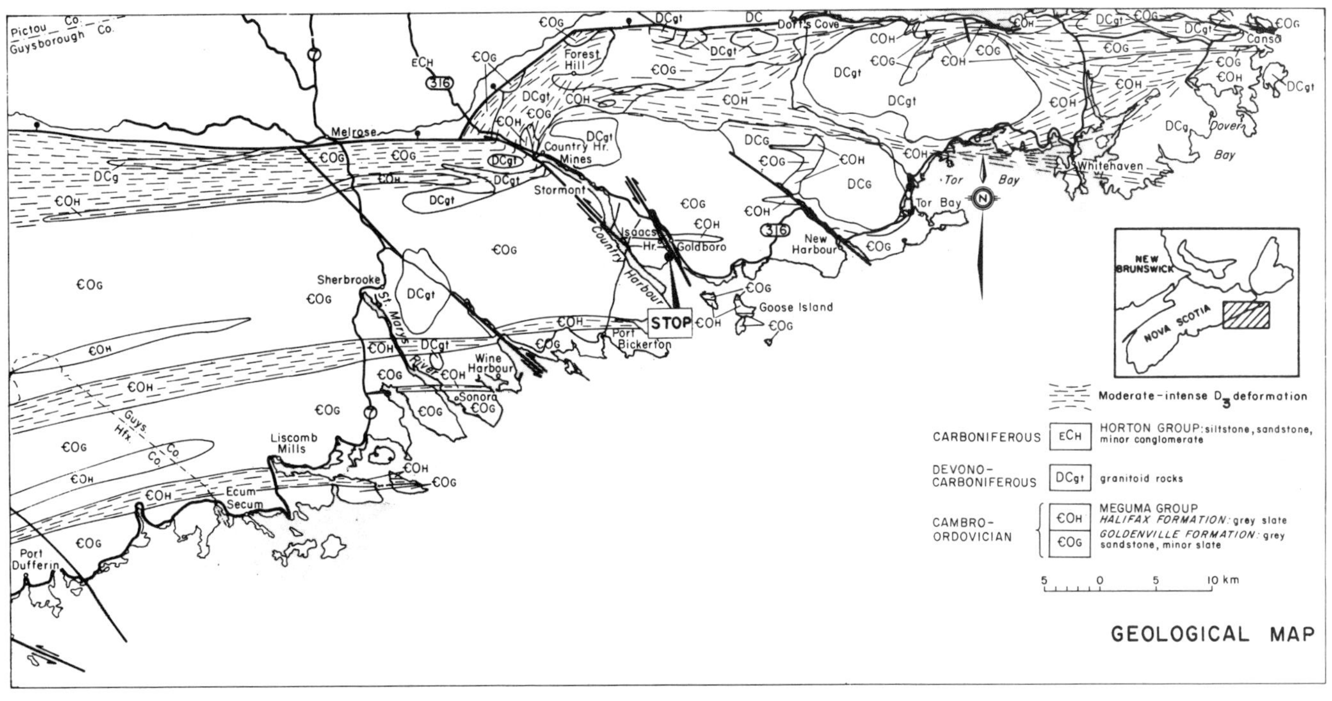

Figure 1. Geological map of the easternmost Meguma Terrane, Nova Scotia, showing field trip site.

LOCATION

Isaacs Harbour is located near the eastern end of the south coast of mainland Nova Scotia (Fig. 1). It can be reached by driving along Nova Scotia 316 to the south coast and taking the branch into Isaacs Harbour. Drive south through Isaacs Harbour to the last house, beyond which the road becomes a narrow track (Fig. 2). Park here so as to allow school buses to turn around. This spot is located immediately above Stop L, which may be reached by going down a steep path to the shore; take great care in the descent. To reach Stops B through K, walk 500 ft (150 m) south along the track, across a small stream, and take the first path to the shore just south of the stream. Stop A may be reached by walking 1,000 ft (300 m) south along the main track and cutting down to the shore about 300 ft (100 m) north of the lighthouse along a very small stream.

SIGNIFICANCE

This locality (Fig. 2) is one of the best sites to observe a wide variety of sedimentological, structural, and economic features in the Cambro-Ordovician Goldenville Formation, the lower unit of the Meguma Group, which forms a large part of the Meguma Terrane.

Sedimentological and structural features (Site 92). The interbedded sandstones and slates of the Goldenville Formation at this locality display many features typical of turbidites, such as partial to complete Bouma sequences (a graded basal unit followed upward by a parallel-laminated sandstone, a cross-laminated sandstone that sometimes has convolute lamination, and a slate horizon), flute and groove casts, and dewatering structures (pipe and sheet structures leading upward to sand volcanoes on the upper surface of the sandstone beds). In contrast, the Halifax Formation is predominantly slate with thin sandstone interbeds. Graptolites recovered from the Halifax Formation indicate an Ordovician age.

This site also displays many of the structures present in the Meguma Terrane. The earliest structures include upright, monoclinal, cylindrical, and conical Acadian folds associated with a spaced and slaty pressure solution cleavage defined by chlorite, muscovite, and biotite, inferred to have formed by sinistral shear during the Early Devonian. These structures were subsequently deformed by a subhorizontal crenulation cleavage preferentially developed in the slaty rocks. Decussate biotite porphyroblasts in the contact aureole of Late Devonian granitic plutons may be observed to have overgrown this crenulation cleavage. This was closely followed by, or was partly synchronous with tightening of the Acadian folds and reactivation of the Acadian cleavage dur-

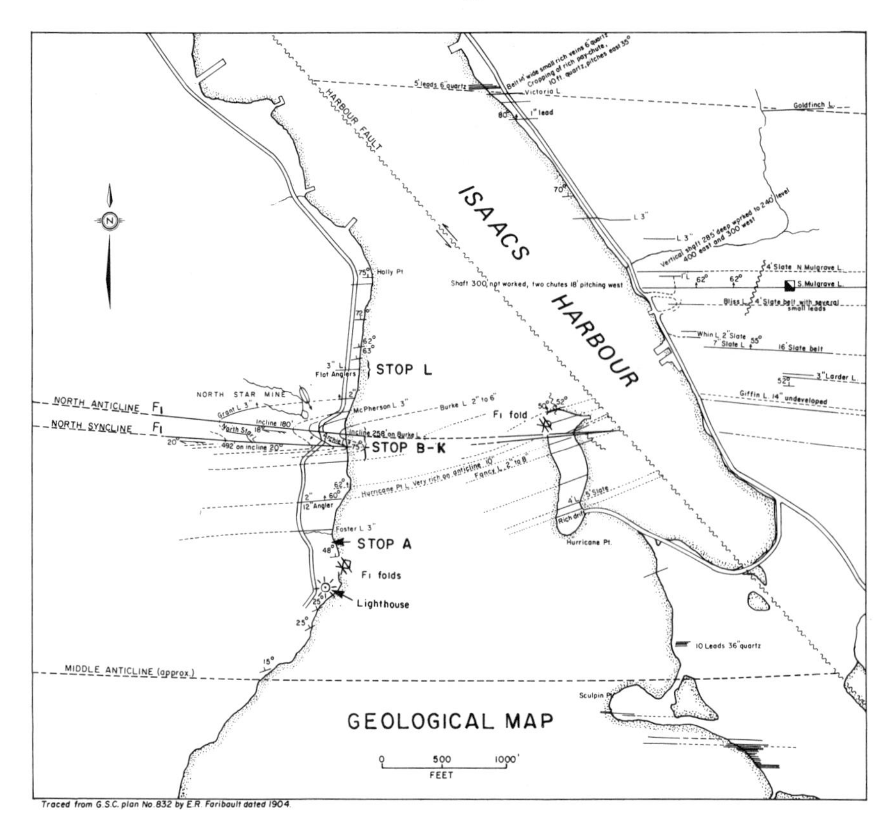

Figure 2. Geological map of the Isaacs Harbour Gold District showing locations of Stops A through L (modified from Faribault, 1904).

ing Late Devonian dextral shear deformation. This was synchronous with the start of westward obduction of the Meguma Terrane, which continued into Permian times (Dallmeyer and Keppie, 1986; Keppie and Dallmeyer, 1986). The Meguma Terrane is bounded along its northern margin by the Minas Geofracture, representing a lateral ramp that merges at depth with the thrust zone beneath the Meguma allochthon. The Minas Geofracture displays structures typical of a dextral shear zone (Keppie and others, 1985). During the Mid to Late Devonian, the Meguma Terrane was eroded and then unconformably overlain by about 1.8 mi (3 km) of Carboniferous rocks. Finally, during the early Mesozoic, the Meguma Terrane was displaced eastward along a listric normal fault that appears to coincide with the thrust at the base of the Meguma Terrane. The NW–SE faults and associated NW–SE, sinistral kink bands along Isaacs Harbour and Country Harbour are believed to have formed at this time. The youngest structures are N–S shear joint zones. Stops A-C, I, K display primarily geological features.

Metallogenic features (Site 93). Five types of veins representative of those present in the Meguma Terrane, including those mined for gold, are present at this locality. They are (1) stratabound, laminated, ribbon veins preferentially developed in the slaty horizons during the Acadian deformation; (2) en echelon veins in the limbs of the main Acadian folds formed by extension during flexural folding; (3) and (4) saddle reef and radial veins formed in tight Acadian fold hinges due to tangential longitudinal strain; and (5) north-south vertical veins, which generally cut across the Acadian structures but were gently folded and cut by the cleavage during the Late Devonian reactivation. Stops D-H, and J display features related to metallogenesis.

SITE INFORMATION

The regional setting of the Meguma Terrane has been published by Keppie (1982, 1985) and Keppie and others (1985), together with an exhaustive bibliography, to which the reader is referred for a more comprehensive treatment than space permits here. A series of maps, including geological, structural, metamorphic, tectonic, and metallogenic, have been published recently (Keppie, 1979a, b, 1982a, b; Chatterjee, 1983) and would be most useful to the reader.

Stop A (Figs. 3–5). At this stop, interbedded sandstones and slates of the Goldenville Formation display a variety of sedimentary structures, including graded bedding, cross stratification, sand volcano pipes, sandstone dikes, and a slumped and disrupted horizon. These beds were first deformed by moderately northwest plunging, upright, F_E folds, which are not associated with any cleavage development (Fig. 3). These early, F_E, NW–SE folds are transected by the E–W vertical, S1 cleavage, which is parallel to the axial planes of minor F1 folds occurring in the middle limb

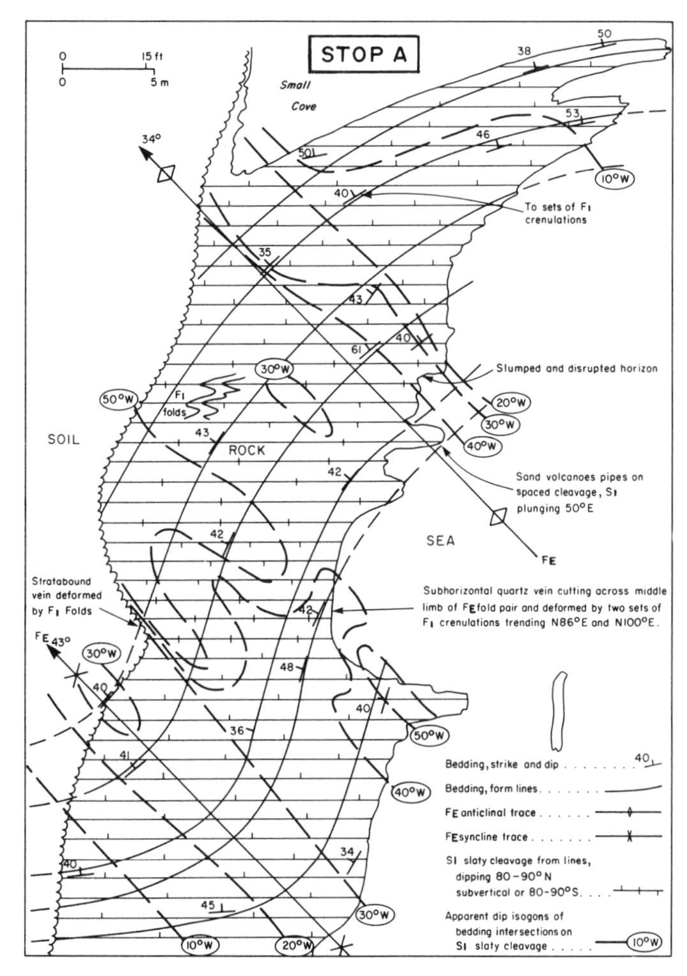

Figure 3. Geological map of Stop A.

Figure 4. Stereoplot of structural data from Stop A.

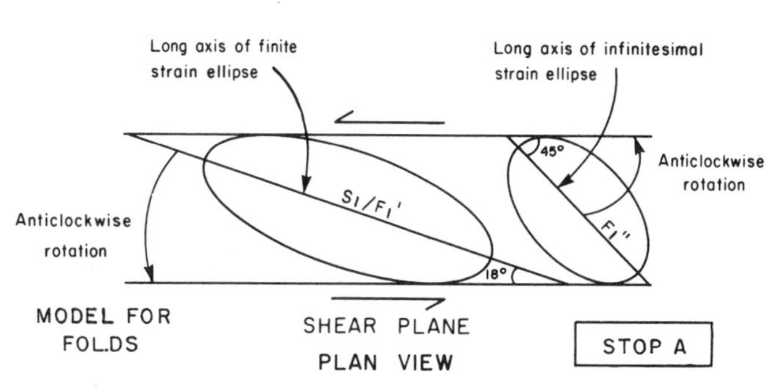

Figure 5. Plan view of sinistral simple shear during D1 to explain the two sets of F1 crenulations at Stop A.

of the NW–SE folds. The lineation formed by the intersection of the bedding on the cleavage appears to be curved because it is an oblique section through the early NW–SE folds (Fig. 4). Note the spaced cleavage in the sandstones in which the cleavage seams are relatively rich in chlorite and micas. Two sets of F1 crenulations are present near the northern margin of the map: one lies in the S1 cleavage, while the other lies slightly clockwise. This relationship, together with the s-shaped symmetry of the minor F1 folds, indicates a sinistral component in the D1 strain (Fig. 5). In this context, it is possible to produce the early, F_E, NW–SE folds during the nascent stage of the same D1 rotational strain regime. The slaty horizons also display a subhorizontal crenulation cleavage, S2, deforming both the bedding and the S1 slaty cleavage. Decussate biotite may be observed enclosing the S2 cleavage here. A few late sinistral, NW–SE, kink bands are also present. In the cliff, stratabound veins deformed by the F1 folds crop out.

Stop B (Figs. 6b and c). Gently dipping S2 crenulation cleavage, preferentially developed in the slaty beds, is cut by several, N–S, vertical quartz veins. However, note that these quartz veins are mildly deformed by the S1 cleavage, a feature interpreted as the result of reactivation during the Late Devonian deformation.

Stop C (Figs. 6a, b, and 7). Cylindrical North Syncline deforming graded sandstone beds with a convergent cleavage fan of S1 spaced cleavage. Note that the cleavage fan axis and the L1 bedding/cleavage intersection lineations are not quite parallel to the F1 fold axis (Fig. 7). A NW–SE sinistral kink band deforms the S1 cleavage here.

Stop D (Figs. 6b and 8). Here, stratabound auriferous veins occur mainly in the slaty horizons. Note that veins parallel to the slaty cleavage cut across and in places join the layer-parallel veins. Early striations on the surface of the layer-parallel veins are deformed by both F1 and F2 crenulations. **Please do not sample the veins.**

Stop E (Figs. 6a-d). Here, a faulted F1 monocline occurs with the excavated Burke vein in the steep limb. A small, folded, sulphide-bearing side vein occurs in the adjacent beds on the southern side of the fault. The fault is inferred from the discordant attitude of the bedding on either side of the excavated area and

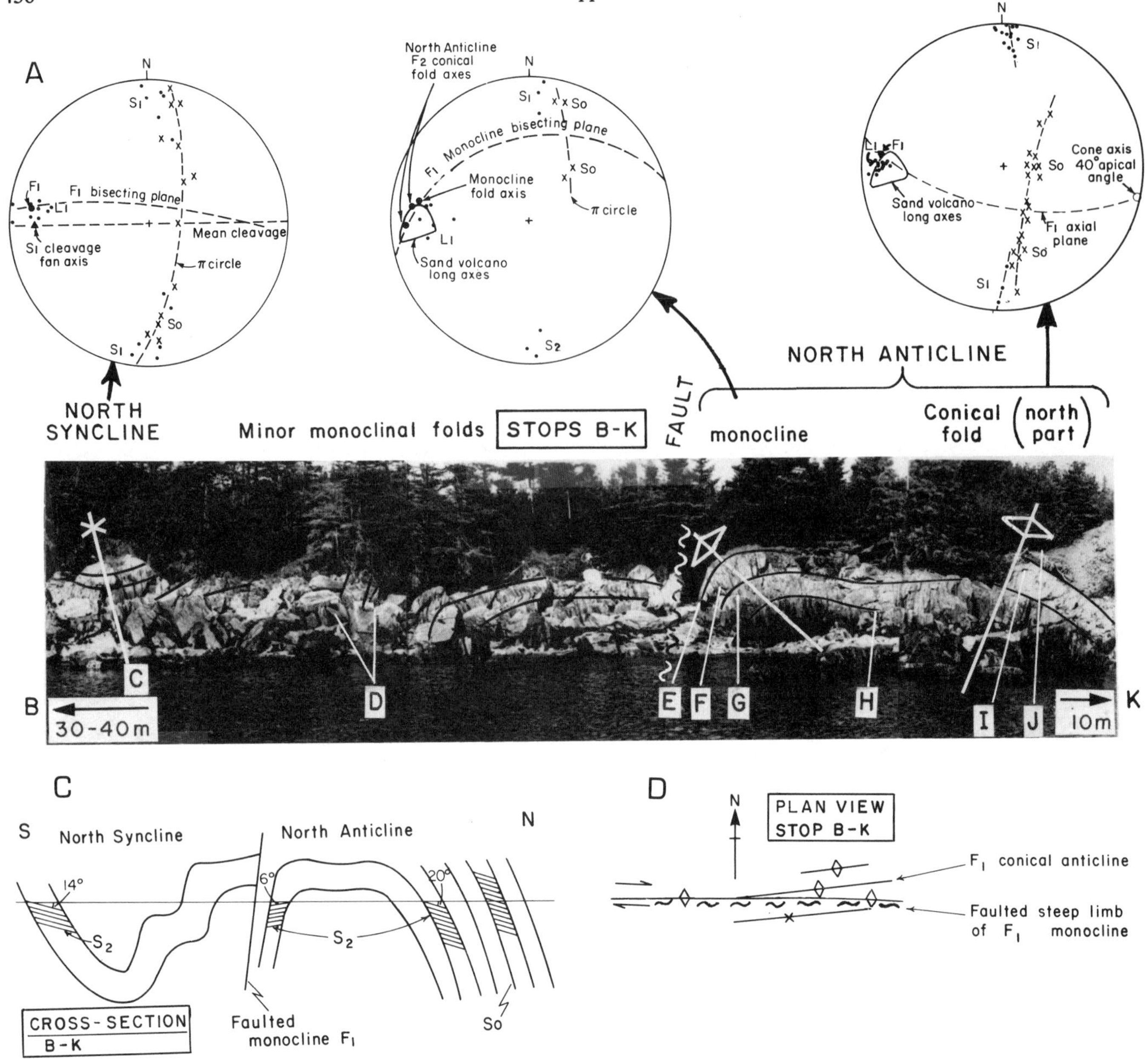

Figure 6. (a) Stereoplots, (b) photograph, (c) cross-section, and (d) plan view of the North Anticline and North Syncline at Stops B through K.

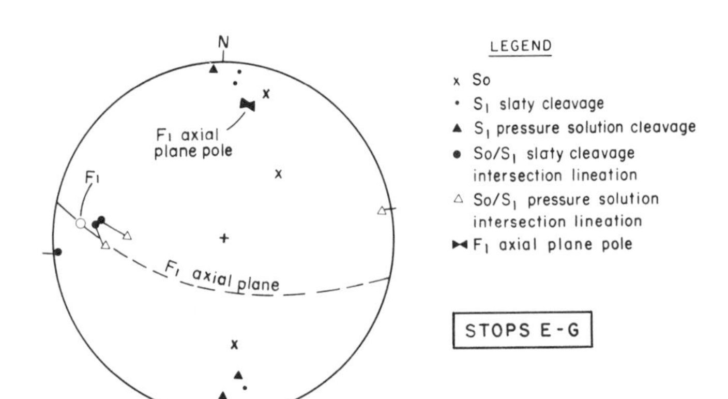

Figure 7. Stereoplot to show nonparallelism of S1 spaced and slaty cleavage and F1 fold axial planes and their respective intersection lineations, L1, with bedding in the North Anticline, Stops E through G. Tie lines join intersection lineations measured at the same place.

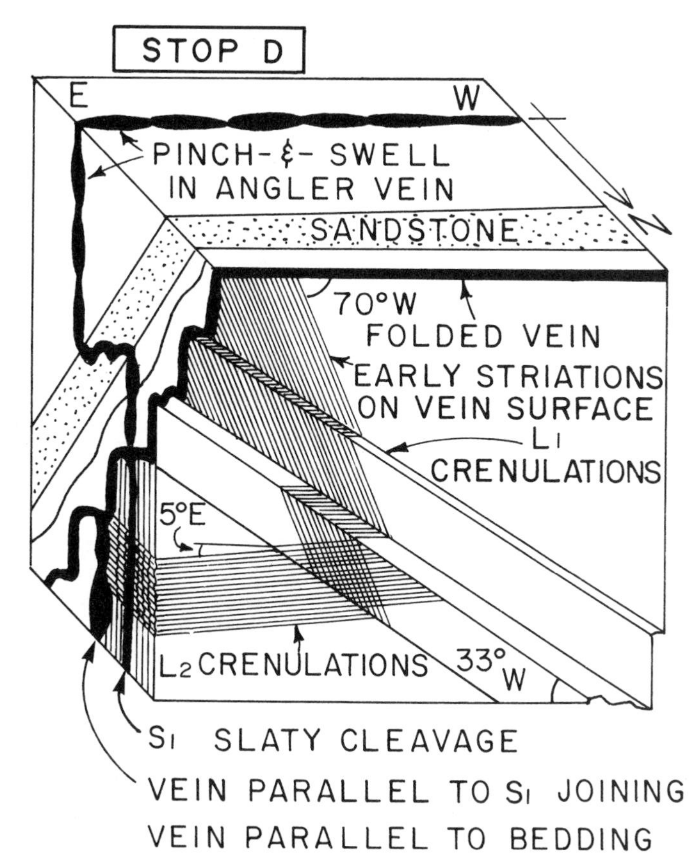

Figure 8. Block diagram of Stop D showing early striations on layer parallel vein deformed by crenulations L1 and L2.

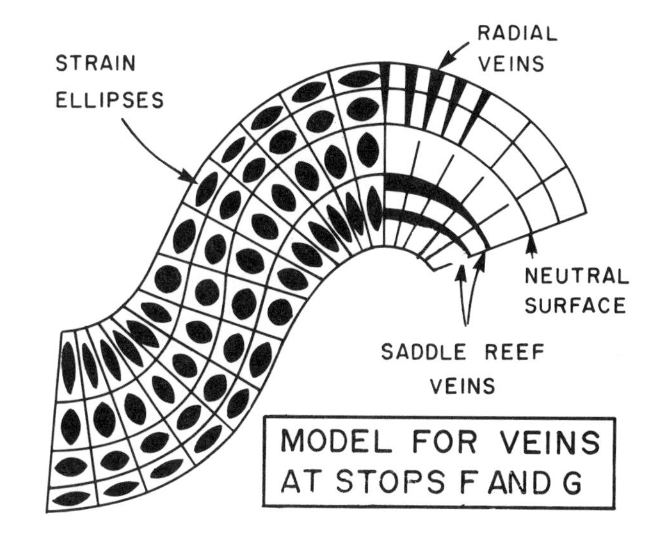

Figure 9. Strain distributions and predicted vein orientations in a fold produced by tangential longitudinal strain; model for veins at Stops F and G.

the lack of stratigraphic correlation. Subhorizontal L2 crenulations may be observed on the north side of the trench. Deformed N–S veins are also present.

Stop F (Figs. 6a, b, and 9). Note the sandstone-hosted, saddle reef quartz vein in the F1 fold hinge, inferred to be the result of extension perpendicular to bedding on the inner side of the neutral surface during tangential longitudinal folding. Also observe the late N–S shear joint zone cutting across the F1 monoclinal hinge.

Stop G (Figs. 6a, b, and 9). Sandstone-hosted radial quartz veins occur in the fold hinge, inferred to be the result of extension parallel to the bedding on the outer side of the neutral surface during tangential longitudinal folding. Note that the veins are cut by the S1 spaced cleavage.

Stop H (Figs. 6a, b and 10). Here, a layer-parallel vein displays early striations on the vein surface, which are deformed by F1 folds. In places, there is a planar structure approximately perpendicular to the striations. The striations are parallel to the long axes of quartz crystals, whereas the planar structure appears to be parallel to the basal pinacoid platelets. This suggests that the vein formed by a crack-seal mechanism presumably perpendicular to the vein wall, followed by bedding-parallel slip, which

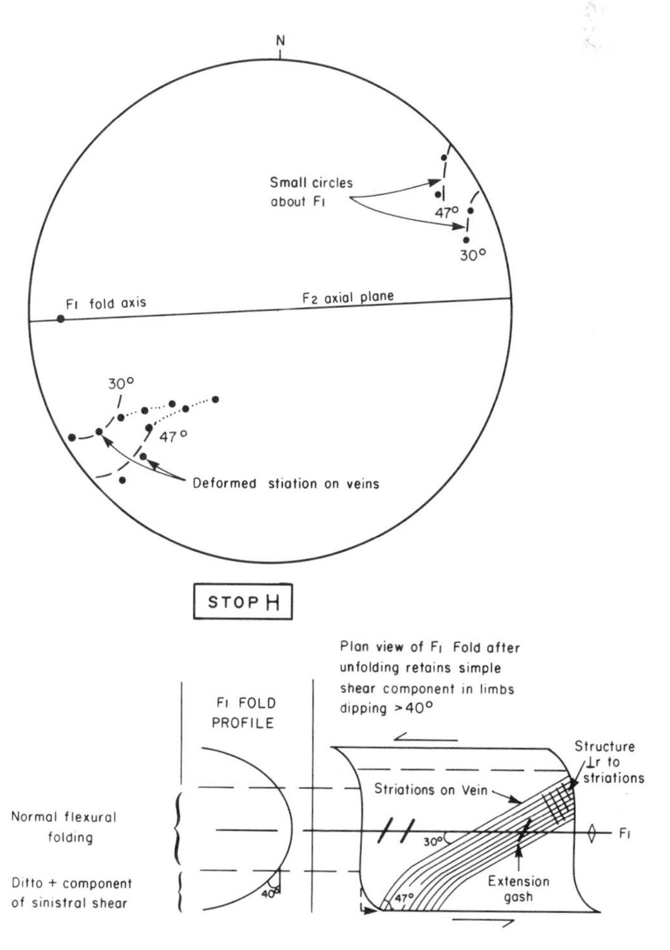

Figure 10. Stereoplot and deformational model of locus of striations on layer-parallel veins deformed by F1 folds at Stop H.

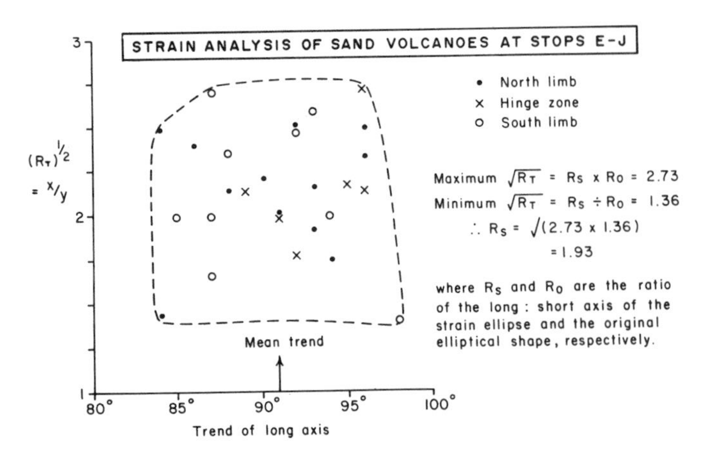

Figure 11. Graph of the long/short axes (X/Y) of sand volcanoes plotted against the orientation of the long axis around the North Anticline, Stops E through J.

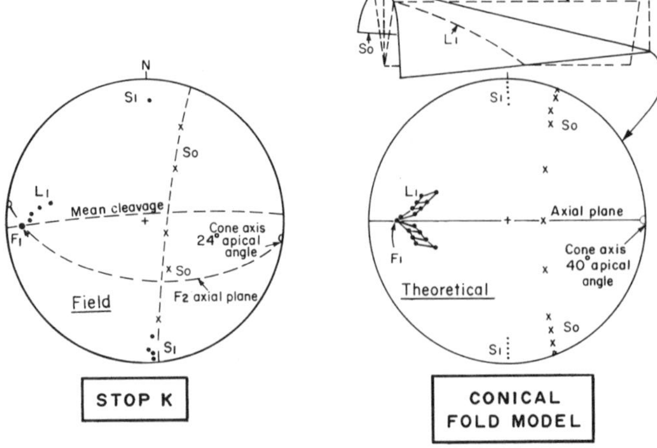

Figure 12. Stereoplot of field data from minor conical fold at Stop K: bedding plane poles define a small circle with a cone apical angle of 24°; S1 slaty cleavage is slightly fanned and the bedding/cleavage intersection lineations, L1, are curvilinear. Compare with sketch and stereoplot of theoretical conical fold with a cleavage fan.

rotated the crystals into the bedding plane. Deformation of these striations by the F1 folds produced a complex lineation locus in which normal flexural folding occurred in the fold hinge; however, a component of sinistral strain becomes evident in limb dips greater than 40°. Also present within these layer-parallel veins are NNE–SSW trending extension gashes inferred to have formed in the same sinistral strain regime.

Stop I (Figs. 6a, b, and 11). Here, deformed sand volcanoes occur on the hinge of an F1 conical fold. Measurements of the sand volcanoes gives a 1.9/1 strain ratio on the bedding planes. Note that the F1 fold axis lies slightly clockwise of the mean long axis of the sand volcanoes, a direction consistent with a component of sinistral strain during the D1 deformation.

Stop J (Fig. 6b). Excavated saddle reef vein (Burke Lead).

Stop K (Figs. 6b and 12). Here, a minor F1 conical fold formed at a facies change from slate to sandstone. Note the curvilinear bedding/cleavage intersection lineations in the fold limbs and compare field and theoretical structural data on the stereoplots. The conical fold is cut by two N–S veins, which in turn are cut by the S1 cleavage reactivated during the Late Devonian deformation. Note that the veins are folded when they strike NNW whereas NNE-striking veins are planar. This is attributed to a component of dextral strain on S1 during the Late Devonian deformation. Also present here are several slump folds transected by the S1 slaty cleavage.

Stop L (Figs. 13 and 14). Here, several en echelon veins occur on the north limb of the North Anticline. These veins cut the sandstones at high angles, whereas they have been rotated towards the bedding in the slaty horizons by flexural folding during the D1 deformation. Boudins with both horizontal and vertical axes formed in the en echelon veins in the slate horizons, but are absent in the sandstones. The en echelon veins are composed of massive quartz, K-feldspar, muscovite, chlorite, carbonates, and sulphides and contain up to 400 ppb gold (Haynes, 1983). Several NW–SE quartz-K-feldspar veins cut the en echelon veins and the country rock. At low tide, a stratabound auriferous quartz-carbonate vein may be observed cutting obliquely across the bedding in a predominantly slaty horizon. Towards the top of the cliff, the stratabound vein is juxtaposed against the en echelon vein, and both veins thin and pinch out, possibly due to boudinage. Haynes (1983) interpreted these veins in terms of a submarine hydrothermal vent–hot spring model in which the en echelon vein is inferred to have been a feeder to a siliceous sinter or geyserite (the quartz-carbonate vein). This hy-

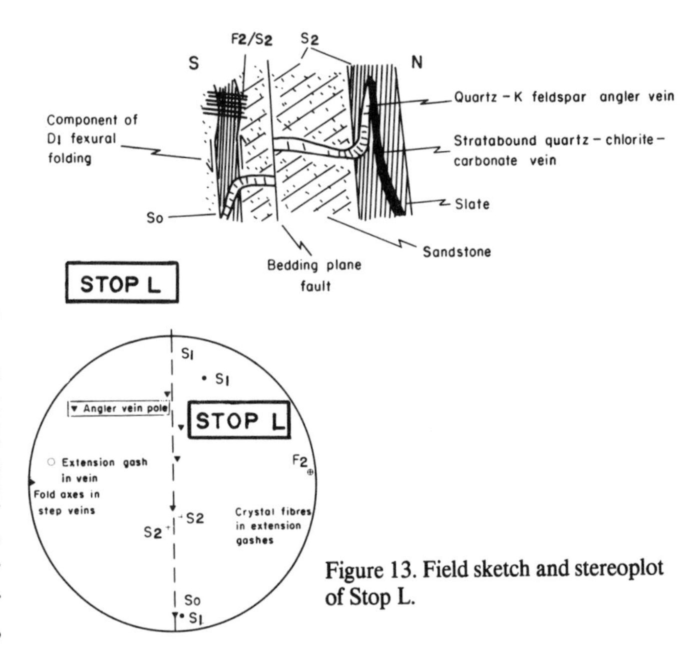

Figure 13. Field sketch and stereoplot of Stop L.

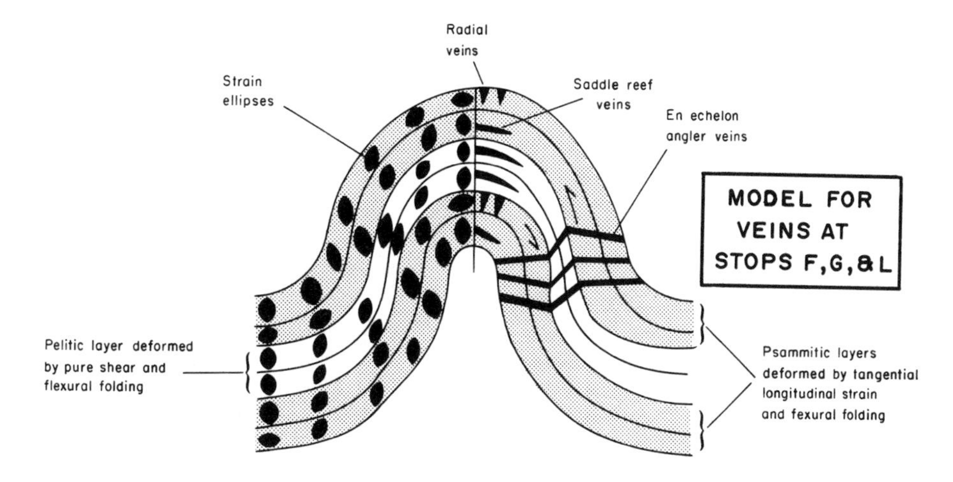

Figure 14. Strain model for the en echelon veins at Stop L and for those at Stops F and G.

pothesis appears untenable because the quartz-carbonate vein cuts across the bedding. Instead, the en echelon veins are inferred to be en echelon extension gashes formed and rotated by flexural folding in the F1 fold limbs. Also note the gently northward dipping S2 crenulation cleavage (slightly fanned) parallel to the axial plane of an F2 fold deforming bedding, S1 and the en echelon vein. Also observe the bedding-parallel fault in the thick sandstone horizon, which displaces the en echelon vein in the same sense as the flexural strain associated with the main F1 folding.

REFERENCES

Chatterjee, A. K., 1983, Metallogenic map of Nova Scotia: Nova Scotia Department of Mines and Energy, Halifax, Nova Scotia, scale 1:500,000.

Dallmeyer, R. D., and Keppie, J. D., 1986, Late Paleozoic tectonothermal evolution of the Meguma Terrane, Nova Scotia (in press).

Faribault, E. R., 1904, Plan and sections, Isaacs Harbour Gold District, Guysborough County, Nova Scotia: Geological Survey of Canada, Map No. 832, scale 1 in = 500 ft.

Graves, M. C., and Zentilli, M., 1982, A review of the geology of gold in Nova Scotia, *in* Hodder, R. W., and Petruk, W., eds., Geology of Canadian Gold Deposits: Canadian Institute of Mining and Metallurgy Special Volume 24, p. 233–242.

Haynes, S. J., 1983, Typomorphism of turbidite-hosted auriferous quartz veins, southern Guysborough County: Nova Scotia Department of Mines and Energy, Report 83-1, p. 183–224.

Keppie, J. D., 1976, Structural model for the saddle reef and associated gold veins in the Meguma Group, Nova Scotia: Canadian Institute of Mining and Metallurgy, v. 69, p. 103–116.

—— , 1979a, Geological map of Nova Scotia: Halifax, Nova Scotia Department of Mines and Energy, scale 1:500,000.

—— , 1979b, Structural map of Nova Scotia: Halifax, Nova Scotia Department of Mines and Energy, scale 1:1,000,000.

—— , 1982, Tectonic map of Nova Scotia: Halifax, Nova Scotia Department of Mines and Energy, scale 1:500,000.

—— , 1982, The Minas Geofracture: Geological Association of Canada Special Paper 24, p. 263–280.

—— , 1985, The Appalachian collage, *in* Gee, D. G., and Sturt, B., eds., The Caledonide Orogen, Scandinavia, and related areas: New York, J. Wiley and Sons Ltd., p. 1217–1226.

Keppie, J. D., and Dallmeyer, R. D., 1986, Geochronological constraints on the accretion of the Meguma Terrane, Nova Scotia (in press).

Keppie, J. D., and Muecke, G. K., 1979, Metamorphic map of Nova Scotia: Halifax, Nova Scotia Department of Mines and Energy, scale 1:1,000,000.

Keppie, J. D., Currie, K., Murphy, J. B., Pickerill, R. K., Fyffe, L. R., and St. Julien, P., 1985, Appalachian Geotraverse (Canadian Mainland): Geological Association of Canada Excursion 1, p. 1–181.

Schenk, P. E., 1970, Regional variation in the flysch-like Meguma Group (Lower Paleozoic) of Nova Scotia, compared to recent sedimentation off the Scotian shelf: Geological Association of Canada Special Paper 7, p. 127–153.

The Annieopsquotch Complex, southwest Newfoundland; An Early Ordovician ophiolite and its unconformable Silurian cover

G. R. Dunning, Department of Mineralogy and Geology, Royal Ontario Museum, 100 Queens Park, Toronto, Ontario M5S 2C6, Canada

LOCATION

The Annieopsquotch Complex is located in southwest Newfoundland on crown land in the southwest corner of the Puddle Pond map area (NTS 12A/5 1:50,000). It is located immediately east of the Burgeo Road (Newfoundland 480), approximately 30 mi (50 km) southeast of its intersection with Trans-Canada 1 near Stephenville (Fig. 1). A short hike due east from point "A" on the road through open country, partly on a moose trail through tuckamoor, will bring you to the rising barren slopes of the ophiolite. The gabbro and sheeted dyke zones of the complex, as well as the Silurian sedimentary-volcanic sequence, can be seen on the traverse. The ground is sufficiently open, with abundant outcrop on the ophiolite, that an exact route need not be specified. The line on Figure 2 indicates a good route, and one should follow valleys and ridges. The use of air photos A19471-92 and A19470-188, 189 (available from the Newfoundland Department of Mines and Energy, Howley Building, St. John's, Newfoundland) in combination with a topographic map is strongly recommended as the country is rugged.

Besides this route east of the road, the site includes outcrops along the Burgeo Road over a distance of approximately 1.2 mi (2 km) (Fig. 2). Gabbro and sheeted dikes of the ophiolite locally cut by granodiorite dikes and elsewhere overlain by a coarse basal conglomerate of the Silurian sedimentary sequence are exposed along the route. A cross section through most of the Silurian red sedimentary and volcanic sequence can be examined by walking along the road north from the quarry.

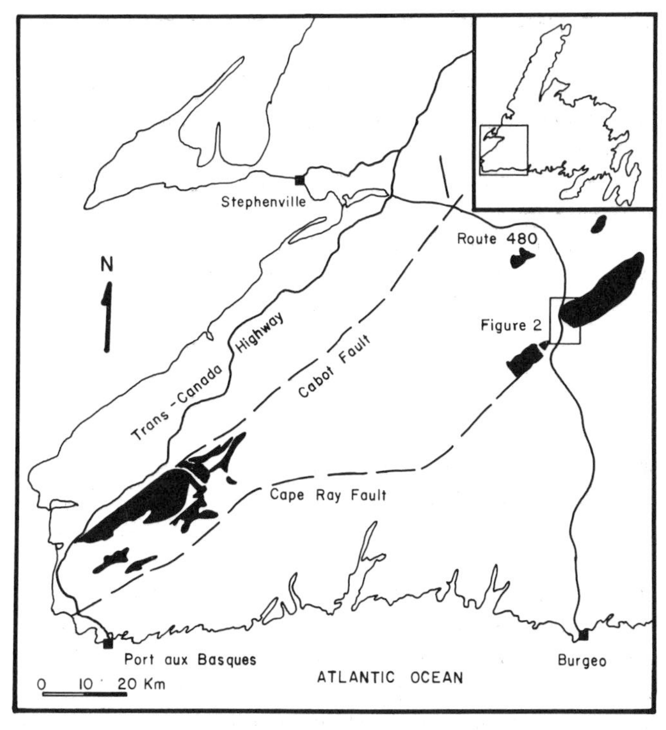

Figure 1. Location map of southwest Newfoundland showing Trans-Canada Highway, ophiolites represented in black, Newfoundland 480 to Burgeo and the area of the site outlined.

SIGNIFICANCE

The Annieopsquotch Complex is a fragment of Ordovician oceanic crust emplaced over the ancient continental margin of North America. It consists of layered olivine-clinopyroxene-plagioclase cumulates, a high-level gabbro zone, sheeted dike zone, and pillow lava zone, which trend northeast and face and dip southeast. The Annieopsquotch Complex is the largest, most complete, and least disrupted of a group of more than 20 ophiolite fragments that occur in a belt from near Buchans southwest to King George IV Lake. Because they show the same field relationships, orientation of sheeted dikes and igneous layering, and MORB geochemical signature (Dunning, 1984), it is suggested that they represent remnants of a major allochthon of Iapetus Ocean crust. This allochthon was disrupted during and after emplacement over the continental margin by intrusion, faulting, and erosion.

The Annieopsquotch Complex is in fault contact to the southeast with an isoclinally folded volcanic arc sequence, the Victoria Lake Group (Fig. 2, unit 5). It is intruded to the west and southwest by Ordovician tonalite to granodiorite (unit 6) that was generated by partial melting of a complex source during arc-continent collision. These intrusions and the ophiolite were in turn cut by small gabbro-diorite intrusions in latest Ordovician time (unit 7). Uplift and erosion of the ophiolite, tonalite, and gabbro and diorite intrusions and felsic volcanism in early Silurian time led to deposition of the red clastic sedimentary sequence with intercalated lahars and rhyolite flows that unconformably overlies the ophiolite at its southwestern end. These sedimentary rocks post-date closure of the Iapetus Ocean and the cessation of arc-related volcanism and plutonism. Subsequently the ophiolite was intruded by a hornblende and biotite-bearing granite of presumed Devonian (Acadian) age exposed east of the area covered by Figure 2. All of these units are disrupted and tilted along faults of regional extent, some of which were active until Carboniferous time.

These rocks, exposed in a small area, display many of the

geological relationships of significance to the interpretation of Newfoundland geology in general. The ophiolite represents either oceanic or marginal basin crust from the Iapetus Ocean; the tonalitic rocks represent the transition from an oceanic to continental regime during ocean closure while the Silurian sequence of sedimentary and volcanic rocks is of continental character.

SITE INFORMATION

The Annieopsquotch Complex and the Silurian sedimentary sequence can be seen at the site. Elsewhere along the road the Victoria Lake Group, tonalite-granodiorite suite, and a late Ordovician gabbro-diorite intrusion can also be seen; however, these are not discussed in this chapter.

Ophiolitic rocks of the Annieopsquotch Complex underlie an area of 56 mi^2 (140 km^2) of the Annieopsquotch Mountains south of the Lloyds River (Fig. 2).

Layered Cumulate Zone

Layered olivine-clinopyroxene-plagioclase cumulates, representing the "critical zone" or MOHO section of Malpas (1976), are exposed in two large areas east of the area of Figure 2. These are beyond the distance that one could reasonably walk from the road in one day. Layers strike 160 to 180 degrees in both these bodies and are steeply dipping, similar to dike orientations in the sheeted dike zone described below. Smaller bodies of layered cumulates shown in Figure 2 (unit 1) exhibit undeformed magmatic structures such as graded layering and trough structures. Elsewhere the rocks have been metamorphosed and show the effects of high temperature ductile shear. The latter are interpreted to be related to the mantle tectonite fabric found in the mantle ultramafic rocks of many ophiolites (e.g., Moores and Vine, 1971; Cann, 1974), although no tectonized harzburgites (depleted mantle) or other mantle lithologies have been identified in the Annieopsquotch Complex.

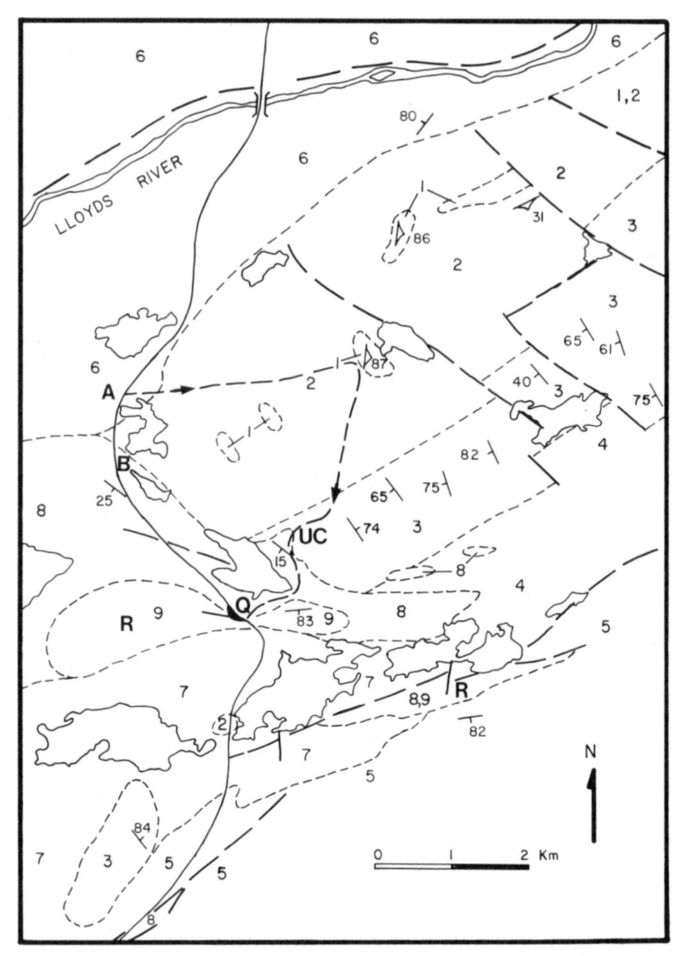

Figure 2. Geology of the Annieopsquotch Complex and associated rocks and overlying Silurian red clastic sedimentary and volcanic sequence. A = starting point, B = greenish pebble conglomerate, Q = Quarry, R = Rhyolite center, UC = site of exposed unconformity. 1 = layered cumulates, 2 = high-level gabbro, 3 = sheeted dykes, 4 = pillow lava, 5 = Victoria Lake Group, 6 = tonalite, granodiorite, 7 = gabbro, diorite, 8 = sandstone, conglomerate, and 9 = rhyolite, lahar. Thick dashed line with arrows is the general suggested route.

High Level Gabbro Zone

The gabbro zone, composed of seven fault-bounded blocks, occurs along the northwestern edge of the mountains (Fig. 1). Most of the gabbro (unit 2) is coarse grained and equigranular, composed of nearly equal amounts of plagioclase and clinopyroxene. It is commonly crosscut by diabase dikes and contains trondhjemite and pegmatitic gabbro pods. Small patches of magnetite gabbro and quartz gabbro are also present along the top (southeast) of the gabbro zone. These details cannot be shown on the map but many of these features will be encountered along the route shown. The top of the zone is offset along faults, many of which cannot be traced through the sheeted dike zone; several of these faults may follow the margins of discrete gabbro intrusions. Juxtaposed fault blocks of the gabbro contrast in the number of crosscutting diabase dikes and trondhjemite and pegmatite pods. In some exposures the gabbro is criss-crossed by hundreds of raised, amphibole-bearing veinlets, which may represent hydrothermal fluid pathways. Rocks near the base of the gabbro zone have been recrystallized locally into two contrasting metamorphic lithologies—amphibolite and granulite. Only amphibolites, commonly associated with shear zones, can be seen near the west end of the complex. They are medium- to coarse-grained and show compositional banding due to variations in the proportion of plagioclase to hornblende. Areas of coarse gabbro cut by a few diabase dikes are preserved within the zones, and relict, coarse plagioclase crystals are present within the amphibolite. Two-pyroxene mafic granulites occur as numerous blocks or irregular zones that grade into regular massive gabbro near outcrops of the underlying critical zone gabbro.

At the top (southeast) of the gabbro zone, gabbro cut by only a few dikes may locally grade within 150 ft (100 m) to 100%

sheeted dikes. Near the very top of the zone, many textural varieties and grain sizes of gabbro and diabase occur together with curved, irregular contacts. East of the area covered by Figure 2, in several excellent exposures in the central part of the complex, the diabase appears to be rapidly solidified magma pools that have intruded and enclosed coarse grained gabbro. The top of the massive gabbro section is therefore interpreted as the root zone for some of the diabase dikes.

Sheeted Dike Zone

The sheeted dike zone is exposed along the southeast side of the gabbro zone for the full strike length of the Complex (Fig. 2, unit 3). In the area of Figure 2 the dikes strike between 160 and 190 degrees, dip steeply, and vary in width from 0.4 in to 33 ft (1 cm to 10 m), with the majority between 1.5 and 10 ft (0.5 and 3 m). Many have chilled margins. Plagioclase phenocrysts occur in about 10% of the dikes, whereas clinopyroxene and hornblende phenocryst and lithic inclusions are very rare, having been identified in only about 5 dikes in the entire ophiolite. A few trondhjemite dikes and even rarer breccia dikes, composed of small fragments of diabase in an aphanitic diabase matrix, are present about 6 mi (10 km) ENE of Figure 2. The latter are undeformed and have smooth contacts with adjacent dikes. In some areas near the contact with the pillow lava zone, many closely spaced aphanitic diabase dikes intrude along the margins and internal fractures of earlier wider diabase dikes.

Pillow Lava Zone

This zone (unit 4) is up to 0.6 mi (1 km) wide, and is composed of basaltic pillow lava, massive flows, pillow breccia, diabase dikes, and sills. Dikes predominate in the lower part of the zone. Some bud to form pillows or pillow breccia. Only a small amount of pillow lava, cut by greater than 50% diabase dikes, is present at the southwestern end of the complex. The only exposures on Figure 2 are approximately 1.2 mi (2 km) east of the recommended traverse line and the best exposures are approximately 6 mi (10 km) farther ENE.

Pillows are round to elongate, 7 in to 3.3 ft (0.2 to 1 m) in diameter and their shapes do not indicate obvious top directions. They are aphanitic, non-vesicular, and some contain plagioclase phenocrysts. Interpillow material is basalt and minor red chert.

Several faults, which cut the top of the pillow lava zone and thus form the southeastern boundary of the Annieopsquotch Complex, are marked by chlorite schist, which contains lenses of the same Silurian red clastic sedimentary rocks that unconformably overlie the complex at its southwestern end. Therefore, the latest movement on this fault is Silurian or younger. Two trondhjemite pods differentiated from the ophiolitic gabbros have been dated by the U/Pb (zircon) method at 481.4 +4.0/−1.9 Ma and 477.5 +2.6/−2.0 Ma (Dunning and Krogh, 1985). These ages are Arenigian by the time scale of van Eysinga (1975) and are among the youngest from any ophiolite recognized in the Appalachian-Caledonide Mountain Belt.

Diabase dikes and pillow lavas analyzed show clear N-type MORB trace element trends and rare earth element patterns with light rare earth depletion and negative Europium anomalies, that fall in the range of modern MORBs (Dunning and Chorlton, 1985).

Clastic Sedimentary and Volcanic Sequence

Red clastic sedimentary rocks (unit 8) with interbedded rhyolite flows, welded tuffs, and a lahar (unit 9) outcrop around the south end of the Annieosquotch Complex and in narrow fault-bounded slivers farther south on the road and along the southeast slope of the mountains above Victoria Lake (not shown on Fig. 2). The unconformity between crudely stratified basal conglomerate beds of the sequence and sheeted dikes of the ophiolite is well exposed at locality "UC" (Fig. 2). Here the beds dip southwestward at about 15 degrees and 90 to 100% of the variably rounded cobbles are of gabbro or diabase (Chandler, 1982). Some thin beds of laminated and trough crossbedded coarse grained sandstone are interbedded with the conglomerates. Imbrication in coarse conglomerate beds indicates sedimentary transport to the northeast (i.e.; toward the present high ground of the ophiolite, because the ophiolite was uplifted after redbed deposition; Chandler, 1982). Two outliers of red conglomerate occur on the ophiolite and these have altered hematized ophiolitic rocks beneath them. Hematization of ophiolitic rocks is present in a valley to the northeast of the present outcrop area of the sedimentary rocks, suggesting that the Silurian rocks once covered a larger area of the ophiolite. A few meters above the unconformity, the abundance of felsic volcanic clasts rapidly increases, and rare granite boulders are present.

The northernmost outcrop of the sedimentary rocks on the road (locality "B") is a greenish pebble conglomerate composed of quartz and feldspar with lesser basalt and rhyolite lithic fragments. It seems to be derived predominantly from granodiorite, which is exposed further north on the road. Farther south the outcrops are composed of gently dipping red, crudely size-graded pebble to cobble conglomerate, which locally contains crossbeds. Clasts are predominantly of rhyolite derived from nearby contemporaneous volcanic centers (hills to E and W of the road; Fig. 2, localities "R"). The occurrence of calcite cement led Chandler (1982) to suggest that the climate was dry during deposition. In outcrops on the road just north of the quarry, fine-grained sandstone predominates and displays fining upward cycles, scoured bases of beds, mud cracks, and cross-bedded and massive beds. At the quarry (locality "Q"), a 330-ft-section (100 m) is exposed with its top to the south, containing at the base a 33-ft-thick (10 m) rhyolite flow faulted against 10 ft (3 m) of red mudstone. Above the mudstone an unsorted unit occurs composed of sand, mud, boulders of rhyolite of various types, gabbro, and diabase, which has been interpreted as a lahar (Dunning and Herd, 1980).

It is interpreted that deposition of the Burgeo Road sequence reflects uplift and erosion of fault-bounded blocks of crust following closure of the Iapetus Ocean. Felsic, and to a lesser extent basaltic, volcanism, and dike intrusion were localized along these faults, building small rhyolite centers from which lavas, tuffs, and lahars flowed over the sediments being shed from the uplifted blocks.

Regionally, this sequence has been correlated (Kean, 1983) with a red sedimentary sequence that crosses the King George IV map area (NTS 12A/6) to the south and with the Windsor Point Group exposed along the Cape Ray Fault in southwestern Newfoundland. However, both of the latter groups contain Emsian spores (early Devonian) and a rhyolite in the Windsor Point Group has been dated at 377 ± 21 Ma by the Rb/Sr whole rock method (Wilton, 1984). In contrast, rhyolite interbedded with the red sandstones near the Burgeo Road is dated at 431 ±5 Ma by the U/Pb (zircon) method (Chandler and Dunning, 1983) indicating that this sequence is early Silurian and therefore not correlative with those to the south. The Silurian sequence here can best be correlated with the Springdale and Botwood Groups of northern and central Newfoundland, respectively.

REFERENCES

Cann, J. R., 1974, A model for oceanic crustal structure developed: Geophysical Journal of the Royal Astronomical Society, v. 3, p. 169–187.

Chandler, F. W., 1982, Sedimentology of two middle Paleozoic terrestrial sequences, King George IV Lake area, Newfoundland, and some comments on regional paleoclimate, *in* Current Research: Canada Geological Survey Paper 82-1A, p. 213–219.

Chandler, F. W., and Dunning, G. R., 1983, Fourfold significance of an early Silurian U/Pb zircon age from rhyolite in redbeds, southwest Newfoundland, NTS 12A/4, *in* Current Research: Canada Geological Survey Paper 83-1B, p. 419–421.

Dunning, G. R., 1984, The geology, geochemistry, geochronology, and regional setting of the Annieopsquotch Complex and related rocks of southwest Newfoundland [Ph.D. thesis]: St. John's, Memorial University of Newfoundland, 403 p.

Dunning, G. R., and Chorlton, L. B., 1985, The Annieopsquotch ophiolite belt of southwest Newfoundland; Geology and tectonic significance: Geological Society of America Bulletin, v. 96, p. 1466–1476.

Dunning, G. R., and Herd, R. K., 1980, The Annieopsquotch ophiolite complex, southwest Newfoundland, and its regional relationships, *in* Current Research, Part A: Canada Geological Survey Paper 80-1A, p. 227–234.

Dunning, G. R., and Krogh, T. E., 1985, Geochronology of ophiolites of the Newfoundland Appalachians: Canadian Journal of Earth Sciences, v. 22, p. 1659–1670.

Kean, B. F., 1983, Geology of the King George IV Lake map area (12A/4), Newfoundland: Newfoundland Department of Mines and Energy Report 83-4, 62 p.

Malpas, J. G., 1976, The geology and petrogenesis of the Bay of Islands ophiolite suite, western Newfoundland (2 parts) [Ph.D. thesis]: St. John's, Memorial University of Newfoundland, 435 p.

Moores, E. M., and Vine, F. J., 1971, The Troodos Massif, Cyprus, and other ophiolites as oceanic crust; Evaluation and implications: Philosophical Transactions of the Royal Society of London, A, v. 268, p. 443–466.

van Eysinga, F.W.B., 1975, Geological time table (3rd edition): New York, Elsevier Publishing Company, 1978.

Wilton, D.H.C., 1984, Metallogenic, tectonic, and geochemical evolution of the Cape Ray Fault Zone with emphasis on electrum mineralization [Ph.D. thesis]: St. John's, Memorial University of Newfoundland, 618 p.

Late Quaternary glaciation and sea-level change southwest Newfoundland, Canada

I. A. ***Brookes,*** *Department of Geography, York University, North York, Ontario M3J 1P3, Canada*

LOCATION

This site is the coast and hinterland of St. George's Bay, southwest Newfoundland, centered on Stephenville (43°, 33′N; 58°, 35′W; Fig. 1). Drift is exposed in 42 mi (70 km) of nearly continuous coastal cliffs westward to Port au Port, and south to Highlands, with important patchy exposures and terrain units on Port au Port Peninsula and along hinterland routes. Access is by air to Stephenville (which has hotel accommodation), and then by auto on paved roads, with short walks from the roads to the sections. Stops 1, 2, and 3 are reached from Newfoundland 460 west of Stephenville. Stops 4, 5, and 6 are reached via Newfoundland 460/461 to Trans-Canada Highway, then via Newfoundland 404 and 405 which branch from it. Consult locality maps for orientation from roads to sections at the stops. No prior permission is required, merely explanation to local people encountered. Useful maps are a provincial road map, the Canada National Topographic System sheet 12B (scale 1:250,000), the 1:50,000 sheets of the NTS 12B series, and GSC Map 15-73 (in Brookes, 1974).

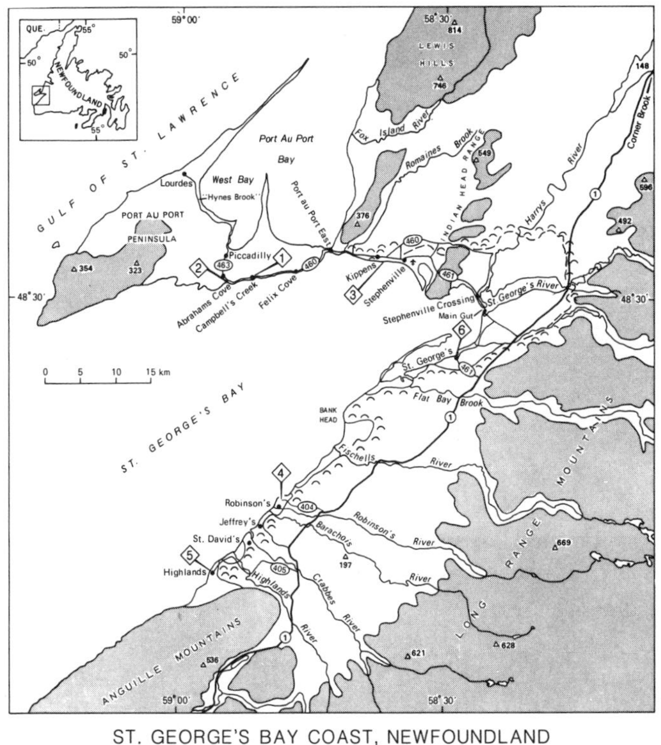

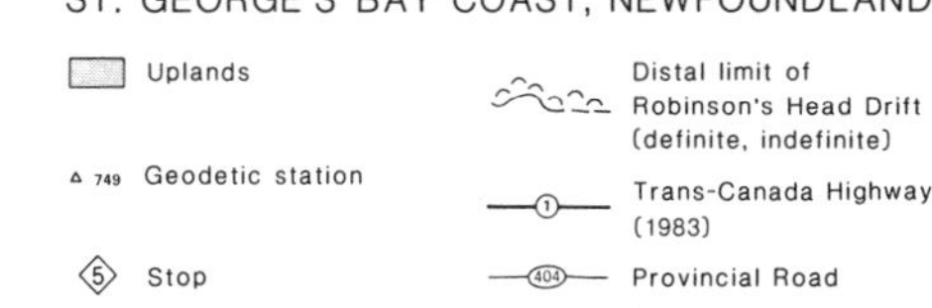

Figure 1. Southwest Newfoundland, showing route and stops of tour, places mentioned in text, upland areas, and Robinson's Head Moraine.

SIGNIFICANCE

Although southwest Newfoundland lies only 120 mi (200 km) southeast of Labrador, across the Gulf of St. Lawrence, striae on bedrock west of Stephenville (with regional till provenance data) show that it was last glaciated by a Late Wisconsinan ice cap flowing west-northwest from Newfoundland interior ice divides. This evidence limits the Late Wisconsinan expansion of the Laurentide Ice Sheet from the Nouveau Québec ice divide. Smooth-topped coastal massifs at 2,640 ft (800 m) north of St. George's Bay, visible from Port au Port Peninsula, stood as nunataks above Newfoundland ice.

Stratigraphic relations between deposits of the maximal and deglacial phases of Late Wisconsinan glaciation and overlying marine sediments demonstrate the mode and pattern of coastal deglaciation and marine submergence. Dates on marine shells provide a chronology. Relations between the marine sediments and overlying glacial deposits demonstrate variable dynamics along an ice margin that readvanced into the sea in some places, but was stabilized by marine influence in others. Coastal landform/sediment associations that are well-defined and often dated, emerged and submerged, provide data for a well-controlled history of relative sea-level change in the area.

SITE INFORMATION

Background

The glacial and sea-level history of southwest Newfoundland was first outlined by MacClintock and Twenhofel (1940) and Flint (1940), thus opening the "modern" era of glacial studies on the island. MacClintock and Twenhofel recognized three drift units around St. George's Bay: St. George's River Drift, Bay St. George Delta, and Robinson's Head Drift. They interpreted them as a glacial-marine-glacial sequence of Wisconsin age, now understood to mean Late Wisconsinan. Flint mapped emerged shore features in the region, relating them to postglacial sea-level change and isostatic crustal warping. Both studies concluded that Wisconsin-age Labradorean ice completely covered at least western, and probably all of, Newfoundland—Flint (1940), from the northwestward rise of reconstructed isobases along the west coast; MacClintock and Twenhofel (1940), from sets of crossing striae near Port au Port. An island-centered ice cap was postulated to explain a radial striation pattern produced during deglaciation.

Studies by Brookes (1969; 1970a,b; 1974) around St.

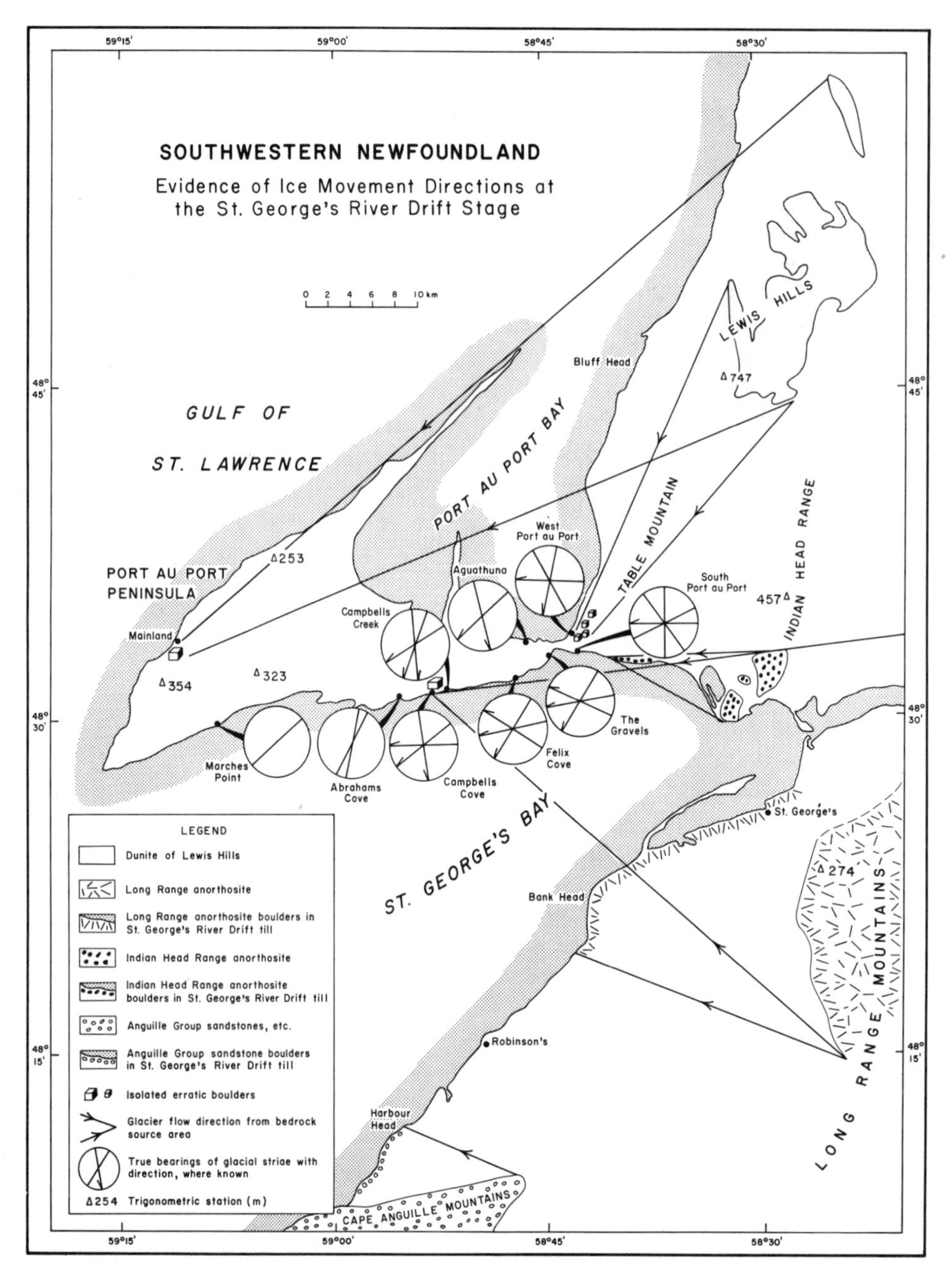

Figure 2. Southwest Newfoundland: evidence of ice movement directions (striae, till provenance, and erratics) at St. George's River Drift Stage (Late Wisconsinan).

George's Bay revealed evidence in glacial erosion microfeatures and till/erratic provenance (Fig. 2) that the till of MacClintock and Twenhofel's (1940) St. George's River Drift was deposited by maximal Late Wisconsinan Newfoundland ice flowing west-northwestward into the Gulf of St. Lawrence, not by Labradorean ice. Deglaciation and instantaneous marine submergence of the present coast occurred 13,500 years B.P., according to eight radiocarbon dates from the region on shells in stony, glaciomarine muds overlying that drift. Submergence of an isostatically depressed coastal fringe attained 144 ft (44 m) above the present sea level at several sites. The marine limit is marked by washed till limits and fragments of a complex delta, MacClintock and Twenhofel's Bay St. George Delta. This is a coarsening-upward complex built by meltwater rivers that were the antecedents of the present centripetal St. George's Bay drainage pattern. At 12,600 years B.P., the margin of the Newfoundland ice cap surged into the sea in several lobes, but was stabilized by calving between them, when the sea level had fallen to a brief stillstand at 89–95 ft (27–29 m). This reactivation produced MacClintock and Twenhofel's Robinson's Head Drift, the seaward limit of which is marked by a prominent kame moraine, a feature with little till, few surface boulders, and a rolling surface topography. Shells in marine sands closely associated with the moraine near Stephenville are dated at 12,600 years B.P. (Brookes, 1977a).

Subsequent rapid deglaciation of the St. George's Bay hinterland left a mantle of coarse till and glaciofluvial deposits, and was accompanied at the coast by continued relative sea-level fall. Sea level fell below its present position about 10,000 years B.P., decelerating to a minimum of 36–46 ft (11–14 m) below present about 6,000 years B.P. From this level it has risen to its present level, forming salt marshes in protected estuaries and spit/bar complexes across more exposed bays (Brookes and others, 1985).

Geomorphic studies in the mountains north and south of St. George's Gay (Grant, 1977; Brookes, 1977b) revealed three systematically arranged geomorphic zones with sharp, seaward-sloping boundaries, in which bedrock and glacial till show step-wise increases in weathering and colluviation with increasing elevation. These boundaries were interpreted as limits of glaciations less extensive with decreasing age. The upper limit of the lowest zone can be traced to Late Wisconsinan glacial deposits at the coast, whereas the two higher zones are tentatively assigned to earlier Wisconsinan and pre-Wisconsinan glaciations. D. R. Grant (in preparation) has extended this interpretation to the approximate 2,600 ft (800 m) level of Lewis Hills and Blow me Down Hills, northwest of Stephenville, which are visible from Port au Port. Terrain above 1650 ft (500 m) in these hills stood above Late Wisconsinan Newfoundland ice.

Stops

En route to Stop 1, Newfoundland 460 from Stephenville traverses rolling terrain of the Robinsons Head Moraine (Fig. 1) and postglacial marine terraces at 95 ft (29 m) and 46 ft (14 m) set into it. The former is prominent as a delta surface best viewed from Port au Port Peninsula after crossing the isthmus. On the peninsula, this road traverses north-dipping, Ordovician carbonate rocks lightly mantled with St. George's River Drift till and again terraced at the above elevations.

Stop 1. Coastal erosion of bedrock and overlying till along this shore is continually exposing glacial microfeatures on abraded rock ledges. At this stop, Campbells Creek (Fig. 1), a broad coastal ledge east of the creek, shows these features in most years. If not exposed, neighboring ledges and those at Campbells Cove to the west may be searched.

It was at this locality that the sequence of striae beneath St. George's River till first became clear (Brookes, 1970b). Strong striae directed westward or southwestward are commonly inscribed by a later, weaker set directed southwestward or southward, respectively (Fig. 2). Ice movements are actually revealed by crescentic fractures (concave down-ice), miniature stoss-and-lee forms, and "rat tails" in the lee of chert nodules. If the latter two are not visible at this stop, they are common all along this coast. St. George's River till shows a pebble fabric more closely related to later than to earlier movements. These ice-flow directions are also indicated by erratic boulders and till provenance, although their sequence is not. The change from generally west to generally south is believed to reflect changing ice margins, first trending parallel to the main northeastward coastal alignment at the maximum, then changing to perpendicular to the south shore of Port au Port Peninsula, as a bathymetrically controlled calving bay developed in the glacier margin in St. George's Bay.

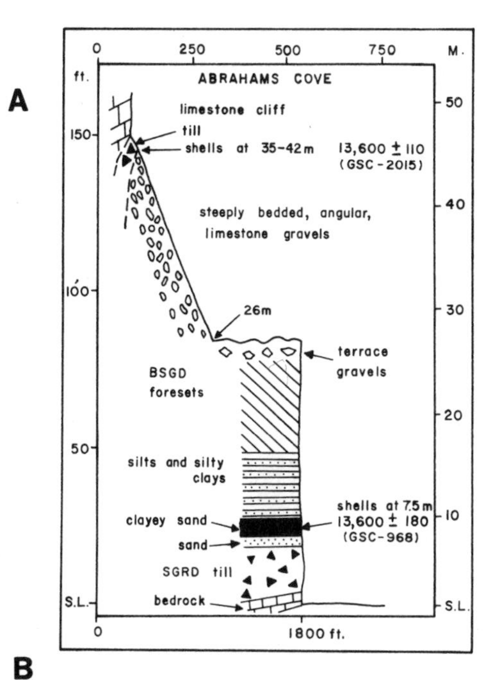

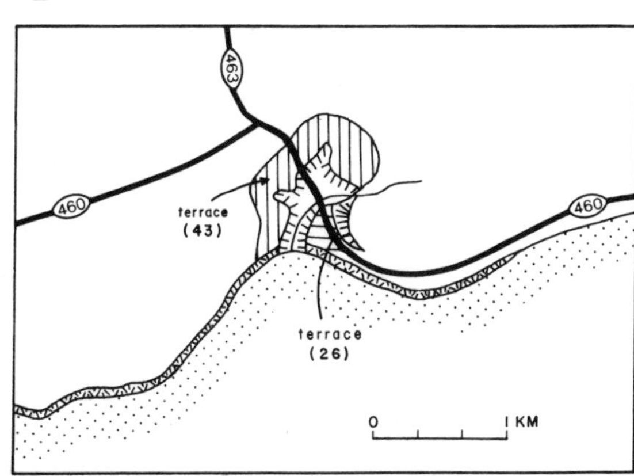

Figure 3. Location, landforms, and section in Quaternary sediments, Abrahams Cove (Stop 2).

Stop 2. Follow Newfoundland 460 to Abrahams Cove. Park on the roadside near the gravel pit and walk to the beach down a track on the east side of the creek (Fig. 3, B).

Figure 3,A shows the generalized composite section from the beach to above the marine limit at Abrahams Cove. Till over southwest-striated bedrock was deposited by maximal Late Wisconsinan Newfoundland ice channelled by the fault-guided valley inland. Shells in overlying glaciomarine stony mud are dated at

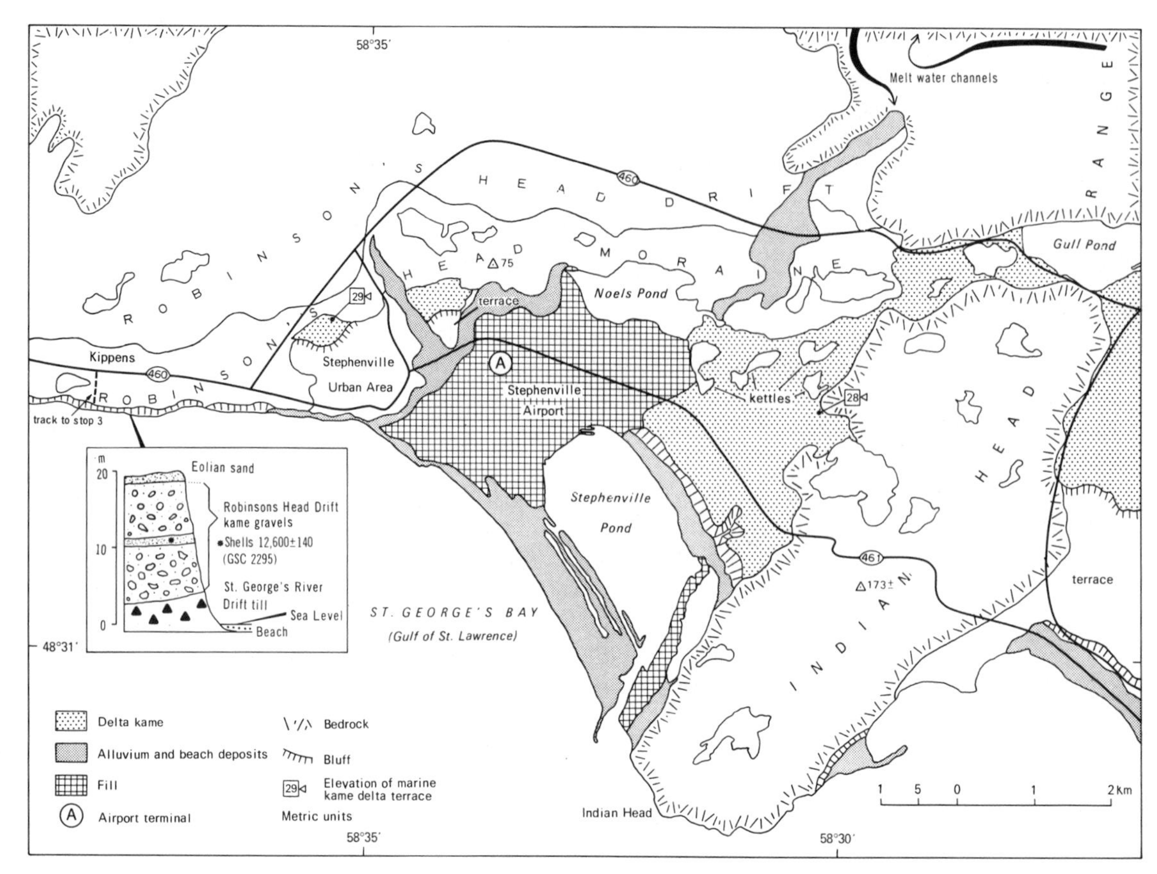

Figure 4. Landforms of Stephenville-Kippens area, with location of access to Kippens cliff section (inset) of Stop 3.

about 13,600 ± 180 years B.P. (GSC-968), and those close to the marine limit are dated at 13,600 ± 110 years B.P. (GSC-2015). Similar dates at elevations approximately 25 ft (7.5 m) and 138 ft (42 m) indicate the penecontemporaneous deposition of a coarsening upward and landward deltaic wedge on deglacial submergence. Regression from a 144-ft (44-m) marine limit was interrupted by a terrace-forming sea-level stand at 89 ft (27 m), correlated with terraces near Stephenville, where it is dated at 12,600 years B.P. (Brookes, 1977a).

Evidence of recent sea-level rise is only poorly seen in marshy bottomland in the creek bed at Stop 2. Better evidence is seen in salt marshes in valleys draining into West Bay, Port au Port Bay. One of these, "Hynes Brook" (Fig. 1) yielded a 3,000-year record of sea-level rise from radiocarbon-dated saltmarsh sediments. These marshes may be reached by branching off Newfoundland 460 onto Newfoundland 463 at Abrahams Cove, and following this to the shore of West Bay (Fig. 1).

Stop 3. This stop is reached by returning on Newfoundland 460 to Kippens (Fig. 1) and branching seaward (south) on Williams Lane. The section shown in the inset in Figure 4 is located about 650 ft (200 m) east of the road's end at the clifftop.

The rolling clifftop profile along this shore intersects the Robinson's Head Moraine, dominated by kame sediments, which in places overlie deep-water stratified marine silts dated at 13,100 ± 180 years B.P. (GSC-4095); in others they overlie the deglacial kame gravels of St. George's River Drift. These complex relations indicate close proximity of ice to the present coast after initial deglaciation about 13,000 years BP, with marine overlap in places, and ice barring the sea in others. The section (inset, Fig. 4) shows a layer of coarse sands with shells intercalated in gravels that underlie the Robinson's Head Moraine surface. Here, initial marine overlap and deep-water sedimentation was barred by remnant coastal ice, which then readvanced into the sea, allowing shallow marine deposition within ice-marginal sediments. Shells from the marine deposits are dated at 12,600 ± 40 years B.P. (GSC-2295) and thus date the moraine. They also date an 89- to 95-ft (27- to 29-m) terrace, often intimately associated with the moraine around St. George's Bay.

To reach Stop 4 (Fig. 1), bypass Stephenville on Newfoundland 460 (east) to join Newfoundland 461, and follow this to the Trans-Canada Highway east of Stephenville Crossing. Follow the highway south for about 24 mi (38 km) to the junction with Newfoundland 404. Turn right on 404, follow across railroad tracks to T-junction. Turn right toward Heatherton, then left on

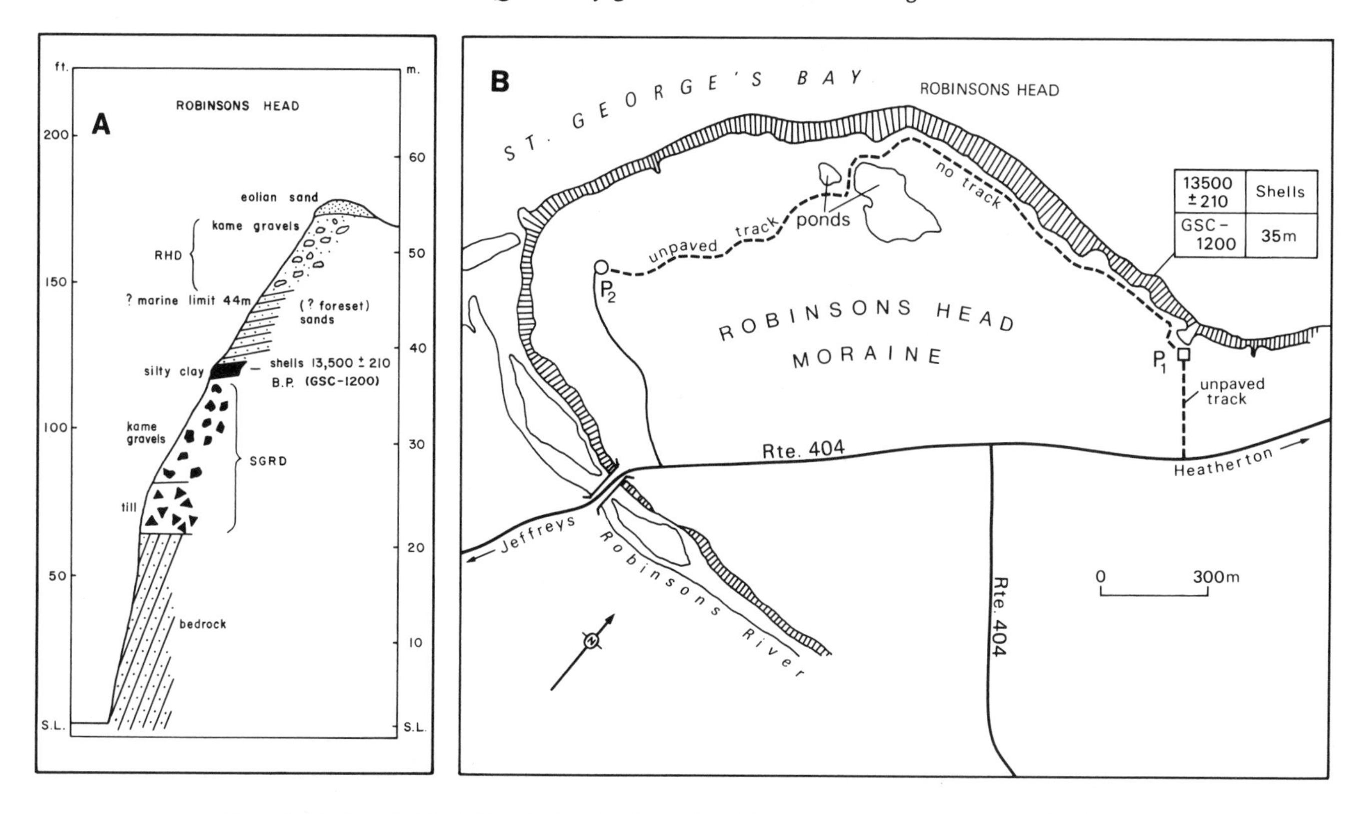

Figure 5. Robinson's Head (Stop 4) showing A, stratigraphic section located on map; and B, landforms, and alternate start and finish points of walk.

an unpaved track to P_1 (Fig. 5) to begin the 1.2-mi (2-km) walk. This is a round-trip walk unless vehicles are driven to point P_2 for pick-up. Buses should not leave pavement at either end; there are no turning places.

Stop 4. A walk along the clifftop at Robinson's Head affords fine views of the prominent Robinson's Head Moraine, also prominent at distant Bank Head, to the northeast, and southwest views over other lobate moraine segments intimately associated with a regressional marine terrace. Careful cliff descents and traverses reveal stratigraphic relations of the complete Late Wisconsinan glacial-marine-glacial sequence at this type section (Fig. 5,A). The section deserves special attention not only because it exposes a complete sequence, but also because basal marine sediments are radiocarbon dated.

Over near-vertically dipping, red, clastic, sedimentary rocks of the Codroy Group (Mississippian), compact, mauve-gray, St. George's River till contains erratics derived from eastward. Loose, sandy, kame gravels overlie the till, indicating deglacial ice stagnation near the coast.

Over the kame gravels at the section located in Figure 5,B, a thin bed of stratified, brown, sandy silt and silty clay contains marine bivalves, at 115 ft (35 m), dated at 13,500 ± 210 years BP (GSC-1200), similar to several other dated shells from comparable sediments around St. George's Bay. As elsewhere, here they represent basal deposition of marine sediments immediately following deglaciation. They are overlain here by seaward-dipping, gravelly sands that rise to 148 ft (45 m). These resemble delta foreset beds. Though they may also be kame sands of the Robinson's Head Moraine above them, assignment to the Bay St. George Delta is preferred here—particularly because they rise to the 148-ft (45-m) marine limit determined on a delta surface 7 mi (11.25 km) to the northeast. In turn, these (?foreset) sands are overlain by coarser, more poorly sorted, structureless, sandy gravels beneath the rolling surface of the moraine. They belong to the laterally variable Robinson's Head Drift, which is dominated by stratified, variably dipping sands and gravels, but which also contains pods of more compact, but still sand-dominated till. Along the cliff the contact between glacial and marine sediments is very irregular and usually lies below marine limit. Ice thus readvanced across deltaic sediments, below the contemporary sea level, plowing them and mixing them with glacial sediments. The view southward from the summit of this headland shows the lobate moraine fronted by a coeval marine terrace at 85 ft (27 m).

From either point P_1 or P_2 indicated in Figure 5, follow Newfoundland 404 westward through Jeffreys to rejoin the Trans-Canada Highway. Turn right and follow this to Newfoundland 405 junction (Fig. 1), turning right toward St. David's for 1.2 mi (2 km). Turn left to Highlands near the railroad tracks and follow road in Figure 6B to Stop 5.

Stop 5. The Highlands coastal section is located and illustrated in Figure 6A, B. The view south along the road is broken by the low, wooded crest of a lobate ridge of the Robinson's

Head Moraine, which is cut by the cliffs. The section near Harbour Head (Fig. 6,A) shows reddish gray, compact Robinson's Head till about 26 ft (8 m) thick, terraced at 92 ft (28 m), overlying truncated foreset sands of Bay St. George Delta. These in turn overlie delta bottomsets, a glaciomarine stony mud (with marine shells dated nearby at 13,420 ± 190 years BP (GSC-598), with St. George's River till over rock at the base. The upper till was again deposited by ice readvancing over a delta to below sea level. The terrace against the moraine flanks shows that this readvance was contemporary with a 92-ft (28-m) sea level, thus indirectly dated at 12,600 years BP by comparison with the Stephenville information.

For Stop 6, return to Trans-Canada Highway. From here an optional tour of Codroy Lowland en route to the Nova Scotia ferry (from Port aux Basques) can be taken, using Brookes (1977b) and Brookes and others (1982) as guides. Turn right for this tour. Otherwise, turn left and follow to junction of Newfoundland 461 to St. George's. In St. George's, locate Turf Point northeast of government wharf.

Stop 6. This is a terrestrial, ombrotrophic bog eroded by rising sea level, showing roots of spruce and fir projecting from fibrous *Sphagnum*-rich peat layers. It could not have begun to develop until sea-level had fallen below its base at modern mean tide level. A radiocarbon date of 7,340 ± 220 years BP (GSC-1145) was obtained on basal organic-rich sands, but a more recent date of 9,350 ± years BP (WAT-883), presumably on lower sands, supersedes it as a minimum age for sea level falling past its present level.

Return to Stephenville on Newfoundland 461, via Stephenville Crossing, taking the southern road over Indian Head Range (Figs. 1, 4). Beside this road, striae on Precambrian gneiss record westward, maximal Late Wisconsinan ice flow, as well as south-southwestward deglacial (or deflected basal) flow toward the sea.

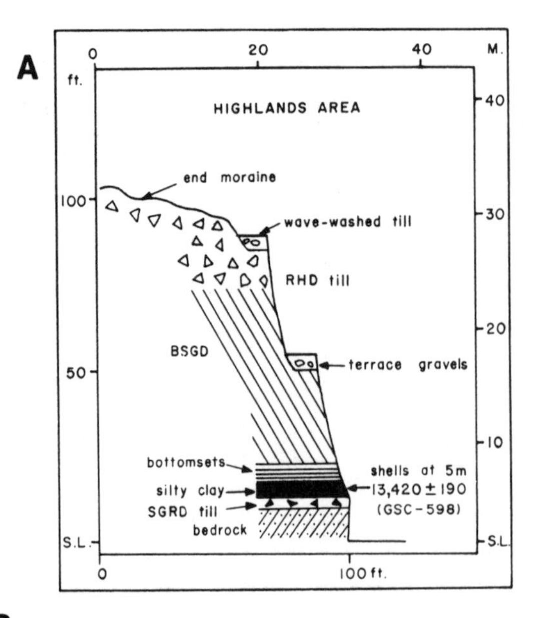

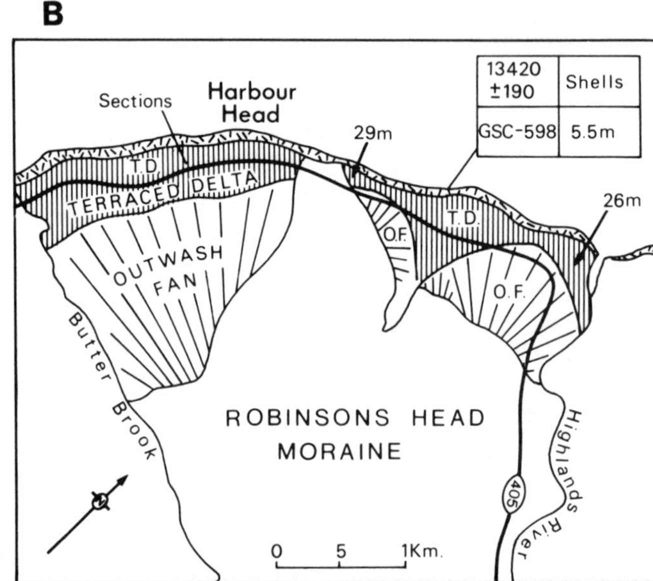

Figure 6. Landforms and section in Quaternary sediments, Highlands (Stop 5).

REFERENCES CITED

Brookes, I. A., 1969, Late-glacial marine overlap in southwestern Newfoundland: Canadian Journal of Earth Sciences, v. 6, p. 1397–1404.

—— , 1970a, The glaciation of southwestern Newfoundland [Ph.D. thesis]: Montreal, McGill University, 208 p.

—— , 1970b, New evidence of an independent Wisconsin-age ice cap over Newfoundland: Canadian Journal of Earth Sciences, v. 7, p. 1374–1382.

—— , 1974, Late-Wisconsin glaciation of southwestern Newfoundland with special reference to the Stephenville map-area: Geological Survey of Canada Paper 74-30, 33 p.

—— , 1977a, Radiocarbon age of Robinson's Head moraine, west Newfoundland and its significance for postglacial sea-level changes: Canadian Journal of Earth Sciences, v. 14, p. 2121–2126.

—— , 1977b, Geomorphology and Quaternary geology of Codroy Lowland and adjacent plateaus, southwest Newfoundland: Canadian Journal of Earth Sciences, v. 14, p. 2101–2120.

Brookes, I. A., McAndrews, J. H., and von Bitter, P. H., 1982, Quaternary interglacial and associated deposits in southwest Newfoundland: Canadian Journal of Earth Sciences, v. 19, p. 410–423.

Brookes, I. A., Scott, D. B., and McAndrews, J. H., 1985, Post-glacial relative sea-level change, Port au Port area, west Newfoundland: Canadian Journal of Earth Sciences, v. 22, p. 1039–1047.

Flint, R. F., 1940, Late Quaternary changes of level in western and southern Newfoundland: Geological Society of America Bulletin, v. 51, p. 1757–1780.

Grant, D. R., 1977, Altitudinal weathering zones and glacial limits in western Newfoundland, with particular reference to Gros Morne National Park: Geological Survey of Canada Paper 77-1A, p. 455–463.

MacClintock, P., and Twenhofel, W. H., 1940, Wisconsin glaciation of Newfoundland: Geological Society of America Bulletin, v. 51, p. 1729–1756.

The Bay of Islands ophiolite; A cross section through Paleozoic crust and mantle in western Newfoundland

J. Malpas, Department of Earth Sciences, Memorial University of Newfoundland, St. John's, Newfoundland A1B 3X5, Canada

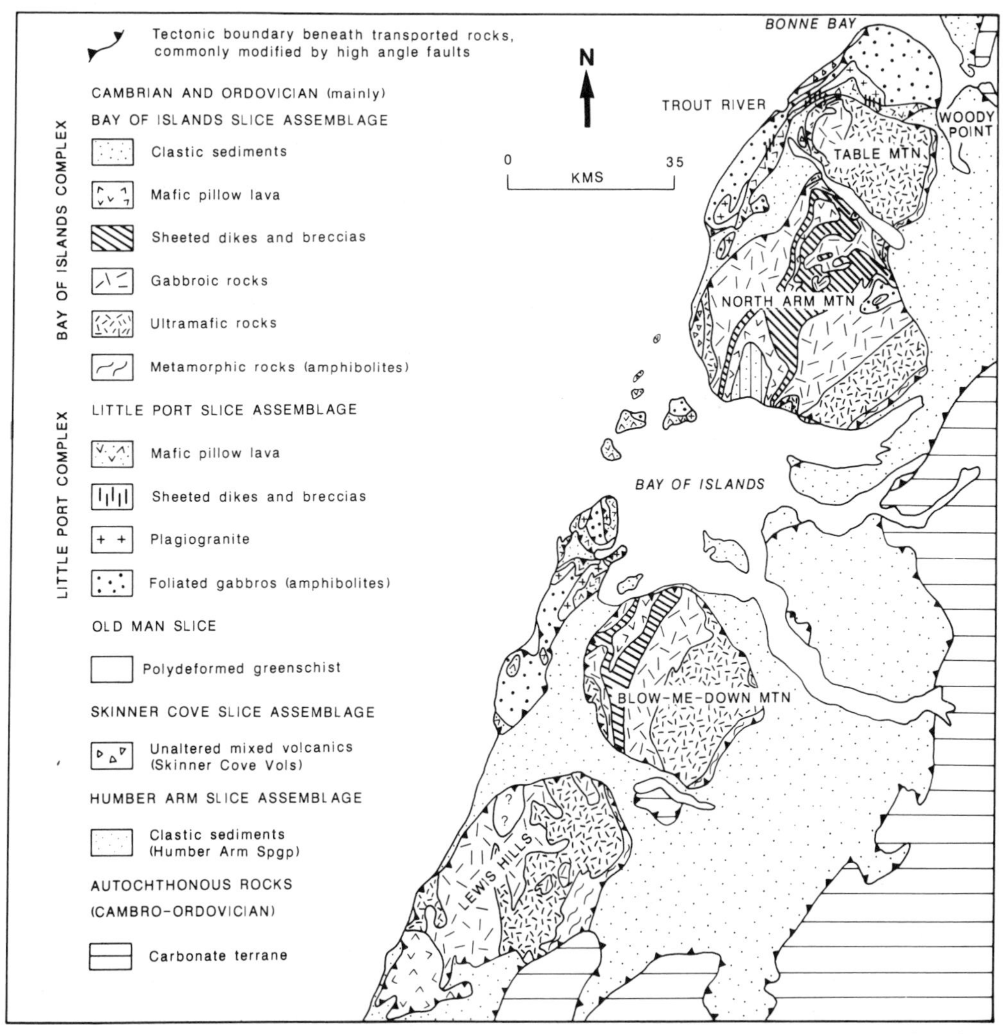

Figure 1. General geology of the Bay of Islands region. Stops 1 and 2 are situated on the northern side of Table Mountain.

LOCATION

The Bay of Islands Complex forms a discontinuous northeasterly trending belt of mafic-ultramafic massifs approximately 60 mi (100 km) long and 15 mi (25 km) wide on the west coast of Newfoundland (Fig. 1). The northernmost of the massifs, Table Mountain, lies within the Gros Morne National Park and can be reached by motor vehicle from the main Northern Peninsula highway (Newfoundland 430) and the Bonne Bay road (Newfoundland 431). Visitors should turn left onto Newfoundland 431 at Woody Point and travel toward Trout River. Exposures described in this guide will be found on the hills to the south of this road between a distance of 2 mi (3 km) and 6 mi (10 km) from the turn-off (Fig. 2). Local access is by foot since many outcrops are on the mountain top, which is, however, flat and easily traversed. Paths of easiest access to specific locations are shown in Figure 3. Park rangers are available to lead field excursions and permits must be obtained if specimens are to be taken. Permits may be applied for by writing to Parks Canada Atlantic Regional Office, D.O.E. Historic Properties, Upper Water Street, Halifax, Nova Scotia, B3J 1S9. It is also advisable to inform Park Headquarters in Rocky Harbour if an extended stay on the mountain is planned. This guide describes two major outcrop sections separated by a distance of approximately 2.5 mi (4 km) driving. Vehicles should be parked on the road side and the traverses completed by foot to each section.

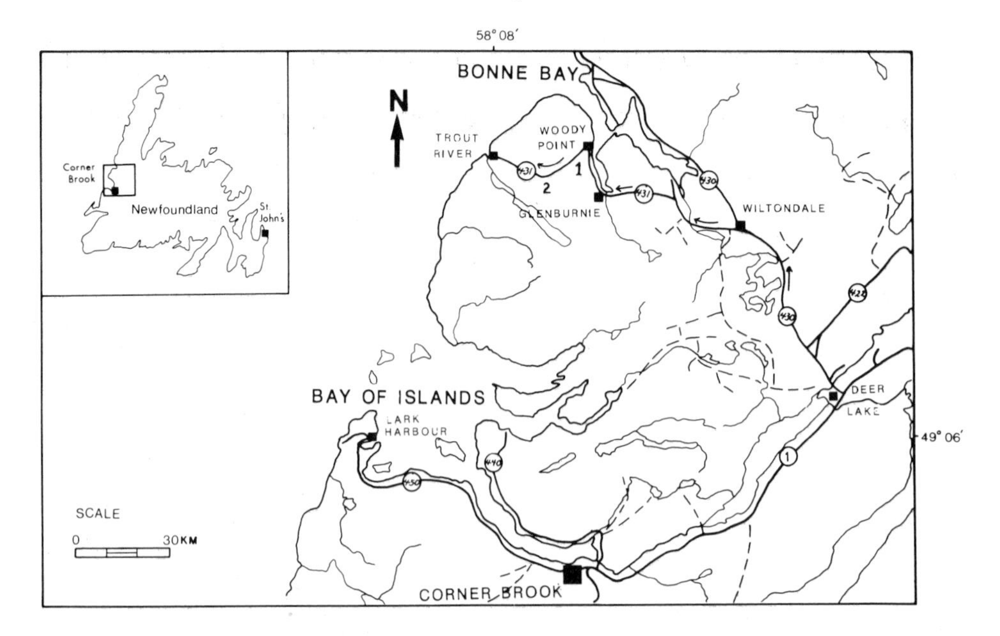

Figure 2. Approach to Table Mountain is along Newfoundland 430 from Deer Lake and then Newfoundland 431 from Wiltondale.

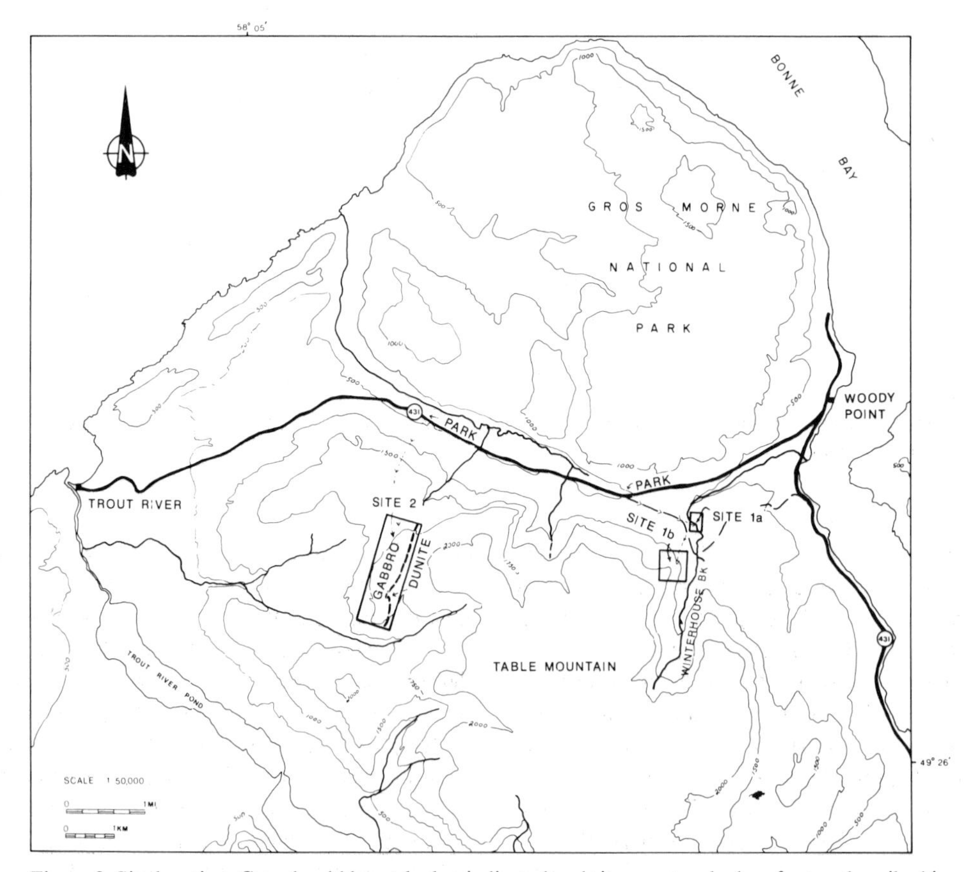

Figure 3. Site location. Cars should be parked as indicated and sites approached on foot as described in text. Gabbro-dunite boundary is indicated at Stop 2.

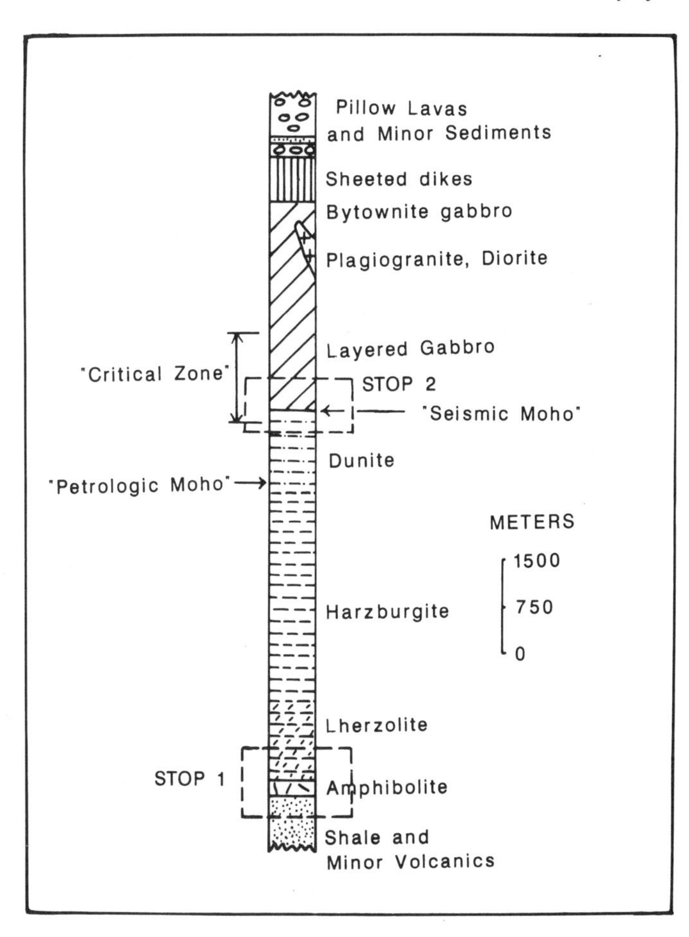

Figure 4. Vertical composite section through the Bay of Islands ophiolite suite. Only ultramafic rocks and gabbros are seen on Table Mountain; the other rock types are seen on the more southerly massifs. Positions of Stops 1 and 2 in the stratigraphy are shown.

Figure 5. Faulted relationship between lherzolites (right) and shales (left) with rodingite alteration along the contact at Stop 1a, Winterhouse Brook.

SIGNIFICANCE

The Bay of Islands Complex forms the highest slice of a series of thrust slices (the Humber Arm Allochthon) emplaced on the continental margin of North America in Middle Ordovician times (Williams, 1973). The complex is a well preserved ophiolite suite comprising rocks of the oceanic mantle and crust (Fig. 4). On two of the four major massifs making up the complex, the higher stratigraphic parts of the suite, sheeted dykes and pillow lavas, are preserved but are relatively inaccessible. On Table Mountain, however, peridotites of the mantle section and gabbros of oceanic layer three are best exhibited and the transition between the two, the Paleozoic Mohorovicic Discontinuity, can be walked upon.

SITE INFORMATION

All of the transported slices of the Allochthon comprise distinct assemblages and are stacked in a consistent order. Facies considerations, combined with structural analyses, imply that the ophiolites were the most easterly of these assemblages when formed and are therefore the farthest travelled. Recent U/Pb determinations produce an age of approximately 485 Ma for the formation of the ophiolite sequence so that it was transported and emplaced soon after. Since emplacement, the ophiolite complex has lain outside the effective realm of mid-Paleozoic orogenesis that was so widespread and intense in central Newfoundland. The sequence of the ophiolite units is disposed in a synclinal form on Table Mountain with a northeast-trending subhorizontal axis and moderately dipping limbs. The tectonic base of the massif is subhorizontal so that the units are truncated at depth in much the same way as they are truncated at the top by the present horizontal topographic surface. Strike-slip faults that bound Table Mountain to the north (e.g., at Winterhouse Brook, Figs. 1 and 3) and south are tear faults that mark the sides of the northwestward transported massif and do not continue westwards to the coast.

Site 1 is the contact between ultramafic rocks of the ophiolite and adjacent shales, with hydrothermal development of rodingite assemblages. Harzburgites of the Table Mountain Massif also outcrop in Winterhouse Brook.

Travel along the Trout River road for 2 mi (3 km). Park at the Y-junction with the old road on left. Leave vehicle and walk back along the old road for approximately 1 mi (1.5 km) to where the road is washed out by Winterhouse Brook. Walk down the Brook for 1.6 ft (500 m) to Stop 1a (Fig. 3). After visiting Stop 1a, Stop 1b is reached by walking back upstream to the base of the steep slopes of Table Mountain. Stop 1b is on the slopes on the west side of the brook (Fig. 3).

At Stop 1a, overlooking Winterhouse Brook, serpentinized peridotites are in fault contact with shales of the Humber Arm Supergroup, part of the lowest thrust slice in the allochthon (Fig. 5). The complete serpentinisation of the ultramafic rocks makes their identification difficult, although lherzolites can be found nearby. Unlike the basal contact of the ophiolite slice, no dynamothermal aureole of amphibolitic or greenschist nature is

Figure 6. Folded tectonite fabric in harzburgites of the mantle sequence at Stop 1b, Winterhouse Canyon.

Figure 7. Layered norites and anorthosites in the Critical Zone on Table Mountain at Stop 2.

seen here although many boulders of these lithologies are seen in the stream bed and are derived from outcrops considerably further upstream (Malpas, 1979). The ultramafic rocks at the faulted contact have been modified by later hydrothermal solutions creating calcium-rich prehnite, calcite, xonotlite, and wollastonite, essentially a rodingite assemblage (Smith, 1958). The contact consists of a resistant, white-weathering layer 12 to 18 in (30 to 45 cm) thick, underlain by green to black shale that grades into red shale. A sandstone lens some 10 ft (3 m) from the contact shows no evidence of thermal metamorphism. The colour changes in the shale and the presence of secondary low temperature silicate or sulphide minerals along the contact zone are also ascribed to the local hydrothermal activity. It cannot be determined whether the hydrothermal metasomatism took place during a late stage of emplacement of the ophiolite, or during later faulting. The only source of calcium existing in the ultramafic rocks is in the small amounts of clinopyroxene. It is therefore more likely that much of the necessary calcium was derived from the sediments rather than from the lherzolites during serpentinization.

At Stop 1b, harzburgite of the mantle tectonic sequence is well exposed. This type of rock forms most of the ultramafic component of the ophiolite and consists of approximately 60% olivine, 30% enstatite, 5% chrome-diopside, and 5% accessory minerals that include a chrome-rich spinel. The olivine averages Fo_{91} in composition with very little deviation from the mean; the enstatite averages Fs_9 in composition. The harzburgites are thought to represent depleted upper mantle produced by the partial melting of more aluminous upper mantle peridotite and subsequent removal of the basaltic liquid (Malpas, 1978).

The harzburgite is characterized by a banded appearance caused mainly by a variation in the enstatite/olivine ratio (Fig. 5). The bands have gradational contacts and range in thickness from 0.5 in (1 cm) or less to 300 ft (100 m). Maximum distances over which the individual bands have been traced are of the order of 1.600 ft (500 m) and no distinct stratigraphy can be assessed. The bands are the result of deformation of veins and dikes of pyroxenite and dunite by plastic flow at high temperatures and high strain rates in the mantle during convective flow and are a manifestation of the mantle tectonite fabric. The banding is cut locally by a number of less deformed veins and dikes that range from olivine pyroxenite to dunite in composition. Pyroxenite veins dominate in this locality, which is near the base of the peridotite sequence. Although best seen crosscutting the mineral layering, some veins also run parallel to the mineral layering for some distance before branching or linking laterally with similarly oriented veins. In this mode they virtually form a part of the banding. Some pyroxenite and dunite veins are folded into open folds (Fig. 5) together with the host peridotite. It is thought that the vein rocks represent early fractionation products of a basic magma produced by the fusion of mantle material. O'Hara (1968) showed that olivine is the first mineral to crystallize at pressures >5 kb from basaltic melts and is succeeded by orthopyroxene and clinopyroxene respectively. Dunitic veins in the peridotites thus represent olivine fractionation from a melt, probably at around 10 kb and pyroxenites result from the coprecipitation of olivine, orthopyroxene, and clinopyroxene. Reequilibration of the pyroxenes at lower pressures and temperatures resulted in exsolution textures.

The features displayed in the rocks at these outcrops—including the pronounced banding and foliation, recrystallization, and isoclinal folding of layers and veins—represent deformation of material in the upper mantle and are comparable to features described first by Nicolas and others (1971) for other ophiolites. Progressive stages of deformation produce the rotation and stretching of the veins with accompanying disaggregation of their minerals so that a banded rock is finally developed.

In order to get to Stop 2, return to the main Trout River

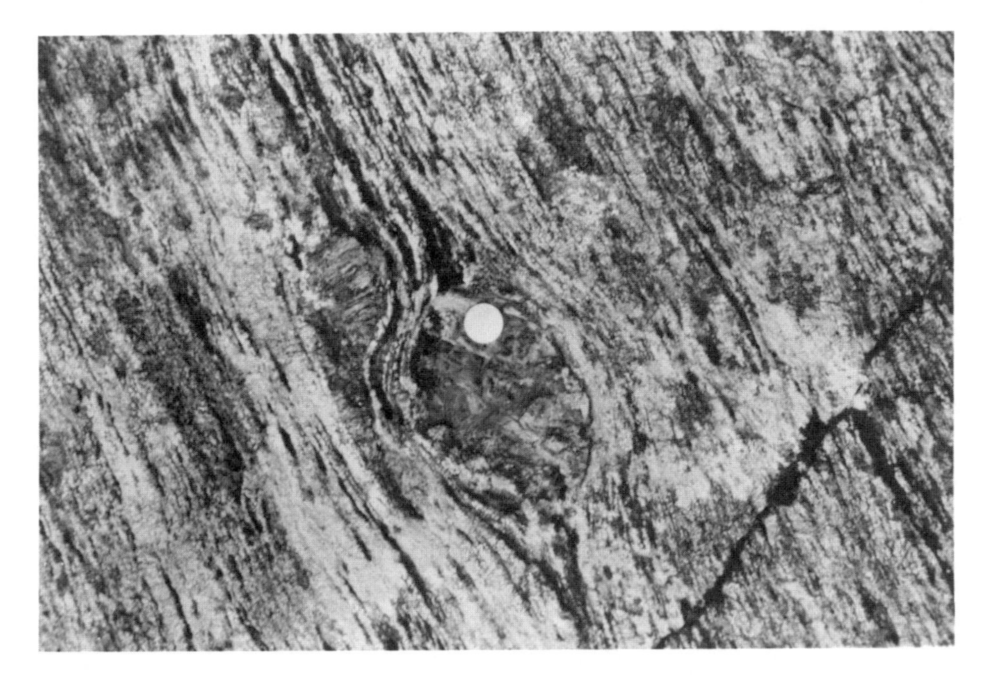

Figure 8. Augen of clinopyroxene in mylonitised gabbro, common on the west side of Stop 2 trending parallel to the ridge. Scale is 10-cent coin.

road and travel westwards towards Trout River for 2.5 mi (4 km). Park at the start of the "Green Gardens" Trail. Leave the vehicle and walk up the hill to the south to the low crest. Upon reaching the crest walk approximately 0.6 mi (1 km) SSE to a low ridge, as shown in Figure 3. Site 2 consists of rocks exposed along the length and breadth of this ridge. Stop 2 is the "Critical Zone" or the transition between peridotites and gabbro. The ultramafic rocks viewed at Stop 1 are represented by dunites higher in the section. At Stop 2 the ultramafic rocks are feldspathic. The "Critical Zone" marks the transition between oceanic mantle and Layer III of the oceanic crust and can be regarded as the Lower Paleozoic Moho.

About 1,600 ft (500 m) to the east of the prominent ridge of interbanded rocks, the harzburgite of the mantle section grade over a short distance to dunitic rocks, locally chromite bearing, which represent a highly deformed cumulate sequence (Fig. 6. As such they have a different origin than the rocks of the mantle immediately beneath them in the stratigraphy. These dunites become more feldspathic higher in the transition zone and on the ridge are interbanded with norite, troctolite, and anorthosite over an interval of about 660 ft (200 m; Fig. 7). Layers in this zone range from inches (centimetres) to several feet (metres) thick and in a few places original cumulate features can be observed. However, for the most part the rocks are affected by syn- and post-accumulation deformation, evidence for which ranges from the recognition of large scale (0.62 ft; 1 km amplitude) isoclinal folds that structurally transpose much of the section, to small scale folds and the development of penetrative schistosities. Large scale ductile simple shear zones are marked by mylonites in which plagioclase has been completely recrystallized while pyroxene remains as augen (Fig. 8). Some layers can be followed around fold hinges as clear marker bands but, for the most part, the hinges of folds have been removed by shearing and layers appear more lensoid than continuous. The complexity of the folding can be ascertained from the variable orientation of the L_1 lineation, which has been clearly affected by a later fold episode. Deformation associated with this later episode also involved the development of lower temperature, greenschist facies, and shear zones (Girardeau and Nicolas, 1981).

The rock types in the "Critical Zone" may be classified according to the relative proportions of olivine (Fo_{85-87}), plagioclase (An_{75}) and clinopyroxene (diopsidic-augite). The sequence of mineral crystallization wehre it is observed in preserved cumulates is olivine-plagioclase-clinopyroxene. Chemical variation within the minerals of the zone seems relatively limited, probably as a result of the large degree of recrystallization, and no original regular cryptic variation has yet been identified. In many places hydrogarnet has replaced plagioclase, its composition lying towards the tri-calcium hexahydrate end member.

Undeformed veinlets of coarse clinopyroxenite occur towards the base of the "Critical Zone," both concordant with, and crosscutting, the foliation. In these, the clinopyroxene is almost pure diopside, $Ca_{47}Mg_{49}Fe_4$, and associated with a small amount of serpentinized olivine. Higher in the zone gabbroic veins and veins and dikes of anthoslite are more common. Some of the anorthosite veins have been rotated into the foliation of the host gabbros so that they appear as layers. Branching and crosscutting features and their sharp margins easily identify them, however, as intrusive features.

The deformation affecting the "Critical Zone" is considered to be in part contemporaneous with the development of the fabrics in the mantle rocks and therefore related to ocean-floor spreading. This ductile deformation has obliterated to a large extent the original plutonic geometry of the complex and it occurred contemporaneously with igneous activity as witnessed by the presence of deformed and undeformed veins and dikes. The deformation has caused the transposition of cumulate fabric elements parallel to mantle flow fabrics in the underlying harzburgites and in the later stages was characterized by displacements of discrete sections of the complex along ductile simple shear zones, becoming progressively cooler, and probably parallel to the horizontal limb of a mantle convection cell.

REFERENCES CITED

Girardeau, J., and Nicolas, A., 1981, The structures of two ophiolite massifs, Bay of Islands, Newfoundland; A model for the oceanic crust and upper mantle: Tectonophysics, v. 77, p. 1–34.

Malpas, J., 1978, Magma generation in the upper mantle; Field evidence from ophiolite suites and application to the generation of oceanic lithosphere: Philosophical Transactions of the Royal Society of London, v. A288, p. 527–546.

—— , 1979, The dynamothermal aureole of the Bay of Islands ophiolite suite: Canadian Journal of Earth Sciences, v. 16, p. 2086–2101.

Nicolas, A., Bouchez, J. L., Boudier, F., Mercier, J. C., 1971, Textures, structures, and fabrics due to solid state flow in some European lherzolites: Tectonophysics, v. 12, p. 55–86.

O'Hara, M. J., 1968, The bearing of phase equilibria studies in synthetic and natural systems on the evolution of basic and ultrabasic rocks: Earth Science Reviews, v. 4, p. 69–133.

Smith, C. H., 1958, Bay of Islands igneous complex, western Newfoundland: Geological Survey of Canada Memoir 290, 132 p.

Williams, H., 1973, Bay of Islands map area, Newfoundland: Geological Survey of Canada Paper 72-34, 7 p.

Geological Society of America Centennial Field Guide—Northeastern Section, 1987

Ophiolitic mélange along the Baie Verte Line, Coachman's Harbour, Newfoundland

James Hibbard, *Department of Geological Sciences, Cornell University, Ithaca, New York 14853*

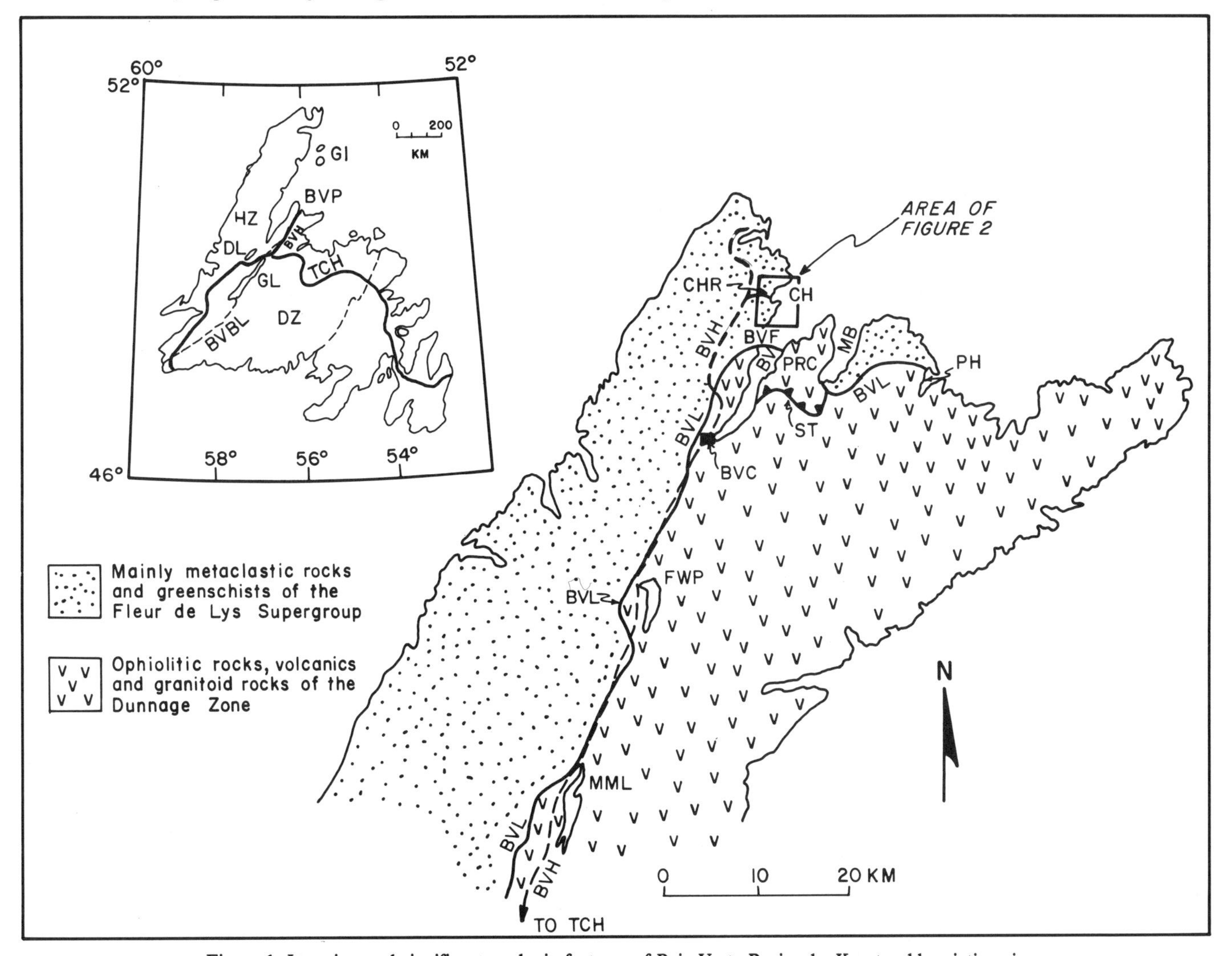

Figure 1. Location and significant geologic features of Baie Verte Peninsula. Key to abbreviations in figure: BV, Baie Verte; BVC, Baie Verte community; BVF, Baie Verte Flexure; BVH, Baie Verte Highway; BVL, Baie Verte Line; CH, Coachman's Harbour; CHR, Coachman's Harbour access road; FWP, Flat Water Pond; MB, Ming's Bight; MML, MicMac Lake; PH, Pacquet Harbour; PRC, Point Rousse Complex; ST, Scrape Thrust. Key to abbreviations in inset: BVBL, Baie Verte–Brompton Line; BVP, Baie Verte Peninsula; BVH, Baie Verte Highway; DL, Deer Lake; DZ, Dunnage Zone; G1, Grey Islands; GL, Grand Lake; HZ, Humber Zone; TCH, Trans Canada Highway.

LOCATION

Coachman's Harbour is located along the northwest coast of Baie Verte, on the Baie Verte Peninsula, Newfoundland (Fig. 1). The peninsula is accessible from the Trans-Canada Highway along Newfoundland 72, the Baie Verte highway. Coachman's Harbour is located at the end of a branch road leading east off the Baie Verte highway, approximately 15 mi (25 km) north of the community of Baie Verte (Fig. 1).

Ophiolitic mélanges are best exposed at two coastal localities on opposite sides of Coachman's Harbour, one near Flat Point and the other at the town picnic site (Fig. 2). For the locality near Flat Point, take the first left off the branch road, near the bottom of the hill leading to the harbour. Proceed for approximately 0.5 mi (0.8 km) to a lane that branches to the right; turn down the lane and park. The locality is immediately to the east-

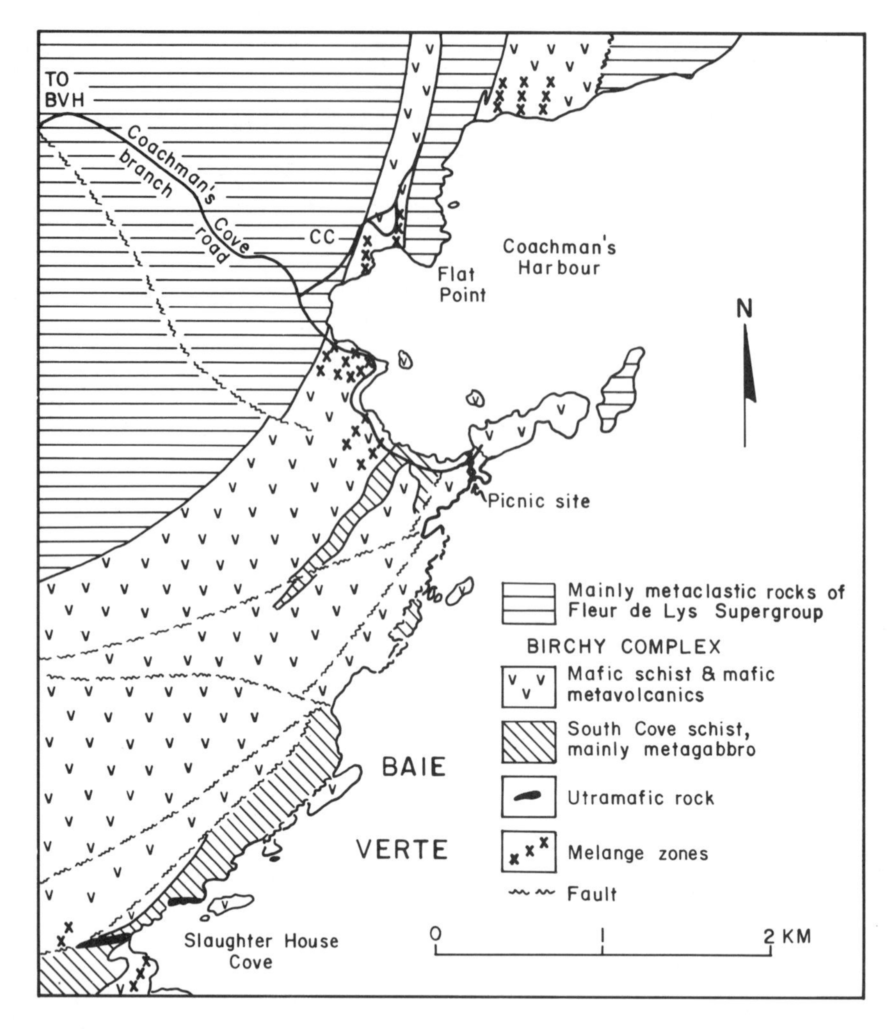

Figure 2. General geology of Coachman's Harbour area (from Kennedy, 1971; Bursnall, 1975; Hibbard, 1983). BVH, Baie Verte Highway; CC, Coachman's Cove community.

southeast along the coast. The picnic site locality is located at the end of the Coachman's Harbour branch road, near the southeast side of South Cove. Park approximately 1,000 ft (300 m) from the end of the road and proceed southeastward on foot to the picnic site; the locality is along the Baie Verte coast, directly below the picnic grounds. No permission is necessary to visit either locality.

SIGNIFICANCE

Ophiolitic mélanges help define the Baie Verte Line; the line is the northern section of the Baie Verte–Brompton Line, a narrow, steeply dipping structural zone that marks an Early Paleozoic continent-ocean interface along the entire length of the Canadian Appalachians (Williams and St. Julien, 1982). It separates North American continental rocks of the Humber Zone from ophiolitic and volcanic rocks of the Dunnage Zone. The ophiolitic mélanges are infolded with multideformed and metamorphosed continental rise and slope rocks at the eastern edge of the ancient continental margin. Since the mélanges are deformed and metamorphosed with the surrounding rocks, they are interpreted to be structural zones marking the early tectonic disruption of the rise and slope deposits. This event has been related to the imbrication and obduction of Taconic allochthonous ophiolites, now located in western Newfoundland, onto the ancient continental margin (Williams, 1977; Williams and others, 1977; Hibbard, 1982, 1983). The ophiolitic mélanges provide critical evidence that that Baie Verte Line most likely represents the root zone for the Taconic allochthonous ophiolites of western Newfoundland (Williams and others, 1977; Williams, 1977).

SITE INFORMATION

Regional Setting

Ophiolitic mélanges at Coachman's Harbour are a part of the recently redefined Fleur de Lys Supergroup (Hibbard, 1983).

The supergroup consists mainly of metaclastic schists with interlayered marble, amphibolite, and greenschist; these components have been multideformed and metamorphosed in the upper greenschist to lower amphibolite facies. The unit forms an extensive orthotectonic belt in western Newfoundland, with a strike length of more than 180 mi (300 km). It extends from the northerly Grey Islands, southward across the Baie Verte Peninsula and into the Deer Lake–Grand Lake area (Fig. 1). This orthotectonic belt forms the deformed eastern margin of the Humber Zone, and abuts against lesser deformed mafic and ultramafic assemblages of the Dunnage Zone, to the east. The boundary between these two zones is the Baie Verte Line, a steep polygenetic structural zone delineated by ophiolitic rocks along much of its length (Fig. 1).

It has long been recognized that parts of the supergroup are the tectonized lithologic equivalents of the relatively undeformed Hadrynian to Early Paleozoic sequence of the western part of the Humber Zone (Neale and Nash, 1963; Williams and Stevens, 1974). This sequence represents a stable carbonate platform along the ancient North American continental margin and overlying allochthonous rocks derived from the east (Williams and Stevens, 1974). Protoliths of the Fleur de Lys Supergroup have been interpreted as the slope-rise prism sequence along this margin (Neale and Nash, 1963; Williams and Stevens, 1974; Bursnall and deWit, 1975; Hibbard, 1983). However, the younger age limit and the age and mechanism of tectonism of the orthotectonic belt have been a topics of major controversy.

The orthotectonic belt contrasts sharply with less deformed Lower Ordovician ophiolitic and mafic volcanic sequences of the adjacent Dunnage Zone. Thus, some workers viewed the tectonism of the Fleur de Lys rocks to be older than the Early Ordovician age of the relatively pristine ophiolitic rocks of the nearby Dunnage Zone (Church, 1969; Dewey and Bird, 1971; Kidd, 1977; Kennedy, 1975). This opinion was further tempered by the recognition of a pre-Ordovician event in the Fleur de Lys correlative Dalradian rocks of Ireland.

Other workers familiar with the regional geology were uneasy with this interpretation because evidence of this pre-Ordovician tectonism is lacking in the bordering western part of the Humber Zone. Bursnall and deWit (1975) first voiced this opinion; they correlated both rocks and structures of the belt with lesser deformed rocks and less intensely developed structures of the flanking Humber and Dunnage Zones. They viewed the deformation and metamorphism of the orthotectonic belt to be of probable Early to Middle Ordovician age and inferred that the tectonism was related to the emplacement of Taconic allochthons in western Newfoundland.

Recognition of ophiolitic mélanges in the supergroup has eased this controvesy and has led to the general consensus that the Fleur de Lys rocks are temporal equivalents of less deformed flanking rocks to both the east and west. The mélanges are compelling evidence for relating the early deformation and metamorphism of the orthotectonic belt to the Ordovician emplacement of ophiolitic rocks onto the Humber Zone carbonate bank.

Relationship of the mélanges to the Baie Verte Line

Ophiolitic mélange is present in many places along the western side of the Baie Verte Line; however, it is best developed north of Baie Verte. South of Baie Verte, the Baie Verte Line is sharply defined by a steeply dipping north-northeast–trending polygenetic fault system that abruptly juxtaposes the Fleur de Lys Supergroup to the west against easterly ophiolitic and mafic volcanic rocks. Large ultramafic bodies are localized along this braided fault system. In part, the fault system is a relatively late feature in the regional tectonic history, as it truncates the east side of a postkinematic Devonian granitoid batholith that is within the Fleur de Lys terrane. Ophiolitic mélange is rare in this area, and, where present, is closely associated with the late fault system. This suggests that mélange may have been more extensive but has been structurally deleted by the later faults along much of this part of the line.

North of the community of Baie Verte, the line fans into a structural zone composed mainly of a heterogeneously deformed imbricate stack of ophiolitic rocks (Bursnall, 1975). The westernmost of these slices, the Birchy Complex (Williams and others, 1977; Hibbard, 1983), consists of greenschist, amphibolite, metagabbro, ultramafic rock, and mélange, which collectively constitute remnants of a dismembered ophiolite (Bursnall, 1975). Since greenschist of the complex is conformably interlayered with typical Fleur de Lys metaclastic rocks and displays all of the deformation and metamorphism of the unit, the complex is considered to be an integral part of the supergroup. Ophiolitic mélange zones as well as slide zones with localized ultramafic bodies are well developed in the complex. (The term "slide" is herein used as defined by Hutton [1979], i.e., essentially a pre- or synmetamorphic fault.) Further imbrication of the Birchy Complex with lesser deformed, more coherent ophiolites to the east took place during later phases of the regional deformation (Bursnall, 1975). Thus in the area north of Baie Verte, the line is marked by a complex zone of ophiolitic rocks that record the nascent stages of the structure, represented by ophiolitic mélange and slide zones, and is accentuated by imbrication along later faults.

East of Baie Verte, the line changes orientation, to an east-west trend, in the Ming's Bight–Pacquet Harbour area. This regional arcuate trend around Baie Verte has been termed the Baie Verte Flexure (Hibbard, 1982, 1983). On the east limb of the flexure the Baie Verte Line is a discrete broad zone of ophiolitic mélange and early slide zones that has not been accentuated by later faulting (Hibbard, 1982, 1983); this zone has been intruded by granite and subjected to the same late tectonism as nearby Fleur de Lys rocks. The eastward continuity of the line from Baie Verte around the flexure to Ming's Bight is apparently blocked by the ophiolitic Point Rousse Complex of the Dunnage Zone. The Point Rousse Complex is interpreted to be emplaced late in the regional tectonic scheme and to structurally overlie this section of the line (Hibbard, 1982).

In summary, ophiolitic mélange and associated early slides form a broad zone of early disruption in the Fleur de Lys Supergroup at its eastern margin. This zone represents the nascent stages of the Baie Verte Line. South of Baie Verte, along the west limb of the Baie Verte Flexure, this zone was accentuated by later faulting that now define the line; on the east limb of the flexure, the line is a broad zone, made more diffuse by intrusion and late deformation.

Description of ophiolitic mélange at Coachman's Harbour

The drive from the Trans-Canada Highway to Coachman's Harbour affords some interesting views of the geology along the Baie Verte Line. From Mic Mac Lake to the area north of Baie Verte, the Baie Verte highway crosses the Baie Verte Line in numerous places (Fig. 1). Roadcuts expose greenschist of the Birchy Complex, ultramafic bodies along the line, and mafic assemblages of the Dunnage Zone. One spectacular outcrop along the highway, at the outlet of Flatwater Pond, exposes a bright green altered ultramafic rock, locally termed "virginite" (quartz-magnesite-fuchsite rock). The virginite is localized along a fault zone in the Flatwater ultramafic body, one of the larger serpentinite bodies defining the Baie Verte Line. Please do not sample this outcrop without permission of the Rock Shop in Springdale, Newfoundland, whose owners have mineral rights to the local area. North of Baie Verte, the highway traverses mainly rocks of the Fleur de Lys Supergroup; outcrops along the highway expose rocks of the Birchy Complex and typical metaclastics of the supergroup.

The Coachman's Harbour access road negotiates a steep hill leading to the harbour; the hill provides a good overview of the local geology. Rocks around the cove are mainly part of the Birchy Complex (Fig. 2). Immediately offshore, to the east, the rusty colored Tin Pot Islands are composed of relatively undeformed ultramafic rocks that are part of the Point Rousse Complex; the complex forms the large peninsula in the distance.

At least eleven localities of ophiolitic mélange are known from around the harbour (Fig. 2); it is obvious that some repetition through folding has occurred, but there is uncertainty as to how many original zones are represented here. The best exposures of the mélanges are located near Flat Point, which is not visible from the access road, and at the town picnic site, located on the peninsula in the foreground of the Tin Pot Islands.

Stop 1. Flat Point locality. Along the northern coast of North Cove, between the community of Coachman's Cove and Flat Point, ophiolitic mélange is infolded with greenschist and minor psammitic schist of the Birchy Complex (Fig. 2). Eastward toward Flat Point, the complex is in conformable contact with Fleur de Lys metaclastic rocks of the Flat Point Formation (Kennedy, 1971; Bursnall, 1975).

The mélange zone in this section is approximately 100 ft (30 m) wide and consists of mainly blackish-gray graphitic pelite containing conspicuous, bright emerald green lenses of recrystallized ultramafic rock and diffuse layers of buff psammitic schist. The recrystallized ultramafic rock is composed chiefly of coarse-grained actinolite that is pale to emerald green, with blades up to 1.6 in (4 cm) long, set in a fine-grained fuchsite-carbonate matrix. These pods range in size from a few inches in diameter to lenses 10 by 3 ft (3 by 1 m). The mineralogy of these rocks indicates an ultramafic protolith (Kennedy, 1971). The psammitic lenses are generally composed of medium-grained albite-quartz-biotite ± muscovite schists that appear to form irregular layers within the pelitic matrix. This particular zone is also characterized by extensive quartz veining.

The mélange displays all of the intense deformation evident in the surrounding Birchy and Flat Point schists. The dominant schisosity in the zone is the regional second-phase fabric, and it is locally associated with second-phase folds that are tight to isoclinal. As in all of the Fleur de Lys terrane, first-phase structures are rare and manifested mainly as microscopic foliations in the hinge areas of second-phase folds. The whole area around Flat Point has been affected by late-phase regional folding that is especially well developed in the mélange.

To the east of the mélange zone, the Birchy schists are intercalated with thin layers of calcareous muscovite–albite pelite and impure dolomitic marble near their contact with the Flat Point Formation. The formation here is composed of medium-layered albite-quartz-muscovite psammite and semipelite; locally, massive muscovite-chlorite-albite semipelite occurs in the unit. The Flat Point Formation exhibits essentially the same structures as seen in the Birchy Complex at this locality. At the eastern part of the Flat Point, small, highly strained pods of actinolite-fuchsite schist may demarcate a pre-D_2 slide zone.

Originally, the Flat Point Formation was interpreted to overlie the Birchy Complex (Kennedy, 1971; Bursnall, 1975). However, primary features indicating younging directions are scarce and commonly not reliable; thus it is doubtful whether there is enough evidence preserved here to reconstruct the original stratigraphic succession.

Stop 2. Picnic site locality. On the narrow neck of land between South Cove and Baie Verte, the Birchy Complex sequence is composed of greenschist, graphitic schist, and ophiolitic mélange (Fig. 2). The ophiolitic mélange forms an irregular zone, up to 60 ft (20 m) wide, that is infolded with the greenschist just below the picnic grounds.

At this locality, the ophiolitic mélange is composed of black graphitic schist and gray muscovite-albite semipelite containing pods of the familiar bright green actinolite-fuchsite schist and dismembered layers of buff psammitic schist and marble. The recrystallized ultramafic pods are less numerous and smaller than those at Flat Point; however, the zone includes a large, equidimensional serpentinite body (165 ft, or 50 m, in diameter). This body locally exhibits a coarse actinolite alteration rind reminiscent of the pods found elsewhere in the mélange. It also displays an internal brecciation that predates serpentinization and incorporation into the mélange (Williams, 1977).

Immediately north of the mélange zone, graphitic schist forms a zone that is interlayered with the Birchy greenschist. This zone may represent either a continuation of the mélange zone or a distinct member in the complex. Farther north on the South Cove shoreline, near an abandoned wharf, the greenschist is interlayered with minor buff psammites and reddish-colored coticules (Bursnall, 1975); this sequence exhibits classic fold interference patterns formed by second- and later phase folds. Also in this area, the greenschists exhibit small epidote-quartz clasts that probably reflect a fragmental protolith for these rocks.

The Point Rousse Complex is well exposed across the waters of Baie Verte to the east of the picnic site. The complex is imbricated along a series of northwest-dipping thrust faults that are discernible even from this distance. The sole thrust to the complex, the Scrape Thrust (Hibbard, 1983) (Fig. 1), is well exposed along the Ming's Bight access road, approximately 10 mi (17 km) east of the Baie Verte community.

The Baie Verte Line is hidden beneath the complex, but emerges in the Ming's Bight–Pacquet Harbour area farther east. Along the coast at Pacquet Harbor, an early slide zone containing rafts of recrystallized gabbro and ultramafic rock is spectacularly exposed. This zone, termed the Big Brook Slide (Hibbard, 1983) for the brook along which it is exposed, contains a metamafic block approximately 30 by 10 ft (10 by 3 m), that displays possible original cumulate layering.

SIGNIFICANCE AND INTERPRETATION

The recrystallized ultramafic pods, such as those at Stops 1 and 2, were originally thought to represent dismembered prekinematic ultramafic dikes (Kennedy, 1971). However, the recognition that they are confined to zones of graphitic pelite and the interpretation that they represent ophiolitic mélanges attach new significance to these zones (Bursnall, 1975; Williams, 1977; Williams and others, 1977; Hibbard, 1982, 1983). They indicate that the initial disruption of the ancient continental margin, represented by the Fleur de Lys Supergroup, preceded the multiphase tectonothermal history of these rocks. The ophiolitic mélanges imply that the early disruption of the margin involved the transport of ophiolites over the continental margin (Bursnall, 1975; Williams, 1977). Similar mélanges, although less deformed and metamorphosed than the Birchy examples, are found between structural slices of Taconic allochthonous ophiolites of western Newfoundland. Most reasonably then, the ophiolitic mélanges are related to the Early Ordovician emplacement of allochthonous ophiolites onto the Lower Paleozoic carbonate shelf of western Newfoundland. Small ultramafic bodies located along early slide zones in the supergroup are also most reasonably related to this event (Bursnall, 1975; Williams, 1977; Hibbard, 1982, 1983).

The ophiolitic mélanges form an integral part of the Baie Verte Line and record the early stages of the juxtaposition of the ancient continental margin with oceanic rocks to the east. Their linkage to the Taconic allochthons of western Newfoundland indicates that the allochthons must have been derived from farther to the east, in the Dunnage Zone. Thus, the Baie Verte Line most likely represents the root zone for the Taconic allochthonous ophiolites of western Newfoundland (Williams and others, 1977; Williams, 1977; Hibbard, 1982, 1983).

Subsequent tectonism has been centered upon these early structural zones, but has been heterogeneously distributed around the Baie Verte Flexure. South of Baie Verte, the mélanges underwent multiphase tectonism that appears to be related to the obduction event, as Fleur de Lys rocks had cooled and uplift had been effected by the Middle Silurian (Dallmeyer, 1977). Subsequently the line has been accentuated by later faults. In contrast, ophiolitic mélanges and early slides along the east limb of the flexure have been involved in Acadian southeast-directed thrusting, folding, and metamorphism (Hibbard, 1982, 1983; Dallmeyer and Hibbard, 1984), and here the Baie Verte Line is a diffuse structural zone.

The significance of Fleur de Lys ophiolitic mélanges with respect to the Baie Verte Line and allochthons of western Newfoundland places important constraints on the tectonothermal history of the western orthotectonic belt of the Northern Appalachians. Geologic relationships similar to those at Baie Verte are consistent along the length of the Baie Verte–Brompton Line, and possibly south into the United States (Williams and St. Julien, 1982). The presence of the mélanges and the consistency of geologic relationships along the belt provides compelling evidence that the early tectonothermal history along the length of the ancient margin was similar to that at Baie Verte.

REFERENCES CITED

Bursnall, J. T., 1975, Stratigraphy, structure, and metamorphism west of Baie Verte, Burlington Peninsula, Newfoundland [Ph.D. thesis]: Cambridge, England, Cambridge University, 337 p.

Bursnall, J. T., and deWit, M. J., 1975, Timing and development of the orthotectonic zone in the Appalachian Orogen of northwest Newfoundland: Canadian Journal of Earth Sciences, v. 12, p. 1712–1722.

Church, W. R., 1969, Metamorphic rocks of Burlington Peninsula and adjoining areas of Newfoundland and their bearing on continental drift in the North Atlantic, *in* Kay, M., ed., North Atlantic; Geology and continental drift: American Association of Petroleum Geologists Memoir 12, p. 212–233.

Dallmeyer, R. D., 1977, $^{40}Ar/^{39}Ar$ age spectra of minerals from the Fleur de Lys terrane in northwest Newfoundland; Their bearing on chronology and metamorphism within the Appalachian orthotectonic zone: Journal of Geology, v. 85, p. 89–103.

Dallmeyer, R. D., and Hibbard, J., 1984, Geochronology of the Baie Verte Peninsula, Newfoundland; Implications for the tectonic evolution of the Humber and Dunnage Zones of the Appalachian Orogen: Journal of Geology, v. 92, p. 489–512.

Dewey, J. F., and Bird, J. M., 1971, Origin and emplacement of the ophiolite suite, Appalachian ophiolites in Newfoundland: Journal of Geophysical Research, v. 76, p. 3179–3206.

Hibbard, J., 1982, Significance of the Baie Verte Flexure, Newfoundland: Geological Society of American Bulletin, v. 93, p. 790–797.

—— 1983, Geology of the Baie Verte Peninsula, Newfoundland: St. John's Newfoundland Department of Mines and Energy Memoir 2, 279 p.

Hutton, I., 1979, Tectonic slides; A review and reappraisal: Earth Science Reviews, v. 15, p. 151–172.

Kennedy, M. J., 1971, Structure and stratigraphy of the Fleur de Lys Supergroup in the Fleur de Lys area, Burlington Peninsula, Newfoundland: Proceedings of the Geological Association of Canada, v. 24, p. 59–71.

—— , 1975, Repetitive orogeny in the northeastern Appalachians; New plate models based upon Newfoundland examples: Tectonophysics, v. 28, p. 39–87.

Kidd, W., 1977, The Baie Verte Lineament; Ophiolite complex floor and mafic volcanic fill of a small Ordovician marginal basin, *in* Talwani, M., and Pitman, N. C., eds., Island arcs, deep sea trenches, and back arc basins: American Geophysical Union Maurice Ewing series, v. 1, p. 407–418.

Neale, E.R.W., and Nash, W. A., 1963, Sandy Lake (east half), Newfoundland: Geological Survey of Canada Paper 62-28, 40 p.

Williams, H., 1977, Ophiolitic mélange and its significance in the Fleur de Lys Supergroup, northern Appalachians: Canadian Journal of Earth Sciences, v. 14, p. 987–1003.

Williams, H., and St. Julien, P., 1982, The Baie Verte–Brompton Line; Early Paleozoic continent-ocean interface in the Canadian Appalachians, *in* St. Julien, P., and Béland, J., eds., Major structural zones and faults of the northern Appalachians: Geological Association of Canada Special Paper 24, p. 177–207.

Williams, H., and Stevens, R. K., 1974, The ancient continental margin of eastern North America, *in* Burk, C. A., and Drake, C. L., eds., The geology of the continental margin: New York, Springer-Verlag, p. 781–796.

Williams, H., Hibbard, J. P., and Bursnall, J. T., 1977, Geological setting of asbestos-bearing ultramafic rocks along the Baie Verte Lineament, Newfoundland: Geological Survey of Canada Paper 77-1A, p. 351–360.

Silurian stratigraphy and mélanges, New World Island, north central Newfoundland

Douglas Reusch, *Apex Geological Consultants Limited, 4 Pennywell Road, St. John's, Newfoundland A1C 2K9, Canada*

LOCATION

This site includes one principal stop at Rogers Cove (Stop 2) southeast of Cobbs Arm, New World Island, and two subordinate stops, one at Clarkes Cove (Stop 1) northwest of Cobbs Arm and another at Reach Run (Stop 3) between Port Albert Peninsula and Chapel Island (Fig. 1). Stop 1 (Silurian unconformity, Clarkes Cove): From Cobbs Arm, continue north on Newfoundland 346. Turn right onto the road to Pikes Arm and then left onto the road to Green Cove. Near the top of the hill before descending toward Green Cove, turn right down a short, steep, side road that ends at Clarkes Cove. Walk east along the rocky beach about 300 ft (100 m) to the unconformity between conglomerates and basalts. Stop 2 (Silurian mélange, Rogers Cove): Proceed to the end of the Rogers Cove road. Park there and walk through the yard of the last house (a white house with green trim belonging to a very amiable Mr. Sydney Mills). Be sure to close the gate in the front yard and take care climbing over the fence in the back yard and descending along the path to Rogers Cove. Stop 3 (Dunnage Mélange, Reach Run): Park in the quarry on the southeast (mainland) side of Reach Run.

SIGNIFICANCE

Marine sedimentary rocks of Middle Ordovician through Silurian age conformably overlie marine volcanic rocks of Middle Ordovician and older age on New World Island in central Newfoundland (Horne, 1970). These conformable marine sections are unique in the Newfoundland Appalachians in that they record continuous sedimentation through the interval of Taconic orogeny. In contrast, Silurian strata in more westerly parts of the orogen lie unconformably on Ordovician marine volcanic rocks, tonalites, and ophiolites.

The conformable marine sections are locally capped by Silurian mélanges, which coincide with the junctures between thrust sheets. These Silurian mélanges are also unique and are critical in models of the mid-Paleozoic history of the Appalachian orogen. They may signify an episode of Silurian thrusting, initially submarine, that further consolidated the orogen.

The extensive and well-known Dunnage Mélange occurs nearby; however, its age, origin, and significance are still conjectural.

SITE INFORMATION

Geological Setting

Ordovician and Silurian rocks crop out in three belts on eastern New World Island between the Dunnage Mélange and

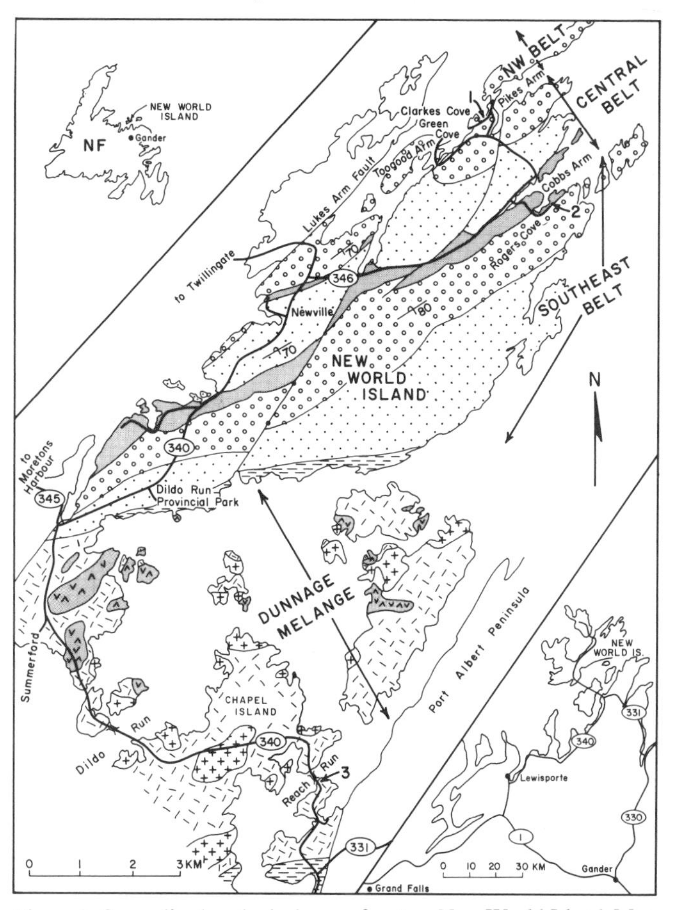

Figure 1. Generalized geological map of eastern New World Island. Map symbols as follows: dark shading, dominantly Middle Ordovician basalts overlain by limestones and black shales, locally includes Silurian sedimentary rocks and mixed assemblages (mélange); horizontal dashed lines, Middle Ordovician shales; stipple, sandstone turbidites; open circles, Silurian conglomerates. Symbols for Dunnage Mélange: random dashed lines, dominantly chaotic shaly sedimentary rocks; dark shading with v, basalts; crosses, dacite porphyries.

the Lukes Arm Fault (Fig. 1; Williams, 1963). In the Southeast Belt (Fig. 2), Middle Ordovician black shales pass conformably upwards through sandstone turbidites into Silurian conglomerates; Silurian mélange caps this section. In the Central Belt, Middle Ordovician and older volcanic rocks, limestones, and black shales also pass conformably upwards through sandstone turbidites into Silurian conglomerates. In the Northwest Belt, however, Ordovician volcanic rocks are overlain unconformably by Silurian conglomerates (Stop 1).

Silurian mélange (Stop 2) occurs within a steep fault zone that separates the Southeast and Central belts. It is composed of a

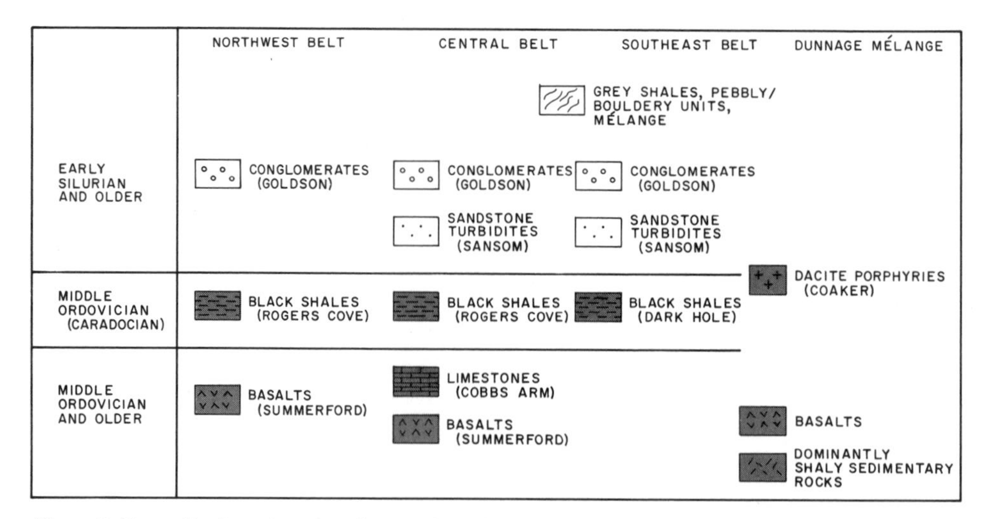

Figure 2. Generalized stratigraphy of eastern New World Island. Commonly cited formation names are in parentheses.

shaly matrix (Silurian) and coarse clasts of black shales, limestones, sandstones, and basalts. These clasts are typical of rocks of the Central Belt.

The Dunnage Mélange (Stop 3) occupies an extensive area adjacent to the Southeast Belt (Kay, 1976). Its matrix is dark shale of Early Ordovician age. Blocks are dominantly clastic sedimentary and volcanic rocks of Cambrian and Early Ordovician age and range up to 3,000 ft (1,000 m) in size.

Beds on eastern New World Island generally dip steeply southeast and face northwest (Fig. 1). Adjacent stratigraphic belts are separated by steep fault zones. In addition, a network of steep, rectilinear faults transects the area.

History of Interpretation

New World Island, including the Dunnage Mélange, is as notorious for interpretational controversy as it is critical to regional analysis of the Appalachian orogen.

Twenhofel and Schrock (1937) viewed the succession of eastern New World Island as a single, continuous northwest-facing section, which implied that the SE Belt has the oldest rocks. Kay and Williams (1963) discovered Silurian fossils in conglomerates of the Southeast Belt and first proposed fault-repeated sections. In a radically different interpretation, McKerrow and Cocks (1978) suggested that Silurian mélange at Rogers Cove is the basal part of an enormous olistostrome (submarine gravity slide deposit) that includes most of the Central and Northwest Belts of eastern New World Island; in their model, which requires no faults, all of the Ordovician rocks northwest of the Southeast Belt are huge clasts (olistoliths). Arnott (1983), however, showed that the mélange is restricted to a narrow zone between the Southeast and Central Belts and confirmed that the three sections are fault-repeated.

Kay and Williams (1963) had also reported an unconformable relationship between Silurian sedimentary rocks and Middle Ordovician limestones and black shales at Rogers Cove (in the fault zone between the Southeast and Central belts). The Silurian sedimentary rocks, which contain clasts of limestone and black shale, were supposed to represent the basal part of the "cover" sequence. However, the Ordovician rocks that were viewed as "basement" are now known to occur as blocks in the mélange (Jacobi and Schweickert, 1976).

The faults that repeat the stratigraphic sections have provoked considerable controversy with respect to their type, polarity, and timing. Kay and Williams (1963) originally suggested northwest-directed thrusts. Kay (1967) later favoured transcurrent faults, although his argument for this switch is not particularly logical. The stratigraphy-parallel nature of the faults and their positions between northwest-facing stratigraphic sections have led most recent workers to interpret them as thrusts. The

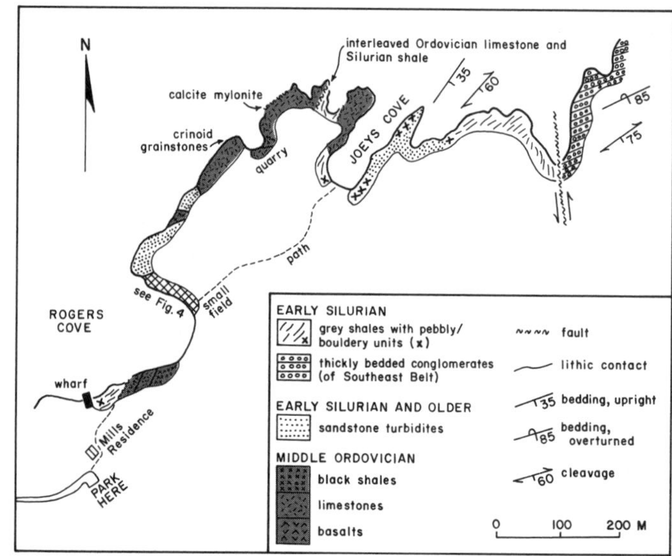

Figure 3. Generalized geological map of the Rogers Cove area (Stop 2: Silurian mélange).

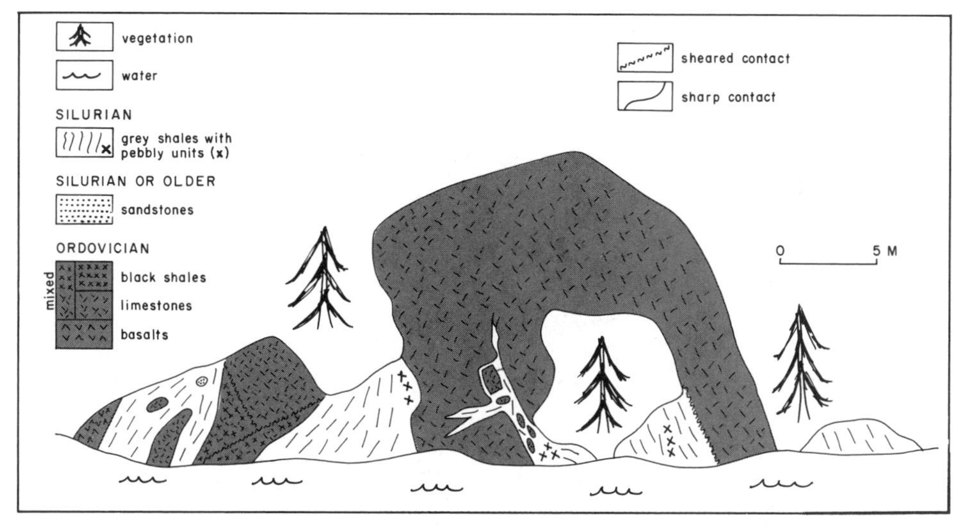

Figure 4. Sketch of cliff exposures of Silurian mélange along the northeastern shoreline of Rogers Cove (looking northeast; see Fig. 3 for location).

thrusting was syn-sedimentary according to the view that the mélange formed along an active submarine fault scarp. Controversy about polarity stems from different interpretations of local structures and restorations of the stratigraphic sections: Dean and Strong (1977), Nelson (1981), and Reusch (1983) have contended that thrusting was southeast-directed, whereas Williams and others (1983) argue that it was northwest-directed. Williams and others (1983) also assert that early thrust zones were reactivated during later widespread strike-slip faulting.

A Silurian unconformity in Newfoundland indicates the Taconic orogeny (Williams, 1967), which is widely believed to have been caused by the collision of a volcanic arc with North America. According to this model, volcanic basement in the Northwest Belt marks the eastern fringe of the collision zone. The marine clastic sedimentary rocks represent the Silurian margin of the newly accreted North American continent. Thrust faults record destruction of this basin in a subsequent accretionary event. Late strike-slip faults mark a transition from a dominantly convergent margin to a dominantly transform one.

The Dunnage Mélange is widely assumed to be Ordovician (e.g., Hibbard and Williams, 1979). However, its boundaries are difficult to define, and the possibility that it includes younger rocks should not be dismissed (Reusch, 1983). It has been interpreted as both a tectonic mélange (Dewey, 1969) and as a sedimentary mélange or olistostrome (Horne, 1969; Hibbard and Williams, 1979; Lorenz, 1982). Until its age and origin are resolved, its regional significance will remain uncertain.

Stop 1. Silurian unconformity, Clarkes Cove. An unconformity separating a Silurian conglomerate unit from Middle Ordovician basalts is well exposed in beach outcrops. In detail, it is a sharp, depositional surface. The overlying conglomerate unit faces northwest. Sandstones near its base contain Silurian corals. Pebbles in the conglomerates are limited in variety and of local derivation. The underlying basalts, which are massive and without bedding, appear to be no more deformed than the conglomerates (although both the basalts and the conglomerates are exceptionally competent and generally do not display pronounced effects of deformation other than tilting). It is therefore difficult to establish the relationship between basement and cover as angular or disconformable.

An unconformable relationship between Ordovician and Silurian rocks is typical of westerly parts of the Newfoundland Appalachians (Dunning, this volume). This stop should prepare the visitor for appreciating the unique nature of the continuous sections exposed in the adjacent thrust sheets to the southeast.

Stop 2. Silurian mélange, Rogers Cove. Amygdaloidal basalts, limestones, and black shales, representing typical Middle Ordovician rocks, and gray shales, which form the matrix of the mélange, may be seen in Rogers Cove directly at the foot of the path that descends from the Mills residence (Fig. 3). Within the gray shales, note the scattered remains of the Silurian brachiopod *Stricklandia lens progressa* and also pebbly horizons, located several feet (meters) to the west near the wharf, that contain angular clasts of Middle Ordovician black shales.

The most spectacular exposure of mélange in the area is located along the northeastern shoreline of the cove (Fig. 4). A large mass of Middle Ordovician limestone is surrounded by Silurian gray shales containing fragments, up to boulder-size, of limestone, black shale, and lesser basalt. Note the penetrations of Silurian gray shale into the Ordovician limestone: the shale appears to have been highly mobile, whereas the limestone was fragmented, locally forming inclusions in the shale matrix (a relationship analogous to xenoliths within an intrusion). Note also faulted contacts and the internally veined and brecciated limestone, which suggest high-level, extremely brittle deformation.

Additional examples of bouldery shales may be seen in Joeys Cove to the northeast of Rogers Cove, accessible by a slightly overgrown path through the woods adjacent to a fence around a small field (Fig. 3). Sandstone turbidites and shales, located along the shoreline to the east of Joeys Cove, contain Silurian brachiopods and corals and are locally complexly deformed. The more coherent beds dip and face southeast and dip less steeply than cleavage. They are structurally anomalous in comparison with thickly bedded Silurian conglomerates of the Southeast Belt that face northwest and that dip *more* steeply than cleavage; these are located still farther east on the far side of a sinistral cross fault. The Silurian conglomerates contain an assemblage of pebbles fairly representative of the Ordovician volcanic-tonalite-ophiolite terrane to the west that was eroded during the Taconic orogeny (Dunning, this volume; Helwig and Sarpi, 1969).

Some scrambling through the woods may be required to arrive at outcrops northwest of Joeys Cove. Along an exposed contact between Ordovician limestones and Silurian shales, the limestone and shale are delicately interleaved, indicating that the contact is tectonic. The limestones in particular display markedly inhomogeneous deformation: in the vicinity of an old quarry that faces Cobbs Arm, crinoid columns with no distortion contrast sharply with finely laminated calcite mylonite along a minor late (?) fault.

Stop 3. Dunnage Mélange, Reach Run. Where Newfoundland 340 crosses high ground north of Dildo Run Provincial Park, it provides a sweeping vista of Dildo Run between New World Island and Port Albert Peninsula. Many of the myriad islands and abrupt hills are discrete clasts of resistant sedimentary and volcanic rocks within the Dunnage Mélange.

The quarry on the south side of Reach Run (Fig. 1) exposes the shaly matrix of the mélange and clasts of various sizes and types set within it. The chaotic texture and lack of bedding are impressive and are fairly representative of the mélange as a whole.

Along the shoreline of Reach Run about 300 ft (100 m) north of the quarry, the mélange matrix includes bands 1 to 4 in (2 to 10 cm) thick of green and black shales. A dike of dacite porphyry exposed nearby is typical of the Coaker Porphyry, the most extensive of several types of intrusive rocks within the Dunnage Mélange (Lorenz, 1982).

REFERENCES CITED

Arnott, R. J., 1983, Sedimentology of Upper Ordovician–Silurian sequences on New World Island, Newfoundland; Separate fault-controlled basins?: Canadian Journal of Earth Sciences, v. 20, p. 345–354.

Dean, P. L., and Strong, D. F., 1977, Folded thrust faults in Notre Dame Bay, Newfoundland: American Journal of Science, v. 277, p. 97–108.

Dewey, J. F., 1969, Evolution of the Appalachian/Caledonian orogen: Nature, v. 222, p. 124–129.

Helwig, J., and Sarpi, E., 1969, Plutonic-pebble conglomerates, New World Island, Newfoundland, and history of eugeosynclines, *in* Kay, M., ed., North Atlantic geology and continental drift: American Association of Petroleum Geologists Memoir 12, p. 443–466.

Hibbard, J. B., and Williams, H., 1979, Regional setting of the Dunnage Mélange in the Newfoundland Appalachians: American Journal of Science, v. 279, p. 993–1021.

Horne, G. S., 1969, Early Ordovician chaotic deposits in the central volcanic belt of northeastern Newfoundland: Geological Society of America Bulletin, v. 80, p. 2451–2464.

—— , 1970, Complex volcanic-sedimentary patterns in the Magog belt of northeastern Newfoundland: Geological Society of America Bulletin, v. 81, p. 1767–1788.

Jacobi, R. D., and Schweickert, R. A., 1976, Implications of new data on stratigraphic and structural relations of Ordovician rocks on New World Island, north-central Newfoundland: Geological Society of America Abstracts with Programs, v. 8, p. 206.

Kay, M., 1967, Stratigraphy and structure of northeastern Newfoundland bearing on drift in North Atlantic: American Association of Petroleum Geologists Bulletin, v. 14, p. 579–600.

—— , 1976, Dunnage Mélange and subduction of the Protacadic Ocean, northeast Newfoundland: Geological Society of America Special Paper 175, 49 p.

Kay, M., and Williams, H., 1963, Ordovician-Silurian relations on New World Island, Notre Dame Bay, northeast Newfoundland [abs.]: Geological Society of America Bulletin, v. 74, p. 807.

Lorenz, B., 1982, Timing of igneous activity in the Dunnage Mélange, Newfoundland: Geological Society of America Abstracts with Programs, v. 14, p. 35.

McKerrow, W. S., and Cocks, L.R.M., 1978, A lower Paleozoic trench-fill sequence, New World Island, Newfoundland: Geological Society of America Bulletin, v. 89, p. 1121–1132.

Nelson, K. D., 1981, Mélange development in the Boones Point Complex, north-central Newfoundland: Canadian Journal of Earth Sciences, v. 18, p. 433–442.

Reusch, D. N., 1983, The New World Island Complex and its relationships to nearby formations, north-central Newfoundland [M.S. thesis]: St. John's, Memorial University of Newfoundland, 235 p.

Twenhofel, W. H., and Schrock, R. R., 1937, Silurian strata of Notre Dame Bay and Exploits Valley, Newfoundland: Geological Society of America Bulletin, v. 48, p. 1743–1771.

Williams, H., 1963, Twillingate map-area, Newfoundland: Geological Survey of Canada Paper 63-36, 30 p.

—— , 1967, Silurian rocks of Newfoundland, *in* Geology of the Atlantic Region: Geological Association of Canada Special Paper 4, p. 93–137.

Williams, P. F., Karlstrom, K. E., and van der Pluijm, B., 1983, Thrusting in the New World Island–Hamilton Sound Area of Newfoundland, *in* Schenk, P. F., ed., Regional trends in the geology of the Appalachian-Caledonian-Hercynian-Mauritanide orogen: D. Reidel Publishing Company, p. 377–378.

Stratigraphy of Cambrian rocks at Bacon Cove, Duffs, and Manuels River, Conception Bay, Avalon Peninsula, eastern Newfoundland

*M. M. **Anderson**, Departments of Biology and Earth Sciences, Memorial University of Newfoundland, St. John's, Newfoundland A1B 3X9, Canada*

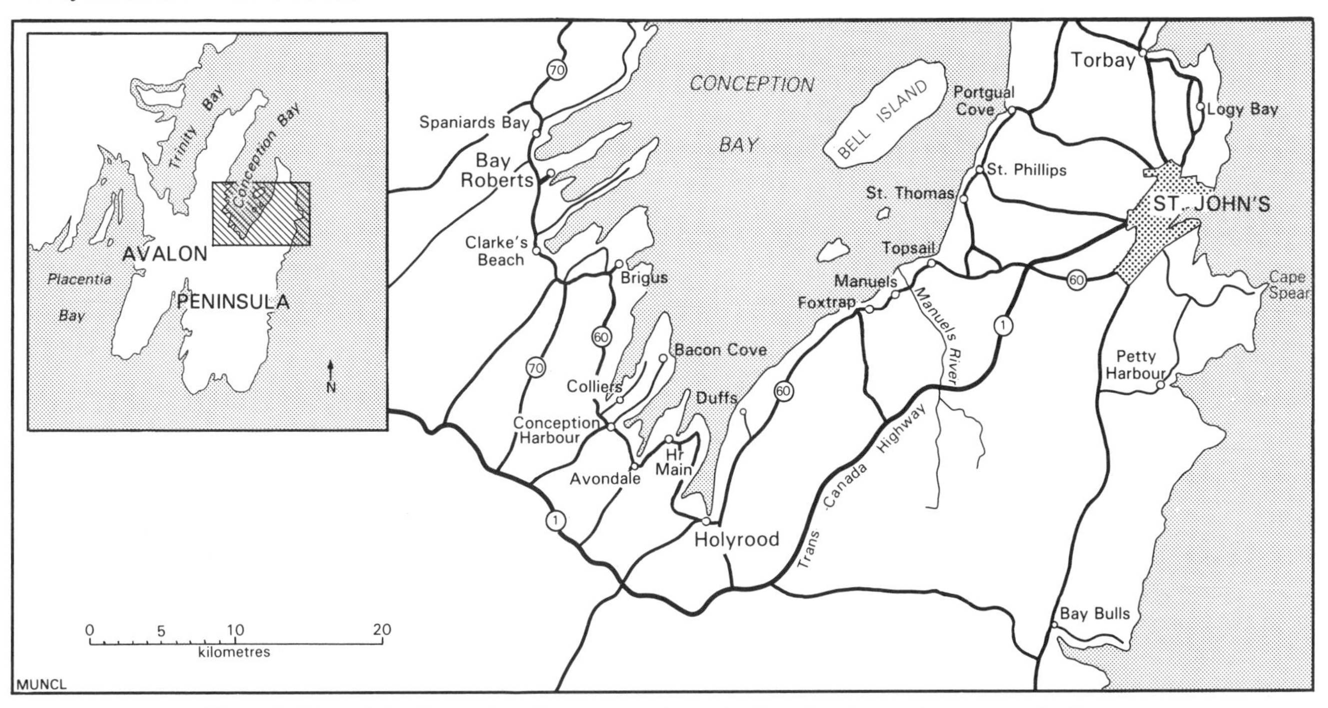

Figure 1. Map of the Conception Bay area, northeast Avalon, showing roads, towns, and villages, including Bacon Cove and Duffs, and, on the southeastern side of the bay, Manuels River.

LOCATIONS

Cambrian rocks bordering Conception Bay crop out in a northeast-trending belt adjacent to the coast on the southeastern side of the bay and in several small isolated areas on its southwestern side. The outcrops of concern here are those on the coast at Bacon Cove and Duffs, and those, just inland, in the lower valley of Manuels River (Fig. 1). Access to them is unrestricted.

Stop 1: Bacon Cove. From St. John's, drive west along the Trans-Canada Highway to the Avondale access road (38 mi; 64 km), turn north to Avondale (6 mi; 10 km), then at the "T" junction turn left at the coastal road (Newfoundland 60) to Conception Harbour (3 mi; 5 km). In that village, turn sharp right (the turn-off is obscured by a building), continue beside the sea until a "Y" junction and bear left for Bacon Cove (5 mi; 8 km from Conception Harbour). The road beside the cove is narrow, so park just beyond the cove on open ground to the left of the road.

Stop 2: Duffs. From Bacon Cove, return to Conception Harbour and turn left onto Newfoundland 60; follow that highway around the head of Conception Bay to 4 mi (6.5 km) beyond the centre of Holyrood where a wide road on the left leads to a Power Station (conspicuous smoke stacks). Turn into that road and, almost immediately, bear left onto a gravel road leading to Duffs (1 mi; 1.6 km). Park at the end of the wide road. Walk along the route shown by arrows on Fig. 2 to where the oil pipeline passes under the railway; turn left and push thick bushes aside to reach the top of a tree-covered slope. Clamber down between trees to the rocks along the shore. Warning: low cliffs are present in places immediately beyond the last row of trees at the foot of the slope.

Stop 3: Manuels River. Return from Duffs to Newfoundland 60 and proceed eastward to a bridge over the Manuels River (10 mi; 16.5 km). Park beside the road just before the bridge (Fig. 3). Examine the rocks immediately upstream and downstream from the bridge. Return to the car and drive across the bridge. Turn left at the light and then, 1,300 ft (400 m) farther on, turn left (at the second entrance to a large school) onto a gravel road leading to a church in the northwest corner of a playground. Park beside the church and walk along the trail on its northern side (Fig. 3) to the riverbank and Cambrian outcrops. To return to St. John's, drive east on Newfoundland 60 to the junction with

Trans-Canada 1. From there, take Trans-Canada 1 northeast to St. John's (Fig. 1).

SIGNIFICANCE

On the west side of Bacon Cove, Lower Cambrian sediments of the Bonavista Formation rest with marked angular unconformity on steeply dipping siltstones of the late Precambrian Conception Group. The irregular erosion surface beneath the Cambrian sediments, with its potholes and veneer of stromatolites, marks the position of an Early Cambrian shoreline. The presence below the unconformity of folded Precambrian rocks is significant because Bacon Cove is one of the very few places in southeastern Newfoundland that provides evidence for folding at the close of the Precambrian (Avalonian orogeny). However, most of the folding in the region is attributed to the Acadian orogeny.

A thick layer of diamictite in a sequence of Conception Group sediments cropping out on the northeast side of Bacon Cove belongs to a glaciomarine unit, the Gaskiers Formation. This unit, widely distributed on the Avalon Peninsula, is evidence for a glacial episode of regional extent in southeastern Newfoundland during late Precambrian time. No other part of eastern North America is known to have been subjected to regional glaciation at that time.

At Duffs, late Precambrian granitic rocks of a composite pluton (Holyrood granite) are overlain nonconformably by Lower Cambrian strata of the Brigus Formation. Remnants of an erosion surface on the granitic rocks are relics of the coastline that existed there in the late Early Cambrian. The nonconformity is significant because it shows that, prior to the advance of the sea into the area, prolonged subaerial erosion resulted in the loss of thousands of metres of late Precambrian sediments represented elsewhere on the Avalon Peninsula, and in the partial unroofing of the Holyrood granite. The intrusive rocks at Duffs are of particular interest because they represent two unusual, intimately associated, facies of the Holyrood granite that are unique to the area: granophyre and pegmatite. The pegmatite contains giant euhedral crystals of quartz up to 20 in (50 cm) across.

Manuels River is one of the most famous Cambrian localities in the world. Gently dipping Lower, Middle, and Upper Cambrian sedimentary rocks, with a total thickness of more than 1,000 ft (300 m), are exposed in its entrenched lower valley. Many of the beds, especially those of the Middle Cambrian part of the section, are richly fossiliferous. Most of the species of trilobites, brachiopods, and other organisms in the Manuels River section are known also from contemporaneous beds in southern Britain, Maritime Canada, and eastern parts of the U.S.: the Avalonian province (Theokritoff, 1979).

SITE INFORMATION

Regional setting. Precambrian rocks underlying the Avalon Peninsula can conveniently be divided into two major assemblages: a lower of late Riphean to middle Vendian age, and an upper of late Vendian age. The lower assemblage (at least 10,000 to 20,000 ft; 3 to 5 km thick) comprises terrestrial and marine, basic to acid, volcanic rocks (Harbour Main Group), and volcaniclastic sedimentary rocks (Conception Group); the Conception Group includes a glaciomarine unit (Gaskiers Formation) in its lower part. The upper assemblage (25,000 to 30,000 ft; 8 to 9 km thick) consists of thick detrital sequences composed mainly of black shales, red sandstones, and conglomerates (St. John's and Signal Hill Groups in the eastern part of the Avalon Peninsula and stratigraphically equivalent "groups" in the central and western parts of the peninsula) (Anderson and King, 1980).

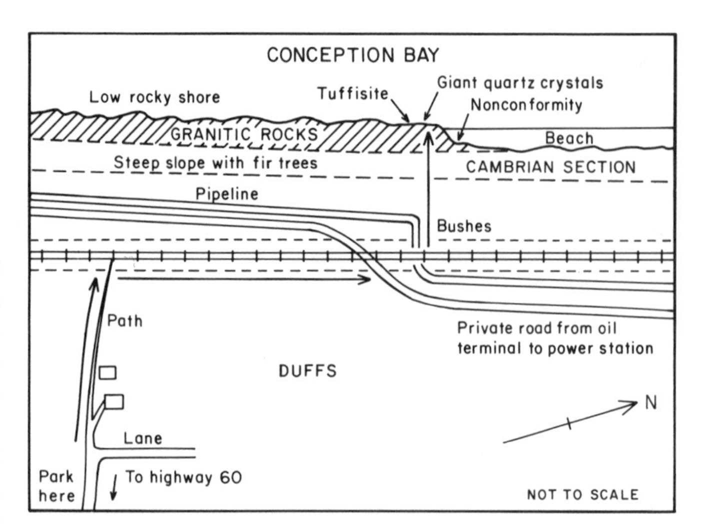

Figure 2. Map of the Duffs area showing the road leading to the area and the outcrops of Precambrian and Cambrian rocks on the coast.

Adjacent to the head of Conception Bay, granitic rocks (Holyrood granite) of late Vendian age (ca. 620 Ma; Krogh and others, 1983) intrude volcanic rocks of the Harbour Main Group; rocks of the upper assemblage are not represented there because they were entirely destroyed by erosion in latest Precambrian time. Hence, south of Conception Bay, Cambrian rocks directly overlie rocks of the lower assemblage and Holyrood granite.

Four formations are recognized in the Lower Cambrian succession on the Avalon Peninsula, namely (oldest to youngest), the Random, Bonavista, Smith Point, and Brigus Formations. In the Conception Bay area, the Lower Cambrian succession is partly to largely incomplete, i.e., on the southwestern side of the bay the Random Formation is missing, and on its southeastern side the only Lower Cambrian sediments present are those of a lower part of the Brigus Formation. The incompleteness of the Lower Cambrian succession on both sides of the bay is accounted for by nondeposition (Anderson, 1981).

Stop 1: Bacon Cove. Along the shore on the western side of Bacon Cove, almost horizontal Lower Cambrian sediments of the Bonavista Formation rest with marked angular unconformity on steeply dipping late Precambrian siltstones of the Conception Group. The erosion surface marking the unconformity is irregu-

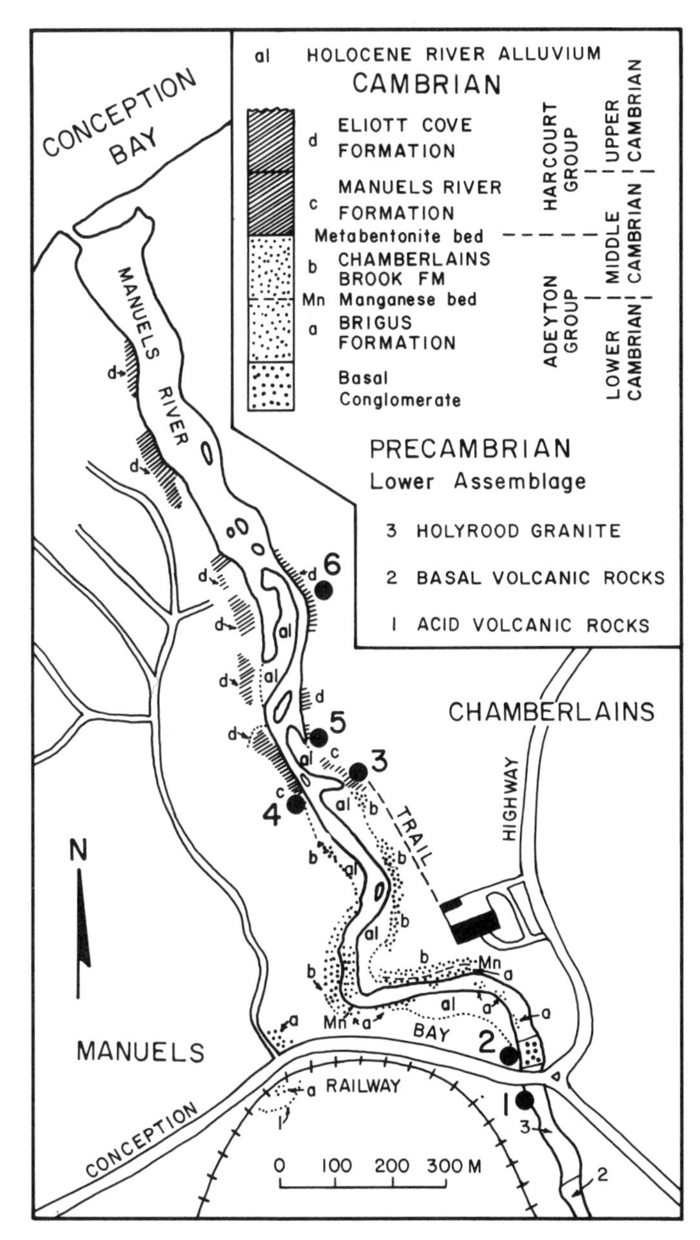

Figure 3. Map of the lower valley of Manuels River showing the distribution of Cambrian outcrops and their ages. The two stations at Manuels Bridge and the four stations in the entrenched part of the valley north of the bridge are indicated; also shown are the school (between the river and the Conception Bay Highway) and the trail leading to Station 3 from the church in the northwest corner of the playground behind the school. (Based on Figure 3 of Howell, 1925.)

lar, i.e., it has a distinct relief that, in places, exceeds 6 ft (2 m). The lower part of the Cambrian sequence thus occupies the low areas or depressions, which vary considerably in areal extent and depth. Small, steep-sided excavations, more or less circular in plan, are potholes; a vertical section through an infilled pothole is present in a cliff near the southern end of the Cambrian outcrop.

A breccia-conglomerate consisting of angular to well-rounded clasts of locally derived sedimentary and volcanic rocks in a calcareous arkosic matrix is present at the base of the Cambrian sequence in the deeper depressions. These pockets of breccia-conglomerate are of variable thickness (rarely exceeding 12 in; 30 cm) and the contained clasts range in size from granules to rare boulders.

The breccia-conglomerate either grades upward into a variable thickness (up to 24 in; 60 cm) of coarse-grained, purple, calcareous arkose, or, more commonly, is separated from the arkose by a thin layer (mm to some 34 in; 9 cm thick) of purplish red stromatolitic limestone. The stromatolites are stratiform with flat to wavy laminae, but in some places they are columnar. The flat to wavy types are also present here and there in the breccia-conglomerate, either in pockets between the clasts or encrusting them.

In places the arkose contains thin, laterally impersistent layers of siltstone (mm to a few cm thick). The arkose infilling of potholes near the northern end of the Cambrian outcrop includes, basally, layers of phosphate and, above them, angular to well-rounded, pebble-size and smaller fragments of phosphate derived from other, previously existing layers.

A widely distributed but discontinuous layer of dark red to purple stromatolitic limestone (a few mm to several cm thick) overlies the arkose. The stromatolites are similar to those occurring below the arkose, and like them contain well-preserved fossil algae (*Epiphyton, Rivularia*) (Edhorn and Anderson, 1977). However, columnar forms (up to 1 in; 3 cm high and a cm or less in diameter) are more common, and some of them exhibit branching.

Above the stromatolitic limestone (or, locally, the arkose) is a bed of pink limestone of variable thickness (at least several cm) that contains oncolites, especially numerous near its top, and a variety of small shelly fossils, mainly hyolithids. In broad, shallow depressions on the erosion surface, the limestone is separated into two, more rarely three, layers by one or two intercalations of red mudstone (up to 2 in; 5 cm thick) that thin to disappearance laterally. In such cases the maximum thickness of the "unit" is between 8 and 10 in (20 and 25 cm).

The pink fossiliferous limestone is overlain by about 15 ft (5 m) of red and green mudstones that are the youngest Cambrian sediments exposed on the west side of Bacon Cove.

Each of the beds above the breccia-conglomerate, in turn, overlaps the bed below it to rest unconformably on Conception Group rocks underlying successively higher parts of the erosion surface, thereby locally becoming the basal bed of the Cambrian sequence; irrespective of lithology, the basal bed contains scattered, pebble-size and smaller, intraclasts.

The irregular surface on the Late Precambrian rocks on the west side of Bacon Cove (remnants are also present on the north side of the cove) marks the position of an Early Cambrian shoreline that was subsequently buried by a fining-upward sedimentary sequence deposited during the continued, predominantly eastward, advance of the sea (Cambrian transgression) over what is now the Conception Bay area.

Lower Cambrian sediments, younger than those of the Bonavista Formation, are exposed on the hillside above the cliffs on the north side of Bacon Cove in a syncline pitching to the south; the syncline is bounded to east and west by faults. Red mudstones of the uppermost part of the Bonavista Formation are overlain by limestones of the Smith Point Formation (15 ft; 5 m), which are overlain by mudstones with interbeds of nodular limestone of the Brigus Formation.

On the small promontory forming the northeast corner of Bacon Cove, two tiny, flat-lying remnants of the Bonavista Formation overlie steeply dipping Conception Group rocks including a thick layer of diamictite of glacial origin (Gaskiers Formation).

Stop 2: Duffs. The granophyre cropping out south of the reentrant in the coastline at Duffs (Fig. 2) is an unusual facies of the Holyrood granite (Hughes, 1971). Intimately associated with it is another facies of the granite unique to Duffs, namely, pegmatite with giant crystals of quartz up to 20 in (50 cm) in diameter. The pegmatite forms blind bodies up to 15 ft (5 m) across within the granophyre, some of these bodies are surrounded by a zone of concentrically banded rock (alternation of fine-grained dark-coloured microgranite with a coarser-grained light-coloured rock consisting mainly of perthite). Numerous veins, up to 4 in (10 cm) wide, of dark tuffisite (fragments of granophyre in a comminuted matrix of the same material) cut the granophyre. The veins thicken and thin, branch, and intersect one another. Hughes (1971) attributed the brecciation of the granophyre to the last event in its cooling history, an explosive escape of gas from lower parts of the intrusion.

North of the reentrant in the coastline at Duffs (Fig. 2), gently northwest-dipping strata of the Brigus Formation (late Early Cambrian) rest nonconformably upon the granophyre. The erosion surface on the granophyre is exposed only in the vicinity of the low, irregular cliff that forms the south side of the reentrant, i.e., for a few metres to the north and south of the foot and the top of the cliff, respectively. Adjacent to the cliff, the erosion surface is irregular and fissured, reflecting the highly fractured nature of the underlying granophyre. A breccia-conglomerate infills the irregularities. It consists of angular blocks and smaller fragments of granophyre, well-rounded cobbles and pebbles of granophyre, and pebbles of quartz in a matrix of black arkosic argillite. Just north of the reentrant, the erosion surface is almost flat, and the basal bed there is a thin layer of black argillite that also infills opened-up joints and other fractures in the underlying granophyre. In some places, the walls of the fractures have thin patches of red stromatolitic limestone adhering to them.

Northward from the exposure of black argillite, for several hundred metres along the coast, successively younger beds of the Brigus Formation crop out in low cliffs behind a beach of cobbles and boulders. The argillite is overlain by a much-disturbed and poorly exposed sequence (possibly 6 ft; 2 m thick) of beds of limetone alternating with beds of hard, dark gray to black mudstone. Two of the limestones are crowded with small shelly fossils, and the uppermost limestone (about 16 in; 40 cm thick) contains numerous oncolites. Above this conspicuous bed (only exposed on the beach), green and red mudstones (in units 3 to 6 ft; 1 to 2 m thick) alternate with beds of gray to pink limestone that are either massive or nodular. Nodules of limestone are also present in two of the mudstone units. Trace fossils ("worm-burrows") are ubiquitous in the mudstones. McCartney (1967) recorded finding the trilobites *Callavia broeggeri* and *Strenuella strenua* in the limestones.

The irregular cliff and the wave-cut surfaces adjacent to it at Duffs are remnants of the coastline that existed there late in Early Cambrian time.

Engelhardt (1975) found an aerodynamically shaped bomb fragment in a boulder of the Cambrian basal breccia-conglomerate at Duffs, and within the bomb, angular quartz crystals containing planar deformation structures highly symptomatic of shock metamorphism. On the basis of this and other evidence from the coastal section at Duffs, he suggested that "the basal breccia and the fragmentation of the granophyre was produced by an extraterrestrial body which impacted the Precambrian surface in Early Cambrian times and formed a crater . . ." the crater subsequently being filled by marine sediments of Early Cambrian age. Engelhardt's evidence was reviewed by Anderson and Hughes (1976) who concluded that there is no *prima facie* case for a meteorite impact having been responsible for the brecciation of the granophyre at Duffs, or for the nature of the unconformable contact there. The source of the bomb, an important find, has yet to be resolved.

Stop 3: Manuels River: Cambrian succession and faunal assemblages. The Cambrian sediments exposed in the valley of Manuels River (Fig. 3) belong to two groups established by Jenness (1963) on the basis of lithology (although they also differ in colour):

(1) The Adeyton Group, here about 300 ft (100 m) thick, consists of a basal conglomerate (15 to 20 ft; 5 to 6 m thick) followed above by a succession of green to gray and, less commonly, red mudstones or shales with some thin, largely nodular limestone interbeds. About 60 ft (20 m) above the base, several of the beds of shale consist mainly of nodules of rhodochrosite (superficially weathered to psilomelane) (Mn of Fig. 3). The beds of the Adeyton Group below and above the base of the manganese beds belong, respectively, to the Brigus Formation (late Lower Cambrian) and the Chamberlain's Brook Formation (early Middle Cambrian). The manganese beds mark an important nonsequence (virtual cessation of deposition) here and elsewhere in the northern part of southeastern Newfoundland where the upper part of the Brigus Formation and the lower part of the Chamberlain's Brook Formation are missing. The missing beds are present on Cape St. Mary's Peninsula, southwest Avalon (Fletcher, 1972).

(2) The Harcourt Group, of which about 600 ft (200 m) are exposed in Manuels River valley, consists of gray to black, commonly pyritiferous shales with thin interbeds and large lens-shaped concretions of dense black limestone in its lower part (90 ft; 26 m), and thin interbeds of sandstone and rare, rusty weathering, black nodules of impure limestone in its upper part

(580 ft; 176+ m). The lower and upper parts belong to the Manuels River Formation (late Middle Cambrian) and the Elliot Cove Formation (highest Middle Cambrian, Upper Cambrian), respectively.

Howell (1925) distinguished seven successive faunal assemblages (predominantly of trilobites) in the Manuels River Cambrian succession; an eighth was recognized by Poulsen and Anderson (1975). The assemblages are characterized from bottom to top by: (1) *Catadoxides magnificus,* in the Brigus Formation (late Early Cambrian) below the manganese beds; (2) *Eccaparadoxides bennetti,* in the Chamberlain's Brook Formation (early Middle Cambrian) between the manganese beds and a layer of metabentonite (see description Station 4); (3) *Hydrocephalus hicksi,* and (4) *Paradoxides davidis,* in the Manuels River Formation (late Middle Cambrian) above the layer of metabentonite and below a conglomerate with phosphatic clasts (see description Station 5); (5) *Lejopyge laevigata,* in 30 ft (10 m) of shales above the conglomerate that represent the lowest part of the Elliot Cove Formation (latest Middle Cambrian); (6) *Agnostus pisiformis,* (7) *Homagnostus obesus, Olenus* spp., (8) *Parabolina spinulosa* and *Orusia lenticularis,* at successively higher levels in the Elliot Cove Formation (late Cambrian).

Manuels River Bridge. Station 1. The bedrock upstream for some distance from the bridge is Holyrood granite with inclusions of acidic and basic volcanic rocks of the Harbour Main Group.

Station 2. Just downstream from the bridge, at the crest of the waterfall, granite and rhyolite are overlain nonconformably by Cambrian basal conglomerate (Lower Cambrian Brigus Formation), 16.5 to 20 ft (5 to 6 m) thick, dipping about 10° to the north. The conglomerate consists dominantly of rhyolite cobbles and pebbles with the interstices between them filled with sand; it represents a local beach deposit that accumulated close to a rocky coastal cliff.

Manuels River Valley. Station 3. Note the cliff on the east side of a conspicuous embayment in Manuels River. Exposed in the cliff is a succession of thinly to thickly bedded dark gray to black shales with numerous lenses and interbeds of gray to black fossiliferous limestone. This succession represents much of the lower part of the Manuels River Formation (the lowest part is concealed by talus) and contains fossils of the *Hydrocephalus hicksi* assemblage. Howell (1925) gave a bed-by-bed description of the Manuels River Middle Cambrian succession and listed the fossils present in each bed.

Cross the river where it is shallowest (just north of an embayment) to the west bank.

Station 4. Note the cliff on the west side of a valley just south of the embayment (depicted in Fig. 2, Plate 2, of Howell, 1925). A layer 1.5 in (3.5 cm) thick of white, plastic, clay-like material, metabentonite, in a recess about 10 ft (3 m) above the base of the cliff marks the boundary between underlying and overlying beds of the Chamberlain's Brook and Manuels River Formations, respectively. (The metabentonite, a postdepositionally altered volcanic ash, is evidence for contemporaneous volcanic activity.) The greater part of the Chamberlain's Brook Formation exposed in the cliff below the boundary consists of beds of hard, olive-gray shale, most of which contain numerous small nodules of gray limestone. However, the uppermost 24 in (60 cm) of the formation is transitional in character and difficult to distinguish lithologically from the Manuels River Formation. It consists (below the metabentonite) of a bed of shaly, nodular, dark gray limestone 12 in (30 cm) thick underlain by a similar thickness of bedded black shales containing many small crystals of barite.

The 10 ft (3 m) of dark gray to black shales with interbeds of dark gray limestone overlying the metabentonite form the lowest part of the Manuels River Formation, i.e., the part that is concealed at Station 3. Return to the east bank of the river.

Station 5. Note the outcrop of shales in the valley side just north of Station 3. The northward-dipping shales at the southern end of the section are separated from those at its northern end by a conglomeratic layer of variable thickness (2 to 6 in; 5 to 15 cm) consisting of phosphatic clasts. The dark gray to black, silty, occasionally pyritiferous shales below the conglomerate form the uppermost part of the Manuels River Formation. Trilobite remains are particularly abundant in the softer shales of the lower part of the sequence. The species present were listed by Bergström and Levi-Setti (1978) who described several new sub-species of *Paradoxides davidis.*

The shales above the conglomerate belong to the Elliot Cove Formation. Fossils found in the 30 ft (10 m) of shales at the base of the formation by Poulsen and Anderson (1975) show that the lowermost part of it (previously considered barren) is of latest Middle Cambrian age, and that the conglomerate marks a break in sedimentation (temporary withdrawal of the sea). Thus, the conglomerate does not coincide with the base of the Upper Cambrian, as formerly believed, and the Middle-Upper Cambrian boundary, which is conformable, lies within the Elliot Cove Formation.

Station 6. Note the low outcrops in the bank of the river at the marked northeastward bend in its course. The black, micaceous, Upper Cambrian shales at this locality contain fragments of *Olenus* and numerous specimens of *Homagnostus obesus,* the inarticulate brachiopod *Lingulella,* and small ostracodes.

REFERENCES

Anderson, M. M., 1981, The Random Formation of southeastern Newfoundland; A discussion aimed at establishing its age and relationship to bounding formations: American Journal of Science, v. 281, p. 807–830.

Anderson, M. M., and Hughes, C. J., 1976, Indications for a meteoritic impact of Early Cambrian age at Conception Bay, Newfoundland; Discussion: Die Naturwissenschaften, v. 63, p. 87–88.

Anderson, M. M., and King, A. F., 1980, Rocks and fossils of Late Precambrian and early Paleozoic age (including Late Precambrian glacial sequence),

Avalon Peninsula, Newfoundland: Geological Association of Canada/Mineralogical Association of Canada, Joint Annual Meeting, Guidebook 12, 28 p.

Bergström, J., and Levi-Setti, R., 1978, Phenotypic variation in the Middle Cambrian trilobite *Paradoxides davidis* Salter at Manuels, SE Newfoundland: Geologica et Palaeontologica, v. 12, p. 1–40.

Edhorn, A., and Anderson, M. M., 1977, Algal remains in the Lower Cambrian Bonavista Formation, Conception Bay, southeastern Newfoundland, *in* Flugel, E., ed., Fossil Algae: Berlin-Heidelberg-New York, Springer-Verlag, p. 113–123.

Engelhardt, W. V., 1975, Indications for a meteoritic impact of Early Cambrian age at Conception Bay, Newfoundland: Die Naturwissenschaften, v. 62, p. 234–235.

Fletcher, T. P., 1972, Geology and Lower to Middle Cambrian trilobite faunas of the southwest Avalon, Newfoundland [Ph.D. thesis]: Cambridge, England, University of Cambridge, 530 p.

Howell, B. F., 1925, The faunas of the Cambrian *Paradoxides* beds at Manuels, Newfoundland: Bulletins of American Paleontology, v. 11, no. 43, 140 p.

Hughes, C. J., 1971, Anatomy of a granophyre intrusion: Lithos, v. 4, p. 403–415.

Jenness, S. E., 1963, Terra Nova and Bonavista map-areas, Newfoundland: Geological Survey of Canada Memoir 327, 184 p.

Krogh, T. E., Strong, D. F., and Papezik, V. S., 1983, Precise U-Pb ages of zircons from volcanic and plutonic units in the Avalon Peninsula: Geological Society of America Abstracts with Programs, v. 15, p. 135.

McCartney, W. D., 1967, Whitbourne map-area, Newfoundland: Geological Survey of Canada Memoir 341, 135 p.

Poulsen, V., and Anderson, M. M., 1975, The Middle–Upper Cambrian transition in southeastern Newfoundland, Canada: Canadian Journal of Earth Sciences, v. 12, p. 2065–2079.

Theokritoff, G., 1979, Early Cambrian provincialism and biogeographic boundaries in the North Atlantic region: Lethaia, v. 12, p. 281–295.

Late Precambrian glaciomarine sequence, Conception Group, Double Road Point, St. Mary's Bay, southeast Newfoundland

*M. M. **Anderson**, Departments of Biology and Earth Sciences, Memorial University of Newfoundland, St. John's, Newfoundland A1B 3X9, Canada*

LOCATION

Double Road Point is a small headland near the village of St. Mary's, a fishing community on the east side of St. Mary's Bay on the Avalon Peninsula, Newfoundland (Fig. 1). From the Trans-Canada Highway (Route 1) west of St. John's, take Newfoundland 90 south to St. Mary's (42 mi; 70 km). At the church in the village (Fig. 1) bear right onto a dirt road beside the harbour. Take the second left turn and drive up a short hill to 'T' junction at its top and turn right. Bear left at the next fork (about 1,000 ft; 300 m from the 'T' junction) onto a narrow track (negotiable with care by car) that ends beside a large fenced field; park on firm ground. The southwestern corner of the field is adjacent to the northern end of the Double Road Point section. Follow an obvious footpath around the eastern (landward) side of the fenced field, which is private. The path continues, close to the shore, past the southern end of the section. No permission is needed to visit the site.

SIGNIFICANCE

The Gaskiers Formation, a glaciomarine unit in the late Precambrian Conception Group, is widely distributed on the Avalon Peninsula of southeastern Newfoundland, particularly in the St. Mary's Bay area. At Double Road Point, the formation consists essentially of a succession of thick layers of diamictite, indistinguishable from tillite, some of which are separated from one another by thin stratified units of mudstone, siltstone, and sandstone that usually contain dropstones.

The presence of the glacigenic Gaskiers Formation in the Conception Group, which otherwise consists of volcaniclastic sedimentary rocks, is significant because it indicates that (1) glaciation occurred in eastern Newfoundland in late Precambrian time and interrupted marine sedimentation there; (2) the whole of eastern Newfoundland presently included in the tectonostratigraphic Avalon Zone (Williams, 1979) was affected by glaciation, that is, glaciation was of regional extent, and not, as formerly believed, of local extent; and (3) the glaciation, because of its regional nature and age, was probably related to a worldwide episode of glaciation in the early Vendian, which suggests that, at that time, eastern Newfoundland was situated far from its present site and formed part of a 'continental block' elsewhere that was also subjected to glaciation in the late Precambrian.

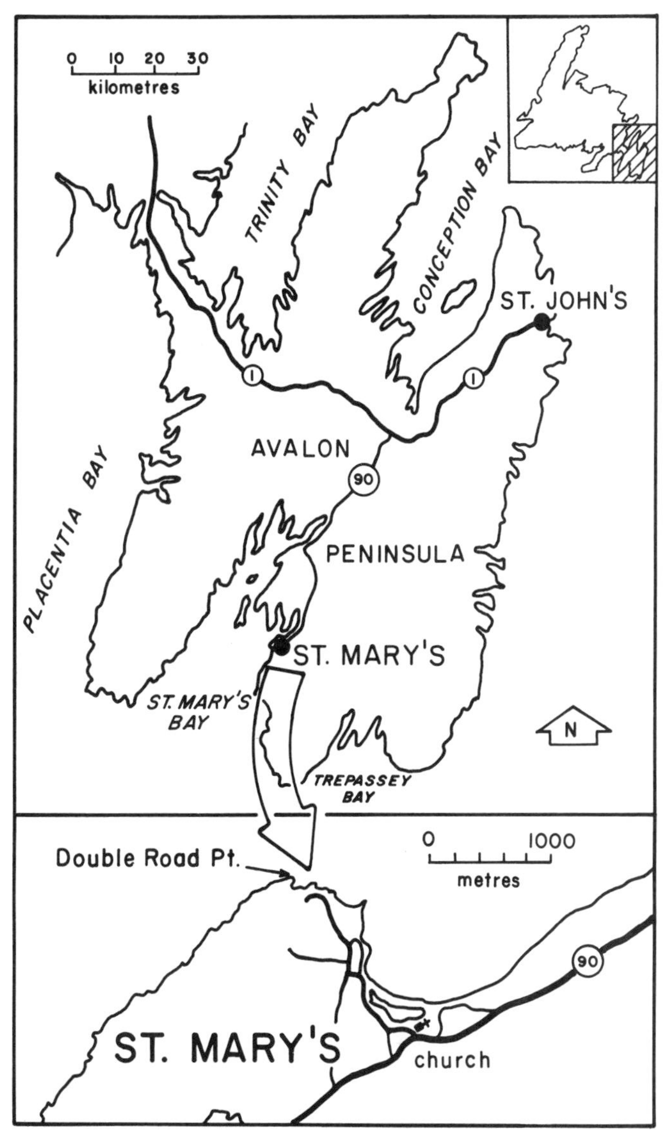

Figure 1. Outline map of Avalon Peninsula showing the route from the city of St. John's to the village of St. Mary's, and (below) the roads in the village between the church on the main road and the coastal section at Double Road Point.

SITE INFORMATION

The Conception Group is a 1.8 to 3 mi (3 to 5 km) thick sequence of late Precambrian marine, flysch-like volcaniclastic sedimentary rocks with intercalations of airfall tuff that includes, in its lower part, a glaciomarine unit. The group is exposed only on the Avalon Peninsula, southeastern Newfoundland, where it is divided, in ascending order, into the Mall Bay, Gaskiers, Drook, Briscal, and Mistaken Point Formations. The Gaskiers Formation, the only one of direct concern here, differs from the others in that it consists predominantly of sediments of glacial origin. The type area of the formation is in the St. Mary's Bay area where it has its thickest development (to about 1,000 ft; 300 m). Figure 2 shows the distribution of outcrops of the Gaskiers Formation in the type area and the location of the field site at Double Road Point. The succession of rock types there is as shown in Figure 3. The rocks are exposed in low coastal cliffs. No significant gaps are apparent in the section. However, there are several narrow, steep-sided clefts in the cliffs along the coast; at least one of them marks the position of a fault. Thus, minor gaps due to faulting are probably present. These are indicated on Figure 3.

The sequence is best studied by beginning at the southern end of the exposure, where the lowermost part of the Gaskiers Formation is repeated by faults and the basal bed (a reddish mudstone with sparse dropstones) overlies rocks of the Mall Bay Formation with apparent conformity, and working northeastward along the coast (up section) to the top of the formation, a massive reddish-purple diamictite overlain by less than 3 ft (1 m) of red mudstone with dropstones in its lower part. Above the mudstone, and conformable with it, are sediments of the Drook Formation.

Beds of diamictite represent about 90% of the total thickness of the Gaskiers Formation. They range from 10 to 18 ft (3 to 55 m) in thickness, and their lower contacts, where observable, are shasrp and conformable. The diamictites are, in general, massive, internally structureless, and made up of a variety of rock fragments of different sizes and shapes and of differing degrees of roundness, set in a groundmass of sand-sized and finer-grained particles. However, some of these otherwise massive beds contain rare, thin intercalations of tuff that provide evidence for ash falls at times of non-deposition, while in others there are thin layers of sandstone or lenses of conglomerate with evidence of basal scour that possibly mark breaks in deposition (Gravenor, 1980). The pyroclastic material was probably derived from an active volcano in the vicinity of Great Colinet Island (Fig. 2), as the sequence of diamictites exposed at the southern end of the island includes not only intercalations of tuff and a significant amount of volcanic debris in its upper part, but also two layers of agglomerate close to its top (Williams and King, 1979).

Two of the massive diamictites in the lower part of the section at Double Road Point (Fig. 3c, d) contain deformed and undeformed slabs of intrabasinal sediments up to several meters in length. About 360 ft (110 m) above the base of the section, a 10-ft-thick (3-m) band of conglomerate (essentially a diamictite with an unusually high proportion of clasts) separates two thick diamictites (Fig. 3e). This separation is unique, as the other diamictites in the section either lie one above another or are separated by stratified units, rarely more than a meter thick, of gray to green (more rarely purple) mudstone, siltstone, and sandstone. Stratified units are largely confined to the lower part of the section. Rhythmites are present in some of these units. They consist of a succession of graded laminae or thin beds (2 mm or less to several cm or more in thickness) each of which is divisible into a lower, dark-coloured layer of coarse sand to silt and an upper, lighter-coloured layer of mud. This alternation in the grain size and colour of the 'couplets' is responsible for the regular banded appearance of the rhythmites. Good examples of rhythmic bedding, both uncontorted and contorted, are well exposed in the lower part of the stratified unit overlying the lowermost massive diamictite (Fig. 3a). The beds in the upper part of that unit (purple mudstones, gray to green sandstones, and diamictite), which has an overall thickness of nearly 30 ft (10 m), are contorted and disrupted (Fig. 3b).

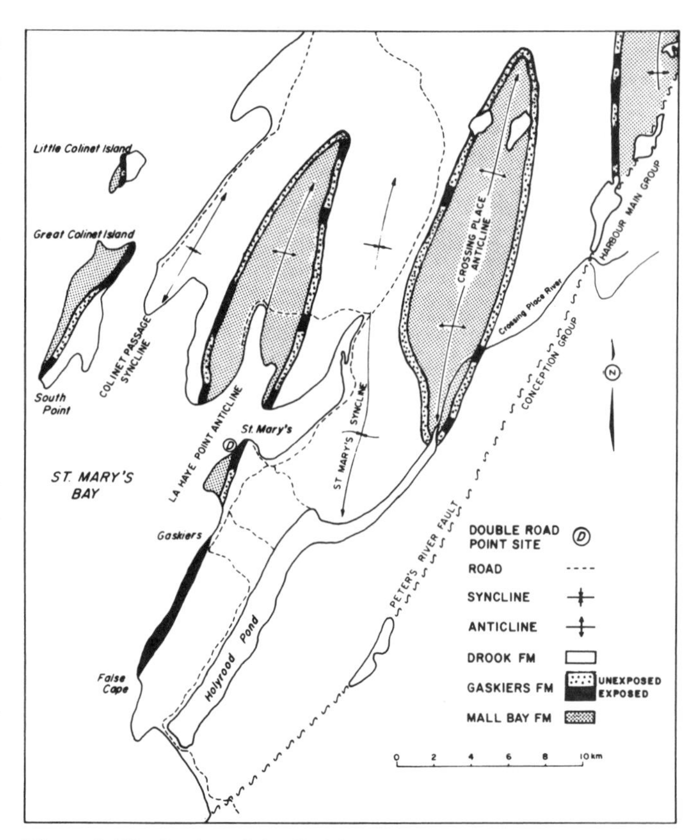

Figure 2. Distribution of the Gaskiers Formation and major folds in the St. Mary's Bay area. (Based on Fig. 24 of Williams and King, 1979.)

Some sandstones in the stratified units are structureless, but others display convolute lamination, ripple-drift lamination, and small-scale slump folds, all indicating transport to the east and southeast (Williams and King, 1979).

Dropstones associated with the stratified units have, in numerous cases, disrupted and deformed the layers into which they have fallen; their diameters commonly exceed the thickness of the layers. Most dropstones are granules or pebbles but larger ones (cobbles, small boulders) are also present.

The matrix of individual diamictites, as seen in outcrop, is generally either a sandy mudstone or a muddy sandstone, gray, green, or red to purple in colour. The clasts are matrix supported, haphazardly distributed, and, with rare local exceptions, are not oriented. Clasts of pebble-size and larger constitute between 2.5% and 12% by volume of the individual diamictites. In general,

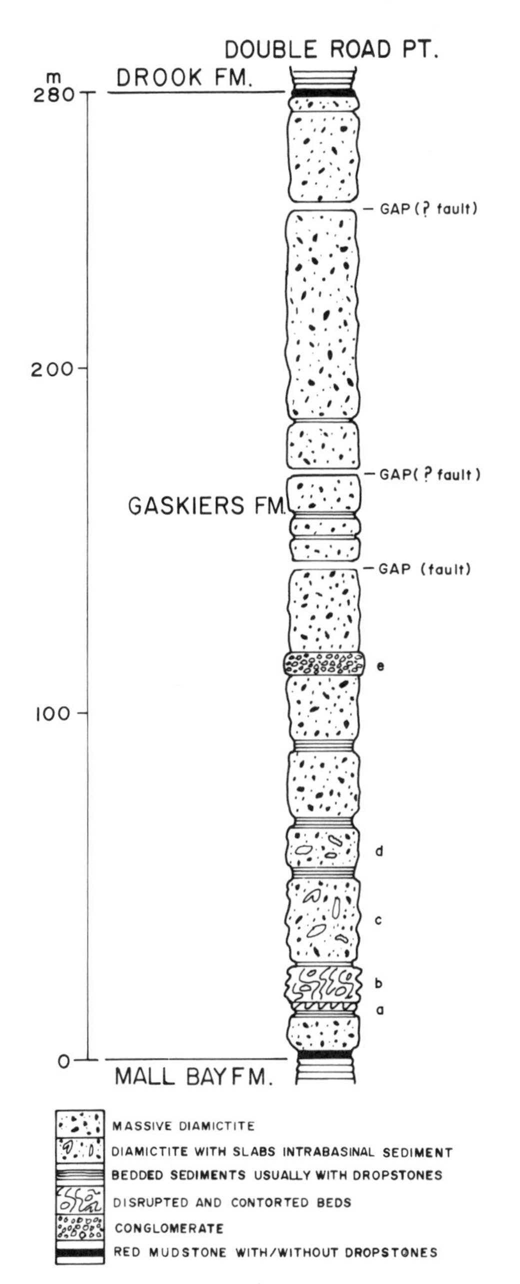

Figure 3. Double Road Point section, St. Mary's Bay. Thickness of stratified units and possible gaps in section exaggerated to show their positions more clearly.

clasts are tabular or equidimensional; those of pebble-size and larger are usually subrounded to well-rounded. Faceted and striated clasts are common, and typical 'flat-irons' can also be found (for illustrations see Brückner and Anderson, 1971).

In decreasing order of abundance, clasts are of igneous (extrusive and intrusive), sedimentary, and metamorphic rocks. Common types present include red to pink silicic volcanic rocks, vesicular to massive basalt, a variety of coarse-grained pink to red granites, porphyritic granite, granophyre, diorite, sandstone, siltstone, limestone, quartz, volcanic pebble-conglomerate, quartzite, and foliated granite.

The source of most of the rock types found as clasts has previously been referred to as intrabasinal, with only the quartzite and foliated granite specifically designated as exotic to the Avalon Peninsula (Williams and King, 1979; Anderson and King, 1981), which implies that nearly all the others were derived by glacial erosion from rocks belonging to formations presently exposed on the Avalon Peninsula. However, at the time of the glacial episode, apart from a volcanic island to the south of Conception Bay and another not far distant from Great Colinet Island, that region was covered by sea and the site of deposition of the products of glaciation. The glacial debris in the Gaskiers Formation in the St. Mary's Bay area was derived (as indicated by the direction of transport noted earlier) from a land area west (although not necessarily due west) of the Avalon Peninsula region.

Sediments of the Gaskiers Formation contain garnets for which there is no exposed source on the Avalon Peninsula; the presence of garnets derived by glacial grinding from garnet-bearing rocks is significant because, (1) some of them bear chattermark trails indicating that they have undergone grain-to-grain grinding during glacial transport, thus providing additional evidence for a glacial origin for the sediments of the Gaskiers Formation and (2) the percentage of garnets bearing such trails increases with distance travelled in englacial transport so that it provides an indication of how far the grains were transported (Gravenor, 1979, 1980). About 9% of the fresh garnets extracted from the Gaskiers Formation by Gravenor (1980) have chattermark trails; he concluded that, "The sample population is large enough to indicate with reasonable certainty that the total transport distance is less than that of the Pleistocene glacial deposits in North America (14%; 360+ mi; 600+ km) but greater than that of the Pleistocene glacial deposits of New Zealand (1%: 30 mi; 50 km)" (Gravenor, 1980, p. 1333). The percentage of fresh garnets with chattermark trails found in the latter is significantly less than that found in the Gaskiers Formation, and therefore it is reasonable to assume that the source of the garnets in the Gaskiers Formation in the St. Mary's Bay area was at least 60 mi (100 km) inland from the coastline at the western margin of the basin of deposition, although a more distant source, perhaps as much as 120 to 180 mi (200 to 300 km) inland from that margin, cannot be ruled out.

Deep water Conception Group sediments (which bound all known occurrences of the Gaskiers Formation) extend as far west as Placentia Bay (Fig. 1). Therefore, the western margin of the basin of deposition, where true till and intertill deposits were laid down in shallow water offshore beneath a grounded ice-sheet, lay well to the west of Placentia Bay. The ice that advanced into the sea descended from an elevated region inland (centre of ice accumulation) that lay still further to the west, that is, in the western part of the Avalon Zone. If the grains of garnet were transported over a distance of more than 60 mi (100 km), they may have come from beyond the western margin of that zone in unknown terrain, present at that time but subsequently lost, probably during the movement of 'Avalon Zone rocks' into their present position adjacent to the rest of the island of Newfoundland in the Silurian or the Devonian (Williams and Hatcher, 1982).

It is apparent from the preceding discussion that the sediments of the Gaskiers Formation preserved in the St. Mary's Bay area (and northeastward beyond it to the southwest side of Conception Bay) were deposited far from the western margin of the basin of deposition, and to the east of the seaward margin of a floating ice-shelf (the source of the icebergs mentioned below), on the lower slope or floor of the basin in deep, open water.

The presence in the Gaskiers formation of penecontemporaneous volcanic detritus as interbeds and also within some of the massive diamictites (air-fall bombs, blocks, and lapilli), notably at the southern end of Great Colinet Island, provides conclusive evidence for the absence of an ice-cover above the site of deposition. The movement of icebergs over the area during the glacial episode would have been impeded had there been an ice-cover. In fact, the only glacial debris in the Gaskiers Formation actually known to have been deposited directly from ice was that released from melting icebergs.

A deep-water intrabasinal site of deposition for the Gaskiers Formation means that the glacial debris in the diamictites and the stratified units must have been derived, by redeposition and resedimentation, from the margin of the basin, the only part of the basin where in situ deposition of glacial deposits was taking place. Thus, from time to time during the glacial episode, glacial debris that had accumulated at the seaward margin of the grounded ice-shelf became unstable and flowed slowly downslope under the pull of gravity as a debris flow. Gravenor was the first to recognize that the diamictites of the Gaskiers Formation represent debris flows; he called them glaciogenic subaqueous debris flows (Gravenor, 1980). However, the diamictites described earlier as containing slabs of intra-basinal sediments may have been emplaced by slumping.

Apart from the contribution made by melting icebergs and contemporaneous volcanic activity mentioned earlier, the remainder of the Gaskiers Formation consists of bedded sediments that separate some of the debris flows. Rhythmites that exhibit graded bedding predominate and represent the distal deposits of turbidity currents. Less common are beds of sandstone, siltstone, and mudstone (lacking grading) that were deposited from bottom meltwater currents (winnowing upslope–rentrainment of sediment–deposition downslope). Least common are conglomerates, coarse residual deposits formed by the loss of finer material as a result of winnowing by strong bottom currents.

The contorted and disrupted beds in the lower part of the Gaskiers Formation owe their structure to post-depositional, gravity-induced lateral movement, probably of a local nature.

The glacial debris of eastern occurrences of the Gaskiers Formation on the Avalon Peninsula (including the Virgin Rocks Shoal area 120 mi (200 km) southeast of St. John's; Brückner and Anderson, 1971) was probably derived from the east so that, in that direction, the land that was glaciated was probably no less distant from the Avalon Peninsula than was the continental area on the western side of the basin. On the northern and southern sides of the peninsula, rocks of the Gaskiers Formation pass beneath the sea (following the regional strike), and there is, therefore, every likelihood of their continuing for some distance beyond the peninsula in both directions. It is now evident that the whole of the Avalon Zone, in eastern Newfoundland, both on land and beneath the sea, was affected directly or indirectly by glaciation in late Precambrian time. Thus, the glaciation recorded by the rocks of the Gaskiers Formation was of regional extent.

No age determinations have been made on rocks of the Conception Group. However, fossils of soft-bodied multicellular organisms present in its upper part (Anderson, 1978) are representatives of the Ediacara fauna, of world-wide distribution, which is recognized elsewhere as being of middle Vendian age. The fossils first appear in the Conception Group some 5,000 ft (1,500 m) above the top of the Gaskiers Formation, which suggests that the latter is of early Vendian age.

Apart from the glacial and periglacial deposits of the Mount Rogers and Grandfather Mountain Formations in the Central Appalachians, considered to be the products of local highland glaciation (Schwab, 1981), glacial deposits comparable in age to those of the Gaskiers Formation are unknown elsewhere in eastern North America. However, glaciation in eastern Newfoundland was of regional extent and, therefore, probably related to a world-wide episode of glaciation, most likely the earlier of the two episodes (pre–650 Ma) that occurred in the Vendian (Hambrey, 1983).

REFERENCES CITED

Anderson, M. M., 1978, Ediacaran fauna, *in* Lapedes, D. N., ed., McGraw–Hill Yearbook of Science and Technology: McGraw–Hill Book Company, Inc., p. 146–149.

Anderson, M. M., and King, A. F., 1981, Precambrian tillites of the Conception Group on the Avalon Peninsula, southeastern Newfoundland, *in* Hambrey, M. J., and Harland, W. B., eds., Earth's Pre-Pleistocene Glacial Record: Cambridge University Press, p. 760–766.

Brückner, W. D., and Anderson, M. M., 1971, Late Precambrian glacial deposits in southeastern Newfoundland; A preliminary note: Proceedings of the Geological Association of Canada, v. 24, p. 95–102.

Gravenor, C. P., 1979, The nature of the Late Paleozoic glaciation in Gondwana as determined from an analysis of garnets and other heavy minerals: Canadian Journal of Earth Sciences, v. 16, p. 1137–1153.

——, 1980, Heavy minerals and sedimentological studies on the glaciogenic Late Precambrian Gaskiers Formation of Newfoundland: Canadian Journal of Earth Sciences, v. 17, p. 1331–1341.

Hambrey, M. J., 1983, Correlation of Late Proterozoic tillites in the North Atlantic region and Europe: Geological Magazine, v. 120, p. 209–232.

Schwab, F. L., 1981, Late Precambrian tillites of the Appalachians, *in* Hambrey, M. J., and Harland, W. B., eds., Earth's Pre-Pleistocene Glacial Record: Cambridge University Press, p. 751–755.

Williams, H., 1979, Appalachian Orogen in Canada: Canadian Journal of Earth Sciences, v. 16, p. 792–807.

Williams, H., and Hatcher, R. D., Jr., 1982, Suspect terranes and accretionary history of the Appalachian orogen: Geology, v. 10, p. 530–536.

Williams, H., and King, A. F., 1979, Trepassey Map Area, Newfoundland: Geological Survey of Canada Memoir 389, 24 p.

Index

Typeset by WESType Publishing Services, Inc., Boulder, Colorado
Printed in U.S.A. by Malloy Lithographing, Inc., Ann Arbor, Michigan

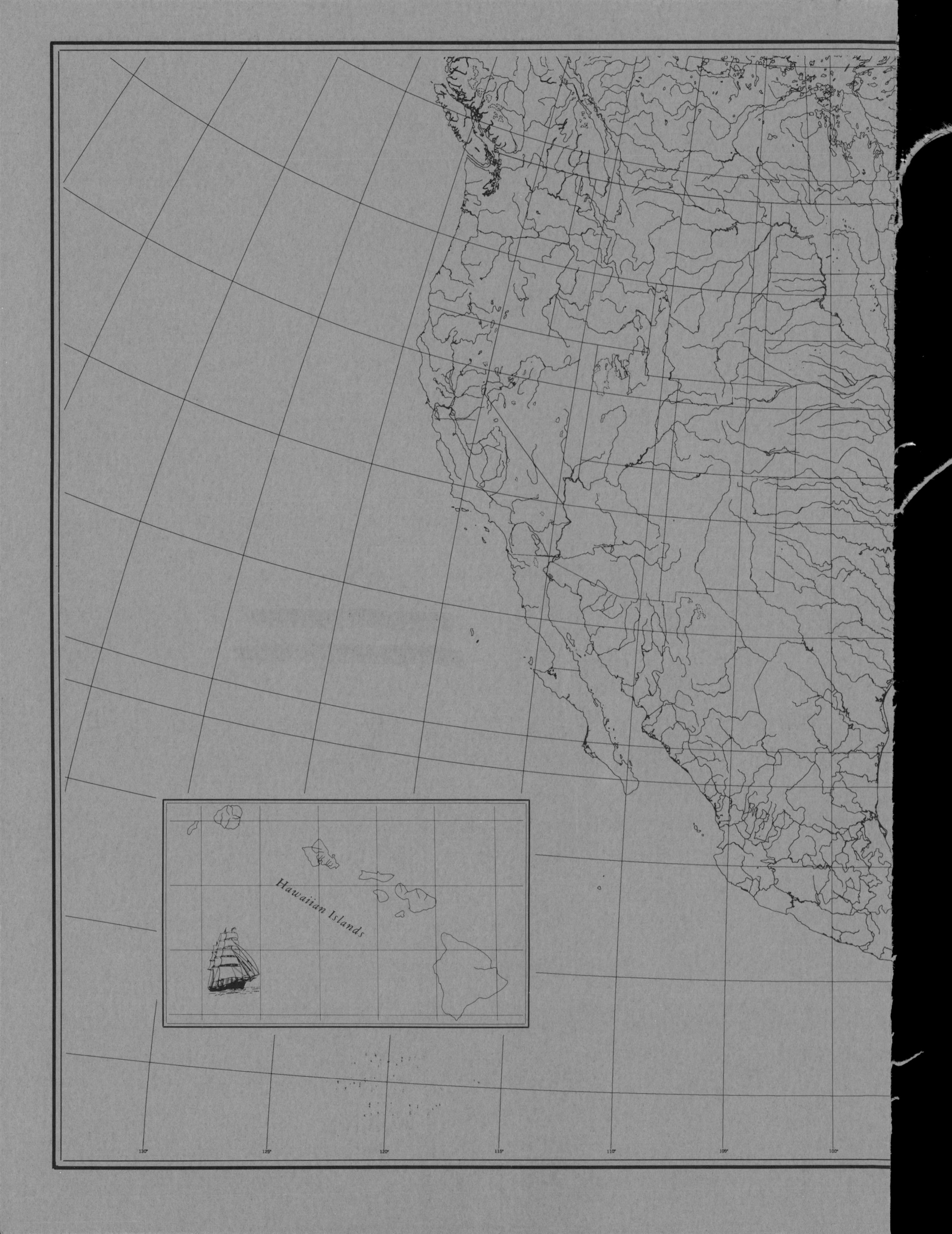
Hawaiian Islands
130°
125°
120°
115°
110°
105°
100°